NONEQUILIBRIUM MANY-BODY THEORY OF QUANTUM SYSTEMS

The Green's function method is among the most powerful and versatile formalisms in physics, and its nonequilibrium version has proved invaluable in many research fields. With entirely new chapters and updated example problems, the second edition of this popular text continues to provide an ideal introduction to nonequilibrium many-body quantum systems and ultrafast phenomena in modern science. Retaining the unique and self-contained style of the original, this new edition has been thoroughly revised to address interacting systems of fermions and bosons, simplified many-body approaches like the GKBA, the Bloch equations, and the Boltzmann equations, and the connection between Green's functions and newly developed time-resolved spectroscopy techniques. Small gaps in the theory have been filled, and frequently overlooked subtleties have been systematically highlighted and clarified. With an abundance of illustrative examples, insightful discussions, and modern applications, this book remains the definitive guide for students and researchers alike.

GIANLUCA STEFANUCCI is Professor of Condensed Matter Theory at the University of Rome Tor Vergata, Italy. His current research interests are in nonequilibrium quantum matter, low-dimensional systems, and time-resolved spectroscopy. He is the author of more than 100 research articles in prestigious, peer-reviewed journals.

ROBERT VAN LEEUWEN is Professor of Physics at the University of Jyväskylä, Finland. His primary area of research is quantum many-body theory, with a focus on nonequilibrium systems. Within this research field he has published more than 100 articles in distinguished, scientific journals.

NONEQUILIBRIUM MANY-BODY THEORY OF QUANTUM SYSTEMS

A Modern Introduction

Second Edition

GIANLUCA STEFANUCCI
Università degli Studi di Roma 'Tor Vergata'

ROBERT VAN LEEUWEN
University of Jyväskylä, Finland

Shaftesbury Road, Cambridge CB2 8EA, United Kingdom

One Liberty Plaza, 20th Floor, New York, NY 10006, USA

477 Williamstown Road, Port Melbourne, VIC 3207, Australia

314–321, 3rd Floor, Plot 3, Splendor Forum, Jasola District Centre,
New Delhi - 110025, India

103 Penang Road, #05–06/07, Visioncrest Commercial, Singapore 238467

Cambridge University Press is part of Cambridge University Press & Assessment,
a department of the University of Cambridge.

We share the University's mission to contribute to society through the pursuit of
education, learning and research at the highest international levels of excellence.

www.cambridge.org
Information on this title: www.cambridge.org/9781009536790

DOI: 10.1017/9781009536776

First published 2013
Second edition 2025

A catalogue record for this publication is available from the British Library

A Cataloging-in-Publication data record for this book is available from the Library of Congress

ISBN 978-1-009-53679-0 Hardback

To my wife Marina
GS

Contents

Preface to the Second Edition

During the past decade, research in physics has considerably drifted toward the nonequilibrium properties of quantum matter. Many different time-resolved spectroscopy techniques (angle-resolved photoemission, streaking, fragmentation chronoscopy, absorption, reflectivity, transport, etc.) have been developed and further refined. It is today possible to "film" the motion of electrons and nuclei with high temporal resolution (about one-billionth of one-millionth of a second) and to investigate a large variety of ultrafast quantum phenomena. The nonequilibrium Green's function formalism remains the most versatile theoretical technique to address the new emerging physics. We therefore felt it necessary to integrate the material of the first edition with a textbook introduction on the connection between Green's functions and the outcomes of the newly developed time-resolved spectroscopy techniques.

We have taken the opportunity of writing this second edition to improve the presentation of the existing material. We have optimized the presentation by rearranging some of the topics and by grouping related topics together. An example is the new chapter on the electron gas, which combines sections that were previously scattered throughout different chapters.

The first edition also lacked a fundamental topic, namely the Green's function treatment of multi-component systems, such as electrons and photons or electrons and phonons. The second edition contains an entirely new chapter entitled "Green's Functions for Nonequilibrium Fermion–Boson Systems" devoted to this topic. Here we also provide a pedagogical introduction to the recently developed ab initio theory of electrons and phonons. Finally, we have included one additional chapter entitled "From Green's Functions to Simplified Many-Body Approaches," where we discuss popular approaches like the Generalized Kadanoff–Baym Ansatz, the semiconductor Bloch equations, the Boltzmann equations, the Redfield equation, and the Lindblad equation. Starting from the nonequilibrium Green's function formalism, we derive all these approaches step by step, highlighting the underlying approximations and limitations.

This second edition maintains the same pedagogical style as the first edition. The book remains a self-contained and self-learning book for masters and PhD students and a textbook for undergraduate courses. As in the first edition, *there is not a single result which is not derived.*

Lastly but not least importantly, we want to express our gratitude to our students over the past 10 years, particularly Simone Latini, Fabio Covito, Francesco Fantini, Simone Manti, Paolo Gazzaneo, Silvia Bianchi, Tommaso Mazzocchi, Kai Wu, Zhenlin Zhang, and Alessandro Moreci, for their valuable suggestions and observations. Their input has greatly contributed to refining the presentation of various topics year after year. Any errors that may remain are solely the responsibility of the authors.

Preface to the First Edition

This textbook contains a pedagogical introduction to the theory of Green's functions *in* and *out* of equilibrium, and is accessible to students with a standard background in basic quantum mechanics and complex analysis. Two main motivations prompted us to write a monograph for beginners on this topic.

The first motivation is research-oriented. With the advent of nanoscale physics and ultrafast lasers it became possible to probe the correlation between particles in excited quantum states. New fields of research (e.g., molecular transport, nanoelectronics, Josephson nanojunctions, attosecond physics, nonequilibrium phase transitions, ultracold atomic gases in optical traps, optimal control theory, kinetics of Bose condensates, quantum computation) added up to the already existing fields in mesoscopic physics and nuclear physics. The Green's function method is probably one of the most powerful and versatile formalisms in physics, and its nonequilibrium version has already proven to be extremely useful in several of the aforementioned contexts. Extending the method to deal with the new emerging nonequilibrium phenomena holds promise to facilitate and quicken our comprehension of the excited state properties of matter. At present, unfortunately, to learn the nonequilibrium Green's function formalism requires more effort than learning the equilibrium (zero-temperature or Matsubara) formalism, despite the fact that *nonequilibrium Green's functions are not more difficult.* This brings us to the second motivation.

The second motivation is educational in nature. As students we had to learn the method of Green's functions at zero temperature, with the normal-orderings and contractions of Wick's theorem, the adiabatic switching-on of the interaction, the Gell-Mann–Low theorem, the Feynman diagrams, etc. Then we had to learn the finite-temperature or Matsubara formalism where there is no need of normal-orderings to prove Wick's theorem, and where it is possible to prove a diagrammatic expansion without the adiabatic switching-on and the Gell-Mann–Low theorem. The Matsubara formalism is often taught as a disconnected topic but the diagrammatic expansion is exactly the same as that of the zero-temperature formalism. Why do the two formalisms look the same? Why do we need more "assumptions" in the zero-temperature formalism? And isn't it enough to study the finite-temperature formalism? After all, zero-temperature is just one possible temperature. When we became postdocs we bumped into yet another version of Green's functions, the nonequilibrium Green's functions or the so-called Keldysh formalism. And again another different way to prove Wick's theorem and the diagrammatic expansion. Furthermore, while several excellent textbooks on the equilibrium formalisms are available, here the learning process is considerably slowed down by the absence of introductory textbooks. There exist a few review articles on the

Keldysh formalism, but they are scattered over the years and the journals. Students have to face different jargon and different notation, dig out original papers (not all downloadable from the web), and have to find the answer to lots of typical newcomer questions: Why is the diagrammatic expansion of the Keldysh formalism again the same as that of the zero-temperature and Matsubara formalisms? How do we see that the Keldysh formalism reduces to the zero-temperature formalism in equilibrium? How to introduce the temperature in the Keldysh formalism? It is easy to imagine the frustration of many students during their early days of study of nonequilibrium Green's functions. In this book we will introduce only *one* formalism, which we can call the *contour formalism*, and we will do it using a very pedagogical style. The contour formalism is not more difficult than the zero-temperature, Matsubara or Keldysh formalisms and we will explicitly show how it reduces to those under special conditions. Furthermore, the contour formalism provides a natural answer to all previous questions. Thus, the message is: *There is no need to learn the same thing three times.*

Starting from basic quantum mechanics, we introduce the contour Green's function formalism step by step. The physical content of the Green's function is discussed with particular attention to the time-dependent aspect and applied to different physical systems ranging from molecules and nanostructures to metals and insulators. With this powerful tool at our disposal we then go through the Feynman diagrams, the theory of conserving approximations, the Kadanoff–Baym equations, the Luttinger–Ward variational functionals, the Bethe–Salpeter equation, and the Hedin equations.

This book is not a collection of chapters on different applications, but a self-contained introduction to mathematical and physical concepts of general use. We made a serious effort in organizing apparently disconnected topics in a *logical* instead of *chronological* way, and in filling up many small gaps. The adjective "modern" in the title refers to the presentation more than to specific applications. The overall goal of the present book is to derive a set of kinetic equations governing the quantum dynamics of many identical particles and to develop perturbative as well as nonperturbative approximation schemes for their solution.

About 600 pages may seem too many for a textbook on Green's functions, so let us justify this voluminousness. First of all, *there is not a single result which is not derived.* This means that we inserted several intermediate steps to guide the reader through every calculation. Second, for every formal development or new mathematical quantity we present carefully selected examples which illustrate the physical content of what we are doing. Without examples and illustrations (more than 250 figures) this book would be half the size but the actual understanding would probably be much less. The large number of examples compensates for the moderate number of exercises. Third, in the effort of writing a comprehensive presentation of the various topics, we came across several small subtleties which, if not addressed and properly explained, could give rise to serious misunderstandings. We therefore added many remarks and clarifying discussions throughout the text.

The structure of the book is illustrated in Fig. 1 and can be roughly partitioned in three parts: mathematical tools, approximation schemes, and applications. For the detailed list of topics the reader can have a look at the table of contents. Of course, the choice of topics reflects our personal background and preferences. However, we feel reasonably confident to have covered all fundamental aspects of Green's function theory in and out of equilibrium. We tried to create a self-contained and self-study book capable of bringing

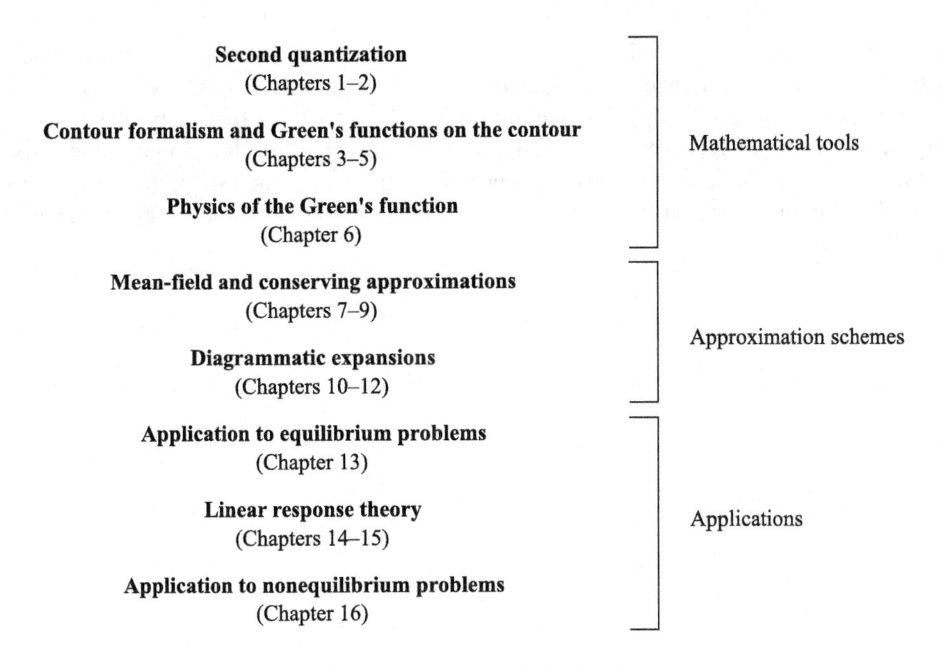

Figure 1 Structure of the book.

the undergraduate or PhD student to the level of approaching the modern literature and enabling him/her to model or solve new problems with physically justified approximations. If we are successful in this endeavor, it will be due to the enthusiastic and motivated students in Rome and Jyväskylä to whom we had the privilege to teach part of this book. We thank them for their feedback, from which we indeed benefited enormously.

Speaking of thanks: Our first and biggest thanks goes to Carl-Olof Almbladh and Ulf von Barth, who introduced us to the wonderful world of many-body perturbation theory and Green's function theory during our postdoc years in Lund. Only now that we were forced to deepen our understanding in order to explain these methods could we fully appreciate all their "of-course-I-don't-need-to-tell-you" or "you-probably-already-know" answers to our questions. We are also thankful to Evert Jan Baerends, Michele Cini, and Hardy Gross, from whom we learned a large part of what today is our background in physics and chemistry and with whom we undertook many exciting research projects. We wish to express our gratitude to our PhD students, postdocs, and local colleagues Klaas Giesbertz, Petri Myöhänen, Enrico Perfetto, Michael Ruggenthaler, Niko Säkkinen, Adrian Stan, Riku Tuovinen, and Anna-Maija Uimonen, for providing us with many valuable suggestions and for helping out in generating several figures. The research on the Kadanoff–Baym equations and their implementation which forms the last chapter of the book would not have been possible without the enthusiasm and the excellent numerical work of Nils Erik Dahlen. We are indebted to Heiko Appel, Karsten Balzer, Michael Bonitz, Raffaele Filosofi, Ari Harju, Maria Hellgren, Stefan Kurth, Matti Manninen, Kristian Thygesen, and Claudio Verdozzi, with whom we had many inspiring and insightful discussions which either directly or indirectly influenced part of the contents of the book. We further thank the Department of Physics and the Nanoscience Center of the University of Jyväskylä and the Department of Physics of the University of Rome Tor Vergata for creating a very pleasant and supportive environment for the writing of the book. Finally, we would like to thank a large number of people, too numerous to mention, in the research community who have shaped our view on many scientific topics in and outside of many-body theory.

Abbreviations and Acronyms

ARPES	angle-resolved photoemission spectroscopy
a.u.	atomic units
BvK	Born–von Karman
GKBA	Generalized Kadanoff–Baym Ansatz
h.c.	Hermitian conjugate
HF	Hartree–Fock
HOMO	highest occupied molecular orbital
HSEX	Hartree plus screened exchange
KMS	Kubo–Martin–Schwinger
LUMO	lowest unoccupied molecular orbital
LW	Luttinger–Ward
MBPT	many-body perturbation theory
NEGF	nonequilibrium Green's function
PES	photoemission spectroscopy
ph	particle–hole
PPP	Pariser–Parr–Pople
QMC	Quantum Monte Carlo
RPA	Random Phase Approximation
Tph	T-matrix approximation in the particle–hole channel
Tpp	T-matrix approximation in the particle–particle channel
WBLA	Wide-Band Limit Approximation
XC	exchange-correlation

Fundamental Constants and Basic Relations

Fundamental Constants

Electron charge: $e = -1$ a.u. $= 1.60217646 \times 10^{-19}$ coulomb

Electron mass: $m_e = 1$ a.u. $= 9.10938188 \times 10^{-31}$ kg

Planck constant: $\hbar = 1$ a.u. $= 1.054571 \times 10^{-34}$ J s = 6.58211×10^{-16} eV s

Speed of light: $c = 137$ a.u. $= 3 \times 10^5$ km/s

Boltzmann constant: $K_\mathrm{B} = 8.3 \times 10^{-5}$ eV/K

Basic Quantities and Relations

Bohr radius: $a_\mathrm{B} = \frac{\hbar^2}{m_e e^2} = 1$ a.u. $= 0.5$ Å

Electron gas density: $n = \frac{(\hbar p_\mathrm{F})^3}{3\pi^2} =$ (p_F being the Fermi momentum)

Electron gas radius: $\frac{1}{n} = \frac{4\pi}{3}(a_\mathrm{B} r_s)^3$, $r_s = \frac{(9\pi/4)^{1/3}}{\hbar a_\mathrm{B} p_\mathrm{F}}$

Plasma frequency: $\omega_\mathrm{p} = \sqrt{\frac{4\pi e^2 n}{m_e}}$ (n being the electron gas density)

Rydberg $R = \frac{e^2}{2a_\mathrm{B}} = 0.5$ a.u. $\simeq 13.6$ eV

Bohr magneton $\mu_\mathrm{B} = \frac{e\hbar}{2m_e c} = 3.649 \times 10^{-3}$ a.u. $= 5.788 \times 10^{-5}$ eV/T

Room temperature ($T \sim 300$ K) energy: $K_B T \sim \frac{1}{40}$ eV

$\hbar c \sim 197$ MeV fm (1 fm = 10^{-15} m)

$m_e c^2 = 0.5447$ MeV

1

Second Quantization

In this chapter we revisit the quantum mechanical description of one-particle systems and many-particle systems. We highlight the differences between distinguishable and indistinguishable, or identical, particles and bring to the front the mathematical complications that arise when dealing with identical particles. We then introduce the second quantization formalism and show how to overcome these complications. The main actors of the second quantization formalism are the field operators, which can be used to represent states and quantum observables in the Hilbert space of identical particles.

1.1 Quantum Mechanics of One Particle

In quantum mechanics the physical state of a particle is described in terms of a *ket* $|\Psi\rangle$. This ket belongs to a *Hilbert space*, which is nothing but a vector space endowed with an inner product. The dimension of the Hilbert space is essentially fixed by our physical intuition; it is us who decide which kets are relevant to the description of the particle. For instance, if we want to describe how a laser works we can choose those energy eigenkets that get populated and depopulated, and discard the rest. This selection of states leads to the well-known description of a laser in terms of a three-level system, four-level system, etc. A fundamental property following from the vector nature of the Hilbert space is that any linear superposition of kets is another ket in the Hilbert space. In other words, we can make a linear superposition of physical states and the result is another physical state. In quantum mechanics, however, it is only the "direction" of the ket that matters, so $|\Psi\rangle$ and $C|\Psi\rangle$ represent the same physical state for all complex numbers C. This redundancy prompts us to work with *normalized* kets. What do we mean by that? We said before that there is an inner product in the Hilbert space. Let us denote by $\langle\Phi|\Psi\rangle = \langle\Psi|\Phi\rangle^*$ the inner product between two kets $|\Psi\rangle$ and $|\Phi\rangle$ of the Hilbert space. Then every ket has a real positive inner product with itself,

$$0 < \langle\Psi|\Psi\rangle < \infty.$$

A ket is said to be normalized if the inner product with itself is 1. Throughout this book we always assume that a ket is normalized unless otherwise stated. Every ket can be normalized by choosing the complex constant $C = e^{\mathrm{i}\alpha}/\sqrt{\langle\Psi|\Psi\rangle}$ with α an arbitrary real number. Thus, the normalization fixes the ket of a physical state only modulo a phase factor. As we see in Section 1.3, this freedom is the basis of a fundamental property about the nature

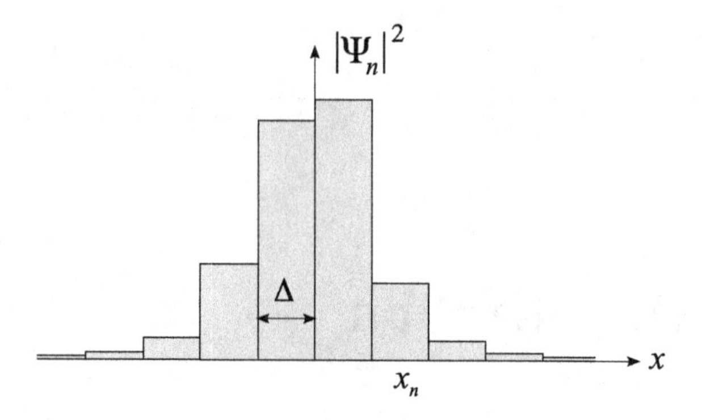

Figure 1.1 Histogram of the normalized number of clicks of the detector in $x_n = n\Delta$. The height of the bars corresponds to the probabilities $|\Psi_n|^2$.

of elementary particles. The notion of inner product also allows us to define the *dual space* as the vector space of linear operators $\langle\Phi|$, which deliver the complex number $\langle\Phi|\Psi\rangle$ when acting on the ket $|\Psi\rangle$. The elements of the dual space are called *bra*, and we can think of the inner product as the action of a bra on a ket. The formulation of quantum mechanics in terms of bras and kets is due to Dirac [1, 2] and turns out to be extremely useful.

According to the basic principles of quantum mechanics [2],

- With every physical observable is associated a Hermitian operator whose eigenvalues λ represent the outcome of an experimental measurement of the observable.
- If the particle is described by the ket $|\Psi\rangle$, then the probability of measuring λ is given by
$$P(\lambda) = |\langle\lambda|\Psi\rangle|^2,$$
where $|\lambda\rangle$ is the eigenket of the operator with eigenvalue λ.
- The experimental measurement is so invasive that just after measurement the particle *collapses* in the ket $|\lambda\rangle$.

Let us discuss the implications of these principles with an example.

Discrete formulation Suppose that we want to measure the position of a particle living in a one-dimensional world. We can construct a detector with the property that it clicks whenever the particle is no further away than, say, $\Delta/2$ from the position of the detector. We distribute these detectors on a uniform grid $x_n = n\Delta$, with n integers, so as to cover the entire one-dimensional world. The experiment consists in preparing the particle in a state $|\Psi\rangle$ and in taking note of which detector clicks. After the click, we know for sure that the particle is in the interval $x_n \pm \Delta/2$, where x_n is the position of the detector that clicked. Repeating the experiment $N \gg 1$ times, counting the number of times that a given detector clicks, and dividing the result by N, we obtain the probability that the particle is in the interval $x_n \pm \Delta/2$, see histogram in Fig. 1.1. Quantum mechanics tells us that this probability is
$$P(n) = |\langle n|\Psi\rangle|^2,$$

where $|n\rangle$ is the ket describing the particle in the interval $x_n \pm \Delta/2$. The experimental setup does not allow us to say where exactly the particle is within this interval. In fact, it does not make sense to speak about the exact position of the particle since it cannot be measured. From the experimental output we could even argue that the one-dimensional world is discrete! What we want to say is that in our experiment the "exact position" of the particle is a mere speculative concept, like the gender, color, or happiness of the particle. These degrees of freedom may also exist, but if they cannot be measured then we should not include them in the description of the physical world. As scientists we can only assign a ket $|n\rangle$ to the state of the particle just after measurement, and we can interpret this ket as describing the particle in some discrete position. The probability of finding the particle in $|n'\rangle$ just after the nth detector has clicked is zero for all $n' \neq n$ and unity for $n' = n$, and hence,

$$\langle n'|n\rangle = \delta_{n'n}. \tag{1.1}$$

The kets $|n\rangle$ are orthonormal and it is easy to show that they form a basis of our Hilbert space. Suppose by absurdum that there exists another ket $|\chi\rangle$ orthogonal to all the $|n\rangle$. If the particle is described by this ket then the probability that the nth detector clicks is $|\langle n|\chi\rangle|^2 = 0$ for all n. This cannot be the case unless the particle is somewhere outside the one-dimensional world – that is, in a state not included in our original description.

Let us continue to elaborate on the example of the particle in a one-dimensional world. We said before, that the kets $|n\rangle$ form a basis. Therefore, any ket $|\Psi\rangle$ can be expanded as

$$|\Psi\rangle = \sum_n \Psi_n |n\rangle. \tag{1.2}$$

Since the basis is orthonormal, the coefficient Ψ_n is simply

$$\Psi_n = \langle n|\Psi\rangle, \tag{1.3}$$

and its square modulus is exactly the probability $P(n)$:

$$|\Psi_n|^2 = \left(\begin{array}{c} \text{probability of finding the particle in} \\ \text{volume element } \Delta \text{ around } x_n \end{array} \right).$$

It is important to appreciate the advantage of working with normalized kets. Since $\langle\Psi|\Psi\rangle = 1$, then

$$\sum_n |\Psi_n|^2 = 1, \tag{1.4}$$

according to which the probability of finding the particle anywhere is unity. The interpretation of the $|\Psi_n|^2$ as probabilities would not be possible if $|\Psi\rangle$ and $|n\rangle$ were not normalized.

Given an orthonormal basis, the inner product of a normalized ket $|\Psi\rangle$ with a basis ket gives the probability amplitude of having the particle in that ket.

Inserting (1.3) back into (1.2), we find the interesting relation

$$|\Psi\rangle = \sum_n \langle n|\Psi\rangle\, |n\rangle = \sum_n |n\rangle\langle n|\Psi\rangle.$$

This relation is interesting because it is true for all $|\Psi\rangle$ and hence

$$\sum_n |n\rangle\langle n| = \hat{\mathbb{1}}, \tag{1.5}$$

with $\hat{\mathbb{1}}$ the identity operator. Equation (1.5) is known as the *completeness relation* and expresses the fact that the set $\{|n\rangle\}$ is an orthonormal basis. Vice versa, any orthonormal basis satisfies the completeness relation.

Continuum formulation We now assume that we can construct more and more precise detectors and hence reduce the range Δ. Then we can also refine the description of our particle by putting the detectors closer and closer. In the limit $\Delta \to 0$, the probability $|\Psi_n|^2$ approaches zero and it makes more sense to reason in terms of the *probability density* $|\Psi_n|^2/\Delta$ of finding the particle in x_n. Let us rewrite (1.2) as

$$|\Psi\rangle = \Delta \sum_n \frac{\Psi_n}{\sqrt{\Delta}} \frac{|n\rangle}{\sqrt{\Delta}}. \tag{1.6}$$

We now define the continuous function $\Psi(x_n)$ and the continuous ket $|x_n\rangle$ as

$$\Psi(x_n) \equiv \lim_{\Delta \to 0} \frac{\Psi_n}{\sqrt{\Delta}}, \qquad |x_n\rangle = \lim_{\Delta \to 0} \frac{|n\rangle}{\sqrt{\Delta}}.$$

In this definition the limiting function $\Psi(x_n)$ is well defined, while the limiting ket $|x_n\rangle$ makes *mathematical sense* only under an integral sign since the norm $\langle x_n|x_n\rangle = \infty$. However, we can still give to $|x_n\rangle$ a precise *physical meaning* since in quantum mechanics only the "direction" of a ket matters.[1] With these definitions (1.6) can be seen as the Riemann sum of $\Psi(x_n)|x_n\rangle$. In the limit $\Delta \to 0$ the sum becomes an integral over x, and we can write

$$|\Psi\rangle = \int dx\, \Psi(x)|x\rangle.$$

The function $\Psi(x)$ is usually called the *wavefunction* or the *probability amplitude,* and its square modulus $|\Psi(x)|^2$ is the probability density of finding the particle in x, or equivalently

$$|\Psi(x)|^2\, dx = \left(\begin{array}{c} \text{probability of finding the particle} \\ \text{in volume element } dx \text{ around } x \end{array} \right).$$

In the continuum formulation the orthonormality relation (1.1) becomes

$$\langle x_{n'}|x_n\rangle = \lim_{\Delta \to 0} \frac{\delta_{n'n}}{\Delta} = \delta(x_{n'} - x_n),$$

where $\delta(x)$ is the Dirac δ-function, see Appendix A. Similarly, the completeness relation becomes

$$\int dx\, |x\rangle\langle x| = \hat{\mathbb{1}}.$$

[1] The formulation of quantum mechanics using nonnormalizable states requires the extension of Hilbert spaces to *rigged Hilbert spaces.* Readers interested in the mathematical foundations of this extension can consult, for example, Ref. [3]. Here we simply note that in a rigged Hilbert space everything works as in the more familiar Hilbert space. We simply have to keep in mind that every divergent quantity comes from some continuous limit and that in all physical quantities the divergency is canceled by an infinitesimally small quantity.

The entire discussion can easily be generalized to particles with spin in three (or any other) dimension. Let us denote by $\mathbf{x} = (\mathbf{r}\sigma)$ the collective index for the position $\mathbf{r}$ and the spin projection (say along the z axis) σ of the particle. If in every point of space we put a spin-polarized detector which clicks only if the particle has spin σ then $|\mathbf{x}\rangle$ is the state of the particle just after the spin-polarized detector in $\mathbf{r}$ has clicked. The *position–spin kets* $|\mathbf{x}\rangle$ are orthonormal

$$\langle \mathbf{x}'|\mathbf{x}\rangle = \delta_{\sigma'\sigma}\delta(\mathbf{r}' - \mathbf{r}) \equiv \delta(\mathbf{x}' - \mathbf{x}), \tag{1.7}$$

and form a basis. Hence they satisfy the completeness relation, which in this case reads

$$\boxed{\int d\mathbf{x}\; |\mathbf{x}\rangle\langle \mathbf{x}| = \hat{\mathbb{1}}} \tag{1.8}$$

Here and in the remainder of the book we use the symbol

$$\int d\mathbf{x} \equiv \sum_{\sigma} \int d\mathbf{r}$$

to signify a sum over spin and an integral over space. The expansion of a ket in this continuous Hilbert space follows directly from the completeness relation

$$|\Psi\rangle = \hat{\mathbb{1}}|\Psi\rangle = \int d\mathbf{x}\; |\mathbf{x}\rangle\langle \mathbf{x}|\Psi\rangle,$$

and the square modulus of the wavefunction $\Psi(\mathbf{x}) \equiv \langle \mathbf{x}|\Psi\rangle$ is the probability density of finding the particle in $\mathbf{x} = (\mathbf{r}\sigma)$:

$$|\Psi(\mathbf{x})|^2\, d\mathbf{r} \;=\; \left(\begin{array}{c} \text{probability of finding the particle with spin } \sigma \\ \text{in volume element } d\mathbf{r} \text{ around } \mathbf{r} \end{array} \right).$$

Operators So far we have only discussed the possible states of the particle, and the physical interpretation of the expansion coefficients. To say something about the dynamics of the particle, we must know the Hamiltonian operator $\hat{h}$. The knowledge of the Hamiltonian in quantum mechanics is analogous to knowledge of the forces in Newtonian mechanics. In Newtonian mechanics the dynamics of the particle is completely determined by the position and velocity at a certain time and by the forces. In quantum mechanics the dynamics of the wavefunction is completely determined by the wavefunction at a certain time and by $\hat{h}$. The Hamiltonian operator $\hat{h} \equiv h(\hat{\boldsymbol{r}}, \hat{\boldsymbol{p}}, \hat{\boldsymbol{S}})$ does, in general, depend on the position operator $\hat{\boldsymbol{r}}$, the momentum operator $\hat{\boldsymbol{p}}$, and the spin operator $\hat{\boldsymbol{S}}$. An example is the Hamiltonian for a particle of mass m, charge q, and gyromagnetic ratio g moving in an external scalar potential ϕ, vector potential $\mathbf{A}$, and whose spin is coupled to the magnetic field $\mathbf{B} = \boldsymbol{\nabla} \times \mathbf{A}$:

$$\hat{h} = \frac{1}{2m}\left(\hat{\boldsymbol{p}} - \frac{q}{c}\mathbf{A}(\hat{\boldsymbol{r}})\right)^2 + q\phi(\hat{\boldsymbol{r}}) - g\mu_{\mathrm{B}}\mathbf{B}(\hat{\boldsymbol{r}}) \cdot \hat{\boldsymbol{S}}, \tag{1.9}$$

with c the speed of light and μ_{B} the Bohr magneton.[2] Unless otherwise stated in this book we use atomic units, so $\hbar = 1$, $c \sim 137$, electron charge $e = -1$, and electron mass

[2] Other relativistic corrections like the spin–orbit interaction can be incorporated without any conceptual complication.

$m_e = 1$. Thus, in (1.9) the Bohr magneton $\mu_B = \frac{e\hbar}{2m_e c} \sim 3.649 \times 10^{-3}$, and charge and mass of the particles are measured in units of e and m_e, respectively. To distinguish operators from scalar or matrix quantities we always put the symbol " ^ " (read "hat") on them. The position–spin kets are eigenstates of the position operator and of the z-component of the spin operator:

$$\hat{\boldsymbol{r}}|\mathbf{x}\rangle = \mathbf{r}|\mathbf{x}\rangle, \qquad \hat{S}_z|\mathbf{x}\rangle = \sigma|\mathbf{x}\rangle,$$

with $\sigma = -S, -S+1, \ldots, S-1, S$ for spin S particles. The eigenstates of the momentum operator are instead the *momentum–spin kets* $|\mathbf{p}\sigma\rangle$:

$$\hat{\boldsymbol{p}}|\mathbf{p}\sigma\rangle = \mathbf{p}|\mathbf{p}\sigma\rangle.$$

These kets are also eigenstates of $\hat{S}_z$ with eigenvalue σ. The momentum–spin kets form an orthonormal basis like the position–spin kets. The inner product between $|\mathbf{x}\rangle = |\mathbf{r}\sigma\rangle$ and $|\mathbf{p}\sigma'\rangle$ is proportional to $\delta_{\sigma\sigma'}$ times the plane wave $e^{\mathrm{i}\mathbf{p}\cdot\mathbf{r}}$. In this book we choose the constant of proportionality to be unity, so that

$$\boxed{\langle\mathbf{x}|\mathbf{p}\sigma'\rangle = \delta_{\sigma\sigma'}\langle\mathbf{r}|\mathbf{p}\rangle \qquad \text{with} \qquad \langle\mathbf{r}|\mathbf{p}\rangle = e^{\mathrm{i}\mathbf{p}\cdot\mathbf{r}}} \tag{1.10}$$

This inner product fixes uniquely the form of the completeness relation for the kets $|\mathbf{p}\sigma\rangle$. We have

$$\begin{aligned}\langle\mathbf{p}'\sigma'|\mathbf{p}\sigma\rangle &= \delta_{\sigma'\sigma}\langle\mathbf{p}'|\mathbf{p}\rangle = \delta_{\sigma'\sigma}\int d\mathbf{r}\,\langle\mathbf{p}'|\mathbf{r}\rangle\langle\mathbf{r}|\mathbf{p}\rangle = \delta_{\sigma'\sigma}\int d\mathbf{r}\, e^{\mathrm{i}(\mathbf{p}-\mathbf{p}')\cdot\mathbf{r}} \\ &= (2\pi)^3\delta_{\sigma'\sigma}\delta(\mathbf{p}'-\mathbf{p}),\end{aligned}$$

and therefore

$$\boxed{\sum_\sigma \int \frac{d\mathbf{p}}{(2\pi)^3}|\mathbf{p}\sigma\rangle\langle\mathbf{p}\sigma| = \hat{\mathbb{1}}} \tag{1.11}$$

as can easily be verified by acting with (1.11) on the ket $|\mathbf{p}'\sigma'\rangle$ or on the bra $\langle\mathbf{p}'\sigma'|$.

Before moving to the quantum mechanical description of many particles, let us briefly recall how to calculate the matrix elements of the Hamiltonian $\hat{h}$ in the position–spin basis. If $|\Psi\rangle$ is the ket of the particle, then

$$\langle\mathbf{x}|\hat{\boldsymbol{p}}|\Psi\rangle = -\mathrm{i}\boldsymbol{\nabla}\langle\mathbf{x}|\Psi\rangle \quad \Rightarrow \quad \langle\Psi|\hat{\boldsymbol{p}}|\mathbf{x}\rangle = \mathrm{i}\langle\Psi|\mathbf{x}\rangle\overleftarrow{\boldsymbol{\nabla}},$$

where the arrow over the gradient specifies that $\boldsymbol{\nabla}$ acts on the quantity to its left. It follows from these identities that

$$\langle\mathbf{x}|\hat{\boldsymbol{p}}|\mathbf{x}'\rangle = -\mathrm{i}\delta_{\sigma\sigma'}\boldsymbol{\nabla}\delta(\mathbf{r}-\mathbf{r}') = \mathrm{i}\delta_{\sigma\sigma'}\delta(\mathbf{r}-\mathbf{r}')\overleftarrow{\boldsymbol{\nabla}}', \tag{1.12}$$

where $\boldsymbol{\nabla}'$ means that the gradient acts on the primed variable. Therefore, the matrix element $\langle\mathbf{x}|\hat{h}|\mathbf{x}'\rangle$ with $\hat{h} = h(\hat{\boldsymbol{r}}, \hat{\boldsymbol{p}}, \hat{\boldsymbol{S}})$ can be written as

$$\boxed{\langle\mathbf{x}|\hat{h}|\mathbf{x}'\rangle = h_{\sigma\sigma'}(\mathbf{r}, -\mathrm{i}\boldsymbol{\nabla}, \mathbf{S})\delta(\mathbf{r}-\mathbf{r}') = \delta(\mathbf{r}-\mathbf{r}')h_{\sigma\sigma'}(\mathbf{r}', \mathrm{i}\overleftarrow{\boldsymbol{\nabla}}', \mathbf{S})} \tag{1.13}$$

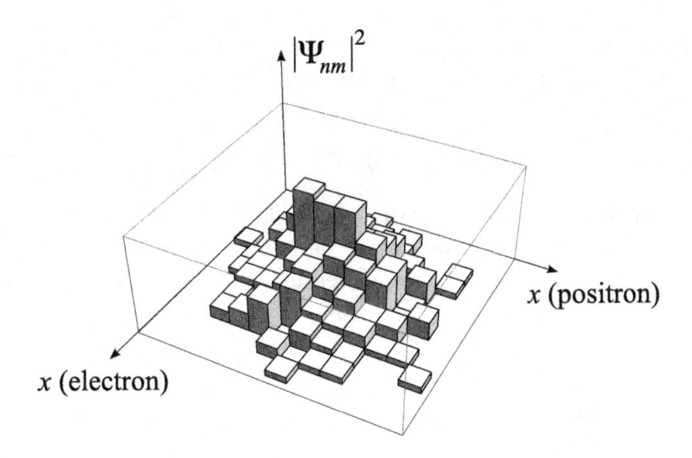

Figure 1.2 Histogram of the normalized number of simultaneous clicks of the electron and positron detectors in $x_n = n\Delta$ and $x_m = m\Delta$, respectively. The height of the parallelepipeds corresponds to the probabilities $|\Psi_{nm}|^2$.

where $\mathbf{S}$ is the matrix of the spin operator with elements $\langle\sigma|\hat{\boldsymbol{S}}|\sigma'\rangle = \mathbf{S}_{\sigma\sigma'}$. For example, for the one-particle Hamiltonian in (1.9) we have

$$h_{\sigma\sigma'}(\mathbf{r}, -\mathrm{i}\boldsymbol{\nabla}, \mathbf{S}) = \frac{\delta_{\sigma\sigma'}}{2m}\left(-\mathrm{i}\boldsymbol{\nabla} - \frac{q}{c}\mathbf{A}(\mathbf{r})\right)^2 + \delta_{\sigma\sigma'}q\phi(\mathbf{r}) - g\mu_{\mathrm{B}}\mathbf{B}(\mathbf{r})\cdot\mathbf{S}_{\sigma\sigma'}.$$

We use (1.13) over and over in the following chapters to recognize the matrix structure of several equations.

1.2 Quantum Mechanics of Many Particles

We want to generalize the concepts of the previous section to many particles. Let us first discuss the case of *distinguishable particles*. Particles are called distinguishable if one or more of their properties, such as mass, charge, spin, etc., are different. Let us consider, for instance, an electron and a positron in one dimension. These particles are distinguishable since the charge of the positron is opposite to the charge of the electron.

Discrete formulation for two particles To measure the position of the electron and the position of the positron at a certain time, we put an electron detector and a positron detector at every point $x_n = n\Delta$ of the real axis and perform a *coincidence experiment*. This means that we take note of the position of the electron detector and of the positron detector only if they click *at the same time*. The result of the experiment is the pair of points (x_n, x_m), where x_n refers to the electron and x_m refers to the positron. Performing the experiment $N \gg 1$ times, counting the number of times that the pair (x_n, x_m) is measured and dividing the result by N, we obtain the probability that the electron is in x_n and the positron in x_m, see the histogram in Fig. 1.2. According to quantum mechanics, the electron–positron pair collapses in the ket $|n\rangle|m\rangle$ just after measurement. This ket describes an electron in the interval $x_n \pm \Delta/2$ and a positron in the interval $x_m \pm \Delta/2$. Therefore, the probability of finding the electron-positron pair in $|n'\rangle|m'\rangle$ is zero unless $n' = n$ and $m' = m$; that is,

$$(\,\langle n'|\langle m'|\,)\,(\,|n\rangle|m\rangle\,) = \delta_{n'n}\delta_{m'm}.$$

The kets $|n\rangle|m\rangle$ are orthonormal and form a basis since if there was a ket $|\chi\rangle$ orthogonal to all of them then the electron-positron pair described by $|\chi\rangle$ would not be on the real axis. The orthonormality of the basis is expressed by the completeness relation

$$\sum_{nm}(\,|n\rangle|m\rangle\,)\,(\,\langle n|\langle m|\,) = \hat{\mathbb{1}}.$$

This relation can be used to expand any ket as

$$|\Psi\rangle = \hat{\mathbb{1}}|\Psi\rangle = \sum_{nm}(\,|n\rangle|m\rangle\,)\,(\,\langle n|\langle m|\,)\,|\Psi\rangle,$$

and if $|\Psi\rangle$ is normalized then the square modulus of the coefficients $\Psi_{nm} \equiv (\,\langle n|\langle m|\,)\,|\Psi\rangle$ is the probability represented in the histogram.

Continuum formulation for two particles As in the previous section, we could refine the experiment by putting the detectors closer and closer. We could also rethink the entire experiment in three (or any other) dimensions and use spin-polarized detectors. We then arrive at the position–spin kets $|\mathbf{x}_1\rangle|\mathbf{x}_2\rangle$ for the electron–positron pair with inner product

$$(\,\langle \mathbf{x}'_1|\langle \mathbf{x}'_2|\,)\,(\,|\mathbf{x}_1\rangle|\mathbf{x}_2\rangle\,) = \delta(\mathbf{x}'_1 - \mathbf{x}_1)\delta(\mathbf{x}'_2 - \mathbf{x}_2),$$

from which we deduce the completeness relation

$$\int d\mathbf{x}_1 d\mathbf{x}_2\,(\,|\mathbf{x}_1\rangle|\mathbf{x}_2\rangle\,)\,(\,\langle \mathbf{x}_1|\langle \mathbf{x}_2|\,) = \hat{\mathbb{1}}.$$

The expansion of a generic ket is

$$|\Psi\rangle = \int d\mathbf{x}_1 d\mathbf{x}_2\,(\,|\mathbf{x}_1\rangle|\mathbf{x}_2\rangle\,)\,(\,\langle \mathbf{x}_1|\langle \mathbf{x}_2|\,)\,|\Psi\rangle,$$

and if $|\Psi\rangle$ is normalized then the square modulus of the wavefunction $\Psi(\mathbf{x}_1, \mathbf{x}_2) \equiv (\,\langle \mathbf{x}_1|\langle \mathbf{x}_2|\,)\,|\Psi\rangle$ yields the probability density of finding the electron in $\mathbf{x}_1 = (\mathbf{r}_1\sigma_1)$ and the positron in $\mathbf{x}_2 = (\mathbf{r}_2\sigma_2)$:

$$|\Psi(\mathbf{x}_1, \mathbf{x}_2)|^2\,d\mathbf{r}_1 d\mathbf{r}_2 = \left(\begin{array}{c}\text{probability of finding the electron with spin } \sigma_1 \\ \text{in volume element } d\mathbf{r}_1 \text{ around } \mathbf{r}_1 \text{ and the positron} \\ \text{with spin } \sigma_2 \text{ in volume element } d\mathbf{r}_2 \text{ around } \mathbf{r}_2\end{array}\right).$$

Continuum formulation for N particles The generalization to N distinguishable particles is straightforward. The position–spin ket $|\mathbf{x}_1\rangle \ldots |\mathbf{x}_N\rangle$ describes the physical state in which the first particle is in $\mathbf{x}_1$, the second particle is in $|\mathbf{x}_2\rangle$, etc. These kets form an orthonormal basis with inner product

$$(\,\langle \mathbf{x}'_1| \ldots \langle \mathbf{x}'_N|\,)\,(\,|\mathbf{x}_1\rangle \ldots |\mathbf{x}_N\rangle\,) = \delta(\mathbf{x}'_1 - \mathbf{x}_1) \ldots \delta(\mathbf{x}'_N - \mathbf{x}_N), \tag{1.14}$$

and therefore the completeness relation reads

$$\int d\mathbf{x}_1 \ldots d\mathbf{x}_N \, (\, |\mathbf{x}_1\rangle \ldots |\mathbf{x}_N\rangle \,) \, (\, \langle \mathbf{x}_1| \ldots \langle \mathbf{x}_N| \,) = \hat{\mathbb{1}}.$$

Operators Having discussed the Hilbert space for N distinguishable particles, we now consider the operators acting on the N-particle kets. We start with an example and consider again the electron–positron pair. Suppose that there is an electric field $\mathbf{E}(\mathbf{r}) = -\boldsymbol{\nabla}\phi(\mathbf{r})$ extending across all of space and that we are interested in measuring the total potential energy. This is an observable quantity and, hence, associated with it there exists an operator $\hat{\mathcal{H}}_{\text{pot}}$. By definition the eigenstates of this operator are the position–spin kets $|\mathbf{x}_1\rangle|\mathbf{x}_2\rangle$ and the corresponding eigenvalues are $-\phi(\mathbf{r}_1)+\phi(\mathbf{r}_2)$, independent of the spin of the particles (in atomic units the charge of the electron is $q=-1$ and hence the charge of the positron is $q=+1$). The operator $\hat{\mathcal{H}}_{\text{pot}}$ is then the sum of the electrostatic potential operator acting on the first particle and doing nothing to the second particle and the electrostatic potential operator acting on the second particle and doing nothing to the first particle:

$$\hat{\mathcal{H}}_{\text{pot}} = -\phi(\hat{\boldsymbol{r}}) \otimes \hat{\mathbb{1}} + \hat{\mathbb{1}} \otimes \phi(\hat{\boldsymbol{r}}). \tag{1.15}$$

The symbol $\otimes$ denotes the *tensor product* of operators acting on different particles:

$$\hat{\mathcal{H}}_{\text{pot}}|\mathbf{x}_1\rangle|\mathbf{x}_2\rangle = -\phi(\hat{\boldsymbol{r}})|\mathbf{x}_1\rangle\hat{\mathbb{1}}|\mathbf{x}_2\rangle + \hat{\mathbb{1}}|\mathbf{x}_1\rangle\phi(\hat{\boldsymbol{r}})|\mathbf{x}_2\rangle = \big[-\phi(\mathbf{r}_1)+\phi(\mathbf{r}_2)\big]|\mathbf{x}_1\rangle|\mathbf{x}_2\rangle.$$

The generalization of the potential energy operator to N particles of charge $q_1,\ldots,q_N$ is rather voluminous

$$\hat{\mathcal{H}}_{\text{pot}} = q_1\phi(\hat{\boldsymbol{r}})\otimes\underbrace{\hat{\mathbb{1}}\otimes\ldots\otimes\hat{\mathbb{1}}}_{N-1 \text{ times}} + q_2\hat{\mathbb{1}}\otimes\phi(\hat{\boldsymbol{r}})\otimes\underbrace{\ldots\otimes\hat{\mathbb{1}}}_{N-2 \text{ times}} + \ldots + q_N\underbrace{\hat{\mathbb{1}}\otimes\hat{\mathbb{1}}\otimes\ldots}_{N-1 \text{ times}}\otimes\phi(\hat{\boldsymbol{r}}), \tag{1.16}$$

and it is typically shortened to

$$\hat{\mathcal{H}}_{\text{pot}} = \sum_{j=1}^{N} q_j\phi(\hat{\boldsymbol{r}}_j),$$

where $\hat{\boldsymbol{r}}_j$ is the position operator acting on the jth particle and doing nothing to the other particles. Similarly, the noninteracting part of the Hamiltonian of N particles is typically written as

$$\hat{\mathcal{H}}_0 = \sum_{j=1}^{N} \hat{h}_j = \sum_{j=1}^{N} h(\hat{\boldsymbol{r}}_j, \hat{\boldsymbol{p}}_j, \hat{\boldsymbol{S}}_j), \tag{1.17}$$

while the interaction part is written as

$$\hat{\mathcal{H}}_{\text{int}} = \frac{1}{2}\sum_{i\neq j}^{N} v(\hat{\boldsymbol{r}}_i, \hat{\boldsymbol{r}}_j), \tag{1.18}$$

with $v(\mathbf{r}_1,\mathbf{r}_2)$ the interparticle interaction. We observe that these operators depend explicitly on the number of particles and are therefore difficult to manipulate in problems where the number of particles can fluctuate, such as in systems at finite temperature. As we see later in this chapter, another disadvantage is that the evaluation of their action on kets describing *identical* particles is very lengthy. Fortunately, an incredible simplification occurs for identical particles and the expressions for operators and kets become much lighter and easier to manipulate. To appreciate this simplification, however, we first have to understand how the quantum-mechanical formulation changes when the particles are identical.

1.3 Quantum Mechanics of Many Identical Particles

Two particles are called *identical particles* or *indistinguishable particles* if they have the same internal properties (i.e., the same mass, charge, spin etc.). For example, two electrons are two identical particles. To understand the qualitative difference between distinguishable and identical particles, let us perform the coincidence experiment of the previous section for two electrons both with spin projection $1/2$ and again in one dimension.

Discrete formulation for two particles At every point $x_n = n\Delta$ we put a spin-polarized electron detector and since the particles are identical we need only one kind of detector. If the detectors in x_n and x_m click at the same time, then we can be sure that just after this time there is one electron around x_n and another electron around x_m. Let us denote by $|nm\rangle$ with $n \geq m$ the ket describing the physical state in which the two electrons collapse after measurement. For mathematical convenience we also define the ket $|nm\rangle$ with $n \leq m$ as the ket describing the *same physical state* as $|mn\rangle$. Notice the different notation with respect to the previous section, where we have used the ket $|n\rangle|m\rangle$ to describe the first particle around x_n and the second particle around x_m. In the case of the electron–positron pair we could make the positron-click louder than the electron-click and hence distinguish the state $|n\rangle|m\rangle$ from the state $|m\rangle|n\rangle$. However, in this case we only have electron detectors and it is impossible to distinguish which electron has made a given detector click.

In Section 1.1 we observed that the normalized ket of a physical state is uniquely defined up to a phase factor. For our mathematical description to make sense, we then must impose that

$$|nm\rangle = e^{i\alpha}|mn\rangle \qquad \text{for all } n, m.$$

Using the above relation twice, we find that $e^{2i\alpha} = 1$, or equivalently $e^{i\alpha} = \pm 1$. Consequently, the ket

$$|nm\rangle = \pm|mn\rangle \tag{1.19}$$

is either symmetric or antisymmetric under the interchange of the electron positions. This is a fundamental property of nature: All particles can be grouped in two main classes. Particles described by a symmetric ket are called *bosons*, while those described by an antisymmetric ket are called *fermions*. The electrons of our example are fermions. Here and in the rest of the book the upper sign always refers to bosons and the lower sign to fermions. In the case of fermions (1.19) implies $|nn\rangle = -|nn\rangle$ and hence $|nn\rangle$ must be the *null ket* $|\emptyset\rangle$ – that is, it is not possible to create two fermions in the same position and with the same spin. This peculiarity of fermions is known as the *Pauli exclusion principle*.

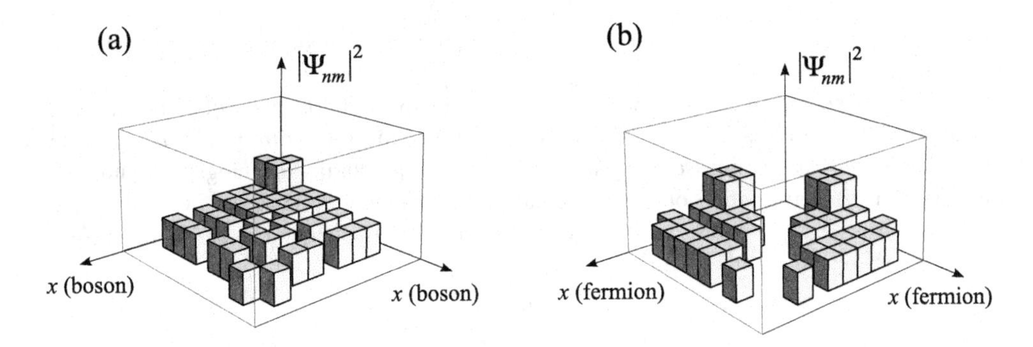

Figure 1.3 Histogram of the normalized number of simultaneous clicks of the detector in $x_n = n\Delta$ and in $x_m = m\Delta$ for (a) two bosons and (b) two fermions. The height of the parallelepipeds corresponds to the probabilities $|\Psi_{nm}|^2$.

If we now repeat the coincidence experiment $N \gg 1$ times, count the number of times that the detectors click simultaneously in x_n and x_m, and divide the result by N we can draw the histograms of Fig. 1.3 for bosons and fermions. The probability is symmetric under the interchange $n \leftrightarrow m$ due to property (1.19). The fermions are easily recognizable since the probability of finding them in the same place is zero.

In this book we learn how to deal with systems of many identical particles, such as molecules or solids, and therefore we do not always repeat that the particles are identical: by *particles* we mean *identical particles* unless otherwise stated. Unlike the case of the electron–positron pair, the probability of measuring a particle in $x_{n'}$ and the other in $x_{m'}$ just after the detectors in x_n and x_m have simultaneously clicked is zero unless $n = n'$ and $m = m'$ or $n = m'$ and $m = n'$, and hence

$$\langle n'm'|nm\rangle = c_1\delta_{n'n}\delta_{m'm} + c_2\delta_{m'n}\delta_{n'm}. \tag{1.20}$$

To fix the constants c_1 and c_2 we observe that

$$\langle n'm'|nm\rangle = \pm\langle n'm'|mn\rangle = \pm c_1\delta_{n'm}\delta_{m'n} \pm c_2\delta_{m'm}\delta_{n'n},$$

from which it follows that $c_1 = \pm c_2$. Furthermore, since the kets are normalized we must have for all $n \neq m$

$$1 = \langle nm|nm\rangle = c_1.$$

For $n = m$ the ket $|nn\rangle$ exists only for bosons, and one finds $\langle nn|nn\rangle = 2c_1$. It is therefore more convenient to work with a nonnormalized ket $|nn\rangle$ so that $c_1 = 1$ in all cases. We choose the normalization of the bosonic ket $|nn\rangle$ to be 2:

$$\langle nn|nn\rangle = 2. \tag{1.21}$$

Putting everything together we can rewrite the inner product (1.20) as

$$\langle n'm'|nm\rangle = \delta_{n'n}\delta_{m'm} \pm \delta_{m'n}\delta_{n'm}.$$

The inner product for the fermionic ket $|nn\rangle$ is automatically zero, in agreement with the fact that $|nn\rangle = |\emptyset\rangle$.

Let us now come to the completeness relation in the Hilbert space of two particles. Since $|nm\rangle = \pm|mn\rangle$, a basis in this space is given by the set $\{|nm\rangle\}$ with $n \geq m$. In other words the basis comprises only *inequivalent* configurations, meaning configurations not related by a permutation of the coordinates. The elements of this set are orthogonal and normalized except for the bosonic ket $|nn\rangle$, whose normalization is 2. Therefore, the completeness relation reads

$$\sum_{n>m} |nm\rangle\langle nm| + \frac{1}{2}\sum_{n} |nn\rangle\langle nn| = \hat{\mathbb{1}},$$

where the second sum does not contribute in the fermionic case. We can rewrite the completeness relation as an unrestricted sum over all n and m using the (anti)symmetry property (1.19). The resulting expression is

$$\frac{1}{2}\sum_{nm} |nm\rangle\langle nm| = \hat{\mathbb{1}},$$

which is much more elegant. The completeness relation can be used to expand any other ket in the same Hilbert space:

$$|\Psi\rangle = \hat{\mathbb{1}}|\Psi\rangle = \frac{1}{2}\sum_{nm} |nm\rangle\langle nm|\Psi\rangle, \tag{1.22}$$

and if $|\Psi\rangle$ is normalized then the square moduli of the coefficients of the expansion $\Psi_{nm} \equiv \langle nm|\Psi\rangle$ have the standard probabilistic interpretation:

$$|\Psi_{nm}|^2 = \left(\begin{array}{c}\text{probability of finding one particle in volume}\\ \text{element } \Delta \text{ around } x_n \text{ and the other particle}\\ \text{in volume element } \Delta \text{ around } x_m\end{array}\right)$$

for all $n \neq m$. For $n = m$ we must remember that the normalization of the ket $|nn\rangle$ is 2 and therefore $|\Psi_{nn}|^2$ gives *twice* the probability of finding two particles in the same place (since the proper normalized ket is $|nn\rangle/\sqrt{2}$). Consequently,

$$\frac{|\Psi_{nn}|^2}{2} = \left(\begin{array}{c}\text{probability of finding two particles in}\\ \text{volume element } \Delta \text{ around } x_n\end{array}\right).$$

Continuum formulation for two particles We can now refine the experiment by putting the detectors closer and closer. The continuum limit works exactly in the same manner as in the previous two sections. We rewrite the expansion (1.22) as

$$|\Psi\rangle = \frac{1}{2}\,\Delta^2\sum_{nm}\frac{|nm\rangle}{\Delta}\frac{\Psi_{nm}}{\Delta}, \tag{1.23}$$

and define the continuous wavefunction $\Psi(x_n, x_m)$ and the continuous ket $|x_n x_m\rangle$ according to

$$\Psi(x_n, x_m) = \lim_{\Delta\to 0}\frac{\Psi_{nm}}{\Delta}, \qquad |x_n x_m\rangle = \lim_{\Delta\to 0}\frac{|nm\rangle}{\Delta}.$$

The expansion (1.23) can then be seen as the Riemann sum of $\Psi(x_n, x_m)|x_n x_m\rangle$ and in the limit $\Delta \to 0$ the sum becomes the integral

$$|\Psi\rangle = \frac{1}{2} \int dx dx' \; \Psi(x, x')|xx'\rangle.$$

We can also derive the continuous representation of the completeness relation and the continuous representation of the inner product between two basis kets. We have

$$\lim_{\Delta \to 0} \frac{1}{2} \Delta^2 \sum_{nm} \frac{|nm\rangle}{\Delta} \frac{\langle nm|}{\Delta} = \frac{1}{2} \int dx dx' |xx'\rangle\langle xx'| = \hat{\mathbb{1}} \tag{1.24}$$

and

$$\begin{aligned} \lim_{\Delta \to 0} \frac{\langle n'm'|nm\rangle}{\Delta^2} &= \langle x_{n'} x_{m'}|x_n x_m\rangle \\ &= \delta(x_{n'} - x_n)\delta(x_{m'} - x_m) \pm \delta(x_{m'} - x_n)\delta(x_{n'} - x_m). \end{aligned} \tag{1.25}$$

The generalization to higher dimensions and to particles with different spin projections is straightforward. We define the position–spin ket $|\mathbf{x}_1\mathbf{x}_2\rangle$ as the ket of the physical state in which the particles collapse after the simultaneous clicking of a spin-polarized detector for particles of spin projection σ_1 placed in $\mathbf{r}_1$ and a spin-polarized detector for particles of spin projection σ_2 placed in $\mathbf{r}_2$. The set of inequivalent configurations $|\mathbf{x}_1\mathbf{x}_2\rangle$ forms a basis of the Hilbert space of two identical particles. In the following we refer to this space as $\mathbb{H}_2$. In analogy with (1.25) the continuous kets have inner product

$$\begin{aligned} \langle \mathbf{x}'_1\mathbf{x}'_2|\mathbf{x}_1\mathbf{x}_2\rangle &= \delta(\mathbf{x}'_1 - \mathbf{x}_1)\delta(\mathbf{x}'_2 - \mathbf{x}_2) \pm \delta(\mathbf{x}'_1 - \mathbf{x}_2)\delta(\mathbf{x}'_2 - \mathbf{x}_1) \\ &= \sum_P (\pm)^P \delta(\mathbf{x}'_1 - \mathbf{x}_{P(1)})\delta(\mathbf{x}'_2 - \mathbf{x}_{P(2)}), \end{aligned} \tag{1.26}$$

where the upper/lower sign refers to bosons/fermions. The second line of this equation is an equivalent way of rewriting the (anti)symmetric product of δ-functions. The sum runs over the permutations P of $(1, 2)$, which are the identity permutation $(P(1), P(2)) = (1, 2)$ and the interchange $(P(1), P(2)) = (2, 1)$. The quantity $(\pm)^P$ is equal to $+1$ if the permutation requires an even number of interchanges and ± 1 if the permutation requires an odd number of interchanges. In the fermionic case, all position–spin kets have the same norm since (1.26) implies

$$\langle \mathbf{x}_1\mathbf{x}_2|\mathbf{x}_1\mathbf{x}_2\rangle = \delta(0)^2 \qquad \text{for fermions.}$$

Due to the possibility in the bosonic case that two coordinates are identical, the norms of the position–spin kets are instead not all the same since

$$\langle \mathbf{x}_1\mathbf{x}_2|\mathbf{x}_1\mathbf{x}_2\rangle = \delta(0)^2 \times \left\{ \begin{array}{ll} 1 & \text{if } \mathbf{x}_1 \neq \mathbf{x}_2 \\ 2 & \text{if } \mathbf{x}_1 = \mathbf{x}_2 \end{array} \right. \qquad \text{for bosons,}$$

in agreement with (1.21).

In complete analogy with (1.24) we can also write the completeness relation according to

$$\frac{1}{2}\int d\mathbf{x}_1 d\mathbf{x}_2 |\mathbf{x}_1\mathbf{x}_2\rangle\langle\mathbf{x}_1\mathbf{x}_2| = \hat{\mathbb{1}}. \tag{1.27}$$

Then, any ket $|\Psi\rangle \in \mathbb{H}_2$ can be expanded in the position–spin basis as

$$|\Psi\rangle = \hat{\mathbb{1}}|\Psi\rangle = \frac{1}{2}\int d\mathbf{x}_1 d\mathbf{x}_2\, |\mathbf{x}_1\mathbf{x}_2\rangle \underbrace{\langle\mathbf{x}_1\mathbf{x}_2|\Psi\rangle}_{\Psi(\mathbf{x}_1,\mathbf{x}_2)}. \tag{1.28}$$

If $|\Psi\rangle$ is normalized, we can give a probability interpretation to the square modulus of the wavefunction $\Psi(\mathbf{x}_1, \mathbf{x}_2)$:

$$|\Psi(\mathbf{x}_1,\mathbf{x}_2)|^2 d\mathbf{r}_1 d\mathbf{r}_2 = \left(\begin{array}{c}\text{probability of finding one particle with spin } \sigma_1 \text{ in} \\ \text{volume element } d\mathbf{r}_1 \text{ around } \mathbf{r}_1 \text{ and the other particle with} \\ \text{spin } \sigma_2 \text{ in volume element } d\mathbf{r}_2 \text{ around a } \textit{different} \text{ point } \mathbf{r}_2\end{array}\right).$$

However, in the case $\mathbf{x}_1 = \mathbf{x}_2$ the above formula needs to be replaced by

$$\frac{|\Psi(\mathbf{x}_1,\mathbf{x}_1)|^2}{2} d\mathbf{r}_1 d\mathbf{r}_2 = \left(\begin{array}{c}\text{probability of finding one particle with spin } \sigma_1 \text{ in} \\ \text{volume element } d\mathbf{r}_1 \text{ around } \mathbf{r}_1 \text{ and the other particle} \\ \text{with the } \textit{same} \text{ spin in volume element } d\mathbf{r}_2 \text{ around} \\ \text{the } \textit{same} \text{ point } \mathbf{r}_1\end{array}\right),$$

due to the different normalization of the diagonal kets. We stress again that the above probability interpretation follows from the normalization $\langle\Psi|\Psi\rangle = 1$, which in the continuum case reads [see (1.28)]

$$1 = \frac{1}{2}\int d\mathbf{x}_1 d\mathbf{x}_2\, |\Psi(\mathbf{x}_1,\mathbf{x}_2)|^2.$$

Continuum formulation for N particles It should now be clear how to extend the above relations to the case of N identical particles. We say that if the detector for a particle of spin projection σ_1 placed in $\mathbf{r}_1$, the detector for a particle of spin projection σ_2 placed in $\mathbf{r}_2$, etc. all click at the same time then the N-particle state collapses in the position–spin ket $|\mathbf{x}_1 \ldots \mathbf{x}_N\rangle$. Due to the nature of identical particles this ket must have the symmetry property (as usual, upper/lower sign refers to bosons/fermions):

$$\boxed{|\mathbf{x}_{P(1)} \ldots \mathbf{x}_{P(N)}\rangle = (\pm)^P |\mathbf{x}_1 \ldots \mathbf{x}_N\rangle} \tag{1.29}$$

where P is a permutation of the labels $(1, \ldots, N)$, and $(\pm)^P = 1$ for even permutations and ± 1 for odd permutations (thus for bosons is always 1). A permutation is even/odd if the number of interchanges is even/odd.[3] Therefore, given the ket $|\mathbf{x}_1 \ldots \mathbf{x}_N\rangle$ with all different coordinates there are $N!$ equivalent configurations that represent the same physical state. More generally, if the ket $|\mathbf{x}_1 \ldots \mathbf{x}_N\rangle$ has m_1 coordinates equal to $\mathbf{y}_1$, m_2 coordinates equal to $\mathbf{y}_2 \neq \mathbf{y}_1$, …, m_M coordinates equal to $\mathbf{y}_M \neq \mathbf{y}_1, \ldots, \mathbf{y}_{M-1}$, with $m_1 + \ldots + m_M = N$,

[3] The reader can learn more on how to calculate the sign of a permutation in Appendix B.

then the number of equivalent configurations is $N!/(m_1! \ldots m_M!)$. In the fermionic case, if two or more coordinates are the same then the ket $|\mathbf{x}_1 \ldots \mathbf{x}_N\rangle$ is the null ket $|\emptyset\rangle$. The set of position–spin kets corresponding to inequivalent configurations form a basis in the Hilbert space of N identical particles; we refer to this space as $\mathbb{H}_N$.

The inner product between two position–spin kets is

$$\langle \mathbf{x}'_1 \ldots \mathbf{x}'_N | \mathbf{x}_1 \ldots \mathbf{x}_N \rangle = \sum_P c_P \prod_{j=1}^N \delta(\mathbf{x}'_j - \mathbf{x}_{P(j)}),$$

where the c_Ps are numbers depending on the permutation P. As in the two-particle case, the (anti)symmetry (1.29) of the position–spin kets requires that $c_P = c(\pm)^P$, and the normalization of $|\mathbf{x}_1 \ldots \mathbf{x}_N\rangle$ with all *different* coordinates fixes the constant $c = 1$. Hence,

$$\boxed{\langle \mathbf{x}'_1 \ldots \mathbf{x}'_N | \mathbf{x}_1 \ldots \mathbf{x}_N \rangle = \sum_P (\pm)^P \prod_{j=1}^N \delta(\mathbf{x}'_j - \mathbf{x}_{P(j)})} \tag{1.30}$$

This is the familiar expression for the permanent/determinant $|A|_\pm$ of a $N \times N$ matrix A (see Appendix B):

$$|A|_\pm \equiv \sum_P (\pm)^P A_{1P(1)} \ldots A_{NP(N)}.$$

Choosing the matrix elements of A to be $A_{ij} = \delta(\mathbf{x}'_i - \mathbf{x}_j)$, we can rewrite (1.30) as

$$\langle \mathbf{x}'_1 \ldots \mathbf{x}'_N | \mathbf{x}_1 \ldots \mathbf{x}_N \rangle = \begin{vmatrix} \delta(\mathbf{x}'_1 - \mathbf{x}_1) & \ldots & \delta(\mathbf{x}'_1 - \mathbf{x}_N) \\ \cdot & \ldots & \cdot \\ \cdot & \ldots & \cdot \\ \delta(\mathbf{x}'_N - \mathbf{x}_1) & \ldots & \delta(\mathbf{x}'_N - \mathbf{x}_N) \end{vmatrix}_\pm . \tag{1.31}$$

As in the two-particle case, these formulas are so elegant because we take the bosonic kets at equal coordinates with a slightly different normalization. Consider N bosons in M *different* coordinates of which m_1 have coordinate $\mathbf{y}_1$, ..., m_M have coordinate $\mathbf{y}_M$ (hence $m_1 + \ldots + m_M = N$). Then the norm is given by

$$\langle \overbrace{\mathbf{y}_1 \ldots \mathbf{y}_1}^{m_1} \ldots \overbrace{\mathbf{y}_M \ldots \mathbf{y}_M}^{m_M} | \overbrace{\mathbf{y}_1 \ldots \mathbf{y}_1}^{m_1} \ldots \overbrace{\mathbf{y}_M \ldots \mathbf{y}_M}^{m_M} \rangle = \delta(0)^N m_1! m_2! \ldots m_M!,$$

as follows directly from (1.30).[4] In the case of fermions, instead, all position–spin kets have norm $\delta(0)^N$ since it is not possible for two or more fermions to have the same coordinate.

Given the norm of the position–spin kets, the completeness relation for N particles is a straightforward generalization of (1.27) and reads

$$\boxed{\frac{1}{N!} \int d\mathbf{x}_1 \ldots d\mathbf{x}_N \, |\mathbf{x}_1 \ldots \mathbf{x}_N\rangle\langle \mathbf{x}_1 \ldots \mathbf{x}_N| = \hat{\mathbb{1}}} \tag{1.32}$$

[4]According to (1.29), the order of the arguments in the inner product does not matter.

Therefore the expansion of a ket $|\Psi\rangle \in \mathbb{H}_N$ can be written as

$$|\Psi\rangle = \hat{\mathbb{1}}|\Psi\rangle = \frac{1}{N!}\int d\mathbf{x}_1 \dots d\mathbf{x}_N |\mathbf{x}_1 \dots \mathbf{x}_N\rangle \underbrace{\langle \mathbf{x}_1 \dots \mathbf{x}_N|\Psi\rangle}_{\Psi(\mathbf{x}_1,\dots,\mathbf{x}_N)},$$

which generalizes the expansion (1.28) to the case of N particles. The wavefunction $\Psi(\mathbf{x}_1,\dots,\mathbf{x}_N)$ is totally symmetric for bosons and totally antisymmetric for fermions due to (1.29). If $|\Psi\rangle$ is normalized then the normalization of the wavefunction reads

$$1 = \langle\Psi|\Psi\rangle = \frac{1}{N!}\int d\mathbf{x}_1 \dots d\mathbf{x}_N |\,\Psi(\mathbf{x}_1,\dots,\mathbf{x}_N)|^2. \tag{1.33}$$

The probabilistic interpretation of the square modulus of the wavefunction can be extracted using the same line of reasoning as for the two-particle case:

$$\frac{|\Psi(\overbrace{\mathbf{y}_1\dots\mathbf{y}_1}^{m_1}\dots\overbrace{\mathbf{y}_M\dots\mathbf{y}_M}^{m_M})|^2}{m_1!\dots m_M!}\prod_{j=1}^{M} d\mathbf{R}_j = \begin{pmatrix} \text{probability of finding} \\ m_1 \text{ particles in } d\mathbf{R}_1 \text{ around } \mathbf{y}_1 \\ \vdots \\ m_M \text{ particles in } d\mathbf{R}_M \text{ around } \mathbf{y}_M \end{pmatrix}, \tag{1.34}$$

where $d\mathbf{R}_j$ is the product of volume elements,

$$d\mathbf{R}_j \equiv \prod_{i=m_1+\dots+m_{j-1}+1}^{m_1+\dots+m_j} d\mathbf{r}_i.$$

When all coordinates are different, (1.34) tells us that the quantity $|\Psi(\mathbf{x}_1,\dots,\mathbf{x}_N)|^2 d\mathbf{r}_1 \dots d\mathbf{r}_N$ is the probability of finding one particle in volume element $d\mathbf{r}_1$ around $\mathbf{x}_1$, ..., and one particle in volume element $d\mathbf{r}_N$ around $\mathbf{x}_N$. We could have absorbed the prefactor $1/N!$ in (1.33) in the wavefunction (as is commonly done) but then we could not interpret the quantity $|\Psi(\mathbf{x}_1,\dots,\mathbf{x}_N)|^2 d\mathbf{r}_1 \dots d\mathbf{r}_N$ as the right-hand side (r.h.s.) of (1.34) since this would amount to regarding equivalent configurations as distinguishable and consequently the probability would be overestimated by a factor of $N!$.

The reader might wonder why we have been so punctilious about the possibility of having more than one boson with the same position–spin coordinate, since these configurations are of zero measure in the space of all configurations. However, such configurations are the physically most relevant in bosonic systems at low temperature. Indeed, bosons can condense in states in which certain (continuum) quantum numbers are *macroscopically* occupied and hence have a finite probability. A common example is the zero momentum state of a free boson gas in three dimensions.

First quantization We close this section by illustrating a practical way to construct the N-particle position–spin kets using the N-particle position–spin kets of distinguishable particles. The procedure simply consists in forming (anti)symmetrized products of one-particle position–spin kets. For instance, we have for the case of two particles

$$|\mathbf{x}_1\mathbf{x}_2\rangle = \frac{|\mathbf{x}_1\rangle|\mathbf{x}_2\rangle \pm |\mathbf{x}_2\rangle|\mathbf{x}_1\rangle}{\sqrt{2}}, \tag{1.35}$$

and more generally for N particles

$$|\mathbf{x}_1 \ldots \mathbf{x}_N\rangle = \frac{1}{\sqrt{N!}} \sum_P (\pm)^P |\mathbf{x}_{P(1)}\rangle \ldots |\mathbf{x}_{P(N)}\rangle. \tag{1.36}$$

Using the inner product (1.14), one can check directly that these states have inner product (1.30). We refer to the above representation of the position–spin kets as kets in *first quantization* since it is the representation usually found in basic quantum mechanics books. Using (1.36) we could proceed to calculate matrix elements of operators such as the potential energy, total energy, spin, angular momentum, density, etc. However, this involves rather cumbersome expressions with a large number of terms differing only in the sign and the order of the coordinates. In the next section we describe a formalism, known as *second quantization*, that makes it easy to do such calculations efficiently, as the position–spin ket is represented by a *single* ket rather than by $N!$ products of one-particle kets as in (1.36). As we see, the merits of second quantization are the compactness of the expressions and an enormous simplification in the calculation of the action of operators over states in $\mathbb{H}_N$. This formalism further treats systems with different numbers of identical particles on the same footing and it is therefore well suited to study of ionization processes, transport phenomena, and finite temperature effects within the grand canonical ensemble of quantum statistical physics.

1.4 Field Operators

The advantage of the bra-and-ket notation invented by Dirac is twofold. First of all, it provides a geometric interpretation of the physical states in Hilbert space as abstract kets independent of the basis in which they are expanded. For example, it does not matter whether we expand $|\Psi\rangle$ in terms of the position–spin kets or momentum–spin kets; $|\Psi\rangle$ remains the same although the expansion coefficients in the two bases are different. The second advantage is that the abstract kets can be systematically generated by repeated applications of a *creation operator* on the empty or zero-particle state. This approach forms the basis of an elegant formalism known as *second quantization*, which we describe in detail in this section.

Fock space To deal with arbitrary many identical particles we define a collection $\mathbb{F}$ of Hilbert spaces, also known as *Fock space*, according to

$$\mathbb{F} = \{\mathbb{H}_0, \mathbb{H}_1, \ldots, \mathbb{H}_N, \ldots\},$$

with $\mathbb{H}_N$ the Hilbert space for N identical particles. An arbitrary element of the Fock space is a ket that can be written as

$$|\Psi\rangle = \sum_{N=0}^{\infty} c_N |\Psi_N\rangle, \tag{1.37}$$

where $|\Psi_N\rangle$ belongs to $\mathbb{H}_N$. The inner product between the ket (1.37) and another element in the Fock space,

$$|\chi\rangle = \sum_{N=0}^{\infty} d_N |\chi_N\rangle,$$

is defined as

$$\langle\chi|\Psi\rangle \equiv \sum_{N=0}^{\infty} d_N^* c_N \langle\chi_N|\Psi_N\rangle,$$

where $\langle\chi_N|\Psi_N\rangle$ is the inner product in $\mathbb{H}_N$. This definition is dictated by common sense: The probability of having $M \neq N$ particles in an N-particle ket is zero and therefore kets with a different number of particles are orthogonal (i.e., have zero overlap).

The Hilbert space $\mathbb{H}_0$ is the space with *zero* particles. Since an empty system has no degrees of freedom, $\mathbb{H}_0$ is a one-dimensional space and we denote by $|0\rangle$ the only normalized ket in $\mathbb{H}_0$,

$$\langle 0|0\rangle = 1.$$

According to the expansion (1.37), the ket $|0\rangle$ has all $c_N = 0$ except for c_0. This state should not be confused with the null ket $|\emptyset\rangle$, which is defined as the state in Fock space with all $c_N = 0$ and, therefore, is *not* a physical state. The *empty ket* $|0\rangle$ is a physical state; indeed the normalization $\langle 0|0\rangle = 1$ means that the probability of finding nothing in an empty space is 1.

Field operators The goal of this section is to find a clever way to construct a basis for each Hilbert space $\mathbb{H}_1, \mathbb{H}_2, \ldots$. To accomplish this goal the central idea of the second quantization formalism is to define a *field operator* $\hat{\psi}^\dagger(\mathbf{x}) = \hat{\psi}^\dagger(\mathbf{r}\sigma)$ that generates the position–spin kets by repeated action on the empty ket:

$$\begin{aligned} |\mathbf{x}_1\rangle &= \hat{\psi}^\dagger(\mathbf{x}_1)|0\rangle \\ |\mathbf{x}_1\mathbf{x}_2\rangle &= \hat{\psi}^\dagger(\mathbf{x}_2)|\mathbf{x}_1\rangle = \hat{\psi}^\dagger(\mathbf{x}_2)\hat{\psi}^\dagger(\mathbf{x}_1)|0\rangle \\ |\mathbf{x}_1\ldots\mathbf{x}_N\rangle &= \hat{\psi}^\dagger(\mathbf{x}_N)|\mathbf{x}_1\ldots\mathbf{x}_{N-1}\rangle = \hat{\psi}^\dagger(\mathbf{x}_N)\ldots\hat{\psi}^\dagger(\mathbf{x}_1)|0\rangle \end{aligned} \tag{1.38}$$

Since an operator is uniquely defined from its action on a complete set of states in the Hilbert space (the Fock space in our case), the above relations *define* the field operator $\hat{\psi}^\dagger(\mathbf{x})$ for all $\mathbf{x}$. The field operator $\hat{\psi}^\dagger(\mathbf{x})$ transforms a ket of $\mathbb{H}_N$ into a ket of $\mathbb{H}_{N+1}$ for all N, see Fig. 1.4(a). We may say that the field operator $\hat{\psi}^\dagger(\mathbf{x})$ creates a particle in $\mathbf{x}$ and it is therefore called the *creation operator.* Since the position–spin kets change a plus or minus sign under interchange of any two particles, it follows that

$$\begin{aligned} \hat{\psi}^\dagger(\mathbf{x})\hat{\psi}^\dagger(\mathbf{y})|\mathbf{x}_1\ldots\mathbf{x}_N\rangle &= |\mathbf{x}_1\ldots\mathbf{x}_N\,\mathbf{y}\,\mathbf{x}\rangle = \pm|\mathbf{x}_1\ldots\mathbf{x}_N\,\mathbf{x}\,\mathbf{y}\rangle \\ &= \pm\hat{\psi}^\dagger(\mathbf{y})\hat{\psi}^\dagger(\mathbf{x})|\mathbf{x}_1\ldots\mathbf{x}_N\rangle, \end{aligned}$$

where we recall that the upper sign in $\pm$ refers to bosons and the lower sign to fermions. This identity is true for all $\mathbf{x}_1, \ldots, \mathbf{x}_N$ and for all N (i.e., for all states in $\mathbb{F}$), and hence

$$\hat{\psi}^\dagger(\mathbf{x})\hat{\psi}^\dagger(\mathbf{y}) = \pm\hat{\psi}^\dagger(\mathbf{y})\hat{\psi}^\dagger(\mathbf{x}).$$

If we define the (anti)commutator between two generic operators $\hat{A}$ and $\hat{B}$ according to

$$\left[\hat{A}, \hat{B}\right]_{\mp} = \hat{A}\hat{B} \mp \hat{B}\hat{A},$$

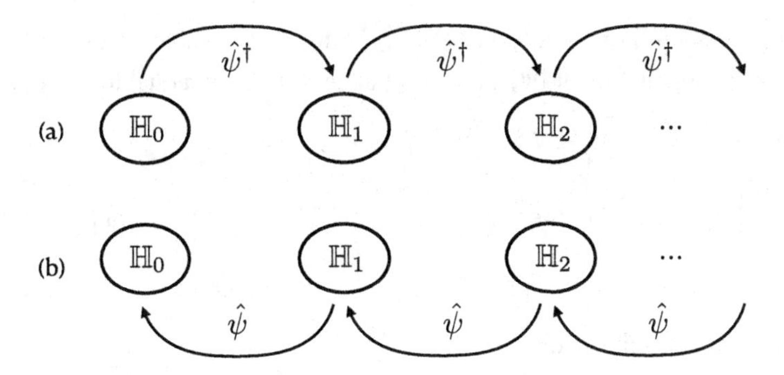

Figure 1.4 Schematic action of the creation operator $\hat{\psi}^\dagger$ in (a) and of the annihilation operator $\hat{\psi}$ in (b).

we can rewrite the above relation as

$$\boxed{\left[\hat{\psi}^\dagger(\mathbf{x}), \hat{\psi}^\dagger(\mathbf{y})\right]_\mp = 0} \tag{1.39}$$

Corresponding to the operator $\hat{\psi}^\dagger(\mathbf{x})$ there is the adjoint operator $\hat{\psi}(\mathbf{x})$ [or equivalently $\hat{\psi}^\dagger(\mathbf{x})$ is the adjoint of $\hat{\psi}(\mathbf{x})$]. Let us remind the reader about the definition of adjoint operators. An operator $\hat{O}^\dagger$ with the superscript "†" (read *dagger*) is the adjoint of the operator $\hat{O}$ if

$$\langle\chi|\hat{O}|\Psi\rangle = \langle\Psi|\hat{O}^\dagger|\chi\rangle^*$$

for all $|\chi\rangle$ and $|\Psi\rangle$, which implies $(\hat{O}^\dagger)^\dagger = \hat{O}$. In particular, when $\hat{O} = \hat{\psi}(\mathbf{x})$ we have

$$\langle\chi|\hat{\psi}(\mathbf{x})|\Psi\rangle = \langle\Psi|\hat{\psi}^\dagger(\mathbf{x})|\chi\rangle^*.$$

Since for any $|\Psi\rangle \in \mathbb{H}_{N+1}$ the quantity $\langle\Psi|\hat{\psi}^\dagger(\mathbf{x})|\chi\rangle$ is zero for all $|\chi\rangle$ with no components in $\mathbb{H}_N$, the above equation implies that $\hat{\psi}(\mathbf{x})|\Psi\rangle \in \mathbb{H}_N$ - that is, the operator $\hat{\psi}(\mathbf{x})$ maps the elements of $\mathbb{H}_{N+1}$ into elements of $\mathbb{H}_N$, see Fig. 1.4(b). Thus, whereas the operator $\hat{\psi}^\dagger(\mathbf{x})$ adds a particle, its adjoint operator $\hat{\psi}(\mathbf{x})$ removes a particle and, for this reason, is called the *annihilation operator*. Below we study its properties and how it acts on the position–spin kets.

By taking the adjoint of the identity (1.39), we immediately obtain the (anti)commutation relation

$$\boxed{\left[\hat{\psi}(\mathbf{x}), \hat{\psi}(\mathbf{y})\right]_\mp = 0} \tag{1.40}$$

The action of $\hat{\psi}(\mathbf{x})$ on any state can be deduced from its definition as the adjoint of $\hat{\psi}^\dagger(\mathbf{x})$ together with the inner product (1.31) between the position–spin kets. Let us illustrate this first for the action on the empty ket $|0\rangle$. For any $|\Psi\rangle \in \mathbb{F}$,

$$\langle\Psi|\hat{\psi}(\mathbf{x})|0\rangle = \langle 0|\hat{\psi}^\dagger(\mathbf{x})|\Psi\rangle^* = 0,$$

since $\hat{\psi}^\dagger(\mathbf{x})|\Psi\rangle$ contains at least one particle and is therefore orthogonal to $|0\rangle$. We conclude that $\hat{\psi}(\mathbf{x})|0\rangle$ is orthogonal to all $|\Psi\rangle$ in $\mathbb{F}$ and hence it must be equal to the null ket

$$\hat{\psi}(\mathbf{x})|0\rangle = |\emptyset\rangle. \tag{1.41}$$

The action of $\hat{\psi}(\mathbf{x})$ on the one-particle ket $|\mathbf{y}\rangle$ can be inferred from (1.30) and (1.38); we have

$$\delta(\mathbf{y}-\mathbf{x}) = \langle \mathbf{y}|\mathbf{x}\rangle = \langle \mathbf{y}|\hat{\psi}^\dagger(\mathbf{x})|0\rangle = \langle 0|\hat{\psi}(\mathbf{x})|\mathbf{y}\rangle^*.$$

Since $\hat{\psi}(\mathbf{x})|\mathbf{y}\rangle \in \mathbb{H}_0$, it follows that

$$\hat{\psi}(\mathbf{x})|\mathbf{y}\rangle = \delta(\mathbf{y}-\mathbf{x})|0\rangle. \tag{1.42}$$

We see from this relation that the operator $\hat{\psi}(\mathbf{x})$ removes a particle from the state $|\mathbf{y}\rangle$ when $\mathbf{x} = \mathbf{y}$ and otherwise yields zero.

The derivation of the action of $\hat{\psi}(\mathbf{x})$ on the empty ket and on the one-particle ket was rather elementary. Let us now derive the action of $\hat{\psi}(\mathbf{x})$ on the general N-particle ket $|\mathbf{y}_1 \ldots \mathbf{y}_N\rangle$. For this purpose we consider the matrix element

$$\langle \mathbf{x}_1 \ldots \mathbf{x}_{N-1}|\hat{\psi}(\mathbf{x}_N)|\mathbf{y}_1 \ldots \mathbf{y}_N\rangle = \langle \mathbf{x}_1 \ldots \mathbf{x}_N|\mathbf{y}_1 \ldots \mathbf{y}_N\rangle. \tag{1.43}$$

The overlap on the r.h.s. is given in (1.31); expanding the permanent/determinant along row N (see Appendix B), we get

$$\langle \mathbf{x}_1 \ldots \mathbf{x}_{N-1}|\hat{\psi}(\mathbf{x}_N)|\mathbf{y}_1 \ldots \mathbf{y}_N\rangle$$
$$= \sum_{k=1}^{N} (\pm)^{N+k} \delta(\mathbf{x}_N - \mathbf{y}_k)\langle \mathbf{x}_1 \ldots \mathbf{x}_{N-1}|\mathbf{y}_1 \ldots \mathbf{y}_{k-1}\mathbf{y}_{k+1} \ldots \mathbf{y}_N\rangle.$$

This expression is valid for any $|\mathbf{x}_1 \ldots \mathbf{x}_{N-1}\rangle$ and since $\hat{\psi}(\mathbf{x})$ maps from $\mathbb{H}_N$ only to $\mathbb{H}_{N-1}$, we conclude that

$$\boxed{\hat{\psi}(\mathbf{x})|\mathbf{y}_1 \ldots \mathbf{y}_N\rangle = \sum_{k=1}^{N} (\pm)^{N+k} \delta(\mathbf{x} - \mathbf{y}_k)\, |\mathbf{y}_1 \ldots \mathbf{y}_{k-1}\mathbf{y}_{k+1} \ldots \mathbf{y}_N\rangle} \tag{1.44}$$

We have just derived an important equation for the action of the annihilation operator on a position-spin ket. It correctly reduces to (1.42) when $N = 1$ and for $N > 1$ yields, for example,

$$\hat{\psi}(\mathbf{x})|\mathbf{y}_1\mathbf{y}_2\rangle = \delta(\mathbf{x}-\mathbf{y}_2)|\mathbf{y}_1\rangle \pm \delta(\mathbf{x}-\mathbf{y}_1)|\mathbf{y}_2\rangle,$$
$$\hat{\psi}(\mathbf{x})|\mathbf{y}_1\mathbf{y}_2\mathbf{y}_3\rangle = \delta(\mathbf{x}-\mathbf{y}_3)|\mathbf{y}_1\mathbf{y}_2\rangle \pm \delta(\mathbf{x}-\mathbf{y}_2)|\mathbf{y}_1\mathbf{y}_3\rangle + \delta(\mathbf{x}-\mathbf{y}_1)|\mathbf{y}_2\mathbf{y}_3\rangle.$$

So the annihilation operator removes subsequently a particle from every position-spin coordinate while keeping the final result totally symmetric or antisymmetric in all $\mathbf{y}$ variables by adjusting the signs of the prefactors.

(Anti)commutation rule With the help of (1.44) we can derive a fundamental (anti)commutation relation between the annihilation and creation operators. Acting on both sides of (1.44) with $\hat{\psi}^\dagger(\mathbf{y})$ and denoting by $|R\rangle$ the ket on the r.h.s., we have

$$\hat{\psi}^\dagger(\mathbf{y})\hat{\psi}(\mathbf{x})|\mathbf{y}_1\ldots\mathbf{y}_N\rangle = \hat{\psi}^\dagger(\mathbf{y})|R\rangle. \tag{1.45}$$

Exchanging the order of the field operators in the left-hand side (l.h.s.) of the above identity and using (1.44), we find

$$\begin{aligned}\hat{\psi}(\mathbf{x})\hat{\psi}^\dagger(\mathbf{y})|\mathbf{y}_1\ldots\mathbf{y}_N\rangle &= \hat{\psi}(\mathbf{x})|\mathbf{y}_1\ldots\mathbf{y}_N\mathbf{y}\rangle = \delta(\mathbf{x}-\mathbf{y})\,|\mathbf{y}_1\ldots\mathbf{y}_N\rangle \\ &+ \sum_{k=1}^{N}(\pm)^{N+1+k}\delta(\mathbf{x}-\mathbf{y}_k)\,|\mathbf{y}_1\ldots\mathbf{y}_{k-1}\mathbf{y}_{k+1}\ldots\mathbf{y}_N\mathbf{y}\rangle \\ &= \delta(\mathbf{x}-\mathbf{y})\,|\mathbf{y}_1\ldots\mathbf{y}_N\rangle \pm \hat{\psi}^\dagger(\mathbf{y})|R\rangle.\end{aligned} \tag{1.46}$$

Subtraction and addition of (1.45) and (1.46) for bosons and fermions respectively then gives

$$\left[\hat{\psi}(\mathbf{x}),\hat{\psi}^\dagger(\mathbf{y})\right]_\mp |\mathbf{y}_1\ldots\mathbf{y}_N\rangle = \delta(\mathbf{x}-\mathbf{y})|\mathbf{y}_1\ldots\mathbf{y}_N\rangle,$$

which must be valid for all position–spin kets and for all N, and therefore

$$\boxed{\left[\hat{\psi}(\mathbf{x}),\hat{\psi}^\dagger(\mathbf{y})\right]_\mp = \delta(\mathbf{x}-\mathbf{y})} \tag{1.47}$$

The (anti)commutation relations (1.39), (1.40), and (1.47) are the main results of this section and form the basis of most derivations in this book. As we see in Section 1.6, all many-particle operators, such as total energy, density, current, spin, etc., consist of simple expressions in terms of the field operators $\hat{\psi}$ and $\hat{\psi}^\dagger$, and the calculation of their averages can easily be performed with the help of the (anti)commutation relations. It is similar to the harmonic oscillator of quantum mechanics: Both the eigenstates and the operators are expressed in terms of the raising and lowering operators $\hat{a}^\dagger$ and $\hat{a}$, and to calculate all sorts of averages it is enough to know the commutation relations $[\hat{a},\hat{a}]_- = [\hat{a}^\dagger,\hat{a}^\dagger]_- = 0$ and $[\hat{a},\hat{a}^\dagger]_- = 1$. The difference with second quantization is that we have a "harmonic oscillator" for every $\mathbf{x}$. Using the (anti)commutation properties we can manipulate directly the kets and never have to deal with the rather cumbersome expressions of the wavefunctions; the field operators take care of the symmetry of the kets automatically. The great achievement of second quantization is comparable to that of a programming language. When we program we use a nice, friendly text editor to write code that tells the computer what operations to do, and we do not worry about whether the instructions given through the text editor are correctly executed by the machine. A bug in the code is an error in the text of the program (the way we manipulate the field operators) and not an erroneous functioning of some logic gate (the violation of the symmetry properties of the many-particle kets).

Exercise 1.1 We define the *density operator*

$$\hat{n}(\mathbf{x}) \equiv \hat{\psi}^\dagger(\mathbf{x})\hat{\psi}(\mathbf{x}).$$

Using the identities $[\hat{A}\hat{B},\hat{C}]_- = \hat{A}[\hat{B},\hat{C}]_- + [\hat{A},\hat{C}]_-\hat{B} = \hat{A}[\hat{B},\hat{C}]_+ - [\hat{A},\hat{C}]_+\hat{B}$, prove the following relations for fermionic and bosonic field operators:

$$[\hat{n}(\mathbf{x}),\hat{\psi}(\mathbf{x}')]_- = -\delta(\mathbf{x}-\mathbf{x}')\hat{\psi}(\mathbf{x}), \tag{1.48}$$

$$[\hat{n}(\mathbf{x}),\hat{\psi}^\dagger(\mathbf{x}')]_- = \delta(\mathbf{x}-\mathbf{x}')\hat{\psi}^\dagger(\mathbf{x}). \tag{1.49}$$

1.5 General Basis States

In the previous section we learned how to construct states of many identical particles with a given spin and position. The position–spin is, however, just one possible choice of quantum numbers to characterize every single particle. We now show how the field operators can be used to construct states of many identical particles in which every particle is labeled by general quantum numbers, such as momentum, energy, etc.

Let us consider a normalized one-particle ket $|n\rangle$. The quantum number $n = (s\tau)$ comprises an orbital quantum number s and the spin projection τ along some quantization axis. Choosing the quantization axis of the spin to be the same as that of the position–spin ket $|\mathbf{x}\rangle = |\mathbf{r}\sigma\rangle$, the overlap between $|n\rangle$ and $|\mathbf{x}\rangle$ is

$$\langle\mathbf{x}|n\rangle \equiv \varphi_n(\mathbf{x}) = \varphi_s(\mathbf{r})\delta_{\tau\sigma}. \tag{1.50}$$

The one-particle ket $|n\rangle$ can be expanded in the position–spin kets using the completeness relation (1.8):

$$|n\rangle = \int d\mathbf{x}\,|\mathbf{x}\rangle\langle\mathbf{x}|n\rangle = \int d\mathbf{x}\,\varphi_n(\mathbf{x})|\mathbf{x}\rangle = \int d\mathbf{x}\,\varphi_n(\mathbf{x})\hat{\psi}^\dagger(\mathbf{x})|0\rangle. \tag{1.51}$$

One can easily check that the normalization $\langle n|n\rangle = 1$ is equivalent to saying that $\int d\mathbf{x}|\varphi_n(\mathbf{x})|^2 = 1$. From (1.51) we see that $|n\rangle$ is obtained by applying to the empty ket $|0\rangle$ the operator

$$\boxed{\hat{d}_n^\dagger \equiv \int d\mathbf{x}\,\varphi_n(\mathbf{x})\hat{\psi}^\dagger(\mathbf{x})} \tag{1.52}$$

that is, $\hat{d}_n^\dagger|0\rangle = |n\rangle$. We may say that $\hat{d}_n^\dagger$ creates a particle with quantum number n. Similarly, if we take the adjoint of (1.52),

$$\boxed{\hat{d}_n \equiv \int d\mathbf{x}\,\varphi_n^*(\mathbf{x})\hat{\psi}(\mathbf{x})} \tag{1.53}$$

we obtain an operator that destroys a particle with quantum number n since

$$\hat{d}_n|n\rangle = \hat{d}_n\hat{d}_n^\dagger|0\rangle = \int d\mathbf{x}d\mathbf{x}'\varphi_n^*(\mathbf{x})\varphi_n(\mathbf{x}')\underbrace{\hat{\psi}(\mathbf{x})\hat{\psi}^\dagger(\mathbf{x}')|0\rangle}_{\delta(\mathbf{x}-\mathbf{x}')|0\rangle} = \int d\mathbf{x}\,|\varphi_n(\mathbf{x})|^2|0\rangle = |0\rangle.$$

The operators $\hat{d}_n$ and $\hat{d}_n^\dagger$, being linear combinations of field operators at different $\mathbf{x}$, can act on states with arbitrary many particles. Below we derive some important relations for the $\hat{d}$-operators when the set $\{|n\rangle\}$ forms an orthonormal basis in the one-particle Hilbert space.

We can easily derive the important (anti)commutation relations using the corresponding relations for the field operators,

$$\left[\hat{d}_n, \hat{d}^\dagger_m\right]_\mp = \int d\mathbf{x}d\mathbf{x}'\, \varphi^*_n(\mathbf{x})\varphi_m(\mathbf{x}') \underbrace{\left[\hat{\psi}(\mathbf{x}), \hat{\psi}^\dagger(\mathbf{x}')\right]_\mp}_{\delta(\mathbf{x}-\mathbf{x}')} = \langle n|m\rangle = \delta_{nm}, \tag{1.54}$$

and the more obvious ones,

$$\left[\hat{d}_n, \hat{d}_m\right]_\mp = \left[\hat{d}^\dagger_n, \hat{d}^\dagger_m\right]_\mp = 0, \tag{1.55}$$

that follow similarly. It is worth noticing that the $\hat{d}$-operators obey the same (anti)commutation relations as the field operators, with the index n playing the role of $\mathbf{x}$. This is a very important observation since the results of the previous section rely only on the (anti)commutation relations of $\hat{\psi}$ and $\hat{\psi}^\dagger$, and hence remain valid in this more general basis. To convince the reader of this fact, we derive some of the results of the previous section directly from the (anti)commutation relations. We define the N-particle ket

$$|n_1 \ldots n_N\rangle \equiv \hat{d}^\dagger_{n_N} \ldots \hat{d}^\dagger_{n_1}|0\rangle = \hat{d}^\dagger_{n_N}|n_1 \ldots n_{N-1}\rangle \tag{1.56}$$

which has the symmetry property

$$|n_{P(1)} \ldots n_{P(N)}\rangle = (\pm)^P |n_1 \ldots n_N\rangle,$$

as follows immediately from (1.55). Like the position–spin kets, the kets $|n_1 \ldots n_N\rangle$ span the N-particle Hilbert space $\mathbb{H}_N$. The action of $\hat{d}_n$ on $|n_1 \ldots n_N\rangle$ is similar to the action of $\hat{\psi}(\mathbf{x})$ on $|\mathbf{x}_1 \ldots \mathbf{x}_N\rangle$. Using the (anti)commutation relation (1.54), we can move the $\hat{d}_n$-operator through the string of $d^\dagger$-operators:[5]

$$\begin{aligned}
\hat{d}_n|n_1 \ldots n_N\rangle &= \left(\left[\hat{d}_n, \hat{d}^\dagger_{n_N}\right]_\mp \pm \hat{d}^\dagger_{n_N}\hat{d}_n\right)|n_1 \ldots n_{N-1}\rangle \\
&= \delta_{nn_N}|n_1 \ldots n_{N-1}\rangle \pm \hat{d}^\dagger_{n_N}\left(\left[\hat{d}_n, \hat{d}^\dagger_{n_{N-1}}\right]_\mp \pm \hat{d}^\dagger_{n_{N-1}}\hat{d}_n\right)|n_1 \ldots n_{N-2}\rangle \\
&= \delta_{nn_N}|n_1 \ldots n_{N-1}\rangle \pm \delta_{nn_{N-1}}|n_1 \ldots n_{N-2}n_N\rangle \\
&\quad (\pm)^2 \hat{d}^\dagger_{n_N}\hat{d}^\dagger_{n_{N-1}}\left(\left[\hat{d}_n, \hat{d}^\dagger_{n_{N-2}}\right]_\mp \pm \hat{d}^\dagger_{n_{N-2}}\hat{d}_n\right)|n_1 \ldots n_{N-3}\rangle \\
&= \sum_{k=1}^{N}(\pm)^{N+k}\delta_{nn_k}|n_1 \ldots n_{k-1}n_{k+1} \ldots n_N\rangle.
\end{aligned} \tag{1.57}$$

This result can also be used to calculate directly the overlap between two states of the general basis. For example, for the case of two particles we have

$$\begin{aligned}
\langle n'_1 n'_2|n_1 n_2\rangle &= \langle n'_1|\hat{d}_{n'_2}|n_1 n_2\ \rangle = \langle n'_1|\left(\delta_{n'_2 n_2}|n_1\rangle \pm \delta_{n'_2 n_1}|n_2\rangle\right) \\
&= \delta_{n'_1 n_1}\delta_{n'_2 n_2} \pm \delta_{n'_1 n_2}\delta_{n'_2 n_1},
\end{aligned}$$

[5] Alternatively (1.57) can be derived from (1.44) together with the definitions of the $\hat{d}$-operators.

which is the analog of (1.26). More generally, for N particles we have

$$\langle n'_1 \dots n'_N | n_1 \dots n_N \rangle = \sum_P (\pm)^P \prod_{j=1}^{N} \delta_{n'_j \, n_{P(j)}}, \tag{1.58}$$

which should be compared with the overlap $\langle \mathbf{x}'_1 \dots \mathbf{x}'_N | \mathbf{x}_1 \dots \mathbf{x}_N \rangle$ in (1.30).

The states $|n_1 \dots n_N\rangle$ are orthonormal (with the exception of the bosonic kets with two or more equal quantum numbers) and can be used to construct a basis. In analogy with (1.32), the completeness relation is

$$\frac{1}{N!} \sum_{n_1, \dots, n_N} |n_1 \dots n_N\rangle\langle n_1 \dots n_N| = \hat{\mathbb{1}},$$

and hence the expansion of a ket $|\Psi\rangle$ belonging to $\mathbb{H}_N$ reads

$$|\Psi\rangle = \hat{\mathbb{1}}|\Psi\rangle = \frac{1}{N!} \sum_{n_1, \dots, n_N} |n_1 \dots n_N\rangle \underbrace{\langle n_1 \dots n_N | \Psi\rangle}_{\Psi(n_1, \dots, n_N)}. \tag{1.59}$$

If $|\Psi\rangle$ is normalized, then the coefficients $\Psi(n_1, \dots, n_N)$ have the following probabilistic interpretation

$$\frac{|\Psi(\overbrace{n_1 \dots n_1}^{m_1} \dots \overbrace{n_M \dots n_M}^{m_M})|^2}{m_1! \dots m_M!} = \begin{pmatrix} \text{probability of finding} \\ m_1 \text{ particles with quantum number } n_1 \\ \vdots \\ m_M \text{ particles with quantum number } n_M \end{pmatrix}.$$

We already observed that the $\hat{d}$-operators obey the same (anti)commutation relations as the field operators provided that $\{|n\rangle\}$ is an orthonormal basis in $\mathbb{H}_1$. Likewise, we can construct linear combinations of the $\hat{d}$-operators that preserve the (anti)commutation relations. It is left as an exercise for the reader to prove that the operators

$$\hat{c}_\alpha = \sum_n U_{\alpha n} \hat{d}_n, \qquad \hat{c}^\dagger_\alpha = \sum_n U^*_{\alpha n} \hat{d}^\dagger_n$$

obey

$$\left[\hat{c}_\alpha, \hat{c}^\dagger_\beta\right]_\mp = \delta_{\alpha\beta},$$

provided that

$$U_{\alpha n} \equiv \langle \alpha | n \rangle$$

is the inner product between the elements of the original orthonormal basis $\{|n\rangle\}$ and the elements of another orthonormal basis $\{|\alpha\rangle\}$. Indeed, in this case the $U_{\alpha n}$ are the matrix elements of a unitary matrix since

$$\sum_n U_{\alpha n} U^\dagger_{n\beta} = \sum_n \langle \alpha | n \rangle \langle n | \beta \rangle = \langle \alpha | \beta \rangle = \delta_{\alpha\beta},$$

where we use the completeness relation. In particular, when $\alpha = \mathbf{x}$ we have $U_{\mathbf{x}n} = \langle \mathbf{x}|n\rangle = \varphi_n(\mathbf{x})$, and we find that $\hat{c}_{\mathbf{x}} = \hat{\psi}(\mathbf{x})$. We thus recover the field operators as

$$\hat{\psi}(\mathbf{x}) = \sum_n \varphi_n(\mathbf{x})\hat{d}_n, \qquad \hat{\psi}^\dagger(\mathbf{x}) = \sum_n \varphi_n^*(\mathbf{x})\hat{d}_n^\dagger. \tag{1.60}$$

These relations tell us that the expansion of the position–spin kets in terms of the kets $|n_1 \ldots n_N\rangle$ is simply

$$\boxed{|\mathbf{x}_1 \ldots \mathbf{x}_N\rangle = \sum_{n_1 \ldots n_N} \varphi_{n_1}^*(\mathbf{x}_1) \ldots \varphi_{n_N}^*(\mathbf{x}_N)|n_1 \ldots n_N\rangle} \tag{1.61}$$

Conversely, using (1.52) we can expand the general basis kets in terms of the position–spin kets as

$$\boxed{|n_1 \ldots n_N\rangle = \int d\mathbf{x}_1 \ldots d\mathbf{x}_N\, \varphi_{n_1}(\mathbf{x}_1) \ldots \varphi_{n_N}(\mathbf{x}_N)|\mathbf{x}_1 \ldots \mathbf{x}_N\rangle} \tag{1.62}$$

Slater determinants If we are given a state $|\Psi\rangle$ that is expanded in a general basis and we subsequently want to calculate properties in position–spin space, such as the particle density or the current density, we need to calculate the overlap between $|n_1 \ldots n_N\rangle$ and $|\mathbf{x}_1 \ldots \mathbf{x}_N\rangle$. This overlap is the wavefunction for N particles with quantum numbers $n_1, \ldots, n_N$:

$$\Psi_{n_1 \ldots n_N}(\mathbf{x}_1, \ldots, \mathbf{x}_N) = \langle \mathbf{x}_1 \ldots \mathbf{x}_N | n_1 \ldots n_N\rangle.$$

The explicit form of the wavefunction follows directly from the inner product (1.30) and from the expansion (1.62), and reads

$$\begin{aligned}\Psi_{n_1 \ldots n_N}(\mathbf{x}_1, \ldots, \mathbf{x}_N) &= \sum_P (\pm)^P \varphi_{n_1}(\mathbf{x}_{P(1)}) \ldots \varphi_{n_N}(\mathbf{x}_{P(N)}) \\ &= \begin{vmatrix} \varphi_{n_1}(\mathbf{x}_1) & \ldots & \varphi_{n_1}(\mathbf{x}_N) \\ \cdot & \ldots & \cdot \\ \cdot & \ldots & \cdot \\ \varphi_{n_N}(\mathbf{x}_1) & \ldots & \varphi_{n_N}(\mathbf{x}_N) \end{vmatrix}_\pm .\end{aligned} \tag{1.63}$$

Since for any matrix A we have $|A|_\mp = |A^T|_\mp$ with A^T the transpose of A, we can equivalently write

$$\Psi_{n_1 \ldots n_N}(\mathbf{x}_1, \ldots, \mathbf{x}_N) = \begin{vmatrix} \varphi_{n_1}(\mathbf{x}_1) & \ldots & \varphi_{n_N}(\mathbf{x}_1) \\ \cdot & \ldots & \cdot \\ \cdot & \ldots & \cdot \\ \varphi_{n_1}(\mathbf{x}_N) & \ldots & \varphi_{n_N}(\mathbf{x}_N) \end{vmatrix}_\pm .$$

In the case of fermions, the determinant is also known as the *Slater determinant.* For those readers already familiar with Slater determinants we note that the absence on the r.h.s. of the prefactor $1/\sqrt{N!}$ is a consequence of forcing on the square modulus of the wavefunction a probability interpretation, as discussed in detail in Section 1.3. The action of the $\hat{d}$-operators has a simple algebraic interpretation in terms of permanents or determinants.

The action of the creation operator $\hat{d}_n^\dagger$ on $|n_1 \dots n_N\rangle$ in the position–spin representation, $\langle \mathbf{x}_1 \dots \mathbf{x}_{N+1}|\hat{d}_n^\dagger|n_1 \dots n_N\rangle$, simply amounts to adding a column with coordinate $\mathbf{x}_{N+1}$ and a row with wavefunction φ_n in (1.63). For the annihilation operator we have a similar algebraic interpretation. Taking the inner product with $\langle \mathbf{x}_1 \dots \mathbf{x}_{N-1}|$ of both sides of (1.57), we get

$$\langle \mathbf{x}_1 \dots \mathbf{x}_{N-1}|\hat{d}_n|n_1 \dots n_N\rangle = \sum_{k=1}^{N} (\pm)^{N+k}\, \delta_{n n_k}$$
$$\times \left| \begin{array}{cccccc} \varphi_{n_1}(\mathbf{x}_1) & \dots & \dots & \varphi_{n_1}(\mathbf{x}_{N-1}) & \varphi_{n_1}(\mathbf{x}_N) \\ \vdots & \vdots & \vdots & \vdots & \\ \varphi_{n_{k-1}}(\mathbf{x}_1) & \dots & \dots & \varphi_{n_{k-1}}(\mathbf{x}_{N-1}) & \varphi_{n_{k-1}}(\mathbf{x}_N) \\ \varphi_{n_k}(\mathbf{x}_1) & \dots & \dots & \varphi_{n_k}(\mathbf{x}_{N-1}) & \varphi_{n_k}(\mathbf{x}_N) \\ \varphi_{n_{k+1}}(\mathbf{x}_1) & \dots & \dots & \varphi_{n_{k+1}}(\mathbf{x}_{N-1}) & \varphi_{n_{k+1}}(\mathbf{x}_N) \\ \vdots & \vdots & \vdots & \vdots & \\ \varphi_{n_N}(\mathbf{x}_1) & \dots & \dots & \varphi_{n_N}(\mathbf{x}_{N-1}) & \varphi_{n_N}(\mathbf{x}_N) \end{array} \right|_{\pm} .$$

That is, the action of the $\hat{d}_n$-operator amounts to deleting the last column and, if present, the row with quantum number n from the permanent/determinant of (1.63), and otherwise yields zero. Already at this stage the reader can appreciate how powerful it is to work with the field operators and not to have anything to do with Slater determinants.

Exercise 1.2 Prove the inverse relations (1.60).

Exercise 1.3 Let $|n\rangle = |\mathbf{p}\tau\rangle$ be a momentum–spin ket so that $\langle \mathbf{x}|\mathbf{p}\tau\rangle = e^{\mathrm{i}\mathbf{p}\cdot\mathbf{r}}\delta_{\sigma\tau}$, see (1.10). Show that the (anti)commutation relation in (1.54) then reads

$$\left[\hat{d}_{\mathbf{p}\tau}, \hat{d}^\dagger_{\mathbf{p}'\tau'}\right]_\mp = (2\pi)^3 \delta(\mathbf{p}-\mathbf{p}')\delta_{\tau\tau'}, \tag{1.64}$$

and that the expansion (1.60) of the field operators in terms of the $\hat{d}$-operators is

$$\hat{\psi}(\mathbf{x}) = \int \frac{d\mathbf{p}}{(2\pi)^3} e^{\mathrm{i}\mathbf{p}\cdot\mathbf{r}} \hat{d}_{\mathbf{p}\sigma}, \qquad \hat{\psi}^\dagger(\mathbf{x}) = \int \frac{d\mathbf{p}}{(2\pi)^3} e^{-\mathrm{i}\mathbf{p}\cdot\mathbf{r}} \hat{d}^\dagger_{\mathbf{p}\sigma}. \tag{1.65}$$

1.6 Hamiltonian in Second Quantization

The field operators are useful not only to construct the kets of N identical particles but also the operators acting on them. Let us consider again two identical particles and the total potential energy (1.16), with $N = 2$ and $q_1 = q_2 = q$. In first quantization the ket $|\mathbf{x}_1\mathbf{x}_2\rangle$ is represented by the (anti)symmetrized product (1.35) of one-particle kets. It is instructive to calculate the action of $\hat{\mathcal{H}}_{\text{pot}}$ on $|\mathbf{x}_1\mathbf{x}_2\rangle$ to later appreciate the advantages of second quantization. We have

$$\begin{aligned}
\hat{\mathcal{H}}_{\text{pot}}\frac{|\mathbf{x}_1\rangle|\mathbf{x}_2\rangle \pm |\mathbf{x}_2\rangle|\mathbf{x}_1\rangle}{\sqrt{2}} &= q\frac{\phi(\mathbf{r}_1)|\mathbf{x}_1\rangle|\mathbf{x}_2\rangle \pm \phi(\mathbf{r}_2)|\mathbf{x}_2\rangle|\mathbf{x}_1\rangle + \phi(\mathbf{r}_2)|\mathbf{x}_1\rangle|\mathbf{x}_2\rangle \pm \phi(\mathbf{r}_1)|\mathbf{x}_2\rangle|\mathbf{x}_1\rangle}{\sqrt{2}} \\
&= q\left[\phi(\mathbf{r}_1) + \phi(\mathbf{r}_2)\right]\frac{|\mathbf{x}_1\rangle|\mathbf{x}_2\rangle \pm |\mathbf{x}_2\rangle|\mathbf{x}_1\rangle}{\sqrt{2}}.
\end{aligned} \tag{1.66}$$

Throughout this book we use calligraphic letters for operators acting on kets written in first quantization as opposed to operators (such as the field operators) acting on kets written in second quantization (e.g., $\hat{\psi}^\dagger(\mathbf{x}_N)\ldots\hat{\psi}^\dagger(\mathbf{x}_1)|0\rangle$). We refer to the former as *operators in first quantization* and to the latter as *operators in second quantization*. We now show that the very same result (1.66) can be obtained if we write the potential energy operator as

$$\hat{H}_{\text{pot}} = q\int d\mathbf{x}\,\phi(\mathbf{r})\,\hat{n}(\mathbf{x}),$$

where

$$\boxed{\hat{n}(\mathbf{x}) = \hat{\psi}^\dagger(\mathbf{x})\hat{\psi}(\mathbf{x})} \tag{1.67}$$

is the so-called *density operator* already introduced in Exercise 1.1. The origin of this name for the operator $\hat{n}(\mathbf{x})$ stems from the fact that $|\mathbf{x}_1\ldots\mathbf{x}_N\rangle$ is an eigenket of the density operator whose eigenvalue is exactly the density of N particles in the position–spin coordinates $\mathbf{x}_1,\ldots,\mathbf{x}_N$. Indeed,

$$\begin{aligned}
\hat{n}(\mathbf{x})\underbrace{\hat{\psi}^\dagger(\mathbf{x}_N)\hat{\psi}^\dagger(\mathbf{x}_{N-1})\ldots\hat{\psi}^\dagger(\mathbf{x}_1)|0\rangle}_{|\mathbf{x}_1\ldots\mathbf{x}_N\rangle} &= \left[\hat{n}(\mathbf{x}),\hat{\psi}^\dagger(\mathbf{x}_N)\right]_-\hat{\psi}^\dagger(\mathbf{x}_{N-1})\ldots\hat{\psi}^\dagger(\mathbf{x}_1)|0\rangle \\
&\quad + \hat{\psi}^\dagger(\mathbf{x}_N)\left[\hat{n}(\mathbf{x}),\hat{\psi}^\dagger(\mathbf{x}_{N-1})\right]_-\ldots\hat{\psi}^\dagger(\mathbf{x}_1)|0\rangle \\
&\quad\ \vdots \\
&\quad + \hat{\psi}^\dagger(\mathbf{x}_N)\hat{\psi}^\dagger(\mathbf{x}_{N-1})\ldots\left[\hat{n}(\mathbf{x}),\hat{\psi}^\dagger(\mathbf{x}_1)\right]_-|0\rangle \\
&= \underbrace{\left(\sum_{i=1}^{N}\delta(\mathbf{x}-\mathbf{x}_i)\right)}_{\substack{\text{density of } N \text{ particles} \\ \text{in } \mathbf{x}_1,\ldots,\mathbf{x}_N}}|\mathbf{x}_1\ldots\mathbf{x}_N\rangle,
\end{aligned} \tag{1.68}$$

where we repeatedly use (1.49). This result tells us that any ket with N particles is an eigenket of the operator

$$\boxed{\hat{N} \equiv \int d\mathbf{x}\,\hat{n}(\mathbf{x})}$$

with eigenvalue N. For this reason $\hat{N}$ is called the *operator of the total number of particles.*

By acting with $\hat{H}_{\rm pot}$ on $|\mathbf{x}_1\mathbf{x}_2\rangle = \hat{\psi}^\dagger(\mathbf{x}_2)\hat{\psi}^\dagger(\mathbf{x}_1)|0\rangle$ and taking into account (1.68), we find

$$\hat{H}_{\rm pot}|\mathbf{x}_1\mathbf{x}_2\rangle = q\int d\mathbf{x}\,\phi(\mathbf{r})\sum_{i=1}^{2}\delta(\mathbf{x}-\mathbf{x}_i)|\mathbf{x}_1\mathbf{x}_2\rangle = q\big[\phi(\mathbf{r}_1+\phi(\mathbf{r}_2)\big]|\mathbf{x}_1\mathbf{x}_2\rangle.$$

Simple and elegant! Both the operator and the ket are easy to manipulate and their expressions are undoubtedly shorter than the corresponding expressions in first quantization. A further advantage of second quantization is that the operator $\hat{H}_{\rm pot}$ keeps the very same form independently of the number of particles; using (1.68) it is straightforward to verify that

$$\hat{H}_{\rm pot}|\mathbf{x}_1\ldots\mathbf{x}_N\rangle = q\left(\sum_{i=1}^{N}\phi(\mathbf{r}_i)\right)|\mathbf{x}_1\ldots\mathbf{x}_N\rangle.$$

To the contrary, $\hat{\mathcal{H}}_{\rm pot}$ in (1.16) acts only on kets belonging to $\mathbb{H}_N$. Thus, when working in Fock space it would be more rigorous to specify on which Hilbert space $\hat{\mathcal{H}}_{\rm pot}$ acts. Denoting by $\hat{\mathcal{H}}_{\rm pot}(N)$ the operator in (1.16), we can write down the relation between operators in first and second quantization as

$$\hat{H}_{\rm pot} = \sum_{N=0}^{\infty}\hat{\mathcal{H}}_{\rm pot}(N),$$

with the extra rule that $\hat{\mathcal{H}}_{\rm pot}(N)$ yields the null ket when acting on a state of $\mathbb{H}_{M\neq N}$. In this book, however, we are not so meticulous with the notation. The Hilbert space on which operators in first quantization act is clear from the context.

The goal of this section is to extend the above example to general operators and in particular to derive an expression for the many-particle Hamiltonian $\hat{H} = \hat{H}_0 + \hat{H}_{\rm int}$. According to (1.17), the matrix element of the noninteracting Hamiltonian $\hat{H}_0$ between a position-spin ket and a generic ket $|\Psi\rangle$ is

$$\langle\mathbf{x}_1\ldots\mathbf{x}_N|\hat{H}_0|\Psi\rangle = \sum_{j=1}^{N}\sum_{\sigma'}h_{\sigma_j\sigma'}(\mathbf{r}_j,-\mathrm{i}\boldsymbol{\nabla}_j,\mathbf{S})\,\Psi(\mathbf{x}_1,\ldots,\mathbf{x}_{j-1},\mathbf{r}_j\sigma',\mathbf{x}_{j+1},\ldots,\mathbf{x}_N). \tag{1.69}$$

It is worth observing that for $N=1$ this expression reduces to

$$\langle\mathbf{x}|\hat{H}_0|\Psi\rangle = \sum_{\sigma'}h_{\sigma\sigma'}(\mathbf{r},-\mathrm{i}\boldsymbol{\nabla},\mathbf{S})\Psi(\mathbf{r}\sigma'),$$

which agrees with (1.13) when $|\Psi\rangle = |\mathbf{x}''\rangle$ since in this case $\Psi(\mathbf{r}\sigma') = \langle\mathbf{r}\sigma'|\mathbf{r}''\sigma''\rangle = \delta(\mathbf{r}-\mathbf{r}'')\delta_{\sigma'\sigma''}$. Similarly, we see from (1.18) that the matrix element of the interaction Hamiltonian $\hat{H}_{\rm int}$ between a position-spin ket and a generic ket $|\Psi\rangle$ is

$$\langle\mathbf{x}_1\ldots\mathbf{x}_N|\hat{H}_{\rm int}|\Psi\rangle = \frac{1}{2}\sum_{i\neq j}^{N}v(\mathbf{x}_i,\mathbf{x}_j)\,\Psi(\mathbf{x}_1,\ldots,\mathbf{x}_N). \tag{1.70}$$

In (1.70) we consider the more general case of spin-dependent interactions $v(\mathbf{x}_1, \mathbf{x}_2)$, according to which the interaction energy between a particle in $\mathbf{r}_1$ and a particle in $\mathbf{r}_2$ depends also on the spin orientation σ_1 and σ_2 of these particles. Now the question is: How do we express $\hat{H}_0$ and $\hat{H}_{\text{int}}$ in terms of field operators?

Noninteracting Hamiltonian We start our discussion with the noninteracting Hamiltonian. For pedagogical purposes we derive the operator $\hat{H}_0$ in second quantization in two different ways.

Derivation I: In first quantization the noninteracting Hamiltonian $\hat{\mathcal{H}}_0$ of a system of N particles each described by $\hat{h}$ is given in (1.17). The first quantization eigenkets of $\hat{\mathcal{H}}_0$ are obtained by forming (anti)symmetrized products of one-particle eigenkets of $\hat{h}$, and look like

$$|n_1 \ldots n_N\rangle = \frac{1}{\sqrt{N!}} \sum_P (\pm)^P |n_{P(1)}\rangle \ldots |n_{P(N)}\rangle, \tag{1.71}$$

with

$$\hat{h}|n\rangle = \epsilon_n |n\rangle.$$

We leave it as an exercise for the reader to show that

$$\hat{\mathcal{H}}_0 |n_1 \ldots n_N\rangle = (\epsilon_{n_1} + \ldots + \epsilon_{n_N}) \, |n_1 \ldots n_N\rangle.$$

The proof of this identity involves the same kind of manipulations used to derive (1.66). To carry them out is useful to appreciate the simplicity of second quantization. We show below that in second quantization the noninteracting Hamiltonian $\hat{H}_0$ takes the compact form

$$\boxed{\hat{H}_0 = \int d\mathbf{x} d\mathbf{x}' \, \hat{\psi}^\dagger(\mathbf{x}) \langle \mathbf{x}|\hat{h}|\mathbf{x}'\rangle \hat{\psi}(\mathbf{x}')} \tag{1.72}$$

independently of the number of particles. We prove (1.72) by showing that the second quantization ket $|n_1 \ldots n_N\rangle$ is an eigenket of $\hat{H}_0$ with eigenvalue $\epsilon_{n_1} + \ldots + \epsilon_{n_N}$. In second quantization $|n_1 \ldots n_N\rangle = \hat{d}^\dagger_{n_N} \ldots \hat{d}^\dagger_{n_1}|0\rangle$, with the $\hat{d}$-operators defined in (1.52) and (1.53). It is then natural to express $\hat{H}_0$ in terms of the $\hat{d}$-operators. Inserting a completeness relation between $\hat{h}$ and $|\mathbf{x}'\rangle$, we find

$$\begin{aligned}
\hat{H}_0 &= \sum_n \int d\mathbf{x} d\mathbf{x}' \, \hat{\psi}^\dagger(\mathbf{x}) \langle \mathbf{x}|\hat{h}|n\rangle \langle n|\mathbf{x}'\rangle \hat{\psi}(\mathbf{x}') \\
&= \sum_n \epsilon_n \int d\mathbf{x} \, \hat{\psi}^\dagger(\mathbf{x}) \underbrace{\langle \mathbf{x}|n\rangle}_{\varphi_n(\mathbf{x})} \int d\mathbf{x}' \underbrace{\langle n|\mathbf{x}'\rangle}_{\varphi_n^*(\mathbf{x}')} \hat{\psi}(\mathbf{x}') = \sum_n \epsilon_n \hat{d}^\dagger_n \hat{d}_n,
\end{aligned} \tag{1.73}$$

where we use $\hat{h}|n\rangle = \epsilon_n|n\rangle$. The $\hat{d}$-operators bring the Hamiltonian into a diagonal form – that is, none of the off-diagonal combinations $\hat{d}^\dagger_n d_m$ with $m \neq n$ appear in $\hat{H}_0$. The *occupation operator*

$$\hat{n}_n \equiv \hat{d}^\dagger_n \hat{d}_n \tag{1.74}$$

is the analog of the density operator $\hat{n}(\mathbf{x})$ in the position–spin basis; it counts how many particles have quantum number n. Using the (anti)commutation relations (1.54) and (1.55), it is easy to prove that

$$\left[\hat{n}_n, \hat{d}^\dagger_m\right]_- = \delta_{nm}\hat{d}^\dagger_m, \qquad \left[\hat{n}_n, \hat{d}_m\right]_- = -\delta_{nm}\hat{d}_m, \tag{1.75}$$

which should be compared with the relations (1.48) and (1.49). The action of $\hat{H}_0$ on $|n_1 \ldots n_N\rangle$ is then

$$\begin{aligned}
\hat{H}_0 \underbrace{\hat{d}^\dagger_{n_N}\hat{d}^\dagger_{n_{N-1}}\ldots\hat{d}^\dagger_{n_1}|0\rangle}_{|n_1\ldots n_N\rangle} &= \sum_n \epsilon_n \Big(\left[\hat{n}_n, \hat{d}^\dagger_{n_N}\right]_- \hat{d}^\dagger_{n_{N-1}}\ldots\hat{d}^\dagger_{n_1}|0\rangle \\
&\quad + \hat{d}^\dagger_{n_N}\left[\hat{n}_n, \hat{d}^\dagger_{n_{N-1}}\right]_- \ldots\hat{d}^\dagger_{n_1}|0\rangle \\
&\quad \vdots \\
&\quad + \hat{d}^\dagger_{n_N}\hat{d}^\dagger_{n_{N-1}}\ldots\left[\hat{n}_n, \hat{d}^\dagger_{n_1}\right]_- |0\rangle \Big) \\
&= (\epsilon_{n_1} + \ldots + \epsilon_{n_N})\,\hat{d}^\dagger_{n_N}\hat{d}^\dagger_{n_{N-1}}\ldots\hat{d}^\dagger_{n_1}|0\rangle.
\end{aligned} \tag{1.76}$$

This is exactly the result we wanted to prove: The Hamiltonian $\hat{H}_0$ is the correct second quantized form of $\hat{\mathcal{H}}_0$. We can write $\hat{H}_0$ in different ways using the matrix elements (1.13) of $\hat{h}$. For instance,

$$\hat{H}_0 = \sum_{\sigma\sigma'} \int d\mathbf{r}\, \hat{\psi}^\dagger(\mathbf{r}\sigma) h_{\sigma\sigma'}(\mathbf{r}, -\mathrm{i}\boldsymbol{\nabla}, \mathbf{S}) \hat{\psi}(\mathbf{r}\sigma'), \tag{1.77}$$

or, equivalently,

$$\hat{H}_0 = \sum_{\sigma\sigma'} \int d\mathbf{r}\, \hat{\psi}^\dagger(\mathbf{r}\sigma) h_{\sigma\sigma'}(\mathbf{r}, \mathrm{i}\overleftarrow{\boldsymbol{\nabla}}, \mathbf{S}) \hat{\psi}(\mathbf{r}\sigma'). \tag{1.78}$$

In these expressions the action of the gradient $\boldsymbol{\nabla}$ on a field operator is a formal expression which makes sense only when we sandwich $\hat{H}_0$ with a bra and a ket. For instance,

$$\langle\chi|\hat{\psi}^\dagger(\mathbf{r}\sigma)\boldsymbol{\nabla}\hat{\psi}(\mathbf{r}\sigma')|\Psi\rangle \equiv \lim_{\mathbf{r}'\to\mathbf{r}} \boldsymbol{\nabla}'\langle\chi|\hat{\psi}^\dagger(\mathbf{r}\sigma)\hat{\psi}(\mathbf{r}'\sigma')|\Psi\rangle, \tag{1.79}$$

where $\boldsymbol{\nabla}'$ is the gradient with respect to the primed variable. It is important to observe that for any arbitrary large but finite system the physical states have no particles at infinity. Therefore, if $|\chi\rangle$ and $|\Psi\rangle$ are physical states, then (1.79) vanishes when $|\mathbf{r}| \to \infty$. More generally, the sandwich of a string of field operators $\hat{\psi}^\dagger(\mathbf{x}_1)\ldots\hat{\psi}^\dagger(\mathbf{x}_N)\hat{\psi}(\mathbf{y}_1)\ldots\hat{\psi}(\mathbf{y}_M)$ with two physical states vanishes when one of the coordinates of the field operators approaches infinity. The equivalence between (1.77) and (1.78) has to be understood as an equivalence between the sandwich of the corresponding r.h.s. with physical states. Consider, for example, $\hat{h} = \hat{p}^2/2m$. Equating the r.h.s. of (1.77) and (1.78) we get

$$\sum_\sigma \int d\mathbf{r}\, \hat{\psi}^\dagger(\mathbf{r}\sigma)\left[-\frac{\nabla^2}{2m}\hat{\psi}(\mathbf{r}\sigma)\right] = \sum_\sigma \int d\mathbf{r} \left[-\frac{\nabla^2}{2m}\hat{\psi}^\dagger(\mathbf{r}\sigma)\right]\hat{\psi}(\mathbf{r}\sigma).$$

This is an equality only provided that the integration by parts produces a vanishing boundary term – that is, only provided that for any two physical states $|\chi\rangle$ and $|\Psi\rangle$ the quantity in (1.79) vanishes when $|\mathbf{r}| \to \infty$.

Derivation 2: The second derivation consists in showing that the matrix elements of (1.77) or (1.78) are given by (1.69). Using (1.44), we find

$$\begin{aligned}
&\langle \mathbf{x}_1 \ldots \mathbf{x}_N | \hat{\psi}^\dagger(\mathbf{r}\sigma) h_{\sigma\sigma'}(\mathbf{r}, -\mathrm{i}\nabla, \mathbf{S}) \hat{\psi}(\mathbf{r}\sigma') | \Psi \rangle \\
&= \lim_{\mathbf{r}' \to \mathbf{r}} h_{\sigma\sigma'}(\mathbf{r}', -\mathrm{i}\nabla', \mathbf{S}) \sum_{j=1}^{N} (\pm)^{N+j} \delta(\mathbf{x}_j - \mathbf{x}) \langle \mathbf{x}_1 \ldots \mathbf{x}_{j-1} \mathbf{x}_{j+1} \ldots \mathbf{x}_N | \hat{\psi}(\mathbf{x}') | \Psi \rangle \\
&= \lim_{\mathbf{r}' \to \mathbf{r}} h_{\sigma\sigma'}(\mathbf{r}', -\mathrm{i}\nabla', \mathbf{S}) \sum_{j=1}^{N} \delta(\mathbf{x}_j - \mathbf{x}) \Psi(\mathbf{x}_1, \ldots \mathbf{x}_{j-1}, \mathbf{x}', \mathbf{x}_{j+1}, \ldots, \mathbf{x}_N),
\end{aligned}$$

where we use that it requires $N - j$ interchanges to put $\mathbf{x}'$ at the position between $\mathbf{x}_{j-1}$ and $\mathbf{x}_{j+1}$. Summing over σ, σ' and integrating over $\mathbf{r}$ we get

$$\langle \mathbf{x}_1 \ldots \mathbf{x}_N | \hat{H}_0 | \Psi \rangle = \sum_{j=1}^{N} \sum_{\sigma'} \lim_{\mathbf{r}' \to \mathbf{r}_j} h_{\sigma_j\sigma'}(\mathbf{r}', -\mathrm{i}\nabla', \mathbf{S}) \Psi(\mathbf{x}_1, \ldots, \mathbf{x}_{j-1}, \mathbf{x}', \mathbf{x}_{j+1}, \ldots, \mathbf{x}_N), \tag{1.80}$$

which coincides with the matrix element (1.69). Here and in the following we call *one-body operators* those operators in second quantization that can be written as a quadratic form of the field operators. The Hamiltonian $\hat{H}_0$ and the potential energy operator $\hat{H}_{\text{pot}}$ are one-body operators.

Interacting Hamiltonian From (1.76) it is evident that one-body Hamiltonians can only describe noninteracting systems since the eigenvalues are the sum of one-particle eigenvalues, and the latter do not depend on the position of the other particles. If there is an interaction $v(\mathbf{x}_1, \mathbf{x}_2)$ between one particle in $\mathbf{x}_1$ and another particle in $\mathbf{x}_2$, the corresponding interaction energy operator $\hat{H}_{\text{int}}$ cannot be a one-body operator. The energy to put a particle in a given point depends on where the other particles are located. Suppose that there is a particle in $\mathbf{x}_1$. Then if we want to put a particle in $\mathbf{x}_2$ we must pay an energy $v(\mathbf{x}_1, \mathbf{x}_2)$. The addition of another particle in $\mathbf{x}_3$ costs an energy $v(\mathbf{x}_1, \mathbf{x}_3) + v(\mathbf{x}_2, \mathbf{x}_3)$. In general, if we have N particles in $\mathbf{x}_1, \ldots, \mathbf{x}_N$ the total interaction energy is $\sum_{i<j} v(\mathbf{x}_i, \mathbf{x}_j) = \frac{1}{2} \sum_{i \neq j} v(\mathbf{x}_i, \mathbf{x}_j)$. To derive the form of $\hat{H}_{\text{int}}$ in second quantization we simply notice that the ket $|\mathbf{x}_1 \ldots \mathbf{x}_N\rangle$ is an eigenket of $\hat{H}_{\text{int}}$ with eigenvalue $\frac{1}{2} \sum_{i \neq j} v(\mathbf{x}_i, \mathbf{x}_j)$:

$$\hat{H}_{\text{int}} |\mathbf{x}_1 \ldots \mathbf{x}_N\rangle = \left(\frac{1}{2} \sum_{i \neq j} v(\mathbf{x}_i, \mathbf{x}_j) \right) |\mathbf{x}_1 \ldots \mathbf{x}_N\rangle. \tag{1.81}$$

Equivalently, (1.81) follows directly from the matrix element (1.70), which is valid for all $|\Psi\rangle$. Due to the presence of a double sum in (1.81) the operator $\hat{H}_{\text{int}}$ must be a *quartic* form in the field operators. In (1.68) we proved that $|\mathbf{x}_1 \ldots \mathbf{x}_N\rangle$ is an eigenket of the density operator

$\hat{n}(\mathbf{x})$ with eigenvalue $\sum_i \delta(\mathbf{x}-\mathbf{x}_i)$. This implies that $|\mathbf{x}_1 \ldots \mathbf{x}_N\rangle$ is also an eigenket of the operator $\hat{n}(\mathbf{x})\hat{n}(\mathbf{x}')$ with eigenvalue $\sum_{i,j} \delta(\mathbf{x}-\mathbf{x}_i)\delta(\mathbf{x}'-\mathbf{x}_j)$. Thus, taking into account that the double sum in (1.81) does not contain terms with $i=j$, the interaction energy operator is given by

$$\begin{aligned}\hat{H}_{\text{int}} &= \frac{1}{2}\int d\mathbf{x}\, d\mathbf{x}'\, v(\mathbf{x},\mathbf{x}')\hat{n}(\mathbf{x})\hat{n}(\mathbf{x}') - \frac{1}{2}\int d\mathbf{x}\, v(\mathbf{x},\mathbf{x})\hat{n}(\mathbf{x}) \\ &= \frac{1}{2}\int d\mathbf{x}\, d\mathbf{x}'\, v(\mathbf{x},\mathbf{x}')\left(\hat{\psi}^\dagger(\mathbf{x})\hat{\psi}(\mathbf{x})\hat{\psi}^\dagger(\mathbf{x}')\hat{\psi}(\mathbf{x}') - \delta(\mathbf{x}-\mathbf{x}')\hat{\psi}^\dagger(\mathbf{x})\hat{\psi}(\mathbf{x})\right) \\ &= \frac{1}{2}\int d\mathbf{x}\, d\mathbf{x}'\, v(\mathbf{x},\mathbf{x}')\hat{\psi}^\dagger(\mathbf{x})\hat{\psi}^\dagger(\mathbf{x}')\hat{\psi}(\mathbf{x}')\hat{\psi}(\mathbf{x}).\end{aligned} \tag{1.82}$$

In the last equality we first use the (anti)commutation relation (1.47) to cancel the term proportional to $\delta(\mathbf{x}-\mathbf{x}')$, and then (1.40) to exchange the operators $\hat{\psi}(\mathbf{x})$ and $\hat{\psi}(\mathbf{x}')$. It is easy to verify that the action of $\hat{H}_{\text{int}}$ on $|\mathbf{x}_1 \ldots \mathbf{x}_N\rangle$ yields (1.81). Like the one-body Hamiltonian $\hat{H}_0$, the interaction energy operator keeps the very same form independently of the number of particles. We call *two-body operators* those operators that can be written as a quartic form of the field operators and, in general, *n-body operators* those operators that contain a string of n field operators $\hat{\psi}^\dagger$ followed by a string of n field operators $\hat{\psi}$.

Total Hamiltonian The total Hamiltonian of a system of interacting identical particles is the sum of $\hat{H}_0$ and $\hat{H}_{\text{int}}$ and reads

$$\boxed{\hat{H} = \int d\mathbf{x} d\mathbf{x}'\, \hat{\psi}^\dagger(\mathbf{x})\langle \mathbf{x}|\hat{h}|\mathbf{x}'\rangle\hat{\psi}(\mathbf{x}') + \frac{1}{2}\int d\mathbf{x}\, d\mathbf{x}' v(\mathbf{x},\mathbf{x}')\hat{\psi}^\dagger(\mathbf{x})\hat{\psi}^\dagger(\mathbf{x}')\hat{\psi}(\mathbf{x}')\hat{\psi}(\mathbf{x})} \tag{1.83}$$

Equation (1.83) is the main result of this section. To calculate the action of $\hat{H}$ on a ket $|\Psi\rangle$ we only need to know the (anti)commutation relations since $|\Psi\rangle$ can always be expanded in terms of $\hat{\psi}^\dagger(\mathbf{x}_1)\ldots\hat{\psi}^\dagger(\mathbf{x}_N)|0\rangle$. Equivalently, given a convenient one-body basis $\{|n\rangle\}$ we may work with the $\hat{d}$-operators. This is done by expressing $\hat{H}$ in terms of the $\hat{d}$-operators, expanding $|\Psi\rangle$ on the basis $\hat{d}^\dagger_{n_1}\ldots\hat{d}^\dagger_{n_N}|0\rangle$, and then using the (anti)commutation relations (1.54) and (1.55). To express $\hat{H}$ in terms of the $\hat{d}$-operators, we simply substitute the expansion (1.60) in (1.83) and find

$$\hat{H} = \underbrace{\sum_{ij} h_{ij}\hat{d}^\dagger_i\hat{d}_j}_{\hat{H}_0} + \underbrace{\frac{1}{2}\sum_{ijmn} v_{ijmn}\hat{d}^\dagger_i\hat{d}^\dagger_j\hat{d}_m\hat{d}_n}_{\hat{H}_{\text{int}}}, \tag{1.84}$$

with

$$h_{ij} = \langle i|\hat{h}|j\rangle = \sum_{\sigma\sigma'}\int d\mathbf{r}\, \varphi^*_i(\mathbf{r}\sigma)h_{\sigma\sigma'}(\mathbf{r},-\mathrm{i}\boldsymbol{\nabla},\mathbf{S})\varphi_j(\mathbf{r}\sigma') = h^*_{ji}, \tag{1.85}$$

and the so-called *Coulomb integrals*[6]

$$v_{ijmn} = \int d\mathbf{x}\, d\mathbf{x}'\, \varphi^*_i(\mathbf{x})\varphi^*_j(\mathbf{x}')v(\mathbf{x},\mathbf{x}')\varphi_m(\mathbf{x}')\varphi_n(\mathbf{x}). \tag{1.86}$$

[6] In fact, the nomenclature Coulomb integral is appropriate only if v is the Coulomb interaction.

In the new basis the single-particle Hamiltonian in first quantization can be written in the ket–bra form

$$\hat{h} = \sum_{ij} h_{ij} |i\rangle\langle j|, \tag{1.87}$$

as can easily be checked by taking the matrix element $\langle i|\hat{h}|j\rangle$ and comparing with (1.85).

We recall that the quantum numbers of the general basis comprise an orbital and a spin quantum number. For later purposes it is instructive to highlight the spin structure in (1.84). We write the quantum numbers i, j, m, n as

$$i = s_1\sigma_1, \quad j = s_2\sigma_2, \quad m = s_3\sigma_3, \quad n = s_4\sigma_4.$$

Then the one-body part reads

$$\hat{H}_0 = \sum_{\substack{s_1 s_2 \\ \sigma_1\sigma_2}} h_{s_1\sigma_1\, s_2\sigma_2} \hat{d}^\dagger_{s_1\sigma_1} \hat{d}_{s_2\sigma_2}.$$

In the absence of magnetic fields or spin–orbit coupling, h does not depend on $\mathbf{S}$ and hence its matrix elements are diagonal in spin space $h_{ij} = \delta_{\sigma_1\sigma_2} h_{s_1 s_2}$. In this case $\hat{H}_0$ takes the simpler form

$$\hat{H}_0 = \sum_{s_1 s_2} \sum_{\sigma} h_{s_1 s_2} \hat{d}^\dagger_{s_1\sigma} \hat{d}_{s_2\sigma}, \tag{1.88}$$

where $h_{s_1 s_2}$ is the spatial integral in (1.85) with the functions $\varphi_s(\mathbf{r})$ defined in (1.50). For interparticle interactions $v(\mathbf{x}_1, \mathbf{x}_2) = v(\mathbf{r}_1, \mathbf{r}_2)$ which are independent of spin the interaction Hamiltonian can be manipulated in a similar manner. From (1.86) we see that v_{ijmn} vanishes if j and m have different spin projection ($\sigma_2 \neq \sigma_3$) or if i and n have different spin projection ($\sigma_1 \neq \sigma_4$):

$$v_{ijmn} = \delta_{\sigma_2\sigma_3} \delta_{\sigma_1\sigma_4} v_{s_1 s_2 s_3 s_4}, \tag{1.89}$$

where $v_{s_1 s_2 s_3 s_4}$ is the spatial integral in (1.86) with the functions $\varphi_s(\mathbf{r})$. Inserting this form of the interaction into $\hat{H}_{\text{int}}$ we find

$$\hat{H}_{\text{int}} = \frac{1}{2} \sum_{\substack{s_1 s_2 s_3 s_4 \\ \sigma\sigma'}} v_{s_1 s_2 s_3 s_4} \hat{d}^\dagger_{s_1\sigma} \hat{d}^\dagger_{s_2\sigma'} \hat{d}_{s_3\sigma'} \hat{d}_{s_4\sigma}. \tag{1.90}$$

We propose below a few simple exercises to practice with operators in second quantization. In the next chapter we illustrate physically relevant examples and use some of the identities from the exercises to acquire familiarity with the formalism of second quantization.

Exercise 1.4 Let $\hat{n}_n \equiv \hat{d}^\dagger_n \hat{d}_n$ be the occupation operator for particles with quantum number n, see (1.74). Prove that in the fermionic case

$$\hat{n}_n^2 = \hat{n}_n, \tag{1.91}$$

and hence that the eigenvalues of $\hat{n}_n$ are either 0 or 1 – that is, it is not possible to create two fermions in the same state $|n\rangle$. This is a direct consequence of the Pauli exclusion principle.

Exercise 1.5 Prove that the total number of particle operators $\hat{N} = \int d\mathbf{x}\, \hat{\psi}^\dagger(\mathbf{x})\hat{\psi}(\mathbf{x})$ can also be written as $\hat{N} = \sum_n \hat{d}_n^\dagger \hat{d}_n$ for any orthonormal basis $|n\rangle$. Calculate the action of $\hat{N}$ on a generic ket $|\Psi_N\rangle$ with N particles ($|\Psi_N\rangle \in \mathbb{H}_N$) and prove that

$$\hat{N}|\Psi_N\rangle = N|\Psi_N\rangle.$$

Exercise 1.6 Prove that $\hat{N}$ commutes with $\hat{H}_0$ and $\hat{H}_{\text{int}}$ - that is,

$$[\hat{N}, \hat{H}_0]_- = [\hat{N}, \hat{H}_{\text{int}}]_- = 0. \tag{1.92}$$

This means that the eigenkets of $\hat{H}$ can be chosen as kets with a fixed number of particles.

Exercise 1.7 Let $n = s\sigma$ and $\sigma = \uparrow, \downarrow$ be the spin projection for fermions of spin $1/2$. We consider the operators

$$\hat{S}^z_s \equiv \frac{1}{2}(\hat{n}_{s\uparrow} - \hat{n}_{s\downarrow}), \qquad \hat{S}^+_s \equiv \hat{d}^\dagger_{s\uparrow}\hat{d}_{s\downarrow}, \qquad \hat{S}^-_s \equiv \hat{d}^\dagger_{s\downarrow}\hat{d}_{s\uparrow} = (\hat{S}^+_s)^\dagger. \tag{1.93}$$

Using the anticommutation relations, prove that the action of the above operators on the kets $|s\sigma\rangle \equiv \hat{d}^\dagger_{s\sigma}|0\rangle$ is

$$\hat{S}^z_s|s\uparrow\rangle = \frac{1}{2}|s\uparrow\rangle, \qquad \hat{S}^+_s|s\uparrow\rangle = |\emptyset\rangle, \qquad \hat{S}^-_s|s\uparrow\rangle = |s\downarrow\rangle,$$

and

$$\hat{S}^z_s|s\downarrow\rangle = -\frac{1}{2}|s\downarrow\rangle, \qquad \hat{S}^+_s|s\downarrow\rangle = |s\uparrow\rangle, \qquad \hat{S}^-_s|s\downarrow\rangle = |\emptyset\rangle.$$

To what operators do $\hat{S}^z_s$, $\hat{S}^+_s$, $\hat{S}^-_s$ correspond?

Exercise 1.8 Let us define the *spin operators* along the x and y directions as

$$\hat{S}^x_s \equiv \frac{1}{2}(\hat{S}^+_s + \hat{S}^-_s), \qquad \hat{S}^y_s \equiv \frac{1}{2\mathrm{i}}(\hat{S}^+_s - \hat{S}^-_s),$$

and the spin operator $\hat{S}^z_s$ along the z direction as in (1.93). Prove that these operators can also be written as

$$\hat{S}^j_s = \frac{1}{2}\sum_{\sigma\sigma'} \hat{d}^\dagger_{s\sigma}\sigma^j_{\sigma\sigma'}\hat{d}_{s\sigma'}, \qquad j = x,\, y,\, z, \tag{1.94}$$

with

$$\sigma^x = \begin{pmatrix} 0 & 1 \\ 1 & 0 \end{pmatrix}, \quad \sigma^y = \begin{pmatrix} 0 & -\mathrm{i} \\ \mathrm{i} & 0 \end{pmatrix}, \quad \sigma^z = \begin{pmatrix} 1 & 0 \\ 0 & -1 \end{pmatrix},$$

the *Pauli matrices.* Using the anticommutation relations, verify that

$$[\hat{S}^i_s, \hat{S}^j_{s'}]_- = \mathrm{i}\delta_{ss'} \sum_{k=x,y,z} \varepsilon_{ijk}\hat{S}^k_s,$$

where ε_{ijk} is the *Levi-Civita tensor.*[7]

---

[7] The Levi-Civita tensor is zero if at least two indices are equal and otherwise

$$\varepsilon_{P(1)P(2)P(3)} = (-)^P,$$

where P is an arbitrary permutation of the indices $1, 2, 3$.

2

Getting Familiar with Second Quantization: Model Hamiltonians

In this chapter we get acquainted with the formalism of second quantization. We discuss how to choose a proper set of basis functions for some relevant physical systems, construct the corresponding Hamiltonians, and derive a few elementary results. We do not aim at providing an exhaustive presentation of these *model Hamiltonians*, something that would require a monograph for each model. We rather aim at warming up the reader with the mathematical operations that are often encountered when working with creation and annihilation operators. The best way to learn how to manipulate these operators is by seeing them at work.

2.1 Model Hamiltonians

In all practical calculations the properties of a many-particle system are extracted by using a finite number of (physically relevant) single-particle basis functions. For instance, in systems such as crystals or molecules the electrons are attracted by the positive charge of the nuclei, and it is reasonable to expect that a few localized orbitals around each nucleus provide a good-enough basis set. If we think of the H_2 molecule, the simplest description consists in taking one basis function, $\varphi_{n=1\sigma}$, for an electron of spin σ localized around the first nucleus and another one, $\varphi_{n=2\sigma}$, for an electron of spin σ localized around the second nucleus, see the schematic representation below.

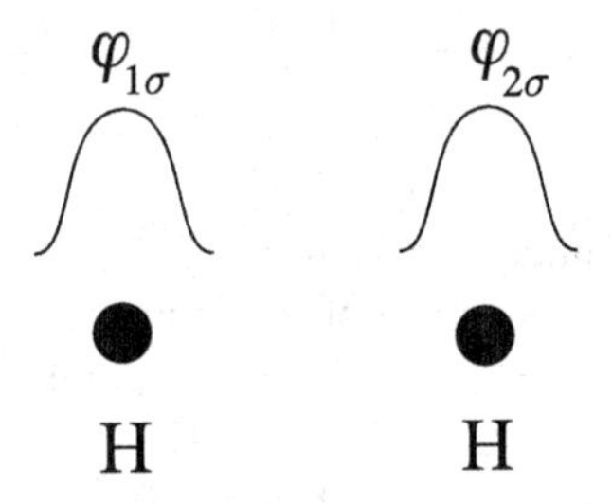

If the set $\{\varphi_n\}$ is not complete in $\mathbb{H}_1$, then the expansion (1.60) is an approximation for the field operators. The approximate field operators satisfy approximate (anti)commutation

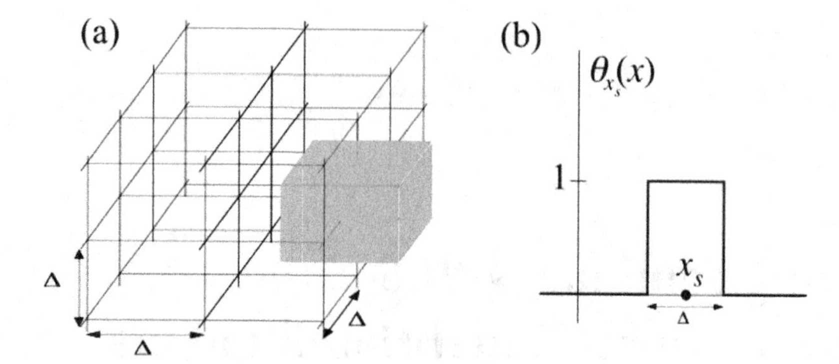

Figure 2.1 (a) Space grid with vertices (grid points) in $\mathbf{r}_s$. To each grid point is assigned a basis function as described in the main text. (b) The piecewise function used to construct the approximate basis.

relations. Let us show it with an example. We consider a space grid in three-dimensional space with uniform grid spacing Δ. To each grid point $\mathbf{r}_s = (x_s, y_s, z_s)$ we assign a basis function that is constant in a cube of linear dimension Δ centered in $\mathbf{r}_s$, see Fig. 2.1(a). These basis functions are orthonormal and have the following mathematical structure:

$$\varphi_{n=s\sigma_s}(\mathbf{r}\sigma) = \delta_{\sigma\sigma_s}\frac{\theta_{x_s}(x)\theta_{y_s}(y)\theta_{z_s}(z)}{\Delta^{3/2}}, \tag{2.1}$$

with the Heaviside function $\theta_a(x) = \theta(\frac{1}{2}\Delta - |a - x|)$, see Fig. 2.1(b), and the prefactor $1/\Delta^{3/2}$, which guarantees the correct normalization. For any finite grid spacing Δ, the set $\{\varphi_n\}$ is not a complete set. However, if the physical properties of the system vary on a length scale much larger than Δ, then the use of the approximate field operators

$$\hat{\psi}^\dagger(\mathbf{r}\sigma) \sim \hat{\psi}^\dagger_\Delta(\mathbf{r}\sigma) = \sum_s \frac{\theta_{x_s}(x)\theta_{y_s}(y)\theta_{z_s}(z)}{\Delta^{3/2}}\hat{d}^\dagger_{s\sigma}$$

is expected to work fine. The (anti)commutation relation for these approximate field operators is

$$\left[\hat{\psi}_\Delta(\mathbf{x}), \hat{\psi}^\dagger_\Delta(\mathbf{x}')\right]_\mp = \delta_{\sigma\sigma'}\sum_{ss'}\frac{\theta_{x_s}(x)\theta_{y_s}(y)\theta_{z_s}(z)\theta_{x_{s'}}(x')\theta_{y_{s'}}(y')\theta_{z_{s'}}(z')}{\Delta^3}\underbrace{\left[\hat{d}_{s\sigma}, \hat{d}^\dagger_{s'\sigma}\right]_\mp}_{\delta_{ss'}}.$$

Thus we see that the (anti)commutator is zero if $\mathbf{r}$ and $\mathbf{r}'$ belong to different cubes and is equal to $\delta_{\sigma\sigma'}/\Delta^3$ otherwise. The accuracy of the results can be checked by increasing the number of basis functions. In our example this corresponds to reducing the spacing Δ. In the limit $\Delta \to 0$ the product $\theta_{x_s}(x)\theta_{x_s}(x')/\Delta \to \delta(x - x')$, and similarly for y and z, and hence the (anti)commutator approaches the exact result

$$\left[\hat{\psi}_\Delta(\mathbf{x}), \hat{\psi}^\dagger_\Delta(\mathbf{x}')\right]_\mp \xrightarrow[\Delta\to 0]{} \delta_{\sigma\sigma'}\delta(x - x')\delta(y - y')\delta(z - z') = \delta(\mathbf{x} - \mathbf{x}').$$

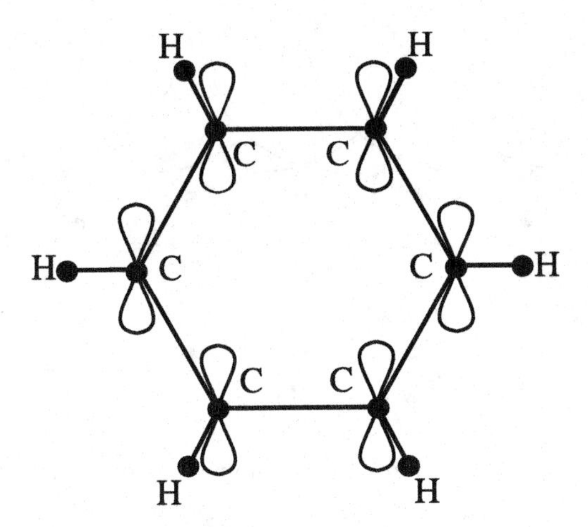

Figure 2.2 Schematic representation of the benzene molecule C_6H_6 with a p_z orbital for each carbon atom. We recall that the electronic configuration of carbon is $1s^2 2s^2 2p^2$ and that in the benzene geometry the $2s$, $2p_x$, $2p_y$ orbitals hybridize to form three sp^2 orbitals [7]. The latter share an electron with the nearest hydrogen as well as with the two nearest-neighbor carbons.

2.2 Pariser–Parr–Pople Model

A popular model Hamiltonian often employed to describe organic molecules is the so-called *Pariser–Parr–Pople model*, or simply the PPP model [4, 5]. We here give an elementary derivation of the PPP model and refer the reader to more specialized textbooks for a careful justification of the simplifications involved [6]. As a concrete example we consider an atomic ring, such as the benzene molecule C_6H_6 in Fig. 2.2, but the basic ideas can be used for other molecular geometries as well. If we are interested in the low-energy physics of the system, such as its ground-state properties, we can assume that the inner shell electrons are "frozen" in their molecular orbitals while the outer shell electrons are free to wander around the molecule. In the case of benzene we may consider as frozen the two 1s electrons of each carbon atom C as well as the three electrons of the in-plane sp^2 orbitals that form σ-bonds with the hydrogen atom H and with the two nearest carbon atoms. For the description of the remaining six electrons (one per C-H unit) we could limit ourselves to use a p_z orbital for each carbon atom. In general, the problem is always to find a minimal set of functions to describe the dynamics of the "free," also called *valence*, electrons responsible for the low-energy excitations.

Let us assign a single orbital to each atomic position $\mathbf{R}_s$ of the ring:

$$\tilde{\varphi}_{s\tau}(\mathbf{r}\sigma) = \delta_{\sigma\tau} f(\mathbf{r} - \mathbf{R}_s),$$

where $s = 1, \ldots, N$ and N is the number of atoms in the ring. The function $f(\mathbf{r})$ is localized around $\mathbf{r} = 0$; an example could be the exponential function $e^{-\alpha|\mathbf{r}|}$, see Fig. 2.3. The set of functions $\{\tilde{\varphi}_{s\tau}\}$ is, in general, not an orthonormal set since the *overlap matrix*

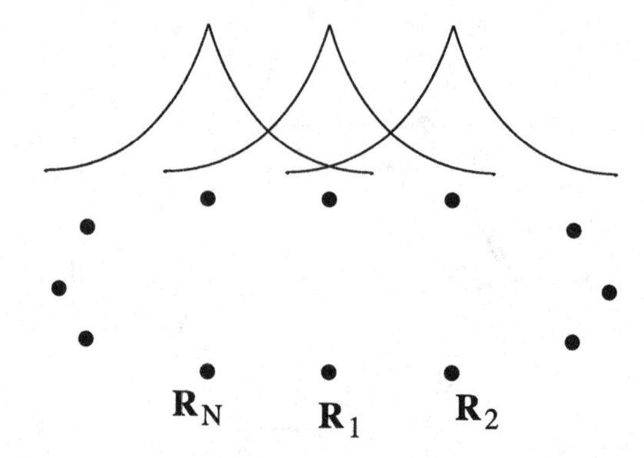

Figure 2.3 Orbitals $\tilde{\varphi}_{s\tau}$ localized around the atomic position $\mathbf{R}_s$ of the atomic ring.

$$S_{ss'} = \int d\mathbf{x}\, \tilde{\varphi}^*_{s\tau}(\mathbf{x})\tilde{\varphi}_{s'\tau}(\mathbf{x}) = \int d\mathbf{r}\, f^*(\mathbf{r}-\mathbf{R}_s)f(\mathbf{r}-\mathbf{R}_{s'})$$

may have nonvanishing off-diagonal elements. A simple way to orthonormalize the set $\{\tilde{\varphi}_{s\tau}\}$ without losing the local character of the functions is the following. The overlap matrix is Hermitian and positive definite,[1] meaning that all its eigenvalues λ_k are larger than zero. Let $D = \mathrm{diag}(\lambda_1, \lambda_2, \ldots)$ and U be the unitary matrix that brings S in its diagonal form – that is, $S = UDU^\dagger$. We define the square root of the matrix S according to

$$S^{1/2} = UD^{1/2}U^\dagger \qquad \text{with} \qquad D^{1/2} = \mathrm{diag}(\sqrt{\lambda_1}, \sqrt{\lambda_2}, \ldots).$$

The matrix $S^{1/2}$ is also Hermitian and positive-definite and can easily be inverted

$$S^{-1/2} = UD^{-1/2}U^\dagger \qquad \text{with} \qquad D^{-1/2} = \mathrm{diag}(\frac{1}{\sqrt{\lambda_1}}, \frac{1}{\sqrt{\lambda_2}}, \ldots).$$

We then construct the new set of functions

$$\varphi_{s\tau}(\mathbf{x}) = \sum_{s'} \tilde{\varphi}_{s'\tau}(\mathbf{x}) S^{-1/2}_{s's},$$

whose overlap is

$$\int d\mathbf{x}\, \varphi^*_{s_1\tau_1}(\mathbf{x})\varphi_{s_2\tau_2}(\mathbf{x}) = \sum_{s's''} \int d\mathbf{x}\, S^{-1/2}_{s_1s'} \tilde{\varphi}^*_{s'\tau_1}(\mathbf{x})\tilde{\varphi}_{s''\tau_2}(\mathbf{x}) S^{-1/2}_{s''s_2}$$
$$= \delta_{\tau_1\tau_2}(S^{-1/2}SS^{-1/2})_{s_1s_2} = \delta_{\tau_1\tau_2}\delta_{s_1s_2}.$$

[1]The positive definiteness of S follows from its definition. Given an arbitrary vector with components v_s, we have

$$\sum_{ss'} v^*_s S_{ss'} v_{s'} = \int d\mathbf{x} \left(\sum_s v^*_s \tilde{\varphi}^*_{s\tau}(\mathbf{x})\right)\left(\sum_{s'} v_{s'}\tilde{\varphi}_{s'\tau}(\mathbf{x})\right) > 0.$$

The set $\{\varphi_{s\tau}\}$ is therefore orthonormal. If the off-diagonal elements of the overlap matrix are small compared to unity, then the new functions are only slightly delocalized. Consider, for instance, an overlap matrix of the form

$$S = \begin{pmatrix} 1 & \delta & 0 & 0 & \dots & \delta \\ \delta & 1 & \delta & 0 & & 0 \\ 0 & \delta & 1 & \ddots & & 0 \\ 0 & 0 & \ddots & \ddots & \ddots & \vdots \\ \vdots & & & \ddots & 1 & \delta \\ \delta & 0 & 0 & \dots & \delta & 1 \end{pmatrix} = \hat{\mathbb{1}} + \Delta,$$

according to which we only have overlap of amount $\delta \ll 1$ between nearest-neighbor atoms (note that the matrix element $S_{1N} = S_{N1} = \delta$ since atom N is the nearest neighbor of atom 1). To first order in δ the inverse of the square root of S is

$$S^{-1/2} = (\hat{\mathbb{1}} + \Delta)^{-1/2} \sim \hat{\mathbb{1}} - \frac{1}{2}\Delta,$$

and therefore the new functions are slightly spread over the nearest-neighbor atoms

$$\varphi_{s\tau}(\mathbf{x}) = \tilde{\varphi}_{s\tau}(\mathbf{x}) - \frac{\delta}{2}\tilde{\varphi}_{s+1\tau}(\mathbf{x}) - \frac{\delta}{2}\tilde{\varphi}_{s-1\tau}(\mathbf{x}), \tag{2.2}$$

where it is understood that the index $s \pm N$ must be identified with s. The orthonormal functions $\{\varphi_{s\tau}\}$ form the set over which to expand the field operators

$$\hat{\psi}^\dagger(\mathbf{x}) \sim \sum_{s\tau} \varphi^*_{s\tau}(\mathbf{x})\hat{d}^\dagger_{s\tau},$$

with

$$\hat{d}^\dagger_{s\tau} = \int d\mathbf{x}\, \varphi_{s\tau}(\mathbf{x})\hat{\psi}^\dagger(\mathbf{x}),$$

and similar relations for the adjoint operators. The $\hat{d}$-operators satisfy the anticommutation relations (1.54) since $\{\varphi_{s\tau}\}$ is an orthonormal set. Inserting the approximate expansion of the field operators into the Hamiltonian $\hat{H}$ we obtain an approximate Hamiltonian. This Hamiltonian looks like (1.84) but the sums are restricted to the incomplete set of quantum numbers s. In this way, the original field operators $\hat{\psi}(\mathbf{x})$ and $\hat{\psi}^\dagger(\mathbf{x})$ get replaced by a finite (or at most countable) number of $\hat{d}$-operators. The parameters h_{ij} and v_{ijmn} of the approximate Hamiltonian depend on the *specific choice of basis functions* and on the *microscopic details of the system*, such as the mass and charge of the particles (we remind the reader that in this book we work in atomic units so that for electrons $m_e = -e = 1$), the scalar and vector potentials, the interparticle interaction, etc. Once these parameters are given, the approximate Hamiltonian is fully specified and we refer to it as the *model Hamiltonian.* Below we estimate the parameters h_{ij} and v_{ijmn} for the set $\{\varphi_{s\tau}\}$ in (2.2) and for an atomic ring like benzene.

Let us write the functions (2.2) as a product of an orbital and a spin part: $\varphi_{s\tau}(\mathbf{r}\sigma) = \delta_{\sigma\tau}\varphi_s(\mathbf{r})$. Since the $\varphi_s(\mathbf{r})$ are localized around the atomic position $\mathbf{R}_s$, the dominant Coulomb integrals $v_{s_1 s_2 s_3 s_4}$ are those with $s_1 = s_4$ and $s_2 = s_3$, see (1.86). We therefore make the approximation

$$v_{s_1 s_2 s_3 s_4} \sim \delta_{s_1 s_4}\delta_{s_2 s_3} \underbrace{\int d\mathbf{r}d\mathbf{r}' |\varphi_{s_1}(\mathbf{r})|^2 v(\mathbf{r},\mathbf{r}')|\varphi_{s_2}(\mathbf{r}')|^2}_{v_{s_1 s_2}}. \tag{2.3}$$

The quantity $v_{s_1 s_2}$ equals the classical interaction energy between the charge distributions $|\varphi_{s_1}|^2$ and $|\varphi_{s_2}|^2$. For the Coulomb interaction $v(\mathbf{r},\mathbf{r}') = 1/|\mathbf{r}-\mathbf{r}'|$ we can further approximate the integral $v_{s_1 s_2}$ when $s_1 \neq s_2$ as

$$v_{s_1 s_2} \sim \frac{1}{|\mathbf{R}_{s_1} - \mathbf{R}_{s_2}|}, \tag{2.4}$$

since in the neighborhood of $|\mathbf{R}_{s_1} - \mathbf{R}_{s_2}|$ the function $1/r$ is slowly varying and can be considered constant. Inserting these results into (1.90), we obtain the model form of the interaction operator

$$\hat{H}_{\text{int}} = \frac{1}{2}\sum_{\substack{ss'\\ \sigma\sigma'}} v_{ss'}\hat{d}^\dagger_{s\sigma}\hat{d}^\dagger_{s'\sigma'}\hat{d}_{s'\sigma'}\hat{d}_{s\sigma} = \frac{1}{2}\sum_{s\neq s'} v_{ss'}\hat{n}_s\hat{n}_{s'} + \sum_s v_{ss}\hat{n}_{s\uparrow}\hat{n}_{s\downarrow}, \tag{2.5}$$

with $n_{s\sigma} = \hat{d}^\dagger_{s\sigma}\hat{d}_{s\sigma}$, the occupation operator that counts how many electrons (0 or 1) are in the spin-orbital $\varphi_{s\sigma}$ and $\hat{n}_s \equiv \hat{n}_{s\uparrow} + \hat{n}_{s\downarrow}$.[2]

By the same overlap argument we can neglect all matrix elements $h_{ss'}$ between atomic sites that are not nearest neighbors (we are implicitly assuming that $h_{s\sigma\, s'\sigma'} = \delta_{\sigma\sigma'}h_{ss'}$). Let $\langle ss'\rangle$ denote the couples of nearest-neighbor atomic sites. Then the noninteracting part (1.88) of the Hamiltonian takes the form

$$\hat{H}_0 = \sum_s h_{ss}\hat{n}_s + \sum_{\langle ss'\rangle}\sum_\sigma h_{ss'}\hat{d}^\dagger_{s\sigma}\hat{d}_{s'\sigma}.$$

In the present case the coefficients $h_{ss'}$ are given by

$$h_{ss'} = \langle s\sigma|\hat{h}|s'\sigma\rangle = \int d\mathbf{r}\, \varphi^*_s(\mathbf{r})\left[-\frac{\nabla^2}{2} - \phi(\mathbf{r})\right]\varphi_{s'}(\mathbf{r}).$$

The potential $\phi(\mathbf{r})$ in this expression is the sum of the electrostatic potentials between an electron in $\mathbf{r}$ and the atomic nuclei in $\mathbf{R}_s$, that is,

$$\phi(\mathbf{r}) = \sum_s \frac{Z_s}{|\mathbf{r} - \mathbf{R}_s|},$$

[2] In writing the first term on the r.h.s. of (2.5) we use that for $s\sigma \neq s'\sigma'$ the operator $\hat{d}_{s\sigma}$ commutes with $\hat{d}^\dagger_{s'\sigma'}\hat{d}_{s'\sigma'}$. For the second term we further take into account that for $s = s'$ and $\sigma = \sigma'$ the product $\hat{d}_{s\sigma}\hat{d}_{s\sigma} = 0$.

where Z_s is the effective nuclear positive charge of atom s - that is, the sum of the bare nuclear charge and the screening charge of the frozen electrons. Let us manipulate a bit the diagonal elements h_{ss}. Using the explicit form of the potential, we write

$$h_{ss} = \epsilon_s + \beta_s,$$

with

$$\epsilon_s = \int d\mathbf{r}\, \varphi_s^*(\mathbf{r}) \left[-\frac{\nabla^2}{2} - \frac{Z_s}{|\mathbf{r} - \mathbf{R}_s|} \right] \varphi_s(\mathbf{r}) \tag{2.6}$$

and

$$\beta_s = -\sum_{s' \neq s} \int d\mathbf{r}\, |\varphi_s(\mathbf{r})|^2 \frac{Z_{s'}}{|\mathbf{r} - \mathbf{R}_{s'}|}.$$

Since $|\varphi_s|^2$ is the charge distribution of an electron localized in $\mathbf{R}_s$ we can, as before, approximately write

$$\beta_s \sim -\sum_{s' \neq s} \frac{Z_{s'}}{|\mathbf{R}_s - \mathbf{R}_{s'}|} = -\sum_{s' \neq s} Z_{s'} v_{ss'}$$

where we use (2.4). Inserting these results into $\hat{H}_0$ and adding $\hat{H}_{\text{int}}$, we find the PPP model Hamiltonian

$$\begin{aligned} \hat{H} = \hat{H}_0 + \hat{H}_{\text{int}} &= \sum_s \epsilon_s \hat{n}_s + \sum_{\langle ss' \rangle} \sum_\sigma h_{ss'} \hat{d}^\dagger_{s\sigma} \hat{d}_{s'\sigma} \\ &+ \frac{1}{2} \sum_{s \neq s'} v_{ss'} (\hat{n}_s - Z_s)(\hat{n}_{s'} - Z_{s'}) + \sum_s v_{ss} \hat{n}_{s\uparrow} \hat{n}_{s\downarrow}, \end{aligned} \tag{2.7}$$

where we also add the constant $\frac{1}{2} \sum_{s \neq s'} v_{ss'} Z_s Z_{s'}$ corresponding to the electrostatic energy of the screened nuclei. If the ground state $|\Psi_0\rangle$ has exactly Z_s electrons on atom s (i.e., $\hat{n}_s |\Psi_0\rangle = Z_s |\Psi_0\rangle$ for all s), then the only interaction energy comes from the last term. In general, however, this is not the case since for $|\Psi_0\rangle$ to be an eigenstate of all $\hat{n}_s$ it must be $|\Psi_0\rangle = |s_1\sigma_1\, s_2\sigma_2 \ldots\rangle$. This state is an eigenstate of $\hat{H}$ only provided that the off-diagonal elements $h_{ss'} = 0$, which is satisfied when the atoms are infinitely far apart from each other.

Every model Hamiltonian in the scientific literature has underlying assumptions and approximations which are similar to those of the PPP model. Most of these models cannot be solved exactly despite the fact that the Hamiltonian is undoubtedly simpler than the original continuum Hamiltonian. In the next sections we discuss other examples of model Hamiltonians suited to describing other physical systems. Before that, however, we use the PPP model to study the smallest possible molecule: the hydrogen molecule.

2.2.1 Hydrogen Molecule

The hydrogen molecule H_2 consists of two protons (the nuclei) and two electrons. Choosing real orbitals to describe the electrons around each nucleus, the PPP Hamiltonian (2.7) reads

$$\begin{aligned} \hat{H} &= \epsilon(\hat{n}_1 + \hat{n}_2) + T \sum_\sigma (\hat{d}^\dagger_{1\sigma} \hat{d}_{2\sigma} + \hat{d}^\dagger_{2\sigma} \hat{d}_{1\sigma}) \\ &+ v_{12}(\hat{n}_1 - 1)(\hat{n}_2 - 1) + v(\hat{n}_{1\uparrow}\hat{n}_{1\downarrow} + \hat{n}_{2\uparrow}\hat{n}_{2\downarrow}), \end{aligned} \tag{2.8}$$

where $\epsilon \equiv \epsilon_1 = \epsilon_2$, $T \equiv h_{12} = h_{21}$, and $v \equiv v_{11} = v_{22}$. The so-called *dissociation limit* corresponds to pulling the protons infinitely far apart, hence $T = 0$ and $v_{12} = 0$. In this limit the ground-state energy with two electrons is 2ϵ, which is the same energy as two isolated hydrogen atoms if the chosen orbitals are the $1s$ eigenfunctions of the hydrogen atom, see (2.6).

Let us manipulate the Hamiltonian. The square of the operator of the total number of electrons $\hat{N} = \hat{n}_1 + \hat{n}_2$ is

$$\hat{N}^2 = (\hat{n}_1 + \hat{n}_2)^2 = \hat{N} + 2(\hat{n}_{1\uparrow}\hat{n}_{1\downarrow} + \hat{n}_{2\uparrow}\hat{n}_{2\downarrow} + \hat{n}_1\hat{n}_2),$$

where we take into account that $\hat{n}_{i\sigma}^2 = \hat{n}_{i\sigma}$, see Exercise 1.4. We can therefore rewrite (2.8) as

$$\hat{H} = \epsilon\hat{N} + \frac{v_{12}}{2}(\hat{N}^2 - 3\hat{N} + 2) + \hat{H}_{\mathrm{Hd}}.$$

The operator

$$\hat{H}_{\mathrm{Hd}} \equiv T\sum_{\sigma}(\hat{d}^{\dagger}_{1\sigma}\hat{d}_{2\sigma} + \hat{d}^{\dagger}_{2\sigma}\hat{d}_{1\sigma}) + U(\hat{n}_{1\uparrow}\hat{n}_{1\downarrow} + \hat{n}_{2\uparrow}\hat{n}_{2\downarrow}), \qquad U \equiv v - v_{12}, \tag{2.9}$$

is known as the Hamiltonian of the *Hubbard dimer*, see Section 2.5. Since $\hat{H}_{\mathrm{Hd}}$ and $\hat{N}$ commute, they can be simultaneously diagonalized. The simultaneous eigenkets of $\hat{H}_{\mathrm{Hd}}$ with eigenvalue E and of $\hat{N}$ with eigenvalue N are also eigenkets of $\hat{H}$ with eigenvalue $E + \epsilon N + \frac{v_{12}}{2}(N^2 - 3N + 2)$. In the following we derive eigenkets and eigenvalues of $\hat{H}_{\mathrm{Hd}}$ for every N. In the charge neutral H_2 molecule we have $N = 2$.

The Fock space $\mathbb{F} = \mathbb{H}_0 \oplus \mathbb{H}_1 \oplus \mathbb{H}_2 \oplus \mathbb{H}_3 \oplus \mathbb{H}_4$ of the Hubbard dimer is finite because we cannot have more than four electrons. The zero-particle Hilbert space $\mathbb{H}_0$ contains only the empty ket, which is an eigenket of $\hat{H}_{\mathrm{Hd}}$ with eigenvalue 0. In $\mathbb{H}_1$ we have four possible kets $|i\sigma\rangle = \hat{d}^{\dagger}_{i\sigma}|0\rangle$ with $i = 1, 2$. They are all eigenkets of the interaction operator with eigenvalue 0. Furthermore, the matrix $\langle i\sigma|\hat{H}_{\mathrm{Hd}}|j\sigma'\rangle = \delta_{\sigma\sigma'}h_{ij}$ is spin diagonal. The eigenvalues of $h = \begin{pmatrix} 0 & T \\ T & 0 \end{pmatrix}$ are $\pm T$ and the corresponding eigenvectors are $\frac{1}{\sqrt{2}}(1, \pm 1)$. Thus, the one-particle eigenkets of the Hubbard dimer are $\frac{1}{\sqrt{2}}(\hat{d}^{\dagger}_{1\sigma} \pm \hat{d}^{\dagger}_{2\sigma})|0\rangle$, see also Table 2.1. Let us now consider the two-particle Hilbert space $\mathbb{H}_2$. If both electrons have spin σ, then the only possible ket is $\hat{d}^{\dagger}_{1\sigma}\hat{d}^{\dagger}_{2\sigma}|0\rangle$ due to the Pauli exclusion principle. The reader can easily verify that this ket is an eigenket of $\hat{H}_{\mathrm{Hd}}$ with eigenvalue 0. In particular, it is also an eigenket of the interaction part (the term proportional to U) with eigenvalue 0 since the electrons have parallel spin. On the other hand, if the electrons have opposite spin then we have four possible kets:

$$|\Psi_1\rangle = \hat{d}^{\dagger}_{1\uparrow}\hat{d}^{\dagger}_{1\downarrow}|0\rangle, \quad |\Psi_2\rangle = \hat{d}^{\dagger}_{2\uparrow}\hat{d}^{\dagger}_{2\downarrow}|0\rangle, \quad |\Psi_3\rangle = \hat{d}^{\dagger}_{1\uparrow}\hat{d}^{\dagger}_{2\downarrow}|0\rangle, \quad |\Psi_4\rangle = \hat{d}^{\dagger}_{1\downarrow}\hat{d}^{\dagger}_{2\uparrow}|0\rangle.$$

The Hamiltonian does not couple these states to those with parallel spin. Therefore, we can calculate the remaining eigenvalues and eigenkets in $\mathbb{H}_2$ by diagonalizing the matrix h_2

Space	Eigenvalues	Eigenkets
$\mathbb{H}_0$	0	$\|0\rangle$
$\mathbb{H}_1$	$\pm T$	$\frac{1}{\sqrt{2}}(\hat{d}^\dagger_{1\sigma} \pm \hat{d}^\dagger_{2\sigma})\|0\rangle$
$\mathbb{H}_2$	0	$\hat{d}^\dagger_{1\uparrow}\hat{d}^\dagger_{2\uparrow}\|0\rangle$
	0	$\hat{d}^\dagger_{1\downarrow}\hat{d}^\dagger_{2\downarrow}\|0\rangle$
	0	$\frac{\hat{d}^\dagger_{1\uparrow}\hat{d}^\dagger_{2\downarrow} + \hat{d}^\dagger_{1\downarrow}\hat{d}^\dagger_{2\uparrow}}{\sqrt{2}}\|0\rangle$
	U	$\frac{\hat{d}^\dagger_{1\uparrow}\hat{d}^\dagger_{1\downarrow} - \hat{d}^\dagger_{2\uparrow}\hat{d}^\dagger_{2\downarrow}}{\sqrt{2}}\|0\rangle$
	$E_3 = \frac{1}{2}(U - \Delta)$	$\frac{E_3\hat{d}^\dagger_{1\uparrow}\hat{d}^\dagger_{1\downarrow} + E_3\hat{d}^\dagger_{2\uparrow}\hat{d}^\dagger_{2\downarrow} + 2T\hat{d}^\dagger_{1\uparrow}\hat{d}^\dagger_{2\downarrow} - 2T\hat{d}^\dagger_{1\downarrow}\hat{d}^\dagger_{2\uparrow}}{\sqrt{2E_3^2 + 8T^2}}\|0\rangle$
	$E_4 = \frac{1}{2}(U + \Delta)$	$\frac{E_4\hat{d}^\dagger_{1\uparrow}\hat{d}^\dagger_{1\downarrow} + E_4\hat{d}^\dagger_{2\uparrow}\hat{d}^\dagger_{2\downarrow} + 2T\hat{d}^\dagger_{1\uparrow}\hat{d}^\dagger_{2\downarrow} - 2T\hat{d}^\dagger_{1\downarrow}\hat{d}^\dagger_{2\uparrow}}{\sqrt{2E_4^2 + 8T^2}}\|0\rangle$
$\mathbb{H}_3$	$\pm T + U$	$\frac{1}{\sqrt{2}}(\hat{d}^\dagger_{1\sigma}\hat{d}^\dagger_{2\uparrow}\hat{d}^\dagger_{2\downarrow} \pm \hat{d}^\dagger_{1\uparrow}\hat{d}^\dagger_{1\downarrow}\hat{d}^\dagger_{2\sigma})\|0\rangle$
$\mathbb{H}_4$	$2U$	$\hat{d}^\dagger_{1\uparrow}\hat{d}^\dagger_{1\downarrow}\hat{d}^\dagger_{2\uparrow}\hat{d}^\dagger_{2\downarrow}\|0\rangle$

Table 2.1 Eigenvalues and normalized eigenkets of the Hubbard dimer in the different Hilbert spaces. In the table, the quantity $\Delta = \sqrt{16T^2 + U^2}$.

with elements $(h_2)_{ij} = \langle\Psi_i|\hat{H}_{\mathrm{Hd}}|\Psi_j\rangle$. After some simple algebra, one finds

$$h_2 = \begin{pmatrix} U & 0 & T & -T \\ 0 & U & T & -T \\ T & T & 0 & 0 \\ -T & -T & 0 & 0 \end{pmatrix}.$$

This matrix has eigenvalues $E_1 = 0$, $E_2 = U$, $E_3 = \frac{1}{2}(U - \Delta)$, and $E_4 = \frac{1}{2}(U + \Delta)$, where $\Delta = \sqrt{16T^2 + U^2}$. Let us take, for example, $U > 0$ so that $E_3 < 0$ and $E_4 > 0$.

The normalized eigenkets corresponding to these eigenvalues are reported in Table 2.1. The occurrence of the zero eigenvalue E_1 should not come as a surprise. The Hamiltonian commutes with the total spin operator $\hat{S}^2 = (\hat{\mathbf{S}}_1 + \hat{\mathbf{S}}_2) \cdot (\hat{\mathbf{S}}_1 + \hat{\mathbf{S}}_2)$, as well as with $\hat{S}^z = \hat{S}_1^z + \hat{S}_2^z$ [the spin operators are defined in (1.94)]. Therefore the degenerate eigenkets of $\hat{H}_{\mathrm{Hd}}$ must belong to spin multiplets. It is easy to see that the three eigenkets with vanishing eigenvalue belong to a triplet, whereas the eigenkets with eigenvalues E_2, E_3, and E_4 are singlets. To conclude our analysis we must calculate the eigenkets and eigenvalues with three and four electrons. This can be done along the same lines as for the other Hilbert spaces, and the final results are reported in Table 2.1.

It is interesting to observe that in the dissociation limit ($T \to 0$ and $v_{12} \to 0$, hence $U \to v$) the eigenvalues of $\hat{H}_{\mathrm{Hd}}$ in $\mathbb{H}_2$ become 0 (four times degenerate) and $U = v$ (two times degenerate), and hence the eigenvalues of $\hat{H}$ become 2ϵ (four times degenerate) and $2\epsilon + v$ (two times degenerate). These results agree with our physical intuition; if two isolated hydrogen atoms have one electron each (either with spin up or down), then the energy is 2ϵ, whereas if both electrons are on the same atom then the energy is $2\epsilon + U$. This simple treatment of the hydrogen molecule describes with rather high accuracy both the ground state and the first excited state [8].

2.3 Bloch Theorem and Band Structure

We discussed in Section 1.6 how to find eigenvalues and eigenvectors of the noninteracting Hamiltonian $\hat{H}_0$. In the absence of interactions, the many-particle problem reduces to a single-particle problem. The interparticle interaction makes our lives much more complicated (and interesting) and we have to resort to approximate methods to make progress. The zeroth-order approximation consists of neglecting the interparticle interaction altogether. How much physics can we capture this way? The answer to this question clearly depends on the system at hand and on the physical properties that interest us. It turns out that in nature *some* physical properties of *some* systems are not so sensitive to interparticle interactions. For instance, a noninteracting treatment of crystals is, in many cases, enough to assess whether the crystal is a metal or an insulator.

A crystal consists of a unit cell repeated periodically in three, two, or one dimensions. Each unit cell contains the same (finite) number of atoms arranged in the same geometry. An example is *graphene*, which is a planar structure of sp^2-bonded carbon atoms arranged in a honeycomb lattice, as illustrated in Fig. 2.4. In this case the unit cell consists of two carbon atoms, a and b in Fig. 2.4, repeated periodically along the directions $\mathbf{v}_+$ and $\mathbf{v}_-$. The unit cells can be labeled with a vector of integers $\mathbf{n} = (n_1, \ldots, n_d)$, where d is the dimension of the crystal, see again Fig. 2.4. The expansion of the vector $\mathbf{n}$ over the orthonormal basis $\{\mathbf{e}_i\}$ with $(\mathbf{e}_i)_j = \delta_{ij}$ reads

$$\mathbf{n} = \sum_{i=1}^{d} n_i \mathbf{e}_i.$$

Two unit cells with labels $\mathbf{n}$ and $\mathbf{n}'$ are nearest neighbors if $|\mathbf{n} - \mathbf{n}'| = 1$. As for the PPP model, we assign to each unit cell a set of localized orbitals $\{\varphi_{\mathbf{n}s\sigma}\}$ which we assume already orthonormal and denote by $\hat{d}^\dagger_{\mathbf{n}s\sigma}$ and $\hat{d}_{\mathbf{n}s\sigma}$ the creation and annihilation operators

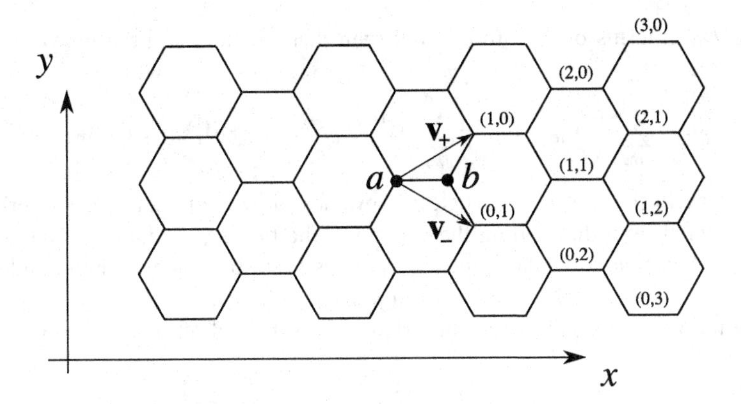

Figure 2.4 The crystal structure of graphene: a unit cell with two carbon atoms (a and b) repeated periodically along the directions $\mathbf{v}_+$ and $\mathbf{v}_-$ to form a honeycomb lattice. For illustrative purposes, some unit cells have been labeled with the two-dimensional vector of integers $\mathbf{n} = (n_1, n_2)$.

for electrons in the orbital $\varphi_{\mathbf{n}s\sigma}$. The index s runs over all orbitals of a given unit cell. In the case of graphene we may assign a single p_z orbital to each carbon, so $s = 1, 2$.

In the absence of magnetic fields or other couplings that break the spin symmetry, the matrix elements of $\hat{h}$ are

$$h_{\mathbf{n}s\sigma\,\mathbf{n}'s'\sigma'} = \delta_{\sigma\sigma'} h_{\mathbf{n}s\,\mathbf{n}'s'}.$$

If we choose orbitals localized around the atomic positions, then the matrix elements $h_{\mathbf{n}s\,\mathbf{n}'s'}$ are very small for $|\mathbf{n} - \mathbf{n}'| \gg 1$. We then discard $h_{\mathbf{n}s\,\mathbf{n}'s'}$ unless $\mathbf{n}$ and $\mathbf{n}'$ are nearest neighbors. The periodicity of a crystal is reflected in the fact that the unit cell Hamiltonian $h_{\mathbf{n}s\,\mathbf{n}s}$ and the Hamiltonian connecting two nearest-neighbor cells $h_{\mathbf{n}s\,\mathbf{n}\pm\mathbf{e}_i s}$ do not depend on $\mathbf{n}$. We therefore define the matrices

$$h_{ss'} \equiv h_{\mathbf{n}s\,\mathbf{n}s'} \tag{2.10}$$

and

$$T_{i,ss'} \equiv h_{\mathbf{n}+\mathbf{e}_i s\,\mathbf{n}s'}, \quad \Rightarrow \quad T^\dagger_{i,ss'} = h^*_{\mathbf{n}+\mathbf{e}_i s'\,\mathbf{n}s} = h_{\mathbf{n}s\,\mathbf{n}+\mathbf{e}_i s'} = h_{\mathbf{n}-\mathbf{e}_i s\,\mathbf{n}s'}. \tag{2.11}$$

With these definitions our model for the noninteracting Hamiltonian of a crystal takes the form

$$\hat{H}_0 = \sum_{\mathbf{n}\sigma} \sum_{ss'} \left(h_{ss'} \hat{d}^\dagger_{\mathbf{n}s\sigma} \hat{d}_{\mathbf{n}s'\sigma} + \sum_{i=1}^{d} T_{i,ss'} \hat{d}^\dagger_{\mathbf{n}+\mathbf{e}_i s\sigma} \hat{d}_{\mathbf{n}s'\sigma} + T^\dagger_{i,ss'} \hat{d}^\dagger_{\mathbf{n}-\mathbf{e}_i s\sigma} \hat{d}_{\mathbf{n}s'\sigma} \right).$$

To better visualize the matrix structure of $\hat{H}_0$, we introduce the vector of $\hat{d}$-operators

$$\hat{\mathbf{d}}^\dagger_{\mathbf{n}\sigma} \equiv (\hat{d}^\dagger_{\mathbf{n}1\sigma}, \hat{d}^\dagger_{\mathbf{n}2\sigma}, \ldots), \qquad \hat{\mathbf{d}}_{\mathbf{n}\sigma} = (\hat{d}^\dagger_{\mathbf{n}1\sigma}, \hat{d}^\dagger_{\mathbf{n}2\sigma}, \ldots)^\dagger,$$

and rewrite $\hat{H}_0$ in terms of the product between these vectors and the matrices h, T_i, and $T_i^\dagger$:

$$\hat{H}_0 = \sum_{\mathbf{n}\sigma} \left(\hat{\mathbf{d}}^\dagger_{\mathbf{n}\sigma} \, h \, \hat{\mathbf{d}}_{\mathbf{n}\sigma} + \sum_{i=1}^{d} \hat{\mathbf{d}}^\dagger_{\mathbf{n}+\mathbf{e}_i\sigma} \, T_i \, \hat{\mathbf{d}}_{\mathbf{n}\sigma} + \hat{\mathbf{d}}^\dagger_{\mathbf{n}-\mathbf{e}_i\sigma} \, T_i^\dagger \, \hat{\mathbf{d}}_{\mathbf{n}\sigma} \right). \tag{2.12}$$

The strategy to find the one-particle eigenvalues of $\hat{H}_0$ consists of considering a finite block of the crystal and then letting the volume of the block go to infinity. This block is, for convenience, chosen as a parallelepiped with edges given by d linearly independent vectors of integers $\mathbf{N}_1, \ldots, \mathbf{N}_d$ each radiating from a given unit cell. Without loss of generality we can choose the radiating unit cell at the origin. The set V_b of all unit cells contained in the block is then

$$\mathsf{V}_b = \left\{ \mathbf{n} : 0 \leq \frac{\mathbf{n} \cdot \mathbf{N}_i}{\mathbf{N}_i \cdot \mathbf{N}_i} < 1 \text{ for all } i \right\}. \tag{2.13}$$

In the case of a one-dimensional crystal we can choose only one vector $\mathbf{N}_1 = N$ and the unit cells of the block $\mathbf{n} = n \in \mathsf{V}_b$ are those with $n = 0, 1, \ldots, N-1$. For the two-dimensional graphene, instead, we may choose, for example, $\mathbf{N}_1 = (N, 0)$ and $\mathbf{N}_2 = (0, M)$ or $\mathbf{N}_1 = (N, N)$ and $\mathbf{N}_2 = (M, -M)$, or any other couple of linearly independent vectors. When the volume of the block tends to infinity, the eigenvalues of the crystal are independent of the choice of the block. Interestingly, however, the procedure below allows us to know the eigenvalues of $\hat{H}_0$ for any *finite* $\mathbf{N}_1, \ldots, \mathbf{N}_d$, and hence to have access to the eigenvalues of blocks of different shapes. We come back to this aspect at the end of this section. To simplify the mathematical treatment, we impose periodic boundary conditions along $\mathbf{N}_1, \ldots, \mathbf{N}_d$ - that is, we "wrap" the block onto itself, forming a ring in $d = 1$, a torus in $d = 2$, etc. This choice of boundary conditions is known as the *Born–von Karman (BvK) boundary condition* and turns out to be very convenient. Other kinds of boundary conditions would lead to the same results in the limit of large V_b. The BvK condition implies that the cells $\mathbf{n}$ and $\mathbf{n} + \mathbf{N}_i$ are actually the same cell and we therefore make the identifications

$$\hat{\mathbf{d}}^\dagger_{\mathbf{n}+\mathbf{N}_i\sigma} \equiv \hat{\mathbf{d}}^\dagger_{\mathbf{n}\sigma}, \qquad \hat{\mathbf{d}}_{\mathbf{n}+\mathbf{N}_i\sigma} \equiv \hat{\mathbf{d}}_{\mathbf{n}\sigma} \tag{2.14}$$

for all $\mathbf{N}_i$.

We are now in the position to show how the diagonalization procedure works. Consider the Hamiltonian (2.12) in which the sum over $\mathbf{n}$ is restricted to unit cells in V_b and the identification (2.14) holds for all boundary terms with one $\hat{d}$-operator outside V_b. To bring $\hat{H}_0$ in a diagonal form, we must find a suitable linear combination of the $\hat{d}$-operators that preserves the anticommutation relations, as discussed in Section 1.6. For this purpose, we construct the matrix U with elements

$$U_{\mathbf{nk}} = \frac{1}{\sqrt{|\mathsf{V}_b|}} e^{i\mathbf{k}\cdot\mathbf{n}}, \tag{2.15}$$

where $|\mathsf{V}_b|$ is the number of unit cells in V_b. In (2.15) the row index runs over all $\mathbf{n} \in \mathsf{V}_b$, whereas the column index runs over all vectors $\mathbf{k} = (k_1, \ldots, k_d)$ with components $-\pi < k_i \leq \pi$ and, more importantly, fulfilling

$$\mathbf{k} \cdot \mathbf{N}_i = 2\pi m_i, \qquad \text{with } m_i \text{ integers}. \tag{2.16}$$

We leave it as an exercise for the reader to prove that the number of $\mathbf{k}$ vectors with these properties is exactly $|\mathrm{V}_b|$ and hence that U is a square matrix. The set of all $\mathbf{k}$ is called the *first Brillouin zone*. Due to property (2.16) the quantity $U_{\mathbf{nk}}$ is periodic in $\mathbf{n}$, with periods $\mathbf{N}_1, \ldots, \mathbf{N}_d$. It is this periodicity that we now exploit to prove that U is unitary. Consider the following set of equalities

$$\begin{aligned} e^{ik'_1} \sum_{\mathbf{n}\in\mathrm{V}_b} U^*_{\mathbf{nk}} U_{\mathbf{nk}'} &= \sum_{\mathbf{n}\in\mathrm{V}_b} U^*_{\mathbf{nk}} U_{\mathbf{n}+\mathbf{e}_1\mathbf{k}'} \\ &= \sum_{\mathbf{n}\in\mathrm{V}_b} U^*_{\mathbf{n}-\mathbf{e}_1\mathbf{k}} U_{\mathbf{nk}'} \\ &= e^{ik_1} \sum_{\mathbf{n}\in\mathrm{V}_b} U^*_{\mathbf{nk}} U_{\mathbf{nk}'}, \end{aligned}$$

and the likes with $k'_2, \ldots, k'_d$. In the second line of the above identities we use that the sum over all $\mathbf{n} \in \mathrm{V}_b$ of a periodic function $f(\mathbf{n})$ is the same as the sum of $f(\mathbf{n} - \mathbf{e}_1)$. For the left- and right-hand sides of these equations to be the same we must either have $\mathbf{k} = \mathbf{k}'$ or

$$\sum_{\mathbf{n}} U^*_{\mathbf{nk}} U_{\mathbf{nk}'} = 0 \qquad \text{for } \mathbf{k} \neq \mathbf{k}'.$$

Since for $\mathbf{k} = \mathbf{k}'$ the sum $\sum_{\mathbf{n}} U^*_{\mathbf{nk}} U_{\mathbf{nk}} = 1$, the matrix U is unitary and hence the operators

$$\hat{\mathbf{c}}_{\mathbf{k}\sigma} = \frac{1}{\sqrt{|\mathrm{V}_b|}} \sum_{\mathbf{n}\in\mathrm{V}_b} e^{-i\mathbf{k}\cdot\mathbf{n}}\, \hat{\mathbf{d}}_{\mathbf{n}\sigma}, \qquad \hat{\mathbf{c}}^\dagger_{\mathbf{k}\sigma} = \frac{1}{\sqrt{|\mathrm{V}_b|}} \sum_{\mathbf{n}\in\mathrm{V}_b} e^{i\mathbf{k}\cdot\mathbf{n}}\, \hat{\mathbf{d}}^\dagger_{\mathbf{n}\sigma},$$

preserve the anticommutation relations. The inverse relations read

$$\hat{\mathbf{d}}_{\mathbf{n}\sigma} = \frac{1}{\sqrt{|\mathrm{V}_b|}} \sum_{\mathbf{k}} e^{i\mathbf{k}\cdot\mathbf{n}}\, \hat{\mathbf{c}}_{\mathbf{k}\sigma}, \qquad \hat{\mathbf{d}}^\dagger_{\mathbf{n}\sigma} = \frac{1}{\sqrt{|\mathrm{V}_b|}} \sum_{\mathbf{k}} e^{-i\mathbf{k}\cdot\mathbf{n}}\, \hat{\mathbf{c}}^\dagger_{\mathbf{k}\sigma}, \tag{2.17}$$

and the reader can easily check that due to property (2.16) the $\hat{d}$-operators satisfy the BvK boundary conditions (2.14). Inserting these inverse relations into (2.12) (in which the sum is restricted to $\mathbf{n} \in \mathrm{V}_b$) we get the Hamiltonian

$$\hat{H}_0 = \sum_{\mathbf{k}\sigma} \hat{\mathbf{c}}^\dagger_{\mathbf{k}\sigma} \underbrace{\left(h + \sum_{i=1}^{d} (T_i e^{-ik_i} + T_i^\dagger e^{ik_i}) \right)}_{h_{\mathbf{k}}} \hat{\mathbf{c}}_{\mathbf{k}\sigma}.$$

In this expression the matrix $h_{\mathbf{k}}$ is Hermitian and can be diagonalized. Let $\epsilon_{\mathbf{k}\nu}$ be the eigenvalues of $h_{\mathbf{k}}$ and $u_{\mathbf{k}}$ be the unitary matrix that brings $h_{\mathbf{k}}$ in the diagonal form – that is, $h_{\mathbf{k}} = u_{\mathbf{k}} \mathrm{diag}(\epsilon_{\mathbf{k}1}, \epsilon_{\mathbf{k}2}, \ldots) u^\dagger_{\mathbf{k}}$. The unitary matrix $u_{\mathbf{k}}$ has the dimension of the number of orbitals in the unit cell and should not be confused with the matrix U, whose dimension is $|\mathrm{V}_b|$. We now perform a further change of basis and construct the following linear combinations of the $\hat{c}$-operators with fixed $\mathbf{k}$ vector

$$\hat{\mathbf{b}}_{\mathbf{k}\sigma} = u^\dagger_{\mathbf{k}} \hat{\mathbf{c}}_{\mathbf{k}\sigma}, \qquad \hat{\mathbf{b}}^\dagger_{\mathbf{k}\sigma} = \hat{\mathbf{c}}^\dagger_{\mathbf{k}\sigma} u_{\mathbf{k}}.$$

Denoting by $u_{\mathbf{k}\nu}(s) = (u_{\mathbf{k}})_{s\nu}$ the (s,ν) matrix element of $u_{\mathbf{k}}$, the explicit form of the $\hat{b}$-operators is

$$\hat{b}_{\mathbf{k}\nu\sigma} = \sum_s u^*_{\mathbf{k}\nu}(s)\hat{c}_{\mathbf{k}s\sigma}, \qquad \hat{b}^\dagger_{\mathbf{k}\nu\sigma} = \sum_s u_{\mathbf{k}\nu}(s)\hat{c}^\dagger_{\mathbf{k}s\sigma}.$$

With these definitions the Hamiltonian $\hat{H}_0$ takes the desired form since

$$\begin{aligned}\hat{H}_0 &= \sum_{\mathbf{k}\sigma} \hat{\mathbf{c}}^\dagger_{\mathbf{k}\sigma} u_{\mathbf{k}} \, \mathrm{diag}(\epsilon_{\mathbf{k}1}, \epsilon_{\mathbf{k}2}, \ldots) \, u^\dagger_{\mathbf{k}} \hat{\mathbf{c}}_{\mathbf{k}\sigma} \\ &= \sum_{\mathbf{k}\nu\sigma} \epsilon_{\mathbf{k}\nu} \hat{b}^\dagger_{\mathbf{k}\nu\sigma} \hat{b}_{\mathbf{k}\nu\sigma}.\end{aligned}$$

We have just derived the *Bloch theorem*: The one-particle eigenvalues of $\hat{H}_0$ are obtained by diagonalizing $h_{\mathbf{k}}$ for all $\mathbf{k}$ and the corresponding one-particle eigenkets $|\mathbf{k}\nu\sigma\rangle = \hat{b}^\dagger_{\mathbf{k}\nu\sigma}|0\rangle$ have overlap with the original basis functions $|\mathbf{n}s\sigma'\rangle = d^\dagger_{\mathbf{n}s\sigma'}|0\rangle$ given by

$$\psi_{\mathbf{k}\nu\sigma}(\mathbf{n}s\sigma') = \langle \mathbf{n}s\sigma'|\mathbf{k}\nu\sigma\rangle = \frac{\delta_{\sigma\sigma'}}{\sqrt{|\mathrm{V}_b|}} e^{\mathrm{i}\mathbf{k}\cdot\mathbf{n}} \, u_{\mathbf{k}\nu}(s), \tag{2.18}$$

which is a plane-wave with different amplitudes on different atoms of the same unit cell. When the volume of the block $|\mathrm{V}_b| \to \infty$, the $\mathbf{k}$ becomes a continuum index, see (2.16), called *quasi-momentum*[3] and the eigenvalues $\epsilon_{\mathbf{k}\nu}$ become a continuum called *band*. We then have a band for each ν and the total number of bands coincides with the number of localized orbitals per unit cell. Each crystal is characterized by its *band structure* and, as we see in Section 6.4, the band structure can be experimentally measured.

A *metal* is a crystal with partially filled bands; therefore, electrons at the Fermi energy can be excited using an infinitesimal amount of energy. It is precisely this property that makes a metal a good conductor of electrical current. In contrast, a *semiconductor* or an *insulator* is a crystal characterized by a finite energy gap ($\sim 1 \div 2$ eV for semiconductors and >2 eV for insulators) between the highest occupied state and the lowest unoccupied one. In this case we have bands that are either completely filled, known as *valence bands*, or completely empty, known as *conduction bands*.

Before applying the Bloch theorem to some simple crystals, one final observation is in order. If we choose the localized basis functions to be of the form (2.1) and subsequently we want to increase the accuracy of the calculations by reducing the spacing Δ, then in the limit $\Delta \to 0$ the quantities $u_{\mathbf{k}\nu}(s) \to u_{\mathbf{k}\nu}(\mathbf{r})$ become continuous functions of the position $\mathbf{r}$ in the cell. These functions can be periodically extended to all space by imposing that they assume the same value in the same point of all unit cells. In the literature the periodic functions $u_{\mathbf{k}\nu}(\mathbf{r})$ are called *Bloch functions*. Below we illustrate some elementary applications of this general framework.

One-band model The simplest example is a one-dimensional crystal with one orbital per unit cell, see Fig. 2.5(a). Then the matrices $h \equiv \epsilon$ and $T \equiv t$ are 1×1 matrices and the one-particle eigenvalues are

$$\epsilon_k = \epsilon + te^{-\mathrm{i}k} + te^{\mathrm{i}k} = \epsilon + 2t\cos k. \tag{2.19}$$

[3] Actually the $\mathbf{k}$ vectors are dimensionless quasi-momenta. The relation between the $\mathbf{k}$ vectors and the true quasi-momenta is discussed in Appendix C.

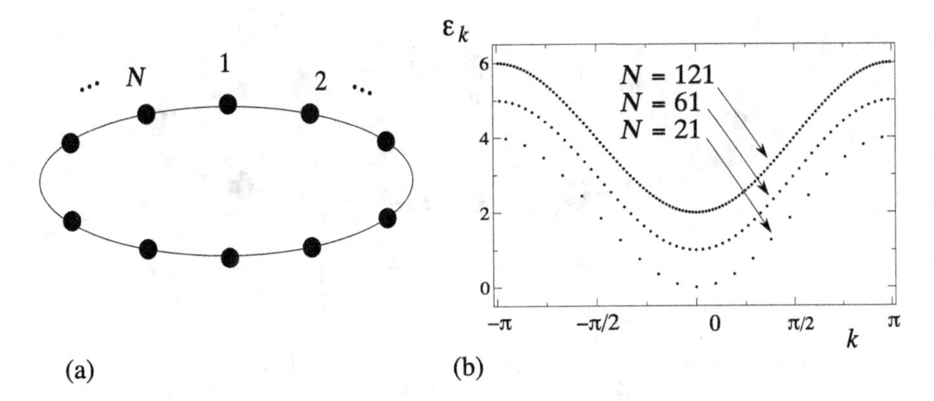

Figure 2.5 (a) Illustration of a ring with N unit cells. (b) Eigenvalues of the ring Hamiltonian with $t = -1$ and $\epsilon = 2$ for different numbers N of unit cells. The eigenvalues with $N = 61$ and 121 are shifted upward by 1 and 2, respectively.

If the number N of unit cells is, for example, odd, then the first Brillouin zone for the one-dimensional k vector is the set $k = 2\pi m/N$ with $m = -\frac{(N-1)}{2}, \ldots, \frac{(N-1)}{2}$, as follows from (2.16). The eigenvalues (2.19) are displayed in Fig. 2.5(b) for different N. It is clear that when $N \to \infty$ they become a continuum and form one band.

Two-band model Another example of a one-dimensional crystal is shown in Fig. 2.6(a). Each unit cell consists of two different atoms, a and b in Fig. 2.6(a). We assign a single orbital per atom and neglect all off-diagonal matrix elements except those connecting atoms of type a to atoms of type b. For simplicity we also consider the case that the distance between two nearest-neighbor atoms is everywhere the same along the crystal. Then the matrices h and T are 2×2 matrices with the following structure:

$$h = \begin{pmatrix} \epsilon + \Delta & t \\ t & \epsilon - \Delta \end{pmatrix}, \qquad T = \begin{pmatrix} 0 & t \\ 0 & 0 \end{pmatrix},$$

where Δ is an energy parameter that takes into account the different nature of the atoms. The structure of $T_{ss'} = h_{n+1s\,ns'}$, see (2.11), can be derived by checking the overlap between atom s in cell $n+1$ and atom s' in cell n. For instance, if we take cell 2 we see from Fig. 2.6(a) that atom a has overlap with atom b in cell 1, hence $T_{ab} = t$, but it does not have overlap with atom a in cell 1, hence $T_{aa} = 0$. On the other hand, atom b in cell 2 does not have overlap with any atom in cell 1, and hence $T_{ba} = T_{bb} = 0$. The eigenvalues of

$$h_k = h + Te^{-ik} + T^{\dagger}e^{ik} = \begin{pmatrix} \epsilon + \Delta & t(1 + e^{-ik}) \\ t(1 + e^{ik}) & \epsilon - \Delta \end{pmatrix}$$

are

$$\epsilon_{k\pm} = \epsilon \pm \sqrt{\Delta^2 + 2t^2(1 + \cos k)}, \tag{2.20}$$

and are displayed in Fig. 2.6(b) for different numbers N of unit cells. Like in the previous example, the first Brillouin zone is the set of $k = 2\pi m/N$ with $m = -\frac{(N-1)}{2}, \ldots, \frac{(N-1)}{2}$

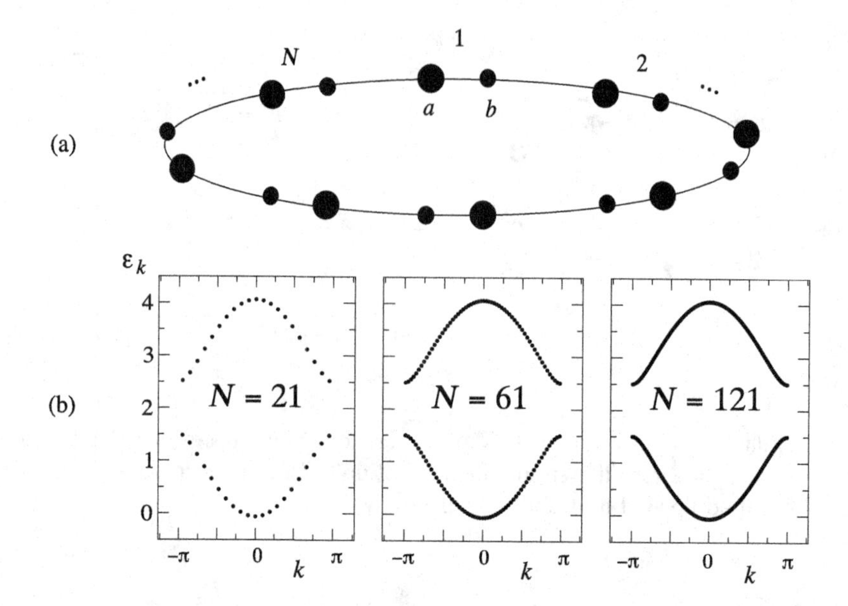

Figure 2.6 (a) Illustration of a ring with N unit cells and two atoms per cell. (b) Eigenvalues of the ring Hamiltonian with $t = -1$, $\epsilon = 2$, and $\Delta = 1/2$ for different numbers N of unit cells.

for odd N. When $N \to \infty$ the $\epsilon_{k\pm}$ become a continuum and form two bands separated by an *energy gap* of width 2Δ. If the crystal contains two electrons per unit cell, then the lower band is fully occupied, whereas the upper band is empty. In this situation we must provide a minimum energy 2Δ to excite one electron in an empty state; the crystal behaves like an insulator if the gap is large and like a semiconductor if the gap is small. It is also worth noting that for $\Delta = 0$ we recover the previous example but with $2N$ atoms rather than N.

Graphene Another important example is the two-dimensional crystal with which we opened this section. Graphene is a *single layer* of graphite and, as such, not easy to isolate. It was not until 2004 that graphene was experimentally realized by transferring a single layer of graphite onto a silicon dioxide substrate [9], an achievement that was awarded with the Nobel Prize in physics in 2010. The coupling between the graphene and the substrate is very weak and does not alter the electrical properties of the graphene. Before 2004, experimentalists were already able to produce tubes of graphene with a diameter of a few nanometers. These *carbon nanotubes* can be considered as an infinitely long graphene strip wrapped onto a cylinder. Due to their mechanical (strong and stiff) and electrical properties, carbon nanotubes have long being investigated. There exist several kind of nanotubes, depending on how the strip is wrapped. The wrapping is specified by a pair of integers (P, Q) so that atoms separated by $P\mathbf{v}_+ + Q\mathbf{v}_-$ are identified. Below we consider the *armchair nanotubes* for which $(P, Q) = (N, N)$ and hence the axis of the tube is parallel to $\mathbf{v}_+ - \mathbf{v}_-$.

Let us define the vector $\mathbf{N}_1 = (N, N)$ and the vector parallel to the tube axis $\mathbf{N}_2 =$

$(M, -M)$. The armchair nanotube is recovered when $M \to \infty$. For simplicity we assign a single p_z orbital to each carbon atom and consider only those off-diagonal matrix elements that connect atoms a to atoms b. From Fig. 2.4 we see that the matrices h, T_1, and T_2 must have the form[4]

$$h = \begin{pmatrix} 0 & t \\ t & 0 \end{pmatrix}, \qquad T_1 = \begin{pmatrix} 0 & t \\ 0 & 0 \end{pmatrix}, \qquad T_2 = \begin{pmatrix} 0 & t \\ 0 & 0 \end{pmatrix}.$$

The eigenvalues of

$$h_{\mathbf{k}} = h + \sum_{i=1}^{2} (T_i e^{-\mathrm{i}k_i} + T_i^{\dagger} e^{\mathrm{i}k_i}) = \begin{pmatrix} 0 & t(1 + e^{-\mathrm{i}k_1} + e^{-\mathrm{i}k_2}) \\ t(1 + e^{\mathrm{i}k_1} + e^{\mathrm{i}k_2}) & 0 \end{pmatrix}$$

can easily be calculated and read

$$\epsilon_{\mathbf{k}\pm} = \pm t \sqrt{1 + 4\cos\frac{k_1 - k_2}{2}\left(\cos\frac{k_1 - k_2}{2} + \cos\frac{k_1 + k_2}{2}\right)}.$$

The first Brillouin zone is a square with vertices in $(\pm\pi, \pm\pi)$ and $\mathbf{k}$ vectors fulfilling (2.16); that is,

$$\begin{aligned} \mathbf{k} \cdot \mathbf{N}_1 &= N(k_1 + k_2) = 2\pi m_1, \\ \mathbf{k} \cdot \mathbf{N}_2 &= M(k_1 - k_2) = 2\pi m_2. \end{aligned} \tag{2.21}$$

From these relations it is evident that an equivalent domain of $\mathbf{k}$ vectors, like the one illustrated in Fig. 2.7(a), is more convenient for our analysis. This equivalent domain is a rectangle tilted by $\pi/4$, with the short edge equal to $\sqrt{2}\pi$ and the long edge equal to $2\sqrt{2}\pi$. If we define $k_x = k_1 + k_2$ and $k_y = k_1 - k_2$, then $-2\pi < k_x \le 2\pi$ and $-\pi < k_y \le \pi$, and hence

$$\begin{aligned} k_x &= 2\pi\frac{m_1}{N}, \quad \text{with } m_1 = -N + 1, \ldots, N, \\ k_y &= 2\pi\frac{m_2}{M}, \quad \text{with } m_2 = -\frac{(M-1)}{2}, \ldots, \frac{M-1}{2}, \end{aligned}$$

for odd M. The reader can easily check that the number of $\mathbf{k}$ is the same as the number of atoms in the graphene block. In the limit $M \to \infty$ the quantum number k_y becomes a continuous index and we obtain $2 \times 2N$ one-dimensional bands corresponding to the number of carbon atoms in the transverse direction:[5]

$$\epsilon_{k_y k_x \pm} = \pm t \sqrt{1 + 4\cos\frac{k_y}{2}\left(\cos\frac{k_y}{2} + \cos\frac{k_x}{2}\right)}.$$

The set of all carbon atoms in the transverse direction can be considered as the unit cell of the one-dimensional crystal, which is the nanotube.

[4]The addition of a constant energy to the diagonal elements of h simply leads to a rigid shift of the eigenvalues.

[5]In this example the quantum number k_x together with the sign $\pm$ of the eigenvalues plays the role of the band index ν.

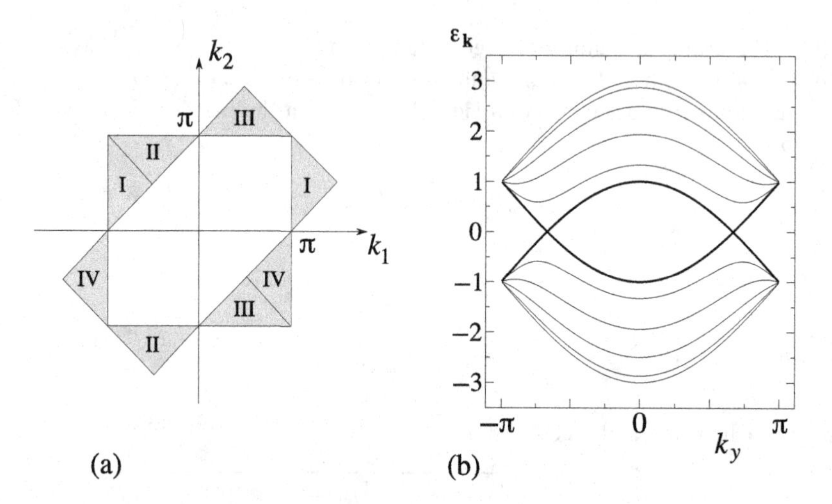

Figure 2.7 (a) Equivalent domain of $\mathbf{k}$ vectors. The gray areas with the same roman numerals are separated either by the vector $(2\pi, 0)$ or by the vector $(0, 2\pi)$. (b) Eigenvalues of a $(5, 5)$ armchair nanotube. The thick lines correspond to the highest valence band (below zero) and the lower conduction band (above zero). All other bands (thin lines) are doubly degenerate.

In Fig. 2.7(b) we plot the eigenvalues $\epsilon_{k_y k_x \pm}$ for the $(N, N) = (5, 5)$ armchair nanotube. Since the neutral nanotube has one electron per atom, all bands with negative energy are fully occupied. It is noteworthy that the highest occupied band (valence band) and the lowest unoccupied band (conduction band) touch *only at two points.* Around these points the energy dispersion is linear and the electrons behave as if they were relativistic [10]. Increasing the diameter of the nanotube, k_x too becomes a continuous index and the $\epsilon_{\mathbf{k}\pm}$ become the eigenvalues of the two-dimensional graphene. In this limit the $\epsilon_{\mathbf{k}\pm}$ form two two-dimensional bands that still touch at only two points. Due to this peculiar band structure, graphene is classified as a *semi-metal* – that is, is a crystal between a metal and a semiconductor.

Exercise 2.1 Show that the eigenvalues (2.20) with $\Delta = 0$ coincide with the eigenvalues (2.19) of the simple ring with $2N$ atoms.

Exercise 2.2 Consider the one-dimensional crystal below,

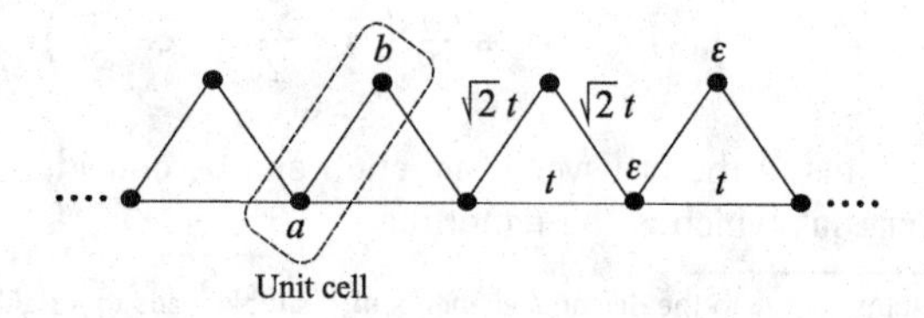

with matrices $h = \begin{pmatrix} \epsilon & \sqrt{2}t \\ \sqrt{2}t & \epsilon \end{pmatrix}$ and $T = \begin{pmatrix} t & \sqrt{2}t \\ 0 & 0 \end{pmatrix}$. Show that for $t > 0$ the two bands are

$$\epsilon_{k1} = \epsilon - 2t,$$
$$\epsilon_{k2} = \epsilon + 2t + 2t\cos k,$$

with $k \in (-\pi, \pi)$. The first band is therefore perfectly flat. If we have half an electron per unit cell, then the ground state is highly degenerate. For instance, the states obtained by occupying each k-level of the flat band with an electron of either spin up or down all have the same energy. This degeneracy is lifted by the electron–electron interaction and the ground state turns out to be the one in which all electrons have parallel spins. The crystal is then a *ferromagnet.* The ferromagnetism in flat-band crystals was proposed by Mielke and Tasaki [11, 12, 13] and is usually called *flat-band ferromagnetism.*

2.4 Anderson Model

The Anderson model is ubiquitous in condensed matter physics as it represents the simplest schematization of a discrete state "interacting" with a continuum of states. It was originally introduced by Anderson to describe the behavior of materials with magnetic impurities [14]. Today the Anderson model is also used to explain the so-called Kondo effect – that is, the formation of a sharp resonance involving a localized electronic spin surrounded by a spin cloud [15, 16], see Section 6.3.3. We introduce it here to model an atom or a molecule adsorbed on the surface of a metal, see the schematic illustration below.

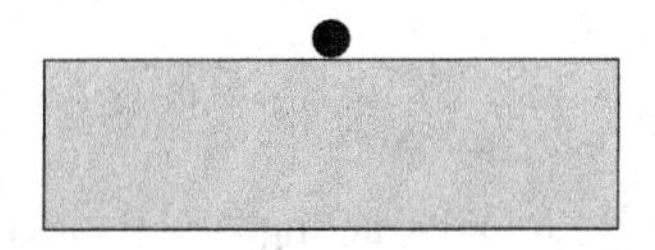

Let $|\epsilon Q\rangle$ be the single-particle energy eigenket of the metal, where Q is a quantum number (continuous or discrete) that accounts for the degeneracy of the eigenvalue ϵ. For the atom we consider only a single localized orbital and denote by $|\epsilon_0\rangle$ the corresponding ket. The atomic Hamiltonian $h = \langle\epsilon_0|\hat{h}|\epsilon_0\rangle \equiv \epsilon_0$ is therefore a 1×1 matrix. In this schematization of the problem, the atom can accommodate at most two electrons (of opposite spin); we denote by $U \equiv v_{\epsilon_0\epsilon_0\epsilon_0\epsilon_0}$ their repulsive energy. For large distances between the atom and the surface, the Hamiltonian of the system is the sum of the Hamiltonian $\hat{H}_{\text{met}}$ of the metal and the Hamiltonian $\hat{H}_{\text{at}}$ of the atom:

$$\hat{H}_{\text{met}} + \hat{H}_{\text{at}} = \sum_\sigma \int d\epsilon\, dQ\, \epsilon\, \hat{d}^\dagger_{\epsilon Q\sigma}\hat{d}_{\epsilon Q\sigma} + \sum_\sigma \epsilon_0\, \hat{d}^\dagger_{0\sigma}\hat{d}_{0\sigma} + U\hat{n}_{0\uparrow}\hat{n}_{0\downarrow},$$

where the $\hat{d}$-operators create or annihilate electrons and hence satisfy the anticommutation relations:

$$\left[\hat{d}_{\epsilon Q\sigma}, \hat{d}^\dagger_{\epsilon' Q'\sigma'}\right]_+ = \delta(\epsilon - \epsilon')\delta(Q - Q')\delta_{\sigma\sigma'}, \qquad \left[\hat{d}_{0\sigma}, \hat{d}^\dagger_{0\sigma'}\right]_+ = \delta_{\sigma\sigma'}.$$

The interaction energy is written in terms of the occupation operators $\hat{n}_{0\sigma} = \hat{d}^{\dagger}_{0\sigma}\hat{d}_{0\sigma}$, see (1.74), as we already did in (2.5). When the atom approaches the surface, the overlap $T_{\epsilon Q} \equiv \langle \epsilon Q|\hat{h}|\epsilon_0\rangle$ increases and cannot be neglected any longer. In this regime the Hamiltonian of the system becomes

$$\hat{H} = \hat{H}_{\text{met}} + \hat{H}_{\text{at}} + \sum_{\sigma} \int d\epsilon\, dQ \left(T_{\epsilon Q} \hat{d}^{\dagger}_{\epsilon Q\sigma} \hat{d}_{0\sigma} + T^{*}_{\epsilon Q} \hat{d}^{\dagger}_{0\sigma} \hat{d}_{\epsilon Q\sigma} \right). \tag{2.22}$$

This is the general structure of the *Anderson model.*

It is common (and for later purposes also instructive) to discretize the continuum of states by retaining only one state $|\epsilon_k Q_k\rangle$ in the volume element $\Delta\epsilon\Delta Q$. Then, for small volume elements, the Anderson Hamiltonian can be approximated as

$$\hat{H} = \sum_{k\sigma} \epsilon_k \hat{d}^{\dagger}_{k\sigma}\hat{d}_{k\sigma} + \sum_{k\sigma} \left(T_k \hat{d}^{\dagger}_{k\sigma}\hat{d}_{0\sigma} + T^{*}_k \hat{d}^{\dagger}_{0\sigma}\hat{d}_{k\sigma} \right) + \sum_{\sigma} \epsilon_0 \hat{d}^{\dagger}_{0\sigma}\hat{d}_{0\sigma} + U \hat{n}_{0\uparrow}\hat{n}_{0\downarrow}, \tag{2.23}$$

where the discrete $\hat{d}$-operators satisfy the anticommutation relations $\left[\hat{d}_k, \hat{d}^{\dagger}_{k'}\right]_+ = \delta_{kk'}$. To recover the continuum limit, we establish the correspondence

$$T_k = \sqrt{\Delta\epsilon\Delta Q}\, T_{\epsilon_k Q_k} \tag{2.24}$$

for the off-diagonal matrix elements of $\hat{h}$ and

$$\hat{d}_{k\sigma} = \sqrt{\Delta\epsilon\Delta Q}\, \hat{d}_{\epsilon_k Q_k \sigma}, \qquad \hat{d}^{\dagger}_{k\sigma} = \sqrt{\Delta\epsilon\Delta Q}\, \hat{d}^{\dagger}_{\epsilon_k Q_k \sigma},$$

for the fermionic operators. Letting $\Delta\epsilon\Delta Q \to 0$, the discrete Hamiltonian (2.23) reduces to the continuum Hamiltonian (2.22) and

$$\left[\hat{d}_{\epsilon_k Q_k \sigma}, \hat{d}^{\dagger}_{\epsilon_{k'} Q_{k'} \sigma'}\right]_+ = \frac{\delta_{kk'}}{\Delta\epsilon\Delta Q}\delta_{\sigma\sigma'} \to \delta(\epsilon - \epsilon')\delta(Q - Q')\delta_{\sigma\sigma'}.$$

In the following we study the noninteracting Anderson model, hence we set $U = 0$. The noninteracting Hamiltonian $\hat{H}_0 = \hat{H}_{0\uparrow} + \hat{H}_{0\downarrow}$ is the sum of two identical Hamiltonians differing only by the spin projection of the fermionic operators. As the spin projections can be treated separately, we consider only one of the two, say $\hat{H}_{0\uparrow}$. Being the spin index of the fermionic operators always $\sigma =\uparrow$ we omit to write the spin index in the remainder of this section: $\hat{d}_k$ and $\hat{d}_0$ stand for $\hat{d}_{k\uparrow}$ and $\hat{d}_{0\uparrow}$.

Atomic occupation We are interested in calculating the atomic occupation n_0 for a given Fermi energy ϵ_F, which is the energy of the highest occupied level of the metal. Let $|\lambda\rangle$ be the single-particle eigenkets of $\hat{H}_0$ with eigenenergies ϵ_λ. The corresponding annihilation and creation operators $\hat{c}_\lambda$ and $\hat{c}^{\dagger}_\lambda$ are a linear combination of the original $\hat{d}$-operators, as discussed in Section 1.5:

$$\hat{c}^{\dagger}_\lambda = \langle \epsilon_0|\lambda\rangle\, \hat{d}^{\dagger}_0 + \sum_k \langle k|\lambda\rangle\, \hat{d}^{\dagger}_k,$$

and with a similar equation for the adjoint. The ground state $|\Phi_0\rangle$ of the system is obtained by occupying all the $|\lambda\rangle$ with energies $\epsilon_\lambda \le \epsilon_F$ and reads

$$|\Phi_0\rangle = \prod_{\lambda:\epsilon_\lambda \le \epsilon_F} \hat{c}^{\dagger}_\lambda\, |0\rangle. \tag{2.25}$$

To calculate the atomic occupation $n_0 = \langle\Phi_0|\hat{d}_0^\dagger\hat{d}_0|\Phi_0\rangle$, we evaluate the ket $\hat{d}_0|\Phi_0\rangle$ by moving the operator $\hat{d}_0$ through the string of $\hat{c}^\dagger$-operators, as we did in (1.57). The difference here is that the anticommutator is $[\hat{d}_0, \hat{c}_\lambda^\dagger]_+ = \langle\epsilon_0|\lambda\rangle$ rather than a Kronecker delta. We then find

$$\hat{d}_0|\Phi_0\rangle = \sum_\lambda (-)^{p_\lambda}\langle\epsilon_0|\lambda\rangle \prod_{\lambda'\neq\lambda} \hat{c}_{\lambda'}^\dagger|0\rangle,$$

where the sum and the product are restricted to states below the Fermi energy. The integer p_λ in the above equation refers to the position of $\hat{c}_\lambda^\dagger$ in the string of operators (2.25), in agreement with (1.57). The atomic occupation n_0 is the inner product of the many-particle state $\hat{d}_0|\Phi_0\rangle$ with itself. In this inner product all cross terms vanish since they are proportional to the inner product of states with different strings of $\hat{c}_\lambda^\dagger$-operators, see (1.58). Taking into account that $\prod_{\lambda'\neq\lambda}\hat{c}_{\lambda'}^\dagger|0\rangle$ is normalized to 1 for every λ, we obtain the intuitive result

$$n_0 = \sum_{\lambda:\epsilon_\lambda\leq\epsilon_F} |\langle\epsilon_0|\lambda\rangle|^2\,; \tag{2.26}$$

the atomic occupation is the sum over all occupied states of the probability of finding an electron in $|\epsilon_0\rangle$.

From (2.26) it seems that we need to know the eigenkets $|\lambda\rangle$ and eigenenergies ϵ_λ in order to determine n_0. We now show that this is not strictly the case. Let us rewrite n_0 as

$$n_0 = \int_{-\infty}^{\epsilon_F}\frac{d\omega}{2\pi}\sum_\lambda 2\pi\delta(\omega-\epsilon_\lambda)|\langle\epsilon_0|\lambda\rangle|^2. \tag{2.27}$$

The eigenkets $|\lambda\rangle$ satisfy $\hat{h}|\lambda\rangle = \epsilon_\lambda|\lambda\rangle$, where $\hat{h}$ is the single-particle Hamiltonian in first quantization. For the Anderson model, $\hat{h}$ has the following ket-bra form:

$$\hat{h} = \sum_k \epsilon_k|k\rangle\langle k| + \epsilon_0|\epsilon_0\rangle\langle\epsilon_0| + \sum_k\left(T_k|k\rangle\langle\epsilon_0| + T_k^*|\epsilon_0\rangle\langle k|\right),$$

where we use (1.87). Then,

$$\sum_\lambda \delta(\omega-\epsilon_\lambda)|\langle\epsilon_0|\lambda\rangle|^2 = \sum_\lambda\langle\epsilon_0|\delta(\omega-\hat{h})|\lambda\rangle\langle\lambda|\epsilon_0\rangle = \langle\epsilon_0|\delta(\omega-\hat{h})|\epsilon_0\rangle,$$

where we use the completeness relation $\sum_\lambda|\lambda\rangle\langle\lambda| = \hat{\mathbb{1}}$. Inserting this result into (2.27), we get

$$n_0 = \int_{-\infty}^{\epsilon_F}\frac{d\omega}{2\pi}\,\langle\epsilon_0|2\pi\delta(\omega-\hat{h})|\epsilon_0\rangle.$$

This is our first encounter with the *spectral function* (first quantization) operator,

$$\hat{\mathcal{A}}(\omega) \equiv 2\pi\delta(\omega-\hat{h}) = \mathrm{i}\left[\frac{1}{\omega-\hat{h}+\mathrm{i}\eta} - \frac{1}{\omega-\hat{h}-\mathrm{i}\eta}\right].$$

In the second equality η is an infinitesimally small positive constant and we use the *Cauchy relation*

$$\boxed{\frac{1}{\omega-\epsilon\pm\mathrm{i}\eta} = P\frac{1}{\omega-\epsilon} \mp \mathrm{i}\pi\delta(\omega-\epsilon)} \tag{2.28}$$

where P denotes the principal part. Thus we can calculate n_0 if we find a way to determine the matrix element $A_{00}(\omega) = \langle\epsilon_0|\hat{\mathcal{A}}(\omega)|\epsilon_0\rangle$ of the spectral function. As we see in Chapter 6, this matrix element can be interpreted as the probability that an electron in $|\epsilon_0\rangle$ has energy ω.[6]

To determine $A_{00}(\omega)$, we separate $\hat{h} = \hat{\mathcal{E}} + \hat{\mathcal{T}}$ into a sum of the metal+atom Hamiltonian $\hat{\mathcal{E}}$ and the off-diagonal part $\hat{\mathcal{T}}$, and use the identity

$$\frac{1}{\zeta - \hat{h}} = \frac{1}{\zeta - \hat{\mathcal{E}}} + \frac{1}{\zeta - \hat{h}}\,\hat{\mathcal{T}}\,\frac{1}{\zeta - \hat{\mathcal{E}}}, \tag{2.29}$$

where ζ is an arbitrary complex number. This identity can easily be verified by multiplying both sides from the left by $(\zeta - \hat{h})$. Bracketing (2.29) with $\langle\epsilon_0|$ and $|\epsilon_0\rangle$ and between $\langle\epsilon_0|$ and $|k\rangle$ we find

$$\langle\epsilon_0|\frac{1}{\zeta - \hat{h}}|\epsilon_0\rangle = \frac{1}{\zeta - \epsilon_0} + \sum_k T_k\,\langle\epsilon_0|\frac{1}{\zeta - \hat{h}}|k\rangle\,\frac{1}{\zeta - \epsilon_0},$$

$$\langle\epsilon_0|\frac{1}{\zeta - \hat{h}}|k\rangle = T_k^*\,\langle\epsilon_0|\frac{1}{\zeta - \hat{h}}|\epsilon_0\rangle\,\frac{1}{\zeta - \epsilon_k}.$$

Substituting the second of these equations into the first, we arrive at the following important result:

$$\langle\epsilon_0|\frac{1}{\zeta - \hat{h}}|\epsilon_0\rangle = \frac{1}{\zeta - \epsilon_0 - \Sigma_{\rm em}(\zeta)}, \qquad \text{with} \quad \Sigma_{\rm em}(\zeta) = \sum_k \frac{|T_k|^2}{\zeta - \epsilon_k}. \tag{2.30}$$

The *embedding self-energy* $\Sigma_{\rm em}(\zeta)$ appears because the atom is not isolated; we could think of it as a correction to the atomic energy ϵ_0 induced by the presence of the metal. Taking $\zeta = \omega + {\rm i}\eta$, we can separate $\Sigma_{\rm em}(\zeta)$ into a real and an imaginary part:

$$\Sigma_{\rm em}(\omega + {\rm i}\eta) = \sum_k \frac{|T_k|^2}{\omega - \epsilon_k + {\rm i}\eta} = \underbrace{P\sum_k \frac{|T_k|^2}{\omega - \epsilon_k}}_{\Lambda_{\rm em}(\omega)} - \frac{\rm i}{2}\underbrace{2\pi\sum_k |T_k|^2\delta(\omega - \epsilon_k)}_{\Gamma_{\rm em}(\omega)}. \tag{2.31}$$

The real and imaginary parts are not independent, but instead related by a *Hilbert transformation*,

$$\Lambda_{\rm em}(\omega) = P\int \frac{d\omega'}{2\pi}\frac{\Gamma_{\rm em}(\omega')}{\omega - \omega'},$$

as can be verified by inserting the explicit expression of $\Gamma_{\rm em}(\omega)$ into the r.h.s. In conclusion, we have obtained an expression for n_0 in terms of the quantity $\Gamma_{\rm em}(\omega)$ only:

$$\begin{aligned} n_0 &= \int_{-\infty}^{\epsilon_{\rm F}} \frac{d\omega}{2\pi} A_{00}(\omega) \\ &= \int_{-\infty}^{\epsilon_{\rm F}} \frac{d\omega}{2\pi}\,{\rm i}\left[\frac{1}{\omega + {\rm i}\eta - \epsilon_0 - \Sigma_{\rm em}(\omega + {\rm i}\eta)} - \frac{1}{\omega - {\rm i}\eta - \epsilon_0 - \Sigma_{\rm em}(\omega - {\rm i}\eta)}\right] \\ &= -2\int_{-\infty}^{\epsilon_{\rm F}} \frac{d\omega}{2\pi}\,{\rm Im}\frac{1}{\omega - \epsilon_0 - \Lambda_{\rm em}(\omega) + \frac{\rm i}{2}\Gamma_{\rm em}(\omega) + {\rm i}\eta}. \end{aligned} \tag{2.32}$$

[6] For the time being we observe that this interpretation is supported by the normalization condition $\int \frac{d\omega}{2\pi} A_{00}(\omega) = 1$ – that is, the probability that the electron has energy between $-\infty$ and ∞ is 1.

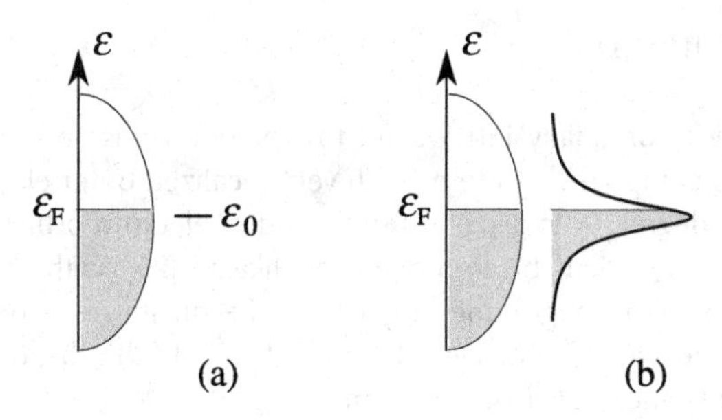

Figure 2.8 (a) The metallic band and the sharp level in the atomic limit. (b) The metallic band and the spectral function A_{00} for finite $\Gamma_{\rm em}$.

If the coupling T_k between the atom and the metal is weak, then $\Lambda_{\rm em}$ and $\Gamma_{\rm em}$ are small and the dominant contribution to the above integral comes from a region around ϵ_0. In the continuum limit the function $\Gamma_{\rm em}(\omega)$ is a smooth function since

$$\Gamma_{\rm em}(\omega) \to 2\pi \int d\epsilon\, dQ\, |T_{\epsilon Q}|^2 \delta(\omega - \epsilon) = 2\pi \int dQ\, |T_{\omega Q}|^2, \tag{2.33}$$

and in the integral (2.32) we can approximate $\Gamma_{\rm em}(\omega)$ by a constant $\Gamma_{\rm em} = \Gamma_{\rm em}(\epsilon_0)$. This approximation is known as the *Wide-Band Limit Approximation* (WBLA) since for the r.h.s. of (2.33) to be ω-independent the spectrum of the metal must extend from $-\infty$ to $+\infty$. In the WBLA the Hilbert transform $\Lambda_{\rm em}(\omega)$ vanishes and the formula for the atomic occupation simplifies to[7]

$$n_0 = \int_{-\infty}^{\epsilon_F} \frac{d\omega}{2\pi} \frac{\Gamma_{\rm em}}{(\omega - \epsilon_0)^2 + \Gamma_{\rm em}^2/4}. \tag{2.34}$$

We thus see that in the WBLA n_0 is the integral up to the Fermi energy of a Lorentzian of width $\Gamma_{\rm em}$ centered at ϵ_0. The atomic level is occupied ($n_0 \sim 1$) if $\epsilon_0 < \epsilon_F - \Gamma_{\rm em}$, whereas it is empty ($n_0 \sim 0$) if $\epsilon_0 > \epsilon_F + \Gamma_{\rm em}$. This result should be compared with the atomic limit, $\Gamma_{\rm em} = 0$, corresponding to the isolated atom. In this limit the Lorentzian becomes a δ-function since

$$\lim_{\Gamma_{\rm em} \to 0} \frac{1}{\pi} \frac{\Gamma_{\rm em}/2}{(\omega - \epsilon_0)^2 + \Gamma_{\rm em}^2/4} = \delta(\omega - \epsilon_0),$$

as follows immediately from the Cauchy relation (2.28). In the atomic limit the atomic occupation is exactly 1 for $\epsilon_0 < \epsilon_F$ and 0 otherwise. We say that the presence of the metal broadens the sharp atomic level and transforms it into a resonance of finite width, as illustrated in Fig. 2.8. This broadening is a general feature of discrete levels "interacting" or "in contact" with a continuum and it is observed in interacting systems as well.

[7]Since $\Gamma_{\rm em}(\omega) = \Gamma_{\rm em}$ for all ω, we can discard the infinitesimal η in the denominator of (2.32).

2.5 Hubbard Model

The *Hubbard model* was originally introduced to describe transition metals and rare earth metals – that is, solids composed by atoms with very localized outer electrons (d or f shells). In these materials the degree of localization of the outer-electron orbitals is so high that the Coulomb integrals $v_{s_1s_2s_3s_4}$ can be approximated like in (2.3) with the diagonal elements $U_s \equiv v_{ss}$ about an order of magnitude larger than the off-diagonal ones. For this reason, Hubbard, in a milestone paper from 1963 [17], included only the diagonal interaction U_s in his treatment and wrote the model Hamiltonian

$$\hat{H} = \hat{H}_0 + \hat{H}_{\rm int} = \sum_\sigma \sum_{ss'} h_{ss'} \hat{d}^\dagger_{s\sigma} \hat{d}_{s'\sigma} + \sum_s U_s \hat{n}_{s\uparrow} \hat{n}_{s\downarrow}, \tag{2.35}$$

that today carries his name. In this Hamiltonian the sums run over the lattice position s of the nuclei, the $\hat{d}$-operators are fermionic annihilation and creation operators for electrons with the usual anticommutation relations

$$\left[\hat{d}_{s\sigma}, \hat{d}^\dagger_{s'\sigma'}\right]_+ = \delta_{\sigma\sigma'}\delta_{ss'} \quad , \quad \left[\hat{d}_{s\sigma}, \hat{d}_{s'\sigma'}\right]_+ = 0,$$

and $\hat{n}_{s\sigma} = \hat{d}^\dagger_{s\sigma}\hat{d}_{s\sigma}$ is the occupation operator for electrons in s with spin σ. The matrix elements $h_{ss'}$ are typically set to zero if the distance between s and s' exceeds a few lattice spacings. In the context of model Hamiltonians, the off-diagonal matrix elements $h_{ss'}$ are also called *hopping integrals* or simply *hoppings* since they multiply the operator $\hat{d}^\dagger_{s\sigma}\hat{d}_{s'\sigma}$ which destroys an electron in s' and creates an electron in s, an operation that can be pictured as the hopping of an electron from s' to s. It is interesting to notice that in the Hubbard model two electrons interact only if they occupy the same atomic site (and of course have opposite spins). This feature allows for writing the interaction part using two different forms of the Coulomb integrals. The first form is the *spin-independent* interaction introduced in (1.89):

$$v^{(1)}_{s_1\sigma_1 s_2\sigma_2 s_3\sigma_3 s_4\sigma_4} = \delta_{\sigma_1\sigma_4}\delta_{\sigma_2\sigma_3} \times \delta_{s_1s_2}\delta_{s_2s_3}\delta_{s_3s_4} U_{s_1}, \tag{2.36}$$

whereas the second form is the *spin-dependent* interaction:

$$v^{(2)}_{s_1\sigma_1 s_2\sigma_2 s_3\sigma_3 s_4\sigma_4} = \delta_{\sigma_1\sigma_4}\delta_{\sigma_2\sigma_3} \times \delta_{\sigma_1\bar{\sigma}_2}\delta_{s_1s_2}\delta_{s_2s_3}\delta_{s_3s_4} U_{s_1}, \tag{2.37}$$

with $\bar{\sigma}_2$ the spin opposite to σ_2. Taking into account the Pauli principle $(\hat{d}_{s\sigma})^2 = 0$, the reader can easily check that

$$\sum_s U_s \hat{n}_{s\uparrow}\hat{n}_{s\downarrow} = \frac{1}{2} \sum_{\substack{s_1s_2s_3s_4 \\ \sigma_1\sigma_2\sigma_3\sigma_4}} v^{(i)}_{s_1\sigma_1 s_2\sigma_2 s_3\sigma_3 s_4\sigma_4} d^\dagger_{s_1\sigma_1} d^\dagger_{s_2\sigma_2} d_{s_3\sigma_3} d_{s_4\sigma_4}. \tag{2.38}$$

To gain some insight into the physics of the Hubbard model we compare two limiting situations: $\hat{H}_{\rm int} = 0$ and $\hat{H}_0 = 0$. We see that electrons behave as "waves" if $\hat{H}_{\rm int} = 0$ and as "particles" if $\hat{H}_0 = 0$; how they behave when both $\hat{H}_0$ and $\hat{H}_{\rm int}$ are nonzero is a fascinating problem intimately related to the wave–particle dualism. To readers interested in the physics of the Hubbard model we suggest the review article in Ref. [18] and the books in Refs. [19, 20].

No Hubbard interaction For $\hat{H}_{\rm int} = 0$ the Hubbard Hamiltonian describes noninteracting electrons on a lattice. Let $|k\sigma\rangle = c^\dagger_{k\sigma}|0\rangle$ be the one-particle eigenkets with energies ϵ_k. If we order the energies $\epsilon_k \le \epsilon_{k+1}$, the state of lowest energy with $N_{\uparrow/\downarrow}$ electrons of spin up/down is

$$|\Phi\rangle = \prod_{k=1}^{N_\uparrow}\prod_{k'=1}^{N_\downarrow} \hat{c}^\dagger_{k\uparrow}\hat{c}^\dagger_{k'\downarrow}|0\rangle \qquad \text{with} \quad \hat{H}_0|\Phi\rangle = \left(\sum_{k=1}^{N_\uparrow}\epsilon_k + \sum_{k'=1}^{N_\downarrow}\epsilon_{k'}\right)|\Phi\rangle. \tag{2.39}$$

We can say that in $|\Phi\rangle$ the electrons behave as "waves" since they have probability 0 or 1 to be in a delocalized state $|k\sigma\rangle$. The N-particle ground state of $\hat{H}_0$ is obtained by minimizing the eigenvalue in (2.39) with respect to $N_\uparrow$ and $N_\downarrow$ under the constraint $N_\uparrow + N_\downarrow = N$. For nondegenerate energies $\epsilon_k < \epsilon_{k+1}$, the ground state of $\hat{H}_0$ with an even number N of electrons is unique and reads

$$|\Phi_0\rangle = \prod_{k=1}^{N/2} \hat{c}^\dagger_{k\uparrow}\hat{c}^\dagger_{k\downarrow}|0\rangle. \tag{2.40}$$

We leave it as an exercise for the reader to show that $|\Phi_0\rangle$ is also an eigenstate of the total spin operators [see (1.94)],

$$\hat{\mathbf{S}} = (\hat{S}^x, \hat{S}^y, \hat{S}^z) \equiv \sum_s \hat{\mathbf{S}}_s = \sum_s (\hat{S}^x_s, \hat{S}^y_s, \hat{S}^z_s)$$

with vanishing eigenvalue, i.e., it is a singlet.

If we perturb the system with a weak external magnetic field $\mathbf{B}$ along, say, the z axis and discard the coupling to the orbital motion, the noninteracting part of the Hamiltonian changes to

$$\hat{H}_0 \to \hat{H}_0 - g\mu_{\rm B}\hat{\mathbf{S}}\cdot\mathbf{B} = \hat{H}_0 - \frac{1}{2}g\mu_{\rm B}B(\hat{N}_\uparrow - \hat{N}_\downarrow), \tag{2.41}$$

with g the electron gyromagnetic ratio, $\mu_{\rm B}$ the Bohr magneton, and $\hat{N}_\sigma = \sum_s \hat{n}_{s\sigma}$ the operator for the total number of particles of spin σ. The eigenkets (2.39) are also eigenkets of this new Hamiltonian, but with a different eigenvalue:

$$(\hat{H}_0 - g\mu_{\rm B}\hat{\mathbf{S}}\cdot\mathbf{B})|\Phi\rangle = \left(\sum_{k=1}^{N_\uparrow}(\epsilon_k - \frac{1}{2}g\mu_{\rm B}B) + \sum_{k'=1}^{N_\downarrow}(\epsilon_{k'} + \frac{1}{2}g\mu_{\rm B}B)\right)|\Phi\rangle.$$

Thus, in the presence of an external magnetic field the state (2.40) is no longer the lowest in energy since for, for example, $B > 0$, it becomes energetically convenient to have more electrons of spin up than of spin down. This is the typical behavior of a *Pauli paramagnet* - that is, a system whose total spin is zero for $B = 0$ and grows parallel to $\mathbf{B}$ for $B \neq 0$.[8]

[8]The Pauli paramagnetic behavior is due to the spin degrees of freedom and it is therefore distinct from the paramagnetic behavior due to the orbital degrees of freedom, see Section 3.4.

Only Hubbard interaction Next we discuss the case $\hat{H}_0 = 0$. The Hamiltonian $\hat{H}_{\text{int}}$ is already in a diagonal form and the generic eigenket can be written as

$$|\Phi_{XY}\rangle = \prod_{s\in X}\prod_{s'\in Y} \hat{d}^\dagger_{s\uparrow}\hat{d}^\dagger_{s'\downarrow}|0\rangle,$$

where X and Y are two arbitrary collections of atomic sites. We can say that in $|\Phi_{XY}\rangle$ the electrons behave like "particles" since they have probability 0 or 1 to be on a given atomic site. The energy eigenvalue of $|\Phi_{XY}\rangle$ is

$$E_{XY} = \sum_{s\in X\cap Y} U_s.$$

The ground state(s) for a given number $N = N_\uparrow + N_\downarrow$ of electrons can be constructed by choosing X and Y that minimize the energy E_{XY}. Denoting by N_V the total number of atomic sites we have that for $N \le N_V$ all states with $X \cap Y = \emptyset$ are ground states (with zero energy) and, again, the system behaves like a Pauli paramagnet.

In conclusion, neither $\hat{H}_0$ nor $\hat{H}_{\text{int}}$ favor any kind of magnetic order. However, their sum $\hat{H}_0 + \hat{H}_{\text{int}}$ sometimes does. In Section 2.6 we illustrate an example of this phenomenon.

2.5.1 Particle-Hole Symmetry

In this section we describe the properties of the Hubbard model defined on *bipartite lattices* - that is, lattices that can be divided into two sets of sites A and B with hoppings only from sites of A to sites of B and vice versa:

$$\hat{H}_{\text{bip}} = \sum_\sigma \sum_{\substack{s\in A\\ s'\in B}} \left(h_{ss'}\hat{d}^\dagger_{s\sigma}\hat{d}_{s'\sigma} + h_{s's}\hat{d}^\dagger_{s'\sigma}\hat{d}_{s\sigma}\right) + \sum_s U_s \hat{n}_{s\uparrow}\hat{n}_{s\downarrow}.$$

The Hubbard dimer in Section 2.2.1 represents the simplest example of bipartite lattice, A being site 1, B being site 2 and $h_{12} = T$. If the lattice is bipartite and if $h_{ss'} = h_{s's}$ and $U_s = U$ is independent of s, the Hubbard Hamiltonian enjoys an interesting *particle-hole symmetry* that can be used to simplify the calculation of eigenvalues and eigenkets. Let us first explain what this symmetry is. The fermionic operators

$$\hat{b}_{s\sigma} \equiv \begin{cases} \hat{d}^\dagger_{s\sigma} & s\in A \\[2ex] -\hat{d}^\dagger_{s\sigma} & s\in B \end{cases}$$

satisfy the same anticommutation relations as the original $\hat{d}$-operators, and in terms of them $\hat{H}_{\text{bip}}$ becomes

$$\begin{aligned}\hat{H}_{\text{bip}} &= -\sum_\sigma \sum_{\substack{s\in A\\ s'\in B}} \left(h_{ss'}\hat{b}_{s\sigma}\hat{b}^\dagger_{s'\sigma} + h_{s's}\hat{b}_{s'\sigma}\hat{b}^\dagger_{s\sigma}\right) + U\sum_s \hat{b}_{s\uparrow}\hat{b}^\dagger_{s\uparrow}\hat{b}_{s\downarrow}\hat{b}^\dagger_{s\downarrow} \\ &= \sum_\sigma \sum_{\substack{s\in A\\ s'\in B}} \left(h_{ss'}\hat{b}^\dagger_{s\sigma}\hat{b}_{s'\sigma} + h_{s's}\hat{b}^\dagger_{s'\sigma}\hat{b}_{s\sigma}\right) + U\sum_s \hat{n}^{(b)}_{s\uparrow}\hat{n}^{(b)}_{s\downarrow} - U\hat{N}^{(b)} + UN_V,\end{aligned}$$

where $\hat{n}^{(b)}_{s\sigma} = \hat{b}^\dagger_{s\sigma}\hat{b}_{s\sigma}$, $\hat{N}^{(b)} = \sum_{s\sigma}\hat{n}^{(b)}_{s\sigma}$, and N_V is the total number of sites. Except for the last two terms, which both commute with the first two terms, the Hamiltonian written with the $\hat{b}$-operators is identical to the Hamiltonian written with the $\hat{d}$-operators. This means that if

$$|\Psi\rangle = \sum_{s_1\ldots s_{N_\uparrow}}\sum_{s'_1\ldots s'_{N_\downarrow}} \Psi(s_1\uparrow,\ldots,s_{N_\uparrow}\uparrow,s'_1\downarrow,\ldots,s'_{N_\downarrow}\downarrow)\,\hat{d}^\dagger_{s_1\uparrow}\ldots\hat{d}^\dagger_{s_{N_\uparrow}\uparrow}\hat{d}^\dagger_{s'_1\downarrow}\ldots\hat{d}^\dagger_{s'_{N_\downarrow}\downarrow}|0\rangle$$

is an eigenstate of $\hat{H}_{\rm bip}$ with N_σ electrons of spin σ and energy E, then the state $|\Psi^{(b)}\rangle$ obtained from $|\Psi\rangle$ by replacing the $\hat{d}$-operators with the $\hat{b}$-operators *and* the empty ket $|0\rangle$ with the empty ket $|0^{(b)}\rangle$ of the $\hat{b}$-operators is also an eigenstate of $\hat{H}_{\rm bip}$ but with energy $E - U(N_\uparrow + N_\downarrow) + UN_V$. The empty ket $|0^{(b)}\rangle$ is by definition the ket for which

$$\hat{b}_{s\sigma}|0^{(b)}\rangle = 0 \qquad \text{for all } s \text{ and } \sigma.$$

Recalling the definition of the $\hat{b}$-operators, this is equivalent to saying that $\hat{d}^\dagger_{s\sigma}|0^{(b)}\rangle = 0$ for all s and σ – that is, the ket $|0^{(b)}\rangle$ is the full ket with one electron of spin up and down on every site. Thus the state $|\Psi^{(b)}\rangle$ can alternatively be written as a linear combination of products of annihilation $\hat{d}$-operators acting on the full ket. The number of electrons of spin σ in $|\Psi^{(b)}\rangle$ is therefore $N_V - N_\sigma$. In conclusion, we can say that if E is an eigenvalue of $\hat{H}_{\rm bip}$ with N_σ electrons of spin σ then $E - U(N_\uparrow + N_\downarrow) + UN_V$ is also an eigenvalue of $\hat{H}_{\rm bip}$ but with $N_V - N_\sigma$ electrons of spin σ. The relation between the corresponding eigenkets is that the empty ket is replaced by the full ket and the creation operators are replaced by the annihilation operators.

Exercise 2.3 Consider a chain of N sites, $s = 1,\ldots,N$, with $T_{ss'} = T$ if s and s' are nearest neighbors, and 0 otherwise. The one-body Hamiltonian is

$$\hat{H}_0 = T\sum_\sigma\sum_{s=1}^{N-1}(\hat{d}^\dagger_{s\sigma}\hat{d}_{s+1\sigma} + \hat{d}^\dagger_{s+1\sigma}\hat{d}_{s\sigma}).$$

Prove that the single-particle eigenkets of $\hat{H}_0$ are

$$|k\sigma\rangle = \hat{c}^\dagger_{k\sigma}|0\rangle = \sum_s\sqrt{\frac{2}{N+1}}\sin(\frac{\pi ks}{N+1})\hat{d}^\dagger_{s\sigma}|0\rangle,$$

with $k = 1,\ldots,N$, and have eigenvalues $\epsilon_k = 2T\cos(\frac{\pi k}{N+1})$.

Exercise 2.4 Let $\hat{N}_\sigma = \sum_s \hat{d}^\dagger_{s\sigma}\hat{d}_{s\sigma}$ be the operator for the total number of particles of spin σ and $\hat{S}^j = \sum_s \hat{S}^j_s$ be the *total spin operators*, where the spin density operators are defined in (1.94). Show that these operators commute with both $\hat{H}_0$ and $\hat{H}_{\rm int}$ in (2.35).

2.6 Heisenberg Model

A model system in which the number N of electrons is identical to the number N_V of space-orbitals basis functions is said to be half-filled, since the maximum possible value of N is $2N_V$. In the half-filled Hubbard model, when the U_s are much larger than the $h_{ss'}$, states with two electrons on the same site have very high energy. It would then be useful to construct an effective theory in the truncated space of states with one single electron per atomic site,

$$|\Phi_{\{\sigma\}}\rangle = \prod_s \hat{d}^\dagger_{s\sigma(s)}|0\rangle, \tag{2.42}$$

where $\sigma(s) = \uparrow, \downarrow$ is a collection of spin indices. The kets (2.42) are eigenkets of $\hat{H}_{\rm int}$ in (2.35) with eigenvalue zero and of $\hat{N}$ with eigenvalue N_V. Their number is equal to the number of possible spin configurations, that is 2^{N_V}. Since $\hat{H}_{\rm int}$ is positive semidefinite, the kets $|\Phi_{\{\sigma\}}\rangle$ form a basis in the ground (lowest energy) subspace of $\hat{H}_{\rm int}$. To construct the effective low-energy theory we need to understand how the states (2.42) are mixed by the one-body part $\hat{H}_0$ of the Hubbard Hamiltonian (2.35). Let us start by separating the matrix h into a diagonal part, $\epsilon_s = h_{ss}$, and an off-diagonal part $T_{ss'} = h_{ss'}$ for $s \neq s'$. For simplicity, in the discussion below we set the diagonal elements $\epsilon_s = 0$. If $|\Psi\rangle$ is an eigenket of the full Hamiltonian $\hat{H} = \hat{H}_0 + \hat{H}_{\rm int}$ with energy E, then the eigenvalue equation can be written as

$$|\Psi\rangle = \frac{1}{E - \hat{H}_{\rm int}}\hat{H}_0|\Psi\rangle. \tag{2.43}$$

We now approximate $|\Psi\rangle$ as a linear combination of the $|\Phi_{\{\sigma\}}\rangle$. Then, the action of $(E - \hat{H}_{\rm int})^{-1}\hat{H}_0$ on $|\Psi\rangle$ yields a linear combination of kets with a doubly occupied site.[9] Since these kets are orthogonal to $|\Psi\rangle$, we iterate (2.43) so as to generate an eigenvalue problem in the subspace of the $|\Phi_{\{\sigma\}}\rangle$,

$$\langle\Phi_{\{\sigma\}}|\Psi\rangle = \langle\Phi_{\{\sigma\}}|\frac{1}{E - \hat{H}_{\rm int}}\hat{H}_0\frac{1}{E - \hat{H}_{\rm int}}\hat{H}_0|\Psi\rangle, \qquad |\Psi\rangle = \sum_{\{\sigma\}}\alpha_{\{\sigma\}}|\Phi_{\{\sigma\}}\rangle. \tag{2.44}$$

This eigenvalue equation tells us how $\hat{H}_0$ lifts the ground-state degeneracy to second order in $\hat{H}_0$. To evaluate the r.h.s. we consider a generic term in $\hat{H}_0|\Psi\rangle$, say, $\hat{d}^\dagger_{s\uparrow}\hat{d}_{s'\uparrow}|\Phi_{\{\sigma\}}\rangle$, with $s \neq s'$ (recall that $\epsilon_s = 0$). This term corresponds to removing an electron of spin up from s' and creating an electron of spin up in s. If the spin configuration $\{\sigma\}$ is such that $\sigma(s') = \uparrow$ and $\sigma(s) = \downarrow$ the ket $\hat{d}^\dagger_{s\uparrow}\hat{d}_{s'\uparrow}|\Phi_{\{\sigma\}}\rangle$ has the site s occupied by two electrons (of opposite spin) and the site s' empty; in all other cases $\hat{d}^\dagger_{s\uparrow}\hat{d}_{s'\uparrow}|\Phi_{\{\sigma\}}\rangle = |\emptyset\rangle$, since we cannot remove an electron of spin up from s' if $\sigma(s') = \downarrow$ and we cannot create an electron of spin up in s if there is already one (i.e., $\sigma(s) = \uparrow$). From this observation we conclude that $\hat{d}^\dagger_{s\uparrow}\hat{d}_{s'\uparrow}|\Phi_{\{\sigma\}}\rangle$ is either an eigenket of $\hat{H}_{\rm int}$ with eigenvalue U_s or the null ket. Similar considerations apply to $\hat{d}^\dagger_{s\downarrow}\hat{d}_{s'\downarrow}|\Phi_{\{\sigma\}}\rangle$. Therefore, we can write

$$\frac{1}{E - \hat{H}_{\rm int}}\hat{H}_0|\Psi\rangle = \sum_\sigma\sum_{ss'}\frac{T_{ss'}}{E - U_s}\hat{d}^\dagger_{s\sigma}\hat{d}_{s'\sigma}|\Psi\rangle \sim -\sum_\sigma\sum_{ss'}\frac{T_{ss'}}{U_s}\hat{d}^\dagger_{s\sigma}\hat{d}_{s'\sigma}|\Psi\rangle,$$

[9] If the reader does not see that this is the case, it becomes clear in a few lines.

where in the last step we used that the U_s are much larger than E.[10] Inserting this result into (2.44) and taking into account that $\langle\Phi_{\{\sigma\}}|(E-\hat{H}_{\rm int})^{-1} = \langle\Phi_{\{\sigma\}}|/E$, we find the eigenvalue equation

$$E\langle\Phi_{\{\sigma\}}|\Psi\rangle = \langle\Phi_{\{\sigma\}}|\underbrace{-\sum_{\sigma\sigma'}\sum_{ss'}\sum_{rr'}\frac{T_{rr'}T_{ss'}}{U_s}\hat{d}^\dagger_{r\sigma'}\hat{d}_{r'\sigma'}\hat{d}^\dagger_{s\sigma}\hat{d}_{s'\sigma}}_{\hat{H}_{\rm eff}}|\Psi\rangle. \tag{2.45}$$

The eigenvalues E of the *effective Hamiltonian* $\hat{H}_{\rm eff}$ in the subspace of the $|\Phi_{\{\sigma\}}\rangle$ are the second-order corrections of the highly degenerate ground-state energy. We now show that in the subspace of the $|\Phi_{\{\sigma\}}\rangle$ the effective Hamiltonian takes a very elegant form.

The terms of $\hat{H}_{\rm eff}|\Psi\rangle$ that have a nonvanishing inner product with the $|\Phi_{\{\sigma\}}\rangle$ are those generated by removing an electron from s' (which then remains empty), creating an electron of the same spin in s (which becomes doubly occupied), and removing the same electron or the one with opposite spin from $r' = s$ and creating it back in $r = s'$, see the illustration below.

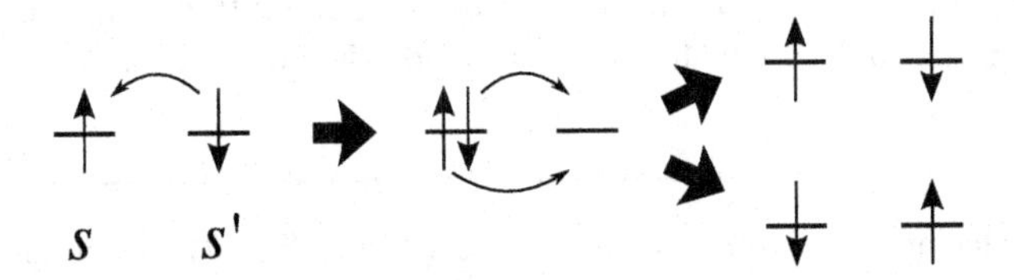

In (2.45) we can therefore restrict the sums to $r' = s$ and $r = s'$. In doing so the effective Hamiltonian simplifies to

$$\begin{aligned}\hat{H}_{\rm eff} &\equiv -\sum_{\sigma\sigma'}\sum_{ss'}\frac{T_{s's}T_{ss'}}{U_s}\hat{d}^\dagger_{s'\sigma'}\hat{d}_{s\sigma'}\hat{d}^\dagger_{s\sigma}\hat{d}_{s'\sigma}\\ &= -\sum_{\sigma\sigma'}\sum_{ss'}\frac{|T_{ss'}|^2}{U_s}(\delta_{\sigma\sigma'}\hat{n}_{s'\sigma} - \hat{d}^\dagger_{s'\sigma'}\hat{d}_{s'\sigma}\hat{d}^\dagger_{s\sigma}\hat{d}_{s\sigma'}).\end{aligned}$$

Using the definition (1.93) of the spin density operators, we can perform the sum over σ, σ' and rewrite the effective Hamiltonian as

$$\hat{H}_{\rm eff} = -\sum_{ss'}\frac{|T_{ss'}|^2}{U_s}\left(\hat{n}_{s'\uparrow} + \hat{n}_{s'\downarrow} - \hat{n}_{s'\uparrow}\hat{n}_{s\uparrow} - \hat{n}_{s'\downarrow}\hat{n}_{s\downarrow} - \hat{S}^+_{s'}\hat{S}^-_s - \hat{S}^-_{s'}\hat{S}^+_s\right).$$

The above equation can be further manipulated to obtain a physically transparent formula. In the subspace of the $|\Phi_{\{\sigma\}}\rangle$, the operator $\hat{n}_{s'\uparrow} + \hat{n}_{s'\downarrow}$ is always 1. This implies that $\hat{S}^z_s = \frac{1}{2}(\hat{n}_{s\uparrow} - \hat{n}_{s\downarrow}) = \hat{n}_{s\uparrow} - \frac{1}{2} = -\hat{n}_{s\downarrow} + \frac{1}{2}$, and hence $\hat{n}_{s'\uparrow}\hat{n}_{s\uparrow} + \hat{n}_{s'\downarrow}\hat{n}_{s\downarrow} = 2\hat{S}^z_{s'}\hat{S}^z_s + \frac{1}{2}$.

[10] We remind the reader that E is the correction to the degenerate ground-state energy due to the presence of $\hat{H}_0$ and therefore $E \to 0$ for $T_{ss'} \to 0$.

In conclusion, $\hat{H}_{\text{eff}}$ can be expressed solely in terms of the spin density operators:

$$\hat{H}_{\text{eff}} = \sum_{ss'} J_{ss'}(\hat{\mathbf{S}}_s \cdot \hat{\mathbf{S}}_{s'} - \frac{1}{4}), \qquad \text{with} \quad J_{ss'} = |T_{ss'}|^2 \left(\frac{1}{U_s} + \frac{1}{U_{s'}} \right).$$

This effective Hamiltonian is known as the *Heisenberg model* [21] and tells us that the interaction between two electrons frozen in their atomic positions has the same form as the interaction between two magnetic moments. There is, however, an important difference. The coupling constants $J_{ss'}$ are not proportional to μ_{B}^2 (the square of the Bohr magneton). The origin of the Heisenberg spin–spin interaction is purely quantum mechanical and has no classical analog; it stems from virtual transitions where an electron hops to another occupied site and then hops back with the same or opposite spin.

Since $J_{ss'} > 0$, the Heisenberg Hamiltonian favors those configurations in which the electron spins in s and s' point in opposite directions. For instance, in a one-dimensional chain with sites $s = -\infty, \ldots, -1, 0, 1, \ldots \infty$ and $T_{ss'} = T$ for $|s-s'| = 1$ and 0 otherwise, the ground-state average $\langle\Psi|\hat{\mathbf{S}}_s \cdot \hat{\mathbf{S}}_{s'}|\Psi\rangle = (-)^{s-s'} f(s-s')$ with $f(s-s') > 0$. This means that the spin magnetization is *staggered* and the system behaves like an *antiferromagnet*. The Heisenberg model is a clear example of how the subtle interplay between the tendency toward localization brought about by $\hat{H}_{\text{int}}$ and delocalization due to $\hat{H}_0$ favors some kind of magnetic order.

Exercise 2.5 Calculate the ground-state average $\langle\Psi|\hat{\mathbf{S}}_s \cdot \hat{\mathbf{S}}_{s'}|\Psi\rangle$ of the Heisenberg dimer $\hat{H}_{\text{eff}} = J\hat{\mathbf{S}}_1 \cdot \hat{\mathbf{S}}_2$ for $s, s' = 1, 2$.

2.7 BCS Model and the Exact Richardson Solution

The BCS model was introduced by Bardeen, Cooper, and Schrieffer in 1957 [22] to explain the phenomenon of superconductivity in metals at low enough temperatures. In the simplest version of the BCS model, two electrons can interact only if they are in the energy eigenstates $\varphi_{k\uparrow}(\mathbf{r}\sigma) = \delta_{\sigma\uparrow}\varphi_k(\mathbf{r})$ and $\varphi_{k\downarrow}(\mathbf{r}\sigma) = \delta_{\sigma\downarrow}\varphi_k^*(\mathbf{r})$, scattering into another couple of states $\varphi_{k'\uparrow}$ and $\varphi_{k'\downarrow}$.[‖] The probability for this process is assumed to be independent of k and k' and all other scattering processes are discarded. We defer the reader to the original paper for the microscopic justification of these assumptions. Denoting by ϵ_k the single-particle eigenenergies, the BCS Hamiltonian reads

$$\hat{H} = \hat{H}_0 + \hat{H}_{\text{int}} = \sum_{k\sigma} \epsilon_k \hat{c}^\dagger_{k\sigma}\hat{c}_{k\sigma} - v \sum_{kk'} \hat{c}^\dagger_{k\uparrow}\hat{c}^\dagger_{k\downarrow}\hat{c}_{k'\downarrow}\hat{c}_{k'\uparrow}, \tag{2.46}$$

where v is the so-called *scattering amplitude* and the sum runs over a set of one-particle states such as the Bloch states of Section 2.3. The operators $\hat{c}^\dagger_{k\sigma}$ create an electron in $\varphi_{k\sigma}$ and satisfy the usual anticommutation relations $\left[\hat{c}_{k\sigma}, \hat{c}^\dagger_{k'\sigma'}\right]_+ = \delta_{\sigma\sigma'}\delta_{kk'}$. The BCS model is usually treated within the so-called BCS approximation, according to which the ground

‖The state $\varphi_{k\downarrow}(\mathbf{r}\sigma)$ is the time reversal of $\varphi_{k\uparrow}(\mathbf{r}\sigma)$.

state is a superposition of states with a different number of electrons. That this is certainly an approximation follows from the fact that $\hat{H}$ commutes with the total number of particle operator $\hat{N}$, and hence the *exact* ground state has a well-defined number of electrons. Nevertheless, in macroscopic systems the fluctuations around the average value of $\hat{N}$ tend to zero in the thermodynamic limit, and the BCS approximation is, in these cases, very accurate. The BCS approximation breaks down in small systems such as superconducting aluminum nanograins [23]. Starting from the mid-1990s the experimental progress in characterizing superconducting nanostructures has renewed the interest in the BCS model beyond the BCS approximation. In the effort to construct new approximation schemes, a series of old papers in nuclear physics containing the *exact solution* of the BCS model came to light. In fact, the Hamiltonian (2.46) also describes a system of nucleons with an effective nucleon–nucleon interaction v; in this context it is known as the *pairing model.* In the mid-1960s Richardson derived a set of algebraic equations to calculate all eigenstates and eigenvalues of the pairing model [24]. Below we derive the Richardson solution of the BCS model.

The first important observation is that $\hat{H}_{\text{int}}$ acts only on doubly occupied k-states. This means that given a many-particle ket $|\Psi\rangle$ in which a k-state is occupied by a single electron (either with spin up or down), also the ket $\hat{H}|\Psi\rangle$ has the k-state occupied by a single electron. In other words, the singly occupied states remain blocked from participating in the dynamics. The labels of these states are therefore good quantum numbers. Let B be a subset of k-states whose number is $|B|$. Then, a generic eigenket of $\hat{H}$ has the form

$$|\Psi_B^{(N)}\rangle = \prod_{k\in B} \hat{c}^\dagger_{k\,\sigma(k)}|\Psi^{(N)}\rangle,$$

with $\sigma(k) = \uparrow, \downarrow$ a collection of spin indices and $|\Psi^{(N)}\rangle$ a $2N$-electron ket whose general form is

$$|\Psi^{(N)}\rangle = \sum_{k_1\ldots k_N \notin B} \alpha_{k_1\ldots k_N} \hat{b}^\dagger_{k_1}\ldots\hat{b}^\dagger_{k_N}|0\rangle, \tag{2.47}$$

where $\hat{b}^\dagger_k \equiv \hat{c}^\dagger_{k\uparrow}\hat{c}^\dagger_{k\downarrow}$ are electron-pair creation operators or simply pair operators. Since $\hat{b}^\dagger_k\hat{b}^\dagger_k = 0$, we can restrict the sum in (2.47) to $k_1 \neq k_2 \neq \ldots \neq k_N$. Furthermore, $\hat{b}^\dagger_k\hat{b}^\dagger_{k'} = \hat{b}^\dagger_{k'}\hat{b}^\dagger_k$ and hence the amplitudes $\alpha_{k_1\ldots k_N}$ are symmetric under a permutation of the indices $k_1, \ldots, k_N$. The eigenket $|\Psi_B^{(N)}\rangle$ describes $2N + |B|$ electrons, $|B|$ of which are in singly occupied k-states and contribute $E_B = \sum_{k\in B}\epsilon_k$ to the energy, and the remaining N pairs of electrons are distributed among the remaining unblocked k-states. Since the dynamics of the blocked electrons is trivial, in the following we assume $B = \emptyset$.

The key to solving the BCS model is the commutation relation between the pair operators

$$\left[\hat{b}_k, \hat{b}^\dagger_{k'}\right]_- = \delta_{kk'}(1 - \hat{n}_{k\uparrow} - \hat{n}_{k\downarrow}), \qquad \left[\hat{b}_k, \hat{b}_{k'}\right]_- = \left[\hat{b}^\dagger_k, \hat{b}^\dagger_{k'}\right]_- = 0,$$

with $\hat{n}_{k\sigma} = \hat{c}^\dagger_{k\sigma}\hat{c}_{k\sigma}$ the occupation operator. In the Hilbert space of states (2.47) one can discard the term $\hat{n}_{k\uparrow} + \hat{n}_{k\downarrow}$ in the commutator since $k_1 \neq k_2 \neq \ldots \neq k_N$ and hence

$$\begin{aligned}
\hat{b}_k \hat{b}^\dagger_{k_1} \hat{b}^\dagger_{k_2} \ldots \hat{b}^\dagger_{k_N} |0\rangle &= \left[\hat{b}_k, \hat{b}^\dagger_{k_1}\right]_- \hat{b}^\dagger_{k_2} \ldots \hat{b}^\dagger_{k_N} |0\rangle + \hat{b}^\dagger_{k_1} \left[\hat{b}_k, \hat{b}^\dagger_{k_2}\right]_- \ldots \hat{b}^\dagger_{k_N} |0\rangle \\
&+ \ldots + \hat{b}^\dagger_{k_1} \hat{b}^\dagger_{k_2} \ldots \left[\hat{b}_k, \hat{b}^\dagger_{k_N}\right]_- |0\rangle \\
&= \delta_{kk_1} \hat{b}^\dagger_{k_2} \hat{b}^\dagger_{k_3} \ldots \hat{b}^\dagger_{k_N} |0\rangle + \delta_{kk_2} \hat{b}^\dagger_{k_1} \hat{b}^\dagger_{k_3} \ldots \hat{b}^\dagger_{k_N} |0\rangle \\
&+ \ldots + \delta_{kk_N} \hat{b}^\dagger_{k_1} \hat{b}^\dagger_{k_2} \ldots \hat{b}^\dagger_{k_{N-1}} |0\rangle,
\end{aligned} \tag{2.48}$$

which is the same result as (1.57) for bosonic operators. This is not, a posteriori, surprising, since $\hat{b}^\dagger_k$ creates two electrons in a *singlet state*. These pair singlets are known as the *Cooper pairs* in honor of Cooper, who first realized the formation of bound states in a fermionic system with attractive interactions [25]. The analogy between Cooper pairs and bosons is, however, not so strict: We cannot create two Cooper pairs in the same state, but we can certainly do it for two bosons. We may say that Cooper pairs are *hardcore bosons* since we cannot find two or more of them in the same state.

To understand the logic of the Richardson solution we start with the simplest nontrivial case: a number $N = 2$ of pairs,

$$|\Psi^{(2)}\rangle = \sum_{p\neq q} \alpha_{pq} \hat{b}^\dagger_p \hat{b}^\dagger_q |0\rangle,$$

with $\alpha_{pq} = \alpha_{qp}$. Using (2.48) we find

$$\hat{H}|\Psi^{(2)}\rangle = \sum_{p\neq q} \alpha_{pq} \left((2\epsilon_p + 2\epsilon_q) \hat{b}^\dagger_p \hat{b}^\dagger_q - v \sum_{k\neq q} \hat{b}^\dagger_k \hat{b}^\dagger_q - v \sum_{k\neq p} \hat{b}^\dagger_p \hat{b}^\dagger_k \right) |0\rangle.$$

Renaming the indices $k \leftrightarrow p$ in the first sum and $k \leftrightarrow q$ in the second sum, we can easily extract from $\hat{H}|\Psi^{(2)}\rangle = E|\Psi^{(2)}\rangle$ an eigenvalue equation for the amplitudes α_{pq}

$$(2\epsilon_p + 2\epsilon_q)\alpha_{pq} - v \sum_{k\neq q} \alpha_{kq} - v \sum_{k\neq p} \alpha_{pk} = E\alpha_{pq}. \tag{2.49}$$

The constraint in the sums is reminiscent of the fact that there cannot be more than one pair in a k-state. It is this constraint that renders the problem complicated since for a pair to stay in k no other pair must be there. Nevertheless, (2.49) still admits a simple solution! The idea is to reduce (2.49) to two coupled eigenvalue equations. We therefore make the ansatz

$$\alpha_{pq} = \alpha^{(1)}_p \alpha^{(2)}_q + \alpha^{(1)}_q \alpha^{(2)}_p,$$

which entails the symmetry property $\alpha_{pq} = \alpha_{qp}$, and write the energy E as the sum of two pair energies $E = E_1 + E_2$. Then, the eigenvalue equation (2.49) becomes

$$\begin{aligned}
&\{\alpha^{(2)}_q (2\epsilon_p - E_1)\alpha^{(1)}_p + \alpha^{(1)}_q (2\epsilon_p - E_2)\alpha^{(2)}_p\} + \{p \leftrightarrow q\} \\
&= v \left(\{\alpha^{(2)}_q \sum_k \alpha^{(1)}_k + \alpha^{(1)}_q \sum_k \alpha^{(2)}_k\} + \{p \leftrightarrow q\} \right) - 2v(\alpha^{(1)}_p \alpha^{(2)}_p + \alpha^{(1)}_q \alpha^{(2)}_q),
\end{aligned} \tag{2.50}$$

where on the r.h.s. we extended the sums to all k and subtracted the extra term. Without this extra term (2.50) is solved by the amplitudes

$$\alpha_p^{(1)} = \frac{1}{2\epsilon_p - E_1}, \qquad \alpha_p^{(2)} = \frac{1}{2\epsilon_p - E_2}, \tag{2.51}$$

with E_i a root of

$$\sum_k \frac{1}{2\epsilon_k - E_i} = \frac{1}{v}.$$

This solution would correspond to two *independent* pairs since $\alpha_p^{(1)}$ does not depend on $\alpha_p^{(2)}$. The inclusion of the extra term does not change the structure (2.51) of the α, but instead changes the equation that determines E_i. Indeed, from (2.51) it follows that

$$\alpha_p^{(1)}\alpha_p^{(2)} = \frac{\alpha_p^{(2)} - \alpha_p^{(1)}}{E_2 - E_1},$$

and therefore (2.50) can be rewritten as

$$\{\alpha_q^{(2)}\left((2\epsilon_p - E_1)\alpha_p^{(1)} - v\sum_k \alpha_k^{(1)} + \frac{2v}{E_2 - E_1}\right)\} + \{p \leftrightarrow q\}$$

$$+\{\alpha_q^{(1)}\left((2\epsilon_p - E_2)\alpha_p^{(2)} - v\sum_k \alpha_k^{(2)} - \frac{2v}{E_2 - E_1}\right)\} + \{p \leftrightarrow q\} = 0.$$

We have just found that (2.51) is a solution provided that the pair energies E_1 and E_2 are the roots of the coupled system of algebraic equations:

$$\sum_k \frac{1}{2\epsilon_k - E_1} = \frac{1}{v} + \frac{2}{E_2 - E_1},$$

$$\sum_k \frac{1}{2\epsilon_k - E_2} = \frac{1}{v} + \frac{2}{E_1 - E_2}.$$

The generalization to N pairs is tedious but straightforward. In the remainder of this section we do not introduce concepts or formulas needed for the following chapters. Therefore, the reader can, if not interested, move forward to the next section. Let $|\Psi^{(N)}\rangle$ be the N-pair eigenket of $\hat{H}$ and $\{k\}_i$ be the set $\{k_1, \ldots, k_{i-1}, k_{i+1}, \ldots, k_N\}$ in which the k_i state is removed. The eigenvalue equation for the symmetric tensor $\alpha_{k_1 \ldots k_N}$ reads

$$\left(\sum_{i=1}^N 2\epsilon_{k_i}\right)\alpha_{k_1\ldots k_N} - v\sum_{i=1}^N \sum_{p\neq\{k\}_i} \alpha_{k_1\ldots k_{i-1}\, p\, k_{i+1}\ldots k_N} = E\alpha_{k_1\ldots k_N}.$$

Like in the case of two pairs, the sum over p on the l.h.s. is constrained and some extra work must be done to make the equation separable. The ansatz is

$$\alpha_{k_1\ldots k_N} = \sum_P \alpha_{k_{P(1)}}^{(1)} \cdots \alpha_{k_{P(N)}}^{(N)} = \sum_P \prod_{i=1}^N \alpha_{k_{P(i)}}^{(i)},$$

where the sum is over all permutations of $(1,\ldots,N)$, and $E = E_1 + \ldots + E_N$. Substitution into the eigenvalue equation leads to

$$\sum_P \sum_{i=1}^N \Bigg\{ \Big(\prod_{l\neq i}^N \alpha_{k_{P(l)}}^{(l)} \Big) \Bigg[(2\epsilon_{k_{P(i)}} - E_i)\alpha_{k_{P(i)}}^{(i)} - v \sum_p \alpha_p^{(i)} \Bigg] \\ + v \sum_{j\neq i}^N \Big(\prod_{l\neq i,j}^N \alpha_{k_{P(l)}}^{(l)} \Big) \alpha_{k_{P(j)}}^{(i)} \alpha_{k_{P(j)}}^{(j)} \Bigg] = 0, \tag{2.52}$$

where, as in the example with $N = 2$, the last term is what we have to subtract to perform the unconstrained sum over p. We look for solutions of the form

$$\alpha_k^{(i)} = \frac{1}{2\epsilon_k - E_i}, \tag{2.53}$$

from which it follows that the product of two pair amplitudes with the same k can be written as

$$\alpha_k^{(i)} \alpha_k^{(j)} = \frac{\alpha_k^{(j)} - \alpha_k^{(i)}}{E_j - E_i}.$$

We use this identity to manipulate a bit the last term in (2.52),

$$v \sum_P \sum_{i=1}^N \sum_{j\neq i}^N \Big(\prod_{l\neq i,j}^N \alpha_{k_{P(l)}}^{(l)} \Big) \frac{\alpha_{k_{P(j)}}^{(j)} - \alpha_{k_{P(j)}}^{(i)}}{E_j - E_i} = 2v \sum_P \sum_{i=1}^N \Big(\prod_{l\neq i}^N \alpha_{k_{P(l)}}^{(l)} \Big) \sum_{j\neq i}^N \frac{1}{E_j - E_i},$$

a result which allows us to decouple the eigenvalue equation

$$\sum_P \sum_{i=1}^N \Big(\prod_{l\neq i}^N \alpha_{k_{P(l)}}^{(l)} \Big) \Bigg[(2\epsilon_{k_{P(i)}} - E_i)\alpha_{k_{P(i)}}^{(i)} - v \sum_p \alpha_p^{(i)} + 2v \sum_{j\neq i} \frac{1}{E_j - E_i} \Bigg] = 0.$$

Therefore the amplitudes (2.53) are solutions provided that the pair energies E_i are the roots of the coupled system of algebraic equations:

$$\boxed{\frac{1}{v_i} \equiv \frac{1}{v} + 2\sum_{j\neq i} \frac{1}{E_j - E_i} = \sum_k \frac{1}{2\epsilon_k - E_i}} \tag{2.54}$$

The system (2.54) can be regarded as a generalization of the eigenvalue equation for a single pair, where the effective scattering amplitude v_i depends on the relative distribution of all other pairs. Numerical solutions show that with increasing v some of the E_i become complex; however, they always occur in complex conjugate pairs, so that the total energy E remains real.

2.8 Holstein Model

The motion of an electron in a crystal (in particular an ionic crystal) causes a displacement of the nuclei and, therefore, differs quite substantially from the motion through a rigid nuclear structure. In Section 16.2 we derive the exact form of the Hamiltonian for electrons and nuclei assuming that the nuclei remain close to their equilibrium positions. Here we provide an intuitive and nonrigorous presentation of a model Hamiltonian that can be derived from the exact result. Consider an electron at some point of the crystal: As a result of the attractive interaction, the nuclei move toward new positions and create a potential well for the electron. If the well is deep enough and if the electron is sufficiently slow, this effect causes the self-trapping of the electron: The electron cannot move unless accompanied by the well, also called *polarization cloud.* The *quasi-particle,* which consists of the electron and of the surrounding polarization cloud, is called the *polaron.*

Depending on the nature of the material, the electron–nuclear interaction gives rise to different kind of polarons, small or large, heavy or light, etc. In ionic crystals (e.g., sodium chloride) the electron–nuclear interaction is long-ranged and a popular model to describe the polaron features is the Fröhlich model [26]. In 1955 Feynman proposed a variational solution in which the polaron was considered as an electron bound to a particle of mass M by a spring of elastic constant K [27]. The Feynman solution is in good agreement with the numerical results for both strong and weak electron–nuclear coupling, and hence it provides a good physical picture of the polaron behavior in these regimes. Other physically relevant models leading to the formation of polarons are the Su–Schrieffer–Heeger model [28] for conducting polymers (originally polyacetylene) and the Holstein model [29] for molecular crystals (originally one-dimensional). Below we discuss the Holstein model.

Let us consider electrons moving along a chain of diatomic molecules whose center of mass and orientations are fixed, whereas the intra-nuclear distance can vary. The Hamiltonian of the system can be regarded as the sum of the molecular chain Hamiltonian $\hat{H}_C$, the electron Hamiltonian $\hat{H}_{\rm el}$, and the Hamiltonian $\hat{H}_{\rm int}$ that describes the interaction between the electrons and the nuclei.

Vibrational Hamiltonian Assuming that the potential energy of a single molecule in the chain is not too different from that of the isolated molecule, and approximating the latter with the energy of a harmonic oscillator of frequency ω_0, the Hamiltonian of the molecular chain takes the form

$$\hat{H}_C = \sum_s \left(\frac{\hat{p}_s^2}{2M} + \frac{1}{2} M \omega_0^2 \hat{x}_s^2 \right),$$

with $\hat{x}_s$ the intra-nuclear distance (measured with respect to the equilibrium position) operator of the sth molecule, $\hat{p}_s$ the corresponding conjugate momentum, and M the relative mass. Introducing the lowering operators $\hat{a}_s = \sqrt{\frac{M\omega_0}{2}}(\hat{x}_s + \frac{\mathrm{i}}{M\omega_0}\hat{p}_s)$ and raising operators $\hat{a}_s^\dagger = \sqrt{\frac{M\omega_0}{2}}(\hat{x}_s - \frac{\mathrm{i}}{M\omega_0}\hat{p}_s)$ with *bosonic* commutation relations $[\hat{a}_s, \hat{a}_{s'}^\dagger]_- = \delta_{ss'}$ and $[\hat{a}_s, \hat{a}_{s'}]_- = 0$, the chain Hamiltonian can be rewritten as

$$\hat{H}_C = \sum_s \omega_0 (\hat{a}_s^\dagger \hat{a}_s + \frac{1}{2}). \tag{2.55}$$

We here wish to spend a few words on the physical meaning of the bosonic operators $\hat{a}_s^\dagger$ and $\hat{a}_s$. First, it is important to stress that *they do not create or destroy the molecules but quanta of vibrations.* These quanta can be labeled with their energy (in our case the energy is the same for all oscillators) or with their position (the center of mass x_s of the molecules) and have, therefore, the *same degrees of freedom as a "particle" of spin zero.* In solid-state physics these "particles" are called *phonons* or, if the molecule is isolated, *vibrons.* Let us elaborate on this "particle" interpretation by considering a one-dimensional harmonic oscillator. In quantum mechanics the energy eigenket $|j\rangle$ describes *one particle* (either a fermion or a boson) in the jth energy level. Within the above interpretation of the vibrational quanta the ket $|j\rangle$ corresponds to a state with j *phonons* since $|j\rangle = \frac{1}{\sqrt{j!}}(\hat{a}^\dagger)^j|0\rangle$; in particular, the ground state $|0\rangle$, which describes *one* particle in the lowest-energy eigenstate, corresponds to a state with *zero* phonons. Even more "exotic" is the new interpretation of ket $|x\rangle$ that in quantum mechanics describes *one* particle in position x. Indeed, $|x\rangle = \sum_j |j\rangle\langle j|x\rangle$ and therefore it is a linear combination of kets *with different numbers of phonons.* Accordingly, an eigenstate of the position operator is obtained by acting with the operator $\hat{A}^\dagger(x) \equiv \sum_j \langle j|x\rangle \frac{1}{\sqrt{j!}}(\hat{a}^\dagger)^j$ on the zero-phonons ket $|0\rangle$. The interpretation of the quantum of vibration as a bosonic particle has been and continues to be a very fruitful idea. After the scientific revolutions of Special Relativity and Quantum Mechanics scientists started to look for a quantum theory consistent with relativistic mechanics.[12] In the attempt to construct such a theory, Dirac proposed a relativistic equation for the wavefunction [30]. The Dirac equation, however, led to a new interpretation of the electrons. The electrons, and more generally the fermions, can be seen as the vibrational quanta of some *fermionic oscillator* in the same way as the phonons, and more generally the bosons, are the vibrational quanta of some *bosonic oscillator.* Of course to describe a particle in the three-dimensional space we need many oscillators, like in the Holstein model. In a continuum description the operator $\hat{a}$ is labeled with the continuous variable $\mathbf{r}$, $\hat{a}_s \to \hat{a}(\mathbf{r})$, and it becomes what is called a *quantum field.* The origin of the name quantum field stems from the fact that to every point in space is assigned a quantity, as is done with the classical electric field $\mathbf{E}(\mathbf{r})$ or magnetic field $\mathbf{B}(\mathbf{r})$, but this quantity is an operator, see Section 16.1. Similarly, the fermions are described by some fermionic quantum field. For these reasons the relativistic generalization of quantum mechanics is called Quantum Field Theory [31, 32, 33, 34, 35, 36].

Interaction Hamiltonian Let us continue with the description of the Holstein model and derive an approximation for the interaction between the electrons and the vibrational quanta of the molecules. We denote by $E(x)$ the ground-state energy of the isolated molecule with one more electron (molecular ion). Due to the presence of the extra electron $E(x)$ is not stationary at the equilibrium position $x = 0$ of the isolated charge-neutral molecule. For small x we can expand the energy to linear order and write $E(x) = -g\sqrt{2M\omega_0}\,x$,[13] where the constant g governs the strength of the coupling. Depending on the system, the coupling g can be either negative or positive. We here choose $g > 0$. Due to the minus sign, the presence of the electron causes a force $F = -dE/dx > 0$, which tends to increase the interatomic distance in the molecule, as shown in the schematic representation below.

[12] We recall that the original Schrödinger equation is invariant under Galilean transformation but not under the more fundamental Lorentz transformations.

[13] The zeroth-order term leads to a trivial shift of the total energy for any fixed number of electrons and is therefore ignored.

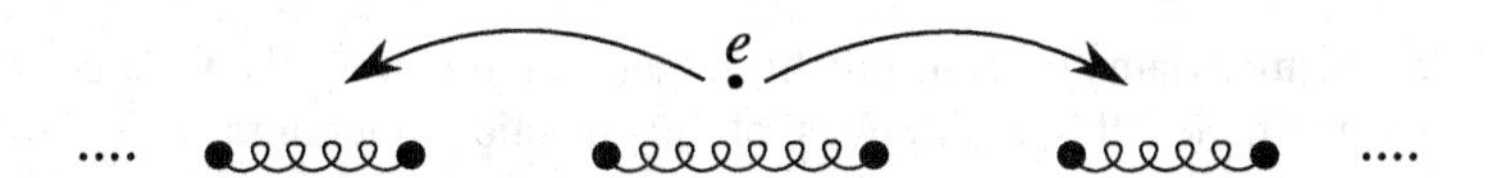

If we discard the corrections induced by the presence of the other molecules, the interaction Hamiltonian can then be modeled as

$$\hat{H}_{\text{int}} = -g\sqrt{2M\omega_0}\sum_{s\sigma} \hat{d}^\dagger_{s\sigma}\hat{d}_{s\sigma}\hat{x}_s = -g\sum_{s\sigma} \hat{d}^\dagger_{s\sigma}\hat{d}_{s\sigma}(\hat{a}_s + \hat{a}^\dagger_s),$$

where the $\hat{d}_{s\sigma}$-operator destroys an electron of spin σ on the sth molecule and therefore $\hat{d}^\dagger_{s\sigma}\hat{d}_{s\sigma}$ is simply the occupation operator that counts how many electrons (zero or one) of spin σ sit on the sth molecule. It is also important to say that the fermionic operators $\hat{d}$ *commute* with the bosonic operators $\hat{a}$.

Holstein Hamiltonian Finally we need to model the Hamiltonian $\hat{H}_{\text{el}}$ that describes "free" electrons moving along the chain. Since the orthonormal functions $\varphi_{s\sigma}$ that define the $\hat{d}$-operators are localized around the sth molecule, we neglect all matrix elements $T_{ss'} \equiv \langle s\sigma|\hat{h}|s'\sigma\rangle$ except those for which $|s - s'| = 1$ (nearest-neighbor molecules). The diagonal matrix elements can be ignored as well since $\epsilon_s \equiv \langle s\sigma|\hat{h}|s\sigma\rangle$ does not depend on s; these matrix elements simply give rise to a constant energy shift for any fixed number of electrons. In conclusion we have

$$\hat{H}_{\text{el}} = T\sum_{s\sigma}(\hat{d}^\dagger_{s\sigma}\hat{d}_{s+1\sigma} + \hat{d}^\dagger_{s+1\sigma}\hat{d}_{s\sigma}),$$

with $T = T_{ss\pm1}$. The Hamiltonian that defines the Holstein model is

$$\hat{H} = \hat{H}_C + \hat{H}_{\text{int}} + \hat{H}_{\text{el}}. \tag{2.56}$$

This Hamiltonian commutes with the operator $\hat{N}_{\text{el}} = \sum_{s\sigma}\hat{d}^\dagger_{s\sigma}\hat{d}_{s\sigma}$ of the total number of electrons but it does not commute with the operator $\hat{N}_{\text{ph}} = \sum_s \hat{a}^\dagger_s\hat{a}_s$ of the total number of phonons due to the presence of $\hat{H}_{\text{int}}$. The eigenstates of $\hat{H}$ are therefore linear combinations of kets with a fixed number of electrons and different number of phonons. Despite its simplicity, the Holstein model cannot be solved exactly and one has to resort to approximations.

2.8.1 Peierls Instability

A common approximation is the so-called *Born–Oppenheimer approximation*, which considers the mass M infinitely large.[14] Nuclei of large mass move so slowly that the electrons have ample time to readjust their wavefunctions in the configuration of minimum energy (i.e., the ground state). In terms of the elastic constant $K = M\omega_0^2$ and the coupling $\tilde{g} = g\sqrt{2M\omega_0}$ the Holstein model in the Born–Oppenheimer approximation reads

$$\hat{H}^{\text{BO}} = T\sum_{s\sigma}(\hat{d}^\dagger_{s\sigma}\hat{d}_{s+1\sigma} + \hat{d}^\dagger_{s+1\sigma}\hat{d}_{s\sigma}) - \tilde{g}\sum_{s\sigma}\hat{d}^\dagger_{s\sigma}\hat{d}_{s\sigma}\hat{x}_s + \frac{1}{2}K\sum_s \hat{x}_s^2.$$

[14] We discuss this approximation in more detail in Section 16.2.2.

The Born–Oppenheimer Hamiltonian is much simpler than the original one since $[\hat{H}^{\rm BO}, \hat{x}_s]_- = 0$ for all s, and hence the eigenvalues of the position operators are good quantum numbers. We then introduce the operators

$$\hat{A}_s^\dagger(x) \equiv \sum_j \langle j|x\rangle \frac{1}{\sqrt{j!}}(\hat{a}_s^\dagger)^j,$$

which create an eigenket of $\hat{x}_s$ with eigenvalue x when acting on the empty ket $|0\rangle$. The general eigenket of $\hat{H}^{\rm BO}$ has the form

$$\hat{A}_1^\dagger(x_1)\hat{A}_2^\dagger(x_2)\ldots|\Phi_{\rm el}, \{x_s\}\rangle,$$

where $|\Phi_{\rm el}, \{x_s\}\rangle$ is a pure electronic ket (obtained by acting with the $\hat{d}^\dagger$-operators over the empty ket) obeying the eigenvalue equation

$$\left(\hat{H}_{\rm el} - \tilde{g}\sum_{s\sigma} x_s \hat{d}_{s\sigma}^\dagger \hat{d}_{s\sigma}\right)|\Phi_{\rm el}, \{x_s\}\rangle = E(\{x_s\})|\Phi_{\rm el}, \{x_s\}\rangle. \tag{2.57}$$

If $|\Phi_{\rm el}, \{x_s\}\rangle$ is the ground state of (2.57) with energy $E_0(\{x_s\})$, the lowest energy of $\hat{H}^{\rm BO}$ is obtained by minimizing

$$E_0(\{x_s\}) + \frac{1}{2}K\sum_s x_s^2$$

over the space of all possible configurations $x_1, x_2, \ldots$. Even though the Hamiltonian in (2.57) is a one-body operator, and hence it does not take long to calculate $E_0(\{x_s\})$ for a given set of x_s, the minimization of a function of many variables is, in general, a complicated task. The best we can do to gain some physical insight is to calculate $E_0(\{x_s\})$ for some "reasonable" configuration and then compare the results.

Let us consider, for instance, a molecular ring with $2N$ molecules and $2N$ electrons (half-filling), N of spin up and N of spin down. It is intuitive to expect that in the ground state the nuclear displacement is uniform, $x_s = x$ for all s, since the ring is invariant under (discrete) rotations. If so, we could calculate the ground-state energy by finding the x that minimize the total energy. For a uniform displacement the Hamiltonian in (2.57) becomes

$$\hat{H}_{\rm el} - \tilde{g}x\sum_{s=1}^{2N}\sum_\sigma \hat{d}_{s\sigma}^\dagger \hat{d}_{s\sigma} = \hat{H}_{\rm el} - \tilde{g}x\hat{N}_{\rm el},$$

and from the result (2.19) we know that the single-particle eigenenergies of this Hamiltonian are $\epsilon_k = -\tilde{g}x + 2T\cos k$ with $k = 2\pi m/2N$. Therefore, the lowest energy of $\hat{H}^{\rm BO}$ (with $2N$ electrons) *in the subspace of uniform displacements* is the minimum of

$$E_{\rm unif}(x) = 2\times\left[\sum_{m=-\frac{N}{2}+1}^{\frac{N}{2}} 2T\cos\frac{2\pi m}{2N}\right] - 2N\tilde{g}x + \frac{2NKx^2}{2},$$

where, for simplicity, we assumed that N is even and that $T < 0$. In this formula the factor of 2 multiplying the first term on the r.h.s. comes from spin.

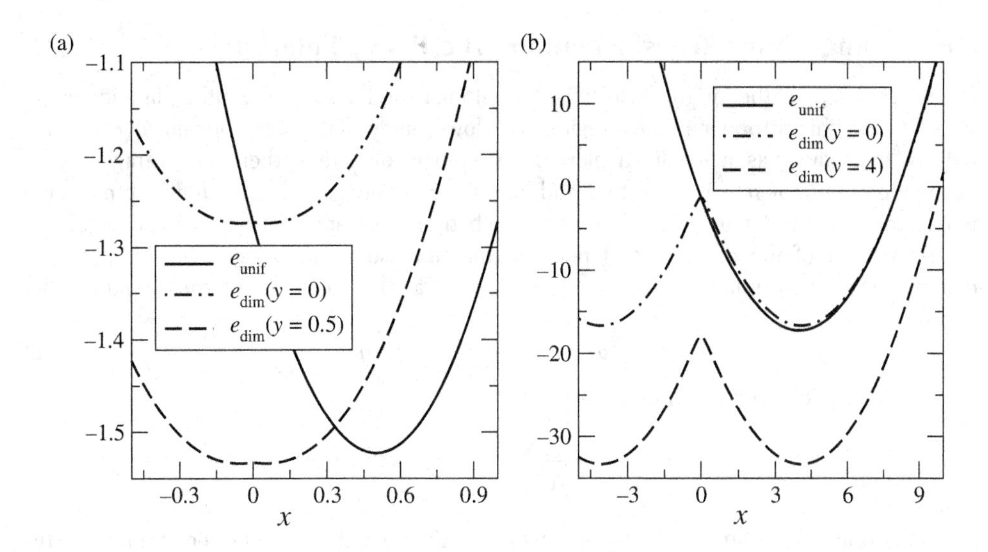

Figure 2.9 Uniform and dimerized energy densities $e_{\text{unif}}(x) = E_{\text{unif}}(x)/2N$ and $e_{\text{dim}}(x,y) = E_{\text{dim}}(x,y)/2N$ for a ring of 100 molecules and $T = -1$, $K = 2$. In (a) $\tilde{g} = 1$, while in (b) $\tilde{g} = 8$. The minimum of the dimerized energy $e_{\text{dim}}(x,y)$ is always the lowest when y is close to the minimum of the uniform energy $e_{\text{unif}}(y)$.

Intuition is not always a good guide. Below we show that the lowest energy of $\hat{H}^{\text{BO}}$ in the subspace of *dimerized configurations* $x_s = y+(-)^s x$ is always lower than the minimum of $E_{\text{unif}}(x)$. In the dimerized case the electronic operator in (2.57) can be written as

$$\hat{H}_{\text{el}} - \tilde{g}y\hat{N}_{\text{el}} - \tilde{g}x\sum_{s=1}^{2N}\sum_{\sigma}(-)^s\hat{d}^{\dagger}_{s\sigma}\hat{d}_{s\sigma}.$$

We already encountered this Hamiltonian in Section 2.3. It describes the ring of Fig. 2.6(a) with $\epsilon = -\tilde{g}y$, $\Delta = \tilde{g}x$, $T = t$, and N unit cells. Using the one-particle eigenvalues given in (2.20), we find that the lowest energy of $\hat{H}^{\text{BO}}$ (with $2N$ electrons) *in the subspace of dimerized configurations* is the minimum of

$$E_{\text{dim}}(x,y) = -2\times\Big[\sum_{m=-\frac{N}{2}+1}^{\frac{N}{2}}\sqrt{(\tilde{g}x)^2 + 2T^2(1+\cos\frac{2\pi m}{N})}\,\Big] - 2N\tilde{g}y + \frac{NK}{2}\left[(y+x)^2 + (y-x)^2\right].$$

In Fig. 2.9 we display $E_{\text{unif}}(x)/2N$ and $E_{\text{dim}}(x,y)/2N$ as a function of x and for different values of y. In all cases there exist values of y for which the *dimerized configuration has lower energy*. This phenomenon is called the *Peierls instability* [37]: A periodic one-dimensional molecular crystal is unstable toward dimerization since the energy gain in opening a gap in the electronic spectrum is always larger than the elastic energy loss.

2.8.2 Lang-Firsov Transformation: The Heavy Polaron

Let us go back to the original Holstein Hamiltonian and consider another limiting case: $g \gg T$. In this strong coupling regime the low-energy states have localized electrons. Indeed, if a state has a localized electron of spin σ on site s then the average of the occupation operator $\hat{n}_{s\sigma}$ is close to 1 and hence the energy gain in stretching the molecule is maximized. In order to treat $\hat{H}_{\rm el}$ as a perturbation it is convenient to perform a unitary transformation of the electron and phonon operators so as to bring $\hat{H}_C + \hat{H}_{\rm int}$ into a diagonal form. This unitary transformation is known as the *Lang-Firsov transformation* [38] and reads

$$\hat{p}_{s\sigma} = e^{{\rm i}\hat{S}}\hat{d}_{s\sigma}e^{-{\rm i}\hat{S}}, \qquad \hat{b}_s = e^{{\rm i}\hat{S}}\hat{a}_s e^{-{\rm i}\hat{S}}, \tag{2.58}$$

where $\hat{S}$ is the following Hermitian operator

$$\hat{S} = -{\rm i}\frac{g}{\omega_0}\sum_{s\sigma}\hat{n}_{s\sigma}(\hat{a}_s^\dagger - \hat{a}_s).$$

Since the transformation is unitary, the new operators obey the same (anti)commutation relations as the old ones, which in our case means $\left[\hat{p}_{s\sigma}, \hat{p}^\dagger_{s'\sigma'}\right]_+ = \delta_{ss'}\delta_{\sigma\sigma'}$ and $\left[\hat{b}_s, \hat{b}^\dagger_{s'}\right]_- = \delta_{ss'}$. To express the old operators in terms of the new ones we use the following trick. Let $\hat{O}$ be a generic operator and $\hat{F}_\alpha(\hat{O}) \equiv e^{{\rm i}\alpha\hat{S}}\hat{O}e^{-{\rm i}\alpha\hat{S}}$. This transformation is unitary for every real α and has the properties

1. $\hat{F}_0(\hat{O}) = \hat{O}$;

2. $\hat{F}_\alpha(\hat{O}^\dagger) = \hat{F}^\dagger_\alpha(\hat{O})$;

3. $\hat{F}_\alpha(\hat{O}_1 + \hat{O}_2) = \hat{F}_\alpha(\hat{O}_1) + \hat{F}_\alpha(\hat{O}_2)$;

4. $\hat{F}_\alpha(\hat{O}_1\hat{O}_2) = \hat{F}_\alpha(\hat{O}_1)\hat{F}_\alpha(\hat{O}_2)$.

Furthermore, the derivative with respect to α is simply

$$\hat{F}'_\alpha(\hat{O}) = {\rm i}\,[\hat{S}, \hat{F}_\alpha(\hat{O})] = {\rm i}\,e^{{\rm i}\alpha\hat{S}}\,[\hat{S}, \hat{O}]\,e^{-{\rm i}\alpha\hat{S}} = {\rm i}F_\alpha([\hat{S}, \hat{O}]).$$

We now show how to invert (2.58) by calculating $\hat{F}'_\alpha(\hat{O})$ with $\hat{O}$ either the fermionic $\hat{d}$-operators or the bosonic $\hat{a}$-operators.

The commutators we are interested in are

$$[\hat{S}, \hat{d}_{s\sigma}] = {\rm i}\frac{g}{\omega_0}\hat{d}_{s\sigma}(\hat{a}_s^\dagger - \hat{a}_s), \quad \Rightarrow \quad [\hat{S}, \hat{d}^\dagger_{s\sigma}] = -{\rm i}\frac{g}{\omega_0}\hat{d}^\dagger_{s\sigma}(\hat{a}_s^\dagger - \hat{a}_s) \tag{2.59}$$

and

$$[\hat{S}, \hat{a}_s] = {\rm i}\frac{g}{\omega_0}\sum_\sigma \hat{n}_{s\sigma}, \quad \Rightarrow \quad [\hat{S}, \hat{a}_s^\dagger] = {\rm i}\frac{g}{\omega_0}\sum_\sigma \hat{n}_{s\sigma}. \tag{2.60}$$

From (2.60) it follows that (property 3) $\hat{F}'_\alpha(\hat{a}_s^\dagger - \hat{a}_s) = 0$ and hence $\hat{F}_\alpha(\hat{a}_s^\dagger - \hat{a}_s) = \hat{F}_0(\hat{a}_s^\dagger - \hat{a}_s) = \hat{a}_s^\dagger - \hat{a}_s$ (property 1) does not depend on α. This independence allows us to integrate the equation for the electron operators. From the first relation in (2.59) we find

$$\hat{F}'_\alpha(\hat{d}_{s\sigma}) = -\frac{g}{\omega_0}\hat{F}_\alpha(\hat{d}_{s\sigma}(\hat{a}^\dagger_s - \hat{a}_s)) = -\frac{g}{\omega_0}\hat{F}_\alpha(\hat{d}_{s\sigma})(\hat{a}^\dagger_s - \hat{a}_s),$$

where in the last equality we use property 4. Taking into account that $\hat{F}_0(\hat{d}_{s\sigma}) = \hat{d}_{s\sigma}$, the solution of the differential equation is

$$\hat{F}_\alpha(\hat{d}_{s\sigma}) = \hat{d}_{s\sigma}e^{-\alpha\frac{g}{\omega_0}(\hat{a}^\dagger_s - \hat{a}_s)} \quad \overset{\text{property 2}}{\Longrightarrow} \quad \hat{F}_\alpha(\hat{d}^\dagger_{s\sigma}) = \hat{d}^\dagger_{s\sigma}e^{\alpha\frac{g}{\omega_0}(\hat{a}^\dagger_s - \hat{a}_s)}.$$

With the transformed electron operators we can calculate the transformed phonon operators. We use the result in (2.60) and the fact that $\hat{F}_\alpha(\hat{n}_{s\sigma}) = \hat{F}_\alpha(\hat{d}^\dagger_{s\sigma})\hat{F}_\alpha(\hat{d}_{s\sigma}) = \hat{n}_{s\sigma}$ does not depend on α; then we find $\hat{F}'_\alpha(\hat{a}_s) = -\frac{g}{\omega_0}\sum_\sigma \hat{n}_{s\sigma}$. Integrating over α, we eventually obtain (property 1)

$$\hat{F}_\alpha(\hat{a}_s) = \hat{a}_s - \alpha\frac{g}{\omega_0}\sum_\sigma \hat{n}_{s\sigma} \quad \overset{\text{property 2}}{\Longrightarrow} \quad \hat{F}_\alpha(\hat{a}^\dagger_s) = \hat{a}^\dagger_s - \alpha\frac{g}{\omega_0}\sum_\sigma \hat{n}_{s\sigma}.$$

The transformed operators in (2.58) follow from the above identities with $\alpha = 1$ and the inverse transformation reads

$$\hat{d}_{s\sigma} = \hat{p}_{s\sigma}e^{\frac{g}{\omega_0}(\hat{b}^\dagger_s - \hat{b}_s)}, \qquad \hat{a}_s = \hat{b}_s + \frac{g}{\omega_0}\sum_\sigma \hat{n}_{s\sigma}. \tag{2.61}$$

These results are physically sound. The unitary transformation $\hat{S}$ is the exponential of an operator proportional to the sum of the momentum operators, $\hat{p}_s \propto (\hat{a}^\dagger_s - \hat{a}_s)$, and therefore it is similar to a translation operator. As a consequence, the new position operator is shifted; this follows from the second relation in (2.61), which implies

$$\hat{F}_1(\hat{x}_s) = \frac{\hat{b}^\dagger_s + \hat{b}_s}{\sqrt{2M\omega_0}} = \underbrace{\frac{\hat{a}^\dagger_s + \hat{a}_s}{\sqrt{2M\omega_0}}}_{\hat{x}_s} - 2\frac{g}{\omega_0}\frac{\sum_\sigma \hat{n}_{s\sigma}}{\sqrt{2M\omega_0}}.$$

Upon substitution of (2.61) in the Holstein Hamiltonian (2.56) we find

$$\hat{H}_C + \hat{H}_{\text{int}} = \omega_0\sum_s(\hat{b}^\dagger_s\hat{b}_s + \frac{1}{2}) - \frac{g^2}{\omega_0}\sum_s\left(\sum_\sigma \hat{n}_{s\sigma}\right)^2,$$

and

$$\hat{H}_{\text{el}} = T\sum_{s\sigma}\left(\hat{B}^\dagger_s\hat{B}_{s+1}\hat{p}^\dagger_{s\sigma}\hat{p}_{s+1\sigma} + \hat{B}^\dagger_{s+1}\hat{B}_s\hat{p}^\dagger_{s+1\sigma}\hat{p}_{s\sigma}\right),$$

where we define $\hat{B}_s \equiv e^{\frac{g}{\omega_0}(\hat{b}^\dagger_s - \hat{b}_s)}$. As anticipated, the zeroth-order Hamiltonian $\hat{H}_C + \hat{H}_{\text{int}}$ is diagonalized by the Lang–Firsov transformation. The eigenstates comprise a given number of transformed electrons and transformed phonons. For instance, if X is the set of molecules hosting a transformed electron of spin $\uparrow$ and Y is the set of molecules hosting a transformed electron of spin $\downarrow$, then the ket

$$|\Phi\rangle = \prod_{s\in X}\hat{p}^\dagger_{s\uparrow}\prod_{s'\in Y}\hat{p}^\dagger_{s'\downarrow}\prod_r(\hat{b}^\dagger_r)^{m_r}|0\rangle$$

is an eigenket of $\hat{H}_C + \hat{H}_{\text{int}}$ with eigenvalue

$$E = \omega_0 \sum_r (m_r + \frac{1}{2}) - \frac{g^2}{\omega_0} \left(|X| + |Y| + 2|X \cap Y| \right),$$

where $|X|$, $|Y|$ is the number of elements in the set X, Y. We refer to the transformed electrons as the *polarons* since the action of $\hat{p}^\dagger_{s\sigma}$ on the empty ket $|0\rangle$ generates an electron of spin σ in the sth molecule surrounded by a cloud of phonons.

In the subspace with only one polaron, say of spin up, all kets with no transformed phonons, $|\Phi_s\rangle \equiv \hat{p}^\dagger_{s\uparrow}|0\rangle$, have the same energy $-g^2/\omega_0$, which is also the lowest possible energy. This means that the ground state is degenerate; a ket of the ground-state multiplet describes a still electron trapped in s by the polarization cloud. The ground-state degeneracy is lifted by the perturbation $\hat{H}_{\text{el}}$. To first order, the splitting is given by the eigenvalues of the matrix $\langle \Phi_s|\hat{H}_{\text{el}}|\Phi_{s'}\rangle$. Since the polarons can only hop to nearest-neighbor molecules, the matrix elements are zero unless $s' = s \pm 1$, in which case

$$\langle \Phi_s|\hat{H}_{\text{el}}|\Phi_{s\pm1}\rangle = T\langle 0|\hat{B}^\dagger_s \hat{B}_{s\pm1}|0\rangle.$$

To evaluate the matrix element on the r.h.s. we make use of the Baker-Campbell-Hausdorff formula.[15] Recalling the definition of the $\hat{B}$-operators we have

$$\begin{aligned}
\hat{B}^\dagger_s \hat{B}_{s\pm1}|0\rangle &= e^{-\frac{g}{\omega_0}\hat{b}^\dagger_s} e^{\frac{g}{\omega_0}\hat{b}_s} e^{-\frac{g^2}{2\omega_0^2}} e^{\frac{g}{\omega_0}\hat{b}^\dagger_{s\pm1}} e^{-\frac{g}{\omega_0}\hat{b}_{s\pm1}} e^{-\frac{g^2}{2\omega_0^2}}|0\rangle \\
&= e^{-\frac{g^2}{\omega_0^2}} e^{-\frac{g}{\omega_0}\hat{b}^\dagger_s} e^{\frac{g}{\omega_0}\hat{b}^\dagger_{s\pm1}}|0\rangle, \qquad (2.62)
\end{aligned}$$

where in the last step we use that the $\hat{b}$-operators with different indices commute and that for any constant α the ket $e^{\alpha\hat{b}_s}|0\rangle = \sum_k \frac{1}{k!}(\alpha\hat{b}_s)^k|0\rangle = |0\rangle$, since $\hat{b}_s|0\rangle = |\emptyset\rangle$. Taylor-expanding the exponentials in (2.62) and multiplying by the bra $\langle 0|$, it is evident that only the zeroth-order terms remain. Hence $\langle 0|\hat{B}^\dagger_s \hat{B}_{s\pm1}|0\rangle = \exp(-g^2/\omega_0^2)$ which, as expected, does not depend on s. We conclude that

$$\langle \Phi_s|\hat{H}_{\text{el}}|\Phi_{s'}\rangle = T_{\text{eff}}(\delta_{s,s'+1} + \delta_{s,s'-1}), \qquad T_{\text{eff}} \equiv T e^{-\frac{g^2}{\omega_0^2}}. \qquad (2.63)$$

For rings with N molecular sites the eigenvalues of the matrix (2.63) are $\epsilon_k = 2T_{\text{eff}} \cos k$, as we already saw in (2.19). In the large N limit the variable $k = 2\pi m/N$ and the eigenvalues ϵ_k become a continuum and we can extract the *effective mass* of the polaron. This is typically done by comparing the energy dispersion ϵ_k (at low energies) with the energy dispersion $p^2/(2m^*)$ of a free particle of mass m^* and momentum p. In our case the momentum is the crystal momentum $p = k/a \in (-\pi/a, \pi/a)$, with a the equilibrium distance between two neighboring molecules. Assuming, for example, $T < 0$, the lowest energies are those

[15] Given two operators $\hat{A}$ and $\hat{B}$ the Baker-Campbell-Hausdorff formula is the solution to $\hat{C} = \ln(e^{\hat{A}}e^{\hat{B}})$ or, equivalently, $e^{\hat{C}} = e^{\hat{A}}e^{\hat{B}}$. There is no close expression for the solution of this problem. However, if the commutator $[\hat{A}, \hat{B}] = c\hat{\mathbb{1}}$, then

$$e^{\hat{A}+\hat{B}} = e^{\hat{A}}e^{\hat{B}}e^{-\frac{c}{2}}.$$

with $k \ll 1$, and hence we can approximate ϵ_k as $\epsilon_{k=ap} \sim 2T_{\text{eff}} + |T_{\text{eff}}|a^2p^2$, from which it follows that the polaron mass is

$$m^* = \frac{e^{g^2/\omega_0^2}}{2|T|a^2}.$$

Thus, in the strong coupling regime the polaron mass increases exponentially with the coupling g. The polaron is heavy and the molecular deformations are localized around it.

3

Time-Dependent Problems and Equations of Motion

In this chapter we discuss the time-dependent Schrödinger equation in the formalism of second quantization. We introduce the concept of the evolution operator, time-order operator, and operators in the Heisenberg picture. We also show how to recover the continuity equation and the Newton equation for systems of charged particles in a time-varying electromagnetic field.

3.1 Time-Dependent Hamiltonian

After the excursion of Chapter 2 on model systems, we now go back to the general second-quantized Hamiltonian (1.83) and specialize the discussion to common physical situations. The operator $\hat{h}$ typically describes a particle of mass m and charge q in an external electromagnetic field and reads

$$\hat{h} = \frac{1}{2m}\left(\hat{\boldsymbol{p}} - \frac{q}{c}\mathbf{A}(\hat{\boldsymbol{r}})\right)^2 + q\phi(\hat{\boldsymbol{r}}) - g\mu_{\mathrm{B}}\mathbf{B}(\hat{\boldsymbol{r}}) \cdot \hat{\boldsymbol{\mathcal{S}}}, \tag{3.1}$$

with ϕ the scalar potential, $\mathbf{A}$ the vector potential, and $\mathbf{B} = \boldsymbol{\nabla} \times \mathbf{A}$ the magnetic field.[1] In (3.1) the constant c is the speed of light while g and μ_{B} are the gyromagnetic ratio and the Bohr magneton, respectively, see also (1.9). The many-body Hamiltonian $\hat{H}$ in (1.83) with $\hat{h}$ from (3.1) describes a system of interacting identical particles in some external *static* field. If we consider a molecule, in the absence of a magnetic field a standard choice for the potentials is $\mathbf{A}(\mathbf{r}) = 0$ and $\phi(\mathbf{r}) = \sum_i Z_i/|\mathbf{r} - \mathbf{R}_i|$, with Z_i being the charge of the ith nucleus at position $\mathbf{R}_i$. In the next chapters we develop approximation schemes to calculate equilibrium properties, such as total energies, ionization energies, spectral functions, and charge and current distributions, of these systems. We are especially interested in situations where at some time t_0 a *time-dependent* perturbation is switched on. For instance, one

[1] Rigorously speaking, the many-body Hamiltonian (1.83) with one-particle Hamiltonian (3.1) can only be derived in the so-called *Coulomb gauge* $\boldsymbol{\nabla} \cdot \mathbf{A} = 0$ (transverse vector potential). In this gauge ϕ corresponds to the *external* scalar potential, whereas $\mathbf{A}$ corresponds to the *total* (and transverse) vector potential, given by the sum of the external and induced vector potentials. We prove this statement in Chapter 16, see Section 16.10.1. For the time being we assume that the induced vector potential can be neglected.

could change the electromagnetic field and study how the particles move under the influence of a time-dependent electric field $\mathbf{E}(\mathbf{r},t)$ and magnetic field $\mathbf{B}(\mathbf{r},t)$ with

$$\mathbf{E}(\mathbf{r},t) = -\boldsymbol{\nabla}\phi(\mathbf{r},t) - \frac{1}{c}\frac{\partial}{\partial t}\mathbf{A}(\mathbf{r},t), \tag{3.2a}$$

$$\mathbf{B}(\mathbf{r},t) = \boldsymbol{\nabla} \times \mathbf{A}(\mathbf{r},t), \tag{3.2b}$$

and, of course, $\phi(\mathbf{r},t \leq t_0) = \phi(\mathbf{r})$ and $\mathbf{A}(\mathbf{r},t \leq t_0) = \mathbf{A}(\mathbf{r})$. In this case $\hat{h} \to \hat{h}(t)$ becomes time-dependent through its dependence on the external potentials. Consequently also the full Hamiltonian (1.83) acquires a time dependence, $\hat{H} \to \hat{H}(t) = \hat{H}_0(t) + \hat{H}_{\text{int}}$. The time evolution of the system is governed by the *Schrödinger equation,*

$$\boxed{\mathrm{i}\frac{d}{dt}|\Psi(t)\rangle = \hat{H}(t)|\Psi(t)\rangle} \tag{3.3}$$

with $|\Psi(t)\rangle$ the ket of the system at time t. The Schrödinger equation is a first-order differential equation in time and, therefore, $|\Psi(t)\rangle$ is uniquely determined once the initial ket $|\Psi(t_0)\rangle$ is given. For time-independent Hamiltonians $\hat{H}(t) = \hat{H}(t_0)$ for all times t, and (3.3) is solved by

$$|\Psi(t)\rangle = e^{-\mathrm{i}\hat{H}(t_0)(t-t_0)}|\Psi(t_0)\rangle. \tag{3.4}$$

How does this solution change when $\hat{H}(t)$ is time-dependent?

3.2 Evolution Operator

To generalize (3.4) we look for an operator $\hat{U}(t,t_0)$ which maps $|\Psi(t_0)\rangle$ into $|\Psi(t)\rangle$:

$$|\Psi(t)\rangle = \hat{U}(t,t_0)|\Psi(t_0)\rangle. \tag{3.5}$$

The operator $\hat{U}(t,t_0)$ must be unitary since the time-dependent Schrödinger equation preserves the norm of the states - that is, $\langle\Psi(t)|\Psi(t)\rangle = \langle\Psi(t_0)|\Psi(t_0)\rangle$.

Forward evolution Let us discuss first the case $t > t_0$. We start by considering a Hamiltonian $\hat{H}(t)$ which is piecewise constant - that is, $\hat{H}(t) = \hat{H}(t_p)$ for $t_p < t \leq t_{p+1}$, where the t_p are times at which the Hamiltonian changes suddenly. If we know the ket at time t_0 then we can calculate the ket at time $t \in (t_n, t_{n+1})$ by using (3.4) repeatedly:

$$\begin{aligned}|\Psi(t)\rangle &= e^{-\mathrm{i}\hat{H}(t_n)(t-t_n)}|\Psi(t_n)\rangle = e^{-\mathrm{i}\hat{H}(t_n)(t-t_n)}e^{-\mathrm{i}\hat{H}(t_{n-1})(t_n-t_{n-1})}|\Psi(t_{n-1})\rangle \\ &= e^{-\mathrm{i}\hat{H}(t_n)(t-t_n)}e^{-\mathrm{i}\hat{H}(t_{n-1})(t_n-t_{n-1})}\ldots e^{-\mathrm{i}\hat{H}(t_0)(t_1-t_0)}|\Psi(t_0)\rangle.\end{aligned} \tag{3.6}$$

As expected, the operator acting on $|\Psi(t_0)\rangle$ is unitary since it is the product of unitary operators. It is important to observe the order of the operators: The exponential calculated with the Hamiltonian $\hat{H}(t_p)$ is on the left of all exponentials calculated with $\hat{H}(t_q)$ if $t_p > t_q$ (t_p later than t_q). Equation (3.6) takes a simpler form when the times t_p are equally spaced

– that is, $t_p = t_0 + p\Delta_t$ with Δ_t some given time interval. Then, $t_p - t_{p-1} = \Delta_t$ for all p and hence

$$|\Psi(t_{n+1})\rangle = e^{-i\hat{H}(t_n)\Delta_t} e^{-i\hat{H}(t_{n-1})\Delta_t} \ldots e^{-i\hat{H}(t_0)\Delta_t} |\Psi(t_0)\rangle. \tag{3.7}$$

At this point it is natural to introduce the so-called *chronological ordering operator* or *time ordering operator* T. Despite its name, T is not an operator in the usual sense,[2] like the Hamiltonian or the density operator, but rather a rule that establishes how to rearrange products of operators. We define the chronologically ordered product of m Hamiltonians at times $t_m \geq \ldots \geq t_1$ as

$$T\left\{\hat{H}(t_{P(m)})\hat{H}(t_{P(m-1)}) \ldots \hat{H}(t_{P(1)})\right\} = \hat{H}(t_m)\hat{H}(t_{m-1}) \ldots \hat{H}(t_1), \tag{3.8}$$

for all permutations P of $(1, \ldots, m-1, m)$. The action of T is to rearrange the Hamiltonians according to the *two ls rule*: *l*ater times go to the *l*eft. Since the order of the Hamiltonians in the T product is irrelevant, under the T sign they can be treated as commuting operators. This observation can be used to rewrite (3.7) in a more compact and useful form. Indeed, in (3.7) the Hamiltonians are already chronologically ordered and hence nothing changes if we act with the T operator on the product of the exponentials:

$$|\Psi(t_{n+1})\rangle = T\left\{e^{-i\hat{H}(t_n)\Delta_t} e^{-i\hat{H}(t_{n-1})\Delta_t} \ldots e^{-i\hat{H}(t_0)\Delta_t}\right\} |\Psi(t_0)\rangle. \tag{3.9}$$

The equivalence between (3.7) and (3.9) can easily be checked by expanding the exponentials in power series and by taking into account that the action of the T operator on Hamiltonians with the same time argument is unambiguously defined since these Hamiltonians commute. For instance, in the case of only two exponentials

$$e^{-i\hat{H}(t_1)\Delta_t} e^{-i\hat{H}(t_0)\Delta_t} = \sum_{m,n=0}^{\infty} \frac{(-i\Delta_t)^n}{n!} \frac{(-i\Delta_t)^m}{m!} \hat{H}(t_1)^n \hat{H}(t_0)^m,$$

and we see that the action of T does not change this expression. We now recall that according to the Baker-Campbell-Hausdorff formula for two commuting operators $\hat{A}$ and $\hat{B}$ the product $\exp(\hat{A})\exp(\hat{B})$ is equal to $\exp(\hat{A}+\hat{B})$ (see also footnote 15 in Section 2.8.2). Under the T sign $\hat{H}(t_p)$ commutes with $\hat{H}(t_q)$ for all p and q, and therefore

$$|\Psi(t_{n+1})\rangle = T\left\{e^{-i\Delta_t \sum_{p=0}^{n} \hat{H}(t_p)}\right\} |\Psi(t_0)\rangle, \tag{3.10}$$

which is a very nice result.

We are now in the position to answer the general question of how to solve (3.3) for time-dependent Hamiltonians. Let $t > t_0$ be the time at which we want to know the time-evolved ket $|\Psi(t)\rangle$. We divide the interval (t_0, t) in $n+1$ equal sub-intervals $(t_p, t_{p+1} = t_p + \Delta_t)$ with $\Delta_t = (t - t_0)/(n+1)$, as shown in the figure below:

$\hat{H}(t_0)$ $\hat{H}(t_1)$ $\hat{H}(t_n)$

$t_0 \longleftrightarrow t_1$ t_2 .... t_n $t_{n+1} = t$

Δ_t

[2]Notice that the symbol T for the chronological ordering operator does not have a "hat."

If n is large enough, the Hamiltonian $\hat{H}(t) \sim \hat{H}(t_p)$ for $t \in (t_p, t_{p+1})$ and $|\Psi(t)\rangle$ is approximately given by (3.10). Increasing n, and hence reducing Δ_t, the approximated ket $|\Psi(t)\rangle$ approaches the exact one and eventually coincides with it when $n \to \infty$. We conclude that the general solution of (3.3) can be written as

$$|\Psi(t)\rangle = \lim_{n\to\infty} T\left\{e^{-\mathrm{i}\Delta_t \sum_{p=0}^{n} \hat{H}(t_p)}\right\} |\Psi(t_0)\rangle = T\left\{e^{-\mathrm{i}\int_{t_0}^{t} d\bar{t}\, \hat{H}(\bar{t})}\right\} |\Psi(t_0)\rangle. \tag{3.11}$$

The operator in (3.5) is therefore

$$\hat{U}(t,t_0) \equiv T\left\{e^{-\mathrm{i}\int_{t_0}^{t} d\bar{t}\, \hat{H}(\bar{t})}\right\}, \tag{3.12}$$

and it is usually called the *evolution operator.*

Equation of motion for the evolution operator The evolution operator obeys a simple differential equation. Inserting solution (3.11) in the time-dependent Schrödinger equation (3.3), we find

$$\mathrm{i}\frac{d}{dt}\hat{U}(t,t_0)|\Psi(t_0)\rangle = \hat{H}(t)\hat{U}(t,t_0)|\Psi(t_0)\rangle,$$

and since this must be true for all initial states $|\Psi(t_0)\rangle$ we conclude that

$$\mathrm{i}\frac{d}{dt}\hat{U}(t,t_0) = \hat{H}(t)\hat{U}(t,t_0), \qquad \hat{U}(t_0,t_0) = \hat{\mathbb{1}}, \tag{3.13}$$

with $\hat{\mathbb{1}}$ the identity operator. The differential equation in (3.13) together with the initial condition at t_0 determines uniquely $\hat{U}(t,t_0)$. We can use (3.13) for an alternative proof of (3.12). We integrate (3.13) between t_0 and t and take into account that $\hat{U}(t_0,t_0) = \hat{\mathbb{1}}$:

$$\hat{U}(t,t_0) = \hat{\mathbb{1}} - \mathrm{i}\int_{t_0}^{t} dt_1\, \hat{H}(t_1)\hat{U}(t_1,t_0).$$

This integral equation contains the same information as (3.13). This is a completely general fact: A first-order differential equation endowed with a boundary condition can always be written as an integral equation which automatically incorporates the boundary condition. We can now replace the evolution operator under the integral sign with the whole r.h.s. and iterate. In doing so we find

$$\begin{aligned}\hat{U}(t,t_0) &= \hat{\mathbb{1}} - \mathrm{i}\int_{t_0}^{t} dt_1\, \hat{H}(t_1) + (-\mathrm{i})^2 \int_{t_0}^{t} dt_1 \int_{t_0}^{t_1} dt_2\, \hat{H}(t_1)\hat{H}(t_2)\hat{U}(t_2,t_0) \\ &= \sum_{k=0}^{\infty} (-\mathrm{i})^k \int_{t_0}^{t} dt_1 \int_{t_0}^{t_1} dt_2 \ldots \int_{t_0}^{t_{k-1}} dt_k\, \hat{H}(t_1)\hat{H}(t_2)\ldots\hat{H}(t_k).\end{aligned}$$

The product of the Hamiltonians is chronologically time-ordered since we integrate t_j between t_0 and t_{j-1} for all j. Therefore nothing changes if we act with T on the r.h.s. We now remind the reader that under the T sign the Hamiltonians can be treated as commuting operators and hence

$$\hat{U}(t,t_0) = \sum_{k=0}^{\infty} \frac{(-\mathrm{i})^k}{k!} \int_{t_0}^{t} dt_1 \int_{t_0}^{t} dt_2 \ldots \int_{t_0}^{t} dt_k \, T\left\{\hat{H}(t_1)\hat{H}(t_2)\ldots\hat{H}(t_k)\right\}$$

$$= T\left\{e^{-\mathrm{i}\int_{t_0}^{t} d\bar{t}\hat{H}(\bar{t})}\right\}, \tag{3.14}$$

where in the first equality we extend the domain of integration from (t_0, t_{j-1}) to (t_0, t) for all t_j and we divide by $k!$ – that is, the number of identical contributions generated by the extension of the domain.[3]

Backward evolution Next we consider the evolution operator $\hat{U}(t, t_0)$ for $t < t_0$, or equivalently $\hat{U}(t_0, t)$ for $t > t_0$ (since t and t_0 are two arbitrary times $\hat{U}(t < t_0, t_0)$ is given by $\hat{U}(t_0, t > t_0)$ after renaming the times $t \to t_0$ and $t_0 \to t$). By definition the evolution operator has the property

$$\hat{U}(t_3, t_2)\hat{U}(t_2, t_1) = \hat{U}(t_3, t_1), \tag{3.15}$$

from which it follows that $\hat{U}(t_0, t)\hat{U}(t, t_0) = \hat{\mathbb{1}}$. This result can be used to find the explicit expression of $\hat{U}(t_0, t)$. Taking into account (3.11), we have

$$\lim_{n\to\infty} \hat{U}(t_0, t) e^{-\mathrm{i}\hat{H}(t_n)\Delta_t} e^{-\mathrm{i}\hat{H}(t_{n-1})\Delta_t} \ldots e^{-\mathrm{i}\hat{H}(t_0)\Delta_t} = \hat{\mathbb{1}}.$$

We then see by inspection that

$$\hat{U}(t_0, t) = \lim_{n\to\infty} e^{\mathrm{i}\hat{H}(t_0)\Delta_t} \ldots e^{\mathrm{i}\hat{H}(t_{n-1})\Delta_t} e^{\mathrm{i}\hat{H}(t_n)\Delta_t}$$

$$= \bar{T}\left\{e^{\mathrm{i}\int_{t_0}^{t} d\bar{t}\,\hat{H}(\bar{t})}\right\},$$

where we introduce the *anti-chronological ordering operator* or *anti-time ordering operator* $\bar{T}$ whose action is to move the operators with later times to the right. The operator $\hat{U}(t_0, t)$ can be interpreted as the operator that evolves a ket backward in time from t to t_0.

To summarize, the main result of this section is

$$\boxed{\hat{U}(t_2, t_1) = \begin{cases} T\left\{e^{-\mathrm{i}\int_{t_1}^{t_2} d\bar{t}\hat{H}(\bar{t})}\right\} & t_2 > t_1 \\ \bar{T}\left\{e^{+\mathrm{i}\int_{t_2}^{t_1} d\bar{t}\hat{H}(\bar{t})}\right\} & t_2 < t_1 \end{cases}} \tag{3.16}$$

[3]For the k-dimensional integral on a hypercube we have

$$\int_{t_0}^{t} dt_1 \int_{t_0}^{t} dt_2 \ldots \int_{t_0}^{t} dt_k \, f(t_1, \ldots, t_k) = \sum_P \int_{t_0}^{t} dt_1 \int_{t_0}^{t_1} dt_2 \ldots \int_{t_0}^{t_{k-1}} dt_k \, f(t_{P(1)}, \ldots, t_{P(k)}),$$

where the sum over P runs over all permutations of $(1, \ldots, k)$. In the special case of a totally symmetric function f[i.e., $f(t_{P(1)}, \ldots, t_{P(k)}) = f(t_1, \ldots, t_k)$], the above identity reduces to

$$\int_{t_0}^{t} dt_1 \int_{t_0}^{t} dt_2 \ldots \int_{t_0}^{t} dt_k \, f(t_1, \ldots, t_k) = k! \int_{t_0}^{t} dt_1 \int_{t_0}^{t_1} dt_2 \ldots \int_{t_0}^{t_{k-1}} dt_k \, f(t_1, \ldots, t_k).$$

This is the case of the time-ordered product of Hamiltonians in (3.14).

Exercise 3.1 Show that

$$\bar{T}\left\{e^{\mathrm{i}\int_{t_0}^{t} d\bar{t}\,\hat{H}(\bar{t})}\right\} = T\left\{e^{-\mathrm{i}\int_{t}^{2t-t_0} d\bar{t}\,\hat{H}'(\bar{t})}\right\}$$

where $\hat{H}'(\bar{t}) = -\hat{H}(2t - \bar{t})$.

Exercise 3.2 Consider the Hamiltonian of a forced harmonic oscillator,

$$\hat{H}(t) = \omega\hat{d}^\dagger\hat{d} + f(t)(\hat{d}^\dagger + \hat{d}),$$

where $f(t)$ is a real function of time and $[\hat{d}, \hat{d}^\dagger]_- = 1$. Let $|\Phi_n\rangle = \frac{(\hat{d}^\dagger)^n}{\sqrt{n!}}|0\rangle$ be the normalized eigenstates of $\hat{d}^\dagger\hat{d}$ with eigenvalues n. Show that the ket

$$|\Psi_n(t)\rangle = e^{-\mathrm{i}\alpha(t)}e^{y(t)\hat{d}^\dagger - y^*(t)\hat{d}}\,|\Phi_n\rangle$$

is normalized to 1 for any real function $\alpha(t)$ and for any complex function $y(t)$. Then show that this ket also satisfies the time-dependent Schrödinger equation $\mathrm{i}\frac{d}{dt}|\Psi_n(t)\rangle = \hat{H}(t)|\Psi_n(t)\rangle$ with boundary condition $|\Psi_n(t_0)\rangle = |\Phi_n\rangle$ provided that the functions $\alpha(t)$ and $y(t)$ satisfy the differential equations

$$\frac{dy}{dt} + \mathrm{i}\omega y = -\mathrm{i}f, \qquad \frac{d\alpha}{dt} = (n + |y|^2)\omega + (y + y^*)f + \frac{\mathrm{i}}{2}\left(y\frac{dy^*}{dt} - \frac{dy}{dt}y^*\right)$$

with boundary conditions $y(t_0) = 0$ and $\alpha(t_0) = 0$. From these results, prove that the evolution operator is given by

$$\hat{U}(t, t_0) = e^{\mathrm{i}\int_{t_0}^{t} dt_1 \int_{t_0}^{t_1} dt_2\, f(t_1)f(t_2)\sin[\omega(t_1 - t_2)]}\, e^{y(t)\hat{d}^\dagger - y^*(t)\hat{d}}\, e^{-\mathrm{i}\omega\hat{d}^\dagger\hat{d}\,(t - t_0)},$$

with $y(t) = -\mathrm{i}e^{-\mathrm{i}\omega t}\int_{t_0}^{t} dt' f(t')e^{\mathrm{i}\omega t'}$. Useful relations to solve this exercise are

$$\hat{d}\,e^{y\hat{d}^\dagger} = e^{y\hat{d}^\dagger}(\hat{d} + y), \qquad \hat{d}^\dagger e^{-y^*\hat{d}} = e^{-y^*\hat{d}}(\hat{d}^\dagger + y^*),$$

and $e^{y\hat{d}^\dagger - y^*\hat{d}} = e^{y\hat{d}^\dagger}e^{-y^*\hat{d}}e^{-|y|^2/2}$.

3.3 Equations of Motion for Operators in the Heisenberg Picture

In quantum mechanics we associate with any observable quantity O a Hermitian operator $\hat{O}$ and an experimental measurement of O yields one of the eigenvalues of $\hat{O}$. The probability to measure one of these eigenvalues depends on the state of the system at the time of measurement. If $|\Psi(t)\rangle$ is the normalized ket describing the system at time t, then the probability of measuring the eigenvalue λ_i is $P_i(t) = |\langle\Psi_i|\Psi(t)\rangle|^2$, with $|\Psi_i\rangle$ the normalized eigenket of $\hat{O}$ with eigenvalue λ_i – that is, $\hat{O}|\Psi_i\rangle = \lambda_i|\Psi_i\rangle$. The same is true if the observable quantity depends explicitly on time, such as the time-dependent Hamiltonian introduced in the previous sections. In this case the eigenvalues and eigenkets of $\hat{O}(t)$ depend on t, $\lambda_i \to \lambda_i(t)$ and $|\Psi_i\rangle \to |\Psi_i(t)\rangle$, and the probability of measuring $\lambda_i(t)$ at time t becomes $P_i(t) = |\langle\Psi_i(t)|\Psi(t)\rangle|^2$. Notice that $|\Psi_i(t)\rangle$ *is not* the time-evolved ket $|\Psi_i\rangle$ but instead the ith eigenket of the operator $\hat{O}(t)$, which can have any time dependence. The knowledge of all probabilities can be used to construct a dynamical *quantum average* of the observable O according to $\sum_i \lambda_i(t)P_i(t)$. This quantity represents the average of the outcomes of an experimental measurement performed in $N \to \infty$ independent systems all in the same state $|\Psi(t)\rangle$. Using the completeness relation $\sum_i |\Psi_i(t)\rangle\langle\Psi_i(t)| = \hat{\mathbb{1}}$ (which is valid for all t), the dynamical quantum average can be rewritten as

$$\begin{aligned}\sum_i \lambda_i(t)P_i(t) &= \sum_i \lambda_i(t)\langle\Psi(t)|\Psi_i(t)\rangle\langle\Psi_i(t)|\Psi(t)\rangle \\ &= \sum_i \langle\Psi(t)|\hat{O}(t)|\Psi_i(t)\rangle\langle\Psi_i(t)|\Psi(t)\rangle \\ &= \langle\Psi(t)|\hat{O}(t)|\Psi(t)\rangle.\end{aligned}$$

The dynamical quantum average is the expectation value of $\hat{O}(t)$ over the state of the system at time t. The enormous amount of information that can be extracted from the dynamical averages prompts us to develop mathematical techniques for their calculation. In the remainder of this chapter we introduce some fundamental concepts and derive a few important identities to lay down the basis of a very powerful mathematical apparatus.

Heisenberg picture From the results of the previous section we know that the time-evolved ket $|\Psi(t)\rangle = \hat{U}(t,t_0)|\Psi(t_0)\rangle$ and therefore the expectation value $\langle\Psi(t)|\hat{O}(t)|\Psi(t)\rangle$ can also be written as $\langle\Psi(t_0)|\hat{U}(t_0,t)\hat{O}(t)\hat{U}(t,t_0)|\Psi(t_0)\rangle$. This leads us to introduce the notion of operators in the *Heisenberg picture*. An operator $\hat{O}(t)$ in the Heisenberg picture is denoted by $\hat{O}_H(t)$ and is defined according to

$$\boxed{\hat{O}_H(t) \equiv \hat{U}(t_0,t)\hat{O}(t)\hat{U}(t,t_0)} \tag{3.17}$$

We apply the above definition not only to those operators associated with observable quantities, but to *all* operators, including the field operators. For instance, the density operator $\hat{n}(\mathbf{x}) = \hat{\psi}^\dagger(\mathbf{x})\hat{\psi}(\mathbf{x})$ is associated with an observable quantity and is written in terms of two field operators which also admit a Heisenberg picture:

$$\begin{aligned}\hat{n}_H(\mathbf{x},t) &= \hat{U}(t_0,t)\hat{\psi}^\dagger(\mathbf{x})\hat{\psi}(\mathbf{x})\hat{U}(t,t_0) = \hat{U}(t_0,t)\hat{\psi}^\dagger(\mathbf{x})\underbrace{\hat{U}(t,t_0)\hat{U}(t_0,t)}_{\hat{\mathbb{1}}}\hat{\psi}(\mathbf{x})\hat{U}(t,t_0) \\ &= \hat{\psi}^\dagger_H(\mathbf{x},t)\hat{\psi}_H(\mathbf{x},t).\end{aligned} \tag{3.18}$$

From (3.18) it is evident that the Heisenberg picture of the product of two operators $\hat{O} = \hat{O}_1\hat{O}_2$ is simply the product of the two operators in the Heisenberg picture – that is, $\hat{O}_H(t) = \hat{O}_{1,H}(t)\hat{O}_{2,H}(t)$. An important consequence of this fact is that when the (anti)commutation relation between two operators is proportional to the identity operator, then operators in the Heisenberg picture at equal times satisfy the *same* (anti)commutation relation as the original operators. In particular, for the field operators we have

$$\left[\hat{\psi}_H(\mathbf{x},t), \hat{\psi}^\dagger_H(\mathbf{x}',t)\right]_\mp = \delta(\mathbf{x}-\mathbf{x}').$$

Operators in the Heisenberg picture obey a simple *equation of motion.* Taking into account (3.13), we find

$$\mathrm{i}\frac{d}{dt}\hat{O}_H(t) = -\hat{U}(t_0,t)\hat{H}(t)\hat{O}(t)\hat{U}(t,t_0) + \hat{U}(t_0,t)\hat{O}(t)\hat{H}(t)\hat{U}(t,t_0) + \hat{U}(t_0,t)\left(\mathrm{i}\frac{d}{dt}\hat{O}(t)\right)\hat{U}(t,t_0). \tag{3.19}$$

For operators independent of time, the last term vanishes. In most textbooks the last term is also written as $\mathrm{i}\frac{\partial}{\partial t}\hat{O}_H(t)$, where the symbol of partial derivative signifies that only the derivative with respect to the explicit time dependence of $\hat{O}(t)$ must be taken. In this book we use the same notation. The first two terms can be rewritten in two equivalent ways. The first, and most obvious, one is $\hat{U}(t_0,t)[\hat{O}(t),\hat{H}(t)]_-\hat{U}(t,t_0)$. We could, alternatively, insert the identity operator $\hat{\mathbb{1}} = \hat{U}(t,t_0)\hat{U}(t_0,t)$ between the Hamiltonian and $\hat{O}(t)$ in order to have both operators in the Heisenberg picture. Thus,

$$\begin{aligned}\mathrm{i}\frac{d}{dt}\hat{O}_H(t) &= \hat{U}(t_0,t)\left[\hat{O}(t),\hat{H}(t)\right]_-\hat{U}(t,t_0) + \mathrm{i}\frac{\partial}{\partial t}\hat{O}_H(t)\\ &= \left[\hat{O}_H(t),\hat{H}_H(t)\right]_- + \mathrm{i}\frac{\partial}{\partial t}\hat{O}_H(t).\end{aligned} \tag{3.20}$$

Equation of motion for field operators In second quantization all operators are expressed in terms of field operators $\hat{\psi}(\mathbf{x})$, $\hat{\psi}^\dagger(\mathbf{x})$, and hence the equations of motion for $\hat{\psi}_H(\mathbf{x},t)$ and $\hat{\psi}^\dagger_H(\mathbf{x},t)$ play a very special role; they actually constitute the seed of the formalism that we develop in the next chapters. Due to their importance, we derive them below. Let us consider the many-body Hamiltonian (1.83) with some time-dependent one-body part $\hat{h}(t)$:

$$\hat{H}(t) = \underbrace{\sum_{\sigma\sigma'}\int d\mathbf{r}\,\hat{\psi}^\dagger(\mathbf{r}\sigma)h_{\sigma\sigma'}(\mathbf{r},-\mathrm{i}\boldsymbol{\nabla},\mathbf{S},t)\hat{\psi}(\mathbf{r}\sigma')}_{\hat{H}_0(t)} + \frac{1}{2}\int d\mathbf{x}\,d\mathbf{x}'\,v(\mathbf{x},\mathbf{x}')\hat{\psi}^\dagger(\mathbf{x})\hat{\psi}^\dagger(\mathbf{x}')\hat{\psi}(\mathbf{x}')\hat{\psi}(\mathbf{x}). \tag{3.21}$$

Since the field operators have no explicit time dependence, the last term in (3.20) vanishes and we only need to evaluate the commutators between $\hat{\psi}(\mathbf{x})$, $\hat{\psi}^\dagger(\mathbf{x})$ and the Hamiltonian. Let us start with the fermionic field operator $\hat{\psi}(\mathbf{x})$. Using the identity

$$\left[\hat{\psi}(\mathbf{x}),\hat{A}\hat{B}\right]_- = \left[\hat{\psi}(\mathbf{x}),\hat{A}\right]_+\hat{B} - \hat{A}\left[\hat{\psi}(\mathbf{x}),\hat{B}\right]_+$$

with

$$\hat{A} = \hat{\psi}^\dagger(\mathbf{r}'\sigma'), \qquad \hat{B} = \sum_{\sigma''}h_{\sigma'\sigma''}(\mathbf{r}',-\mathrm{i}\boldsymbol{\nabla}',\mathbf{S},t)\hat{\psi}(\mathbf{r}'\sigma''),$$

and taking into account that $\left[\hat{\psi}(\mathbf{x}),\hat{B}\right]_+ = 0$ since $\left[\hat{\psi}(\mathbf{x}),\hat{\psi}(\mathbf{x}')\right]_+ = 0$, we find

$$\left[\hat{\psi}(\mathbf{r}\sigma),\hat{H}_0(t)\right]_- = \sum_{\sigma'}h_{\sigma\sigma'}(\mathbf{r},-\mathrm{i}\boldsymbol{\nabla},\mathbf{S},t)\hat{\psi}(\mathbf{r}\sigma').$$

The evaluation of $\left[\hat{\psi}(\mathbf{x}), \hat{H}_{\text{int}}\right]_-$ is also a simple exercise. Indeed,

$$\left[\hat{\psi}(\mathbf{x}), \hat{\psi}^\dagger(\mathbf{x}')\hat{\psi}^\dagger(\mathbf{x}'')\hat{\psi}(\mathbf{x}'')\hat{\psi}(\mathbf{x}')\right]_- = \left[\hat{\psi}(\mathbf{x}), \hat{\psi}^\dagger(\mathbf{x}')\hat{\psi}^\dagger(\mathbf{x}'')\right]_- \hat{\psi}(\mathbf{x}'')\hat{\psi}(\mathbf{x}')$$
$$= \left(\delta(\mathbf{x}-\mathbf{x}')\hat{\psi}^\dagger(\mathbf{x}'') - \delta(\mathbf{x}-\mathbf{x}'')\hat{\psi}^\dagger(\mathbf{x}')\right)\hat{\psi}(\mathbf{x}'')\hat{\psi}(\mathbf{x}'),$$

and therefore

$$\left[\hat{\psi}(\mathbf{x}), \hat{H}_{\text{int}}\right]_- = \frac{1}{2}\int d\mathbf{x}'' v(\mathbf{x},\mathbf{x}'')\hat{\psi}^\dagger(\mathbf{x}'')\hat{\psi}(\mathbf{x}'')\hat{\psi}(\mathbf{x}) - \frac{1}{2}\int d\mathbf{x}' v(\mathbf{x}',\mathbf{x})\hat{\psi}^\dagger(\mathbf{x}')\hat{\psi}(\mathbf{x})\hat{\psi}(\mathbf{x}')$$
$$= \int d\mathbf{x}' v(\mathbf{x},\mathbf{x}')\hat{n}(\mathbf{x}')\hat{\psi}(\mathbf{x}),$$

where in the last step we use the symmetry property $v(\mathbf{x},\mathbf{x}') = v(\mathbf{x}',\mathbf{x})$. Substituting these results in (3.20), the equation of motion for the field operator reads

$$\boxed{\mathrm{i}\frac{d}{dt}\hat{\psi}_H(\mathbf{x},t) = \sum_{\sigma'} h_{\sigma\sigma'}(\mathbf{r},-\mathrm{i}\boldsymbol{\nabla},\mathbf{S},t)\hat{\psi}_H(\mathbf{r}\sigma',t) + \int d\mathbf{x}' v(\mathbf{x},\mathbf{x}')\hat{n}_H(\mathbf{x}',t)\hat{\psi}_H(\mathbf{x},t)} \tag{3.22}$$

The equation of motion for $\hat{\psi}^\dagger_H(\mathbf{x},t)$ can be obtained from the adjoint of the equation above and reads

$$\boxed{\mathrm{i}\frac{d}{dt}\hat{\psi}^\dagger_H(\mathbf{x},t) = -\sum_{\sigma'} \hat{\psi}^\dagger_H(\mathbf{r}\sigma',t)h_{\sigma'\sigma}(\mathbf{r},\mathrm{i}\overleftarrow{\boldsymbol{\nabla}},\mathbf{S},t) - \int d\mathbf{x}' v(\mathbf{x},\mathbf{x}')\hat{\psi}^\dagger_H(\mathbf{x},t)\hat{n}_H(\mathbf{x}',t)} \tag{3.23}$$

In the next sections we use (3.22) and (3.23) to derive the equation of motion for other physically relevant operators such as the density and the total momentum of the system. These results are exact and hence provide important benchmarks to check the quality of an approximation.

Exercise 3.3 Show that the same equations of motion (3.22) and (3.23) are valid for the bosonic field operators.

Exercise 3.4 Show that for a system of identical particles in an external time-dependent electromagnetic field with

$$h_{\sigma\sigma'}(\mathbf{r},-\mathrm{i}\boldsymbol{\nabla},\mathbf{S},t) = \delta_{\sigma\sigma'}\left[\frac{1}{2m}\left(-\mathrm{i}\boldsymbol{\nabla} - \frac{q}{c}\mathbf{A}(\mathbf{r},t)\right)^2 + q\phi(\mathbf{r},t)\right], \tag{3.24}$$

the equations of motion of the field operators are invariant under the gauge transformation

$$\boxed{\begin{aligned} \mathbf{A}(\mathbf{r},t) &\to \mathbf{A}(\mathbf{r},t) + \boldsymbol{\nabla}\Lambda(\mathbf{r},t) \\ \phi(\mathbf{r},t) &\to \phi(\mathbf{r},t) - \frac{1}{c}\frac{\partial}{\partial t}\Lambda(\mathbf{r},t) \\ \hat{\psi}_H(\mathbf{x},t) &\to \hat{\psi}_H(\mathbf{x},t)\exp\left[\mathrm{i}\frac{q}{c}\Lambda(\mathbf{r},t)\right] \end{aligned}} \tag{3.25}$$

3.4 Continuity Equation: Paramagnetic and Diamagnetic Currents

Let us consider a system of interacting and identical particles under the influence of an external time-dependent electromagnetic field. The one-body part of the Hamiltonian is given by the first term of (3.21), with $h(\mathbf{r}, -\mathrm{i}\boldsymbol{\nabla}, \mathbf{S}, t)$ as in (3.24). The equation of motion for the density operator $\hat{n}_H(\mathbf{x}, t) = \hat{\psi}^\dagger_H(\mathbf{x}, t)\hat{\psi}_H(\mathbf{x}, t)$ can easily be obtained using (3.22) and (3.23) since[4]

$$\mathrm{i}\frac{d}{dt}\hat{n}_H = \left(\mathrm{i}\frac{d}{dt}\hat{\psi}^\dagger_H\right)\hat{\psi}_H + \hat{\psi}^\dagger_H\left(\mathrm{i}\frac{d}{dt}\hat{\psi}_H\right).$$

On the r.h.s. the terms containing the scalar potential ϕ and the interaction v cancel. This cancellation is a direct consequence of the fact that the operator $\int d\mathbf{x}'\phi(\mathbf{r}', t)\hat{n}(\mathbf{x}')$ in $\hat{H}_0(t)$ and $\hat{H}_{\mathrm{int}}$ are expressed in terms of the density only, see (1.82), and therefore commute with $\hat{n}(\mathbf{x})$. To calculate the remaining terms, we write

$$\left(\pm\mathrm{i}\boldsymbol{\nabla} - \frac{q}{c}\mathbf{A}\right)^2 = -\nabla^2 \mp \frac{\mathrm{i}q}{c}(\boldsymbol{\nabla}\cdot\mathbf{A}) \mp \frac{2\mathrm{i}q}{c}\mathbf{A}\cdot\boldsymbol{\nabla} + \frac{q^2}{c^2}A^2, \tag{3.26}$$

from which it follows that also the term proportional to A^2 cancels (like the scalar potential this term is coupled to the density operator).[5] Collecting the remaining terms, we find

$$\begin{aligned}\frac{d}{dt}\hat{n}_H &= \frac{1}{2m\mathrm{i}}\left[\left(\nabla^2\hat{\psi}^\dagger_H\right)\hat{\psi}_H - \hat{\psi}^\dagger_H\left(\nabla^2\hat{\psi}_H\right)\right] \\ &+ \frac{q}{mc}\left[\hat{n}_H\boldsymbol{\nabla} + \left(\boldsymbol{\nabla}\hat{\psi}^\dagger_H\right)\hat{\psi}_H + \hat{\psi}^\dagger_H\left(\boldsymbol{\nabla}\hat{\psi}_H\right)\right]\cdot\mathbf{A}.\end{aligned}$$

The first term on the r.h.s. can be written as minus the divergence of the *paramagnetic current density operator,*

$$\boxed{\hat{\mathbf{j}}(\mathbf{x}) \equiv \frac{1}{2m\mathrm{i}}\left[\hat{\psi}^\dagger(\mathbf{x})\left(\boldsymbol{\nabla}\hat{\psi}(\mathbf{x})\right) - \left(\boldsymbol{\nabla}\hat{\psi}^\dagger(\mathbf{x})\right)\hat{\psi}(\mathbf{x})\right]} \tag{3.27}$$

in the Heisenberg picture, while the second term can be written as minus the divergence of the *diamagnetic current density operator,*

$$\boxed{\hat{\mathbf{j}}_d(\mathbf{x}, t) \equiv -\frac{q}{mc}\hat{n}(\mathbf{x})\mathbf{A}(\mathbf{r}, t)} \tag{3.28}$$

also in the Heisenberg picture. Notice that the diamagnetic current depends explicitly on time through the time dependence of the vector potential. The resulting equation of motion for the density $\hat{n}_H(\mathbf{x}, t)$ is known as the *continuity equation*:

$$\boxed{\frac{d}{dt}\hat{n}_H(\mathbf{x}, t) = -\boldsymbol{\nabla}\cdot\left[\hat{\mathbf{j}}_H(\mathbf{x}, t) + \hat{\mathbf{j}}_{d,H}(\mathbf{x}, t)\right]} \tag{3.29}$$

[4] In the rest of this section we omit the arguments of the operators as well as of the scalar and vector potentials if there is no ambiguity.

[5] Equation (3.26) can easily be verified by applying the differential operator to a test function.

Paramagnetic and diamagnetic currents The origin of the names paramagnetic and diamagnetic currents stems from the behavior of the orbital magnetic moment that these currents generate when the system is exposed to a magnetic field. To make these definitions less abstract, let us consider a system of noninteracting particles and rewrite the Hamiltonian $\hat{H}_0$ using the identity (3.26) together with the rearrangement

$$\frac{1}{mi}\hat{\psi}^\dagger(\nabla\hat{\psi})\cdot\mathbf{A} = \hat{\mathbf{j}}\cdot\mathbf{A} + \frac{1}{2mi}(\nabla\hat{n})\cdot\mathbf{A}.$$

We find

$$\hat{H}_0 = -\frac{1}{2m}\int\hat{\psi}^\dagger\nabla^2\hat{\psi} - \frac{q}{c}\int\hat{\mathbf{j}}\cdot\mathbf{A} + \int\hat{n}\left(q\phi + \frac{q^2}{2mc^2}A^2\right) + \frac{iq}{2mc}\int\nabla\cdot(\hat{n}\mathbf{A}). \quad (3.30)$$

Equation (3.30) is an exact manipulation of the original $\hat{H}_0$.[6] We now specialize the discussion to situations with no electric field, $\mathbf{E} = 0$, and with a static and uniform magnetic field $\mathbf{B}$. A possible gauge for the scalar and vector potentials is $\phi = 0$ and $\mathbf{A}(\mathbf{r}) = -\frac{1}{2}\mathbf{r}\times\mathbf{B}$. Substituting these potentials in (3.30), we find a coupling between $\mathbf{B}$ and the currents

$$\hat{H}_0 = -\frac{1}{2m}\int\hat{\psi}^\dagger\nabla^2\hat{\psi} - \int\left(\hat{\mathbf{m}} + \frac{1}{2}\hat{\mathbf{m}}_d\right)\cdot\mathbf{B} + \frac{iq}{2mc}\int\nabla\cdot(\hat{n}\mathbf{A}), \quad (3.31)$$

where we define the magnetic moments

$$\hat{\mathbf{m}}(\mathbf{x}) \equiv \frac{q}{2c}\mathbf{r}\times\hat{\mathbf{j}}(\mathbf{x}), \qquad \hat{\mathbf{m}}_d(\mathbf{x},t) \equiv \frac{q}{2c}\mathbf{r}\times\hat{\mathbf{j}}_d(\mathbf{x},t),$$

generated by the paramagnetic and diamagnetic currents. In (3.31) we use the identity $\mathbf{a}\cdot(\mathbf{b}\times\mathbf{c}) = -(\mathbf{b}\times\mathbf{a})\cdot\mathbf{c}$ with $\mathbf{a}$, $\mathbf{b}$, and $\mathbf{c}$ three arbitrary vectors. The negative sign in front of the coupling $\hat{\mathbf{m}}\cdot\mathbf{B}$ energetically favors configurations in which the paramagnetic moment is large and aligned along $\mathbf{B}$, a behavior similar to the total spin magnetization in a Pauli paramagnet. This explain the name "paramagnetic current." To the contrary, the diamagnetic moment tends, by definition, to be aligned in the opposite direction since

$$\begin{aligned}\hat{\mathbf{m}}_d = \frac{q}{2c}\mathbf{r}\times\hat{\mathbf{j}}_d &= -\frac{q^2}{2mc^2}\hat{n}\,[\mathbf{r}\times\mathbf{A}] \\ &= \frac{q^2}{4mc^2}\hat{n}\,[\mathbf{r}\times(\mathbf{r}\times\mathbf{B})] \\ &= \frac{q^2}{4mc^2}\hat{n}\,[\mathbf{r}(\mathbf{r}\cdot\mathbf{B}) - r^2\mathbf{B}],\end{aligned}$$

and hence, denoting by θ the angle between $\mathbf{r}$ and $\mathbf{B}$,

$$\hat{\mathbf{m}}_d\cdot\mathbf{B} = -\frac{q^2}{4mc^2}\hat{n}\,(rB)^2(1-\cos^2\theta) \leq 0.$$

[6] A brief comment about the last term is in order. For any arbitrary large but finite system, all many-body kets $|\Psi_i\rangle$ relevant to its description yield matrix elements of the density operator $\langle\Psi_i|\hat{n}(\mathbf{r})|\Psi_j\rangle$ that vanish at large $\mathbf{r}$ (no particles at infinity). Then, the total divergence in the last term of (3.30) does not contribute to the quantum averages and can be discarded.

This explain the name "diamagnetic current." The minus sign in front of the coupling $\hat{\mathbf{m}}_d \cdot \mathbf{B}$ in (3.31) favors configurations in which the diamagnetic moment is small. It should be said, however, that there are important physical situations in which the diamagnetic contribution is the dominant one. For instance, in solids composed by atoms with filled shells the average of $\hat{\mathbf{m}}$ over the ground state is zero and the system is called *Larmor diamagnet*. Furthermore, interactions among particles may change drastically the relative contribution between the paramagnetic and diamagnetic moments. A system is said to exhibit perfect diamagnetism if the magnetic field generated by the total magnetic moment $\mathbf{M} = \mathbf{m} + \mathbf{m}_d$ cancels exactly the external magnetic field $\mathbf{B}$. An example of a perfect diamagnet is a bulk superconductor in which $\mathbf{B}$ is expelled outside the bulk by the diamagnetic currents, a phenomenon known as the *Meissner effect*. As a final remark, we observe that neither the paramagnetic nor the diamagnetic current density operators are invariant under the gauge transformation (3.25). On the contrary, the *current density operator*

$$\boxed{\hat{\mathbf{J}}(\mathbf{x}, t) = \hat{\mathbf{j}}(\mathbf{x}) + \hat{\mathbf{j}}_d(\mathbf{x}, t)} \tag{3.32}$$

in the Heisenberg picture, $\hat{\mathbf{J}}_H(\mathbf{x}, t)$, is gauge-invariant and hence observable.

Exercise 3.5 Show that

$$\left[\hat{\mathbf{J}}(\mathbf{x}, t), \hat{n}(\mathbf{x}')\right]_- = -\frac{\mathrm{i}}{m}\hat{n}(\mathbf{x})\boldsymbol{\nabla}\delta(\mathbf{x} - \mathbf{x}').$$

Exercise 3.6 Show that under the gauge transformation (3.25)

$$\hat{\mathbf{j}}_H(\mathbf{x}, t) \to \hat{\mathbf{j}}_H(\mathbf{x}, t) + \frac{q}{mc}\hat{n}_H(\mathbf{x}, t)\boldsymbol{\nabla}\Lambda(\mathbf{r}, t),$$
$$\hat{\mathbf{j}}_{d,H}(\mathbf{x}, t) \to \hat{\mathbf{j}}_{d,H}(\mathbf{x}, t) - \frac{q}{mc}\hat{n}_H(\mathbf{x}, t)\boldsymbol{\nabla}\Lambda(\mathbf{r}, t).$$

3.5 Lorentz Force

From the continuity equation it follows that the dynamical quantum average of the current density operator $\hat{\mathbf{J}}(\mathbf{x}, t)$ is the particle flux, and hence the operator

$$\boxed{\hat{\mathbf{P}}(t) \equiv m \int d\mathbf{x}\, \hat{\mathbf{J}}(\mathbf{x}, t)} \tag{3.33}$$

is the *total momentum operator*. Consequently, the time derivative of $\hat{\mathbf{P}}_H(t)$ is the total force acting on the system. For particles of charge q in an external electromagnetic field this force is the Lorentz force. In this section we see how the Lorentz force comes out from our equations.

The calculation of $\mathrm{i}\frac{d}{dt}\hat{\mathbf{P}}_H$ can be performed along the same lines as the continuity equation. We proceed by first evaluating $\mathrm{i}\frac{d}{dt}\hat{\mathbf{j}}_{d,H}$, then $\mathrm{i}\frac{d}{dt}\hat{\mathbf{j}}_H$, and finally integrating over all space and spin the sum of the two. The equation of motion for the diamagnetic current density follows directly from the continuity equation and reads

$$\mathrm{i}\frac{d}{dt}\hat{\mathbf{j}}_{d,H} = \frac{\mathrm{i}q}{mc}\mathbf{A}(\boldsymbol{\nabla}\cdot\hat{\mathbf{J}}_H) - \frac{\mathrm{i}q}{mc}\hat{n}_H\frac{d}{dt}\mathbf{A}. \tag{3.34}$$

The equation of motion for the paramagnetic current density is slightly more complicated. It is convenient to work in components and define $f_{k=x,y,x}$ as the components of a generic vector function $\mathbf{f}$ and $\partial_{k=x,y,z}$ as the partial derivative with respect to x, y, and z. The kth component of the paramagnetic current density (3.27) in the Heisenberg picture is $\hat{j}_{k,H} = \frac{1}{2m\mathrm{i}}[\hat{\psi}^\dagger_H(\partial_k\hat{\psi}_H) - (\partial_k\hat{\psi}^\dagger_H)\hat{\psi}_H]$. To calculate its time derivative we rewrite the equations of motion (3.22) and (3.23) for the field operators as

$$\mathrm{i}\frac{d}{dt}\hat{\psi}_H = \mathrm{i}\frac{d}{dt}\hat{\psi}_H\bigg|_{\phi=\mathbf{A}=0} + \frac{\mathrm{i}q}{mc}\sum_p\left[A_p\partial_p + \frac{1}{2}(\partial_p A_p)\right]\hat{\psi}_H + w\,\hat{\psi}_H,$$

$$\mathrm{i}\frac{d}{dt}\hat{\psi}^\dagger_H = \mathrm{i}\frac{d}{dt}\hat{\psi}^\dagger_H\bigg|_{\phi=\mathbf{A}=0} + \frac{\mathrm{i}q}{mc}\sum_p\left[A_p\partial_p + \frac{1}{2}(\partial_p A_p)\right]\hat{\psi}^\dagger_H - w\,\hat{\psi}^\dagger_H,$$

where the first term is defined as the contribution to the derivative which does not *explicitly* depend on the scalar and vector potentials (there is, of course, an implicit dependence through the evolution operator $\hat{U}$ in the field operators $\hat{\psi}_H$ and $\hat{\psi}^\dagger_H$). The function w is defined as

$$w(\mathbf{r},t) = q\phi(\mathbf{r},t) + \frac{q^2}{2mc^2}A^2(\mathbf{r},t). \tag{3.35}$$

We can evaluate $\mathrm{i}\frac{d}{dt}\hat{j}_{k,H}$ in a systematic way by collecting the terms that do not depend on ϕ and $\mathbf{A}$, the terms linear in $\mathbf{A}$, and the terms proportional to w. We find

$$\mathrm{i}\frac{d}{dt}\hat{j}_{k,H} = \mathrm{i}\frac{d}{dt}\hat{j}_{k,H}\bigg|_{\phi=\mathbf{A}=0} + \frac{\mathrm{i}q}{mc}\sum_p\left[\partial_p(A_p\,\hat{j}_{k,H}) + (\partial_k A_p)\hat{j}_{p,H}\right] - \frac{\mathrm{i}}{m}\hat{n}_H\partial_k w. \tag{3.36}$$

It is not difficult to show that

$$\mathrm{i}\frac{d}{dt}\hat{j}_{k,H}\bigg|_{\phi=\mathbf{A}=0} = -\mathrm{i}\sum_p\partial_p\hat{T}_{pk,H} - \mathrm{i}\hat{W}_{k,H},$$

where the *momentum-stress tensor operator* $\hat{T}_{pk} = \hat{T}_{kp}$ reads

$$\hat{T}_{pk} = \frac{1}{2m^2}\left[(\partial_k\hat{\psi}^\dagger)(\partial_p\hat{\psi}) + (\partial_p\hat{\psi}^\dagger)(\partial_k\hat{\psi}) - \frac{1}{2}\partial_k\partial_p\hat{n}\right], \tag{3.37}$$

while the operator $\hat{W}_k(\mathbf{x})$ is given by

$$\begin{aligned}\hat{W}_k(\mathbf{x},t) &= \frac{1}{m}\int d\mathbf{x}'\hat{\psi}^\dagger(\mathbf{x})\hat{\psi}^\dagger(\mathbf{x}')(\partial_k v(\mathbf{x},\mathbf{x}'))\hat{\psi}(\mathbf{x}')\hat{\psi}(\mathbf{x}) \\ &= \mathrm{i}\left[\hat{j}_k(\mathbf{x}),\hat{H}_{\mathrm{int}}\right]_-.\end{aligned}$$

The equation of motion for the current density operator $\hat{\mathbf{J}}_H$ follows by adding (3.34) to (3.36). After some algebra one finds

$$\mathrm{i}\frac{d}{dt}\hat{J}_{k,H} = \frac{\mathrm{i}q}{m}\hat{n}_H\left[-\partial_k\phi - \frac{1}{c}\frac{d}{dt}A_k\right] + \frac{\mathrm{i}q}{mc}\sum_p \hat{J}_{p,H}\left[\partial_k A_p - \partial_p A_k\right]$$
$$- \mathrm{i}\sum_p \partial_p\left[\hat{T}_{pk,H} - \frac{q}{mc}(A_k\hat{J}_{p,H} + A_p\hat{j}_{k,H})\right] - \mathrm{i}\hat{W}_{k,H}. \tag{3.38}$$

In the first row we recognize the kth component of the electric field $\mathbf{E}$ as well as the kth component of the vector product $\hat{\mathbf{J}}_H \times (\boldsymbol{\nabla} \times \mathbf{A}) = \hat{\mathbf{J}}_H \times \mathbf{B}$.[7] The equation of motion for $\hat{\mathbf{P}}_H$ is obtained by integrating (3.38) over all space and spin. The first two terms give $\int(q\,\hat{n}_H\mathbf{E}+\frac{q}{c}\hat{\mathbf{J}}_H\times\mathbf{B})$, which is exactly the operator of the Lorentz force, as expected. What about the remaining terms? Let us start by discussing $\hat{W}_{k,H}$. The interparticle interaction $v(\mathbf{x},\mathbf{x}')$ is symmetric under the interchange $\mathbf{x} \leftrightarrow \mathbf{x}'$. Moreover, to not exert a net force on the system v must depend only on the difference $\mathbf{r} - \mathbf{r}'$, a requirement that guarantees the conservation of $\hat{\mathbf{P}}_H$ in the absence of external fields.[8] Taking into account these properties, we have $\partial_k v(\mathbf{x},\mathbf{x}') = -\partial'_k v(\mathbf{x},\mathbf{x}') = -\partial'_k v(\mathbf{x}',\mathbf{x})$, where ∂'_k is the partial derivative with respect to x', y', z'. Therefore $\partial_k v(\mathbf{x},\mathbf{x}')$ is antisymmetric under the interchange $\mathbf{x} \leftrightarrow \mathbf{x}'$ and consequently the integral over all space of $\hat{W}_{k,H}$ is zero. No further simplifications occur upon integration and hence the Lorentz force operator is recovered only modulo the total divergence of another operator [the first term in the second row of (3.38)]. This fact should not worry the reader. Operators themselves are not measurable; it is the quantum average of an operator that can be compared to an experimental result. As already mentioned in footnote 6, for any large but finite system the probability of finding a particle far away from the system is vanishingly small for all physical states. Therefore, *in an average sense* the total divergence does not contribute and we can write

$$\boxed{\frac{d}{dt}\langle\hat{\mathbf{P}}_H\rangle = \int\left(q\langle\hat{n}_H\rangle\mathbf{E} + \frac{q}{c}\langle\hat{\mathbf{J}}_H\rangle \times \mathbf{B}\right)} \tag{3.39}$$

where $\langle\hat{O}\rangle$ denotes the quantum average of the operator $\hat{O}$ over some physical state.

[7] Given three arbitrary vectors $\mathbf{a}$, $\mathbf{b}$, and $\mathbf{c}$, the kth component of $\mathbf{a} \times (\mathbf{b} \times \mathbf{c})$ can be obtained as follows:

$$\begin{aligned}[\mathbf{a} \times (\mathbf{b} \times \mathbf{c})]_k &= \sum_{pq} \varepsilon_{kpq} a_p (\mathbf{b} \times \mathbf{c})_q \\ &= \sum_{pq}\sum_{lm} \varepsilon_{kpq}\varepsilon_{qlm} a_p b_l c_m = \sum_{plm}(\delta_{kl}\delta_{pm} - \delta_{km}\delta_{pl}) a_p b_l c_m \\ &= \sum_p (a_p b_k c_p - a_p b_p c_k),\end{aligned}$$

where we use the identity $\sum_q \varepsilon_{kpq}\varepsilon_{qlm} = (\delta_{kl}\delta_{pm} - \delta_{km}\delta_{pl})$ for the contraction of two Levi-Civita tensors.

[8] These properties do not imply a dependence on $|\mathbf{r} - \mathbf{r}'|$ only. For instance, $v(\mathbf{x},\mathbf{x}') = \delta_{\sigma\sigma'}/(y - y')^2$ is symmetric and depends only on $\mathbf{r} - \mathbf{r}'$, but cannot be written as a function of $|\mathbf{r} - \mathbf{r}'|$ only. In the special (and physically relevant) case that $v(\mathbf{x},\mathbf{x}')$ depends only on $|\mathbf{r} - \mathbf{r}'|$, then the internal torque is zero and the angular momentum is conserved.

4

The Contour Idea

In this chapter we introduce the idea of the *contour* to calculate the quantum average of any quantum observable. Depending on the physical system and on the physical situation (nonequilibrium or equilibrium, finite or zero temperature), we show how to simplify the calculations by deforming the contour. Finally, we introduce the operator correlators at the heart of the nonequilibrium Green's function formalism, and derive their equation of motion.

4.1 Time-Dependent Quantum Averages

In the previous chapter we discussed how to calculate the time-dependent quantum average of an operator $\hat{O}(t)$ at time t when the system is prepared in the state $|\Psi(t_0)\rangle \equiv |\Psi_0\rangle$ at time t_0: If the time evolution is governed by the Schrödinger equation (3.3), the expectation value $O(t)$ is given by

$$O(t) = \langle\Psi(t)|\hat{O}(t)|\Psi(t)\rangle = \langle\Psi_0|\hat{U}(t_0,t)\hat{O}(t)\hat{U}(t,t_0)|\Psi_0\rangle,$$

with $\hat{U}$ the evolution operator (3.16). We may say that $O(t)$ is the overlap between the initial bra $\langle\Psi_0|$ and a ket obtained by evolving $|\Psi_0\rangle$ from t_0 to t, after which the operator $\hat{O}(t)$ acts, and then evolving the ket backward from t to t_0.

For $t > t_0$ the evolution operator $\hat{U}(t,t_0)$ is expressed in terms of the chronological ordering operator, while $\hat{U}(t_0,t)$ is in terms of the anti-chronological ordering operator. Inserting their explicit expressions in $O(t)$, we find

$$O(t) = \langle\Psi_0|\bar{T}\left\{e^{-\mathrm{i}\int_t^{t_0} d\bar{t}\,\hat{H}(\bar{t})}\right\}\,\hat{O}(t)\,T\left\{e^{-\mathrm{i}\int_{t_0}^{t} d\bar{t}\,\hat{H}(\bar{t})}\right\}|\Psi_0\rangle. \tag{4.1}$$

The structure of the r.h.s. is particularly interesting. Reading the operators from right to left we note that inside the T operator all Hamiltonians are ordered with later time arguments to the left, the latest time being t. Then there is the operator $\hat{O}(t)$, and to its left all Hamiltonians inside the $\bar{T}$ operator ordered with earlier time arguments to the left, the earliest time being t_0. The main purpose of this section is to elucidate the mathematical structure of (4.1) and to introduce a convenient notation to manipulate chronologically and anti-chronologically ordered products of operators.

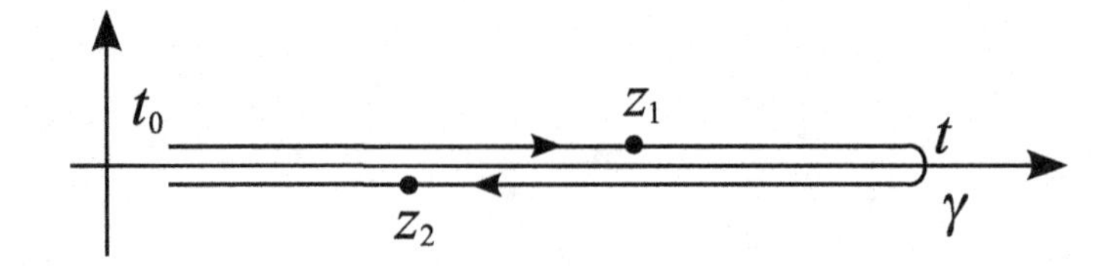

Figure 4.1 The oriented contour γ in the complex time plane as described in the main text. The contour consists of a forward and a backward branch along the real axis between t_0 and t. The branches are displaced from the real axis only for graphical purposes. According to the orientation, point z_2 is later than point z_1.

If we expand the exponentials in (4.1) in powers of the Hamiltonian, then a generic term of the expansion consists of integrals over time of operators such as

$$\bar{T}\left\{\hat{H}(t_1)\dots\hat{H}(t_n)\right\}\hat{O}(t)\,T\left\{\hat{H}(t_1')\dots\hat{H}(t_m')\right\}, \tag{4.2}$$

where all $\{t_i\}$ and $\{t_i'\}$ have values between t_0 and t.[1] This quantity can be rewritten in a more convenient way by introducing a few new definitions. We start by defining the oriented "contour,"

$$\gamma \equiv \underbrace{(t_0,t)}_{\gamma_-}\oplus\underbrace{(t,t_0)}_{\gamma_+}, \tag{4.3}$$

which goes from t_0 to t and then back to t_0. The contour γ consists of two paths: a forward branch γ_- and a backward branch γ_+, as shown in Fig. 4.1. A generic point z' of γ can lie either on γ_- or on γ_+ and once the branch is specified it can assume any value between t_0 and t. We denote by $z' = t'_-$ the point of γ lying on the branch γ_- with value t' and by $z' = t'_+$ the point of γ lying on the branch γ_+ with value t'. Having defined γ, we introduce operators with arguments on the contour according to

$$\hat{A}(z') \equiv \begin{cases} \hat{A}_-(t') & \text{if } z' = t'_- \\ \hat{A}_+(t') & \text{if } z' = t'_+ \end{cases}. \tag{4.4}$$

In general, the operator $\hat{A}(z')$ on the forward branch ($\hat{A}_-(t')$) can be different from the operator on the backward branch ($\hat{A}_+(t')$). We further define a suitable ordering operator for the product of many operators with arguments on γ. Let $\mathcal{T}$ be the *contour-ordering operator* that moves operators with "later" contour arguments to the left. Then for every permutation P of the times z_m later than z_{m-1} later than z_{m-2} ... later than z_1 we have

$$\mathcal{T}\left\{\hat{A}_{P(m)}(z_{P(m)})\dots\hat{A}_{P(2)}(z_{P(2)})\hat{A}_{P(1)}(z_{P(1)})\right\} = \hat{A}_m(z_m)\dots\hat{A}_2(z_2)\hat{A}_1(z_1),$$

which should be compared with (3.8). A point z_2 is later than a point z_1 if z_1 is closer to the starting point, see again Fig. 4.1. In particular, a point on the backward branch is always

[1] We remind the reader that *by construction* the (anti-)time-ordered operator acting on the multiple integral of Hamiltonians is the multiple integral of the (anti-)time-ordered operator acting on the Hamiltonians; see, for instance, (3.14). In other words, $T\int\dots \equiv \int T\dots$ and $\bar{T}\int\dots \equiv \int\bar{T}\dots$.

later than a point on the forward branch. Furthermore, due to the orientation, if $t_1 > t_2$ then t_{1-} is later than t_{2-} while t_{1+} is earlier than t_{2+}. Thus, $\mathcal{T}$ acts like the chronological ordering operator for arguments on γ_- and like the anti-chronological ordering operator for arguments on γ_+. The definition of $\mathcal{T}$, however, allows us to consider also other cases. For example, given two operators $\hat{A}(z_1)$ and $\hat{B}(z_2)$ with argument on the contour, we have the following possibilities:

$$\mathcal{T}\left\{\hat{A}(z_1)\hat{B}(z_2)\right\} = \begin{cases} T\left\{\hat{A}_-(t_1)\hat{B}_-(t_2)\right\} & \text{if } z_1 = t_{1-} \text{ and } z_2 = t_{2-} \\ \hat{A}_+(t_1)\hat{B}_-(t_2) & \text{if } z_1 = t_{1+} \text{ and } z_2 = t_{2-} \\ \hat{B}_+(t_2)\hat{A}_-(t_1) & \text{if } z_1 = t_{1-} \text{ and } z_2 = t_{2+} \\ \bar{T}\left\{\hat{A}_+(t_1)\hat{B}_+(t_2)\right\} & \text{if } z_1 = t_{1+} \text{ and } z_2 = t_{2+} \end{cases}. \tag{4.5}$$

Operators on the contour and the contour-ordering operator can be used to rewrite (4.2) in a compact form. We define the Hamiltonian and the operator $\hat{O}$ with arguments on γ according to

$$\hat{H}(z' = t'_\pm) \equiv \hat{H}(t'), \qquad \hat{O}(z' = t'_\pm) \equiv \hat{O}(t'). \tag{4.6}$$

Both the Hamiltonian and $\hat{O}$ are the same on the forward and backward branches, and they equal the corresponding operators with real-time argument; this is a special case of (4.4) with $\hat{A}_- = \hat{A}_+$. In this book, the field operators – and hence all operators associated with observable quantities (e.g., density, current, energy) – with argument on the contour are defined as in (4.6) – that is, they are the same on the two branches of γ and they equal the corresponding operators with real-time argument. Since the field operators carry no dependence on time, we have

$$\boxed{\hat{\psi}(\mathbf{x}, z = t_\pm) \equiv \hat{\psi}(\mathbf{x}), \qquad \hat{\psi}^\dagger(\mathbf{x}, z = t_\pm) \equiv \hat{\psi}^\dagger(\mathbf{x})} \tag{4.7}$$

and, hence, the density with argument on the contour is

$$\hat{n}(\mathbf{x}, z) = \hat{\psi}^\dagger(\mathbf{x}, z)\hat{\psi}(\mathbf{x}, z) = \hat{\psi}^\dagger(\mathbf{x})\hat{\psi}(\mathbf{x}) = \hat{n}(\mathbf{x}),$$

the diamagnetic current with arguments on the contour is

$$\hat{\mathbf{j}}_d(\mathbf{x}, z) = -\frac{q}{mc}\hat{n}(\mathbf{x}, z)\mathbf{A}(\mathbf{x}, z) = -\frac{q}{mc}\hat{n}(\mathbf{x})\mathbf{A}(\mathbf{x}, t),$$

and so on. Examples of operators that take different values on the forward and backward branches can be constructed as in (4.5): Let $\hat{O}_1(z_1)$ and $\hat{O}_2(z_2)$ be such that $\hat{O}_1(z_1 = t_{1\pm}) = \hat{O}_1(t_1)$ and $\hat{O}_2(z_2 = t_{2\pm}) = \hat{O}_2(t_2)$. Then, for any fixed value of z_2 the operator

$$\hat{A}(z_1) \equiv \mathcal{T}\left\{\hat{O}_1(z_1)\hat{O}_2(z_2)\right\}$$

is, in general, different on the two branches, see again (4.5). Notice the slight abuse of notation in (4.6) and (4.7). The same symbol $\hat{H}$ (or $\hat{O}$) is used for the operator with argument on the contour and for the operator with real times. There is, however, no risk of ambiguity as long as we always specify the argument: From now on we use the letter z for variables on γ.

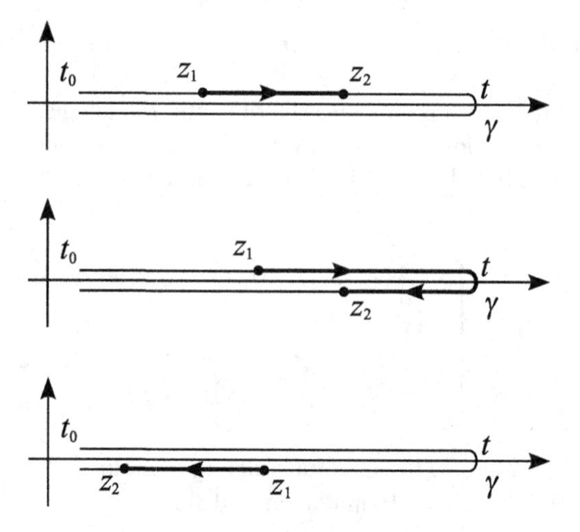

Figure 4.2 Illustration of the three possible locations of z_2 and z_1 for z_2 later than z_1. The domain of integration is highlighted with bold lines.

With definition (4.6) we can rewrite (4.2) as

$$\mathcal{T}\left\{\hat{H}(t_{1+})\ldots\hat{H}(t_{n+})\hat{O}(t_{\pm})\hat{H}(t'_{1-})\ldots\hat{H}(t'_{m-})\right\}, \tag{4.8}$$

where the argument of the operator $\hat{O}$ can be either t_+ or t_-. This result is only the first of a series of simplifications entailed by our new definitions. We can proceed further by introducing the contour integral between two points z_1 and z_2 on γ in the same way as the standard integral along any contour. If z_2 is later than z_1, see Fig. 4.2, then we have

$$\int_{z_1}^{z_2} d\bar{z}\,\hat{A}(\bar{z}) = \begin{cases} \int_{t_1}^{t_2} d\bar{t}\,\hat{A}_-(\bar{t}) & \text{if } z_1 = t_{1-} \text{ and } z_2 = t_{2-} \\ \int_{t_1}^{t} d\bar{t}\,\hat{A}_-(\bar{t}) + \int_{t}^{t_2} d\bar{t}\,\hat{A}_+(\bar{t}) & \text{if } z_1 = t_{1-} \text{ and } z_2 = t_{2+}\,, \\ \int_{t_1}^{t_2} d\bar{t}\,\hat{A}_+(\bar{t}) & \text{if } z_1 = t_{1+} \text{ and } z_2 = t_{2+} \end{cases}$$

while if z_2 is earlier than z_1

$$\int_{z_1}^{z_2} d\bar{z}\,\hat{A}(\bar{z}) = -\int_{z_2}^{z_1} d\bar{z}\,\hat{A}(\bar{z}).$$

In this definition $\bar{z}$ is the integration variable along γ and *not* the complex conjugate of z. The latter is denoted by z^*. The generic term of the expansion (4.1) is obtained by integrating over all $\{t_i\}$ between t and t_0 and over all $\{t'_i\}$ between t_0 and t the operator (4.2). Taking into account that (4.2) is equivalent to (4.8) and using the definition of the contour integral,

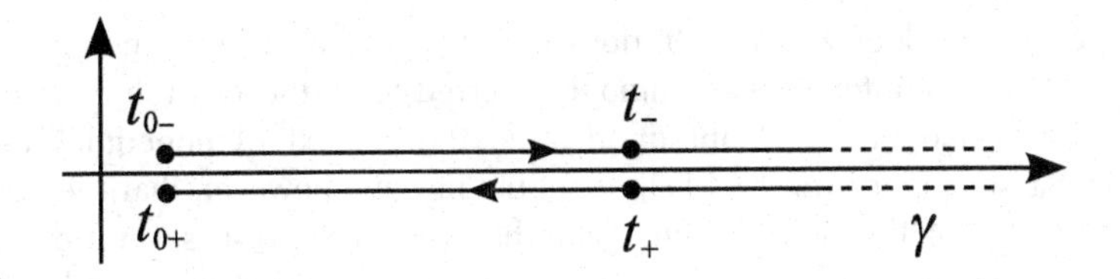

Figure 4.3 The extended oriented contour γ described in the main text with a forward and a backward branch between t_0 and ∞. For any physical time t we have two points $t_\pm$ on γ at the same distance from t_0.

we can write

$$\int_t^{t_0} dt_1 \dots dt_n \int_{t_0}^{t} dt'_1 \dots dt'_m \, \bar{T}\left\{\hat{H}(t_1)\dots\hat{H}(t_n)\right\} \hat{O}(t)\, T\left\{\hat{H}(t'_1)\dots\hat{H}(t'_m)\right\}$$
$$= \int_{\gamma+} dz_1 \dots dz_n \int_{\gamma-} dz'_1 \dots dz'_m \, \mathcal{T}\left\{\hat{H}(z_1)\dots\hat{H}(z_n)\hat{O}(t_\pm)\hat{H}(z'_1)\dots\hat{H}(z'_m)\right\},$$

where the symbol $\int_{\gamma+}$ signifies that the integral is between t_+ and t_{0+} and the symbol $\int_{\gamma-}$ signifies that the integral is between t_{0-} and t_-.[2] Using this general result for all the terms of the expansion, we can rewrite the time-dependent quantum average (4.1) as

$$O(t) = \langle\Psi_0|\mathcal{T}\left\{e^{-\mathrm{i}\int_{\gamma+} d\bar{z}\,\hat{H}(\bar{z})}\hat{O}(t_\pm)e^{-\mathrm{i}\int_{\gamma-} d\bar{z}\,\hat{H}(\bar{z})}\right\}|\Psi_0\rangle.$$

Next we use that operators inside the $\mathcal{T}$ sign can be treated as commuting operators (like for the chronological or anti-chronological ordering operators) and hence

$$O(t) = \langle\Psi_0|\mathcal{T}\left\{e^{-\mathrm{i}\int_\gamma d\bar{z}\,\hat{H}(\bar{z})}\hat{O}(t_\pm)\right\}|\Psi_0\rangle, \tag{4.9}$$

where $\int_\gamma = \int_{\gamma-} + \int_{\gamma+}$ is the contour integral between t_{0-} and t_{0+}. Equation (4.9) is, at the moment, no more than a compact way to rewrite $O(t)$. As we shall see, however, the new notation is extremely useful to manipulate more complicated quantities. We emphasize that in (4.9) $\hat{O}(t)$ is *not* the operator in the Heisenberg picture [the latter is denoted by $\hat{O}_H(t)$].

The contour γ has an aesthetically unpleasant feature: its length depends on t. It would be desirable to have a formula similar to (4.9) but in terms of a universal contour that does not change when we vary the time. Let us explore the implications of extending γ up to infinity, as shown in Fig. 4.3. We evaluate the contour-ordered product in (4.9), with γ the extended contour, when the operator $\hat{O}$ is placed in the position $t_\pm$,

$$\mathcal{T}\left\{e^{-\mathrm{i}\int_\gamma d\bar{z}\,\hat{H}(\bar{z})}\,\hat{O}(t_-)\right\} = \hat{U}(t_0,\infty)\hat{U}(\infty,t)\hat{O}(t)\hat{U}(t,t_0) = \hat{U}(t_0,t)\hat{O}(t)\hat{U}(t,t_0),$$

and similarly,

$$\mathcal{T}\left\{e^{-\mathrm{i}\int_\gamma d\bar{z}\,\hat{H}(\bar{z})}\,\hat{O}(t_+)\right\} = \hat{U}(t_0,t)\hat{O}(t)\hat{U}(t,\infty)\hat{U}(\infty,t_0) = \hat{U}(t_0,t)\hat{O}(t)\hat{U}(t,t_0).$$

[2] Considering the orientation of the contour, the notation is very intuitive.

Thus, the expectation value $O(t)$ in (4.9) does not change if we extend the contour γ as in Fig. 4.3. The extended contour γ is commonly referred to as the *Keldysh contour* in honor of Keldysh, who developed the "contour idea" in the context of nonequilibrium Green's functions in a classic paper from 1964 [39]. It should be said, however, that the idea of using the contour γ of Fig. 4.3 (which over the years has been named in several different ways, including "round trip contour" and "closed loop contour") was first presented by Schwinger a few years earlier, in 1961 [40]. For this reason the contour γ is sometimes called the *Schwinger–Keldysh contour.* Schwinger's paper [40] deals with the study of the Brownian motion of a quantum oscillator in external fields, and even though his idea is completely general (like most of his ideas), the modern nonequilibrium Green's functions formalism is much closer to that developed independently by Keldysh.[3] In the following we simply use the name *contour* since, as we see in the next section, other authors also came up with the "contour idea" independently and more or less at the same time. Furthermore, we do not give a *chronological* presentation of the formalism but, rather, a *logical* presentation. As we shall see, *all Green's function formalisms usually treated as independent naturally follow from a single one.*

Let us continue and go back to (4.9). An important remark about this equation concerns the explicit time dependence of $\hat{O}(t)$. If the operator does not depend on time, we can safely write $\hat{O}(t) = \hat{O}$ in (4.1). However, if we do so in (4.9) it is not clear where to place the operator $\hat{O}$ when acted upon by $\mathcal{T}$. The reason to keep the contour argument even for operators that do not have an explicit time dependence (like the field operators) stems from the need to specify their position along the contour, thus rendering unambiguous the action of $\mathcal{T}$. Once the operators are ordered, we can omit the time arguments if there is no time dependence.

Finally, we observe that the l.h.s. in (4.9) contains the physical time t, while the r.h.s. contains operators with arguments on γ. We can transform (4.9) into an identity between quantities on the contour if we define $O(t_\pm) \equiv O(t)$. In this way, (4.9) takes the elegant form

$$\boxed{O(z) = \langle\Psi_0|\mathcal{T}\left\{e^{-\mathrm{i}\int_\gamma d\bar{z}\,\hat{H}(\bar{z})}\,\hat{O}(z)\right\}|\Psi_0\rangle} \tag{4.10}$$

In (4.10) the contour argument z can be either t_- or t_+, and according to (4.9) our definition is consistent since $O(t_-) = O(t_+) = O(t)$.

4.2 Time-Dependent Ensemble Averages

So far we have used the term *system* to denote an isolated system of particles. In reality, however, it is not possible to completely isolate the system from the surrounding environment; the isolated system is an idealization. The interaction, no matter how weak, between the system and the environment renders a description in terms of one single many-body state impossible. The approach of quantum statistical physics to this problem consists in assigning a probability $w_n \in [0,1]$ of finding the system at time t_0 in the state $|\chi_n\rangle$, with $\sum_n w_n = 1$. The states $|\chi_n\rangle$ are normalized, $\langle\chi_n|\chi_n\rangle = 1$, but they may not be orthogonal

[3] For an interesting historical review on the status of the Russian science in nonequilibrium physics in the 1950s and early 1960s, see the article by Keldysh in Ref. [41].

and they may have different energies, momentum, spin, and also different number of particles. The underlying idea is to describe the system+environment in terms of the isolated system only and to account for the interaction with the environment through the probability distribution w_n. The latter, of course, depends on the features of the environment itself. It is important to stress that such probabilistic description is not a consequence of the quantum limitations imposed by the Heisenberg principle. There are no *theoretical* limitations to how well a system can be isolated. Furthermore, for a perfectly isolated system there are no *theoretical* limitations to the accuracy with which one can determine its quantum state. In quantum mechanics a state is uniquely characterized by a complete set of quantum numbers and these quantum numbers are the eigenvalues of a complete set of *commuting operators.* Thus, it is in principle possible to measure *at the same time* the value of all these operators and determine the exact state of the system. In the language of statistical physics we then say that the system is in a *pure state* since the probabilities w_n are all zero except for a single w_n that is 1.

The *ensemble average* of an operator $\hat{O}(t)$ at time t_0 is defined in the most natural way as

$$O(t_0) = \sum_n w_n \langle \chi_n | \hat{O}(t_0) | \chi_n \rangle, \tag{4.11}$$

and reduces to the quantum average previously introduced in the case of pure states. If we imagine an ensemble of identical *and* isolated systems each in a different pure state $|\chi_n\rangle$, then the ensemble average is the result of calculating the weighted sum of the quantum averages $\langle \chi_n | \hat{O}(t_0) | \chi_n \rangle$ with weights w_n. Ensemble averages incorporate the interaction between the system and the environment.

Density matrix operator The ensemble average leads us to introduce an extremely useful quantity called the *density matrix operator* $\hat{\rho}$, which contains all the statistical information

$$\hat{\rho} = \sum_n w_n |\chi_n\rangle\langle\chi_n|. \tag{4.12}$$

The density matrix operator is self-adjoint, $\hat{\rho} = \hat{\rho}^\dagger$, and positive semidefinite since

$$\langle \Psi | \hat{\rho} | \Psi \rangle = \sum_n w_n |\langle \Psi | \chi_n \rangle|^2 \geq 0,$$

for all states $|\Psi\rangle$. Denoting by $|\Psi_k\rangle$ a generic basis of orthonormal states, we can rewrite the ensemble average (4.11) in terms of $\hat{\rho}$ as[4]

$$\begin{aligned} O(t_0) &= \sum_k \sum_n w_n \langle \chi_n | \Psi_k \rangle \langle \Psi_k | \hat{O}(t_0) | \chi_n \rangle = \sum_k \langle \Psi_k | \hat{O}(t_0) \hat{\rho} | \Psi_k \rangle \\ &= \mathrm{Tr}\left[\hat{O}(t_0)\, \hat{\rho} \right] = \mathrm{Tr}\left[\hat{\rho}\, \hat{O}(t_0) \right], \end{aligned} \tag{4.13}$$

where the symbol Tr denotes a trace over all many-body states – that is, a trace in the Fock space $\mathbb{F}$. If $\hat{O}(t_0) = \hat{\mathbb{1}}$ then from (4.11) $O(t_0) = 1$ since the $|\chi_n\rangle$ are normalized and we have the property

$$\mathrm{Tr}\,[\hat{\rho}] = 1.$$

[4]We recall that the states $|\chi_n\rangle$ may not be orthogonal.

Choosing the kets $|\Psi_k\rangle$ to be the eigenkets of $\hat{\rho}$ with eigenvalues ρ_k, we can write $\hat{\rho} = \sum_k \rho_k |\Psi_k\rangle\langle\Psi_k|$. The eigenvalues ρ_k are nonnegative (since $\hat{\rho}$ is positive semidefinite) and sum up to 1, meaning that $\rho_k \in [0,1]$ and hence that $\mathrm{Tr}\left[\hat{\rho}^2\right] \leq 1$. The most general expression for the ρ_k which incorporates the above constraints is

$$\rho_k = \frac{e^{-x_k}}{\sum_p e^{-x_p}},$$

where x_k are real (positive or negative) numbers. In particular, if $\rho_k = 0$ then $x_k = \infty$. For reasons that will soon become clear, we write $x_k = \beta E_k^{\mathrm{M}}$, where β is a real positive constant, and construct the operator $\hat{H}^{\mathrm{M}}$ according to[5]

$$\hat{H}^{\mathrm{M}} = \sum_k E_k^{\mathrm{M}} |\Psi_k\rangle\langle\Psi_k|.$$

The density matrix operator can then be written as

$$\boxed{\hat{\rho} = \sum_k \frac{e^{-\beta E_k^{\mathrm{M}}}}{Z} |\Psi_k\rangle\langle\Psi_k| = \frac{e^{-\beta \hat{H}^{\mathrm{M}}}}{Z}} \tag{4.14}$$

with the *partition function*

$$\boxed{Z \equiv \sum_k e^{-\beta E_k^{\mathrm{M}}} = \mathrm{Tr}\left[e^{-\beta \hat{H}^{\mathrm{M}}}\right]} \tag{4.15}$$

For example, if we number the $|\Psi_k\rangle$ with an integer $k = 0, 1, 2, \ldots$ and if $\hat{\rho} = \rho_1|\Psi_1\rangle\langle\Psi_1| + \rho_5|\Psi_5\rangle\langle\Psi_5|$, then

$$\hat{H}^{\mathrm{M}} = E_1^{\mathrm{M}}|\Psi_1\rangle\langle\Psi_1| + E_5^{\mathrm{M}}|\Psi_5\rangle\langle\Psi_5| + \lim_{E\to\infty} E \sum_{k\neq 1,5} |\Psi_k\rangle\langle\Psi_k|, \qquad \begin{cases} E_1^{\mathrm{M}} = -\frac{1}{\beta}\ln(\rho_1 Z) \\ E_5^{\mathrm{M}} = -\frac{1}{\beta}\ln(\rho_5 Z) \end{cases}.$$

In the special case of pure states, $\hat{\rho} = |\Psi_0\rangle\langle\Psi_0|$, we could either take $E_k^{\mathrm{M}} \to \infty$ for all $k \neq 0$ or alternatively we could take E_k^{M} finite but larger than E_0^{M} and $\beta \to \infty$ since

$$\lim_{\beta\to\infty} \hat{\rho} = \lim_{\beta\to\infty} \frac{\sum_k e^{-\beta E_k^{\mathrm{M}}} |\Psi_k\rangle\langle\Psi_k|}{\sum_k e^{-\beta E_k^{\mathrm{M}}}} = |\Psi_0\rangle\langle\Psi_0|. \tag{4.16}$$

For operators $\hat{H}^{\mathrm{M}}$ with degenerate ground states, (4.16) reduces instead to an equally weighted ensemble of degenerate ground states.

In general, the expression of $\hat{H}^{\mathrm{M}}$ in terms of field operators is a complicated linear combination of one-body, two-body, three-body, etc. operators. However, in most physical

[5]The superscript "M" stands for "Matsubara" since quantities with this superscript have to do with the initial preparation of the system. Matsubara put forward a perturbative formalism, described in the next chapter, to evaluate the ensemble averages of operators at the initial time t_0.

situations the density matrix $\hat{\rho}$ is chosen to describe a system in thermal equilibrium at a given temperature T and chemical potential μ. This density matrix can be determined by maximizing the entropy of the system with the constraints that the average energy and number of particles are fixed, see Appendix D. The resulting $\hat{\rho}$ can be written as in (4.14) with

$$\hat{H}^{\mathrm{M}} = \hat{H} - \mu\hat{N} \qquad \text{and} \qquad \beta = \frac{1}{K_{\mathrm{B}}T},$$

where $\hat{H}$ is the Hamiltonian of the system and K_{B} is the Boltzmann constant. Thus, for Hamiltonians as in (1.83) the operator $\hat{H}^{\mathrm{M}}$ is the sum of one-body and two-body operators.

Konstantinov–Perel' contour Let us now ask the question of how the ensemble averages evolve in time. According to the statistical picture outlined above we have to evolve each system of the ensemble and then calculate the weighted sum of the time-dependent quantum averages $\langle\chi_n(t)|\hat{O}(t)|\chi_n(t)\rangle$ with weights w_n. The systems of the ensemble are all identical and hence described by the same Hamiltonian $\hat{H}(z)$. Using the same logic that led to (4.10), we find

$$\begin{aligned} O(z) &= \sum_n w_n \langle\chi_n|\hat{U}(t_0,t)\hat{O}(t)\hat{U}(t,t_0)|\chi_n\rangle = \mathrm{Tr}\left[\hat{\rho}\,\hat{U}(t_0,t)\hat{O}(t)\hat{U}(t,t_0)\right] \\ &= \mathrm{Tr}\left[\hat{\rho}\,\mathcal{T}\left\{e^{-\mathrm{i}\int_\gamma d\bar{z}\,\hat{H}(\bar{z})}\,\hat{O}(z)\right\}\right], \end{aligned} \tag{4.17}$$

and taking into account the representation (4.14) for the density matrix operator,

$$O(z) = \frac{\mathrm{Tr}\left[e^{-\beta\hat{H}^{\mathrm{M}}}\,\mathcal{T}\left\{e^{-\mathrm{i}\int_\gamma d\bar{z}\,\hat{H}(\bar{z})}\,\hat{O}(z)\right\}\right]}{\mathrm{Tr}\left[e^{-\beta\hat{H}^{\mathrm{M}}}\right]}. \tag{4.18}$$

Two observations are now in order:

1. The contour-ordered product

$$\mathcal{T}\left\{e^{-\mathrm{i}\int_\gamma d\bar{z}\,\hat{H}(\bar{z})}\right\} = \hat{U}(t_0,\infty)\hat{U}(\infty,t_0) = \hat{\mathbb{1}}, \tag{4.19}$$

and can therefore be inserted inside the trace in the denominator of (4.18).

2. The exponential of $\hat{H}^{\mathrm{M}}$ can be written as

$$e^{-\beta\hat{H}^{\mathrm{M}}} = e^{-\mathrm{i}\int_{\gamma^{\mathrm{M}}} d\bar{z}\,\hat{H}^{\mathrm{M}}},$$

where γ^{M} is any contour in the complex plane starting in z_a and ending in z_b, with the only constraint that

$$\boxed{z_b - z_a = -\mathrm{i}\beta}$$

Using observations (1) and (2) in (4.18) we find

$$O(z) = \frac{\mathrm{Tr}\left[e^{-\mathrm{i}\int_{\gamma^{\mathrm{M}}} d\bar{z}\hat{H}^{\mathrm{M}}}\,\mathcal{T}\left\{e^{-\mathrm{i}\int_\gamma d\bar{z}\,\hat{H}(\bar{z})}\,\hat{O}(z)\right\}\right]}{\mathrm{Tr}\left[e^{-\mathrm{i}\int_{\gamma^{\mathrm{M}}} d\bar{z}\hat{H}^{\mathrm{M}}}\,\mathcal{T}\left\{e^{-\mathrm{i}\int_\gamma d\bar{z}\,\hat{H}(\bar{z})}\right\}\right]}. \tag{4.20}$$

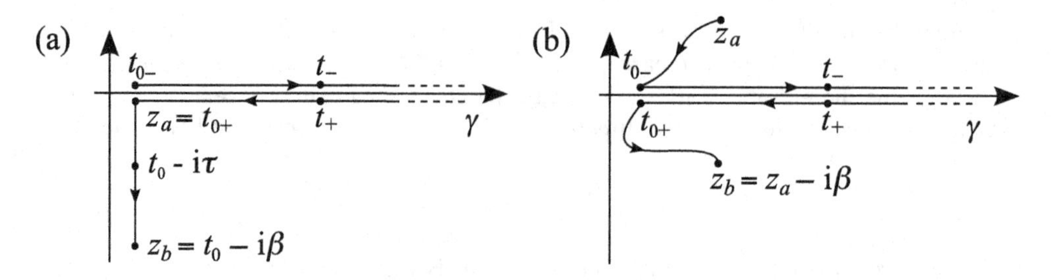

Figure 4.4 Two examples of the extension of the original contour. In (a) a vertical track going from t_0 to $t_0 - \mathrm{i}\beta$ has been added and, according to the orientation, any point on this track is later than a point on the forward or backward branch. In (b) any point between z_a and t_{0-} (t_{0+} and z_b) is earlier (later) than a point on the forward or backward branch.

This is a very interesting formula. Performing a statistical average is similar to performing a time propagation, as they are both described by the exponential of a "Hamiltonian" operator. In particular, the statical average is equivalent to a time propagation along the complex path γ^{M}. The complex time evolution can be incorporated inside the contour-ordering operator provided that we connect γ^{M} to the original contour and define $\hat{H}(z) = \hat{H}^{\mathrm{M}}$ for any z on γ^{M}. Two examples of such contours are given in Fig. 4.4.[6] The idea of a contour with a complex path γ^{M} which incorporates the information on how the system is initially prepared was proposed for the first time in 1960 by Konstantinov and Perel' [42], and subsequently developed by several authors, see, e.g., Refs. [43, 44]. According to the orientation displayed in the figure, a point on γ^{M} lying between z_a and t_{0-} is *earlier* than any point lying on the forward or backward branch (there are no such points for the contour of Fig. 4.4(a)). Similarly, a point on γ^{M} lying between t_{0+} and z_b is *later* than any point lying on the forward or backward branch. We use this observation and the cyclic property of the trace to rewrite the numerator of (4.20) as

$$\mathrm{Tr}\left[e^{-\mathrm{i}\int_{t_{0+}}^{z_b} d\bar{z}\hat{H}^{\mathrm{M}}}\mathcal{T}\left\{e^{-\mathrm{i}\int_{\gamma} d\bar{z}\,\hat{H}(\bar{z})}\hat{O}(z)\right\}e^{-\mathrm{i}\int_{z_a}^{t_{0-}} d\bar{z}\hat{H}^{\mathrm{M}}}\right] = \mathrm{Tr}\left[\mathcal{T}\left\{e^{-\mathrm{i}\int_{\gamma^{\mathrm{M}}\oplus\gamma} d\bar{z}\,\hat{H}(\bar{z})}\hat{O}(z)\right\}\right],$$

where $\gamma^{\mathrm{M}}\oplus\gamma$ denotes a Konstantinov–Perel' contour of Fig. 4.4 and $\mathcal{T}$ is the contour-ordering operator along $\gamma^{\mathrm{M}}\oplus\gamma$. From now on we simply denote by γ the Konstantinov–Perel' contour - that is,

$$\gamma = \gamma_- \oplus \gamma_+ \oplus \gamma^{\mathrm{M}},$$

with γ_- the forward branch and γ_+ the backward branch. Furthermore, in the remainder of the book we simply refer to γ as the *contour*. Performing the same manipulations for the

[6] Strictly speaking it is not necessary to connect γ^{M} to the horizontal branches since we can define an ordering also for disconnected contours by saying, for example, that all points on a piece are earlier or later than all points on the other piece. The fundamental motivation for us to connect γ^{M} is explained at the end of Section 5.1. For the time being, let us say that it is an aesthetically appealing choice.

denominator of (4.20), we get

$$O(z) = \frac{\mathrm{Tr}\left[\mathcal{T}\left\{e^{-\mathrm{i}\int_\gamma d\bar{z}\,\hat{H}(\bar{z})}\,\hat{O}(z)\right\}\right]}{\mathrm{Tr}\left[\mathcal{T}\left\{e^{-\mathrm{i}\int_\gamma d\bar{z}\,\hat{H}(\bar{z})}\right\}\right]} \tag{4.21}$$

Equation (4.21) is the main result of this section.

We have already shown that if z lies on the forward/backward branch then (4.21) yields the time-dependent ensemble average of the observable $O(t)$. Does (4.21) make any sense if z lies on γ^{M}? In general it does not since the operator $\hat{O}(z)$ itself is not defined in this case. In the absence of a definition for $\hat{O}(z)$ when $z \in \gamma^{\mathrm{M}}$ we can always invent one. A natural definition would be

$$\hat{O}(z \in \gamma^{\mathrm{M}}) \equiv \hat{O}^{\mathrm{M}}, \tag{4.22}$$

which is the same operator for every point $z \in \gamma^{\mathrm{M}}$ and hence it is compatible with our definition of $\hat{H}(z \in \gamma^{\mathrm{M}}) = \hat{H}^{\mathrm{M}}$. Throughout this book we use the superscript "M" to indicate the constant value of the Hamiltonian or of any other operator (both in first and second quantization) along the path γ^{M}:

$$\hat{H}^{\mathrm{M}} \equiv \hat{H}(z \in \gamma^{\mathrm{M}}), \qquad \hat{h}^{\mathrm{M}} \equiv \hat{h}(z \in \gamma^{\mathrm{M}}), \qquad \hat{O}^{\mathrm{M}} \equiv \hat{O}(z \in \gamma^{\mathrm{M}}). \tag{4.23}$$

We often consider the case $\hat{O}^{\mathrm{M}} = \hat{O}(t_0)$, except for the Hamiltonian which we take as $\hat{H}^{\mathrm{M}} = \hat{H}(t_0) - \mu\hat{N}$. However, the formalism itself is not restricted to these situations. In some of the applications of the following chapters we use this freedom for studying systems initially prepared in a stationary ensemble. It is important to stress again that definition (4.22) is operative only after the operators have been ordered along the contour. Inside the $\mathcal{T}$ product all operators must have a contour argument, as must those operators with no explicit time dependence. Having a definition for $\hat{O}(z)$ for all z on the contour, we can calculate what (4.21) yields for $z \in \gamma^{\mathrm{M}}$. Taking into account (4.19) we find

$$O(z \in \gamma^{\mathrm{M}}) = \frac{\mathrm{Tr}\left[e^{-\mathrm{i}\int_z^{z_b} d\bar{z}\hat{H}(\bar{z})}\hat{O}^{\mathrm{M}}e^{-\mathrm{i}\int_{z_a}^{z} d\bar{z}\hat{H}(\bar{z})}\right]}{\mathrm{Tr}\left[e^{-\beta\hat{H}^{\mathrm{M}}}\right]} = \frac{\mathrm{Tr}\left[e^{-\beta\hat{H}^{\mathrm{M}}}\hat{O}^{\mathrm{M}}\right]}{Z}, \tag{4.24}$$

where we use the cyclic property of the trace. The r.h.s. is independent of z and for systems in thermal equilibrium it coincides with the thermal average of the observable O^{M}.

Let us summarize what we derived so far. In (4.21) the variable z lies on the contour γ of Fig. 4.4; the r.h.s. gives the time-dependent ensemble average of $\hat{O}(t)$ when $z = t_\pm$ lies on the forward or backward branch, and the ensemble average of $\hat{O}^{\mathrm{M}}$ when z lies on γ^{M}.

We conclude this section with an observation. We already explained that the system–environment interaction is taken into account by assigning a density matrix $\hat{\rho}$ or equivalently an operator $\hat{H}^{\mathrm{M}}$. This corresponds to having an ensemble of identical and *isolated* systems with probabilities w_n. In calculating a time-dependent ensemble average, however, each of these systems evolves as an *isolated* system. To understand the implications of this procedure, we consider an environment at zero temperature, so that the density matrix is simply $\hat{\rho} = |\Psi_0\rangle\langle\Psi_0|$, with $|\Psi_0\rangle$ the ground state of the system. Suppose that we switch on a perturbation which we switch off after time T. In the real world the average of any observable quantity goes back to its original value for $t \gg T$ due to the interaction with the environment. Strictly speaking this is *not* what our definition of time-dependent ensemble average predicts. The average $\langle\Psi_0(t)|\hat{O}|\Psi_0(t)\rangle$ corresponds to the following thought experiment: The system is initially in contact with the environment at zero temperature (and hence in its ground state), then it is *disconnected* from the environment and it is allowed to evolve as an isolated system [42]. In other words, the effects of the system–environment interaction are discarded during the time propagation. In particular, if we inject energy into the system (like we do when we perturb it) there is no way to dissipate it. It is therefore important to keep in mind that the results of our calculations are valid up to times much shorter than the typical relaxation time of the system–environment interaction. To overcome this limitation one should explicitly include the system–environment interaction in the formalism. This leads to a dissipative formulation of the problem, see Section 17.7 and Refs. [45, 46].

4.3 Initial Equilibrium and Adiabatic Switching

The actual evaluation of (4.21) is, in general, a very difficult task. The formula involves the trace over the full Fock space $\mathbb{F}$ of the product of several operators.[7] As we shall see, we can make progress whenever $\hat{H}(z)$ is the sum of a one-body operator $\hat{H}_0(z)$, which is typically easy to deal with, and an interaction energy operator $\hat{H}_{\rm int}(z)$ which is, in some sense, "small" and can therefore be treated perturbatively. The derivation of the perturbative scheme to calculate (4.21) is the topic of the next chapter. In this section we look for alternative ways of including $\hat{H}_{\rm int}(z)$ along the contour without altering the exact result. Depending on the problem at hand it can be advantageous to use one formula or the other when dealing with $\hat{H}_{\rm int}$ perturbatively. The alternative formulas, however, are valid only under some extra assumptions.

We consider a system initially in equilibrium at a given temperature and chemical potential so that

$$\hat{H}^{\rm M} = \hat{H}_0^{\rm M} + \hat{H}_{\rm int} \qquad \text{with} \qquad \hat{H}_0^{\rm M} = \hat{H}_0 - \mu\hat{N}.$$

The curious reader can generalize the following discussion to more exotic initial preparations. As usual we take t_0 to be the time at which the Hamiltonian,

$$\hat{H} \to \hat{H}(t) = \hat{H}_0(t) + \hat{H}_{\rm int},$$

acquires some time dependence. To be concrete we also choose the contour of Fig. 4.4(a). In Fig. 4.5(a) we illustrate how the Hamiltonian $\hat{H}(z)$ appearing in (4.21) changes along the contour.

[7]The product of several operators stems from the Taylor expansion of the exponential.

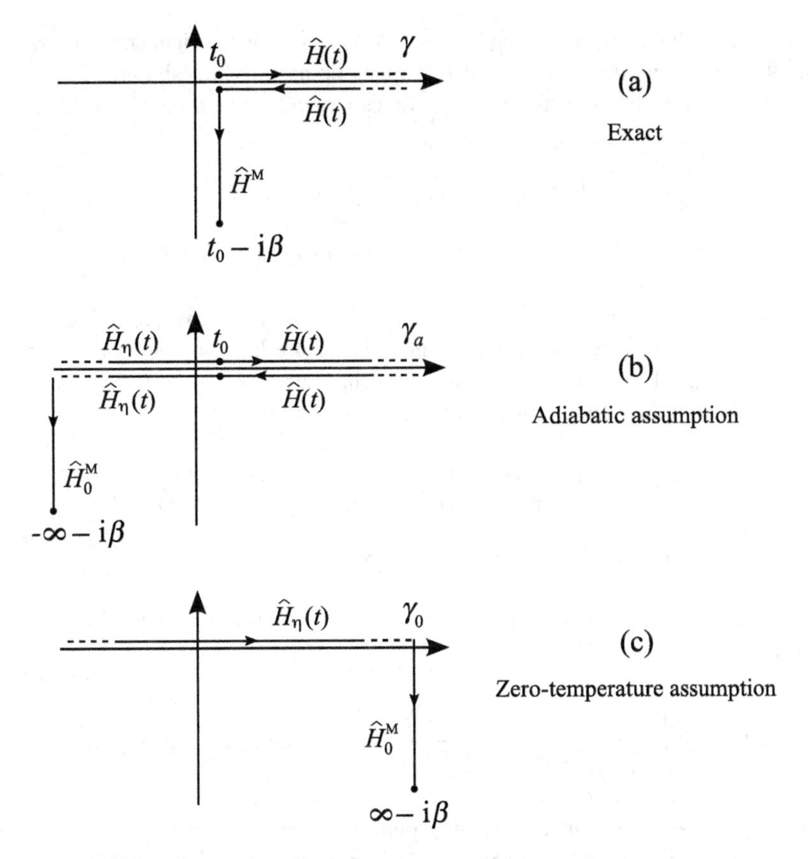

Figure 4.5 Contours and Hamiltonian $\hat{H}(z)$ for (a) the exact formula, (b) the adiabatic formula, and (c) the zero-temperature formula with $\beta \to \infty$.

Adiabatic assumption The *adiabatic assumption* is based on the idea that one can generate the density matrix $\hat{\rho}$ with Hamiltonian $\hat{H}^{\mathrm{M}}$ starting from the density matrix $\hat{\rho}_0$ with Hamiltonian $\hat{H}_0^{\mathrm{M}}$ and then switching on the interaction adiabatically:

$$\hat{\rho} = \frac{e^{-\beta \hat{H}^{\mathrm{M}}}}{Z} = \hat{U}_\eta(t_0, -\infty)\, \frac{e^{-\beta \hat{H}_0^{\mathrm{M}}}}{Z_0}\, \hat{U}_\eta(-\infty, t_0) = \hat{U}_\eta(t_0, -\infty)\, \hat{\rho}_0\, \hat{U}_\eta(-\infty, t_0), \tag{4.25}$$

where $\hat{U}_\eta$ is the real-time evolution operator with Hamiltonian

$$\hat{H}_\eta(t) = \hat{H}_0 + e^{-\eta|t-t_0|}\hat{H}_{\mathrm{int}},$$

and η is an infinitesimally small positive constant. This Hamiltonian coincides with the noninteracting Hamiltonian when $t \to -\infty$ and with the full interacting Hamiltonian when $t = t_0$. Mathematically the adiabatic assumption is supported by the *Gell-Mann-Low theorem* [47, 48], according to which if $\beta \to \infty$ and $\hat{H}_0^{\mathrm{M}}$ has a nondegenerate ground state

$|\Phi_0\rangle$ (hence $\hat{\rho}_0 = |\Phi_0\rangle\langle\Phi_0|$), then $\hat{U}_\eta(t_0, -\infty)|\Phi_0\rangle \equiv |\Psi_0\rangle$ is an eigenstate[8] of $\hat{H}^{\rm M}$ (hence $\hat{\rho} = |\Psi_0\rangle\langle\Psi_0|$). In general the validity of the adiabatic assumption should be checked case by case. Under the adiabatic assumption we can rewrite the time-dependent ensemble average in (4.17) as

$$\begin{aligned} O(z = t_\pm) &= \mathrm{Tr}\left[\hat{\rho}\,\hat{U}(t_0,t)\hat{O}(t)\hat{U}(t,t_0)\right] \\ &= \mathrm{Tr}\left[\hat{\rho}_0\,\hat{U}_\eta(-\infty,t_0)\,\hat{U}(t_0,t)\hat{O}(t)\hat{U}(t,t_0)\hat{U}_\eta(t_0,-\infty)\right], \end{aligned} \tag{4.26}$$

where the cyclic property of the trace has been used. Similarly to what we did in the previous sections, we cast (4.26) in terms of a contour-ordered product of operators. Consider the contour γ_a of Fig. 4.5(b), which is essentially the same contour as Fig. 4.5(a), where $t_0 \to -\infty$. If the Hamiltonian changes along the contour as

$$\hat{H}(t_\pm) = \begin{cases} \hat{H}_\eta(t) = \hat{H}_0 + e^{-\eta|t-t_0|}\hat{H}_{\rm int} & \text{for } t < t_0 \\ \hat{H}(t) = \hat{H}_0(t) + \hat{H}_{\rm int} & \text{for } t > t_0 \end{cases}$$

$$\hat{H}(z \in \gamma^{\rm M}) = \hat{H}_0^{\rm M} = \hat{H}_0 - \mu\hat{N},$$

then (4.26) takes the same form as (4.21) in which $\gamma \to \gamma_a$ and the Hamiltonian goes into the Hamiltonian of Fig. 4.5(b):

$$O(z) = \frac{\mathrm{Tr}\left[\mathcal{T}\left\{e^{-\mathrm{i}\int_{\gamma_a} d\bar{z}\,\hat{H}(\bar{z})}\,\hat{O}(z)\right\}\right]}{\mathrm{Tr}\left[\mathcal{T}\left\{e^{-\mathrm{i}\int_{\gamma_a} d\bar{z}\,\hat{H}(\bar{z})}\right\}\right]} \qquad \text{(adiabatic assumption)}. \tag{4.27}$$

We refer to this way of calculating time-dependent ensemble averages as the *adiabatic formula*. This is exactly the formula used by Keldysh in his original paper [39]. The adiabatic formula is correct only provided that the adiabatic assumption is fulfilled. The adiabatic formula gives the noninteracting ensemble average of the operator $\hat{O}$ if $z \in \gamma^{\rm M}$.

Zero-temperature assumption We can derive yet another expression of the ensemble average of the operator $\hat{O}$ for systems unperturbed by external driving fields and hence described by a Hamiltonian $\hat{H}(t > t_0) = \hat{H}$ independent of time. In this case, for any finite time t we can approximate $\hat{U}(t,t_0)$ with $\hat{U}_\eta(t,t_0)$ since we can always choose $\eta \ll 1/|t - t_0|$. If we do so in (4.26) we get

$$O(z = t_\pm) = \mathrm{Tr}\left[\hat{\rho}_0\,\hat{U}_\eta(-\infty,t)\hat{O}(t)\hat{U}_\eta(t,-\infty)\right]. \tag{4.28}$$

According to the adiabatic assumption, we can generate the interacting $\hat{\rho}$ starting from the noninteracting $\hat{\rho}_0$ and then propagating *forward* in time from $-\infty$ to t_0 using the evolution operator $\hat{U}_\eta$. If so, then ρ could also be generated starting from $\hat{\rho}_0$ and then propagating *backward* in time from ∞ to t_0 using the same evolution operator $\hat{U}_\eta$ since $\hat{H}_\eta(t_0 - \Delta_t) = \hat{H}_\eta(t_0 + \Delta_t)$. In other words,

$$\hat{\rho} = \hat{U}_\eta(t_0,\infty)\,\hat{\rho}_0\,\hat{U}_\eta(\infty,t_0).$$

[8] The state $|\Psi_0\rangle$ is not necessarily the ground state.

Comparing this equation with (4.25), we conclude that

$$\hat{\rho}_0 = \hat{U}_\eta(-\infty, \infty)\,\hat{\rho}_0\,\hat{U}_\eta(\infty, -\infty). \tag{4.29}$$

It may be reasonable to expect that this identity is fulfilled because every eigenstate $|\Phi_k\rangle$ of $\hat{\rho}_0$ goes back to $|\Phi_k\rangle$ by switching on and off the interaction adiabatically:

$$\langle\Phi_k|\hat{U}_\eta(\infty, -\infty) = e^{\mathrm{i}\alpha_k}\langle\Phi_k|. \tag{4.30}$$

This expectation is generally wrong due to the occurrence of level crossings and degeneracies in the spectrum of $\hat{H}_\eta$.[9] However, if $\hat{\rho}_0 = |\Phi_0\rangle\langle\Phi_0|$ is a pure state then (4.29) implies that (4.30) with $k = 0$ is satisfied.[10] For $\hat{\rho}_0$ to be a pure state we must take the zero-temperature limit $\beta \to \infty$ and have a nondegenerate ground state $|\Phi_0\rangle$ of $\hat{H}_0^{\mathrm{M}}$. We refer to the adiabatic assumption in combination with equilibrium at zero temperature and with the condition of no ground-state degeneracy as the *zero-temperature assumption*. The zero-temperature assumption can be used to manipulate (4.28) a little more. Since $|\Phi_0\rangle$ is nondegenerate we have

$$\lim_{\beta\to\infty}\hat{\rho}_0 = |\Phi_0\rangle\langle\Phi_0| = \frac{|\Phi_0\rangle\langle\Phi_0|\hat{U}_\eta(\infty, -\infty)}{\langle\Phi_0|\hat{U}_\eta(\infty, -\infty)|\Phi_0\rangle} = \lim_{\beta\to\infty}\frac{e^{-\beta\hat{H}_0^{\mathrm{M}}}\hat{U}_\eta(\infty, -\infty)}{\mathrm{Tr}\left[e^{-\beta\hat{H}_0^{\mathrm{M}}}\hat{U}_\eta(\infty, -\infty)\right]}.$$

Inserting this result into (4.28), we find

$$\lim_{\beta\to\infty} O(z = t_\pm) = \lim_{\beta\to\infty}\frac{\mathrm{Tr}\left[e^{-\beta\hat{H}_0^{\mathrm{M}}}\hat{U}_\eta(\infty, t)\hat{O}(t)\hat{U}_\eta(t, -\infty)\right]}{\mathrm{Tr}\left[e^{-\beta\hat{H}_0^{\mathrm{M}}}\hat{U}_\eta(\infty, -\infty)\right]}. \tag{4.31}$$

If we now rewrite the exponential $e^{-\beta\hat{H}_0^{\mathrm{M}}}$ as $\exp[-\mathrm{i}\int_{\gamma^{\mathrm{M}}} d\bar{z}\,\hat{H}_0^{\mathrm{M}}]$ and construct the contour γ_0 which starts at $-\infty$ and goes all the way to ∞ and then down to $\infty - \mathrm{i}\beta$, we see that (4.31) has again the same mathematical structure as (4.21), in which $\gamma \to \gamma_0$ and the Hamiltonian along the contour changes as illustrated in Fig. 4.5(c). It is worth noticing that the contour γ_0 has the special property of having only a forward branch. We refer to this way of calculating ensemble averages as the *zero-temperature formula*. The zero-temperature formula is correct only provided that the zero-temperature assumption is fulfilled. This is certainly not the case if the Hamiltonian of the system is time-dependent. There is indeed no reason to expect that by switching on and off the interaction the system goes back to the same state in the presence of external driving fields. Finally we observe that the zero-temperature formula,

[9] Consider, for instance, a density matrix $\hat{\rho}_0 = \frac{1}{2}(|\Phi_1\rangle\langle\Phi_1| + |\Phi_2\rangle\langle\Phi_2|)$. Then the most general solution of (4.29) is not (4.30) but

$$\begin{aligned}\langle\Phi_1|\hat{U}_\eta(\infty, -\infty) &= e^{\mathrm{i}\alpha}\cos\theta\langle\Phi_1| + \sin\theta\langle\Phi_2|,\\ \langle\Phi_2|\hat{U}_\eta(\infty, -\infty) &= \sin\theta\langle\Phi_1| - e^{-\mathrm{i}\alpha}\cos\theta\langle\Phi_2|,\end{aligned}$$

with α and θ two arbitrary real numbers.

[10] For completeness we mention that the phase factor $\alpha_0 \sim 1/\eta$. This has no consequence in the calculation of observable quantities, but it may lead to instabilities when solving the time-dependent Schrödinger equation numerically, see also Exercise 4.1.

like the adiabatic formula, gives the noninteracting ensemble average of the operator $\hat{O}$ if $z \in \gamma^{\mathrm{M}}$.

To summarize, the average of the operator $\hat{O}$ can be calculated using the formula (4.21) in which the contour and the Hamiltonian along the contour are one of those illustrated in Fig. 4.5. The equivalence between these different flavors of (4.21) relies on the validity of the adiabatic assumption or the zero-temperature assumption.

Exercise 4.1 Consider the Hamiltonian of a bosonic harmonic oscillator, $\hat{H} = \omega \hat{d}^\dagger \hat{d}$, with eigenstates $|\Phi_n\rangle = \frac{(\hat{d}^\dagger)^n}{\sqrt{n!}}|0\rangle$ and eigenvalues $n\omega$. Suppose to switch on adiabatically the perturbation $\hat{H}' = \lambda(\hat{d}^\dagger + \hat{d})$ so that the total adiabatic Hamiltonian reads

$$\hat{H}_\eta(t) = \omega \hat{d}^\dagger \hat{d} + e^{-\eta|t|}\lambda(\hat{d}^\dagger + \hat{d}),$$

where η is an infinitesimally small energy. Using the results of Exercise 3.2, show that the state $\hat{U}_\eta(0, \pm\infty)|\Phi_n\rangle$ is, up to an infinite phase factor, the normalized eigenstate of the shifted harmonic oscillator, $\hat{H}_\eta(0) = \omega \hat{d}^\dagger \hat{d} + \lambda(\hat{d}^\dagger + \hat{d})$, with eigenvalue $n\omega - \lambda^2/\omega$. We remind the reader that the eigenstates of the shifted harmonic oscillator are $|\Psi_n\rangle = e^{-\frac{\lambda}{\omega}(\hat{d}^\dagger - \hat{d})}|\Phi_n\rangle$.

Exercise 4.2 Consider the noninteracting Hubbard model at zero temperature in the presence of a magnetic field $\mathbf{B}$ along the z axis. The Hamiltonian is therefore $\hat{H}_0 - \frac{1}{2} g\mu_{\mathrm{B}} B(\hat{N}_\uparrow - \hat{N}_\downarrow)$, see (2.41). Suppose that $B > 0$ is large enough so that the ground state has more electrons of spin up than electrons of spin down. Show that starting from the ground state of $\hat{H}_0$, hence with the same number of spin up and down electrons, and then switching on the magnetic field adiabatically, we never generate the spin-polarized ground state.

4.4 Equations of Motion for Operators in the Contour Heisenberg Picture

The interpretation of the density matrix operator as an evolution operator along a complex path prompts us to extend the definition of the evolution operator. We define the *contour evolution operator* $\hat{U}(z_2, z_1)$, with z_1, z_2 belonging to some of the contours discussed in the previous section, as

$$\hat{U}(z_2, z_1) = \begin{cases} \mathcal{T}\left\{e^{-\mathrm{i}\int_{z_1}^{z_2} d\bar{z}\, \hat{H}(\bar{z})}\right\} & z_2 \text{ later than } z_1 \\ \bar{\mathcal{T}}\left\{e^{+\mathrm{i}\int_{z_2}^{z_1} d\bar{z}\, \hat{H}(\bar{z})}\right\} & z_2 \text{ earlier than } z_1 \end{cases}, \tag{4.32}$$

where we introduce the anti-chronological contour-ordering operator $\bar{\mathcal{T}}$, which rearranges operators with later contour variables to the right. The contour evolution operator is unitary only for z_1 and z_2 on the horizontal branches: If z_1 and/or z_2 lie on γ^{M} then $\hat{U}$ is proportional to the exponential of a Hermitian operator (like the density matrix operator). In particular, $\hat{U}(z_b, z_a) = e^{-\beta \hat{H}^{\mathrm{M}}}$. The contour evolution operator has properties very similar to the real-time evolution operator, namely

1. $\hat{U}(z, z) = \hat{\mathbb{1}}$;

2. $\hat{U}(z_3, z_2)\hat{U}(z_2, z_1) = \hat{U}(z_3, z_1)$;

3. $\hat{U}$ satisfies a simple differential equation: If z is later than z_0,

$$\mathrm{i}\frac{d}{dz}\hat{U}(z, z_0) = \mathcal{T}\left\{\mathrm{i}\frac{d}{dz}e^{-\mathrm{i}\int_{z_0}^{z} d\bar{z}\,\hat{H}(\bar{z})}\right\} = \mathcal{T}\left\{\hat{H}(z)e^{-\mathrm{i}\int_{z_0}^{z} d\bar{z}\,\hat{H}(\bar{z})}\right\} = \hat{H}(z)\hat{U}(z, z_0) \tag{4.33}$$

and

$$\mathrm{i}\frac{d}{dz}\hat{U}(z_0, z) = \bar{\mathcal{T}}\left\{\mathrm{i}\frac{d}{dz}e^{+\mathrm{i}\int_{z_0}^{z} d\bar{z}\,\hat{H}(\bar{z})}\right\} = -\bar{\mathcal{T}}\left\{\hat{H}(z)e^{+\mathrm{i}\int_{z_0}^{z} d\bar{z}\,\hat{H}(\bar{z})}\right\} = -\hat{U}(z_0, z)\hat{H}(z). \tag{4.34}$$

In (4.33) we use that $\mathcal{T}\{\hat{H}(z)\ldots\} = \hat{H}(z)\mathcal{T}\{\ldots\}$ since the operators in $\{\ldots\}$ are calculated at earlier times. A similar property has been used to obtain (4.34).

It goes without saying that the derivative with respect to z of an operator $\hat{A}(z)$ with argument on the contour is defined in the same way as the standard contour derivative. Let z' be a point on γ infinitesimally later than z. If $z = t_-$ (forward branch), we can write $z' = (t+\varepsilon)_-$ and the distance between these two points is simply $(z' - z) = \varepsilon$. Then

$$\frac{d}{dz}\hat{A}(z) = \lim_{z'\to z}\frac{\hat{A}(z') - \hat{A}(z)}{z' - z} = \lim_{\varepsilon\to 0}\frac{\hat{A}_-(t+\varepsilon) - \hat{A}_-(t)}{\varepsilon} = \frac{d}{dt}\hat{A}_-(t). \tag{4.35}$$

On the other hand, if $z = t_+$ (backward branch), then a point infinitesimally later than z can be written as $z' = (t-\varepsilon)_+$ and the distance is in this case $(z' - z) = -\varepsilon$. Therefore,

$$\frac{d}{dz}\hat{A}(z) = \lim_{z'\to z}\frac{\hat{A}(z') - \hat{A}(z)}{z' - z} = \lim_{\varepsilon\to 0}\frac{\hat{A}_+(t-\varepsilon) - \hat{A}_+(t)}{-\varepsilon} = \frac{d}{dt}\hat{A}_+(t). \tag{4.36}$$

From this result it follows that operators that are the same on the forward and backward branches have a derivative which is also the same on the forward and backward branches. Finally we consider the case $z \in \gamma^{\mathrm{M}}$. For simplicity let γ^{M} be the vertical track of Fig. 4.4(a), so that $z = t_0 - \mathrm{i}\tau$ (no extra complications arise from more general paths). A point infinitesimally later than z can be written as $z' = t_0 - \mathrm{i}(\tau + \varepsilon)$ so that the distance between these two points is $(z' - z) = -\mathrm{i}\varepsilon$. The derivative along the imaginary track is then

$$\frac{d}{dz}\hat{A}(z) = \lim_{z'\to z}\frac{\hat{A}(z') - \hat{A}(z)}{z' - z} = \lim_{\varepsilon\to 0}\frac{\hat{A}(t_0 - \mathrm{i}(\tau+\varepsilon)) - \hat{A}(t_0 - \mathrm{i}\tau)}{-\mathrm{i}\varepsilon} = \mathrm{i}\frac{d}{d\tau}\hat{A}(t_0 - \mathrm{i}\tau).$$

Equations (4.33) and (4.34) are differential equations along a contour. They have been derived without using the property of unitarity and they should be compared with the differential equation (3.13) and its adjoint for the real-time evolution operator. The conclusion is that $\hat{U}(t_2, t_1)$ and $\hat{U}(z_2, z_1)$ are closely related. In particular, taking into account that on the forward/backward branch $\mathcal{T}$ orders the operators as the chronological/anti-chronological ordering operator while $\bar{\mathcal{T}}$ orders them as the anti-chronological/chronological ordering operator, the reader can easily verify that

$$\boxed{\hat{U}(t_2, t_1) = \hat{U}(t_{2-}, t_{1-}) = \hat{U}(t_{2+}, t_{1+})} \tag{4.37}$$

Contour Heisenberg picture The contour evolution operator can be used to rewrite the ensemble average (4.21) in an alternative way. Let us denote by z_i the initial point of the contour and by z_f the final point of the contour.[||] Then

$$O(z) = \frac{\mathrm{Tr}\left[\hat{U}(z_\mathrm{f}, z)\,\hat{O}(z)\,\hat{U}(z, z_\mathrm{i})\right]}{\mathrm{Tr}\left[\hat{U}(z_\mathrm{f}, z_\mathrm{i})\right]} = \frac{\mathrm{Tr}\left[\hat{U}(z_\mathrm{f}, z_\mathrm{i})\hat{U}(z_\mathrm{i}, z)\,\hat{O}(z)\,\hat{U}(z, z_\mathrm{i})\right]}{\mathrm{Tr}\left[\hat{U}(z_\mathrm{f}, z_\mathrm{i})\right]}.$$

Looking at this result, it is natural to introduce the *contour Heisenberg picture* according to

$$\boxed{\hat{O}_H(z) \equiv \hat{U}(z_\mathrm{i}, z)\,\hat{O}(z)\,\hat{U}(z, z_\mathrm{i})} \tag{4.38}$$

If z lies on the horizontal branches, then the property (4.37) implies a simple relation between the contour Heisenberg picture and the standard Heisenberg picture,

$$\hat{O}_H(t_+) = \hat{O}_H(t_-) = \hat{O}_H(t),$$

where $\hat{O}_H(t)$ is the operator in the standard Heisenberg picture.

Equation of motion for field operators on the contour The equation of motion for an operator in the contour Heisenberg picture is easily derived from (4.33) and (4.34), and reads

$$\begin{aligned} \mathrm{i}\frac{d}{dz}\hat{O}_H(z) &= \hat{U}(z_\mathrm{i}, z)\,[\hat{O}(z), \hat{H}(z)]\,\hat{U}(z, z_\mathrm{i}) + \mathrm{i}\frac{\partial}{\partial z}\hat{O}_H(z) \\ &= [\hat{O}_H(z), \hat{H}_H(z)] + \mathrm{i}\frac{\partial}{\partial z}\hat{O}_H(z), \end{aligned} \tag{4.39}$$

where the partial derivative is with respect to the explicit z-dependence of the operator $\hat{O}(z)$. Equation (4.39) has exactly the same structure as the real-time equation of motion (3.20). In particular, for $z = t_\pm$ the r.h.s. of (4.39) is identical to the r.h.s. of (3.20), in which t is replaced by z. This means that the equation of motion for the field operators on the contour is given by (3.22) and (3.23), in which $\hat{\psi}_H(\mathbf{x}, t) \to \hat{\psi}_H(\mathbf{x}, z)$, $\hat{\psi}^\dagger_H(\mathbf{x}, t) \to \hat{\psi}^\dagger_H(\mathbf{x}, z)$. For $z \in \gamma^\mathrm{M}$ the r.h.s. of (4.39) contains the commutator $[\hat{O}^\mathrm{M}, \hat{H}^\mathrm{M}]$, see (4.22) and (4.23). We define the field operators with argument on γ^M as

$$\boxed{\hat{\psi}(\mathbf{x}, z \in \gamma^\mathrm{M}) \equiv \hat{\psi}(\mathbf{x}), \qquad \hat{\psi}^\dagger(\mathbf{x}, z \in \gamma^\mathrm{M}) \equiv \hat{\psi}^\dagger(\mathbf{x})} \tag{4.40}$$

[||]The contour can be, for example, that of Fig. 4.4(a), in which case $z_\mathrm{i} = t_{0-}$ and $z_\mathrm{f} = t_0 - \mathrm{i}\beta$, or that of Fig. 4.4(b), in which case $z_\mathrm{i} = z_a$ and $z_\mathrm{f} = z_b$, or also that of Fig. 4.5(c), in which case $z_\mathrm{i} = -\infty$ and $z_\mathrm{f} = \infty - \mathrm{i}\beta$ with $\beta \to \infty$.

which together with (4.7) implies that the field operators are constant along the entire contour. In order to keep the presentation suitable to that of an introductory book we only consider systems prepared with $\hat{H}^{\mathrm{M}}$ of the form

$$\hat{H}^{\mathrm{M}} = \underbrace{\int d\mathbf{x}d\mathbf{x}'\hat{\psi}^{\dagger}(\mathbf{x})\langle\mathbf{x}|\hat{h}^{\mathrm{M}}|\mathbf{x}'\rangle\hat{\psi}(\mathbf{x}')}_{\hat{H}_0^{\mathrm{M}}} + \underbrace{\frac{1}{2}\int d\mathbf{x}d\mathbf{x}'v^{\mathrm{M}}(\mathbf{x},\mathbf{x}')\hat{\psi}^{\dagger}(\mathbf{x})\hat{\psi}^{\dagger}(\mathbf{x}')\hat{\psi}(\mathbf{x}')\hat{\psi}(\mathbf{x})}_{\hat{H}_{\mathrm{int}}^{\mathrm{M}}}. \tag{4.41}$$

The generalization to more complicated $\hat{H}^{\mathrm{M}}$ with three-body or higher-order interactions is simply more tedious, but it does not require more advanced mathematical tools. For our purposes it is instructive enough to show that no complication arises when $v^{\mathrm{M}} \neq v$ and $\hat{h}^{\mathrm{M}} \neq \hat{h} - \mu$.[12] With $\hat{H}^{\mathrm{M}}$ of the form (4.41), the equation of motion for the field operators on the entire contour can be written as [compare with (3.22) and (3.23)]

$$\mathrm{i}\frac{d}{dz}\hat{\psi}_H(\mathbf{x},z) = \sum_{\sigma'} h_{\sigma\sigma'}(\mathbf{r},-\mathrm{i}\boldsymbol{\nabla},\mathbf{S},z)\hat{\psi}_H(\mathbf{r}\sigma',z) + \int d\mathbf{x}'v(\mathbf{x},\mathbf{x}',z)\hat{n}_H(\mathbf{x}',z)\hat{\psi}_H(\mathbf{x},z), \tag{4.42}$$

$$-\mathrm{i}\frac{d}{dz}\hat{\psi}^{\dagger}_H(\mathbf{x},z) = \sum_{\sigma'} \hat{\psi}^{\dagger}_H(\mathbf{r}\sigma',z)h_{\sigma'\sigma}(\mathbf{r},\mathrm{i}\overleftarrow{\boldsymbol{\nabla}},\mathbf{S},z) + \int d\mathbf{x}'v(\mathbf{x},\mathbf{x}',z)\hat{\psi}^{\dagger}_H(\mathbf{x},z)\hat{n}_H(\mathbf{x}',z), \tag{4.43}$$

with

$$\begin{cases} \hat{h}(z=t_{\pm}) = \hat{h}(t) \\ \hat{h}(z\in\gamma^{\mathrm{M}}) = \hat{h}^{\mathrm{M}} \end{cases} \qquad \begin{cases} v(\mathbf{x},\mathbf{x}',t_{\pm}) = v(\mathbf{x},\mathbf{x}',t) \\ v(\mathbf{x},\mathbf{x}',z\in\gamma^{\mathrm{M}}) = v^{\mathrm{M}}(\mathbf{x},\mathbf{x}') \end{cases}.$$

In this equation we include an explicit time dependence in the interparticle interaction $v(\mathbf{x},\mathbf{x}',t_{\pm}) = v(\mathbf{x},\mathbf{x}',t)$. The reader can easily verify that in the derivation of (3.22) and (3.23) we did not make any use of the time independence of v. This extension is useful to deal with situations like those of the previous section, in which the interaction was switched on adiabatically.

Even though simple, the equations of motion for the field operators still look a bit complicated. We can unravel the underlying "matrix structure" using (1.13), which we rewrite here for convenience:

$$\langle\mathbf{x}|\hat{h}(z)|\mathbf{x}'\rangle = h_{\sigma\sigma'}(\mathbf{r},-\mathrm{i}\boldsymbol{\nabla},\mathbf{S},z)\delta(\mathbf{r}-\mathbf{r}') = \delta(\mathbf{r}-\mathbf{r}')h_{\sigma\sigma'}(\mathbf{r}',\mathrm{i}\overleftarrow{\boldsymbol{\nabla}}',\mathbf{S},z).$$

By inspection we see that (4.42) and (4.43) are equivalent to

$$\boxed{\mathrm{i}\frac{d}{dz}\hat{\psi}_H(\mathbf{x},z) = \int d\mathbf{x}'\langle\mathbf{x}|\hat{h}(z)|\mathbf{x}'\rangle\hat{\psi}_H(\mathbf{x}',z) + \int d\mathbf{x}'v(\mathbf{x},\mathbf{x}',z)\hat{n}_H(\mathbf{x}',z)\hat{\psi}_H(\mathbf{x},z)} \tag{4.44}$$

[12] If $v^{\mathrm{M}} = v$ and $\hat{h}^{\mathrm{M}} = \hat{h} - \mu$, then $\hat{H}^{\mathrm{M}} = \hat{H}_0 + \hat{H}_{\mathrm{int}} - \mu\hat{N}$ yields the density matrix of thermal equilibrium.

$$\boxed{-\mathrm{i}\frac{d}{dz}\hat{\psi}^{\dagger}_{H}(\mathbf{x},z) = \int d\mathbf{x}'\hat{\psi}^{\dagger}_{H}(\mathbf{x}',z)\langle\mathbf{x}'|\hat{h}(z)|\mathbf{x}\rangle + \int d\mathbf{x}' v(\mathbf{x},\mathbf{x}',z)\hat{\psi}^{\dagger}_{H}(\mathbf{x},z)\hat{n}_{H}(\mathbf{x}',z)} \tag{4.45}$$

These equations are valid for systems initially in equilibrium and then perturbed by external fields, as well as for more exotic situations such as an interacting system ($v \neq 0$) initially described by a noninteracting ensemble ($v^{\mathrm{M}} = 0$) or a noninteracting system ($v = 0$) initially described by an interacting ensemble ($v^{\mathrm{M}} \neq 0$) and then perturbed by external fields. Finally, we wish to emphasize that the operator $\hat{\psi}^{\dagger}_{H}(\mathbf{x},z)$ is not the adjoint of $\hat{\psi}_{H}(\mathbf{x},z)$ if $z \in \gamma^{\mathrm{M}}$, since $\hat{U}(z,z_{\mathrm{i}})$ is not unitary in this case.

Exercise 4.3 Prove (4.37).

Exercise 4.4 Suppose that the interaction Hamiltonian contains also a three-body operator,

$$\begin{aligned}\hat{H}_{\mathrm{int}}(z) = &\frac{1}{2}\int d\mathbf{x}_1 d\mathbf{x}_2\, v(\mathbf{x}_1,\mathbf{x}_2,z)\hat{\psi}^{\dagger}(\mathbf{x}_1)\hat{\psi}^{\dagger}(\mathbf{x}_2)\hat{\psi}(\mathbf{x}_2)\hat{\psi}(\mathbf{x}_1)\\ &+\frac{1}{3}\int d\mathbf{x}_1 d\mathbf{x}_2 d\mathbf{x}_3\, v(\mathbf{x}_1,\mathbf{x}_2,\mathbf{x}_3,z)\hat{\psi}^{\dagger}(\mathbf{x}_1)\hat{\psi}^{\dagger}(\mathbf{x}_2)\hat{\psi}^{\dagger}(\mathbf{x}_3)\hat{\psi}(\mathbf{x}_3)\hat{\psi}(\mathbf{x}_2)\hat{\psi}(\mathbf{x}_1),\end{aligned} \tag{4.46}$$

where $v(\mathbf{x}_1,\mathbf{x}_2,\mathbf{x}_3,z)$ is totally symmetric under a permutation of $\mathbf{x}_1,\mathbf{x}_2,\mathbf{x}_3$. Show that the equation of motion for the field operator is

$$\begin{aligned}\mathrm{i}\frac{d}{dz}\hat{\psi}_{H}(\mathbf{x},z) = &\int d\mathbf{x}_1\langle\mathbf{x}|\hat{h}(z)|\mathbf{x}_1\rangle\hat{\psi}_{H}(\mathbf{x}_1,z) + \int d\mathbf{x}_1\, v(\mathbf{x},\mathbf{x}_1,z)\hat{n}_{H}(\mathbf{x}_1,z)\hat{\psi}_{H}(\mathbf{x},z)\\ &+\int d\mathbf{x}_1 d\mathbf{x}_2\, v(\mathbf{x}_1,\mathbf{x}_2,\mathbf{x},z)\hat{\psi}^{\dagger}_{H}(\mathbf{x}_1,z)\hat{\psi}^{\dagger}_{H}(\mathbf{x}_2,z)\hat{\psi}_{H}(\mathbf{x}_2,z)\hat{\psi}_{H}(\mathbf{x}_1,z)\hat{\psi}_{H}(\mathbf{x},z).\end{aligned}$$

4.5 Operator Correlators on the Contour

In the previous sections we have shown how to rewrite the time-dependent ensemble average of an operator $\hat{O}(t)$ and derived the equation of motion for operators in the contour Heisenberg picture. The main question that remains to be answered is: How can we calculate $O(t)$? We have already mentioned that the difficulty lies in evaluating the exponential in (4.21) and in taking the trace over the Fock space. For the evaluation of the exponential, a natural way to proceed is to expand it in a Taylor series. This would lead to traces of time-ordered strings of operators on the contour. These strings are of the general form

$$\hat{k}(z_1,\ldots,z_n) = \mathcal{T}\left\{\hat{O}_1(z_1)\ldots\hat{O}_n(z_n)\right\}, \tag{4.47}$$

in which $\hat{O}_k(z_k)$ are operators located at position z_k on the contour. Alternatively, we could calculate $O(t)$ by tracing with $\hat{\rho}$ the equation of motion for $\hat{O}_H(z)$ and then solving the resulting differential equation. However, as clearly shown in (4.44) and (4.45), the time derivative generates new operators whose time derivative generates yet other and more complex operators, and so on. As a result we are again back to calculate the trace of a string of operators like (4.47), but this time with operators in the contour Heisenberg picture. For instance, the r.h.s. of (4.44) contains the operator $\hat{n}_H(\mathbf{x}', z)\hat{\psi}_H(\mathbf{x}, z) = \mathcal{T}\left\{n_H(\mathbf{x}', z^+)\hat{\psi}_H(\mathbf{x}, z)\right\}$, where z^+ is a contour time infinitesimally later than z.

From the above discussion we conclude that if we want to calculate $O(t)$ we must be able to manipulate objects like (4.47). We refer to these strings of operators as the *operator correlators*. As we shall see, they play an important part in the subsequent development of the perturbative scheme. The aim of this section is to discuss some of their basic properties. At this stage it is not important what the origin of the time-dependence of the operators is. We can take operators in the contour Heisenberg picture or consider any other arbitrary time-dependence. It is not even important what the actual shape of the contour is. The only thing that matters in the derivations that follow is that the operators are under the contour-ordered sign $\mathcal{T}$.

Contour ordering for fermionic operators The most natural way to find relations for the operator correlators is to differentiate them with respect to their contour arguments. As we shall see, this leads to a very useful set of hierarchy equations. In order not to overcrowd our equations we introduce the abbreviation

$$\hat{O}_j \equiv O_j(z_j).$$

The simplest example of an operator correlator is the contour-ordered product of just two operators [see also (4.5)],

$$\mathcal{T}\left\{\hat{O}_1\hat{O}_2\right\} = \theta(z_1, z_2)\hat{O}_1\hat{O}_2 + \theta(z_2, z_1)\hat{O}_2\hat{O}_1, \tag{4.48}$$

where $\theta(z_1, z_2) = 1$ if z_1 is later that z_2 on the contour and 0 otherwise. This θ-function can be thought of as the Heaviside function on the contour. If we differentiate (4.48) with respect to the contour variable z_1 we obtain

$$\frac{d}{dz_1}\mathcal{T}\left\{\hat{O}_1\hat{O}_2\right\} = \delta(z_1, z_2)\left[\hat{O}_1, \hat{O}_2\right]_- + \mathcal{T}\left\{\left(\frac{d}{dz_1}\hat{O}_1\right)\hat{O}_2\right\}. \tag{4.49}$$

In this formula we define the Dirac δ-function on the contour in the obvious way:

$$\delta(z_1, z_2) \equiv \frac{d}{dz_1}\theta(z_1, z_2) = -\frac{d}{dz_2}\theta(z_1, z_2).$$

By definition, $\delta(z_1, z_2)$ is zero everywhere except that in $z_1 = z_2$, where it is infinite.

Furthermore, for any operator $\hat{A}(z)$ we have[13]

$$\int_{z_i}^{z_f} d\bar{z}\, \delta(z,\bar{z})\hat{A}(\bar{z}) = \hat{A}(z).$$

A nice feature of (4.49) is the equal time commutator between the two operators. Since we can build any many-body operator from the field operators, one of the most important cases to consider is when $\hat{O}_1$ and $\hat{O}_2$ are field operators. In this case the commutator has a simple value for bosons [either 0 or a δ-function, see (1.39), (1.40), and (1.47)] but has no simple expression for fermions. A simple expression for fermions would be obtained if we could replace the commutator with the anticommutator. This can be readily achieved by defining the time ordering of two fermionic field operators to be given by

$$\mathcal{T}\left\{\hat{O}_1\hat{O}_2\right\} \equiv \theta(z_1,z_2)\hat{O}_1\hat{O}_2 - \theta(z_2,z_1)\hat{O}_2\hat{O}_1,$$

where we introduced a minus sign in front of the last term. If we differentiate this expression with respect to the time z_1 we get

$$\frac{d}{dz_1}\mathcal{T}\left\{\hat{O}_1\hat{O}_2\right\} = \delta(z_1,z_2)\left[\hat{O}_1,\hat{O}_2\right]_+ + \mathcal{T}\left\{\left(\frac{d}{dz_1}\hat{O}_1\right)\hat{O}_2\right\},$$

where now the anticommutator appears. In order for this nice property also to be present in general n-point correlators we introduce the following generalized definition of the contour-ordered product:

$$\mathcal{T}\left\{\hat{O}_1\ldots\hat{O}_n\right\} = \sum_P (\pm)^P \theta_n(z_{P(1)},\ldots,z_{P(n)})\hat{O}_{P(1)}\ldots\hat{O}_{P(n)}, \tag{4.50}$$

where we sum over all permutations P of n variables and where the $+$ sign refers to a string of bosonic field operators and the $-$ sign refers to a string of fermionic field operators. In (4.50) we further define the n-time θ-function θ_n to be

$$\theta_n(z_1,\ldots,z_n) = \begin{cases} 1 & \text{if } z_1 > z_2 > \ldots > z_n \\ 0 & \text{otherwise} \end{cases},$$

or equivalently

$$\theta_n(z_1,\ldots,z_n) = \theta(z_1,z_2)\theta(z_2,z_3)\ldots\theta(z_{n-1},z_n).$$

From now on we interchangeably write "$z_1 > z_2$" or "z_1 later than z_2" as well as "$z_1 < z_2$" or "z_1 earlier than z_2." Definition (4.50) considers all possible orderings of the contour

[13]This identity can easily be checked with, for example, an integration by part. We have

$$\begin{aligned}\int_{z_i}^{z_f} d\bar{z}\,\delta(z,\bar{z})\hat{A}(\bar{z}) &= -\int_{z_i}^{z_f} d\bar{z}\,[\frac{d}{d\bar{z}}\theta(z,\bar{z})]\hat{A}(\bar{z}) \\ &= \hat{A}(z_i) + \int_{z_i}^{z_f} d\bar{z}\,\theta(z,\bar{z})\frac{d}{d\bar{z}}\hat{A}(\bar{z}) = \hat{A}(z_i) + \hat{A}(z) - \hat{A}(z_i).\end{aligned}$$

times through the sum over P, and for a given set of times $z_1, \ldots, z_n$ only one term of the sum survives. It follows from the definition that

$$\mathcal{T}\left\{\hat{O}_1 \ldots \hat{O}_n\right\} = (\pm)^P \mathcal{T}\left\{\hat{O}_{P(1)} \ldots \hat{O}_{P(n)}\right\}.$$

In particular, it follows that bosonic field operators commute within the contour-ordered product, whereas fermionic field operators anticommute.

At this point we should observe that the contour-ordering operator is ambiguously defined if the operators have the same time variable. To cure this problem we introduce the further rule that operators at equal times do not change their relative order after the contour ordering. Thus, for instance,

$$\mathcal{T}\left\{\hat{\psi}(\mathbf{x}_1, z)\hat{\psi}(\mathbf{x}_2, z)\right\} = \hat{\psi}(\mathbf{x}_1)\hat{\psi}(\mathbf{x}_2)$$

or

$$\mathcal{T}\left\{\hat{\psi}(\mathbf{x}_1, t_-)\hat{\psi}(\mathbf{x}_2, t'_+)\hat{\psi}^\dagger(\mathbf{x}_3, t_-)\right\} = \pm\hat{\psi}(\mathbf{x}_2)\hat{\psi}(\mathbf{x}_1)\hat{\psi}^\dagger(\mathbf{x}_3)$$

or

$$\mathcal{T}\left\{\hat{\psi}(\mathbf{x}_1, z)\hat{\psi}(\mathbf{x}_2, z)\hat{\psi}^\dagger(\mathbf{x}_2, z^+)\hat{\psi}^\dagger(\mathbf{x}_1, z^+)\right\} = \hat{\psi}^\dagger(\mathbf{x}_2)\hat{\psi}^\dagger(\mathbf{x}_1)\hat{\psi}(\mathbf{x}_1)\hat{\psi}(\mathbf{x}_2),$$

where z^+ is a time infinitesimally later than z. From the last example we infer that a composite operator which consists of M equal-time *fermionic* field operators, such as the density or the current or $\hat{n}(\mathbf{x})\hat{\psi}(\mathbf{y})$, behaves like a bosonic/fermionic field operator for even/odd M under the $\mathcal{T}$ sign. We use the general nomenclature bosonic/fermionic operators for these kind of composite operators. In particular, a string of Hamiltonians behaves like a set of bosonic operators. Therefore, our generalized definition (4.50) is consistent with the earlier definition in Section 4.1.

According to our new definition the derivative of the two-operator correlator is given by

$$\frac{d}{dz_1}\mathcal{T}\left\{\hat{O}_1\hat{O}_2\right\} = \delta(z_1, z_2)\left[\hat{O}_1, \hat{O}_2\right]_\mp + \mathcal{T}\left\{\left(\frac{d}{dz_1}\hat{O}_1\right)\hat{O}_2\right\}, \tag{4.51}$$

where the upper/lower sign refers to bosonic/fermionic operators.

Equations of motion for high-order operator correlators Let us generalize this result to higher-order operator correlators. For long strings of fermionic operators, finding the sign of the prefactor in (4.50) can be awkward. However, there exists a very nice and elegant graphical way to find the sign by the simple drawing of one diagram. Let us illustrate it with an example. Consider the case of five fermionic operators $\hat{O}_1 \ldots \hat{O}_5$ with the contour variables $z_2 > z_1 > z_4 > z_5 > z_3$. Then,

$$\mathcal{T}\left\{\hat{O}_1\hat{O}_2\hat{O}_3\hat{O}_4\hat{O}_5\right\} = -\hat{O}_2\hat{O}_1\hat{O}_4\hat{O}_5\hat{O}_3.$$

The reordering of these operators corresponds to the permutation

$$P(1, 2, 3, 4, 5) = (2, 1, 4, 5, 3)$$

and has sign -1. This permutation can be drawn graphically as in Fig. 4.6. The operators $\hat{O}_1 \ldots \hat{O}_5$ are denoted by dots and are ordered from the top downward on a vertical line on

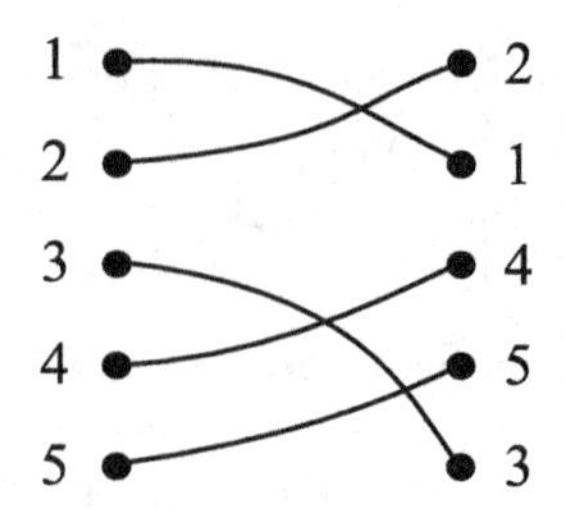

Figure 4.6 A graph to calculate the sign of a permutation.

the left of the figure. The permuted operators $\hat{O}_{P(1)} \dots \hat{O}_{P(5)}$ are similarly drawn on a vertical line on the right of the figure, with the latest time on the top and the earliest time on the bottom. We then connect the left dot i with the right dot i by a line. This line does not have to be a straight line, it can be any curve as long as it does not extend beyond the left and right boundaries of the graph. If we count the number of crossings n_c then the sign of the permutation is simply given by $(-1)^{n_c}$. Since in our case there are three crossings, the sign of the permutation is $(-1)^3 = -1$. This graphical trick is explained in more detail in Appendix B, where it is also used to derive some useful relations for determinants and permanents. It is readily seen that any interchange of neighboring operators on the left or right vertical line increases or decreases the number of crossings by one. An interchange on the left vertical line corresponds to transposing operators under the contour-ordering sign. For example, if we interchange the operators $\hat{O}_3$ and $\hat{O}_4$,

$$\mathcal{T}\left\{\hat{O}_1\hat{O}_2\hat{O}_3\hat{O}_4\hat{O}_5\right\} = -\mathcal{T}\left\{\hat{O}_1\hat{O}_2\hat{O}_4\hat{O}_3\hat{O}_5\right\}, \tag{4.52}$$

in agreement with the fact that the number of crossings for the operator correlator on the r.h.s. is 2. On the other hand, if we interchange two operators on the right vertical line, we change the contour ordering. For instance, interchanging operators $\hat{O}_1$ and $\hat{O}_2$ on the right changes the contour ordering from $z_2 > z_1$ to $z_1 > z_2$. These operations and graphs such as that in Fig. 4.6 are very useful in proving relations involving contour-ordered products. We use these pictures to derive a generalization of (4.51).

We consider strings of only fermionic or only bosonic operators as the generalization to mixed strings is straightforward. From (4.50) we see that the derivative of a time-ordered product consists of a piece where the Heaviside functions are differentiated and a piece where an operator is differentiated:

$$\frac{d}{dz_k}\mathcal{T}\left\{\hat{O}_1\dots\hat{O}_n\right\} = \partial^{\theta}_{z_k}\mathcal{T}\left\{\hat{O}_1\dots\hat{O}_n\right\} + \mathcal{T}\left\{\hat{O}_1\dots\hat{O}_{k-1}\left(\frac{d}{dz_k}\hat{O}_k\right)\hat{O}_{k+1}\dots\hat{O}_n\right\}, \tag{4.53}$$

where we define [49]

$$\partial^{\theta}_{z_k}\mathcal{T}\left\{\hat{O}_1\dots\hat{O}_n\right\} \equiv \sum_P (\pm)^P \left(\frac{d}{dz_k}\theta_n(z_{P(1)},\dots,z_{P(n)})\right)\hat{O}_{P(1)}\dots\hat{O}_{P(n)}. \tag{4.54}$$

It remains to give a more explicit form to (4.54). Imagine that we have a given contour ordering of the operators as in Fig. 4.6 and that we subsequently change the time z_k. As previously explained, this corresponds to moving a dot on the right vertical line. When z_k moves along the contour, the Heaviside function leads to a sudden change in the time-ordering whenever z_k passes another time z_l. In such a case the derivative of the Heaviside function gives a contribution proportional to $\delta(z_k, z_l)$. We only need to find what the prefactor of this δ-function is and for this purpose we need to know how the correlator behaves when z_k is very close to z_l. Let us start by moving the operator $\hat{O}_l$ directly after $\hat{O}_k$ inside the contour ordering. This requires $l-k-1$ interchanges when $l > k$ and $k-l$ interchanges when $k > l$, and therefore

$$\mathcal{T}\left\{\hat{O}_1 \ldots \hat{O}_n\right\} = (\pm)^{l-k-1}\mathcal{T}\left\{\hat{O}_1 \ldots \hat{O}_k\hat{O}_l\hat{O}_{k+1} \ldots \hat{O}_{l-1}\hat{O}_{l+1} \ldots \hat{O}_n\right\}$$

when $l > k$ and

$$\mathcal{T}\left\{\hat{O}_1 \ldots \hat{O}_n\right\} = (\pm)^{k-l}\mathcal{T}\left\{\hat{O}_1 \ldots \hat{O}_{l-1}\hat{O}_{l+1} \ldots \hat{O}_k\hat{O}_l\hat{O}_{k+1} \ldots \hat{O}_n\right\}$$

when $k > l$. The key observation is now that the operators $\hat{O}_k$ and $\hat{O}_l$ stay next to each other after the contour ordering since z_k has been assumed to be very close to z_l, and therefore there is no other time z_j between them. The sign of the permutation that achieves this contour ordering is therefore equal to the sign of the permutation Q of the subset of operators not including $\hat{O}_k$ and $\hat{O}_l$. This is easily seen graphically since the pair of lines connecting the operators $\hat{O}_k$ and $\hat{O}_l$ to their images is always crossed an even number of times. For instance, if we move 4 just below 1 in the left vertical line of Fig. 4.6 and then shift the pair 1 and 4 in the right vertical line upward or downward, we see that the number of crossings n_c always changes by an even number. Then for $l > k$ and for z_k just above and below z_l we can write

$$\begin{aligned}\mathcal{T}\left\{\hat{O}_1 \ldots \hat{O}_n\right\} &= (\pm)^{l-k-1}(\pm)^Q\Big[\theta(z_k, z_l)\hat{O}_{Q(1)} \ldots \hat{O}_k\hat{O}_l \ldots \hat{O}_{Q(n)} \\ &\quad \pm\ \theta(z_l, z_k)\hat{O}_{Q(1)} \ldots \hat{O}_l\hat{O}_k \ldots \hat{O}_{Q(n)}\Big].\end{aligned}$$

We can now differentiate the Heaviside function with respect to z_k and find that

$$\begin{aligned}\partial^\theta_{z_k}\mathcal{T}\left\{\hat{O}_1 \ldots \hat{O}_n\right\} &= (\pm)^{l-k-1}\delta(z_k, z_l)(\pm)^Q\hat{O}_{Q(1)} \ldots \left[\hat{O}_k, \hat{O}_l\right]_\mp \ldots \hat{O}_{Q(n)} \\ &= (\pm)^{l-k-1}\delta(z_k, z_l)\mathcal{T}\left\{\hat{O}_1 \ldots \left[\hat{O}_k, \hat{O}_l\right]_\mp \hat{O}_{k+1} \ldots \hat{O}_{l-1}\hat{O}_{l+1} \ldots \hat{O}_n\right\}.\end{aligned}$$

Due to the presence of the δ-function the (anti)commutator under the contour-ordering sign can be regarded as a function of one time variable only and behaves like a bosonic operator. Similarly for $k > l$ we can, in exactly the same way, derive the equation

$$\partial^\theta_{z_k}\mathcal{T}\left\{\hat{O}_1 \ldots \hat{O}_n\right\} = (\pm)^{k-l}\delta(z_k, z_l)\mathcal{T}\left\{\hat{O}_1 \ldots \hat{O}_{l-1}\hat{O}_{l+1} \ldots \left[\hat{O}_k, \hat{O}_l\right]_\mp \hat{O}_{k+1} \ldots \hat{O}_n\right\}.$$

These two equations are valid only for z_k very close to z_l. The general result for the derivative (4.54) is obtained by summing over all possible values of $l \neq k$. This yields the final expression

$$
\begin{aligned}
&\partial^\theta_{z_k} \mathcal{T}\left\{\hat{O}_1 \ldots \hat{O}_n\right\} \\
&\quad = \sum_{l=1}^{k-1} (\pm)^{k-l} \delta(z_k, z_l) \mathcal{T}\left\{\hat{O}_1 \ldots \hat{O}_{l-1}\hat{O}_{l+1} \ldots \left[\hat{O}_k, \hat{O}_l\right]_\mp \hat{O}_{k+1} \ldots \hat{O}_n\right\} \\
&\quad + \sum_{l=k+1}^{n} (\pm)^{l-k-1} \delta(z_k, z_l) \mathcal{T}\left\{\hat{O}_1 \ldots \left[\hat{O}_k, \hat{O}_l\right]_\mp \hat{O}_{k+1} \ldots \hat{O}_{l-1}\hat{O}_{l+1} \ldots \hat{O}_n\right\}. \qquad (4.55)
\end{aligned}
$$

This expression can be rewritten in various other ways since the (anti)commutator behaves as a bosonic operator and can therefore be placed anywhere we like under the $\mathcal{T}$ sign. Equation (4.53) together with (4.55) represents the generalization of (4.51). For example, if we differentiate the five-operator correlator in (4.52) with respect to z_3 we get

$$
\begin{aligned}
\frac{d}{dz_3} \mathcal{T}\left\{\hat{O}_1\hat{O}_2\hat{O}_3\hat{O}_4\hat{O}_5\right\} &= \delta(z_3, z_1) \mathcal{T}\left\{\hat{O}_2 \left[\hat{O}_3, \hat{O}_1\right]_\mp \hat{O}_4\hat{O}_5\right\} \\
&\pm \delta(z_3, z_2) \mathcal{T}\left\{\hat{O}_1 \left[\hat{O}_3, \hat{O}_2\right]_\mp \hat{O}_4\hat{O}_5\right\} + \delta(z_3, z_4) \mathcal{T}\left\{\hat{O}_1\hat{O}_2 \left[\hat{O}_3, \hat{O}_4\right]_\mp \hat{O}_5\right\} \\
&\pm \delta(z_3, z_5) \mathcal{T}\left\{\hat{O}_1\hat{O}_2 \left[\hat{O}_3, \hat{O}_5\right]_\mp \hat{O}_4\right\} + \mathcal{T}\left\{\hat{O}_1\hat{O}_2 \left(\frac{d}{dz_3}\hat{O}_3\right) \hat{O}_4\hat{O}_5\right\},
\end{aligned}
$$

where the sign in front of each δ-function is simply given by the number of interchanges required to shifting the operators $\hat{O}_l$ for $l = 1, 2, 4, 5$ directly after $\hat{O}_3$ inside the $\mathcal{T}$ product.

We already mentioned that an important application of (4.55) is when the operators $\hat{O}_l$ are the field operators in the contour Heisenberg picture. In this case the equal-time (anti)commutator is a real number equal to either zero or a δ-function. Let us therefore investigate how (4.55) simplifies if

$$
\left[\hat{O}_k(z), \hat{O}_l(z)\right]_\mp = c_{kl}(z)\hat{\mathbb{1}}, \qquad (4.56)
$$

where $c_{kl}(z)$ is a scalar function of z and where on the l.h.s. we reinsert the explicit time dependence of the operators to stress that (4.56) is valid only for the equal-time (anti)commutator. Since the unit operator $\hat{\mathbb{1}}$ commutes with all operators in Fock space, the (anti)commutators in (4.55) can be moved outside the contour-ordered product and we obtain

$$
\begin{aligned}
\partial^\theta_{z_k} \mathcal{T}\left\{\hat{O}_1 \ldots \hat{O}_n\right\} &= \sum_{l=1}^{k-1} (\pm)^{k-l} \delta(z_k, z_l) \left[\hat{O}_k, \hat{O}_l\right]_\mp \mathcal{T}\{\hat{O}_1 \ldots \overset{\sqcap}{\hat{O}_l} \ldots \overset{\sqcap}{\hat{O}_k} \ldots \hat{O}_n\} \\
&+ \sum_{l=k+1}^{n} (\pm)^{l-k-1} \delta(z_k, z_l) \left[\hat{O}_k, \hat{O}_l\right]_\mp \mathcal{T}\{\hat{O}_1 \ldots \overset{\sqcap}{\hat{O}_k} \ldots \overset{\sqcap}{\hat{O}_l} \ldots \hat{O}_n\}, \qquad (4.57)
\end{aligned}
$$

where the symbol $\sqcap$ above an operator $\hat{O}_k$ signifies that this operator is missing from the list; for example,

$$
\hat{O}_1 \overset{\sqcap}{\hat{O}_2} \hat{O}_3 \overset{\sqcap}{\hat{O}_4} \hat{O}_5 = \hat{O}_1\hat{O}_3\hat{O}_5.
$$

Equation (4.57) describes very clearly why it was useful to introduce definition (4.50) of the contour ordering. If we had stuck to our original version of contour ordering, then (4.55) would still have been valid but with all prefactors $+$ and, more importantly, with a commutator also for fermionic operators. The unpleasant consequence of this fact would be that for the fermionic field operators ψ and $\psi^\dagger$ no simplification like in (4.57) can be made. This is a nice example of something to keep in mind when learning new topics: definitions are always introduced to simplify the calculations and never fall from the sky. To appreciate the simplification entailed by definition (4.50), we work out the contour-time derivative of a string of four field operators in the contour Heisenberg picture. Let us introduce a notation which is used throughout the book:

$$i = \mathbf{x}_i, z_i, \quad j = \mathbf{x}_j, z_j, \quad \ldots$$

and

$$i' = \mathbf{x}'_i, z'_i, \quad j' = \mathbf{x}'_j, z'_j, \quad \ldots$$

and

$$\bar{i} = \bar{\mathbf{x}}_i, \bar{z}_i, \quad \bar{j} = \bar{\mathbf{x}}_j, \bar{z}_j, \quad \ldots$$

and so on, to denote the collective position–spin–time coordinate and

$$\delta(j;k) \equiv \delta(z_j, z_k)\delta(\mathbf{x}_j - \mathbf{x}_k)$$

to denote the space–spin–time δ-function. Then, (4.57) together with (4.53) tells us, for instance, that

$$\frac{d}{dz_2}\mathcal{T}\left\{\hat{\psi}_H(1)\hat{\psi}_H(2)\hat{\psi}^\dagger_H(3)\hat{\psi}^\dagger_H(4)\right\} = \mathcal{T}\left\{\hat{\psi}_H(1)\left(\frac{d}{dz_2}\hat{\psi}_H(2)\right)\hat{\psi}^\dagger_H(3)\hat{\psi}^\dagger_H(4)\right\}$$
$$+\delta(2;3)\mathcal{T}\left\{\hat{\psi}_H(1)\hat{\psi}^\dagger_H(4)\right\} \pm \delta(2;4)\mathcal{T}\left\{\hat{\psi}_H(1)\hat{\psi}^\dagger_H(3)\right\},$$

where we take into account that the equal-time (anti)commutator between $\hat{\psi}_H(2)$ and $\hat{\psi}_H(1)$ vanishes.

Equations of motion for Green's function correlators At the beginning of the section we motivated the importance of studying the operator correlators k with their natural appearance in the expansion of the contour-ordered exponential and in the equations of motion. Thus, the operators $\hat{O}_k$ are typically composite operators corresponding to observable quantities and, as such, with an equal number of creation and annihilation field operators. Let us therefore define the *Green's function correlators*

$$\hat{G}_n(1,\ldots,n;1',\ldots,n') \equiv \frac{1}{\mathrm{i}^n}\mathcal{T}\left\{\hat{\psi}_H(1)\ldots\hat{\psi}_H(n)\hat{\psi}^\dagger_H(n')\ldots\hat{\psi}^\dagger_H(1')\right\}, \tag{4.58}$$

where the primed $j' = \mathbf{x}'_j, z'_j$ and unprimed $j = \mathbf{x}_j, z_j$ coordinates label creation and annihilation field operators, respectively (which can be either all bosonic or all fermionic). As we shall see, the prefactor $1/\mathrm{i}^n$ in this equation is a useful convention. For the case $n = 0$ we define $\hat{G}_0 \equiv \hat{\mathbb{1}}$. We can derive an extremely useful set of coupled equations for these operator correlators using (4.53) and (4.57) with the identification

$$\hat{O}_j = \begin{cases} \hat{\psi}_H(j) & \text{for } j = 1,\ldots,n \\ \hat{\psi}^\dagger_H((2n-j+1)') & \text{for } j = n+1,\ldots,2n \end{cases}.$$

After some relabeling we easily find

$$\mathrm{i}\frac{d}{dz_k}\hat{G}_n(1,\ldots,n;1',\ldots,n')$$
$$= \frac{1}{\mathrm{i}^n}\mathcal{T}\left\{\hat{\psi}_H(1)\ldots\left(\mathrm{i}\frac{d}{dz_k}\hat{\psi}_H(k)\right)\ldots\hat{\psi}_H(n)\hat{\psi}^\dagger_H(n')\ldots\hat{\psi}^\dagger_H(1')\right\}$$
$$+\sum_{j=1}^{n}(\pm)^{k+j}\,\delta(k;j')\,\hat{G}_{n-1}(1,\ldots\overset{\sqcap}{k}\ldots,n;1',\ldots\overset{\sqcap}{j'}\ldots,n') \tag{4.59}$$

and

$$-\mathrm{i}\frac{d}{dz'_k}\hat{G}_n(1,\ldots,n;1',\ldots,n')$$
$$= \frac{1}{\mathrm{i}^n}\mathcal{T}\left\{\hat{\psi}_H(1)\ldots\hat{\psi}_H(n)\hat{\psi}^\dagger_H(n')\ldots\left(-\mathrm{i}\frac{d}{dz'_k}\hat{\psi}^\dagger_H(k')\right)\ldots\hat{\psi}^\dagger_H(1')\right\}$$
$$+\sum_{j=1}^{n}(\pm)^{k+j}\,\delta(j;k')\,\hat{G}_{n-1}(1,\ldots\overset{\sqcap}{j}\ldots,n;1',\ldots\overset{\sqcap}{k'}\ldots,n'). \tag{4.60}$$

Indeed the shift of, for example, $\hat{\psi}^\dagger_H(j')$ just after $\hat{\psi}_H(k)$ in (4.59) requires $(n-j)+(n-k)$ interchanges and $(\pm)^{(n-j)+(n-k)} = (\pm)^{k+j}$.

The terms involving the time derivatives of the field operators can be worked out from (4.44) and (4.45). These equations can be rewritten in the following compact form:

$$\int d\bar{1}\,\overrightarrow{G}_0^{-1}(k;\bar{1})\hat{\psi}_H(\bar{1}) = \int d\bar{1}\,v(k,\bar{1})\hat{n}_H(\bar{1})\hat{\psi}_H(k), \tag{4.61a}$$

$$\int d\bar{1}\hat{\psi}^\dagger_H(\bar{1})\overleftarrow{G}_0^{-1}(\bar{1};k') = \int d\bar{1}\,v(k',\bar{1})\hat{\psi}^\dagger_H(k')\hat{n}_H(\bar{1}), \tag{4.61b}$$

where we introduce the definition

$$\overrightarrow{G}_0^{-1}(1;1') \equiv \langle\mathbf{x}_1|\,\mathrm{i}\frac{d}{dz_1} - \hat{h}(z_1)|\mathbf{x}'_1\rangle\delta(z_1,z'_1), \tag{4.62a}$$

$$\overleftarrow{G}_0^{-1}(1;1') \equiv \delta(z_1,z'_1)\langle\mathbf{x}_1| - \mathrm{i}\frac{\overleftarrow{d}}{dz'_1} - \hat{h}(z'_1)|\mathbf{x}'_1\rangle, \tag{4.62b}$$

as well as the short-hand notation

$$v(i;j) \equiv \delta(z_i,z_j)v(\mathbf{x}_i,\mathbf{x}_j,z_i). \tag{4.63}$$

The arrow over the contour-time derivative simply indicates that the derivative acts on the left. The r.h.s. of (4.61) contains three field operators with the same time argument. For instance, in (4.61a) we have

$$\hat{n}_H(\bar{1})\hat{\psi}_H(k) = \hat{\psi}^\dagger_H(\bar{1})\hat{\psi}_H(\bar{1})\hat{\psi}_H(k) = \pm\hat{\psi}^\dagger_H(\bar{1})\hat{\psi}_H(k)\hat{\psi}_H(\bar{1}),$$

where we use the fact that field operators in the Heisenberg picture at equal times satisfy the same (anti)commutation relations as the field operators. Inserting this composite operator inside the contour ordering, we would like to move the field operator $\hat{\psi}_H^\dagger(\bar{1})$ to the right in order to form a $\hat{G}_n$. To make sure that after the reordering $\hat{\psi}_H^\dagger(\bar{1})$ ends up to the left of $\hat{\psi}_H(\bar{1})$ and $\hat{\psi}_H(k)$, we calculate it in $\bar{1}^+ = \bar{\mathbf{x}}_1, \bar{z}_1^+$, where $\bar{z}_1^+$ is infinitesimally later than $\bar{z}_1$. Inside the $\mathcal{T}$ sign we can then write

$$\mathcal{T}\Big\{\ldots \hat{n}_H(\bar{1})\hat{\psi}_H(k)\ldots\Big\} = \pm\mathcal{T}\Big\{\ldots \hat{\psi}_H(k)\hat{\psi}_H(\bar{1})\hat{\psi}_H^\dagger(\bar{1}^+)\ldots\Big\}.$$

As $\hat{\psi}_H(\bar{1})\hat{\psi}_H^\dagger(\bar{1}^+)$ is composed by an even number of field operators it behaves like a bosonic operator inside the contour ordering and can therefore be placed wherever we like without caring about the sign. With this trick, (4.59) can be written as

$$\begin{aligned}\int d\bar{1}\, \overrightarrow{G}_0^{-1}(k;\bar{1})\, \hat{G}_n(1,..,\bar{1},..,n;1',..,n') \\ = \pm\mathrm{i}\int d\bar{1}\, v(k;\bar{1})\, \hat{G}_{n+1}(1,..,n,\bar{1};1',..,n',\bar{1}^+) \\ + \sum_{j=1}^{n}(\pm)^{k+j}\,\delta(k;j')\,\hat{G}_{n-1}(1,..,\sqcap\!\!\!\!k,..,n;1',..,\sqcap\!\!\!\!j',..,n')\end{aligned} \tag{4.64}$$

where the argument $\bar{1}$ in the l.h.s. is at position k.

A similar equation can be derived in a completely analogous way for the contour-ordered product in (4.60). This time we write

$$\mathcal{T}\Big\{\ldots \hat{\psi}_H^\dagger(k')\hat{n}_H(\bar{1})\ldots\Big\} = \pm\mathcal{T}\Big\{\ldots \hat{\psi}_H(\bar{1}^-)\hat{\psi}_H^\dagger(\bar{1})\hat{\psi}_H^\dagger(k')\ldots\Big\},$$

where the time coordinate $\bar{z}_1^-$ is infinitesimally earlier than $\bar{z}_1$. We again use that $\hat{\psi}_H(\bar{1}^-)\hat{\psi}_H^\dagger(\bar{1})$ is a bosonic operator, and hence it can be placed between $\hat{\psi}_H(n)$ and $\hat{\psi}_H^\dagger(n')$ without caring about the sign. Inserting this result into the equation of motion, we find

$$\begin{aligned}\int d\bar{1}\, \hat{G}_n(1,..,n;1',..,\bar{1},..,n')\overleftarrow{G}_0^{-1}(\bar{1};k') \\ = \pm\mathrm{i}\int d\bar{1}\, v(k';\bar{1})\, \hat{G}_{n+1}(1,..,n,\bar{1}^-;1',..,n',\bar{1}) \\ + \sum_{j=1}^{n}(\pm)^{k+j}\,\delta(j;k')\,\hat{G}_{n-1}(1,..,\sqcap\!\!\!\!j,..,n;1',..,\sqcap\!\!\!\!k',..,n')\end{aligned} \tag{4.65}$$

where the argument $\bar{1}$ in the l.h.s. is at position k'.

We have thus derived an infinite hierarchy of operator equations in Fock space in which the derivative of $\hat{G}_n$ is expressed in terms of $\hat{G}_{n-1}$ and $\hat{G}_{n+1}$. These equations are very general as no assumptions have been made on the particular shape of the contour, which can be any of the contours encountered in the previous sections. We have only used an identity for the contour derivative of a string of contour ordered-operators and the equations of motion for $\hat{\psi}_H$ and $\hat{\psi}_H^\dagger$. Equations (4.64) and (4.65) play a pivotal role in the development of the theory. As we explain in the following chapters, the whole structure of diagrammatic perturbation theory is encoded in them.

Exercise 4.5 Let $\hat{O}_j \equiv \hat{O}_j(z_j)$ be a set of composite operators consisting of an even or odd number of fermion field operators. Show that a bosonic operator (even number) always commutes within the contour-ordered product and that a fermionic operator (odd number) anticommutes with another fermionic operator. Further, let $z_1 > z_2 > \ldots > z_n$. Show that

$$\mathcal{T}\left\{\hat{O}_{P(1)} \ldots \hat{O}_{P(n)}\right\} = (-1)^F \hat{O}_1 \ldots \hat{O}_n,$$

where P is a permutation of the labels and where F is the number of interchanges of fermion operators in the permutation P that puts all the operators in the right order.

Exercise 4.6 How do (4.64) and (4.65) change if the interaction Hamiltonian contains a three-body operator, like in (4.46)?

Exercise 4.7 Consider the Hamiltonian on the contour

$$\hat{H}(z) = \sum_{i=1}^{N} \left(\omega_i \hat{d}_i^\dagger \hat{d}_i + f_i(z)\hat{x}_i\right),$$

where $\hat{x}_i = \frac{1}{\sqrt{2}}(\hat{d}_i^\dagger + \hat{d}_i)$ and the $\hat{d}$-operators satisfy the bosonic commutation relations $[\hat{d}_i, \hat{d}_j^\dagger]_- = \delta_{ij}$ and $[\hat{d}_i, \hat{d}_j]_- = [\hat{d}_i^\dagger, \hat{d}_j^\dagger]_- = 0$. Show that the operator correlator

$$\mathcal{T}\left\{\hat{x}_{1,H}(z_1) \ldots \hat{x}_{n,H}(z_n)\right\}$$

satisfies the equation of motion

$$\begin{aligned}
&\left[\frac{d^2}{dz_k^2} + \omega_k^2\right] \mathcal{T}\left\{\hat{x}_{1,H}(z_1) \ldots \hat{x}_{n,H}(z_n)\right\} \\
&\quad = -\omega_k f_k(z) \mathcal{T}\left\{\hat{x}_{1,H}(z_1) \ldots \overset{\sqcap}{\hat{x}_{k,H}}(z_k) \ldots \hat{x}_{n,H}(z_n)\right\} \\
&\quad\quad - \mathrm{i}\omega_k \sum_{j \neq k}^{n} \delta(j;k) \mathcal{T}\left\{\hat{x}_{1,H}(z_1) \ldots \overset{\sqcap}{\hat{x}_{k,H}}(z_k) \ldots \overset{\sqcap}{\hat{x}_{j,H}}(z_j) \ldots \hat{x}_{n,H}(z_n)\right\},
\end{aligned}$$

where $\delta(j;k) = \delta_{jk}\delta(z_j, z_k)$. Why do we need a second-order equation to obtain a closed set of equations for these types of operator correlator?

5

Nonequilibrium Green's Functions

In this chapter we lay down the basis of many-body perturbation theory (MBPT). We define the n-particle nonequilibrium Green's functions (NEGF), derive the Martin–Schwinger hierarchy for these quantities, and show how to solve the hierarchy using Wick's theorem. We close the chapter with the derivation of an extremely useful set of rules, known as the Langreth rules, needed to convert equations on the contour into coupled equations on the real-time axis.

5.1 Martin–Schwinger Hierarchy

In Chapter 4 we derive the differential equations (4.64) and (4.65) for the operator correlators $\hat{G}_n$ which form the building blocks to construct any other operator correlator. This set of operator equations can be turned into a coupled set of differential equations by taking the average with $\hat{\rho} = e^{-\beta \hat{H}^{\mathrm{M}}}/Z$. If we discuss a system that is initially in thermodynamic equilibrium, then $\hat{\rho}$ is the grand canonical density matrix. More generally we define the *n-particle nonequilibrium Green's function* (NEGF) G_n to be

$$G_n(1,\ldots,n;1',\ldots,n') \equiv \frac{\mathrm{Tr}\left[e^{-\beta \hat{H}^{\mathrm{M}}}\hat{G}_n(1,\ldots,n;1',\ldots,n')\right]}{\mathrm{Tr}\left[e^{-\beta \hat{H}^{\mathrm{M}}}\right]}$$
$$= \frac{1}{\mathrm{i}^n}\frac{\mathrm{Tr}\left[\mathcal{T}\left\{ e^{-\mathrm{i}\int_\gamma d\bar{z}\hat{H}(\bar{z})}\hat{\psi}(1)\ldots\hat{\psi}(n)\hat{\psi}^\dagger(n')\ldots\hat{\psi}^\dagger(1')\right\}\right]}{\mathrm{Tr}\left[\mathcal{T}\left\{e^{-\mathrm{i}\int_\gamma d\bar{z}\hat{H}(\bar{z})}\right\}\right]}. \tag{5.1}$$

In the second equality we use the definition of operators in the contour Heisenberg picture. Consider, for instance, G_1 with $z_1 < z_1'$. Then,

$$\begin{aligned} e^{-\beta \hat{H}^{\mathrm{M}}}\mathcal{T}\left\{\hat{\psi}_H(1)\hat{\psi}_H^\dagger(1')\right\} &= \pm\hat{U}(z_{\mathrm{f}},z_{\mathrm{i}})\hat{U}(z_{\mathrm{i}},z_1')\hat{\psi}^\dagger(1')\hat{U}(z_1',z_1)\hat{\psi}(1)\hat{U}(z_1,z_{\mathrm{i}}) \\ &= \pm\mathcal{T}\left\{e^{-\mathrm{i}\int_\gamma d\bar{z}\hat{H}(\bar{z})}\hat{\psi}^\dagger(1')\hat{\psi}(1)\right\} \\ &= \mathcal{T}\left\{e^{-\mathrm{i}\int_\gamma d\bar{z}\hat{H}(\bar{z})}\hat{\psi}(1)\hat{\psi}^\dagger(1')\right\}, \end{aligned}$$

where we take into account that $e^{-\beta\hat{H}^{\rm M}} = \hat{U}(z_{\rm f}, z_{\rm i})$ and that under the $\mathcal{T}$ sign the field operators (anti)commute. Similar considerations apply to G_n. The Green's function G_1 at times $z_1 = z$ and $z_1' = z^+$ infinitesimally later than z_1 is the time-dependent ensemble average of the operator $\hat{\psi}^\dagger(\mathbf{x}_1')\hat{\psi}(\mathbf{x}_1)$ (modulo a factor of i) from which we can calculate the time-dependent ensemble average of *any* one-body operator, see also Appendix E. More generally, by choosing the contour arguments of G_n to be $z_i = z$ and $z_i' = z^+$ for all i, we obtain the time-dependent ensemble average of the operator $\hat{\psi}^\dagger(\mathbf{x}_n')...\hat{\psi}^\dagger(\mathbf{x}_1')\hat{\psi}(\mathbf{x}_1)...\hat{\psi}(\mathbf{x}_n)$ (modulo a factor i^n), from which we can calculate the time-dependent ensemble average of *any* n-body operator.

It is worth observing that under the adiabatic assumption, see Section 4.3, we can calculate the Green's functions G_n from the second line of (5.1) in which $\gamma \to \gamma_a$ is the adiabatic contour of Fig. 4.5(b) and the "adiabatic Hamiltonian" changes along the contour, as illustrated in the same figure. This is the same as saying that in the first line of (5.1) we replace $\hat{H}^{\rm M} \to \hat{H}_0^{\rm M}$, and take the arguments of $\hat{G}_n$ on γ_a and the field operators in the Heisenberg picture with respect to the adiabatic Hamiltonian. These adiabatic Green's functions coincide with the exact ones when all contour arguments are larger than t_0. The proof of this statement is the same as the proof of the adiabatic formula (4.27). If we further have a system in equilibrium at zero temperature, then we can use the zero-temperature assumption. The Green's functions can then be calculated from the second line of (5.1) in which $\gamma \to \gamma_0$ is the zero-temperature contour of Fig. 4.5(c) and the "zero-temperature Hamiltonian" changes along the contour, as illustrated in the same figure. Equivalently, the zero-temperature Green's functions can be calculated from the first line of (5.1) in which $\hat{H}^{\rm M} \to \hat{H}_0^{\rm M}$, the arguments of $\hat{G}_n$ are on the contour γ_0, and the field operators are in the Heisenberg picture with respect to the zero-temperature Hamiltonian.[1] These zero-temperature Green's functions coincide with the exact ones for contour arguments on the forward branch, as can readily be verified by following the same logic of Section 4.3. Using the (equal-time) adiabatic or zero-temperature Green's functions corresponds to using the adiabatic or zero-temperature formula for the time-dependent averages. The reader is strongly encouraged to pause a while and get convinced of this fact. A useful exercise is to write the explicit expression of $G_1(\mathbf{x}, z; \mathbf{x}', z^+)$ from both the first and second line of (5.1) with γ one of the three contours of Fig. 4.5 [and, of course, with the Hamiltonian $\hat{H}(z)$ as illustrated in these contours] and to show that they are the same under the corresponding assumptions. The special appealing feature of writing the Green's functions as in the second line of (5.1) is that this formula nicely embodies all possible cases (exact, adiabatic, and zero-temperature ones) once we assign the form of γ and the way the Hamiltonian changes along the contour.

Let us now come back to the differential equations (4.64) and (4.65). They have been derived without any assumption on the shape of the contour and without any assumption on the time-dependence of $\hat{h}(z)$ and $v(\mathbf{x}, \mathbf{x}', z)$ along the contour. Therefore, no matter how we define the Green's functions G_n (to be the exact, the adiabatic, or the zero-temperature), they satisfy the same kind of differential equations. To show it we simply multiply (4.64) and (4.65) by the appropriate density matrix $\hat{\rho}$ (equal to $e^{-\beta\hat{H}^{\rm M}}/Z$ in the exact case and equal to

[1] We recall that under the zero-temperature assumption $\hat{U}_\eta(-\infty, \infty)|\Phi_0\rangle = e^{-\mathrm{i}\alpha_0}|\Phi_0\rangle$, where $|\Phi_0\rangle$ is the ground state of $\hat{H}_0^{\rm M}$.

$e^{-\beta \hat{H}_0^{\mathrm{M}}}/Z_0$ in the adiabatic and zero-temperature cases), take the trace, and use definition (5.1) to obtain the following system of coupled differential equations:

$$\begin{aligned}\int d\bar{1}\, \overrightarrow{G}_0^{-1}(k;\bar{1})\, G_n(1,..,\bar{1},..,n;1',..,n') \\ = \pm \mathrm{i} \int d\bar{1}\, v(k;\bar{1})\, G_{n+1}(1,..,n,\bar{1};1',..,n',\bar{1}^+) \\ + \sum_{j=1}^{n} (\pm)^{k+j}\, \delta(k;j')\, G_{n-1}(1,..,\overset{\sqcap}{k},..,n;1',..,\overset{\sqcap}{j'},..,n')\end{aligned} \tag{5.2}$$

and

$$\begin{aligned}\int d\bar{1}\, G_n(1,..,n;1',..,\bar{1},..,n')\overleftarrow{G}_0^{-1}(\bar{1};k') \\ = \pm \mathrm{i} \int d\bar{1}\, v(k';\bar{1})\, G_{n+1}(1,..,n,\bar{1}^-;1',..,n',\bar{1}) \\ + \sum_{j=1}^{n} (\pm)^{k+j}\, \delta(j;k')\, G_{n-1}(1,..,\overset{\sqcap}{j},..,n;1',..,\overset{\sqcap}{k'},..,n')\end{aligned} \tag{5.3}$$

This set of equations is known as the *Martin–Schwinger hierarchy* [50] for the nonequilibrium Green's functions.[2] If we could solve it we would be able to calculate any time-dependent ensemble average of any observable that we want. For example, the average of the density operator and paramagnetic current density operator are given in terms of the *one-particle Green's function* G_1 as

$$n(\mathbf{x},z) = \frac{\mathrm{Tr}\left[e^{-\beta \hat{H}^{\mathrm{M}}}\hat{\psi}_H^\dagger(\mathbf{x},z)\hat{\psi}_H(\mathbf{x},z)\right]}{\mathrm{Tr}\left[e^{-\beta \hat{H}^{\mathrm{M}}}\right]} = \pm \mathrm{i} G_1(\mathbf{x},z;\mathbf{x},z^+) \tag{5.4}$$

and [see (3.27)]

$$\begin{aligned}\mathbf{j}(\mathbf{x},z) &= \frac{1}{2m\mathrm{i}}\frac{\mathrm{Tr}\left[e^{-\beta \hat{H}^{\mathrm{M}}}\left(\hat{\psi}_H^\dagger(\mathbf{x},z)(\boldsymbol{\nabla}\hat{\psi}_H(\mathbf{x},z)) - (\boldsymbol{\nabla}\hat{\psi}_H^\dagger(\mathbf{x},z))\hat{\psi}_H(\mathbf{x},z)\right)\right]}{\mathrm{Tr}\left[e^{-\beta \hat{H}^{\mathrm{M}}}\right]} \\ &= \pm\left(\frac{\boldsymbol{\nabla}-\boldsymbol{\nabla}'}{2m}G(\mathbf{x},z;\mathbf{x}',z^+)\right)_{\mathbf{x}'=\mathbf{x}},\end{aligned} \tag{5.5}$$

while, for instance, the interaction energy is given in terms of the *two-particle Green's*

[2] By definition $G_n = 1$ for $n = 0$.

function as[3]

$$E_{\text{int}}(z) = \frac{1}{2}\int d\mathbf{x}d\mathbf{x}'\, v(\mathbf{x},\mathbf{x}';z)\frac{\text{Tr}\left[e^{-\beta\hat{H}^{\text{M}}}\hat{\psi}^{\dagger}_{H}(\mathbf{x},z)\hat{\psi}^{\dagger}_{H}(\mathbf{x}',z)\hat{\psi}_{H}(\mathbf{x}',z)\hat{\psi}_{H}(\mathbf{x},z)\right]}{\text{Tr}\left[e^{-\beta\hat{H}^{\text{M}}}\right]}$$

$$= -\frac{1}{2}\int d\mathbf{x}d\mathbf{x}'\, v(\mathbf{x},\mathbf{x}';z)G_2(\mathbf{x}',z,\mathbf{x},z;\mathbf{x}',z^+,\mathbf{x},z^+). \quad (5.6)$$

Choosing z on the vertical track, we have the initial value of these observable quantities while for $z = t_{\pm}$ on the horizontal branches we have their ensemble average at time t. Moreover, according to the previous discussion, we can state that:

The exact, adiabatic, and zero-temperature formulas correspond to using the G_n which solve the Martin–Schwinger hierarchy on the contours of Fig. 4.5 with a single-particle Hamiltonian h and interaction v as specified in the same figure.

Kubo–Martin–Schwinger boundary conditions In all cases the task is therefore to solve the hierarchy. As with any set of differential equations, their solution is unique provided we pose appropriate spatial and temporal boundary conditions. These boundary conditions obviously depend on the physical problem at hand. The spatial boundary conditions on the Green's functions are determined directly by the corresponding spatial boundary conditions on the many-body states of the system. For instance, if we describe a finite system such as an isolated molecule, we know that the many-body wavefunctions vanish at spatial infinity and hence we require the same for the Green's functions. For an infinite periodic crystal the many-body wavefunctions are invariant (modulo a phase factor) under translations over a lattice vector. For this case we thus demand that the Green's functions obey the same lattice-periodic symmetry in each of their spatial coordinates, since the physics of adding or removing a particle cannot depend on the particular unit cell where we add or remove the particle, see Appendix C. Since the differential equations are first order in the time derivatives, we also need one condition per time-argument for each Green's function G_n. From definition (5.1) it follows that (we derive these relations below)

$$\begin{aligned} G_n(1,\ldots,\mathbf{x}_k,z_{\text{i}},\ldots,n;1',\ldots,n') &= \pm G_n(1,\ldots,\mathbf{x}_k,z_{\text{f}},\ldots,n;1',\ldots,n') \\ G_n(1,\ldots,n;1',\ldots,\mathbf{x}'_k,z_{\text{i}},\ldots,n') &= \pm G_n(1,\ldots,n;1',\ldots,\mathbf{x}'_k,z_{\text{f}},\ldots,n') \end{aligned} \quad (5.7)$$

with sign $+/-$ for the bosonic/fermionic case. The Green's functions are therefore (anti)periodic along the contour γ. The boundary conditions (5.7) are known as the *Kubo–Martin–Schwinger* (KMS) *relations* [50, 51]. To derive these relations we consider only the numerator of (5.1). Let us insert the value z_{f} in the kth contour argument. Since z_{f} is the latest time on the contour we have, using the cyclic property of the trace,[4] the set of identities illustrated

[3]We remind the reader that under the $\mathcal{T}$ sign the order of operators with the same contour argument is preserved, see also examples in Section 4.5.

[4]Since in the bosonic case $[\psi(\mathbf{x}),\psi^{\dagger}(\mathbf{x}')]_- = \delta(\mathbf{x}-\mathbf{x}')$, it seems dubious to use the property $\text{Tr}[\hat{A}\hat{B}] = \text{Tr}[\hat{B}\hat{A}]$ for field operators. However, under the trace the field operators are always multiplied by $e^{-\beta\hat{H}^{\text{M}}}$ and the use of the cyclic property is allowed. For a mathematical discussion, see Ref. [52].

$$\mathrm{Tr}\left[\mathcal{T}\left\{e^{-\mathrm{i}\int_\gamma d\bar{z}\hat{H}(\bar{z})}\hat{\psi}(1)\ldots\hat{\psi}(k-1)\hat{\psi}(\mathbf{x}_k,z_\mathrm{f})\hat{\psi}(k+1)\ldots\hat{\psi}^\dagger(1')\right\}\right]$$

$$=(\pm)^{k-1}\mathrm{Tr}\left[\hat{\psi}(\mathbf{x}_k)\mathcal{T}\left\{e^{-\mathrm{i}\int_\gamma d\bar{z}\hat{H}(\bar{z})}\hat{\psi}(1)\ldots\hat{\psi}(k-1)\hat{\psi}(k+1)\ldots\hat{\psi}^\dagger(1')\right\}\right]$$

$$=(\pm)^{k-1}\mathrm{Tr}\left[\mathcal{T}\left\{e^{-\mathrm{i}\int_\gamma d\bar{z}\hat{H}(\bar{z})}\hat{\psi}(1)\ldots\hat{\psi}(k-1)\hat{\psi}(k+1)\ldots\hat{\psi}^\dagger(1')\right\}\hat{\psi}(\mathbf{x}_k)\right]$$

$$=(\pm)^{k-1}\mathrm{Tr}\left[\mathcal{T}\left\{e^{-\mathrm{i}\int_\gamma d\bar{z}\hat{H}(\bar{z})}\hat{\psi}(1)\ldots\hat{\psi}(k-1)\hat{\psi}(k+1)\ldots\hat{\psi}^\dagger(1')\hat{\psi}(\mathbf{x}_k,z_\mathrm{i})\right\}\right]$$

$$=(\pm)^{k-1}(\pm)^{2n-k}\mathrm{Tr}\left[\mathcal{T}\left\{e^{-\mathrm{i}\int_\gamma d\bar{z}\hat{H}(\bar{z})}\hat{\psi}(1)\ldots\hat{\psi}(k-1)\hat{\psi}(\mathbf{x}_k,z_\mathrm{i})\hat{\psi}(k+1)\ldots\hat{\psi}^\dagger(1')\right\}\right]$$

Figure 5.1 Proof of the KMS relations.

in Fig. 5.1. In the figure the arrows indicate how the operator $\hat{\psi}(\mathbf{x}_k)$ moves from one step to the next. The final result is that we have replaced the contour-argument z_f by z_i and gained a sign $(\pm)^{k-1}(\pm)^{2n-k}=\pm1$. One can similarly derive the second KMS relation for a time-argument z'_k. As we see in Section 5.3, these boundary conditions are sufficient to determine a unique solution for the differential equations (5.2) and (5.3), provided that the boundary points belong to the same connected contour on which the differential equations are solved. It is indeed very important that the Martin–Schwinger hierarchy is valid on the vertical track and that the vertical track is connected to the horizontal branches. If the vertical track was not attached, then there would be a unique solution on the vertical track, but there would be no unique solution on the horizontal branches, see Exercise 5.1. This provides a posteriori yet another and more fundamental reason to attach the vertical track. In Section 5.3 we give an exact formal solution of the hierarchy for the one- and two-particle Green's function as an infinite expansion in powers of the interparticle interaction v.

Exercise 5.1 In order to illustrate the importance of the KMS boundary conditions and of connected domains when solving the Martin–Schwinger hierarchy, consider the differential equation

$$\mathrm{i}\frac{df(z)}{dz}-\epsilon f(z)=g(z),$$

valid for $z\in(0,-\mathrm{i}\beta)$ and $z\in(1,\infty)$ and with boundary condition $f(0)=\pm f(-\mathrm{i}\beta)$. This equation has the same structure of the Martin–Schwinger hierarchy but the real domain $(1,\infty)$ is disconnected from the imaginary domain $(0,-\mathrm{i}\beta)$. Show that for any complex ϵ with a nonvanishing real part the solution is unique in $(0,-\mathrm{i}\beta)$ but not in $(1,\infty)$. Are there imaginary values of ϵ for which the solution is not unique in $(0,-\mathrm{i}\beta)$?

5.2 Wick's Theorem

The main difficulty in solving the hierarchy is due to the coupling of the equation for G_n to those for G_{n+1} and G_{n-1}. However, we note that in the noninteracting case, $v = 0$, we have only a coupling to lower-order Green's functions given by the equations

$$\int d\bar{1}\, \overrightarrow{G}_0^{-1}(k;\bar{1})\, G_n(1,\ldots,\bar{1},\ldots,n;1',\ldots,n') \\ = \sum_{j=1}^{n} (\pm)^{k+j}\, \delta(k;j')\, G_{n-1}(1,\ldots \overset{\sqcap}{k} \ldots,n;1',\ldots \overset{\sqcap}{j'} \ldots,n'), \tag{5.8}$$

$$\int d\bar{1}\, G_n(1,\ldots,n;1',\ldots,\bar{1},\ldots,n')\overleftarrow{G}_0^{-1}(\bar{1};k') \\ = \sum_{j=1}^{n} (\pm)^{k+j}\, \delta(j;k')\, G_{n-1}(1,\ldots \overset{\sqcap}{j} \ldots,n;1',\ldots \overset{\sqcap}{k'} \ldots,n'). \tag{5.9}$$

In this case the hierarchy can be solved exactly. Let us denote by $G_{0,n}$ the solution of (5.8) and (5.9). We first observe that the noninteracting one-particle Green's function $G_0 \equiv G_{0,1}$ satisfies the first equations of the hierarchy:

$$\int d\bar{1}\, \overrightarrow{G}_0^{-1}(1;\bar{1})\, G_0(\bar{1};1') = \delta(1;1'), \tag{5.10a}$$

$$\int d\bar{1}\, G_0(1;\bar{1})\, \overleftarrow{G}_0^{-1}(\bar{1};1') = \delta(1;1'). \tag{5.10b}$$

These equations are decoupled from the others and can be used to determine G_0 uniquely. We now show that if G_0 satisfies (5.10) then the solution of the hierarchy is simply [49, 50]

$$\boxed{G_{0,n}(1,\ldots,n;1',\ldots,n') = \begin{vmatrix} G_0(1;1') & \ldots & G_0(1;n') \\ \vdots & & \vdots \\ G_0(n;1') & \ldots & G_0(n;n') \end{vmatrix}_{\pm}} \tag{5.11}$$

As in Section 1.5, the quantity $|A|_\pm$, where A is a $n \times n$ matrix with elements A_{ij}, is the permanent/determinant for bosons/fermions:

$$|A|_\pm \equiv \sum_P (\pm)^P A_{1P(1)} \ldots A_{nP(n)}.$$

In (5.11) matrix A has elements $A_{ij} = G_0(i;j')$.

Let us expand the permanent/determinant in (5.11) along row k (see Appendix B):

$$G_{0,n}(1,\ldots,n;1',\ldots,n') = \sum_{j=1}^{n} (\pm)^{k+j} G_0(k;j')\, G_{0,n-1}(1,\ldots \overset{\sqcap}{k} \ldots n;1',\ldots \overset{\sqcap}{j'} \ldots,n').$$

Setting $k = \bar{1}$ and acting from the left with $\int d\bar{1} \overrightarrow{G}_0^{-1}(k;\bar{1})$, we see immediately that this $G_{0,n}$ satisfies (5.8) since the first equation of the hierarchy tells us that G_0 satisfies (5.10a).

On the other hand, expanding along column k we find

$$G_{0,n}(1,\ldots,n;1',\ldots,n') = \sum_{j=1}^{n} (\pm)^{k+j} G_0(j;k')\, G_{0,n-1}(1,\ldots \overset{\sqcap}{j} \ldots,n;1',\ldots \overset{\sqcap}{k'} \ldots,n'),$$

which can similarly be seen to satisfy the Martin–Schwinger hierarchy equations (5.9) using (5.10b). Therefore we conclude that (5.11) is a solution to the noninteracting Martin–Schwinger hierarchy. However, since this is just a particular solution of the set of differential equations, we still need to check that it satisfies the KMS boundary conditions (5.7). This is readily done. From (5.11), $G_{0,n}$ satisfies these conditions whenever G_0 satisfies them as multiplying G_0 by a ± 1 in a row or a column the permanent/determinant is multiplied by the same factor. It is therefore sufficient to solve (5.10a) and (5.10b) with KMS boundary conditions to obtain all $G_{0,n}$!

The result (5.11) is also known as *Wick's theorem* [53], and it is exceedingly useful in deriving diagrammatic perturbation theory. It seems somewhat of an exaggeration to give the simple statement summarized in (5.11) the status of a theorem. Wick's theorem, however, is usually derived in textbooks in a different way that requires much more effort and a separate discussion, depending on whether we are working with the exact Green's function in equilibrium (also known as finite temperature Matsubara formalism) or in nonequilibrium or with the adiabatic Green's function or with the zero-temperature Green's function. The above derivation, apart from being shorter, is valid in all cases and highlights the physical content of Wick's theorem as a solution to a boundary value problem for the Martin–Schwinger hierarchy (or equations of motion for the nonequilibrium Green's functions). To recover the Wick theorem of preference it is enough to assign the corresponding contour and the Hamiltonian along the contour. We comment more on how to recover the various formalisms in Section 5.4.

5.3 Exact Solution of the Hierarchy from Wick's Theorem

Wick's theorem suggests that we can calculate the interacting n-particle Green's function by a *brute force* expansion in powers of the interaction. From definition (5.1) it follows that (omitting the arguments of the Green's function)

$$G_n = \frac{1}{\mathrm{i}^n} \frac{\mathrm{Tr}\left[\mathcal{T}\left\{ e^{-\mathrm{i}\int_\gamma d\bar{z}\hat{H}_0(\bar{z})} e^{-\mathrm{i}\int_\gamma d\bar{z}\hat{H}_{\mathrm{int}}(\bar{z})} \hat{\psi}(1)\ldots\hat{\psi}^\dagger(1')\right\}\right]}{\mathrm{Tr}\left[\mathcal{T}\left\{e^{-\mathrm{i}\int_\gamma d\bar{z}\hat{H}_0(\bar{z})} e^{-\mathrm{i}\int_\gamma d\bar{z}\hat{H}_{\mathrm{int}}(\bar{z})}\right\}\right]},$$

where we split the Hamiltonian into a one-body part $\hat{H}_0$ and a two-body part $\hat{H}_{\mathrm{int}}$ and we use the property that $\hat{H}_0$ and $\hat{H}_{\mathrm{int}}$ commute under the $\mathcal{T}$ sign. Expanding in powers of $\hat{H}_{\mathrm{int}}$ we get

$$G_n = \frac{1}{\mathrm{i}^n} \frac{\sum_{k=0}^{\infty} \frac{(-\mathrm{i})^k}{k!} \int_\gamma d\bar{z}_1 \ldots d\bar{z}_k \, \langle \mathcal{T}\left\{\hat{H}_{\mathrm{int}}(\bar{z}_1)\ldots\hat{H}_{\mathrm{int}}(\bar{z}_k)\hat{\psi}(1)\ldots\hat{\psi}^\dagger(1')\right\}\rangle_0}{\sum_{k=0}^{\infty} \frac{(-\mathrm{i})^k}{k!} \int_\gamma d\bar{z}_1 \ldots d\bar{z}_k \, \langle \mathcal{T}\left\{\hat{H}_{\mathrm{int}}(\bar{z}_1)\ldots\hat{H}_{\mathrm{int}}(\bar{z}_k)\right\}\rangle_0}, \tag{5.12}$$

where we introduce the short-hand notation

$$\mathrm{Tr}\left[\mathcal{T}\left\{e^{-\mathrm{i}\int_\gamma d\bar{z}\hat{H}_0(\bar{z})}\ldots\right\}\right] = \langle\mathcal{T}\{\ldots\}\rangle_0,$$

according to which any string of operators can be inserted for the dots. We now recall that $\hat{H}_{\mathrm{int}}(z)$ is a two-body operator both on the horizontal branches and on the vertical track of the contour, see Section 4.4. Therefore we can write

$$\hat{H}_{\mathrm{int}}(z) = \frac{1}{2}\int dz'\int d\mathbf{x}d\mathbf{x}' v(\mathbf{x},z;\mathbf{x}',z')\hat{\psi}^\dagger(\mathbf{x},z^+)\hat{\psi}^\dagger(\mathbf{x}',z'^+)\hat{\psi}(\mathbf{x}',z')\hat{\psi}(\mathbf{x},z), \tag{5.13}$$

where

$$v(\mathbf{x},z;\mathbf{x}',z') = \delta(z,z')\left\{\begin{array}{ll} v(\mathbf{x},\mathbf{x}',t) & \text{if } z = t_\pm \text{ is on the horizontal branches of } \gamma \\ v^{\mathrm{M}}(\mathbf{x},\mathbf{x}') & \text{if } z \text{ is on the vertical track of } \gamma \end{array}\right. .$$

The reason for shifting the contour arguments of the $\hat{\psi}^\dagger$ operators from z, z' to z^+, z'^+ in (5.13) stems from the possibility of moving the field operators freely under the $\mathcal{T}$ sign without losing the information that after the reordering the $\hat{\psi}^\dagger$ operators must be placed to the left of the $\hat{\psi}$ operators, see also below. Taking into account (5.13) and reordering the field operators, (5.12) provides an expansion of G_n in terms of noninteracting Green's functions $G_{0,m}$. Below we work out explicitly the case of the one- and two-particle Green's functions $G \equiv G_1$ and G_2, which are the ones most commonly used.

One-particle Green's function Let us start with the one-particle Green's function. For notational convenience we denote by $a = (\mathbf{x}_a, z_a)$ and $b = (\mathbf{x}_b, z_b)$ and rename all time-integration variables $\bar{z}_k$ to z_k. Then (5.12) for $n = 1$ yields

$$G(a;b) = \frac{1}{\mathrm{i}}\frac{\sum\limits_{k=0}^{\infty}\frac{(-\mathrm{i})^k}{k!}\int_\gamma dz_1\ldots dz_k\,\langle\mathcal{T}\left\{\hat{H}_{\mathrm{int}}(z_1)\ldots\hat{H}_{\mathrm{int}}(z_k)\hat{\psi}(a)\hat{\psi}^\dagger(b)\right\}\rangle_0}{\sum\limits_{k=0}^{\infty}\frac{(-\mathrm{i})^k}{k!}\int_\gamma dz_1\ldots dz_k\,\langle\mathcal{T}\left\{\hat{H}_{\mathrm{int}}(z_1)\ldots\hat{H}_{\mathrm{int}}(z_k)\right\}\rangle_0}. \tag{5.14}$$

Using the explicit form (5.13) of the interaction Hamiltonian, the numerator of this equation can be rewritten as

$$\sum_{k=0}^{\infty}\frac{1}{k!}\left(-\frac{\mathrm{i}}{2}\right)^k\int d1\ldots dk\,d1'\ldots dk' v(1;1')\ldots v(k;k')$$
$$\times\langle\mathcal{T}\left\{\hat{\psi}^\dagger(1^+)\hat{\psi}^\dagger(1'^+)\hat{\psi}(1')\hat{\psi}(1)\ldots\hat{\psi}^\dagger(k^+)\hat{\psi}^\dagger(k'^+)\hat{\psi}(k')\hat{\psi}(k)\hat{\psi}(a)\hat{\psi}^\dagger(b)\right\}\rangle_0.$$

We reorder the quantity in the bracket as follows:

$$\langle\mathcal{T}\left\{\hat{\psi}(a)\hat{\psi}(1)\hat{\psi}(1')\ldots\hat{\psi}(k)\hat{\psi}(k')\hat{\psi}^\dagger(k'^+)\hat{\psi}^\dagger(k^+)\ldots\hat{\psi}^\dagger(1'^+)\hat{\psi}^\dagger(1^+)\hat{\psi}^\dagger(b)\right\}\rangle_0,$$

which requires an even number of interchanges so there is no sign change. In this expression we recognize the noninteracting $(2k+1)$-particle Green's function

$$G_{0,2k+1}(a,1,1',\ldots,k,k';b,1^+,1'^+,\ldots k^+,k'^+)$$

multiplied by $\mathrm{i}^{2k+1}Z_0$.[5] Similarly, the kth-order bracket of the denominator can be written as the noninteracting $2k$-particle Green's function

$$G_{0,2k}(1,1',\ldots,k,k';1^+,1'^+,\ldots k^+,k'^+)$$

multiplied by $\mathrm{i}^{2k}Z_0$. We therefore find that (5.14) is equivalent to

$$G(a;b)=\frac{\sum_{k=0}^{\infty}\frac{1}{k!}\left(\frac{\mathrm{i}}{2}\right)^k\int v(1;1')\ldots v(k;k')G_{0,2k+1}(a,1,1',\ldots;b,1^+,1'^+,\ldots)}{\sum_{k=0}^{\infty}\frac{1}{k!}\left(\frac{\mathrm{i}}{2}\right)^k\int v(1;1')\ldots v(k;k')G_{0,2k}(1,1',\ldots;1^+,1'^+,\ldots)},\tag{5.15}$$

where the integrals are over $1,1',\ldots,k,k'$. Next we observe that the $G_{0,n}$ can be decomposed in products of noninteracting one-particle Green's functions G_0 using the Wick's theorem (5.11). We thus arrive at the following important formula:

$$\boxed{G(a;b)=\frac{\sum_{k=0}^{\infty}\frac{1}{k!}\left(\frac{\mathrm{i}}{2}\right)^k\int v(1;1')\ldots v(k;k')\begin{vmatrix}G_0(a;b) & G_0(a;1^+) & \ldots & G_0(a;k'^+)\\ G_0(1;b) & G_0(1;1^+) & \ldots & G_0(1;k'^+)\\ \vdots & \vdots & \ddots & \vdots\\ G_0(k';b) & G_0(k';1^+) & \ldots & G_0(k';k'^+)\end{vmatrix}_{\pm}}{\sum_{k=0}^{\infty}\frac{1}{k!}\left(\frac{\mathrm{i}}{2}\right)^k\int v(1;1')\ldots v(k;k')\begin{vmatrix}G_0(1;1^+) & G_0(1;1'^+) & \ldots & G_0(1;k'^+)\\ G_0(1';1^+) & G_0(1';1'^+) & \ldots & G_0(1';k'^+)\\ \vdots & \vdots & \ddots & \vdots\\ G_0(k';1^+) & G_0(k';1'^+) & \ldots & G_0(k';k'^+)\end{vmatrix}_{\pm}}}\tag{5.16}$$

which is an exact expansion of the interacting G in terms of the noninteracting G_0.

Two-particle Green's function We can derive the expansion for the two-particle Green's function in a similar way. We have

$$G_2(a,b;c,d)=\frac{1}{\mathrm{i}^2}\frac{\sum_{k=0}^{\infty}\frac{(-\mathrm{i})^k}{k!}\int_\gamma dz_1\ldots dz_k\langle\mathcal{T}\left\{\hat{H}_{\rm int}(z_1)\ldots\hat{H}_{\rm int}(z_k)\hat{\psi}(a)\hat{\psi}(b)\hat{\psi}^\dagger(d)\hat{\psi}^\dagger(c)\right\}\rangle_0}{\sum_{k=0}^{\infty}\frac{(-\mathrm{i})^k}{k!}\int_\gamma dz_1\ldots dz_k\,\langle\mathcal{T}\left\{\hat{H}_{\rm int}(z_1)\ldots\hat{H}_{\rm int}(z_k)\right\}\rangle_0}.\tag{5.17}$$

Using again the explicit form (5.13) of the interaction Hamiltonian, the numerator of (5.17) can be rewritten as

$$\sum_{k=0}^{\infty}\frac{1}{k!}\left(-\frac{\mathrm{i}}{2}\right)^2\int d1\ldots dkd1'\ldots dk'v(1;1')\ldots v(k;k')$$
$$\times\langle\mathcal{T}\left\{\hat{\psi}^\dagger(1^+)\hat{\psi}^\dagger(1'^+)\hat{\psi}(1')\hat{\psi}(1)\ldots\hat{\psi}^\dagger(k^+)\hat{\psi}^\dagger(k'^+)\hat{\psi}(k')\hat{\psi}(k)\right.$$
$$\times\left.\hat{\psi}(a)\hat{\psi}(b)\hat{\psi}^\dagger(d)\hat{\psi}^\dagger(c)\right\}\rangle_0.$$

[5]We remind the reader that $Z_0=e^{-\beta\hat{H}_0^{\rm M}}=\mathrm{Tr}[\mathcal{T}\{e^{-\mathrm{i}\int_\gamma d\bar{z}\hat{H}_0(\bar{z})}\}]$ is the noninteracting partition function.

Reordering the quantity in the bracket as

$$\langle \mathcal{T}\left\{\hat{\psi}(a)\hat{\psi}(b)\hat{\psi}(1)\hat{\psi}(1')\ldots\hat{\psi}(k)\hat{\psi}(k')\hat{\psi}^\dagger(k'^+)\hat{\psi}^\dagger(k^+)\ldots\hat{\psi}^\dagger(1'^+)\hat{\psi}^\dagger(1^+)\hat{\psi}^\dagger(d)\hat{\psi}^\dagger(c)\right\}\rangle_0$$

requires an even number of interchanges and we recognize the noninteracting $(2k+2)$-particle Green's function

$$G_{0,2k+2}(a,b,1,1'\ldots,k,k';c,d,1^+,1'^+,\ldots,k^+,k'^+)$$

multiplied by $\mathrm{i}^{2k+2}Z_0$. The denominator in (5.17) is the same as that of the one-particle Green's function in (5.14) and therefore G_2 becomes

$$G_2(a,b;c,d)=\frac{\sum\limits_{k=0}^{\infty}\frac{1}{k!}\left(\frac{\mathrm{i}}{2}\right)^k\int v(1;1')\ldots v(k;k')G_{0,2k+2}(a,b,1,1'\ldots;c,d,1^+,1'^+,\ldots)}{\sum\limits_{k=0}^{\infty}\frac{1}{k!}\left(\frac{\mathrm{i}}{2}\right)^k\int v(1;1')\ldots v(k;k')G_{0,2k}(1,1',\ldots;1^+,1'^+,\ldots)},$$

where the integrals are over $1,1',\ldots,k,k'$. Using Wick's theorem, we can now transform G_2 into an exact perturbative expansion in terms of the noninteracting Green's function G_0:

$$G_2(a,b;c,d)=\frac{\sum\limits_{k=0}^{\infty}\frac{1}{k!}\left(\frac{\mathrm{i}}{2}\right)^k\int v(1;1')\ldots v(k;k')\begin{vmatrix}G_0(a;c) & G_0(a;d) & \ldots & G_0(a;k'^+)\\ G_0(b;c) & G_0(b;d) & \ldots & G_0(b;k'^+)\\ \vdots & \vdots & \ddots & \vdots\\ G_0(k';c) & G_0(k';d) & \ldots & G_0(k';k'^+)\end{vmatrix}_\pm}{\sum\limits_{k=0}^{\infty}\frac{1}{k!}\left(\frac{\mathrm{i}}{2}\right)^k\int v(1;1')\ldots v(k;k')\begin{vmatrix}G_0(1;1^+) & G_0(1;1'^+) & \ldots & G_0(1;k'^+)\\ G_0(1';1^+) & G_0(1';1'^+) & \ldots & G_0(1';k'^+)\\ \vdots & \vdots & \ddots & \vdots\\ G_0(k';1^+) & G_0(k';1'^+) & \ldots & G_0(k';k'^+)\end{vmatrix}_\pm} \tag{5.18}$$

Partition function In a similar way, the reader can work out the equations for the higher-order Green's functions as well as for the partition function Z, which we write below:[6]

$$\frac{Z}{Z_0}=\sum_{k=0}^{\infty}\frac{1}{k!}\left(\frac{\mathrm{i}}{2}\right)^k\int v(1;1')\ldots v(k;k')\begin{vmatrix}G_0(1;1^+) & G_0(1;1'^+) & \ldots & G_0(1;k'^+)\\ G_0(1';1^+) & G_0(1';1'^+) & \ldots & G_0(1';k'^+)\\ \vdots & \vdots & \ddots & \vdots\\ G_0(k';1^+) & G_0(k';1'^+) & \ldots & G_0(k';k'^+)\end{vmatrix}_\pm \tag{5.19}$$

[6]The ratio Z/Z_0 is simply the denominator of the formulas (5.16) and (5.18) for G and G_2.

Many-body perturbation theory for the NEGF is now completely defined. In Section 6.2 we learn how to calculate the noninteracting Green's function G_0 for any physical system and we show that it is as easy as solving a one-particle problem in quantum mechanics. The evaluation of (5.16), (5.18), or (5.19) is therefore a well-defined mathematical problem and there exist many useful tricks to carry on the calculations in an efficient way. In his memorable talk at the Pocono Manor Inn in 1948, Feynman showed how to represent the cumbersome MBPT expansion in terms of physical insightful diagrams [54], and since then the Feynman diagrams have been an invaluable tool in many areas of physics. The Feynman diagrams give us a deep insight into the microscopic scattering processes, help us to evaluate (5.16), (5.18), or (5.19) more efficiently, and also provide us with techniques to resum certain classes of diagrams to infinite order in the interaction strength. These resummations are particularly important in bulk systems, where diagrams can be divergent (such as in the case of the long-range Coulomb interaction), as well as in finite systems. The expansion in powers of the interaction is indeed an *asymptotic* expansion, meaning that the inclusion of diagrams beyond a certain order brings the sum further and further away from the exact result - see, for example, Ref. [55]. For a brief introduction to asymptotic expansions, see Appendix F.

5.4 Finite and Zero-Temperature Formalism from the Exact Solution

In this section we discuss how different versions of MBPT that are commonly used follow as special cases of the perturbation theory developed in the previous section. Examples of these formalisms are the finite-temperature Matsubara formalism, the zero-temperature formalism, the Keldysh formalism, and the Konstantinov–Perel' formalism. In most textbooks these different flavors of MBPT appear as disconnected topics even though the perturbative terms have the same mathematical structure. This is not a coincidence, but has its origin in the Martin–Schwinger hierarchy as defined on the contour. We now demonstrate this fact by discussing these formalisms one by one.

Konstantinov–Perel' formalism In the Konstantinov–Perel' formalism the Green's functions satisfy the Martin–Schwinger hierarchy on the contour of Fig. 4.5(a) and, therefore, they are given by (5.16) and (5.18), where the z-integrals run on the same contour. The averages calculated from these Green's functions at equal time (i.e., $z_i = z$ and $z_i' = z^+$ for all i) correspond to the exact formula (4.21). We remind the reader that this formula yields the initial ensemble average for z on the vertical track and the time-dependent ensemble average for z on the horizontal branches. It is therefore clear that if we are interested in calculating time-dependent ensemble averages up to a maximum time T, we only need the Green's functions with real-time contour arguments up to T. *In this case it is sufficient to solve the Martin–Schwinger hierarchy over a shrunken contour like the one illustrated below:*

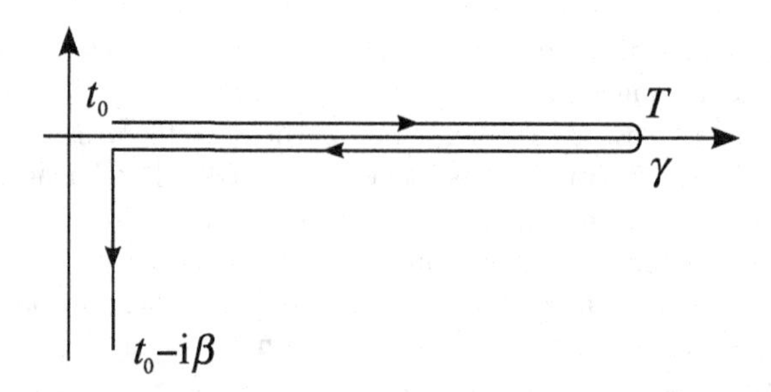

The solution is again given by (5.16) and (5.18), but the z-integrals run over the shrunken contour. This observation tells us that if the contour is longer than T, then the terms with integrals after T cancel off. The cancellation is analogous to the one discussed in Section 4.1, where we showed that nothing changes in (4.9) if the contour extends all the way to $+\infty$.

Matsubara formalism The Matsubara formalism is used to calculate the initial ensemble average (4.24) and it is typically applied to systems in equilibrium at finite temperature. For this reason the Matsubara formalism is sometimes called the "finite-temperature formalism." The crucial observation here is that in order to calculate the initial averages, we only need the Green's function with times on the vertical track. Then, according to the previous discussion, we can take the time $T = t_0$ and shrink the horizontal branches to a point leaving only the vertical track. The corresponding Matsubara Green's functions are given by the solution of the Martin–Schwinger hierarchy on γ^{M}. The Matsubara formalism therefore consists of expanding the Green's functions as in (5.16) and (5.18) with the z-integrals restricted to the vertical track. It is important to realize that no assumptions, like the adiabatic or the zero-temperature assumption, are made in this formalism. The Matsubara formalism is exact but limited to initial (or equilibrium) averages. Notice that not all equilibrium properties can be extracted from the Matsubara Green's functions. As we shall see, quantities like photo-emission currents, hyper-polarizabilities, and more generally high-order response properties require knowledge of equilibrium Green's functions with *real*-time contour arguments.

Keldysh formalism The formalism originally used by Keldysh was based on the adiabatic assumption according to which the interacting density matrix $\hat{\rho}$ can be obtained from the noninteracting $\hat{\rho}_0$ by an adiabatic switch-on of the interaction.[7] Under this assumption we can calculate time-dependent ensemble averages from the adiabatic Green's functions – that is, from the solution of the Martin–Schwinger hierarchy on the contour γ_a of Fig. 4.5(b) with the adiabatic Hamiltonian shown in the same figure. These adiabatic Green's functions are again given by (5.16) and (5.18), but the z-integrals run over γ_a and the Hamiltonian is the adiabatic Hamiltonian. The important simplification entailed by the adiabatic assumption is that the interaction v is zero on the vertical track, see again Fig. 4.5(b). Consequently, in (5.16) and (5.18) we can restrict the z-integrals to the horizontal branches. Like the exact formalism, the adiabatic formalism can be used to deal with nonequilibrium situations in which the external perturbing fields are switched on after time t_0. In the special case of no external fields we can calculate interacting *equilibrium* Green's functions at any finite temperature with *real*-time contour arguments. Different to the exact and Matsubara formalisms,

[7]As already observed, this assumption is supported by the Gell-Mann–Low theorem [47].

however, the Green's functions with imaginary time contour arguments are noninteracting since on the vertical track we have $\hat{H}_0^{\mathrm{M}}$ and not $\hat{H}^{\mathrm{M}}$.

To summarize: *The adiabatic Green's functions can be expanded as in (5.16) or (5.18) where the z-integrals run over the contour below*:

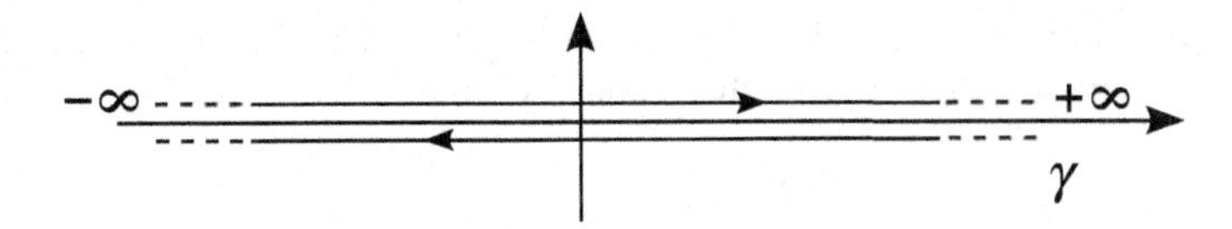

Zero-temperature formalism The zero-temperature formalism relies on the zero-temperature assumption, according to which the ground state of $\hat{H}_0^{\mathrm{M}}$ evolves into the ground state of $\hat{H}^{\mathrm{M}}$ by adiabatically switching on the interaction in the remote past or in the distant future. As we already discussed, this assumption makes sense only *in the absence of external fields.* The corresponding zero-temperature Green's functions are the solution of the Martin–Schwinger hierarchy on the contour γ_0 of Fig. 4.5(c), where the Hamiltonian changes along the contour as shown in the same figure. Like in the adiabatic case, the interaction v of the zero-temperature Hamiltonian vanishes along the vertical track and hence the z-integrals in the expansions (5.16) and (5.18) can be restricted to a contour that goes from $-\infty$ to ∞. The contour ordering on the horizontal branch of γ_0 is the same as the standard time ordering. For this reason the zero-temperature Green's functions are also called time-ordered Green's functions. The zero-temperature formalism allows us to calculate interacting Green's functions in equilibrium at zero temperature with *real*-time contour arguments. It cannot, however, be used to study systems out of equilibrium and/or at finite temperature. In some cases, however, the zero-temperature formalism is used also at finite temperatures (finite β) as the finite temperature corrections are small.[8] This approximated formalism is sometimes referred to as the *real-time finite temperature formalism* [48]. It is worth remarking that in the real-time finite-temperature formalism (as in the Keldysh formalism) the temperature enters in (5.16) and (5.18) only through G_0, which satisfies the KMS relations. In the Konstantinov–Perel' formalism, on the other hand, the temperature enters through G_0 and through the contour integrals since the interaction is nonvanishing along the vertical track.

To summarize: *The zero-temperature Green's functions can be expanded as in (5.16) and (5.18) where the z-integrals run over the contour below:*

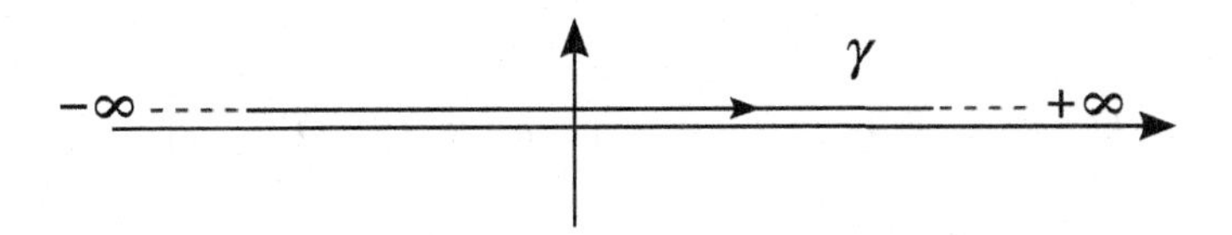

This contour is the same as the real-time axis and therefore the contour ordering is the

[8] At finite temperature the use of the contour γ_0 implies the assumption that *all* eigenstates of $\hat{H}_0^{\mathrm{M}}$ evolve into themselves by adiabatically switching on and then off the interaction – that is, the satisfaction of (4.30).

same as the time ordering. For this reason the Green's functions of the zero-temperature formalism are also called *time-ordered Green's functions.*

In conclusion, if one does not want to make the assumptions implicit in the last two formalisms, then there is no alternative but to use the full contour of Fig. 4.5(a). This is the contour that we use in the remainder of the book. However, all derivations in the book rely exclusively on the solution of the Martin–Schwinger hierarchy and therefore can be directly adapted to any formalism preferred by the reader by simply changing

$$\int_{\gamma_{\rm book}} \rightarrow \int_{\gamma_{\rm preferred}}$$

in every integral that you see with a contour symbol attached.

As a final remark, we point out that if we are only interested in the equal-time G_n then we can easily derive an exact hierarchy for these quantities starting from the Martin–Schwinger hierarchy. This is the so-called BBGKY hierarchy, and we defer the reader to Appendix G for its derivation and further discussion.

5.5 Langreth Rules

In the expansion of the interacting Green's functions, see (5.16) or (5.18), convolutions such as $\int d\bar{1} G_0(1,\bar{1})G_0(\bar{1},2)$ or products such as $G_0(1;2)G_0(2;1)$ appear. In order to evaluate these formal expressions we must convert contour integrals into standard real-time integrals and products of functions on the contour into products of functions with real-time arguments. The purpose of this section is to derive a set of identities to transform contour convolutions and contour products into convolutions and products that can be calculated analytically or implemented numerically. Some of these identities are known as the *analytic continuation rules*, or *Langreth rules* [56], while others, equally important, are often ignored.[9] We think that the name "analytic continuation rules" misleads the reader to think that these rules are applicable only to analytic functions of the complex time. The analyticity property is never used in the derivations below and, therefore, we prefer to use the name "generalized Langreth rules" or simply "Langreth rules." In this section we present a comprehensive and self-contained derivation of all of them. A table at the end of the section summarizes the main results. For those readers who are not already familiar with the formalism, we strongly recommend following this section with pencil and paper.

Keldysh space We specialize the discussion to two-point correlators (like the one-particle Green's function),

$$k(z,z') = \mathrm{Tr}\left[\hat{\rho}\,\hat{k}(z,z')\right] = \mathrm{Tr}\left[\hat{\rho}\,\mathcal{T}\left\{\hat{O}_1(z)\hat{O}_2(z')\right\}\right],$$

which are the ensemble average of the two-point operator correlators (4.47). Higher order correlators are more involved, but conceptually no more complicated and can be treated in a similar manner [57, 58]. As in Section 4.5 we do not specify the origin of the z-dependence of the operators; they could be operators with an explicit z-dependence or operators in the

[9]This is due to the fact that MBPT is often used within the Keldysh (adiabatic) formalism, see Section 5.4.

Heisenberg picture. The important thing is that $\hat{O}(t_+) = \hat{O}(t_-)$.[10] Unless otherwise stated, here and in the remainder of the book we always consider the contour of Fig. 4.5(a).

Due to the contour ordering, $k(z,z')$ has the following structure:

$$k(z,z') = \theta(z,z')k^>(z,z') + \theta(z',z)k^<(z,z'),$$

with $\theta(z,z')$ the Heaviside function on the contour and

$$k^>(z,z') = \mathrm{Tr}\left[\hat{\rho}\,\hat{O}_1(z)\hat{O}_2(z')\right], \qquad k^<(z,z') = \pm\mathrm{Tr}\left[\hat{\rho}\,\hat{O}_2(z')\hat{O}_1(z)\right]$$

where the $\pm$ sign in $k^<$ is for bosonic/fermionic operators $\hat{O}_1$ and $\hat{O}_2$. It is important to observe that the functions $k^>$ and $k^<$ are well defined for *all* z and z' on the contour. Furthermore, these functions have the important property that their value is independent of whether z, z' lie on the forward or backward branch:

$$\boxed{k^{\lessgtr}(t_+,z') = k^{\lessgtr}(t_-,z'), \qquad k^{\lessgtr}(z,t'_+) = k^{\lessgtr}(z,t'_-)} \tag{5.20}$$

since $\hat{O}_i(t_+) = \hat{O}_i(t_-)$ for both $i = 1,2$. We say that a function $k(z,z')$ belongs to the *Keldysh space* if it can be written as

$$k(z,z') = k^\delta(z)\delta(z,z') + \theta(z,z')k^>(z,z') + \theta(z',z)k^<(z,z'),$$

with $k^{\lessgtr}(z,z')$ satisfying the properties (5.20), $k^\delta(t_+) = k^\delta(t_-) \equiv k^\delta(t)$, and $\delta(z,z')$ the δ-function on the contour. We here observe that the δ-function on the contour is zero if z and z' lie on different branches, $\delta(t_\pm,t_\mp) = 0$, and that, due to the orientation of the contour, $\delta(t_-,t'_-) = \delta(t-t')$, whereas $\delta(t_+,t'_+) = -\delta(t-t')$, with $\delta(t-t')$ the δ-function on the real axis. The precise meaning of these identities is the following. From the definition of δ-function on the contour [i.e., $\int_\gamma dz'\delta(z,z')f(z') = f(z)$] and from the definition (4.4) of functions on the contour, we have for $z = t_-$:

$$f_-(t) = f(t_-) = \int_\gamma dz'\delta(t_-,z')f(z') = \int_{t_0}^{\infty} dt'\,\delta(t_-,t'_-)f(t'_-) = \int_{t_0}^{\infty} dt'\,\delta(t-t')f_-(t'),$$

whereas for $z = t_+$ we have:

$$f_+(t) = f(t_+) = \int_\gamma dz'\delta(t_+,z')f(z') = \int_{\infty}^{t_0} dt'\,\delta(t_+,t'_+)f(t'_+) = -\int_{t_0}^{\infty} dt'\,\delta(t-t')f_+(t').$$

Keldysh components In order to extract physical information from $k(z,z')$ we must evaluate this function for all possible positions of z and z' on the contour: both arguments on the horizontal branches, one argument on the vertical track and the other on the horizontal branches, or both arguments on the vertical track. We define the *greater* and *lesser*

[10] This property is, by definition, satisfied by the field operators $\hat{\psi}$ and $\hat{\psi}^\dagger$ (and hence by arbitrary products of field operators), as well as by the field operators $\hat{\psi}_H$ and $\hat{\psi}^\dagger_H$ in the contour Heisenberg picture. See also the discussion around (4.6) and (4.7).

Keldysh components as the following functions on the real-time axis:

$$\boxed{\begin{aligned} k^{>}(t,t') &\equiv k(t_+, t'_-) \\ k^{<}(t,t') &\equiv k(t_-, t'_+) \end{aligned}} \tag{5.21}$$

In other words, $k^{>}(t,t')$ is the real-time function with the values of the contour function $k^{>}(z,z')$.[||] Similarly, $k^{<}(t,t') = k^{<}(z,z')$. From these considerations we see that the *equal-time* greater and lesser functions can also be written as

$$k^{>}(t,t) = k(z^+, z) = k(z, z^-), \qquad k^{<}(t,t) = k(z, z^+) = k(z^-, z), \tag{5.22}$$

where z^+ (z^-) is a contour point infinitesimally later (earlier) than the contour point $z = t_\pm$ which can lie either on the forward or backward branch. We sometime use these alternative expressions to calculate time-dependent ensemble averages.

We also define the *left* and *right* Keldysh components from $k(z,z')$ with one real-time t and one imaginary-time $t_0 - \mathrm{i}\tau$:

$$\boxed{\begin{aligned} k^{\lceil}(\tau,t) &\equiv k(t_0 - \mathrm{i}\tau, t_\pm) \\ k^{\rceil}(t,\tau) &\equiv k(t_\pm, t_0 - \mathrm{i}\tau) \end{aligned}} \tag{5.23}$$

The names "left" and "right" refer to the position of the vertical segment of the hooks "$\lceil$" and "$\rceil$" with respect to the horizontal segment. In the definition of $k^{\lceil}$ and $k^{\rceil}$ we can arbitrarily choose t_+ or t_- since $t_0 - \mathrm{i}\tau$ is later than both of them and $k^{\lessgtr}(z,z')$ fulfills (5.20). In (5.23) the notation has been chosen to help the visualization of the contour arguments [59]. For instance, the symbol "$\rceil$" has a horizontal segment followed by a vertical one; accordingly, $k^{\rceil}$ has a first argument which is real (and thus lies on the horizontal axis) and a second argument which is imaginary (and lies on the vertical track). We can explain the use of the left symbol "$\lceil$" in a similar way. In the definition of the left and right functions we also introduced a convention of denoting with Latin letters the real times and with Greek letters the imaginary times. This convention is adopted throughout the book.

As we shall see, it is also useful to define the *Matsubara* component $k^{\mathrm{M}}(\tau,\tau')$ with both contour arguments on the vertical track:

$$\boxed{\begin{aligned} k^{\mathrm{M}}(\tau,\tau') &\equiv k(t_0 - \mathrm{i}\tau, t_0 - \mathrm{i}\tau') \\ &= \delta(t_0 - \mathrm{i}\tau, t_0 - \mathrm{i}\tau') k^{\delta}(t_0 - \mathrm{i}\tau) + k_r^{\mathrm{M}}(\tau,\tau') \end{aligned}} \tag{5.24}$$

where

$$\boxed{k_r^{\mathrm{M}}(\tau,\tau') = \theta(\tau - \tau') k^{>}(t_0 - \mathrm{i}\tau, t_0 - \mathrm{i}\tau') + \theta(\tau' - \tau) k^{<}(t_0 - \mathrm{i}\tau, t_0 - \mathrm{i}\tau')}$$

[||]As we have already emphasized, $k^{>}(z,z')$ is independent of whether z and z' are on the forward or backward branch.

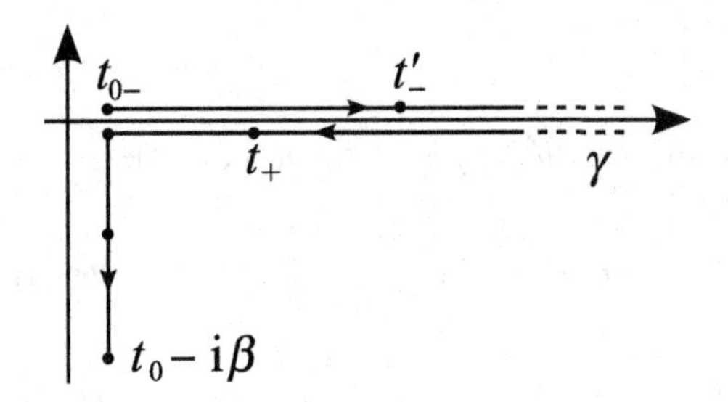

Figure 5.2 Contour with t on the backward branch and t' on the forward branch.

is the regular part of the function.[12] To convert the δ-function on the vertical track into a standard δ-function we use again the definition

$$f(t_0 - \mathrm{i}\tau) = \int_\gamma dz' \delta(t_0 - \mathrm{i}\tau, z') f(z') \qquad \text{(setting } z' = t_0 - \mathrm{i}\tau'\text{)}$$
$$= -\mathrm{i} \int_0^\beta d\tau' \delta(t_0 - \mathrm{i}\tau, t_0 - \mathrm{i}\tau') f(t_0 - \mathrm{i}\tau'),$$

from which it follows that $\delta(t_0 - \mathrm{i}\tau, t_0 - \mathrm{i}\tau') = \mathrm{i}\delta(\tau - \tau')$. Thus, if we introduce the short-hand notation $k^\delta(\tau) \equiv k^\delta(t_0 - \mathrm{i}\tau)$, we can rewrite the Matsubara component as

$$k^{\mathrm{M}}(\tau, \tau') = \mathrm{i}\delta(\tau - \tau') k^\delta(\tau) + k_r^{\mathrm{M}}(\tau, \tau'). \tag{5.25}$$

The advantage of working with the Keldysh components is that $k^{\lessgtr}$, $k^{\lceil,\rceil}$, and k^{M} are functions of real variables (as opposed to contour variables) and therefore can be numerically implemented, visualized, plotted, etc.; in other words, the Keldysh components are easier to handle. Furthermore, they encode all the information to reconstruct $k^{\lessgtr}(z, z')$.

Convolutions We now consider the convolution

$$c(z, z') = \int_\gamma d\bar{z}\, a(z, \bar{z}) b(\bar{z}, z') \tag{5.26}$$

between two functions $a(z, z')$ and $b(z, z')$ in Keldysh space. It is easy to check that $c(z, z')$ belongs to the Keldysh space as well. The question we ask is how to express the Keldysh components of c in terms of the Keldysh components of a and b. We start by evaluating, for example, the greater component. From definition (5.21) and with the help of Fig. 5.2 we

[12] In writing k_r^{M} we take into account that $\theta(t_0 - \mathrm{i}\tau, t_0 - \mathrm{i}\tau')$ is 1 if $\tau > \tau'$ and 0 otherwise; hence it is equal to $\theta(\tau - \tau')$.

can write

$$c^>(t,t') = c(t_+,t'_-) = a(t_+,t'_-)b^\delta(t'_-) + a^\delta(t_+)b(t_+,t'_-) + \int_{t_{0-}}^{t'_-} d\bar{z}\, a^>(t_+,\bar{z})b^<(\bar{z},t'_-)$$
$$+ \int_{t'_-}^{t_+} d\bar{z}\, a^>(t_+,\bar{z})b^>(\bar{z},t'_-) + \int_{t_+}^{t_0-\mathrm{i}\beta} d\bar{z}\, a^<(t_+,\bar{z})b^>(\bar{z},t'_-)$$
$$= a^>(t,t')b^\delta(t') + a^\delta(t)b^>(t,t') + \int_{t_0}^{t'} d\bar{t}\, a^>(t,\bar{t})b^<(\bar{t},t')$$
$$+ \int_{t'}^{t} d\bar{t}\, a^>(t,\bar{t})b^>(\bar{t},t') + \int_{t}^{t_0} d\bar{t}\, a^<(t,\bar{t})b^>(\bar{t},t') - \mathrm{i}\int_0^\beta d\bar{\tau}\, a^\rceil(t,\bar{\tau})b^\lceil(\bar{\tau},t').$$

The first integral in the last line is an ordinary integral on the real axis and can be rewritten as

$$\int_{t'}^{t} d\bar{t}\, a^>(t,\bar{t})b^>(\bar{t},t') = \int_{t'}^{t_0} d\bar{t}\, a^>(t,\bar{t})b^>(\bar{t},t') + \int_{t_0}^{t} d\bar{t}\, a^>(t,\bar{t})b^>(\bar{t},t'),$$

which inserted in the previous expression leads to

$$c^>(t,t') = a^>(t,t')b^\delta(t') + a^\delta(t)b^>(t,t') - \int_{t_0}^{t'} d\bar{t}\, a^>(t,\bar{t})[b^>(\bar{t},t') - b^<(\bar{t},t')]$$
$$+ \int_{t_0}^{t} d\bar{t}\, [a^>(t,\bar{t}) - a^<(t,\bar{t})]b^>(\bar{t},t') - \mathrm{i}\int_0^\beta d\bar{\tau}\, a^\rceil(t,\bar{\tau})b^\lceil(\bar{\tau},t'). \tag{5.27}$$

We see that it is convenient to define two more Keldysh components with real-time arguments:

$$\boxed{k^{\mathrm{R}}(t,t') \equiv k^\delta(t)\delta(t-t') + \theta(t-t')[k^>(t,t') - k^<(t,t')]} \tag{5.28}$$

$$\boxed{k^{\mathrm{A}}(t,t') \equiv k^\delta(t)\delta(t-t') - \theta(t'-t)[k^>(t,t') - k^<(t,t')]} \tag{5.29}$$

with the real-time Heaviside function $\theta(t) = 1$ for $t > 0$ and 0 otherwise. The *retarded* component $k^{\mathrm{R}}(t,t')$ vanishes for $t < t'$, while the *advanced* component $k^{\mathrm{A}}(t,t')$ vanishes for $t > t'$. The retarded and advanced functions can be used to rewrite (5.27) in a very elegant form:

$$c^>(t,t') = \int_{t_0}^{\infty} d\bar{t}\, [a^>(t,\bar{t})b^{\mathrm{A}}(\bar{t},t') + a^{\mathrm{R}}(t,\bar{t})b^>(\bar{t},t')] - \mathrm{i}\int_0^\beta d\bar{\tau}\, a^\rceil(t,\bar{\tau})b^\lceil(\bar{\tau},t').$$

This formula can be made even more compact if we introduce a short-hand notation for convolutions between t_0 and ∞, and for convolutions between 0 and β. For two arbitrary functions f and g we define

$$f \cdot g \equiv \int_{t_0}^{\infty} d\bar{t}\, f(\bar{t})g(\bar{t}),$$
$$f \star g \equiv -\mathrm{i}\int_0^\beta d\bar{\tau}\, f(\bar{\tau})g(\bar{\tau}).$$

Then, the formula for $c^>$ becomes

$$\boxed{c^> = a^> \cdot b^{\mathrm{A}} + a^{\mathrm{R}} \cdot b^> + a^\rceil \star b^\lceil} \tag{5.30}$$

In a similar way, we can extract the lesser component and find that

$$\boxed{c^< = a^< \cdot b^{\mathrm{A}} + a^{\mathrm{R}} \cdot b^< + a^\rceil \star b^\lceil} \tag{5.31}$$

Equations (5.30) and (5.31) can be used to extract the retarded and advanced components of c. Taking into account that the singular part c^δ is simply the product $a^\delta b^\delta$, and using the definition (5.28), we find

$$\begin{aligned} c^{\mathrm{R}}(t,t') &= c^\delta(t)\delta(t-t') + \theta(t-t')[c^>(t,t') - c^<(t,t')] \\ &= a^\delta(t)b^\delta(t)\delta(t-t') + \theta(t-t')\int_{t_0}^{\infty} d\bar{t}\, a^{\mathrm{R}}(t,\bar{t})[b^>(\bar{t},t') - b^<(\bar{t},t')] \\ &+ \theta(t-t')\int_{t_0}^{\infty} d\bar{t}\, [a^>(t,\bar{t}) - a^<(t,\bar{t})]b^{\mathrm{A}}(\bar{t},t'). \end{aligned}$$

Let us manipulate this expression a bit. Similarly to the Matsubara component, we separate out the regular contribution from the retarded/advanced components: $k^{\mathrm{R/A}}(t,t') = k^\delta(t)\delta(t-t') + k_r^{\mathrm{R/A}}(t,t')$. Then we observe that the last term on the r.h.s. vanishes unless $t \geq t'$ and that $b^{\mathrm{A}}(\bar{t},t')$ vanishes unless $t' \geq \bar{t}$. We can, therefore, replace $[a^>(t,\bar{t}) - a^<(t,\bar{t})]$ with $\theta(t-\bar{t})[a^>(t,\bar{t}) - a^<(t,\bar{t})] = a_r^{\mathrm{R}}(t,\bar{t})$ without changing the result. Next we consider the second term on the r.h.s. Writing a^{R} as the sum of the singular and regular contributions and taking into account that $b^> - b^< = b_r^{\mathrm{R}} - b_r^{\mathrm{A}}$, we find

$$\begin{aligned} \theta(t-t')\int_{t_0}^{\infty} d\bar{t}\, a^{\mathrm{R}}(t,\bar{t})[b^>(\bar{t},t') - b^<(\bar{t},t')] &= a^\delta(t)b_r^{\mathrm{R}}(t,t') \\ &+\theta(t-t')\int_{t_0}^{\infty} d\bar{t}\, a_r^{\mathrm{R}}(t,\bar{t})[b_r^{\mathrm{R}}(\bar{t},t') - b_r^{\mathrm{A}}(\bar{t},t')]. \end{aligned}$$

Collecting all these results, it is a matter of simple algebra to show that

$$\boxed{c^{\mathrm{R}} = a^{\mathrm{R}} \cdot b^{\mathrm{R}}} \tag{5.32}$$

Using similar manipulations it is possible to show that

$$\boxed{c^{\mathrm{A}} = a^{\mathrm{A}} \cdot b^{\mathrm{A}}} \tag{5.33}$$

It is worth noting that neither c^{R} nor c^{A} contain an integral along the vertical track of the contour.

Another important component that can be constructed from the greater and lesser functions is the *time-ordered* component. This is obtained by choosing both contour arguments in $k(z,z')$ on the forward branch:

$$\boxed{k^{\mathrm{T}}(t,t') \equiv k(t_-,t'_-) = k^\delta(t)\delta(t-t') + \theta(t-t')k^>(t,t') + \theta(t'-t)k^<(t,t')} \tag{5.34}$$

Similarly, the *anti-time-ordered* component is obtained by choosing both z and z' on the backward branch:

$$\boxed{k^{\bar{\mathrm{T}}}(t,t') \equiv k(t_+,t'_+) = -k^\delta(t)\delta(t-t') + \theta(t'-t)k^>(t,t') + \theta(t-t')k^<(t,t')} \tag{5.35}$$

The reader can easily verify that

$$\begin{aligned} k^{\mathrm{T}} &= k^< + k^{\mathrm{R}} = k^> + k^{\mathrm{A}}, \\ k^{\bar{\mathrm{T}}} &= k^> - k^{\mathrm{R}} = k^< - k^{\mathrm{A}} . \end{aligned}$$

Consequently, the time-ordered component of the convolution of two functions in Keldysh space can be extracted using the results for $c^{\lessgtr}$ and $c^{\mathrm{R,A}}$:

$$\begin{aligned} c^{\mathrm{T}} &= c^> + c^{\mathrm{A}} = a^{\mathrm{R}} \cdot b^> + a^> \cdot b^{\mathrm{A}} + a^{\mathrm{A}} \cdot b^{\mathrm{A}} + a^\rceil \star b^\lceil \\ &= a^{\mathrm{R}} \cdot b^> + a^{\mathrm{T}} \cdot b^{\mathrm{A}} + a^\rceil \star b^\lceil \\ &= [a^{\mathrm{T}} - a^<] \cdot b^> + a^{\mathrm{T}} \cdot [b^{\mathrm{T}} - b^>] + a^\rceil \star b^\lceil \\ &= a^{\mathrm{T}} \cdot b^{\mathrm{T}} - a^< \cdot b^> + a^\rceil \star b^\lceil . \end{aligned}$$

Similar manipulations lead to

$$c^{\bar{\mathrm{T}}} = a^> \cdot b^< - a^{\bar{\mathrm{T}}} \cdot b^{\bar{\mathrm{T}}} + a^\rceil \star b^\lceil . \tag{5.36}$$

Next we show how to express the right and left components $c^\rceil$ and $c^\lceil$. Let us start with $c^\rceil$. In definition (5.23) we choose, for example, t_- as the first argument of c. Then we find

$$\begin{aligned} c^\rceil(t,\tau) &= a^\delta(t_-)b(t_-,t_0-\mathrm{i}\tau) + a(t_-,t_0-\mathrm{i}\tau)b^\delta(t_0-\mathrm{i}\tau) \\ &+ \int_{t_{0-}}^{t_-} d\bar{z}\, a^>(t_-,\bar{z})b^<(\bar{z},t_0-\mathrm{i}\tau) + \int_{t_-}^{t_{0+}} d\bar{z}\, a^<(t_-,\bar{z})b^<(\bar{z},t_0-\mathrm{i}\tau) \\ &+ \int_{t_0}^{t_0-\mathrm{i}\tau} d\bar{z}\, a^<(t_-,\bar{z})b^<(\bar{z},t_0-\mathrm{i}\tau) + \int_{t_0-\mathrm{i}\tau}^{t_0-\mathrm{i}\beta} d\bar{z}\, a^<(t_-,\bar{z})b^>(\bar{z},t_0-\mathrm{i}\tau) \\ &= a^\delta(t)b^\rceil(t,\tau) + a^\rceil(t,\tau)b^\delta(\tau) + \int_{t_0}^{t} d\bar{t}\, [a^>(t,\bar{t}) - a^<(t,\bar{t})]b^\rceil(\bar{t},\tau) \\ &- \mathrm{i}\int_0^\beta d\bar{\tau}\, a^\rceil(t,\bar{\tau})b^{\mathrm{M}}_\tau(\bar{\tau},\tau). \end{aligned}$$

Recalling definition (5.28) of the retarded component and taking into account (5.25), we arrive at the following compact formula:

$$\boxed{c^\rceil = a^{\mathrm{R}} \cdot b^\rceil + a^\rceil \star b^{\mathrm{M}}} \tag{5.37}$$

The formula for $c^\lceil$ can be derived similarly and reads

$$\boxed{c^\lceil = a^\lceil \cdot b^{\mathrm{A}} + a^{\mathrm{M}} \star b^\lceil} \tag{5.38}$$

Definition	$c(z,z') = \int_\gamma d\bar{z}\, a(z,\bar{z})b(\bar{z},z')$	$c(z,z') = a(z,z')b(z',z)$ $[a^\delta = b^\delta = 0]$
$k^>(t,t') = k(t_+,t'_-)$	$c^> = a^> \cdot b^A + a^R \cdot b^> + a^\rceil \star b^\lceil$	$c^> = a^> b^<$
$k^<(t,t') = k(t_-,t'_+)$	$c^< = a^< \cdot b^A + a^R \cdot b^< + a^\rceil \star b^\lceil$	$c^< = a^< b^>$
$k^R(t,t') = k^\delta(t)\delta(t-t') + \theta(t-t')[k^>(t,t') - k^<(t,t')]$	$c^R = a^R \cdot b^R$	$c^R = \begin{cases} a^R b^< + a^< b^A \\ a^R b^> + a^> b^A \end{cases}$
$k^A(t,t') = k^\delta(t)\delta(t-t') - \theta(t'-t)[k^>(t,t') - k^<(t,t')]$	$c^A = a^A \cdot b^A$	$c^A = \begin{cases} a^A b^< + a^< b^R \\ a^A b^> + a^> b^R \end{cases}$
$k^T(t,t') = k(t_-,t'_-)$	$c^T = a^T \cdot b^T - a^< \cdot b^> + a^\rceil \star b^\lceil$	$c^T = \begin{cases} a^< b^T + a^R b^< \\ a^T b^> + a^> b^A \end{cases}$
$k^{\bar{T}}(t,t') = k(t_+,t'_+)$	$c^{\bar{T}} = a^> \cdot b^< - a^{\bar{T}} \cdot b^{\bar{T}} + a^\rceil \star b^\lceil$	$c^{\bar{T}} = \begin{cases} a^{\bar{T}} b^< - a^< b^A \\ a^> b^{\bar{T}} - a^R b^> \end{cases}$
$k^\rceil(t,\tau) = k(t_\pm, t_0 - i\tau)$	$c^\rceil = a^R \cdot b^\rceil + a^\rceil \star b^M$	$c^\rceil = a^\rceil b^\lceil$
$k^\lceil(\tau,t) = k(t_0 - i\tau, t_\pm)$	$c^\lceil = a^\lceil \cdot b^A + a^M \star b^\lceil$	$c^\lceil = a^\lceil b^\rceil$
$k^M(\tau,\tau') = k(t_0 - i\tau, t_0 - i\tau')$	$c^M = a^M \star b^M$	$c^M = a^M b^M$

Table 5.1 Definitions of Keldysh components (first column) and identities for the convolution (second column) and the product (third column) of two functions in Keldysh space.

Finally, it is straightforward to prove that the Matsubara component of c is simply given by

$$\boxed{c^M = a^M \star b^M} \tag{5.39}$$

since the integral along the forward branch cancels exactly the integral along the backward branch. This last identity exhausts the Langreth rules for the convolutions of two functions in Keldysh space. The results are summarized in the second column of Table 5.1.

Products There is another class of important identities regarding the product of two functions in Keldysh space:

$$c(z,z') = a(z,z')b(z',z).$$

Unlike the convolution, the product does not always belong to the Keldysh space. If, for instance, the singular part a^δ is nonvanishing, then the product c contains the term $\delta(z,z')a^\delta(z)b(z',z)$. The problems with this term are (1) if b^δ is also nonvanishing then c contains the square of a δ-function and hence it does not belong to the Keldysh space; and (2) the function $b(z',z)$ is defined for $z' \to z$ but exactly in $z' = z$ it can be discontinuous and hence $\delta(z,z')b(z',z)$ is ill-defined. The product c is well defined only provided that the singular parts a^δ and b^δ are identically zero, in which case c belongs to the Keldysh space with $c^\delta = 0$. Below we consider only this sub-class of functions in Keldysh space. In practice we never have to deal with product of functions with a singular part.

For nonsingular functions, the Keldysh components of c can easily be extracted in terms of those of a and b. The reader can check that

$$c^>(t,t') = a^>(t,t')b^<(t',t), \quad c^<(t,t') = a^<(t,t')b^>(t',t),$$
$$c^\rceil(t,\tau) = a^\rceil(t,\tau)b^\lceil(\tau,t), \qquad c^\lceil(\tau,t) = a^\lceil(\tau,t)b^\rceil(t,\tau),$$
$$c^{\rm M}(\tau,\tau') = a^{\rm M}(\tau,\tau')b^{\rm M}(\tau',\tau).$$

The retarded/advanced component is then obtained using the above identities. Taking into account that $c^\delta = 0$, we have

$$c^{\rm R}(t,t') = \theta(t-t')[a^>(t,t')b^<(t',t) - a^<(t,t')b^>(t',t)].$$

We could eliminate the θ-function by adding and subtracting either $a^<b^<$ or $a^>b^>$ and rearranging the terms. The final result is

$$\begin{aligned} c^{\rm R}(t,t') &= a^{\rm R}(t,t')b^<(t',t) + a^<(t,t')b^{\rm A}(t',t) \\ &= a^{\rm R}(t,t')b^>(t',t) + a^>(t,t')b^{\rm A}(t',t). \end{aligned}$$

Similarly, one finds

$$\begin{aligned} c^{\rm A}(t,t') &= a^{\rm A}(t,t')b^<(t',t) + a^<(t,t')b^{\rm R}(t',t) \\ &= a^{\rm A}(t,t')b^>(t',t) + a^>(t,t')b^{\rm R}(t',t). \end{aligned}$$

The time-ordered and anti-time-ordered functions can be derived in a similar way, and the reader can consult the third column of Table 5.1 for the complete list of identities.

Exercise 5.2 Let $a(z,z') = a^\delta(z)\delta(z,z')$ be a function in Keldysh space with only a singular part. Show that the convolution $c(z,z') = \int_\gamma d\bar{z}\, a(z,\bar{z})b(\bar{z},z')$ has the following components:

$$c^\lessgtr(t,t') = a^\delta(t)b^\lessgtr(t,t'), \qquad c^\rceil(t,\tau) = a^\delta(t)b^\rceil(t,\tau),$$
$$c^\lceil(\tau,t) = a^\delta(\tau)b^\lceil(\tau,t), \qquad c^{\rm M}(\tau,\tau') = a^\delta(\tau)b^{\rm M}(\tau,\tau').$$

Find also the components for the convolution $c(z,z') = \int_\gamma d\bar{z}\, b(z,\bar{z})a(\bar{z},z')$.

Exercise 5.3 Let a, b, c, be three functions in Keldysh space with $b(z,z') = b^\delta(z)\delta(z,z')$. Denoting by

$$d(z,z') = \int_\gamma d\bar{z}d\bar{z}' a(z,\bar{z})b(\bar{z},\bar{z}')c(\bar{z}',z')$$

the convolution between the three functions, show that

$$\begin{aligned} d^\lessgtr(t,t') = &\int_{t_0}^\infty d\bar{t}\, \left[a^\lessgtr(t,\bar{t})b^\delta(\bar{t})c^{\rm A}(\bar{t},t') + a^{\rm R}(t,\bar{t})b^\delta(\bar{t})c^\lessgtr(\bar{t},t')\right] \\ &- {\rm i}\int_0^\beta d\bar{\tau} a^\rceil(t,\bar{\tau})b^\delta(\bar{\tau})c^\lceil(\bar{\tau},t'). \end{aligned}$$

Exercise 5.4 Let c be the convolution between two functions a and b in Keldysh space. Show that

$$\int_\gamma dz\; c(z, z^\pm) = (-\mathrm{i})^2 \int_0^\beta d\tau d\tau' a^{\mathrm{M}}(\tau, \tau') b^{\mathrm{M}}(\tau', \tau^\pm),$$

and hence the result involves only integrals along the imaginary track. Hint: Use that in accordance with (5.22) $c(z, z^\pm) = c^{\gtrless}(t, t)$ for both $z = t_-$ and $z = t_+$.

Exercise 5.5 Let $a(z, z')$ be a function in Keldysh space and $f(z)$ a function on the contour with $f(t_+) = f(t_-)$ and $f(t_0 - \mathrm{i}\tau) = 0$. Show that

$$\int_\gamma dz' a(t_\pm, z') f(z') = \int_{t_0}^\infty dt' a^{\mathrm{R}}(t, t') f(t')$$

and

$$\int_\gamma dz' f(z') a(z', t_\pm) = \int_{t_0}^\infty dt' f(t') a^{\mathrm{A}}(t', t).$$

6

One-Particle Green's Function: Exact Results

Before delving into the Wick expansions (5.16), (5.18), or (5.19), we get acquainted with the one-particle Green's function G or simply the Green's function. This chapter is divided into three parts. In Section 6.1 we introduce some general concepts and illustrate what kind of physical information can be directly extracted from the different Keldysh components of G. In Section 6.2 we discuss the noninteracting Green's function G_0, which is the fundamental quantity of the Wick expansion. Finally, we derive several exact properties of the interacting G in Section 6.3, and discuss other physical (and measurable) quantities that can be calculated from it in Sections 6.4 and 6.5.

6.1 What Can We Learn from G?

We start our overview with a preliminary discussion on the different character of the space-spin and time dependence in $G(1;2)$. In the Dirac formalism the time-dependent wavefunction $\Psi(\mathbf{x},t)$ of a single particle is the inner product between the position–spin ket $|\mathbf{x}\rangle$ and the time-evolved ket $|\Psi(t)\rangle$. In other words, the wavefunction $\Psi(\mathbf{x},t)$ is the representation of the ket $|\Psi(t)\rangle$ in the position–spin basis. Likewise, the Green's function $G(1;2)$ can be thought of as the representation in the position–spin basis of a (single-particle) operator in first quantization,

$$\hat{\mathcal{G}}(z_1, z_2) = \int d\mathbf{x}_1 d\mathbf{x}_2 \, |\mathbf{x}_1\rangle G(1;2) \langle \mathbf{x}_2|,$$

with matrix elements

$$\langle \mathbf{x}_1 | \hat{\mathcal{G}}(z_1, z_2) | \mathbf{x}_2 \rangle = G(1;2). \tag{6.1}$$

It is important to appreciate the difference between $\hat{\mathcal{G}}$ and the operator correlator $\hat{G}_1$ of (4.58). The former is an operator in first quantization (and according to our notation is denoted by a calligraphic letter), while the latter is an operator in Fock space. The Green's function operator $\hat{\mathcal{G}}$ is a very useful quantity. Consider, for instance, the expansion (1.60) of the field operators

$$\hat{\psi}(\mathbf{x}) = \sum_i \varphi_i(\mathbf{x}) \hat{d}_i, \qquad \hat{\psi}^\dagger(\mathbf{x}) = \sum_i \varphi_i^*(\mathbf{x}) \hat{d}_i^\dagger$$

over some more convenient basis for the problem at hand (if the new basis is not complete then the expansion is an approximated expansion). Then, it is also more convenient to work with the Green's function

$$G_{ji}(z_1, z_2) = \frac{1}{\mathrm{i}} \frac{\mathrm{Tr}\left[e^{-\beta \hat{H}^{\mathrm{M}}} \mathcal{T}\left\{\hat{d}_{j,H}(z_1)\hat{d}^{\dagger}_{i,H}(z_2)\right\}\right]}{\mathrm{Tr}\left[e^{-\beta \hat{H}^{\mathrm{M}}}\right]}$$

rather than with $G(1;2)$. Now the point is that $G_{ji}(z_1, z_2)$ and $G(1;2)$ are just different matrix elements of the same Green's function operator. Inserting the expansion of the field operators in $G(1;2)$, we find

$$G(1;2) = \sum_{ji} \varphi_j(\mathbf{x}_1) G_{ji}(z_1, z_2) \varphi_i^*(\mathbf{x}_2) = \sum_{ji} \langle \mathbf{x}_1 | j \rangle G_{ji}(z_1, z_2) \langle i | \mathbf{x}_2 \rangle,$$

and comparing this equation with (6.1) we conclude that

$$\hat{\mathcal{G}}(z_1, z_2) = \sum_{ji} |j\rangle G_{ji}(z_1, z_2) \langle i|,$$

from which the result

$$\langle j | \hat{\mathcal{G}}(z_1, z_2) | i \rangle = G_{ji}(z_1, z_2)$$

follows directly. Another advantage of working with the Green's function operator rather than with its matrix elements is that one can cast the equations of motion in an invariant form – that is, in a form that is independent of the basis. The analog in quantum mechanics would be to work with kets rather than with wavefunctions. For simplicity, let us consider the equations of motion of G (first equations of the Martin–Schwinger hierarchy) for a system of noninteracting particles: $v = 0$ and hence $\hat{H}_{\mathrm{int}} = 0$. These are given in (5.10); we write them below in a more explicit form:

$$\mathrm{i}\frac{d}{dz_1} G(1;2) - \int d3\, h(1;3) G(3;2) = \delta(1;2), \tag{6.2}$$

$$-\mathrm{i}\frac{d}{dz_2} G(1;2) - \int d3\, G(1;3) h(3;2) = \delta(1;2), \tag{6.3}$$

where we define

$$h(1;2) \equiv \langle \mathbf{x}_1 | \hat{h}(z_1) | \mathbf{x}_2 \rangle\, \delta(z_1, z_2). \tag{6.4}$$

We see by inspection that (6.2) and (6.3) are obtained by sandwiching with $\langle \mathbf{x}_1 |$ and $| \mathbf{x}_2 \rangle$ the following equations for operators in first quantization:

$$\left[\mathrm{i}\frac{d}{dz_1} - \hat{h}(z_1)\right] \hat{\mathcal{G}}(z_1, z_2) = \delta(z_1, z_2), \tag{6.5}$$

$$\hat{\mathcal{G}}(z_1, z_2) \left[-\mathrm{i}\frac{\overleftarrow{d}}{dz_2} - \hat{h}(z_2)\right] = \delta(z_1, z_2). \tag{6.6}$$

As usual, the arrow over d/dz_2 specifies that the derivative acts on the left. The operator formulation helps us to visualize the structure of the equations of motion. In particular we see that (6.6) resembles the adjoint of (6.5). From now on we refer to (6.6) as the adjoint equation of motion.[1] In a similar way, one can construct the (n-particle) operator in first quantization for the n-particle Green's function and cast the whole Martin–Schwinger hierarchy in an operator form. This, however, goes beyond the scope of the section; let us go back to the Green's function.

6.1.1 The Inevitable Emergence of Memory

All the fundamental equations encountered so far are linear *differential equations* in time. An example is the time-dependent Schrödinger equation, according to which the state of the system $|\Psi(t+\Delta_t)\rangle$ infinitesimally after time t can be calculated from the state of the system $|\Psi(t)\rangle$ at time t, see Section 3.2. This means we do not need to "remember" $|\Psi(t')\rangle$ for all $t' < t$ in order to evolve the state from t to $t+\Delta_t$; the time-dependent Schrödinger equation has no memory. Likewise, the Martin–Schwinger hierarchy for the many-particle Green's functions is a set of coupled *differential equations* in the (contour) time that can be used to calculate all the G_n from their values at an infinitesimally earlier (contour) time. Again, there is no memory involved. The *brute force* solution of the Schrödinger equation or of the Martin–Schwinger hierarchy is, however, not a viable route to make progress. In the presence of many particles the huge amount of degrees of freedom of the state of the system (for the Schrödinger equation) or of the many-particle Green's functions G, G_2, $G_3, \ldots$ (for the Martin–Schwinger hierarchy) renders these equations practically unsolvable. Luckily, we are typically not interested in the full knowledge of these quantities. For instance, to calculate the density $n(\mathbf{x}, t)$ we only need to know the Green's function G, see (5.4). Then the question arises whether it is possible to construct an exact effective equation for the quantities we are interested in by "embedding" all the other degrees of freedom into such an effective equation. As we shall see below and later on in this book, the answer is affirmative, but the embedding procedure inevitably leads to the appearance of memory. This is a very profound concept and deserves a careful explanation.

Wavefunction perspective Let us consider a system consisting of two subsystems coupled to each other. A possible realization is the Anderson model, where the first subsystem is the atom and the second subsystem is the metal, and the Hamiltonian for a single particle in first quantization is, see Section 2.4,

$$\hat{h} = \underbrace{\sum_k \epsilon_k |k\rangle\langle k|}_{\text{metal}} + \underbrace{\epsilon_0 |\epsilon_0\rangle\langle\epsilon_0|}_{\text{atom}} + \underbrace{\sum_k (T_k |k\rangle\langle\epsilon_0| + T_k^* |\epsilon_0\rangle\langle k|)}_{\text{coupling}} . \tag{6.7}$$

The evolution of a single-particle ket $|\Psi\rangle$ is determined by the time-dependent Schrödinger equation $\mathrm{i}\frac{d}{dt}|\Psi(t)\rangle = \hat{h}|\Psi(t)\rangle$. Taking the inner product of the Schrödinger equation with

[1] In fact, (6.6) is the adjoint of (6.5) for any z_1 and z_2 on the horizontal branches of the contour. This is proved in Chapter 12.

$\langle \epsilon_0|$ and $\langle k|$, we find a coupled system of equations for the amplitudes $\varphi_0(t) = \langle \epsilon_0|\Psi(t)\rangle$ and $\varphi_k(t) = \langle k|\Psi(t)\rangle$:

$$\mathrm{i}\frac{d}{dt}\varphi_0(t) = \epsilon_0\varphi_0(t) + \sum_k T_k^* \varphi_k(t),$$

$$\mathrm{i}\frac{d}{dt}\varphi_k(t) = \epsilon_k\varphi_k(t) + T_k\varphi_0(t).$$

The second equation can easily be solved for $\varphi_k(t)$. The function

$$g_k^{\mathrm{R}}(t, t_0) = -\mathrm{i}\theta(t - t_0)e^{-\mathrm{i}\epsilon_k(t-t_0)} \tag{6.8}$$

obeys[2]

$$\left[\mathrm{i}\frac{d}{dt} - \epsilon_k\right] g_k^{\mathrm{R}}(t, t_0) = \delta(t - t_0),$$

and therefore for any $t > t_0$

$$\varphi_k(t) = \mathrm{i}g_k^{\mathrm{R}}(t, t_0)\varphi_k(t_0) + \int_{t_0}^{\infty} dt'\, g_k^{\mathrm{R}}(t, t')T_k\varphi_0(t'). \tag{6.9}$$

The first term on the r.h.s. is the solution of the homogeneous equation and correctly depends on the boundary condition. Taking into account that for $t \to t_0^+$ the integral vanishes, ($g_k^{\mathrm{R}}(t, t')$ is zero for $t' > t$), whereas $g_k^{\mathrm{R}}(t, t_0) \to -\mathrm{i}$, we see that (6.9) is an equality in this limit. For times $t > t_0$ the reader can verify that by acting with $\left[\mathrm{i}\frac{d}{dt} - \epsilon_k\right]$ on both sides of (6.9) we recover the differential equation for φ_k. Substitution of (6.9) into the equation for $\varphi_0(t)$ gives

$$\left[\mathrm{i}\frac{d}{dt} - \epsilon_0\right] \varphi_0(t) = \mathrm{i}\sum_k T_k^* g_k^{\mathrm{R}}(t, t_0)\varphi_k(t_0) + \int_{t_0}^{\infty} dt'\, \Sigma_{\mathrm{em}}^{\mathrm{R}}(t, t')\varphi_0(t'), \tag{6.10}$$

where we define the so-called (retarded) *embedding self-energy*,

$$\Sigma_{\mathrm{em}}^{\mathrm{R}}(t, t') = \sum_k T_k^* g_k^{\mathrm{R}}(t, t')T_k. \tag{6.11}$$

The embedding self-energy takes into account that the particle can escape from the atom at time t', wander in the metal, and then come back to the atom at time t. For the isolated atom all the T_k vanish and the solution of (6.10) reduces to $\varphi_0(t) = e^{-\mathrm{i}\epsilon_0(t-t_0)}\varphi_0(t_0)$. The important message carried by (6.10) is that we can propagate in time the atomic amplitude φ_0 without knowing the metallic amplitudes φ_k (except for their value at the initial time). The price to pay, however, is that (6.10) is an *integro-differential equation* with a *memory kernel* given by the embedding self-energy: To calculate $\varphi_0(t)$ we must know $\varphi_0(t')$ at all previous times. In other words, memory appears because we embedded the degrees of freedom of one subsystem so to have an exact effective equation for the degrees of

[2] As we shall see, $g_k^{\mathrm{R}}(t, t')$ is the retarded component of the Green's function of a noninteracting system with single-particle Hamiltonian $\hat{h}_k = \epsilon_k|k\rangle\langle k|$.

freedom of the other subsystem. This result is completely general: The consequence of the embedding is the appearance of a memory kernel.

Green's function perspective It is instructive (and mandatory for a book on Green's functions) to show how memory appears in the Green's function language. For this purpose we consider a system described by the noninteracting Anderson Hamiltonian $\hat{h}$ in (6.7) and initially in equilibrium at a given temperature and chemical potential. The Green's function can be calculated from (6.5) and (6.6) with

$$\hat{h}(z) = \begin{cases} \hat{h}^{\mathrm{M}} = \hat{h} - \mu & \text{for } z = t_0 - \mathrm{i}\tau \\ \hat{h} & \text{for } z = t_\pm, \end{cases}$$

and by imposing the KMS boundary conditions. By sandwiching (6.5) with $\langle \epsilon_0 |$ and $| \epsilon_0 \rangle$ and with $\langle k |$ and $| \epsilon_0 \rangle$, we find the following coupled system of equations:

$$\left[\mathrm{i}\frac{d}{dz_1} - h_{00}(z_1) \right] G_{00}(z_1, z_2) - \sum_k h_{0k}(z_1) G_{k0}(z_1, z_2) = \delta(z_1, z_2),$$

$$\left[\mathrm{i}\frac{d}{dz_1} - h_{kk}(z_1) \right] G_{k0}(z_1, z_2) - h_{k0}(z_1) G_{00}(z_1, z_2) = 0,$$

with the obvious notation $h_{00}(z) = \langle \epsilon_0 | \hat{h}(z) | \epsilon_0 \rangle$, $h_{0k}(z) = \langle \epsilon_0 | \hat{h}(z) | k \rangle$, etc., and similarly for the Green's function. Like in the example on the single-particle wavefunction, we solve the second of these equations by introducing the Green's function g of the isolated metal whose matrix elements $g_{kk'} = \delta_{kk'} g_k$ obey

$$\left[\mathrm{i}\frac{d}{dz_1} - h_{kk}(z_1) \right] g_k(z_1, z_2) = \delta(z_1, z_2),$$

with KMS boundary condition $g_k(t_{0-}, z_2) = \pm g_k(t_0 - \mathrm{i}\beta, z_2)$ (upper/lower sign for bosons/fermions). As we see in the next section, the retarded component of g_k is exactly the function in (6.8). We then have

$$G_{k0}(z_1, z_2) = \int_\gamma d\bar{z}\, g_k(z_1, \bar{z}) T_k G_{00}(\bar{z}, z_2),$$

where we take into account that $h_{k0}(z) = T_k$ for all z. From this result it should be clear the reason for imposing the KMS boundary condition on g_k: Any other boundary condition would have generated a G_{k0} which would not fulfill the KMS relations. Substitution of G_{k0} into the equation for G_{00} leads to an exact effective equation for the atomic Green's function

$$\left[\mathrm{i}\frac{d}{dz_1} - h_{00}(z_1) \right] G_{00}(z_1, z_2) - \int_\gamma d\bar{z}\, \Sigma_{\mathrm{em}}(z_1, \bar{z}) G_{00}(\bar{z}, z_2) = \delta(z_1, z_2), \tag{6.12}$$

where we define the embedding self-energy on the contour according to

$$\Sigma_{\mathrm{em}}(z_1, z_2) = \sum_k T_k^*\, g_k(z_1, z_2)\, T_k.$$

The adjoint equation of (6.12) can be derived similarly. Once again, the embedding of the metallic degrees of freedom leads to an equation for the atomic Green's function that contains a memory kernel. Equation (6.12) can be solved without knowing G_{k0} or $G_{kk'}$; we may say that memory emerges when leaving out some information. The equations of motion (6.5) and (6.6) clearly show that memory cannot emerge if *all* the matrix elements of the Green's function are taken into account.

Self-energy Interacting systems are more complicated since the equations of motion for G involve the two-particle Green's function G_2. Therefore, even considering all the matrix elements of G, the equations of motion do not form a closed set of equations for G. We may use the equations of motion for G_2 to express G_2 in terms of G, but these equations also involve the three-particle Green's function G_3, so we should first determine G_3. More generally, the equations of motion couple G_n to $G_{n\pm1}$ and the problem of finding an exact effective equation for G is rather complicated. Nevertheless, a formal solution exists and it has the same mathematical structure as (6.12):

$$\left[\mathrm{i}\frac{d}{dz_1} - \hat{h}(z_1)\right]\hat{\mathcal{G}}(z_1,z_2) - \int_\gamma d\bar{z}\,\hat{\Sigma}(z_1,\bar{z})\hat{\mathcal{G}}(\bar{z},z_2) = \delta(z_1,z_2). \tag{6.13}$$

In this equation the memory kernel $\hat{\Sigma}$ (a single-particle operator like $\hat{\mathcal{G}}$) depends only on the Green's function and is known as the many-body *self-energy*. We prove (6.13) in Chapter 7, see (7.18) and (7.19). For the moment it is important to appreciate that once more and in a completely different context the embedding of degrees of freedom (all the G_n with $n \geq 2$) leads to an exact effective *integro-differential equation* for the Green's function that contains a memory kernel. Due to the same mathematical structure of (6.12) and (6.13), the solution of the noninteracting Anderson model provides us with interesting physical information on the behavior of an interacting many-body system.

6.1.2 Matsubara Green's Function and Initial Preparations

The Matsubara component of the Green's function follows from $\hat{\mathcal{G}}(z_1,z_2)$ by setting $z_1 = t_0 - \mathrm{i}\tau_1$ and $z_2 = t_0 - \mathrm{i}\tau_2$. The generic matrix element then reads

$$G^{\mathrm{M}}_{ji}(\tau_1,\tau_2) = \frac{1}{\mathrm{i}}\left\{\theta(\tau_1-\tau_2)\frac{\mathrm{Tr}\left[e^{(\tau_1-\tau_2-\beta)\hat{H}^{\mathrm{M}}}\hat{d}_j e^{(\tau_2-\tau_1)\hat{H}^{\mathrm{M}}}\hat{d}^\dagger_i\right]}{\mathrm{Tr}\left[e^{-\beta\hat{H}^{\mathrm{M}}}\right]}\right.$$
$$\left.\pm\theta(\tau_2-\tau_1)\frac{\mathrm{Tr}\left[e^{(\tau_2-\tau_1-\beta)\hat{H}^{\mathrm{M}}}\hat{d}^\dagger_i e^{(\tau_1-\tau_2)\hat{H}^{\mathrm{M}}}\hat{d}_j\right]}{\mathrm{Tr}\left[e^{-\beta\hat{H}^{\mathrm{M}}}\right]}\right\}, \tag{6.14}$$

and does not contain any information on how the system evolves in time. Instead, it contains information on how the system is initially prepared. As already pointed out in Chapter 4, the initial state of the system can be the thermal equilibrium state (in which case $\hat{H}^{\mathrm{M}} = \hat{H}(t_0) - \mu\hat{N}$) or any other state. It is easy to show that *from the equal-time Matsubara Green's function we can calculate the initial ensemble average of any one-body operator*

$$\hat{O} = \sum_{ij} O_{ij} \hat{d}_i^\dagger \hat{d}_j, \tag{6.15}$$

since

$$O = \frac{\mathrm{Tr}\left[e^{-\beta \hat{H}^{\mathrm{M}}} \hat{O}\right]}{\mathrm{Tr}\left[e^{-\beta \hat{H}^{\mathrm{M}}}\right]} = \sum_{ij} O_{ij} \frac{\mathrm{Tr}\left[e^{-\beta \hat{H}^{\mathrm{M}}} \hat{d}_i^\dagger \hat{d}_j\right]}{\mathrm{Tr}\left[e^{-\beta \hat{H}^{\mathrm{M}}}\right]} = \pm \mathrm{i} \sum_{ij} O_{ij} G_{ji}^{\mathrm{M}}(\tau, \tau^+), \tag{6.16}$$

with τ^+ a time infinitesimally larger than τ. The r.h.s. is independent of τ since all matrix elements of $\hat{\mathcal{G}}^{\mathrm{M}}(\tau_1, \tau_2)$ depend on the time difference $\tau_1 - \tau_2$ only. We further observe that the KMS relations imply

$$\hat{\mathcal{G}}^{\mathrm{M}}(0, \tau) = +\hat{\mathcal{G}}^{\mathrm{M}}(\beta, \tau), \qquad \hat{\mathcal{G}}^{\mathrm{M}}(\tau, 0) = \pm \hat{\mathcal{G}}^{\mathrm{M}}(\tau, \beta), \tag{6.17}$$

i.e., the Matsubara Green's function is a periodic function for bosons and an antiperiodic function for fermions and the period is given by the inverse temperature β. We can therefore expand the Matsubara Green's function in a Fourier series according to

$$\boxed{\hat{\mathcal{G}}^{\mathrm{M}}(\tau_1, \tau_2) = \frac{1}{-\mathrm{i}\beta} \sum_{m=-\infty}^{\infty} e^{-\omega_m(\tau_1 - \tau_2)} \hat{\mathcal{G}}^{\mathrm{M}}(\omega_m)} \tag{6.18}$$

with *Matsubara frequencies*

$$\omega_m = \left\{ \begin{array}{ll} \frac{2m\pi}{-\mathrm{i}\beta} & \text{for bosons} \\ \\ \frac{(2m+1)\pi}{-\mathrm{i}\beta} & \text{for fermions} \end{array} \right. . \tag{6.19}$$

Noninteracting Matsubara Green's function Let us calculate the coefficients of the expansion for a noninteracting density matrix – that is, for a density matrix with a one-body operator

$$\hat{H}^{\mathrm{M}} = \sum_{ij} h_{ij}^{\mathrm{M}} \hat{d}_i^\dagger \hat{d}_j.$$

Setting $z_1 = t_0 - \mathrm{i}\tau_1$ and $z_2 = t_0 - \mathrm{i}\tau_2$ in the equation of motion (6.5), we find

$$\left[-\frac{d}{d\tau_1} - \hat{h}^{\mathrm{M}} \right] \hat{\mathcal{G}}^{\mathrm{M}}(\tau_1, \tau_2) = \delta(-\mathrm{i}\tau_1 + \mathrm{i}\tau_2) = \mathrm{i}\delta(\tau_1 - \tau_2),$$

where in the last step we use $\delta(-\mathrm{i}\tau) = \mathrm{i}\delta(\tau)$, see the discussion before (5.25). Inserting (6.18) into the above equation and exploiting the identities (see Appendix A),

$$\delta(\tau) = \frac{1}{\beta} \sum_{m=-\infty}^{\infty} \left\{ \begin{array}{l} e^{-\mathrm{i}\frac{2m\pi}{\beta}\tau} \\ e^{-\mathrm{i}\frac{(2m+1)\pi}{\beta}\tau} \end{array} \right. = \frac{1}{\beta} \sum_{m=-\infty}^{\infty} e^{-\omega_m \tau}, \tag{6.20}$$

we can easily extract the coefficients

$$\hat{\mathcal{G}}^{\mathrm{M}}(\omega_m) = \frac{1}{\omega_m - \hat{h}^{\mathrm{M}}}. \tag{6.21}$$

To familiarize ourselves with these new formulas we now calculate the occupation n_0 of the noninteracting Anderson model and show that the result agrees with (2.32).

Occupation in the noninteracting Anderson model The occupation operator for the atomic site in the Anderson model is $\hat{n}_0 = \hat{d}_0^\dagger \hat{d}_0$ and from (6.16) (the lower sign for fermions) the ensemble average of $\hat{n}_0$ is

$$n_0 = -\mathrm{i}G_{00}^{\mathrm{M}}(\tau, \tau^+) = \frac{-\mathrm{i}}{-\mathrm{i}\beta} \sum_m e^{\eta\omega_m} G_{00}^{\mathrm{M}}(\omega_m). \tag{6.22}$$

The second equality follows from (6.18), in which we set $\tau_1 = \tau$ and $\tau_2 = \tau^+ = \tau + \eta$, where η is an infinitesimal positive constant. We consider the system in thermal equilibrium at a given temperature and chemical potential so that $\hat{h}^{\mathrm{M}} = \hat{h} - \mu$. The expansion coefficients can be calculated from (6.21) following the same steps leading to (2.30) and read

$$G_{00}^{\mathrm{M}}(\omega_m) = \langle \epsilon_0 | \frac{1}{\omega_m - \hat{h}^{\mathrm{M}}} | \epsilon_0 \rangle = \frac{1}{\omega_m + \mu - \epsilon_0 - \Sigma_{\mathrm{em}}(\omega_m + \mu)}.$$

To evaluate the sum in (6.22) we use the following trick. We first observe that the function $Q(\zeta) \equiv G_{00}^{\mathrm{M}}(\zeta)$ of the complex variable ζ is analytic everywhere except along the real axis,[3] where it has poles (and/or branch cuts when the spectrum ϵ_k becomes a continuum). For any such function we can write

$$\frac{1}{-\mathrm{i}\beta} \sum_{m=-\infty}^{\infty} e^{\eta\omega_m} Q(\omega_m) = \int_{\Gamma_a} \frac{d\zeta}{2\pi} f(\zeta) e^{\eta\zeta} Q(\zeta), \tag{6.23}$$

where $f(\zeta) = 1/(e^{\beta\zeta} + 1)$ is the Fermi function with simple poles in $\zeta = \omega_m$ and residues $-1/\beta$, while Γ_a is the contour of Fig. 6.1(a) that encircles all Matsubara frequencies clockwisely.[4] Taking into account that $\eta > 0$, we can deform the contour Γ_a into the contour Γ_b

[3]Suppose that the denominator of $G_{00}^{\mathrm{M}}(\zeta)$ vanishes for some complex $\zeta = x + \mathrm{i}y$. Then, taking the imaginary part of the denominator, we find

$$y \left(1 + \sum_k \frac{|T_k|^2}{(x + \mu - \epsilon_k)^2 + y^2} \right) = 0,$$

which can be satisfied only for $y = 0$.

[4]We remind the reader that the Cauchy residue theorem for a meromorphic function $f(z)$ in a domain D states that

$$\oint_\gamma dz f(z) = 2\pi\mathrm{i} \sum_j \lim_{z \to z_j} (z - z_j) f(z),$$

where γ is an anticlockwise-oriented contour in D and the sum runs over the simple poles z_j of $f(z)$ contained in γ (a minus sign in front of the formula appears for clockwise-oriented contours like Γ_a).

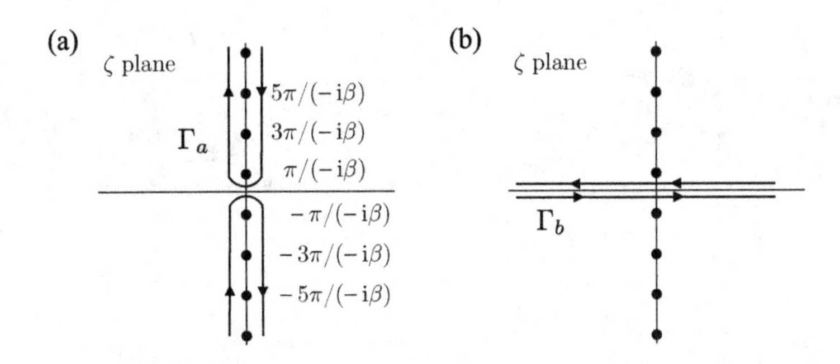

Figure 6.1 Contours for the evaluation of the sum $\sum_m e^{\eta\omega_m} G^{\rm M}_{00}(\omega_m)$. The points displayed are the fermionic Matsubara frequencies $\omega_m = (2m+1)\pi/(-{\rm i}\beta)$.

of Fig. 6.1(b), since $\lim_{\zeta\to\pm\infty} e^{\eta\zeta} f(\zeta) = 0$, thus obtaining

$$\frac{1}{-{\rm i}\beta}\sum_{m=-\infty}^{\infty} e^{\eta\omega_m} Q(\omega_m) = \lim_{\delta\to 0^+}\left[\int_{-\infty}^{\infty}\frac{d\omega}{2\pi} f(\omega) Q(\omega - {\rm i}\delta) + \int_{\infty}^{-\infty}\frac{d\omega}{2\pi} f(\omega) Q(\omega + {\rm i}\delta)\right]$$

$$= \lim_{\delta\to 0^+}\int_{-\infty}^{\infty}\frac{d\omega}{2\pi} f(\omega)\left[Q(\omega - {\rm i}\delta) - Q(\omega + {\rm i}\delta)\right].$$

Using the same notation as in (2.31) we have

$$\lim_{\delta\to 0^+}\left[G^{\rm M}_{00}(\omega - {\rm i}\delta) - G^{\rm M}_{00}(\omega + {\rm i}\delta)\right] = -2{\rm i}\,{\rm Im}\frac{1}{\omega+\mu-\epsilon_0-\Lambda_{\rm em}(\omega+\mu)+\frac{\rm i}{2}\Gamma_{\rm em}(\omega+\mu)+{\rm i}\delta},$$

and hence

$$n_0 = -2\int_{-\infty}^{\infty}\frac{d\omega}{2\pi} f(\omega-\mu)\,{\rm Im}\frac{1}{\omega-\epsilon_0-\Lambda_{\rm em}(\omega)+\frac{\rm i}{2}\Gamma_{\rm em}(\omega)+{\rm i}\delta}. \tag{6.24}$$

This result generalizes (2.32) to finite temperatures and correctly reduces to (2.32) at zero temperature.

6.1.3 Lesser/Greater Green's Function and the Concept of Quasi-particles

To access the dynamical properties of a system it is necessary to know (1) how the system is initially prepared and (2) how the system evolves in time. If we restrict ourselves to the time-dependent ensemble average of one-body operators as in (6.15), both (1) and (2) are encoded in the lesser and greater Green's functions, which by definition read

$$G^<_{ji}(t,t') = \mp{\rm i}\frac{{\rm Tr}\left[e^{-\beta\hat{H}^{\rm M}}\hat{d}^\dagger_{i,H}(t')\hat{d}_{j,H}(t)\right]}{{\rm Tr}\left[e^{-\beta\hat{H}^{\rm M}}\right]} = \mp{\rm i}\sum_k \rho_k\langle\Psi_k|\hat{d}^\dagger_{i,H}(t')\hat{d}_{j,H}(t)|\Psi_k\rangle, \tag{6.25}$$

$$G^>_{ji}(t,t') = -{\rm i}\frac{{\rm Tr}\left[e^{-\beta\hat{H}^{\rm M}}\hat{d}_{j,H}(t)\hat{d}^\dagger_{i,H}(t')\right]}{{\rm Tr}\left[e^{-\beta\hat{H}^{\rm M}}\right]} = -{\rm i}\sum_k \rho_k\langle\Psi_k|\hat{d}_{j,H}(t)\hat{d}^\dagger_{i,H}(t')|\Psi_k\rangle. \tag{6.26}$$

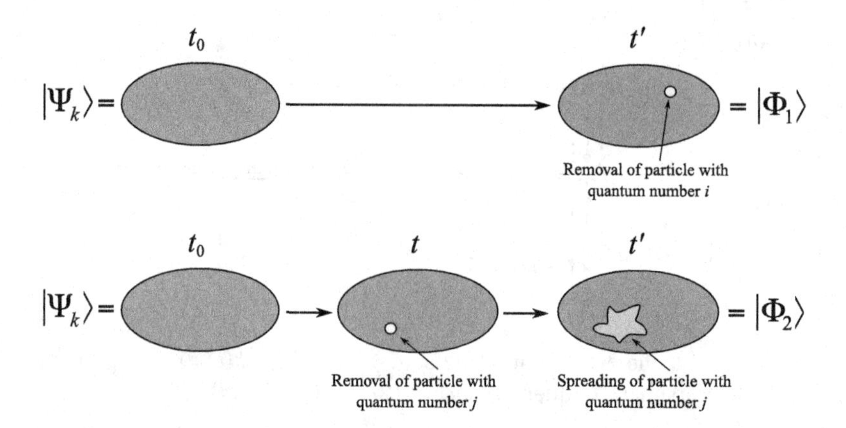

Figure 6.2 Schematic representation of the states $|\Phi_1\rangle$ and $|\Phi_2\rangle$ appearing in (6.29).

The $\hat{d}$-operators in these formulas are in the standard (as opposed to contour) Heisenberg picture (3.17). In the second equalities we simply introduce a complete set of eigenkets $|\Psi_k\rangle$ of $\hat{H}^{\mathrm{M}}$ with eigenvalue E_k^{M} so that $\rho_k = e^{-\beta E_k^{\mathrm{M}}}/Z$. It is straightforward to verify that the greater/lesser Green's function has the anti-hermiticity property

$$\boxed{[G^{>}_{ji}(t,t')]^* = -G^{>}_{ij}(t',t)} \quad , \quad \boxed{[G^{<}_{ji}(t,t')]^* = -G^{<}_{ij}(t',t)} \tag{6.27}$$

and consequently the retarded and advanced Green's functions are related by

$$\boxed{\hat{\mathcal{G}}^{\mathrm{R}}(t,t') = \theta(t-t')\left[\hat{\mathcal{G}}^{>}(t,t') - \hat{\mathcal{G}}^{<}(t,t')\right] = \left[\hat{\mathcal{G}}^{\mathrm{A}}(t',t)\right]^{\dagger}} \tag{6.28}$$

Below we discuss the lesser Green's function and leave it as an exercise for the reader to go through the same logical and mathematical steps in the case of the greater Green's function.

A generic term of the sum in (6.25) contains the quantity

$$\langle\Psi_k|\hat{d}^{\dagger}_{i,H}(t')\hat{d}_{j,H}(t)|\Psi_k\rangle = \underbrace{\langle\Psi_k|\hat{U}(t_0,t')\hat{d}^{\dagger}_i}_{\langle\Phi_1|}\ \underbrace{\hat{U}(t',t)\,\hat{d}_j\,\hat{U}(t,t_0)|\Psi_k\rangle}_{|\Phi_2\rangle}, \tag{6.29}$$

with $\hat{U}$ the evolution operator (3.16). This quantity is proportional to the probability amplitude that evolving $|\Psi_k\rangle$ from t_0 to t, then removing a particle with quantum number j and letting the new state evolve from t to t' (this state is $|\Phi_2\rangle$) we find the same state as evolving $|\Psi_k\rangle$ from t_0 to t', at which time a particle with quantum number i is removed (this state is $|\Phi_1\rangle$), see Fig. 6.2. As suggested by the figure, when time passes the disturbance (removal of a particle) "spreads" and, therefore, it is reasonable to expect that the probability amplitude vanishes for $|t-t'| \to \infty$ independently of the quantum number i. Of course, this expectation makes sense provided that the system has infinitely many degrees of freedom coupled to each other. If the system has only a finite number of degrees of freedom (such as in the PPP model for benzene or in the hydrogen molecule discussed in Chapter 2), then the probability amplitude exhibits an oscillatory behavior.

Quasi-particles In systems with infinitely many degrees of freedom, a particle with quantum number i is said to be an *independent* or *free particle* and i is said to be a *good quantum number* if $G^{\lessgtr}_{ii}(t,t')$ does not vanish for $|t-t'| \to \infty$. Independent particles behave as elementary particles in an empty space; they do not decay or, equivalently, have an infinitely long lifetime. Depending on the system there may be zero, one, a few, or even a complete set of good quantum numbers. The latter situation is always the case in noninteracting systems and, vice versa, any noninteracting system admits a complete set of good quantum numbers, see Section 6.2.2. In interacting systems a useful concept is the "quasi-good" quantum number - that is, a quantum number for which $G^{\lessgtr}_{ii}(t,t')$ decays to zero (as $|t-t'| \to \infty$) over a timescale that is much longer than the timescale of the dynamical process we are interested in. Particles with these quantum numbers are called *quasi-particles*. Quasi-particles decay after a long time and their existence allows us to develop many useful and insightful approximations. We come back to this concept in Sections 6.3.2 and 10.2.

Time-dependent ensemble averages Let us go back to (6.29). We note that for $t = t'$ the disturbance has no time to spread. In this special (but relevant) case the probability amplitude reduces to the overlap between $\hat{d}_j|\Psi_k(t)\rangle$ and $\hat{d}_i|\Psi_k(t)\rangle$, where $|\Psi_k(t)\rangle = \hat{U}(t,t_0)|\Psi_k\rangle$. For $i = j$ the overlap becomes the probability of finding a particle with quantum number i when the system is in $|\Psi_k(t)\rangle$ - that is, the overlap becomes the quantum average of the occupation operator $\hat{n}_i = \hat{d}^\dagger_i \hat{d}_i$.[5] This is not surprising since the time-dependent ensemble average of $\hat{n}_i$ is $n_i(t) = \pm\mathrm{i}\, G_{ii}(z,z^+) = \pm\mathrm{i}\, G^<_{ii}(t,t)$. More generally, *the lesser Green's function allows us to calculate the time-dependent ensemble average of any one-body operator* since, given an operator $\hat{O}$ as in (6.15), we have

$$O(t) = \pm\mathrm{i}\sum_{ij} O_{ij} G^<_{ji}(t,t). \tag{6.30}$$

For $t = t_0$ the average $O(t_0)$ must be equal to the initial average (6.16), and therefore

$$\hat{\mathcal{G}}^<(t_0,t_0) = \hat{\mathcal{G}}^{\mathrm{M}}(\tau,\tau^+) \quad \text{for all } \tau. \tag{6.31}$$

This identity can also be directly deduced from the definition of the Matsubara and lesser components of the Green's function.

A special case that is worth discussing in more detail is a system with Hamiltonian $\hat{H}(t) = \hat{H}(t_0) = \hat{H}$ constant in time. The corresponding Green's function can describe either: (1) a system with Hamiltonian $\hat{H}' \equiv \hat{H}^{\mathrm{M}} + \mu\hat{N}$ initially in thermal equilibrium[6] and then driven out of equilibrium by the sudden switch-on of a perturbation $\Delta\hat{H} = \hat{H} - \hat{H}'$, so that $\hat{H}(t > t_0) = \hat{H}' + \Delta\hat{H} = \hat{H}$; or (2) a system with Hamiltonian $\hat{H}$ initially prepared in an excited (not necessarily stationary) ensemble described by a density matrix $\hat{\rho} = e^{-\beta\hat{H}^{\mathrm{M}}}/Z$.

[5]It is maybe worth commenting on how a probability amplitude can become a probability. We show it with an example. Given an unnormalized state $|\Psi\rangle$ [like $|\Phi_1\rangle$ or $|\Phi_2\rangle$ in (6.29)], the quantity $\langle\Psi|\Psi\rangle$ is proportional to the probability amplitude of finding $|\Psi\rangle$ when the system is in $|\Psi\rangle$. Now if we write $|\Psi\rangle = |\Psi_1\rangle\langle\Psi_2|\Phi\rangle$, with $|\Psi_1\rangle$, $|\Psi_2\rangle$, and $|\Phi\rangle$ normalized to 1, then $\langle\Psi|\Psi\rangle = |\langle\Psi_2|\Phi\rangle|^2$ equals the probability of finding $|\Psi_2\rangle$ when the system is in $|\Phi\rangle$. The reader can generalize this argument to the case in which $|\Psi_1\rangle\langle\Psi_2|$ is replaced by the operator $\hat{d}_i$.

[6]The equilibrium density matrix of a system with Hamiltonian $\hat{H}'$ is $\hat{\rho} = e^{-\beta\hat{H}^{\mathrm{M}}}/Z$, with $\hat{H}^{\mathrm{M}} = \hat{H}' - \mu\hat{N}$.

No matter what the physical situation is, the evolution operator is a simple exponential and (6.25) becomes

$$G^{<}_{ji}(t,t') = \mp \mathrm{i} \sum_k \rho_k \langle \Psi_k | e^{\mathrm{i}\hat{H}(t'-t_0)} \hat{d}^\dagger_i e^{-\mathrm{i}\hat{H}(t'-t)} \hat{d}_j e^{-\mathrm{i}\hat{H}(t-t_0)} | \Psi_k \rangle. \tag{6.32}$$

In general, the lesser Green's function is *not* a function of the time difference $t-t'$ only. This is a consequence of the fact that if the $|\Psi_k\rangle$ are not eigenstates of $\hat{H}$ then their evolution is not given by a multiplicative (time-dependent) phase factor.

6.2 Noninteracting Green's Function

In this section we calculate the Green's function for a system of noninteracting particles that is initially described by a noninteracting density matrix. This means that the Hamiltonian along the contour is

$$\hat{H}(z) = \sum_{ij} h_{ij}(z) \hat{d}^\dagger_i \hat{d}_j = \sum_{ij} \langle i | \hat{h}(z) | j \rangle \, \hat{d}^\dagger_i \hat{d}_j,$$

with $\hat{h}(z = t_0 - \mathrm{i}\tau) = \hat{h}^{\mathrm{M}}$ the constant single-particle Hamiltonian along the imaginary track and $\hat{h}(z = t_\pm) = \hat{h}(t)$ the single-particle Hamiltonian along the horizontal branches. The noninteracting Green's function operator satisfies the equations of motion (6.5) and (6.6). To solve them, we write $\hat{\mathcal{G}}$ as

$$\hat{\mathcal{G}}(z_1, z_2) = \hat{\mathcal{U}}_L(z_1) \hat{\mathcal{F}}(z_1, z_2) \hat{\mathcal{U}}_R(z_2),$$

where the (first quantization) operators $\hat{\mathcal{U}}_{L/R}(z)$ fulfill

$$\mathrm{i}\frac{d}{dz}\hat{\mathcal{U}}_L(z) = \hat{h}(z)\hat{\mathcal{U}}_L(z), \qquad -\mathrm{i}\frac{d}{dz}\hat{\mathcal{U}}_R(z) = \hat{\mathcal{U}}_R(z)\hat{h}(z),$$

with boundary conditions $\hat{\mathcal{U}}_L(t_{0-}) = \hat{\mathcal{U}}_R(t_{0-}) = \hat{\mathbb{1}}$. The structure of these equations is the same as that of the evolution operator on the contour - compare with (4.33) and (4.34). The explicit form of $\hat{\mathcal{U}}_{L/R}(z)$ is therefore the same as that in (4.32):

$$\hat{\mathcal{U}}_L(z) = \mathcal{T}\left\{ e^{-\mathrm{i}\int_{t_{0-}}^{z} d\bar{z}\, \hat{h}(\bar{z})} \right\}, \qquad \hat{\mathcal{U}}_R(z) = \bar{\mathcal{T}}\left\{ e^{+\mathrm{i}\int_{t_{0-}}^{z} d\bar{z}\, \hat{h}(\bar{z})} \right\}. \tag{6.33}$$

The operator $\hat{\mathcal{U}}_L(z)$ can be seen as the single-particle forward evolution operator on the contour (from t_{0-} to z), whereas $\hat{\mathcal{U}}_R(z)$ can be seen as the single-particle backward evolution operator on the contour (from z to t_{0-}). This is also in agreement with the fact that $\hat{\mathcal{U}}_L(z)\hat{\mathcal{U}}_R(z) = \hat{\mathcal{U}}_R(z)\hat{\mathcal{U}}_L(z) = \hat{\mathbb{1}}$, which follows directly from their definition. Substituting $\hat{\mathcal{G}}$ into the equations of motion (6.5) and (6.6), we obtain the following differential equations for $\hat{\mathcal{F}}$

$$\mathrm{i}\frac{d}{dz_1}\hat{\mathcal{F}}(z_1, z_2) = \delta(z_1, z_2), \qquad -\mathrm{i}\frac{d}{dz_2}\hat{\mathcal{F}}(z_1, z_2) = \delta(z_1, z_2).$$

The most general solution of these differential equations is

$$\hat{\mathcal{F}}(z_1, z_2) = \theta(z_1, z_2)\hat{\mathcal{F}}^{>} + \theta(z_2, z_1)\hat{\mathcal{F}}^{<},$$

where the constant operators $\hat{\mathcal{F}}^{>}$ and $\hat{\mathcal{F}}^{<}$ are constrained by

$$\hat{\mathcal{F}}^{>} - \hat{\mathcal{F}}^{<} = -\mathrm{i}\,\hat{\mathbb{1}}. \tag{6.34}$$

To determine $\hat{\mathcal{F}}^{>}$ (or $\hat{\mathcal{F}}^{<}$) we can use one of the two KMS relations. Choosing, for example,

$$\hat{\mathcal{G}}(t_{0-}, z') = \pm\hat{\mathcal{G}}(t_0 - \mathrm{i}\beta, z') \quad \begin{cases} + \text{ for bosons} \\ - \text{ for fermions,} \end{cases}$$

it is straightforward to find

$$\hat{\mathcal{F}}^{<} = \pm\hat{\mathcal{U}}_L(t_0 - \mathrm{i}\beta)\hat{\mathcal{F}}^{>} = \pm e^{-\beta\hat{h}^{\mathrm{M}}}\hat{\mathcal{F}}^{>} \quad \begin{cases} + \text{ for bosons} \\ - \text{ for fermions.} \end{cases}$$

Solving this equation for $\hat{\mathcal{F}}^{>}$ and inserting the result into (6.34), we get

$$\hat{\mathcal{F}}^{<} = \mp\mathrm{i}\frac{1}{e^{\beta\hat{h}^{\mathrm{M}}} \mp \hat{\mathbb{1}}} = \mp\mathrm{i}f(\hat{h}^{\mathrm{M}}),$$

where $f(\omega) = 1/[e^{\beta\omega} \mp 1]$ is the Bose/Fermi function. Consequently, the operator $\hat{\mathcal{F}}^{>}$ reads

$$\hat{\mathcal{F}}^{>} = \pm\mathrm{i}\frac{1}{e^{-\beta\hat{h}^{\mathrm{M}}} \mp 1} = -\mathrm{i}\bar{f}(\hat{h}^{\mathrm{M}}),$$

with $\bar{f}(\omega) = 1 \pm f(\omega) = e^{\beta\omega}f(\omega)$. It is left as an exercise for the reader to show that we would have got the same results using the other KMS relation – that is, $\hat{\mathcal{G}}(z, t_{0-}) = \pm\hat{\mathcal{G}}(z, t_0 - \mathrm{i}\beta)$. To summarize, the noninteracting Green's function can be written as

$$\boxed{\hat{\mathcal{G}}(z_1, z_2) = -\mathrm{i}\hat{\mathcal{U}}_L(z_1)\left[\theta(z_1, z_2)\bar{f}(\hat{h}^{\mathrm{M}}) \pm \theta(z_2, z_1)f(\hat{h}^{\mathrm{M}})\right]\hat{\mathcal{U}}_R(z_2)} \tag{6.35}$$

with $\hat{\mathcal{U}}_{L/R}$ given in (6.33). Having the Green's function on the contour, we can now extract all its Keldysh components.

6.2.1 Matsubara Component

The Matsubara Green's function follows from (6.35) by setting $z_1 = t_0 - \mathrm{i}\tau_1$ and $z_2 = t_0 - \mathrm{i}\tau_2$ and reads

$$\boxed{\hat{\mathcal{G}}^{\mathrm{M}}(\tau_1, \tau_2) = -\mathrm{i}\left[\theta(\tau_1 - \tau_2)\bar{f}(\hat{h}^{\mathrm{M}}) \pm \theta(\tau_2 - \tau_1)f(\hat{h}^{\mathrm{M}})\right]e^{-(\tau_1 - \tau_2)\hat{h}^{\mathrm{M}}}} \tag{6.36}$$

where we make use of the fact that

$$\hat{\mathcal{U}}_L(t_0 - \mathrm{i}\tau) = e^{-\tau\hat{h}^{\mathrm{M}}} \quad , \quad \hat{\mathcal{U}}_R(t_0 - \mathrm{i}\tau) = e^{\tau\hat{h}^{\mathrm{M}}},$$

which also implies that $\hat{\mathcal{U}}_{L/R}(t_0 - \mathrm{i}\tau)$ commutes with $\hat{h}^{\mathrm{M}}$. Inserting a complete set of eigenkets $|\lambda^{\mathrm{M}}\rangle$ of $\hat{h}^{\mathrm{M}}$ with eigenvalues $\epsilon_\lambda^{\mathrm{M}}$, the Matsubara Green's function can also be written as

$$\hat{\mathcal{G}}^{\mathrm{M}}(\tau_1,\tau_2) = -\mathrm{i}\sum_\lambda \left[\theta(\tau_1-\tau_2)\bar{f}(\epsilon_\lambda^{\mathrm{M}}) \pm \theta(\tau_2-\tau_1)f(\epsilon_\lambda^{\mathrm{M}})\right] e^{-(\tau_1-\tau_2)\epsilon_\lambda^{\mathrm{M}}}|\lambda^{\mathrm{M}}\rangle\langle\lambda^{\mathrm{M}}|.$$

This result should be compared with (6.18). For noninteracting density matrices the coefficients of the expansion are given by (6.21) and hence

$$\boxed{\hat{\mathcal{G}}^{\mathrm{M}}(\tau_1,\tau_2) = \frac{1}{-\mathrm{i}\beta}\sum_{m=-\infty}^{\infty}\frac{e^{-\omega_m(\tau_1-\tau_2)}}{\omega_m - \hat{h}^{\mathrm{M}}}} \tag{6.37}$$

We thus have two different ways, (6.36) and (6.37), of writing the same quantity. We close this section by proving the equivalence between them.

The strategy is to perform the sum over the Matsubara frequencies using a generalization of (6.23). The Bose/Fermi function $f(\zeta) = 1/(e^{\beta\zeta}\mp 1)$ of the complex variable ζ has simple poles in $\zeta = \omega_m$ with residues $\pm 1/\beta$ (as usual upper/lower sign for bosons/fermions). Therefore, given a function $Q(\zeta)$ analytic around all Matsubara frequencies we can write the following two equivalent identities:

$$\frac{1}{-\mathrm{i}\beta}\sum_{m=-\infty}^{\infty} Q(\omega_m)e^{-\omega_m\tau} = \int_{\Gamma_a}\frac{d\zeta}{2\pi}Q(\zeta)e^{-\zeta\tau}\times\begin{cases}\mp f(\zeta)\\ \\ -e^{\beta\zeta}f(\zeta).\end{cases} \tag{6.38}$$

The functions in the first and second rows of the curly bracket yield the same value in $\zeta = \omega_m$, since $e^{\beta\omega_m} = \pm 1$. The contour Γ_a must encircle all the Matsubara frequencies clockwise without including any singularity (pole or branch point) of Q - that is, Q must be analytic inside Γ_a. In particular, we consider functions Q with singularities only in $\zeta = \epsilon_\lambda^{\mathrm{M}}$. Then, for fermionic systems we can choose the contour Γ_a as shown in Fig. 6.1(a). The bosonic case is a bit more subtle since the Matsubara frequency $\omega_0 = 0$ lies on the real axis and hence the contour Γ_a does not exist if Q has a singularity in $\zeta = 0$. At any finite temperature, however, the lowest eigenvalue of $\hat{h}^{\mathrm{M}}$ must be strictly positive, otherwise both (6.36) and (6.37) are ill-defined.[7] Thus, $Q(\zeta)$ has no poles in $\zeta = 0$ and a possible choice of contour Γ_a is illustrated in Fig. 6.3(a). Next, we observe that if $\beta > \tau > 0$, we could use the bottom identity in (6.38) and deform the contour as shown in Fig. 6.1(b) (for fermions) and Fig. 6.3(b) for bosons, since $\lim_{\zeta\to\pm\infty} e^{(\beta-\tau)\zeta}f(\zeta) = 0$. On the contrary, if $0 > \tau > -\beta$, we

[7] In bosonic systems the lowest eigenvalue $\epsilon_0^{\mathrm{M}} = \epsilon_0 - \mu$ of $\hat{h}^{\mathrm{M}}$ approaches zero when $\beta\to\infty$ as $\epsilon_0^{\mathrm{M}}\sim 1/\beta$ so that the product $\beta\epsilon_0^{\mathrm{M}}$ remains finite. If the energy spectrum of $\hat{h}^{\mathrm{M}}$ is continuous and if the density of single-particle states $D(\omega) = 2\pi\sum_\lambda\delta(\omega-\epsilon_\lambda^{\mathrm{M}})$ vanishes when $\omega\to 0$, then all bosons have the same quantum number ϵ_0^{M} in the zero-temperature limit. This phenomenon is known as the *Bose condensation*. Thus, in physical bosonic systems the density matrix $\hat{\rho}$ can be of the form $|\Phi_0\rangle\langle\Phi_0|$ (pure state) only at zero temperature and only in the presence of Bose condensation. Without Bose condensation $\hat{\rho}$ is a mixture also at zero temperature, see Exercise 6.3.

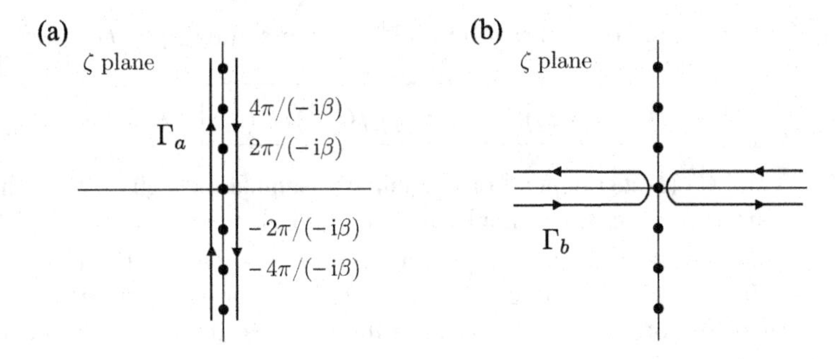

Figure 6.3 Contours for the evaluation of the bosonic sum in (6.38). The points displayed are the bosonic Matsubara frequencies $\omega_m = 2m\pi/(-\mathrm{i}\beta)$. The contour Γ_b contains all the poles of the function $Q(\zeta)$.

could use the top identity and still deform the contour as shown in Fig. 6.1(b) (for fermions) and Fig. 6.3(b) (for bosons) since $\lim_{\zeta\to\pm\infty} e^{-\tau\zeta} f(\zeta) = 0$. We then conclude that

$$\boxed{\frac{1}{-\mathrm{i}\beta}\sum_{m=-\infty}^{\infty} Q(\omega_m)e^{-\omega_m\tau} = \int_{\Gamma_b}\frac{d\zeta}{2\pi}Q(\zeta)e^{-\zeta\tau}\left[-\theta(\tau)e^{\beta\zeta}f(\zeta)\mp\theta(-\tau)f(\zeta)\right]} \tag{6.39}$$

We take $Q(\zeta) = 1/(\zeta - \epsilon_\lambda^{\mathrm{M}})$ - that is, $Q(\zeta)$ is a meromorphic function with one simple pole in $\zeta = \epsilon_\lambda^{\mathrm{M}}$ inside Γ_b of residue 1. The Cauchy residue theorem gives

$$\frac{1}{-\mathrm{i}\beta}\sum_{m=-\infty}^{\infty}\frac{e^{-\omega_m\tau}}{\omega_m - \epsilon_\lambda^{\mathrm{M}}} = -\mathrm{i}\left[\theta(\tau)\bar{f}(\epsilon_\lambda^{\mathrm{M}}) \pm \theta(-\tau)f(\epsilon_\lambda^{\mathrm{M}})\right]e^{-\tau\epsilon_\lambda^{\mathrm{M}}}.$$

Multiplying both sides by $|\lambda^{\mathrm{M}}\rangle\langle\lambda^{\mathrm{M}}|$ and summing over λ we find the equivalence between (6.36) and (6.37) we were looking for.

6.2.2 Lesser and Greater Components

The operators $\hat{\mathcal{U}}_L(z)$ and $\hat{\mathcal{U}}_R(z)$ evaluated for z on the forward/backward branch of the contour reduce to the single-particle real-time evolution operators

$$\hat{\mathcal{U}}_L(t_\pm) \equiv \hat{\mathcal{U}}(t) = T\left\{e^{-\mathrm{i}\int_{t_0}^{t}d\bar{t}\,\hat{h}(\bar{t})}\right\},$$
$$\hat{\mathcal{U}}_R(t_\pm) = \hat{\mathcal{U}}^\dagger(t),$$

with T the chronological ordering operator introduced in (3.8). The action of $\hat{\mathcal{U}}(t)$ on a generic single-particle ket $|\Psi\rangle$ yields the time-evolved ket $|\Psi(t)\rangle$, which obeys the Schrödinger equation $\mathrm{i}\frac{d}{dt}|\Psi(t)\rangle = \hat{h}(t)|\Psi(t)\rangle$. The lesser component of the noninteracting Green's function follows from (6.35) when setting $z_1 = t_{1-}$ and $z_2 = t_{2+}$. We find

$$\boxed{\hat{\mathcal{G}}^<(t_1,t_2) = \mp\mathrm{i}\hat{\mathcal{U}}(t_1)\,f(\hat{h}^{\mathrm{M}})\hat{\mathcal{U}}^\dagger(t_2)} \tag{6.40}$$

Similarly, the greater component follows from (6.35) when setting $z_1 = t_{1+}$ and $z_2 = t_{2-}$, and reads

$$\boxed{\hat{\mathcal{G}}^{>}(t_1, t_2) = -\,\mathrm{i}\hat{\mathcal{U}}(t_1)\,\bar{f}(\hat{h}^{\mathrm{M}})\hat{\mathcal{U}}^{\dagger}(t_2)} \tag{6.41}$$

Both $\hat{\mathcal{G}}^{>}(t_1, t_2)$ and $\hat{\mathcal{G}}^{<}(t_1, t_2)$ depend on the initial ensemble through $f(\hat{h}^{\mathrm{M}})$. This should not come as a surprise since, for example, the diagonal element $\pm\mathrm{i}\, G^{<}_{ii}(t,t)$ is the time-dependent ensemble average of the occupation operator $\hat{n}_i = \hat{d}^{\dagger}_i \hat{d}_i$. The physical content of the lesser/greater Green's function can be more easily visualized by inserting a complete set of eigenstates of $\hat{h}^{\mathrm{M}}$ between the Bose/Fermi function and the evolution operators:

$$\begin{aligned}\hat{\mathcal{G}}^{<}(t_1, t_2) &= \mp\mathrm{i}\sum_{\lambda} f(\epsilon^{\mathrm{M}}_{\lambda})\hat{\mathcal{U}}(t_1)\,|\lambda^{\mathrm{M}}\rangle\langle\lambda^{\mathrm{M}}|\hat{\mathcal{U}}^{\dagger}(t_2)\\ &= \mp\mathrm{i}\sum_{\lambda} f(\epsilon^{\mathrm{M}}_{\lambda})\,|\lambda^{\mathrm{M}}(t_1)\rangle\,\langle\lambda^{\mathrm{M}}(t_2)|.\end{aligned}$$

In particular, the generic matrix element in the position–spin basis reads

$$G^{<}(1;2) = \mp\mathrm{i}\sum_{\lambda} f(\epsilon^{\mathrm{M}}_{\lambda})\,\varphi^{\mathrm{M}}_{\lambda}(\mathbf{x}_1, t_1)\,\varphi^{\mathrm{M}*}_{\lambda}(\mathbf{x}_2, t_2),$$

with $\varphi^{\mathrm{M}}_{\lambda}(\mathbf{x}, t) = \langle\mathbf{x}|\lambda^{\mathrm{M}}(t)\rangle$ the time-evolved eigenfunction of $\hat{h}^{\mathrm{M}}$ (a similar formula can be derived for the greater component). Thus, the lesser (greater) Green's function of a noninteracting system can be constructed by populating the single-particle eigenfunctions of $\hat{h}^{\mathrm{M}}$ according to the Bose/Fermi function f ($\bar{f}$) and then evolving them according to the time-dependent Schrödinger equation with Hamiltonian $\hat{h}(t)$. The familiar result for the time-dependent density is readily recovered:

$$n(\mathbf{x}, t) = \pm\mathrm{i}\, G^{<}(\mathbf{x}, t; \mathbf{x}, t) = \sum_{\lambda} f(\epsilon^{\mathrm{M}}_{\lambda})\,|\varphi^{\mathrm{M}}_{\lambda}(\mathbf{x}, t)|^2. \tag{6.42}$$

Time-independent Hamiltonian The general dependence of $\hat{h}(t)$ on time prevents us from doing more analytic manipulations. Below we discuss the case in which $\hat{h}(t) = \hat{h}$ is time-independent. Then, the evolution operator is simply $\hat{\mathcal{U}}(t) = \exp[-\mathrm{i}\hat{h}(t - t_0)]$ and (6.40) and (6.41) simplify to

$$\hat{\mathcal{G}}^{<}(t_1, t_2) = \mp\mathrm{i}\, e^{-\mathrm{i}\hat{h}(t_1 - t_0)}\, f(\hat{h}^{\mathrm{M}})\, e^{\mathrm{i}\hat{h}(t_2 - t_0)}, \tag{6.43a}$$

$$\hat{\mathcal{G}}^{>}(t_1, t_2) = -\,\mathrm{i}\, e^{-\mathrm{i}\hat{h}(t_1 - t_0)}\, \bar{f}(\hat{h}^{\mathrm{M}})\, e^{\mathrm{i}\hat{h}(t_2 - t_0)}. \tag{6.43b}$$

Denoting by $|\lambda\rangle$ the eigenkets of $\hat{h}$ with eigenvalues ϵ_λ, we then see that

$$\begin{aligned}G^{<}_{\lambda\lambda}(t_1, t_2) &\equiv \langle\lambda|\hat{\mathcal{G}}^{<}(t_1, t_2)|\lambda\rangle = \mp\mathrm{i}\, e^{-\mathrm{i}\epsilon_\lambda(t_1 - t_2)}\langle\lambda|f(\hat{h}^{\mathrm{M}})|\lambda\rangle,\\ G^{>}_{\lambda\lambda}(t_1, t_2) &\equiv \langle\lambda|\hat{\mathcal{G}}^{>}(t_1, t_2)|\lambda\rangle = -\,\mathrm{i}\, e^{-\mathrm{i}\epsilon_\lambda(t_1 - t_2)}\langle\lambda|\bar{f}(\hat{h}^{\mathrm{M}})|\lambda\rangle.\end{aligned}$$

Neither $G^{<}_{\lambda\lambda}$ nor $G^{>}_{\lambda\lambda}$ vanish as $(t_1 - t_2) \to \infty$. Noninteracting systems therefore admit a complete set of good quantum numbers; particles with quantum numbers λ behave as independent particles.

We observe that the lesser and greater Green's functions in (6.43) are not functions of the time difference $t_1 - t_2$, in agreement with the discussion below (6.32). The invariance under time translations requires that the system is prepared in an eigenstate (or in a mixture of eigenstates) of $\hat{h}$ - that is, that $\hat{h}^{\mathrm{M}}$ commutes with $\hat{h}$. In this case,

$$\hat{\mathcal{G}}^{<}(t_1,t_2) = \mp \mathrm{i}\, f(\hat{h}^{\mathrm{M}})\, e^{-\mathrm{i}\hat{h}(t_1-t_2)},$$
$$\hat{\mathcal{G}}^{>}(t_1,t_2) = -\,\mathrm{i}\, \bar{f}(\hat{h}^{\mathrm{M}})\, e^{-\mathrm{i}\hat{h}(t_1-t_2)},$$

and the dependence on t_0 disappears. The time translational invariance allows us to define the Fourier transform,

$$\hat{\mathcal{G}}^{\lessgtr}(t_1,t_2) = \int \frac{d\omega}{2\pi} e^{-\mathrm{i}\omega(t_1-t_2)} \hat{\mathcal{G}}^{\lessgtr}(\omega), \tag{6.44}$$

and we see by inspection that

$$\boxed{\hat{\mathcal{G}}^{<}(\omega) = \mp 2\pi \mathrm{i}\, f(\hat{h}^{\mathrm{M}})\, \delta(\omega - \hat{h})} \tag{6.45}$$

$$\boxed{\hat{\mathcal{G}}^{>}(\omega) = -2\pi \mathrm{i}\, \bar{f}(\hat{h}^{\mathrm{M}})\, \delta(\omega - \hat{h})} = \pm e^{\beta \hat{h}^{\mathrm{M}}}\, \hat{\mathcal{G}}^{<}(\omega). \tag{6.46}$$

6.2.3 All Other Components

From the knowledge of the greater and lesser Green's functions we can extract all the remaining Keldysh components, see Table 5.1.

Retard and advanced Green's functions By definition the retarded Green's function is

$$\hat{\mathcal{G}}^{\mathrm{R}}(t_1,t_2) = \theta(t_1-t_2)[\hat{\mathcal{G}}^{>}(t_1,t_2) - \hat{\mathcal{G}}^{<}(t_1,t_2)] = -\mathrm{i}\,\theta(t_1-t_2)\hat{\mathcal{U}}(t_1)\hat{\mathcal{U}}^{\dagger}(t_2)$$
$$= -\mathrm{i}\,\theta(t_1-t_2) T\left\{e^{-\mathrm{i}\int_{t_2}^{t_1} d\bar{t}\, \hat{h}(\bar{t})}\right\}, \tag{6.47}$$

whereas the advanced Green's function reads

$$\hat{\mathcal{G}}^{\mathrm{A}}(t_1,t_2) = \mathrm{i}\,\theta(t_2-t_1)\bar{T}\left\{e^{\mathrm{i}\int_{t_1}^{t_2} d\bar{t}\, \hat{h}(\bar{t})}\right\} = [\hat{\mathcal{G}}^{\mathrm{R}}(t_2,t_1)]^{\dagger}. \tag{6.48}$$

It is interesting to observe that the retarded/advanced noninteracting Green's function does not depend on the initial density matrix. This means that $\hat{\mathcal{G}}^{\mathrm{R/A}}$ does not change by varying the initial number of particles or the distribution of the particles among the different energy levels. The information carried by $\hat{\mathcal{G}}^{\mathrm{R,A}}$ is the same as the information carried by the single-particle evolution operator $\hat{\mathcal{U}}$. We use this observation to rewrite $\hat{\mathcal{G}}^{\lessgtr}$ in (6.40) and (6.41) in terms of the retarded/advanced Green's function:

$$\boxed{\hat{\mathcal{G}}^{\lessgtr}(t_1,t_2) = \hat{\mathcal{G}}^{\mathrm{R}}(t_1,t_0)\hat{\mathcal{G}}^{\lessgtr}(t_0,t_0)\hat{\mathcal{G}}^{\mathrm{A}}(t_0,t_2)} \tag{6.49}$$

Left and right Green's functions: Analogous relations can be derived for the left/right Green's functions:

$$\hat{\mathcal{G}}^{\rceil}(t,\tau) = \mp \mathrm{i}\hat{\mathcal{U}}(t) f(\hat{h}^{\mathrm{M}}) e^{\tau\hat{h}^{\mathrm{M}}} = \mathrm{i}\hat{\mathcal{G}}^{\mathrm{R}}(t,t_0)\hat{\mathcal{G}}^{\mathrm{M}}(0,\tau), \tag{6.50}$$

$$\hat{\mathcal{G}}^{\lceil}(\tau,t) = -\mathrm{i}\, e^{-\tau\hat{h}^{\mathrm{M}}} \bar{f}(\hat{h}^{\mathrm{M}})\hat{\mathcal{U}}^{\dagger}(t) = -\mathrm{i}\hat{\mathcal{G}}^{\mathrm{M}}(\tau,0)\hat{\mathcal{G}}^{\mathrm{A}}(t_0,t). \tag{6.51}$$

Time-ordered Green's functions The time-ordered Green's function is

$$\hat{\mathcal{G}}^{\mathrm{T}}(t_1,t_2) = -\mathrm{i}\hat{\mathcal{U}}(t_1)\left[\theta(t_1-t_2)\bar{f}(\hat{h}^{\mathrm{M}}) \pm \theta(t_2-t_1) f(\hat{h}^{\mathrm{M}})\right]\hat{\mathcal{U}}^{\dagger}(t_2).$$

Time-independent Hamiltonian We conclude this section by considering again a system with Hamiltonian $\hat{h}(t) = \hat{h}$ constant in time. Then, from (6.47) and (6.48) the retarded/advanced Green's functions become

$$\hat{\mathcal{G}}^{\mathrm{R}}(t_1,t_2) = -\mathrm{i}\,\theta(t_1-t_2)e^{-\mathrm{i}\hat{h}(t_1-t_2)}, \tag{6.52a}$$

$$\hat{\mathcal{G}}^{\mathrm{A}}(t_1,t_2) = +\mathrm{i}\,\theta(t_2-t_1)e^{+\mathrm{i}\hat{h}(t_2-t_1)}, \tag{6.52b}$$

which, as expected, depend only on the time difference $t_1 - t_2$. This allows us to define the Fourier transform of $\hat{\mathcal{G}}^{\mathrm{R/A}}$ according to

$$\hat{\mathcal{G}}^{\mathrm{R,A}}(t_1,t_2) = \int \frac{d\omega}{2\pi} e^{-\mathrm{i}\omega(t_1-t_2)}\hat{\mathcal{G}}^{\mathrm{R,A}}(\omega).$$

The calculation of the Fourier transform $\hat{\mathcal{G}}^{\mathrm{R,A}}(\omega)$ is most easily done by using the representation of the Heaviside function,

$$\theta(t_1-t_2) = \mathrm{i}\int \frac{d\omega}{2\pi}\frac{e^{-\mathrm{i}\omega(t_1-t_2)}}{\omega+\mathrm{i}\eta}, \tag{6.53}$$

with η an infinitesimal positive constant. We find

$$\hat{\mathcal{G}}^{\mathrm{R}}(t_1,t_2) = \int \frac{d\omega}{2\pi}\frac{e^{-\mathrm{i}(\omega+\hat{h})(t_1-t_2)}}{\omega+\mathrm{i}\eta},$$

and changing the variable of integration $\omega \to \omega - \hat{h}$ we can identify the Fourier transform with

$$\boxed{\hat{\mathcal{G}}^{\mathrm{R}}(\omega) = \frac{1}{\omega-\hat{h}+\mathrm{i}\eta} = \sum_{\lambda}\frac{|\lambda\rangle\langle\lambda|}{\omega-\epsilon_\lambda+\mathrm{i}\eta}} \tag{6.54}$$

where the sum runs over a complete set of eigenkets $|\lambda\rangle$ of $\hat{h}$ with eigenvalues ϵ_λ. To calculate the Fourier transform of the advanced Green's function, we notice that the property (6.48) implies

$$\hat{\mathcal{G}}^{\mathrm{A}}(\omega) = [\hat{\mathcal{G}}^{\mathrm{R}}(\omega)]^{\dagger},$$

and hence

$$\boxed{\hat{\mathcal{G}}^{\mathrm{A}}(\omega) = \frac{1}{\omega - \hat{h} - \mathrm{i}\eta} = \sum_{\lambda} \frac{|\lambda\rangle\langle\lambda|}{\omega - \epsilon_\lambda - \mathrm{i}\eta}} \tag{6.55}$$

The retarded Green's function is analytic in the upper half of the complex ω plane, whereas the advanced Green's function is analytic in the lower half. As we see in Section 6.3, this analytic structure is completely general since it is a direct consequence of *causality*: $\hat{\mathcal{G}}^{\mathrm{R}}(t_1, t_2)$ vanishes for $t_1 < t_2$, whereas $\hat{\mathcal{G}}^{\mathrm{A}}(t_1, t_2)$ vanishes for $t_1 > t_2$.

Fluctuation–dissipation theorem The results (6.54) and (6.55) are interesting also for another reason. From the definition of the retarded/advanced component of a Keldysh function we have the general relation (omitting the time arguments)

$$\hat{\mathcal{G}}^{\mathrm{R}} - \hat{\mathcal{G}}^{\mathrm{A}} = \hat{\mathcal{G}}^{>} - \hat{\mathcal{G}}^{<},$$

which tells us that the *difference* between the lesser and greater Green's function can be expressed in terms of the difference between the retarded and advanced Green's function. We now show that a system prepared in a stationary excited state has lesser and greater Green's functions that can *separately* be written in terms of $\hat{\mathcal{G}}^{\mathrm{R}} - \hat{\mathcal{G}}^{\mathrm{A}}$. For the system to be in a stationary excited state the Hamiltonian $\hat{h}^{\mathrm{M}}$ must commute with $\hat{h}$. Then, $\hat{\mathcal{G}}^{\lessgtr}$ is given by (6.45) and (6.46) which, taking into account (6.54) and (6.55) together with the Cauchy relation (2.28), can also be written as

$$\boxed{\hat{\mathcal{G}}^{<}(\omega) = \pm f(\hat{h}^{\mathrm{M}})[\hat{\mathcal{G}}^{\mathrm{R}}(\omega) - \hat{\mathcal{G}}^{\mathrm{A}}(\omega)]}$$

$$\boxed{\hat{\mathcal{G}}^{>}(\omega) = \bar{f}(\hat{h}^{\mathrm{M}})[\hat{\mathcal{G}}^{\mathrm{R}}(\omega) - \hat{\mathcal{G}}^{\mathrm{A}}(\omega)]}$$

In the special case that $\hat{h}^{\mathrm{M}} = \hat{h} - \mu$ describes the system in thermal equilibrium, the above relations reduce to the so-called *fluctuation–dissipation theorem*,

$$\hat{\mathcal{G}}^{<}(\omega) = \pm f(\omega - \mu)[\hat{\mathcal{G}}^{\mathrm{R}}(\omega) - \hat{\mathcal{G}}^{\mathrm{A}}(\omega)],$$
$$\hat{\mathcal{G}}^{>}(\omega) = \bar{f}(\omega - \mu)[\hat{\mathcal{G}}^{\mathrm{R}}(\omega) - \hat{\mathcal{G}}^{\mathrm{A}}(\omega)].$$

As we see in the next section, the fluctuation–dissipation theorem is valid in interacting systems as well.

Time-ordered Green's function for stationary states Let us also calculate the time-ordered Green's function for the case in which $\hat{h}^{\mathrm{M}}$ commutes with $\hat{h}$. Using the results (6.45) and (6.46) as well as the representation (6.53) of the Heaviside function, we find

$$\hat{\mathcal{G}}^{\mathrm{T}}(t_1, t_2) = \int \frac{d\omega}{2\pi} e^{-\mathrm{i}\omega(t_1 - t_2)} \underbrace{\left[\frac{\bar{f}(\hat{h}^{\mathrm{M}})}{\omega - \hat{h} + \mathrm{i}\eta} \mp \frac{f(\hat{h}^{\mathrm{M}})}{\omega - \hat{h} - \mathrm{i}\eta} \right]}_{\hat{\mathcal{G}}^{\mathrm{T}}(\omega)}.$$

The Fourier transform of the time-ordered Green's function has poles on both sides of the complex plane. For fermions in equilibrium at zero temperature $\hat{h}^{\mathrm{M}} = \hat{h} - \epsilon_{\mathrm{F}}$, with

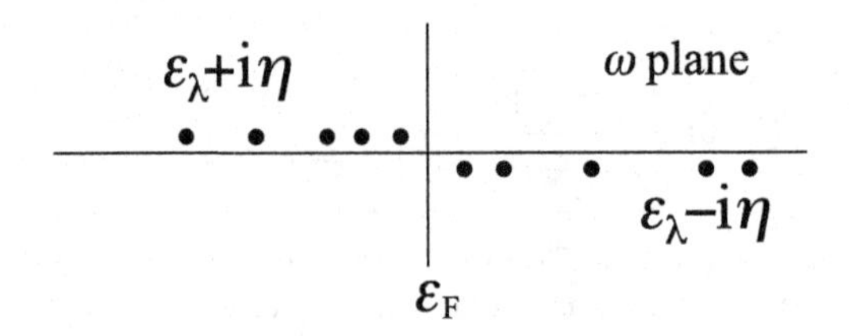

Figure 6.4 Location of the poles of the time-ordered Green's function for a system of fermions at zero temperature.

$\epsilon_F = \lim_{\beta\to\infty}\mu$ the Fermi energy, and $f(\epsilon_\lambda - \mu) = \theta(\epsilon_F - \epsilon_\lambda)$. The equilibrium ensemble is a pure state where all eigenkets $|\lambda\rangle$ of $\hat{h}$ with energy $\epsilon_\lambda < \epsilon_F$ are populated. Then

$$\hat{\mathcal{G}}^{\mathrm{T}}(\omega) = \sum_{\epsilon_\lambda > \epsilon_F} \frac{|\lambda\rangle\langle\lambda|}{\omega - \epsilon_\lambda + \mathrm{i}\eta} + \sum_{\epsilon_\lambda < \epsilon_F} \frac{|\lambda\rangle\langle\lambda|}{\omega - \epsilon_\lambda - \mathrm{i}\eta}. \qquad \text{Fermions at zero temperature}$$

As illustrated in Fig. 6.4, the poles of $\hat{\mathcal{G}}^{\mathrm{T}}(\omega)$ with real part smaller than ϵ_F lie on the upper half of the complex ω-plane, whereas those with real part larger than ϵ_F lie on the lower half of the complex ω-plane. This property is maintained in interacting systems as well.

Exercise 6.1 Consider a system of particles in one dimension with single-particle Hamiltonian $\hat{h} = \hat{p}^2/2$. Taking into account that the single-particle eigenkets are the momentum–spin kets $|p\sigma\rangle$ with eigenvalue $p^2/2$, use (6.54) to show that

$$\langle x\sigma|\hat{\mathcal{G}}^{\mathrm{R}}(\omega)|x'\sigma'\rangle = \delta_{\sigma\sigma'} G^{\mathrm{R}}(x, x'; \omega),$$

with

$$G^{\mathrm{R}}(x, x'; \omega) = -\frac{1}{\sqrt{2|\omega|}} \begin{cases} \mathrm{i}\, e^{\mathrm{i}\sqrt{2|\omega|}\,|x-x'|} & \omega > 0 \\ e^{-\sqrt{2|\omega|}\,|x-x'|} & \omega < 0 \end{cases}.$$

Exercise 6.2 Consider the same system as Exercise 6.1, in which a δ-like potential $\hat{\phi} = \lambda\delta(\hat{x})$, $\lambda > 0$, is added to the free Hamiltonian. Show that the new Green's function is given by

$$G^{\mathrm{R}}(x, x'; \omega) = -\frac{1}{\sqrt{2|\omega|}} \begin{cases} \mathrm{i}\, e^{\mathrm{i}\sqrt{2|\omega|}\,|x-x'|} + \dfrac{\lambda e^{\mathrm{i}\sqrt{2|\omega|}\,(|x|+|x'|)}}{\sqrt{2|\omega|} + \mathrm{i}\lambda} & \omega > 0 \\[2ex] e^{-\sqrt{2|\omega|}\,|x-x'|} - \dfrac{\lambda e^{-\sqrt{2|\omega|}\,(|x|+|x'|)}}{\sqrt{2|\omega|} + \lambda} & \omega < 0 \end{cases}.$$

How does this result change if $\lambda < 0$? (Think about the formation of bound states.)

Exercise 6.3 Consider the single-level noninteracting Hamiltonian $\hat{H} = \epsilon\, \hat{d}^\dagger \hat{d}$ with eigenkets $|\Psi_k\rangle = \frac{(\hat{d}^\dagger)^k}{\sqrt{k!}}|0\rangle$ and eigenvalues $E_k = k\epsilon$. Show that if the average occupation $\mathrm{Tr}[\,\hat{\rho}\,\hat{d}^\dagger \hat{d}\,] =$

n, then the density matrix in thermal equilibrium reads

$$\hat{\rho} = \frac{e^{-\beta(\hat{H}-\mu\hat{N})}}{\mathrm{Tr}\left[e^{-\beta(\hat{H}-\mu\hat{N})}\right]} = \frac{\sum_k \left(\frac{n}{n\pm1}\right)^k |\Psi_k\rangle\langle\Psi_k|}{\sum_k \left(\frac{n}{n\pm1}\right)^k},$$

where the upper/lower sign applies to bosons/fermions and the sum over k runs between 0 and ∞ in the case of bosons and between 0 and 1 in the case of fermions. We then see that in fermionic systems $\hat{\rho}$ is a pure state for $n \to 0$ or $n \to 1$, whereas in bosonic systems $\hat{\rho}$ is a mixture of states for all values of $n \neq 0$.

6.3 Interacting Green's Function and Lehmann Representation

In this section we discuss the Green's function of an interacting system with Hamiltonian $\hat{H}(t) = \hat{H}$ constant in time. The *Lehmann representation* of the n-particle Green's function G_n is simply a rewriting of the definition of G_n in which every evolution operator is expanded over a complete set of eigenstates of $\hat{H}$. The resulting expression is, in general, rather cumbersome already for the one-particle Green's function G; for instance, in (6.32) one should expand three evolution operators. Below we confine the discussion to systems initially prepared in a stationary ensemble. This case includes the one treated in most textbooks: thermal equilibrium.

6.3.1 Fluctuation-Dissipation Theorem and Other Exact Properties

Systems prepared in a stationary ensemble of $\hat{H}$ are described by a Matsubara operator $\hat{H}^{\mathrm{M}}$ that commutes with $\hat{H}$: $[\hat{H}^{\mathrm{M}}, \hat{H}]_- = 0$. To this class of initial states belongs the equilibrium state for which $\hat{H}^{\mathrm{M}} = \hat{H} - \mu\hat{N}$. The eigenkets $|\Psi_k\rangle$ of $\hat{H}^{\mathrm{M}}$ in (6.25) and (6.26) can then be chosen to be also eigenkets of $\hat{H}$ with eigenvalue E_k, and the general formula for the lesser and greater Green's function simplifies to

$$G^<_{ji}(t,t') = \mp\mathrm{i}\sum_k \rho_k \langle\Psi_k|\hat{d}^\dagger_i e^{-\mathrm{i}(\hat{H}-E_k)(t'-t)}\hat{d}_j|\Psi_k\rangle, \tag{6.56}$$

$$G^>_{ji}(t,t') = -\mathrm{i}\sum_k \rho_k \langle\Psi_k|\hat{d}_j e^{-\mathrm{i}(\hat{H}-E_k)(t-t')}\hat{d}^\dagger_i|\Psi_k\rangle. \tag{6.57}$$

As expected, all matrix elements of $\hat{\mathcal{G}}^{\lessgtr}(t,t')$ depend only on the difference $t-t'$ and, therefore, can be Fourier transformed as in (6.44). In Fourier space the relation between $\hat{\mathcal{G}}^{\lessgtr}(\omega)$ and $\hat{\mathcal{G}}^{\mathrm{R/A}}(\omega)$ is particularly elegant. Using the representation (6.53) of the Heaviside function, we have

$$\hat{\mathcal{G}}^{\mathrm{R}}(t,t') = \mathrm{i}\int\frac{d\omega}{2\pi}\frac{e^{-\mathrm{i}\omega(t-t')}}{\omega+\mathrm{i}\eta}\int\frac{d\omega'}{2\pi}e^{-\mathrm{i}\omega'(t-t')}\left[\hat{\mathcal{G}}^>(\omega') - \hat{\mathcal{G}}^<(\omega')\right],$$

from which it follows that

$$\hat{\mathcal{G}}^{\mathrm{R}}(\omega) = \mathrm{i} \int \frac{d\omega'}{2\pi} \frac{\hat{\mathcal{G}}^{>}(\omega') - \hat{\mathcal{G}}^{<}(\omega')}{\omega - \omega' + \mathrm{i}\eta}. \tag{6.58}$$

For the Fourier transform of the advanced Green's function we observe that (6.28) implies $\hat{\mathcal{G}}^{\mathrm{A}}(\omega) = [\hat{\mathcal{G}}^{\mathrm{R}}(\omega)]^{\dagger}$ and hence

$$\hat{\mathcal{G}}^{\mathrm{A}}(\omega) = \mathrm{i} \int \frac{d\omega'}{2\pi} \frac{\hat{\mathcal{G}}^{>}(\omega') - \hat{\mathcal{G}}^{<}(\omega')}{\omega - \omega' - \mathrm{i}\eta}, \tag{6.59}$$

where we take into account that the operator $\mathrm{i}[\hat{\mathcal{G}}^{>}(\omega') - \hat{\mathcal{G}}^{<}(\omega')]$ is self-adjoint, see (6.27).

Going back to (6.56) and (6.57) we expand the evolution operator over a complete set of eigenstates of $\hat{H}$ and obtain the Lehmann representation,

$$\boxed{G^{<}_{ji}(t,t') = \mp \mathrm{i} \sum_{pk} \rho_k \, \Phi^*_{pk}(i) \Phi_{pk}(j) e^{-\mathrm{i}(E_p - E_k)(t'-t)}} \tag{6.60}$$

$$\boxed{G^{>}_{ji}(t,t') = -\mathrm{i} \sum_{pk} \rho_k \, \Phi_{kp}(j) \Phi^*_{kp}(i) e^{-\mathrm{i}(E_p - E_k)(t-t')}} \tag{6.61}$$

where we define the amplitudes

$$\Phi_{kp}(i) = \langle \Psi_k | \hat{d}_i | \Psi_p \rangle. \tag{6.62}$$

For these amplitudes to be different from zero, $|\Psi_p\rangle$ must contain one particle more than $|\Psi_k\rangle$.

Quasi-particle wavefunctions To gain some insight into the physical meaning of the Φ_{kp}, let us consider a noninteracting Hamiltonian $\hat{H} = \sum_\lambda \hat{c}^\dagger_\lambda \hat{c}_\lambda$. Then, the generic eigenket with N particles is

$$|\Psi_k\rangle = \hat{c}^\dagger_{\lambda_1} \ldots \hat{c}^\dagger_{\lambda_N} |0\rangle,$$

and the only nonvanishing amplitudes are those with $|\Psi_p\rangle = \hat{c}^\dagger_\lambda |\Psi_k\rangle$ - that is, $|\Psi_p\rangle$ must be an eigenket with $N+1$ particles, N of which are in the energy levels $\lambda_1, \ldots, \lambda_N$. In this case,

$$\Phi_{kp}(i) = \langle \Psi_k | \hat{d}_i \hat{c}^\dagger_\lambda | \Psi_k \rangle = C \varphi_\lambda(i)$$

is proportional to the wavefunction of the λth single-particle eigenstate;[8] in particular, if $\hat{d}_i$ is the field operator $\hat{\psi}(\mathbf{x})$ then $\Phi_{kp}(\mathbf{x}) = C\varphi_\lambda(\mathbf{x})$. In interacting systems the Φ_{kp} cannot be identified with a single-particle eigenfunction. Nevertheless, they satisfy a single-particle Schrödinger-like equation that we now derive. Let $\hat{H} = \sum_{ij} h_{ij} \hat{d}^\dagger_i \hat{d}_j + \hat{H}_{\mathrm{int}}$ be the Hamiltonian of the interacting system and consider the sandwich of the commutator

$$\left[\hat{d}_i, \hat{H}\right]_- = \sum_j h_{ij} \hat{d}_j + \left[\hat{d}_i, \hat{H}_{\mathrm{int}}\right]_-$$

[8] The constant of proportionality C is 1 for fermions and $n_\lambda!$ for bosons, where n_λ is the number of bosons in the λth energy level, see Chapter 1.

with the states $\langle\Psi_k|$ and $|\Psi_p\rangle$. Taking into account definition (6.62), we find

$$\sum_j h_{ij}\Phi_{kp}(j) + \sum_q \left[\Phi_{kq}(i)I_{qp} - I_{kq}\Phi_{qp}(i)\right] = (E_p - E_k)\Phi_{kp}(i),$$

where

$$I_{qp} = \langle\Psi_q|\hat{H}_{\text{int}}|\Psi_p\rangle.$$

For $\hat{H}_{\text{int}} = 0$, the single-particle Schrödinger-like equation is solved by $E_p - E_k = \epsilon_\lambda$ and $\Phi_{kp}(i) = \varphi_\lambda(i)$, in agreement with the discussion above. In the general case we refer to $\Phi_{kp}(i)$ as the *quasi-particle wavefunction.*

Fluctuation-dissipation theorem From (6.60) and (6.61) we can immediately read the Fourier transform of the lesser and greater Green's functions:

$$\boxed{G^<_{ji}(\omega) = \mp 2\pi\text{i}\sum_{pk}\rho_k\,\Phi^*_{pk}(i)\Phi_{pk}(j)\delta(\omega - E_k + E_p) = -G^{<*}_{ij}(\omega)} \tag{6.63}$$

$$\boxed{G^>_{ji}(\omega) = -2\pi\text{i}\sum_{pk}\rho_k\,\Phi_{kp}(j)\Phi^*_{kp}(i)\delta(\omega - E_p + E_k) = -G^{>*}_{ij}(\omega)} \tag{6.64}$$

The diagonal elements of $\text{i}G^{\lessgtr}$ have a well-defined sign for all frequencies ω,

$$\text{i}G^>_{jj}(\omega) \geq 0 \quad , \quad \text{i}G^<_{jj}(\omega) \begin{array}{ll} \geq 0 & \text{for bosons} \\ \leq 0 & \text{for fermions.} \end{array} \tag{6.65}$$

Substituting these results into (6.58) and (6.59), we find

$$\boxed{G^{\text{R/A}}_{ji}(\omega) = \sum_{pk}\frac{\Phi_{kp}(j)\Phi^*_{kp}(i)}{\omega - E_p + E_k \pm \text{i}\eta}\left[\,\rho_k \mp \rho_p\,\right]} \tag{6.66}$$

where in $G^<$ we have renamed the summation indices $k \leftrightarrow p$. We see that the Fourier transform of the retarded Green's function is analytic in the upper half of the complex ω plane, whereas the Fourier transform of the advanced Green's function is analytic in the lower half of the complex ω plane, in agreement with the results of Section 6.2.3.

The Fourier transforms can be used to derive an important result for systems in thermal equilibrium, $\hat{H}^{\text{M}} = \hat{H} - \mu\hat{N}$. In this case,

$$\rho_k = \frac{e^{-\beta(E_k - \mu N_k)}}{\text{Tr}\left[e^{-\beta(\hat{H} - \mu\hat{N})}\right]} = e^{-\beta(E_k - E_p) + \beta\mu(N_k - N_p)}\rho_p,$$

with N_k the number of particles in state $|\Psi_k\rangle$. Substituting this result into $G^>$, renaming the summation indices $k \leftrightarrow p$, and taking into account that the only nonvanishing Φ_{kp} are those for which $N_k - N_p = -1$, we find

$$\boxed{\hat{\mathcal{G}}^>(\omega) = \pm e^{\beta(\omega-\mu)}\hat{\mathcal{G}}^<(\omega)} \tag{6.67}$$

Next we recall that by the very same definition of retarded/advanced functions,

$$\hat{\mathcal{G}}^{>}(\omega) - \hat{\mathcal{G}}^{<}(\omega) = \hat{\mathcal{G}}^{\mathrm{R}}(\omega) - \hat{\mathcal{G}}^{\mathrm{A}}(\omega).$$

The combination of these last two identities leads to the *fluctuation–dissipation theorem,*

$$\boxed{\hat{\mathcal{G}}^{<}(\omega) = \pm f(\omega - \mu)[\hat{\mathcal{G}}^{\mathrm{R}}(\omega) - \hat{\mathcal{G}}^{\mathrm{A}}(\omega)]} \tag{6.68}$$

$$\boxed{\hat{\mathcal{G}}^{>}(\omega) = \bar{f}(\omega - \mu)[\hat{\mathcal{G}}^{\mathrm{R}}(\omega) - \hat{\mathcal{G}}^{\mathrm{A}}(\omega)]} \tag{6.69}$$

which was previously demonstrated only for noninteracting particles.

There exists also another way of deriving (6.67) and hence the fluctuation–dissipation theorem. By definition the left Green's function is

$$\begin{aligned} G^{\lceil}_{ji}(\tau,t') &= -\mathrm{i}\sum_k \rho_k \langle \Psi_k | \underbrace{e^{\mathrm{i}(\hat{H}-\mu\hat{N})(-\mathrm{i}\tau)}\hat{d}_j e^{-\mathrm{i}(\hat{H}-\mu\hat{N})(-\mathrm{i}\tau)}}_{\hat{d}_{j,H}(t_0-\mathrm{i}\tau)} \underbrace{e^{\mathrm{i}\hat{H}(t'-t_0)}\hat{d}^{\dagger}_i e^{-\mathrm{i}\hat{H}(t'-t_0)}}_{\hat{d}^{\dagger}_{i,H}(t')} |\Psi_k\rangle \\ &= -\mathrm{i}e^{\mu\tau}\sum_k \rho_k \langle \Psi_k | \hat{d}_j e^{-\mathrm{i}(\hat{H}-E_k)(t_0-\mathrm{i}\tau-t')}\hat{d}^{\dagger}_i |\Psi_k\rangle, \end{aligned}$$

and comparing this result with (6.57) we conclude that

$$\hat{\mathcal{G}}^{\lceil}(\tau,t') = e^{\mu\tau}\hat{\mathcal{G}}^{>}(t_0 - \mathrm{i}\tau, t').$$

A similar relation can be derived for the lesser and right Green's function,

$$\hat{\mathcal{G}}^{\rceil}(t,\tau) = e^{-\mu\tau}\hat{\mathcal{G}}^{<}(t, t_0 - \mathrm{i}\tau).$$

Combining these results with the KMS relations we find

$$\begin{aligned} \hat{\mathcal{G}}^{<}(t_0,t') &= \hat{\mathcal{G}}(t_{0-},t'_{+}) \\ &= \pm\hat{\mathcal{G}}(t_0 - \mathrm{i}\beta, t'_{+}) \\ &= \pm\hat{\mathcal{G}}^{\lceil}(\beta,t') \\ &= \pm e^{\mu\beta}\hat{\mathcal{G}}^{>}(t_0 - \mathrm{i}\beta, t'). \end{aligned}$$

Taking the Fourier transform of both sides, we recover (6.67).

Relation between retarded/advanced and Matsubara components For systems in thermal equilibrium we can also derive an important relation between the Matsubara Green's function and the retarded/advanced Green's function. The starting point is (6.14). Expanding the exponentials in a complete set of eigenstates of $\hat{H}^{\mathrm{M}} = \hat{H} - \mu\hat{N}$, we find

$$G^{\mathrm{M}}_{ji}(\tau_1,\tau_2) = \frac{1}{\mathrm{i}}\sum_{kp}\rho_p\Big[\theta(\tau_1-\tau_2)e^{\beta(E^{\mathrm{M}}_p - E^{\mathrm{M}}_k)} \pm \theta(\tau_2-\tau_1)\Big]e^{-(\tau_1-\tau_2)(E^{\mathrm{M}}_p - E^{\mathrm{M}}_k)}\Phi_{kp}(j)\Phi^{*}_{kp}(i),$$

where $E_p^{\mathrm{M}} = E_p - \mu N_p$. To obtain this result we have renamed the summation indices $k \leftrightarrow p$ in the first term and used that $\rho_k = e^{\beta(E_p^{\mathrm{M}} - E_k^{\mathrm{M}})}\rho_p$. In Section 6.2.1 we have proven the identity [recall that $\bar{f}(E) = e^{\beta E} f(E)$]

$$\frac{1}{-\mathrm{i}\beta}\sum_{m=-\infty}^{\infty}\frac{e^{-\omega_m \tau}}{\omega_m - E} = \frac{1}{\mathrm{i}}\left[\theta(\tau)e^{\beta E} \pm \theta(-\tau)\right] f(E)e^{-\tau E}.$$

The r.h.s. of this equation has precisely the same structure appearing in G^{M}. Therefore we can rewrite the Matsubara Green's function as

$$G_{ji}^{\mathrm{M}}(\tau_1,\tau_2) = \frac{1}{-\mathrm{i}\beta}\sum_{m=-\infty}^{\infty} e^{-\omega_m(\tau_1-\tau_2)} \underbrace{\sum_{kp}\frac{\rho_p}{f(E_p^{\mathrm{M}} - E_k^{\mathrm{M}})}\frac{\Phi_{kp}(j)\Phi_{kp}^*(i)}{\omega_m - E_p^{\mathrm{M}} + E_k^{\mathrm{M}}}}_{G_{ji}^{\mathrm{M}}(\omega_m)}.$$

Let us manipulate this formula. We have

$$\frac{\rho_p}{f(E_p^{\mathrm{M}} - E_k^{\mathrm{M}})} = \rho_p\left(e^{\beta(E_p^{\mathrm{M}} - E_k^{\mathrm{M}})} \mp 1\right) = \rho_k \mp \rho_p.$$

Furthermore, the only nonvanishing terms in the sum over p and k are those for which $N_p - N_k = 1$, and hence

$$E_p^{\mathrm{M}} - E_k^{\mathrm{M}} = E_p - E_k - \mu(N_p - N_k) = E_p - E_k - \mu.$$

Inserting these results into the formula for G^{M}, we find

$$\boxed{G_{ji}^{\mathrm{M}}(\omega_m) = \sum_{kp}\frac{\Phi_{kp}(j)\Phi_{kp}^*(i)}{\omega_m + \mu - E_p + E_k}\left[\,\rho_k \mp \rho_p\,\right]}$$

and a comparison with (6.66) leads to the important relation

$$\hat{\mathcal{G}}^{\mathrm{M}}(\zeta) = \begin{cases} \hat{\mathcal{G}}^{\mathrm{R}}(\zeta + \mu) & \text{for } \mathrm{Im}[\zeta] > 0 \\ \hat{\mathcal{G}}^{\mathrm{A}}(\zeta + \mu) & \text{for } \mathrm{Im}[\zeta] < 0. \end{cases} \tag{6.70}$$

Thus, $\hat{\mathcal{G}}^{\mathrm{M}}(\zeta)$ is analytic everywhere except along the real axis where it can have poles or branch points. The reader can easily check that (6.70) agrees with formulas (6.21), (6.54), and (6.55) for noninteracting Green's functions. In particular, (6.70) implies that for $\zeta = \omega \pm \mathrm{i}\eta$,

$$\boxed{\hat{\mathcal{G}}^{\mathrm{M}}(\omega \pm \mathrm{i}\eta) = \hat{\mathcal{G}}^{\mathrm{R/A}}(\omega + \mu)} \tag{6.71}$$

according to which $\hat{\mathcal{G}}^{\mathrm{M}}$ has a discontinuity given by the difference $\hat{\mathcal{G}}^{\mathrm{R}} - \hat{\mathcal{G}}^{\mathrm{A}}$ when the complex frequency crosses the real axis.

6.3.2 Spectral Function and Probability Interpretation

The Lehmann representation (6.60) and (6.61) simplifies further for systems that are initially in a pure state, and hence $\hat{\rho} = |\Psi_{N,0}\rangle\langle\Psi_{N,0}|$. In the discussion that follows the eigenstate $|\Psi_{N,0}\rangle$ with eigenenergy $E_{N,0}$ can be either the ground state or an excited state of $\hat{H}$ with N particles. We denote by $|\Psi_{N\pm1,m}\rangle$ the eigenstates of $\hat{H}$ with $N \pm 1$ particles and define the *quasi-particle wavefunctions* P_m and the *quasi-hole wavefunctions* Q_m according to

$$P_m(i) = \langle\Psi_{N,0}|\hat{d}_i|\Psi_{N+1,m}\rangle, \qquad Q_m(i) = \langle\Psi_{N-1,m}|\hat{d}_i|\Psi_{N,0}\rangle. \tag{6.72}$$

Then, the lesser and greater Green's functions become

$$G^<_{ji}(t,t') = \mp\mathrm{i}\sum_m Q_m(j)Q^*_m(i)e^{-\mathrm{i}(E_{N-1,m}-E_{N,0})(t'-t)}, \tag{6.73}$$

$$G^>_{ji}(t,t') = -\mathrm{i}\sum_m P_m(j)P^*_m(i)e^{-\mathrm{i}(E_{N+1,m}-E_{N,0})(t-t')}, \tag{6.74}$$

where $E_{N\pm1,m}$ is the energy eigenvalue of $|\Psi_{N\pm1,m}\rangle$. From (6.73) we see that the probability amplitude illustrated pictorially in Fig. 6.2 can be written as the sum of oscillatory functions whose frequency $(E_{N,0} - E_{N-1,m})$ corresponds to a possible *ionization energy* (also called *removal energy*) of the system. Similarly, the probability amplitude (described by $G^>$) that by adding a particle with quantum number i at time t' and then evolving until t we find the same state as evolving until t and then adding a particle with quantum number j, can be written as the sum of oscillatory functions whose frequency $(E_{N+1,m} - E_{N,0})$ corresponds to a possible *affinity* (also called *addition energy*) of the system.

It is especially interesting to look at the Fourier transforms of these functions:

$$\boxed{G^<_{ji}(\omega) = \mp2\pi\mathrm{i}\sum_m Q_m(j)Q^*_m(i)\,\delta(\omega - [E_{N,0} - E_{N-1,m}])} \tag{6.75}$$

$$\boxed{G^>_{ji}(\omega) = -2\pi\mathrm{i}\sum_m P_m(j)P^*_m(i)\,\delta(\omega - [E_{N+1,m} - E_{N,0}])} \tag{6.76}$$

The Fourier transform of $G^<$ is peaked at the removal energies, whereas the Fourier transform of $G^>$ is peaked at the addition energies. We can say that by removing (adding) a particle, the system gets excited in a combination of eigenstates with one particle fewer (one particle more). In noninteracting systems the matrix elements $G^{\lessgtr}_{\lambda\lambda'}(\omega)$ in the basis that diagonalizes $\hat{h}$ can be extracted from (6.45) and (6.46) and read

$$G^<_{\lambda\lambda'}(\omega) = \mp\delta_{\lambda\lambda'}2\pi\mathrm{i}f(\epsilon^{\mathrm{M}}_\lambda)\delta(\omega - \epsilon_\lambda), \qquad G^>_{\lambda\lambda'}(\omega) = -\delta_{\lambda\lambda'}2\pi\mathrm{i}\bar{f}(\epsilon^{\mathrm{M}}_\lambda)\delta(\omega - \epsilon_\lambda).$$

Thus, the removal (addition) of a particle with quantum number λ excites the system in only one way since the lesser (greater) Green's function is peaked at the removal (addition) energy ϵ_λ and is zero otherwise. This property reflects the fact that in a noninteracting system there exists a complete set of *good quantum numbers* – that is, a complete set of states in which the particle is not scattered by any other particle. Particles in these states are the *independent particles* introduced in Section 6.1.3; they have a well-defined energy and hence an infinitely long lifetime. In interacting systems the good quantum numbers, if any, form a subset of zero measure in the one-particle basis set. In general, a particle scatters with all the other particles and its energy cannot be sharply defined.

Spectral function It would be useful to construct a frequency-dependent operator $\hat{\mathcal{A}}(\omega)$ whose average $A_{jj}(\omega) = \langle j|\hat{\mathcal{A}}(\omega)|j\rangle$ contains information about the probability for an added/removed particle with quantum number j to have energy ω. From the above discussion a natural proposal for this operator is

$$\boxed{\hat{\mathcal{A}}(\omega) = \mathrm{i}\,[\hat{\mathcal{G}}^{>}(\omega) - \hat{\mathcal{G}}^{<}(\omega)] = \mathrm{i}\,[\hat{\mathcal{G}}^{\mathrm{R}}(\omega) - \hat{\mathcal{G}}^{\mathrm{A}}(\omega)]} \tag{6.77}$$

Taking into account that $\hat{\mathcal{G}}^{\lessgtr}(\omega)$ is anti-Hermitian, see (6.63) and (6.64), we infer that $\hat{\mathcal{A}}(\omega)$ is a self-adjoint operator for all ω:

$$\hat{\mathcal{A}}(\omega) = \hat{\mathcal{A}}^{\dagger}(\omega). \tag{6.78}$$

In noninteracting systems[9] we have $\hat{\mathcal{A}}(\omega) = 2\pi\delta(\omega - \hat{h})$ and hence

$$A_{jj}(\omega) = 2\pi \sum_{\lambda} |\langle j|\lambda\rangle|^2 \delta(\omega - \epsilon_\lambda) \geq 0. \tag{6.79}$$

The standard interpretation of this result is that the probability for a particle with quantum number j to have energy ω is zero unless ω is one of the single-particle energies, in which case the probability is proportional to $|\langle j|\lambda\rangle|^2$. This interpretation is sound also for another reason. The probability that the particle has energy ω in the range $(-\infty, \infty)$ should be 1 for any j, and indeed from (6.79) we have

$$\int \frac{d\omega}{2\pi} A_{jj}(\omega) = \sum_{\lambda} |\langle j|\lambda\rangle|^2 = 1.$$

Furthermore, the sum over the complete set j of $A_{jj}(\omega)$ should correspond to the density of states $D(\omega)$, and indeed

$$D(\omega) = \sum_{j} A_{jj}(\omega) = 2\pi \sum_{\lambda} \delta(\omega - \epsilon_\lambda). \tag{6.80}$$

A suitable name for the operator $\hat{\mathcal{A}}(\omega)$ is *spectral function* operator since it contains information on the energy spectrum of a single particle. The spectral function is a very useful mathematical quantity, but the above probabilistic interpretation is questionable. Is $A_{jj}(\omega)$ the probability that a removed particle or an added particle with quantum number j has energy ω? Are we sure that $A_{jj}(\omega) \geq 0$ also in the interacting case?

---

[9]See also the discussion in Section 2.4.

In the interacting case, the matrix element $A_{jj}(\omega)$ can be derived from (6.75) and (6.76) and reads

$$A_{jj}(\omega) = 2\pi \left[\sum_m |P_m(j)|^2 \delta(\omega - [E_{N+1,m} - E_{N,0}]) \mp \sum_m |Q_m(j)|^2 \delta(\omega - [E_{N,0} - E_{N-1,m}]) \right]. \tag{6.81}$$

A comparison with (6.79) shows how natural it is to interpret the functions P_m and Q_m as quasi-particle and quasi-hole wavefunctions. In fermionic systems $A_{jj}(\omega) \geq 0$, but in bosonic systems $A_{jj}(\omega)$ can be negative since the second term on the r.h.s. of (6.81) is nonpositive. For instance, one can show that the spectral function of the bosonic Hubbard model is negative for some ω [60, 61], see also Exercise 6.7. As we see in the next section, the particle and hole contributions to $A(\omega)$ can be measured separately and therefore the quantities $\mathrm{i}G^{>}(\omega) > 0$ and $\pm \mathrm{i}G^{<}(\omega) > 0$, see (6.65), are more fundamental than $A(\omega)$. Even though these quantities do not integrate to unity we can always normalize them and interpret $\mathrm{i}G^{>}(\omega)$ as the probability that an added particle has energy ω, and $\pm \mathrm{i}G^{<}(\omega)$ as the probability that a removed particle has energy ω.

Despite the nonpositivity of the bosonic spectral function, the normalization condition is fulfilled for both fermions and bosons. Upon integration of (6.81) over ω we end up with the sums $\sum_m |P_m(j)|^2$ and $\sum_m |Q_m(j)|^2$. From definition (6.72) it is easy to see that

$$\sum_m |P_m(j)|^2 = \langle \Psi_{N,0} | \hat{d}_j \hat{d}_j^\dagger | \Psi_{N,0} \rangle, \quad \sum_m |Q_m(j)|^2 = \langle \Psi_{N,0} | \hat{d}_j^\dagger \hat{d}_j | \Psi_{N,0} \rangle,$$

and hence

$$\int \frac{d\omega}{2\pi} A_{jj}(\omega) = \langle \Psi_{N,0} | \left[\hat{d}_j, \hat{d}_j^\dagger \right]_\mp | \Psi_{N,0} \rangle = 1.$$

More generally, the integral of the matrix elements of the spectral function operator satisfies the sum rule,

$$\boxed{\int \frac{d\omega}{2\pi} A_{ji}(\omega) = \delta_{ji}} \tag{6.82}$$

which can be verified similarly. The reader can also derive this results in the position–spin basis; using the field operators instead of the $\hat{d}$-operators, the matrix elements of $\hat{\mathcal{A}}$ are functions $A(\mathbf{x}, \mathbf{x}'; \omega)$ of the position–spin coordinates, and the r.h.s. of the sum rule (6.82) is replaced by $\delta(\mathbf{x} - \mathbf{x}')$.

We have discussed the physical interpretation of $\hat{\mathcal{A}}$ only for systems in a pure state. However, definition (6.77) makes sense for any initial ensemble such that $[\hat{H}^{\mathrm{M}}, \hat{H}]_- = 0$, such as for the equilibrium ensemble at finite temperature. In this case, $G^<$ and $G^>$ are given by (6.63) and (6.64), and the curious reader can easily generalize the previous discussion as well as check that (6.82) is still satisfied. It is important to stress that for systems in thermal equilibrium the knowledge of $\hat{\mathcal{A}}$ is enough to calculate all Keldysh components of $\hat{\mathcal{G}}$ with real-time arguments. From (6.58) and (6.59) we have

$$\boxed{\hat{\mathcal{G}}^{\mathrm{R}}(\omega) = \int \frac{d\omega'}{2\pi} \frac{\hat{\mathcal{A}}(\omega')}{\omega - \omega' + \mathrm{i}\eta}} \qquad \boxed{\hat{\mathcal{G}}^{\mathrm{A}}(\omega) = \int \frac{d\omega'}{2\pi} \frac{\hat{\mathcal{A}}(\omega')}{\omega - \omega' - \mathrm{i}\eta}} \tag{6.83}$$

and from the fluctuation–dissipation theorem (6.68) and (6.69) we have

$$\boxed{\hat{\mathcal{G}}^{<}(\omega) = \mp \mathrm{i} f(\omega-\mu)\hat{\mathcal{A}}(\omega)} \qquad \boxed{\hat{\mathcal{G}}^{>}(\omega) = -\mathrm{i}\bar{f}(\omega-\mu)\hat{\mathcal{A}}(\omega)} \tag{6.84}$$

In a *zero-temperature* fermionic system the function $f(\omega-\mu)=\theta(\mu-\omega)$, and hence the spectral function has peaks only at the addition energies for $\omega > \mu$ and peaks only at the removal energies for $\omega < \mu$. This nice separation is not possible at finite temperature. We further observe that if we write the retarded Green's functions as the sum of a Hermitian and anti-Hermitian quantity,

$$\hat{\mathcal{G}}^{\mathrm{R}}(\omega) = \hat{\mathcal{C}}(\omega) - \frac{\mathrm{i}}{2}\hat{\mathcal{A}}(\omega) = \left[\hat{\mathcal{G}}^{\mathrm{A}}(\omega)\right]^{\dagger}, \tag{6.85}$$

then (6.83) implies that $\hat{\mathcal{C}}$ and $\hat{\mathcal{A}}$ are connected by a Hilbert transformation,

$$\hat{\mathcal{C}}(\omega) = P\int \frac{d\omega'}{2\pi}\frac{\hat{\mathcal{A}}(\omega')}{\omega-\omega'}, \tag{6.86}$$

similarly to the real and imaginary part of the embedding self-energy in (2.31). This is a general property of several many-body quantities, see Section 10.1.

Finally, we would like to point out that the shape of the spectral function can be used to predict the time-dependent behavior of the Green's function and vice versa. If a matrix element of the spectral function has δ-like peaks, like the $A_{\lambda\lambda}(\omega) = 2\pi\delta(\omega-\epsilon_\lambda)$ of a noninteracting system or the $A_{ji}(\omega)$ of a finite system, then the corresponding matrix element of the Green's function oscillates in time. As already discussed in Section 6.1.3, for the Green's function $G^{\lessgtr}_{ji}(t,t')$ to decay as $|t-t'|$ increases, the quantum numbers j and i must be coupled to infinitely many degrees of freedom. This occurs both in *macroscopic noninteracting* systems if j and i are not good quantum numbers or in *macroscopic interacting* systems for almost all quantum numbers. The effect of the coupling is to broaden the δ-like peaks, thus giving a finite lifetime to the particles. This can be seen in (6.79) (for noninteracting systems) and (6.81) (for interacting systems); $A_{jj}(\omega)$ becomes a continuous function of ω since the sum over discrete states becomes an integral over a *continuum* of states. In Section 10.2 we deepen this aspect further. The transition from sharp δ-like peaks to a continuous function often generates some consternation. In Appendix H we show that there is nothing mysterious in this transition. Our first example of a continuous, broadened spectral function was the Lorentzian $A_{00}(\omega)$ of the noninteracting Anderson model, see (2.34):

$$A_{00}(\omega) = \langle\epsilon_0|\hat{\mathcal{A}}(\omega)|\epsilon_0\rangle = \frac{\Gamma_{\mathrm{em}}}{[(\omega-\epsilon_0)^2+\Gamma^2_{\mathrm{em}}/4]}. \tag{6.87}$$

The state $|\epsilon_0\rangle$ is not an eigenstate of the single-particle Hamiltonian and hence it does not have a well-defined energy (equivalently, ϵ_0 is not a good quantum number): A particle in $|\epsilon_0\rangle$ has a finite probability of having any energy, the most probable energy being ϵ_0. In the limit $\Gamma_{\mathrm{em}} \to 0$ (no contacts between the metal and the atom), the Lorentzian approaches a δ-function in agreement with the fact that $|\epsilon_0\rangle$ becomes an eigenstate with eigenvalue ϵ_0 (equivalently, ϵ_0 becomes a good quantum number). The effect of the interparticle interaction is to broaden (almost) *all* matrix elements of the spectral function and hence to destroy the independent-particle picture.

6.3.3 Anderson Model in the Coulomb Blockade Regime

As an appetizer of the effects of the interparticle interaction on the shape of the spectral function, we consider the Anderson model at finite U. How does $A_{00}(\omega)$ in (6.87) change? To get some insight into the problem, we begin with the isolated atom – that is, we set $T_k = 0$ in (2.23). The atomic Hamiltonian is equivalent to the Hubbard model with only one site. The Fock space of the isolated atom consists of the empty ket $|0\rangle$, the singly occupied kets $|\sigma\rangle = \hat{d}^\dagger_{0\sigma}|0\rangle$, $\sigma = \uparrow, \downarrow$, and the doubly occupied ket $|\uparrow\downarrow\rangle = \hat{d}^\dagger_{0\uparrow}\hat{d}^\dagger_{0\downarrow}|0\rangle$. These kets are eigenkets of the occupation number operators $\hat{n}_{0\sigma}$ and therefore they are also eigenkets of the atomic Hamiltonian $\hat{H}_{\rm at}$, with eigenvalues $E_0 = 0$, $E_\sigma = \epsilon_0$, and $E_{\uparrow\downarrow} = 2\epsilon_0 + U$, and of the total number operator $\hat{N} = \hat{n}_{0\uparrow} + \hat{n}_{0\downarrow}$, with eigenvalues $N_0 = 0$, $N_\sigma = 1$, and $N_{\uparrow\downarrow} = 2$.

If the atom is in thermal equilibrium at chemical potential μ and inverse temperature β, then the atomic occupation is

$$n_{0\sigma} = \mathrm{Tr}[\hat{\rho}\,\hat{n}_{0\sigma}] = \rho_\sigma + \rho_{\uparrow\downarrow}, \tag{6.88}$$

where we define the weights $\rho_i = e^{-\beta(E_i - \mu N_i)}/Z$ with $Z = \mathrm{Tr}[e^{-\beta(\hat{H}_{\rm at} - \mu\hat{N})}]$ the partition function. Since $\rho_\uparrow = \rho_\downarrow$, the result is independent of the spin σ, so we define $n_0 = n_{0\sigma}$. The lesser/greater Green's function can be calculated from (6.63) and (6.64). The only nonvanishing amplitudes are

$$\Phi_{0,\sigma}(\sigma') = \langle 0|\hat{d}_{0\sigma'}|\sigma\rangle = \delta_{\sigma\sigma'} \quad , \quad \Phi_{\sigma,\uparrow\downarrow}(\sigma') = \langle\sigma|\hat{d}_{0\sigma'}|\uparrow\downarrow\rangle = 1 - \delta_{\sigma\sigma'}.$$

Therefore,

$$G^<_{0\uparrow 0\uparrow}(\omega) = 2\pi\mathrm{i}\Big[\rho_\uparrow\,\delta(\omega-\epsilon_0) + \rho_{\uparrow\downarrow}\,\delta(\omega-\epsilon_0-U)\Big],$$

$$G^>_{0\uparrow 0\uparrow}(\omega) = -2\pi\mathrm{i}\Big[\rho_0\,\delta(\omega-\epsilon_0) + \rho_\downarrow\,\delta(\omega-\epsilon_0-U)\Big].$$

It is left as an exercise for the reader to show that $G^\lessgtr_{0\downarrow 0\downarrow}(\omega) = G^\lessgtr_{0\uparrow 0\uparrow}(\omega)$, whereas the off-diagonal elements $G^\lessgtr_{0\uparrow 0\downarrow}(\omega) = G^\lessgtr_{0\downarrow 0\uparrow}(\omega) = 0$. Henceforth we omit the spin indices and write simply $G^\lessgtr_{00}(\omega)$ to denote either the $\uparrow\uparrow$ or the $\downarrow\downarrow$ component.

The spectral function follows directly from (6.77):

$$A_{00}(\omega) = 2\pi\Big[(\rho_0 + \rho_\uparrow)\,\delta(\omega-\epsilon_0) + (\rho_\downarrow + \rho_{\uparrow\downarrow})\,\delta(\omega-\epsilon_0-U)\Big].$$

Taking into account (6.88) and the normalization condition $\mathrm{Tr}[\hat{\rho}] = \sum_i \rho_i = 1$, we can express the spectral function in terms of the atomic occupation according to

$$A_{00}(\omega) = 2\pi\Big[(1-n_0)\,\delta(\omega-\epsilon_0) + n_0\,\delta(\omega-\epsilon_0-U)\Big]. \tag{6.89}$$

In the noninteracting ($U = 0$) case, this result correctly reduces to the $\Gamma_{\rm em} \to 0$ limit of (6.87). Another exact property satisfied by (6.89) is the sum rule (6.82): $\int\frac{d\omega}{2\pi}A_{00}(\omega) = 1$.

In the noninteracting case, the couplings T_k between the atom and the metal broadens the sharp δ-like peak of the spectral function and transforms it into a Lorentzian of width

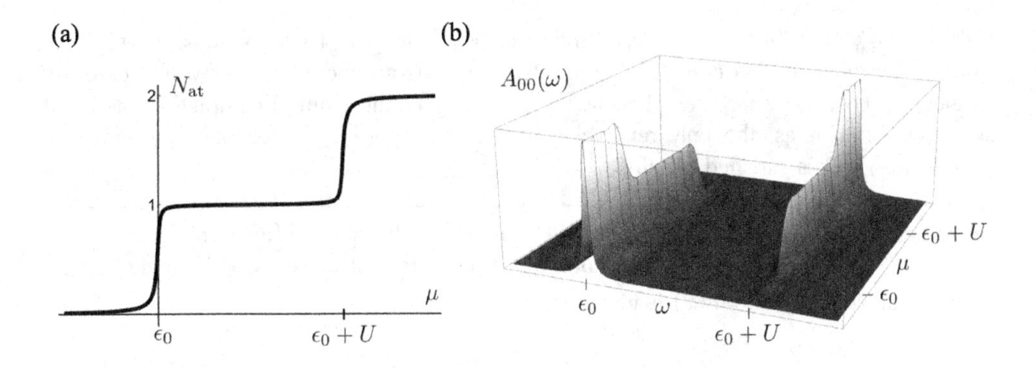

Figure 6.5 The Anderson model in the Coulomb blockade regime. (a) Total atomic occupation versus the chemical potential. (b) Spectral function for different chemical potentials.

$\Gamma_{\rm em}$. In the interacting case this picture remains correct provided that $\Gamma_{\rm em} \lesssim 1/\beta = K_{\rm B}T$ [62, 63]. This is the so-called *Coulomb blockade regime*, and the spectral function reads

$$A_{00}(\omega) = (1 - n_0)\frac{\Gamma_{\rm em}}{[(\omega - \epsilon_0)^2 + \Gamma_{\rm em}^2/4]} + n_0\frac{\Gamma_{\rm em}}{[(\omega - \epsilon_0 - U)^2 + \Gamma_{\rm em}^2/4]}, \tag{6.90}$$

which correctly reduces to (6.89) for $\Gamma_{\rm em} \to 0$. The atomic occupation n_0 can be calculated from the fluctuation–dissipation theorem:

$$n_0 = -{\rm i}G^<_{00}(t,t) = -{\rm i}\int \frac{d\omega}{2\pi} G^<_{00}(\omega) = \int \frac{d\omega}{2\pi} f(\omega - \mu)A_{00}(\omega).$$

Defining the integral

$$I(\mu - x) \equiv \int \frac{d\omega}{2\pi} f(\omega - \mu)\frac{\Gamma_{\rm em}}{[(\omega - x)^2 + \Gamma_{\rm em}^2/4]},$$

we can write the atomic occupation as

$$n_0 = \frac{I(\mu - \epsilon_0)}{1 + I(\mu - \epsilon_0) - I(\mu - \epsilon_0 - U)}. \tag{6.91}$$

In Fig. 6.5 (a) we show the total atomic occupation $N_{\rm at} = 2n_0$ versus the chemical potential for (in arbitrary units) $\epsilon_0 = 0$, $\Gamma_{\rm em} = 0.02$ and $K_{\rm B}T = 1/\Gamma_{\rm em}$, and $U = 1$. Unlike the noninteracting case $N_{\rm at}$ does not jump from 0 to 2 as μ crosses ϵ_0. Adding one electron to the empty atom costs an energy ϵ_0. However, once the first electron is in, the addition of a second electron costs an energy $\epsilon_0 + U$. Therefore the atomic occupation remains pinned at $N_{\rm at} = 1$ until the chemical potential is large enough for the second electron to overcome the repulsion barrier; this occurs for $\mu > \epsilon_0 + U$. We can say that the Coulomb repulsion U blocks the second electron, an effect known as the *Coulomb blockade*. In Fig. 6.5(b) we show the spectral function for different chemical potentials. For $\mu < \epsilon_0$ the atom is empty and only $G^{>}_{00}$ contributes, the only addition energy being ϵ_0. For $\epsilon_0 < \mu < \epsilon_0 + U$ the atom is half-filled. In this case, adding an electron costs an energy $\epsilon_0 + U$ while removing an electron costs an energy ϵ_0. Finally, for $\mu > \epsilon_0 + U$ the atom is completely filled and only $G^{<}_{00}$ contributes, the only removal energy being $\epsilon_0 + U$. Notice that the sum rule $\int \frac{d\omega}{2\pi} A_{00}(\omega) = 1$ is satisfied for all μ.

Exercise 6.4 Using the results of Exercise 6.2, show that the spectral function $A(x,x';\omega) = \mathrm{i}\left[G^{\mathrm{R}}(x,x';\omega) - G^{\mathrm{A}}(x,x';\omega)\right]$ of a noninteracting system of particles with single-particle Hamiltonian $\hat{h} = \hat{p}^2/2 + \lambda\delta(\hat{x})$ is given by

$$A(x,x';\omega) = \frac{2}{\sqrt{2\omega}}\left\{\cos(\sqrt{2\omega}|x-x'|) + \frac{\lambda}{2\omega+\lambda^2}\mathrm{Im}\left[(\sqrt{2\omega} - \mathrm{i}\lambda)e^{\mathrm{i}\sqrt{2\omega}\,(|x|+|x'|)}\right]\right\}$$

for $\omega > 0$ and $A(x,x';\omega) = 0$ for $\omega < 0$. Show further that $A(x,x;\omega) \geq 0$.

Exercise 6.5 Consider the spectral function of Exercise 6.4. Use the fluctuation–dissipation theorem to show that the density $n(x\sigma) = \pm\mathrm{i}\int \frac{d\omega}{2\pi} G^{<}(x,x;\omega)$ for particles in position x with spin σ is given by

$$n(x\sigma) = 2\int_0^\infty \frac{dq}{2\pi}\frac{1}{e^{\beta(\frac{q^2}{2}-\mu)} \mp 1}\left\{1 + \lambda\,\frac{q\sin(2q|x|) - \lambda\cos(2q|x|)}{q^2+\lambda^2}\right\}.$$

Further show that

$$\lim_{x\to\pm\infty} n(x\sigma) = \int_{-\infty}^{\infty}\frac{dq}{2\pi}\frac{1}{e^{\beta(\frac{q^2}{2}-\mu)} \mp 1} = n_0,$$

where n_0 is the density per spin of the homogeneous system – that is, with $\lambda = 0$.

Exercise 6.6 Consider the density of Exercise 6.5 for fermions at zero temperature. In this case, the Fermi function $\lim_{\beta\to\infty}(e^{\beta(\frac{q^2}{2}-\mu)}+1)^{-1}$ is unity for $|q| < p_{\mathrm{F}}$ and zero otherwise, where the Fermi momentum $p_{\mathrm{F}} = \sqrt{2\mu}$. Show that $p_{\mathrm{F}} = \pi n_0$ and that the density at the origin is given by

$$n(0\sigma) = n_0\left[1 - \frac{\lambda}{p_{\mathrm{F}}}\arctan(\frac{p_{\mathrm{F}}}{\lambda})\right].$$

Show further that for $\lambda \to \infty$ the density profile becomes

$$n(x\sigma) = n_0\left[1 - \frac{\sin(2p_{\mathrm{F}}|x|)}{2p_{\mathrm{F}}|x|}\right].$$

Therefore the density exhibits spatial oscillations with wavevector $2p_{\mathrm{F}}$. These are known as the *Friedel oscillations.* How does this result change for $\lambda \to -\infty$?

Exercise 6.7 Consider a single-level bosonic Hamiltonian $\hat{H} = E(\hat{n})$, where $\hat{n} = \hat{d}^\dagger\hat{d}$ is the occupation operator and $E(x)$ is an arbitrary real function of x. Show that in thermal equilibrium the spectral function is

$$A(\omega) = 2\pi\sum_{k=0}^{\infty}(k+1)\left[\rho_k - \rho_{k+1}\right]\delta(\omega - E(k+1) + E(k)),$$

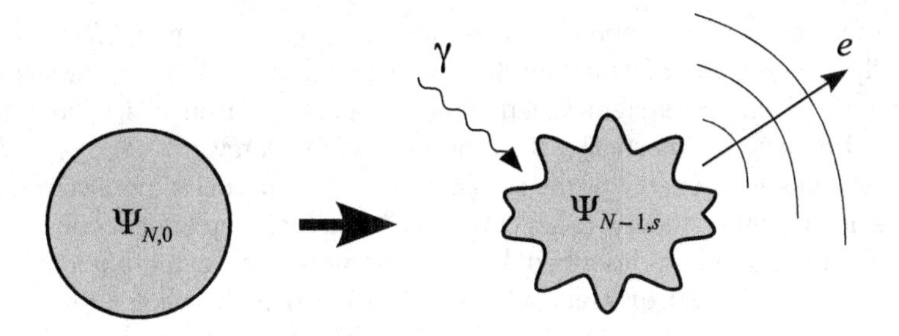

Figure 6.6 Schematic representation of a photoemission experiment. After the absorption of the photon γ of energy ω_0, the system is left in an excited state formed by the photoelectron and $N-1$ electrons in the sth excited state.

with $\rho_k = e^{-\beta(E(k)-\mu k)}/Z$ and Z the partition function. From this result we see that if $E(k+1)-\mu < E(k)$ for some k, then $[\rho_k - \rho_{k+1}] < 0$ and hence the spectral function can be negative for some frequency.

Exercise 6.8 Show that for a system of fermions the diagonal matrix element $G^{\mathrm{R}}_{jj}(\zeta)$ is nonvanishing in the upper half of the complex ζ plane. Similarly, $G^{\mathrm{A}}_{jj}(\zeta)$ is nonvanishing in the lower half of the complex ζ plane. Hint: Calculate the imaginary part from (6.83) and use the fact that $A_{jj}(\omega) \geq 0$.

Exercise 6.9 Show that the eigenvalues of the self-adjoint matrices $\mathrm{i}G^{>}_{ij}(\omega)$ and $\pm\mathrm{i}G^{<}_{ij}(\omega)$ are nonnegative for all ω. This is a generalization of property (6.65). Show also that for fermionic systems the eigenvalues of the spectral function are nonnegative. Hint: Use that if the expectation value of a Hermitian operator is nonnegative for all states, then its eigenvalues are nonnegative.

6.4 Time-Resolved and Angle-Resolved Photoemission Spectroscopy

At the moment the spectral function is a pure mathematical object providing information on the underlying physical system (we have seen that it has δ-like peaks at the removal and addition energies). Can we also relate it to some measurable quantity? In this section we show that the *spectral function can be measured with a photoemission experiment*. In Section 9.5 we discuss an alternative strategy based on quantum transport experiments.

6.4.1 Heuristic Derivation

Let us start by considering a system of electrons (fermionic system). A photoemission experiment consists in irradiating the system with light of frequency $\omega_0 > 0$ and then measuring the number of ejected electrons (or photoelectrons) with energy ϵ, see Fig. 6.6. This experimental technique is known as *photoemission spectroscopy* (PES). Without loss of generality we can set the threshold of the continuum (scattering) states to zero so that

$\epsilon > 0$. Due to energy conservation $E_{N,0} + \omega_0 = E_{N-1,s} + \epsilon$, where $E_{N,0}$ is the energy of the initially unperturbed system with N particles and $E_{N-1,s}$ is the energy of the sth excited state in which the system is left after the photoelectron has been kicked out.[10] Clearly, if the frequency ω_0 is smaller than the *ionization energy* $I \equiv E_{N-1,0} - E_{N,0}$, with $E_{N-1,0}$ the ground-state energy of the system with $N-1$ particles, no electron is ejected. Let us develop a simple theory to calculate the result of a photoemission experiment. The electromagnetic field is described by a monochromatic vector potential $\mathbf{A}(\mathbf{r},t) = \mathbf{A}(\mathbf{r})e^{\mathrm{i}\omega_0 t} + \mathbf{A}^*(\mathbf{r})e^{-\mathrm{i}\omega_0 t}$. Experimentally one observes that the photocurrent (number of ejected electrons per unit time) is proportional to the intensity of the electromagnetic field, and hence we can discard the A^2 term appearing in the single-particle Hamiltonian (3.1). The time-dependent perturbation which couples the light to the electrons is then proportional to $\hat{\boldsymbol{p}} \cdot \mathbf{A}(\hat{\boldsymbol{r}},t) + \mathbf{A}(\hat{\boldsymbol{r}},t) \cdot \hat{\boldsymbol{p}}$. Expanding the field operators over some convenient basis, the most general form of this perturbation in second quantization reads

$$\hat{H}_{\text{light-matter}}(t) = \sum_{ij} (h_{ij} e^{\mathrm{i}\omega_0 t} + h^*_{ij} e^{-\mathrm{i}\omega_0 t}) \hat{d}^\dagger_i \hat{d}_j.$$

According to the *Fermi golden rule*, see Appendix I, the probability per unit time of the transition from the initial state $|\Psi_{N,0}\rangle$ with energy $E_{N,0}$ to an excited state $|\Psi_{N,m}\rangle$ with energy $E_{N,m}$ is given by

$$P_{0\to m} = 2\pi \left| \langle \Psi_{N,m}| \sum_{ij} h^*_{ij} \hat{d}^\dagger_i \hat{d}_j |\Psi_{N,0}\rangle \right|^2 \delta(\omega_0 - [E_{N,m} - E_{N,0}]).$$

The photocurrent $I_{ph}(\epsilon)$ with electrons of energy ϵ is proportional to the sum of all the transition probabilities with excited states $|\Psi_{N,m}\rangle = \hat{d}^\dagger_\epsilon |\Psi_{N-1,s}\rangle$. These states describe an electron outside the system with energy ϵ and $N-1$ electrons inside the system in the sth excited state, see again Fig. 6.6. In fact, for energies ϵ sufficiently large, the photoelectron does not feel the presence of the $N-1$ electrons left behind and therefore $|\Psi_{N,m}\rangle$ is simply obtained by creating the photoelectron over the interacting $(N-1)$-particle excited state. If we restrict ourselves to these kinds of excited states and rename the corresponding transition probability $P_{0\to m} = P_s(\epsilon)$, we see that the index i of the perturbation must be equal to ϵ for $P_s(\epsilon)$ not to vanish.[11] We then have

$$P_s(\epsilon) = 2\pi \left| \langle \Psi_{N-1,s}| \sum_j h^*_{\epsilon j} \hat{d}_j |\Psi_{N,0}\rangle \right|^2 \delta(\omega_0 - \epsilon - [E_{N-1,s} - E_{N,0}]).$$

Summing over all the excited states of the $N-1$ particle system and comparing the result with (6.75), we obtain

$$I_{ph}(\epsilon) \propto \sum_s P_s(\epsilon) = -\mathrm{i} \sum_{jj'} h^*_{\epsilon j} h_{\epsilon j'} G^<_{jj'}(\epsilon - \omega_0). \tag{6.92}$$

Thus, the photocurrent is proportional to $-\mathrm{i}G^<(\epsilon - \omega_0)$. If the system is initially in equilibrium at zero temperature, then $E_{N,0}$ is the energy of the ground state, and the photocurrent

[10] Even though in most experiments the system is initially in its ground state, our treatment is applicable also to situations in which $E_{N,0}$ refers to some excited state – read again the beginning of Section 6.3.2.

[11] The ground state $|\Psi_{N,0}\rangle$ does not contain photoelectrons.

is proportional to the spectral function at energies $\epsilon - \omega_0 < \mu$, see the observation below (6.84).[12]

Bosonic photoemission If we had started from a system of bosons we could have gone through the same logical and mathematical steps as before to then get the same formula for the photocurrent in terms of a sum of transition probabilities $P_s(\epsilon)$. In the bosonic case, however, the sum over all s of $P_s(\epsilon)$ gives a linear combination of $+\mathrm{i}G^<_{jj'}(\epsilon - \omega_0)$. Thus the sign of the bosonic photocurrent is opposite to the sign of the hole part of $A_{jj'}(\epsilon - \omega_0)$. This argument provides a physical explanation of the possible nonpositivity of the bosonic spectral function.

Inverse photoemission The outcome of a photoemission experiment relates the photocurrent to $G^<$ independently of the statistics of the particles. To have access to $G^>$, we must perform an *inverse photoemission experiment.* This experiment consists in directing a beam of particles of well-defined energy at the sample. The incident particles penetrate the surface and decay in low-energy unoccupied states by emitting photons. With considerations similar to those of a photoemission experiment, we can develop a simple theory to determine the number of emitted photons of energy ω_0. Not surprisingly, the result is proportional to $G^>$. Thus, $G^>$ and $G^<$ can be separately measured and interpreted.

Exercise 6.10 Show that the current of photons of energy ω_0 in an inverse photoemission experiment can be expressed solely in terms of $G^>$.

6.4.2 Formula of the Photocurrent

In an *angle-resolved photoemission spectroscopy* (ARPES) experiment, it is possible to measure the energy as well as the angle of emission of the photoelectrons. The *time-resolved* extension (TR-ARPES) of this idea can be used to reveal the electron dynamics of out-of-equilibrium systems. In TR-ARPES, a *pump* pulse excites electrons from occupied states to unoccupied states without causing ionization. The photocurrent is generated by a second *probe* pulse which impinges the excited system after a tunable pump–probe delay. If the probed system is a crystal, then the time-, angle-, and energy-dependent photocurrent contains information on the band structure, on the time-dependent occupation of the Bloch states, and much more. In this section we derive the formula of the TR-ARPES photocurrent.

LEED states Without any loss of generality we assume that the crystal surface lies on the x–y plane, see Fig. 6.7. The surface breaks the periodicity along the z direction but it preserves that along the plane. We can think of the system like a two-dimensional crystal with an infinitely long unit cell extending from $z = -\infty$ to $z = +\infty$, and repeating in the x and y directions (for positive z the unit cell is pure vacuum). According to the Bloch theorem, see Section 2.3, the single-particle Hamiltonian $\hat{h}$ has eigenvalues $\epsilon_{\mathbf{k}_\parallel \nu}$ and eigenkets $|\mathbf{k}_\parallel \nu\rangle$, where $\mathbf{k}_\parallel$ is the quasi-momentum parallel to the surface and ν is a collective index for the band and spin degrees of freedom. Setting to zero the minimum energy of free electrons (these electrons are not bounded to the crystal), we can use the *degenerate* eigenkets of positive energy to construct the free-electron states $|f_{\mathbf{k}\sigma}\rangle$ describing an outgoing electron of spin σ and momentum $\mathbf{k} = (\mathbf{k}_\parallel, k_\perp)$, $k_\perp > 0$ being the component perpendicular to the

---

[12] At zero temperature $\mu = E_{N,0} - E_{N-1,0} \equiv -I$ is the ionization energy, and therefore $I_{ph}(\epsilon) = 0$ for $\omega_0 < \epsilon + I$, as expected.

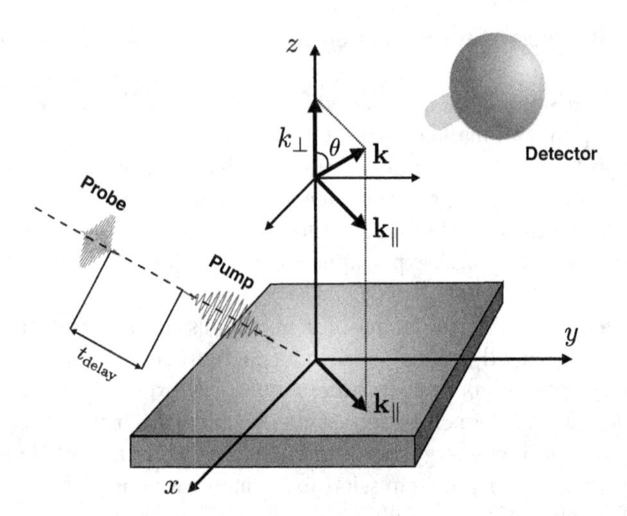

Figure 6.7 Illustration of TR-ARPES experiments in crystals. A pump pulse brings the crystal in an excited or nonstationary state. After a delay time t_{delay}, a probe pulse ionizes the crystal by expelling an electron with quasi-momentum $\mathbf{k}_\parallel$; the electron is then detected as a photoelectron of momentum $\mathbf{k} = (\mathbf{k}_\parallel, k_\perp)$.

surface, see Fig 6.7. The free-electron states have energy $\epsilon_{\mathbf{k}} = k^2/2 = (k_\perp^2 + k_\parallel^2)/2$ and can be written as a linear combination of the form

$$|f_{\mathbf{k}\sigma}\rangle = |f_{(\mathbf{k}_\parallel, k_\perp)\sigma}\rangle = \sum_{\nu:\ \epsilon_{\mathbf{k}_\parallel \nu} = \epsilon_{\mathbf{k}}} \alpha_{k_\perp \sigma, \nu}(\mathbf{k}_\parallel)\, |\mathbf{k}_\parallel \nu\rangle.$$

In technical jargon, the states $|f_{\mathbf{k}\sigma}\rangle$ are known as time-reversed *low-energy electron-diffraction* (LEED) states [64, 65, 66]. How to obtain the coefficients $\alpha_{k_\perp \sigma,\nu}(\mathbf{k}_\parallel)$ of the expansion goes beyond the purpose of this book; we refer the interested reader to Refs. [67, 68, 69]. As in the heuristic derivation, we discard the Coulomb interaction between free electrons and electrons bound to the crystal; the higher is the energy of the probe-photons, and hence of the photoelectrons, the more accurate is this approximation. Taking into account that the pump pulse excites the electrons without expelling them from the crystal, we can write the Hamiltonian of the full system *without* the probe pulse as

$$\hat{H}(t) = \hat{H}_{\text{crystal}}(t) + \sum_{\mathbf{k}\sigma} \epsilon_{f\mathbf{k}} \hat{f}^\dagger_{\mathbf{k}\sigma} \hat{f}_{\mathbf{k}\sigma}.$$

Henceforth, we denote the fermionic annihilation (creation) operators for bound electrons by $\hat{d}_{\mathbf{k}_\parallel \nu}$ ($\hat{d}^\dagger_{\mathbf{k}_\parallel \nu}$) and those for free electrons by $\hat{f}_{\mathbf{k}\sigma}$ ($\hat{f}^\dagger_{\mathbf{k}\sigma}$). The crystal Hamiltonian then contains only operators $\hat{d}_{\mathbf{k}_\parallel \nu}$ and $\hat{d}^\dagger_{\mathbf{k}_\parallel \nu}$, and it depends on time through the coupling between the bound electrons and the (time-dependent) pump pulse.

Probe-matter interaction To describe the coupling of the system with the probe pulse, we work in the dipole approximation (which is accurate for photon energies below 10 keV) and consider a vector potential $\mathbf{A}(t) = \boldsymbol{\eta} A(t)$ parallel to some unit vector $\boldsymbol{\eta}$. Let

$$D_{\mathbf{k}\sigma\nu} \equiv \frac{1}{c}\langle f_{\mathbf{k}\sigma}|(\hat{\mathbf{p}}\cdot\boldsymbol{\eta})|\mathbf{k}_{\parallel}\nu\rangle \tag{6.93}$$

be the matrix element of the light-matter interaction operator between a free-electron state and a bound state, see the Hamiltonian (3.1). Notice that $\langle f_{\mathbf{k}\sigma}|\hat{\mathbf{p}}|\mathbf{k}'_{\parallel}\nu\rangle \propto \delta_{\mathbf{k}_{\parallel},\mathbf{k}'_{\parallel}}$, reflecting the fact that parallel momentum is conserved in a photoemission process. The probe-system coupling can then be written as

$$\hat{H}_{\text{probe-matter}}(t) = A(t)\sum_{\mathbf{k}\sigma,\nu}\left(D_{\mathbf{k}\sigma\nu}\hat{f}^{\dagger}_{\mathbf{k}\sigma}\hat{d}_{\mathbf{k}_{\parallel}\nu} + D^{*}_{\mathbf{k}\sigma\nu}\hat{d}^{\dagger}_{\mathbf{k}_{\parallel}\nu}\hat{f}_{\mathbf{k}\sigma}\right).$$

Photocurrent The photocurrent of electrons with momentum $\mathbf{k}$ at a certain time t is given by the rate of change of the occupation of the free-electron state with momentum $\mathbf{k}$:

$$I_{ph}(\mathbf{k},t) \equiv \frac{d}{dt}\sum_{\sigma}\langle\hat{f}^{\dagger}_{\mathbf{k}\sigma,H}(t)\hat{f}_{\mathbf{k}\sigma,H}(t)\rangle = -\mathrm{i}\frac{d}{dt}\sum_{\sigma}G^{<}_{\mathbf{k}\sigma}(t,t), \tag{6.94}$$

where operators in the Heisenberg picture evolve with the total Hamiltonian $\hat{H}_{\text{tot}}(t) \equiv \hat{H}(t) + \hat{H}_{\text{light-matter}}(t)$; hence, both pump and probe pulses are contained in $\hat{H}_{\text{tot}}(t)$. In (6.94) appears the lesser component of the free-electron Green's function, which is defined according to

$$G_{\mathbf{k}\sigma}(z,z') \equiv \frac{1}{\mathrm{i}}\langle\mathcal{T}\left\{\hat{f}_{\mathbf{k}\sigma,H}(z)\hat{f}^{\dagger}_{\mathbf{k}\sigma,H}(z')\right\}\rangle,$$

where the short-hand notation $\langle\ldots\rangle$ stands for the ensemble average. Taking into account that the operators $\hat{f}_{\mathbf{k}\sigma}$ commute with $\hat{H}_{\text{crystal}}(t)$, the equations of motion for $G_{\mathbf{k}\sigma}$ simply read

$$\left[\mathrm{i}\frac{d}{dz} - \epsilon_{\mathbf{k}}\right]G_{\mathbf{k}\sigma}(z,z') - \sum_{\nu}D_{\mathbf{k}\sigma\nu}\,A(z)\,G_{\mathbf{k}\nu\sigma}(z,z') = \delta(z,z'), \tag{6.95}$$

$$\left[-\mathrm{i}\frac{d}{dz'} - \epsilon_{\mathbf{k}}\right]G_{\mathbf{k}\sigma}(z,z') - \sum_{\nu}D^{*}_{\mathbf{k}\sigma\nu}\,A(z')\,G_{\mathbf{k}\sigma\nu}(z,z') = \delta(z,z'), \tag{6.96}$$

where $A(t_0 - \mathrm{i}\beta) = 0$, $A(t_{\pm}) = A(t)$, and

$$G_{\mathbf{k}\nu\sigma}(z,z') \equiv \frac{1}{\mathrm{i}}\langle\mathcal{T}\left\{\hat{d}_{\mathbf{k}_{\parallel}\nu,H}(z)\hat{f}^{\dagger}_{\mathbf{k}\sigma,H}(z')\right\}\rangle,$$
$$G_{\mathbf{k}\sigma\nu}(z,z') \equiv \frac{1}{\mathrm{i}}\langle\mathcal{T}\left\{\hat{f}_{\mathbf{k}\sigma,H}(z)\hat{d}^{\dagger}_{\mathbf{k}_{\parallel}\nu,H}(z')\right\}\rangle.$$

Equations (6.95) and (6.96) and all subsequent equations of motion have to be solved with KMS boundary conditions, see Section 5.1.

The equation for the equal-time lesser component $G^<_{\mathbf{k}\sigma}(t,t)$ follows when setting $z = t_-$ and $z' = t_+$, and then subtracting (6.96) from (6.95). We find

$$\mathrm{i}\frac{d}{dt}\sum_{\sigma} G^<_{\mathbf{k}\sigma}(t,t) = -2\sum_{\sigma\nu} \mathrm{Re}\left[D^*_{\mathbf{k}\sigma\nu} A(t) G^<_{\mathbf{k}\sigma\nu}(t,t)\right], \tag{6.97}$$

where we take into account the anti-hermiticity property (6.27), which in our case reads $G^<_{\mathbf{k}\nu\sigma}(t,t) = -[G^<_{\mathbf{k}\sigma\nu}(t,t)]^*$. We can now express the right-hand side of (6.97) in terms of the crystal Green's function,

$$G_{\mathbf{k}_{\parallel}\nu\nu'}(z,z') \equiv \frac{1}{\mathrm{i}}\langle \mathcal{T}\left\{\hat{d}_{\mathbf{k}_{\parallel}\nu}(z)\hat{d}^{\dagger}_{\mathbf{k}_{\parallel}\nu'}(z')\right\}\rangle.$$

The equation of motion for $G_{\mathbf{k}\sigma\nu}$ reads

$$\left[\mathrm{i}\frac{d}{dz} - \epsilon_{\mathbf{k}}\right] G_{\mathbf{k}\sigma\nu}(z,z') - \sum_{\nu'} D_{\mathbf{k}\sigma\nu'}\, A(z)\, G_{\mathbf{k}_{\parallel}\nu'\nu}(z,z') = 0. \tag{6.98}$$

If we define the unperturbed (probe-free) Green's function as the solution of

$$\left[\mathrm{i}\frac{d}{dz} - \epsilon_{\mathbf{k}}\right] g_{\mathbf{k}}(z,z') = \delta(z,z'),$$

then (6.98) can be solved for $G_{\mathbf{k}\sigma\nu}$, yielding

$$G_{\mathbf{k}\sigma\nu}(z,z') = \sum_{\nu'}\int d\bar{z}\, g_{\mathbf{k}}(z,\bar{z})\, D_{\mathbf{k}\sigma\nu'}\, A(\bar{z})\, G_{\mathbf{k}_{\parallel}\nu'\nu}(\bar{z},z').$$

Substituting this result into (6.97), we find it convenient to define the *ionization self-energy*:

$$\Sigma^{\mathrm{ion}}_{\mathbf{k}\nu\nu'}(z,z') \equiv \sum_{\sigma} D^*_{\mathbf{k}\sigma\nu}\, A(z)\, g_{\mathbf{k}}(z,z')\, A(z')\, D_{\mathbf{k}\sigma\nu'}. \tag{6.99}$$

In fact, the square bracket in (6.97) is nothing but the lesser component of the contour-time convolution of the product $\Sigma^{\mathrm{ion}}_{\mathbf{k}\nu\nu'}(z,\bar{z})G_{\mathbf{k}_{\parallel}\nu'\nu}(\bar{z},z')$. Like the embedding self-energy discussed in Section 6.1.1, the ionization self-energy accounts for the fact that an electron can escape from the crystal at time z', wander in free space, and then come back to the crystal at time z.

Using the Langreth rules of Section 5.5 and taking into account that $\Sigma^{\mathrm{ion},<}_{\mathbf{k}\nu\nu'} \propto g^<_{\mathbf{k}} \propto f(\epsilon_{\mathbf{k}} - \mu) = 0$, see (6.40), which agrees with the fact that the initial occupation of the free-electron states vanishes, we can write the photocurrent in (6.94) as

$$I_{ph}(\mathbf{k},t) = 2\sum_{\nu\nu'}\int_{t_0}^{\infty} d\bar{t}\; \mathrm{Re}\left[\Sigma^{\mathrm{ion,R}}_{\mathbf{k}\nu\nu'}(t,\bar{t})G^<_{\mathbf{k}_{\parallel}\nu'\nu}(\bar{t},t)\right]. \tag{6.100}$$

The retarded component of the ionization self-energy follows directly from (6.99) when the contour function $g_{\mathbf{k}}$ is replaced by the retarded function $g^{\mathrm{R}}_{\mathbf{k}}$. Taking into account result (6.52), we find

$$\Sigma^{\mathrm{ion,R}}_{\mathbf{k}\nu\nu'}(t,\bar{t}) = -\mathrm{i}\theta(t-\bar{t})\sum_{\sigma} D^*_{\mathbf{k}\sigma\nu} D_{\mathbf{k}\sigma\nu'} A(t)A(\bar{t})e^{-\mathrm{i}\epsilon_{\mathbf{k}}(t-\bar{t})}. \tag{6.101}$$

Equation (6.100) is the NEGF formula for the *time-dependent photocurrent* [70]. The formula is valid for systems in arbitrary nonequilibrium states and for any temporal shape of the probe field, the only approximation being that the free electrons do not interact with the crystal electrons. As we have already pointed out, the probe field is usually extremely weak and we can ignore its effects on $G_{\mathbf{k}_\parallel \nu'\nu}$. In fact, the ionization self-energy, and hence the photocurrent, is already quadratic in the probe field. We also remark that (6.100) agrees with another formula for the photocurrent, derived in Ref. [71], since the time derivative $\frac{d}{dt}\langle \hat{f}^\dagger_{\mathbf{k}\sigma,H}(t)\hat{f}_{\mathbf{k}\sigma,H}(t)\rangle$ is proportional to the current carried by photoelectrons of momentum $\mathbf{k}$ - that is, $|\mathbf{k}|\langle \hat{f}^\dagger_{\mathbf{k}\sigma,H}(t)\hat{f}_{\mathbf{k}\sigma,H}(t)\rangle$.

ARPES for stationary states To make contact with the heuristic derivation in Section 6.4.1, we consider the special case of a system left in a stationary excited state after the action of the pump pulse, hence $G^<_{\mathbf{k}_\parallel \nu'\nu}(\bar{t},t)$ depends only on the time difference $(\bar{t}-t)$, and take a probe pulse sharply peaked at frequency ω_0 - that is, $A(t) = \theta(t-t_0)\left(A_0 e^{\mathrm{i}\omega_0 t} + A_0^* e^{-\mathrm{i}\omega_0 t}\right)$. If we are interested in the photocurrent for $t\to\infty$, then the only terms of the ionization self-energy that contribute in (6.100) are those depending on the time difference:

$$\Sigma^{\mathrm{ion,R}}_{\mathbf{k}\nu\nu'}(t,\bar{t}) \simeq -\mathrm{i}\theta(t-\bar{t})\sum_\sigma |A_0|^2 D^*_{\mathbf{k}\sigma\nu} D_{\mathbf{k}\sigma\nu'} e^{-\mathrm{i}\epsilon_{\mathbf{k}}(t-\bar{t})}\left(e^{\mathrm{i}\omega_0(t-\bar{t})} + e^{-\mathrm{i}\omega_0(t-\bar{t})}\right). \quad (6.102)$$

Fourier transforming the lesser Green's function in (6.100) and using (6.102), we find

$$I_{ph}(\mathbf{k},t) = -2\mathrm{i}\,|A_0|^2 \sum_{\sigma\nu\nu'} D^*_{\mathbf{k}\sigma\nu} D_{\mathbf{k}\sigma\nu'} \int \frac{d\omega}{2\pi}\, G^<_{\mathbf{k}_\parallel \nu'\nu}(\omega)$$
$$\times\, \mathrm{Re}\left[\int_{t_0}^{t} d\bar{t}\, \left(e^{-\mathrm{i}(\epsilon_{\mathbf{k}}-\omega_0-\omega)(t-\bar{t})} + e^{-\mathrm{i}(\epsilon_{\mathbf{k}}+\omega_0-\omega)(t-\bar{t})}\right)\right].$$

To derive this formula, we use that $\mathrm{i}\,G^<_{\mathbf{k}_\parallel \nu'\nu}(\omega)$ is self-adjoint, see (6.63). We observe that $\mathrm{i}\,G^<_{\mathbf{k}_\parallel \nu'\nu}(\omega)$ is also negative definite, see Exercise 6.9, and therefore the sum over ν and ν' of the product $D^*_{\mathbf{k}\sigma\nu}\,\mathrm{i}\,G^<_{\mathbf{k}_\parallel \nu'\nu}(\omega)\,D_{\mathbf{k}\sigma\nu'}$ is real and negative for all $\mathbf{k}$. Performing the time integral and taking into account that $\lim_{t\to\infty}\sin(\Omega t)/\Omega = \pi\delta(\Omega)$, the long-time limit of the photocurrent is given by

$$I_{ph}(\mathbf{k}) \equiv \lim_{t\to\infty} I(\mathbf{k},t)$$
$$= -2\mathrm{i}\,|A_0|^2 \sum_{\sigma\nu\nu'} D^*_{\mathbf{k}\sigma\nu} D_{\mathbf{k}\sigma\nu'} \left[G^<_{\mathbf{k}_\parallel \nu'\nu}(\epsilon_{\mathbf{k}}-\omega_0) + G^<_{\mathbf{k}_\parallel \nu'\nu}(\epsilon_{\mathbf{k}}+\omega_0)\right]. \quad (6.103)$$

Assuming $\omega_0 > 0$ and thermal equilibrium at zero temperature, meaning that the stationary state is the ground state, only the first term in the square bracket is nonvanishing. This can be inferred from the fluctuation-dissipation theorem (6.68), according to which $G^<_{\mathbf{k}_\parallel \nu'\nu}(\omega) \propto f(\omega-\mu)$ vanishes (at zero temperature) for $\omega > \mu$.[13] The result (6.103) generalizes the photocurrent in (6.92) to arbitrary stationary conditions (including thermal equilibrium at

[13] As the minimum energy of free electrons has been set to zero, the chemical potential must be negative.

finite temperature) and it allows for resolving the angle and the energy of the emitted electrons.

Band structure from ARPES Let us comment on the ARPES photocurrent formula (6.103). Consider a noninteracting crystal in thermal equilibrium at a certain inverse temperature β and chemical potential μ. Then the lesser Green's function can be directly deduced from (6.45):

$$G^{<}_{\mathbf{k}_{\parallel}\nu'\nu}(\omega) = \delta_{\nu\nu'} 2\pi \mathrm{i}\, f(\epsilon_{\mathbf{k}_{\parallel}\nu} - \mu)\delta(\omega - \epsilon_{\mathbf{k}_{\parallel}\nu}).$$

Inserting this expression into (6.103), we find

$$I_{ph}(\mathbf{k}) = 4\pi \sum_{\sigma\nu} |A_0 D_{\mathbf{k}\sigma\nu}|^2 f(\epsilon_{\mathbf{k}_{\parallel}\nu} - \mu)\delta(\epsilon_{\mathbf{k}} - \omega_0 - \epsilon_{\mathbf{k}_{\parallel}\nu}). \tag{6.104}$$

Suppose we place the detector so that only electrons with parallel momentum pointing in some direction $\mathbf{n}_{\parallel} = \mathbf{k}_{\parallel}/k_{\parallel}$ can enter into it, see Fig. 6.7. By measuring the photocurrent for all angles $\theta = \arcsin(k_{\parallel}/k)$ and by resolving (for each angle θ) the photocurrent in the energy ϵ of the photoelectrons, we can construct the surface $I_{ph}(\epsilon, \theta)$. Cutting the surface along the line $\sqrt{2m\epsilon}\sin\theta = C$, we obtain the photocurrent due to electrons with parallel momentum $k_{\parallel} = C$, since $k_{\parallel} = k\sin\theta$ and $k = \sqrt{2m\epsilon}$. This photocurrent is nonvanishing only for some discrete energies ϵ_{ν}. From (6.104) we conclude that $\epsilon_{\nu} - \omega_0$ is an eigenvalue of the crystal Hamiltonian with parallel momentum $\mathbf{k}_{\parallel}$ - that is, $\epsilon_{\mathbf{k}_{\parallel}\nu} = \epsilon_{\nu} - \omega_0$. Repeating the experiment for all directions $\mathbf{n}_{\parallel}$, we can reconstruct the occupied band structure of the crystal [72, 73].

In the real world, electrons interact with each other but the concept of band structure remains a very useful concept since the removal of electrons with a given quasi-momentum generates one main excitation with a long lifetime.[14] Consequently, the photocurrent $I_{ph}(\mathbf{k})$ remains a peaked function of ω_0, although the position of the peak is in general different from the one of the noninteracting crystal. According to the discussion in Section 6.1.3, we can say that electrons in Bloch states behave like quasi-particles. In Section 10.2 we deepen the concept of quasi-particles and show how to estimate the lifetime from the width of the peak in $I_{ph}(\mathbf{k})$. The development of perturbative methods for including interaction effects in the Green's function is crucial to improve the results of a noninteracting treatment [74, 75, 76, 77, 78, 79].

6.4.3 Excitons in Photoemission

Excitons are bound electron–hole pairs that may form in an insulator when electrons are excited from the valence (v) bands to the conduction (c) bands [80]. To illustrate the signature of excitons in TR-ARPES, we consider a simple one-dimensional model with one valence band and one conduction band separated by a direct gap of magnitude Δ, see also the example in Section 2.3. Since the formation of excitons is due to the attraction between the valence hole and the conduction electron, we simplify the presentation and discard the Coulomb

[14]In particular, the excitation generated by the removal of an electron with momentum on the Fermi surface has an infinitely long lifetime and hence the corresponding matrix element of the spectral function has a sharp δ-peak, see also Section 15.6.

interaction between electrons in the same band. For simplicity we also discard spin. Let $k = \mathbf{k}_{\parallel}$ the one-dimensional quasi-momentum and $\hat{d}_{k\nu}$ be the operators that annihilate an electron of quasi-momentum k in band $\nu = c, v$ and satisfying the usual anticommutation rules $[\hat{d}_{k\nu}, \hat{d}_{k'\nu'}]_+ = \delta_{\nu\nu'}\delta_{kk'}$. The Hamiltonian of our one-dimensional two-band insulator then reads

$$\hat{H}_{\rm ins} = \sum_k \left(\epsilon_{kv}\hat{d}^\dagger_{kv}\hat{d}_{kv} + \epsilon_{kc}\hat{d}^\dagger_{kc}\hat{d}_{kc}\right) + \frac{1}{N}\sum_{k_1k_2q} v_q\, \hat{d}^\dagger_{k_1+qv}\hat{d}^\dagger_{k_2-qc}\hat{d}_{k_2c}\hat{d}_{k_1v}, \tag{6.105}$$

where $v_q > 0$ is the Coulomb repulsion and N is the number of discretized quasi-momenta (which is the same as the number of electrons in the valence band).[15] The physical interpretation of the second term – that is, the interaction Hamiltonian – is transparent: An electron with quasi-momentum k_1 in the v-band scatters against an electron of quasi-momentum k_2 in the c-band, and after the scattering their quasi-momenta change into $k_1 + q$ and $k_2 - q$. The total quasi-momentum is correctly conserved due to the discrete translational symmetry, see also Appendix C. Choosing the chemical potential inside the gap, the lowest-energy state (in Fock space) of $\hat{H}_{\rm ins} - \mu\hat{N}$ is precisely the ground state $|\Phi_0\rangle$ of the insulator; it is characterized by a filled v-band and an empty c-band (hence it belongs to the Hilbert space $\mathbb{H}_N$ of N identical particles), see Exercise 6.11. Without any loss of generality we take the ground-state energy $E_0 = 0$.

Excitons To form excitons we drive the insulator, initially in the ground state, out of equilibrium using an ultrafast and low-intensity laser pulse (pump) that excites electrons from the valence band to the conduction band. To the lowest order in the light intensity the state of the system at the end of the pump pulse, say at time $t = 0$, can be written as

$$|\Psi\rangle = a|\Phi_0\rangle + b|\Phi_{\rm eh}\rangle, \tag{6.106}$$

with $b = \sqrt{1 - a^2} \ll 1$ and

$$|\Phi_{\rm eh}\rangle = \sum_k Y_k \hat{d}^\dagger_{kc}\hat{d}_{kv}|\Phi_0\rangle$$

the component of the full many-body state $|\Psi\rangle$ having one electron in the conduction band and one hole in the valence band.[16] The coefficients Y_k depend on the laser pulse parameters, such as duration and frequency. The state $|\Phi_{\rm eh}\rangle$ can be expanded in eigenstates $|\Phi_j\rangle$ of $\hat{H}_{\rm ins}$ that fulfill $\hat{N}_c|\Phi_j\rangle = |\Phi_j\rangle$ (one electron in the conduction band) and $\hat{N}_v|\Phi_j\rangle =$

[15] The prefactor $1/N$ in the interaction Hamiltonian guarantees the correct scaling in the continuum limit $N \to \infty$. According to (2.16) the spacing between two consecutive quasi-momenta is $dk = 2\pi/N$, and therefore $\lim_{N\to\infty}\sum_k = \lim_{N\to\infty}\int \frac{dk}{2\pi}N$. If we define the continuum field operators $\hat{d}_\nu(k) = \lim_{N\to\infty}\sqrt{N}\,\hat{d}_{k\nu}$, $\nu = c, v$, having anticommutation rules $[\hat{d}_\nu(k), \hat{d}_{\nu'}(k')]_+ = \lim_{N\to\infty} N\delta_{\nu\nu'}\delta_{kk'} = 2\pi\delta_{\nu\nu'}\delta(k - k')$, then we see that

$$\lim_{N\to\infty}\hat{H}_{\rm ins} = \int\frac{dk}{2\pi}\left(\epsilon_{kv}\hat{d}^\dagger_v(k)\hat{d}_v(k) + \epsilon_{kc}\hat{d}^\dagger_c(k)\hat{d}_c(k)\right) + \int\frac{dk_1dk_2dq}{(2\pi)^3}v_q\,\hat{d}^\dagger_v(k_1+q)\hat{d}^\dagger_c(k_2-q)\hat{d}_c(k_2)\hat{d}_v(k_1).$$

[16] The momentum of light for pump pulses in the $1 \div 10$ eV range is negligible and therefore we do not include particle-hole states of the form $\hat{d}^\dagger_{kc}\hat{d}_{k'v}|\Phi_0\rangle$ with $k \neq k'$ in the expansion of $|\Phi_{\rm eh}\rangle$.

$(N-1)|\Phi_j\rangle$ (one hole in the valence band). Writing these eigenstates as

$$|\Phi_j\rangle = \sum_k Y_k^{(j)} |\chi_k\rangle,$$

where $|\chi_k\rangle \equiv \hat{d}^\dagger_{kc}\hat{d}_{kv}|\Phi_0\rangle$ is an orthonormal basis in the considered subspace, the eigenvalue equation for the coefficients $Y_k^{(j)}$ reads

$$(\epsilon_{kc} - \epsilon_{kv} + v_0 - \Omega_j)Y_k^{(j)} = \frac{1}{N}\sum_q v_q Y_{k-q}^{(j)}. \tag{6.107}$$

The reader can easily derive this result by multiplying $\hat{H}_{\rm ins}|\Phi_j\rangle = \Omega_j|\Phi_j\rangle$ with the bra $\langle\chi_k|$. The eigenvalues Ω_j are the possible excitation energies of the system with one electron–hole pair. Let us discuss the eigenvalue equation.

If the Coulomb interaction $v_q = v_0$ is q-independent, then the r.h.s. of (6.107) is independent of k since all functions are periodic in the first Brillouin zone and therefore $\sum_q Y_{k-q}^{(j)} = \sum_q Y_{-q}^{(j)} = \sum_q Y_q^{(j)}$. Consequently, all solutions have the form $Y_k^{(j)} = 1/(\epsilon_{kc} - \epsilon_{kv} + v_0 - \Omega_j)$, where Ω_j is a solution of

$$1 = \frac{v_0}{N}\sum_q \frac{1}{\epsilon_{qc} - \epsilon_{qv} + v_0 - \Omega}.$$

Let us define $\Delta \equiv \min_q(\epsilon_{qc} - \epsilon_{qv}) + v_0$. In the range $\Omega \in (-\infty, \Delta)$, the r.h.s. is a positive function that increases monotonically from 0 (at $\Omega = -\infty$) to $+\infty$ (at $\Omega = \Delta$). On the contrary, for $\Omega > \Delta$ the r.h.s. jumps from $-\infty$ and $+\infty$ every time Ω crosses one of the energies $(\epsilon_{qc} - \epsilon_{qv}) + v_0$. We infer that there is an isolated solution at energy $\Omega = \Omega_{\rm x} < \Delta$ and a continuum of solutions (when $N \to \infty$) for energies $\Omega > \Delta$. The quantity Δ is called the *optical gap* and it corresponds to the minimum energy of a particle–hole pair in the continuum. The isolated solution at $\Omega_{\rm x}$ (below the optical gap) instead corresponds to a bound electron–hole pair – that is, an *exciton*. In a more realistic scenario (two- or three-dimensional systems, multiple conduction and valence bands, intraband and interband interaction) the electron–hole spectrum consists of a continuum of energies above a certain minimum (the optical gap) and a certain number of isolated solutions at energies below the optical gap. The isolated solutions correspond to excitons and their number depends on the considered material (it can be zero, finite, or infinite but countable).

Green's function for resonant pumping Let $\Omega_{\rm x}$ be the lowest exciton energy and let us denote by $|\Phi_{\rm x}\rangle$ and $Y_k^{({\rm x})}$ the corresponding exciton eigenstate and amplitudes. In the so-called *resonant pumping* experiments the laser has frequency $\lesssim \Omega_{\rm x}$, so to excite only the lowest-energy exciton. The state of the system at the end of the pump is then given by (6.106) with $|\Phi_{\rm eh}\rangle = |\Phi_{\rm x}\rangle$. Consequently, at later times $t > 0$ we have

$$|\Psi(t)\rangle = e^{-{\rm i}\hat{H}_{\rm ins}t}|\Psi\rangle = a|\Phi_0\rangle + b\,e^{-{\rm i}\Omega_{\rm x}t}|\Phi_{\rm x}\rangle.$$

Let us calculate the lesser Green's function to the lowest nonvanishing order in b. Taking into account that we are in a pure state – that is, the density matrix operator $\hat{\rho} = |\Psi\rangle\langle\Psi|$ –

we have from (6.29)

$$G^<_{kvv}(t,t') = \mathrm{i}\,\langle\Psi(t')|\hat{d}^\dagger_{kv} e^{-\mathrm{i}\hat{H}_{\rm ins}(t'-t)}\hat{d}_{kv}|\Psi(t)\rangle$$
$$\simeq \mathrm{i}\,|a|^2\,\langle\Phi_0|\hat{d}^\dagger_{kv} e^{-\mathrm{i}\hat{H}_{\rm ins}(t'-t)}\hat{d}_{kv}|\Phi_0\rangle = \mathrm{i}\,|a|^2 e^{\mathrm{i}\epsilon_{kv}(t'-t)}, \tag{6.108}$$

where we observe that $\hat{d}_{kv}|\Phi_0\rangle$ is an eigenstate of $\hat{H}_{\rm ins}$ with eigenvalue $-\epsilon_{kv}$. The vv component is therefore of order zero in b. For the cv component we find

$$G^<_{kcv}(t,t') = \mathrm{i}\,\langle\Psi(t')|\hat{d}^\dagger_{kv} e^{-\mathrm{i}\hat{H}_{\rm ins}(t'-t)}\hat{d}_{kc}|\Psi(t)\rangle$$
$$= \mathrm{i}\,a^* b\,\langle\Phi_0|\hat{d}^\dagger_{kv} e^{-\mathrm{i}\hat{H}_{\rm ins}(t'-t)}\hat{d}_{kc}|\Phi_{\rm x}\rangle\, e^{-\mathrm{i}\Omega_{\rm x}t} = \mathrm{i}\,a^* b\,Y^{({\rm x})}_k e^{\mathrm{i}\epsilon_{kv}(t'-t)} e^{-\mathrm{i}\Omega_{\rm x}t}, \tag{6.109}$$

which is of first order in b. Notice that this component does not depend on the time difference only. Finally, the cc component yields

$$G^<_{kcc}(t,t') = \mathrm{i}\,\langle\Psi(t')|\hat{d}^\dagger_{kc} e^{-\mathrm{i}\hat{H}_{\rm ins}(t'-t)}\hat{d}_{kc}|\Psi(t)\rangle$$
$$= \mathrm{i}\,|b|^2\langle\Phi_{\rm x}|\hat{d}^\dagger_{kc} e^{-\mathrm{i}\hat{H}_{\rm ins}(t'-t)}\hat{d}_{kc}|\Phi_{\rm x}\rangle\, e^{\mathrm{i}\Omega_{\rm x}(t'-t)} = \mathrm{i}\,|bY^{({\rm x})}_k|^2 e^{\mathrm{i}(\epsilon_{kv}+\Omega_{\rm x})(t'-t)}, \tag{6.110}$$

which is of second order in b. We use these results to calculate the photocurrent for two different kind of probe pulses. For simplicity we take a light-matter coupling $D_{k\nu} = D$ independent of the quasi-momentum and band index.

Everlasting monochromatic probes The first kind of pulse is the everlasting monochromatic probe $A(t) = \theta(t)(A_0 e^{\mathrm{i}\omega_0 t} + A_0^* e^{-\mathrm{i}\omega_0 t})$, $\omega_0 > 0$, already studied in the previous section. As pointed out above (6.102), only terms depending on the time difference contribute to the long-time limit of the photocurrent. This holds for the ionization self-energy as well as for the Green's function, and it is a direct consequence of the Riemann–Lebesgue theorem. We can then ignore $G^<_{kcv}$ in the calculation of $I_{ph}(k)$. Inserting the Fourier transform of $G^<_{kvv}$ and $G^<_{kcc}$ in (6.103), we find

$$I_{ph}(\mathbf{k}) = 4\pi|A_0 D|^2\Big[|a|^2\delta(\epsilon_{\mathbf{k}} - \omega_0 - \epsilon_{kv}) + |b|^2|Y^{({\rm x})}_k|^2\delta(\epsilon_{\mathbf{k}} - \omega_0 - \epsilon_{kv} - \Omega_{\rm x})\Big]. \tag{6.111}$$

Let us evaluate this formula for a valence band $\epsilon_{kv} = -\frac{\Delta}{2} - \Delta(1 - \cos k)$, a conduction band $\epsilon_{kc} = -\epsilon_{kv} - v_0$, and a q-independent interaction $v_q = v_0 = 0.65\Delta$. For these parameters $\hat{H}_{\rm ins}$ has one single excitonic solution of energy $\Omega_{\rm x} \simeq 0.81\Delta$. In Fig. 6.8 we show the density plot of $I_{ph}(\mathbf{k})$ as a function of the energy variable $\epsilon \equiv \epsilon_{\mathbf{k}} - \omega_0$ (vertical axis) and parallel quasi-momentum k (horizontal axis). For $\epsilon < 0$ we recognize the shape of the valence band; the signal comes from the first term in (6.111) and it would be the only signal if the insulator were in the ground state. No signal is instead detected from the conduction band, whose shape is given by the faint white line. The exciton gives rise to a sideband located at energy $\Omega_{\rm x}$ above the valence-band maximum, see second term in (6.111). It is a replica of the valence band with spectral weight proportional to the square of the exciton wavefunction in k-space: $|Y^{({\rm x})}_k|^2$.

The possibility of measuring the exciton wavefunction using TR-ARPES was realized in 2016 [70]. Theoretical investigations [81, 82, 83], refinements [84, 85, 86], and implications

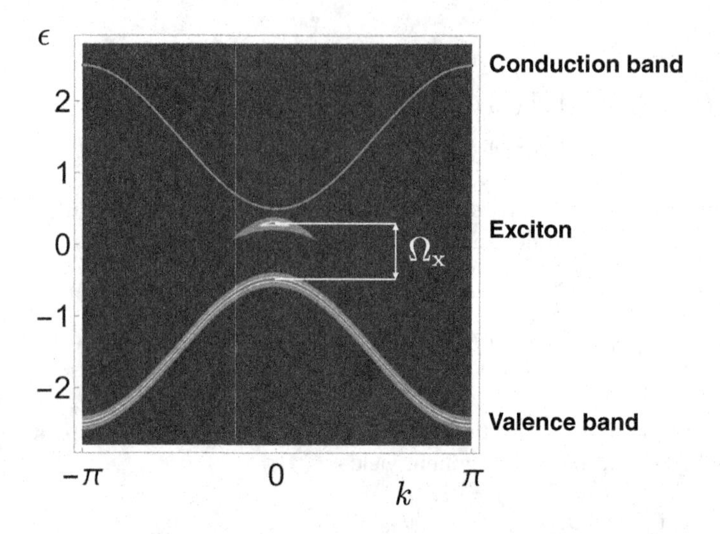

Figure 6.8 Density plot of the photocurrent (6.111) versus the energy $\epsilon \equiv \epsilon_{\mathbf{k}} - \omega_0$ (vertical axis) and parallel quasi-momentum k (horizontal axis). We have chosen $a \simeq 1$ and scale up the photocurrent for $\epsilon > 0$ to improve visibility. Energies are measured in units of the bare gap Δ.

[87, 88, 89] then followed, but it was not until 2021 that the first experimental evidence was finally obtained [90, 91].

Ultrafast probes The second kind of pulse that we consider is an ultrafast probe $A(t) = A_0\,\delta(t - t_{\text{delay}})$, where $t_{\text{delay}} > 0$ is the time delay between the pump, ending at time $t = 0$, and the probe. The ionization self-energy (6.101) for our two-band insulator becomes

$$\Sigma^{\text{ion,R}}_{\mathbf{k}\nu\nu'}(t,\bar{t}) = -\frac{\mathrm{i}}{2}\delta(t - t_{\text{delay}})\delta(\bar{t} - t_{\text{delay}})|A_0 D|^2,$$

which inserted into (6.100) gives for the photocurrent

$$I_{ph}(\mathbf{k},t) = \delta(t - t_{\text{delay}})|A_0 D|^2 \sum_{\nu\nu'} \text{Re}\left[-\mathrm{i}G^{<}_{k\nu'\nu}(t,t)\right].$$

Thus the photocurrent is nonzero only at time $t = t_{\text{delay}}$ and, noteworthy, its value depends on t_{delay}. The dependence on t_{delay} is brought by the cv and vc matrix elements of the Green's function. Inserting the results in (6.108), (6.109), and (6.110), we obtain

$$I_{ph}(\mathbf{k},t) = \delta(t - t_{\text{delay}})|A_0 D|^2\Big(|a|^2 + |bY^{(\text{x})}_k|^2 + 2\text{Re}\Big[a^* bY^{(\text{x})}_k e^{-\mathrm{i}\Omega_{\text{x}} t}\Big]\Big). \tag{6.112}$$

Thus the photocurrent consists of a delay-independent part and a monochromatic oscillatory contribution of frequency Ω_{x}. We conclude that ultrafast probes can be used to reveal excitonic coherence in resonantly pumped insulators. We also observe that for this kind of probe pulses the information on band structure and exciton sidebands is lost. Numerical

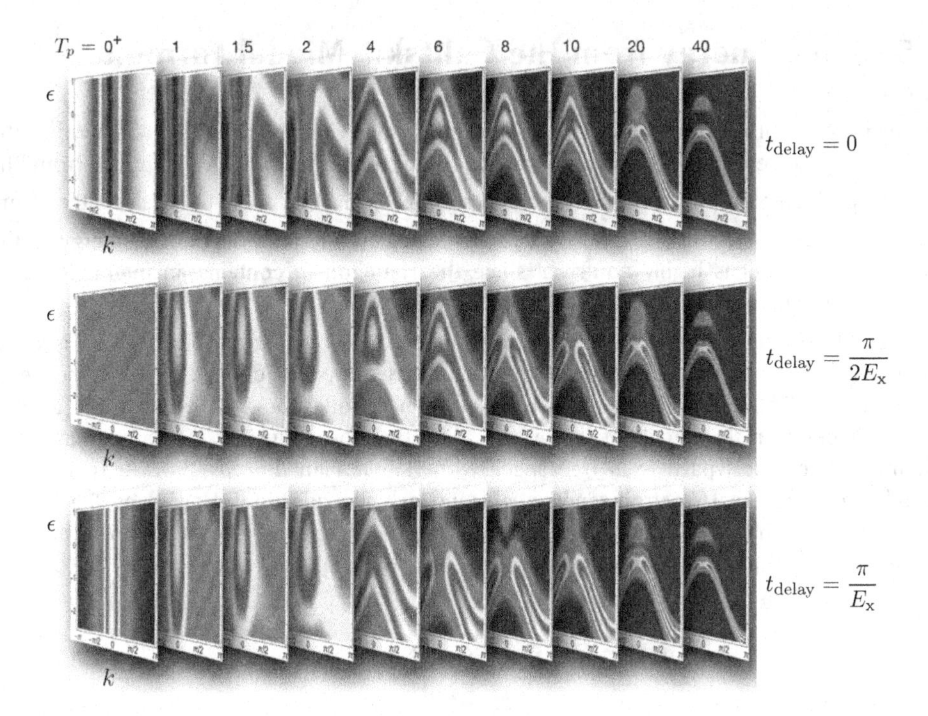

Figure 6.9 Density plot of the integrated TR-ARPES photocurrent $\bar{I}_{ph}$ versus $\epsilon = \epsilon_{\mathbf{k}} - \omega_0$ and k for probe pulses of different duration T_p impinging the system at different times $t_{\rm delay}$. Energies are measured in units of the gap Δ and times in units of $1/\Delta$.

simulations with probes of decreasing duration nicely show that the spectral features in the time-independent photocurrent of Fig. 6.8 gradually fade away and, at the same time, the photocurrent acquires a dependence on $t_{\rm delay}$ [89]. In Fig. 6.9 we report the outcome of these simulations for a model Hamiltonian similar to (6.105). The density plots show the integrated photocurrent $\bar{I}_{ph}(\mathbf{k}) = \int_0^\infty dt\, I_{ph}(\mathbf{k}, t)$ versus $\epsilon = \epsilon_{\mathbf{k}} - \omega_0$ and parallel quasi-momentum k for a probing pulse

$$A(t) = A_0 \sin\left(\frac{\pi(t - t_{\rm delay})}{T_p}\right)^2 \sin\left(\omega_0(t - t_{\rm delay})\right)$$

active from $t = t_{\rm delay}$ until $t = t_{\rm delay} + T_p$ and zero otherwise. The limiting cases in (6.111) and (6.112) are recovered for $T_p \to \infty$ and $T_p \to 0^+$, respectively.

Exercise 6.11 Let N be the number of k-points in the Hamiltonian (6.105). Prove that the lowest-energy state with N electrons is obtained by filling all the states of the v-band. Prove also that this state is the lowest-energy state of $\hat{H}_{\rm ins} - \mu\hat{N}$ in Fock space when μ lies between the valence and the conduction bands.

Exercise 6.12 Prove (6.107).

6.5 Total Energy from the Galitskii–Migdal Formula

In our introductory discussion on the lesser and greater Green's function we show that the time-dependent ensemble average of any one-body operator can be computed from the equal-time lesser Green's function $\hat{\mathcal{G}}^<(t,t)$, see (6.30). In this section we show that from the knowledge of the full $\hat{\mathcal{G}}^<(t,t')$ we can also calculate the time-dependent energy of the system. This result is highly nontrivial since the Hamiltonian contains an interaction part that is a two-body operator.

How to define the energy operator Let us start by clarifying a point which is often the source of some confusion: How is the total energy operator defined in the presence of a time-dependent external field? This is a very general question and, as such, the answer cannot depend on the details of the system. We, therefore, consider the simple case of a quantum mechanical particle in free space with Hamiltonian $\hat{h} = \hat{p}^2/2m$. At time t_0 we switch on an external electromagnetic field and ask the question: How does the energy of the particle change in time? Let $|\Psi(t)\rangle$ be the ket of the particle at time t. The time evolution is governed by the Schrödinger equation $\mathrm{i}\frac{d}{dt}|\Psi(t)\rangle = \hat{h}(t)|\Psi(t)\rangle$ with

$$\hat{h}(t) = \frac{1}{2m}\left(\hat{\boldsymbol{p}} - \frac{q}{c}\mathbf{A}(\hat{\boldsymbol{r}},t)\right)^2 + q\phi(\hat{\boldsymbol{r}},t).$$

Is the energy of the particle given by the average over $|\Psi(t)\rangle$ of $\hat{p}^2/2m$ (which describes the original system with no external fields) or of the full Hamiltonian $\hat{h}(t)$? The correct answer is neither. The average of $\hat{h}(t)$ must be ruled out since it contains the coupling energy between the particle and the *external* field $q\phi(\hat{\boldsymbol{r}},t)$. Our system is a free particle and therefore its energy is simply the average of the kinetic energy operator – that is, the velocity operator squared times $m/2$. In the presence of a vector potential, however, the velocity operator is $[\hat{\boldsymbol{p}} - \frac{q}{c}\mathbf{A}(\hat{\boldsymbol{r}},t)]/m$ and not just $\hat{\boldsymbol{p}}/m$. We conclude that the energy of the system at time t must be calculated by averaging the operator $\frac{1}{2m}\left(\hat{\boldsymbol{p}} - \frac{q}{c}\mathbf{A}(\hat{\boldsymbol{r}},t)\right)^2 = \hat{h}(t) - q\phi(\hat{\boldsymbol{r}},t)$ over $|\Psi(t)\rangle$. Further evidence in favor of the above choice comes from the fact that all physical quantities must be invariant under a gauge transformation. For the case of a single particle, the gauge transformation $\mathbf{A} \to \mathbf{A} + \boldsymbol{\nabla}\Lambda$, $\phi \to \phi - \frac{1}{c}\frac{\partial}{\partial t}\Lambda$ implies that the ket changes according to $|\Psi(t)\rangle \to \exp\left[\mathrm{i}\frac{q}{c}\Lambda(\hat{\boldsymbol{r}},t)\right]|\Psi(t)\rangle$, see (3.25). The reader can easily verify that neither $\langle\Psi(t)|\hat{p}^2/2m|\Psi(t)\rangle$ nor $\langle\Psi(t)|\hat{h}(t)|\Psi(t)\rangle$ are gauge invariant, while $\langle\Psi(t)|\hat{h}(t) - q\phi(\hat{\boldsymbol{r}},t)|\Psi(t)\rangle$ is.

Let us now apply the same reasoning to a system of interacting identical particles with Hamiltonian $\hat{H}(t) = \hat{H}_0(t) + \hat{H}_{\rm int}$. According to the previous discussion, the energy of the system at a generic time t_1, $E_{\rm S}(t_1)$, is the time-dependent ensemble average of the operator

$$\hat{H}_{\rm S}(t_1) \equiv \hat{H}(t_1) - q\int d\mathbf{x}_1\, \hat{n}(\mathbf{x}_1)\phi_\Delta(1),$$

where $\phi_\Delta(1) = \phi(\mathbf{r}_1,t_1) - \phi(\mathbf{r}_1)$ is the difference between the scalar potential at time t_1

and the scalar potential in equilibrium. Then we have

$$E_S(z_1) = \sum_k \rho_k \langle \Psi_k | \hat{U}(t_{0-}, z_1) \left[\int d\mathbf{x}_1 d\mathbf{x}_2 \, \hat{\psi}^\dagger(\mathbf{x}_1) \langle \mathbf{x}_1 | \hat{h}_S(z_1) | \mathbf{x}_2 \rangle \hat{\psi}(\mathbf{x}_2) \right.$$
$$\left. + \frac{1}{2} \int d\mathbf{x}_1 d\mathbf{x}_2 \, v(\mathbf{x}_1, \mathbf{x}_2) \, \hat{\psi}^\dagger(\mathbf{x}_1) \hat{\psi}^\dagger(\mathbf{x}_2) \hat{\psi}(\mathbf{x}_2) \hat{\psi}(\mathbf{x}_1) \right] \hat{U}(z_1, t_{0-}) | \Psi_k \rangle,$$

where $\hat{h}_S(z_1) = \hat{h}(z_1) - q\phi_\Delta(\hat{\mathbf{r}}_1, z_1)$ is the single-particle Hamiltonian of the system.

Energy with NEGF We remind the reader that the energy $E_S(t_1)$ as a function of the physical time t_1 is given by $E_S(t_{1\pm})$, see again (4.10). The reason for introducing the contour time z_1 is that we now recognize on the r.h.s. the one- and two-particle Green's functions with a precise order of the contour-time variables. We have, see also (5.6),

$$E_S(z_1) = \underbrace{\pm \mathrm{i} \int d\mathbf{x}_1 d2 \, h_S(1;2) G(2;1^+)}_{E_{\text{one}}} \underbrace{- \frac{1}{2} \int d\mathbf{x}_1 d2 \, v(1;2) G_2(1,2;1^+,2^+)}_{E_{\text{int}}} . \qquad (6.113)$$

The energy E_{one} is the one-body part of the total energy – that is, the sum of the kinetic and potential energy. The particular form of the interaction energy allows us to express E_S in terms of G only. Adding the equation of motion for G to its adjoint [see the first equations of the Martin–Schwinger hierarchy (5.2) and (5.3)] and then setting $2 = 1^+$, we find

$$\left[\left(\mathrm{i}\frac{d}{dz_1} - \mathrm{i}\frac{d}{dz_2} \right) G(1;2) \right]_{2=1^+} - \int d3 \left[h(1;3) G(3;1^+) + G(1;3^+) h(3;1) \right]$$
$$= \pm 2\mathrm{i} \int d3 \, v(1;3) G_2(1,3;1^+,3^+),$$

where $h(1;3)$ is defined in (6.4). Inserting this result into $E_S(z_1)$, we arrive at the very interesting formula

$$E_S(z_1) = \pm \, \mathrm{i} \int d\mathbf{x}_1 \langle \mathbf{x}_1 | \left[\hat{h}_S(z_1) - \frac{1}{2}\hat{h}(z_1) \right] \hat{\mathcal{G}}(z_1, z_1^+) | \mathbf{x}_1 \rangle$$
$$\pm \frac{\mathrm{i}}{4} \int d\mathbf{x}_1 \left[\left(\mathrm{i}\frac{d}{dz_1} - \mathrm{i}\frac{d}{dz_2} \right) \langle \mathbf{x}_1 | \hat{\mathcal{G}}(z_1, z_2) | \mathbf{x}_1 \rangle \right]_{z_2 = z_1^+} . \qquad (6.114)$$

This formula yields the initial energy of the system E_S^M when $z_1 = t_0 - \mathrm{i}\tau_1$ and the time-dependent energy $E_S(t_1)$ when $z_1 = t_{1\pm}$. In the former case $\hat{h}_S(z_1) = \hat{h}(z_1) = \hat{h}^M$ and expanding $\hat{\mathcal{G}}$ in the Matsubara series (6.18) we get

$$\boxed{E_S^M = \pm \frac{\mathrm{i}}{2} \frac{1}{-\mathrm{i}\beta} \sum_m e^{\eta\omega_m} \int d\mathbf{x} \, \langle \mathbf{x} | (\omega_m + \hat{h}^M) \hat{\mathcal{G}}^M(\omega_m) | \mathbf{x} \rangle} \qquad (6.115)$$

For the time-dependent energy $E_S(t)$ we recall that $\hat{h}_S(t) = \hat{h}(t) - q\phi_\Delta(\hat{\mathbf{r}}, t)$, and hence from (6.114) with $z_1 = t_\pm$

$$E_S(t) = \pm \frac{\mathrm{i}}{4} \int d\mathbf{x} \, \langle \mathbf{x} | \left(\mathrm{i}\frac{d}{dt} - \mathrm{i}\frac{d}{dt'} + 2\hat{h}(t) \right) \hat{\mathcal{G}}^<(t,t') | \mathbf{x} \rangle \Big|_{t'=t} - q \int d\mathbf{x} \, n(\mathbf{x},t) \phi_\Delta(\mathbf{r},t).$$

Galitskii–Migdal formula In the special case that the Hamiltonian $\hat{H}(t) = \hat{H}$ does not depend on time (hence $\phi_\Delta = 0$) and the system is initially in a stationary ensemble of $\hat{H}$,[17] the lesser Green's function depends on the time difference $t - t'$ only. Then, Fourier transforming $\hat{\mathcal{G}}^<$ as in (6.44), the energy simplifies to

$$\boxed{E_{\rm S} = \pm\frac{\rm i}{2}\int\frac{d\omega}{2\pi}\int d\mathbf{x}\,\langle\mathbf{x}|(\omega+\hat{h})\hat{\mathcal{G}}^<(\omega)|\mathbf{x}\rangle} \tag{6.116}$$

This result represents a generalization of the so-called *Galitskii–Migdal formula* since it is valid not only for systems initially in thermal equilibrium, but also for systems in a stationary ensemble. In the case of thermal equilibrium, $E_{\rm S}^{\rm M} = E_{\rm S} - \mu N$, and hence (6.116) provides an alternative formula to (6.115) for the calculation of the initial energy.

As an example of stationary ensemble we consider the case of noninteracting systems. Then the lesser Green's function is given by (6.45), and (6.116) yields

$$E_{\rm S} = \sum_\lambda f(\epsilon_\lambda^{\rm M})\epsilon_\lambda. \tag{6.117}$$

The energy of the system is the weighted sum of the single-particle energies. As expected, the equilibrium energy is recovered for $\hat{h}^{\rm M} = \hat{h} - \mu$. In general, however, we can calculate the energy of an arbitrary stationary ensemble by a proper choice of $\hat{h}^{\rm M}$. We use (6.117) in Section 8.3.2 to study the spin-polarized ground state of an electron gas.

[17] This means that $\hat{\rho} = \sum_k \rho_k |\Psi_k\rangle\langle\Psi_k|$ where all $|\Psi_k\rangle$ are eigenkets of $\hat{H}$.

7

One-Particle Green's Function: Diagrammatic Expansion

In Chapter 6 we discussed the physical information contained in the one-particle Green's function and derived several exact properties. In this chapter we go back to the formula (5.16) for G and present a practical and efficient method to collect, group, and reorganize all the terms of Wick's expansion. The method is graphical and consists in representing every term of the expansion with a diagram. The idea is due to Feynman and has two main appealing features: (1) it is much easier to manipulate diagrams rather than lengthy and intricate mathematical expressions; and (2) the Feynman diagrams explicitly unravel the underlying physical content of the various terms of the expansion, thus allowing for constructing physical and controllable approximations to G.

7.1 Getting Started with Feynman Diagrams

We write (5.16) again for convenience:

$$G(a;b) = \frac{\sum_{k=0}^{\infty} \frac{1}{k!}\left(\frac{\mathrm{i}}{2}\right)^k \int v(1;1')\,..\,v(k;k') \begin{vmatrix} G_0(a;b) & G_0(a;1^+) & \dots & G_0(a;k'^+) \\ G_0(1;b) & G_0(1;1^+) & \dots & G_0(1;k'^+) \\ \vdots & \vdots & \ddots & \vdots \\ G_0(k';b) & G_0(k';1^+) & \dots & G_0(k';k'^+) \end{vmatrix}_{\pm}}{\sum_{k=0}^{\infty} \frac{1}{k!}\left(\frac{\mathrm{i}}{2}\right)^k \int v(1;1')\,..\,v(k;k') \begin{vmatrix} G_0(1;1^+) & G_0(1;1'^+) & \dots & G_0(1;k'^+) \\ G_0(1';1^+) & G_0(1';1'^+) & \dots & G_0(1';k'^+) \\ \vdots & \vdots & \ddots & \vdots \\ G_0(k';1^+) & G_0(k';1'^+) & \dots & G_0(k';k'^+) \end{vmatrix}_{\pm}}. \tag{7.1}$$

This equation gives explicitly all the terms needed to calculate G to all orders in the interaction strength v. What we need to do is find an efficient way to collect them. To get some experience, let us work out some low-order terms explicitly.

Vacuum diagrams We start with the denominator, which is the ratio Z/Z_0 between the partition function of the interacting and noninteracting systems, see (5.19). To first order we have

$$\left(\frac{Z}{Z_0}\right)^{(1)} = \frac{\mathrm{i}}{2}\int d1d1'\,v(1;1') \begin{vmatrix} G_0(1;1^+) & G_0(1;1'^+) \\ G_0(1';1^+) & G_0(1';1'^+) \end{vmatrix}_{\pm}$$
$$= \frac{\mathrm{i}}{2}\int d1d1'\,v(1;1')\left[G_0(1;1^+)G_0(1';1'^+) \pm G_0(1;1'^+)G_0(1';1^+)\right]. \tag{7.2}$$

The basic idea of the Feynman diagrams is to provide a simple set of rules to convert a drawing into a well-defined mathematical quantity, like (7.2). Since (7.2) contains only Green's functions and interparticle interactions, we must assign to G_0 and v a graphical object. We represent a Green's function $G_0(1;2^+)$ by an oriented line going from 2 to 1:

$G_0(1;2^+)$ = 1 ——◀—— 2

The Green's function line is oriented to distinguish $G_0(1;2^+)$ from $G_0(2;1^+)$. The orientation is, of course, a pure convention. We could have chosen the opposite orientation as long as we consistently use the same orientation for all Green's functions. The convention above is the standard one. It stems from the intuitive picture that in $G_0(1;2^+)$ we create a particle in 2 and destroy it back in 1. Thus the particle "moves" from 2 to 1. The interaction $v(1;2)$ is represented by a wiggly line:

$v(1;2)$ = 1 ∿∿∿∿ 2

which has no direction since $v(1;2) = v(2;1)$. Then, the two terms in (7.2) have the graphical form

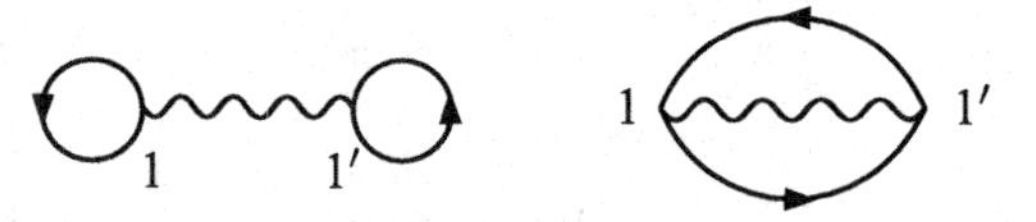

where integration over all internal vertices (in this case 1 and 1′) is understood. It is important to observe that the infinitesimal contour-time shift in $G_0(1;2^+)$ plays a role only when the start and end points of G are the same (as in the first diagram above) or when points 1 and 2 are joined by an interaction line $v(1;2) = \delta(z_1,z_2)v(\mathbf{x}_1,\mathbf{x}_2)$ (as in the second diagram above). In all other cases we can safely discard the shift since $z_1 = z_2$ is a set of zero measure in the integration domain.

> *In the remainder of the book we sometimes omit the infinitesimal shift. If a diagram contains a Green's function with the same contour-time arguments, then the second argument is understood to be infinitesimally later than the first.*

It is also worth noting that the prefactor of the diagrams is determined by (7.2); it is (i/2) for the first diagram and (±i/2) for the second diagram. More generally, to any order in the

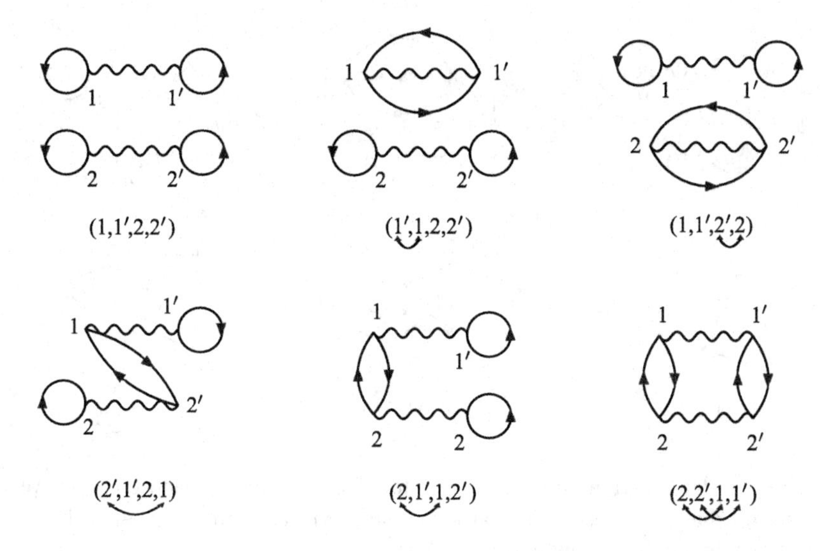

Figure 7.1 Some of the second-order diagrams of the MBPT expansion of Z/Z_0.

interaction strength the prefactors follow directly from (7.1). From now on the prefactor of a diagram is incorporated in the diagram itself – that is, to each diagram corresponds an integral of Green's functions and interactions with the appropriate prefactor.

To evaluate the second-order contribution to Z/Z_0 we must expand the permanent/determinant of a 4×4 matrix, which yields $4! = 24$ terms:

$$\begin{aligned} \left(\frac{Z}{Z_0}\right)^{(2)} &= \frac{1}{2!}\frac{\mathrm{i}^2}{2^2}\int d1d1'd2d2'\, v(1;1')v(2;2') \\ &\times \sum_P (\pm)^P G_0(1;P(1))G_0(1';P(1'))G_0(2;P(2))G_0(2';P(2')). \end{aligned} \tag{7.3}$$

In Fig. 7.1 we show some of the diagrams originating from this expansion. Below each diagram we indicate the permutation that generates it. The prefactor is simply $\frac{1}{2!}(\frac{\mathrm{i}}{2})^2$ times the sign of the permutation. Going to higher order in v, the number of diagrams grows and their topology becomes increasingly more complicated. However, they all have a common feature: Their mathematical expression contains an integral over all vertices. We refer to these diagrams, as the *vacuum diagrams*. Thus, given a vacuum diagram, the rules to convert it into a mathematical expression are:

- Number all vertices and assign an interaction $v(i;j)$ to a wiggly line connecting i and j and a Green's function $G_0(i;j^+)$ to an oriented line from j to i.
- Integrate over all vertices and multiply by $[(\pm)^P \frac{1}{k!}(\frac{\mathrm{i}}{2})^k]$, where $(\pm)^P$ is the sign of the permutation and k is the number of interaction lines.

Green's function diagrams Let us now turn our attention to the numerator, $N(a;b)$, of (7.1). To first order in the interaction strength we must evaluate the permanent/determinant

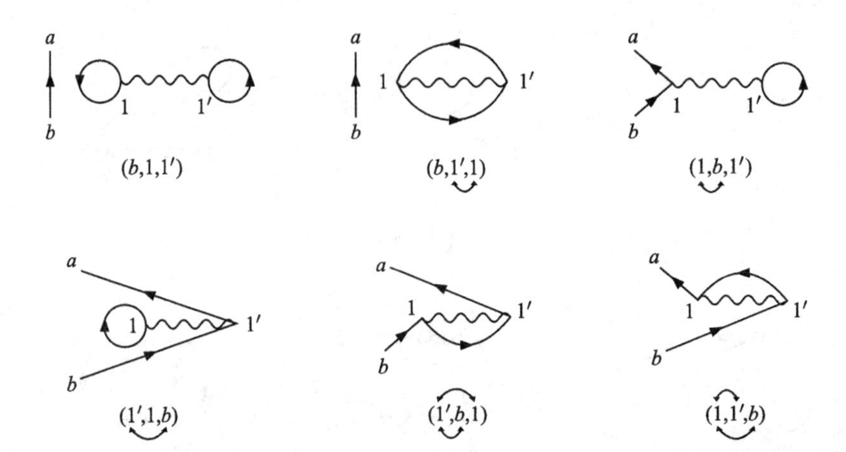

Figure 7.2 First-order diagrams for the numerator $N(a;b)$. Under each diagram we report the permutation that generates it. The arrows help to visualize the number of interchanges (the first interchange is the arrow under the tern).

of a 3×3 matrix. The expansion along the first column leads to

$$\begin{aligned} N^{(1)}(a;b) &= \frac{\mathrm{i}}{2} G_0(a;b) \int d1d1' v(1;1') \begin{vmatrix} G_0(1;1^+) & G_0(1;1'^+) \\ G_0(1';1^+) & G_0(1';1'^+) \end{vmatrix}_{\pm} \\ &\pm \frac{\mathrm{i}}{2} \int d1d1' v(1;1') G_0(1;b) \begin{vmatrix} G_0(a;1^+) & G_0(a;1'^+) \\ G_0(1';1^+) & G_0(1';1'^+) \end{vmatrix}_{\pm} \\ &+ \frac{\mathrm{i}}{2} \int d1d1' v(1;1') G_0(1';b) \begin{vmatrix} G_0(a;1^+) & G_0(a;1'^+) \\ G_0(1;1^+) & G_0(1;1'^+) \end{vmatrix}_{\pm} . \end{aligned} \quad (7.4)$$

To each term we can easily give a diagrammatic representation. In the first line of (7.4) we recognize the ratio $(Z/Z_0)^{(1)}$ of (7.2) multiplied by $G_0(a;b)$. The corresponding diagrams are simply those of $(Z/Z_0)^{(1)}$ with an extra line going from b to a. The remaining terms can be drawn in a similar manner and the full set of diagrams (together with the corresponding permutations) is shown in Fig. 7.2. The prefactor is simply $(\mathrm{i}/2)$ times the sign of the permutation. The diagrams for $N(a;b)$ are different from the vacuum diagrams since there are two *external* vertices (a and b) over which we do not integrate. We refer to these diagrams as the *Green's function diagrams.* The rules to convert a Green's function diagram into a mathematical expression are the same as those for the vacuum diagrams, with the exception that there is no integration over the external vertices. In contrast, a vacuum diagram contains only *internal* vertices.

At this point the only quantity which is a bit awkward to determine is the sign of the permutation. It would be useful to have a simple rule to fix the sign by giving the diagram a cursory glance. This is the topic of the next section.

Exercise 7.1 Calculate all the second-order terms of Z/Z_0 and draw the corresponding diagrams.

Exercise 7.2 Show that the diagrammatic representation of the last two lines of (7.4) are the last four diagrams of Fig. 7.2.

7.2 Loop Rule

From (7.1) or (7.3) we see that the sign of a diagram is determined by the sign of the permutation that changes the second argument of the Green's functions. In graphical terms this amounts to a permutation of the starting points of the Green's function lines of a diagram. Since every permutation can be obtained by successive interchanges of pairs of labels (i, j), we only need to investigate how such interchanges modify a diagram. A vacuum diagram consists of a certain number of loops and, therefore, an interchange can occur either between two starting points of the same loop or between two starting points of different loops. In the former case we have the generic situation

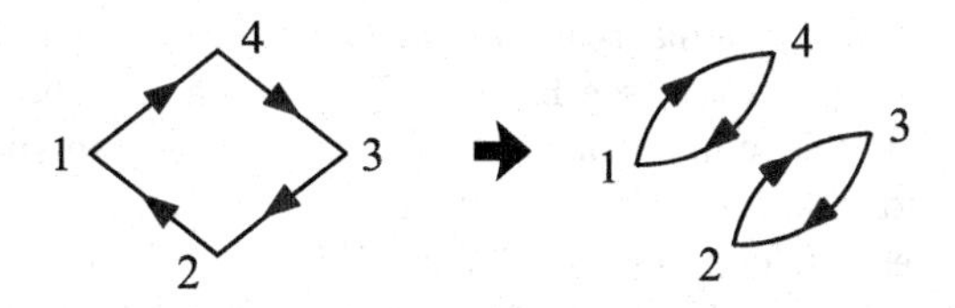

in which we interchange the starting points 2 and 4 so that $G_0(1;2)G_0(2;3)G_0(3;4)$ $G_0(4;1) \to G_0(1;4)G_0(2;3)G_0(3;2)G_0(4;1)$ and hence the number of loops increases by one. In the latter case we have the generic situation

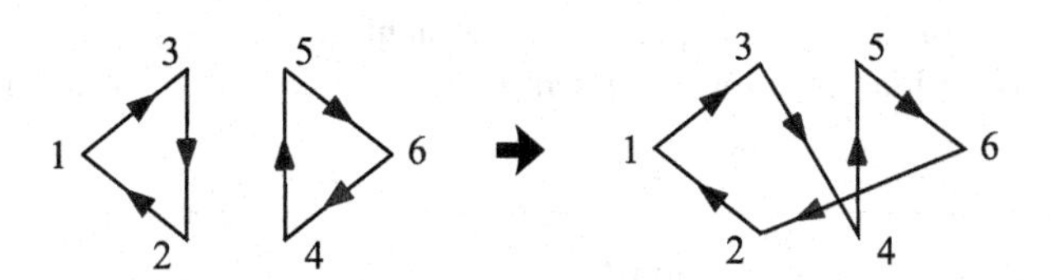

in which we interchange the starting points 3 and 6 so that $G_0(1;2)G_0(2;3)G_0(3;1)$ $G_0(5;4)G_0(4;6)G_0(6;5) \to G_0(1;2)G_0(2;6)G_0(3;1)G_0(5;4)G_0(4;3)G_0(6;5)$ and hence the number of loops decreases by one. It is not difficult to convince ourselves that this is a general rule: An interchange of starting points changes the number of loops by one. For a Green's function diagram, in addition to the interchanges just considered, we have two more possibilities: The interchange occurs either between two starting points on the path connecting b to a or between a starting point on the path and a starting point on a loop. In the first case we have the generic situation

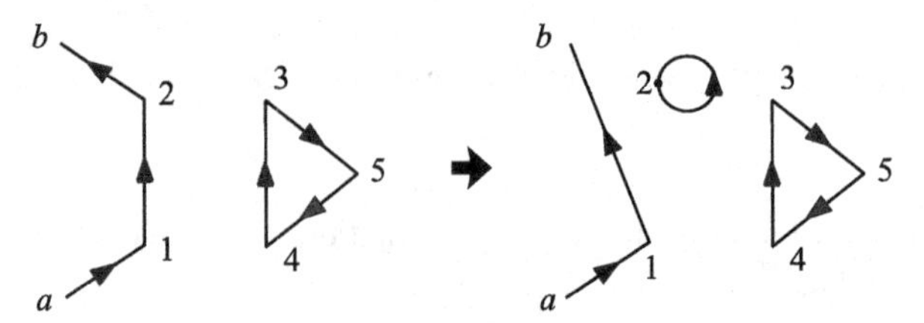

in which we interchange the starting points 1 and 2, and the number of loops increases by one. The reader can easily verify that the number of loops would have increased by one also by an interchange of a and 2 or of a and 1. In the second case an interchange of, for example, the starting points a and 3 leads to

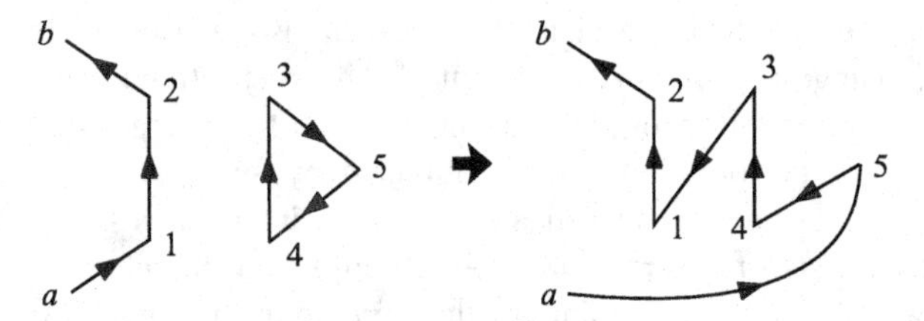

and the number of loops decreases by one. Again, this is completely general: An interchange of starting points changes the number of loops by one. Taking into account that for the identity permutation the sign of a diagram, be it a vacuum or a Green's function diagram, is + and the number of loops is even, see Exercise 7.3, we can state the so-called *loop rule*: $(\pm)^P = (\pm)^l$ where l is the number of loops! As an example we show in Fig. 7.3 some of the $5! = 120$ second-order Green's function diagrams together with the corresponding permutation; in all cases the loop rule is fulfilled. This example also shows that there are several diagrams [(a) to (e)] which are products of a *connected diagram* (connected to a and b) and a vacuum diagram. It turns out that the vacuum diagrams are divided out by the denominator of the Green's function in (7.1), thus leading to a large reduction of the number of diagrams to be considered. Furthermore, there are diagrams that have the same numerical value [e.g., (a)–(b)–(e), (c)–(d), and (f)–(g)]. This is due to a permutation and mirror symmetry of the interaction lines $v(j; j')$. We can achieve a large reduction in the number of diagrams by taking into account these symmetries. In the next two sections we discuss how to do it.

Exercise 7.3 Draw the G-diagram and the vacuum diagram of order n – that is, with n interaction lines – corresponding to the identity permutation.

7.3 Cancellation of Disconnected Diagrams

We already observed that the number of terms generated by (7.1) grows very rapidly when going to higher order in the interaction strength. In this section we show that the disconnected vacuum diagrams of the numerator are exactly canceled by the vacuum diagrams

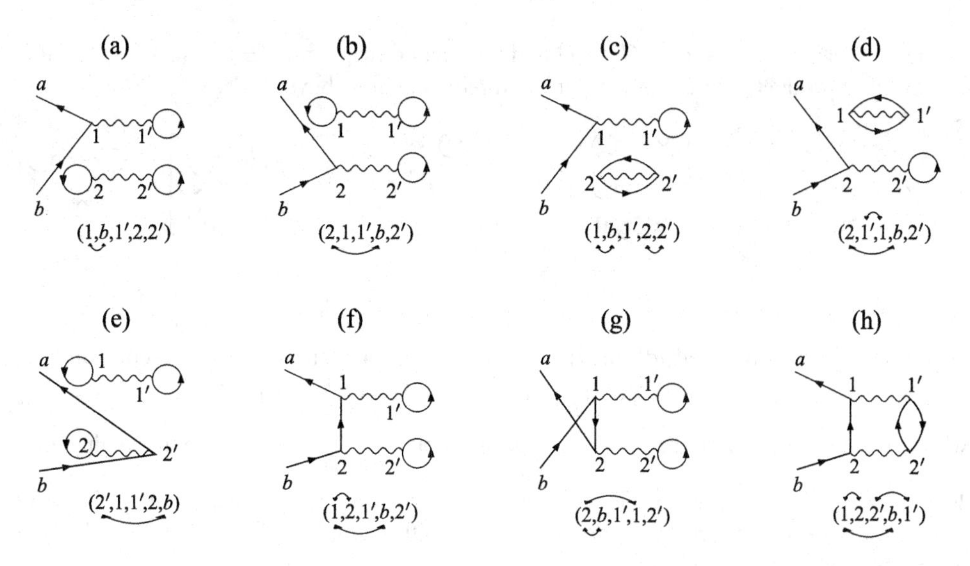

Figure 7.3 Some of the second-order Green's function diagrams and the corresponding permutations.

of the denominator. Let us start with an example. Consider the following three Green's function diagrams that are part of the expansion of the numerator of (7.1) to third order:

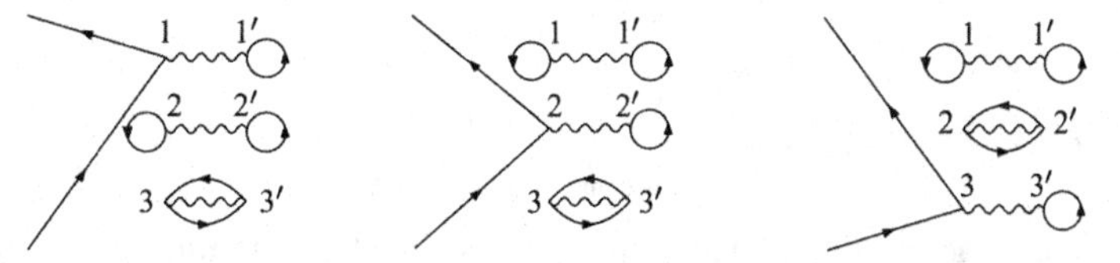

They are all related by a permutation of the interaction lines $v(i;i') \leftrightarrow v(j;j')$ that preserves the structure of two disjoint pieces, which are: (piece 1) the connected Green's function diagram corresponding to the third diagram of Fig. 7.2 and (piece 2) the vacuum diagram corresponding to the third diagram of Fig. 7.1. Furthermore, they all have the same prefactor $(\pm)^4 \frac{1}{3!}(\frac{\mathrm{i}}{2})^3$ so that their total contribution is

$$3 \times \frac{1}{3!}\left(\frac{\mathrm{i}}{2}\right)^3 \int [G_0 G_0 G_0 v] \int [G_0 G_0 G_0 G_0 v v], \tag{7.5}$$

with the obvious notation that the first factor $\int [G_0 G_0 G_0 v]$ refers to the Green's function diagram and the second factor $\int [G_0 G_0 G_0 G_0 v v]$ refers to the vacuum diagram. Let us now consider the product of diagrams:

As the first diagram is $\pm\frac{\mathrm{i}}{2}\int[G_0G_0G_0v]$ and the second diagram is $(\pm)^3\frac{1}{2!}\left(\frac{\mathrm{i}}{2}\right)^2\int[G_0G_0G_0 G_0vv]$, their product equals the sum of the three diagrams above:

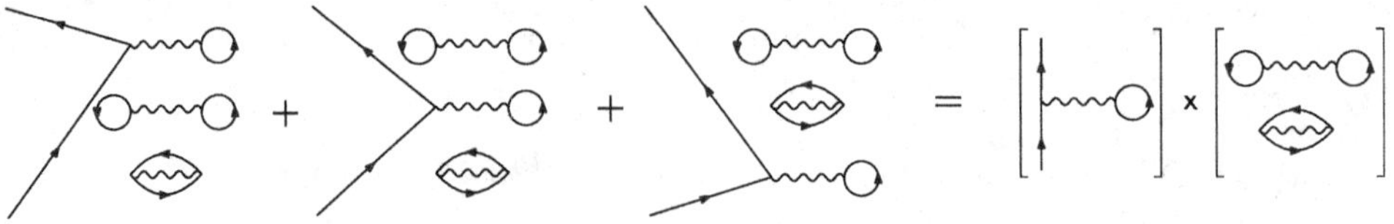

This result is readily seen to be generally valid. To kth order there are $\binom{k}{n}$ ways to construct a given connected nth-order Green's function diagram out of the interaction lines $v(1;1'),\ldots,v(k;k')$, which all give the same contribution $G^{(n)}_{c,i}(a;b)=\int[G_0\ldots G_0v\ldots v]_i$, where i labels the given diagram. In our example $\binom{3}{1}=3$. The remaining part $V_j^{(k-n)}=\int[G_0\ldots G_0v\ldots v]_j$ is proportional to the jth vacuum diagram (consisting of one or more disjoint pieces) of order $k-n$. Thus, the total contribution of the kth-order term of the numerator of (7.1) is

$$
\begin{aligned}
N^{(k)}(a;b) &= \frac{1}{k!}\left(\frac{\mathrm{i}}{2}\right)^k \sum_{n=0}^{k}\binom{k}{n} \sum_{i=\substack{G\text{-connected}\\ \text{diagrams}}}\ \sum_{j=\substack{\text{vacuum}\\ \text{diagrams}}} (\pm)^{l_i+l_j} G^{(n)}_{c,i}(a;b)V_j^{(k-n)} \\
&= \sum_{n=0}^{k}\frac{1}{n!}\left(\frac{\mathrm{i}}{2}\right)^n \sum_{i=\substack{G\text{-connected}\\ \text{diagrams}}} (\pm)^{l_i} G^{(n)}_{c,i}(a;b) \\
&\quad\times \frac{1}{(k-n)!}\left(\frac{\mathrm{i}}{2}\right)^{k-n} \sum_{j=\substack{\text{vacuum}\\ \text{diagrams}}} (\pm)^{l_j} V_j^{(k-n)},
\end{aligned}
$$

where l_i and l_j are the number of loops in diagrams i and j. The third line in this equation is exactly $(Z/Z_0)^{(k-n)}$. Therefore, if we denote by

$$
G^{(n)}_{c}(a;b) = \frac{1}{n!}\left(\frac{\mathrm{i}}{2}\right)^n \sum_{i=\substack{G\text{-connected}\\ \text{diagrams}}} (\pm)^{l_i} G^{(n)}_{c,i}(a;b)
$$

the sum of all nth-order connected diagrams of $N(a;b)$, we have

$$
\begin{aligned}
N(a;b) &= \sum_{k=0}^{\infty}\sum_{n=0}^{k} G^{(n)}_{c}(a;b)\left(\frac{Z}{Z_0}\right)^{(k-n)} = \sum_{n=0}^{\infty}\sum_{k=n}^{\infty} G^{(n)}_{c}(a;b)\left(\frac{Z}{Z_0}\right)^{(k-n)} \\
&= \left(\frac{Z}{Z_0}\right)\sum_{n=0}^{\infty} G^{(n)}_{c}(a;b).
\end{aligned}
$$

We have just found the important and beautiful result that all vacuum diagrams of the denominator of (7.1) are canceled out by the disconnected part of the numerator! The MBPT formula (7.1) simplifies to

$$G(a;b) = \sum_{n=0}^{\infty} \frac{1}{n!}\left(\frac{\mathrm{i}}{2}\right)^n \int v(1;1')\ldots v(n;n') \begin{vmatrix} G_0(a;b) & G_0(a;1^+) & \ldots & G_0(a;n'^+) \\ G_0(1;b) & G_0(1;1^+) & \ldots & G_0(1;n'^+) \\ \vdots & \vdots & \ddots & \vdots \\ G_0(n';b) & G_0(n';1^+) & \ldots & G_0(n';n'^+) \end{vmatrix}_{\substack{\pm \\ c}} \tag{7.6}$$

where the symbol $|\ldots|_{\substack{\pm \\ c}}$ signifies that in the expansion of the permanent/determinant only the terms represented by connected diagrams are retained.

7.4 Summing Only the Topologically Inequivalent Diagrams

The cancellation of disconnected diagrams reduces drastically the number of terms in the MBPT expansion of G, but we can do even better. If we write down the diagrams for G, we realize that there are still many connected diagrams with the same value. In first order, for instance, the third and fourth diagram, as well as the fifth and sixth diagram, of Fig. 7.2, clearly lead to the same integrals, for only the labels 1 and $1'$ are interchanged. In second order each connected diagram comes in eight variants, all with the same value. An example is the diagrams of Fig. 7.4, in which (b) is obtained from (a) by mirroring the interaction line $v(1;1')$, (c) is obtained from (a) by mirroring the interaction line $v(2;2')$, and (d) is obtained from (a) by mirroring both interaction lines. For an nth-order diagram we thus have 2^n such mirroring operations ($2^2 = 4$ in Fig. 7.4). The second row of the figure is obtained by interchanging the interaction lines $v(1;1') \leftrightarrow v(2;2')$ in the diagrams of the first row. If we have n interaction lines, then there are $n!$ possible permutations ($2! = 2$ in Fig. 7.4). We conclude that there exist $2^n n!$ diagrams with the same value to order n. Since these diagrams are obtained by mirroring and permutations of interaction lines, they are also *topologically equivalent* – that is, they are obtained from one another by a continuous deformation. Therefore we only need to consider diagrams with different topology and multiply by $2^n n!$ where n is the number of interaction lines. For these diagrams the new value of the prefactor becomes

$$2^n n! \frac{1}{n!}\left(\frac{\mathrm{i}}{2}\right)^n (\pm)^l = \mathrm{i}^n (\pm)^l,$$

and (7.6) can be rewritten as

$$G(a;b) = \sum_{n=0}^{\infty} \mathrm{i}^n \int v(1;1')\ldots v(n;n') \begin{vmatrix} G_0(a;b) & G_0(a;1^+) & \ldots & G_0(a;n'^+) \\ G_0(1;b) & G_0(1;1^+) & \ldots & G_0(1;n'^+) \\ \vdots & \vdots & \ddots & \vdots \\ G_0(n';b) & G_0(n';1^+) & \ldots & G_0(n';n'^+) \end{vmatrix}_{\substack{\pm \\ c \\ t.i.}} \tag{7.7}$$

where the symbol $|\ldots|_{\substack{\pm \\ c \\ t.i.}}$ signifies that in the expansion of the permanent/determinant only the terms represented by connected and topologically inequivalent diagrams are retained.

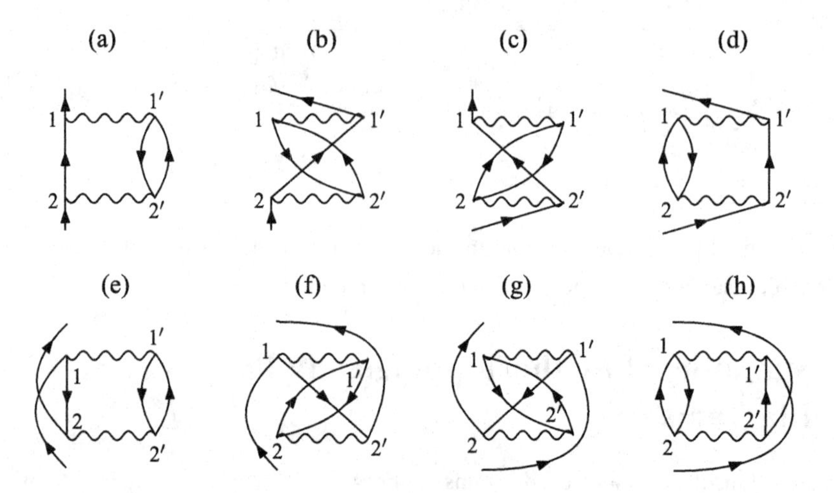

Figure 7.4 A class of eight topologically equivalent second-order connected diagrams for the Green's function.

From now on we work only with these diagrams. Thus, the new rules to convert a Green's function diagram into a mathematical expression are:

- Number all vertices and assign an interaction $v(i;j)$ to a wiggly line connecting i and j and a Green's function $G_0(i;j^+)$ to an oriented line from j to i.
- Integrate over all internal vertices and multiply by $\mathrm{i}^n(\pm)^l$, where l is the number of loops and n is the number of interaction lines.

Using (7.7) we can easily expand the Green's function to second order; the result is shown in Fig. 7.5. We find 2 first-order diagrams and 10 second-order diagrams. This is an enormous simplification! Without using the cancellation of disconnected diagrams and the resummation of topologically equivalent diagrams, the Green's function to second-order $G^{(2)} = (N^{(0)} + N^{(1)} + N^{(2)}) - (Z/Z_0)^{(1)}(N^{(0)} + N^{(1)}) + \{[(Z/Z_0)^{(1)}]^2 - (Z/Z_0)^{(2)}\}N^{(0)}$ would consist of $(1 + 3! + 5!) + 2!(1 + 3!) + (2!)^2 + 4! = 169$ diagrams.

Exercise 7.4 Prove that G to second order in v is given by Fig. 7.5.

7.5 Self-Energy and the Dyson Equation

To reduce further the number of diagrams it is necessary to introduce a new quantity: the *self-energy* Σ. It is clear from the diagrammatic structure that the Green's function has the general form

$$G = \quad + \quad + \quad + \ldots \tag{7.8}$$

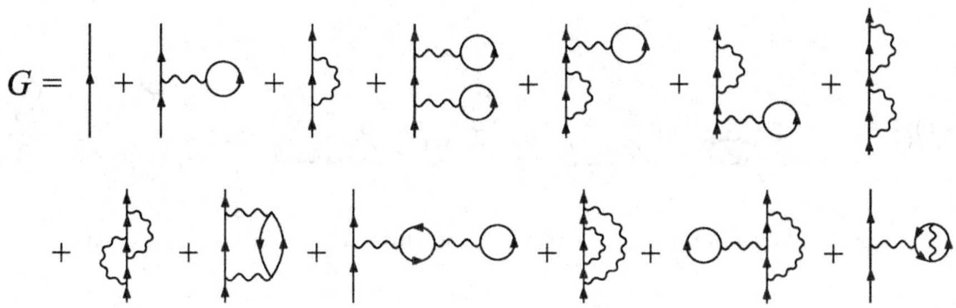

$G = \; + \; + \; + \; + \; + \; + \;$

$+ \; + \; + \; + \; + \; + \;$

Figure 7.5 Diagrammatic MBPT expansion of the Green's function to second order using the new Feynman rules.

where the self-energy

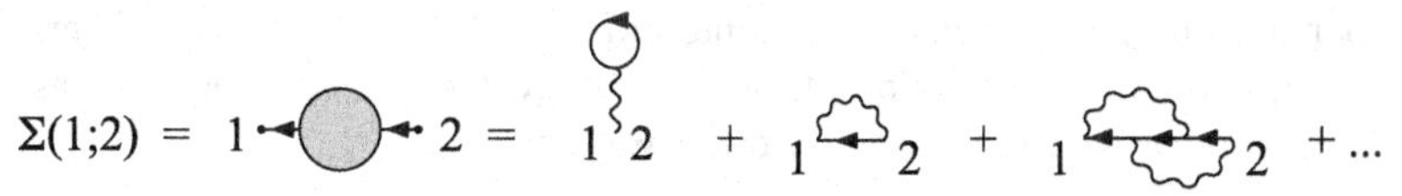

$\Sigma(1;2) = 1 \; 2 = 1 \, 2 + 1 \, 2 + 1 \, 2 + \ldots$

The self-energy consists of all diagrams that do not break up into two disjoint pieces by cutting a single G_0-line. So, for instance, the fourth, fifth, sixth, and seventh diagrams of Fig. 7.5 belong to the third term on the r.h.s. of (7.8), while all diagrams in the second row of the same figure belong to the second term. The self-energy diagrams are called *one-particle irreducible diagrams*, or simply *irreducible diagrams*. By construction, $\Sigma = \Sigma[G_0, v]$ depends on the noninteracting Green's function G_0 and on the interaction v. If we represent the interacting Green's function by an oriented double line,

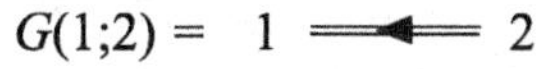

$G(1;2) = 1 \; 2$

then we can rewrite (7.8) as

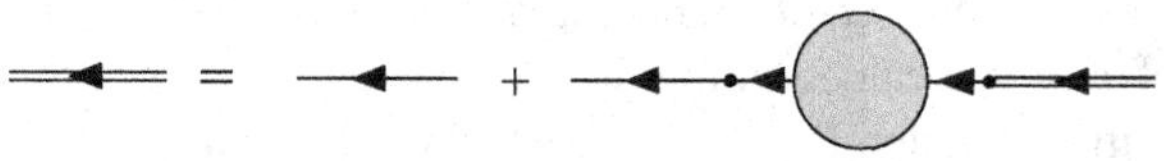

$= \; + \;$ (7.9)

or, equivalently,

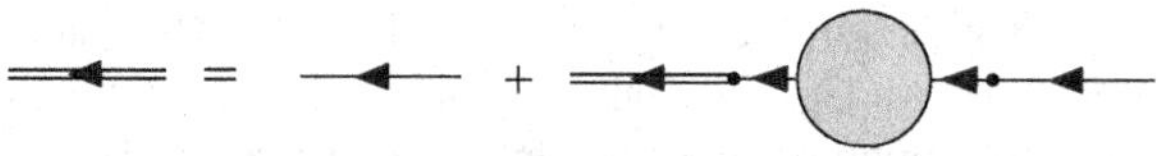

$= \; + \;$ (7.10)

Equations (7.9) and (7.10) are readily seen to generate (7.8) by iteration. The mathematical expression of these equations is

$$\begin{aligned} G(1;2) &= G_0(1;2) + \int d3d4\, G_0(1;3)\Sigma(3;4)G(4;2) \\ &= G_0(1;2) + \int d3d4\, G(1;3)\Sigma(3;4)G_0(4;2), \end{aligned} \tag{7.11}$$

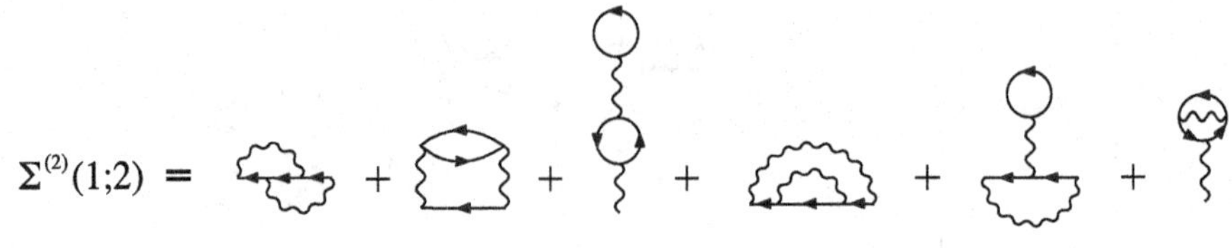

$$\Sigma^{(2)}(1;2) = \quad + \quad + \quad + \quad + \quad +$$

Figure 7.6 Self-energy diagrams to second order in the interaction strength.

which is known as the *Dyson equation.* Since Σ is expressed in terms of G_0 and v, the Dyson equation (7.11) is a linear integral equation for the Green's function G

Thanks to the introduction of the self-energy, we can reduce the number of diagrams even further since we only need to consider topologically inequivalent Σ-diagrams (these diagrams are, by definition, connected and one-particle irreducible). Through the Dyson equation (7.11) these diagrams are summed to *infinite order* in the Green's function. Hence a finite number of Σ-diagrams implies an infinite number of Green's function diagrams.

For the diagrammatic construction of the self-energy, the rules are the same as those for the Green's function. For instance, to first order we have

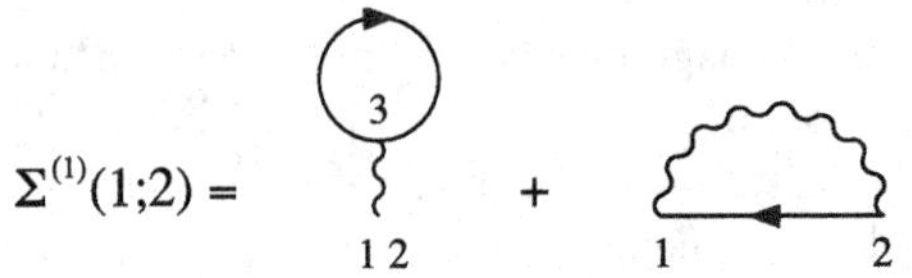

$$\Sigma^{(1)}(1;2) = \quad + $$

3

1 2 1 2

or as a formula:

$$\Sigma^{(1)}(1;2) = \pm\mathrm{i}\,\delta(1;2)\int d3\, v(1;3)G_0(3;3^+) + \mathrm{i}\, v(1;2)G_0(1;2^+). \tag{7.12}$$

The first-order self-energy is also called the *Hartree–Fock self-energy.* It consists of a tadpole diagram, the Hartree contribution, and of an oyster diagram, the Fock (or exchange) contribution. We deepen the physical content of this approximation in the next chapter. Notice that the prefactor is included in the diagrams, in accordance with our convention. In the Hartree contribution, the first term in (7.12), we have also added a δ-function since in the corresponding Green's function diagram (the second diagram of Fig. 7.5) we have $1 = 2$. This applies to all self-energy diagrams that start and end with the *same* interaction vertex (see, e.g., the third and sixth diagrams in the second row of Fig. 7.5). The full set of second-order self-energy diagrams is shown in Fig. 7.6. There are only 6 diagrams to be considered against the 10 second-order Green's function diagrams. In the next two sections we achieve another reduction in the number of diagrams by introducing a very useful topological concept.

7.6 G-skeleton Diagrams

A *G-skeleton diagram* for the self-energy is obtained by removing all self-energy insertions from a given diagram. A self-energy insertion is a piece that can be cut away from a diagram by cutting two Green's function lines. For example, the diagram

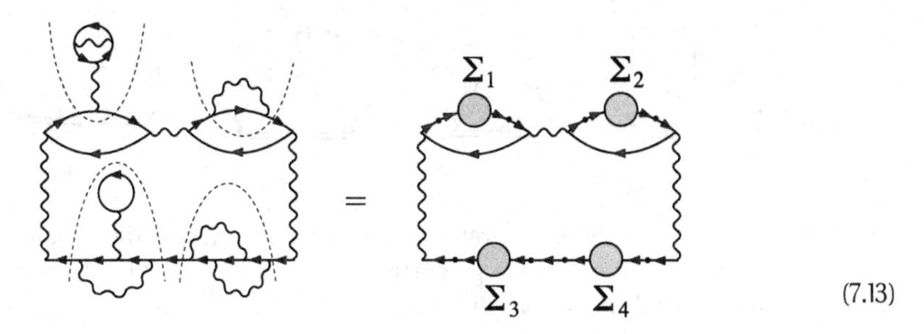

(7.13)

has four self-energy insertions residing inside the thin-dashed parabolic lines. The r.h.s. highlights the structure of the diagram and implicitly defines the self-energy insertions Σ_i, $i = 1, 2, 3, 4$. The G-skeleton diagram corresponding to (7.13) is therefore

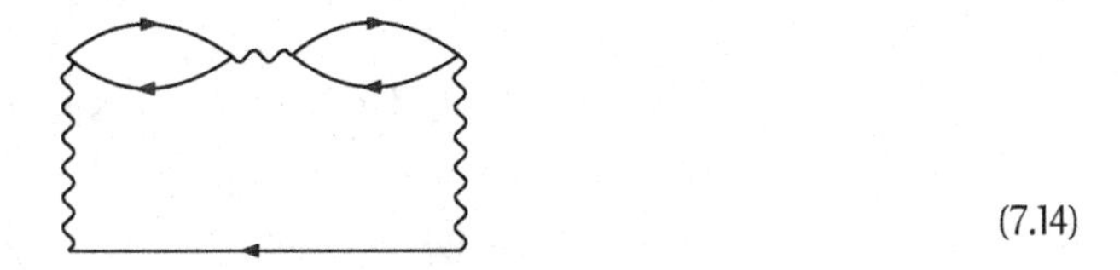

(7.14)

The G-skeleton diagrams allow us to express the self-energy in terms of the interacting (dressed) Green's function G rather than the noninteracting Green's function G_0. Consider again the example (7.14). If we sum over all possible self-energy insertions, we find

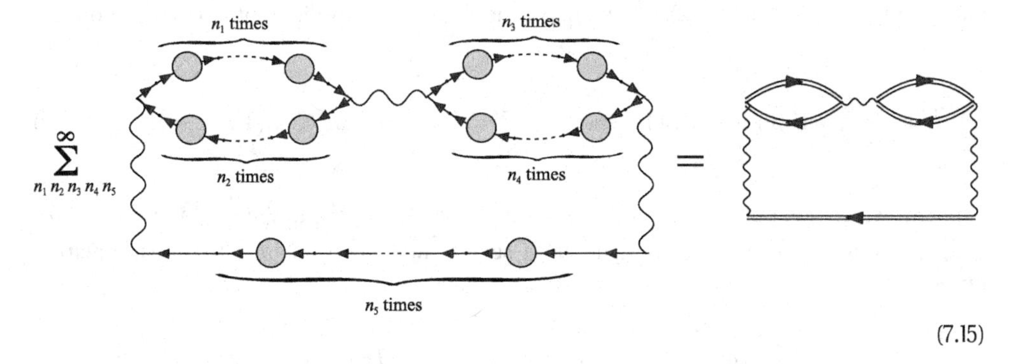

(7.15)

where each gray circle represents the exact self-energy. Thus, the sum over n_i on the l.h.s. gives the dressed G and the result is the G-skeleton diagram (7.14) in which G_0 is replaced by G. This procedure yields the self-energy $\Sigma = \Sigma_s[G, v]$ as a functional of the interaction v and of the dressed Green's function G. The subscript "s" specifies that the functional is constructed by taking only G-skeleton diagrams from the functional $\Sigma = \Sigma[G_0, v]$ and then replacing G_0 with G: $\Sigma[G_0, v] = \Sigma_s[G, v]$. In Chapter 11 we show how to use this result to construct a functional $\Phi[G, v]$ whose functional derivative with respect to G is the exact self-energy.

Using the G-skeleton diagrams, we can write the self-energy up to second order in the interaction as

$\Sigma_s[G,v]$ = 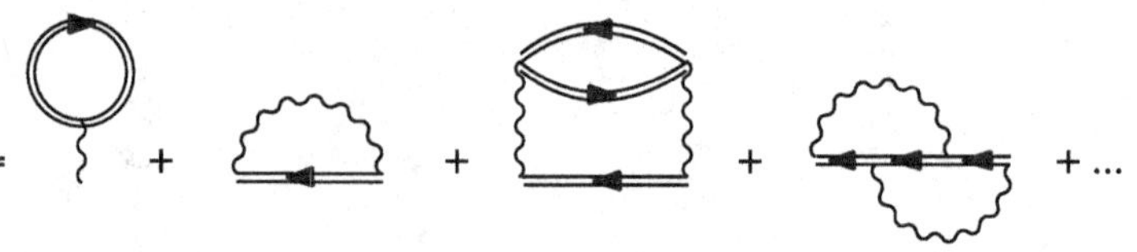 (7.16)

There are only two G-skeleton diagrams of second order against the six diagrams in Fig. 7.6. The approximation for the self-energy corresponding to the four G-skeleton diagrams illustrated in (7.16) is called the *second Born approximation*.

More generally, for any given approximation to $\Sigma_s[G,v]$, we can calculate an approximate Green's function from the Dyson equation:

$$\begin{aligned} G(1;2) &= G_0(1;2) + \int d3d4\, G_0(1;3)\Sigma_s[G,v](3;4)G(4;2) \\ &= G_0(1;2) + \int d3d4\, G(1;3)\Sigma_s[G,v](3;4)G_0(4;2), \end{aligned} \tag{7.17}$$

which in this form is a *nonlinear* integral equation for G on the contour. Of course, the solution of the Dyson equation with the exact self-energy (obtained by summing all the G-skeleton self-energy diagrams) yields the exact Green's function.

Integro-differential equations for G An alternative strategy to calculate the Green's function consists in applying $\overrightarrow{G}_0^{-1}(1';1) = [\mathrm{i}\frac{d}{dz_1'}\delta(1';1) - h(1';1)]$ to the first row of (7.17), integrate over 1, and use (6.2) to generate a nonlinear integro-differential equation on the contour:

$$\int d1 \left[\mathrm{i}\frac{d}{dz_1'}\delta(1';1) - h(1';1)\right] G(1;2) = \delta(1';2) + \int d4\, \Sigma_s[G,v](1';4)G(4;2). \tag{7.18}$$

Similarly, we could apply $\overleftarrow{G}_0^{-1}(2;2') = [-\mathrm{i}\delta(2;2')\frac{\overleftarrow{d}}{dz_2'} - h(2;2')]$ to the second row of (7.17), integrate over 2, and use (6.3) to generate a second nonlinear integro-differential equation on the contour:

$$\int d2\, G(1;2) \left[-\mathrm{i}\delta(2;2')\frac{\overleftarrow{d}}{dz_2'} - h(2;2')\right] = \delta(1;2') + \int d3\, G(1;3)\Sigma_s[G,v](3;2'). \tag{7.19}$$

Both equations must be solved with KMS boundary conditions. Equations (7.18) and (7.19) have exactly the mathematical form anticipated in (6.13): The self-energy stems from the fact that all the degrees of freedom in $G_2, G_3, \ldots$ have been embedded into one single effective equation for the one-particle Green's function G.

We observe that (7.18) and (7.19) are equations for the contour Green's function G. In Chapter 9 we show how to transform them into a coupled system of equations, known as the *Kadanoff–Baym equations*, for the Keldysh components of G.

7.7 W-skeleton Diagrams

The topological concept of G-skeleton diagrams can also be applied to the interaction lines and leads to a further reduction of the number of diagrams. Let us call a piece of diagram a *polarization* insertion if it can be cut away by cutting two interaction lines. For example,

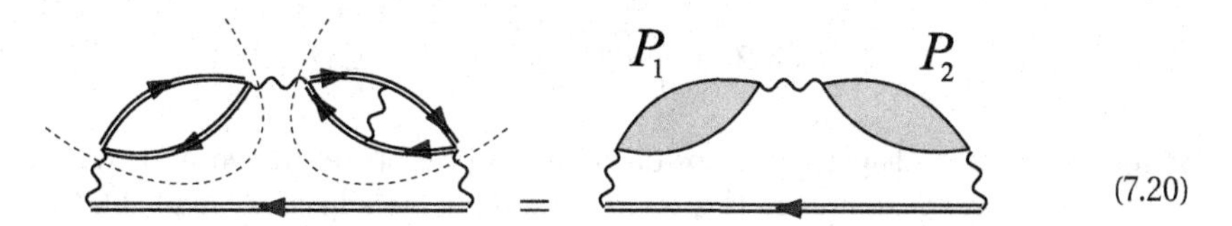

(7.20)

is a G-skeleton diagram with two polarization insertions,

$P_1 =$ $\qquad$ $P_2 =$ (7.21)

The polarization diagrams, like the ones above, must be one-interaction-line irreducible – that is, they cannot break into two disjoint pieces by cutting an interaction line. Thus, even though we could cut away the piece $P_1 v P_2$ by cutting two interaction lines on the l.h.s. of (7.20), the diagram $P_1 v P_2$ is not a polarization diagram since it breaks into the disjoint pieces P_1 and P_2 by cutting the interaction line in the middle.

Screened interaction The polarization diagrams can be used to define the *dressed* or *screened interaction* W according to

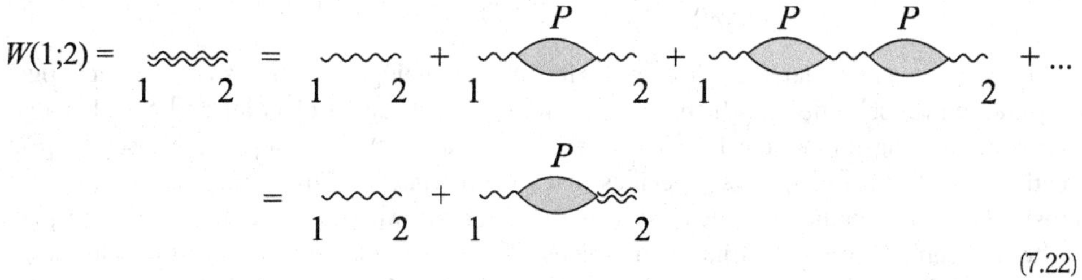

$W(1;2) =$ 1 2 $=$ 1 2 $+$ 1 P 2 $+$ 1 P P 2 $+ \ldots$

$=$ 1 2 $+$ 1 P 2 (7.22)

In this diagrammatic equation, P is the sum of all possible polarization diagrams. We refer to P as the *polarization*. Similarly to the self-energy, the polarization can be considered either as a functional of the noninteracting Green's function G_0 and the interaction v, $P = P[G_0, v]$, or as a functional of the dressed G and v, $P = P_s[G, v]$. Equation (7.22) has the form of the Dyson equation (7.8) for the Green's function. In formulas it reads

$$\begin{aligned} W(1;2) &= v(1;2) + \int v(1;3)P(3;4)v(4;2) + \int v(1;3)P(3;4)v(4;5)P(5;6)v(6;2) + \ldots \\ &= v(1;2) + \int v(1;3)P(3;4)W(4;2), \end{aligned} \quad (7.23)$$

where the integral is over all repeated variables.

Alternatively we can express the screened interaction in terms of the so-called *reducible polarization* or *density response function*, which is defined diagrammatically by

$$\chi(1;2) = P(1;2) + \int d3d4\, P(1;3)v(3;4)\chi(4;2). \tag{7.24}$$

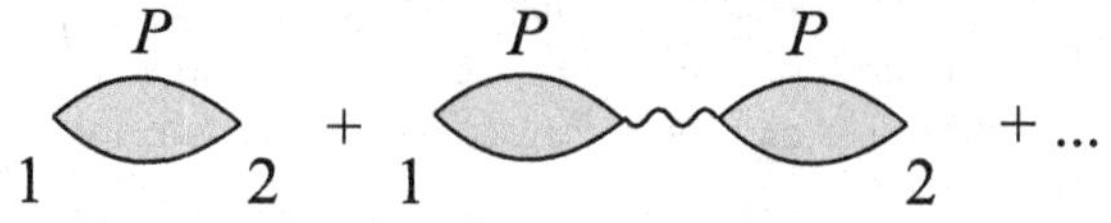

Different to the P-diagrams, the χ-diagrams can be broken down into two disjoint pieces by cutting an interaction line. This is the reason for the adjective "reducible" in the name of χ. Using (7.24) in the Dyson equation for the screened interaction, see (7.23), we obtain

$$W(1;2) = v(1;2) + \int d3d4\, v(1;3)\chi(3;4)v(4;2). \tag{7.25}$$

Let us express $\Sigma_s[G, v]$ in terms of the screened interaction W. We say that a diagram is a W*-skeleton diagram* if it does not contain polarization insertions. Then the desired expression for Σ is obtained by discarding all those diagrams which are not W-skeletonic in the expansion (7.16), and then replacing v with W. The only diagram for which we should not proceed with the replacement is the Hartree diagram [the first diagram in (7.16)] since here every polarization insertion is equivalent to a self-energy insertion – see, for instance, the third diagram of Fig. 7.6. In the Hartree diagram the replacement $v \to W$ would lead to double counting. Therefore,

$$\Sigma = \Sigma_{ss}[G, W] = \Sigma_{\rm H}[G, v] + \Sigma_{ss,\rm xc}[G, W], \tag{7.26}$$

where $\Sigma_{\rm H}[G, v]$ is the Hartree self-energy while the remaining part is the so-called *exchange-correlation* (XC) self-energy, which includes the Fock (exchange) diagram and all other diagrams accounting for nonlocal effects in time (correlations) – see again the discussion in Section 6.1.1. The subscript "ss" specifies that to construct the functional Σ_{ss} or $\Sigma_{ss,\rm xc}$ we must take into account only those self-energy diagrams which are skeletonic with respect to both Green's function and interaction lines. The skeletonic expansion of the self-energy in terms of G and W (up to third order in W) is shown in Fig. 7.7(a). Similarly, the MBPT expansion of the polarization in terms of G and W is obtained by taking only G- and W-skeleton diagrams and then replacing $G_0 \to G$ and $v \to W$. This operation leads to a polarization $P = P_{ss}[G, W]$, which can be regarded as a functional of G and W and whose expansion (up to second order in W) is shown in Fig. 7.7(b). The number of self-energy and polarization diagrams to a given order in the interaction has been derived in Ref. [92] for the nonskeletonic, skeletonic in G, and skeletonic in G and W expansions.

The diagrammatic expansion in terms of W instead of v does not only have the *mathematical* advantage of reducing the number of diagrams. There is also a *physical* advantage [93]. In bulk systems and for long-range interactions, like the Coulomb interaction, many self-energy diagrams are *divergent.* The effect of the polarization insertions is to cut off the long-range nature of v. Replacing v with W makes these diagrams finite and physically interpretable. We see an example of divergent self-energy diagrams in Section 15.2.

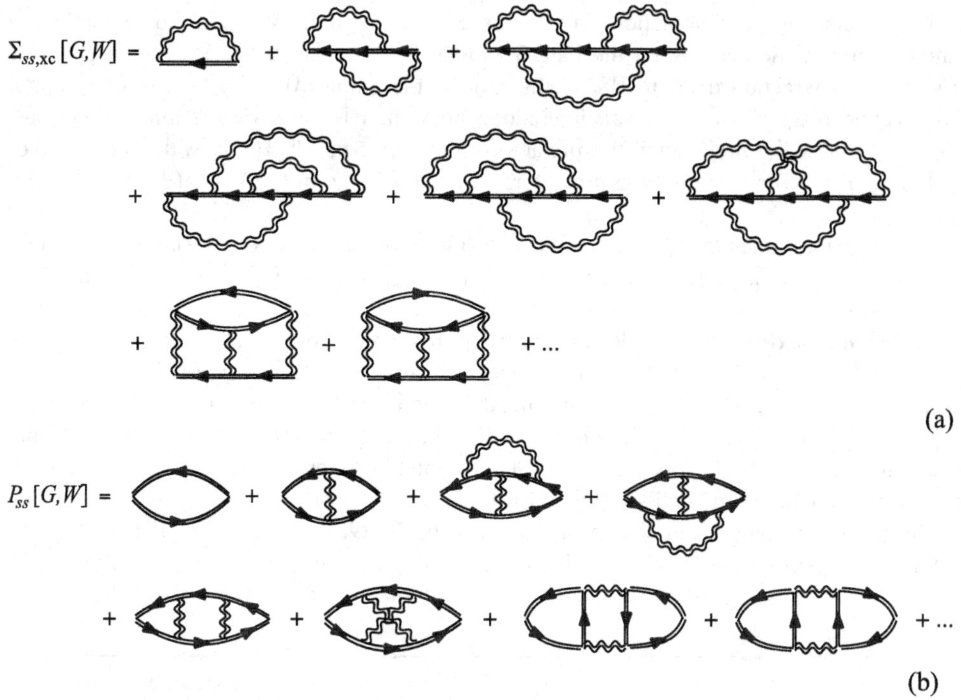

Figure 7.7 Diagrammatic MBPT expansion in terms of G and W of (a) the self-energy up to third order and (b) the polarization up to second order.

From the diagrammatic structure of the polarization diagrams we see that

$$\boxed{P(1;2) = P(2;1)} \tag{7.27}$$

and, as a consequence, also

$$\boxed{W(1;2) = W(2;1)} \tag{7.28}$$

which tells us that the screened interaction between the particles is symmetric, as one would expect. It is worth noticing, however, that this symmetry property is not fulfilled by every single diagram. For instance, the third and fourth diagrams of Fig. 7.7(b) are, separately, not symmetric. The symmetry is recovered only when they are summed. This is a general property of the polarization diagrams: either they are symmetric or they come in pairs of mutually *reversed* diagrams. By a reversed polarization diagram we mean the same diagram in which the end points are interchanged. If we label the left and right vertices of the third and fourth diagrams of Fig. 7.7(b) with 1 and 2, then the fourth diagram with relabeled vertices 2 and 1 (reversed diagram) becomes identical to the third diagram.

Diagrammatic rules for the polarization Let us derive the diagrammatic rules for the polarization diagrams. In a nth-order Σ-diagram with n interaction lines, the prefactor is $\mathrm{i}^n(\pm)^l$ since every interaction line comes with a factor i. We want to maintain the same diagrammatic rules when the bare interaction v is replaced by W. Let us consider a screened interaction diagram W_k of the form, for example, $W_k = vP_kv$, where P_k is a polarization diagram. If P_k has m interaction lines then W_k has $m+2$ interaction lines. For W_k to have the prefactor i, the prefactor of P_k must be $\mathrm{i}^{m+1}(\pm)^l$, with l the number of loops in P_k. For example, the prefactors of the diagrams in (7.21) are $\mathrm{i}(\pm)^1$ for P_1 and $\mathrm{i}^2(\pm)^1$ for P_2.

The Feynman rules to convert a χ-diagram into a mathematical expression are the same as those for the polarization: a prefactor $\mathrm{i}^{m+1}(\pm)^l$ for a diagram with m interaction lines and l loops.

GW approximation The lowest-order approximation one can make in Fig. 7.7 is $\Sigma_{ss,\mathrm{xc}}(1;2) = \mathrm{i}G(1;2^+)\ W(1;2)$ (first diagram) and $P(1;2) = \pm\mathrm{i}G(1;2)G(2;1)$ (first diagram). This approximation was introduced by Hedin in 1965 and is today known as the GW approximation [93]. The GW approximation has been rather successful in describing spectral properties [94, 95, 96, 97, 98, 99, 100] and total energies [101, 102, 103]. We come back to this approximation in Sections 12.6 and 15.6.

In general, for any given choice of diagrams in $\Sigma_{ss}[G,W]$ and $P_{ss}[G,W]$ we have to solve simultaneously the Dyson equation (7.17) and (7.23) to find an approximation to G and W which, we stress again, corresponds to the resummation of an infinite set of diagrams.

Exercise 7.5 Show that the G- and W-skeleton diagrams for Σ up to third order in W are those of Fig. 7.7(a) and that the G- and W-skeleton diagrams for P up to second order in W are those of Fig. 7.7(b).

7.8 Summary and Feynman Rules

We finally summarize the achievements of this chapter. Rather than working out all diagrams for the Green's function, it is more advantageous to work out the diagrams for the self-energy. In this way, many diagrams are already summed to infinite order by means of the Dyson equation. The Feynman rules to construct the self-energy are

1. ***Undressed case:*** $\Sigma = \Sigma[G_0, v]$:
   - Draw all topologically inequivalent self-energy diagrams using G_0 and v. By definition a self-energy diagram is connected and one-particle irreducible.
   - If the diagram has n interaction lines and l loops, then the prefactor is $\mathrm{i}^n(\pm)^l$.
   - Integrate over all internal vertices of the diagram.
2. ***Partially dressed case:*** $\Sigma = \Sigma_s[G, v]$:
   - The same as in (1), but in the first point we should consider only the G-skeleton diagrams (no self-energy insertions) and replace G_0 with G.

3. ***Fully dressed case:*** $\Sigma = \Sigma_{ss}[G, W]$, $P = P_{ss}[G, W]$:

- The same as in (1), but in the first point we should consider only diagrams that are both G-skeletonic (no self-energy insertions) and W-skeletonic (no polarization insertion), and replace G_0 with G and v with W.
- In this case we also need to construct the polarization diagrams. For m interaction lines and l loops, the prefactor of these diagrams is $\mathrm{i}^{m+1}(\pm)^l$.

Feynman rules for general bases The presentation of the diagrammatic expansion has been done with (1.83) as the reference Hamiltonian. For this Hamiltonian the building blocks of a diagram are the Green's function and the interaction, and they combine in the following manner

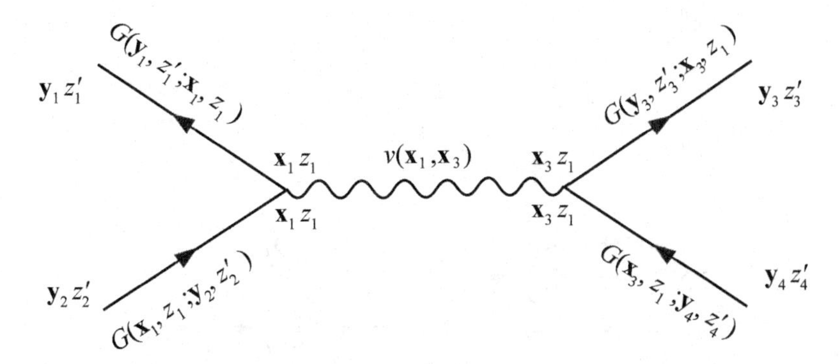

With the exception of special systems such as the electron gas, the evaluation of a generic diagram with pencil and paper is essentially impossible and we must resort to computer programs. In order to evaluate a diagram numerically, however, we must expand the field operators over some discrete (and incomplete) basis. In this case the Green's function $G(\mathbf{x}, z; \mathbf{x}', z')$ becomes a matrix $G_{ii'}(z, z')$, the self-energy $\Sigma(\mathbf{x}, z; \mathbf{x}', z')$ becomes a matrix $\Sigma_{ii'}(z, z')$, the interaction $v(\mathbf{x}, \mathbf{x}')$ becomes a four-index tensor (Coulomb integrals) v_{ijmn}, the screened interaction $W(\mathbf{x}, z; \mathbf{x}', z')$ becomes a four-index tensor $W_{ijmn}(z, z')$, and the polarization $P(\mathbf{x}, z; \mathbf{x}', z')$ becomes a four-index tensor $P_{ijmn}(z, z')$. The diagrammatic rules to construct these quantities are exactly the same as those already derived, and the entire discussion of this and the following chapters remains valid provided that the building blocks (Green's function and interaction) are combined according to

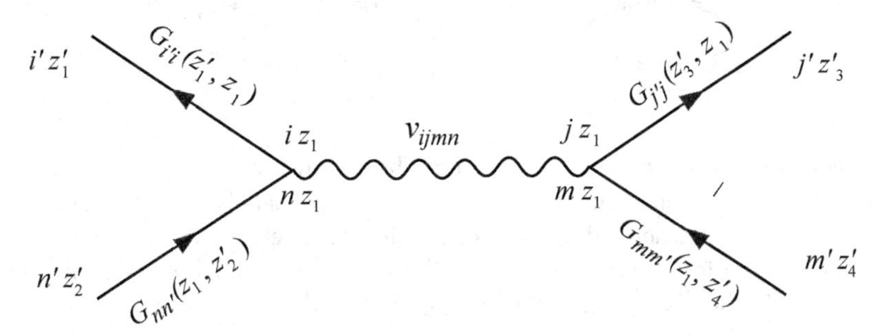

We see that there is no difference in the time-label of a vertex since the interaction remains local in time. On the other hand, the orbital-spin-label of every vertex duplicates since the orbital-spin-label of the G that enters a vertex is, in general, different from the orbital-spin-label of the G that exits the same vertex. The matrix elements v_{ijmn} join the outgoing Green's functions with entries i and j to the ingoing Green's functions with entries m and n. The labels (i, n) and (j, m) in v_{ijmn} correspond to basis functions calculated in the same position–spin coordinate, see (1.86), and therefore must be placed next to each other. For instance, the self-energy up to second order in the interaction is (omitting the time coordinates)

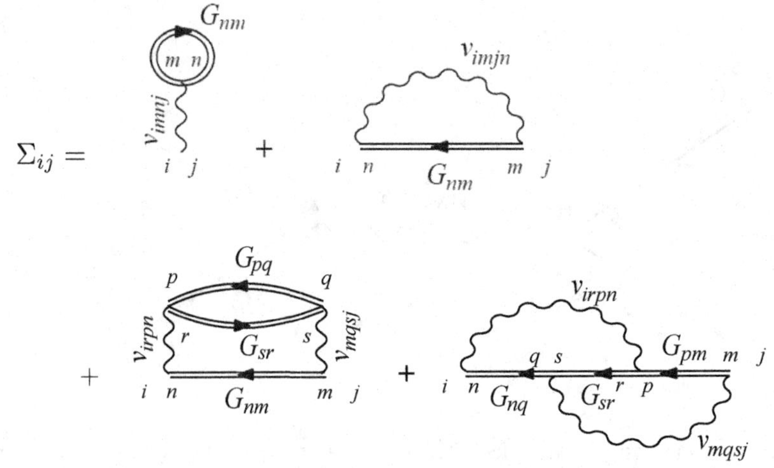

$$\Sigma_{ij} = \quad (7.29)$$

while the diagrammatic expansion of the screened interaction looks like:

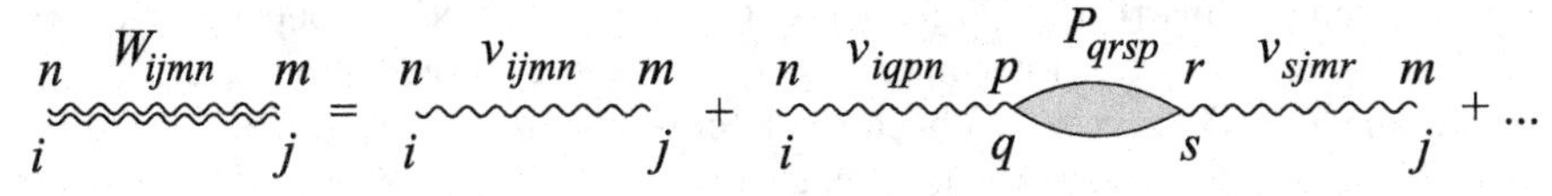

$$W_{ijmn} = v_{ijmn} + v_{iqpn} P_{qrsp} v_{sjmr} + \ldots \quad (7.30)$$

where the diagrammatic expansion of the polarization has the structure:

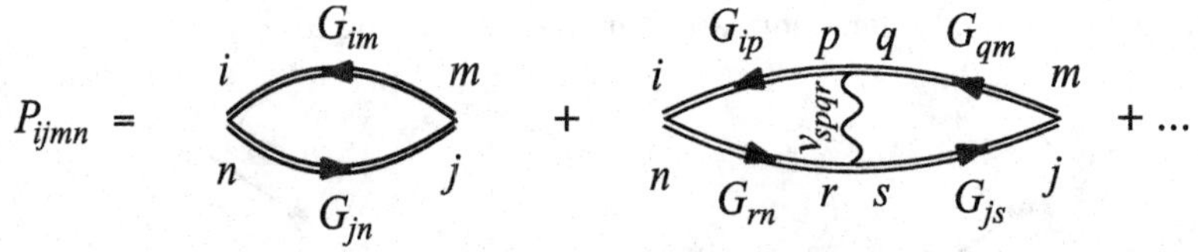

$$P_{ijmn} = \ldots + \ldots \quad (7.31)$$

As in the continuum case, once a diagram has been drawn we must integrate the internal times over the contour and sum the internal orbital-spin-labels over the discrete basis. Remember that for spin-independent interactions the matrix elements v_{ijmn} are zero unless the spin of i is equal to the spin of n and/or the spin of j is equal to the spin of m. This implies that also in the discrete case the spin is conserved at every vertex, as it should be.

8

Hartree-Fock Approximation

In this chapter we analyze the self-energy to first order in the interaction strength. This is the so-called *Hartree-Fock approximation.* We derive some general properties of the Hartree-Fock approximation and then apply it to a gas of interacting particles, highlighting the merits and limitations.

8.1 Time Locality

The Hartree-Fock self-energy is the sum of the tadpole diagram (Hartree) and the oyster (or Fock or exchange) diagram, see (7.12). In the G-skeleton framework, see Section 7.6, these diagrams must be calculated with G rather than G_0; hence, (7.12) becomes

$$\boxed{\Sigma_{\rm HF}(1;2) = \pm \mathrm{i}\,\delta(1;2)\int d3\,v(1;3)G(3;3^+) + \mathrm{i}\,v(1;2)G(1;2^+)} \tag{8.1}$$

This self-energy is local in contour times z_1 and z_2 since the Hartree term contains $\delta(1;2) = \delta(\mathbf{x}_1 - \mathbf{x}_2)\delta(z_1, z_2)$ and the Fock term contains $v(1;2) = v(\mathbf{x}_1, \mathbf{x}_2, z_1)\delta(z_1, z_2)$. Therefore, (8.1) has the following mathematical structure:

$$\Sigma_{\rm HF}(1;2) = \delta(z_1, z_2)E(\mathbf{x}_1, \mathbf{x}_2, z_1). \tag{8.2}$$

Inserting this approximate self-energy into (7.18), we find

$$\int d\mathbf{x}_1 \left[\delta(\mathbf{x}_1' - \mathbf{x}_1)\,\mathrm{i}\frac{d}{dz_1'} - \langle \mathbf{x}_1'|\hat{h}(z_1')|\mathbf{x}_1\rangle - E(\mathbf{x}_1', \mathbf{x}_1, z_1')\right] G(\mathbf{x}_1, z_1'; 2) = \delta(1';2). \tag{8.3}$$

If we define the (first-quantization) operator

$$\hat{\mathcal{E}}(z) = \int d\mathbf{x}_1 d\mathbf{x}_2\, |\mathbf{x}_1\rangle E(\mathbf{x}_1, \mathbf{x}_2, z)\langle \mathbf{x}_2|, \tag{8.4}$$

then we recognize that (8.3) is the sandwich between $\langle \mathbf{x}_1'|$ and $|\mathbf{x}_2\rangle$ of

$$\left[\mathrm{i}\frac{d}{dz_1} - \hat{h}(z_1) - \hat{\mathcal{E}}(z_1)\right] \hat{\mathcal{G}}(z_1, z_2) = \delta(z_1, z_2).$$

Similarly, inserting the Hartree-Fock self-energy into (7.19), we find

$$\hat{\mathcal{G}}(z_1,z_2)\left[-\mathrm{i}\frac{\overleftarrow{d}}{dz_2}-\hat{h}(z_2)-\hat{\mathcal{E}}(z_1)\right]=\delta(z_1,z_2).$$

The time locality of the Hartree-Fock self-energy generates equations of motion that have the same mathematical structure as the equations of motion of the noninteracting Green's function, see (6.5) and (6.6). These kind of approximations are also known as *mean-field approximations.* The physical idea behind a mean-field approximation is to include the effects of the interaction through an effective potential. In bulk systems this idea makes sense if there are so many particles, or equivalently if the density is so high, that we can treat the interaction of a particle with all the other particles as an effective average interaction – that is, as an effective (or mean) field. As we shall see, the property of time locality implies the existence of *good quantum numbers* and hence of one-particle states with an infinitely long lifetime, see again Section 6.3.2. In Section 9.1 we show that any approximation beyond Hartree-Fock leads to a self-energy that is nonlocal in the contour time.

8.2 Hartree Approximation

Let us start by discussing the Hartree contribution. The Hartree self-energy is given by the first term in the r.h.s. of (8.1) and is usually written as

$$\Sigma_{\rm H}(1;2)=\delta(1;2)V_{\rm H}(1),$$

where

$$V_{\rm H}(1)=V_{\rm H}(\mathbf{x}_1,z_1)=\pm\mathrm{i}\int d3\,v(1;3)G(3;3^+)=\int d3\,v(1;3)n(3) \tag{8.5}$$

is the effective potential of the Hartree approximation, also known as the *Hartree potential.* For the Coulomb interaction $v(1;2)=\delta(z_1,z_2)/|\mathbf{r}_1-\mathbf{r}_2|$ the Hartree potential coincides with the classical potential generated by a density distribution $n(1)=n(\mathbf{x}_1,z_1)$:

$$V_{\rm H}(1)=\int d\mathbf{x}\,\frac{n(\mathbf{x},z_1)}{|\mathbf{r}_1-\mathbf{r}|}\qquad\text{(for Coulombic interactions).} \tag{8.6}$$

Equations (8.3), with $E(\mathbf{x}_1,\mathbf{x}_2,z_1)=\delta(\mathbf{x}_1-\mathbf{x}_2)V_{\rm H}(\mathbf{x}_1,z_1)$, and (8.5) form a system of coupled equations for the Hartree Green's function. In practical calculations they are solved by making an initial (and reasonable) guess for the Green's function, $G^{(0)}$, which is then used to determine an initial guess for the Hartree potential, $V_{\rm H}^{(0)}$, using (8.5); $V_{\rm H}^{(0)}$ is then inserted in (8.4) to obtain a new Green's function, $G^{(1)}$, which is then used in (8.5) to determine a new Hartree potential $V_{\rm H}^{(1)}$, and so on until convergence is achieved. In other words, the Green's function must be calculated *self-consistently* since the operator acting on G depends on G itself.

For the evaluation of both the Green's function and the Hartree potential we describe a method based on solving a system of coupled equations for single-particle wavefunctions. Except for the fact that $V_{\rm H}$ depends on G, the equations of motion are formally identical to the equations of motion of noninteracting particles. The general solution is therefore obtained as described in Section 6.2, as long as we replace $\hat{h}(z)$ with

$$\hat{h}_{\mathrm{H}}(z) \equiv \hat{h}(z) + \hat{\mathcal{V}}_{\mathrm{H}}(z),$$

where, see (8.4),

$$\begin{aligned}\hat{\mathcal{V}}_{\mathrm{H}}(z) &= \int d\mathbf{x}_1 d\mathbf{x}_2\, |\mathbf{x}_1\rangle \delta(\mathbf{x}_1 - \mathbf{x}_2) V_{\mathrm{H}}(\mathbf{x}_1, z) \langle \mathbf{x}_2| \\ &= \int d\mathbf{x}\, |\mathbf{x}\rangle V_{\mathrm{H}}(\mathbf{x}, z) \langle \mathbf{x}| \end{aligned} \tag{8.7}$$

is the Hartree potential operator in first quantization. By definition, $\hat{\mathcal{V}}_{\mathrm{H}}(z)$ is a constant operator along the imaginary track ($z = t_0 - \mathrm{i}\tau$), which we denote by $\hat{\mathcal{V}}^{\mathrm{M}}_{\mathrm{H}}$, consistently with our notation.

8.2.1 Hartree Equations

Let us consider an interacting system initially in equilibrium in some external potential $\phi(\mathbf{x} = \mathbf{r}\sigma)$, depending on both the position and the spin projection along the z axis (the inclusion of a vector potential is straightforward). For particles of mass m and charge q the single-particle Hamiltonian $\hat{h}^{\mathrm{M}}$ is

$$\hat{h}^{\mathrm{M}} = \hat{h} - \mu = \frac{\hat{p}^2}{2m} + q\phi(\hat{\boldsymbol{r}}, \hat{S}_z) - \mu,$$

and the interaction $v^{\mathrm{M}} = v$. To calculate the Green's function we first need to solve the eigenvalue problem

$$\left(\hat{h}^{\mathrm{M}} + \hat{\mathcal{V}}^{\mathrm{M}}_{\mathrm{H}}\right)|\lambda\rangle = (\epsilon^{\mathrm{H,M}}_{\lambda} - \mu)|\lambda\rangle,$$

where $(\epsilon^{\mathrm{H,M}}_{\lambda} - \mu)$ are the one-particle eigenvalues of $\hat{h}^{\mathrm{M}}_{\mathrm{H}} \equiv (\hat{h}^{\mathrm{M}} + \hat{\mathcal{V}}^{\mathrm{M}}_{\mathrm{H}})$ and $|\lambda\rangle = |\lambda^{\mathrm{M}}\rangle$ are the corresponding eigenkets. In order to lighten the notation we here omit the superscript "M" in the eigenvalues and eigenkets. By sandwiching with the bra $\langle \mathbf{x}|$ the eigenvalue equation, we find

$$\boxed{\left[-\frac{\nabla^2}{2m} + q\phi(\mathbf{x}) + V^{\mathrm{M}}_{\mathrm{H}}(\mathbf{x})\right]\varphi_\lambda(\mathbf{x}) = \epsilon^{\mathrm{H}}_{\lambda}\varphi_\lambda(\mathbf{x})} \tag{8.8}$$

where according to (8.5) and to the formula (6.42) for the density of noninteracting particles,

$$\boxed{V^{\mathrm{M}}_{\mathrm{H}}(\mathbf{x}) = \int d\mathbf{x}'\, v(\mathbf{x}, \mathbf{x}') \overbrace{\sum_{\nu} f(\epsilon^{\mathrm{H}}_{\nu} - \mu)|\varphi_\nu(\mathbf{x}')|^2}^{n(\mathbf{x}')}} \tag{8.9}$$

The coupled equations (8.8) and (8.9) for the single-particle wavefunctions $\{\varphi_\lambda\}$ are known as the *Hartree equations* [104]. Notice that the Hartree wavefunctions $\{\varphi_\lambda\}$ form an orthonormal basis since they are the eigenfunctions of the Hermitian operator $\hat{h}^{\mathrm{M}} + \hat{\mathcal{V}}^{\mathrm{M}}_{\mathrm{H}}$.

Zero-temperature fermions In a system of fermions at zero temperature, only states with energy $\epsilon^{\mathrm{H}}_{\lambda} < \mu$ contribute to the sum over λ in (8.9). The number of these states, say N, is the number of fermions in the system. For $N = 1$ the *exact* solution of the problem is the same as the noninteracting solution, since a single fermion cannot interact with itself. Does the Hartree approximation give the exact result for $N = 1$? The Hartree equation for the occupied wavefunction of a single fermion reads

$$\left[-\frac{\nabla^2}{2m} + q\phi(\mathbf{x}) + \int d\mathbf{x}'\, v(\mathbf{x},\mathbf{x}')|\varphi(\mathbf{x}')|^2\right]\varphi(\mathbf{x}) = \epsilon^{\mathrm{H}}\varphi(\mathbf{x}), \tag{8.10}$$

which differs from the noninteracting eigenvalue equation. Thus the Hartree approximation is not exact even for $N = 1$ since the Hartree potential is equal to the classical potential generated by the fermion itself. This is an intrinsic feature of the Hartree approximation known as the *self-interaction error*. Each fermion feels the potential generated by itself. The self-interaction error goes like $1/N$ and is therefore vanishingly small in bulk systems, but it can be rather large in finite systems. As we see in Section 8.3 the Hartree–Fock approximation cures this problem for the occupied states.

Zero-temperature bosons In bosonic systems at zero temperature, bosons can condense in the lowest energy level φ, meaning that all occupied φ_{λ} in (8.8) collapse in the same wavefunction φ. Then, as for a single fermion at zero temperature, the Hartree equations reduce to a single nonlinear equation:

$$\left[-\frac{\nabla^2}{2m} + q\phi(\mathbf{x}) + N\int d\mathbf{x}'\, v(\mathbf{x},\mathbf{x}')|\varphi(\mathbf{x}')|^2\right]\varphi(\mathbf{x}) = \epsilon^{\mathrm{H}}\varphi(\mathbf{x}), \tag{8.11}$$

where N is the number of bosons in the system. The bosonic case also suffers from the self-interaction error. In particular, for $N = 1$ the above equation does not reduce to the eigenvalue equation for one single boson. As we see in Section 8.3 the Hartree–Fock approximation does *not* cure this problem in bosonic systems.

If we multiply (8.11) by $\sqrt{N}$ and define $\tilde{\varphi} = \sqrt{N}\varphi$, we obtain exactly (8.10). This kind of equation is called a *nonlinear Schrödinger equation*. A renowned physical example of a nonlinear Schrödinger equation is that of a system of hardcore bosons with interparticle interaction $v(\mathbf{x},\mathbf{x}') = v_0\delta(\mathbf{x}-\mathbf{x}')$. In this case, (8.11) becomes

$$\left[-\frac{\nabla^2}{2m} + q\phi(\mathbf{x}) + v_0|\tilde{\varphi}(\mathbf{x})|^2\right]\tilde{\varphi}(\mathbf{x}) = \epsilon^{\mathrm{H}}\tilde{\varphi}(\mathbf{x}),$$

which is called the *Gross–Pitaevskii equation* [105, 106]. The Gross–Pitaevskii equation is enjoying increasing popularity due to experimental advances in trapping and cooling weakly interacting atoms using lasers. The first ever observation of a Bose condensate dates back to 1995, and since then many groups have been able to reproduce such a remarkable phenomenon. For a review on Bose condensation in these systems, see Ref. [107].

Total energy in the Hartree approximation We already pointed out that in the Hartree approximation the particles behave as free particles. Then we would (naively) expect that the total energy of the system in thermal equilibrium is the noninteracting energy (6.117) with single-particle Hartree energies $\epsilon^{\mathrm{H}}_{\lambda}$ and occupations $f_{\lambda} \equiv f(\epsilon^{\mathrm{H}}_{\lambda} - \mu)$. This is, however, not so! Let us clarify this subtle point. If we multiply (8.8) by $\varphi_\lambda^*(\mathbf{x})$ and integrate over $\mathbf{x}$, we can express the generic eigenvalue in terms of the matrix elements of $\hat{h} + \hat{\mathcal{V}}_{\rm H}^{\rm M}$:

$$\begin{aligned}\epsilon_\lambda^{\rm H} &= h_{\lambda\lambda} + \int d\mathbf{x}d\mathbf{x}'\, v(\mathbf{x},\mathbf{x}')n(\mathbf{x}')|\varphi_\lambda(\mathbf{x})|^2 \\ &= h_{\lambda\lambda} + \sum_\nu f_\nu \int d\mathbf{x}d\mathbf{x}'\, \varphi_\lambda^*(\mathbf{x})\varphi_\nu^*(\mathbf{x}')v(\mathbf{x},\mathbf{x}')\varphi_\nu(\mathbf{x}')\varphi_\lambda(\mathbf{x}) \\ &= h_{\lambda\lambda} + \sum_\nu f_\nu v_{\lambda\nu\nu\lambda},\end{aligned}$$

where we use the definitions (1.85) and (1.86). Let us now evaluate the total energy. In (6.116), $\hat{h}$ is the time-independent single-particle Hamiltonian and *not* the single-particle Hartree Hamiltonian. The latter is $\hat{h}_{\rm H} \equiv \hat{h} + \hat{\mathcal{V}}_{\rm H} = \hat{h} + \hat{\mathcal{V}}_{\rm H}^{\rm M}$, where we take into account that in equilibrium $\hat{\mathcal{V}}_{\rm H}(t) = \hat{\mathcal{V}}_{\rm H}^{\rm M}$. In the Hartree approximation,

$$\hat{\mathcal{G}}^<(\omega) = \mp 2\pi {\rm i}\, f(\hat{h}_{\rm H}^{\rm M})\,\delta(\omega - \hat{h}_{\rm H}),$$

and using that $\int d\mathbf{x}\langle\mathbf{x}|\ldots|\mathbf{x}\rangle = \sum_\lambda\langle\lambda|\ldots|\lambda\rangle$ (invariance of the trace under unitary transformations) we find

$$\begin{aligned}E &= \frac{1}{2}\int\frac{d\omega}{2\pi}\sum_\lambda\langle\lambda|(\hat{h}+\omega)f(\underbrace{\hat{h}+\hat{\mathcal{V}}_{\rm H}^{\rm M}}_{\hat{h}_{\rm H}^{\rm M}}-\mu)\,2\pi\,\delta(\omega-[\underbrace{\hat{h}+\hat{\mathcal{V}}_{\rm H}^{\rm M}}_{\hat{h}_{\rm H}}])|\lambda\rangle \\ &= \frac{1}{2}\sum_\lambda f_\lambda\langle\lambda|\hat{h}+\epsilon_\lambda^{\rm H}|\lambda\rangle = \sum_\lambda f_\lambda\epsilon_\lambda^{\rm H} - \frac{1}{2}\sum_{\lambda\nu} f_\lambda f_\nu v_{\lambda\nu\nu\lambda},\end{aligned} \tag{8.12}$$

where in the last equality we express $h_{\lambda\lambda}$ in terms of $\epsilon_\lambda^{\rm H}$ and the Coulomb integrals. Why does this energy differ from the energy of a truly noninteracting system? The explanation of the apparent paradox is simple. The eigenvalue $\epsilon_\lambda^{\rm H}$ contains the interaction energy $v_{\lambda\nu\nu\lambda}$ between a particle in φ_λ and a particle in φ_ν. If we sum over all λ this interaction energy is counted twice. In (8.12) the double counting is correctly removed by subtracting the last term.

Time-dependent Hartree equations Once the equilibrium problem is solved, we can construct the Matsubara Green's function as, for example, in (6.36). For all other components, however, we need to propagate the eigenstates of $\hat{h}_{\rm H}^{\rm M}$ in time. If the interacting system is exposed to a time-dependent electromagnetic field, then the single-particle Hamiltonian $\hat{h}(t)$ is time-dependent and so is the Hartree Hamiltonian $\hat{h}_{\rm H}(t) = \hat{h}(t) + \hat{\mathcal{V}}_{\rm H}(t)$. The evolution is governed by the Schrödinger equation ${\rm i}\frac{d}{dt}|\lambda(t)\rangle = \hat{h}_{\rm H}(t)|\lambda(t)\rangle$ with $|\lambda(t_0)\rangle = |\lambda\rangle$ the eigenkets of $\hat{h}_{\rm H}^{\rm M}$. By sandwiching the time-dependent Schrödinger equation with the bra $\langle\mathbf{x}|$, we get

$${\rm i}\frac{d}{dt}\varphi_\lambda(\mathbf{x},t) = \left[\frac{1}{2m}\left(-{\rm i}\boldsymbol{\nabla} - \frac{q}{c}\mathbf{A}(\mathbf{x},t)\right)^2 + q\phi(\mathbf{x},t) + V_{\rm H}(\mathbf{x},t)\right]\varphi_\lambda(\mathbf{x},t), \tag{8.13}$$

where $\mathbf{A}$ and ϕ are the external time-dependent vector and scalar potentials.[1] The time-dependent Hartree potential is given in (8.5); expressing the density in terms of the time-evolved eigenfunctions, we have

$$V_{\mathrm{H}}(\mathbf{x},t) = \int d\mathbf{x}'\, v(\mathbf{x},\mathbf{x}') \overbrace{\sum_{\lambda} f(\epsilon_{\lambda}^{\mathrm{H}} - \mu)|\varphi_{\lambda}(\mathbf{x}',t)|^2}^{n(\mathbf{x}',t)}. \tag{8.14}$$

The coupled equations (8.13) and (8.14) for the wavefunctions $\varphi_{\lambda}(\mathbf{x},t)$ are known as the *time-dependent Hartree equations.* For the unperturbed system the Hartree equations are solved by $\varphi_{\lambda}(\mathbf{x},t) = e^{-\mathrm{i}\epsilon_{\lambda}^{\mathrm{H}} t}\varphi_{\lambda}(\mathbf{x})$, as it should.

8.2.2 Equation of State of the Electron Gas

The *electron gas* is a system of interacting electrons (spin $1/2$ fermions) with single-particle Hamiltonian $\hat{h}(t) = \hat{h} = \hat{p}^2/2$ and with a spin-independent interparticle interaction $v(\mathbf{x}_1,\mathbf{x}_2) = v(\mathbf{r}_1 - \mathbf{r}_2)$ that we leave general for the time being.[2] The reader can consult Refs. [48, 108] for a detailed and thorough presentation of the physics of the electron gas. Below we consider the electron gas in thermal equilibrium so that $\hat{h}^{\mathrm{M}} = \hat{h} - \mu$ and $v^{\mathrm{M}} = v$. The full Hamiltonian $\hat{H} = \hat{H}_0 + \hat{H}_{\mathrm{int}}$ is invariant under space and time translations and consequently all physical quantities have the same property. In particular, the electron density per spin $n(\mathbf{x},z) = n/2$ is independent of $\mathbf{x}$ and z and so is the Hartree potential:[3]

$$V_{\mathrm{H}}(1) = \int d\mathbf{x}\, v(\mathbf{x}_1,\mathbf{x})\frac{n}{2} = \sum_{\sigma}\int d\mathbf{r}\, v(\mathbf{r}_1 - \mathbf{r})\frac{n}{2} \equiv n\tilde{v}_0, \tag{8.15}$$

where $\tilde{v}_0$ is the Fourier transform $\tilde{v}_{\mathbf{p}} = \int d\mathbf{r}\, e^{-\mathrm{i}\mathbf{p}\cdot\mathbf{r}}\, v(\mathbf{r})$ with $\mathbf{p} = 0$. The eigenkets of the Hartree Hamiltonian $\hat{h}_{\mathrm{H}}^{\mathrm{M}} = \hat{h} + \hat{V}_{\mathrm{H}}^{\mathrm{M}} - \mu$ are the momentum-spin kets $|\mathbf{p}\sigma\rangle$ with eigenenergy $p^2/2 + n\tilde{v}_0 - \mu$. The lesser Green's function can then be calculated using (6.45) and reads

$$\hat{\mathcal{G}}^{<}(\omega) = 2\pi\mathrm{i}\sum_{\sigma}\int \frac{d\mathbf{p}}{(2\pi)^3}\, |\mathbf{p}\sigma\rangle\langle\mathbf{p}\sigma| \frac{\delta(\omega - \frac{p^2}{2} - n\tilde{v}_0)}{e^{\beta(\frac{p^2}{2} + n\tilde{v}_0 - \mu)} + 1},$$

from which we can extract the value n of the density

$$\begin{aligned}\frac{n}{2} &= -\mathrm{i}\,\langle\mathbf{x}|\hat{\mathcal{G}}^{<}(t,t)|\mathbf{x}\rangle = -\mathrm{i}\int\frac{d\omega}{2\pi}\langle\mathbf{r}\sigma|\hat{\mathcal{G}}^{<}(\omega)|\mathbf{r}\sigma\rangle \\ &= \int\frac{d\mathbf{p}}{(2\pi)^3}\frac{1}{e^{\beta(\frac{p^2}{2} + n\tilde{v}_0 - \mu)} + 1},\end{aligned} \tag{8.16}$$

where we use that $\langle\mathbf{r}|\mathbf{p}\rangle = e^{\mathrm{i}\mathbf{p}\cdot\mathbf{r}}$, see (1.10). For any given initial temperature $T = 1/(K_{\mathrm{B}}\beta)$ and chemical potential μ, (8.16) provides a self-consistent equation for the density. Thus,

[1]For simplicity, in (8.13) the coupling between the time-dependent magnetic field and the spin degrees of freedom has not been included.

[2]The case of Coulomb interactions is considered in Section 8.3.2.

[3]We recall that in our units the electron charge $q = e = -1$, while the electron mass $m = m_e = 1$.

for an electron gas the solution of the Hartree equations reduces to the solution of (8.16): Once n is known, the Hartree problem is solved. Let us comment on the solution of (8.16) when $\tilde{v}_0 > 0$ (repulsive interaction). The r.h.s. is a decreasing function of n as can easily be checked by taking the derivative with respect to n. Consequently the self-consistent density is always smaller than the noninteracting density. For the interacting system to have the same density as the noninteracting system we have to increase μ - that is, the free energy per electron. This agrees with our intuitive picture that the energy needed to put together interacting particles increases with the strength of the repulsion. Of course, the opposite is true for attractive ($\tilde{v}_0 < 0$) interactions.

Equation of state From the above result we also expect that the pressure of the interacting system is larger or smaller than that of the noninteracting system, depending on whether the interaction is repulsive or attractive [109]. We recall that the pressure P is obtained from the density n by performing the integral (see Appendix D)

$$P(\beta, \mu) = \int_{-\infty}^{\mu} d\mu' \, n(\beta, \mu') \quad \Rightarrow \quad dP = n d\mu, \tag{8.17}$$

where $n(\beta, \mu)$ is the density given by the solution of (8.16). In the low-density limit $\beta\mu \to -\infty$, and the self-consistent equation (8.16) simplifies to

$$n(\beta, \mu) = 2 \int \frac{d\mathbf{p}}{(2\pi)^3} e^{-\beta(\frac{p^2}{2} + n\tilde{v}_0 - \mu)} = 2 \frac{e^{-\beta(n\tilde{v}_0 - \mu)}}{\sqrt{(2\pi\beta)^3}}.$$

Differentiating the above equation at fixed β, we get $dn = -\beta\tilde{v}_0 n dn + \beta n d\mu$, from which we find $d\mu = \tilde{v}_0 dn + \frac{1}{\beta}\frac{dn}{n}$. Substituting this result into (8.17) and integrating over n between 0 and n, we deduce the equation of state for an electron gas in the Hartree approximation:

$$P = nK_{\mathrm{B}}T + \frac{1}{2}\tilde{v}_0 n^2, \tag{8.18}$$

where we use that the pressure at zero density vanishes. The correction to the equation of state $P = nK_{\mathrm{B}}T$ (noninteracting system) is positive in the repulsive case and negative otherwise, as expected. However, in the attractive case the result does not make sense for too low temperatures. The derivative of P with respect to n should always be positive since

$$\left(\frac{\partial P}{\partial n}\right)_T = \left(\frac{\partial P}{\partial \mu}\right)_T \left(\frac{\partial \mu}{\partial n}\right)_T = n \left(\frac{\partial \mu}{\partial n}\right)_T > 0.$$

The inequality follows from the positivity of the density n and of the derivative

$$\left(\frac{\partial \mu}{\partial n}\right)_T^{-1} = \left(\frac{\partial n}{\partial \mu}\right)_T = \frac{1}{\mathrm{V}} \frac{\partial}{\partial \mu} \frac{\mathrm{Tr}\left[e^{-\beta(\hat{H} - \mu\hat{N})}\hat{N}\right]}{\mathrm{Tr}\left[e^{-\beta(\hat{H} - \mu\hat{N})}\right]} = \frac{\beta}{\mathrm{V}} \langle \left(\hat{N} - N\right)^2 \rangle,$$

where V is the volume of the system and $\langle \ldots \rangle$ denotes the ensemble average. If we take the derivative of (8.18), we find

$$\left(\frac{\partial P}{\partial n}\right)_T = K_{\mathrm{B}}T + \tilde{v}_0 n,$$

which becomes negative for temperatures $T < T_c \equiv -\tilde{v}_0 n/K_{\rm B}$. This meaningless result does actually contain some physical information. As the temperature decreases from values above T_c, the derivative $(\partial P/\partial n)_T$ approaches zero from positive values and hence the fluctuation in the number of particles, $\langle(\hat{N}-N)^2\rangle$, diverges in this limit. Typically the occurrence of large fluctuations in physical quantities signals the occurrence of an instability. In fact, the electron gas with attractive interparticle interaction undergoes a phase transition at sufficiently low temperatures, turning into a superconductor.

van der Waals equation It is also interesting to observe that (8.18) has the form of the van der Waals equation $(P - \alpha n^2)(\mathrm{V} - \mathrm{V}_{\rm exc}) = n\mathrm{V}K_{\rm B}T$ with α a constant and $\mathrm{V}_{\rm exc}$ the exclusion volume – that is, the hardcore impenetrable volume of the particles. In the Hartree approximation $\alpha = -\frac{1}{2}\tilde{v}_0$ while $\mathrm{V}_{\rm exc} = 0$. The fact that $\mathrm{V}_{\rm exc} = 0$ is somehow expected. As already pointed out, the Hartree approximation treats the particles as effectively free, meaning that two particles can get close to each other without paying energy. To have a nonvanishing exclusion volume we must go beyond the Hartree–Fock approximation and introduce correlations in the self-energy.[4]

The free nature of the electrons in the Hartree approximation is most evident from the spectral function. From definition (6.77) we find

$$\langle \mathbf{p}\sigma|\hat{\mathcal{A}}(\omega)|\mathbf{p}'\sigma'\rangle = 2\pi\delta_{\sigma\sigma'}\delta(\mathbf{p}-\mathbf{p}')\delta(\omega - \frac{p^2}{2} - n\tilde{v}_0).$$

According to our interpretation of the spectral function, the above result tells us that an electron with definite momentum and spin has a well-defined energy, $\epsilon^{\rm H}_{\mathbf{p}} = \frac{p^2}{2} + n\tilde{v}_0$, and hence an infinitely long lifetime. This is possible only provided that there is no scattering between the electrons.

Exercise 8.1 Consider the time-dependent Gross–Pitaevskii equation in one dimension with $\phi = \mathbf{A} = 0$:

$$\mathrm{i}\frac{d}{dt}\varphi(x,t) = -\frac{1}{2m}\frac{d^2}{dx^2}\varphi(x,t) + v_0|\varphi(x,t)|^2\varphi(x,t),$$

where for simplicity we omitted the spin index. Show that for $v_0 > 0$ a possible solution of the Gross–Pitaevskii equation is

$$\varphi(x,t) = \tanh\left[\sqrt{mv_0}(x - \mathrm{v}t)\right] e^{\mathrm{i}\left[m\mathrm{v}x - \left(m\frac{\mathrm{v}^2}{2} + v_0\right)t\right]},$$

whereas for $v_0 < 0$ a possible solution is

$$\varphi(x,t) = \mathrm{sech}\left[\sqrt{-mv_0}(x - \mathrm{v}t)\right] e^{\mathrm{i}\left[m\mathrm{v}x - \left(m\frac{\mathrm{v}^2}{2} + \frac{v_0}{2}\right)t\right]},$$

where v is an arbitrary velocity. These solutions are solitary waves, or *solitons*. Indeed the function $|\varphi(x,t)|^2$ propagates in the medium with velocity v without changing its profile. The reader interested in the theory of solitons can consult Ref. [110].

---

[4] This is nicely discussed in Ref. [109].

Exercise 8.2 Using (6.116), show that the energy of the electron gas in the Hartree approximation is given by

$$E = \mathsf{V}\left[\frac{1}{2}\tilde{v}_0 n^2 + 2\int \frac{d\mathbf{p}}{(2\pi)^3}\frac{p^2/2}{e^{\beta(\frac{p^2}{2}+\tilde{v}_0 n-\mu)}+1}\right],$$

where $\mathsf{V} \equiv \int d\mathbf{r}$ is the volume of the system.

Exercise 8.3 Using (8.16), plot μ as a function of temperature T for different values of density n and check that the plots are consistent with the result $\beta\mu \to -\infty$ for $n \to 0$.

Exercise 8.4 Consider the Hubbard model with only one site: $\hat{H} = \epsilon_0 \sum_\sigma \hat{d}^\dagger_\sigma \hat{d}_s + U\hat{n}_\uparrow \hat{n}_\downarrow$. Assuming that the Green's function is diagonal in the spin indices, show that the Hartree potential is $V_{\mathrm{H}}(\sigma, z) = U(n_\uparrow(z) + n_\downarrow(z))$ for interaction (2.36) and $V_{\mathrm{H}}(\sigma, z) = Un_{\bar{\sigma}}(z)$ for interaction (2.37). Here, $n_\sigma(z) = -\mathrm{i}G_{\sigma\sigma}(z, z^+)$ is the number of electrons with spin σ, and $\bar{\sigma}$ is the spin opposite to σ.

8.3 Hartree-Fock Approximation

The main advance of the Hartree-Fock approximation over the Hartree approximation is the incorporation of exchange effects. This is most easily understood in terms of the two-particle Green's function, and we indeed come back to this point in Chapter 14. In this section the exchange effects brought about by the Fock term become evident in the formula for the total energy.

The Hartree-Fock self-energy is given in (8.2), where $E(\mathbf{x}_1, \mathbf{x}_2, z) = V_{\mathrm{HF}}(\mathbf{x}_1, \mathbf{x}_2, z)$ is the *Hartree-Fock potential,* see again (8.1),

$$\begin{aligned} V_{\mathrm{HF}}(\mathbf{x}_1, \mathbf{x}_2, z) &= \left[\delta(\mathbf{x}_1 - \mathbf{x}_2)V_{\mathrm{H}}(\mathbf{x}_1, z) + \mathrm{i}\, v(\mathbf{x}_1, \mathbf{x}_2)G(\mathbf{x}_1, z; \mathbf{x}_2, z^+)\right] \\ &= \delta(\mathbf{x}_1 - \mathbf{x}_2)\int d\mathbf{x}\, v(\mathbf{x}_1, \mathbf{x})n(\mathbf{x}, z) \pm v(\mathbf{x}_1, \mathbf{x}_2)n(\mathbf{x}_1, \mathbf{x}_2, z). \end{aligned} \tag{8.19}$$

In (8.19) we define

$$n(\mathbf{x}_1, \mathbf{x}_2, z) \equiv \pm \mathrm{i}\, G(\mathbf{x}_1, z; \mathbf{x}_2, z^+) \tag{8.20}$$

as the time-dependent *one-particle density matrix.*[5] Introducing the Hartree-Fock potential operator in first quantization, see (8.4),

$$\hat{\mathcal{V}}_{\mathrm{HF}}(z) = \int d\mathbf{x}_1 d\mathbf{x}_2\, |\mathbf{x}_1\rangle V_{\mathrm{HF}}(\mathbf{x}_1, \mathbf{x}_2, z)\langle \mathbf{x}_2|, \tag{8.21}$$

the Hartree-Fock Green's function is defined as the solution of

$$\left[\mathrm{i}\frac{d}{dz_1} - \hat{h}(z_1) - \hat{\mathcal{V}}_{\mathrm{HF}}(z_1)\right]\hat{\mathcal{G}}(z_1, z_2) = \delta(z_1, z_2). \tag{8.22}$$

[5]On the diagonal the one-particle density matrix is the same as the density: $n(\mathbf{x}_1, \mathbf{x}_1, z) = n(\mathbf{x}_1, z)$. More about the one-particle and the n-particle density matrices can be found in Appendix G.

The Green's function in (8.22) must be solved self-consistently and, like in the Hartree approximation, this can be done by solving a set of coupled equations for one-particle wavefunctions. These equations are known as the *Hartree–Fock equations* and are derived below.

8.3.1 Hartree–Fock Equations

The Hartree–Fock G can be calculated as in Section 6.2, as long as we replace $\hat{h}(z)$ with

$$\hat{h}_{\mathrm{HF}}(z) = \hat{h}(z) + \hat{\mathcal{V}}_{\mathrm{HF}}(z). \tag{8.23}$$

Along the imaginary track the Hartree–Fock potential operator is a constant operator that we denote by $\hat{\mathcal{V}}^{\mathrm{M}}_{\mathrm{HF}}$. We again specialize the discussion to particles of mass m and charge q initially in equilibrium in some external potential $\phi(\mathbf{x})$.[6] The single-particle Hamiltonian which describes the system is therefore $\hat{h}^{\mathrm{M}} = \hat{p}^2/(2m) + q\phi(\hat{\boldsymbol{r}}, \hat{S}_z) - \mu$ and the interaction is $v^{\mathrm{M}} = v$. The first step consists in finding the kets $|\lambda\rangle$ which solve the eigenvalue problem

$$\left[\hat{h}^{\mathrm{M}} + \hat{\mathcal{V}}^{\mathrm{M}}_{\mathrm{HF}}\right] |\lambda\rangle = (\epsilon^{\mathrm{HF}}_{\lambda} - \mu)|\lambda\rangle. \tag{8.24}$$

By sandwiching with the bra $\langle\mathbf{x}|$ and using the explicit form of $\hat{h}^{\mathrm{M}}$, the eigenvalue equation reads

$$\boxed{\left[-\frac{\nabla^2}{2m} + q\phi(\mathbf{x})\right]\varphi_\lambda(\mathbf{x}) + \int d\mathbf{x}'\, V^{\mathrm{M}}_{\mathrm{HF}}(\mathbf{x},\mathbf{x}')\varphi_\lambda(\mathbf{x}') = \epsilon^{\mathrm{HF}}_{\lambda}\varphi_\lambda(\mathbf{x})} \tag{8.25}$$

with

$$\boxed{V^{\mathrm{M}}_{\mathrm{HF}}(\mathbf{x},\mathbf{x}') = \delta(\mathbf{x}-\mathbf{x}')V^{\mathrm{M}}_{\mathrm{H}}(\mathbf{x}) \pm v(\mathbf{x},\mathbf{x}')\sum_{\nu} f(\epsilon^{\mathrm{HF}}_{\nu} - \mu)\varphi_\nu(\mathbf{x})\varphi^*_\nu(\mathbf{x}')} \tag{8.26}$$

and the Hartree potential $V^{\mathrm{M}}_{\mathrm{H}}$ given in (8.9). In (8.26) we use result (6.36), according to which

$$G(\mathbf{x}, t_0 - \mathrm{i}\tau; \mathbf{x}', t_0 - \mathrm{i}\tau^+) = G^{\mathrm{M}}(\mathbf{x}, \tau; \mathbf{x}', \tau^+) = \mp\mathrm{i}\sum_{\nu} f(\epsilon^{\mathrm{HF}}_{\nu} - \mu)\langle\mathbf{x}|\varphi_\nu\rangle\langle\varphi_\nu|\mathbf{x}'\rangle.$$

We thus obtain a coupled system of nonlinear equations for the eigenfunctions φ_λ. These equations are known as the *Hartree–Fock equations* [111, 112]. As usual, the upper/lower sign in (8.26) refers to bosons/fermions.

We have already mentioned that the Hartree–Fock approximation cures the self-interaction problem in fermionic systems. Using (8.26), the second term on the l.h.s. of (8.25) becomes

$$\int d\mathbf{x}'\, V^{\mathrm{M}}_{\mathrm{HF}}(\mathbf{x},\mathbf{x}')\varphi_\lambda(\mathbf{x}') = \sum_{\nu} f(\epsilon^{\mathrm{HF}}_{\nu} - \mu)\int d\mathbf{x}'\, v(\mathbf{x},\mathbf{x}') \times \left[|\varphi_\nu(\mathbf{x}')|^2\varphi_\lambda(\mathbf{x}) \pm \varphi_\nu(\mathbf{x})\varphi^*_\nu(\mathbf{x}')\varphi_\lambda(\mathbf{x}')\right]. \tag{8.27}$$

[6] The inclusion of a vector potential and spin-flip interactions is straightforward.

The term $\nu = \lambda$ in the above sum vanishes for fermions while it yields twice the Hartree contribution for bosons. Thus, for a system of bosons at zero temperature the Hartree-Fock equations are identical to the Hartree equations (8.11) with $N \to 2N$. It is also noteworthy that if the interaction is spin-independent - that is, $v(\mathbf{x}_1, \mathbf{x}_2) = v(\mathbf{r}_1 - \mathbf{r}_2)$ - the second term in the square brackets vanishes unless ν and λ have the same spin projection. In other words, there are no exchange contributions coming from particles of different spin projection.

Hartree-Fock equations in a general basis The position-spin basis is not always the most convenient basis to solve the Hartree-Fock equations. Given a general basis $\{|i\rangle\}$, the sandwich of (8.24) with the bra $\langle i|$ yields

$$\sum_j (h^{\rm M}_{ij} + V^{\rm M}_{{\rm HF},ij})\langle j|\lambda\rangle = (\epsilon^{\rm HF}_\lambda - \mu)\langle i|\lambda\rangle, \tag{8.28}$$

where $h^{\rm M}_{ij} = \langle i|\hat{h}^{\rm M}|j\rangle$ and

$$\begin{aligned} V^{\rm M}_{{\rm HF},ij} = \langle i|\hat{\mathcal{V}}^{\rm M}_{\rm HF}|j\rangle &\underbrace{=}_{(8.21)} \int d\mathbf{x}_1 d\mathbf{x}_2\, \varphi^*_i(\mathbf{x}_1) V_{\rm HF}(\mathbf{x}_1,\mathbf{x}_2,z)\varphi_j(\mathbf{x}_2) \\ &\underbrace{=}_{(8.19)} \int d\mathbf{x}_1 d\mathbf{x}_2 \Big[\varphi^*_i(\mathbf{x}_1)\varphi_j(\mathbf{x}_1)v(\mathbf{x}_1,\mathbf{x}_2)n(\mathbf{x}_2) \pm \varphi^*_i(\mathbf{x}_1)\varphi_j(\mathbf{x}_2)v(\mathbf{x}_1,\mathbf{x}_2)n(\mathbf{x}_1,\mathbf{x}_2)\Big]. \end{aligned} \tag{8.29}$$

The one-particle density matrix can be expanded in the general basis according to

$$\begin{aligned} n(\mathbf{x}_1,\mathbf{x}_2) &= \pm {\rm i}\langle \mathbf{x}_1|\hat{\mathcal{G}}^{\rm M}(0,0^+)|\mathbf{x}_2\rangle \underbrace{=}_{(6.36)} \langle \mathbf{x}_1|f(\hat{h}^{\rm M} + \hat{\mathcal{V}}^{\rm M}_{\rm HF})|\mathbf{x}_2\rangle \\ &= \sum_{pq} \underbrace{\sum_\nu f(\epsilon^{\rm HF}_\nu - \mu)\langle p|\nu\rangle\langle \nu|q\rangle}_{n_{pq}} \varphi_p(\mathbf{x}_1)\varphi^*_q(\mathbf{x}_2) = \sum_{pq} \varphi_p(\mathbf{x}_1)\varphi^*_q(\mathbf{x}_2) n_{pq}, \end{aligned} \tag{8.30}$$

where n_{pq} is the one-particle density matrix in the general basis. Substituting (8.30) into (8.29) and recalling definition (1.86) of the Coulomb integrals, we get

$$V^{\rm M}_{{\rm HF},ij} = \sum_{pq} (v_{iqpj} \pm v_{iqjp}) n_{pq}. \tag{8.31}$$

Equations (8.28) and (8.31) form a coupled system of nonlinear equations for the amplitudes $\langle i|\lambda\rangle$ and the eigenvalues $\epsilon^{\rm HF}_\lambda$. It is easy to verify that these equations correctly reduce to (8.25) and (8.26) if the general basis is the position-spin basis.

Time-dependent Hartree-Fock equations Once the equilibrium problem is solved, we can construct the Matsubara Green's function. For all other Keldysh components we must propagate the $|\lambda\rangle$ in time according to the Schrödinger equation ${\rm i}\frac{d}{dt}|\lambda(t)\rangle = [\hat{h}(t) + \hat{\mathcal{V}}_{\rm HF}(t)]|\lambda(t)\rangle$ with initial conditions $|\lambda(t_0)\rangle = |\lambda\rangle$. By sandwiching again with $\langle \mathbf{x}|$ we find the so-called *time-dependent Hartree–Fock equations*. They are simply obtained from the static equations (8.25) by replacing $[-\nabla^2/(2m) + q\phi(\mathbf{x})]$ with the time-dependent single-particle Hamiltonian $h(\mathbf{r}, -\mathrm{i}\boldsymbol{\nabla}, \mathbf{S}, t)$, the static wavefunctions $\varphi_\lambda(\mathbf{x})$ with the time-evolved wavefunctions $\varphi_\lambda(\mathbf{x}, t)$, the static Hartree–Fock potential $V^{\mathrm{M}}_{\mathrm{HF}}(\mathbf{x}, \mathbf{x}')$ with the time-dependent Hartree–Fock potential $V_{\mathrm{HF}}(\mathbf{x}, \mathbf{x}', t)$, and $\epsilon^{\mathrm{HF}}_\lambda$ with $\mathrm{i}\frac{d}{dt}$.

Total energy in the Hartree–Fock approximation In thermal equilibrium ($\hat{h}^{\mathrm{M}} = \hat{h} - \mu$ and $v^{\mathrm{M}} = v$) the single-particle Hamiltonian $\hat{h}(t) = \hat{h}$ is independent of time. As a consequence, the Hartree–Fock Hamiltonian $\hat{h}_{\mathrm{HF}} = \hat{h} + \hat{\mathcal{V}}_{\mathrm{HF}} = \hat{h} + \hat{\mathcal{V}}^{\mathrm{M}}_{\mathrm{HF}}$ is also independent of time and the lesser Green's function reads

$$\hat{\mathcal{G}}^<(\omega) = \mp 2\pi\mathrm{i}\, f(\underbrace{\hat{h} + \hat{\mathcal{V}}^{\mathrm{M}}_{\mathrm{HF}} - \mu}_{\hat{h}^{\mathrm{M}}_{\mathrm{HF}}})\, \delta(\omega - [\underbrace{\hat{h} + \hat{\mathcal{V}}^{\mathrm{M}}_{\mathrm{HF}}}_{\hat{h}_{\mathrm{HF}}}]).$$

Substituting this $\hat{\mathcal{G}}^<$ in (6.116) and inserting a complete set of eigenstates of $\hat{h} + \hat{\mathcal{V}}^{\mathrm{M}}_{\mathrm{HF}}$, we get

$$\begin{aligned} E &= \sum_\lambda f_\lambda \left[\epsilon^{\mathrm{HF}}_\lambda - \frac{1}{2} \int d\mathbf{x} \langle \mathbf{x}|\hat{\mathcal{V}}^{\mathrm{M}}_{\mathrm{HF}}|\lambda\rangle\langle\lambda|\mathbf{x}\rangle \right] \\ &= \sum_\lambda f_\lambda \left[\epsilon^{\mathrm{HF}}_\lambda - \frac{1}{2} \sum_\nu f_\nu (v_{\lambda\nu\nu\lambda} \pm v_{\lambda\nu\lambda\nu}) \right], \qquad f_\lambda \equiv f(\epsilon^{\mathrm{HF}}_\lambda - \mu), \end{aligned} \tag{8.32}$$

where in the last equality we use (8.27). As expected, the total energy is not the weighted sum of the one-particle Hartree–Fock energies, see discussion below (8.12). It is instructive to express the $\epsilon^{\mathrm{HF}}_\lambda$ in terms of $h_{\lambda\lambda}$ and Coulomb integrals. If we multiply (8.25) by $\varphi^*_\lambda(\mathbf{x})$ and integrate over $\mathbf{x}$, we obtain

$$\epsilon^{\mathrm{HF}}_\lambda = h_{\lambda\lambda} + \sum_\nu f_\nu (v_{\lambda\nu\nu\lambda} \pm v_{\lambda\nu\lambda\nu}), \tag{8.33}$$

from which it is evident that the self-interaction energy - that is, the contribution $\lambda = \nu$ in the sum - vanishes in the case of fermions.

Koopmans' theorem Since the total energy $E \neq \sum_\lambda f_\lambda \epsilon^{\mathrm{HF}}_\lambda$ we cannot interpret the $\epsilon^{\mathrm{HF}}_\lambda$ as the energy of a particle. Can we still give a physical interpretation to the $\epsilon^{\mathrm{HF}}_\lambda$? To answer this question, we insert (8.33) into (8.32) and find

$$E = \sum_\lambda f_\lambda h_{\lambda\lambda} + \frac{1}{2} \sum_{\lambda\nu} f_\lambda f_\nu (v_{\lambda\nu\nu\lambda} \pm v_{\lambda\nu\lambda\nu}). \tag{8.34}$$

We now consider an ultrafast ionization process in which the particle in the ρth level is suddenly brought infinitely far away from the system. If we measure the energy of the ionized system before it has time to relax in some lower-energy state, we find

$$E_\rho = \sum_{\lambda \neq \rho} f_\lambda h_{\lambda\lambda} + \frac{1}{2} \sum_{\lambda\nu \neq \rho} f_\lambda f_\nu (v_{\lambda\nu\nu\lambda} \pm v_{\lambda\nu\lambda\nu}).$$

The difference between the initial energy E and the energy E_ρ is

$$E - E_\rho = f_\rho h_{\rho\rho} + \frac{1}{2} f_\rho f_\rho (v_{\rho\rho\rho\rho} \pm v_{\rho\rho\rho\rho})$$
$$+ \frac{1}{2} \sum_{\nu \neq \rho} f_\rho f_\nu (v_{\rho\nu\nu\rho} \pm v_{\rho\nu\rho\nu}) + \frac{1}{2} \sum_{\lambda \neq \rho} f_\lambda f_\rho (v_{\lambda\rho\rho\lambda} \pm v_{\lambda\rho\lambda\rho}).$$

Using the symmetry $v_{ijkl} = v_{jilk}$ of the Coulomb integrals, we can rewrite $E - E_\rho$ in the following compact form:

$$E - E_\rho = f_\rho \left[\epsilon_\rho^{\mathrm{HF}} - \frac{1}{2} f_\rho (v_{\rho\rho\rho\rho} \pm v_{\rho\rho\rho\rho}) \right].$$

Thus, for a system of fermions the eigenvalue $\epsilon_\rho^{\mathrm{HF}}$ multiplied by the occupation f_ρ is the difference between the energy of the initial system and the energy of the ionized and unrelaxed system. At zero temperature we expect that the removal of a particle from the highest occupied level (HOMO) does not cause a dramatic relaxation, and hence $\epsilon_{\mathrm{HOMO}}^{\mathrm{HF}}$ should provide a good estimate of the ionization energy. This result is known as *Koopmans' theorem.* A similar interpretation is not possible for a system of bosons.

8.3.2 Electron Gas and Spin-Polarized Solutions

Let us consider again an electron gas in equilibrium with single-particle Hamiltonian $\hat{h}^{\mathrm{M}} = \hat{p}^2/2 - \phi_0 - \mu$, where ϕ_0 is a constant energy shift (for electrons the charge $q = -1$). Due to translational invariance the eigenkets of $\hat{h}^{\mathrm{M}} + \hat{\mathcal{V}}_{\mathrm{HF}}^{\mathrm{M}}$ are the momentum–spin kets $|\mathbf{p}\sigma\rangle$:

$$\left[\hat{h}^{\mathrm{M}} + \hat{\mathcal{V}}_{\mathrm{HF}}^{\mathrm{M}} \right] |\mathbf{p}\sigma\rangle = (\epsilon_{\mathbf{p}}^{\mathrm{HF}} - \mu) |\mathbf{p}\sigma\rangle. \tag{8.35}$$

The Matsubara Green's function (needed to evaluate $\hat{\mathcal{V}}_{\mathrm{HF}}^{\mathrm{M}}$) is given in (6.36) and reads

$$G^{\mathrm{M}}(\mathbf{x}_1, \tau; \mathbf{x}_2, \tau^+) = \mathrm{i} \sum_\sigma \int \frac{d\mathbf{k}}{(2\pi)^3} f_{\mathbf{k}} \langle \mathbf{x}_1 | \mathbf{k}\sigma \rangle \langle \mathbf{k}\sigma | \mathbf{x}_2 \rangle, \tag{8.36}$$

where we use the short-hand notation $f_{\mathbf{k}} \equiv f(\epsilon_{\mathbf{k}}^{\mathrm{HF}} - \mu)$ and insert the completeness relation (1.11). Fourier transforming the interparticle interaction,

$$v(\mathbf{r}_1 - \mathbf{r}_2) = \int \frac{d\mathbf{q}}{(2\pi)^3} e^{\mathrm{i}\mathbf{q}\cdot(\mathbf{r}_1 - \mathbf{r}_2)} \tilde{v}_{\mathbf{q}} = \int \frac{d\mathbf{q}}{(2\pi)^3} \tilde{v}_{\mathbf{q}} \langle \mathbf{r}_1 | \mathbf{q} \rangle \langle \mathbf{q} | \mathbf{r}_2 \rangle,$$

and using the identity $\langle \mathbf{r}_1 | \mathbf{k} \rangle \langle \mathbf{r}_1 | \mathbf{q} \rangle = \langle \mathbf{r}_1 | \mathbf{k} + \mathbf{q} \rangle$ we can rewrite the Hartree–Fock operator (8.21), with $V_{\mathrm{HF}}^{\mathrm{M}}(\mathbf{x}_1, \mathbf{x}_2)$ from (8.19), in a diagonal form:

$$\hat{\mathcal{V}}_{\mathrm{HF}}^{\mathrm{M}} = \int d\mathbf{x}_1 d\mathbf{x}_2 \, |\mathbf{x}_1\rangle \underbrace{\langle \mathbf{x}_1 | \left[n\tilde{v}_0 - \sum_\sigma \int \frac{d\mathbf{k} d\mathbf{q}}{(2\pi)^6} |\mathbf{q}\sigma\rangle f_{\mathbf{k}} \tilde{v}_{\mathbf{q}-\mathbf{k}} \langle \mathbf{q}\sigma | \right] |\mathbf{x}_2\rangle}_{V_{\mathrm{HF}}^{\mathrm{M}}(\mathbf{x}_1, \mathbf{x}_2)} \langle \mathbf{x}_2 |$$
$$= \sum_\sigma \int \frac{d\mathbf{q}}{(2\pi)^3} |\mathbf{q}\sigma\rangle \left[n\tilde{v}_0 - \int \frac{d\mathbf{k}}{(2\pi)^3} f_{\mathbf{k}} \tilde{v}_{\mathbf{q}-\mathbf{k}} \right] \langle \mathbf{q}\sigma |. \tag{8.37}$$

Substituting this result into the eigenvalue equation (8.35), we obtain the following expression for $\epsilon_{\mathbf{p}}^{\mathrm{HF}}$:

$$\epsilon_{\mathbf{p}}^{\mathrm{HF}} = \frac{p^2}{2} - \phi_0 + n\tilde{v}_0 - \int \frac{d\mathbf{k}}{(2\pi)^3} f_{\mathbf{k}} \tilde{v}_{\mathbf{p}-\mathbf{k}}. \tag{8.38}$$

The total energy is the integral over all momenta of $\epsilon_{\mathbf{p}}^{\mathrm{HF}}$ minus one-half the interaction energy weighted with the Fermi function [see (8.32)]:

$$E = 2\mathrm{V} \int \frac{d\mathbf{p}}{(2\pi)^3} f_{\mathbf{p}} \left[\frac{p^2}{2} - \phi_0 + \frac{1}{2} n\tilde{v}_0 - \frac{1}{2} \int \frac{d\mathbf{k}}{(2\pi)^3} f_{\mathbf{k}} \tilde{v}_{\mathbf{p}-\mathbf{k}} \right], \tag{8.39}$$

where the factor of 2 comes from spin and where $\mathrm{V} = \int d\mathbf{r}$ is the volume of the system.

Let us specialize these formulas to the Coulomb interaction $v(\mathbf{r}_1 - \mathbf{r}_2) = 1/|\mathbf{r}_1 - \mathbf{r}_2|$. The Fourier transform is in this case $\tilde{v}_{\mathbf{p}} = 4\pi/p^2$ and hence $\tilde{v}_0$ diverges! Is this physical? The answer is yes. In the absence of any positive charge that attracts and bounds the electrons together the energy that we must spend to put an extra electron in a box containing a macroscopic number of electrons is infinite. We therefore consider the physical situation in which the space is permeated by a uniform density $n_b = n$ of positive charge, in such a way that the whole system is charge-neutral. Then the potential ϕ_0 felt by an electron in $\mathbf{r}$ is $\phi_0 = n_b \int d\mathbf{r}_1\, v(\mathbf{r} - \mathbf{r}_1) = n\tilde{v}_0$, which is also divergent, and the one-particle eigenvalues turn out to be finite:

$$\epsilon_{\mathbf{p}}^{\mathrm{HF}} = \frac{p^2}{2} - 4\pi \int \frac{d\mathbf{k}}{(2\pi)^3} \frac{f_{\mathbf{k}}}{|\mathbf{p} - \mathbf{k}|^2}. \tag{8.40}$$

The energy E is, however, still divergent. This is due to the fact that E contains only the electronic energy. Adding to E the energy of the positive background $E_b = \frac{1}{2} \int d\mathbf{r}_1 d\mathbf{r}_2 n_b^2/|\mathbf{r}_1 - \mathbf{r}_2| = \frac{1}{2} \mathrm{V} n^2 \tilde{v}_0$, the total energy of the system electrons+background is finite and reads[7]

$$E_{\mathrm{tot}} = E + E_b = \mathrm{V} \int \frac{d\mathbf{p}}{(2\pi)^3} f_{\mathbf{p}} \left[2\frac{p^2}{2} - 4\pi \int \frac{d\mathbf{k}}{(2\pi)^3} \frac{f_{\mathbf{k}}}{|\mathbf{p} - \mathbf{k}|^2} \right]. \tag{8.41}$$

Let us evaluate (8.40) and (8.41) at zero temperature.

Quasi-particles and total energy At zero temperature the chemical potential μ coincides with the Fermi energy ϵ_{F} and we can write $f_{\mathbf{p}} = \theta(\epsilon_{\mathrm{F}} - \epsilon_{\mathbf{p}}^{\mathrm{HF}})$. The eigenstates with energy below ϵ_{F} are filled while the others are empty. The density (per spin) is therefore

$$\frac{n}{2} = \int \frac{d\mathbf{p}}{(2\pi)^3} \theta(\epsilon_{\mathrm{F}} - \epsilon_{\mathbf{p}}^{\mathrm{HF}}) = p_{\mathrm{F}}^3/(6\pi^2), \tag{8.42}$$

where p_{F} is the Fermi momentum – that is, the value of the modulus of $\mathbf{p}$ for which $\epsilon_{\mathbf{p}}^{\mathrm{HF}} = \epsilon_{\mathrm{F}}$. Notice that the Fermi momentum is the same as that of a noninteracting gas with the same density. As we see in Section 11.6, this result is a general consequence of the conserving nature of the Hartree–Fock approximation. We also observe that the eigenvalue $\epsilon_{\mathbf{p}}^{\mathrm{HF}}$ in (8.40) depends only on $p = |\mathbf{p}|$. This is true for all interactions $v(\mathbf{r}_1 - \mathbf{r}_2)$ that are invariant under rotations. The calculation of the integral in (8.40) can be performed by

[7] Take into account that the density of the electron gas is given by $n = 2 \int \frac{d\mathbf{p}}{(2\pi)^3} f_{\mathbf{p}}$.

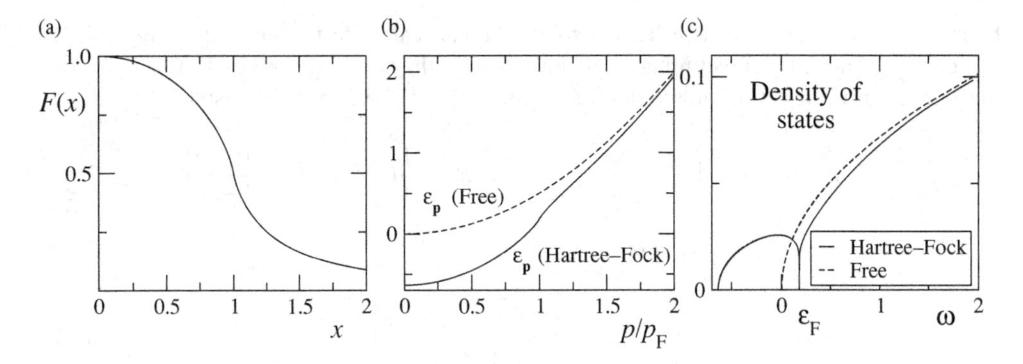

Figure 8.1 (a) The function $F(x)$ whose derivatives diverge in $x = 1$. (b) Comparison between the free-electron eigenvalues and the Hartree–Fock eigenvalues in units of p_F^2. (c) Density of states $D(\omega)/2\pi$ in units of p_F (energy is in units of p_F^2). In (b) and (c) $p_F = 1$, which corresponds to $r_s \sim 1.92$.

expanding $|\mathbf{p}-\mathbf{k}|^2 = p^2 + k^2 - 2kp\cos\theta$ (with θ the angle between $\mathbf{p}$ and $\mathbf{k}$) and changing the measure from Cartesian to polar coordinates $d\mathbf{k} = d\varphi d(\cos\theta)dk\,k^2$. The integral of the angular part yields

$$4\pi\int\frac{d\mathbf{k}}{(2\pi)^3}\frac{f_{\mathbf{k}}}{|\mathbf{p}-\mathbf{k}|^2} = \frac{1}{\pi p}\int_0^{p_F} dk\,k\ln\left|\frac{p+k}{p-k}\right| = \frac{p_F}{\pi x}\int_0^1 dy\,y\ln\left|\frac{x+y}{x-y}\right|,$$

where in the last equality we introduce the dimensionless variables $x = p/p_F$ and $y = k/p_F$. The remaining integral is a standard logarithmic integral,

$$I(a,b) \equiv \int_0^1 dy\,y\ln\left|\frac{b+ay}{b-ay}\right| = \frac{b}{a} + \frac{1}{2}(1-\frac{b^2}{a^2})\ln\left|\frac{b+a}{b-a}\right|, \tag{8.43}$$

and we thus arrive at the result

$$4\pi\int\frac{d\mathbf{k}}{(2\pi)^3}\frac{f_{\mathbf{k}}}{|\mathbf{p}-\mathbf{k}|^2} = \frac{2p_F}{\pi}F(\frac{p}{p_F}), \tag{8.44}$$

where

$$F(x) = \frac{1}{2} + \frac{1-x^2}{4x}\ln\left|\frac{1+x}{1-x}\right|. \tag{8.45}$$

The function $F(x)$ and a comparison between the free-electron eigenvalue $p^2/2$ and the Hartree–Fock eigenvalue $\epsilon_{\mathbf{p}}^{\mathrm{HF}}$ is shown in Fig. 8.1 for $p_F = 1$. The exchange contribution lowers the free-electron eigenvalues:

$$\underbrace{\epsilon_{\mathbf{p}} = \frac{p^2}{2}}_{\text{free}} \quad\to\quad \underbrace{\epsilon_{\mathbf{p}}^{\mathrm{HF}} = \frac{p^2}{2} - \frac{2p_F}{\pi}F(\frac{p}{p_F})}_{\text{Hartree-Fock}}. \tag{8.46}$$

This is a very important result since it provides a (partial) explanation of the stability of matter. Without the exchange correction the electrons would behave as free particles and hence they could easily escape from a solid. We can calculate the binding energy due to the exchange term by substituting (8.44) into (8.41) with $f_{\mathbf{p}} = \theta(\epsilon_{\mathrm{F}} - \epsilon^{\mathrm{HF}}_{\mathbf{p}})$. Changing the measure from Cartesian to polar coordinates and exploiting the analytic result,

$$\int dx\, x(1-x^2)\ln\left|\frac{1+x}{1-x}\right| = \frac{1}{2}x - \frac{1}{6}x^3 - \frac{1}{4}(1-x^2)^2\ln\left|\frac{1+x}{1-x}\right|,$$

we find that the total energy at zero temperature is

$$\boxed{E_{\mathrm{tot}} = n\mathrm{V}\left[\frac{3}{5}\frac{p_{\mathrm{F}}^2}{2} - \frac{3}{4}\frac{p_{\mathrm{F}}}{\pi}\right]} \tag{8.47}$$

It is common to express this result in terms of the "average distance" between the electrons. This simply means that if there is one electron in a sphere of radius R, then the density of the gas is $n^{-1} = \frac{4}{3}\pi R^3$. Writing the radius $R = r_s a_{\mathrm{B}}$ in units of the Bohr radius $a_{\mathrm{B}} = 1$ a.u. and taking into account that $p_{\mathrm{F}} = (3\pi^2 n)^{1/3}$ [see (8.42)], we find $p_{\mathrm{F}} = (\frac{9\pi}{4})^{\frac{1}{3}}\frac{1}{r_s}$ and hence

$$\boxed{\frac{E_{\mathrm{tot}}}{n\mathrm{V}} = \frac{1}{2}\left[\frac{2.21}{r_s^2} - \frac{0.916}{r_s}\right] \quad \text{(in atomic units)}} \tag{8.48}$$

The dimensionless radius r_s is also called the *Wigner–Seitz radius.* Most metals have r_s in the range $2 \div 6$ and therefore the exchange energy gives an important contribution to the binding energy.

Density of states So far, good news. The Hartree–Fock approximation, however, does also suffer serious problems. For instance, the density of states $D(\omega)$ defined in (6.80) gives

$$D(\omega) = 2\pi\int\frac{d\mathbf{p}}{(2\pi)^3}\delta(\omega - \epsilon^{\mathrm{HF}}_{\mathbf{p}}) = \frac{1}{\pi}\int dp\, p^2\frac{\delta(p - p(\omega))}{|\partial\epsilon^{\mathrm{HF}}_{\mathbf{p}}/\partial p|_{p(\omega)}} = \frac{p^2(\omega)}{\pi}\left|\frac{d\epsilon^{\mathrm{HF}}_{\mathbf{p}}}{dp}\right|^{-1}_{p(\omega)},$$

with $p(\omega)$ the value of the momentum p for which $\epsilon^{\mathrm{HF}}_{\mathbf{p}} = \omega$. In the noninteracting gas $p(\omega) = \sqrt{2\omega}$ and the density of states is proportional to $\sqrt{\omega}$. In the Hartree–Fock approximation $D(\omega)$ is zero at the Fermi energy, see Fig. 8.1, due to the divergence of the derivative of $F(x)$ in $x = 1$. This is in neat contrast to the experimentally observed density of states of metals which are smooth and finite across ϵ_{F}. Another problem of the Hartree–Fock approximation is the overestimation of the bandwidths of metals and of the band-gaps of semiconductors and insulators. These deficiencies are due to the lack of screening effects. As the interparticle interaction is repulsive, an electron pushes away electrons around it and remains surrounded by a positive charge (coming from the positive background charge). The consequence of the screening is that the effective repulsion between two electrons is reduced (and, as we shall see, retarded).

Spin-polarized solution A further problem of the Hartree–Fock approximation that we wish to discuss is related to the stability of the solution. We prepare the system in a different initial state by changing the single-particle Hamiltonian $\hat{h}^{\mathrm{M}}$ under the constraint that $\hat{h}^{\rm M}_{\rm HF}$ and $\hat{h}_{\rm HF}(t_0)$ commute.[8] Then the eigenkets $|\lambda\rangle$ of $\hat{h}^{\rm M}_{\rm HF}$ are also eigenkets of $\hat{h}_{\rm HF}(t_0)$ with eigenvalues, say, $\epsilon^{\rm M}_\lambda$ and $\epsilon^{\rm HF}_\lambda$ respectively. The time-evolved kets $|\lambda(t)\rangle = e^{-{\rm i}\epsilon^{\rm HF}_\lambda(t-t_0)}|\lambda\rangle$ are the self-consistent solution of the time-dependent Hartree-Fock equations, since $\hat{h}_{\rm HF}(t)$, evaluated with $|\lambda(t)\rangle$, is time-independent and hence ${\rm i}\frac{d}{dt}|\lambda(t)\rangle = \hat{h}_{\rm HF}(t)|\lambda(t)\rangle = \hat{h}_{\rm HF}(t_0)|\lambda(t)\rangle$ is satisfied. In this case, the real-time Green's functions depend only on the time difference, meaning that $\hat{h}^{\rm M}$ describes an excited stationary ensemble.

According to (6.45) the lesser Green's function is obtained by populating the kets $|\lambda\rangle$ with occupations $f(\epsilon^{\rm M}_\lambda)$ and, in principle, may give a total energy which is lower than (8.47). If so, the Hartree-Fock approximation is unstable and this instability indicates either that the approximation is not a good approximation or that the system would like to rearrange the electrons in a different way. Changing the Hamiltonian $\hat{h}^{\rm M}$ and checking the stability of an approximation is a very common procedure to assess the quality of the approximation and/or to understand if the ground state of the system has the same symmetry as the noninteracting ground state. Let us illustrate how it works with an example.

As a trial single-particle Hamiltonian we consider

$$\hat{h}^{\rm M} = \hat{h} + \sum_\sigma \int \frac{d\mathbf{p}}{(2\pi)^3}\, \epsilon_\sigma |\mathbf{p}\sigma\rangle\langle\mathbf{p}\sigma| - \mu, \qquad \hat{h} = \hat{p}^2/2 - \phi_0. \tag{8.49}$$

As we see shortly, this choice of $\hat{h}^{\rm M}$ corresponds to an initial configuration with different number of spin up and down electrons. We are therefore exploring the possibility that a *spin-polarized* electron gas has lower energy than the spin-unpolarized one.

Having broken the spin symmetry, the one-particle eigenvalues of the Hartree-Fock Hamiltonian $\hat{h}^{\rm M}_{\rm HF} = \hat{h}^{\rm M} + \hat{\mathcal{V}}^{\rm M}_{\rm HF}$ are spin-dependent. The extra term in (8.49), however, does not break the translational invariance and therefore the momentum-spin kets $|\mathbf{p}\sigma\rangle$ are still good eigenkets. At zero temperature $\mu = \epsilon_{\rm F}$, and the Matsubara Green's function needed to calculate $\hat{\mathcal{V}}^{\rm M}_{\rm HF}$ reads

$$G^{\rm M}(\mathbf{x}_1,\tau;\mathbf{x}_2,\tau^+) = {\rm i}\sum_\sigma \int \frac{d\mathbf{k}}{(2\pi)^3}\, f_{\mathbf{k}\sigma}\langle\mathbf{x}_1|\mathbf{k}\sigma\rangle\langle\mathbf{k}\sigma|\mathbf{x}_2\rangle,$$

where $f_{\mathbf{k}\sigma} = \theta(\epsilon_{\rm F} - \epsilon^{\rm HF}_{\mathbf{k}\sigma} - \epsilon_\sigma)$ and $\epsilon^{\rm HF}_{\mathbf{k}\sigma} + \epsilon_\sigma - \epsilon_{\rm F}$ are the eigenvalues of $\hat{h}^{\rm M}_{\rm HF}$. Inserting this result into the Hartree-Fock potential (8.19), we obtain the following Hartree-Fock potential operator [compare with (8.37)]:

$$\hat{\mathcal{V}}^{\rm M}_{\rm HF} = \sum_\sigma \int \frac{d\mathbf{q}}{(2\pi)^3}\, |\mathbf{q}\sigma\rangle \left[(n_\uparrow + n_\downarrow)\tilde{v}_0 - \int \frac{d\mathbf{k}}{(2\pi)^3} f_{\mathbf{k}\sigma}\tilde{v}_{\mathbf{q}-\mathbf{k}}\right]\langle\mathbf{q}\sigma|,$$

with $n_\sigma = \int \frac{d\mathbf{p}}{(2\pi)^3} f_{\mathbf{p}\sigma}$ the uniform density for electrons of spin σ. Notice that $\hat{\mathcal{V}}^{\rm M}_{\rm HF}$ remains diagonal in the basis of the momentum-spin kets, a property that guarantees $[\hat{h}^{\rm M}_{\rm HF}, \hat{h}_{\rm HF}(t_0)] = 0$. We also observe that at zero temperature $n_\sigma = p^3_{{\rm F}\sigma}/(6\pi^2)$, where the spin-dependent Fermi momentum is defined as the solution of $\epsilon^{\rm HF}_{\mathbf{p}\sigma} = \epsilon_{\rm F} - \epsilon_\sigma$; thus we can treat either ϵ_σ or n_σ or $p_{{\rm F}\sigma}$ as the independent variable.

[8] We remind the reader that $\hat{h}^{\rm M}$ specifies the initial preparation and does not have to represent the physical Hamiltonian $\hat{h}$ of the system. In thermal equilibrium $\hat{h}^{\rm M} = \hat{h} - \mu$. Any other choice of $\hat{h}^{\rm M}$ describes an excited ensemble.

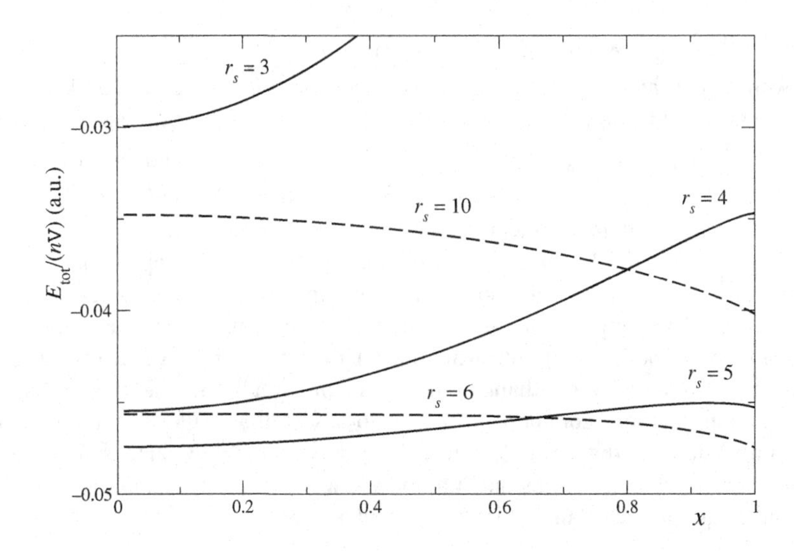

Figure 8.2 Energy density of the spin-polarized electron gas in Hartree–Fock approximation.

The Hartree–Fock Hamiltonian at positive times is time-independent and given by

$$\hat{h}_{\mathrm{HF}} = \frac{\hat{p}^2}{2} - \phi_0 + \hat{\mathcal{V}}^{\mathrm{M}}_{\mathrm{HF}},$$

with eigenvalues $\epsilon^{\mathrm{HF}}_{\mathbf{p}\sigma}$. We consider again Coulombic interactions and neutralize the system with a positive background charge $n_b = n_\uparrow + n_\downarrow$. Then the eigenvalues of $\hat{h}_{\mathrm{HF}}$ are given by

$$\epsilon^{\mathrm{HF}}_{\mathbf{p}\sigma} = \frac{p^2}{2} - \frac{2p_{\mathrm{F}\sigma}}{\pi} F(\frac{p}{p_{\mathrm{F}\sigma}}),$$

and we find

$$\begin{aligned}\hat{\mathcal{G}}^{<}(\omega) &= 2\pi \mathrm{i} f(\hat{h}^{\mathrm{M}}_{\mathrm{HF}})\delta(\omega - \hat{h}_{\mathrm{HF}}) \\ &= 2\pi\mathrm{i}\sum_\sigma \int \frac{d\mathbf{p}}{(2\pi)^3}\, |\mathbf{p}\sigma\rangle f_{\mathbf{p}\sigma}\, \underbrace{\delta\left(\omega - \frac{p^2}{2} + \frac{2p_{\mathrm{F}\sigma}}{\pi}F(\frac{p}{p_{\mathrm{F}\sigma}})\right)}_{\delta(\omega-\epsilon^{\mathrm{HF}}_{\mathbf{p}\sigma})}\langle \mathbf{p}\sigma|.\end{aligned}$$

The total energy of the spin-polarized electron gas can readily be obtained using (6.116), and reads

$$E_{\mathrm{tot}} = \mathrm{V}\sum_\sigma \int \frac{d\mathbf{p}}{(2\pi)^3} f_{\mathbf{p}\sigma}\left[\frac{p^2}{2} - \frac{p_{\mathrm{F}\sigma}}{\pi}F(\frac{p}{p_{\mathrm{F}\sigma}})\right] = \mathrm{V}\sum_\sigma n_\sigma \left[\frac{3}{5}\frac{p^2_{\mathrm{F}\sigma}}{2} - \frac{3}{4}\frac{p_{\mathrm{F}\sigma}}{\pi}\right].$$

Let us study what this formula predicts at fixed $n_\uparrow + n_\downarrow \equiv n$. If we define the spin polarization $x = (n_\uparrow - n_\downarrow)/n$, we can write $n_\uparrow = \frac{n}{2}(1+x)$ and $n_\downarrow = \frac{n}{2}(1-x)$, and hence we can express E_{tot} in terms of n and x only (recall that $p_{\mathrm{F}\sigma} = (6\pi^2 n_\sigma)^{1/3}$). In Fig. 8.2 we plot

$E_{\text{tot}}/(n\text{V})$ as a function of the spin polarization x for different values of $r_s = \left(\frac{3}{4\pi n}\right)^{1/3}$. For small r_s the state of minimum energy is spin-unpolarized since the energy has a minimum in $x = 0$. However, by increasing r_s, or equivalently by decreasing the density, the system prefers to break the spin symmetry and to relax in a spin-polarized state. The tendency toward spin-polarized solutions at small densities can easily be understood. The exchange interaction between two electrons is negative if the electrons have the same spin and it is zero otherwise [see the discussion below (8.27)]. Consider a spin-unpolarized electron gas and imagine flipping the spin of one electron. Due to the Pauli exclusion principle, particles with the same spin cannot occupy the same level and therefore after the spin flip we must put the electron in a state with momentum greater than p_{F}. If the density is small, this increase in kinetic energy is lower than the exchange energy gain. Then we can lower the energy further by flipping the spin of another electron and then of another one and so on until we start to pay too much kinetic energy. If the fraction of flipped spin is finite, then the more stable solution is spin-polarized. The results in Fig. 8.2 predict that the transition from spin-unpolarized to spin-polarized solutions occurs at a critical value of r_s between 5 and 6 (the precise value is $r_s = 5.45$). This critical value is unphysical since it is in the range of the r_s of ordinary metals (ferromagnetic materials have r_s an order of magnitude larger).

Exercise 8.5 Consider the Hubbard model with only one site of Exercise 8.4. Assuming that the Green's function is diagonal in the spin indices, show that the HF potential is $V_{\text{HF}}(\sigma, z) = U n_{\bar{\sigma}}(z)$, independent of whether one uses interaction (2.36) or (2.37).

9

Kadanoff–Baym Equations

In this chapter we learn how to calculate the Green's function for self-energy approximations beyond Hartree–Fock. We first discuss important properties of the self-energy and then derive the so-called Kadanoff–Baym equations to propagate the Green's function in time. We also generalize the Kadanoff–Baym equations to open systems, thereby extending the range of applicability to the field of quantum transport.

9.1 Self-Energy as a Function in Keldysh Space

To begin with we rewrite (7.18) and (7.19) in a slightly different way:

$$\mathrm{i}\frac{d}{dz_1}G(1;2) - \int d3\; [h(1;3) + \Sigma(1;3)]\, G(3;2) = \delta(1;2), \tag{9.1}$$

$$-\mathrm{i}\frac{d}{dz_2}G(1;2) - \int d3\, G(1;3)\, [h(3;2) + \Sigma(3;2)] = \delta(1;2), \tag{9.2}$$

with $h(1;3) = \langle \mathbf{x}_1|\hat{h}(z_1)|\mathbf{x}_3\rangle\delta(z_1, z_3)$, see (6.4). The self-energy is the most natural object to describe the propagation of a particle in an interacting medium. In fact, Σ adds up to h and their sum defines a sort of *self*-consistent Hamiltonian (or *energy* operator) which accounts for the effects of the interparticle interaction. Since the exact self-energy is nonlocal in time, the energy operator $h+\Sigma$ does not only consider the instantaneous position of the particle, but also where the particle was. These kinds of nonlocal (in time) effects are usually referred to as *retardation effects.*

Let us define the self-energy operator in first quantization, $\hat{\Sigma}$, in a similar way as we have already done for the Green's function and the Hartree–Fock potential:

$$\hat{\Sigma}(z_1, z_2) = \int d\mathbf{x}_1 d\mathbf{x}_2\, |\mathbf{x}_1\rangle\Sigma(1;2)\langle\mathbf{x}_2|.$$

Then, the equations of motion (9.1) and (9.2) are the representation in the position–spin basis of the following equations for operators in first quantization,

$$\left[\mathrm{i}\frac{d}{dz_1} - \hat{h}(z_1)\right]\hat{\mathcal{G}}(z_1, z_2) = \delta(z_1, z_2) + \int_\gamma dz\; \hat{\Sigma}(z_1, z)\hat{\mathcal{G}}(z, z_2), \tag{9.3}$$

$$\hat{\mathcal{G}}(z_1, z_2)\left[-\mathrm{i}\frac{\overleftarrow{d}}{dz_2} - \hat{h}(z_2)\right] = \delta(z_1, z_2) + \int_\gamma dz\, \hat{\mathcal{G}}(z_1, z)\hat{\Sigma}(z, z_2). \tag{9.4}$$

They correctly reduce to the equations of motion for the noninteracting Green's function operator (6.5) and (6.6) when $\hat{\Sigma} = 0$. Equations (9.3) and (9.4) are first-order integro-differential equations (in the contour times) to be solved with the KMS boundary conditions, see (5.7). They contain a convolution on the contour between two functions, G and Σ, in Keldysh space. To extract an equation for the lesser, greater, etc. components, we have to use the Langreth rules. The resulting equations are known as the *Kadanoff–Baym equations*, and we derive them in Section 9.2.

Self-energy and two-particle Green's function Comparing (9.1) and (9.2) with the first equations of the Martin–Schwinger hierarchy, see (5.2) and (5.3), we obtain an important relation between the self-energy and the two-particle Green's function:

$$\int d3\, \Sigma(1;3)G(3;2) = \pm \mathrm{i} \int d3\, v(1;3)G_2(1,3;2,3^+), \tag{9.5}$$

and

$$\int d3\, G(1;3)\Sigma(3;2) = \pm \mathrm{i} \int d3\, G_2(1,3^-;2,3)v(3;2). \tag{9.6}$$

Thus, to any approximate Σ corresponds an approximate G_2, and vice versa, through one of the two equivalent equations (9.5) and (9.6). In Chapter 14 we deepen this connection further.

KMS for the self-energy From (9.5) and (9.6) it also follows that if G_2 satisfies the KMS relations, then the self-energy satisfies the same KMS relations, as G:

$$\hat{\Sigma}(z_1, t_{0-}) = \pm\hat{\Sigma}(z_1, t_0 - \mathrm{i}\beta), \tag{9.7}$$

$$\hat{\Sigma}(t_{0-}, z_2) = \pm\hat{\Sigma}(t_0 - \mathrm{i}\beta, z_2). \tag{9.8}$$

Mathematical structure in Keldysh space The two-particle Green's function with the second and fourth arguments in 3 and 3^+ can be written as

$$G_2(1,3;2,3^+) = \mp\langle\mathcal{T}\left\{\hat{n}_H(3)\hat{\psi}_H(1)\hat{\psi}_H^\dagger(2)\right\}\rangle,$$

where the short-hand notation $\langle\ldots\rangle \equiv \mathrm{Tr}[e^{-\beta\hat{H}^{\mathrm{M}}}\ldots]/\mathrm{Tr}[e^{-\beta\hat{H}^{\mathrm{M}}}]$ denotes the ensemble average. If we define the operators[1]

$$\hat{\gamma}_H(1) \equiv \int d3\, v(1;3)\hat{n}_H(3)\hat{\psi}_H(1),$$

$$\hat{\gamma}_H^\dagger(1) \equiv \int d3\, v(1;3)\hat{\psi}_H^\dagger(1)\hat{n}_H(3),$$

then (9.5) and (9.6) can be rewritten as

$$\int d3\, \Sigma(1;3)G(3;2) = -\mathrm{i}\,\langle\mathcal{T}\left\{\hat{\gamma}_H(1)\hat{\psi}_H^\dagger(2)\right\}\rangle, \tag{9.9}$$

[1]The operator $\hat{\gamma}_H^\dagger$ is the adjoint of $\hat{\gamma}_H$ only for contour arguments on the horizontal branches.

$$\int d3\, G(1;3)\Sigma(3;2) = -\mathrm{i}\,\langle \mathcal{T}\left\{\hat{\psi}_H(1)\hat{\gamma}_H^\dagger(2)\right\}\rangle, \tag{9.10}$$

where we use that in the $\mathcal{T}$ product the field operators (anti)commute. The operators $\hat{\gamma}_H$ and $\hat{\gamma}_H^\dagger$ should not be new to the reader; they also appear in the equations of motion of the field operators and, of course, this is not a coincidence. Let us cast (4.61) in terms of $\hat{\gamma}_H$ and $\hat{\gamma}_H^\dagger$:

$$\int d1 \left[\mathrm{i}\frac{d}{dz_4}\delta(4;1) - h(4;1)\right]\hat{\psi}_H(1) = \hat{\gamma}_H(4), \tag{9.11}$$

$$\int d2\, \hat{\psi}_H^\dagger(2)\left[-\mathrm{i}\delta(2;4)\frac{\overleftarrow{d}}{dz_4} - h(2;4)\right] = \hat{\gamma}_H^\dagger(4). \tag{9.12}$$

We now act on (9.9) from the right with $\overleftarrow{G}_0^{-1}(2;4) = [-\mathrm{i}\delta(2;4)\frac{\overleftarrow{d}}{dz_4} - h(2;4)]$, and integrate over 2.[2] Using (9.2) and (9.12) we get

$$\begin{aligned}\Sigma(1;4) + \int d2d3\, \Sigma(1;3)G(3;2)\Sigma(2;4) &= \delta(z_1,z_4)\langle\left[\hat{\gamma}_H(\mathbf{x}_1,z_1),\hat{\psi}_H^\dagger(\mathbf{x}_4,z_1)\right]_\mp\rangle \\ &\quad - \mathrm{i}\,\langle\mathcal{T}\left\{\hat{\gamma}_H(1)\hat{\gamma}_H^\dagger(4)\right\}\rangle.\end{aligned} \tag{9.13}$$

The (anti)commutator in the first term on the r.h.s. originates from the derivative of the contour θ-functions implicit in the $\mathcal{T}$ product (as always, the upper/lower sign refers to bosons/fermions, respectively). A similar equation can be derived by acting on (9.10) from the left with $\overrightarrow{G}_0^{-1}(4;1) = [\mathrm{i}\frac{d}{dz_4}\delta(4;1) - h(4;1)]$ and integrating over 1. Taking into account (9.1) and (9.11), we get

$$\begin{aligned}\Sigma(4;2) + \int d1d3\, \Sigma(4;1)G(1;3)\Sigma(3;2) &= \delta(z_2,z_4)\langle\left[\hat{\psi}_H(\mathbf{x}_4,z_2),\hat{\gamma}_H^\dagger(\mathbf{x}_2,z_2)\right]_\mp\rangle \\ &\quad - \mathrm{i}\,\langle\mathcal{T}\left\{\hat{\gamma}_H(4)\hat{\gamma}_H^\dagger(2)\right\}\rangle.\end{aligned} \tag{9.14}$$

Comparing (9.13) with (9.14), we find

$$\langle\left[\hat{\gamma}_H(\mathbf{x}_1,z_1),\hat{\psi}_H^\dagger(\mathbf{x}_2,z_1)\right]_\mp - \left[\hat{\psi}_H(\mathbf{x}_1,z_1),\hat{\gamma}_H^\dagger(\mathbf{x}_2,z_1)\right]_\mp\rangle = 0. \tag{9.15}$$

It is instructive to verify this identity independently. Let us calculate the first (anti)commutator on the l.h.s. From definition (4.38) of operators in the contour Heisenberg picture we have

$$\langle\left[\hat{\gamma}_H(\mathbf{x}_1,z_1),\hat{\psi}_H^\dagger(\mathbf{x}_2,z_1)\right]_\mp\rangle = \langle\hat{U}(t_{0-},z_1)\left[\hat{\gamma}(\mathbf{x}_1),\hat{\psi}^\dagger(\mathbf{x}_2)\right]_\mp\hat{U}(z_1,t_{0-})\rangle.$$

Using the identity

$$\left[\hat{A}\hat{B},\hat{C}\right]_\mp = \hat{A}\left[\hat{B},\hat{C}\right]_\mp \pm \left[\hat{A},\hat{C}\right]_-\hat{B},$$

[2] We are here assuming that we can interchange integration and differentiation.

it is a matter of simple algebra to derive

$$\left[\hat{\gamma}(\mathbf{x}_1), \hat{\psi}^\dagger(\mathbf{x}_2)\right]_\mp = \delta(\mathbf{x}_1 - \mathbf{x}_2) \int d\mathbf{x}_3\, v(\mathbf{x}_1, \mathbf{x}_3)\hat{n}(\mathbf{x}_3) \pm v(\mathbf{x}_1, \mathbf{x}_2)\hat{\psi}^\dagger(\mathbf{x}_2)\hat{\psi}(\mathbf{x}_1). \quad (9.16)$$

We leave as an exercise for the reader to prove that the second (anti)commutator in (9.15) yields exactly the same result.

Equation (9.13) is particularly interesting since it shows that the self-energy, as a function in Keldysh space, has the following structure [43]:

$$\hat{\Sigma}(z_1, z_2) = \delta(z_1, z_2)\hat{\Sigma}^\delta(z_1) + \theta(z_1, z_2)\hat{\Sigma}^>(z_1, z_2) + \theta(z_2, z_1)\hat{\Sigma}^<(z_1, z_2), \quad (9.17)$$

where the singular part $\langle \mathbf{x}_1|\hat{\Sigma}^\delta(z_1)|\mathbf{x}_4\rangle$ is given by the first term on the r.h.s. of (9.13). Taking into account definition (8.19) of the Hartree–Fock potential and comparing it with (9.16), we also see that $\hat{\Sigma}^\delta = \hat{V}_{\mathrm{HF}}$. This is the result anticipated at the end of Section 8.1: Any approximation beyond the Hartree–Fock approximation leads to a nonlocal (in the contour times) self-energy.

Correlation self-energy We refer to the nonlocal part of the total self-energy as the *correlation self-energy,*

$$\boxed{\hat{\Sigma}_{\mathrm{c}}(z_1, z_2) \equiv \theta(z_1, z_2)\hat{\Sigma}^>(z_1, z_2) + \theta(z_2, z_1)\hat{\Sigma}^<(z_1, z_2)} \quad (9.18)$$

The correlation self-energy takes into account all effects beyond the Hartree–Fock theory.

Is the self-energy a correlator? Unlike the Green's function, the self-energy does not have a mathematical expression in terms of a contour-ordered average of operators. The mathematical expression for Σ which is closest to a contour-ordered average can be deduced from (9.13). The first term on the r.h.s. of (9.13) is the singular (Hartree–Fock) part of the self-energy. The structure $\Sigma G \Sigma$ is reducible and hence does not belong to Σ. Consequently, the correlation self-energy can be written as

$$\boxed{\Sigma_{\mathrm{c}}(1;2) = -\mathrm{i}\,\langle \mathcal{T}\left\{\hat{\gamma}_H(1)\hat{\gamma}_H^\dagger(2)\right\}\rangle_{\mathrm{irr}}} \quad (9.19)$$

where $\langle \ldots \rangle_{\mathrm{irr}}$ signifies that in the Wick's expansion of the average we only retain those terms whose diagrammatic representation is an irreducible diagram [43].

Exercise 9.1 Evaluate the r.h.s. of (9.19) to second order in the interaction v and show that it is given by the sum of the first and second diagrams in Fig. 7.6 (all other diagrams in the same figure contain a $\delta(z_1, z_2)$ and hence they are part of the singular (Hartree–Fock) self-energy).

9.2 Kadanoff–Baym Equations

To solve the equations of motion (9.3) and (9.4) in practice we need to transform them into ordinary integro-differential equations for quantities with real-time arguments (as opposed to contour-time arguments). This can be done by taking the contour times of the Green's function and self-energy on different branches of the contour and then using the Langreth rules of Table 5.1. The Green's function has no singular contribution (i.e., $G^\delta = 0$), while the self-energy has a singular contribution given by the Hartree–Fock part, see Section 9.1.

Taking both arguments of $\hat{\mathcal{G}}$ on the vertical track, $z_{1,2} = t_0 - \mathrm{i}\tau_{1,2}$, we obtain a pair of integro-differential equations for the Matsubara Green's function:

$$\left[-\frac{d}{d\tau_1} - \hat{h}^{\mathrm{M}}\right] \hat{\mathcal{G}}^{\mathrm{M}}(\tau_1, \tau_2) = \mathrm{i}\delta(\tau_1 - \tau_2) + \left[\hat{\Sigma}^{\mathrm{M}} \star \hat{\mathcal{G}}^{\mathrm{M}}\right](\tau_1, \tau_2), \tag{9.20}$$

$$\hat{\mathcal{G}}^{\mathrm{M}}(\tau_1, \tau_2) \left[\frac{\overleftarrow{d}}{d\tau_2} - \hat{h}^{\mathrm{M}}\right] = \mathrm{i}\delta(\tau_1 - \tau_2) + \left[\hat{\mathcal{G}}^{\mathrm{M}} \star \hat{\Sigma}^{\mathrm{M}}\right](\tau_1, \tau_2). \tag{9.21}$$

At this point it is important to recall the observation we made in Section 5.3. If we are interested in calculating the Green's function with real-time arguments up to a maximum time T, then we do not need a contour γ that goes all the way to ∞. It is enough that γ reaches T. This implies that *if the external vertices of a self-energy diagram are smaller than T, then all integrals over the internal vertices that go from T to ∞ cancel with those from ∞ to T.* Consequently, to calculate $\hat{\Sigma}(z_1, z_2)$ with z_1 and z_2 up to a maximum real-time T, we only need the Green's function $\hat{\mathcal{G}}(z_1, z_2)$ with real times up to T. In particular, to calculate $\hat{\Sigma}^{\mathrm{M}}$ we can set $T = t_0$ and therefore we only need $\hat{\mathcal{G}}(z_1, z_2)$ with arguments on the vertical track – that is, we only need $\hat{\mathcal{G}}^{\mathrm{M}}$. Let us consider, for instance, the diagrams in Fig. 7.7(a); the diagrams in the second row contain the convolution of five Green's functions and for z_1 and z_2 on γ^{M} these diagrams can be written as the convolution along γ^{M} of five Matsubara Green's functions, see (5.39). Another example is the second Born bubble diagram in (7.16) (third diagram). In this case we have to integrate over the entire contour the vertices of the bubble; however, since the interaction (wiggly lines) is local in time, this integral reduces to $G(\mathbf{x}_3, z_1; \mathbf{x}_4, z_2)G(\mathbf{x}_4, z_2; \mathbf{x}_3, z_1)$, which is the product of two Matsubara Green's function when $z_{1,2} = t_0 - \mathrm{i}\tau_{1,2}$. We then conclude that (9.20) and (9.21) constitute a closed system of equations for $\hat{\mathcal{G}}^{\mathrm{M}}$ – that is, there is no mixing between $\hat{\mathcal{G}}^{\mathrm{M}}$ and the other components of the Green's function. We also infer that the two equations are not independent. In fact, the single-particle Hamiltonian $\hat{h}(z)$ is constant along the vertical track and therefore $\hat{\mathcal{G}}^{\mathrm{M}}(\tau_1, \tau_2)$, and consequently $\hat{\Sigma}^{\mathrm{M}}(\tau_1, \tau_2)$, depend only on the time difference $\tau_1 - \tau_2$. Taking into account the KMS relations, we can expand $\hat{\mathcal{G}}^{\mathrm{M}}$ and $\hat{\Sigma}^{\mathrm{M}}$ in Matsubara frequencies:

$$\hat{\mathcal{G}}^{\mathrm{M}}(\tau_1, \tau_2) = \frac{1}{-\mathrm{i}\beta} \sum_{m=-\infty}^{\infty} e^{-\omega_m(\tau_1 - \tau_2)} \hat{\mathcal{G}}^{\mathrm{M}}(\omega_m), \tag{9.22a}$$

$$\hat{\Sigma}^{\mathrm{M}}(\tau_1, \tau_2) = \frac{1}{-\mathrm{i}\beta} \sum_{m=-\infty}^{\infty} e^{-\omega_m(\tau_1 - \tau_2)} \hat{\Sigma}^{\mathrm{M}}(\omega_m), \tag{9.22b}$$

use (6.20) and find[3]

$$\hat{\mathcal{G}}^{\mathrm{M}}(\omega_m) = \frac{1}{\omega_m - \hat{h}^{\mathrm{M}} - \hat{\Sigma}^{\mathrm{M}}(\omega_m)}. \tag{9.23}$$

Therefore, (9.20) and (9.21) can be converted into a system of algebraic equations for the $\hat{\mathcal{G}}^{\mathrm{M}}(\omega_m)$. In general, $\hat{\Sigma}^{\mathrm{M}}(\omega_m)$ depends on all the Matsubara coefficients $\{\hat{\mathcal{G}}^{\mathrm{M}}(\omega_n)\}$. Thus (9.23) is a coupled system of equations for the unknown $\{\hat{\mathcal{G}}^{\mathrm{M}}(\omega_n)\}$. The solution of this system amounts to determining the initial ensemble and it constitutes the preliminary step to solving the equations of motion.

With the Matsubara component at our disposal, we can calculate all other components by time propagation. The equation for the right Green's function is obtained by taking $z_1 = t_-$ or t_+ and $z_2 = t_0 - \mathrm{i}\tau$ in (9.3), and reads

$$\left[\mathrm{i}\frac{d}{dt} - \hat{h}(t)\right]\hat{\mathcal{G}}^{\rceil}(t,\tau) = \left[\hat{\Sigma}^{\mathrm{R}} \cdot \hat{\mathcal{G}}^{\rceil} + \hat{\Sigma}^{\rceil} \star \hat{\mathcal{G}}^{\mathrm{M}}\right](t,\tau) \tag{9.24}$$

Similarly, for $z_1 = t_0 - \mathrm{i}\tau$ and $z_2 = t_-$ or t_+, the equation of motion (9.4) yields

$$\hat{\mathcal{G}}^{\lceil}(\tau,t)\left[-\mathrm{i}\frac{\overleftarrow{d}}{dt} - \hat{h}(t)\right] = \left[\hat{\mathcal{G}}^{\lceil} \cdot \hat{\Sigma}^{\mathrm{A}} + \hat{\mathcal{G}}^{\mathrm{M}} \star \hat{\Sigma}^{\lceil}\right](\tau,t) \tag{9.25}$$

At fixed τ, (9.24) and (9.25) are first-order integro-differential equations in t which must be solved with initial conditions

$$\hat{\mathcal{G}}^{\rceil}(t_0,\tau) = \hat{\mathcal{G}}^{\mathrm{M}}(0,\tau), \qquad \hat{\mathcal{G}}^{\lceil}(\tau,t_0) = \hat{\mathcal{G}}^{\mathrm{M}}(\tau,0). \tag{9.26}$$

The retarded/advanced and the left/right components of the self-energy in (9.24) and (9.25) depend not only on $G^{\rceil}$ and $G^{\lceil}$, but also on the lesser and greater Green's functions. Therefore, (9.24) and (9.25) do not form a closed set of equations for $G^{\rceil}$ and $G^{\lceil}$. To close the set we need the equations of motion for $G^{\lessgtr}$. These can easily be obtained by setting $z_1 = t_{1\pm}$ and $z_2 = t_{2\mp}$ in (9.3) and (9.4):

$$\left[\mathrm{i}\frac{d}{dt_1} - \hat{h}(t_1)\right]\hat{\mathcal{G}}^{\lessgtr}(t_1,t_2) = \left[\hat{\Sigma}^{\lessgtr} \cdot \hat{\mathcal{G}}^{\mathrm{A}} + \hat{\Sigma}^{\mathrm{R}} \cdot \hat{\mathcal{G}}^{\lessgtr} + \hat{\Sigma}^{\rceil} \star \hat{\mathcal{G}}^{\lceil}\right](t_1,t_2) \tag{9.27}$$

$$\hat{\mathcal{G}}^{\lessgtr}(t_1,t_2)\left[-\mathrm{i}\frac{\overleftarrow{d}}{dt_2} - \hat{h}(t_2)\right] = \left[\hat{\mathcal{G}}^{\lessgtr} \cdot \hat{\Sigma}^{\mathrm{A}} + \hat{\mathcal{G}}^{\mathrm{R}} \cdot \hat{\Sigma}^{\lessgtr} + \hat{\mathcal{G}}^{\rceil} \star \hat{\Sigma}^{\lceil}\right](t_1,t_2) \tag{9.28}$$

which must be solved with initial conditions

$$\hat{\mathcal{G}}^{<}(t_0,t_0) = \hat{\mathcal{G}}^{\mathrm{M}}(0,0^+), \qquad \hat{\mathcal{G}}^{>}(t_0,t_0) = \hat{\mathcal{G}}^{\mathrm{M}}(0^+,0). \tag{9.29}$$

[3]Use that for convolutions along the imaginary track γ^{M} of the contour $-\mathrm{i}\int_0^\beta d\tau\, e^{(\omega_m - \omega_{m'})\tau} = -\mathrm{i}\beta\delta_{mm'}$, see also Appendix A.

The set of equations (9.24), (9.25), (9.27), and (9.28) are known as the *Kadanoff–Baym equations* [43, 109]. The Kadanoff–Baym equations, together with the initial conditions (9.26) and (9.29), completely determine the Green's function with one and two real times once a choice for the self-energy is made. Below we derive some exact relations that are of great help in the actual implementation, see Appendix J. Different techniques have been used to solve the Kadanoff–Baym equations; among them we mention time-stepping techniques [113, 114, 115, 116, 118, 119, 120], low-rank compression of the Green's functions and self-energies [121], adaptive time-stepping techniques [122], or piecewise high-order orthogonal-polynomial expansions of the Green's function [123].

From the definition of the Green's function we have derived relation (6.27), which in operator form reads

$$\boxed{\hat{\mathcal{G}}^{\lessgtr}(t_1, t_2) = -\left[\hat{\mathcal{G}}^{\lessgtr}(t_2, t_1)\right]^{\dagger}} \tag{9.30}$$

In a similar way, it is easy to show that (as always, the upper/lower sign refers to bosons/fermions, respectively)

$$\boxed{\hat{\mathcal{G}}^{\lceil}(\tau, t) = \mp\left[\hat{\mathcal{G}}^{\rceil}(t, \beta - \tau)\right]^{\dagger}} \tag{9.31}$$

and

$$\boxed{\hat{\mathcal{G}}^{\mathrm{M}}(\tau_1, \tau_2) = -\left[\hat{\mathcal{G}}^{\mathrm{M}}(\tau_1, \tau_2)\right]^{\dagger}} \tag{9.32}$$

These properties of the Green's function can be transferred directly to the self-energy. Consider, for instance, the adjoint of (9.20). Taking into account that the derivative of $\hat{\mathcal{G}}^{\mathrm{M}}(\tau_1, \tau_2)$ with respect to τ_1 is *minus* the derivative with respect to τ_2, using (9.32) we find

$$-\hat{\mathcal{G}}^{\mathrm{M}}(\tau_1, \tau_2)\left[\frac{\overleftarrow{d}}{d\tau_2} - \hat{h}^{\mathrm{M}}\right] = -\mathrm{i}\delta(\tau_1 - \tau_2) + \mathrm{i}\int_0^{\beta} d\tau \left(-\hat{\mathcal{G}}^{\mathrm{M}}(\tau, \tau_2)\right)\left[\hat{\Sigma}^{\mathrm{M}}(\tau_1, \tau)\right]^{\dagger}.$$

We rename the integration variable $\tau \to \tau_1 + \tau_2 - \tau$. Then the argument of the Green's function becomes $\tau_1 - \tau$ while the argument of the self-energy becomes $\tau - \tau_2$ (recall that these quantities depend only on the time difference). The domain of integration remains $(0, \beta)$ since G and Σ are (anti)periodic. Comparing the resulting equation with (9.21), we conclude that

$$\int_0^{\beta} d\tau\, \hat{\mathcal{G}}^{\mathrm{M}}(\tau_1, \tau)\hat{\Sigma}^{\mathrm{M}}(\tau, \tau_2) = -\int_0^{\beta} d\tau\, \hat{\mathcal{G}}^{\mathrm{M}}(\tau_1, \tau)\left[\hat{\Sigma}^{\mathrm{M}}(\tau, \tau_2)\right]^{\dagger}.$$

This identity must be true for all τ_1 and τ_2 and since $\hat{\mathcal{G}}^{\mathrm{M}}$ is invertible (i.e., $\hat{\mathcal{G}}^{\mathrm{M}} \star \hat{\Sigma}^{\mathrm{M}} = 0$ only for $\hat{\Sigma}^{\mathrm{M}} = 0$), it follows that[4]

$$\boxed{\hat{\Sigma}^{\mathrm{M}}(\tau_1, \tau_2) = -\left[\hat{\Sigma}^{\mathrm{M}}(\tau_1, \tau_2)\right]^{\dagger}} \tag{9.33}$$

[4]We could have also derived (9.33) directly from (9.23).

Vice versa, if the self-energy satisfies (9.33), then the Matsubara Green's function satisfies (9.32). In a similar way we can deduce the relations between the left, right, lesser, and greater self-energy and the corresponding adjoint quantities. In this case, however, we must use the four coupled equations (9.24), (9.25), (9.27), and (9.28) at the same time. The derivation is a bit lengthier, but the final result is predictable. We therefore leave it as an exercise for the reader to prove that

$$\boxed{\hat{\Sigma}^{\lessgtr}(t_1, t_2) = -\left[\hat{\Sigma}^{\lessgtr}(t_2, t_1)\right]^{\dagger}} \tag{9.34}$$

$$\boxed{\hat{\Sigma}^{\lceil}(\tau, t) = \mp\left[\hat{\Sigma}^{\rceil}(t, \beta - \tau)\right]^{\dagger}} \tag{9.35}$$

In particular, (9.34) implies

$$\boxed{\hat{\Sigma}^{\mathrm{R}}(t_1, t_2) = \left[\hat{\Sigma}^{\mathrm{A}}(t_2, t_1)\right]^{\dagger}} \tag{9.36}$$

This last relation could have been deduced also from the retarded/advanced component of the Dyson equation (7.11) together with property (6.28). To summarize, the self-energy has the same properties as the Green's function under complex conjugation.

The r.h.s. of the Kadanoff–Baym equations is a Keldysh component of either the convolution

$$\hat{\mathcal{I}}_L(z_1, z_2) \equiv \int_{\gamma} dz\, \hat{\Sigma}(z_1, z)\hat{\mathcal{G}}(z, z_2),$$

or the convolution

$$\hat{\mathcal{I}}_R(z_1, z_2) \equiv \int_{\gamma} dz\, \hat{\mathcal{G}}(z_1, z)\hat{\Sigma}(z, z_2).$$

The quantities $\hat{\mathcal{I}}_L$ and $\hat{\mathcal{I}}_R$ belong to the Keldysh space and are usually referred to as the *collision integrals* since they contain information on how the particles scatter. From the above relations for G and Σ it follows that

$$\hat{\mathcal{I}}_L^{\lceil}(\tau, t) = \mp\left[\hat{\mathcal{I}}_R^{\rceil}(t, \beta - \tau)\right]^{\dagger}, \tag{9.37}$$

$$\hat{\mathcal{I}}_L^{\lessgtr}(t_1, t_2) = -\left[\hat{\mathcal{I}}_R^{\lessgtr}(t_2, t_1)\right]^{\dagger}. \tag{9.38}$$

Therefore, it is sufficient to calculate, say, $\hat{\mathcal{I}}_L^{<}$, $\hat{\mathcal{I}}_R^{>}$, and $\hat{\mathcal{I}}_L^{\lceil}$ in order to solve the Kadanoff–Baym equations.

A final remark pertains to the solution of the Kadanoff–Baym equations for systems in equilibrium. We know that in this case $\hat{\mathcal{G}}^{\lessgtr}$, and hence also $\hat{\Sigma}^{\lessgtr}$, depends on the time difference only. However, it is not obvious that $\hat{\mathcal{G}}^{\lessgtr}(t_1 - t_2)$ can be a solution of (9.27) and (9.28) since neither the two real-time convolutions between t_0 and ∞ nor the convolution along the vertical track depend on $t_1 - t_2$. Is there anything wrong with these equations? In Section 10.5 we answer this question and show how to recover the equilibrium results.

Exercise 9.2 Prove (9.31) and (9.32).

Exercise 9.3 Prove (9.34) and (9.35).

9.3 Formal Solution of the Kadanoff–Baym Equations

The scope of solving the Kadanoff–Baym equations is to obtain the lesser and greater Green's functions from which to calculate, for example, the time-dependent ensemble average of any one-body operator, the time-dependent total energy, the addition and removal energies, etc. In this section we generalize the formal solution (6.49) to interacting systems using the Dyson equation (i.e., the integral form of the Kadanoff–Baym equations). From a practical point of view it is much more advantageous to solve the Kadanoff–Baym equations rather than the integral Dyson equation. However, there exist situations for which the formal solution simplifies considerably. As we shall see, the simplified solution is extremely useful for deriving analytic results and/or setting up numerical algorithms for calculating $\hat{\mathcal{G}}^{\lessgtr}$ without explicitly propagating the Green's function in time.

We follow the derivation of Ref. [59]. The starting point is the Dyson equation (7.11) for the Green's function that we rewrite below in the operator form:

$$\hat{\mathcal{G}}(z,z') = \hat{\mathcal{G}}_0(z,z') + \int_\gamma d\bar{z}d\bar{z}'\,\hat{\mathcal{G}}(z,\bar{z})\hat{\Sigma}(\bar{z},\bar{z}')\hat{\mathcal{G}}_0(\bar{z}',z'), \tag{9.39}$$

where $\hat{\mathcal{G}}_0$ is the noninteracting Green's function. We separate the self-energy $\hat{\Sigma} = \hat{\Sigma}_{\mathrm{HF}} + \hat{\Sigma}_{\mathrm{c}}$ into the Hartree–Fock self-energy and the correlation self-energy (9.18). Then, the Dyson equation can be rewritten as

$$\hat{\mathcal{G}}(z,z') = \hat{\mathcal{G}}_{\mathrm{HF}}(z,z') + \int_\gamma d\bar{z}d\bar{z}'\,\hat{\mathcal{G}}(z,\bar{z})\hat{\Sigma}_{\mathrm{c}}(\bar{z},\bar{z}')\hat{\mathcal{G}}_{\mathrm{HF}}(\bar{z}',z'), \tag{9.40}$$

with $\hat{\mathcal{G}}_{\mathrm{HF}}$ the Hartree-Fock Green's function. The equivalence between these two forms of the Dyson equation can easily be verified by acting on (9.39) from the right with $[-\mathrm{i}\frac{\overleftarrow{d}}{dz'} - \hat{h}(z')]$ and on (9.40) from the right with $[-\mathrm{i}\frac{\overleftarrow{d}}{dz'} - \hat{h}_{\mathrm{HF}}(z')]$. In both cases the result is the equation of motion (9.4). The advantage of using (9.40) is that the correlation self-energy is nonlocal in time and, in macroscopic interacting systems, often decays to zero when the separation between its time arguments approaches infinity. This circumstance is at the basis of an important simplification in the long-time limit.

Using the Langreth rules, the lesser Green's function reads

$$\hat{\mathcal{G}}^< = \left[\delta + \hat{\mathcal{G}}^{\mathrm{R}} \cdot \hat{\Sigma}_{\mathrm{c}}^{\mathrm{R}}\right] \cdot \hat{\mathcal{G}}_{\mathrm{HF}}^< + \hat{\mathcal{G}}^< \cdot \hat{\Sigma}_{\mathrm{c}}^{\mathrm{A}} \cdot \hat{\mathcal{G}}_{\mathrm{HF}}^{\mathrm{A}} + \left[\hat{\mathcal{G}}^{\mathrm{R}} \cdot \hat{\Sigma}_{\mathrm{c}}^< + \hat{\mathcal{G}}^{\rceil} \star \hat{\Sigma}_{\mathrm{c}}^{\lceil}\right] \cdot \hat{\mathcal{G}}_{\mathrm{HF}}^{\mathrm{A}}$$
$$+ \hat{\mathcal{G}}^{\mathrm{R}} \cdot \hat{\Sigma}_{\mathrm{c}}^{\rceil} \star \hat{\mathcal{G}}_{\mathrm{HF}}^{\lceil} + \hat{\mathcal{G}}^{\rceil} \star \hat{\Sigma}_{\mathrm{c}}^{\mathrm{M}} \star \hat{\mathcal{G}}_{\mathrm{HF}}^{\lceil},$$

and solving for $\hat{\mathcal{G}}^<$

$$\hat{\mathcal{G}}^< = \left[\delta + \hat{\mathcal{G}}^{\mathrm{R}} \cdot \hat{\Sigma}_{\mathrm{c}}^{\mathrm{R}}\right] \cdot \hat{\mathcal{G}}_{\mathrm{HF}}^< \cdot \left[\delta + \hat{\Sigma}_{\mathrm{c}}^{\mathrm{A}} \cdot \hat{\mathcal{G}}^{\mathrm{A}}\right] + \left[\hat{\mathcal{G}}^{\mathrm{R}} \cdot \hat{\Sigma}_{\mathrm{c}}^< + \hat{\mathcal{G}}^{\rceil} \star \hat{\Sigma}_{\mathrm{c}}^{\lceil}\right] \cdot \hat{\mathcal{G}}^{\mathrm{A}}$$
$$+ \left[\hat{\mathcal{G}}^{\mathrm{R}} \cdot \hat{\Sigma}_{\mathrm{c}}^{\rceil} \star \hat{\mathcal{G}}_{\mathrm{HF}}^{\lceil} + \hat{\mathcal{G}}^{\rceil} \star \hat{\Sigma}_{\mathrm{c}}^{\mathrm{M}} \star \hat{\mathcal{G}}_{\mathrm{HF}}^{\lceil}\right] \cdot \left[\delta + \hat{\Sigma}_{\mathrm{c}}^{\mathrm{A}} \cdot \hat{\mathcal{G}}^{\mathrm{A}}\right].$$

In obtaining this result we use the identity

$$\left[\delta - \hat{\Sigma}_{\mathrm{c}}^{\mathrm{A}} \cdot \hat{\mathcal{G}}^{\mathrm{A}}\right] \cdot \left[\delta + \hat{\Sigma}_{\mathrm{c}}^{\mathrm{A}} \cdot \hat{\mathcal{G}}_{\mathrm{HF}}^{\mathrm{A}}\right] = \delta,$$

which is a direct consequence of the advanced Dyson equation. Next we observe that (6.49) is valid for both noninteracting and mean-field Green's functions, and hence

$$\hat{\mathcal{G}}_{\mathrm{HF}}^<(t,t') = \hat{\mathcal{G}}_{\mathrm{HF}}^{\mathrm{R}}(t,t_0)\hat{\mathcal{G}}_{\mathrm{HF}}^<(t_0,t_0)\hat{\mathcal{G}}_{\mathrm{HF}}^{\mathrm{A}}(t_0,t'), \tag{9.41}$$

and similarly from (6.51),

$$\hat{\mathcal{G}}_{\mathrm{HF}}^{\lceil}(\tau,t') = -\mathrm{i}\hat{\mathcal{G}}_{\mathrm{HF}}^{\mathrm{M}}(\tau,0)\hat{\mathcal{G}}_{\mathrm{HF}}^{\mathrm{A}}(t_0,t').$$

Therefore,

$$\hat{\mathcal{G}}^<(t,t') = \hat{\mathcal{G}}^{\mathrm{R}}(t,t_0)\hat{\mathcal{G}}_{\mathrm{HF}}^<(t_0,t_0)\hat{\mathcal{G}}^{\mathrm{A}}(t_0,t') + \left[\hat{\mathcal{G}}^{\mathrm{R}} \cdot \hat{\Sigma}_{\mathrm{c}}^< \cdot \hat{\mathcal{G}}^{\mathrm{A}}\right](t,t')$$
$$+ \left[\hat{\mathcal{G}}^{\rceil} \star \hat{\Sigma}_{\mathrm{c}}^{\lceil} \cdot \hat{\mathcal{G}}^{\mathrm{A}}\right](t,t') - \mathrm{i}\left[\hat{\mathcal{G}}^{\mathrm{R}} \cdot \hat{\Sigma}_{\mathrm{c}}^{\rceil} \star \hat{\mathcal{G}}_{\mathrm{HF}}^{\mathrm{M}} + \hat{\mathcal{G}}^{\rceil} \star \hat{\Sigma}_{\mathrm{c}}^{\mathrm{M}} \star \hat{\mathcal{G}}_{\mathrm{HF}}^{\mathrm{M}}\right](t,t_0)\hat{\mathcal{G}}^{\mathrm{A}}(t_0,t').$$

In order to eliminate the right Green's function from this equation we use again the Dyson equation. The right component of (9.40) yields

$$\hat{\mathcal{G}}^{\rceil} \star \left[\delta - \hat{\Sigma}_{\mathrm{c}}^{\mathrm{M}} \star \hat{\mathcal{G}}_{\mathrm{HF}}^{\mathrm{M}}\right] = \left[\delta + \hat{\mathcal{G}}^{\mathrm{R}} \cdot \hat{\Sigma}_{\mathrm{c}}^{\mathrm{R}}\right] \cdot \hat{\mathcal{G}}_{\mathrm{HF}}^{\rceil} + \hat{\mathcal{G}}^{\mathrm{R}} \cdot \hat{\Sigma}_{\mathrm{c}}^{\rceil} \star \hat{\mathcal{G}}_{\mathrm{HF}}^{\mathrm{M}}.$$

Using the identity

$$\left[\delta - \hat{\Sigma}_{\mathrm{c}}^{\mathrm{M}} \star \hat{\mathcal{G}}^{\mathrm{M}}\right] \cdot \left[\delta + \hat{\Sigma}_{\mathrm{c}}^{\mathrm{M}} \star \hat{\mathcal{G}}_{\mathrm{HF}}^{\mathrm{M}}\right] = \delta,$$

which follows from the Matsubara–Dyson equation, as well as (6.50) for Hartree–Fock Green's functions,

$$\hat{\mathcal{G}}_{\mathrm{HF}}^{\rceil}(t,\tau) = \mathrm{i}\hat{\mathcal{G}}_{\mathrm{HF}}^{\mathrm{R}}(t,t_0)\hat{\mathcal{G}}_{\mathrm{HF}}^{\mathrm{M}}(0,\tau),$$

we find

$$\hat{\mathcal{G}}^{\rceil}(t,\tau) = \mathrm{i}\hat{\mathcal{G}}^{\mathrm{R}}(t,t_0)\hat{\mathcal{G}}^{\mathrm{M}}(0,\tau) + \left[\hat{\mathcal{G}}^{\mathrm{R}} \cdot \hat{\Sigma}_{\mathrm{c}}^{\rceil} \star \hat{\mathcal{G}}^{\mathrm{M}}\right](t,\tau). \tag{9.42}$$

Substituting this result into the equation for $\hat{\mathcal{G}}^<$, we obtain the generalization of (6.49):

$$\boxed{\begin{aligned}\hat{\mathcal{G}}^<(t,t') &= \hat{\mathcal{G}}^{\mathrm{R}}(t,t_0)\hat{\mathcal{G}}^<(t_0,t_0)\hat{\mathcal{G}}^{\mathrm{A}}(t_0,t') + \left[\hat{\mathcal{G}}^{\mathrm{R}} \cdot \left(\hat{\Sigma}_{\mathrm{c}}^< + \hat{\Sigma}_{\mathrm{c}}^{\rceil} \star \hat{\mathcal{G}}^{\mathrm{M}} \star \hat{\Sigma}_{\mathrm{c}}^{\lceil}\right) \cdot \hat{\mathcal{G}}^{\mathrm{A}}\right] \\ &+ \mathrm{i}\hat{\mathcal{G}}^{\mathrm{R}}(t,t_0)\left[\hat{\mathcal{G}}^{\mathrm{M}} \star \hat{\Sigma}_{\mathrm{c}}^{\lceil} \cdot \hat{\mathcal{G}}^{\mathrm{A}}\right](t_0,t') - \mathrm{i}\left[\hat{\mathcal{G}}^{\mathrm{R}} \cdot \hat{\Sigma}_{\mathrm{c}}^{\rceil} \star \hat{\mathcal{G}}^{\mathrm{M}}\right](t,t_0)\hat{\mathcal{G}}^{\mathrm{A}}(t_0,t')\end{aligned}} \tag{9.43}$$

The formula for the greater Green's function is identical to (9.43), but with all the superscripts "<" replaced by ">". For $\hat{\Sigma}_c = 0$ only the first term survives, in agreement with the fact that in this case the Green's function is a mean-field Green's function, see identity (9.41). For nonvanishing correlation self-energies all terms must be retained.

It is important to clarify a point about the formal solution (9.43). Suppose that we have solved the equilibrium problem and hence that we know $\hat{\mathcal{G}}^{\mathrm{M}}$. Taking into account that $\hat{\mathcal{G}}^{\mathrm{R/A}}$ is defined in terms of $\hat{\mathcal{G}}^{\lessgtr}$, couldn't we use (9.42), (9.43), and the analogous equations for $\hat{\mathcal{G}}^{\lceil}$ and $\hat{\mathcal{G}}^{>}$ to determine $\hat{\mathcal{G}}^{\lessgtr}$, $\hat{\mathcal{G}}^{\rceil}$, and $\hat{\mathcal{G}}^{\lceil}$? If the answer was positive we would bump into a serious conundrum since these equations do not know anything about the Hamiltonian at positive times! In other words, we would have the same Green's function independently of the external fields. Thus the answer to the above question must be negative and (9.42), (9.43), and the like for $\hat{\mathcal{G}}^{\lceil}$ and $\hat{\mathcal{G}}^{>}$ cannot be independent from one another. We can easily show that this is the case in a system of noninteracting particles. Here we have

$$\hat{\mathcal{G}}^{\lessgtr}(t,t') = -[\hat{\mathcal{G}}^{>}(t,t_0) - \hat{\mathcal{G}}^{<}(t,t_0)]\,\hat{\mathcal{G}}^{\lessgtr}(t_0,t_0)\,[\hat{\mathcal{G}}^{>}(t_0,t') - \hat{\mathcal{G}}^{<}(t_0,t')],$$

where we use $\hat{\mathcal{G}}^{\mathrm{R}}(t,t_0) = \hat{\mathcal{G}}^{>}(t,t_0) - \hat{\mathcal{G}}^{<}(t,t_0)$ and $\hat{\mathcal{G}}^{\mathrm{A}}(t_0,t') = -[\hat{\mathcal{G}}^{>}(t_0,t') - \hat{\mathcal{G}}^{<}(t_0,t')]$. Now consider the above equation with, for example, $t' = t_0$:

$$\hat{\mathcal{G}}^{<}(t,t_0) = \mathrm{i}\,[\hat{\mathcal{G}}^{>}(t,t_0) - \hat{\mathcal{G}}^{<}(t,t_0)]\,\hat{\mathcal{G}}^{<}(t_0,t_0),$$

$$\hat{\mathcal{G}}^{>}(t,t_0) = \mathrm{i}\,[\hat{\mathcal{G}}^{>}(t,t_0) - \hat{\mathcal{G}}^{<}(t,t_0)]\,\hat{\mathcal{G}}^{>}(t_0,t_0),$$

where we take into account that $\hat{\mathcal{G}}^{>}(t_0,t_0) - \hat{\mathcal{G}}^{<}(t_0,t_0) = -\mathrm{i}$. This is a system of coupled equations for $\hat{\mathcal{G}}^{<}(t,t_0)$ and $\hat{\mathcal{G}}^{>}(t,t_0)$. We see, however, that the system admits infinitely many solutions since subtracting the first equation from the second one we get a trivial identity. We conclude that the set of equations (9.42), (9.43), and the like for $\hat{\mathcal{G}}^{\lceil}$ and $\hat{\mathcal{G}}^{>}$ is not sufficient to calculate the Green's function. To form a complete system of equations we must include the retarded or advanced Dyson equation: $\hat{\mathcal{G}}^{\mathrm{R/A}} = \hat{\mathcal{G}}^{\mathrm{R/A}}_{\mathrm{HF}} + \hat{\mathcal{G}}^{\mathrm{R/A}}_{\mathrm{HF}} \cdot \hat{\Sigma}^{\mathrm{R/A}}_{\mathrm{c}} \cdot \hat{\mathcal{G}}^{\mathrm{R/A}}$. This equation depends on the external fields through the Hartree–Fock Green's function and, therefore, it is certainly independent from the other equations.

Long-time limit There exist special circumstances in which a considerable simplification occurs. Suppose that we are interested in the behavior of $\hat{\mathcal{G}}^{<}$ for times t, t' much larger than t_0. In most macroscopic interacting systems the memory carried by $\hat{\Sigma}_c$ is lost when the separation between the time arguments increases. If so, then also the Green's function vanishes in the same limit. This means that initial correlations and initial-state dependencies are washed out in the long-time limit, and only one term remains in (9.43):

$$\lim_{t,t'\to\infty} \hat{\mathcal{G}}^{\lessgtr}(t,t') = \left[\hat{\mathcal{G}}^{\mathrm{R}} \cdot \hat{\Sigma}^{\lessgtr}_{\mathrm{c}} \cdot \hat{\mathcal{G}}^{\mathrm{A}}\right](t,t'). \tag{9.44}$$

We shall demonstrate that this is an *exact* result *for all* times t, t' and for both finite and macroscopic systems provided we are in thermal equilibrium. In other words, for systems in thermal equilibrium, all terms in (9.43) vanish except for the convolution $\hat{\mathcal{G}}^{\mathrm{R}} \cdot \hat{\Sigma}^{\lessgtr}_{\mathrm{c}} \cdot \hat{\mathcal{G}}^{\mathrm{A}}$.

Out of equilibrium this simplification occurs only for long times and only for systems with a decaying memory kernel (self-energy).[5] We further observe that if the Hamiltonian $\hat{H}(t)$ becomes independent of time when $t \to \infty$, then it is reasonable to expect that the Green's function and self-energy depend only on the time difference. In this case the nonequilibrium $\hat{\mathcal{G}}$ and $\hat{\Sigma}$ can be Fourier transformed and the convolution in (9.44) becomes a simple product in frequency space:

$$\hat{\mathcal{G}}^{\lessgtr}(\omega) = \hat{\mathcal{G}}^{\rm R}(\omega)\hat{\Sigma}_{\rm c}^{\lessgtr}(\omega)\hat{\mathcal{G}}^{\rm A}(\omega). \tag{9.45}$$

This result allows us to calculate steady-state quantities without solving the Kadanoff–Baym equations. Indeed, under the above simplifying assumptions the correlation self-energy with real times can be written solely in terms of the Green's function with real times. Consider a generic self-energy diagram with external vertices on the forward or backward branches at a distance t and t' from the origin. If both t and t' tend to infinity, then all internal convolutions along the imaginary track tend to zero. In this limit, $\hat{\Sigma}_{\rm c}^{\lessgtr}$ depends only on $\hat{\mathcal{G}}^{<}$ and $\hat{\mathcal{G}}^{>}$.[6] Then (9.45) together with the retarded or advanced Dyson equation,

$$\hat{\mathcal{G}}^{\rm R/A}(\omega) = \hat{\mathcal{G}}_{\rm HF}^{\rm R/A}(\omega) + \hat{\mathcal{G}}_{\rm HF}^{\rm R/A}(\omega)\hat{\Sigma}_{\rm c}^{\rm R/A}(\omega)\hat{\mathcal{G}}^{\rm R/A}(\omega), \tag{9.46}$$

constitute a closed set of coupled equations for $\hat{\mathcal{G}}^{<}(\omega)$ and $\hat{\mathcal{G}}^{>}(\omega)$ to be solved self-consistently.

9.4 Kadanoff–Baym Equations for Open Systems

During the last decades, the size of electronic circuits has continuously been reduced. Today, systems such as quantum wires and quantum dots are routinely produced on the nanometer scale. The seemingly ultimate limit of miniaturization has been achieved by several experimental groups who were able to place single molecules between two macroscopic electrodes [124, 125, 126, 127, 128]. In this section we apply the Kadanoff–Baym equations to finite *interacting* systems possibly connected to *noninteracting* macroscopic reservoirs. An example of this kind of system is the Anderson model discussed in Section 2.4. No complication arises if the interaction in the reservoirs is treated at the Hartree level, see below.

The general Hamiltonian that we have in mind is

$$\hat{H} = \hat{H}_0 + \hat{H}_{\rm int},$$

where the noninteracting part reads

$$\hat{H}_0 = \underbrace{\sum_{\substack{k\alpha\\ \sigma}} \epsilon_{k\alpha}\hat{n}_{k\alpha\sigma}}_{\text{reservoirs}} + \underbrace{\sum_{\substack{mn\\ \sigma}} T_{mn}\hat{d}^\dagger_{m\sigma}\hat{d}_{n\sigma}}_{\text{molecule}} + \underbrace{\sum_{\substack{m,k\alpha\\ \sigma}} \left(T_{m\,k\alpha}\hat{d}^\dagger_{m\sigma}\hat{d}_{k\alpha\sigma} + T_{k\alpha\,m}\hat{d}^\dagger_{k\alpha\sigma}\hat{d}_{m\sigma}\right)}_{\text{coupling}}. \tag{9.47}$$

---

[5] The same reasoning can also be applied to special matrix elements of the $\hat{\mathcal{G}}^{\lessgtr}$ of macroscopic noninteracting systems like the $G_{00}^{\lessgtr}$ of the noninteracting Anderson model, where the embedding self-energy plays the role of the correlation self-energy, see Section 6.1.1.

[6] The dependence is, of course, determined by the diagrammatic approximation to the self-energy.

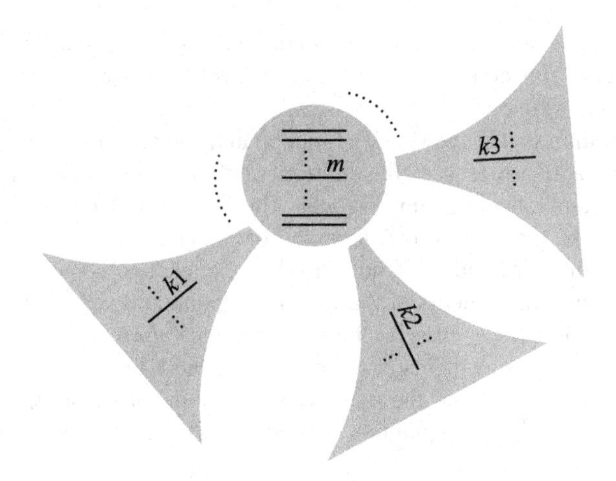

Figure 9.1 Schematic representation of the molecule in contact with many reservoirs.

The index $k\alpha$ refers to the kth eigenfunction of the αth reservoir, whereas the indices m, n refer to basis functions of the finite system which from now on we call "the molecule," see Fig. 9.1; the index σ refers to the spin degree of freedom. As usual, $\hat{n}_{s\sigma} = \hat{d}^\dagger_{s\sigma}\hat{d}_{s\sigma}$ is the occupation number operator for the basis function $s = k\alpha$ or $s = m$. The interacting part of the Hamiltonian is taken to be of the general form

$$\hat{H}_{\text{int}} = \frac{1}{2}\sum_{\substack{ijmn\\ \sigma\sigma'}} v_{ijmn}\hat{d}^\dagger_{i\sigma}\hat{d}^\dagger_{j\sigma'}\hat{d}_{m\sigma'}\hat{d}_{n\sigma}. \tag{9.48}$$

The sum is restricted to indices in the molecule so that there is no interaction between a particle in the molecule and a particle in the reservoirs or between two particles in the reservoirs.[7]

The system in Fig. 9.1 is initially in thermal equilibrium at inverse temperature β and chemical potential μ. The initial density matrix is then $\hat{\rho} = \exp[-\beta(\hat{H} - \mu\hat{N})]/Z$, which corresponds to having $\hat{H}^{\text{M}} = \hat{H} - \mu\hat{N}$ on the vertical track of the contour. At a certain time t_0 we drive the system out of equilibrium by changing the parameters of the Hamiltonian. We could change the interaction or the couplings between the molecule and the reservoirs or the energies $\epsilon_{k\alpha}$. The formalism is so general that we can deal with all sorts of combinations. The Hamiltonian at times $t > t_0$ is therefore given by $\hat{H}$, in which the parameters are time-dependent – that is, $\epsilon_{k\alpha} \to \epsilon_{k\alpha}(t)$, $T_{m\,k\alpha} \to T_{m\,k\alpha}(t)$, $v_{ijmn} \to v_{ijmn}(t)$, etc. The goal is to calculate the Green's function from the Kadanoff–Baym equations and then to extract quantities of physical interest, such as the density, current, screened interaction, polarization, nonequilibrium spectral functions, etc. This is in general a very difficult task since the system of Fig. 9.1 has infinitely many degrees of freedom and does not have

[7]The form (9.48) of the interaction is identical to (1.90), and it should not be confused with the interaction in (1.84) where the spin indices are not spelled out. Thus, the i, j, m, n indices in this section are orbital indices, as opposed to those in Section 1.6, which were orbital-spin indices.

any special symmetry.[8] However, for our particular choice of the interaction an important simplification occurs: The equations to be solved involve only a finite number of matrix elements of the Green's function.

The first observation is that the Green's function and the self-energy are diagonal in spin space since the interaction preserves the spin orientation at every vertex and the noninteracting part of the Hamiltonian is spin diagonal. The inclusion of terms in the Hamiltonian which flip the spin, like the spin–orbit interaction or the coupling with a noncollinear magnetic field, is straightforward and does not introduce conceptual complications. We denote by G_{sr} and Σ_{sr} the matrix elements of the Green's function and self-energy for a given spin projection, where the indices s, r belong to either the reservoirs or the molecule. Similarly, we denote by h_{sr} the matrix elements of the first quantized Hamiltonian $\hat{h}$. The matrices G, Σ, and h are the representation of the operators (in first quantization) $\hat{\mathcal{G}}$, $\hat{\Sigma}$, and $\hat{h}$, in the one-particle basis $\{k\alpha, m\}$. The equations of motion (9.3) and (9.4) for the Green's function matrix G read

$$\left[\mathrm{i}\frac{d}{dz} - h(z)\right] G(z,z') = \delta(z,z') + \int_\gamma d\bar{z}\; \Sigma(z,\bar{z})G(\bar{z},z'), \tag{9.49a}$$

$$G(z,z')\left[-\mathrm{i}\frac{\overleftarrow{d}}{dz'} - h(z')\right] = \delta(z,z') + \int_\gamma d\bar{z}\; G(z,\bar{z})\Sigma(\bar{z},z'), \tag{9.49b}$$

and must be solved with the KMS boundary conditions $G(t_{0-}, z') = -G(t_0 - \mathrm{i}\beta, z')$ and the like for the second argument.

At this point we find it convenient to introduce the "blocks" forming the matrices h, G, and Σ. A block of h, G, and Σ is the projection of these matrices onto the subspace of the reservoirs or the molecule. Thus, for instance, $h_{\alpha\alpha'}$ is the block of h with matrix elements $[h_{\alpha\alpha'}]_{kk'} = \delta_{\alpha\alpha'}\delta_{kk'}\epsilon_{k\alpha}$, $h_{\alpha M}$ is the block of h with matrix elements $[h_{\alpha M}]_{km} = T_{k\alpha\, m}$, and h_{MM} is the block of h with matrix elements $[h_{MM}]_{mn} = T_{mn}$. If we number the reservoirs as $\alpha = 1, 2, 3 \ldots$, the block form of matrix h is

$$h = \begin{pmatrix} h_{11} & 0 & 0 & \ldots & h_{1M} \\ 0 & h_{22} & 0 & \ldots & h_{2M} \\ 0 & 0 & h_{33} & \ldots & h_{3M} \\ \vdots & \vdots & \vdots & \vdots & \vdots \\ h_{M1} & h_{M2} & h_{M3} & \ldots & h_{MM} \end{pmatrix},$$

where each entry (including the 0s) is a block matrix. For the time being we do not specify the number of reservoirs. Similarly to h, we can write the matrices G and Σ in a block form. Matrix G has nonvanishing blocks everywhere. Physically there is no reason to expect that $G_{\alpha\alpha'} \propto \delta_{\alpha\alpha'}$, since an electron is free to go from reservoir α to reservoir α' (passing through the molecule). To the contrary, the self-energy Σ *has only one nonvanishing block,* which is Σ_{MM}. This is a direct consequence of the fact that the diagrammatic expansion of the self-energy starts and ends with an interaction line, which in our case is confined

[8] Continuous symmetries of the Hamiltonian can simplify the calculations considerably. In the electron gas the invariance under translations allows for deriving several analytic results.

to the molecular region. This also implies that $\Sigma_{MM}[G_{MM}]$ is a functional of G_{MM} only. The fact that Σ_{MM} depends only on G_{MM} is very important since it allows us to close the equations of motion for the molecular Green's function.

Let us project the equation of motion (9.49a) onto regions MM and αM:

$$\left[\mathrm{i}\frac{d}{dz} - h_{MM}(z)\right] G_{MM}(z, z') = \delta(z, z') + \sum_{\alpha} h_{M\alpha}(z) G_{\alpha M}(z, z') + \int_{\gamma} d\bar{z}\, \Sigma_{MM}(z, \bar{z}) G_{MM}(\bar{z}, z'), \tag{9.50}$$

$$\left[\mathrm{i}\frac{d}{dz} - h_{\alpha\alpha}(z)\right] G_{\alpha M}(z, z') = h_{\alpha M}(z) G_{MM}(z, z').$$

The latter equation can be solved for $G_{\alpha M}$ and yields

$$G_{\alpha M}(z, z') = \int_{\gamma} d\bar{z}\, g_{\alpha\alpha}(z, \bar{z}) h_{\alpha M}(\bar{z}) G_{MM}(\bar{z}, z'), \tag{9.51}$$

where we define the Green's function $g_{\alpha\alpha}$ of the isolated αth reservoir as the solution of

$$\left[\mathrm{i}\frac{d}{dz} - h_{\alpha\alpha}(z)\right] g_{\alpha\alpha}(z, z') = \delta(z, z'),$$

with KMS boundary conditions. Again we stress the importance of solving the equation for $g_{\alpha\alpha}$ with KMS boundary conditions so that $G_{\alpha M}$ too fulfills the KMS relations. Any other boundary conditions for $g_{\alpha\alpha}$ would lead to unphysical time-dependent results. It is also important to realize that $g_{\alpha\alpha}$ is the Green's function of a noninteracting system and is therefore very easy to calculate. In particular, since the block Hamiltonian $h_{\alpha\alpha}(z)$ is diagonal with matrix elements $\epsilon_{k\alpha}(z)$, we simply have $[g_{\alpha\alpha'}]_{kk'}(z, z') = \delta_{\alpha\alpha'}\delta_{kk'} g_{k\alpha}(z, z')$ with

$$g_{k\alpha}(z, z') = -\mathrm{i}\left[\theta(z, z')\bar{f}(\epsilon^{\mathrm{M}}_{k\alpha}) - \theta(z', z) f(\epsilon^{\mathrm{M}}_{k\alpha})\right] e^{-\mathrm{i}\int_{z'}^{z} d\bar{z}\, \epsilon_{k\alpha}(\bar{z})},$$

where in accordance with our notation $\epsilon^{\mathrm{M}}_{k\alpha} = \epsilon_{k\alpha}(t_0 - \mathrm{i}\tau) = \epsilon_{k\alpha} - \mu$. The reader can verify the correctness of this result using (6.35).

Inserting the solution for $G_{\alpha M}$ into the second term on the r.h.s. of (9.50), we get

$$\sum_{\alpha} h_{M\alpha}(z) G_{\alpha M}(z, z') = \int d\bar{z}\, \Sigma_{\mathrm{em}}(z, \bar{z}) G_{MM}(\bar{z}, z'),$$

where we define the *embedding self-energy* in the usual way:

$$\Sigma_{\mathrm{em}}(z, z') = \sum_{\alpha} \Sigma_{\alpha}(z, z'), \qquad \Sigma_{\alpha}(z, z') = h_{M\alpha}(z) g_{\alpha\alpha}(z, z') h_{\alpha M}(z'). \tag{9.52}$$

Like $g_{\alpha\alpha}$, the embedding self-energy is independent of the electronic interaction, and hence of G_{MM}, and it is completely specified by the reservoir Hamiltonian $h_{\alpha\alpha}$ and by the contact

Hamiltonian $h_{\alpha M}$. In conclusion, the equation of motion for the molecular Green's function becomes

$$\left[\mathrm{i}\frac{d}{dz} - h_{MM}(z)\right] G_{MM}(z, z') = \delta(z, z') + \int_\gamma d\bar{z}\, \left[\Sigma_{MM}(z, \bar{z}) + \Sigma_{\rm em}(z, \bar{z})\right] G_{MM}(\bar{z}, z'). \tag{9.53}$$

The adjoint equation of motion can be derived similarly. Interestingly, the reservoirs modify the equations of motion for the Green's function of the isolated molecule by simply adding the embedding self-energy to the many-body self-energy. Therefore, all results in Sections 9.2 remain valid, provided we interpret Σ as the sum of the many-body and embedding self-energies. We also emphasize that (9.53) is a closed equation for G_{MM} for any diagrammatic approximation to the many-body self-energy Σ_{MM}.[9] This is a very nice result since G_{MM} is the Green's function of a finite system and therefore (9.53) can be implemented numerically using a finite basis.

The equations of motion can be formally integrated to obtain a Dyson equation for the molecular Green's function. Defining the Green's function of the noninteracting and noncontacted molecule as the solution of

$$\left[\mathrm{i}\frac{d}{dz} - h_{MM}(z)\right] g_{MM}(z, z') = \delta(z, z')$$

with KMS boundary conditions, we get

$$G_{MM}(z, z') = g_{MM}(z, z') + \int_\gamma d\bar{z} d\bar{z}' g_{MM}(z, \bar{z}) \Sigma_{\rm tot}(\bar{z}, \bar{z}') G_{MM}(\bar{z}', z'), \tag{9.54}$$

where

$$\Sigma_{\rm tot}(\bar{z}, \bar{z}') \equiv \Sigma_{MM}(\bar{z}, \bar{z}') + \Sigma_{\rm em}(\bar{z}, \bar{z}'). \tag{9.55}$$

The formal solution for $G^{\lessgtr}_{MM}(t, t')$ can be worked out like in Section 9.3. Notice that $\Sigma_{\rm em}$ has no singular part and it should therefore be added to the correlation part of Σ_{MM}.

Exercise 9.4 Consider a one-dimensional reservoir α described by the tight-binding Hamiltonian

$$\hat{H}_\alpha = \sum_{j=0}^{\infty} \sum_\sigma \left[\epsilon_\alpha \hat{d}^\dagger_{j\alpha\sigma} \hat{d}^\dagger_{j\alpha\sigma} + T_\alpha \hat{d}^\dagger_{j+1\alpha\sigma} \hat{d}^\dagger_{j\alpha\sigma} + T^*_\alpha \hat{d}^\dagger_{j\alpha\sigma} \hat{d}^\dagger_{j+1\alpha\sigma})\right].$$

The reservoir is connected only to the m_αth molecular orbital through the site $j = 0$, with hopping integral $T_{0\,m_\alpha}$. Show that the embedding self-energy of reservoir α has the form $[\Sigma_\alpha(z, z')]_{mm'} = \delta_{mm_\alpha} \delta_{m'm_\alpha} \sigma_\alpha(z, z')$ and that the retarded component of the Keldysh function $\sigma_\alpha(z, z')$ in frequency space is

$$\sigma^{\rm R}_\alpha(\omega) = \frac{|T_{0\,m_\alpha}|^2}{2|T_\alpha|^2} \begin{cases} (\omega - \epsilon_\alpha) - \sqrt{(\omega - \epsilon_\alpha)^2 - 4|T_\alpha|^2} & (\omega - \epsilon_\alpha) > 2|T_\alpha| \\ (\omega - \epsilon_\alpha) - \mathrm{i}\sqrt{4|T_\alpha|^2 - (\omega - \epsilon_\alpha)^2} & |\omega - \epsilon_\alpha| < 2|T_\alpha| \\ (\omega - \epsilon_\alpha) + \sqrt{(\omega - \epsilon_\alpha)^2 - 4|T_\alpha|^2} & (\omega - \epsilon_\alpha) < 2|T_\alpha| \end{cases}.$$

[9] If the exact many-body self-energy were used then the resulting G_{MM} would be exact.

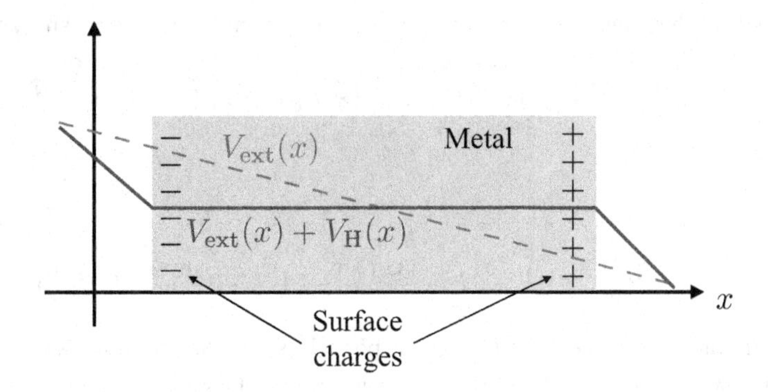

Figure 9.2 Profile of the external potential and total effective potential in the interior of a metal.

9.5 Quantum Transport

We consider the open system of Fig. 9.1 populated by electrons of charge $q = -1$ and initially in thermal equilibrium at inverse temperature β and chemical potential μ. At times $t > t_0$ we switch on an electric field that mimics an applied bias, $\epsilon_{k\alpha}(t) = \epsilon_{k\alpha} + V_\alpha(t)$, where $V_\alpha(t) = q\phi_\alpha(t)$ is the potential felt by the fermions of charge q in reservoir α. In fact, when a metal is exposed to an external electric field $\mathbf{E}_{\text{ext}} = -\boldsymbol{\nabla}\phi$, the effect of the Hartree potential V_{H} is to screen $V_{\text{ext}} = q\phi$ in such a way that the classical (external plus Hartree) potential is uniform in the interior of the metal. Consequently, the potential drop is entirely confined to the molecular region, as shown in the schematic representation of Fig. 9.2. The uniform potential $V_\alpha(t)$ in reservoir α can therefore be interpreted as the sum of the external and Hartree potential (this effectively means that the reservoirs are treated at the Hartree level). Of course, for this potential to be uniform *at all times* we must vary the external potential on a timescale much longer than the characteristic time for the formation of surface charges. A reasonable estimate of this characteristic time is the inverse of the plasma frequency, which in a bulk metal is of the order of $\sim$10 eV, corresponding to a (sub-femtosecond) timescale of $\sim 10^{-15} \div 10^{-16}$ seconds, see Section 15.5. Below we consider the case $V_\alpha(t \gg t_0) = V_\alpha$ independent of time.

The described physical situation is the one of a typical quantum transport setup. In this context the reservoirs play the role of electrodes and we refer to V_α as the applied voltage. From the knowledge of G_{MM}, we have access to the time-dependent ensemble average of any one-body operator acting on the molecular degrees of freedom. We also have access to the total current I_α flowing between the molecule and the electrode α. This current is given by the time derivative of the total number of electrons N_α in electrode α multiplied by the charge q: $I_\alpha(t) = q\frac{d}{dt}N_\alpha(t)$. Since $N_\alpha(t)$ is the ensemble average of the number operator in the Heisenberg picture, $\hat{N}_{\alpha,H}(t)$, its time derivative is the ensemble average of the commutator $[\hat{N}_{\alpha,H}(t), \hat{H}_H(t)]_-$, see (3.20). The operator $\hat{N}_\alpha$ commutes with $\hat{H}_{\text{int}}$ and with all the terms of $\hat{H}_0$ except the one describing the coupling between the molecule and the reservoirs [the third term in (9.47)]. It is a matter of simple algebra to show that

$$\left[\hat{N}_\alpha, \hat{H}(t)\right]_- = \sum_{k\sigma} \left[\hat{n}_{k\alpha\sigma}, \hat{H}(t)\right]_- = \sum_{\substack{mk \\ \sigma}} \left(-T_{m\,k\alpha}\hat{d}^\dagger_{m\sigma}\hat{d}_{k\alpha\sigma} + T_{k\alpha\,m}\hat{d}^\dagger_{k\alpha\sigma}\hat{d}_{m\sigma}\right),$$

and therefore

$$I_\alpha(t) = q\frac{d}{dt}N_\alpha(t) = 2q \times 2\,\mathrm{Re}\left\{\mathrm{Tr}_M\left[h_{M\alpha}(t)G^<_{\alpha M}(t,t)\right]\right\}, \tag{9.56}$$

where Tr_M denotes the trace over the single-particle states of the molecular basis and the factor of 2 comes from spin. In deriving (9.56) we also take into account that the lesser Green's function is anti-Hermitian, see (6.27).[10] Extracting the lesser component of $G_{\alpha M}$ from (9.51), we can rewrite I_α in terms of the Green's function of the molecular region only:

$$\boxed{I_\alpha(t) = 4q\,\mathrm{Re}\left\{\mathrm{Tr}_M\left[\Sigma^<_\alpha \cdot G^{\mathrm{A}}_{MM} + \Sigma^{\mathrm{R}}_\alpha \cdot G^<_{MM} + \Sigma^\rceil_\alpha \star G^\lceil_{MM}\right](t,t)\right\}} \tag{9.57}$$

where the αth contribution to the embedding self-energy (i.e., Σ_α) has been defined in (9.52). The last term in (9.57) explicitly accounts for the initial correlations and the initial-state dependence. In Appendix K we solve analytically the equation of motion (9.53) for noninteracting electrons and derive the explicit time-dependent form of $I_\alpha(t)$, see also Ref. [129].

9.5.1 Meir–Wingreen Formula

If one assumes that the initial correlations and the initial-state dependence are washed out in the long-time limit ($t \to \infty$), then for large times we can discard the imaginary time convolution. The resulting formula is known as the *Meir–Wingreen formula* [130, 131]. Equation (9.57) provides a generalization of the Meir–Wingreen formula to the transient time domain. If in the same limit ($t \to \infty$) we reach a steady state, then the Green's function, and hence the self-energy, becomes a function of the time difference only. In this case we can Fourier transform with respect to the relative time and write the steady-state value of the current as

$$\boxed{I^{(S)}_\alpha \equiv \lim_{t\to\infty} I_\alpha(t) = 2\mathrm{i}q\int\frac{d\omega}{2\pi}\mathrm{Tr}_M\left[\Sigma^<_\alpha(\omega)A_{MM}(\omega) - \Gamma_\alpha(\omega)G^<_{MM}(\omega)\right]} \tag{9.58}$$

where $\Gamma_\alpha(\omega) = \mathrm{i}\left[\Sigma^{\mathrm{R}}_\alpha(\omega) - \Sigma^{\mathrm{A}}_\alpha(\omega)\right]$ is the nonequilibrium quasi-particle width due to the embedding and $A_{MM}(\omega) = \mathrm{i}\left[G^{\mathrm{R}}_{MM}(\omega) - G^{\mathrm{A}}_{MM}(\omega)\right]$ is the nonequilibrium spectral function.[11] These are nonequilibrium quantities since they carry a dependence on the applied voltages V_α. Physically we expect that a necessary condition for the development of a steady state is that the Hamiltonian $\hat{H}(t)$ is independent of t in the long-time limit. In general,

[10] Notice the similarity between (9.56) and the formula (6.97) for the photocurrent.

[11] To derive (9.58), one uses that $\mathrm{Re}\{\mathrm{Tr}[B]\} = \frac{1}{2}\mathrm{Tr}[B + B^\dagger]$ for any matrix B and then the properties of Section 9.2 for the adjoint Green's function and self-energy as well as the cyclic property of the trace.

however, this condition is not sufficient. The presence of bound states in noninteracting systems usually prevents the formation of a steady state [132, 134]. In interacting systems the issue is more difficult to address due to the absence of exact solutions. The numerical simulations so far performed suggest that the current reaches a steady value provided that the self-energy contains some correlations. On the contrary, if Σ_{MM} is approximated at the Hartree or Hartree–Fock level, then the development of a steady state is not guaranteed [135, 136].

In Fourier space the nonequilibrium retarded/advanced embedding self-energy reads

$$\begin{aligned}\Sigma^{\mathrm{R}}_{\alpha}(\omega) &= h_{M\alpha}\frac{1}{\omega - h_{\alpha\alpha} - V_\alpha \pm \mathrm{i}\eta}h_{\alpha M} \\ &= \underbrace{P\, h_{M\alpha}\frac{1}{\omega - h_{\alpha\alpha} - V_\alpha}h_{\alpha M}}_{\Lambda_\alpha(\omega)} \mp \mathrm{i}\underbrace{\pi\, h_{M\alpha}\delta(\omega - h_{\alpha\alpha} - V_\alpha)h_{\alpha M}}_{\Gamma_\alpha(\omega)/2}.\end{aligned}$$

The nonequilibrium lesser embedding self-energy is proportional to the quasi-particle width, since

$$\begin{aligned}\Sigma^{<}_{\alpha}(\omega) &= 2\pi\mathrm{i}\, h_{M\alpha}f(h_{\alpha\alpha} - \mu)\delta(\omega - h_{\alpha\alpha} - V_\alpha)h_{\alpha M} \\ &= \mathrm{i}\, f(\omega - V_\alpha - \mu)\Gamma_\alpha(\omega).\end{aligned} \tag{9.59}$$

Substituting this result into (9.58), we find the equivalent formula

$$\boxed{I^{(S)}_{\alpha} = 2\mathrm{i}q\int\frac{d\omega}{2\pi}\mathrm{Tr}_M\left[\Gamma_\alpha(\omega)\Big(\mathrm{i}\, f(\omega - V_\alpha - \mu)A_{MM}(\omega) - G^{<}_{MM}(\omega)\Big)\right]} \tag{9.60}$$

Proportionate couplings Let us now consider a quantum transport setup with only two electrodes, a left ($\alpha = L$) and a right ($\alpha = R$) one. For *proportionate couplings* to the electrodes, $\Gamma_L(\omega) = \lambda\Gamma_R(\omega)$, we can get rid of the term containing $G^{<}_{MM}$ as follows [130, 131]. We have

$$I^{(S)}_{L} - \lambda I^{(S)}_{R} = -2q\int\frac{d\omega}{2\pi}\left[f(\omega - V_L - \mu) - f(\omega - V_R - \mu)\right]\mathrm{Tr}_M\left[\Gamma_L(\omega)A_{MM}(\omega)\right].$$

Charge conservation implies that $I^{(S)}_{L} = -I^{(S)}_{R}$, and therefore

$$I^{(S)}_{L} = \frac{2q}{1+\lambda}\int\frac{d\omega}{2\pi}\left[f(\omega - V_R - \mu) - f(\omega - V_L - \mu)\right]\mathrm{Tr}_M\left[\Gamma_L(\omega)A_{MM}(\omega)\right]. \tag{9.61}$$

The condition of proportionate couplings is a rather strong requirement for real molecules. Nonetheless, the derived formula has the merit of being physically transparent. In particular, (9.61) makes it evident that the current vanishes for $V_L = V_R$. In Section 11.6.2 we use this formula to derive an exact result in the Anderson model.

9.5.2 Landauer–Büttiker Formula

For noninteracting systems the steady-state formula (9.60) simplifies considerably and its physical meaning becomes more transparent. We recall that (9.60) was derived under the assumptions that (1) both initial correlations and initial-state dependence are washed out in the limit $t \to \infty$; and (2) in the same limit the invariance under time translations is recovered – that is, the Green's function and the self-energy depend only on the time difference.

The lesser Green's function of the molecular region can be calculated from (9.45) where, for the noninteracting case considered here, the self-energy is only the embedding self-energy. To lighten the notation we remove the subscript MM for matrices projected onto the molecular region. Taking into account the result (9.59) for $\Sigma^<_\alpha$, we find

$$G^<(\omega) = G^{\rm R}(\omega)\Sigma^<_{\rm em}(\omega)G^{\rm A}(\omega) = {\rm i}\sum_\alpha f(\omega - V_\alpha - \mu)G^{\rm R}(\omega)\Gamma_\alpha(\omega)G^{\rm A}(\omega).$$

At difference with the WBLA, see Section 2.4, the quasi-particle width matrix $\Gamma_\alpha(\omega)$ is frequency-dependent and $\Lambda_\alpha(\omega)$ is nonvanishing.[12] The retarded Green's function is easily obtained from the Dyson equation (9.54), $G^{\rm R}(\omega) = 1/(\omega - h - \Sigma^{\rm R}_{\rm em}(\omega))$, and from it we can calculate the nonequilibrium spectral function:

$$A(\omega) = {\rm i}[G^{\rm R}(\omega) - G^{\rm A}(\omega)] = \sum_\alpha G^{\rm R}(\omega)\Gamma_\alpha(\omega)G^{\rm A}(\omega).$$

Inserting these results into (9.60), we obtain the *Landauer–Büttiker* formula [137, 138, 139] for the steady-state current:

$$\boxed{I^{(S)}_\alpha = 2q\int\frac{d\omega}{2\pi}\sum_\beta[f(\omega - V_\beta - \mu) - f(\omega - V_\alpha - \mu)]\ {\rm Tr}_M\left[\Gamma_\alpha(\omega)G^{\rm R}(\omega)\Gamma_\beta(\omega)G^{\rm A}(\omega)\right]} \tag{9.62}$$

The quantities

$$\mathcal{T}_{\alpha\beta}(\omega) \equiv {\rm Tr}_M\left[\Gamma_\alpha(\omega)G^{\rm R}(\omega)\Gamma_\beta(\omega)G^{\rm A}(\omega)\right]$$

are related to the probability for an electron of energy ω to be transmitted from electrode α to electrode β. We then see that the current $I^{(S)}_\alpha$ is the sum of the probabilities for electrons in α (and hence with energy below $\mu + V_\alpha$) to go from α to β minus the sum of the probabilities for electrons in β (and hence with energy below $\mu + V_\beta$) to go from β to α. As expected, the steady-state current vanishes if the potentials $V_\alpha = V$ are all the same. It is important to realize that the derivation of the Landauer–Büttiker formula does not apply in the interacting case since $G^< = G^{\rm R}(\Sigma^< + \Sigma^<_{\rm em})G^{\rm A}$ and there is no simple relation between the *nonequilibrium* many-body self-energy $\Sigma^<$ and the *nonequilibrium* many-body quasi-particle width $\Gamma = {\rm i}[\Sigma^> - \Sigma^<]$.

For the case of only two electrodes $\alpha = L$ (left) and $\alpha = R$ (right) the Landauer–Büttiker formula simplifies to

$$I^{(S)}_L = 2q\int\frac{d\omega}{2\pi}[f(\omega - V_R - \mu) - f(\omega - V_L - \mu)]\ {\rm Tr}_M\left[\Gamma_L(\omega)G^{\rm R}(\omega)\Gamma_R(\omega)G^{\rm A}(\omega)\right]$$

and $I^{(S)}_R = -I^{(S)}_L$. Let us apply this formula to the noninteracting Anderson model – that is, a molecule described by a single orbital (hence the dimension of the molecular matrices

[12] We remind the reader that Λ_α and Γ_α are related by a Hilbert transformation.

becomes 1×1). The Hamiltonian of the noninteracting Anderson model is identical to that of Section 2.4; the only difference is that the atomic level (the molecular region in the context of quantum transport) is in contact with more than one metallic reservoir. We also work in the WBLA – that is, $\Gamma_\alpha(\omega) = \Gamma_\alpha$ is independent of ω and consequently $\Lambda_\alpha(\omega) = 0$. In this case,

$$\mathrm{Tr}_M \left[\Gamma_L(\omega) G^{\mathrm{R}}(\omega) \Gamma_R(\omega) G^{\mathrm{A}}(\omega)\right] = \frac{\Gamma_L \Gamma_R}{(\omega - \epsilon_0)^2 + (\Gamma_{\mathrm{em}}/2)^2},$$

where ϵ_0 is the only element of the 1×1 molecular single-particle Hamiltonian h and $\Gamma_{\mathrm{em}} = \Gamma_L + \Gamma_R$. For a small bias between the left and right reservoirs, $V_R = -V_L = q\phi/2$, we can Taylor expand $f(\omega - V_\alpha - \mu) \sim f(\omega - \mu) - f'(\omega - \mu) V_\alpha$. Then,

$$I_L^{(S)} = -2q^2\phi \int \frac{d\omega}{2\pi} f'(\omega - \mu) \frac{\Gamma_L \Gamma_R}{(\omega - \epsilon_0)^2 + (\Gamma_{\mathrm{em}}/2)^2}. \tag{9.63}$$

Notice that the steady-state current is finite despite the absence of dissipative mechanisms leading to a finite resistance. The "resistance" is here only caused by quantum effects: The microscopic cross section of the molecule prevents a macroscopic number of fermions passing through it [59, 140].

In the limit of zero temperature $f'(\omega - \mu) \to -\delta(\omega - \mu)$, the steady-state current simplifies further:

$$I_L^{(S)} = 2 \times \frac{q^2}{2\pi} \frac{\Gamma_L \Gamma_R}{(\epsilon_0 - \mu)^2 + (\Gamma_{\mathrm{em}}/2)^2} \phi. \tag{9.64}$$

The resistance $\mathrm{R} = \phi / I_L^{(S)}$ is minimum when the on-site energy ϵ_0 is aligned to the Fermi energy μ and the molecule is symmetrically coupled to the electrodes – that is, $\Gamma_L = \Gamma_R = \Gamma_{\mathrm{em}}/2$:

$$\mathrm{R}_{\mathrm{min}} = \frac{1}{2} \times \frac{2\pi}{q^2}, \tag{9.65}$$

where the factor $1/2$ comes from spin. The minimum of R at $\epsilon_0 = \mu$ is physically sound. Indeed, for small biases only fermions close to the Fermi energy contribute to the current, and the probability of being reflected is minimum when their energy matches the height of the molecular barrier (i.e., ϵ_0). The inverse of the minimum resistance for particles with unit charge and for a given spin projection is denoted by $\mathcal{C}_0$ and is called the *quantum of conductance*. From (9.65) we have $\mathcal{C}_0 = \frac{1}{2\pi}$, or reinserting the fundamental constants $\mathcal{C}_0 = e^2/h = 1/(25\ \mathrm{kOhm})$, with $-|e|$ the electric charge and h the Planck constant.

Exercise 9.5 Prove (9.58)

Exercise 9.6 Consider a molecule described by a one-dimensional tight-binding wire of N sites, hence $m = 1, \ldots, N$, with on-site energy ϵ_M and nearest-neighbor hopping T_M. The molecule is connected to left and right one-dimensional electrodes described by the Hamiltonian of Exercise 9.4. In particular, the left electrode is connected only to the molecular site $m_L = 1$, whereas the right electrode is connected only to the molecular site $m_R = N$.

Show that the steady-state current (9.62) can be written as

$$I_R^{(S)} = 2q \int \frac{d\omega}{2\pi} \left[f(\omega - V_L - \mu) - f(\omega - V_R - \mu) \right] \sigma_R(\omega - V_R) \sigma_L(\omega - V_L) \times |T_M|^{2N} \left| \mathrm{Det}[G^{\mathrm{R}}(\omega)] \right|^2$$

10

Self-Energy and Screened Interaction: Exact Results

In this chapter we derive several exact properties of the self-energy and screened interaction. We also discuss the physical information carried by these quantities.

10.1 Fluctuation–Dissipation Theorem for Σ

Let us consider a system in equilibrium and hence with $\hat{H}^{\mathrm{M}} = \hat{H} - \mu\hat{N}$. In Section 6.3.1 we saw that the equilibrium Green's function satisfies

$$\hat{\mathcal{G}}^{\lceil}(\tau, t') = e^{\mu\tau}\hat{\mathcal{G}}^{>}(t_0 - \mathrm{i}\tau, t') \quad , \quad \hat{\mathcal{G}}^{\rceil}(t, \tau') = \hat{\mathcal{G}}^{<}(t, t_0 - \mathrm{i}\tau')e^{-\mu\tau'}. \tag{10.1}$$

In a similar way one can also prove that

$$\hat{\mathcal{G}}^{\mathrm{M}}(\tau, \tau') = \begin{cases} e^{\mu\tau}\hat{\mathcal{G}}^{>}(t_0 - \mathrm{i}\tau, t_0 - \mathrm{i}\tau')e^{-\mu\tau'} & \tau > \tau' \\ e^{\mu\tau}\hat{\mathcal{G}}^{<}(t_0 - \mathrm{i}\tau, t_0 - \mathrm{i}\tau')e^{-\mu\tau'} & \tau < \tau' \end{cases}. \tag{10.2}$$

An alternative strategy to derive (10.1) and (10.2) consists of considering the Green's function $\hat{\mathcal{G}}_\mu$ of a system with constant Hamiltonian $\hat{H}(z) = \hat{H}^{\mathrm{M}} = \hat{H} - \mu\hat{N}$ along the entire contour. The Green's function $\hat{\mathcal{G}}_\mu$ differs from $\hat{\mathcal{G}}$ since in $\hat{\mathcal{G}}$ the Hamiltonian $\hat{H}(z) = \hat{H} - \mu\hat{N}$ for z on the vertical track and $\hat{H}(z) = \hat{H}$ for z on the horizontal branches. The relation between $\hat{\mathcal{G}}$ and $\hat{\mathcal{G}}_\mu$ is

$$\hat{\mathcal{G}}(z, z') = \hat{\mathcal{G}}_\mu(z, z') \times \begin{cases} e^{-\mathrm{i}\mu(z - z')} & z, z' \text{ on the horizontal branches} \\ e^{-\mathrm{i}\mu(z - t_0)} & z \text{ on the horizontal branches and } z' \text{ on } \gamma^{\mathrm{M}}, \\ e^{\mathrm{i}\mu(z' - t_0)} & z \text{ on } \gamma^{\mathrm{M}} \text{ and } z' \text{ on the horizontal branches} \\ 1 & z, z' \text{ on } \gamma^{\mathrm{M}} \end{cases}$$

as it follows directly from

$$e^{\mathrm{i}(\hat{H} - \mu\hat{N})(t - t_0)}\,\hat{\psi}(\mathbf{x})\,e^{-\mathrm{i}(\hat{H} - \mu\hat{N})(t - t_0)} = e^{\mathrm{i}\mu(t - t_0)}\,e^{\mathrm{i}\hat{H}(t - t_0)}\,\hat{\psi}(\mathbf{x})\,e^{-\mathrm{i}\hat{H}(t - t_0)}$$

and the like for the creation operator. The crucial observation is that $\hat{\mathcal{G}}^{\lessgtr}_{\mu}(z,z')$ depends only on $z-z'$ for *all* z and z' since the Hamiltonian is independent of z. This implies that if we calculate $\hat{\mathcal{G}}^{>}_{\mu}(t_0,t')$ and then replace $t_0 \to t_0 - \mathrm{i}\tau$, we get

$$\hat{\mathcal{G}}^{>}_{\mu}(t_0-\mathrm{i}\tau,t') = \hat{\mathcal{G}}^{>}_{\mu}(z=t_0-\mathrm{i}\tau,t'_{\pm}) = \hat{\mathcal{G}}^{\lceil}_{\mu}(\tau,t').$$

Using the relation between $\hat{\mathcal{G}}$ and $\hat{\mathcal{G}}_{\mu}$, the first identity in (10.1) follows. The second identity in (10.1) and the identity in (10.2) can be deduced in a similar manner. This alternative derivation allows us to prove a fluctuation–dissipation theorem for several many-body quantities. We here show how it works for the self-energy. Like the Green's function $\hat{\mathcal{G}}^{\lessgtr}_{\mu}(z,z')$, the self-energy $\hat{\Sigma}^{\lessgtr}_{\mu}(z,z')$ also depends only on $z-z'$. Therefore, if we calculate $\hat{\Sigma}^{>}_{\mu}(t_0,t')$ and then replace $t_0 \to t_0 - \mathrm{i}\tau$, we get

$$\hat{\Sigma}^{>}_{\mu}(t_0-\mathrm{i}\tau,t') = \hat{\Sigma}^{>}_{\mu}(z=t_0-\mathrm{i}\tau,t'_{\pm}) = \hat{\Sigma}^{\lceil}_{\mu}(\tau,t').$$

Consider now a generic self-energy diagram for $\hat{\Sigma}_{\mu}$. Expressing every $\hat{\mathcal{G}}_{\mu}$ in terms of $\hat{\mathcal{G}}$, the phase factors cancel out in all internal vertices. Consequently, the relation between $\hat{\Sigma}_{\mu}$ and $\hat{\Sigma}$ is the same as the relation between $\hat{\mathcal{G}}_{\mu}$ and $\hat{\mathcal{G}}$. Combining this relation with the above result, we get $\hat{\Sigma}^{\lceil}(\tau,t') = e^{\mu\tau}\hat{\Sigma}^{>}(t_0-\mathrm{i}\tau,t')$. In a similar way we can work out the other combinations of contour arguments. In conclusion, we have

$$\hat{\Sigma}^{\lceil}(\tau,t') = e^{\mu\tau}\hat{\Sigma}^{>}(t_0-\mathrm{i}\tau,t') \quad , \quad \hat{\Sigma}^{\rceil}(t,\tau') = \hat{\Sigma}^{<}(t,t_0-\mathrm{i}\tau')e^{-\mu\tau'}, \tag{10.3}$$

and

$$\hat{\Sigma}^{\mathrm{M}}(\tau,\tau') = \begin{cases} e^{\mu\tau}\hat{\Sigma}^{>}(t_0-\mathrm{i}\tau,t_0-\mathrm{i}\tau')e^{-\mu\tau'} & \tau>\tau' \\ e^{\mu\tau}\hat{\Sigma}^{<}(t_0-\mathrm{i}\tau,t_0-\mathrm{i}\tau')e^{-\mu\tau'} & \tau<\tau' \end{cases}. \tag{10.4}$$

We can combine these relations with the KMS boundary conditions (9.7) and (9.8) to get

$$\begin{aligned} \hat{\Sigma}^{<}(t_0,t') &= \hat{\Sigma}(t_{0-},t'_{+}) \\ &= \pm\,\hat{\Sigma}(t_0-\mathrm{i}\beta,t'_{+}) \\ &= \pm\hat{\Sigma}^{\lceil}(\beta,t') \\ &= \pm e^{\mu\beta}\hat{\Sigma}^{>}(t_0-\mathrm{i}\beta,t'). \end{aligned}$$

Fourier transforming both sides of this equation, we find

$$\hat{\Sigma}^{>}(\omega) = \pm e^{\beta(\omega-\mu)}\hat{\Sigma}^{<}(\omega), \tag{10.5}$$

which allows us to prove a fluctuation–dissipation theorem for the self-energy.

Quasi-particle width operator If we define the *quasi-particle width operator* as[1]

$$\boxed{\hat{\Gamma}(\omega) \equiv \mathrm{i}[\hat{\Sigma}^{>}(\omega) - \hat{\Sigma}^{<}(\omega)] = \mathrm{i}[\hat{\Sigma}^{\mathrm{R}}(\omega) - \hat{\Sigma}^{\mathrm{A}}(\omega)]} \tag{10.6}$$

[1]The motivation for the name "quasi-particle width operator" becomes clear in Section 10.2.

then we can express the lesser and greater self-energy in terms of $\hat{\Gamma}$ as follows:

$$\boxed{\begin{aligned}\hat{\Sigma}^{<}(\omega) &= \mp\mathrm{i}f(\omega-\mu)\hat{\Gamma}(\omega)\\ \hat{\Sigma}^{>}(\omega) &= -\mathrm{i}\bar{f}(\omega-\mu)\hat{\Gamma}(\omega)\end{aligned}} \tag{10.7}$$

We observe that in (10.6) we could replace $\hat{\Sigma} \to \hat{\Sigma}_{\rm c}$ since $\hat{\Sigma}^{\lessgtr}_{\rm HF} = 0$ (or, equivalently, $\hat{\Sigma}^{\rm R}_{\rm HF} - \hat{\Sigma}^{\rm A}_{\rm HF} = 0$). The quasi-particle width operator for the self-energy is the analog of the spectral function operator for the Green's function. From a knowledge of $\hat{\Gamma}$ we can determine all Keldysh components of $\hat{\Sigma}$ with real-time arguments. The lesser and greater self-energies are obtained from (10.7). The retarded and advanced correlation self-energies follow from the Fourier transform of $\hat{\Sigma}^{\rm R}_{\rm c}(t,t') = \theta(t-t')[\hat{\Sigma}^{>}(t,t') - \hat{\Sigma}^{<}(t,t')]$ and the like for $\hat{\Sigma}^{\rm A}_{\rm c}$, and read

$$\boxed{\hat{\Sigma}^{\rm R/A}_{\rm c}(\omega) = \int \frac{d\omega'}{2\pi}\frac{\hat{\Gamma}(\omega')}{\omega-\omega'\pm\mathrm{i}\eta}} \tag{10.8}$$

Hermiticity properties The Keldysh components of the self-energy have the same hermiticity properties as the Keldysh components of the Green's function. We can easily prove this statement by considering the Fourier transform of the retarded/advanced component of the Dyson equation (9.39).[2] In the operator form it reads

$$\hat{\mathcal{G}}^{\rm R/A}(\omega) = \hat{\mathcal{G}}^{\rm R/A}_0(\omega) + \hat{\mathcal{G}}^{\rm R/A}_0(\omega)\hat{\Sigma}^{\rm R/A}(\omega)\hat{\mathcal{G}}^{\rm R/A}(\omega),$$

where the noninteracting Green's function is $\hat{\mathcal{G}}^{\rm R/A}_0(\omega) = 1/(\omega - \hat{h} \pm \mathrm{i}\eta)$, see (6.54) and (6.55). Solving for $\hat{\mathcal{G}}^{\rm R/A}(\omega)$, we get

$$\hat{\mathcal{G}}^{\rm R/A}(\omega) = \frac{1}{\omega - \hat{h} - \hat{\Sigma}^{\rm R/A}(\omega) \pm \mathrm{i}\eta}. \tag{10.9}$$

From (10.9) and relation (6.85) we infer that

$$\hat{\Sigma}^{\rm R}(\omega) = \left[\hat{\Sigma}^{\rm A}(\omega)\right]^{\dagger}; \tag{10.10}$$

hence the quasi-particle width operator (like the spectral function operator) is self-adjoint,

$$\hat{\Gamma}(\omega) = \left[\hat{\Gamma}(\omega)\right]^{\dagger}. \tag{10.11}$$

Property (10.11) also implies that $\hat{\Sigma}^{\lessgtr}(\omega)$ is an anti-Hermitian operator, see again (10.7), and that $\hat{\Sigma}^{\rm R}(\omega)$ is analytic in the upper half of the complex ω plane, while $\hat{\Sigma}^{\rm A}(\omega)$ is analytic in the lower half of the complex ω plane, see (10.8).

[2] The Dyson equation contains a double convolution on the contour. According to the Langreth rules in Table 5.1 the retarded/advanced component of this double convolution contains only retarded/advanced components of the convoluted functions.

We finally observe that if we write the self-energy $\hat{\Sigma}_{\rm c}^{\rm R/A}$ as the sum of a Hermitian operator $\hat{\Lambda}$ and an anti-Hermitian operator $\mp\frac{\rm i}{2}\hat{\Gamma}$ [compare with the embedding self-energy in (2.31)],

$$\hat{\Sigma}_{\rm c}^{\rm R/A}(\omega) = \hat{\Lambda}(\omega) \mp \frac{\rm i}{2}\hat{\Gamma}(\omega), \tag{10.12}$$

then (10.8) implies that $\hat{\Lambda}$ is the Hilbert transform of $\hat{\Gamma}$

$$\hat{\Lambda}(\omega) = P\int \frac{d\omega'}{2\pi}\frac{\hat{\Gamma}(\omega')}{\omega-\omega'}. \tag{10.13}$$

An analogous relation was found for the retarded/advanced Green's function, see (6.86).

Relation between retarded/advanced and Matsubara components Another important relation that we can derive in equilibrium is the one that establishes a connection between $\hat{\Sigma}^{\rm M}$ and $\hat{\Sigma}^{\rm R/A}$. We consider the Matsubara component of the Dyson equation (9.39) and expand the Green's function and the self-energy in Matsubara frequencies, see (9.22). The Matsubara–Dyson equation in the operator form then reads

$$\hat{\mathcal{G}}^{\rm M}(\omega_m) = \hat{\mathcal{G}}_0^{\rm M}(\omega_m) + \hat{\mathcal{G}}_0^{\rm M}(\omega_m)\hat{\Sigma}^{\rm M}(\omega_m)\hat{\mathcal{G}}^{\rm M}(\omega_m),$$

where the noninteracting Green's function $\hat{\mathcal{G}}_0^{\rm M}(\omega_m) = 1/(\omega_m - \hat{h}^{\rm M})$, see (6.21). Solving for $\hat{\mathcal{G}}^{\rm M}(\omega_m)$, we get

$$\hat{\mathcal{G}}^{\rm M}(\omega_m) = \frac{1}{\omega_m - \hat{h}^{\rm M} - \hat{\Sigma}^{\rm M}(\omega_m)}, \tag{10.14}$$

which agrees with (9.23), as it should. Comparing (10.9) and (10.14) and using (6.70), we conclude that

$$\hat{\Sigma}^{\rm M}(\zeta) = \begin{cases} \hat{\Sigma}^{\rm R}(\zeta+\mu) & \text{for } {\rm Im}[\zeta]>0 \\ \hat{\Sigma}^{\rm A}(\zeta+\mu) & \text{for } {\rm Im}[\zeta]<0 \end{cases}, \tag{10.15}$$

which is the same relation satisfied by the Green's function. Thus, $\hat{\Sigma}^{\rm M}(\zeta)$ is analytic everywhere except along the real axis, where it can have poles or branch points. In particular, for $\zeta = \omega \pm {\rm i}\eta$ we find

$$\boxed{\hat{\Sigma}^{\rm M}(\omega \pm {\rm i}\eta) = \hat{\Sigma}^{\rm R/A}(\omega+\mu)} \tag{10.16}$$

Zero-temperature fermions Let us explore the consequences of these relations in a system of fermions in equilibrium at zero temperature. Assuming that $\hat{\Sigma}^{\lessgtr}(\omega)$ is a continuous function of ω, the fluctuation–dissipation relation (10.7) implies $\hat{\Sigma}^{\lessgtr}(\mu) = 0$ since $f(\omega) = \theta(-\omega)$. This in turn implies that

$$\hat{\Gamma}(\mu) = 0. \tag{10.17}$$

Consequently the retarded self-energy is equal to the advanced self-energy for $\omega = \mu$. From (10.16) we conclude that the Matsubara Green's function is continuous when the complex frequency crosses the real axis in $\omega = 0$.

10.2 Quasi-particles and Lifetimes

We here deepen a concept introduced in Sections 6.1.3 and 6.3.2: quasi-particles in interacting systems. We have seen that the matrix elements $G^<_{ij}(t,t')$ vanish when $|t-t'|\to\infty$ in the absence of good quantum numbers. Physically this means that the probability of finding the system unchanged when a particle is removed from state i at time t and put back in state j at time t' approaches zero for $|t-t'|\to\infty$, see also Fig. 6.2. The Hartree-Fock approximation cannot account for this relaxation since

$$\hat{\Sigma}_{\rm HF}(z_1,z_2)=\delta(z_1,z_2)\hat{\mathcal{V}}_{\rm HF}(z_1)$$

is local in time and hence it does not carry memory. Accordingly, there exist single-particle states [the Hartree-Fock eigenstates in (8.25)] with an infinitely long lifetime. An equivalent way to express the same concept is through the spectral function. In the Hartree-Fock approximation,

$$\hat{\mathcal{A}}(\omega)=2\pi\delta(\omega-\hat{h}_{\rm HF}) \tag{10.18}$$

and using the fluctuation-dissipation theorem (6.84),

$$\begin{aligned}G^<_{ij}(t,t')&=\mp{\rm i}\int\frac{d\omega}{2\pi}e^{-{\rm i}\omega(t-t')}f(\omega-\mu)\,2\pi\,\langle i|\delta(\omega-\hat{h}_{\rm HF})|j\rangle\\&=\mp{\rm i}\,\langle i|f(\hat{h}_{\rm HF}-\mu)e^{-{\rm i}\hat{h}_{\rm HF}(t-t')}|j\rangle.\end{aligned}$$

If $i=j$ is an eigenstate of $\hat{h}_{\rm HF}$ with eigenvalue $\epsilon_i^{\rm HF}$, then

$$G^<_{ii}(t,t')=\mp{\rm i}\,f(\epsilon_i^{\rm HF}-\mu)e^{-{\rm i}\epsilon_i^{\rm HF}(t-t')},$$

which does not decay when $|t-t'|\to\infty$. *The δ-like structure of the spectral function is the signature of the absence of relaxation mechanisms.* Going beyond the Hartree-Fock approximation, we expect the spectral function to become a smooth integrable function (of course, for systems with infinitely many degrees of freedom). Indeed, in this case

$$G^<_{ii}(t,t')=\mp{\rm i}\int\frac{d\omega}{2\pi}e^{-{\rm i}\omega(t-t')}f(\omega-\mu)\langle i|\hat{\mathcal{A}}(\omega)|i\rangle\;\xrightarrow[|t-t'|\to\infty]{}\;0 \tag{10.19}$$

due to the Riemann-Lebesgue theorem.[3]

[3]According to the Riemann-Lebesgue theorem,

$$\lim_{t\to\infty}\int d\omega\,e^{{\rm i}\omega t}f(\omega)=0,$$

provided that f is an integrable function. An intuitive proof of this result consists in splitting the integral over ω into the sum of integrals over a window $\Delta\omega$ around $\omega_n=n\Delta\omega$. If $f(\omega)$ is finite and $\Delta\omega$ is small enough, we can approximate $f(\omega)\sim f(\omega_n)$ in each window and write

$$\begin{aligned}\lim_{t\to\infty}\int d\omega\,e^{{\rm i}\omega t}f(\omega)&\sim\sum_{n=-\infty}^{\infty}f(\omega_n)\lim_{t\to\infty}\int_{\omega_n-\Delta\omega/2}^{\omega_n+\Delta\omega/2}d\omega\,e^{{\rm i}\omega t}\\&=\sum_{n=-\infty}^{\infty}f(\omega_n)\lim_{t\to\infty}\frac{2e^{{\rm i}\omega_n t}\sin(\Delta\omega t/2)}{t}=0.\end{aligned}$$

Relation between spectral function and self-energy These considerations prompt us to look for a transparent relation between the spectral function and the self-energy. We split $\hat{\Sigma} = \hat{\Sigma}_{\rm HF} + \hat{\Sigma}_{\rm c}$ into the Hartree–Fock self-energy and the correlation self-energy, see (9.17) and (9.18). As functions in Keldysh space, $\hat{\Sigma}_{\rm HF}$ has only a singular part, whereas $\hat{\Sigma}_{\rm c}$ has no singular part. Therefore,[4]

$$\hat{\Sigma}^{\rm R/A}(\omega) = \hat{\mathcal{V}}_{\rm HF} + \hat{\Sigma}_{\rm c}^{\rm R/A}(\omega).$$

Recalling that the Hartree–Fock Hamiltonian is $\hat{h}_{\rm HF} = \hat{h} + \hat{\mathcal{V}}_{\rm HF}$, see (8.23), we can then write (10.9) as

$$\hat{\mathcal{G}}^{\rm R/A}(\omega) = \frac{1}{\omega - \hat{h}_{\rm HF} - \hat{\Sigma}_{\rm c}^{\rm R/A}(\omega) \pm {\rm i}\eta}.$$

A remark on the infinitesimal constant η appearing in the denominator is in order. If the imaginary part of $\hat{\Sigma}_{\rm c}^{\rm R/A}(\omega)$ is nonzero we can safely discard η. However, if ${\rm Im}[\hat{\Sigma}_{\rm c}^{\rm R/A}(\omega)] = 0$ for some ω, then η must absolutely be present for otherwise the retarded/advanced Green's function does not have the correct analytic properties, see the discussion below (6.66). Having said that, from now on we incorporate this infinitesimal constant into the retarded/advanced correlation self-energy.

From definition (6.77) of the spectral function we have

$$\begin{aligned}\hat{\mathcal{A}}(\omega) &= {\rm i}\big[\hat{\mathcal{G}}^{\rm R}(\omega) - \hat{\mathcal{G}}^{\rm A}(\omega)\big] = {\rm i}\hat{\mathcal{G}}^{\rm R}(\omega)\left[\frac{1}{\hat{\mathcal{G}}^{\rm A}(\omega)} - \frac{1}{\hat{\mathcal{G}}^{\rm R}(\omega)}\right]\hat{\mathcal{G}}^{\rm A}(\omega)\\ &= {\rm i}\hat{\mathcal{G}}^{\rm R}(\omega)\left[\hat{\Sigma}_{\rm c}^{\rm R}(\omega) - \hat{\Sigma}_{\rm c}^{\rm A}(\omega)\right]\hat{\mathcal{G}}^{\rm A}(\omega).\end{aligned}$$

The difference between the retarded and advanced correlation self-energy is the same as the difference $\hat{\Sigma}^{\rm R}(\omega) - \hat{\Sigma}^{\rm A}(\omega)$, since the Hartree–Fock part cancels. Then, taking into account definition (10.6) of the quasi-particle width operator, we can write the following exact relation for systems in equilibrium:

$$\boxed{\hat{\mathcal{A}}(\omega) = \hat{\mathcal{G}}^{\rm R}(\omega)\hat{\Gamma}(\omega)\hat{\mathcal{G}}^{\rm A}(\omega) = \frac{1}{\omega - \hat{h}_{\rm HF} - \hat{\Sigma}_{\rm c}^{\rm R}(\omega)}\hat{\Gamma}(\omega)\frac{1}{\omega - \hat{h}_{\rm HF} - \hat{\Sigma}_{\rm c}^{\rm A}(\omega)}} \tag{10.20}$$

This is the relation between the spectral function and the self-energy that we were looking for. In fermionic systems the spectral function operator is positive semidefinite (all eigenvalues are nonnegative, see Exercise 6.9) and therefore also the quasi-particle width operator is positive semidefinite, see Exercise 10.1. Combining the fluctuation–dissipation theorem for G and Σ with (10.20), we also find

$$\hat{\mathcal{G}}^{\lessgtr}(\omega) = \hat{\mathcal{G}}^{\rm R}(\omega)\hat{\Sigma}^{\lessgtr}(\omega)\hat{\mathcal{G}}^{\rm A}(\omega). \tag{10.21}$$

Under certain conditions (10.21) is valid even out of equilibrium, see Section 9.3.

[4]In equilibrium the Hartree–Fock potential does not depend on time and is equal to $\hat{\mathcal{V}}_{\rm HF}^{\rm M}$.

Let us explain how to recover the Hartree–Fock result. In this case, $\hat{\Sigma}_c^{R/A}(\omega) = \mp i\eta$ and hence the quasi-particle width operator is $\hat{\Gamma}(\omega) = 2\eta$. Inserting these values into (10.20),

$$\hat{\mathcal{A}}(\omega) = 2\frac{\eta}{(\omega - \hat{h}_{\rm HF})^2 + \eta^2} \underset{\eta\to 0}{\longrightarrow} 2\pi\,\delta(\omega - \hat{h}_{\rm HF}),$$

which correctly agrees with (10.18). We conclude that for $\hat{\mathcal{A}}(\omega)$ to be nonsingular $\hat{\Gamma}(\omega)$ has to be finite.

Quasi-particle lifetime We can provide a justification for the name "quasi-particle width operator" for $\hat{\Gamma}$ by estimating how fast $G^<_{ij}(t,t')$ decays with $|t-t'|$. Let us consider a system invariant under translations, such as the electron gas. In the absence of magnetic fields the momentum–spin kets are eigenkets of $\hat{h}_{\rm HF}$, with the same eigenvalue for spin up and down:

$$\hat{h}_{\rm HF}|\mathbf{p}\sigma\rangle = \epsilon^{\rm HF}_{\mathbf{p}}|\mathbf{p}\sigma\rangle \qquad \left(\text{in the electron gas: } \epsilon^{\rm HF}_{\mathbf{p}} = \frac{p^2}{2} - \phi_0 + \Sigma_{\rm HF}(\mathbf{p})\right).$$

Furthermore, the matrix elements of $\hat{\Gamma}$ in the momentum–spin basis are nonvanishing only along the diagonal,

$$\langle\mathbf{p}\sigma|\hat{\Gamma}(\omega)|\mathbf{p}'\sigma'\rangle = (2\pi)^3\,\delta_{\sigma\sigma'}\delta(\mathbf{p}-\mathbf{p}')\Gamma(\mathbf{p},\omega),$$

and, of course, the same holds true for $\hat{\Lambda}$ and $\hat{\mathcal{A}}$. Denoting by $\Lambda(\mathbf{p},\omega)$ and $A(\mathbf{p},\omega)$ the corresponding diagonal matrix elements (10.20) gives

$$A(\mathbf{p},\omega) = \frac{\Gamma(\mathbf{p},\omega)}{\left(\omega - \epsilon^{\rm HF}_{\mathbf{p}} - \Lambda(\mathbf{p},\omega)\right)^2 + \left(\frac{\Gamma(\mathbf{p},\omega)}{2}\right)^2}. \tag{10.22}$$

We use this result to calculate the matrix elements of $G^<$ in the momentum–spin basis:

$$G^<_{\mathbf{p}\sigma\,\mathbf{p}'\sigma'}(t,t') = (2\pi)^3\delta_{\sigma\sigma'}\delta(\mathbf{p}-\mathbf{p}')\,G^<(\mathbf{p},t-t').$$

From (10.19) we have

$$G^<(\mathbf{p},t-t') = \mp i\int\frac{d\omega}{2\pi}e^{-i\omega(t-t')}f(\omega-\mu)A(\mathbf{p},\omega). \tag{10.23}$$

To estimate the integral, we define the so-called *quasi-particle energy* $E_{\mathbf{p}}$ to be the solution of $\omega - \epsilon^{\rm HF}_{\mathbf{p}} - \Lambda(\mathbf{p},\omega) = 0$. To first order in $(\omega - E_{\mathbf{p}})$ we can write

$$\omega - \epsilon^{\rm HF}_{\mathbf{p}} - \Lambda(\mathbf{p},\omega) = \left(1 - \frac{\partial\Lambda}{\partial\omega}\right)_{\omega=E_{\mathbf{p}}}(\omega - E_{\mathbf{p}}).$$

If the function $\Gamma(\mathbf{p},\omega)$ is small and slowly varying in ω for $\omega\sim E_{\mathbf{p}}$, then the main contribution to the integral (10.23) comes from a region around $E_{\mathbf{p}}$. In this case we can approximate $f(\omega-\mu)$ with $f(E_{\mathbf{p}}-\mu)$ and

$$A(\mathbf{p},\omega) \sim Z_{\mathbf{p}}\frac{1/\tau_{\mathbf{p}}}{(\omega - E_{\mathbf{p}})^2 + (1/2\tau_{\mathbf{p}})^2}, \tag{10.24}$$

where

$$Z_{\mathbf{p}} = \frac{1}{1 - \frac{\partial \Lambda}{\partial \omega}\Big|_{\omega = E_{\mathbf{p}}}} \tag{10.25}$$

is the so-called *quasi-particle renormalization factor,* and

$$\frac{1}{\tau_{\mathbf{p}}} = Z_{\mathbf{p}}\,\Gamma(\mathbf{p}, E_{\mathbf{p}}) \tag{10.26}$$

is the inverse of the so-called *quasi-particle lifetime.* Thus the spectral function becomes a Lorentzian of width proportional to $\Gamma(\mathbf{p}, E_{\mathbf{p}})$ centered at the quasi-particle energy $E_{\mathbf{p}}$. This provides a justification for the name "quasi-particle width operator" for $\hat{\Gamma}$.

Approximating the spectral function as in (10.24), the integral (10.23) for $t > t'$ yields

$$G^{<}(\mathbf{p}, t - t') = \mp \mathrm{i} Z_{\mathbf{p}} f(E_{\mathbf{p}} - \mu) e^{-\mathrm{i}E_{\mathbf{p}}(t-t')} e^{-(t-t')/(2\tau_{\mathbf{p}})}.$$

The probability of finding the system unchanged is the modulus square of the above quantity and decays like $e^{-(t-t')/\tau_{\mathbf{p}}}$. The quasi-particle width $\Gamma(\mathbf{p}, E_{\mathbf{p}})$ is therefore proportional to the inverse of the quasi-particle lifetime of a removed particle with momentum $\mathbf{p}$. In a similar way, one can prove that $\tau_{\mathbf{p}}$ is also the quasi-particle lifetime of an added particle with momentum $\mathbf{p}$. We then conclude that *the lifetime of a single-particle excitation can be estimated from the width of the corresponding peak in the spectral function.* The smaller is $\tau_{\mathbf{p}}^{-1}$, the longer is the lifetime and the more "free particle" is the behavior of the excitation. This is the reason for calling *quasi-particles* those excitations with a long lifetime. Of course, this interpretation makes sense only if $\Gamma(\mathbf{p}, \omega)$ is small for $\omega \sim E_{\mathbf{p}}$.

Renormalization of quasi-particle energy One more observation is about $\hat{\Lambda}$. While the imaginary part of $\hat{\Sigma}_{\mathrm{c}}^{\mathrm{R}}$ gives the width of the spectral peak, the real part of $\hat{\Sigma}_{\mathrm{c}}^{\mathrm{R}}$ gives the energy shift in the position of this peak. The quantity $\hat{\Lambda}$ renormalizes the Hartree–Fock single-particle energy by taking into account the effects of collisions. The real and imaginary parts of $\hat{\Sigma}_{\mathrm{c}}^{\mathrm{R}}$ are not independent, but related through a Hilbert transformation, see (10.13). This is a general aspect of all many-body quantities of equilibrium systems and follows directly from the definition of retarded/advanced functions: Real and imaginary parts of the Fourier transform of the retarded Green's function, self-energy, polarization, screened interaction, etc. are all related through a Hilbert transformation.[5]

Exercise 10.1 The spectral function operator (10.20) has the structure $\hat{\mathcal{A}} = \hat{\mathcal{G}}\,\hat{\Gamma}\,\hat{\mathcal{G}}^{\dagger}$ with $\hat{\mathcal{G}} = \hat{\mathcal{G}}^{\mathrm{R}}$. Therefore, $\hat{\Gamma} = (1/\hat{\mathcal{G}})\ \hat{\mathcal{A}}\ (1/\hat{\mathcal{G}}^{\dagger})$. Show that if $\hat{\mathcal{A}}$ is positive semidefinite then $\langle i|\hat{\Gamma}|i\rangle \geq 0$ for all quantum numbers i and therefore also $\hat{\Gamma}$ is positive semidefinite.

10.3 Total Energy, Interaction Energy, and Correlation Energy with G and Σ

Let us consider a system of identical and interacting particles in equilibrium at a given temperature and chemical potential. In this section we derive a formula for the total energy

[5] And, of course, the same is true for the advanced quantities.

which involves Σ and G. The main advantage of this formula over the Galitskii–Migdal formula derived in Section 6.5 lies in the possibility of separating the total energy into a noncorrelated and correlated part, thus highlighting the dependence on the many-body approximation.

Let $\hat{\rho}$ be the equilibrium density matrix of $\hat{H} = \hat{H}_0 + \hat{H}_{\rm int}$. Then the energy of the system is

$$E = \mathrm{Tr}\left[\hat{\rho}\hat{H}_0\right] + \mathrm{Tr}\left[\hat{\rho}\hat{H}_{\rm int}\right] = E_{\rm one} + E_{\rm int}. \tag{10.27}$$

The one-body part of the total energy, $E_{\rm one}$, is given by the first term in (6.113). Since the system is in equilibrium, the single-particle Hamiltonian $\hat{h}_{\rm S}(t) = \hat{h}$ is time-independent and the Green's function depends only on the time difference. Therefore,

$$\boxed{E_{\rm one} = \pm \mathrm{i} \int d\mathbf{x}\, \langle \mathbf{x}|\hat{h}\,\hat{\mathcal{G}}^<(t,t)|\mathbf{x}\rangle = \pm \mathrm{i} \int \frac{d\omega}{2\pi} \int d\mathbf{x}\, \langle \mathbf{x}|\hat{h}\,\hat{\mathcal{G}}^<(\omega)|\mathbf{x}\rangle} \tag{10.28}$$

where in the last equality we write $\hat{\mathcal{G}}^<(t,t)$ in terms of its Fourier transform.

Interaction energy from the self-energy An important consequence of (9.5) and (9.6) is that they can be used to express the interaction energy in (6.113) in terms of the self-energy,

$$\boxed{E_{\rm int}(z_1) = \pm \frac{\mathrm{i}}{2} \int d\mathbf{x}_1 d3\, \Sigma(1;3)G(3;1^+) = \pm \frac{\mathrm{i}}{2} \int d\mathbf{x}_1 d3\, G(1;3)\Sigma(3;1^+)} \tag{10.29}$$

or, equivalently,

$$\boxed{E_{\rm int}(z_1) = \pm \frac{\mathrm{i}}{2} \int d\bar{\mathbf{x}} d\bar{z}\, \langle \bar{\mathbf{x}}|\hat{\Sigma}(z_1,\bar{z})\hat{\mathcal{G}}(\bar{z},z_1^+)|\bar{\mathbf{x}}\rangle = \pm \frac{\mathrm{i}}{2} \int d\mathbf{x}_1 d\bar{z}\, \langle \bar{\mathbf{x}}|\hat{\mathcal{G}}(z_1,\bar{z})\hat{\Sigma}(\bar{z},z_1^+)|\bar{\mathbf{x}}\rangle} \tag{10.30}$$

It is convenient to separate the singular Hartree–Fock self-energy from the correlation self-energy, see (9.17). We write

$$\hat{\Sigma}(z_1,z_2) = \delta(z_1,z_2)\hat{\mathcal{V}}_{\rm HF} + \hat{\Sigma}_{\rm c}(z_1,z_2), \tag{10.31}$$

where the Hartree–Fock potential operator has matrix elements [see (8.19)]

$$\langle \mathbf{x}_1|\hat{\mathcal{V}}_{\rm HF}|\mathbf{x}_2\rangle = \delta(\mathbf{x}_1 - \mathbf{x}_2)\int d\mathbf{x}\, v(\mathbf{x}_1,\mathbf{x})n(\mathbf{x}) \pm v(\mathbf{x}_1,\mathbf{x}_2)n(\mathbf{x}_1,\mathbf{x}_2),$$

with $n(\mathbf{x}_1,\mathbf{x}_2)$ the one-particle density matrix and $n(\mathbf{x}) = n(\mathbf{x},\mathbf{x})$ the density. Inserting (10.31) into (10.29), we get

$$E_{\rm int} = E_{\rm int,HF} + E_{\rm int,c}.$$

In this equation the Hartree–Fock part of the interaction energy is given by

$$\boxed{E_{\rm int,HF} = \underbrace{\frac{1}{2}\int d\mathbf{x}d\mathbf{x}'\, v(\mathbf{x},\mathbf{x}')n(\mathbf{x})n(\mathbf{x}')}_{E_{\rm int,H}} \underbrace{\pm\frac{1}{2}\int d\mathbf{x}d\mathbf{x}'\, v(\mathbf{x},\mathbf{x}')n(\mathbf{x},\mathbf{x}')n(\mathbf{x}',\mathbf{x})}_{E_{\rm int,x}}} \tag{10.32}$$

which is the sum of a Hartree term $E_{\mathrm{int,H}}$ and a Fock (or exchange) term $E_{\mathrm{int,x}}$. The correlation part reads

$$
\begin{aligned}
E_{\mathrm{int,c}} &= \pm\frac{\mathrm{i}}{2}\int d\mathbf{x}_1 \langle \mathbf{x}_1 | \left[\hat{\Sigma}_{\mathrm{c}}^{\mathrm{M}} \star \hat{\mathcal{G}}^{\mathrm{M}}\right](0,0^+)\,|\mathbf{x}_1\rangle \\
&= \pm\frac{\mathrm{i}}{2}\int d\mathbf{x}_1 \langle \mathbf{x}_1 | \frac{1}{-\mathrm{i}\beta}\sum_m e^{\eta\omega_m}\,\hat{\Sigma}_{\mathrm{c}}^{\mathrm{M}}(\omega_m)\hat{\mathcal{G}}^{\mathrm{M}}(\omega_m)|\mathbf{x}_1\rangle,
\end{aligned}
\tag{10.33}
$$

as follows directly from (10.30) with, for example, $z_1 = t_{0-}$ ($E_{\mathrm{int,c}}$ is independent of time and hence we can choose z_1 anywhere along the contour).

To perform the sum over all Matsubara frequencies we can use the trick in (6.39) since both $\hat{\mathcal{G}}^{\mathrm{M}}(\zeta)$ and $\hat{\Sigma}_{\mathrm{c}}^{\mathrm{M}}(\zeta)$ (as functions of the complex variable ζ) are analytic functions everywhere except that along the real axis, see the discussion below (6.70) and (10.15). We have

$$
\begin{aligned}
\frac{1}{-\mathrm{i}\beta}\sum_m e^{\eta\omega_m}\,\hat{\Sigma}_{\mathrm{c}}^{\mathrm{M}}(\omega_m)\hat{\mathcal{G}}^{\mathrm{M}}(\omega_m) = \mp\int\frac{d\omega}{2\pi}f(\omega)\Big[&\hat{\Sigma}_{\mathrm{c}}^{\mathrm{M}}(\omega-\mathrm{i}\eta)\hat{\mathcal{G}}^{\mathrm{M}}(\omega-\mathrm{i}\eta) \\
&-\hat{\Sigma}_{\mathrm{c}}^{\mathrm{M}}(\omega+\mathrm{i}\eta)\hat{\mathcal{G}}^{\mathrm{M}}(\omega+\mathrm{i}\eta)\Big].
\end{aligned}
$$

Next we use (6.71) for $\hat{\mathcal{G}}^{\mathrm{M}}$ and (10.16) for $\hat{\Sigma}_{\mathrm{c}}^{\mathrm{M}}$. Shifting the integration variable $\omega \to \omega - \mu$, we get

$$
\frac{1}{-\mathrm{i}\beta}\sum_m e^{\eta\omega_m}\,\hat{\Sigma}_{\mathrm{c}}^{\mathrm{M}}(\omega_m)\hat{\mathcal{G}}^{\mathrm{M}}(\omega_m) = \mp\int\frac{d\omega}{2\pi}f(\omega-\mu)\Big[\hat{\Sigma}_{\mathrm{c}}^{\mathrm{A}}(\omega)\hat{\mathcal{G}}^{\mathrm{A}}(\omega) - \hat{\Sigma}_{\mathrm{c}}^{\mathrm{R}}(\omega)\hat{\mathcal{G}}^{\mathrm{R}}(\omega)\Big].
$$

Adding and subtracting $\hat{\Sigma}_{\mathrm{c}}^{\mathrm{A}}(\omega)\hat{\mathcal{G}}^{\mathrm{R}}(\omega)$ and using the fluctuation–dissipation theorem (6.68) for the Green's function and (10.7) for the self-energy, we can rewrite the correlation energy in (10.33) as

$$
E_{\mathrm{int,c}} = \pm\frac{\mathrm{i}}{2}\int\frac{d\omega}{2\pi}\int d\mathbf{x}_1\langle\mathbf{x}_1|\hat{\Sigma}_{\mathrm{c}}^{<}(\omega)\hat{\mathcal{G}}^{\mathrm{A}}(\omega) + \hat{\Sigma}_{\mathrm{c}}^{\mathrm{R}}(\omega)\hat{\mathcal{G}}^{<}(\omega)\,|\mathbf{x}_1\rangle,
\tag{10.34}
$$

where the cyclic property of the trace has been used. This formula does not change if we replace the lesser components with the greater components. To show it, the reader can use the relations

$$
\hat{\mathcal{G}}^{\mathrm{R/A}}(\omega) = \mathrm{i}\int\frac{d\omega'}{2\pi}\frac{\hat{\mathcal{G}}^{>}(\omega')-\hat{\mathcal{G}}^{<}(\omega')}{\omega-\omega'\pm\mathrm{i}\eta},\qquad \hat{\Sigma}_{\mathrm{c}}^{\mathrm{R/A}}(\omega) = \mathrm{i}\int\frac{d\omega'}{2\pi}\frac{\hat{\Sigma}_{\mathrm{c}}^{>}(\omega')-\hat{\Sigma}_{\mathrm{c}}^{<}(\omega')}{\omega-\omega'\pm\mathrm{i}\eta},
$$

already derived in (6.83) and (10.8). Independently of the Keldysh component, lesser or greater, the insertion of the above relations in (10.34) yields the following unique result:

$$
\boxed{E_{\mathrm{int,c}} = \mp\frac{1}{2}\int\frac{d\omega\,d\omega'}{2\pi\;2\pi}\int d\mathbf{x}_1\frac{\langle\mathbf{x}_1|\hat{\Sigma}_{\mathrm{c}}^{<}(\omega)\hat{\mathcal{G}}^{>}(\omega')-\hat{\Sigma}_{\mathrm{c}}^{>}(\omega)\hat{\mathcal{G}}^{<}(\omega')\,|\mathbf{x}_1\rangle}{\omega-\omega'}}
\tag{10.35}
$$

In this formula we remove the infinitesimal imaginary part in the denominator since for $\omega = \omega'$ the numerator vanishes.[6]

Correlation energy Equations (10.28), (10.32), and (10.35) allow us to calculate the energy of an interacting system in equilibrium at a certain temperature for any given many-body approximation to the self-energy. While it is easy to separate the Hartree–Fock part from the correlation part in E_{int}, the same cannot be said of E_{one}. It is in general quite useful to write the *total energy* E as the sum of the Hartree–Fock (noncorrelated) energy plus a correlation energy. The latter provides a measure of how much the system is correlated. A standard trick to perform this separation consists of using the Hellmann–Feynman theorem. The Hellmann–Feynman theorem, however, applies only to density matrices $\hat{\rho}$, which are a mixture of eigenstates with *fixed* coefficients. We therefore specialize the discussion to zero temperature and assume that the degeneracy of the ground-state multiplet of $\hat{H}^{\text{M}}_{\lambda} \equiv \hat{H}^{\text{M}}_{0} + \lambda \hat{H}_{\text{int}}$ does not change with λ. Then the zero-temperature density matrix reads

$$\hat{\rho}(\lambda) = \frac{1}{d} \sum_{g=1}^{d} |\Psi_g(\lambda)\rangle\langle\Psi_g(\lambda)|,$$

where d is the degeneracy and the sum runs over the components of the ground-state multiplet. As the weights in $\hat{\rho}(\lambda)$ are λ-independent, we can use the Hellmann–Feynman theorem and find

$$\frac{d}{d\lambda} \text{Tr}\left[\hat{\rho}(\lambda)\hat{H}_{\lambda}\right] = \text{Tr}\left[\hat{\rho}(\lambda)\hat{H}_{\text{int}}\right]. \tag{10.36}$$

Integrating this equation between $\lambda = 0$ and $\lambda = 1$ we get

$$E = E_0 + \int_0^1 \frac{d\lambda}{\lambda} \text{Tr}\left[\hat{\rho}(\lambda)\lambda\hat{H}_{\text{int}}\right] = E_0 + \int_0^1 \frac{d\lambda}{\lambda} E_{\text{int}}(\lambda),$$

with E the interacting energy and E_0 the energy of the noninteracting system. The *noncorrelated* part of the total energy is defined from the above equation when the interaction energy is evaluated in the Hartree–Fock approximation, $E_{\text{int}}[\Sigma] \to E_{\text{int}}[\Sigma_{\text{HF}}] \equiv E^{\text{HF}}_{\text{int}}$:

$$E^{\text{HF}} = E_0 + \int_0^1 \frac{d\lambda}{\lambda} E^{\text{HF}}_{\text{int}}(\lambda).$$

The correlation energy is therefore the difference between the total energy and the Hartree–Fock energy,

$$\boxed{E_{\text{corr}} \equiv E - E^{\text{HF}} = \int_0^1 \frac{d\lambda}{\lambda} \left(E_{\text{int}}(\lambda) - E^{\text{HF}}_{\text{int}}(\lambda)\right)} \tag{10.37}$$

It is important to appreciate the difference between $E^{\text{HF}}_{\text{int}} = E_{\text{int}}[\Sigma_{\text{HF}}]$ and $E_{\text{int,HF}} = E_{\text{int,HF}}[\Sigma]$. The quantity $E_{\text{int,HF}}[\Sigma]$ given in (10.32) is evaluated with a one-particle density matrix $n(\mathbf{x}_1, \mathbf{x}_2)$ [and density $n(\mathbf{x}) = n(\mathbf{x}, \mathbf{x})$], which comes from a Green's function

[6] From the fluctuation-dissipation theorem for Σ and G, the numerator is proportional to $[f(\omega)\bar{f}(\omega') - \bar{f}(\omega)f(\omega')]\hat{\Gamma}(\omega)\hat{A}(\omega')$, which vanishes for $\omega = \omega'$.

$G = G_0 + G_0\Sigma G$, where Σ can be the second Born, GW, or any other approximate self-energy. As such, $E_{\rm int,HF}[\Sigma]$ contains some correlation as well. Instead, $E_{\rm int}[\Sigma_{\rm HF}]$ is the interaction energy evaluated with a Hartree–Fock Green's function. From (10.35) we see that $E_{\rm int,c}[\Sigma_{\rm HF}] = 0$ and hence $E^{\rm HF}_{\rm int} = E_{\rm int}[\Sigma_{\rm HF}] = E_{\rm int,HF}[\Sigma_{\rm HF}]$. In Section 15.2 we evaluate the correlation energy of an electron gas in the GW approximation.

Exercise 10.2 Verify that $E_{\rm one} + E_{\rm int,HF}$ evaluated with the Hartree–Fock Green's function yields exactly the result in (8.34).

Exercise 10.3 Prove (10.36).

Exercise 10.4 Calculate $E_{\rm int,HF}$ for the Hubbard model.

10.4 Fluctuation–Dissipation Theorem for P and W

Like the self-energy, the polarization and the screened interaction fulfill a fluctuation–dissipation theorem. Let us first consider the polarization. Every diagram for $P(1;2)$ starts with a couple of Green's functions $G(\mathbf{x}_1, z_1; \ldots)G(\ldots; \mathbf{x}_1, z_1)$ and ends with a couple of Green's functions $G(\ldots; \mathbf{x}_2, z_2)G(\mathbf{x}_2, z_2; \ldots)$. Introducing the operator (in first quantization) $\hat{\mathcal{P}}(z_1, z_2)$ with matrix elements

$$\langle \mathbf{x}_1|\hat{\mathcal{P}}(z_1, z_2)|\mathbf{x}_2\rangle = P(1;2),$$

the identities (10.1) and (10.2) imply that

$$\hat{\mathcal{P}}^{\rceil}(\tau, t') = \hat{\mathcal{P}}^{>}(t_0 - \mathrm{i}\tau, t') \quad , \quad \hat{\mathcal{P}}^{\lceil}(t, \tau') = \hat{\mathcal{P}}^{<}(t, t_0 - \mathrm{i}\tau')$$

and

$$\hat{\mathcal{P}}^{\rm M}(\tau, \tau') = \begin{cases} \hat{\mathcal{P}}^{>}(t_0 - \mathrm{i}\tau, t_0 - \mathrm{i}\tau') & \tau > \tau' \\ \hat{\mathcal{P}}^{<}(t_0 - \mathrm{i}\tau, t_0 - \mathrm{i}\tau') & \tau < \tau' \end{cases}.$$

To derive a fluctuation–dissipation theorem for P we need the boundary conditions for this quantity. The polarization $P(1;2)$ contains two Green's functions with argument 1 and two Green's functions with argument 2. Since the Green's function satisfies the KMS relations (5.7), the polarization satisfies the KMS relations below:

$$\hat{\mathcal{P}}(z_1, t_{0-}) = \hat{\mathcal{P}}(z_1, t_0 - \mathrm{i}\beta) \quad , \quad \hat{\mathcal{P}}(t_{0-}, z_2) = \hat{\mathcal{P}}(t_0 - \mathrm{i}\beta, z_2).$$

Therefore we can write

$$\begin{aligned} \hat{\mathcal{P}}^{<}(t_0, t') &= \hat{\mathcal{P}}(t_{0-}, t'_+) \\ &= \hat{\mathcal{P}}(t_0 - \mathrm{i}\beta, t'_+) \\ &= \hat{\mathcal{P}}^{\rceil}(\beta, t') \\ &= \hat{\mathcal{P}}^{>}(t_0 - \mathrm{i}\beta, t'). \end{aligned}$$

Fourier transforming both sides of this equation, we find the important relation

$$\hat{\mathcal{P}}^{>}(\omega) = e^{\beta\omega}\hat{\mathcal{P}}^{<}(\omega).$$

Then the fluctuation–dissipation theorem for the polarization reads

$$\boxed{\hat{\Pi}(\omega) \equiv \mathrm{i}[\hat{\mathcal{P}}^{>}(\omega) - \hat{\mathcal{P}}^{<}(\omega)] \quad \Rightarrow \quad \left\{ \begin{array}{l} \hat{\mathcal{P}}^{<}(\omega) = -\mathrm{i}f(\omega)\hat{\Pi}(\omega) \\[2ex] \hat{\mathcal{P}}^{>}(\omega) = -\mathrm{i}\bar{f}(\omega)\hat{\Pi}(\omega) \end{array} \right.} \tag{10.38}$$

where $f(\omega) = 1/(e^{\beta\omega} - 1)$ is the Bose function and $\bar{f}(\omega) = 1 + f(\omega)$. Clearly the density response function χ defined in (7.24) fulfills the same fluctuation–dissipation theorem as P, since the topology of a χ-diagram is the same as the topology of a P-diagram. In Section 13.2.1 we give an alternative proof of the fluctuation–dissipation theorem for χ based on the Lehmann representation of this quantity.

The derivation of the fluctuation–dissipation theorem for W goes along the same lines. Let us write W in terms of χ according to $W = v + v\chi v$, see (7.25). Since the interaction has only a singular part, we have $v^{\mathrm{R}} = v^{\mathrm{A}} = v$ and $v^{\lessgtr} = 0$. Therefore the lesser and greater screened interaction is simply $W^{\lessgtr} = v\chi^{\lessgtr}v$, and similarly $W^{\lceil} = v\chi^{\lceil}v$ and $W^{\rceil} = v\chi^{\rceil}v$. The interaction v acts like a simple multiplicative factor for the time variable[7] and W fulfills the same relations as χ, which are identical to the relations for P. Therefore, introducing the operator (in first quantization) $\hat{\mathcal{W}}(z_1, z_2)$ with matrix elements

$$\langle \mathbf{x}_1|\hat{\mathcal{W}}(z_1, z_2)|\mathbf{x}_2\rangle = W(1;2),$$

we can write

$$\hat{\mathcal{W}}^{>}(\omega) = e^{\beta\omega}\hat{\mathcal{W}}^{<}(\omega).$$

The fluctuation–dissipation theorem for W reads

$$\boxed{\hat{\Omega}(\omega) \equiv \mathrm{i}[\hat{\mathcal{W}}^{>}(\omega) - \hat{\mathcal{W}}^{<}(\omega)] \quad \Rightarrow \quad \left\{ \begin{array}{l} \hat{\mathcal{W}}^{<}(\omega) = -\mathrm{i}f(\omega)\hat{\Omega}(\omega) \\[2ex] \hat{\mathcal{W}}^{>}(\omega) = -\mathrm{i}\bar{f}(\omega)\hat{\Omega}(\omega) \end{array} \right.} \tag{10.39}$$

with f the Bose function.

Relation between retarded/advanced and Matsubara components Let us write the screened interaction as the sum of the singular part v (local in the contour times) and a rest

$$W(1;2) = v(1;2) + \delta W(1;2).$$

Since $v^{\lessgtr} = 0$, we can equivalently write

$$\hat{\Omega}(\omega) \equiv \mathrm{i}[\delta\hat{\mathcal{W}}^{>}(\omega) - \delta\hat{\mathcal{W}}^{<}(\omega)].$$

[7]This is not true for the space variable since $v\chi v$ involves two space convolutions.

Proceeding along the same line of reasoning as in Section 10.1, one can show that

$$\boxed{\hat{\mathcal{P}}^{\mathrm{R/A}}(\omega) = \int \frac{d\omega'}{2\pi} \frac{\hat{\Pi}(\omega')}{\omega - \omega' \pm \mathrm{i}\eta}} \qquad \boxed{\delta\hat{\mathcal{W}}^{\mathrm{R/A}}(\omega) = \int \frac{d\omega'}{2\pi} \frac{\hat{\Omega}(\omega')}{\omega - \omega' \pm \mathrm{i}\eta}} \tag{10.40}$$

The relation between the retarded/advanced components and the Matsubara component is similar to that for the Green's function and self-energy:

$$\hat{\mathcal{P}}^{\mathrm{M}}(\zeta) = \begin{cases} \hat{\mathcal{P}}^{\mathrm{R}}(\zeta) & \text{for } \mathrm{Im}[\zeta] > 0 \\ \hat{\mathcal{P}}^{\mathrm{A}}(\zeta) & \text{for } \mathrm{Im}[\zeta] < 0 \end{cases},$$

and

$$\delta\hat{\mathcal{W}}^{\mathrm{M}}(\zeta) = \begin{cases} \delta\hat{\mathcal{W}}^{\mathrm{R}}(\zeta) & \text{for } \mathrm{Im}[\zeta] > 0 \\ \delta\hat{\mathcal{W}}^{\mathrm{A}}(\zeta) & \text{for } \mathrm{Im}[\zeta] < 0 \end{cases}.$$

Thus, $\hat{\mathcal{P}}^{\mathrm{M}}(\zeta)$ and $\delta\hat{\mathcal{W}}^{\mathrm{M}}(\zeta)$ are analytic everywhere except along the real axis. In particular, for $\zeta = \omega \pm \mathrm{i}\eta$, we find

$$\boxed{\hat{\mathcal{P}}^{\mathrm{M}}(\omega \pm \mathrm{i}\eta) = \hat{\mathcal{P}}^{\mathrm{R/A}}(\omega)} \qquad \boxed{\delta\hat{\mathcal{W}}^{\mathrm{M}}(\omega \pm \mathrm{i}\eta) = \delta\hat{\mathcal{W}}^{\mathrm{R/A}}(\omega)} \tag{10.41}$$

In Section 11.8.1 we use the fluctuation-dissipation theorems for the polarization and the screened interaction to derive an alternative formula for the correlated part of the interaction energy (10.35)

10.5 Recovering Equilibrium from the Kadanoff-Baym Equations

At the end of Section 9.2 we saw that the Kadanoff-Baym equations for systems in equilibrium are not obviously solved by a Green's function depending on the time difference. In this section we show that the fluctuation-dissipation theorem for the Green's function and the self-energy is all what we need to show that the r.h.s. of (9.27) and (9.28) depends only on $t_1 - t_2$ *and* is independent of t_0. In particular, we prove that

$$\left[\hat{\Sigma}_{\mathrm{c}}^{\lessgtr}\cdot \hat{\mathcal{G}}^{\mathrm{A}} + \hat{\Sigma}_{\mathrm{c}}^{\mathrm{R}}\cdot \hat{\mathcal{G}}^{\lessgtr} + \hat{\Sigma}_{\mathrm{c}}^{\rceil}\star \hat{\mathcal{G}}^{\lceil}\right](t_1,t_2) = \int \frac{d\omega}{2\pi} e^{-\mathrm{i}\omega(t_1-t_2)}\left[\hat{\Sigma}_{\mathrm{c}}^{\lessgtr}(\omega)\hat{\mathcal{G}}^{\mathrm{A}}(\omega) + \hat{\Sigma}_{\mathrm{c}}^{\mathrm{R}}(\omega)\hat{\mathcal{G}}^{\lessgtr}(\omega)\right], \tag{10.42}$$

$$\left[\hat{\mathcal{G}}^{\lessgtr}\cdot \hat{\Sigma}_{\mathrm{c}}^{\mathrm{A}} + \hat{\mathcal{G}}^{\mathrm{R}}\cdot \hat{\Sigma}_{\mathrm{c}}^{\lessgtr} + \hat{\mathcal{G}}^{\rceil}\star \hat{\Sigma}_{\mathrm{c}}^{\lceil}\right](t_1,t_2) = \int \frac{d\omega}{2\pi} e^{-\mathrm{i}\omega(t_1-t_2)}\left[\hat{\mathcal{G}}^{\lessgtr}(\omega)\hat{\Sigma}_{\mathrm{c}}^{\mathrm{A}}(\omega) + \hat{\mathcal{G}}^{\mathrm{R}}(\omega)\hat{\Sigma}_{\mathrm{c}}^{\lessgtr}(\omega)\right], \tag{10.43}$$

for all t_0. In these equations we replaced $\hat{\Sigma} \to \hat{\Sigma}_{\mathrm{c}}$ since

$$\hat{\Sigma}^{\lessgtr,\lceil,\rceil} = \hat{\Sigma}_{\mathrm{c}}^{\lessgtr,\lceil,\rceil}, \qquad \hat{\Sigma}^{\mathrm{R/A}} = \hat{\Sigma}_{\mathrm{HF}}^{\mathrm{R/A}} + \hat{\Sigma}_{\mathrm{c}}^{\mathrm{R/A}},$$

and the products $\hat{\Sigma}_{\mathrm{HF}}^{\mathrm{R}} \cdot \hat{\mathcal{G}}^{\lessgtr}$ and $\hat{\mathcal{G}}^{\lessgtr} \cdot \hat{\Sigma}_{\mathrm{HF}}^{\mathrm{A}}$ depend only on the time difference when $\hat{\mathcal{G}}^{\lessgtr}$ depends only on the time difference ($\hat{\Sigma}_{\mathrm{HF}}^{\mathrm{R/A}}$ is local in time). Thus, for our purposes it is enough to prove (10.42) and (10.43). We would like to point out that standard derivations of (10.42) and (10.43) typically require a few extra (but superfluous) assumptions. Either one uses the adiabatic assumption, in which case $\hat{\Sigma}_c^{\rceil} = \hat{\Sigma}_c^{\lceil} = 0$ (since the self-energy vanishes along the imaginary track) while the convolutions along the real-time axis become the r.h.s. of (10.42) and (10.43) (since $t_0 \to -\infty$). Or, alternatively, one assumes that

$$\lim_{t_0 \to -\infty} \hat{\Sigma}_c^{\rceil}(t, \tau) = \lim_{t_0 \to -\infty} \hat{\Sigma}_c^{\lceil}(\tau, t) = 0, \tag{10.44}$$

which is often satisfied in systems with infinitely many degrees of freedom. However, in systems with an arbitrary large but finite single-particle basis the Green's function is an oscillatory function and so is the self-energy, see again the discussion in Section 6.1.3. Therefore, (10.44) is not satisfied in real or model systems with a discrete spectrum. It is, however, more important to realize that the standard derivations "prove" (10.42) and (10.43) only for $t_0 \to -\infty$. In the following we prove them correct for all t_0 and without any extra assumptions. The derivation nicely illustrates how the apparent dependence on t_0 disappears. Equations (10.42) and (10.43) are therefore much more fundamental than what is commonly thought.

Let us consider the lesser version of (10.42). The l.h.s. is the lesser component of the collision integral $\hat{\mathcal{I}}_L^<(t_1, t_2)$. We use the identity (10.1) for $\hat{\mathcal{G}}^{\lceil}$ and the identity (10.3) for $\hat{\Sigma}^{\rceil} = \hat{\Sigma}_c^{\rceil}$. Expanding all Green's functions and self-energies in Fourier integrals, we get

$$\hat{\mathcal{I}}_L^<(t_1, t_2) = \int \frac{d\omega_1}{2\pi}\frac{d\omega_2}{2\pi} e^{-\mathrm{i}\omega_1 t_1 + \mathrm{i}\omega_2 t_2} \left[\int_{t_0}^{t_2} dt\, e^{\mathrm{i}(\omega_1 - \omega_2)t}\, \hat{\Sigma}_c^<(\omega_1)\hat{\mathcal{G}}^A(\omega_2) \right.$$
$$\left. + \int_{t_0}^{t_1} dt\, e^{\mathrm{i}(\omega_1 - \omega_2)t}\, \hat{\Sigma}_c^R(\omega_1)\hat{\mathcal{G}}^<(\omega_2) - \mathrm{i}\int_0^{\beta} d\tau\, e^{(\omega_1 - \omega_2)(\mathrm{i}t_0 + \tau)}\, \hat{\Sigma}_c^<(\omega_1)\hat{\mathcal{G}}^>(\omega_2) \right].$$

For all these integrals to be well behaved when $t_0 \to -\infty$, we give to ω_1 a small negative imaginary part, $\omega_1 \to \omega_1 - \mathrm{i}\eta/2$, and to ω_2 a small positive imaginary part, $\omega_2 \to \omega_2 + \mathrm{i}\eta/2$. This is just a regularization and it has nothing to do with the adiabatic assumption.[8] With this regularization (10.42) is easily recovered in the limit $t_0 \to -\infty$. In the following we keep

[8]Suppose that we want to recover the result

$$\int_{-\infty}^{\infty} dt\, e^{\mathrm{i}\omega t} = 2\pi\delta(\omega)$$

from the integral $I(T) = \int_{-T}^{T} dt\, e^{\mathrm{i}\omega t}$ when $T \to \infty$. Then we can write

$$I(T) = \int_{-T}^{0} dt\, e^{\mathrm{i}\omega t} + \int_0^T dt\, e^{\mathrm{i}\omega t} = \int_{-T}^{0} dt\, e^{\mathrm{i}(\omega - \mathrm{i}\eta)t} + \int_0^T dt\, e^{\mathrm{i}(\omega + \mathrm{i}\eta)t},$$

where in the last step we simply regularize the integral so that it is well behaved for $T \to \infty$. At the end of the calculation we send $\eta \to 0$. Performing the integral, we find

$$I(T) = \frac{1 - e^{-\mathrm{i}(\omega - \mathrm{i}\eta)T}}{\mathrm{i}(\omega - \mathrm{i}\eta)} + \frac{e^{\mathrm{i}(\omega + \mathrm{i}\eta)T} - 1}{\mathrm{i}(\omega + \mathrm{i}\eta)} \xrightarrow[T \to \infty]{} \frac{1}{\mathrm{i}(\omega - \mathrm{i}\eta)} - \frac{1}{\mathrm{i}(\omega + \mathrm{i}\eta)}.$$

Using the Cauchy relation $1/(\omega \pm \mathrm{i}\eta) = P(1/\omega) \mp \mathrm{i}\pi\delta(\omega)$, with P the principal part, we recover the δ-function result as the regularized limit of $I(T)$.

t_0 finite and show that (10.42) is still true. Performing the integrals over time and using the fluctuation–dissipation theorem to express everything in terms of retarded and advanced quantities, we find

$$
\begin{aligned}
\hat{\mathcal{I}}_L^<(t_1,t_2) = \int \frac{d\omega_1}{2\pi}\frac{d\omega_2}{2\pi}\frac{e^{-\mathrm{i}\omega_1 t_1+\mathrm{i}\omega_2 t_2}}{\mathrm{i}(\omega_1-\omega_2-\mathrm{i}\eta)} \\
\times \Big[\pm f_1 \left(e^{\mathrm{i}(\omega_1-\omega_2)t_2} - e^{\mathrm{i}(\omega_1-\omega_2)t_0}\right)\left(\hat{\Sigma}^{\mathrm{R}}_{\mathrm{c},1} - \hat{\Sigma}^{\mathrm{A}}_{\mathrm{c},1}\right)\hat{\mathcal{G}}^{\mathrm{A}}_2 \\
\pm f_2 \left(e^{\mathrm{i}(\omega_1-\omega_2)t_1} - e^{\mathrm{i}(\omega_1-\omega_2)t_0}\right)\hat{\Sigma}^{\mathrm{R}}_{\mathrm{c},1}\left(\hat{\mathcal{G}}^{\mathrm{R}}_2 - \hat{\mathcal{G}}^{\mathrm{A}}_2\right) \\
\pm f_1\bar{f}_2 \left(e^{(\omega_1-\omega_2)\beta} - 1\right) e^{\mathrm{i}(\omega_1-\omega_2)t_0}\left(\hat{\Sigma}^{\mathrm{R}}_{\mathrm{c},1} - \hat{\Sigma}^{\mathrm{A}}_{\mathrm{c},1}\right)\left(\hat{\mathcal{G}}^{\mathrm{R}}_2 - \hat{\mathcal{G}}^{\mathrm{A}}_2\right)\Big], \quad (10.45)
\end{aligned}
$$

where we introduce the short-hand notation f_1 to denote the Bose/Fermi function $f(\omega_1)$ and similarly $f_2 = f(\omega_2)$ and *mutatis mutandis* the self-energy and the Green's function. Next we observe that

$$f_1\bar{f}_2\left(e^{(\omega_1-\omega_2)\beta} - 1\right) = f_2 - f_1.$$

Using this relation in (10.45) we achieve a considerable simplification, since many terms cancel out and we are left with

$$
\begin{aligned}
\hat{\mathcal{I}}_L^<(t_1,t_2) = \int \frac{d\omega_1}{2\pi}\frac{d\omega_2}{2\pi}\frac{e^{-\mathrm{i}\omega_1 t_1+\mathrm{i}\omega_2 t_2}}{\mathrm{i}(\omega_1-\omega_2-\mathrm{i}\eta)}\Big[e^{\mathrm{i}(\omega_1-\omega_2)t_2}\hat{\Sigma}^<_{\mathrm{c},1}\hat{\mathcal{G}}^{\mathrm{A}}_2 + e^{\mathrm{i}(\omega_1-\omega_2)t_1}\hat{\Sigma}^{\mathrm{R}}_{\mathrm{c},1}\hat{\mathcal{G}}^<_2 \\
- e^{\mathrm{i}(\omega_1-\omega_2)t_0}\hat{\Sigma}^<_{\mathrm{c},1}\hat{\mathcal{G}}^{\mathrm{R}}_2 - e^{\mathrm{i}(\omega_1-\omega_2)t_0}\hat{\Sigma}^{\mathrm{A}}_{\mathrm{c},1}\hat{\mathcal{G}}^<_2\Big]. \quad (10.46)
\end{aligned}
$$

To get rid of the t_0-dependence we exploit the identity

$$0 = \int_{t_0}^{t_2} dt\, \hat{\Sigma}^<_{\mathrm{c}}(t_1,t)\hat{\mathcal{G}}^{\mathrm{R}}(t,t_2),$$

which follows from the fact that the retarded Green's function vanishes whenever its first argument is smaller than the second. In Fourier space this identity looks much more interesting since

$$0 = \int \frac{d\omega_1}{2\pi}\frac{d\omega_2}{2\pi}\frac{e^{-\mathrm{i}\omega_1 t_1+\mathrm{i}\omega_2 t_2}}{\mathrm{i}(\omega_1-\omega_2-\mathrm{i}\eta)}\left(e^{\mathrm{i}(\omega_1-\omega_2)t_2} - e^{\mathrm{i}(\omega_1-\omega_2)t_0}\right)\hat{\Sigma}^<_{\mathrm{c},1}\hat{\mathcal{G}}^{\mathrm{R}}_2.$$

Thus, we see that we can replace t_0 with t_2 in the first term of the second line of (10.46). In a similar way we can show that the second t_0 in (10.46) can be replaced with t_1 and hence the collision integral can be rewritten as

$$
\begin{aligned}
\hat{\mathcal{I}}_L^<(t_1,t_2) = \int \frac{d\omega_1}{2\pi}e^{-\mathrm{i}\omega_1(t_1-t_2)}\hat{\Sigma}^<_{\mathrm{c}}(\omega_1)\underbrace{\int \frac{d\omega_2}{2\pi}\frac{\mathrm{i}[\hat{\mathcal{G}}^{\mathrm{R}}(\omega_2) - \hat{\mathcal{G}}^{\mathrm{A}}(\omega_2)]}{\omega_1-\omega_2-\mathrm{i}\eta}}_{\hat{\mathcal{G}}^{\mathrm{A}}(\omega_1)} \\
+ \int \frac{d\omega_2}{2\pi}e^{-\mathrm{i}\omega_2(t_1-t_2)}\underbrace{\int \frac{d\omega_1}{2\pi}\frac{\mathrm{i}[\hat{\Sigma}^{\mathrm{R}}_{\mathrm{c}}(\omega_1) - \hat{\Sigma}^{\mathrm{A}}_{\mathrm{c}}(\omega_1)]}{\omega_2-\omega_1+\mathrm{i}\eta}}_{\hat{\Sigma}^{\mathrm{R}}_{\mathrm{c}}(\omega_2)}\hat{\mathcal{G}}^<(\omega_2).
\end{aligned}
$$

The quantities below the underbraces are the result of the frequency integral, see (6.83) and (10.8). We have thus proven the lesser version of (10.42) for all t_0. The reader can verify that the greater version of (10.42) and (10.43) can be derived in a similar manner. We observe that for $t_1 = t_2 = t_0$ the l.h.s. of (10.42) and (10.43) reduces to the convolution along the vertical track. This convolution is identical to the one appearing in the interaction energy (10.33). The reader can verify that (10.34) is indeed consistent with the r.h.s. of (10.42) and (10.43).

Recovering the equilibrium solution for P and W Analogous arguments can be used to recover the equilibrium solution from the Dyson equation $W = v + vPW$. Let us be more precise. We define the operator (in first quantization) of the bare interaction $\hat{v}(z_1, z_2) = \delta(z_1, z_2)\hat{v}$ with matrix elements

$$\langle \mathbf{x}_1|\hat{v}(z_1, z_2)|\mathbf{x}_2\rangle = \delta(z_1, z_2)v(\mathbf{x}_1, \mathbf{x}_2).$$

Then the lesser/greater component of the Dyson equation $W = v + vPW$ can be written as

$$\hat{\mathcal{W}}^{\lessgtr}(t, t') = \hat{v}\left[\hat{\mathcal{P}}^{\lessgtr} \cdot \hat{\mathcal{W}}^{\mathrm{A}} + \hat{\mathcal{P}}^{\mathrm{R}} \cdot \hat{\mathcal{W}}^{\lessgtr} + \hat{\mathcal{P}}^{\rceil} \cdot \hat{\mathcal{W}}^{\lceil}\right](t, t'). \tag{10.47}$$

For systems in equilibrium, the l.h.s. depends only on the time difference. However, it is not obvious that this is true also for the r.h.s. since the time convolutions are either from t_0 to ∞ or from 0 to β. In order to prove that the r.h.s. depends only on $t - t'$ and is independent of the initial time t_0 we must use the fluctuation–dissipation theorem derived in Section 10.4, as well as that real and imaginary parts of the retarded/advanced P and $\delta W = W - v$ are related by a Hilbert transformation (here δW is the regular part of the screened interaction). Proceeding along the same lines as for the proof of (10.42) and (10.43), we find the (expected) result

$$\begin{aligned}&\left[\hat{\mathcal{P}}^{\lessgtr} \cdot \delta\hat{\mathcal{W}}^{\mathrm{A}} + \hat{\mathcal{P}}^{\mathrm{R}} \cdot \delta\hat{\mathcal{W}}^{\lessgtr} + \hat{\mathcal{P}}^{\rceil} \cdot \delta\hat{\mathcal{W}}^{\lceil}\right](t, t') \\ &\qquad = \int \frac{d\omega}{2\pi} e^{-\mathrm{i}\omega(t-t')}\left[\hat{\mathcal{P}}^{\lessgtr}(\omega)\delta\hat{\mathcal{W}}^{\mathrm{A}}(\omega) + \hat{\mathcal{P}}^{\mathrm{R}}(\omega)\delta\hat{\mathcal{W}}^{\lessgtr}(\omega)\right].\end{aligned} \tag{10.48}$$

More generally, the following is always true:

> *If the system is in thermal equilibrium, we can simplify the Langreth rules of Table 5.1 by taking $t_0 \to -\infty$ and discarding the vertical track. In this way, real-time convolutions become simple products in frequency space.*

Exercise 10.5 Prove (10.48).

11

Grand Potential: Diagrammatic Expansion and Variational Forms

In this chapter we continue to explore the diagrammatic expansions and focus on the grand potential $\Omega = -\frac{1}{\beta}\ln Z$, with β the inverse temperature and Z the interacting partition function. The analysis leads us to define a functional of the Green's function G having the special property that its functional derivative with respect to G is the self-energy. The same functional can also be used to construct variational forms for Ω that are stationary at the solution of the Dyson equation and that equal the grand potential at the stationary point.

11.1 Linked Cluster Theorem

In Chapter 7 we briefly discussed the diagrammatic expansion of the partition function Z/Z_0 in terms of G_0 and v. We saw that this expansion involves the sum of vacuum diagrams and we also enunciated the diagrammatic rules in Section 7.1. The vacuum diagrams are either connected or consist of disjoint pieces which are, therefore, proportional to the product of connected vacuum diagrams. It would be nice to get rid of the disjoint pieces and to derive a formula for Z/Z_0 in terms of connected diagrams only. To show how this can be done, let us start with an example. Consider the diagram

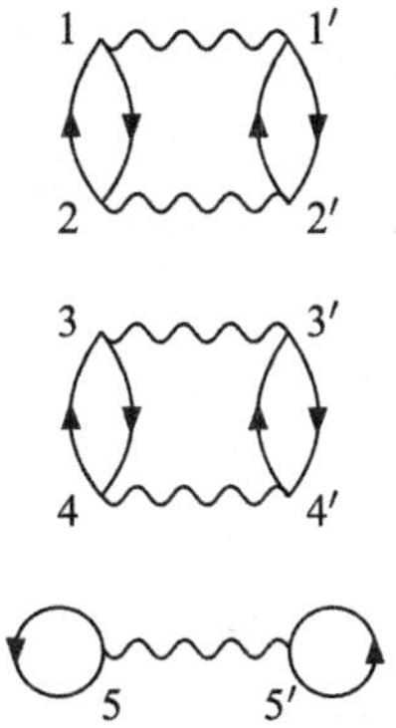

$$= \frac{1}{5!}\left(\frac{\mathrm{i}}{2}\right)^5 (\pm)^6 \int [\underbrace{v \ldots\ldots v}_{5\text{ interactions}} \times \underbrace{G_0 \ldots\ldots G_0}_{10\text{ Green's functions}}], \tag{11.1}$$

where we use the Feynman rules of Section 7.1 to convert the diagram into the mathematical expression on the r.h.s. Corresponding to this diagram are several others that yield the same

value. For instance,

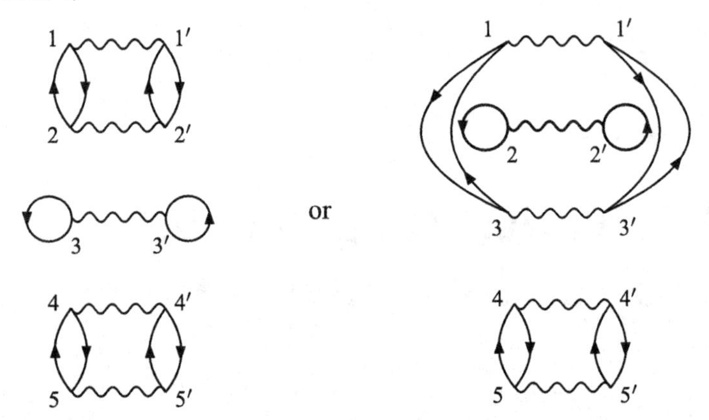

These diagrams simply correspond to different choices of the interaction lines $v(k;k')$ used to draw the connected vacuum diagrams. Let us calculate how many of these diagrams there are. To draw the top diagram in (11.1) we have to choose two interaction lines out of five, and this can be done in $\binom{5}{2} = 10$ ways. The middle diagram in (11.1) can then be drawn by choosing two interaction lines out of the remaining three, and hence we have $\binom{3}{2} = 3$ possibilities. Finally, the bottom diagram requires the choice of one interaction line out of the only one remaining, which can be done in only one way. Since the top and middle diagrams have the same form and the order in which we construct them does not matter, we still have to divide by $2!$. Indeed, in our construction the step

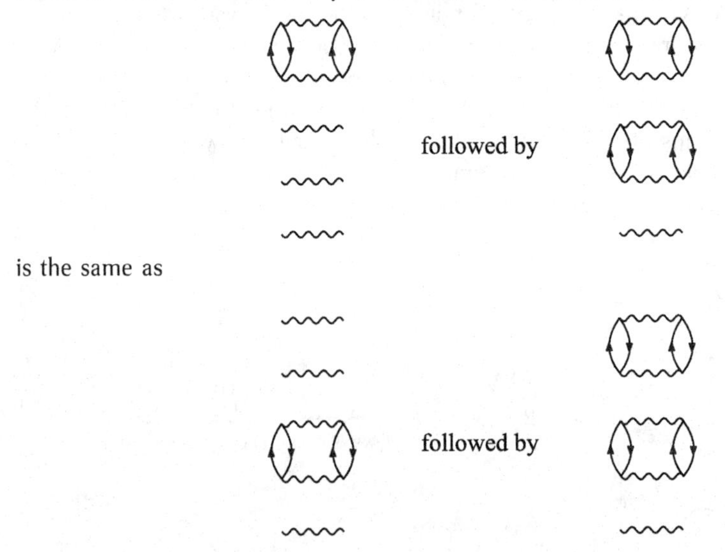

We thus find $\frac{1}{2!}\binom{5}{2}\binom{3}{2} = 15$ diagrams that yield the same value (11.1). It is now easy to make this argument general.

Let us label all the connected vacuum diagrams with an index $i = 1, \ldots, \infty$ and denote by $V_{c,i} = \int [v \ldots v G_0 \ldots G_0]_i$ the integral corresponding to the ith diagram. The number n_i of interaction lines in $V_{c,i}$ is the order of the diagram. A general (and hence not necessarily connected) vacuum diagram contains k_1 times the diagram $V_{c,1}$, k_2 times the diagram $V_{c,2}$, and so on, and therefore its order is

$$n = n_1 k_1 + n_2 k_2 + \ldots,$$

with $k_1,\ k_2, \ldots$ integers between 0 and ∞. There are $\binom{n}{n_1}$ ways to construct the first diagram $V_{c,1}$, $\binom{n-n_1}{n_1}$ ways to construct the second diagram $V_{c,1}$, etc. When we exhaust the diagrams of type $V_{c,1}$ we can start drawing the diagrams $V_{c,2}$. The first diagram $V_{c,2}$ can be constructed in $\binom{n-n_1k_1}{n_2}$ ways, etc. Finally, we have to divide by $k_1!k_2!\ldots$ in order to compensate for the different order in which the same diagrams can be drawn. We thus find that the considered vacuum diagram appears

$$\frac{1}{k_1!k_2!\ldots}\binom{n}{n_1}\binom{n-n_1}{n_1}\cdots\binom{n-n_1k_1}{n_2}\cdots = \frac{1}{k_1!k_2!\ldots}\,\frac{n!}{(n_1!)^{k_1}(n_2!)^{k_2}\ldots}$$

times in the perturbative expansion of Z/Z_0. The total contribution of this type of diagram to Z/Z_0 is therefore

$$\begin{aligned}
&\frac{1}{k_1!k_2!\ldots}\,\frac{n!}{(n_1!)^{k_1}(n_2!)^{k_2}\ldots}\,\frac{1}{n!}\left(\frac{\mathrm{i}}{2}\right)^n (\pm)^l \int [\underbrace{v\ldots v}_{n \text{ times}} \times \underbrace{G_0\ldots G_0}_{2n \text{ times}}] \\
&= \frac{1}{k_1!}\left[\frac{1}{n_1!}\left(\frac{\mathrm{i}}{2}\right)^{n_1}(\pm)^{l_1} V_{c,1}\right]^{k_1} \frac{1}{k_2!}\left[\frac{1}{n_2!}\left(\frac{\mathrm{i}}{2}\right)^{n_2}(\pm)^{l_2} V_{c,2}\right]^{k_2}\ldots
\end{aligned}$$

where l_i is the number of loops in diagram $V_{c,i}$ and $l = k_1 l_1 + k_2 l_2 + \ldots$ is the total number of loops. We have obtained a product of separate connected vacuum diagrams

$$D_{c,i} \equiv \frac{1}{n_i!}\left(\frac{\mathrm{i}}{2}\right)^{n_i}(\pm)^{l_i} V_{c,i}$$

with the right prefactors. To calculate Z/Z_0 we simply have to sum the above expression over all k_i between 0 and ∞

$$\frac{Z}{Z_0} = \sum_{k_1=0}^{\infty}\frac{1}{k_1!}\left(D_{c,1}\right)^{k_1}\sum_{k_2=0}^{\infty}\frac{1}{k_2!}\left(D_{c,2}\right)^{k_2}\ldots = e^{D_{c,1}+D_{c,2}+\ldots}\ . \tag{11.2}$$

This elegant result is known as the *linked cluster theorem,* since it shows that for calculating the partition function it is enough to consider connected vacuum diagrams.

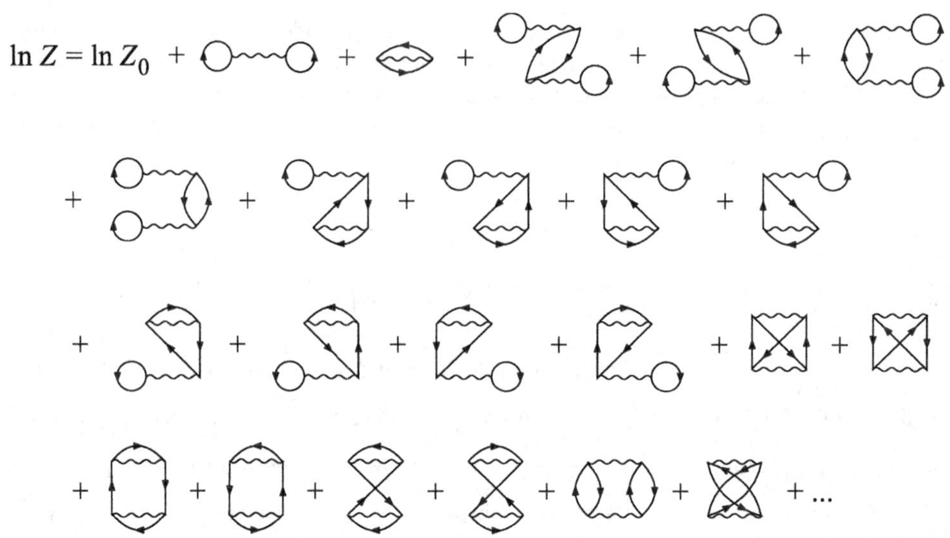

Figure 11.1 Diagrammatic expansion of $\ln Z$ up to second order in the interaction.

Another way to write (11.2) is

$$
\begin{aligned}
\ln\frac{Z}{Z_0} &= \sum_i D_{c,i} \\
&= \sum_{k=0}^{\infty} \frac{1}{k!}\left(\frac{\mathrm{i}}{2}\right)^k \int v(1;1')\dots v(k;k')
\begin{vmatrix}
G_0(1;1^+) & G_0(1;1'^+) & \dots & G_0(1;k'^+) \\
G_0(1';1^+) & G_0(1';1'^+) & \dots & G_0(1';k'^+) \\
\vdots & \vdots & \ddots & \vdots \\
G_0(k';1^+) & G_0(k';1'^+) & \dots & G_0(k';k'^+)
\end{vmatrix}_{\substack{\pm \\ c}}
\end{aligned}
\tag{11.3}
$$

where the integral is over all variables and the symbol $|\dots|_{\substack{\pm \\ c}}$ signifies that in the expansion of the permanent/determinant only the terms represented by connected diagrams are retained. As the logarithm of the partition function is related to the grand potential Ω through the relation $\Omega = -\frac{1}{\beta}\ln Z$, see Appendix D, the linked cluster theorem is an MBPT expansion of Ω in terms of connected vacuum diagrams. In Fig. 11.1 we show the diagrammatic expansion of $\ln Z$ up to second order in the interaction. There are 2 first-order diagrams and 20 second-order diagrams against the $4! = 24$ second-order diagrams resulting from the expansion of the permanent/determinant in (5.19). Thus, the achieved reduction is rather modest.[1] Nevertheless, there is still much symmetry in these diagrams. For instance, the first four second-order diagrams have the same numerical value; the same is true for the next eight diagrams, for the last two diagrams of the third row, for the first four diagrams of the

[1] It should be noticed, however, that 20 diagrams for $\ln Z$ correspond to infinitely many diagrams for $Z = e^{\ln Z}$.

last row, and for the last two diagrams. We can therefore reduce the number of diagrams further if we understand how many vacuum diagrams have the same topology. As we see in the next section, this is less straightforward than for the Green's function diagrams (see Section 7.4) due to the absence of external vertices.

Exercise 11.1 Show that the expansion of $\ln Z$ to second order in the interaction yields the diagrams of Fig. 11.1.

11.2 Summing Only the Topologically Inequivalent Diagrams

In Section 7.4 we showed that each nth-order connected diagram for the Green's function comes in $2^n n!$ variants. An example was given in Fig. 7.4 where we have drawn eight topologically equivalent second-order diagrams. Unfortunately, counting the number of variants in which a vacuum diagram can appear is not as easy. For example, the last two diagrams of Fig. 11.1 are topologically equivalent; since there are no other diagrams with the same topology, the number of variants is, in this case, only 2. For the first diagram in the last row of Fig. 11.1 the number of variants is 4, for the first diagram in the third row of Fig. 11.1 the number of variants is 8, etc. The smaller number of variants for some of the vacuum diagrams is due to the fact that not all permutations and/or mirrorings of the interaction lines lead to a different vacuum diagram – that is, to a different term in the expansion (11.3). If we label the internal vertices of the last two diagrams of Fig. 11.1,

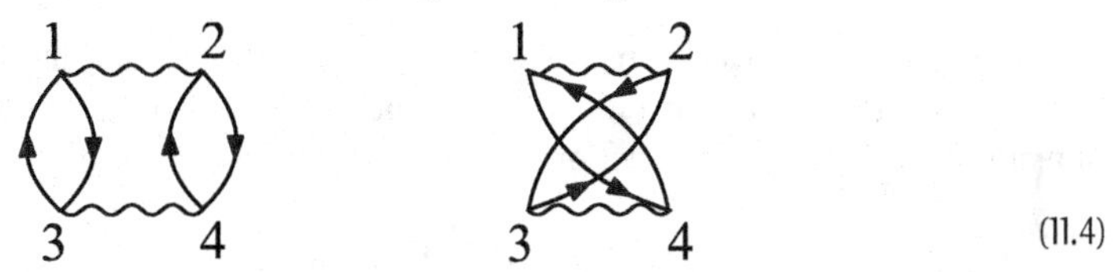

(11.4)

we see that the simultaneous mirroring $v(1;2) \to v(2;1)$ and $v(3;4) \to v(4;3)$ maps each diagram into itself, while the single mirroring $v(1;2) \to v(2;1)$ or $v(3;4) \to v(4;3)$ maps each diagram into the other. It is then clear that to solve the counting problem we must count the symmetries of a given diagram.

Consider a generic vacuum diagram D of order k, and label all its internal vertices from 1 to $2k$. We define $\mathcal{G}$ to be the set of oriented Green's function lines and $\mathcal{V}$ to be the set of interaction lines. For the left diagram in (11.4) we have $\mathcal{G} = \{(1;3),(3;1),(2;4),(4;2)\}$ and $\mathcal{V} = \{(1;2),(3;4)\}$. Notice that for a v-line $(i;j) = (j;i)$, while for a G_0-line we must consider $(i;j)$ and $(j;i)$ as two different elements. The mirrorings and permutations of the v-lines give rise to $2^k k!$ relabelings of the diagram D, which map $\mathcal{V}$ to itself. If such a relabeling also maps $\mathcal{G}$ into $\mathcal{G}$ we call it a symmetry, and the total number of symmetries is denoted by N_S. For the left diagram in (11.4) we have the following symmetries:

$$
\begin{array}{lll}
s_1(1,2,3,4) = (1,2,3,4) & \text{identity} \\
s_2(1,2,3,4) = (2,1,4,3) & \text{mirroring both interaction lines} \\
s_3(1,2,3,4) = (3,4,1,2) & \text{permutation of interaction lines}' \\
s_4(1,2,3,4) = (4,3,2,1) & s_3 \text{ after } s_2
\end{array}
\tag{11.5}
$$

and hence $N_S = 4$. It is easy to see that the set of relabeling forms a group of order $2^k k!$ and the set of symmetries form a subgroup of order N_S. Let s_i, $i = 1, \ldots, N_S$ be the symmetries of D. If we now take a relabeling g different from these s_i we obtain a new vacuum diagram D'. For instance, $g(1,2,3,4) = (2,1,3,4)$ maps the left diagram of (11.4) into the right diagram and vice versa. The diagram D' has also N_S symmetries s'_i given by

$$s'_i = g \circ s_i \circ g^{-1},$$

which map D' into itself. Clearly $s'_i \neq s'_j$ if $s_i \neq s_j$. By taking another relabeling h different from s_i, s'_i, $i = 1, \ldots, N_S$, we obtain another vacuum diagram D'' that also has N_S symmetries $s''_i = h \circ s_i \circ h^{-1}$. Continuing in this way we finally obtain

$$N = \frac{2^k k!}{N_S}$$

different vacuum diagrams with the same topology. Thus, to know the number N of variants we must determine the number N_S of symmetries. Once N_S is known, the diagrammatic expansion of $\ln Z$ can be performed by including only connected and topologically inequivalent vacuum diagrams $V_c = \int [v \ldots v G_0 \ldots G_0]$ with prefactor

$$\frac{2^k k!}{N_S} \frac{1}{k!} \left(\frac{\mathrm{i}}{2}\right)^k (\pm)^l = \frac{\mathrm{i}^k}{N_S} (\pm)^l.$$

From now on we work only with connected and topologically inequivalent vacuum diagrams. It is therefore convenient to change the Feynman rules for the vacuum diagrams similarly to what we did in Section 7.4 for the Green's function:

- Number all vertices and assign an interaction line $v(i;j)$ to a wiggly line between j and i and a Green's function $G_0(i;j^+)$ to an oriented line from j to i.
- Integrate over all vertices and multiply by $\mathrm{i}^k(\pm)^l$, where l is the number of loops and k is the number of interaction lines.

These are the same Feynman rules as for the Green's function diagrams. Since the symmetry factor N_S is not included in the new rules, each diagram is explicitly multiplied by $1/N_S$. For example, the diagrammatic expansion of $\ln Z$ up to second order in the interaction is represented as in Fig. 11.2. Notice the drastic reduction of second-order diagrams: from 20 in Fig. 11.1 to 5 in Fig. 11.2. The diagrammatic expansion of Fig. 11.2 is still an expansion in noninteracting Green's functions G_0. In analogy to what we did in Chapter 7, we may try to expand $\ln Z$ in terms of the dressed Green's function G. As we see, however, the dressed expansion of $\ln Z$ is a bit more complicated than the dressed expansion of Σ. In the next section we illustrate where the problem lies, while in Section 11.4 we show how to overcome it.

Figure 11.2 Diagrammatic expansion of $\ln Z$ up to second order in the interaction with the new Feynman rules.

11.3 The Φ Functional

Let us consider the vacuum diagram

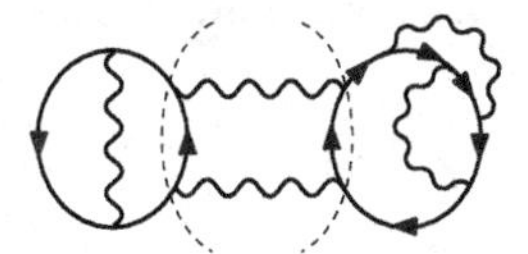

where the thin dashed lines help to visualize the self-energy insertions. This diagram can be regarded either as

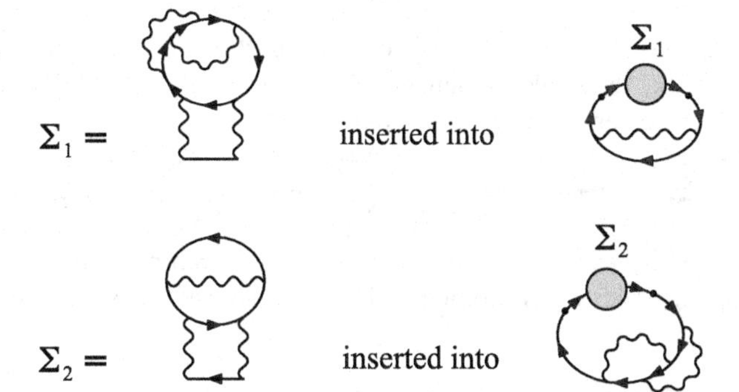

or as

so we could choose the G-skeletonic diagram to be one of the following two diagrams:

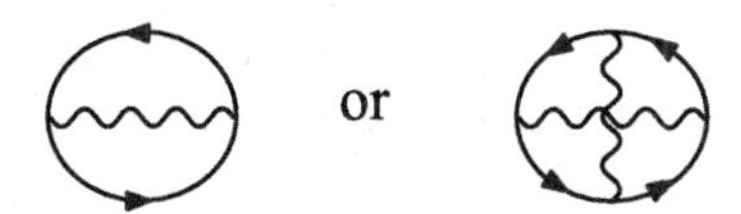

We remind the reader that a diagram is G-skeletonic if it cannot be broken into two disjoint pieces by cutting two Green's function lines, see Section 7.6. The above example shows that a naive sum of skeletonic vacuum diagrams leads to a (at least) double-counting of contributions that should be counted only once. In the case of the Green's function diagrams this problem does not arise since there is a unique G-skeletonic diagram with one ingoing line and one outgoing line. Due to this difficulty it is more convenient to proceed along a new line of argument. We derive the dressed expansion of $\ln Z$ from the self-energy. For this purpose we must establish the relation between a vacuum diagram and a self-energy diagram. This relation also provides us with a different way of determining the number of symmetries N_S of a vacuum diagram.

Classes and symmetries Let us start with some examples. The nonskeletonic vacuum diagram

consists of the set of G_0-lines $\mathcal{G} = \{(1;1),(3;2),(2;3),(4;4)\}$ and the set of v-lines $\mathcal{V} = \{(1;2),(3;4)\}$. Except for the identity, the only other symmetry of this diagram is

$$s(1,2,3,4) = (4,3,2,1),$$

which corresponds to a permutation followed by a simultaneous mirroring of the two v-lines. We now study what happens when we remove a G_0-line from the diagram above. If we remove the G_0-lines $(2;3)$ or $(3;2)$, we produce two topologically equivalent diagrams with the structure below:

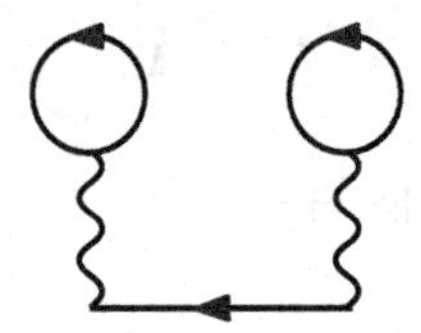

This is a *reducible* self-energy diagram since it can be broken into two disjoint pieces by cutting a G_0-line. The reducible self-energy Σ_r is simply the series (7.8) for the Green's function, in which the external G_0-lines are removed:

$$\Sigma_r = \quad + \quad + \ldots$$

Thus Σ_r contains all Σ-diagrams plus all diagrams that we can make by joining an arbitrary number of Σ-diagrams with Green's function lines. In terms of Σ_r we can rewrite the Dyson equations (7.11) as

$$G(1;2) = G_0(1;2) + \int d3d4\, G_0(1;3)\Sigma_r(3;4)G_0(4;2), \tag{11.6}$$

which implicitly defines the reducible self-energy. If we instead cut the G_0-lines $(1;1)$ or $(4;4)$, we obtain the third diagram of Fig. 7.6. This is a nonskeletonic self-energy diagram since it is one-particle irreducible but contains a self-energy insertion.

A second example is the G-skeletonic vacuum diagram

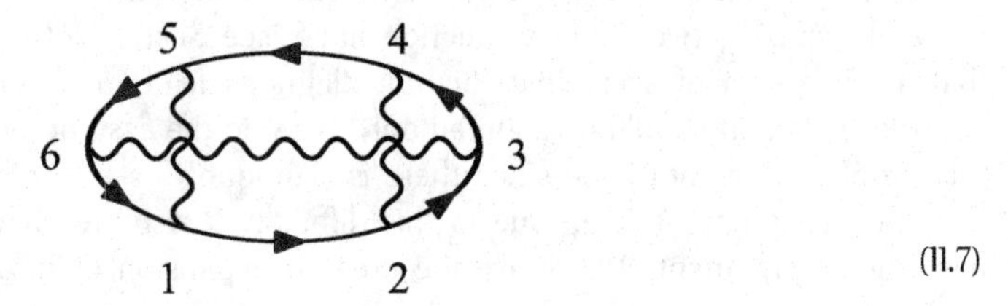

(11.7)

which consists of the set of G_0-lines $\mathcal{G} = \{(2;1),(3;2),(4;3),(5;4),(6;5),(1;6)\}$ and the set of v-lines $\mathcal{V} = \{(1;5),(2;4),(3;6)\}$. The symmetries of this diagram are the identity and

$$s(1,2,3,4,5,6) = (4,5,6,1,2,3), \tag{11.8}$$

which corresponds to a permutation of the vertical v-lines followed by a mirroring of all the v-lines.[2] If we remove $G_0(5;4)$, we obtain the diagram

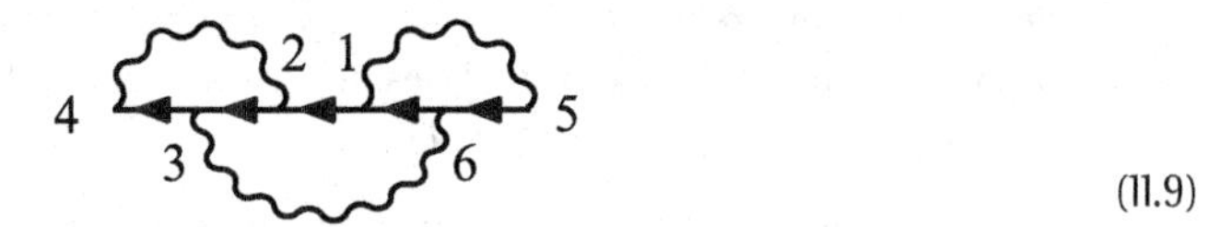

(11.9)

which is a (G-skeletonic) self-energy diagram. A topologically equivalent diagram is obtained if we remove the Green's function $G_0(2;1)$:

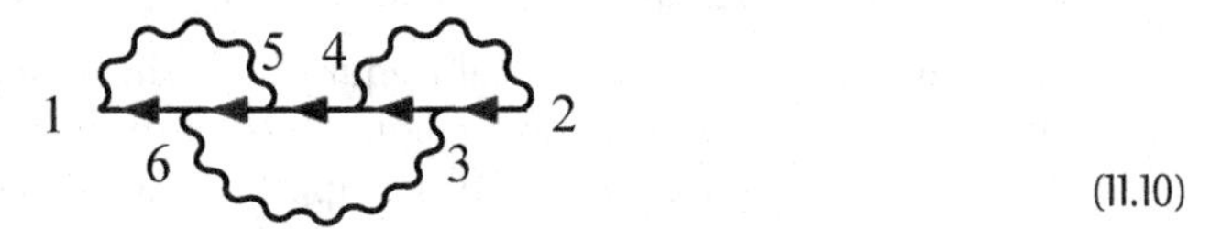

(11.10)

On the other hand, if we remove the Green's function $G_0(4;3)$, we obtain the diagram

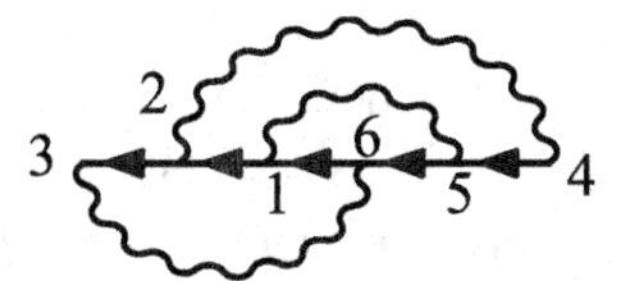

which is again a (G-skeletonic) self-energy diagram but with a different topology.

From these examples we see that by removing G_0-lines from a vacuum diagram we generate reducible self-energy diagrams of the same or different topology. We call two G_0-lines *equivalent* if their removal leads to topologically equivalent Σ_r-diagrams. Since the diagrams (11.9) and (11.10) have the same topology, the G_0-lines $(2;1)$ and $(5;4)$ are equivalent. It is easy to verify that the diagram (11.7) splits into three classes of equivalent lines

$$C_1 = \{(2;1),(5;4)\}, \quad C_2 = \{(3;2),(6;5)\}, \quad C_3 = \{(4;3),(1;6)\}$$

which correspond to the self-energy diagrams

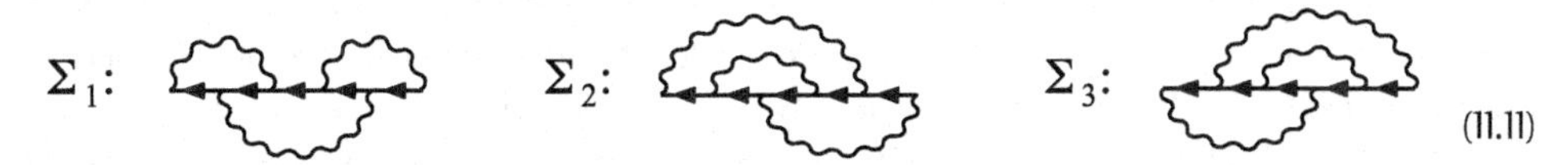

(11.11)

We observe that all classes C_i contain a number of elements equal to the number of symmetry operations. Furthermore, an element of class C_i can be generated from one in

[2] Looking at (11.7), the simultaneous mirroring of the vertical v-lines may look like a symmetry since it corresponds to rotate the diagram by 180 degrees along the axis passing through 3 and 6. However, this rotation also changes the orientation of the G-lines. One can check that the set $\mathcal{G}$ is not mapped to itself under the mirrorings $v(1;5) \to v(5;1)$ and $v(2;4) \to v(4;2)$.

the same class by applying the symmetry operation (11.8). We now show that this fact is generally true: every vacuum diagram with N_S symmetry operations splits into classes of N_S equivalent lines. This statement stems from the following three properties:

- If s is a symmetry of a vacuum diagram D, then the G_0-lines $(i;j)$ and $(s(i);s(j))$ are equivalent.

  Proof. By the symmetry s the diagram D is mapped onto a topologically equivalent diagram D' with relabeled vertices. Superimposing D and D', it is evident that the removal of line $(i;j)$ from D gives the same reducible self-energy diagram as the removal of line $(s(i);s(j))$ from D'.

- If two G_0-lines $(i;j)$ and $(k;l)$ in a vacuum diagram D are equivalent, then there exists a symmetry operation s of D with the property $(s(i);s(j)) = (k;l)$.

  Proof. Let $\Sigma_r^{(i;j)}$ and $\Sigma_r^{(k;l)}$ be the reducible self-energy diagrams obtained by removing from D the G_0-lines $(i;j)$ and $(k;l)$, respectively. By hypothesis the diagrams $\Sigma_r^{(i;j)}$ and $\Sigma_r^{(k;l)}$ are topologically equivalent. Therefore, superimposing the two self-energies, we find a one-to-one mapping between the vertex labels that preserves the topological structure. Hence, this mapping is a symmetry operation. For example, by superimposing the diagrams (11.9) and (11.10) and comparing the labels, we find the one-to-one relation $(4,3,2,1,6,5) \leftrightarrow (1,6,5,4,3,2)$, which is exactly the symmetry operation (11.8).

- If a symmetry operation s of a vacuum diagram maps a G_0-line to itself – that is, $(s(i);s(j)) = (i;j)$ – then s must be the identity operation.

  Proof. A symmetry must preserve the connectivity of the diagram and therefore it is completely determined by the mapping of one vertex. For example, the symmetry s_2 in (11.5) maps vertex 1 into 2: $s(1) = 2$. Since in 1 arrives a G_0-line from 3 and 1 is mapped to 2, in which arrives a G_0-line from 4, then $s(3) = 4$. Furthermore, 1 is connected to 2 by a v-line and hence $s(2) = 1$. Finally, in 2 arrives a G_0-line from 4 and 2 is mapped to 1, in which arrives a G_0-line from 3, and hence $s(4) = 3$. Similarly, for the diagram in (11.7) the symmetry (11.8) maps vertex 1 into 4: $s(1) = 4$. Since in 1 arrives a G_0-line from 6 and 1 is mapped to 4, in which arrives a G_0-line from 3, then $s(6) = 3$. In a similar way one can reconstruct the mapping of all the remaining vertices. Thus, if a G_0-line is mapped onto itself, the symmetry operation must be the identity.

From these properties it follows that a symmetry maps the classes into themselves and that the elements of the same class are related by a symmetry operation. Since for every symmetry $(s(i);s(j)) \neq (i;j)$ unless s is the identity operation, we must have $(s(i);s(j)) \neq (s'(i);s'(j))$ for $s \neq s'$, otherwise $s^{-1} \circ s' = 1$ and hence $s = s'$. Thus, the application of two different symmetry operations to a G_0-line leads to different G_0-lines in the same class. Consequently, all classes must contain the same number of elements and this number equals the number of symmetries N_S. Taking into account that in an nth-order diagram the number of G_0-lines is $2n$, the number of classes is given by

$$N_C = \frac{2n}{N_S}. \tag{11.12}$$

For example, in diagram (11.7) we have $n = 3$ and $N_S = 2$, so there are $N_C = (2 \times 3)/2 = 3$ classes.

Functional derivatives Our analysis has led to a different way of calculating the symmetry number N_S. Rather than finding the permutations and mirrorings of interaction lines that do not change the vacuum diagram, we can count how many G_0-lines yield the same Σ_r-diagram. Mathematically the act of removing a G_0-line corresponds to taking the functional derivative of the vacuum diagram with respect to G_0. Let us clarify this point. Below we show the only two connected vacuum diagrams with one interaction line - see the first two terms in Fig. 11.2:

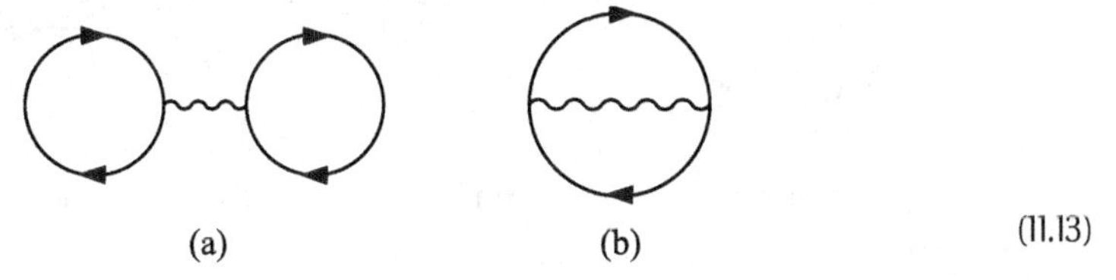

(11.13)

For both diagrams $N_S = 2$. The mathematical expression corresponding to diagrams (a) and (b) are

$$D_a[G_0] = \mathrm{i} \int d1d2\, G_0(1;1^+)v(1;2)G_0(2;2^+)$$

and

$$D_b[G_0] = \pm\mathrm{i} \int d1d2\, G_0(1;2^+)v(1;2)G_0(2;1^+).$$

It is important to realize that if G_0 belongs to the Keldysh space, then the contour integrals reduce to integrals along the imaginary track since the forward branch is exactly canceled by the backward branch (see, for instance, Exercise 5.4). The functionals D_a and D_b, however, are defined for *any* G_0. When we take the functional derivative we consider D_a and D_b as functionals of arbitrary functions G_0, including those that do not belong to the Keldysh space. In this way, a variation of $G_0(1;2)$ with z_1 and/or z_2 on the horizontal branches is, in general, different from zero. After the differentiation we evaluate the result at the physical G_0. That said, let us take the functional derivative of D_a and D_b:

$$\frac{\delta D_a[G_0]}{\delta G_0(2;1^+)} = 2\mathrm{i}\delta(1;2) \int d3\, v(1;3)G_0(3;3^+), \tag{11.14a}$$

$$\frac{\delta D_b[G_0]}{\delta G_0(2;1^+)} = \pm 2\mathrm{i}v(1;2)G_0(1;2^+). \tag{11.14b}$$

Comparing these results with (8.1), we see that (11.14a) is plus/minus the Hartree self-energy multiplied by the symmetry factor $N_S = 2$ of diagram D_a and (11.14b) is plus/minus the Fock (or exchange) self-energy multiplied by the symmetry factor $N_S = 2$ of diagram D_b.

Examples of connected vacuum diagrams with two interaction lines are displayed on the left of Fig. 11.3. For the top diagram the symmetry factor is $N_S = 4$ and the mathematical expression reads

$$D = \mathrm{i}^2(\pm)^2 \int d1d2d3d4\, v(1;2)v(3;4)G_0(1;3)G_0(3;1)G_0(2;4)G_0(4;2). \tag{11.15}$$

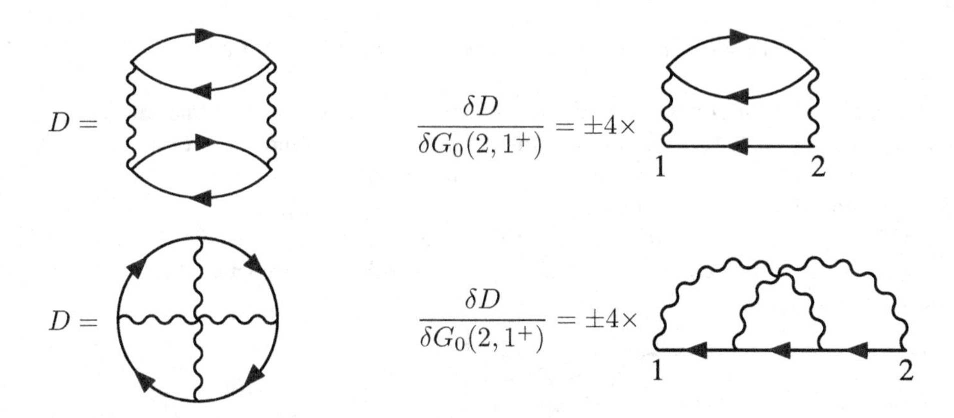

Figure 11.3 Connected vacuum diagrams with two interaction lines and the corresponding self-energy diagrams.

Its functional derivative with respect to $G_0(6;5^+)$ gives

$$\frac{\delta D}{\delta G_0(6;5^+)} = 4\mathrm{i}^2(\pm)^2 \int d2d4\; v(5;2)v(6;4)G_0(5;6)G_0(2;4)G_0(4;2) = \pm 4\Sigma_r(5;6),$$

where the self-energy Σ_r is displayed on the top-right of the same figure. Similar considerations apply to the bottom diagrams of Fig. 11.3. Notice that the sum of the two self-energy diagrams in Fig. 11.3 yields the second-order diagrams of the second Born approximation, see (7.16). From the diagrammatic representation we see that the action of differentiation is the same as removing a G_0-line in all possible ways.

Reducible Φ_r functional Since the removal of a G_0-line always reduces the number of loops by one, we conclude that if D_c is a connected vacuum diagram then

$$\frac{\delta}{\delta G_0(2;1^+)} \frac{1}{N_S} D_c = \pm \Sigma_r^{(D_c)}(1;2),$$

where $\Sigma_r^{(D_c)}$ is a contribution to the reducible self-energy consisting of N_C diagrams. Thus, if we write

$$\ln Z = \ln Z_0 \pm \Phi_r[G_0, v], \tag{11.16}$$

where Φ_r is plus/minus (for bosons/fermions) the sum of all connected and topologically inequivalent vacuum diagrams, each divided by the symmetry factor N_S, then we have

$$\Sigma_r(1;2) = \Sigma_r[G_0, v](1;2) = \frac{\delta \Phi_r}{\delta G_0(2;1^+)}. \tag{11.17}$$

In the second equality we have emphasized that Σ_r is a functional of G_0 and v. It is worth stressing that (11.17) is an identity because the functional derivative of Φ_r contains *all possible reducible self-energy diagrams only once.* If one diagram were missing, then the vacuum diagram obtained by closing the missing Σ_r-diagram with a G_0-line would be connected and

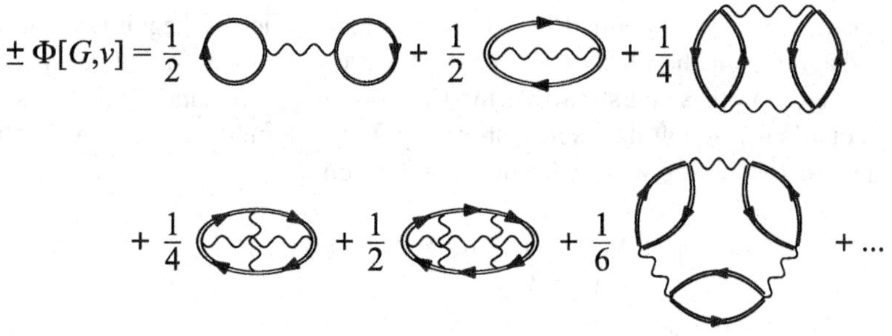

Figure 11.4 Diagrammatic expansion of the Φ functional.

topologically different from all other diagrams in Φ_r, in contradiction with the hypothesis that Φ_r contains *all* connected vacuum diagrams. Furthermore, a Σ_r-diagram cannot be generated by two different vacuum diagrams and hence there is no multiple counting in (11.17).

Equation (11.17) establishes that the reducible self-energy can be obtained as the functional derivative of a functional Φ_r with respect to G_0. We can also go in the opposite direction and "integrate" (11.17) to express Φ_r in terms of reducible self-energy diagrams. We know that there are always $N_{\rm C}$ topologically inequivalent Σ_r-diagrams originating from the same vacuum diagram. Therefore, if we close these diagrams with a G_0-line we obtain the same vacuum diagram $N_{\rm C}$ times multiplied by $(\pm)$ due to the creation of a loop. From (11.12) we have $1/(N_{\rm C}N_{\rm S}) = 1/2n$, and hence

$$\Phi_r[G_0, v] = \sum_{n=1}^{\infty} \frac{1}{2n} \int d1d2\, \Sigma_r^{(n)}[G_0, v](1;2) G_0(2;1^+), \tag{11.18}$$

where $\Sigma_r^{(n)}$ denotes the sum of all reducible and topologically inequivalent self-energy diagrams of order n.

Dressed Φ functional An important consequence of this result is that it tells us how to construct the functional $\Phi = \Phi[G, v]$, whose functional derivative with respect to the *dressed* Green's function G is the self-energy $\Sigma = \Sigma_s[G, v]$ defined in Section 7.6:

$$\Sigma(1;2) = \Sigma_s[G, v](1;2) = \frac{\delta \Phi}{\delta G(2;1^+)}. \tag{11.19}$$

By the same reasoning, $\Phi[G, v]$ is plus/minus the sum of all connected, topologically inequivalent, and G-skeletonic vacuum diagrams, each divided by the corresponding symmetry factor $N_{\rm S}$ and in which $G_0 \to G$, see Fig 11.4. This is so because a nonskeletonic vacuum diagram can only generate nonskeletonic and/or one-particle reducible Σ_r-diagrams, while a G-skeletonic vacuum diagram generates only G-skeletonic irreducible Σ-diagrams. Henceforth, we refer to a self-energy as in (11.19) as a *Φ-derivable self-energy.* An example of Φ-derivable self-energy is the Hartree–Fock self-energy, the Φ-diagrams being the first two diagrams on the r.h.s. of Fig. 11.4, see the discussion below (11.14). A second example is the second Born approximation defined in (7.16), the second-order Φ-diagrams being the third and fourth diagrams on the r.h.s. of Fig. 11.4, see the discussion below (11.15). In Chapter 12 we discuss other examples and show that the Green's function obtained from the solution of the Dyson equation with a Φ-derivable self-energy fulfills all fundamental conservation laws.

The functional Φ can be written in the same fashion as Φ_r:

$$\Phi[G,v] = \sum_{n=1}^{\infty} \frac{1}{2n} \int d1d2\, \Sigma_s^{(n)}[G,v](1;2)G(2;1^+), \tag{11.20}$$

where $\Sigma_s^{(n)}$ is the sum of all topologically inequivalent and G-skeletonic self-energy diagrams of order n in which $G_0 \to G$. It is convenient to introduce the short-hand notation

$$\mathrm{tr}_\gamma[fg] \equiv \int d1d2\, f(1;2)g(2;1) = \mathrm{tr}_\gamma[gf] \tag{11.21}$$

for the trace over space and spin and for the convolution along the contour. Then the expansion of Φ takes the compact form[3]

$$\boxed{\Phi[G,v] = \sum_{n=1}^{\infty} \frac{1}{2n} \mathrm{tr}_\gamma \left[\Sigma_s^{(n)}[G,v]G \right]} \tag{11.22}$$

In order to obtain an expression for $\ln Z$ in terms of the dressed Green's function G rather than G_0, it is tempting to replace $\Phi_r[G_0,v]$ with $\Phi[G,v]$ in (11.16). Unfortunately the value of $\Phi[G,v]$ is not equal to $\Phi_r[G_0,v]$ due to the double-counting problem mentioned at the beginning of the section. Nonetheless, it is possible to derive an exact and simple formula for the correction $\Phi_r[G_0,v] - \Phi[G,v]$; this is the topic of the next section.

Exercise 11.2 Draw all possible Φ-diagrams with three interaction lines and calculate the corresponding self-energies.

11.4 Dressed Expansion of the Grand Potential

Let us consider a system described by the Hamiltonian $\hat{H}_\lambda(z)$ with rescaled interaction $v \to \lambda v$: $\hat{H}_\lambda(z) = \hat{H}_0(z) + \lambda \hat{H}_{\mathrm{int}}(z)$. The grand potential for this Hamiltonian is

$$\Omega_\lambda = -\frac{1}{\beta}\ln Z_\lambda = -\frac{1}{\beta}\ln \mathrm{Tr}\left[e^{-\beta \hat{H}_\lambda^{\mathrm{M}}}\right] = -\frac{1}{\beta}\ln \mathrm{Tr}\left[\mathcal{T}\left\{e^{-\mathrm{i}\int_\gamma dz \hat{H}_\lambda(z)}\right\}\right]. \tag{11.23}$$

The derivative of (11.23) with respect to λ is

$$\frac{d\Omega_\lambda}{d\lambda} = \frac{\mathrm{i}}{\beta}\int_\gamma dz_1 \frac{\mathrm{Tr}\left[\mathcal{T}\left\{e^{-\mathrm{i}\int_\gamma dz \hat{H}_\lambda(z)} \hat{H}_{\mathrm{int}}(z_1)\right\}\right]}{\mathrm{Tr}\left[\mathcal{T}\left\{e^{-\mathrm{i}\int_\gamma dz \hat{H}_\lambda(z)}\right\}\right]} = \frac{\mathrm{i}}{\beta}\int_\gamma dz_1 \langle \hat{H}_{\mathrm{int}}(z_1)\rangle_\lambda, \tag{11.24}$$

---

[3] We observe that definition (11.21) with $f = \Sigma$ and $g = G$ is ambiguous only for the singular Hartree–Fock self-energy since the Green's function should then be evaluated at equal times. The MBPT tells us how to interpret (11.21) in this case: The second time argument of the Green's function must be infinitesimally later than the first time argument.

where we introduce the short-hand notation

$$\langle \ldots \rangle_\lambda = \mathrm{Tr}\left[\mathcal{T}\left\{e^{-\mathrm{i}\int_\gamma dz \hat{H}_\lambda(z)} \ldots\right\}\right] / Z_\lambda,$$

in which any string of operators can be inserted in the dots. The integrand on the r.h.s. of (11.24) multiplied by λ is the interaction energy $E_{\mathrm{int},\lambda}(z_1)$ with rescaled interaction. This energy can be expressed in terms of the rescaled Green's function G_λ and self-energy $\Sigma_\lambda = \Sigma_s[G_\lambda, \lambda v]$ as

$$E_{\mathrm{int},\lambda}(z_1) = \lambda \langle \hat{H}_{\mathrm{int}}(z_1)\rangle_\lambda = \pm\frac{\mathrm{i}}{2}\int d\mathbf{x}_1 d2\, \Sigma_\lambda(1;2)G_\lambda(2;1^+), \tag{11.25}$$

in accordance with (10.29). Inserting this result into (11.24) and integrating over λ between 0 and 1, we find

$$\begin{aligned}\Omega &= \Omega_0 + \frac{\mathrm{i}}{\beta}\int_0^1 \frac{d\lambda}{\lambda}\int_\gamma dz_1 \left(\pm\frac{\mathrm{i}}{2}\right)\int d\mathbf{x}_1 d2\, \Sigma_\lambda(1;2)G_\lambda(2;1^+) \\ &= \Omega_0 \mp \frac{1}{2\beta}\int_0^1 \frac{d\lambda}{\lambda}\, \mathrm{tr}_\gamma\left[\, \Sigma_\lambda G_\lambda \,\right].\end{aligned} \tag{11.26}$$

Equation (11.26) can be used to derive (11.18) in an alternative way. Since (omitting arguments and integrals)

$$\begin{aligned}\Sigma G &= \Sigma[G_0 + G_0\Sigma G_0 + G_0\Sigma G_0\Sigma G_0 + \ldots] = \\ &= [\Sigma + \Sigma G_0\Sigma + \Sigma G_0\Sigma G_0\Sigma + \ldots]G_0 = \Sigma_r G_0,\end{aligned}$$

we can rewrite (11.26) as

$$\begin{aligned}\Omega &= \Omega_0 \mp \frac{1}{2\beta}\int_0^1 \frac{d\lambda}{\lambda}\, \mathrm{tr}_\gamma\left[\, \Sigma_r[G_0, \lambda v]G_0 \,\right] \\ &= \Omega_0 \mp \frac{1}{2\beta}\sum_{n=1}^{\infty}\int_0^1 \frac{d\lambda}{\lambda}\lambda^n \mathrm{tr}_\gamma\left[\, \Sigma_r^{(n)}[G_0, v]G_0 \,\right] \\ &= \Omega_0 \mp \frac{1}{\beta}\sum_{n=1}^{\infty}\frac{1}{2n}\, \mathrm{tr}_\gamma\left[\, \Sigma_r^{(n)}[G_0, v]G_0\right].\end{aligned}$$

Multiplying both sides by $-\beta$ and comparing with the definition of Φ_r in (11.16), we reobtain (11.18).

Let us now consider the functional Φ in (11.20) for a system with rescaled interaction λv:

$$\begin{aligned}\Phi[G_\lambda, \lambda v] &= \sum_{n=1}^{\infty}\frac{1}{2n}\, \mathrm{tr}_\gamma\left[\, \Sigma_s^{(n)}[G_\lambda, \lambda v]G_\lambda \,\right] \\ &= \sum_{n=1}^{\infty}\frac{\lambda^n}{2n}\, \mathrm{tr}_\gamma\left[\, \Sigma_s^{(n)}[G_\lambda, v]G_\lambda \,\right].\end{aligned}$$

The derivative of this expression with respect to λ gives

$$\frac{d}{d\lambda}\Phi[G_\lambda,\lambda v] = \sum_{n=1}^{\infty}\frac{\lambda^{n-1}}{2}\,\mathrm{tr}_\gamma\left[\,\Sigma_s^{(n)}[G_\lambda,v]G_\lambda\,\right] + \mathrm{tr}_\gamma\left[\frac{\delta\Phi[G_\lambda,\lambda v]}{\delta G_\lambda}\frac{dG_\lambda}{d\lambda}\right]$$
$$= \mathrm{tr}_\gamma\left[\frac{1}{2\lambda}\Sigma_\lambda G_\lambda + \Sigma_\lambda\frac{dG_\lambda}{d\lambda}\right], \tag{11.27}$$

where the first term originates from the variation of the interaction lines and the second term from the variation of the Green's function lines. The first term in (11.27) appears also in (11.26), which can therefore be rewritten as

$$\Omega = \Omega_0 \mp \frac{1}{\beta}\int_0^1 d\lambda\left(\frac{d}{d\lambda}\Phi[G_\lambda,\lambda v] - \mathrm{tr}_\gamma\left[\Sigma_\lambda\frac{dG_\lambda}{d\lambda}\right]\right)$$
$$= \Omega_0 \mp \frac{1}{\beta}\left(\Phi - \mathrm{tr}_\gamma\left[\,\Sigma G\,\right]\right) \mp \frac{1}{\beta}\int_0^1 d\lambda\,\mathrm{tr}_\gamma\left[\frac{d\Sigma_\lambda}{d\lambda}G_\lambda\right], \tag{11.28}$$

where in the last equality we perform an integration by parts and take into account that $\Phi[G_\lambda,\lambda v]$ and Σ_λ vanish for $\lambda = 0$, while they equal $\Phi = \Phi[G,v]$ and $\Sigma = \Sigma_s[G,v]$ for $\lambda = 1$. To calculate the last term, we observe that[4]

$$\frac{d}{d\lambda}\mathrm{tr}_\gamma\left[\ln(1-G_0\Sigma_\lambda)\right] = -\frac{d}{d\lambda}\mathrm{tr}_\gamma\left[G_0\Sigma_\lambda + \frac{1}{2}G_0\Sigma_\lambda G_0\Sigma_\lambda + \frac{1}{3}G_0\Sigma_\lambda G_0\Sigma_\lambda G_0\Sigma_\lambda + \ldots\right]$$
$$= -\mathrm{tr}_\gamma\left[G_0\frac{d\Sigma_\lambda}{d\lambda} + G_0\Sigma_\lambda G_0\frac{d\Sigma_\lambda}{d\lambda} + G_0\Sigma_\lambda G_0\Sigma_\lambda G_0\frac{d\Sigma_\lambda}{d\lambda} + \ldots\right]$$
$$= -\mathrm{tr}_\gamma\left[G_\lambda\frac{d\Sigma_\lambda}{d\lambda}\right]. \tag{11.29}$$

Inserting this result in (11.28), we obtain an elegant formula for the grand potential

$$\beta(\Omega - \Omega_0) = \mp\left\{\,\Phi - \mathrm{tr}_\gamma\left[\,\Sigma G + \ln(1 - G_0\Sigma)\,\right]\,\right\}. \tag{11.30}$$

This formula provides the MBPT expansion of Ω (and hence of $\ln Z$) in terms of the dressed Green's function G since we know how to expand the functional $\Phi = \Phi[G,v]$ and the self-energy $\Sigma = \Sigma_s[G,v]$ in G-skeletonic diagrams.

11.5 Luttinger–Ward and Klein Functionals

Equation (11.30) can also be regarded as the definition of a functional $\Omega[G,v]$ which takes the value of the grand potential when G is the Green's function of the underlying physical system: $G = G_0 + G_0\Sigma G$. Any physical Green's function belongs to the Keldysh space and hence the contour integrals in (11.30) reduce to integrals along the imaginary track γ^{M}. If, on the other hand, we evaluate $\Omega[G,v]$ at a G which does not belong to the Keldysh space, then the contour integrals cannot be reduced to integrals along γ^{M} and the full contour

[4]The function of an operator $\hat{A}$ can be defined by its Taylor expansion, and therefore $-\ln(1-\hat{A}) = \sum_n \frac{1}{n}\hat{A}^n$.

must be considered. The functional $\Omega[G, v]$ was first introduced by Luttinger and Ward [141] and we therefore refer to it as the *Luttinger–Ward functional*. To distinguish the functional from the grand potential (which is a number), we denote the former by $\Omega_{\rm LW}[G, v]$:

$$\boxed{\Omega_{\rm LW}[G,v] = \Omega_0 \mp \frac{1}{\beta}\left\{ \Phi[G,v] - \mathrm{tr}_\gamma\left[\Sigma_s[G,v]G + \ln(1 - G_0\Sigma_s[G,v])\right]\right\}} \tag{11.31}$$

A remarkable feature of the Luttinger–Ward (LW) functional is its variational property. If we change the Green's function from G to $G + \delta G$, then the change of $\Omega_{\rm LW}$ reads

$$\delta\Omega_{\rm LW} = \mp\frac{1}{\beta}\left\{ \delta\Phi - \mathrm{tr}_\gamma\left[\delta\Sigma G + \Sigma\delta G - (G_0 + G_0\Sigma G_0 + \ldots)\delta\Sigma\right]\right\}.$$

Since $\delta\Phi = \mathrm{tr}_\gamma[\Sigma\delta G]$, we conclude that the variation $\delta\Omega_{\rm LW}$ vanishes when (omitting arguments and integrals)

$$G = G_0 + G_0\Sigma G_0 + G_0\Sigma G_0\Sigma G_0 + \ldots, \tag{11.32}$$

that is, when G is the self-consistent solution of $G = G_0 + G_0\Sigma G$ with $\Sigma = \Sigma_s[G, v]$. Therefore, $\Omega_{\rm LW}$ equals the grand potential at the stationary point. We further observe that the LW functional preserves the variational property for any approximate Φ provided that the self-energy $\Sigma_s[G, v]$ is calculated as the functional derivative of such Φ. In other words, $\Omega_{\rm LW}$ is stationary at the approximate G which satisfies the Dyson equation (11.32) with $\Sigma_s[G, v] = \delta\Phi[G, v]/\delta G$. It is also worth stressing that in the functional $\Omega_{\rm LW}[G, v]$ there is an explicit reference to the underlying physical system through the noninteracting Green's function G_0, see the last term in (11.31). This is not the case for the Φ functional (and hence neither for the self-energy $\Sigma_s[G, v]$) for which a stationary principle would not make sense.[5]

The variational property of the LW functional naturally introduces a new concept to tackle the equilibrium many-body problem. This concept is based on variational principles instead of diagrammatic MBPT expansions. Since $\Omega_{\rm LW}$ is stationary with respect to changes in G, the value of $\Omega_{\rm LW}$ at a Green's function that deviates from the stationary G by δG leads to an error in $\Omega_{\rm LW}$ which is only of second order in δG. Therefore the quality of the results depend primarily on the chosen approximation for Φ. Notice that although $\Omega_{\rm LW}$ is stationary at the self-consistent G, the stationary point does not have to be a minimum.[6] Finally we observe that the advantage of calculating Ω from variational expressions such as the LW functional is that in this way we avoid solving the Dyson equation. If the self-consistent solution gives accurate Ω, then the LW functional produces approximations to these grand potentials with much less computational effort [142, 144].

Klein functional At this point we would like to draw the attention of the reader to an interesting fact. The variational schemes are by no means unique [145]. By adding to $\Omega_{\rm LW}$ any functional $F[D]$, where

$$D[G,v](1;2) = G(1;2) - G_0(1;2) - \int d3d4\, G(1;3)\Sigma_s[G,v](3;4)G_0(4;2), \tag{11.33}$$

[5]There is no reason for Φ to be stationary at the Green's function of the physical system since, except for the interparticle interaction, Φ does not know anything about it!

[6]In Ref. [142] it was shown that $\Omega_{\rm LW}$ is minimum at the stationary point when Φ is approximated by the infinite sum of ring diagrams. This is the so-called RPA approximation and is discussed in detail in Section 15.4.

obeying

$$F[D=0] = \left(\frac{\delta F}{\delta D}\right)_{D=0} = 0, \tag{11.34}$$

one obtains a new variational functional having the same stationary point and the same value at the stationary point. It might, however, be designed to give a second derivative which also vanishes at the stationary point (something that would be of utmost practical value).

Choosing to add inside the curly brackets of (11.31) the functional

$$F[D] = \mathrm{tr}_\gamma\big[\ln(1 + D\overleftarrow{G}_0^{-1}) - D\overleftarrow{G}_0^{-1}\big], \tag{11.35}$$

with $\overleftarrow{G}_0^{-1}(1;2) = -\mathrm{i}\delta(1;2)\frac{\overleftarrow{d}}{dz_2} - h(1;2)$, the differential operator (acting on quantities to its left) for which $\int d3\, G_0(1;3)\, \overleftarrow{G}_0^{-1}(3;2) = \delta(1;2)$, leads to the functional[7]

$$\boxed{\Omega_{\mathrm{K}}[G,v] = \Omega_0 \mp \frac{1}{\beta}\Big\{\Phi[G,v] + \mathrm{tr}_\gamma\Big[\ln(G\overleftarrow{G}_0^{-1}) + 1 - G\overleftarrow{G}_0^{-1}\Big]\Big\}} \tag{11.36}$$

As the functional F in (11.35) has the desired properties (11.34), this new functional is stationary at the Dyson equation and equals the grand potential at the stationary point. The functional (11.36) was first proposed by Klein [146] and we refer to it as the *Klein functional.* The Klein functional is much easier to evaluate and manipulate as compared to the LW functional, but, unfortunately, it is less stable (large second derivative) at the stationary point [144].

Noninteracting grand potential We conclude this section by showing that Ω_0 can be written as

$$\boxed{\Omega_0 = \mp\frac{1}{\beta}\mathrm{tr}_\gamma\left[\ln(-G_0)\right]} \tag{11.37}$$

according to which the Klein functional can also be written as

$$\Omega_{\mathrm{K}}[G,v] = \mp\frac{1}{\beta}\Big\{\Phi[G,v] + \mathrm{tr}_\gamma\Big[\ln(-G) + 1 - G\overleftarrow{G}_0^{-1}\Big]\Big\}. \tag{11.38}$$

We start by calculating the tr_γ of an arbitrary power of G_0:

$$\begin{aligned}\mathrm{tr}_\gamma\left[G_0^m\right] &= \int d\mathbf{x}\langle\mathbf{x}|\int_\gamma dz_1\ldots dz_m\, \hat{\mathcal{G}}_0(z_1,z_2)\ldots\hat{\mathcal{G}}_0(z_m,z_1^+)|\mathbf{x}\rangle \\ &= \int d\mathbf{x}\langle\mathbf{x}|(-\mathrm{i})^m\int_0^\beta d\tau_1\ldots d\tau_m\, \hat{\mathcal{G}}_0^{\mathrm{M}}(\tau_1,\tau_2)\ldots\hat{\mathcal{G}}_0^{\mathrm{M}}(\tau_m,\tau_1^+)|\mathbf{x}\rangle,\end{aligned}$$

where in the last step we use that for any G_0 in Keldysh space the integral along the forward branch cancels the integral along the backward branch, see also Exercise 5.4. Expanding the

[7] Use that $\mathrm{tr}_\gamma\big[\ln(G\overleftarrow{G}_0^{-1} - G\Sigma)\big] = \mathrm{tr}_\gamma\big[\ln(G\overleftarrow{G}_0^{-1} - G\overrightarrow{G}_0^{-1}G_0\Sigma)\big] = \mathrm{tr}_\gamma\big[\ln(G\overleftarrow{G}_0^{-1} - G\overleftarrow{G}_0^{-1}G_0\Sigma)\big] = \mathrm{tr}_\gamma\big[\ln(G\overleftarrow{G}_0^{-1}) + \ln(1 - G_0\Sigma)\big]$.

Matsubara Green's function as in (6.18) and using the identity $-\mathrm{i}\int_0^\beta d\tau\, e^{(\omega_p-\omega_q)\tau} = -\mathrm{i}\beta\delta_{pq}$, see also Appendix A, we can rewrite the trace as

$$\mathrm{tr}_\gamma\left[G_0^m\right] = \sum_{p=-\infty}^{\infty} e^{\eta\omega_p}\int d\mathbf{x}\langle\mathbf{x}|\frac{1}{(\omega_p-\hat{h}^{\mathrm{M}})^m}|\mathbf{x}\rangle = \sum_\lambda\sum_{p=-\infty}^{\infty} e^{\eta\omega_p}\frac{1}{(\omega_p-\epsilon_\lambda^{\mathrm{M}})^m},$$

with $\epsilon_\lambda^{\mathrm{M}}$ the eigenvalues of $\hat{h}^{\mathrm{M}}$. Therefore we define

$$\mathrm{tr}_\gamma\left[\ln(-G_0)\right] = \sum_\lambda\sum_{p=-\infty}^{\infty} e^{\eta\omega_p}\ln\frac{1}{\epsilon_\lambda^{\mathrm{M}}-\omega_p}.$$

To evaluate the sum over the Matsubara frequencies, we use the trick (6.39) and get

$$\begin{aligned}\mathrm{tr}_\gamma\left[\ln(-G_0)\right] &= \mp(-\mathrm{i}\beta)\sum_\lambda\int_{\Gamma_b}\frac{d\zeta}{2\pi}\,e^{\eta\zeta}f(\zeta)\ln\frac{1}{\epsilon_\lambda^{\mathrm{M}}-\zeta}\\ &= \mathrm{i}\sum_\lambda\int_{\Gamma_b}\frac{d\zeta}{2\pi}\,\frac{e^{\eta\zeta}\ln(1\mp e^{-\beta\zeta})}{\zeta-\epsilon_\lambda^{\mathrm{M}}} = -\sum_\lambda\ln(1\mp e^{-\beta\epsilon_\lambda^{\mathrm{M}}}),\end{aligned} \tag{11.39}$$

where we first perform an integration by parts and then use the Cauchy residue theorem. To relate this result to Ω_0 we observe that

$$Z_0 = \mathrm{Tr}\left[e^{-\beta\hat{H}_0^{\mathrm{M}}}\right] = \prod_\lambda\sum_n e^{-\beta\epsilon_\lambda^{\mathrm{M}}n} = \prod_\lambda\left(1\mp e^{-\beta\epsilon_\lambda^{\mathrm{M}}}\right)^{\mp1},$$

where in the last equality we take into account that the sum over n runs between 0 and ∞ in the case of bosons and between 0 and 1 in the case of fermions. Therefore,

$$\Omega_0 = -\frac{1}{\beta}\ln Z_0 = \pm\frac{1}{\beta}\sum_\lambda\ln(1\mp e^{-\beta\epsilon_\lambda^{\mathrm{M}}}).$$

Comparing this result with (11.39), we find (11.37).

Exercise 11.3 Evaluate the Luttinger–Ward functional (11.31) and the Klein functional (11.36) in the Hartree–Fock approximation.

Exercise 11.4 Use the relation $E-\mu N = \frac{\partial}{\partial\beta}(\beta\Omega)$ to show that both the Luttinger–Ward functional (11.31) and the Klein functional (11.36) yield the energy (8.32) in the Hartree–Fock approximation.

11.6 Luttinger–Ward Theorem

In this section we derive an important consequence of the variational idea in combination with Φ-derivable self-energies. Using (11.37), the Luttinger–Ward functional can be rewritten as[8]

$$\Omega_{\mathrm{LW}} = \mp\frac{1}{\beta}\left\{\Phi - \mathrm{tr}_\gamma\left[\Sigma G + \ln(\Sigma - \overleftarrow{G}_0^{-1})\right]\right\},$$

[8] Use that $\mathrm{tr}_\gamma\left[\ln(1-G_0\Sigma)\right] = \mathrm{tr}_\gamma\left[\ln\left(G_0(\overleftarrow{G}_0^{-1}-\Sigma)\right)\right] = \mathrm{tr}_\gamma\left[\ln(-G_0)+\ln(\Sigma-\overleftarrow{G}_0^{-1})\right]$.

where we omit the explicit dependence of Φ and Σ on G and v. If we evaluate this functional at a physical Green's function (hence belonging to the Keldysh space), then the contour integrals reduce to integrals along the imaginary track $\gamma^{\rm M}$. Expanding G and Σ in Matsubara frequencies, we find

$$\Omega_{\rm LW} = \mp\frac{1}{\beta}\Phi \pm \frac{1}{\beta}\sum_{p=-\infty}^{\infty} e^{\eta\omega_p}\int d\mathbf{x}\,\langle\mathbf{x}|\hat{\Sigma}^{\rm M}(\omega_p)\hat{\mathcal{G}}^{\rm M}(\omega_p) + \ln\left(\hat{\Sigma}^{\rm M}(\omega_p) - \omega_p + \hat{h}^{\rm M}\right)|\mathbf{x}\rangle,$$

where in the last term we use (6.21). From quantum statistical mechanics we know that the total number of particles in the system is given by minus the derivative of the grand potential Ω with respect to the chemical potential μ, see Appendix D. Is it true that for an approximate self-energy (and hence for an approximate G)

$$N = -\frac{\partial\Omega_{\rm LW}}{\partial\mu}? \tag{11.40}$$

The answer is positive provided that Σ is Φ-derivable and that G is self-consistently calculated from the Dyson equation. In this case, $\Omega_{\rm LW}$ is stationary with respect to changes in G and therefore in (11.40) we only need to differentiate with respect to the explicit dependence on μ. This dependence is contained in $\hat{h}^{\rm M} = \hat{h} - \mu$, and therefore

$$\begin{aligned}
-\frac{\partial\Omega_{\rm LW}}{\partial\mu} &= \mp\frac{1}{\beta}\sum_{p=-\infty}^{\infty} e^{\eta\omega_p}\int d\mathbf{x}\,\langle\mathbf{x}|\frac{-1}{\hat{\Sigma}^{\rm M}(\omega_p) - \omega_p + \hat{h}^{\rm M}}|\mathbf{x}\rangle \\
&= \mp\frac{1}{\beta}\sum_{p=-\infty}^{\infty} e^{\eta\omega_p}\int d\mathbf{x}\,\langle\mathbf{x}|\hat{\mathcal{G}}^{\rm M}(\omega_p)|\mathbf{x}\rangle \\
&= \pm\mathrm{i}\int d\mathbf{x}\, G^{\rm M}(\mathbf{x},\tau;\mathbf{x},\tau^+).
\end{aligned} \tag{11.41}$$

The r.h.s. of (11.41) is exactly the total number of particles N calculated using the G for which $\Omega_{\rm LW}$ is stationary. This result was obtained by Baym [147] and shows that Φ-derivable self-energies preserve (11.40).

Fermions at zero temperature Equation (11.40) has an interesting consequence in systems of fermions at zero temperature. Let us rewrite the first line of (11.41) in a slightly different form (lower sign for fermions):

$$\begin{aligned}
N = \frac{1}{\beta}\sum_{p=-\infty}^{\infty} e^{\eta\omega_p}\Big[&\frac{\partial}{\partial\omega_p}\int d\mathbf{x}\,\langle\mathbf{x}|\ln\left(\hat{\Sigma}^{\rm M}(\omega_p) - \omega_p + \hat{h}^{\rm M}\right)|\mathbf{x}\rangle \\
&+ \int d\mathbf{x}\,\langle\mathbf{x}|\hat{\mathcal{G}}^{\rm M}(\omega_p)\frac{\partial\hat{\Sigma}^{\rm M}(\omega_p)}{\partial\omega_p}|\mathbf{x}\rangle\Big].
\end{aligned} \tag{11.42}$$

We now prove that at zero temperature the last term of this equation vanishes. For $\beta\to\infty$, the Matsubara frequency $\omega_p = (2p+1)\pi/(-\mathrm{i}\beta)$ becomes a continuous variable ζ and the sum over p becomes an integral over ζ. Since $d\zeta \equiv \omega_{p+1} - \omega_p = 2\pi/(-\mathrm{i}\beta)$, we have

$$\lim_{\beta\to\infty} \frac{1}{\beta} \sum_{p=-\infty}^{\infty} \int d\mathbf{x}\, \langle \mathbf{x}|\hat{\mathcal{G}}^{\mathrm{M}}(\omega_p) \frac{\partial \hat{\Sigma}^{\mathrm{M}}(\omega_p)}{\partial \omega_p}|\mathbf{x}\rangle = \frac{1}{2\pi\mathrm{i}} \int_{-\mathrm{i}\infty}^{\mathrm{i}\infty} d\zeta \int d\mathbf{x}\, \langle \mathbf{x}|\hat{\mathcal{G}}^{\mathrm{M}}(\zeta) \frac{\partial \hat{\Sigma}^{\mathrm{M}}(\zeta)}{\partial \zeta}|\mathbf{x}\rangle$$

$$= \frac{1}{2\pi\mathrm{i}} \int d\mathbf{x} \int_{-\mathrm{i}\infty}^{\mathrm{i}\infty} d\zeta\, \langle \mathbf{x}| \underbrace{\frac{\partial}{\partial \zeta}\left(\hat{\mathcal{G}}^{\mathrm{M}}(\zeta)\hat{\Sigma}^{\mathrm{M}}(\zeta)\right)}_{\text{total derivative}} - \hat{\Sigma}^{\mathrm{M}}(\zeta) \frac{\partial \hat{\mathcal{G}}^{\mathrm{M}}(\zeta)}{\partial \zeta}|\mathbf{x}\rangle$$

$$= -\frac{1}{2\pi\mathrm{i}} \int d\mathbf{x} \int_{-\mathrm{i}\infty}^{\mathrm{i}\infty} d\zeta\, \langle \mathbf{x}| \frac{\delta\Phi}{\delta\hat{\mathcal{G}}^{\mathrm{M}}(\zeta)} \frac{\partial \hat{\mathcal{G}}^{\mathrm{M}}(\zeta)}{\partial \zeta}|\mathbf{x}\rangle, \tag{11.43}$$

where we take into account that the integral over ζ of the total derivative yields zero and that the self-energy is Φ-derivable. We also set to unity the exponential factor $e^{\eta\zeta}$ since the integral along the imaginary axis is convergent. Due to energy conservation, every Φ-diagram is invariant under the change $\hat{\mathcal{G}}^{\mathrm{M}}(\zeta) \to \hat{\mathcal{G}}^{\mathrm{M}}(\zeta + \delta\zeta)$. Thus, the variation $\delta\Phi = 0$ under this transformation. However, the variation $\delta\Phi$ is formally given by the last line of (11.43), which must therefore vanish.

To evaluate N we have to perform a sum over Matsubara frequencies. The argument of the logarithm in (11.42) is $-[\hat{\mathcal{G}}^{\mathrm{M}}(\omega_p)]^{-1}$. Then we can use again the trick (6.39), where Γ_b is the contour in Fig. 6.1, since $\hat{\mathcal{G}}^{\mathrm{M}}(\zeta)$ is analytic in the complex ζ plane except along the real axis, see (6.70), and the eigenvalues of $\hat{\mathcal{G}}^{\mathrm{M}}(\zeta)$ are nonzero, see Exercise 6.8. We find

$$N = \int_{-\infty}^{\infty} \frac{d\omega}{2\pi\mathrm{i}} f(\omega) e^{\eta\omega} \frac{\partial}{\partial \omega} \int d\mathbf{x}\, \langle \mathbf{x}| \ln\left(-\hat{\mathcal{G}}^{\mathrm{M}}(\omega - \mathrm{i}\delta)\right) - \ln\left(-\hat{\mathcal{G}}^{\mathrm{M}}(\omega + \mathrm{i}\delta)\right)|\mathbf{x}\rangle.$$

Integrating by parts and taking into account that in the zero-temperature limit the derivative of the Fermi function $\partial f(\omega)/\partial\omega = -\delta(\omega)$, we get

$$N = \frac{1}{2\pi\mathrm{i}} \int d\mathbf{x}\, \langle \mathbf{x}| \ln\left(\hat{\Sigma}^{\mathrm{M}}(-\mathrm{i}\delta) + \mathrm{i}\delta + \hat{h} - \mu\right) - \ln\left(\hat{\Sigma}^{\mathrm{M}}(\mathrm{i}\delta) - \mathrm{i}\delta + \hat{h} - \mu\right)|\mathbf{x}\rangle, \tag{11.44}$$

where we insert back the explicit expression of $\hat{\mathcal{G}}^{\mathrm{M}}$ in terms of $\hat{\Sigma}^{\mathrm{M}}$ and use that $\hat{h}^{\mathrm{M}} = \hat{h} - \mu$. According to (10.16) and the discussion below it, we have $\hat{\Sigma}^{\mathrm{M}}(\pm\mathrm{i}\delta) = \hat{\Sigma}^{\mathrm{R}}(\mu) = \hat{\Sigma}^{\mathrm{A}}(\mu)$ since the quasi-particle width operator vanishes at $\omega = \mu$. This also implies that $\hat{\Sigma}^{\mathrm{R}}(\mu)$ is self-adjoint. Let us calculate (11.44) for an electron gas.

Electron gas at zero temperature In the electron gas, $\hat{h}$ and $\hat{\Sigma}^{\mathrm{R}}$ are diagonal in the momentum–spin basis – that is, $|\mathbf{p}\sigma\rangle = \frac{p^2}{2}|\mathbf{p}\sigma\rangle$ and $\hat{\Sigma}^{\mathrm{R}}(\mu)|\mathbf{p}\sigma\rangle = \Sigma^{\mathrm{R}}(\mathbf{p},\mu)|\mathbf{p}\sigma\rangle$, see also Section 10.2. Inserting the completeness relation (1.11) in the bracket $\langle\mathbf{x}|\ldots|\mathbf{x}\rangle$, we find

$$N = 2\frac{\mathrm{V}}{2\pi\mathrm{i}} \int \frac{d\mathbf{p}}{(2\pi)^3} \left[\ln\left(\Sigma^{\mathrm{R}}(\mathbf{p},\mu) + \mathrm{i}\delta + \frac{p^2}{2} - \mu\right) - \ln\left(\Sigma^{\mathrm{R}}(\mathbf{p},\mu) - \mathrm{i}\delta + \frac{p^2}{2} - \mu\right)\right],$$

where the factor of 2 comes from spin and $\mathrm{V} = \int d\mathbf{r}$ is the volume of the system. Since

$$\ln(a \pm \mathrm{i}\delta) = \begin{cases} \ln a & a > 0 \\ \ln|a| \pm \mathrm{i}\pi & a < 0 \end{cases},$$

we can rewrite the total number of particles at zero temperature as

$$\boxed{N = 2\mathrm{V}\int \frac{d\mathbf{p}}{(2\pi)^3}\,\theta(\mu - \frac{p^2}{2} - \Sigma^{\mathrm{R}}(\mathbf{p},\mu))} \tag{11.45}$$

This result is known as the *Luttinger–Ward theorem* and is valid *for any Φ-derivable self-energy* provided that the Green's function is self-consistently calculated using the Dyson equation and the self-energy has the correct analytic properties.[9] Equation (8.42) is an example of the Luttinger–Ward theorem in the Hartree–Fock approximation (remember that the Fermi energy ϵ_{F} is the zero-temperature limit of the chemical potential). In fact, according to the discussion below (11.14), the Hartree–Fock self-energy is Φ-derivable.

Due to the rotational invariance, $\epsilon_{\mathbf{p}}$ and $\Sigma^{\mathrm{R}}(\mathbf{p},\mu)$ depend only on the modulus $p = |\mathbf{p}|$ of the momentum. If we define the Fermi momentum p_{F} as the solution of

$$\mu - \frac{p_{\mathrm{F}}^2}{2} - \Sigma^{\mathrm{R}}(p_{\mathrm{F}},\mu) = 0, \tag{11.46}$$

then the equation for the total number of particles reduces to

$$\frac{N}{\mathrm{V}} = n = 2\int_{p<p_{\mathrm{F}}} \frac{d\mathbf{p}}{(2\pi)^3} = \frac{p_{\mathrm{F}}^3}{3\pi^2}, \tag{11.47}$$

which is the same relation as in the noninteracting case.

In the remainder of the section we discuss a couple of nice applications of the Luttinger–Ward theorem.

11.6.1 Momentum Distribution and Sharpness of the Fermi Surface

We show that in the electron gas the momentum distribution $n_{\mathbf{p}}$ (average number of electrons with momentum $\mathbf{p}$) is discontinuous at $|\mathbf{p}| = p_{\mathrm{F}}$ - that is, at the same Fermi momentum of the noninteracting gas.

The *momentum distribution* $n_{\mathbf{p}}$ of the electron gas at zero temperature is given by the equation

$$n_{\mathbf{p}} = \int_{-\infty}^{0} \frac{d\omega}{2\pi} A(\mathbf{p},\mu+\omega), \tag{11.48}$$

where $A(\mathbf{p},\mu+\omega)$ is the spectral function centered at the chemical potential $\mu = \epsilon_{\mathrm{F}}$.

Interestingly, in terms of this quantity the total energy (6.116) for an electron gas takes the form

$$\begin{aligned} E_{\mathrm{S}} &= \frac{\mathrm{V}}{2}\int \frac{d\omega}{2\pi}\int \frac{d\mathbf{p}}{(2\pi)^3}\langle \mathbf{p}|(\omega+\hat{h})f(\omega-\mu)\hat{\mathcal{A}}(\omega)|\mathbf{p}\rangle \\ &= \frac{\mathrm{V}}{2}\int_{-\infty}^{\mu} \frac{d\omega}{2\pi}\int \frac{d\mathbf{p}}{(2\pi)^3}(\omega+\frac{p^2}{2})A(\mathbf{p},\omega) = \frac{\mathrm{V}}{2}\left[\int \frac{d\mathbf{p}}{(2\pi)^3}\frac{p^2}{2}n_{\mathbf{p}} + \int_{-\infty}^{\mu}\frac{d\omega}{2\pi}\,\omega D(\omega)\right], \end{aligned}$$

where $D(\omega)$ is the density of states. It is easy to show that in the noninteracting case the two integrals in the square brackets are identical.

[9] The class of Φ-derivable approximations which also preserve the analytic properties are discussed in Ref. [148].

Equation (11.48) is difficult to use in practical numerical calculations since it has a very sharp quasi-particle peak close to the Fermi surface, the volume of which is hard to converge by integration. We therefore derive a different formula that also allows us to prove that the momentum distribution is sharp. First we note that since $G^{\rm A}(\mathbf{p},\omega) = [G^{\rm R}(\mathbf{p},\omega)]^*$ we have $A(\mathbf{p},\omega) = -2\,{\rm Im}[G^{\rm R}(\mathbf{p},\omega)]$. The retarded Green's function is analytic in the upper half of the complex ω-plane and therefore

$$0 = \int_{-\infty}^{0} d\omega\, G^{\rm R}(\mathbf{p},\mu+\omega) + {\rm i}\int_{0}^{\infty} d\omega\, G^{\rm R}(\mathbf{p},\mu+{\rm i}\omega) + \lim_{R\to\infty} {\rm i}\int_{\frac{\pi}{2}}^{\pi} d\theta\, Re^{{\rm i}\theta} G^{\rm R}(\mathbf{p},\mu+Re^{{\rm i}\theta}),$$

where we first integrate along the real axis from $-\infty$ to 0, then along the positive imaginary axis all the way to ${\rm i}\infty$, and finally along a quarter-circle from ${\rm i}\infty$ to $-\infty$. Since for $|\omega| \to \infty$ we have that $G^{\rm R}(\mathbf{p},\mu+\omega) \sim 1/\omega$, see (6.66), the integral along the quarter-circle becomes

$$\lim_{R\to\infty} {\rm i}\int_{\frac{\pi}{2}}^{\pi} d\theta\, Re^{{\rm i}\theta} G^{\rm R}(\mathbf{p},\mu+Re^{{\rm i}\theta}) = {\rm i}\frac{\pi}{2},$$

and hence

$$\int_{-\infty}^{0} d\omega\, G^{\rm R}(\mathbf{p},\mu+\omega) = -{\rm i}\frac{\pi}{2} - {\rm i}\int_{0}^{\infty} d\omega\, G^{\rm R}(\mathbf{p},\mu+{\rm i}\omega).$$

Inserting this result into (11.48) then yields

$$n_{\mathbf{p}} = \frac{1}{2} + \frac{1}{\pi}{\rm Re}\int_{0}^{\infty} d\omega\, G^{\rm R}(\mathbf{p},\mu+{\rm i}\omega). \tag{11.49}$$

This formula [149] is the basis for our subsequent derivations. We first write for the retarded Green's function centered at μ:

$$G^{\rm R}(\mathbf{p},\mu+\omega) = \frac{1}{\omega+\mu-\frac{p^2}{2}-\Sigma^{\rm R}(\mathbf{p},\mu+\omega)} = \frac{1}{\omega+a_p-\sigma_p^{\rm R}(\omega)},$$

where we define

$$a_p = \mu - \frac{p^2}{2} - \Sigma_{\rm x}(p),$$

with the exchange self-energy $\Sigma_{\rm x}(p) = -\frac{2p_{\rm F}}{\pi}F(p/p_{\rm F})$, see (8.46), and $\sigma_p^{\rm R}(\omega) = \Sigma_{\rm c}^{\rm R}(\mathbf{p},\mu+\omega)$ the correlation self-energy centered at μ.[10] Of course, these quantities depend only on the modulus $p = |\mathbf{p}|$ due to rotational invariance. Let us work out the integral in (11.49):

$${\rm Re}\int_{0}^{\infty} d\omega\, G^{\rm R}(\mathbf{p},\mu+{\rm i}\omega) = {\rm Re}\int_{0}^{\infty} \frac{d\omega}{{\rm i}\omega+a_p-\sigma_p^{\rm R}({\rm i}\omega)}.$$

Since the denominator vanishes when $\omega \to 0$ and $p = p_{\rm F}$ according to the Luttinger–Ward

[10] The Hartree self-energy cancels with the potential of the uniform positive background charge.

theorem, we add and subtract a term in which σ^{R} is expanded to first order in ω:

$$\frac{1}{\mathrm{i}\omega + a_p - \sigma_p^{\mathrm{R}}(\mathrm{i}\omega)} = \underbrace{\left[\frac{1}{\mathrm{i}\omega + a_p - \sigma_p^{\mathrm{R}}(\mathrm{i}\omega)} - \frac{1}{\mathrm{i}\omega + a_p - \sigma_p^{\mathrm{R}}(0) - \mathrm{i}\omega\frac{\partial\sigma_p^{\mathrm{R}}(0)}{\partial\omega}}\right]}_{F_p(\omega)}$$

$$+ \frac{1}{\mathrm{i}\omega + a_p - \sigma_p^{\mathrm{R}}(0) - \mathrm{i}\omega\frac{\partial\sigma_p^{\mathrm{R}}(0)}{\partial\omega}}.$$

The last term can be integrated analytically. By definition,

$$\sigma_p^{\mathrm{R}}(\omega) = \Lambda(\mathbf{p}, \mu + \omega) - \frac{\mathrm{i}}{2}\Gamma(\mathbf{p}, \mu + \omega). \tag{11.50}$$

The imaginary part of $\sigma_p^{\mathrm{R}}(\omega)$ vanishes quadratically as $\omega \to 0$,[11] and therefore $\sigma_p^{\mathrm{R}}(0) = \Lambda(\mathbf{p}, \mu)$ and $\partial\sigma_p^{\mathrm{R}}(0)/\partial\omega = \partial\Lambda(\mathbf{p}, \mu)/\partial\omega$ are real numbers. In particular, from the Hilbert transform relation (10.13) we also deduce that

$$\frac{\partial\sigma_p^{\mathrm{R}}(0)}{\partial\omega} = -\int \frac{d\omega'}{2\pi}\frac{\Gamma(\mathbf{p}, \omega')}{(\omega' - \mu)^2} < 0, \tag{11.51}$$

since the quasi-particle width $\Gamma(\mathbf{p}, \omega') = -2\mathrm{Im}[\sigma_p^{\mathrm{R}}(\omega' - \mu)]$ is nonnegative, see Exercise 10.1, and vanishes quadratically as $\omega' \to \mu$ (hence the integral is well defined even without the principal part). In conclusion, we have

$$\begin{aligned}
\mathrm{Re}\int_0^\infty \frac{d\omega}{\mathrm{i}\omega + a_p - \sigma_p^{\mathrm{R}}(0) - \mathrm{i}\omega\frac{\partial\sigma_p^{\mathrm{R}}(0)}{\partial\omega}} &= \int_0^\infty d\omega \frac{a_p - \sigma_p^{\mathrm{R}}(0)}{(1 - \frac{\partial\sigma_p^{\mathrm{R}}(0)}{\partial\omega})^2\omega^2 + (a_p - \sigma_p^{\mathrm{R}}(0))^2} \\
&= \frac{1}{1 - \frac{\partial\sigma_p^{\mathrm{R}}(0)}{\partial\omega}} \arctan\left(\frac{1 - \frac{\partial\sigma_p^{\mathrm{R}}(0)}{\partial\omega}}{a_p - \sigma_p^{\mathrm{R}}(0)}\,\omega\right)\Bigg|_0^\infty \\
&= \frac{1}{1 - \frac{\partial\sigma_p^{\mathrm{R}}(0)}{\partial\omega}}\,\frac{\pi}{2}\,\mathrm{sgn}(a_p - \sigma_p^{\mathrm{R}}(0)),
\end{aligned}$$

where in the last equality we take into account that $1 - \partial\sigma_p^{\mathrm{R}}(0)/\partial\omega$ is positive and larger than unity, see (11.51). The prefactor

$$\tilde{Z}_p = \frac{1}{1 - \frac{\partial\sigma_p^{\mathrm{R}}(0)}{\partial\omega}} < 1$$

coincides with the quasi-particle renormalization factor $Z_{\mathbf{p}}$ of (10.25) for $p = p_{\mathrm{F}}$, since in this case the quasi-particle energy $E_{\mathbf{p}} = \mu$. Inserting these results into (11.49) we get

$$n_{\mathbf{p}} = \frac{1}{2}\left[1 + \tilde{Z}_p\,\mathrm{sgn}(\mu - \frac{p^2}{2} - \Sigma^{\mathrm{R}}(\mathbf{p}, \mu))\right] + \frac{1}{\pi}\mathrm{Re}\int_0^\infty d\omega F_p(\omega). \tag{11.52}$$

[11] We prove this fact rigorously in (15.9).

According to the Luttinger–Ward theorem (11.45), the first term in this expression jumps with a value $\tilde{Z}_{p_{\rm F}} = Z$ at the Fermi surface. The function $F_p(\omega)$ is the sum of two terms that are equally singular when $\omega \to 0$ and $p = p_{\rm F}$ in such a way that their difference is finite in this limit. Therefore $F_p(\omega)$ is suitable for numerical treatments. Thus the momentum distribution (11.52) is the sum of a discontinuous function and a smooth function. This allows us to define a Fermi surface as the set of $\mathbf{p}$ points, where $n_{\mathbf{p}}$ is discontinuous (a sphere in the electron gas). In the noninteracting case, $F_p(\omega) = 0$, $Z = 1$, and $n_{\mathbf{p}}$ becomes the standard Heaviside function of a Fermi gas. The effect of the interaction is to reduce the size of the discontinuity and to promote electrons below the Fermi surface to states above the Fermi surface in such a way that $n_{\mathbf{p}} < 1$ for $p < p_{\rm F}$ and $n_{\mathbf{p}} > 0$ for $p > p_{\rm F}$, see Section 15.6. We emphasize that although the discontinuity occurs at the same Fermi momentum of the noninteracting electron gas, the value of the chemical potential (or Fermi energy) yielding the same number of particles is different. This fact has already been pointed out in the context of the Hartree approximation, see discussion below (8.16).

We conclude by observing that in the one-dimensional electron gas $\partial\sigma_p^{\rm R}(0)/\partial\omega$ is not real since the rate function $\Gamma(\mathbf{p}, \omega)$ vanishes as $|\omega - \mu|$, see the discussion in Section 15.1. In this system there is no Fermi surface – that is, the momentum distribution is not discontinuous as p crosses $p_{\rm F}$. The interacting electron gas in one dimension is called *Tomonaga–Luttinger liquid* [150, 151, 152].

11.6.2 Anderson Model: Friedel Sum Rule and Kondo Effect

We consider a quantum transport set up described by the Anderson model, see Section 9.5. We also work in the WBLA – that is, $\Gamma_\alpha(\omega) = \Gamma_\alpha$ is independent of ω. Since all matrices in the molecular region are 1×1 matrices, the condition of proportionate coupling is fulfilled. Taking into account that $\lambda = \Gamma_L/\Gamma_R$, (9.61) simplifies to

$$I_L^{(S)} = 2q\frac{\Gamma_L\Gamma_R}{\Gamma_L + \Gamma_R}\int\frac{d\omega}{2\pi}\left[f(\omega - V_R - \mu) - f(\omega - V_L - \mu)\right]A_{00}(\omega),$$

where $A_{00}(\omega) = \langle\epsilon_0|\hat{\mathcal{A}}(\omega)|\epsilon_0\rangle$ is the nonequilibrium spectral function projected onto the only basis function of the molecular region.

Conductance For a small bias between the left and right reservoirs $V_R = -V_L = q\phi/2$, we can Taylor expand $f(\omega - V_\alpha - \mu) \sim f(\omega - \mu) - f'(\omega - \mu)V_\alpha$. Then, to first order in ϕ,

$$I_L^{(S)} = -2q^2\frac{\Gamma_L\Gamma_R}{\Gamma_L + \Gamma_R}\phi\int\frac{d\omega}{2\pi}f'(\omega - \mu)A_{00}(\omega), \tag{11.53}$$

where $A_{00}(\omega)$ is the *equilibrium* spectral function of the Anderson model – that is, the spectral function evaluated at $\phi = 0$. Equation (11.53) correctly reduces to (9.63) for $U = 0$ since in this case $A_{00}(\omega)$ is given by (6.87), with $\Gamma_{\rm em} = \Gamma_L + \Gamma_R$. The *conductance* $\mathcal{C}$ is defined as the derivative of the steady current with respect to the potential drop ϕ:

$$\mathcal{C} \equiv \frac{dI_L^{(S)}}{d\phi} = -2q^2\frac{\Gamma_L\Gamma_R}{\Gamma_L + \Gamma_R}\int\frac{d\omega}{2\pi}f'(\omega - \mu)A_{00}(\omega). \tag{11.54}$$

In the limit of zero temperature, $f'(\omega-\mu)\to-\delta(\omega-\mu)$ and the conductance becomes

$$\mathcal{C}=2\times\frac{q^2}{2\pi}\frac{\Gamma_L\Gamma_R}{\Gamma_L+\Gamma_R}A_{00}(\mu). \tag{11.55}$$

Quantum transport experiments therefore provide an alternative to photoemission experiments for measuring the spectral function.

Friedel sum rule The spectral function evaluated at the chemical potential is particularly interesting. We have

$$\begin{aligned}A_{00}(\mu)&=\mathrm{i}[G^{\mathrm{R}}_{00}(\mu)-G^{\mathrm{A}}_{00}(\mu)]\\&=\mathrm{i}\left[\frac{1}{\mu-\epsilon_0-\Sigma^{\mathrm{R}}_{00}(\mu)-\Sigma^{\mathrm{R}}_{\mathrm{em}}(\mu)}-\frac{1}{\mu-\epsilon_0-\Sigma^{\mathrm{A}}_{00}(\mu)-\Sigma^{\mathrm{A}}_{\mathrm{em}}(\mu)}\right],\end{aligned}$$

where ϵ_0 is the only entry of the 1×1 Hamiltonian h_{MM}. We separate the many-body self-energy into a real and an imaginary part, $\Sigma^{\mathrm{R/A}}_{00}(\omega)=\Lambda_{00}(\omega)\mp\mathrm{i}\Gamma_{00}(\omega)/2$, see (10.12), and use that for fermions at zero temperature $\Gamma_{00}(\mu)=0$, see (10.17). In the WBLA we have $\Sigma^{\mathrm{R/A}}_{\alpha}(\omega)=\mp\mathrm{i}\Gamma_\alpha/2$ (no real part, see Section 2.4) and therefore $\Sigma^{\mathrm{R/A}}_{\mathrm{em}}=\mp\mathrm{i}\,(\Gamma_L+\Gamma_R)/2$. Thus the spectral function reads

$$A_{00}(\mu)=\frac{\Gamma_{\mathrm{em}}}{\big(\mu-\epsilon_0-\Lambda_{00}(\mu)\big)^2+\Gamma^2_{\mathrm{em}}/4},\qquad\Gamma_{\mathrm{em}}\equiv\Gamma_L+\Gamma_R. \tag{11.56}$$

We now show that $A_{00}(\mu)$ can be directly related to the atomic occupation $N_{\mathrm{at}}=2n_0$.

We calculate the atomic occupation $N_{\mathrm{at}}=2n_0$ using (11.42), where in our case $\hat{\Sigma}^{\mathrm{M}}$ is the sum of the many-body self-energy *and* embedding self-energy. For 1×1 matrices we have

$$\begin{aligned}n_0=\frac{1}{\beta}\sum_{p=-\infty}^{\infty}e^{\eta\omega_p}\Big[&\frac{\partial}{\partial\omega_p}\ln\left(\Sigma^{\mathrm{M}}_{00}(\omega_p)+\Sigma^{\mathrm{M}}_{\mathrm{em}}(\omega_p)-\omega_p+\epsilon^{\mathrm{M}}_0\right)\\&+G^{\mathrm{M}}_{00}(\omega_p)\frac{\partial}{\partial\omega_p}\Big(\Sigma^{\mathrm{M}}_{00}(\omega_p)+\Sigma^{\mathrm{M}}_{\mathrm{em}}(\omega_p)\Big)\Big],\end{aligned} \tag{11.57}$$

where $\epsilon^{\mathrm{M}}_0=\epsilon_0-\mu$ is the value of $h_{MM}(z)$ along the imaginary track of the contour. Using the property (6.70) for the Green's function $g_{\alpha\alpha}$ appearing in definition (9.52) of the embedding self-energy, we find

$$\Sigma^{\mathrm{M}}_{\mathrm{emb}}(\omega_p)=\begin{cases}-\mathrm{i}\Gamma_{\mathrm{em}}/2 & \mathrm{Im}[\omega_p]>0\\+\mathrm{i}\Gamma_{\mathrm{em}}/2 & \mathrm{Im}[\omega_p]<0\end{cases}.$$

Therefore $\partial\Sigma^{\mathrm{M}}_{\mathrm{emb}}(\omega_p)/\partial\omega_p\propto\delta(\mathrm{Im}[\omega_p])$. Since the fermionic Matsubara frequencies are all different from zero, see (6.19), this derivative does not contribute to the r.h.s. of (11.57). In the zero-temperature limit the derivative of the many-body self-energy does not contribute either. This is true for the exact as well as for any Φ-derivable approximation, as was demonstrated in Section 11.6. In the limit of zero temperature we can follow the strategy that leads to (11.44), and find

$$n_0 = \frac{1}{2\pi i}\Big[\ln\big(i\Gamma_{em}/2 + \Lambda_{00}(\mu) + \epsilon_0 - \mu\big) - \ln\big(-i\Gamma_{em}/2 + \Lambda_{00}(\mu) + \epsilon_0 - \mu\big)\Big],$$

where we use (10.15) and (10.17). Recalling that for any $y > 0$,

$$\ln(x + iy) - \ln(x - iy) = i\pi - 2i\,\text{atan}(x/y), \tag{11.58}$$

we eventually obtain

$$n_0 = \frac{1}{2} - \frac{1}{\pi}\text{atan}\left(\frac{\epsilon_0 + \Lambda_{00}(\mu) - \mu}{\Gamma_{em}/2}\right).$$

Inverting this equation, $\epsilon_0 + \Lambda_{00}(\mu) - \mu = -(\Gamma_{em}/2)\tan\left[\frac{\pi}{2}(2n_0 - 1)\right]$, and substituting into (11.56), we find the *Friedel sum rule* [153, 154, 155, 156]:

$$A_{00}(\mu) = \frac{4}{\Gamma_{em}}\frac{1}{\tan^2\left[\frac{\pi}{2}(2n_0 - 1)\right] + 1} = \frac{4}{\Gamma_{em}}\cos^2\left[\frac{\pi}{2}(2n_0 - 1)\right] = \frac{4}{\Gamma_{em}}\sin^2\left[\pi n_0\right]. \tag{11.59}$$

According to this result we can get the exact conductance, see (11.55), from the exact atomic occupation. This observation is the basis of conductance calculations within Density Functional Theory [157, 158, 159, 160].

Kondo effect At half-filling (i.e., $n_0 = 1/2$) the Friedel sum rule predicts that $A_{00}(\mu) = 4/\Gamma_{em}$, a value that is independent of the interaction strength U and that can be very large for weak couplings between the molecular region and the electrodes. In this weak coupling regime but for temperatures larger than Γ_{em} (Coulomb blockade regime), the atomic level is half-filled for $\mu = \epsilon_0 + U/2$, see Fig. 6.5(a), and the corresponding spectral function $A_{00}(\omega)$ is vanishingly small for $\omega = \mu$, see Fig. 6.5(b). We conclude that no matter how small is Γ_{em}, the spectral function must change shape at sufficiently low temperatures ($K_B T < \Gamma_{em}$) in order to approach the value $A_{00}(\mu) = 4/\Gamma_{em}$ at zero temperature. In fact, there exists a critical temperature $T_K = \sqrt{2U\Gamma_{em}}\exp[-\pi U/16\Gamma_{em}]$ [161], known as the *Kondo temperature*, below which a sharp peak of width $\sim T_K$ develops at $\omega = \mu$. This phenomenon is called the *Kondo effect* [15, 16] and, according to our probability interpretation of the spectral function, it implies that an electron added to or removed from the atomic level has a very long lifetime. Physically, this is due to the formation of a stable state characterized by a cloud of electrons in the electrodes having spin opposite to the spin of the electron on the atomic level (Kondo singlet). In Ref. [128] the authors show that the Kondo effect can be captured already at the level of the second Born approximation.

In Fig. 11.5 we show the effects of temperature and interaction on the atomic occupation n_0 and conductance $\mathcal{C}$. We take $\epsilon_0 = 0$ and $\Gamma_L = \Gamma_R = \Gamma_{em}/2$. In Fig. 11.5(a) we plot n_0 as obtained from (6.91). For $U = 0$ the atomic occupation "jumps" from zero to unity when the chemical potential crosses the energy ϵ_0 of the isolated atom. At low temperature the jump is smeared due to the coupling to the electrodes. When the temperature becomes greater than Γ_{em} the smearing is mainly due to thermal broadening. As we have already

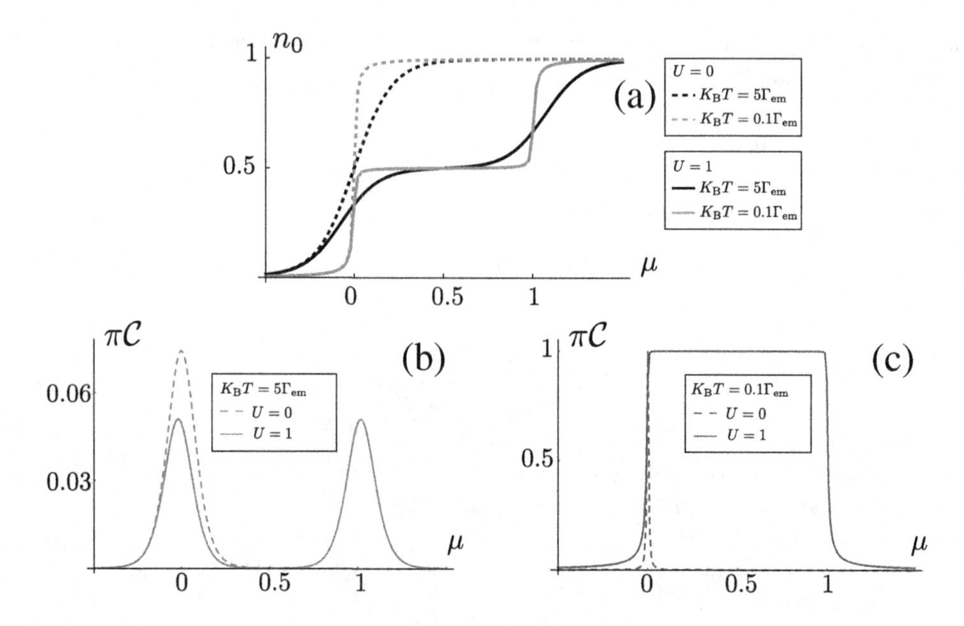

Figure 11.5 (a) Atomic occupation for $\Gamma_{\text{em}} = 0.02$ (in arbitrary units) at temperatures $K_\text{B}T = 0.1\Gamma_{\text{em}}$ and $K_\text{B}T = 5\Gamma_{\text{em}}$ for $U = 0$ (dashed) and $U = 1$ (solid). (b) Conductance at temperature $K_\text{B}T = 5\Gamma_{\text{em}}$. (c) Conductance at temperature $K_\text{B}T = 0.1\Gamma_{\text{em}}$.

seen in Section 6.3.3, the interaction U is responsible for the Coulomb blockaded effect, the atomic site remaining half-filled until the chemical potential is large enough to overcome the Coulomb barrier. The effect of a finite temperature for $U \neq 0$ is to smear the double jump in a similar manner as for $U = 0$. Fig. 11.5(b) we show the conductance at finite temperature for the noninteracting and interacting case. This is obtained from (11.54) with the spectral function $A_{00}(\omega)$ in (6.90). For $U = 0$ we observe a single peak at $\mu = \epsilon_0 = 0$, in agreement with the fact that the noninteracting spectral function peaks at $\omega = \epsilon_0$. In the interacting case the conductance exhibits a double peak, in agreement with the fact that in the Coulomb blockade regime the spectral function has a double-peak structure, see Fig. 6.5. Physically this means that current flows whenever the chemical potential is aligned to either the energy required to fill the atomic site with one electron or with two electrons. The conductance at low temperature is shown in Fig. 11.5(c). From (11.59) and (11.55) it can be approximated as (for $\Gamma_L = \Gamma_R = \Gamma_{\text{em}}/2$)

$$\mathcal{C} = \frac{1}{\pi}\sin^2[\pi n_0] = 2\mathcal{C}_0 \sin^2[\pi n_0], \tag{11.60}$$

where $\mathcal{C}_0 = \frac{1}{2\pi}$ is the quantum of conductance, see the end of Section 9.5.2. The single peak at $U = 0$ of Fig. 11.5(b) becomes sharper and its height reaches the value of twice the quantum of conductance, in agreement with (9.65). The effect of correlations at zero temperature is dramatic. The conductance remains pinned at its maximum value in the whole range of chemical potentials for which $n_0 = 1/2$ (Kondo plateau).

Exercise 11.5 Consider a molecular region described by the Hamiltonian of the so-called *constant interaction model*:

$$\hat{H} = \sum_{m=1}^{M} \epsilon_m \hat{n}_m + \frac{U}{2} \sum_{m \neq m'} \hat{n}_m \hat{n}_{m'}.$$

Assume that each molecular level is coupled to its own left and right WBLA electrode, so that the matrices Γ_L and Γ_R are diagonal. Express the conductance at zero temperature in terms of the occupations n_m of the molecular levels.

11.7 Relation between the Density Response Function and the Φ Functional

The LW functional is based on the MBPT expansion in terms of the dressed Green's function G and interaction v. For long-range interactions like the Coulomb interaction this is not necessarily the best choice of variables. For instance, in bulk systems the most natural and meaningful variable is the screened interaction W [93]. The aim of the next sections is to construct a variational many-body scheme in terms of G and W. To achieve this goal we need a preliminary discussion on the relation between the density response function (or reducible polarization) χ, defined in Section 7.7, and the Φ functional.

It is evident that if we close a χ-diagram with a v-line we obtain a vacuum diagram. Vice versa, cutting off a v-line in a vacuum diagram we obtain a χ-diagram, with only one exception. The diagram that generates the Hartree self-energy (first diagram of Fig. 11.4) is the only vacuum diagram to be one-interaction-line reducible. Therefore, cutting off the v-line from the Hartree diagram does not generate a χ-diagram and, vice versa, no χ-diagram can generate the Hartree diagram if closed with a v-line. We point out that the "exceptional" vacuum diagrams would be infinitely many in a formulation in terms of G_0 rather than G since there would be infinitely many one-interaction-line reducible diagrams (consider, e.g., the first diagram in the second row of Fig. 11.2). This is the main reason for us to present a formulation in terms of dressed vacuum diagrams.

As the operation of cutting off a v-line corresponds to taking a functional derivative with respect to v, we expect the existence of a simple relation between χ and $\delta\Phi/\delta v$. This is indeed the case even though we must be careful in defining the functional derivative with respect to v. The difficulty that arises here, and that does not arise for the functional derivatives with respect to G, has to do with the symmetry $v(i;j) = v(j;i)$. When we take a functional derivative we must allow for arbitrary variations of the interaction, including *nonsymmetric variations*. To clarify this point, consider the vacuum diagram (11.7). The integral associated with it is

$$V_1 = \int v(1;5)v(2;4)v(3;6)G(1;6)G(6;5)G(5;4)G(4;3)G(3;2)G(2;1),$$

where the integral is over all variables. Since v is symmetric we could also write the same integral as

$$V_2 = \int v(5;1)v(2;4)v(3;6)G(1;6)G(6;5)G(5;4)G(4;3)G(3;2)G(2;1),$$

where in V_2 we simply replaced $v(1;5) \to v(5;1)$. For a symmetric interaction $V_1 = V_2$, but otherwise V_1 is different from V_2. If we now take the functional derivative of V_1 with respect to $v(a;b)$ and then evaluate the result for a symmetric interaction, we get

$$\begin{aligned}\frac{\delta V_1}{\delta v(a;b)} &= \int v(2;4)v(3;6)G(a;6)G(6;b)G(b;4)G(4;3)G(3;2)G(2;a) \\ &+ \int v(1;5)v(3;6)G(1;6)G(6;5)G(5;b)G(b;3)G(3;a)G(a;1) \\ &+ \int v(1;5)v(2;4)G(1;b)G(b;5)G(5;4)G(4;a)G(a;2)G(2;1) \\ &= I_{\chi,1}(a;b) + I_{\chi,2}(a;b) + I_{\chi,3}(a;b), \end{aligned} \tag{11.61}$$

which is proportional to the following sum of χ-diagrams:

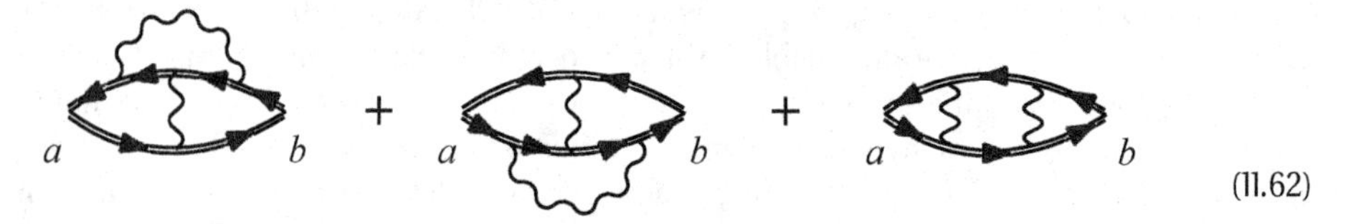

(11.62)

On the other hand, if we take the functional derivative of V_2 with respect to $v(a;b)$ and then evaluate the result for a symmetric interaction, we get

$$\frac{\delta V_2}{\delta v(a;b)} = I_{\chi,1}(b;a) + I_{\chi,2}(a;b) + I_{\chi,3}(a;b) = 2I_{\chi,2}(a;b) + I_{\chi,3}(a;b).$$

In the last equality we take into account that under the interchange $a \leftrightarrow b$ the first diagram in (11.62) [representing $I_{\chi,1}(a;b)$] becomes identical to the second diagram. We thus conclude that $\delta V_1/\delta v(a;b) \neq \delta V_2/\delta v(a;b)$. To remove this ambiguity we define the symmetric derivative,

$$\left.\frac{\delta}{\delta v(a;b)}\cdots\right|_S = \frac{1}{2}\left[\frac{\delta}{\delta v(a;b)}\cdots + \frac{\delta}{\delta v(b;a)}\cdots\right],$$

where any diagram can be inserted in the dots. The reader can easily check that $\left.\frac{\delta V_1}{\delta v(a;b)}\right|_S = \left.\frac{\delta V_2}{\delta v(a;b)}\right|_S$. Clearly, there is not such a problem for $\delta/\delta G$ since G is not symmetric and hence there is only one way to choose the arguments of G in a diagram.[12]

We say that two χ-diagrams are related by a *reversal* symmetry if one diagram can be obtained from the other by interchanging the external vertices. The first two diagrams

[12]We remind the reader that even though the operation $\delta/\delta G$ is unambiguous, also in this case we must allow for arbitrary variations of G including those variations that bring G away from the Keldysh space, see the discussion before (11.14).

in (11.62) are clearly one the reversal of the other. Not all χ-diagrams become another diagram after a reversal operation. For instance, the third diagram in (11.62) or the first two polarization diagrams in Fig. 7.7 are the reversal of themselves.

In graphical terms the symmetric derivative corresponds to cutting a v-line, adding the reversed diagram and dividing the result by 2. The operation of cutting and symmetrizing allows us to define two v-lines as *equivalent* if they lead to the same χ-diagram(s). For instance, cutting a v-line from (11.7) and then symmetrizing yields (up to a prefactor):

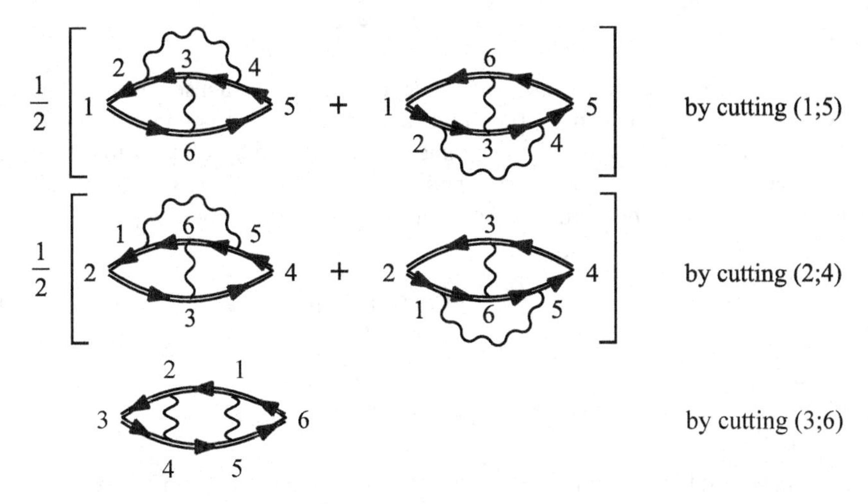

Thus, the v-lines $(1;5)$ and $(2;4)$ are equivalent, while $(3;6)$ is not equivalent to any other line.

Similarly to the G-lines, the v-lines have the following properties:

- If s is a symmetry of a vacuum diagram, then the v-lines $(i;j)$ and $(s(i);s(j))$ are equivalent.

- If two v-lines $(i;j)$ and $(k;l)$ in a vacuum diagram are equivalent, then there exists a symmetry such that $(s(i),s(j)) = (k;l)$ [we remind the reader that for a v-line $(i;j) = (j;i)$].

The proof of both statements is identical to the proof for the Green's function lines (see Section 11.3) and it is left as an exercise for the reader. Let us explore the implications of the above two properties. Consider a vacuum diagram with N_{S} symmetry operations. Given a v-line $(i;j)$, there are two possibilities: (1) there exists a symmetry s_R that reverses the interaction line [i.e., $s_R(i) = j$ and $s_R(j) = i$]; or (2) the symmetry s_R does not exist. In case (1), by cutting $(i;j)$ and symmetrizing, we get a single χ-diagram which is the reversal of itself. Then, by applying all N_{S} symmetry operations to $(i;j)$, we obtain only $N_{\mathrm{S}}/2$ topologically equivalent χ-diagrams since the symmetry s and the symmetry $s \circ s_R$ generate the same χ-diagram and, obviously, $s_1 \circ s_R \neq s_2 \circ s_R$ for $s_1 \neq s_2$. For instance, the only symmetry of (11.7) is a reversal symmetry of the v-line $(3;6)$ and consequently cutting $(3;6)$ and symmetrizing leads to only $N_{\mathrm{S}}/2 = 2/2 = 1$ diagram. In case (2), by cutting $(i;j)$ and symmetrizing, we get two χ-diagrams which are the reversal of each other, and hence there must be a symmetry s_R that maps a diagram into the other. By applying the N_S symmetry operations to $(i;j)$ we then obtain $N_S/2$ couples of topologically equivalent χ-diagrams.

From this analysis it follows that if D_c is a connected vacuum diagram of order n with l loops and V_c is the corresponding integral, $D_c = \mathrm{i}^n(\pm)^l V_c$, we have

$$\frac{1}{N_S}\frac{\delta D_c}{\delta v(1;2)}\bigg|_S = \mathrm{i}^n(\pm)^l \frac{1}{N_S}\frac{\delta V_c}{\delta v(1;2)}\bigg|_S = \frac{1}{2}\chi^{(D_c)}(1;2),$$

where $\chi^{(D_c)}$ is a contribution of order $n-1$ to the density response function consisting of $n/(N_S/2) = 2n/N_S$ diagrams. Therefore, summing over all connected, topologically inequivalent, and G-skeletonic vacuum diagrams each multiplied by $1/N_S$, with the exclusion of the Hartree diagram, we generate a functional from which to calculate the full χ by a functional derivative. This functional is exactly $\pm(\Phi - \Phi_H)$ where Φ_H is the Hartree functional,

$$\Phi_H[G,v] = \pm\frac{1}{2}\mathrm{i}\,(\pm)^2 \int d1d2\, G(1;1^+)v(1;2)G(2;2^+). \tag{11.63}$$

The difference

$$\Phi_{xc}[G,v] = \Phi[G,v] - \Phi_H[G,v]$$

is the exchange-correlation (XC) part of the Φ functional (whose functional derivative yields the XC self-energy). We can then write the important and elegant formula

$$\boxed{\frac{\delta \Phi_{xc}}{\delta v(1;2)}\bigg|_S = \pm\frac{1}{2}\chi(1;2)} \tag{11.64}$$

11.8 The Ψ Functional

Result (11.64) constitutes the starting point to switch from the variables (G,v) to the variables (G,W). There is only one ingredient missing, and that is the Ψ functional. Let us construct the functional $\Psi[G,W]$ by removing from Φ_{xc} all diagrams that contain polarization insertions and then replacing $v \to W$. By the very same reasoning that led to (11.19) and (11.64), the functional derivatives of Ψ with respect to G and W give the self-energy and the polarization:

$$\boxed{\Sigma_{xc}(1;2) = \Sigma_{ss,xc}[G,W](1;2) = \frac{\delta \Psi}{\delta G(2;1^+)}} \tag{11.65}$$

$$\boxed{P(1;2) = P_{ss}[G,W](1;2) = \pm 2\,\frac{\delta \Psi}{\delta W(1;2)}\bigg|_S} \tag{11.66}$$

Similarly to (11.20), the Ψ functional can be expanded in G- and W-skeleton *self-energy* diagrams as

$$\Psi[G, W] = \sum_{n=1}^{\infty} \frac{1}{2n} \int d1d2\, \Sigma^{(n)}_{ss,\mathrm{xc}}[G, W](1; 2)G(2, 1^+)$$
$$= \sum_{n=1}^{\infty} \frac{1}{2n} \mathrm{tr}_\gamma \left[\Sigma^{(n)}_{\mathrm{xc}} G\right]. \tag{11.67}$$

An alternative expression in terms of G- and W-skeleton *polarization* diagrams can be deduced from the obvious invariance $\Psi[G, W] = \Psi[\alpha^{-1/2}G, \alpha W]$. Taking the derivative with respect to α and evaluating the result in $\alpha = 1$, we find

$$0 = \left.\frac{\partial \Psi}{\partial \alpha}\right|_{\alpha=1} = -\frac{1}{2}\mathrm{tr}_\gamma\left[\Sigma_{\mathrm{xc}} G\right] \pm \frac{1}{2}\mathrm{tr}_\gamma\left[PW\right].$$

By equating orders in W, we then obtain

$$\mathrm{tr}_\gamma \left[\Sigma^{(n)}_{\mathrm{xc}} G\right] = \pm \mathrm{tr}_\gamma \left[P^{(n-1)} W\right], \tag{11.68}$$

and substitution into (11.67) gives

$$\boxed{\Psi[G, W] = \pm \sum_{n=1}^{\infty} \frac{1}{2n} \mathrm{tr}_\gamma \left[P^{(n-1)} W\right]} \tag{11.69}$$

A word of caution about equality (11.68) with $n = 1$ must be given. The l.h.s. is simply the Fock (or exchange) diagram and reads

$$\mathrm{i} \int d1d2\, W(1; 2)G(1; 2^+)G(2; 1^+). \tag{11.70}$$

Since the zeroth-order polarization is $P^{(0)}(1; 2) = \pm \mathrm{i}G(1; 2)G(2; 1)$, the r.h.s. of (11.68) reads

$$\mathrm{i} \int d1d2\, W(1; 2)G(1; 2)G(2; 1). \tag{11.71}$$

These two expressions are the same only provided that W is not proportional to $\delta(z_1, z_2)$. If W is approximated with, for example, $W(1; 2) \sim \delta(z_1, z_2)W(\mathbf{x}_1, \mathbf{x}_2)$, then (11.71) is not well defined as it contains two Green's functions with the same time arguments. The ambiguity is resolved by (11.70), which tells us to evaluate the Green's function with the second time argument infinitesimally later than the first: $G(1; 2)G(2; 1) \to G(1; 2^+)G(2; 1^+)$. This agrees with the general remark on notation in Section 7.1.

Relation between Ψ and Φ There exists a simple relation between the Ψ functional and the Φ functional. Let $W = W[G, v]$ be the solution of $W = v + vPW$ [Dyson equation (7.23) for the screened interaction] with $P = P_{ss}[G, W]$. At the self-consistent W we have $P_{ss}[G, W] = P_s[G, v]$ and hence $\Sigma_{ss}[G, W] = \Sigma_s[G, v]$. If we rescale the interaction $v \to \lambda v$ *without changing* G, the self-consistent W changes to $W[G, \lambda v] \equiv \lambda W_\lambda$. Then, the derivative of $\Psi[G, \lambda W_\lambda]$ with respect to λ is the sum of two contributions: one comes from W_λ and the other comes from the explicit dependence on λ. The latter can be calculated in the same way as (11.27), and therefore

$$\frac{d}{d\lambda}\Psi[G, \lambda W_\lambda] = \text{tr}_\gamma \left[\frac{1}{2\lambda}\Sigma_{s,\text{xc}}[G, \lambda v]G \pm \frac{1}{2}\lambda P_s[G, \lambda v]\frac{dW_\lambda}{d\lambda}\right], \tag{11.72}$$

where we use (11.66) for the last term. Next we separate out the Hartree contribution from Φ and calculate the derivative of $\Phi[G, \lambda v]$ with respect to λ:

$$\frac{d}{d\lambda}\Phi[G, \lambda v] = \frac{d}{d\lambda}\Phi_{\text{H}}[G, \lambda v] + \text{tr}_\gamma \left[\frac{1}{2\lambda}\Sigma_{s,\text{xc}}[G, \lambda v]G\right].$$

Comparing with (11.72), we get

$$\frac{d}{d\lambda}\Phi[G, \lambda v] = \frac{d}{d\lambda}\Phi_{\text{H}}[G, \lambda v] + \frac{d}{d\lambda}\Psi[G, \lambda W_\lambda] \mp \frac{1}{2}\text{tr}_\gamma \left[\lambda P_s[G, \lambda v]\frac{dW_\lambda}{d\lambda}\right]. \tag{11.73}$$

The last term in this equation can be written as a total derivative with respect to λ using the same trick leading to (11.29). For convenience we define $P_\lambda \equiv P_s[G, \lambda v]$. Then we have

$$\begin{aligned}
\frac{d}{d\lambda}\text{tr}_\gamma \left[\ln(1 - v\lambda P_\lambda)\right] &= -\frac{d}{d\lambda}\text{tr}_\gamma \left[v(\lambda P_\lambda) + \frac{1}{2}v(\lambda P_\lambda)v(\lambda P_\lambda) + \ldots\right] \\
&= -\text{tr}_\gamma \left[v\frac{d(\lambda P_\lambda)}{d\lambda} + v(\lambda P_\lambda)v\frac{d(\lambda P_\lambda)}{d\lambda} + \ldots\right] \\
&= -\text{tr}_\gamma \left[W_\lambda \frac{d(\lambda P_\lambda)}{d\lambda}\right] = -\frac{d}{d\lambda}\text{tr}_\gamma \left[W_\lambda \lambda P_\lambda\right] + \text{tr}_\gamma \left[\lambda P_\lambda \frac{dW_\lambda}{d\lambda}\right].
\end{aligned}$$

Inserting this result into (11.73) and integrating over λ between 0 and 1, we eventually arrive at the result

$$\boxed{\Phi[G, v] = \Phi_{\text{H}}[G, v] + \Psi[G, W] \mp \frac{1}{2}\text{tr}_\gamma\left[\, WP + \ln(1 - vP)\,\right]} \tag{11.74}$$

with $W = W[G, v]$ and $P = P_s[G, v]$.

11.8.1 Interaction Energy with P and W

An alternative formula for the interaction energy of systems in thermal equilibrium involves the polarization P and the screened interaction W. The starting point is the identity (11.68), which implies

$$\boxed{\text{tr}_\gamma\left[\Sigma_{\text{xc}}G\right] = \pm\text{tr}_\gamma\left[PW\right]} \tag{11.75}$$

Separating out the Fock (exchange) part,

$$\Sigma_{\text{x}}(1; 2) = \text{i}v(1; 2)G(1; 2^+) = \pm\delta(z_1, z_2)v(\mathbf{x}_1, \mathbf{x}_2)n(\mathbf{x}_1, \mathbf{x}_2),$$

from $\Sigma_{\rm xc}$, we can rewrite (11.75) as

$$\mathrm{tr}_\gamma\big[\Sigma_{\rm c}G\big] \mp 2\beta\, E_{\rm int,x} = \pm\mathrm{tr}_\gamma\big[PW\big], \tag{11.76}$$

where β is the inverse temperature[13] and $E_{\rm int,x}$ is the exchange part of the Hartree–Fock interaction energy (10.32). Since the system is in equilibrium,

$$\begin{aligned}\mathrm{tr}_\gamma\big[\Sigma_{\rm c}G\big] &= -\mathrm{i}\beta\int d\mathbf{x}_1\langle\mathbf{x}_1|\left[\hat{\Sigma}_{\rm c}^{\rm M}\star\hat{\mathcal{G}}^{\rm M}\right](0,0^+)\,|\mathbf{x}_1\rangle\\ &= \mp 2\beta E_{\rm int,c}\,,\end{aligned} \tag{11.77}$$

see (10.33). We can evaluate the r.h.s. of (11.76) similarly. We separate out the singular part v from W by defining $W = v + \delta W$ so that

$$\mathrm{tr}_\gamma\big[PW\big] = -\mathrm{i}\beta\int d\mathbf{x}_1\langle\mathbf{x}_1|\hat{\mathcal{P}}(z,z)\hat{v}|\mathbf{x}_1\rangle + \mathrm{tr}_\gamma\big[P\delta W\big].$$

The first term on the r.h.s. requires the calculation of the polarization at equal times. We remind the reader that the many-body expansion of $P(1;2)$ starts with $\pm\mathrm{i}G(1;2)G(2;1)$. For $z_1 = z_2$ this is the only ambiguous term of the expansion. The correct way of removing the ambiguity consists in using $\pm\mathrm{i}G(1;2^+)G(2;1^+)$, in accordance with the discussion below (11.71). If we write

$$P(1;2) = \pm\mathrm{i}G(1;2^+)G(2;1^+) + \delta P(1;2),$$

then

$$-\mathrm{i}\beta\int d\mathbf{x}_1\langle\mathbf{x}_1|\hat{\mathcal{P}}(z,z)\hat{v}|\mathbf{x}_1\rangle = -2\beta\, E_{\rm int,x} - \mathrm{i}\beta\int\frac{d\omega}{2\pi}\int d\mathbf{x}_1\langle\mathbf{x}_1|\delta\hat{\mathcal{P}}^{\lessgtr}(\omega)\hat{v}|\mathbf{x}_1\rangle.$$

As expected, the zeroth-order polarization generates the exchange energy. The second term is independent of whether we choose the greater or lesser component, since $v(\mathbf{x}_1,\mathbf{x}_2)$ is symmetric under $\mathbf{x}_1\leftrightarrow\mathbf{x}_2$ and from (7.27) we have $\delta P^>(1;2) = \delta P^<(2;1)$.[14] Taking into account (11.77) we can express the correlated part of the interaction energy as

$$E_{\rm int,c} = \frac{\mathrm{i}}{2}\int\frac{d\omega}{2\pi}\int d\mathbf{x}_1\langle\mathbf{x}_1|\delta\hat{\mathcal{P}}^{\lessgtr}(\omega)\hat{v}|\mathbf{x}_1\rangle - \frac{1}{2\beta}\mathrm{tr}_\gamma\big[P\delta W\big]. \tag{11.78}$$

[13] We take into account that $\int_\gamma dz = -\mathrm{i}\beta$.

[14] More explicitly we have

$$\begin{aligned}\int d\mathbf{x}_1\langle\mathbf{x}_1|\hat{\mathcal{P}}^<(t,t)\hat{v}|\mathbf{x}_1\rangle &= \int d\mathbf{x}_1 d\mathbf{x}_2\langle\mathbf{x}_1|\hat{\mathcal{P}}^<(t,t)|\mathbf{x}_2\rangle v(\mathbf{x}_2,\mathbf{x}_1)\\ &= \int d\mathbf{x}_1 d\mathbf{x}_2\langle\mathbf{x}_2|\hat{\mathcal{P}}^>(t,t)|\mathbf{x}_1\rangle v(\mathbf{x}_2,\mathbf{x}_1)\\ &= \int d\mathbf{x}_1 d\mathbf{x}_2\langle\mathbf{x}_1|\hat{\mathcal{P}}^>(t,t)|\mathbf{x}_2\rangle v(\mathbf{x}_2,\mathbf{x}_1) = \int d\mathbf{x}_1\langle\mathbf{x}_1|\hat{\mathcal{P}}^>(t,t)\hat{v}|\mathbf{x}_1\rangle,\end{aligned}$$

where in the third line we rename $\mathbf{x}_1\leftrightarrow\mathbf{x}_2$ and use the symmetry of v.

Let us now manipulate the last term of this equation. With the same trick used to derive (10.34), we can write

$$\begin{aligned}\mathrm{tr}_\gamma\left[P\delta W\right] &= -\mathrm{i}\beta\int d\mathbf{x}_1\langle\mathbf{x}_1|\left[\hat{\mathcal{P}}^{\mathrm{M}}\star\delta\hat{\mathcal{W}}^{\mathrm{M}}\right](0,0^+)\,|\mathbf{x}_1\rangle.\\ &= -\mathrm{i}\beta\int\frac{d\omega}{2\pi}\int d\mathbf{x}_1\langle\mathbf{x}_1|\hat{\mathcal{P}}^{\lessgtr}(\omega)\delta\hat{\mathcal{W}}^{\mathrm{A}}(\omega)+\hat{\mathcal{P}}^{\mathrm{R}}(\omega)\delta\hat{\mathcal{W}}^{\lessgtr}(\omega)|\mathbf{x}_1\rangle,\end{aligned} \tag{11.79}$$

where in the last equality we use (10.41) and the fluctuation–dissipation theorems (10.38) and (10.39). Inserting the identities (10.40) for the retarded/advanced quantities, we get[15]

$$\mathrm{tr}_\gamma\left[P\delta W\right] = \beta\int\frac{d\omega\,d\omega'}{2\pi\;2\pi}\int d\mathbf{x}_1\langle\mathbf{x}_1|\frac{\hat{\mathcal{P}}^{<}(\omega)\hat{\mathcal{W}}^{>}(\omega')-\hat{\mathcal{P}}^{>}(\omega)\hat{\mathcal{W}}^{<}(\omega')}{\omega-\omega'}|\mathbf{x}_1\rangle.$$

This result can now be inserted in (11.78) to obtain a formula for the interaction energy in terms of the polarization and the screened interaction:

$$\boxed{\begin{aligned}E_{\mathrm{int,c}} &= \frac{\mathrm{i}}{2}\int\frac{d\omega}{2\pi}\int d\mathbf{x}_1\langle\mathbf{x}_1|\delta\hat{\mathcal{P}}^{\lessgtr}(\omega)\hat{v}|\mathbf{x}_1\rangle\\ &\quad-\frac{1}{2}\int\frac{d\omega\,d\omega'}{2\pi\;2\pi}\int d\mathbf{x}_1\langle\mathbf{x}_1|\frac{\hat{\mathcal{P}}^{<}(\omega)\hat{\mathcal{W}}^{>}(\omega')-\hat{\mathcal{P}}^{>}(\omega)\hat{\mathcal{W}}^{<}(\omega')}{\omega-\omega'}|\mathbf{x}_1\rangle\end{aligned}} \tag{11.80}$$

11.9 Screened Functionals

We are now ready to formulate a variational many-body theory of the grand potential in terms of the independent quantities G and W. First, we observe that by inserting (11.74) into (11.30) we obtain the doubly dressed MBPT expansion of the grand potential:

$$\beta(\Omega-\Omega_0) = \mp\left\{\Phi_{\mathrm{H}}+\Psi-\mathrm{tr}_\gamma\left[\Sigma G+\ln(1-G_0\Sigma)\right]\mp\frac{1}{2}\mathrm{tr}_\gamma\left[PW+\ln(1-vP)\right]\right\}. \tag{11.81}$$

This formula provides the diagrammatic expansion of Ω in the variables G and W since we know how to expand the quantities Ψ, Σ, and P using skeletonic diagrams in G and W.

We can regard (11.81) as the definition of a functional $\Omega[G,W]$ that equals the grand potential when G and W are the Green's function and the screened interaction of the underlying physical system. This functional was first proposed in Ref. [101] and is denoted by $\Omega_{\mathrm{sLW}}[G,W]$ to distinguish it from the grand potential Ω, which is a number (the acronym sLW stands for screened Luttinger–Ward). We then have

$$\begin{aligned}\Omega_{\mathrm{sLW}}[G,W] = \Omega_0\mp\frac{1}{\beta}\Big\{&\Phi_{\mathrm{H}}[G,v]+\Psi[G,W]\\ &-\mathrm{tr}_\gamma\left[\Sigma_{ss}[G,W]G+\ln(1-G_0\Sigma_{ss}[G,W])\right]\\ &\mp\frac{1}{2}\mathrm{tr}_\gamma\left[P_{ss}[G,W]W+\ln(1-vP_{ss}[G,W])\right]\Big\}.\end{aligned} \tag{11.82}$$

[15] The small imaginary part in the denominator can be discarded since the numerator vanishes for $\omega=\omega'$.

A precursor of this functional can be found in Appendix B of Hedin's work [93]. The possibility of constructing W-based variational schemes is also implicit in earlier works by De Dominicis and Martin [162, 164]. The variation of the sLW functional induced by a variation δG and δW is

$$\delta\Omega_{\mathrm{sLW}} = \mp\frac{1}{\beta}\Big\{\,\delta\Phi_{\mathrm{H}} + \delta\Psi - \mathrm{tr}_\gamma\big[\Sigma\delta G + G\delta\Sigma - (G_0 + G_0\Sigma G_0 + \ldots)\delta\Sigma\big] \mp\frac{1}{2}\mathrm{tr}_\gamma\big[P\delta W + W\delta P - (v + vPv + \ldots)\delta P\big]\,\Big\}.$$

Taking into account that $\delta\Phi_{\mathrm{H}} = \mathrm{tr}_\gamma[\Sigma_{\mathrm{H}}\delta G]$ and that $\delta\Psi = \mathrm{tr}_\gamma[\Sigma_{\mathrm{xc}}\delta G \pm \frac{1}{2}P\delta W]$, we see that the sLW functional is stationary when

$$G = G_0 + G_0\Sigma G_0 + G_0\Sigma G_0\Sigma G_0 + \ldots$$

and

$$W = v + vPv + vPvPv + \ldots,$$

that is, when G is the self-consistent Green's function $G = G_0 + G_0\Sigma G$, and W is the self-consistent screened interaction $W = v + vPW$. At the stationary point $\Omega_{\mathrm{sLW}} = \Omega$.

Just as in the case of the LW functional, we can add to the sLW functional any $F[D]$ which fulfills (11.34). The resulting functional is stationary in the same point and it takes the same value at the stationary point. If we add the functional (11.35), with $D = D[G, W]$ defined in (11.33) in which $\Sigma_s[G, v] \to \Sigma_{ss}[G, W]$, we obtain the Klein version of the sLW functional:

$$\Omega_{\mathrm{sK}} = \Omega_0 \mp \frac{1}{\beta}\Big\{\,\Phi_{\mathrm{H}} + \Psi + \mathrm{tr}_\gamma\big[\,\ln(G\overleftarrow{G}_0^{-1}) + 1 - G\overleftarrow{G}_0^{-1}\,\big] \mp \frac{1}{2}\mathrm{tr}_\gamma\big[\,PW + \ln(1 - vP)\,\big]\,\Big\},$$

where the functional dependence of the various quantities is understood. There is, however, an additional freedom which we can use to design more stable functionals [145]. We can add an arbitrary functional $K[Q]$ of a functional $Q[G, W]$, defined by

$$Q[G, W](1;2) = W(1;2) - v(1;2) - \int d3d4\, W(1;2)P_{ss}[G, W](3;4)v(4;2),$$

with the properties

$$K[Q = 0] = \left(\frac{\delta K}{\delta Q}\right)_{Q=0} = 0.$$

We then obtain a new functional for the grand potential with the same stationary point and the same value at the stationary point. Choosing to add inside the curly bracket of (11.82) the functional

$$K[Q] = \pm\frac{1}{2}\mathrm{tr}_\gamma\big[\,\ln(1 + Q\overleftarrow{v}^{-1}) - Q\overleftarrow{v}^{-1}\,\big],$$

with $\overleftarrow{v}^{-1}(1;2)$ the differential operator (acting on quantities to its left) for which $\int d3\, v(1;3)\, \overleftarrow{v}^{-1}(3;2) = \delta(1;2)$,[16] we find

$$\Omega_{\mathrm{ssK}} = \Omega_0 \mp \frac{1}{\beta}\Big\{\,\Phi_{\mathrm{H}} + \Psi - \mathrm{tr}_\gamma\big[\Sigma G + \ln(1 - G_0\Sigma)\,\big] \mp \frac{1}{2}\mathrm{tr}_\gamma\big[\,\ln(Wv^{-1}) + 1 - Wv^{-1}\,\big]\,\Big\}.$$

[16] For instance, for the Coulomb interaction $v(1,2) = \delta(z_1, z_2)/|\mathbf{r}_1 - \mathbf{r}_2|$ and for particles of spin 0 we have $\overleftarrow{v}^{-1}(1;2) = \frac{1}{4\pi}\delta(z_1, z_2)\delta(\mathbf{r}_1 - \mathbf{r}_2)\overleftarrow{\nabla}_2^2$.

Here, ssK stands for the simple version of the screened Klein functional based on the construction of Ref. [101], see also Refs. [145, 165].

12

Conserving Approximations

A major difficulty in the theory of nonequilibrium processes consists in generating approximate Green's functions yielding a particle density $n(\mathbf{x}, t)$ and a current density $\mathbf{J}(\mathbf{x}, t)$ which satisfy the continuity equation, or a total momentum $\mathbf{P}(t)$ and a Lorentz force $\mathbf{F}(t)$ which is the time derivative of $\mathbf{P}(t)$, or a total energy $E(t)$ and a total power which is the time derivative of $E(t)$, etc. All these fundamental relations express the conservation of some physical quantity. In equilibrium the current is zero and hence the density is conserved – that is, it is the same at all times; the force is zero and hence the momentum is conserved; the power fed into the system is zero and hence the energy is conserved, etc. In this chapter we show that the Green's function obtained from the solution of the Dyson equation with a Φ-derivable self-energy satisfies all fundamental conservation laws.

12.1 Φ-derivable Self-Energies

We consider a system of interacting particles with mass m and charge q under the influence of an external electromagnetic field. The single-particle Hamiltonian is

$$h(1; 2) = \langle \mathbf{x}_1 | \hat{h}(z_1) | \mathbf{x}_2 \rangle \delta(z_1, z_2) = \left[\frac{1}{2m} (-\mathrm{i}\,\mathbf{D}_1)^2 + q\phi(1) \right] \delta(1; 2), \tag{12.1}$$

with the gauge-invariant derivative

$$\mathbf{D}_1 \equiv \boldsymbol{\nabla}_1 - \mathrm{i}\frac{q}{c}\mathbf{A}(1). \tag{12.2}$$

The fields $\mathbf{A}(1)$ and $\phi(1)$ are the vector and scalar potentials. The single-particle Hamiltonian $\hat{h}$ is the same on the forward and backward branch, and it is time-independent along the imaginary track of the contour. Therefore, the scalar and vector potentials with arguments on the contour are defined as

$$\mathbf{A}(\mathbf{x}, z = t_\pm) = \mathbf{A}(\mathbf{x}, t), \qquad \mathbf{A}(\mathbf{x}, z = t_0 - \mathrm{i}\tau) = \mathbf{A}(\mathbf{x}), \tag{12.3}$$

and

$$\phi(\mathbf{x}, z = t_\pm) = \phi(\mathbf{x}, t), \qquad \phi(\mathbf{x}, z = t_0 - \mathrm{i}\tau) = \phi(\mathbf{x}). \tag{12.4}$$

We observe that neither $\mathbf{A}$ nor ϕ have an actual dependence on spin. In the following we use either $\phi(\mathbf{x}, z)$ or $\phi(\mathbf{r}, z)$ to represent the scalar potential and either $\mathbf{A}(\mathbf{x}, z)$ or $\mathbf{A}(\mathbf{r}, z)$ to represent the vector potential. The choice of $\mathbf{r}$ or $\mathbf{x}$ is dictated by the principle of having the notation as light as possible.

The equations of motion for the Green's function, see (7.18) and (7.19), with single-particle Hamiltonian $h(1;2)$ given by (12.1) read

$$\Big[\mathrm{i}\frac{d}{dz_1} + \overbrace{\frac{\nabla_1^2}{2m} - \frac{\mathrm{i}q}{2mc}\big[(\boldsymbol{\nabla}_1\cdot\mathbf{A}(1)) + 2\mathbf{A}(1)\cdot\boldsymbol{\nabla}_1\big] - w(1)}^{\frac{1}{2m}D_1^2 - q\phi(1)}\Big]G(1;2) = \delta(1;2) + \int d3\,\Sigma(1;3)G(3;2), \tag{12.5}$$

and

$$\Big[-\mathrm{i}\frac{d}{dz_2} + \overbrace{\frac{\nabla_2^2}{2m} + \frac{\mathrm{i}q}{2mc}\big[(\boldsymbol{\nabla}_2\cdot\mathbf{A}(2)) + 2\mathbf{A}(2)\cdot\boldsymbol{\nabla}_2\big] - w(2)}^{\frac{1}{2m}(D_2^2)^* - q\phi(2)}\Big]G(1;2) = \delta(1;2) + \int d3\,G(1;3)\Sigma(3;2), \tag{12.6}$$

where we introduce the short-hand notation [see also (3.35)]

$$w(1) = q\phi(1) + \frac{q^2}{2mc^2}A^2(1).$$

Subtracting (12.6) from (12.5) and writing $\nabla_1^2 - \nabla_2^2 = (\boldsymbol{\nabla}_1 + \boldsymbol{\nabla}_2)\cdot(\boldsymbol{\nabla}_1 - \boldsymbol{\nabla}_2)$, we find

$$\begin{aligned}\Big[\mathrm{i}\frac{d}{dz_1} + \mathrm{i}\frac{d}{dz_2}\Big]G(1;2) &+ (\boldsymbol{\nabla}_1 + \boldsymbol{\nabla}_2)\cdot\frac{\boldsymbol{\nabla}_1 - \boldsymbol{\nabla}_2}{2m}G(1;2)\\ &- \frac{\mathrm{i}q}{2mc}\big[(\boldsymbol{\nabla}_1\cdot\mathbf{A}(1)) + (\boldsymbol{\nabla}_2\cdot\mathbf{A}(2)) + 2\mathbf{A}(1)\cdot\boldsymbol{\nabla}_1 + 2\mathbf{A}(2)\cdot\boldsymbol{\nabla}_2\big]G(1;2)\\ &- \big(w(1) - w(2)\big)G(1;2) = \int d3\,\big[\Sigma(1;3)G(3;2) - G(1;3)\Sigma(3;2)\big].\end{aligned} \tag{12.7}$$

This is an important equation as it constitutes the starting point to prove that the Green's function obtained from the solution of (12.5) and (12.6) [or equivalently from the solution of the Dyson equation (7.17)] satisfies all conservation laws provided that the self-energy is Φ-derivable:

$$\boxed{\Sigma(1;2) = \frac{\delta\Phi[G]}{\delta G(2;1^+)}} \tag{12.8}$$

In this way, any approximate Φ-derivable self-energy generates an approximate conserving Green's function. This class of approximations is called *conserving approximations*; the underlying theory was developed by Baym in 1962 [147]. As we see, particle, momentum, angular momentum, and energy conservation follow from the invariance of the Φ functional under gauge transformations, spatial translations, rotations, and time translations, respectively.

Conserving approximations do not only have the merit of preserving basic conservation laws. At zero temperature the grand potential equals the ground-state energy $E_{\mathrm{S}}^{\mathrm{M}}$ of a system with Hamiltonian $\hat{H}^{\mathrm{M}}$. A further advantage of conserving approximations is that the energy calculated from $E_{\mathrm{S}}^{\mathrm{M}} = \lim_{\beta\to\infty}\Omega$, with Ω *any* of the grand potential functionals discussed in Chapter 11, or calculated from the Galitskii–Migdal formula (6.115) is the same provided that Σ is Φ-derivable. Indeed, the derivation of the Galitskii–Migdal formula is based on the sole assumption that G satisfies the Dyson equation for some Σ. Therefore, if $\Sigma = \delta\Phi/\delta G$ and if G is calculated self-consistently from $G = G_0 + G_0\Sigma G$, then the two methods yield the same result. Other interesting properties of conserving approximations are, for example, the satisfaction of the Luttinger–Ward theorem, see Section 11.6, the discontinuity of the momentum distribution in a zero-temperature electron gas, see Section 11.6.1, the satisfaction of the virial theorem, see Appendix L, and the Ward identities, see Section 14.10.

We mention that it is also possible to develop partially self-consistent approaches which still guarantee the satisfaction of all conservation laws [166] or only some of them, like the continuity equation [167]. Such approaches are of interest when full self-consistency is hard to achieve, such as due to computational limitations. For a discussion, see Section 17.2.

Conserving approximations for the two-particle Green's function We remark that the Φ-derivability of the self-energy is a sufficient but not necessary condition for the Green's function to be conserving. In 1961, a year before the seminal paper by Baym, Kadanoff and Baym [168] showed that the Green's function that solves the first equations of the Martin–Schwinger hierarchy, see (5.2) and (5.3), is conserving, provided that the approximate two-particle Green's function fulfills:

(C1) the symmetry condition[1]

$$G_2(1,2;1^+,2^+) = G_2(2,1;2^+,1^+);$$

(C2) the KMS boundary conditions (this condition is actually needed only to prove the energy conservation law).

In Section 14.5 we prove that the Φ-derivability of the self-energy guarantees that the two-particle Green's function [obtained from the self-energy through (9.5) or (9.6)] is conserving – that is, conditions (C1) and (C2) are fulfilled. The opposite is not true: If G_2 fulfills conditions (C1) and (C2) then the self-energy obtained from (9.5) or (9.6) may not be Φ-derivable. In this book we do not present the conserving formulation based on the two-particle Green's function and refer the interested reader to the original paper by Kadanoff and Baym [168] (or to the first edition of this book).

[1]This condition is a special case of the more general symmetry property of the two-particle Green's function $G_2(1,2;3,4) = G_2(2,1;4,3)$, which follows directly from definition (5.1).

12.2 Continuity Equation

We can extract an equation for the density by setting $2 = 1^+ = (\mathbf{x}_1, z_1^+)$ in (12.7). Then the first term on the l.h.s. becomes the derivative of the density:

$$\mathrm{i}\frac{d}{dz_1}G(1;1^+) = \pm\frac{d}{dz_1}n(1).$$

The second and the third terms can be grouped to form a total divergence:

$$\text{2nd} + \text{3rd term} = \boldsymbol{\nabla}_1 \cdot \left[\left(\frac{\boldsymbol{\nabla}_1 - \boldsymbol{\nabla}_2}{2m}G(1;2)\right)_{2=1^+} - \frac{\mathrm{i}q}{mc}\mathbf{A}(1)G(1;1^+)\right]. \tag{12.9}$$

The time-dependent ensemble average of the paramagnetic and diamagnetic current density operators are expressed in terms of the Green's function according to [see (5.4) and (5.5)]

$$\mathbf{j}(1) = \pm\left(\frac{\boldsymbol{\nabla}_1 - \boldsymbol{\nabla}_2}{2m}G(1;2)\right)_{2=1^+}, \tag{12.10}$$

and

$$\mathbf{j}_d(1) = -\frac{q}{mc}\mathbf{A}(1)n(1) = \mp\frac{\mathrm{i}q}{mc}\mathbf{A}(1)G(1;1^+).$$

Hence, the r.h.s. of (12.9) is the divergence of the total current density $\mathbf{J}(1) = \mathbf{j}(1) + \mathbf{j}_d(1)$ already defined in (3.32). We thus obtain

$$\frac{d}{dz_1}n(1) + \boldsymbol{\nabla}_1 \cdot \mathbf{J}(1) = \pm\int d3\,\left[\Sigma(1;3)G(3;1^+) - G(1;3)\Sigma(3;1^+)\right]. \tag{12.11}$$

The continuity equation follows provided that the r.h.s. vanishes. To prove that this is the case for any Φ-derivable self-energy, consider the variation of G induced by the infinitesimal gauge transformation

$$G(1;2) \to e^{\mathrm{i}\Lambda(1)}G(1;2)e^{-\mathrm{i}\Lambda(2)} \quad \Rightarrow \quad \delta G(1;2) = \mathrm{i}\,[\Lambda(1) - \Lambda(2)]G(1;2), \tag{12.12}$$

with $\Lambda(\mathbf{x}, t_{0-}) = \Lambda(\mathbf{x}, t_0 - \mathrm{i}\beta)$ to preserve the KMS relations.[2] This transformation leaves Φ unchanged since with every vertex j of a Φ-diagram is associated an ingoing Green's function, $G(\ldots;j^+)$, and an outgoing Green's function, $G(j;\ldots)$. Thus,[3]

$$\begin{aligned} 0 = \delta\Phi &= \int d1d2\frac{\delta\Phi}{\delta G(2;1^+)}\delta G(2;1^+) = \mathrm{i}\int d1d2\;\Sigma(1;2)[\Lambda(2) - \Lambda(1)]G(2;1^+) \\ &= -\mathrm{i}\int d1d2\;\left[\Sigma(1;2)G(2;1^+) - G(1;2)\Sigma(2;1^+)\right]\Lambda(1), \end{aligned} \tag{12.13}$$

[2]It is worth noticing that the Kadanoff–Baym equations are invariant under a gauge transformation for any Φ-derivable self-energy. Indeed, if $G(1;2) \to e^{\mathrm{i}\Lambda(1)}G(1;2)e^{-\mathrm{i}\Lambda(2)}$ then the self-energy $\Sigma(1;2) \to e^{\mathrm{i}\Lambda(1)}\Sigma(1;2)e^{-\mathrm{i}\Lambda(2)}$ as follows directly from the structure of the self-energy diagrams.

[3]The arguments of G and Σ in the second term of the second line of (12.13) can be understood as follows. Renaming the integration variables $1 \leftrightarrow 2$ in the first line of (12.13), we obtain the term $G(1;2^+)\Sigma(2;1)$. The self-energy has, in general, a singular part $\delta(z_2,z_1)\Sigma^\delta(\mathbf{x}_2,\mathbf{x}_1,z_1)$ and a regular part (the correlation self-energy) $\Sigma_c(2;1)$. The product $G(1;2^+)\delta(z_2,z_1)\Sigma^\delta(\mathbf{x}_2,\mathbf{x}_1,z_1) = G(1;2)\delta(z_2,z_1^+)\Sigma^\delta(\mathbf{x}_2,\mathbf{x}_1,z_1)$ while the product of the Green's function with the correlation self-energy gives the same result with or without the superscript "+": $\int d2\,G(1;2^+)\Sigma_c(2;1) = \int d2\,G(1;2)\Sigma_c(2;1) = \int d2\,G(1;2)\Sigma_c(2;1^+)$ since the point $1 = 2$ has zero measure in the domain of integration. Therefore, $\int d2\,G(1;2^+)\Sigma(2;1) = \int d2\,G(1;2)\Sigma(2;1^+)$.

and due to the arbitrariness of Λ

$$\int d2\ \left[\Sigma(1;2)G(2;1^+) - G(1;2)\Sigma(2;1^+)\right] = 0, \tag{12.14}$$

as we wanted to prove. The continuity equation tells us that accumulation of charge in a certain region of space is related to current flow. This is certainly an important relation that one wants to have satisfied in nonequilibrium systems.

The reader may react to seeing that the continuity equation (12.11) is an equation on the contour instead of an equation on the real-time axis. Does (12.11) contain the same information as the continuity equation on the real-time axis? Strictly speaking, (12.11) contains some more information even though the extra information is rather obvious. It is instructive to analyze (12.11) in detail since the same analysis is valid for other physical quantities. To extract physical information from (12.11), we let the contour variable z_1 lie on a different part of the contour. For $z_1 = t_{1-}$ on the forward branch we have

$$\begin{aligned}\frac{d}{dz_1}n(\mathbf{x}_1, z_1) &= \lim_{\epsilon\to 0}\frac{n(\mathbf{x}_1, (t_1+\epsilon)_-) - n(\mathbf{x}_1, t_{1-})}{\epsilon} = \lim_{\epsilon\to 0}\frac{n(\mathbf{x}_1, t_1+\epsilon) - n(\mathbf{x}_1, t_1)}{\epsilon}\\ &= \frac{d}{dt_1}n(\mathbf{x}_1, t_1) = -\boldsymbol{\nabla}_1\cdot\mathbf{J}(\mathbf{x}_1, t_1),\end{aligned}$$

and similarly for $z_1 = t_{1+}$ on the backward branch we have

$$\begin{aligned}\frac{d}{dz_1}n(\mathbf{x}_1, z_1) &= \lim_{\epsilon\to 0}\frac{n(\mathbf{x}_1, (t_1-\epsilon)_+) - n(\mathbf{x}_1, t_{1+})}{-\epsilon} = \lim_{\epsilon\to 0}\frac{n(\mathbf{x}_1, t_1-\epsilon) - n(\mathbf{x}_1, t_1)}{-\epsilon}\\ &= \frac{d}{dt_1}n(\mathbf{x}_1, t_1) = -\boldsymbol{\nabla}_1\cdot\mathbf{J}(\mathbf{x}_1, t_1).\end{aligned}$$

In these two equations we use the definition of the derivative on the contour, see Section 4.4, as well as that for all operators $\hat{O}(t)$ built from the field operators, such as the density and the current density, the time-dependent ensemble average $O(t_\pm) = O(t)$. The continuity equation on the contour contains the same information as the continuity equation on the real-time axis when z_1 lies on the horizontal branches. The extra information contained in (12.11) comes from z_1 on the vertical track. In this case, both the density and the current density are independent of z_1 and the continuity equation reduces to $\boldsymbol{\nabla}_1\cdot\mathbf{J}(1) = 0$. This result tells us that for a system in equilibrium, the current density is divergenceless, and hence there can be no density fluctuations in any portion of the system.

The continuity equation expresses the *local* conservation of particles since it is a differential equation for the density $n(\mathbf{x}_1, z_1)$ in point $\mathbf{x}_1$. The time derivative of the local momentum, angular momentum, or energy is not a total divergence. For instance, the time derivative of the local momentum $\mathbf{P}(1) = m\mathbf{J}(1)$ is a total divergence plus the local Lorentz force, see Section 3.5. Below we prove that momentum, angular momentum, and energy are conserved in a *global* sense – that is, we prove the conservation laws for the integral of the corresponding local quantities.

12.3 Momentum Conservation Law

We saw in (3.33) that the total momentum of a system is the integral over all space and spin of the current density $\mathbf{J}$ multiplied by the mass of the particles,

$$\mathbf{P}(t) \equiv m \int d\mathbf{x}\, \mathbf{J}(\mathbf{x}, t).$$

The total momentum of a system is conserved if the Hamiltonian is invariant under translations. In the absence of external electromagnetic fields ($\mathbf{A} = \phi = 0$) this invariance requires that the interparticle interaction $v(\mathbf{x}_1, \mathbf{x}_2)$ depends only on the difference $\mathbf{r}_1 - \mathbf{r}_2$. Below we show that for such interparticle interactions the time derivative of the total momentum is the total (Lorentz) force, see (3.39).

To calculate the time derivative of $\mathbf{P}$ we calculate separately the time derivative of $\mathbf{j}$ (paramagnetic current density) and $\mathbf{j}_d$ (diamagnetic current density). We introduce the shorthand notation $f_1 = f(1)$ for a scalar function and $f_{1,p} = f_p(1)$ for the pth component of a vector function $\mathbf{f}(1)$. Then, using the Einstein convention of summing of repeated indices, we can rewrite (12.7) as

$$\begin{aligned}\left[\mathrm{i}\frac{d}{dz_1} + \mathrm{i}\frac{d}{dz_2}\right] G(1;2) &+ (\partial_{1,p} + \partial_{2,p})\frac{\partial_{1,p} - \partial_{2,p}}{2m} G(1;2) \\ &- \frac{\mathrm{i}q}{2mc}\big[(\partial_{1,p}A_{1,p}) + (\partial_{2,p}A_{2,p}) + 2A_{1,p}\partial_{1,p} + 2A_{2,p}\partial_{2,p}\big] G(1;2) \\ &- (w_1 - w_2) G(1;2) = \int d3\, \big[\Sigma(1;3)G(3;2) - G(1;3)\Sigma(3;2)\big], \end{aligned} \tag{12.15}$$

where $\partial_{1,p}$ is the pth component of $\boldsymbol{\nabla}_1$. To obtain an equation for the paramagnetic current density, we act on (12.15) with $(\partial_{1,k} - \partial_{2,k})/2m$, and evaluate the result in $2 = 1^+$. The first term in (12.15) yields

$$\frac{\partial_{1,k} - \partial_{2,k}}{2m}\left[\left(\mathrm{i}\frac{d}{dz_1} + \mathrm{i}\frac{d}{dz_2}\right) G(1;2)\right]_{2=1^+} = \mathrm{i}\frac{d}{dz_1}\left[\frac{\partial_{1,k} - \partial_{2,k}}{2m} G(1;2)\right]_{2=1^+} = \pm\mathrm{i}\frac{d}{dz_1} j_{1,k},$$

where we use (12.10). The second term of (12.15) gives a total divergence since

$$\begin{aligned}\frac{\partial_{1,k} - \partial_{2,k}}{2m}(\partial_{1,p} + \partial_{2,p})\frac{\partial_{1,p} - \partial_{2,p}}{2m} G(1;2)\Big|_{2=1^+} \\ = \partial_{1,p}\left[\frac{\partial_{1,k} - \partial_{2,k}}{2m}\frac{\partial_{1,p} - \partial_{2,p}}{2m} G(1;2)\right]_{2=1^+}.\end{aligned}$$

It is easy to show that the quantity in the square bracket is proportional to the time-dependent ensemble average of the momentum stress-tensor operator defined in (3.37). Next we consider the second row of (12.15). For any continuous function f, the quantity $(\partial_{1,k} - \partial_{2,k})(f_1 + f_2)$ vanishes in $2 = 1^+$. Therefore the first two terms give

$$-\frac{\mathrm{i}q}{2mc}\left[\big[(\partial_{1,p}A_{1,p}) + (\partial_{2,p}A_{2,p})\big]\frac{\partial_{1,k} - \partial_{2,k}}{2m} G(1;2)\right]_{2=1^+} = \mp\frac{\mathrm{i}q}{mc}(\partial_{1,p}A_{1,p}) j_{1,k}. \tag{12.16}$$

For the last two terms we use the identity

$$(\partial_{1,k} - \partial_{2,k})(A_{1,p}\partial_{1,p} + A_{2,p}\partial_{2,p}) = (\partial_{1,k}A_{1,p})\partial_{1,p} - (\partial_{2,k}A_{2,p})\partial_{2,p} \\ +(A_{1,p}\partial_{1,p} + A_{2,p}\partial_{2,p})(\partial_{1,k} - \partial_{2,k}),$$

and find

$$-\frac{\mathrm{i}q}{mc}\left[\frac{\partial_{1,k} - \partial_{2,k}}{2m}(A_{1,p}\partial_{1,p} + A_{2,p}\partial_{2,p})G(1;2)\right]_{2=1^+} \\ = \mp\frac{\mathrm{i}q}{mc}\big[j_{1,p}(\partial_{1,k}A_{1,p}) + A_{1,p}(\partial_{1,p}j_{1,k})\big].$$

Notice that, adding (12.16) to the second term on the r.h.s. of this equation, we get a total divergence. Finally, the last two terms on the l.h.s. of (12.15) yield

$$-\left[\frac{\partial_{1,k} - \partial_{2,k}}{2m}(w_1 - w_2)G(1;2)\right]_{2=1^+} = \pm\frac{\mathrm{i}}{m}n_1\partial_{1,k}w_1 \\ = \pm\frac{\mathrm{i}q}{m}n_1\left(\partial_{1,k}\phi_1 + \frac{q}{mc^2}A_{1,p}\partial_{1,k}A_{1,p}\right).$$

Putting together all these results, we obtain the equation for the paramagnetic current density:

$$\frac{d}{dz_1}j_{1,k} = \frac{q}{mc}J_{1,p}\partial_{1,k}A_{1,p} - \frac{q}{m}n_1\partial_{1,k}\phi_1 + \text{total divergence} \\ \mp\mathrm{i}\int d3\left[\frac{\partial_{1,k} - \partial_{2,k}}{2m}\left[\Sigma(1;3)G(3;2) - G(1;3)\Sigma(3;2)\right]\right]_{2=1^+}.$$

The derivative of the diamagnetic current density follows directly from the continuity equation and reads:

$$\frac{d}{dz_1}(j_d)_{1,k} = -\frac{q}{mc}n_1\frac{d}{dz_1}A_{1,k} + \frac{q}{mc}A_{1,k}\partial_{1,p}J_{1,p} \\ = -\frac{q}{mc}n_1\frac{d}{dz_1}A_{1,k} - \frac{q}{mc}J_{1,p}\partial_{1,p}A_{1,k} + \frac{q}{mc}\underbrace{\partial_{1,p}(A_{1,k}J_{1,p})}_{\text{total divergence}}.$$

Summing the last two equations, we find the equation for the current density

$$\frac{d}{dz_1}J_{1,k} = \frac{q}{m}n_1\left(-\partial_{1,k}\phi_1 - \frac{1}{c}\frac{d}{dz_1}A_{1,k}\right) + \frac{q}{mc}J_{1,p}\left(\partial_{1,k}A_{1,p} - \partial_{1,p}A_{1,k}\right) \\ + \text{total divergence} \mp\mathrm{i}\frac{1}{m}\int d3\left[G(1;3)\partial_{1,k}\Sigma(3;1^+) - \Sigma(1;3)\partial_{1,k}G(3;1^+)\right]. \tag{12.17}$$

In the first row we recognize the kth component of the electric field and the kth component of the vector product between $\mathbf{J}$ and the magnetic field, see (3.38) and the discussion below.

Upon integration over $\mathbf{x}_1$, the total divergence does not contribute and we end up with

$$\frac{d}{dz_1}\mathbf{P}(z_1) = \int d\mathbf{x}_1 \Big(qn(1)\mathbf{E}(1) + \frac{q}{c}\mathbf{J}(1) \times \mathbf{B}(1)\Big)$$
$$\mp \mathrm{i}\int d\mathbf{x}_1 \int d3\, \big[G(1;3)\boldsymbol{\nabla}_1\Sigma(3;1^+) - \Sigma(1;3)\boldsymbol{\nabla}_1 G(3;1^+)\big]. \quad (12.18)$$

We thus see that the last term on the r.h.s. must vanish for the approximation to be conserving. This is a consequence of the invariance of the Φ functional under a *time-dependent* rigid shift of the reference frame. Consider the change

$$G(1;2) \to G((\mathbf{r}_1 + \mathbf{R}(z_1))\sigma_1, z_1; (\mathbf{r}_2 + \mathbf{R}(z_2))\sigma_2, z_2),$$

with $\mathbf{R}(t_{0-}) = \mathbf{R}(t_0 - \mathrm{i}\beta)$. A generic Φ-diagram contains an integral over all spatial coordinates and hence the shift $\mathbf{r}_i \to \mathbf{r}_i - \mathbf{R}(z_i)$ brings the transformed G back to the original G without changing $v(i;j)$ since $v(i;j) \to \delta(z_i, z_j)v(\mathbf{r}_i - \mathbf{R}(z_i) - \mathbf{r}_j + \mathbf{R}(z_j)) = v(i;j)$ due to the locality in time. To first order in $\mathbf{R}$ the variation in G is

$$\delta G(2;1^+) = \big[\mathbf{R}(z_2)\cdot\boldsymbol{\nabla}_2 + \mathbf{R}(z_1)\cdot\boldsymbol{\nabla}_1\big]G(2;1^+),$$

and therefore

$$0 = \delta\Phi = \int d1d2 \frac{\delta\Phi}{\delta G(2;1^+)}\delta G(2;1^+)$$
$$= \int d1d2\; \Sigma(1;2)\left[\mathbf{R}(z_2)\cdot\boldsymbol{\nabla}_2 + \mathbf{R}(z_1)\cdot\boldsymbol{\nabla}_1\right]G(2;1^+)$$
$$= \int d1d2\; \left[\Sigma(1;2)\boldsymbol{\nabla}_1 G(2;1^+) - G(1;2)\boldsymbol{\nabla}_1\Sigma(2;1^+)\right]\cdot\mathbf{R}(z_1), \quad (12.19)$$

where in the last equality we perform an integration by parts and rename $1 \leftrightarrow 2$. Since (12.19) is true for all vector functions $\mathbf{R}(z)$, the last term on the r.h.s. of (12.18) vanishes.

12.4 Angular Momentum Conservation Law

The angular momentum of a system is conserved if the Hamiltonian is invariant under rotations. In the absence of external electromagnetic fields ($\mathbf{A} = \phi = 0$) this invariance requires that the interparticle interaction $v(\mathbf{x}_1, \mathbf{x}_2)$ depends only on the distance $|\mathbf{r}_1 - \mathbf{r}_2|$. Below we show that for such interparticle interactions the time derivative of the total angular momentum

$$\mathbf{L}(t) \equiv m\int d\mathbf{x}\; \mathbf{r} \times \mathbf{J}(\mathbf{x}, t)$$

is the total applied torque provided that the approximation is conserving.

We start by rewriting (12.17) in terms of the local Lorentz force $\mathbf{F}(1) \equiv qn(1)\mathbf{E}(1) + \frac{q}{c}\mathbf{J}(1) \times \mathbf{B}(1)$ given by the first two terms on the r.h.s.:

$$m\frac{d}{dz_1}J_{1,k} = F_{1,k}$$
$$\pm \frac{\mathrm{i}}{m}\partial_{1,p}\left[\frac{\partial_{1,k}-\partial_{2,k}}{2}\frac{\partial_{1,p}-\partial_{2,p}}{2}G(1;2)\right]_{2=1^+} + \frac{q}{c}\,\partial_{1,p}\left[A_{1,p}j_{1,k} + A_{1,k}J_{1,p}\right]$$
$$\mp \mathrm{i}\int d3\left[G(1;3)\partial_{1,k}\Sigma(3;1^+) - \Sigma(1;3)\partial_{1,k}G(3;1^+)\right], \tag{12.20}$$

where we have explicitly written down the total divergence (second row). The lth component of the angular momentum is obtained by integrating over space and spin the lth component of $[\mathbf{r}_1 \times \mathbf{J}(1)]_l = \varepsilon_{lik}r_{1,i}J_{1,k}$, where ε_{lik} is the Levi-Civita tensor and summation over repeated indices is understood. Let us consider the total divergence (second row). The first term is the derivative of a symmetric function of k and p. The second term is also the derivative of a symmetric function of k and p since

$$A_{1,p}j_{1,k} + A_{1,k}J_{1,p} = A_{1,p}j_{1,k} + A_{1,k}j_{1,p} - \frac{q}{mc}n_1 A_{1,k}A_{1,p}.$$

We now show that the quantity $\varepsilon_{lik}r_{1,i}\partial_{1,p}S_{1,kp}$ is a total divergence for every symmetric function $S_{1,kp} = S_{1,pk}$, and therefore it vanishes upon integration over space. We have

$$\varepsilon_{lik}r_{1,i}\partial_{1,p}S_{1,kp} = \partial_{1,p}\left[\varepsilon_{lik}r_{1,i}S_{1,kp}\right] - \varepsilon_{lik}S_{1,kp}\underbrace{\partial_{1,p}r_{1,i}}_{\delta_{pi}}$$
$$= \partial_{1,p}\left[\varepsilon_{lik}r_{1,i}S_{1,kp}\right] - \varepsilon_{lik}S_{1,ki},$$

and the last term is zero due to the symmetry of S. Therefore, multiplying (12.20) by $\varepsilon_{lik}r_{1,i}$, integrating over $\mathbf{x}_1$ and reintroducing the vector notation, we find

$$\frac{d}{dz_1}\mathbf{L}(z_1) = \int d\mathbf{x}_1\,(\mathbf{r}_1 \times \mathbf{F}(1))$$
$$\mp \mathrm{i}\int d\mathbf{x}_1 \int d3\,\mathbf{r}_1 \times \left[G(1;3)\boldsymbol{\nabla}_1\Sigma(3;1^+) - \Sigma(1;3)\boldsymbol{\nabla}_1 G(3;1^+)\right]. \tag{12.21}$$

The first term on the r.h.s. is the total torque applied to the system. Hence the angular momentum is conserved provided that the last term vanishes. For rotationally invariant interparticle interactions we can exploit the invariance of Φ under the transformation

$$G(1;2) \to G((R[\boldsymbol{\alpha}(z_1)]\mathbf{r}_1)\sigma_1, z_1; (R[\boldsymbol{\alpha}(z_2)]\mathbf{r}_2)\sigma_2, z_2), \tag{12.22}$$

where $R[\boldsymbol{\alpha}]$ is the 3×3 matrix that rotates a vector by an angle $\boldsymbol{\alpha}$, and $\boldsymbol{\alpha}(t_{0-}) = \boldsymbol{\alpha}(t_0 - \mathrm{i}\beta)$. To first order in $\boldsymbol{\alpha}$ the variation of the spatial coordinates is $\delta\mathbf{r}_i = \boldsymbol{\alpha} \times \mathbf{r}_i$ and hence the variation in G is

$$\delta G(2;1^+) = \left[(\boldsymbol{\alpha}(z_2) \times \mathbf{r}_2)\cdot\boldsymbol{\nabla}_2 + (\boldsymbol{\alpha}(z_1) \times \mathbf{r}_1)\cdot\boldsymbol{\nabla}_1\right]G(2;1^+). \tag{12.23}$$

The invariance of the Φ functional implies

$$\begin{aligned}0=\delta\Phi &= \int d1d2\frac{\delta\Phi}{\delta G(2;1^+)}\delta G(2;1^+)\\ &= \int d1d2\ \Sigma(1;2)\left[(\boldsymbol{\alpha}(z_2)\times\mathbf{r}_2)\cdot\boldsymbol{\nabla}_2+(\boldsymbol{\alpha}(z_1)\times\mathbf{r}_1)\cdot\boldsymbol{\nabla}_1\right]G(2;1^+)\\ &= \int d1d2\ \mathbf{r}_1\times\left[\Sigma(1;2)\boldsymbol{\nabla}_1 G(2;1^+)-G(1;2)\boldsymbol{\nabla}_1\Sigma(2;1^+)\right]\cdot\boldsymbol{\alpha}(z_1),\end{aligned} \tag{12.24}$$

where in the last equality we perform an integration by parts, use the identity $(\mathbf{a}\times\mathbf{b})\cdot\mathbf{c}=\mathbf{a}\cdot(\mathbf{b}\times\mathbf{c})=(\mathbf{b}\times\mathbf{c})\cdot\mathbf{a}$ for any triple of vectors $\mathbf{a}$, $\mathbf{b}$, and $\mathbf{c}$, as well as $\boldsymbol{\nabla}\cdot(\boldsymbol{\alpha}\times\mathbf{r})=0$, and finally rename $1\leftrightarrow 2$. Since (12.24) is true for all vector functions $\boldsymbol{\alpha}(z)$, the last term on the r.h.s. of (12.21) vanishes.

12.5 Energy Conservation Law

According to the discussion in Section 6.5, the energy of the system at a generic time t_1, $E_{\rm S}(t_1)$, is the time-dependent ensemble average of the operator

$$\hat{H}_{\rm S}(t_1)\equiv\hat{H}(t_1)-q\int d\mathbf{x}_1\ \hat{n}(\mathbf{x}_1)\phi_\Delta(1),$$

where $\phi_\Delta(1)=\phi(\mathbf{r}_1,t_1)-\phi(\mathbf{r}_1)$ is the change in the external potential at time t_1. In order to show that a Φ-derivable self-energy leads to a conserving approximation for the total energy, we first have to derive the exact equation for $\frac{d}{dt_1}E_{\rm S}(t_1)$.

We write the density matrix operator as $\hat{\rho}=\sum_k\rho_k|\Psi_k\rangle\langle\Psi_k|$. The time-dependent ensemble average of $\hat{H}_{\rm S}(t_1)$ is then

$$\begin{aligned}E_{\rm S}(t_1)&=\sum_k\rho_k\langle\Psi_k(t_1)|\hat{H}(t_1)-q\int d\mathbf{x}_1\ \hat{n}(\mathbf{x}_1)\phi_\Delta(1)|\Psi_k(t_1)\rangle\\ &=\sum_k\rho_k\langle\Psi_k(t_1)|\hat{H}(t_1)|\Psi_k(t_1)\rangle-q\int d\mathbf{x}_1\ n(1)\phi_\Delta(1).\end{aligned}$$

To calculate the time derivative of the first term we must differentiate the bras $\langle\Psi_k(t_1)|$, the kets $|\Psi_k(t_1)\rangle$, and the Hamiltonian operator. Since all kets $|\Psi_k(t_1)\rangle$ evolve according to the same Schrödinger equation, $\mathrm{i}\frac{d}{dt}|\Psi_k(t)\rangle=\hat{H}(t)|\Psi_k(t)\rangle$, the term coming from $\frac{d}{dt_1}\langle\Psi_k(t_1)|$ cancels the term coming from $\frac{d}{dt_1}|\Psi_k(t_1)\rangle$. The time derivative of $\hat{H}(t_1)$ is the same as the time derivative of the noninteracting Hamiltonian $\hat{H}_0(t_1)$ (the physical interaction is independent of time). Writing $\hat{H}_0(t_1)$ as in (3.30), we find

$$\begin{aligned}\frac{d}{dt_1}E_{\rm S}(t_1)&=\int d\mathbf{x}_1\left[-\frac{q}{c}\mathbf{j}(1)\cdot\frac{d\mathbf{A}(1)}{dt_1}+qn(1)\frac{d}{dt_1}\left(\phi_\Delta(1)+\frac{q}{2mc^2}A^2(1)\right)\right]\\ &\quad-q\frac{d}{dt_1}\int d\mathbf{x}_1\ n(1)\phi_\Delta(1).\\ &=\int d\mathbf{x}_1\left[-\frac{q}{c}\mathbf{J}(1)\cdot\frac{d\mathbf{A}(1)}{dt_1}-q\phi_\Delta(1)\frac{d}{dt_1}n(1)\right].\end{aligned}$$

The last term can be further manipulated using the continuity equation and integrating by parts; we thus arrive at the following important result:

$$\frac{d}{dt_1}E_{\rm S}(t_1) = q\int d\mathbf{x}_1\, \mathbf{J}(1)\cdot\mathbf{E}_\Delta(1), \tag{12.25}$$

with $\mathbf{E}_\Delta = -\boldsymbol{\nabla}\phi_\Delta - \frac{1}{c}\frac{d}{dt}\mathbf{A}$ the change of the electric field. The r.h.s. of (12.25) is the scalar product between the current density and the Lorentz force - that is, it is the power fed into the system (the magnetic field generates a force orthogonal to $\mathbf{J}$ and hence does not contribute to the power).

The proof of energy conservation is a nice example of the importance of using the most convenient notation when carrying out lengthy calculations. Let us write the total energy in terms of the Green's function and the self-energy. The relations (9.5) and (9.6) between the self-energy and the two-particle Green's function lead to two equivalent formulas for the interaction energy $E_{\rm int}(z_1)$, see (10.29). These two formulas certainly give the same result when evaluated with the *exact* self-energy (and hence with the exact G). For an approximate self-energy, however, the equivalence between the two formulas is not guaranteed. The set of approximate self-energies for which the two formulas are equivalent contains the set of self-energies that preserve the continuity equation, see (12.14). We can then write the interaction energy as the arithmetic average

$$E_{\rm int}(z_1) = \pm\frac{\rm i}{4}\int d\mathbf{x}_1 d3\, \left[\Sigma(1;3)G(3;1^+) + G(1;3)\Sigma(3;1^+)\right].$$

According to (6.113) the total energy of the system is

$$E_{\rm S}(z_1) = E_{\rm one}(z_1) + E_{\rm int}(z_1),$$

where

$$E_{\rm one}(z_1) = \pm{\rm i}\int d\mathbf{x}_1 \langle\mathbf{x}_1|\hat{h}_{\rm S}(z_1)\hat{\mathcal{G}}(z_1,z_1^+)|\mathbf{x}_1\rangle \tag{12.26}$$

and $\hat{h}_{\rm S}(z_1)\equiv\hat{h}(z_1)-q\phi_\Delta(\hat{\boldsymbol{r}},z_1)$. For systems in equilibrium, both $E_{\rm one}$ and $E_{\rm int}$ have been thoroughly discussed in Section 10.3. Since (12.26) involves a trace in the one-particle Hilbert space, below we shall often use the cyclic property of the trace to rearrange expressions in the most convenient way.

The derivative of $E_{\rm one}(z_1)$ is the sum of two contributions: one containing the derivative of $\hat{h}_{\rm S}(z_1)$ and another one containing the derivative of $\hat{\mathcal{G}}(z_1,z_1^+)$. To calculate the derivative of the latter we use the equations of motion (9.3) and (9.4). Subtracting the second equation from the first and setting $z_2 = z_1^+$, we find

$$\begin{aligned}\frac{d}{dz_1}\langle\mathbf{x}_2|\hat{\mathcal{G}}(z_1,z_1^+)|\mathbf{x}_1\rangle &= \langle\mathbf{x}_2|\left(\frac{d}{dz_1}+\frac{d}{dz_2}\right)\hat{\mathcal{G}}(z_1,z_2)|\mathbf{x}_1\rangle\Big|_{z_2=z_1^+}\\ &= -{\rm i}\,\langle\mathbf{x}_2|\left[\hat{h}(z_1),\hat{\mathcal{G}}(z_1,z_1^+)\right]|\mathbf{x}_1\rangle\\ &\quad -{\rm i}\int_\gamma dz_3\langle\mathbf{x}_2|\hat{\Sigma}(z_1,z_3)\hat{\mathcal{G}}(z_3,z_1^+) - \hat{\mathcal{G}}(z_1,z_3)\hat{\Sigma}(z_3,z_1^+)|\mathbf{x}_1\rangle.\end{aligned}$$

The derivative of $E_{\rm one}(z_1)$ can then be rewritten as

$$\frac{d}{dz_1}E_{\rm one}(z_1) = P(z_1) + W_{\rm one}(z_1),$$

with

$$P(z_1) = \pm \mathrm{i}\int d\mathbf{x}_1 \langle \mathbf{x}_1| \left(\frac{d\hat{h}_{\rm S}(z_1)}{dz_1} - \mathrm{i}\left[\hat{h}_{\rm S}(z_1), \hat{h}(z_1)\right]_- \right) \hat{\mathcal{G}}(z_1, z_1^+)|\mathbf{x}_1\rangle, \tag{12.27}$$

and

$$\begin{aligned} W_{\rm one}(z_1) &= \pm \int d\mathbf{x}_1 d\mathbf{x}_2 d3\, \langle \mathbf{x}_1|\hat{h}_{\rm S}(z_1)|\mathbf{x}_2\rangle \left[\Sigma(2;3)G(3;1^+) - G(2;3)\Sigma(3;1^+)\right]\Big|_{z_2=z_1} \\ &= \pm \int d\mathbf{x}_1 d3 \left[\left(-\frac{D_2^2}{2m} + q\phi(\mathbf{x}_2)\right)\left[\Sigma(2;3)G(3;1^+) - G(2;3)\Sigma(3;1^+)\right]\right]_{2=1}. \end{aligned}$$

In (12.27) the cyclic property of the trace has been used, $\mathrm{Tr}\left[\hat{A}\,[\hat{B},\hat{C}]_-\right] = \mathrm{Tr}\left[[\hat{A},\hat{B}]_-\hat{C}\right]$. We now show that $P(z_1)$ is exactly the power fed into the system. Let us manipulate the operator to the left of $\hat{\mathcal{G}}(z_1, z_1^+)$. We have

$$\begin{aligned} \frac{d\hat{h}_{\rm S}(z_1)}{dz_1} &= \frac{d}{dz_1}\left[\frac{1}{2m}\left(\hat{p}^2 - \frac{q}{c}\hat{\boldsymbol{p}}\cdot\mathbf{A}(\hat{\boldsymbol{r}}, z_1) - \frac{q}{c}\mathbf{A}(\hat{\boldsymbol{r}}, z_1)\cdot\hat{\boldsymbol{p}} + \frac{q^2}{c^2}A^2(\hat{\boldsymbol{r}}, z_1)\right) + q\phi(\hat{\boldsymbol{r}})\right] \\ &= -\frac{q}{2mc}\hat{\boldsymbol{p}}\cdot\frac{d\mathbf{A}(\hat{\boldsymbol{r}}, z_1)}{dz_1} - \frac{q}{2mc}\frac{d\mathbf{A}(\hat{\boldsymbol{r}}, z_1)}{dz_1}\cdot\hat{\boldsymbol{p}} + \frac{q^2}{mc^2}\mathbf{A}(\hat{\boldsymbol{r}}, z_1)\cdot\frac{d\mathbf{A}(\hat{\boldsymbol{r}}, z_1)}{dz_1}. \end{aligned}$$

Taking into account that the commutator between the momentum operator $\hat{\boldsymbol{p}}$ and an arbitrary function $f(\hat{\boldsymbol{r}})$ of the position operator is $[\hat{\boldsymbol{p}}, f(\hat{\boldsymbol{r}})] = -\mathrm{i}\boldsymbol{\nabla} f(\hat{\boldsymbol{r}})$, we also have

$$\begin{aligned} -\mathrm{i}\left[\hat{h}_{\rm S}(z_1), \hat{h}(z_1)\right]_- &= -\mathrm{i}q\left[\hat{h}_{\rm S}(z_1), \phi_\Delta(\hat{\boldsymbol{r}}, z_1)\right]_- \\ &= -\frac{q}{2m}\hat{\boldsymbol{p}}\cdot\boldsymbol{\nabla}\phi_\Delta(\hat{\boldsymbol{r}}, z_1) - \frac{q}{2m}\boldsymbol{\nabla}\phi_\Delta(\hat{\boldsymbol{r}}, z_1)\cdot\hat{\boldsymbol{p}} + \frac{q^2}{mc}\mathbf{A}(\hat{\boldsymbol{r}}, z_1)\cdot\boldsymbol{\nabla}\phi_\Delta(\hat{\boldsymbol{r}}, z_1). \end{aligned}$$

Putting these results together, we find

$$\begin{aligned} &\frac{d\hat{h}_{\rm S}(z_1)}{dz_1} - \mathrm{i}\left[\hat{h}_{\rm S}(z_1), \hat{h}(z_1)\right]_- \\ &\qquad = q\left(\frac{\hat{\boldsymbol{p}}\cdot\mathbf{E}_\Delta(\hat{\boldsymbol{r}}, z_1) + \mathbf{E}_\Delta(\hat{\boldsymbol{r}}, z_1)\cdot\hat{\boldsymbol{p}}}{2m} - \frac{q}{mc}\mathbf{A}(\hat{\boldsymbol{r}}, z_1)\cdot\mathbf{E}_\Delta(\hat{\boldsymbol{r}}, z_1)\right), \end{aligned} \tag{12.28}$$

where $\mathbf{E}_\Delta(\hat{\boldsymbol{r}}, z_1)$ is the operator for the change of the electric field. Inserting (12.28) into (12.27), we get

$$\begin{aligned} P(z_1) &= \pm \mathrm{i}q\int d\mathbf{x}_1 \langle \mathbf{x}_1|\frac{\left[\hat{\boldsymbol{p}}, \hat{\mathcal{G}}(z_1, z_1^+)\right]_+}{2m} - \frac{q}{mc}\mathbf{A}(\hat{\boldsymbol{r}}, z_1)\hat{\mathcal{G}}(z_1, z_1^+)|\mathbf{x}_1\rangle\cdot\mathbf{E}_\Delta(1) \\ &= q\int d\mathbf{x}_1\, \mathbf{J}(1)\cdot\mathbf{E}_\Delta(1), \end{aligned}$$

where we first use the cyclic property of the trace and then, in the last equality, the identities (1.12) along with (12.10) to express the anticommutator in terms of the paramagnetic current density:[4]

$$\pm\mathrm{i}\,\langle\mathbf{x}_1|\frac{\hat{\mathbf{p}}\hat{\mathcal{G}}(z_1,z_1^+)+\hat{\mathcal{G}}(z_1,z_1^+)\hat{\mathbf{p}}}{2m}|\mathbf{x}_1\rangle=\pm\left(\frac{\nabla_1-\nabla_2}{2m}G(1;2)\right)_{2=1^+}=\mathbf{j}(1).$$

The proof of energy conservation is then reduced to show that the sum of $W_{\mathrm{one}}(z_1)$ and

$$W_{\mathrm{int}}(z_1)\equiv dE_{\mathrm{int}}(z_1)/dz_1$$

vanishes. Let us work out a simpler expression for W_{one}. In the definition of W_{one}, the term with the *static* scalar potential, $q\phi(\mathbf{x}_2)$, vanishes due to the fulfillment of the continuity equation, see (12.14). The remaining term contains the integral of the difference between $[D_1^2\Sigma(1;3)]G(3;1^+)$ and $[D_1^2G(1;3)]\Sigma(3;1^+)$. The gauge-invariant derivative $\mathbf{D}_1=\nabla_1-\mathrm{i}\frac{q}{c}\mathbf{A}(1)$ can be treated similarly to the normal gradient ∇_1 when integrating by parts. Indeed, given two functions $f(1)$ and $g(1)$ that vanish for $|\mathbf{r}_1|\to\infty$, we have

$$\int d\mathbf{r}_1 f(1)D_1^2g(1)=-\int d\mathbf{r}_1[\mathbf{D}_1^*f(1)]\cdot\mathbf{D}_1g(1)=\int d\mathbf{r}_1 g(1)(D_1^2)^*f(1). \tag{12.29}$$

Integrating by part $[D_1^2\Sigma(1;3)]G(3;1^+)$ with the help of (12.29), $W_{\mathrm{one}}(z_1)$ can be rewritten as

$$W_{\mathrm{one}}(z_1)=\pm\int d\mathbf{x}_1d3\left[\left(\frac{D_1^2}{2m}G(1;3)\right)\Sigma(3;1^+)-\Sigma(1;3)\left(\frac{(D_1^2)^*}{2m}G(3;1^+)\right)\right].$$

This result can be further manipulated. We multiply the equation of motion (12.5) by $[-\mathrm{i}\frac{d}{dz_2}+\frac{1}{2m}(D_2^2)^*-q\phi(2)]$, the adjoint equation (12.6) by $[\mathrm{i}\frac{d}{dz_1}+\frac{1}{2m}(D_1^2)-q\phi(1)]$, subtract one from the other and set $2=1^+$. Then, taking into account that Σ satisfies (12.14), we find

$$\int d3\left\{\left[\left(\mathrm{i}\frac{d}{dz_1}+\frac{D_1^2}{2m}\right)G(1;3)\right]\Sigma(3;1^+)+\Sigma(1;3)\left[\left(\mathrm{i}\frac{d}{dz_1}-\frac{(D_1^2)^*}{2m}\right)G(3;1^+)\right]\right\}=0.$$

Thus we see that $W_{\mathrm{one}}(z_1)$ can entirely be expressed in terms of the time derivatives of G. The condition for the energy conservation law can be formulated as

$$\begin{aligned}W_{\mathrm{one}}(z_1)+\frac{dE_{\mathrm{int}}(z_1)}{dz_1}&=\mp\mathrm{i}\int d\mathbf{x}_1d3\left\{\Sigma(1;3)\Big(\frac{d}{dz_1}G(3;1^+)\Big)+\Big(\frac{d}{dz_1}G(1;3)\Big)\Sigma(3;1^+)\right.\\&\left.-\frac{1}{4}\frac{d}{dz_1}\left[\Sigma(1;3)G(3;1^+)+G(1;3)\Sigma(3;1^+)\right]\right\}=0.\end{aligned} \tag{12.30}$$

[4]To prove this equation, consider for instance the first term on the r.h.s. We have

$$\begin{aligned}\langle\mathbf{x}_1|\hat{\mathbf{p}}\hat{\mathcal{G}}(z_1,z_1^+)|\mathbf{x}_1\rangle&=\lim_{\mathbf{x}_2\to\mathbf{x}_1}\int d\mathbf{x}_3\langle\mathbf{x}_1|\hat{\mathbf{p}}|\mathbf{x}_3\rangle\langle\mathbf{x}_3|\hat{\mathcal{G}}(z_1,z_1^+)|\mathbf{x}_2\rangle\\&=-\mathrm{i}\lim_{\mathbf{x}_2\to\mathbf{x}_1}\int d\mathbf{x}_3\nabla_1\delta(\mathbf{x}_1-\mathbf{x}_3)\langle\mathbf{x}_3|\hat{\mathcal{G}}(z_1,z_1^+)|\mathbf{x}_2\rangle\\&=-\mathrm{i}\lim_{\mathbf{x}_2\to\mathbf{x}_1}\nabla_1\langle\mathbf{x}_1|\hat{\mathcal{G}}(z_1,z_1^+)|\mathbf{x}_2\rangle=-\mathrm{i}\,\nabla_1G(1;2)|_{2=1^+}.\end{aligned}$$

The condition (12.30) is satisfied for all Φ-derivable self-energy as it follows from the invariance of Φ under the transformation

$$G(1;2) \to \left(\frac{dw(z_1)}{dz_1}\right)^{1/4} G(\mathbf{r}_1\sigma_1, w(z_1); \mathbf{r}_2\sigma_2, w(z_2)) \left(\frac{dw(z_2)}{dz_2}\right)^{1/4}, \tag{12.31}$$

where $w(z)$ is an invertible function for z on the contour with $w(t_{0-}) = t_{0-}$ and $w(t_0 - \mathrm{i}\beta) = t_0 - \mathrm{i}\beta$. This is an invariance since for every interaction line $v(i;j)$ there are four G that have the integration variables $z_i = z_j$ in common.[5] These four G supply a net factor dw/dz that changes the measure from dz to dw. For the infinitesimal transformation $w(z) = z + \varepsilon(z)$ the variation in the Green's function is, to first order,

$$\delta G(2;1^+) = \left\{\frac{1}{4}\left[\frac{d\varepsilon(z_2)}{dz_2} + \frac{d\varepsilon(z_1)}{dz_1}\right] + \left[\varepsilon(z_2)\frac{d}{dz_2} + \varepsilon(z_1)\frac{d}{dz_1}\right]\right\} G(2;1^+). \tag{12.32}$$

Therefore,

$$\begin{aligned} 0 = \delta\Phi &= \int d1d2 \frac{\delta\Phi}{\delta G(2;1^+)} \delta G(2;1^+) \\ &= \int d1d2\, \Sigma(1;2) \left\{\frac{1}{4}\left[\frac{d\varepsilon(z_2)}{dz_2} + \frac{d\varepsilon(z_1)}{dz_1}\right] + \left[\varepsilon(z_2)\frac{d}{dz_2} + \varepsilon(z_1)\frac{d}{dz_1}\right]\right\} G(2;1^+). \end{aligned} \tag{12.33}$$

Integrating by parts and taking into account that the self-energy satisfies the KMS relations, we recover (12.30)

Exercise 12.1 Using (12.22), prove (12.23).

Exercise 12.2 Using (12.31), prove (12.32) and then (12.33).

12.6 Examples of Φ-derivable Self-Energies: GW and T-matrices

We have already shown that the Hartree–Fock self-energy is Φ-derivable, the Φ-diagrams being the first two diagrams on the r.h.s. of Fig. 11.4. A second example of Φ-derivable self-energy is the second Born approximation, defined in (7.16). It consists of the Hartree–Fock diagrams and the two second-order diagrams in the G-skeletonic expansion. The second-order diagrams are the functional derivative of two distinct Φ-diagrams, the third and fourth diagrams on the r.h.s. of Fig. 11.4, each having symmetry factor $N_S = 4$. Both the Hartree–Fock and the second Born self-energies contain a finite number of diagrams. In this section we present other popular approximations containing an infinite number of self-energy diagrams.

GW self-energy The first approximation is illustrated in Fig. 12.1(a). It consists of an infinite sum of *ring* of *bubble diagrams.* The symmetry factor of the nth-order diagram is

[5] We recall that the interparticle interaction is local in time.

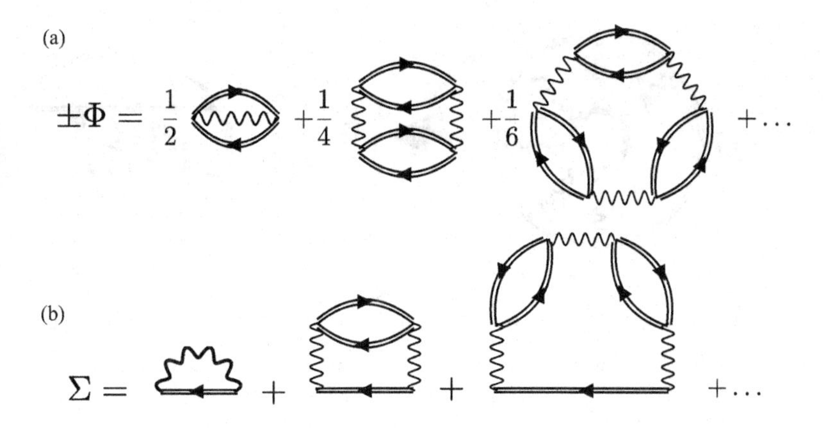

Figure 12.1 GW diagrams for the Φ-functional (a) and the corresponding self-energy (b).

$N_S = 2n$; therefore each Φ-diagram generates only one self-energy diagram. The self-energy obtained by functional differentiation is illustrated in Fig. 12.1(b). It consists of the Fock (or exchange) diagram, the second-order bubble diagram of the second Born approximation, and an infinite series of higher-order bubble diagrams.[6] This self-energy can also be written as the Fock (or exchange) self-energy in which the interparticle interaction v is replaced by the screened interaction W, where W is calculated from (7.22) using the bubble approximation for the polarization:

$$\Sigma(1;2) = \mathrm{i}G(1;2)W(1;2) \quad , \quad P(1;2) = \pm\mathrm{i}G(1;2)G(2;1). \tag{12.34}$$

The careful reader may have recognized this self-energy; it is the GW approximation [93] introduced in Section 7.7. Thus the GW self-energy is also Φ-derivable. The conserving properties of the GW approximation have been numerically verified in Hubbard-like models in Refs. [114, 115, 120]. The implementation of the GW approximation in real systems is still rather burdensome for current computational facilities. Furthermore, the fully self-consistent GW approximation is known to have a number of deficiencies like the washing out of plasmon features and broadened bandwidths in electron gas-like metals [149, 169]. For this reason, partially self-consistent or non-self-consistent treatments are often employed. The partially self-consistent treatment consists in solving the Dyson equation with a self-energy $\Sigma(1;2) = \mathrm{i}G(1;2)W_0(1;2)$. The screened interaction W_0 is calculated using a "quasi-particle" polarization $P(1;2) = \pm\mathrm{i}G_0(1;2)G_0(2;1)$, where G_0 has the structure of a noninteracting Green's function with poles at the quasi-particle energies, and it is kept fixed in the self-consistent iteration cycles. Non-self-consistent treatments consist in solving the Dyson equation with a self-energy $\Sigma(1;2) = \mathrm{i}G_0(1;2)W_0(1;2)$. It should be said that many spectral properties can already be described in partially self-consistent or non-self-consistent treatments (and sometimes even more accurately). In the electron gas the G_0W_0 approximation produces better quasi-particle renormalization factors than the fully

[6]As the Hartree diagram is not present, we have $\Phi = \Phi_{\mathrm{xc}}$ in the GW approximation.

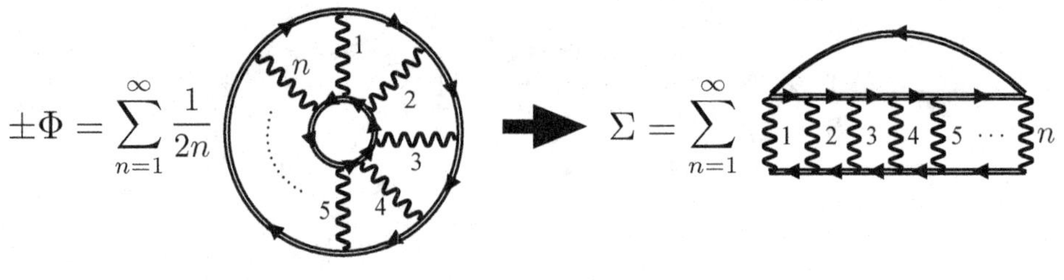

Figure 12.2 T-matrix diagrams in the particle–hole channel for the Φ-functional (left) and the corresponding self-energy (right).

self-consistent GW approximation [93, 148, 169, 170, 171], see also Section 15.6. Good performances of the G_0W_0 approximation have also been reported in a large class of molecular systems [172].

T-matrix self-energy in the particle–hole channel A second important Φ-derivable approximation is the T-matrix approximation in the particle–hole (ph) channel, or simply the Tph approximation. The Φ-diagrams are illustrated on the left of Fig. 12.2. Notice that the arrows of the inner circle flow in the opposite direction to the arrows of the outer circle. As for the GW-diagrams, the symmetry factor of the nth-order Tph-diagram is $N_S = 2n$ and, therefore, the functional differentiation gives rise to only one self-energy diagram. The Tph self-energy is illustrated on the right of the same figure. The first ($n = 1$) diagram is the Hartree, or tadpole, diagram, whereas the second ($n = 2$) diagram is the bubble diagram of the second Born approximation. The Tph self-energy has been heavily used to find the equation of state of an exciton fluid [173, 174, 175, 176], as well as to investigate carrier [177, 178, 179, 180] and exciton scatterings [84] in highly excited semiconductors. The interaction lines of the Tph self-energy do indeed connect Green's functions pointing in opposite directions, thus accounting for the multiple scattering between a particle and a hole. This physical interpretation naturally emerges from the connection between Σ and G_2, see Section 9.1, and from the analysis of the two-particle Green's function in Section 14.7.

T-matrix self-energy in the particle–particle channel with exchange The third approximation that we wish to introduce is the T-matrix approximation in the particle–particle (pp) channel, or simply the Tpp approximation. The Φ-diagrams are those in the first sum on the r.h.s. of Fig. 12.3. The difference with the Φ-diagrams in the Tph approximation is that the arrows of the inner and outer circles flow in the same direction. The symmetry factor of the nth-order diagram is $N_S = 2n$, and therefore each Φ-diagram leads to only one self-energy diagram. The self-energy is illustrated on the bottom part of the same figure (first sum on the r.h.s.). We can add to the diagrams of the Tpp approximation the so-called *exchange* diagrams. These are given by the second sum on the r.h.s. (for both Φ and Σ). The symmetry factor of the nth-order exchange Φ-diagram is again $N_S = 2n$. The reason for calling these diagrams "exchange diagrams" stems from the fact that the two-particle Green's function giving rise to the self-energy in Fig. 12.3(b) has the correct symmetry properties under the exchange of the first two or the last two arguments – that is,

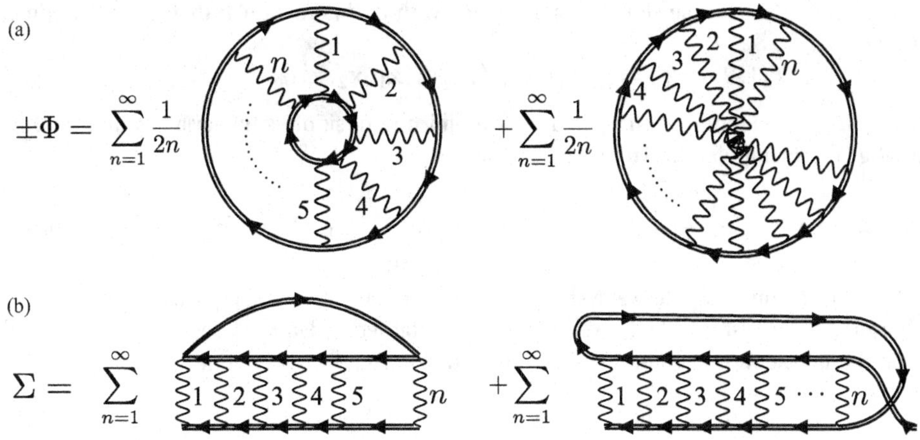

(a)

$$\pm\Phi = \sum_{n=1}^{\infty}\frac{1}{2n} \qquad + \sum_{n=1}^{\infty}\frac{1}{2n}$$

(b)

$$\Sigma = \sum_{n=1}^{\infty} \qquad + \sum_{n=1}^{\infty}$$

Figure 12.3 T-matrix diagrams in the particle–particle channel (the first term on the r.h.s.) and exchange diagrams in the same approximation (the second term on the r.h.s.) for the Φ-functional (a) and the corresponding self-energy (b).

$G_2(1,2;3,4) = \pm G_2(2,1;3,4) = \pm G_2(1,2;4,3)$, see also Section 14.8. Without exchange diagrams these properties would not be fulfilled. We refer to the Tpp approximation with the addition of exchange diagrams as the Tpp+X approximation. Let us discuss some of its properties.

We begin by observing that truncating the Tpp+X approximation to second order in v we recover the second Born approximation. The self-energy in the Tpp+X approximation can be conveniently written as

$$\boxed{\Sigma(1;1') = \pm\mathrm{i}\int d2d2'\,[T(1,2;1',2') \pm T(1,2;2',1')]\,G(2';2^+)} \tag{12.35}$$

where the so-called *transfer matrix* is defined according to

$$T(1,2;1',2') = \delta(1;1')\delta(2;2')v(1';2') + \mathrm{i}\int d3d4\,T(1,2;3,4)G(3;1')G(4;2')v(1';2'). \tag{12.36}$$

We can represent the transfer matrix with the following diagrammatic series:

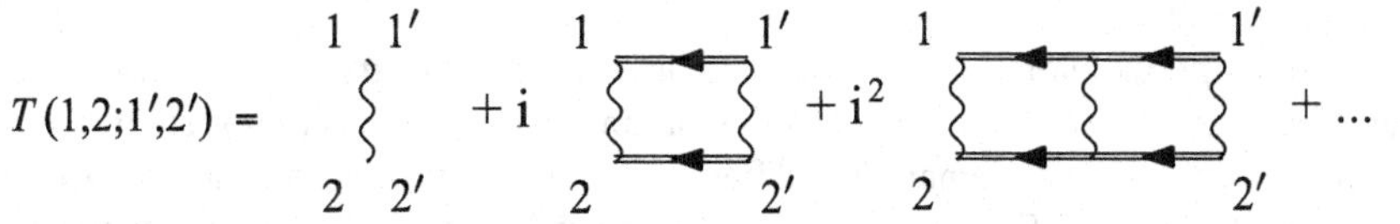

$$T(1,2;1',2') = \quad + \mathrm{i} \quad + \mathrm{i}^2 \quad + \dots \tag{12.37}$$

The transfer matrix is a difficult object to manipulate since it depends on four variables. Nevertheless, the dependence on the time arguments is rather simple. From (12.37) we see that each term of the expansion starts and ends with an interaction v that is local in time. Therefore,

$$T(1,2;1',2') = \delta(z_1,z_2)\delta(z_1',z_2')T(\mathbf{x}_1,\mathbf{x}_2,\mathbf{x}_1',\mathbf{x}_2';z_1,z_1').$$

To highlight the dependence on z_1 and z_1', we omit the position–spin variables and reinsert them after the manipulation. From (12.37) we have

$$T(z_1,z_1') = \delta(z_1,z_1')v + \mathrm{i}v\,G^2(z_1,z_1')\,v + \mathrm{i}^2 v\int d\bar{z}\,G^2(z_1,\bar{z})\,v\,G^2(\bar{z},z_1')\,v + \ldots. \quad (12.38)$$

This function belongs to the Keldysh space and contains a singular part. For systems in equilibrium, the real-time Keldysh components (retarded, advanced, lesser, etc.) depend only on the time difference and can be Fourier transformed. If we define

$$\mathcal{T}(\omega) \equiv \mathrm{i}\left[T^>(\omega) - T^<(\omega)\right] = \mathrm{i}\left[T^{\mathrm{R}}(\omega) - T^{\mathrm{A}}(\omega)\right], \quad (12.39)$$

then, by the definition of retarded and advanced functions, we can write

$$T^{\mathrm{R/A}}(\omega) = v + \int\frac{d\omega'}{2\pi}\frac{\mathcal{T}(\omega')}{\omega-\omega'\pm\mathrm{i}\eta},$$

or, reinserting the dependence on the position–spin variables,

$$T^{\mathrm{R/A}}(\mathbf{x}_1,\mathbf{x}_2,\mathbf{x}_1',\mathbf{x}_2';\omega) = v(\mathbf{x}_1,\mathbf{x}_2)\delta(\mathbf{x}_1,\mathbf{x}_1')\delta(\mathbf{x}_2,\mathbf{x}_2') + \int\frac{d\omega'}{2\pi}\frac{\mathcal{T}(\mathbf{x}_1,\mathbf{x}_2,\mathbf{x}_1',\mathbf{x}_2';\omega')}{\omega-\omega'\pm\mathrm{i}\eta}.$$

The transfer matrix also obeys its own fluctuation–dissipation theorem. The proof goes along the same lines as for the self-energy, polarization, and screened interaction. Since $T(z_1,z_1')$ starts with $G^2(z_1,\ldots)$ and ends with $G^2(\ldots,z_1')$, the starting external vertices produce the factor $e^{2\mu\beta}$, whereas the ending external vertices produce the factor $e^{-2\mu\beta}$. Thus, it is easy to derive

$$T^>(\omega) = e^{\beta(\omega-2\mu)}T^<(\omega). \quad (12.40)$$

Combining (12.40) with definition (12.39), we can determine $T^{\lessgtr}$ from the sole knowledge of $\mathcal{T}$ since

$$\boxed{\begin{aligned} T^<(\omega) &= -\mathrm{i}f(\omega-2\mu)\,\mathcal{T}(\omega) \\ T^>(\omega) &= -\mathrm{i}\bar{f}(\omega-2\mu)\,\mathcal{T}(\omega)\end{aligned}} \quad (12.41)$$

with $f(\omega) = 1/(e^{\beta\omega}-1)$ the Bose function and $\bar{f}(\omega) = 1 + f(\omega)$. This is the fluctuation-dissipation theorem for T.

The T-matrix approximation has been used in several contexts of condensed matter physics and nuclear physics. As the implementation of the fully self-consistent T-matrix self-energy is computationally demanding, different flavors of this approximation have been studied over the years. From (12.36) we see that the transfer matrix satisfies an equation with two Green's functions. If we here set $G = G_0$, we get the so-called T_0-matrix approximation. The transfer matrix T_0 can then be multiplied by G_0 to generate a self-energy according to (12.35). This T_0G_0 approximation was used by Thouless to describe the superconducting instability of a normal metal with attractive interactions [181]. The T_0G_0 approximation, however, gives rise to unphysical results in two dimensions [182]. A different approximation consists in constructing the transfer matrix T from the self-consistent G, which solves the Dyson equation with self-energy $\Sigma = \pm iTG_0$ [183]. Also this scheme is not fully self-consistent and has some problems. The advantage of working with the self-consistent $\Sigma = \pm iTG$ is that several exact properties, like the conservation laws, are preserved [116, 184].

Exercise 12.3 Consider the Hubbard model with only one site of Exercise 8.4. Assuming that the Green's function is diagonal in the spin indices, show that the self-energy in the second Born approximation is spin diagonal and given by

$$\Sigma_\sigma(z_1, z_2) = \delta(z_1, z_2)Un_{\bar{\sigma}}(z_1) - i^2U^2G_\sigma(z_1, z_2)G_{\bar{\sigma}}(z_2, z_1)G_{\bar{\sigma}}(z_1, z_2), \tag{12.42}$$

independent of whether one uses the interaction (2.36) or (2.37).

Exercise 12.4 Consider the system of Exercise 12.3 in thermal equilibrium. Evaluating the lesser and greater second Born self-energy at the Hartree–Fock Green's function, show that

$$\begin{aligned}\Sigma_\sigma^<(\omega) &= U^2\int\frac{d\omega_1}{2\pi}\frac{d\omega_2}{2\pi}G_\sigma^<(\omega+\omega_1-\omega_2)G_{\bar{\sigma}}^>(\omega_1)G_{\bar{\sigma}}^<(\omega_2)\\ &= 2\pi iU^2f_\sigma\bar{f}_{\bar{\sigma}}f_{\bar{\sigma}}\delta(\omega-\epsilon_0-Un_{\bar{\sigma}}),\end{aligned} \tag{12.43}$$

$$\begin{aligned}\Sigma_\sigma^>(\omega) &= U^2\int\frac{d\omega_1}{2\pi}\frac{d\omega_2}{2\pi}G_\sigma^>(\omega+\omega_1-\omega_2)G_{\bar{\sigma}}^<(\omega_1)G_{\bar{\sigma}}^>(\omega_2)\\ &= -2\pi iU^2\bar{f}_\sigma\bar{f}_{\bar{\sigma}}f_{\bar{\sigma}}\delta(\omega-\epsilon_0-Un_{\bar{\sigma}}),\end{aligned} \tag{12.44}$$

where $f_\sigma = f(\epsilon_0 + Un_{\bar{\sigma}} - \mu) = n_\sigma$ and $\bar{f}_\sigma = 1 - f_\sigma$.

Exercise 12.5 Show that the number of symmetries N_S of the nth-order Φ-diagram of the Tpp+X approximation is $N_S = 2n$.

Exercise 12.6 Extract the greater/lesser component of the generic term of expansion (12.38), taking into account that for systems in equilibrium the Langreth rules can be modified – that is, the vertical track can be ignored provided that $t_0 \to -\infty$. Fourier transforming, show that each term of the expansion fulfills (12.40).

13

Linear Response Theory

In the previous chapters we developed a method to study many-particle systems in both equilibrium and nonequilibrium situations. In this chapter we show how the general problem can be simplified by treating the external time-dependent fields perturbatively. To first order, the corresponding theory is known as *linear response theory* (or simply *linear response*) and, of course, it is an approximate theory even for noninteracting systems. Linear response theory naturally leads to introducing the two-particle Green's function $G_2(1,2;1',2')$ calculated in $z'_1 = z_1^+$ and $z'_2 = z_2^+$. With these time constraints the two-particle Green's function becomes a two-times correlator in Keldysh space, known as the *response function.* We derive relevant exact properties of the response function and then relate it to measurable quantities like the photoabsorption spectrum and the absorbed energy.

13.1 Kubo Formula

The problem that we want to solve can be formulated as follows. Consider a system of interacting particles described by the time-dependent Hamiltonian $\hat{H}(t)$ and subject to a further, but "small," time-dependent perturbation $\lambda\hat{H}'(t)$, with $\lambda \ll 1$ a small dimensionless parameter. The full evolution operator obeys the differential equation (3.13), which in our case reads:

$$\mathrm{i}\frac{d}{dt}\hat{U}_{\mathrm{tot}}(t,t_0) = \left[\hat{H}(t) + \lambda\hat{H}'(t)\right]\hat{U}_{\mathrm{tot}}(t,t_0), \tag{13.1}$$

with boundary condition $\hat{U}_{\mathrm{tot}}(t_0,t_0) = \hat{\mathbb{1}}$. We look for a solution of the form

$$\hat{U}_{\mathrm{tot}}(t,t_0) = \hat{U}(t,t_0)\hat{F}(t), \tag{13.2}$$

with $\hat{U}(t,t_0)$ the evolution operator for the system with Hamiltonian $\hat{H}(t)$. Substituting (13.2) into (13.1), we find a differential equation for $\hat{F}$:

$$\mathrm{i}\frac{d}{dt}\hat{F}(t) = \lambda\hat{H}'_H(t)\hat{F}(t), \tag{13.3}$$

where the subscript "H" denotes the operator in the Heisenberg picture: $\hat{H}'_H(t) = \hat{U}(t_0,t)\hat{H}'(t)\hat{U}(t,t_0)$. The operator $\hat{H}'_H(t)$ is self-adjoint for all times t. Equation (13.3) must be solved with boundary condition $\hat{F}(t_0) = \hat{\mathbb{1}}$ and hence an integration between t_0 and t leads to

$$\hat{F}(t) = \hat{\mathbb{1}} - \mathrm{i}\lambda \int_{t_0}^{t} dt' \hat{H}'_H(t')\hat{F}(t') = \hat{\mathbb{1}} - \mathrm{i}\lambda \int_{t_0}^{t} dt' \hat{H}'_H(t') + \mathcal{O}(\lambda^2),$$

where in the last equality we replace $\hat{F}(t')$ with the whole r.h.s. (iterative solution). We conclude that to first order in λ the full evolution operator reads:

$$\hat{U}_{\mathrm{tot}}(t, t_0) = \hat{U}(t, t_0) \left[\hat{\mathbb{1}} - \mathrm{i}\lambda \int_{t_0}^{t} dt' \hat{H}'_H(t')\right]. \tag{13.4}$$

This result must be used with care. We see at once that the operator (13.4) is not unitary. This implies that if we start from a normalized ket $|\Psi\rangle$, then the time-evolved ket $|\Psi(t)\rangle = \hat{U}_{\mathrm{tot}}(t, t_0)|\Psi\rangle$ is no longer normalized. The correction is, however, of second order in λ since

$$\langle\Psi(t)|\Psi(t)\rangle = 1 - \mathrm{i}\lambda \int_{t_0}^{t} dt' \langle\Psi|\hat{H}'_H(t') - \hat{H}'_H(t')|\Psi\rangle + \mathcal{O}(\lambda^2) = 1 + \mathcal{O}(\lambda^2).$$

Nevertheless, the nonunitarity of the evolution operator represents a warning sign that we should not ignore, especially if we are interested in knowing the long-time behavior of the system. For instance, in the simplest case $\hat{H}'(t) = \theta(t - t_0)E\hat{\mathbb{1}}$ is a uniform shift, the evolution operator (13.4) becomes $\hat{U}_{\mathrm{tot}}(t, t_0) = e^{-\mathrm{i}\hat{H}(t-t_0)}[1 - \mathrm{i}\lambda E(t - t_0)]$, and for times $t \gg t_0 + 1/(\lambda E)$ the correction is far from being small. In Appendix I we discuss the subtleties and possible shortcomings of linear response theory, and clarify its domain of applicability.

Let us go back to (13.4). The time-dependent ensemble average of an operator $\hat{O}(t)$ is to first order in λ

$$\begin{aligned}
O(t) &= \langle \hat{U}_{\mathrm{tot}}(t_0, t)\hat{O}(t)\hat{U}_{\mathrm{tot}}(t, t_0)\rangle \\
&\sim \langle \left[\hat{\mathbb{1}} + \mathrm{i}\lambda \int_{t_0}^{t} dt' \hat{H}'_H(t')\right] \hat{U}(t_0, t)\hat{O}(t)\hat{U}(t, t_0) \left[\hat{\mathbb{1}} - \mathrm{i}\lambda \int_{t_0}^{t} dt' \hat{H}'_H(t')\right]\rangle \\
&= \langle \hat{O}_H(t)\rangle - \mathrm{i}\lambda \int_{t_0}^{t} dt' \, \langle \left[\hat{O}_H(t), \hat{H}'_H(t')\right]_- \rangle + \mathcal{O}(\lambda^2),
\end{aligned}$$

where $\langle \ldots \rangle \equiv \mathrm{Tr}[\hat{\rho} \ldots]$ denotes the ensemble average with a density matrix $\hat{\rho}$, see (4.13). Letting $\delta O(t) = O(t) - \langle \hat{O}_H(t)\rangle$ be the change of the time-dependent average of $\hat{O}(t)$ to first order in λ, we find the *Kubo formula*:

$$\boxed{\delta O(t) = -\mathrm{i}\lambda \int_{t_0}^{t} dt' \, \langle \left[\hat{O}_H(t), \hat{H}'_H(t')\right]_- \rangle} \tag{13.5}$$

The Kubo formula is the starting point for our subsequent derivations.

Before concluding this introductory section we would like to make an important remark about the possibility of expanding in powers of λ the change of a time-dependent average.

To first order we see that (13.5) involves an average of operators in the Heisenberg picture with the Hamiltonian $\hat{H}(t)$. For systems in equilibrium at zero temperature, one might be tempted to use the zero-temperature formalism, see Section 5.4, to evaluate this average. However, the zero-temperature formalism gives us direct access to *time-ordered* averages and not to the average of a commutator. Nonetheless, there exists a "simple trick" (see Ref. [185]) to extract the average of a commutator from the time-ordered average $\mathcal{T}\{\hat{O}_H(t)\hat{H}'_H(t')\}$. Going beyond linear response or at finite temperature or for time-dependent Hamiltonians $\hat{H}(t)$, the simple trick can no longer be used. For instance, to second order in λ the change of the time-dependent average involves a double commutator $\left[\left[\hat{O}_H(t), \hat{H}'_H(t')\right]_-, \hat{H}'_H(t'')\right]_-$ and there is no trick to calculate this quantity from the time-ordered average of $\mathcal{T}\{\hat{O}_H(t)\hat{H}'_H(t')\hat{H}'_H(t'')\}$. Second-order changes are relevant every time the linear order change vanishes. This is, for instance, the case of photoemission spectra, where the current is quadratic in the light–matter coupling, see Section 6.4.

13.2 Current and Density Response Functions

We use here the Kubo formula to study a system of interacting particles with mass m and charge q moving under the influence of a vector potential $\mathbf{A}(\mathbf{x}, t)$ and scalar potential $\phi(\mathbf{x}, t)$.[1] The time-dependent Hamiltonian is therefore $\hat{H}(t) = \hat{H}_0(t) + \hat{H}_{\text{int}}$, with

$$\hat{H}_0(t) = \frac{1}{2m}\int d\mathbf{x}\, \hat{\psi}^\dagger(\mathbf{x})\left(-\mathrm{i}\nabla - \frac{q}{c}\mathbf{A}(\mathbf{x}, t)\right)^2 \hat{\psi}(\mathbf{x}) + q\int d\mathbf{x}\, \phi(\mathbf{x}, t)\hat{n}(\mathbf{x}), \tag{13.6}$$

and $\hat{n}(\mathbf{x}) = \hat{\psi}^\dagger(\mathbf{x})\hat{\psi}(\mathbf{x})$ the density operator.[2] To be concrete, we take the system in thermal equilibrium at times $t < t_0$. The equilibrium Hamiltonian $\hat{H}^{\mathrm{M}} = \hat{H}_0 + \hat{H}_{\text{int}} - \mu\hat{N}$ has the one-body part $\hat{H}_0$ given by (13.6), but with static vector and scalar potentials $\mathbf{A}(\mathbf{x})$ and $\phi(\mathbf{x})$. The question we ask is: How does the time evolution change under a change of the external potentials $\mathbf{A} \to \mathbf{A} + \delta\mathbf{A}$ and $\phi \to \phi + \delta\phi$ for times $t > t_0$? To first order in $\delta\mathbf{A}$ and $\delta\phi$ the change of $\hat{H}_0$ can most easily be worked out from (3.30) and reads:[3]

$$\hat{H}_0(t) \to \hat{H}_0(t) - \frac{q}{c}\int d\mathbf{x}\, \hat{\mathbf{j}}(\mathbf{x})\cdot\delta\mathbf{A}(\mathbf{x}, t) + \int d\mathbf{x}\, \hat{n}(\mathbf{x})\left(q\delta\phi(\mathbf{x}, t) + \frac{q^2}{mc^2}\mathbf{A}(\mathbf{x}, t)\cdot\delta\mathbf{A}(\mathbf{x}, t)\right).$$

The perturbation on the r.h.s. of this equation is the explicit form of $\lambda\hat{H}'(t)$. Let us introduce a few definitions and rewrite the perturbation in a more transparent way. We first note that the terms linear in $\delta\mathbf{A}$ can be grouped to form the (gauge-invariant) current density operator $\hat{\mathbf{J}}$ defined in (3.32). It is therefore natural to define the four-dimensional vector of operators $(\hat{n}, \hat{\mathbf{J}}) = (\hat{n}, \hat{J}_x, \hat{J}_y, \hat{J}_z)$ with components $\hat{J}_\mu$, $\mu = 0, 1, 2, 3$, so that $\hat{J}_0 = \hat{n}$ and $\hat{J}_1 = \hat{J}_x$, etc. Similarly, we define the four-dimensional vector $(\delta\phi, -\delta\mathbf{A}/c)$

[1] Even though the vector and the scalar potentials do not depend on spin, we use the variable $\mathbf{x}$ for notational convenience.

[2] For simplicity we do not include the Pauli coupling between the spin of the particles and the magnetic field. We also recall that in (13.6) ϕ is the *external* scalar potential, whereas $\mathbf{A}$ is the *total* (and transverse) vector potential. This fact has been already observed in footnote 1 of Chapter 3, and is proven in Chapter 16.

[3] The last term of (3.30) has been ignored since it is a total divergence.

with components δA^μ, $\mu = 0, 1, 2, 3$, so that $\delta A^0 = \phi$ and $\delta A^1 = -\delta A_x/c$, etc. Then the perturbation takes the following compact form:

$$\lambda \hat{H}'(t) = q \int d\mathbf{x}\, \hat{J}_\mu(\mathbf{x}, t)\, \delta A^\mu(\mathbf{x}, t),$$

with the Einstein convention of summing over repeated indices.

With the explicit form of $\lambda \hat{H}'(t)$ we can try to calculate some physical quantity, such as the change in the time-dependent density n and current density $\mathbf{J}$. Taking into account that $\delta\mathbf{J} = \delta\mathbf{j} - \frac{q}{mc}\delta n\mathbf{A} - \frac{q}{mc} n\delta\mathbf{A}$, see (3.32), the Kubo formula (13.5) yields

$$\begin{aligned}\delta J_\mu(\mathbf{x}, t) =& -\mathrm{i}q \int_{t_0}^{t} dt' \int d\mathbf{x}' \left\langle \left[\hat{J}_{\mu,H}(\mathbf{x}, t), \hat{J}_{\nu,H}(\mathbf{x}', t')\right]_- \right\rangle \delta A^\nu(\mathbf{x}', t') \\ &+ \bar{\delta}_{0\mu} \frac{q}{m} n(\mathbf{x}, t) \delta A^\mu(\mathbf{x}, t),\end{aligned} \tag{13.7}$$

where $\bar{\delta}_{0\mu} = 1 - \delta_{0\mu}$. We now replace the current operators in the above commutator with the fluctuation current operators,

$$\Delta \hat{J}_{\mu,H}(\mathbf{x}, t) \equiv \hat{J}_{\mu,H}(\mathbf{x}, t) - \langle \hat{J}_{\mu,H}(\mathbf{x}, t) \rangle,$$

and rewrite (13.7) as

$$\begin{aligned}\delta J_\mu(\mathbf{x}, t) =& -\mathrm{i}q \int_{t_0}^{t} dt' \int d\mathbf{x}' \left\langle \left[\Delta\hat{J}_{\mu,H}(\mathbf{x}, t), \Delta\hat{J}_{\nu,H}(\mathbf{x}', t')\right]_- \right\rangle \delta A^\nu(\mathbf{x}', t') \\ &+ \bar{\delta}_{0\mu} \frac{q}{m} n(\mathbf{x}, t) \delta A^\mu(\mathbf{x}, t).\end{aligned} \tag{13.8}$$

It is clear that (13.8) is equivalent to (13.7) since a number commutes with all operators.

Response functions The structure of (13.8) prompts us to define the following correlator on the contour[4]

$$\boxed{\chi_{\mu\nu}(\mathbf{x}, z; \mathbf{x}', z') \equiv -\mathrm{i} \left\langle \mathcal{T} \left\{ \Delta\hat{J}_{\mu,H}(\mathbf{x}, z) \Delta\hat{J}_{\nu,H}(\mathbf{x}', z') \right\} \right\rangle} \tag{13.9}$$

which has the symmetry property

$$\boxed{\chi_{\mu\nu}(1; 2) = \chi_{\nu\mu}(2; 1)} \tag{13.10}$$

The correlator $\chi_{\mu\nu}$ has different names depending on the values of μ and ν. For $\mu = \nu = 0$ we have the component $\chi_{00} = \chi$, which we show in Chapter 14 to be the *density response function* defined in (7.24) and analyzed diagrammatically in Section 11.7.[5] The components χ_{0j} and χ_{j0}, with $j = 1, 2, 3$, are called the *current-density response functions*, and the components χ_{jk} with $j, k = 1, 2, 3$ are called the *current response functions*.

[4]In (13.9) the operators are in the contour Heisenberg picture, see (4.38).

[5]In Appendix E we also discuss the relation between the density response function and the so-called *pair correlation function*.

The response function $\chi_{\mu\nu}$ belongs to the Keldysh space and it can be converted into standard time-dependent averages by choosing the contour arguments on different branches. For example, the greater/lesser components of $\chi_{\mu\nu}$ are

$$\chi^{>}_{\mu\nu}(\mathbf{x},t;\mathbf{x}',t') = \chi_{\mu\nu}(\mathbf{x},t_+;\mathbf{x}',t'_-) = -\mathrm{i}\,\langle \Delta\hat{J}_{\mu,H}(\mathbf{x},t)\Delta\hat{J}_{\nu,H}(\mathbf{x}',t')\rangle, \tag{13.11}$$

$$\begin{aligned}\chi^{<}_{\mu\nu}(\mathbf{x},t;\mathbf{x}',t') = \chi_{\mu\nu}(\mathbf{x},t_-;\mathbf{x}',t'_+) &= -\mathrm{i}\,\langle \Delta\hat{J}_{\nu,H}(\mathbf{x}',t')\Delta\hat{J}_{\mu,H}(\mathbf{x},t)\rangle \\ &= -\left[\chi^{>}_{\mu\nu}(\mathbf{x},t;\mathbf{x}',t')\right]^*.\end{aligned} \tag{13.12}$$

Since $\chi_{\mu\nu}$ has no singular part (i.e., $\chi^{\delta}_{\mu\nu} = 0$), the retarded component reads:

$$\begin{aligned}\chi^{\mathrm{R}}_{\mu\nu}(\mathbf{x},t;\mathbf{x}',t') &= \theta(t-t')\left[\chi^{>}_{\mu\nu}(\mathbf{x},t;\mathbf{x}',t') - \chi^{<}_{\mu\nu}(\mathbf{x},t;\mathbf{x}',t')\right] \\ &= -\mathrm{i}\,\theta(t-t')\langle\left[\Delta\hat{J}_{\mu,H}(\mathbf{x},t),\Delta\hat{J}_{\nu,H}(\mathbf{x}',t')\right]_-\rangle,\end{aligned} \tag{13.13}$$

which is exactly the kernel of (13.8). Thus, we can rewrite (13.8) as

$$\boxed{\delta J_\mu(\mathbf{x},t) = q\int_{-\infty}^{\infty} dt' \int d\mathbf{x}'\, \chi^{\mathrm{R}}_{\mu\nu}(\mathbf{x},t;\mathbf{x}',t')\,\delta A^\nu(\mathbf{x}',t') + \bar{\delta}_{0\mu}\frac{q}{m}n(\mathbf{x},t)\delta A^\mu(\mathbf{x},t)} \tag{13.14}$$

where we take into account that δA^ν vanishes for times smaller than t_0. The function $\chi^{\mathrm{R}}_{\mu\nu}(\mathbf{x},t;\mathbf{x}',t')$ is a real function. This can be seen directly from the second line of (13.13) or, equivalently, from the first line of (13.13) using the relation (13.12) between $\chi^<$ and the complex conjugate of $\chi^>$.

Relation between the response functions and G_2 All these components can be expressed in terms of the so-called *two-particle XC function,*[6]

$$\begin{aligned}L(1,2;1',2') &= \pm\left[G_2(1,2;1',2') - G(1;1')G(2;2')\right] \\ &= \mp\langle\mathcal{T}\left\{\hat{\psi}_H(1)\hat{\psi}_H(2)\hat{\psi}^\dagger_H(2')\hat{\psi}^\dagger_H(1')\right\}\rangle \\ &\quad \pm\langle\mathcal{T}\left\{\hat{\psi}_H(1)\hat{\psi}^\dagger_H(1')\right\}\rangle\langle\mathcal{T}\left\{\hat{\psi}_H(2)\hat{\psi}^\dagger_H(2')\right\}\rangle,\end{aligned} \tag{13.15}$$

where, as usual, the upper sign is for bosons and the lower sign for fermions. For instance, the density response function in (13.9) is explicitly given by

$$\chi_{00}(1;2) = -\mathrm{i}\,\langle\mathcal{T}\left\{\hat{n}_H(1)\hat{n}_H(2)\right\}\rangle + \mathrm{i}n(1)n(2) = \pm\mathrm{i}L(1,2;1^+,2^+). \tag{13.16}$$

Similarly, the current-density response function and the current response function are

[6] XC stands for exchange and correlation.

related to L by

$$\chi_{0j}(1;2) = \pm \mathrm{i} \left[\left(\frac{\partial_{2,j} - \partial_{2',j}}{2m\mathrm{i}} - \frac{q}{mc} A_j(2) \right) L(1,2;1^+,2') \right]_{2'=2^+} \tag{13.17a}$$

$$= \pm \mathrm{i} \left[\left(\frac{D_{2,j} - D^*_{2',j}}{2m\mathrm{i}} \right) L(1,2;1^+,2') \right]_{2'=2^+},$$

$$\chi_{j0}(1;2) = \pm \mathrm{i} \left[\left(\frac{D_{1,j} - D^*_{1',j}}{2m\mathrm{i}} \right) L(1,2;1',2^+) \right]_{1'=1^+}, \tag{13.17b}$$

$$\chi_{jk}(1;2) = \pm \mathrm{i} \left[\left(\frac{D_{1,j} - D^*_{1',j}}{2m\mathrm{i}} \right) \left(\frac{D_{2,k} - D^*_{2',k}}{2m\mathrm{i}} \right) L(1,2;1',2') \right]_{\substack{1'=1^+ \\ 2'=2^+}}, \tag{13.17c}$$

with $\mathbf{D}_1 = \boldsymbol{\nabla}_1 - \mathrm{i}\frac{q}{c}\mathbf{A}(1)$ the (gauge-invariant) derivative already introduced in (12.2). Due to the symmetry property $L(1,2;1',2') = L(2,1;2',1')$ of the two-particle XC function, the symmetry property (13.10) is satisfied.

The result (13.14) is an interesting and elegant result, but what are the advantages of using this formula? Ultimately it only gives us access to the linear response change δJ_μ. Why don't we instead calculate the *full* J_μ using the Green's function G, an object much easier to manipulate than $\chi_{\mu\nu}$ or L? In the general case there is indeed no advantage in using (13.14). However, if we are interested in changes with respect to *equilibrium* or *stationary* conditions then it may be more advantageous to work with the correlators $\chi_{\mu\nu}$ since they would depend only on the time difference, whereas G depends (to first order in the external fields) on both time coordinates.

13.2.1 Thermal Equilibrium and Lehmann Representation

When the system is in equilibrium, the time evolution operator is simply the exponential $\hat{U}(t,t_0) = \exp[-\mathrm{i}\hat{H}(t-t_0)]$. Let us see what this simplification leads to. Consider the greater component of the response function (13.11),

$$\chi^>_{\mu\nu}(\mathbf{x},t;\mathbf{x}',t') = -\mathrm{i}\frac{\mathrm{Tr}\left[e^{-\beta(\hat{H}-\mu\hat{N})}e^{\mathrm{i}\hat{H}(t-t_0)}\Delta\hat{J}_\mu(\mathbf{x})e^{-\mathrm{i}\hat{H}(t-t')}\Delta\hat{J}_\nu(\mathbf{x}')e^{-\mathrm{i}\hat{H}(t'-t_0)}\right]}{\mathrm{Tr}\left[e^{-\beta(\hat{H}-\mu\hat{N})}\right]}$$

$$= -\mathrm{i}\sum_a \rho_a \langle\Psi_a|\Delta\hat{J}_\mu(\mathbf{x})e^{-\mathrm{i}(\hat{H}-E_a)(t-t')}\Delta\hat{J}_\nu(\mathbf{x}')|\Psi_a\rangle, \tag{13.18}$$

where the sum is over the eigenkets $|\Psi_a\rangle$ of $\hat{H}$ with energy E_a and number of particles N_a. The weights ρ_a are therefore

$$\rho_a = \frac{e^{-\beta(E_a-\mu N_a)}}{\mathrm{Tr}\left[e^{-\beta(\hat{H}-\mu\hat{N})}\right]}. \tag{13.19}$$

As expected, $\chi^>$ depends only on the time difference $t-t'$. A similar result can be worked out for the lesser component and hence we can define the Fourier transform according to

$$\chi^{\gtrless}_{\mu\nu}(\mathbf{x},t;\mathbf{x}',t') = \int \frac{d\omega}{2\pi} e^{-\mathrm{i}\omega(t-t')}\chi^{\gtrless}_{\mu\nu}(\mathbf{x},\mathbf{x}';\omega).$$

Using the same trick that has led to (6.67), it is easy to find the following exact relation between the greater and lesser components:[7]

$$\chi^{>}_{\mu\nu}(\mathbf{x},\mathbf{x}';\omega) = e^{\beta\omega}\chi^{<}_{\mu\nu}(\mathbf{x},\mathbf{x}';\omega).$$

Like for the equilibrium Green's function, self-energy, polarization, etc., we then have a fluctuation–dissipation theorem for the equilibrium $\chi_{\mu\nu}$. Omitting the position–spin variables, we can write:

$$\boxed{\begin{aligned}\chi^{>}_{\mu\nu}(\omega) &= \bar{f}(\omega)\left[\chi^{\mathrm{R}}_{\mu\nu}(\omega) - \chi^{\mathrm{A}}_{\mu\nu}(\omega)\right]\\ \chi^{<}_{\mu\nu}(\omega) &= f(\omega)\left[\chi^{\mathrm{R}}_{\mu\nu}(\omega) - \chi^{\mathrm{A}}_{\mu\nu}(\omega)\right]\end{aligned}} \tag{13.20}$$

with $f(\omega) = 1/(e^{\beta\omega} - 1)$ the Bose function and $\bar{f}(\omega) = 1 + f(\omega)$.

The Fourier transform of $\chi^{\lessgtr}_{\mu\nu}$ can be used to calculate the Fourier transform of $\chi^{\mathrm{R}}_{\mu\nu}$ since from (13.13)

$$\begin{aligned}\chi^{\mathrm{R}}_{\mu\nu}(\mathbf{x},t;\mathbf{x}',t') &= \int \frac{d\omega}{2\pi} e^{-\mathrm{i}\omega(t-t')}\chi^{\mathrm{R}}_{\mu\nu}(\mathbf{x},\mathbf{x}';\omega)\\ &= \theta(t-t')\int \frac{d\omega'}{2\pi} e^{-\mathrm{i}\omega'(t-t')}\left[\chi^{>}_{\mu\nu}(\mathbf{x},\mathbf{x}';\omega') - \chi^{<}_{\mu\nu}(\mathbf{x},\mathbf{x}';\omega')\right].\end{aligned} \tag{13.21}$$

Taking into account the representation (6.53) of the Heaviside function, we get

$$\chi^{\mathrm{R}}_{\mu\nu}(\mathbf{x},\mathbf{x}';\omega) = \mathrm{i}\int \frac{d\omega'}{2\pi}\frac{\chi^{>}_{\mu\nu}(\mathbf{x},\mathbf{x}';\omega') - \chi^{<}_{\mu\nu}(\mathbf{x},\mathbf{x}';\omega')}{\omega - \omega' + \mathrm{i}\eta} \tag{13.22}$$

with η an infinitesimal positive constant. Similarly, one can show that

$$\chi^{\mathrm{A}}_{\mu\nu}(\mathbf{x},\mathbf{x}';\omega) = \mathrm{i}\int \frac{d\omega'}{2\pi}\frac{\chi^{>}_{\mu\nu}(\mathbf{x},\mathbf{x}';\omega') - \chi^{<}_{\mu\nu}(\mathbf{x},\mathbf{x}';\omega')}{\omega - \omega' - \mathrm{i}\eta}. \tag{13.23}$$

The Fourier transforms (13.22) or (13.23) contain plenty of physical information and play a central role in this chapter.

The explicit form of the response functions at finite temperature are a bit more cumbersome than those at zero temperature. Thus we here discuss only the zero-temperature case and leave the generalization to finite temperature as an exercise for the curious reader. At zero temperature only the ground state $|\Psi_0\rangle$ contributes to the sum over a in (13.18) and, again for simplicity, we assume that $|\Psi_0\rangle$ is nondegenerate and normalized to unity. Then (13.18) simplifies to

$$\begin{aligned}\chi^{>}_{\mu\nu}(\mathbf{x},t;\mathbf{x}',t') &= -\mathrm{i}\,\langle\Psi_0|\Delta\hat{J}_\mu(\mathbf{x})e^{-\mathrm{i}(\hat{H}-E_0)(t-t')}\Delta\hat{J}_\nu(\mathbf{x}')|\Psi_0\rangle\\ &= -\mathrm{i}\sum_b e^{-\mathrm{i}(E_b-E_0)(t-t')}f_{\mu,b}(\mathbf{x})f^*_{\nu,b}(\mathbf{x}'),\end{aligned} \tag{13.24}$$

[7] This identity for $\chi_{00} = \chi$ was already proved in Section 10.4 using the diagrammatic expansion.

where in the second equality we insert the completeness relation $\sum_b |\Psi_b\rangle\langle\Psi_b| = \hat{\mathbb{1}}$ and define the so-called *excitation amplitudes*,

$$\boxed{f_{\mu,b}(\mathbf{x}) \equiv \langle\Psi_0|\Delta\hat{J}_\mu(\mathbf{x})|\Psi_b\rangle} \tag{13.25}$$

The excitation amplitudes $f_{\mu,b}$ vanish if the number of particles in the excited state $|\Psi_b\rangle$ differs from the number of particles in the ground state $|\Psi_0\rangle$. From (13.24) and the like for $\chi^<_{\mu\nu}$ we can readily extract the Fourier transforms:

$$\chi^>_{\mu\nu}(\mathbf{x},\mathbf{x}';\omega) = -2\pi\mathrm{i}\sum_b \delta(\omega-\Omega_b) f_{\mu,b}(\mathbf{x}) f^*_{\nu,b}(\mathbf{x}') = -[\chi^>_{\nu\mu}(\mathbf{x}',\mathbf{x};\omega)]^* \, ,$$

$$\chi^<_{\mu\nu}(\mathbf{x},\mathbf{x}';\omega) = -2\pi\mathrm{i}\sum_b \delta(\omega+\Omega_b) f^*_{\mu,b}(\mathbf{x}) f_{\nu,b}(\mathbf{x}') = -[\chi^<_{\nu\mu}(\mathbf{x}',\mathbf{x};\omega)]^* \, ,$$

with $\Omega_b = E_b - E_0 > 0$ the excitation energies. Substituting these expressions into (13.22), we find the *Lehmann representation* of the retarded response function:

$$\chi^{\mathrm{R}}_{\mu\nu}(\mathbf{x},\mathbf{x}';\omega) = \int db \left[\frac{f_{\mu,b}(\mathbf{x}) f^*_{\nu,b}(\mathbf{x}')}{\omega-\Omega_b+\mathrm{i}\eta} - \frac{f^*_{\mu,b}(\mathbf{x}) f_{\nu,b}(\mathbf{x}')}{\omega+\Omega_b+\mathrm{i}\eta}\right] = [\chi^{\mathrm{A}}_{\nu\mu}(\mathbf{x}',\mathbf{x};\omega)]^* \, . \tag{13.26}$$

The Lehmann representation allows us to understand what physical information is contained in χ^{R}. In the remainder of this section we explore and discuss this result.

13.2.2 Analytic Structure

For real frequencies ω the retarded response function (13.26) has the property

$$\boxed{\chi^{\mathrm{R}}_{\mu\nu}(\mathbf{x},\mathbf{x}';-\omega) = \chi^{\mathrm{R}}_{\mu\nu}(\mathbf{x},\mathbf{x}';\omega)^*} \tag{13.27}$$

which implies that the real part is an even function of ω, whereas the imaginary part is an odd function of ω.[8] For complex ω we see from (13.26) that χ^{R} is analytic in the upper half-plane. This analyticity property together with the fact that $\chi^{\mathrm{R}}_{\mu\nu} \to 0$ for large ω guarantees that $\chi^{\mathrm{R}}(\mathbf{x},t;\mathbf{x}',t')$ vanishes for t' larger than t due to the Cauchy residue theorem. The densities $\delta J_\mu(\mathbf{x},t)$ are sensitive to changes in the potentials $\delta A^\nu(\mathbf{x}',t')$ only if these changes occur at times $t' < t$. We can alternatively say that χ^{R} has the *causality property* or that χ^{R} is causal. In the lower half-plane, instead, χ^{R} has either simple poles in $\pm\Omega_b - \mathrm{i}\eta$ when b is a discrete quantum number or branch cuts along the real axis when b is a continuum quantum number. Consequently the plot of the imaginary part of χ^{R} as a function of the real frequency ω exhibits δ-like peaks at the discrete excitation energies, and it is otherwise a smooth curve in the continuum of excitations and zero everywhere else.

For positive frequencies, the Lehmann representation (13.26) yields

$$\mathrm{Im}\left[\chi^{\mathrm{R}}_{\mu\nu}(\mathbf{x},\mathbf{x}';\omega)\right] = -\pi\sum_b f_{\mu,b}(\mathbf{x}) f^*_{\nu,b}(\mathbf{x}')\delta(\omega-\Omega_b), \qquad \omega > 0.$$

[8] This property is also a direct consequence of the fact that $\chi^{\mathrm{R}}_{\mu\nu}(\mathbf{x},t;\mathbf{x}',t')$ is a real function.

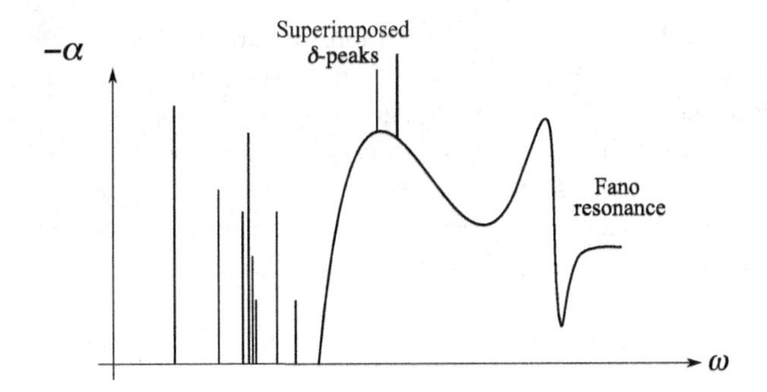

Figure 13.1 General form of the function $\alpha(\omega)$ defined in (13.28).

If we think of $\text{Im}\left[\chi^{\text{R}}_{\mu\nu}\right]$ as a matrix with indices $(\mu\mathbf{x}, \nu\mathbf{x}')$, the above result implies that this matrix is negative definite for $\omega > 0$ since

$$\alpha(\omega) \equiv \sum_{\mu\nu} \int d\mathbf{x}d\mathbf{x}'\alpha^*_\mu(\mathbf{x})\,\text{Im}\left[\chi^{\text{R}}_{\mu\nu}(\mathbf{x}, \mathbf{x}'; \omega)\right] \alpha_\nu(\mathbf{x}') < 0, \tag{13.28}$$

for any set of complex functions $\alpha_\mu(\mathbf{x})$. The general form of the function $\alpha(\omega)$ looks like that shown in Fig. 13.1, where the sharp vertical lines represent δ-like peaks. In principle the δ-like peaks can also be superimposed on the continuum; this is actually a common feature of noninteracting systems. In interacting systems the discrete and continuum non-interacting excitations are coupled by the interaction (unless prohibited by symmetry) and the superimposed δ-like peaks transform into resonances with a characteristic asymmetric lineshape. This effect was pointed out by Fano [186] and is pictorially illustrated in Fig. 13.1. In general the continuum part can have different shapes depending on the nature of the elementary excitations. These excitations are all charge-neutral (or particle-conserving) since the energies E_b that contribute to (13.26) correspond to eigenstates with the same number of particles as in the ground state. This should be contrasted with the excitations of the Green's function, which are not charge-conserving since they involve eigenstates with one particle more or less, see Section 6.3.1.

Discrete excitations and time evolution From the Kubo formula (13.14) we can establish a very nice link between the physics contained in $\chi^{\text{R}}_{\mu\nu}$ and the results of a time evolution. This link is particularly relevant when there are discrete excitations in the spectrum. We therefore analyze the contribution to $\chi^{\text{R}}_{\mu\nu}$ coming from the discrete excitations $b = j$ in detail. To be concrete let us consider the density response function $\chi_{00} = \chi$. Substitution of (13.26) into (13.21) yields

$$\chi^{\text{R}}(\mathbf{x}, t; \mathbf{x}', t') = -\text{i}\theta(t - t') \sum_j \left[e^{-\text{i}\Omega_j(t-t')} f_j(\mathbf{x}) f^*_j(\mathbf{x}') - e^{\text{i}\Omega_j(t-t')} f^*_j(\mathbf{x}) f_j(\mathbf{x}')\right]$$
$$+ \chi^{\text{R}}_{\text{cont}}(\mathbf{x}, t; \mathbf{x}', t'), \tag{13.29}$$

where we drop the subscript 0 in the excitation amplitudes and where $\Omega_j = E_j - E_0$ are the discrete excitation energies. The last term on the r.h.s. of (13.29) is the part of the response function that comes from the integral over the continuum of excitations. If the vector potential $\mathbf{A}(\mathbf{x})$ (of the equilibrium system) vanishes, then the excitation amplitudes can be chosen real. We here assume that this is the case - that is, $f_j = f_j^*$, even though the general conclusion remains valid regardless of this simplification. We now perturb the system with, for example, a scalar potential,

$$\delta\phi(\mathbf{x}, t) = \theta(t - t_0) \int \frac{d\omega}{2\pi} e^{-i\omega t} \delta\phi(\mathbf{x}, \omega),$$

and calculate the time-dependent density induced by this perturbation. Without loss of generality, we take $t_0 = 0$. From the Kubo formula (13.14), the first-order change in the density can be written as

$$\delta n(\mathbf{x}, t) = \int \frac{d\omega}{2\pi} \delta n_\omega(\mathbf{x}, t) + \delta n_{\text{cont}}(\mathbf{x}, t), \tag{13.30}$$

with δn_{cont} the contribution due to $\chi^{\mathrm{R}}_{\text{cont}}$ and

$$\delta n_\omega(\mathbf{x}, t) = -\mathrm{i} \sum_j f_j(\mathbf{x}) \delta\phi_j(\omega) \left[e^{-\mathrm{i}\frac{(\omega+\Omega_j)}{2}t} \frac{\sin\left(\frac{\omega-\Omega_j}{2}t\right)}{\frac{\omega-\Omega_j}{2}} - e^{-\mathrm{i}\frac{(\omega-\Omega_j)}{2}t} \frac{\sin\left(\frac{\omega+\Omega_j}{2}t\right)}{\frac{\omega+\Omega_j}{2}} \right],$$

where the quantities $\delta\phi_j(\omega) \equiv \int d\mathbf{x}' \, f_j(\mathbf{x}') \delta\phi(\mathbf{x}', \omega)$. The linear response density $\delta n_\omega(\mathbf{x}, t)$ is peaked at $\omega \simeq \pm\Omega_j$ [recall that $\lim_{t\to\infty} \sin(\Omega t)/\Omega = \pi\delta(\Omega)$] and therefore the density change in (13.30) oscillates with frequencies $\simeq \pm\Omega_j$. A view from a different perspective of this result is that we can extract the particle-conserving excitation energies of the *unperturbed* system from the frequency of the oscillations of the time-dependent density.

13.2.3 The f-sum Rule

From the Lehmann representation (13.26) we can derive an important sum rule for the density response function. The *Thomas–Reiche–Kuhn sum rule* [187, 188, 189], or simply the f-sum rule, relates the first momentum of the retarded density response function χ^{R} of a system in equilibrium to the corresponding equilibrium density. To prove the f-sum rule, we consider the large ω limit of (13.26) with $\mu = \nu = 0$. Dropping, as before, the subscript 0 from the excitation functions, we have

$$\begin{aligned}\chi^{\mathrm{R}}(\mathbf{x}, \mathbf{x}'; \omega) &= \frac{1}{\omega} \sum_b [f_b(\mathbf{x}) f_b^*(\mathbf{x}') - f_b^*(\mathbf{x}) f_b(\mathbf{x}')] \\ &+ \frac{1}{\omega^2} \sum_b \Omega_b [f_b(\mathbf{x}) f_b^*(\mathbf{x}') + f_b^*(\mathbf{x}) f_b(\mathbf{x}')] + \mathcal{O}(\frac{1}{\omega^3}).\end{aligned} \tag{13.31}$$

The definition $f_b(\mathbf{x}) = \langle\Psi_0|\hat{n}(\mathbf{x})|\Psi_b\rangle$ of the excitation functions and the completeness relation $\hat{\mathbb{1}} = \sum_b |\Psi_b\rangle\langle\Psi_b|$ allows us to recognize in the first term on the r.h.s. the ground-state average of the commutator $[\hat{n}(\mathbf{x}), \hat{n}(\mathbf{x}')]_-$, which is zero. Similarly, recalling that

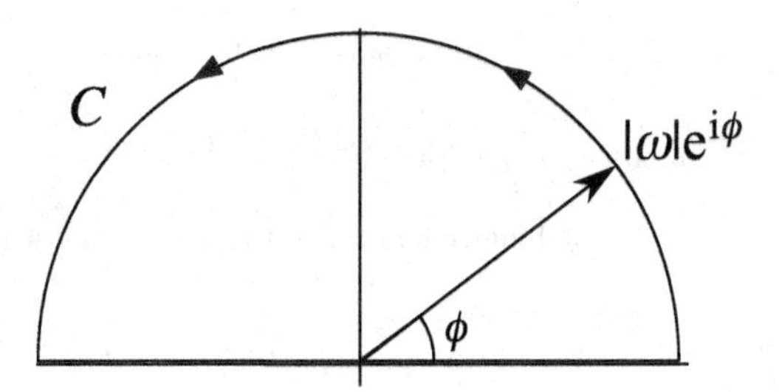

Figure 13.2 The closed contour C going from $-\infty$ to $+\infty$ along a line just above the real axis and from $+\infty$ to $-\infty$ along a semicircle of infinite radius in the upper half-plane.

$\Omega_b = E_b - E_0$, the second term on the r.h.s. of (13.31) can be written as the ground-state average of a double commutator:

$$\begin{aligned}\chi^{\mathrm{R}}(\mathbf{x},\mathbf{x}';\omega) &= \frac{1}{\omega^2}\langle\Psi_0|\big[[\hat{n}(\mathbf{x}),\hat{H}]_-,\hat{n}(\mathbf{x}')\big]_-|\Psi_0\rangle + \mathcal{O}(\frac{1}{\omega^3}) \\ &= \frac{1}{\omega^2}\left[-\frac{1}{m}\boldsymbol{\nabla}\cdot\big(n(\mathbf{x})\boldsymbol{\nabla}\delta(\mathbf{x}-\mathbf{x}')\big)\right] + \mathcal{O}(\frac{1}{\omega^3}).\end{aligned} \tag{13.32}$$

In the second equality we use that $\big[\hat{n}(\mathbf{x}),\hat{H}\big]_- = -\mathrm{i}\boldsymbol{\nabla}\cdot\hat{\mathbf{J}}(\mathbf{x})$ and that the commutator between the current density and the density is (see Exercise 3.5)

$$\big[\hat{\mathbf{J}}(\mathbf{x}),\hat{n}(\mathbf{x}')\big]_- = -\frac{\mathrm{i}}{m}\hat{n}(\mathbf{x})\boldsymbol{\nabla}\delta(\mathbf{x}-\mathbf{x}'). \tag{13.33}$$

Since χ^{R} is analytic in the upper half of the complex ω-plane, the integral of $\omega\chi^{\mathrm{R}}$ along the contour C of Fig. 13.2 is zero. Then we can write

$$0 = \oint_C d\omega\,\omega\chi^{\mathrm{R}}(\mathbf{x},\mathbf{x}';\omega) = \int_{-\infty}^{\infty} d\omega\,\omega\chi^{\mathrm{R}}(\mathbf{x},\mathbf{x}';\omega) + \mathrm{i}\lim_{|\omega|\to\infty}\int_0^{\pi} d\phi\,|\omega|^2 e^{2\mathrm{i}\phi}\chi^{\mathrm{R}}(\mathbf{x},\mathbf{x}';|\omega|e^{\mathrm{i}\phi}).$$

Using (13.32) to evaluate the integral over ϕ, we obtain a relation between the first momentum of χ^{R} and the equilibrium density

$$\boxed{\int_{-\infty}^{\infty} d\omega\,\omega\,\mathrm{Im}[\chi^{\mathrm{R}}(\mathbf{x},\mathbf{x}';\omega)] = \frac{\pi}{m}\boldsymbol{\nabla}\cdot\big(n(\mathbf{x})\boldsymbol{\nabla}\delta(\mathbf{x}-\mathbf{x}')\big)} \tag{13.34}$$

where we take into account that only the imaginary part contributes to the integral since $\mathrm{Re}[\chi^{\mathrm{R}}]$ is even in ω. Equation (13.34) is known as the frequency- or f-sum rule for χ^{R}. It is easy to show that the f-sum rule is valid also at finite temperature.

The f-sum rule is alternatively written in terms of the density operator in momentum space,

$$\hat{n}_{\mathbf{p}\sigma} \equiv \int d\mathbf{r}\,e^{-\mathrm{i}\mathbf{p}\cdot\mathbf{r}}\,\hat{n}(\mathbf{x}) \qquad \Rightarrow \qquad \hat{n}(\mathbf{x}) = \int \frac{d\mathbf{p}}{(2\pi)^3} e^{\mathrm{i}\mathbf{p}\cdot\mathbf{r}}\,\hat{n}_{\mathbf{p}\sigma}.$$

The average of $\hat{n}_{\mathbf{p}\sigma}$ is simply the Fourier transform of the density $n(\mathbf{x})$. The correlator between two of these operators is defined as

$$\Xi_{\sigma\sigma'}(\mathbf{p},z,z') \equiv -\mathrm{i}\langle\mathcal{T}\{\hat{n}_{\mathbf{p}\sigma,H}(z)\hat{n}_{-\mathbf{p}\sigma',H}(z')\}\rangle.$$

The retarded component is directly related to the retarded component of the density response function via[9]

$$\Xi^{\mathrm{R}}_{\sigma\sigma'}(\mathbf{p},\omega) = \int d\mathbf{r}d\mathbf{r}'\, e^{-\mathrm{i}\mathbf{p}\cdot(\mathbf{r}-\mathbf{r}')}\chi^{\mathrm{R}}(\mathbf{x},\mathbf{x}';\omega). \tag{13.35}$$

Multiplying (13.34) by $e^{-\mathrm{i}\mathbf{p}\cdot(\mathbf{r}-\mathbf{r}')}$ and integrating over all $\mathbf{r}$ and $\mathbf{r}'$, we then find a sum rule for $\Xi^{\mathrm{R}}_{\sigma\sigma'}$:

$$\int_{-\infty}^{\infty} d\omega\, \omega\, \Xi^{\mathrm{R}}_{\sigma\sigma'}(\mathbf{p},\omega) = -\mathrm{i}\delta_{\sigma\sigma'}\frac{\pi p^2}{m}\int d\mathbf{r}\, n(\mathbf{x}) = -\mathrm{i}\delta_{\sigma\sigma'}\frac{\pi p^2}{m}N_\sigma,$$

with N_σ the total number of particles of spin σ.

Exercise 13.1 Show that the response function (13.26) can also be written as

$$\chi^{\mathrm{R}}_{\mu\nu}(\mathbf{x},\mathbf{x}';\omega) = \langle\Psi_0|\Delta\hat{J}_\mu(\mathbf{x})\frac{1}{\omega-\hat{H}+E_0+\mathrm{i}\eta}\Delta\hat{J}_\nu(\mathbf{x}')|\Psi_0\rangle$$
$$- \langle\Psi_0|\Delta\hat{J}_\nu(\mathbf{x}')\frac{1}{\omega+\hat{H}-E_0+\mathrm{i}\eta}\Delta\hat{J}_\mu(\mathbf{x})|\Psi_0\rangle.$$

Exercise 13.2 Consider a system for which the action of the Hamiltonian over a complete set of orthonormal states with N particles is

$$\hat{H}|\Psi_0\rangle = E_0|\Psi_0\rangle,$$
$$\hat{H}|\Psi_1\rangle = E_1|\Psi_1\rangle + \int_I^\infty dE\, T_E|\Psi_E\rangle,$$
$$\hat{H}|\Psi_E\rangle = E|\Psi_E\rangle + T^*_E|E_1\rangle,$$

where $|\Psi_0\rangle$ and $|\Psi_1\rangle$ are normalized to unity and $\langle\Psi_E|\Psi_{E'}\rangle = \delta(E-E')$. Assuming that $E_1 > I$, calculate the retarded density–density response function $\chi^{\mathrm{R}}_{00}(\mathbf{x},\mathbf{x}';\omega)$ and study its behavior for ω close to E_1.

Exercise 13.3 In a general one-particle orthonormal basis we define the response function at finite temperature as

$$\chi^{\mathrm{R}}(i,j;\omega) = \sum_a \rho_a\Big[\langle\Psi_a|\Delta\hat{n}_i\frac{1}{\omega-\hat{H}+E_0+\mathrm{i}\eta}\Delta\hat{n}_j|\Psi_a\rangle$$
$$- \langle\Psi_a|\Delta\hat{n}_j\frac{1}{\omega+\hat{H}-E_0+\mathrm{i}\eta}\Delta\hat{n}_i|\Psi_a\rangle\Big],$$

where $\rho_a = e^{-\beta(E_a-\mu N_a)}/\mathrm{Tr}[e^{-\beta(E_a-\mu N_a)}]$, $\hat{n}_i = \sum_\sigma \hat{d}^\dagger_{i\sigma}\hat{d}_{i\sigma}$, and $\Delta\hat{n}_i = \hat{n}_i - \langle\hat{n}_i\rangle$. Calculate $\chi^{\mathrm{R}}(i,j;\omega)$ for the Hamiltonian of the Hubbard dimer, see (2.9), and show that the imaginary part is negative definite for positive frequencies.

---

[9] For a system invariant under translations, the sum over σ,σ' of the r.h.s. of (13.35) yields $\mathrm{V}\chi^{\mathrm{R}}(\mathbf{p},\omega)$, where $\mathrm{V} = \int d\mathbf{r}$ is the volume of the system and $\chi^{\mathrm{R}}(\mathbf{p},\omega)$ is the response function in momentum space, see (13.71).

13.3 Noninteracting Response Functions for Stationary States

We now discuss the response functions of a system of noninteracting particles prepared in a stationary ensemble. There are three motivations for choosing this special case. First, we can acquire some more familiarity with the physics contained in the response function. Second, the noninteracting χ, henceforth denoted by χ^0, is needed to calculate the interacting χ from the Bethe–Salpeter equation, see Section 14.7. And third, it is possible to derive an important analytic formula for χ^0 which is used later on in our examples.

From Wick's theorem in Section 5.2 we know that the two-particle Green's function of a noninteracting system is $G_{0,2}(1,2;1',2') = G_0(1;1')G_0(2;2') \pm G_0(1;2')G_0(2;1')$. Therefore the *noninteracting* two-particle XC function in (13.15) simplifies to

$$L^0(1,2;1',2') = G_0(1;2')G_0(2;1'). \tag{13.36}$$

Let $\hat{d}_n$, $\hat{d}_n^\dagger$ be the annihilation and creation operators that diagonalize $\hat{H} = \hat{H}_0 = \sum_n \epsilon_n \hat{d}_n^\dagger \hat{d}_n$. Expanding the field operators as in (1.60), we can rewrite G_0 as

$$G_0(1;2') = \sum_{nm} \varphi_n(\mathbf{x}_1)\varphi_m^*(\mathbf{x}_2') g_{nm}(z_1, z_2'). \tag{13.37}$$

In the r.h.s., g_{nm} is the matrix element $\langle n|\hat{\mathcal{G}}|m\rangle$, where $\hat{\mathcal{G}}$ is the noninteracting Green's function operator calculated in (6.35) and $|n\rangle = \hat{d}_n^\dagger|0\rangle$. For stationary states the Matsubara Hamiltonian $\hat{h}^{\mathrm{M}}$ commutes with the stationary Hamiltonian $\hat{h}$. The kets $|n\rangle$ are therefore the common eigenvectors of $\hat{h}^{\mathrm{M}}$ and $\hat{h}$ with eigenvalues ϵ_n^{M} and ϵ_n, respectively, and we have

$$g_{nm}(z_1, z_2') = \delta_{nm} g_n(z_1, z_2'), \tag{13.38}$$

where

$$g_n(z_1, z_2') = -\mathrm{i}\left[\theta(z_1, z_2')\bar{f}(\epsilon_n^{\mathrm{M}}) \pm \theta(z_2', z_1) f(\epsilon_n^{\mathrm{M}})\right] e^{-\mathrm{i}\int_{z_2'}^{z_1} d\bar{z}\, \epsilon_n(\bar{z})}, \tag{13.39}$$

and $\epsilon_n(t_0 - \mathrm{i}\tau) = \epsilon_n^{\mathrm{M}}$ while $\epsilon_n(t_\pm) = \epsilon_n$.

We insert the expansion (13.37) into (13.36) to rewrite the two-particle XC function with $z_1' = z_1^+$ and $z_2' = z_2^+$ as

$$L^0(\mathbf{x}_1, z_1, \mathbf{x}_2, z_2; \mathbf{x}_1', z_1^+, \mathbf{x}_2', z_2^+) = \sum_{\substack{ij\\mn}} L^0_{\substack{ij\\mn}}(z_1, z_2)\varphi_i(\mathbf{x}_1)\varphi_j^*(\mathbf{x}_1')\varphi_m(\mathbf{x}_2)\varphi_n^*(\mathbf{x}_2'), \tag{13.40}$$

where $L^0_{\substack{ij\\mn}}(z_1, z_2) = \delta_{in}\delta_{jm}\ell_{nm}(z_1, z_2)$ and

$$\ell_{nm}(z_1, z_2) \equiv g_n(z_1, z_2) g_m(z_2, z_1). \tag{13.41}$$

The noninteracting response function $\chi^0_{\mu\nu}$ can be expressed in terms of ℓ_{nm}. We define the functions $f_{\mu,mn}(\mathbf{x}_1) = f^*_{\mu,nm}(\mathbf{x}_1)$ according to

$$f_{0,mn}(\mathbf{x}_1) = \varphi_n(\mathbf{x}_1)\varphi_m^*(\mathbf{x}_1), \qquad f_{k,mn}(\mathbf{x}_1) = \left(\frac{D_{1,k} - D^*_{1',k}}{2m\mathrm{i}}\right)\varphi_n(\mathbf{x}_1)\varphi_m^*(\mathbf{x}_1')\bigg|_{\mathbf{x}_1 = \mathbf{x}_1'}. \tag{13.42}$$

Using the identities in (13.17) we find

$$\chi^0_{\mu\nu}(1;2) = \pm\mathrm{i}\sum_{\substack{ij\\mn}} L^0_{\substack{ij\\mn}}(z_1,z_2) f_{\mu,ji}(\mathbf{x}_1) f^*_{\nu,mn}(\mathbf{x}_2)$$
$$= \pm\mathrm{i}\sum_{nm} \ell_{nm}(z_1,z_2) f_{\mu,mn}(\mathbf{x}_1) f^*_{\nu,mn}(\mathbf{x}_2). \tag{13.43}$$

We see below that the functions f_μ are precisely the excitation amplitudes (13.25) of noninteracting systems. Let us derive the lesser, greater, retarded, and advanced components of the Keldysh function ℓ_{nm}. We work directly in frequency space as the system is invariant under time translations.

The lesser and greater components of the Green's function g_n follow from (6.45) and (6.46):

$$g^<_n(\omega) = \mp 2\pi i f_n \delta(\omega - \epsilon_n), \qquad f_n \equiv f(\epsilon^{\mathrm{M}}_n), \tag{13.44a}$$
$$g^>_n(\omega) = -2\pi i \bar{f}_n \delta(\omega - \epsilon_n), \qquad \bar{f}_n \equiv \bar{f}(\epsilon^{\mathrm{M}}_n). \tag{13.44b}$$

The function ℓ is a product of functions in Keldysh space; using the Langreth rules in Table 5.1 we find

$$\ell^>_{nm}(\omega) = \int \frac{d\omega'}{2\pi} g^>_n(\omega+\omega') g^<_m(\omega')$$
$$= \mp 2\pi \bar{f}_n f_m \delta(\omega - \epsilon_n + \epsilon_m), \tag{13.45}$$

and similarly

$$\ell^<_{nm}(\omega) = \mp 2\pi f_n \bar{f}_m \delta(\omega - \epsilon_n + \epsilon_m). \tag{13.46}$$

Therefore,

$$\ell^{\mathrm{R/A}}_{nm}(\omega) = \mathrm{i}\int \frac{d\omega'}{2\pi} \frac{\ell^>_{nm}(\omega') - \ell^<_{nm}(\omega')}{\omega - \omega' \pm i\eta}$$
$$= \mp\mathrm{i}\frac{f_m - f_n}{\omega - \epsilon_n + \epsilon_m \pm \mathrm{i}\eta}, \tag{13.47}$$

where we take into account that $\bar{f}_n f_m - f_n \bar{f}_m = f_m - f_n$ for both fermions and bosons.

Derivation based on the Lehmann representation It is instructive to derive the same result using the Lehmann representation. For simplicity we consider a system of fermions in thermal equilibrium at zero temperature. In this case the retarded response function is given by (13.26). For a given chemical potential μ the ground state $|\Psi_0\rangle$ has all levels with energy smaller than μ occupied, whereas those with energy larger than μ are empty. The charge-neutral excited states $|\Psi_b\rangle$ (with the same number of fermions as in $|\Psi_0\rangle$) are obtained by "moving" one, two, three, etc. fermions from the occupied ground-state levels to some empty levels or, in other words, by creating *electron–hole pairs.* If we introduce the convention of labeling the occupied levels with indices n and the unoccupied levels with barred indices $\bar{n}$, then a generic charge-neutral excited state has the form

$$|\Psi_b\rangle = |\Psi_{\bar{n}_1\ldots\bar{n}_N n_1\ldots n_N}\rangle = \hat{d}^\dagger_{\bar{n}_1}\ldots\hat{d}^\dagger_{\bar{n}_N}\hat{d}_{n_1}\ldots\hat{d}_{n_N}|\Psi_0\rangle. \tag{13.48}$$

This state describes a system in which N fermions have been excited from their original levels $n_1, \ldots, n_N$ to the empty levels $\bar{n}_1, \ldots, \bar{n}_N$ and its energy is

$$E_b = E_0 + \sum_{j=1}^{N} (\epsilon_{\bar{n}_j} - \epsilon_{n_j}),$$

with E_0 the ground-state energy. The excitation amplitude $f_{\mu,b}$ is the matrix element of the one-body operator $\Delta \hat{J}_\mu$ between $|\Psi_0\rangle$ and $|\Psi_b\rangle$. It is then clear that the excitation amplitudes vanish if b contains more than one electron–hole pair. Consequently the zero-temperature response function (13.26) for a system of noninteracting fermions simplifies to

$$\chi^{0,\mathrm{R}}_{\mu\nu}(\mathbf{x}, \mathbf{x}'; \omega) = \sum_{n\bar{n}} \left[\frac{f_{\mu,n\bar{n}}(\mathbf{x}) f^*_{\nu,n\bar{n}}(\mathbf{x}')}{\omega - (\epsilon_{\bar{n}} - \epsilon_n) + \mathrm{i}\eta} - \frac{f^*_{\mu,n\bar{n}}(\mathbf{x}) f_{\nu,n\bar{n}}(\mathbf{x}')}{\omega + (\epsilon_{\bar{n}} - \epsilon_n) + \mathrm{i}\eta} \right], \tag{13.49}$$

where only excitations with one electron–hole pair contribute. Taking into account that the Fermi function $f(\epsilon^{\mathrm{M}}_n) = f_n$ is zero for $\epsilon^{\mathrm{M}}_n > 0$ and unity otherwise, we can rewrite $\chi^{0,\mathrm{R}}$ in terms of an unconstrained sum over all one-particle levels:

$$\chi^{0,\mathrm{R}}_{\mu\nu}(\mathbf{x}, \mathbf{x}'; \omega) = \sum_{nm} \bar{f}_n f_m \left[\frac{f_{\mu,mn}(\mathbf{x}) f^*_{\nu,mn}(\mathbf{x}')}{\omega - (\epsilon_n - \epsilon_m) + \mathrm{i}\eta} - \frac{f^*_{\mu,mn}(\mathbf{x}) f_{\nu,mn}(\mathbf{x}')}{\omega + (\epsilon_n - \epsilon_m) + \mathrm{i}\eta} \right]. \tag{13.50}$$

The excitation amplitude for the density response function

$$\begin{aligned} f_{0,mn}(\mathbf{x}) &\equiv \langle \Psi_0 | \Delta \hat{n}(\mathbf{x}) | \Psi_{nm} \rangle = \langle \Psi_0 | \hat{n}(\mathbf{x}) | \Psi_{nm} \rangle = \langle \Psi_0 | \hat{\psi}^\dagger(\mathbf{x}) \hat{\psi}(\mathbf{x}) \hat{d}^\dagger_n \hat{d}_m | \Psi_0 \rangle \\ &= \langle \mathbf{x} | n \rangle \langle m | \mathbf{x} \rangle = \varphi_n(\mathbf{x}) \varphi^*_m(\mathbf{x}) \end{aligned} \tag{13.51}$$

coincides with the function $f_{0,mn}(\mathbf{x})$ in (13.42). Similarly one can show that the excitation amplitudes $f_{k,mn}(\mathbf{x}) \equiv \langle \Psi_0 | \Delta \hat{J}_k(\mathbf{x}) | \Psi_{nm} \rangle$ coincide with the functions $f_{k,mn}(\mathbf{x})$ in (13.42). The second term in the square bracket of (13.50) is the opposite of the first term with $n \leftrightarrow m$. We can then keep only the first term and replace $\bar{f}_n f_m$ with $\bar{f}_n f_m - \bar{f}_n f_m = f_m - f_n$, yielding

$$\chi^{0,\mathrm{R}}_{\mu\nu}(\mathbf{x}, \mathbf{x}'; \omega) = \sum_{nm} f_{\mu,mn}(\mathbf{x}) f^*_{\nu,mn}(\mathbf{x}') \frac{f_m - f_n}{\omega - (\epsilon_n - \epsilon_m) + \mathrm{i}\eta}. \tag{13.52}$$

This result agrees with (13.43) and (13.47).

The derivation based on the Lehmann representation makes us appreciate the advantage of working with Green's functions. In fact, the analogous derivation for a system at finite temperature or in a stationary state would be much more laborious. The response function (13.43) with ℓ from (13.47) generalizes result (13.52) to arbitrary stationary ensembles and is valid for fermions as well as bosons.

Multiple excitations We have said that in the noninteracting case only excited states with one electron–hole pair contribute to the response functions. We can try to understand how and why things change in the interacting case at least for small interactions. If the interaction is weak, then the interacting ground state $|\Psi_0\rangle$ differs from the noninteracting ground state by a small correction. This correction can be written as a linear combination of the states in (13.48). Similarly, the interacting excited eigenstates $|\Psi_b\rangle$ have a dominant term of the form (13.48) plus a small correction; we say that the interacting $|\Psi_b\rangle$ is a *double excited state* if the dominant term of the form (13.48) has $N = 2$, a *triple excited state* if the dominant term has $N = 3$, and so on. In this complicated situation the excitation amplitude $f_{\mu,b}$ - with b a double excitation, a triple excitation, etc. - is generally nonzero. For instance, if $|\Psi_0\rangle$ has a component proportional to $|\Psi_{\bar{n}n}\rangle$, then this component contributes to $f_{0,b}(\mathbf{x})$ with the overlap $\langle\Psi_{\bar{n}n}|\Delta\hat{n}(\mathbf{x})|\Psi_b\rangle$, and this overlap is large when the dominant term of $|\Psi_b\rangle$ has the form $|\Psi_{\bar{n}_1\bar{n}_2n_1n_2}\rangle$ - that is, when b is a double excitation. The interacting response function changes by developing new poles in the discrete part of the spectrum and/or by modifying the continuum lineshape, see Section 14.11.

Exercise 13.4 Calculate the noninteracting retarded density–density response function of a one-dimensional system with single-particle Hamiltonian $\hat{h} = \frac{\hat{p}^2}{2m} + q\phi(\hat{x})$, with $\phi(x) = 0$ for $x \in (0, L)$ and $\phi(x) = \infty$ otherwise, at zero temperature and chemical potential $\mu = \frac{\pi^2}{mL^2}$.

Exercise 13.5 Consider the system of Exercise 13.4 and let $|\varphi_n\rangle$ be the eigenkets of $\hat{h}$. At time $t = 0$ the system is perturbed by the Hamiltonian $\lambda\hat{H}'(t) = \lambda(\hat{d}_1^\dagger\hat{d}_2 + \hat{d}_2^\dagger\hat{d}_1)$ where $\hat{d}_n = \int dx\langle\varphi_n|x\rangle\hat{\psi}(x)$. Calculate the linear response change in the density and compare it with the exact solution.

13.4 Time-Resolved Photoabsorption Spectroscopy

The response functions are fundamental quantities to interpret the results of a time-resolved (or transient) photoabsorption experiment. In this section we develop the theory of time-resolved photoabsorption spectroscopy for finite systems such as atoms, molecules, and nanostructures [190, 191, 192, 193, 194, 195] and refer the reader to fundamental papers [185, 196, 197, 198] and introductory books [199, 200, 201] for extended systems like crystals.

We consider the situation of Fig. 13.3. A finite system is driven out of equilibrium by an intense laser pulse described by the electromagnetic fields $\mathbf{E}(t)$ and $\mathbf{B}(t)$, and subsequently probed with the electromagnetic fields $\mathbf{e}(t)$ and $\mathbf{b}(t)$. We refer to $\mathbf{E}(t)$ and $\mathbf{B}(t)$ as the *pump* fields and to $\mathbf{e}(t)$ and $\mathbf{b}(t)$ as the *probe* fields. To extract as much information as possible, a spectrometer placed between the system and the detector splits the transmitted fields $\mathbf{e}'(t)$ and $\mathbf{b}'(t)$ into two halves, and it generates a tunable time-shift δ for one of the halves. The resulting electric field at the detector is therefore $\mathbf{e}'_{\text{detector}}(t) = \frac{1}{2}(\mathbf{e}'(t) + \mathbf{e}'(t-\delta))$, and likewise for the magnetic field. The detector measures the total transmitted energy $E_{\text{trans}}(\delta)$. This quantity is given by the Poynting vector integrated over time (the duration of the experiment) and surface. Denoting by S the cross section of the detector, we have (here and in the following integrals with no upper and lower limits extend from $-\infty$ to $+\infty$)

$$E_{\text{trans}}(\delta) = S\frac{c}{4\pi}\int dt\,|\mathbf{e}'_{\text{detector}}(t) \times \mathbf{b}'_{\text{detector}}(t)|\,. \tag{13.53}$$

The integral in (13.53) is finite since the probe pulse has a finite duration. In vacuum, the electric and magnetic fields are perpendicular to each other and their cross product is

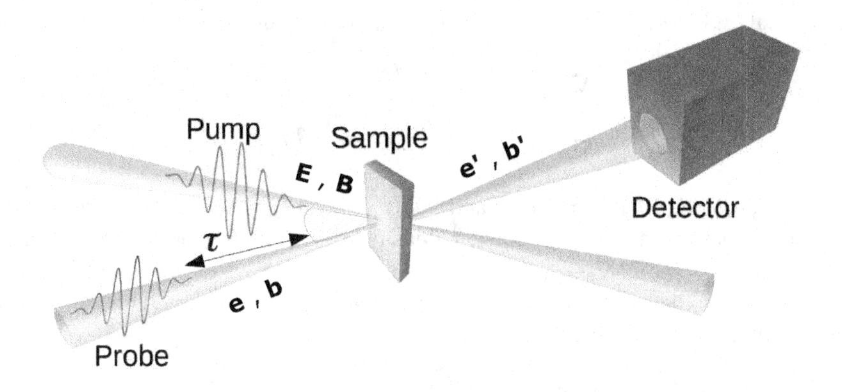

Figure 13.3 Experimental setup for a time-resolved photoabsorption experiment. A pump pulse drives the system out of equilibrium. After a tunable delay a probe pulse is send on the same system and the intensity of the transmitted probe field is measured.

parallel to the direction of propagation. Taking into account that $|\mathbf{b}'| = |\mathbf{e}'|$, the transmitted energy in (13.53) simplifies to

$$E_{\text{trans}}(\delta) = S\frac{c}{4\pi}\int dt\, |\mathbf{e}'_{\text{detector}}(t)|^2 = S\frac{c}{4\pi}\int dt \left|\frac{\mathbf{e}'(t) + \mathbf{e}'(t-\delta)}{2}\right|^2. \tag{13.54}$$

This result can alternatively be written in terms of the Fourier transform of the transmitted field $\mathbf{e}'(t) = \int \frac{d\omega}{2\pi}\, \tilde{\mathbf{e}}'(\omega)e^{-\mathrm{i}\omega t}$, where $\tilde{\mathbf{e}}'^{*}(\omega) = \tilde{\mathbf{e}}'(-\omega)$ since the field is real. We find

$$E_{\text{trans}}(\delta) = S\frac{c}{4\pi}\int \frac{d\omega}{2\pi}\, |\tilde{\mathbf{e}}'(\omega)|^2\, \frac{1+\cos(\omega\delta)}{2}. \tag{13.55}$$

The photoabsorption experiment is repeated for different time-shifts δ, and the results are collected to perform a cosine transform:

$$\tilde{E}_{\text{trans}}(\omega) \equiv \int d\delta\; E_{\text{trans}}(\delta)\cos(\omega\delta). \tag{13.56}$$

Inserting (13.55) into (13.56) and taking into account the identity $\int d\delta \cos(\omega\delta)\cos(\omega'\delta) = \pi\left[\delta(\omega+\omega') + \delta(\omega-\omega')\right]$, we find

$$\tilde{E}_{\text{trans}}(\omega) = 2\pi E_{\text{trans}}(\delta = 0)\delta(\omega) + S\frac{c}{8\pi}\,|\tilde{\mathbf{e}}'(\omega)|^2. \tag{13.57}$$

To interpret the experimental results, we must therefore relate the transmitted probe pulse to the microscopic properties of the system.

Relation between the transmitted field and the dipole moment In a nonmagnetic medium the *total* macroscopic electric and magnetic fields $\boldsymbol{\mathcal{E}}_{\text{tot}}$ and $\boldsymbol{\mathcal{B}}_{\text{tot}}$ – that is, the sum of the external and induced fields - satisfy the Maxwell equations

$$\nabla \times \boldsymbol{\mathcal{E}}_{\text{tot}} = -\frac{1}{c}\frac{\partial \boldsymbol{\mathcal{B}}_{\text{tot}}}{\partial t},$$
$$\nabla \times \boldsymbol{\mathcal{B}}_{\text{tot}} = \frac{1}{c}\frac{\partial \boldsymbol{\mathcal{E}}_{\text{tot}}}{\partial t} + \frac{4\pi q}{c}\langle \mathbf{J}\rangle,$$

where q is the charge of the particles and $\langle \mathbf{J}\rangle(\mathbf{r},t)$ is the macroscopic current density - that is, the spatial average of the current density over a small volume around $\mathbf{r}$. Using that $\nabla \times \nabla \times \boldsymbol{\mathcal{E}}_{\text{tot}} = \nabla(\nabla \cdot \boldsymbol{\mathcal{E}}_{\text{tot}}) - \nabla^2 \boldsymbol{\mathcal{E}}_{\text{tot}}$ and that the system is charge-neutral (in a macroscopic sense), hence $\nabla \cdot \boldsymbol{\mathcal{E}}_{\text{tot}} = 0$, we can combine the Maxwell equations to obtain a wave equation for the electric field:

$$\nabla^2 \boldsymbol{\mathcal{E}}_{\text{tot}} - \frac{1}{c^2}\frac{\partial^2 \boldsymbol{\mathcal{E}}_{\text{tot}}}{\partial t^2} = \frac{4\pi}{c^2} q \frac{\partial \langle \mathbf{J}\rangle}{\partial t}. \tag{13.58}$$

Let $\langle \mathbf{J}\rangle_P(\mathbf{r}t)$ be the current density generated by the pump only, and $\mathbf{E}_{\text{tot}}$ be the solution of (13.58) with $\langle \mathbf{J}\rangle = \langle \mathbf{J}\rangle_P$. The weak probe pulse is responsible for the small change $\langle \mathbf{J}\rangle_P \to \langle \mathbf{J}\rangle = \langle \mathbf{J}\rangle_P + \langle \mathbf{J}\rangle_p$. Correspondingly, $\mathbf{E}_{\text{tot}} \to \boldsymbol{\mathcal{E}}_{\text{tot}} = \mathbf{E}_{\text{tot}} + \mathbf{e}_{\text{tot}}$, where $\mathbf{e}_{\text{tot}}$ must satisfy (13.58) with $\langle \mathbf{J}\rangle = \langle \mathbf{J}\rangle_p$. The quantity $\langle \mathbf{J}\rangle_p$ is the probe-induced change of the macroscopic current while $\mathbf{e}_{\text{tot}}$ is the probe-induced change of the total electric field. We take x as the propagation direction of the incident probe pulse $\mathbf{e}(\mathbf{r},t)$, and we assume that the cross section S is large enough to discard the dependence of $\mathbf{e}(\mathbf{r},t)$ on y and z. We also assume that the transmitted probe pulse $\mathbf{e}'(\mathbf{r},t)$ remains focused and keeps propagating along the same direction x as the incident field. This implies that also the probe-induced macroscopic current $\langle \mathbf{J}\rangle_p$ has a weak dependence on y and z. We then search for solutions $\mathbf{e}_{\text{tot}}(\mathbf{r},t) = \mathbf{e}_{\text{tot}}(x,t)$, where

$$\frac{\partial^2 \mathbf{e}_{\text{tot}}}{\partial x^2} - \frac{1}{c^2}\frac{\partial^2 \mathbf{e}_{\text{tot}}}{\partial t^2} = \frac{4\pi q}{c^2}\frac{\partial \langle \mathbf{J}\rangle_p}{\partial t}. \tag{13.59}$$

The most general solution of (13.59) is the sum of an arbitrary solution $\mathbf{h}(t - x/c)$ of the homogeneous equation and of a special solution $\mathbf{s}(x,t)$: $\mathbf{e}_{\text{tot}}(x,t) = \mathbf{h}(t - x/c) + \mathbf{s}(x,t)$. We choose our reference frame with origin $x = 0$ at the left boundary of the system. We also set the time $t = 0$ as the time when the incident probe pulse $\mathbf{e}(t - x/c)$ is localized somewhere on the left of the system. Imposing the boundary condition $\mathbf{e}_{\text{tot}}(x,0) = \mathbf{e}(-x/c)$, we obtain $\mathbf{h}(-x/c) = \mathbf{e}(-x/c) - \mathbf{s}(x,0)$ and hence

$$\mathbf{e}_{\text{tot}}(x,t) = \mathbf{e}(t - x/c) - \mathbf{s}(x - ct, 0) + \mathbf{s}(x,t).$$

The special solution $\mathbf{s}(x,t)$ is found by inverting the one-dimensional d'Alambertian $\Box \equiv \frac{\partial^2}{\partial x^2} - \frac{1}{c^2}\frac{\partial^2}{\partial t^2}$. The Green's function $\mathfrak{g}(x,t)$ solution of $\Box \mathfrak{g}(x,t) = \delta(x)\delta(t)$ is

$$\mathfrak{g}(x,t) = -\frac{c}{2}\theta(t)\chi_{[-ct,ct]}(x),$$

where $\chi_{[a,b]}(x) = 1$ if $x \in (a,b)$ and 0 otherwise. Therefore, the special solution reads:

$$\mathbf{s}(x,t) = \frac{4\pi q}{c^2}\int dt' dx' \mathfrak{g}(x - x', t - t')\frac{\partial \langle \mathbf{J}\rangle_p}{\partial t'} = -\frac{2\pi q}{c}\int_{-\infty}^{t} dt' \int_{x-c(t-t')}^{x+c(t-t')} dx' \frac{\partial \mathbf{J}_p}{\partial t'}. \tag{13.60}$$

In the last equality we remove the brackets around $\mathbf{J}_p$ since the spatial integral of the macroscopic average is the same as the integral of the microscopic average. Without any loss of generality we can choose the time $t = 0$ as the time before which neither the pump nor the probe have reached the system. Then $\mathbf{J}_p = 0$ for $t < 0$ and hence $\mathbf{s}(x, 0) = 0$ for all x. We conclude that

$$\mathbf{e}_{\text{tot}}(x,t) = \mathbf{e}(t - x/c) + \mathbf{s}(x,t), \tag{13.61}$$

with $\mathbf{s}(x,t)$ given in (13.60).

We are interested in the total electric field on the right of the system - that is, in $x = L$ where L is the system length, since $\mathbf{e}_{\text{tot}}(L,t) = \mathbf{e}'(L,t)$. Taking into account that $\mathbf{J}_p(x',t')$ is nonvanishing only for $x' \in (0, L)$ and $t' > 0$, we have

$$\mathbf{s}(L,t) = -\frac{2\pi q}{c}\int_0^t dt' \int_{L-c(t-t')}^{L} dx' \, \frac{\partial \mathbf{J}_p}{\partial t'}.$$

The lower integration limit of the spatial integral is positive for $t' \in (t - L/c, t)$ and negative for $t' \in (0, t - L/c)$. For thin systems we can discard the contribution coming from $t' \in (t - L/c, t)$ and set the lower integration limit to zero for all $t' \in (0, t)$ [we recall that $\mathbf{J}_p(x', t' < 0) = 0$)]. Thus,

$$\mathbf{s}(L,t) = -\frac{2\pi q}{c}\int_0^t dt' \int_0^L dx' \, \frac{\partial \mathbf{J}_p}{\partial t'} = -\frac{2\pi q}{Sc}\int_V d\mathbf{r}\, \mathbf{J}_p(\mathbf{r},t), \tag{13.62}$$

where in the second equality we integrate over the volume $V = SL$ containing the system. Using the identity $\mathbf{J}_p = \boldsymbol{\nabla} \cdot (\mathbf{J}_p \mathbf{r}) - \mathbf{r}(\boldsymbol{\nabla} \cdot \mathbf{J}_p)$ and extending the integral over all space (outside V the current density vanishes), we can rewrite (13.62) as $\mathbf{s}(L,t) = \frac{2\pi q}{Sc}\int d\mathbf{r}\, \mathbf{r}\,(\boldsymbol{\nabla} \cdot \mathbf{J}_p)$. Further, using the continuity equation $\boldsymbol{\nabla} \cdot \mathbf{J}_p = -\partial n_p/\partial t$, where n_p is the probe-induced change of the density, we can eventually write the transmitted electric field as

$$\mathbf{e}'(L,t) = \mathbf{e}_{\text{tot}}(L,t) = \mathbf{e}(t - L/c) - \frac{2\pi q}{Sc}\frac{d}{dt}\int d\mathbf{r}\, \mathbf{r}\, n_p(\mathbf{r},t). \tag{13.63}$$

The integral on the r.h.s. is the probe-induced dipole moment. Equation (13.63) relates the transmitted probe field to a time-dependent quantum-mechanical average, and it represents the fundamental bridge between theory and experiment.

Absorption energy and response functions The absorption energy is defined as the difference between the incident energy and the transmitted energy. From (13.57) we have for frequencies $\omega \neq 0$

$$\tilde{E}_{\text{abs}}(\omega) \equiv \tilde{E}_{\text{inc}}(\omega) - \tilde{E}_{\text{trans}}(\omega) = S\frac{c}{8\pi}\left(|\tilde{\mathbf{e}}(\omega)|^2 - |\tilde{\mathbf{e}}'(\omega)|^2\right). \tag{13.64}$$

If we define the probe-induced time-dependent dipole moment as

$$\mathbf{d}_p(t) = q\int d\mathbf{r}\, \mathbf{r}\, n_p(\mathbf{r},t), \tag{13.65}$$

then the Fourier transform of the transmitted probe pulse in (13.63) reads (omitting the dependence on L and using the assumption of thin systems):

$$\tilde{\mathbf{e}}'(\omega) = \tilde{\mathbf{e}}(\omega) + \mathrm{i}\frac{2\pi\omega}{Sc}\int dt\, e^{\mathrm{i}\omega t}\mathbf{d}_p(t). \tag{13.66}$$

Inserting this result in (13.64) and keeping only terms of first order in $\mathbf{d}_p(t)$, we find

$$\tilde{E}_{\mathrm{abs}}(\omega) = \frac{\omega}{2}\,\mathrm{Im}\Big[\tilde{\mathbf{e}}^*(\omega)\cdot\int dt\, e^{\mathrm{i}\omega t}\mathbf{d}_p(t)\Big]. \tag{13.67}$$

The probe-induced change of the electronic density can be calculated from linear response theory. For transverse electric fields the scalar potential vanishes and (13.14) yields

$$n_p(\mathbf{r},t) = \sum_\sigma \delta J_0(\mathbf{x},t) = -\frac{q}{c}\int dt' \int d\mathbf{r}' \sum_{\sigma\sigma'}\chi^{\mathrm{R}}_{0k}(\mathbf{x},t;\mathbf{x}',t')\; a_k(\mathbf{r}',t'), \tag{13.68}$$

where a_k is the kth component of the vector potential of the probe pulse.[10] The response function χ_{0k} is written in terms of $\Delta\hat{J}_{k,H}(\mathbf{x}',t')$. We rewrite this operator using the identity below (13.62):

$$\Delta\hat{J}_{k,H}(\mathbf{x}',t') = \sum_j \frac{\partial}{\partial r'_j}\left[r'_k\Delta\hat{J}_{j,H}(\mathbf{x}',t')\right] - r'_k\sum_j\frac{\partial}{\partial r'_j}\Delta\hat{J}_{j,H}(\mathbf{x}',t') \tag{13.69}$$

(the total derivative does not contribute to the integral over all space), and then the continuity equation (3.29) for operators in the Heisenberg picture: $\sum_j \frac{\partial}{\partial r'_j}\Delta\hat{J}_{j,H}(\mathbf{x}',t') = -\frac{\partial}{\partial t'}\Delta\hat{n}_H(\mathbf{x}',t')$. In this way we express the density-current response function in (13.68) in terms of the density-density response function. We find

$$n_p(\mathbf{r},t) = -\frac{q}{c}\int dt' \int d\mathbf{r}'\,\frac{\partial}{\partial t'}\sum_{\sigma\sigma'}\chi^{\mathrm{R}}_{00}(\mathbf{x},t;\mathbf{x}',t')\;\mathbf{r}'\cdot\mathbf{a}(\mathbf{r}',t').$$

Integrating by parts over t' and taking into account that the retarded response function vanishes for $t'=\infty$ while the vector potential vanishes for $t'=-\infty$, we obtain

$$n_p(\mathbf{r},t) = -q\int dt' \int d\mathbf{r}'\;\chi^{\mathrm{R}}_{00}(\mathbf{r},t;\mathbf{r}',t')\;\mathbf{r}'\cdot\mathbf{e}(\mathbf{r}',t'), \tag{13.70}$$

where $\mathbf{e} = -\frac{1}{c}\partial\mathbf{a}/\partial t$ is the external probe field, see (3.2a), and

$$\chi^{\mathrm{R}}_{00}(\mathbf{r},t;\mathbf{r}',t') \equiv \sum_{\sigma\sigma'}\chi^{\mathrm{R}}_{00}(\mathbf{x},t;\mathbf{x}',t') \tag{13.71}$$

is the sum over spins of the response function. The probe-induced change of the density can now be inserted in the equation for $\mathbf{d}_p$, see (13.65). The assumption of thin systems allows us to approximate $\mathbf{e}(\mathbf{r}',t') \simeq \mathbf{e}(L,t')$. Hence (omitting again the dependence on L),

$$\mathbf{d}_p(t) = -q^2\int dt' \int d\mathbf{r}d\mathbf{r}'\;\mathbf{r}\;\chi^{\mathrm{R}}_{00}(\mathbf{r},t;\mathbf{r}',t')\;\mathbf{r}'\cdot\mathbf{e}(t'). \tag{13.72}$$

[10] We are here assuming that the induced vector potential is much smaller than the external vector potential: $|\mathbf{a}_{\mathrm{tot}} - \mathbf{a}|/|\mathbf{a}| \ll 1$.

Therefore the absorption energy (13.67) becomes

$$\boxed{\tilde{E}_{\text{abs}}(\omega) = -\frac{q^2\,\omega}{2}\int dt\, dt' e^{i\omega t}\int d\mathbf{r}d\mathbf{r}'\, \text{Im}\Big[\tilde{\mathbf{e}}^*(\omega)\cdot\mathbf{r}\;\chi^{\text{R}}_{00}(\mathbf{r},t;\mathbf{r}',t')\;\mathbf{r}'\cdot\mathbf{e}(t')\Big]} \tag{13.73}$$

We emphasize that this formula is valid for positive and negative delays between pump and probe pulses, as well as for situations in which the pump and probe pulses overlap in time or even for more exotic situations in which the pump is entirely contained in the time window of the probe.

Ultrafast probe pulse The response function of a system that is either driven by a pump or is left in a nonequilibrium state after the action of a pump depends on t and t' separately. This implies that $\tilde{E}_{\text{abs}}(\omega)$ changes by varying the pump–probe delay. The easiest way to appreciate this fact consists of taking an ultrafast probe pulse. Suppose that the pump impinges on the system at time $t = 0$. After a delay τ we probe the system with $\mathbf{e}(t) = \mathbf{e}_0\delta(t-\tau)$, implying that $\tilde{\mathbf{e}}(\omega) = \mathbf{e}_0\, e^{i\omega\tau}$. Performing the integral over t', the absorption energy reduces to

$$\tilde{E}_{\text{abs}}(\tau,\omega) = -\frac{q^2\,\omega}{2}\int dt\, e^{i\omega(t-\tau)}\int d\mathbf{r}d\mathbf{r}'\, \text{Im}\Big[\mathbf{e}_0\cdot\mathbf{r}\;\chi^{\text{R}}_{00}(\mathbf{r},t;\mathbf{r}',\tau)\;\mathbf{r}'\cdot\mathbf{e}_0\Big], \tag{13.74}$$

where we have emphasized the dependence of $\tilde{E}_{\text{abs}}$ on τ. It is important to realize that the absorption energy depends on τ even for negative τ – that is, before the action of the pump, since the integral over t extends from $-\infty$ to $+\infty$. Interpreting $\tilde{E}_{\text{abs}}(\tau,\omega)$ as the absorption energy of the system at the time of the probe impact is not strictly speaking correct.

Photoabsorption for stationary ensemble Next we consider the system in some stationary ensemble and probe it without using any pumping.[11] Then the response function in (13.73) depends on the time difference only. The inverse Fourier transform of the response function reads, see (13.21),

$$\chi^{\text{R}}_{00}(\mathbf{r},\mathbf{r}';\omega) = \int dt\; e^{i\omega(t-t')}\chi^{\text{R}}_{00}(\mathbf{r},t;\mathbf{r}',t'). \tag{13.75}$$

Taking into account that the inverse Fourier transform of the probe pulse is $\tilde{\mathbf{e}}(\omega) = \int dt' e^{i\omega t'}\mathbf{e}(t')$, the absorption energy (13.73) simplifies to

$$\tilde{E}_{\text{abs}}(\omega) = -\frac{q^2\,\omega}{2}\int d\mathbf{r}d\mathbf{r}'\, \text{Im}\Big[\tilde{\mathbf{e}}^*(\omega)\cdot\mathbf{r}\;\chi^{\text{R}}_{00}(\mathbf{r},\mathbf{r}';\omega)\;\mathbf{r}'\cdot\tilde{\mathbf{e}}(\omega)\Big]. \tag{13.76}$$

This result is clearly independent of the probe impinging time. We observe that for systems in thermal equilibrium at zero temperature the absorption energy is positive for all frequencies, see (13.28), as it should be. Any approximation to χ_{00} that violates this sign property also violates the energy conservation law.

We conclude that time-resolved photoabsorption experiments provide information on the charge-neutral excitation energies and excitation amplitudes as well as on how these quantities change over time when the system is in a nonequilibrium state.

[11]Alternatively we can think of a system that after pumping settles on a stationary ensemble and is subsequently probed after a long time.

13.5 Absorbed Energy and Response Functions

Let us consider a system of particles with charge q driven out of equilibrium by a weak electromagnetic field. No assumptions on the size of the system are made; the derivation in this section applies to atoms, molecules, and bulk materials like crystals. The total absorbed energy $E_{\rm abs}$ is the integral over time of (12.25):

$$E_{\rm abs} = q\int_{-\infty}^{\infty} dt \int d\mathbf{x}\, \mathbf{J}(\mathbf{x},t)\cdot\mathbf{E}_\Delta(\mathbf{r},t) = q\int\frac{d\omega}{2\pi}\int d\mathbf{x}\, \tilde{\mathbf{E}}^*_\Delta(\mathbf{r},\omega)\cdot\tilde{\mathbf{J}}(\mathbf{x},\omega), \tag{13.77}$$

where the change in the electric field is defined according to $\mathbf{E}_\Delta = -\boldsymbol{\nabla}\phi_\Delta - (1/c)\partial\mathbf{A}/\partial t$ see Section 6.5. In the last equality of (13.77) we express $E_{\rm abs}$ in terms of the Fourier transforms of the current and the electric field.[12] To lowest order in the external fields the current is given by (13.14). For simplicity we specialize the discussion to systems prepared in a stationary ensemble – it could be a thermal equilibrium ensemble or a mixture of excited states. Then the response functions depend on the time difference only, and in Fourier space (13.14) becomes

$$\begin{aligned}\tilde{J}_k(\mathbf{x},\omega) = q\int\frac{d\omega}{2\pi}\int d\mathbf{x}'\left[\chi^{\rm R}_{k0}(\mathbf{x},\mathbf{x}';\omega)\tilde{\phi}_\Delta(\mathbf{r}',\omega) - \sum_{j=1}^{3}\chi^{\rm R}_{kj}(\mathbf{x},\mathbf{x}';\omega)\frac{\tilde{A}_j(\mathbf{r}',\omega)}{c}\right]\\ - \frac{q}{mc}n(\mathbf{x})\tilde{A}_k(\mathbf{r},\omega).\end{aligned} \tag{13.78}$$

In Fourier space we also have $\tilde{\mathbf{E}}_\Delta = -\boldsymbol{\nabla}\tilde{\phi}_\Delta + \frac{\mathrm{i}\omega}{c}\tilde{\mathbf{A}}$, and hence $\tilde{\mathbf{A}} = \frac{c}{\mathrm{i}\omega}(\tilde{\mathbf{E}}_\Delta + \boldsymbol{\nabla}\tilde{\phi}_\Delta)$. Using this expression of the vector potential for the second term of (13.78) and substituting the resulting current in (13.77), we get

$$\begin{aligned}E_{\rm abs} = q^2\int\frac{d\omega}{2\pi}\int d\mathbf{x}d\mathbf{x}'\sum_{k=1}^{3}\tilde{E}^*_{\Delta,k}(\mathbf{r},\omega)\\ \times\left\{\left[\chi^{\rm R}_{k0}(\mathbf{x},\mathbf{x}';\omega) - \frac{1}{\mathrm{i}\omega}\sum_{j=1}^{3}\chi^{\rm R}_{kj}(\mathbf{x},\mathbf{x}';\omega)\frac{\partial}{\partial r'_k}\right]\tilde{\phi}_\Delta(\mathbf{r}',\omega)\right.\\ \left. - \frac{1}{\mathrm{i}\omega}\sum_{j=1}^{3}\chi^{\rm R}_{kj}(\mathbf{x},\mathbf{x}';\omega)\tilde{E}_{\Delta,j}(\mathbf{r}',\omega)\right\} - \frac{q^2}{mc}\int\frac{d\omega}{2\pi}\int d\mathbf{x}\, n(\mathbf{x})\tilde{\mathbf{E}}^*_\Delta(\mathbf{r},\omega)\cdot\tilde{\mathbf{A}}(\mathbf{r},\omega).\end{aligned} \tag{13.79}$$

The second line of this equation can be further manipulated. As pointed out below (3.32), the current $\mathbf{J}$ is invariant under the gauge transformations (3.25). In Fourier space the gauge transformation of the scalar and vector potentials reads:

$$\tilde{\phi}_\Delta(\mathbf{r},\omega) \to \tilde{\phi}_\Delta(\mathbf{r},\omega) + \frac{\mathrm{i}\omega}{c}\tilde{\Lambda}(\mathbf{r},\omega) \quad ; \quad \tilde{\mathbf{A}}(\mathbf{r},\omega) \to \tilde{\mathbf{A}}(\mathbf{r},\omega) + \boldsymbol{\nabla}\tilde{\Lambda}(\mathbf{r},\omega). \tag{13.80}$$

[12] For real functions like $\mathbf{E}_\Delta$ and $\mathbf{J}$, the Fourier transform at frequency $-\omega$ is the complex conjugate of the Fourier transform at frequency ω.

The invariance of (13.78) implies that

$$\int \frac{d\omega}{2\pi} \int d\mathbf{x}' \Big[\mathrm{i}\omega \chi^{\mathrm{R}}_{k0}(\mathbf{x},\mathbf{x}';\omega) - \sum_{j=1}^{3} \chi^{\mathrm{R}}_{kj}(\mathbf{x},\mathbf{x}';\omega) \frac{\partial}{\partial r'_j} - \frac{n(\mathbf{x})}{m} \delta(\mathbf{x}-\mathbf{x}') \frac{\partial}{\partial r'_k} \Big] \tilde{\Lambda}(\mathbf{r}',\omega) = 0. \tag{13.81}$$

Integrating by parts and discarding the contribution of the total derivatives, the arbitrariness of the gauge function $\tilde{\Lambda}$ leads to the important identity

$$\Big[\mathrm{i}\omega \chi^{\mathrm{R}}_{k0}(\mathbf{x},\mathbf{x}';\omega) + \sum_{j=1}^{3} \frac{\partial}{\partial r'_j} \chi^{\mathrm{R}}_{kj}(\mathbf{x},\mathbf{x}';\omega) \Big] = -\frac{n(\mathbf{x})}{m} \frac{\partial}{\partial r'_k} \delta(\mathbf{x}-\mathbf{x}'). \tag{13.82}$$

This exact relation between response functions is a consequence of a more general result known as *Ward identities*; we derive the Ward identities in Section 14.10. For the time being we use the Ward identity to rewrite (13.79) in a more compact form. Integrating by parts the term containing the gradient of the external scalar potential and using (13.82), we get

$$\begin{aligned} E_{\mathrm{abs}} = \mathrm{i}\, q^2 \int \frac{d\omega}{2\pi} \frac{1}{\omega} \int d\mathbf{x} d\mathbf{x}' \sum_{kj=1}^{3} \tilde{E}^*_{\Delta,k}(\mathbf{r},\omega) \chi^{\mathrm{R}}_{kj}(\mathbf{x},\mathbf{x}';\omega) \tilde{E}_{\Delta,j}(\mathbf{r}',\omega) \\ + \mathrm{i}\, q^2 \int \frac{d\omega}{2\pi} \frac{1}{\omega} \int d\mathbf{x}\, \frac{n(\mathbf{x})}{m} |\tilde{\mathbf{E}}_{\Delta}(\mathbf{r},\omega)|^2 . \end{aligned} \tag{13.83}$$

The last term does not contribute since $|\tilde{\mathbf{E}}_{\Delta}(\mathbf{x},\omega)|^2$ is an even function of the frequency. Taking into account property (13.27), we also realize that only the imaginary part of the integrand contributes to the integral of the first term. Therefore,

$$\boxed{E_{\mathrm{abs}} = -\, q^2 \int \frac{d\omega}{2\pi} \frac{1}{\omega} \int d\mathbf{r} d\mathbf{r}'\ \mathrm{Im}\Big[\sum_{kj=1}^{3} \tilde{E}^*_{\Delta,k}(\mathbf{r},\omega) \chi^{\mathrm{R}}_{kj}(\mathbf{r},\mathbf{r}';\omega) \tilde{E}_{\Delta,j}(\mathbf{r}',\omega) \Big]} \tag{13.84}$$

We emphasize that this result is valid for systems of any size and for both transverse and longitudinal electric fields. We also observe that for systems in thermal equilibrium at zero temperature the absorbed energy is always positive, as follows directly from (13.28) and as it should be.

Connection to the photoabsorption spectrum Equation (13.84) agrees with (13.76). Using identity (13.69) along with the continuity equation, and discarding the total derivatives (which is appropriate for finite systems), the current–current response function in (13.84) can be replaced by $\omega^2 r_k \chi^{\mathrm{R}}_{00}(\mathbf{r},\mathbf{r}';\omega) r'_k$. Thus, the absorbed energy can equivalently be written as

$$\boxed{E_{\mathrm{abs}} = -q^2 \int \frac{d\omega}{2\pi}\, \omega \int d\mathbf{r} d\mathbf{r}'\ \mathrm{Im}\Big[\tilde{\mathbf{E}}^*_{\Delta}(\mathbf{r},\omega) \cdot \mathbf{r}\ \chi^{\mathrm{R}}_{00}(\mathbf{r},\mathbf{r}';\omega)\ \mathbf{r}' \cdot \tilde{\mathbf{E}}_{\Delta}(\mathbf{r}',\omega) \Big]} \tag{13.85}$$

Equation (13.85) coincides with the integral over all frequencies of $\tilde{E}_{\mathrm{abs}}(\omega)$ in (13.76), except that for a factor of 2. This factor stems from the fact that $\tilde{E}_{\mathrm{abs}}(\omega)$ is the *cosine* transform of the energy flux of the transmitted *and* split field $\mathbf{e}'_{\text{detector}}(t) = \frac{1}{2}(\mathbf{e}'(t) + \mathbf{e}'(t-\delta))$. We remark that the relation between the transmitted energy flux and the absorbed energy per unit frequency is not straightforward for thick or bulky materials. In fact, approximating $\mathbf{e}'$ as a transverse wave propagating in the same direction as the incident field is no longer justified.

Electron gas The simplest bulk system is the electron gas. Let us consider a change in the longitudinal electric field $\mathbf{E}_\Delta(\mathbf{r},t) = -\boldsymbol{\nabla}\phi_\Delta(\mathbf{x},t)$ or, Fourier transforming with respect to space and time, $\tilde{\mathbf{E}}_\Delta(-\mathbf{p},-\omega) = \tilde{\mathbf{E}}^*_\Delta(\mathbf{p},\omega) = \mathrm{i}\mathbf{p}\tilde{\phi}^*_\Delta(\mathbf{p},\omega)$.[13] Substitution of this result into (13.77) generates the scalar product $\mathbf{p}\cdot\tilde{\mathbf{J}}(\mathbf{p},\omega)$, where $\tilde{\mathbf{J}}(\mathbf{p},\omega)$ is the Fourier transform of the current with respect to space and time. The continuity equation in Fourier space reads $-\mathrm{i}\omega\,\tilde{n}(\mathbf{p},\omega) + \mathrm{i}\mathbf{p}\cdot\tilde{\mathbf{J}}(\mathbf{p},\omega) = 0$, and to first order in the electric field the change in the density is given by (13.14):

$$\tilde{n}(\mathbf{p},\omega) = q\,\chi^{\mathrm{R}}_{00}(\mathbf{p},\omega)\tilde{\phi}_\Delta(\mathbf{p},\omega), \tag{13.86}$$

since $\mathbf{A} = 0$. In (13.86) we take advantage of the invariance of the electron gas under translations and define the Fourier transform of the response function according to

$$\chi^{\mathrm{R}}_{00}(\mathbf{r},\mathbf{r}';\omega) = \sum_{\sigma\sigma'}\chi^{\mathrm{R}}_{00}(\mathbf{x},\mathbf{x}';\omega) = \int\frac{d\mathbf{p}}{(2\pi)^3}\,e^{\mathrm{i}\mathbf{p}\cdot(\mathbf{r}-\mathbf{r}')}\chi^{\mathrm{R}}_{00}(\mathbf{p},\omega). \tag{13.87}$$

Collecting these results, we find the following expression for the absorbed energy:

$$E_{\text{abs}} = q^2\int\frac{d\omega d\mathbf{p}}{(2\pi)^4}\frac{\mathrm{i}\omega}{p^2}\chi^{\mathrm{R}}_{00}(\mathbf{p},\omega)|\tilde{\mathbf{E}}_\Delta(\mathbf{p},\omega)|^2. \tag{13.88}$$

Of course, only the imaginary part of χ^{R}_{00} contributes in (13.88) since E_{abs} is a real quantity.[14] We point out that no integration by parts has been performed to derive this result.[15] The absorbed energy is usually expressed in terms of the so-called *energy-loss function* $\mathcal{L}$:

$$\mathcal{L}(\mathbf{p},\omega) \equiv -\frac{4\pi}{p^2}\mathrm{Im}[\chi^{\mathrm{R}}_{00}(\mathbf{p},\omega)] \quad \Rightarrow \quad E_{\text{abs}} = \frac{q^2}{4\pi}\int\frac{d\omega d\mathbf{p}}{(2\pi)^4}\omega\mathcal{L}(\mathbf{p},\omega)|\tilde{\mathbf{E}}_\Delta(\mathbf{p},\omega)|^2.$$

The energy-loss function tells us for which frequencies and momenta the electron gas can dissipate energy. At zero temperature the dissipated energy must be positive for positive frequencies (absorption) and negative for negative frequencies (emission). This implies that

$$\mathrm{Im}[\chi^{\mathrm{R}}_{00}(\mathbf{p},\omega)] \begin{array}{ll} <0 & \text{for } \omega>0 \\ >0 & \text{for } \omega<0 \end{array}, \tag{13.89}$$

in agreement with (13.28).

[13] For real functions $f(\mathbf{r},t)$ the Fourier transform $f(\mathbf{p},\omega) = f^*(-\mathbf{p},-\omega)$.
[14] The real part is an even function of ω – see the discussion below (13.27).
[15] For infinitely large systems the integral of a total divergence cannot be discarded.

14

Two-Particle Green's Function: Diagrammatic Expansion

The effect of the interparticle interaction is to correlate the motion of a particle to the motion of all the other particles. For the two-body interactions considered so far, the action of $\hat{H}_{\text{int}}$ on a many-body ket $|\Phi\rangle = \hat{d}^{\dagger}_{n_1}\hat{d}^{\dagger}_{n_2}\ldots|0\rangle$ (one particle in φ_{n_1}, another one in φ_{n_2}, and so on) yields a linear combination of kets in which at most two particles have changed their state. This implies that the matrix element $\langle\Phi'|\hat{H}_{\text{int}}|\Phi\rangle$ is zero for all those $|\Phi'\rangle$ differing from $|\Phi\rangle$ by more than two $\hat{d}^{\dagger}$ operators. We may say that a particle can scatter at most with another particle and after the scattering the two particles end up in new states. Therefore, if we know how two particles propagate in the system (i.e., if we know G_2), then we can deduce how a single particle propagates in the system (i.e., we can determine G). This is another way to understand the appearance of G_2 in the first equation of the Martin–Schwinger hierarchy. We emphasize that this is true only for two-body interactions. For an interaction Hamiltonian that is a n-body operator – that is, an interaction Hamiltonian that is a linear combination of products of $2n$ field operators – the scattering involves n-particles and the equations of motion for G contain the n-particle Green's function G_n.

In this chapter we develop the diagrammatic theory for the two-particle Green's function G_2. The knowledge of G_2 is not only useful to calculate the one-particle Green's function G. On one hand there exists a simple diagrammatic relation between G_2 and the self-energy Σ, see Section 9.1. Since the G_2-diagrams are easier to interpret than the Σ-diagrams, working with G_2 allows for a better understanding of the physics contained in a self-energy diagram; in turn, this facilitates the construction of suitable self-energy approximations. On the other hand, G_2 is directly related to physically measurable quantities that are not within reach of the one-particle Green's function G.

At the end of this chapter the reader will have the necessary tools to extend the diagrammatic technique to any n-particle Green's function. In most cases the knowledge of G and G_2 is sufficient to determine most of the physical quantities of interest. There exist situations, however, which necessitate the calculation of higher-order Green's functions. For instance, G_3 is needed to study the nucleon–nucleon interaction in deuterium [202] or the Coster–Kronig preceded Auger processes in solids [203]. The expansion of G_3 can be cast in the form of a recursive relation known as the *Faddeev equation* [204, 205].

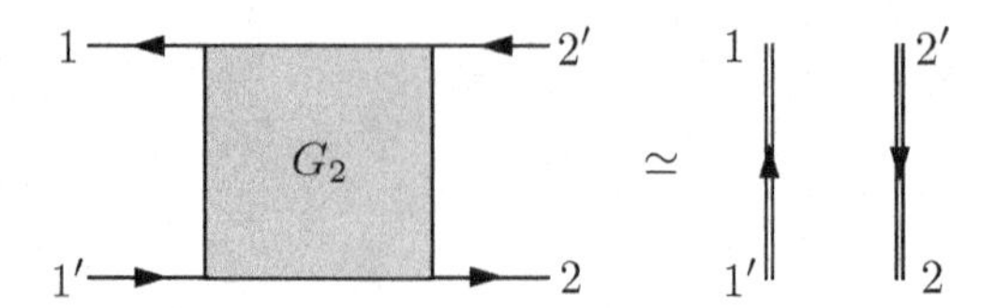

Figure 14.1 Hartree approximation to the two-particle Green's function.

14.1 A Physical Appetizer on the Diagrammatic Expansion

Let us begin by showing how one can generate reasonable approximations to G_2 using nothing more than a bit of physical intuition. Suppose that in our system we could identify two "special" particles that do not interact directly. These two special particles feel the presence of all other particles, but they are insensitive to their mutual position. Then, the probability amplitude for the first particle to go from $1'$ to 1 *and* the second particle to go from $2'$ to 2 is simply the product of the probability amplitudes of the two separate events, or in the Green's function language,

$$G_2(1,2;1',2') \sim G_{2,\mathrm{H}}(1,2;1',2') \equiv e^{\mathrm{i}\alpha} G(1;1')G(2;2'). \tag{14.1}$$

If we represent the two-particle Green's function $G_2(1,2;1',2')$ with two lines that start in $1'$ and $2'$, enter a square where processes of arbitrary complexity can occur and go out in 1 and 2, then (14.1) can be represented diagrammatically as in Fig. 14.1. The phase factor $e^{\mathrm{i}\alpha}$ in (14.1) can be determined by observing that $G_2(1,2;1^+,2^+)$ is the ensemble average of $-\hat{n}_H(1)\hat{n}_H(2)$, and hence

$$\lim_{2\to 1} G_2(1,2;1^+,2^+) = \text{real negative number}.$$

Evaluating the r.h.s. of (14.1) in the same point and taking into account that $G(1;1^+) = \mp \mathrm{i}n(1)$ (upper/lower sign for bosons/fermions) is a negative/positive imaginary number we conclude that $e^{\mathrm{i}\alpha} = 1$. In the next section we show that the approximation (14.1) can be derived from the expansion (5.18), which naturally fixes the phase factor $e^{\mathrm{i}\alpha}$ to be unity. The approximation (14.1) is known as the *Hartree approximation*, and it neglects the direct interaction between two particles.

The Hartree approximation also ignores that the particles are identical. In the exact case, for two identical particles initially in $(1',2')$, it is not possible to distinguish the event in which they are detected in $(1,2)$ from the event in which they are detected in $(2,1)$. Quantum mechanics teaches us that if an event can occur through two different paths then the total probability amplitude is the sum of the probability amplitudes of each path. In the Green's function language this leads to the so-called *Hartree–Fock approximation* for the two-particle Green's function,

$$G_2(1,2;1',2') \sim G_{2,\mathrm{HF}}(1,2;1',2') \equiv G(1;1')G(2;2') + e^{\mathrm{i}\beta} G(1;2')G(2;1'), \tag{14.2}$$

whose diagrammatic representation is illustrated in Fig. 14.2. The phase factor $e^{\mathrm{i}\beta}$ can be determined using the symmetry properties of G_2. From the general definition (5.1) we have

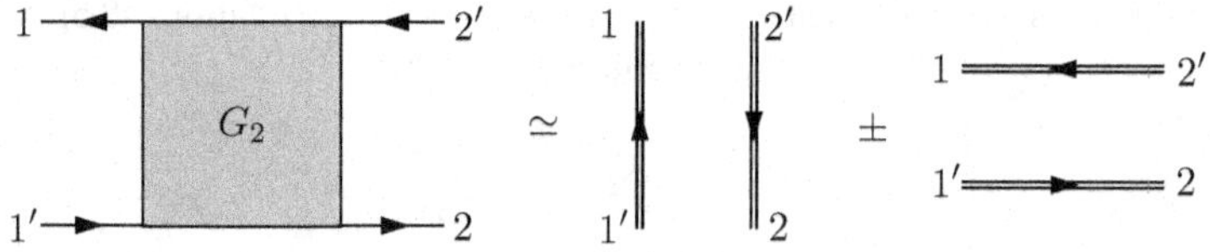

Figure 14.2 Hartree-Fock approximation to the two-particle Green's function.

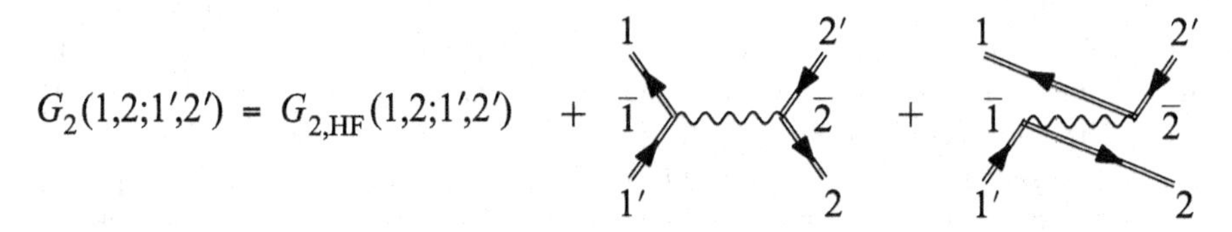

Figure 14.3 Second Born approximation to the two-particle Green's function.

$G_2(1,2;1',2') = \pm G_2(1,2;2',1')$, where the upper/lower sign applies to bosons/fermions, and hence $e^{i\beta} = \pm 1$. The Hartree-Fock approximation (like the Hartree approximation) can also be deduced from the expansion (5.18).

Both the Hartree and Hartree-Fock approximations neglect the direct interaction between two particles, see also the discussion in Section 8.1. In these approximations a particle moves like a free particle under the influence of an effective potential which depends on the position of all the other particles. The reader can easily verify that inserting the Hartree and Hartree-Fock approximation to G_2 in the relations (9.5) or (9.6) one finds the Hartree and Hartree-Fock approximation for the self-energy.

The diagrammatic representation of G_2 is extremely useful for visualizing scattering processes and hence generating suitable approximations for the problem at hand. Without being too rigorous we could already give a taste of how it works. Suppose that there is a region of space where the particle density is smaller than everywhere else. Then we do not expect a mean-field approximation to work in this region as the concept of effective potential has no physical meaning at low densities. We are forced to construct an approximation that takes into account the direct interaction between the particles. In other words, the propagation from $(1', 2')$ to $(1, 2)$ must take into account possible intermediate scatterings in $(\bar{1}, \bar{2})$ as illustrated in Fig. 14.3. In the first diagram two particles propagate from $(1', 2')$ to $(\bar{1}, \bar{2})$ and then scatter. The scattering is caused by the interaction (wiggly line) $v(\bar{1}, \bar{2})$ and after the scattering the propagation continues until the particles arrive in $(1, 2)$. The second diagram contains the same physical information, but the final state $(1, 2)$ is exchanged with $(2, 1)$. Like in the Hartree-Fock case the second diagram guarantees that G_2 has the correct symmetry under the interchange of its arguments. From these considerations we may try to improve over the Hartree-Fock approximation by adding to $G_{2,\mathrm{HF}}$ a term proportional to the diagrams in Fig. 14.3:

$$\int d\bar{1}d\bar{2}\, v(\bar{1};\bar{2})\, [G(1;\bar{1})G(\bar{1};1')G(2;\bar{2})G(\bar{2};2') \pm G(1;\bar{2})G(\bar{2};2')G(2;\bar{1})G(\bar{1};1')]\,,$$

whose dominant contribution comes from the integral over the low-density region. The approximation in Fig. 14.3 is known as the *second Born approximation.* The inclusion of the new scattering processes in the self-energy is possible through relations (9.5) or (9.6), and it leads to the second Born self-energy encountered in (7.16):

$$\Sigma(1;\bar{1}) = \Sigma_{\rm HF}(1;\bar{1}) + \; [2'{=}2,\ \bar{2},\ 1,\ \bar{1}] \; + \; [1,\ \bar{2},\ 2,\ \bar{1}] \tag{14.3}$$

In the second term on the r.h.s. we recognize the process previously described. A particle coming from somewhere reaches $\bar{1}$ and interacts with a particle coming from $2'$ in $\bar{2}$. After the interaction the first particle propagates to 1 and the second particle propagates to $2 = 2'$. The third term in (14.3) stems from the exchange diagram in Fig. 14.3.

This is, in essence, how new physical mechanisms can be incorporated in an approximation within the diagrammatic approach. However, physical intuition alone is not enough to construct a systematic and rigorous expansion. There are several issues that remain to be addressed. What is the prefactor of a diagram for G_2? How should we choose the diagrams so as to have an approximation that is conserving? Furthermore, physical intuition may leave aside exotic diagrams like the one depicted below, where a particle in $4'$ interacts "with itself" in $3'$ and during the propagation between $4'$ and $3'$ scatters with the particle from $1'$. This process seems a bit bizarre but, as we shall see, it exists.

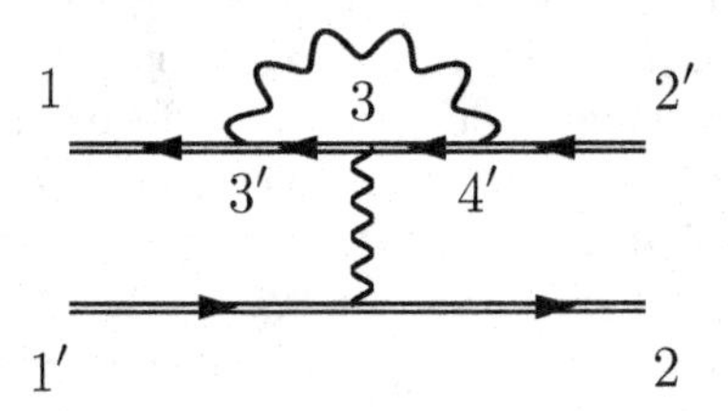

14.2 Diagrams for G_2 and Loop Rule

The expansion of G_2 in terms of the noninteracting Green's function G_0 and interaction v is given in (5.18), which we write again below:

$$G_2(a,b;c,d) = \frac{\sum_{k=0}^{\infty} \frac{1}{k!}\left(\frac{\mathrm{i}}{2}\right)^k \int v(1;1')\ldots v(k;k') \begin{vmatrix} G_0(a;c) & G_0(a;d) & \ldots & G_0(a;k'^+) \\ G_0(b;c) & G_0(b;d) & \ldots & G_0(b;k'^+) \\ \vdots & \vdots & \ddots & \vdots \\ G_0(k';c) & G_0(k';d) & \ldots & G_0(k';k'^+) \end{vmatrix}_\pm}{\sum_{k=0}^{\infty} \frac{1}{k!}\left(\frac{\mathrm{i}}{2}\right)^k \int v(1;1')\ldots v(k;k') \begin{vmatrix} G_0(1;1^+) & G_0(1;1'^+) & \ldots & G_0(1;k'^+) \\ G_0(1';1^+) & G_0(1';1'^+) & \ldots & G_0(1';k'^+) \\ \vdots & \vdots & \ddots & \vdots \\ G_0(k';1^+) & G_0(k';1'^+) & \ldots & G_0(k';k'^+) \end{vmatrix}_\pm}. \tag{14.4}$$

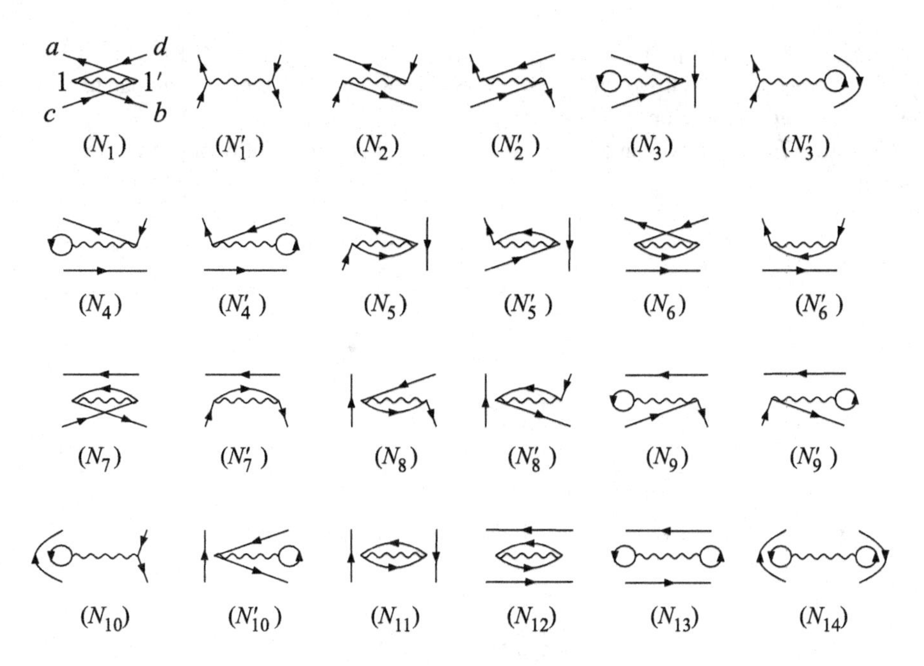

Figure 14.4 The 24 first-order diagrams of the numerator of G_2. The diagrams N_j and N'_j are obtained from one another by a mirroring of the interaction lines. The last four diagrams can be written as the product of a G_2-diagram and a vacuum diagram.

The MBPT for the two-particle Green's function is completely defined by (14.4). To see what kind of simplifications we can make, let us familiarize ourselves with the diagrammatic representation of the various terms. Let $N(a,b;c,d)$ be the numerator of (14.4). To zeroth order in the interaction, N is

$$N^{(0)}(a,b;c,d) = \begin{vmatrix} G_0(a;c) & G_0(a;d) \\ G_0(b;c) & G_0(b;d) \end{vmatrix} = G_0(a;c)G_0(b;d) \pm G_0(a;d)G_0(b;c). \quad (14.5)$$

The two terms in the r.h.s. of (14.5) are represented by

$N^{(0)}(a,b;c,d)$ = [a ↑ c] [d ↓ b] + [a ← d] [c → b] (14.6)

where the prefactor is incorporated in the diagrams and it is given by the sign of the permutation. The calculation of $N(a,b;c,d)$ to first order requires the expansion of the permanent/determinant of a 4×4 matrix. This yields $4! = 24$ terms whose diagrammatic representation is displayed in Fig. 14.4. The prefactor of each diagram is $(\mathrm{i}/2)$ times the sign of the permutation and $N^{(1)}(a,b;c,d)$ is simply the sum of all of them. Going to higher order, we generate diagrams with four external vertices a, b, c, d and a certain number of interaction lines. The prefactor of an nth-order diagram is $\frac{1}{n!}(\mathrm{i}/2)^n(\pm)^P$ with $(\pm)^P$ the sign of the permutation. It would be useful to have a graphical method (similar to the loop rule for the Green's function) to determine the sign of a diagram.

Loop rule for G_2 If we have n interaction lines, then the identity permutation gives the diagram

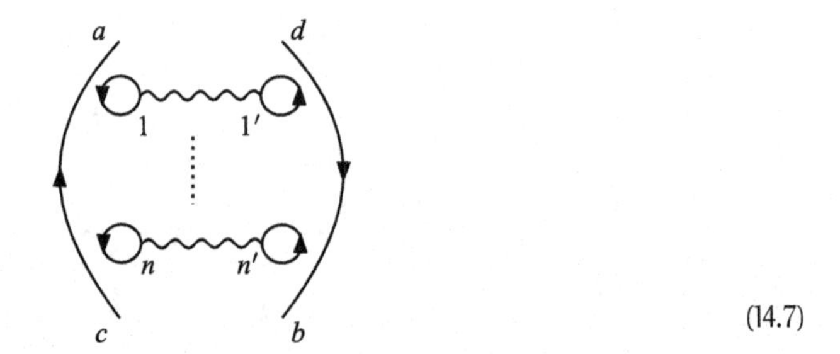

(14.7)

whose sign is $+$. The permutation that interchanges only c and d gives the diagram

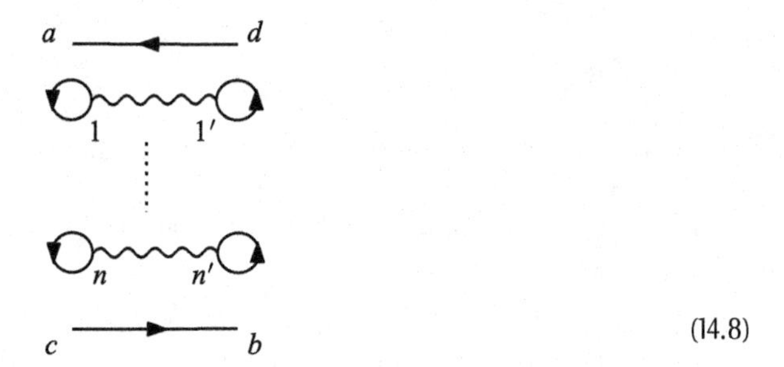

(14.8)

whose sign is $(\pm)$. Any subsequent interchange that keeps connected the vertices (a, c) and (b, d) in (14.7) or (a, d) and (b, c) in (14.8) changes the number of loops by one We then consider a G_2-diagram for $G_2(a, b; c, d)$ with l' loops. By closing this diagram with the Green's function lines $G_0(c; a)$ and $G_0(d; b)$,

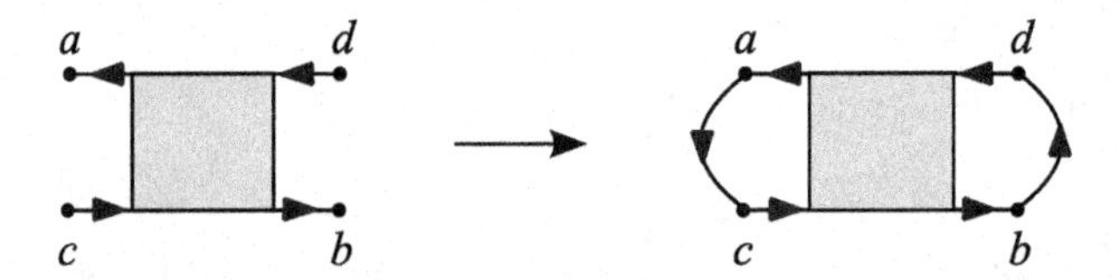

we obtain a diagram with a number of loops $l = l' + 2$ if the G_2-diagram originates from (14.7) [and hence (a, c) and (b, d) are connected] and $l = l' + 1$ if the G_2-diagram originates from (14.8) [and hence (a, d) and (b, c) are connected].[1] Therefore, the sign of a diagram can easily be fixed according to the loop rule for the two-particle Green's function

$$(\pm)^P = (\pm)^l,$$

[1]Notice that these diagrams are not vacuum diagrams. The closing G-lines are only used for graphical purposes to count the number of loops.

where l is the number of loops of the closed G_2-diagram. In the next section we show how to reduce the number of diagrams by exploiting their symmetry properties, as well as the topological notion of skeleton diagrams.

Exercise 14.1 Show that the diagrams in Fig. 14.4 correspond to the 24 terms of the expansion of $N^{(1)}(a, b; c, d)$.

14.3 Skeletonic Expansion in G and W

In analogy with the one-particle Green's function, the vacuum diagrams appearing in the numerator $N(a, b; c, d)$ are canceled by the denominator of (14.4). For instance, the last four diagrams in Fig. 14.4 can be discarded. Then we can rewrite (14.4) without the denominator, provided that we retain only connected G_2-diagrams in the expansion of N (a G_2-diagram is connected if it does not contain vacuum diagrams). To first order, G_2 is the sum of the two diagrams in (14.6) and the first 20 diagrams in Fig. 14.4. The number of diagrams can be further reduced by taking into account that any permutation or mirroring of the interaction lines leads to a topologically equivalent diagram. Thus, to order n each diagram appears in $2^n n!$ variants. For $n = 1$ the number of variants is 2, as it is clearly illustrated in Fig. 14.4. Here the diagrams N_j and N'_j with $j = 1, \ldots, 10$ are obtained from one another by mirroring the interaction line $v(1; 1') \to v(1'; 1)$. In conclusion, the MBPT formula (14.4) for G_2 simplifies to

$$G_2(a, b; c, d) = \sum_{k=0}^{\infty} \mathrm{i}^n \int v(1; 1') \ldots v(k; k') \begin{vmatrix} G_0(a; c) & G_0(a; d) & \ldots & G_0(a; k'^+) \\ G_0(b; c) & G_0(b; d) & \ldots & G_0(b; k'^+) \\ \vdots & \vdots & \ddots & \vdots \\ G_0(k'; c) & G_0(k'; d) & \ldots & G_0(k'; k'^+) \end{vmatrix}_{\substack{\pm \\ c \\ t.i.}}, \tag{14.9}$$

where the symbol $|\ldots|_{\substack{\pm \\ c \\ t.i.}}$ signifies that in the expansion of the permanent/determinant only the terms represented by connected and topologically inequivalent diagrams are retained. In this way the number of first-order diagrams reduces from 20 to 10.

From (14.9) we see that it is convenient to change the Feynman rules for calculating the prefactors. From now on we use the following Feynman rules for the G_2-diagrams:

- Number all vertices and assign an interaction line $v(i; j)$ to a wiggly line between j and i and a Green's function $G_0(i; j^+)$ to an oriented line from j to i.
- Integrate over all internal vertices and multiply by $\mathrm{i}^n(\pm)^l$ where n is the number of interaction lines and l is the number of loops in the closed G_2-diagram.

The expansion (14.9) contains also diagrams with self-energy insertions, see Fig. 14.4. A further reduction in the number of diagrams can be achieved by expanding G_2 in terms of the dressed Green's function G. This means to remove from (14.9) all diagrams with self-energy insertions and then replace G_0 with G. In this way the expansion of G_2 to first order in the interaction contains only four diagrams and reads

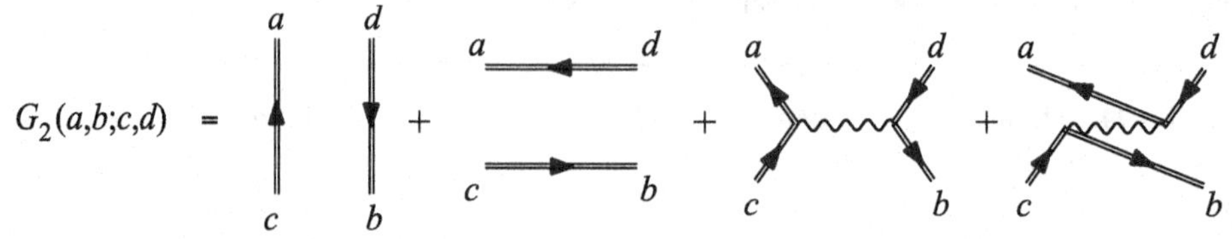

$$G_2(a,b;c,d) = \quad + \quad + \quad + \tag{14.10}$$

which is the second Born approximation to G_2. Finally we could remove from (14.9) all diagrams with polarization insertions and then replace v with W. Keeping only G- and W-skeleton diagrams, (14.9) provides us with an expansion of G_2 in terms of the dressed Green's function G and interaction W. Below we deepen the structure of the expansion of G_2 in G-skeleton diagrams and postpone to Section 14.9 the discussion in terms of G- and W-skeleton diagrams.

14.4 Bethe-Salpeter Equation for the Kernel and the Two-Particle XC Function

We represent $G_2(1,2;3,4)$ as a gray square with two outgoing lines starting from opposite vertices of the square and ending in 1 and 2, and two ingoing lines starting from 3 and 4 and ending in the remaining vertices of the square:

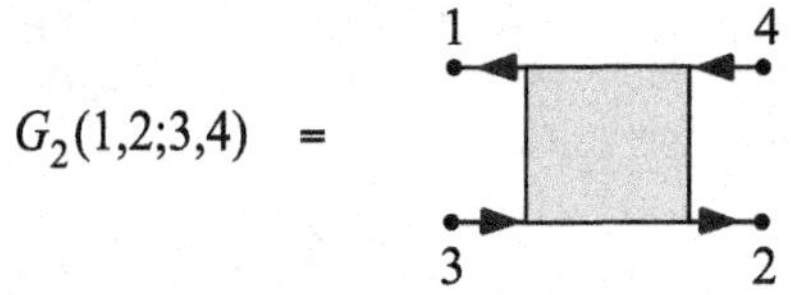

$$G_2(1,2;3,4) = \tag{14.11}$$

This diagrammatic representation is unambiguous if we specify that the third variable (which is 3 in this case) is located to the right of an imaginary *oriented* line connecting diagonally the first to the second variable (which are 1 and 2 in this case). Indeed, the only other possible source of ambiguity is which variable is first; looking at (14.11) we could say that it represents either $G_2(1,2;3,4)$ or $G_2(2;1;4,3)$. However, the exact as well as any conserving approximation to G_2 is such that

$$G_2(1,2;3,4) = G_2(2;1;4,3), \tag{14.12}$$

and therefore we do not need to bother about which variable is first.

Irreducible kernel for G_2 From the G-skeletonic expansion it follows that the general structure of G_2 is

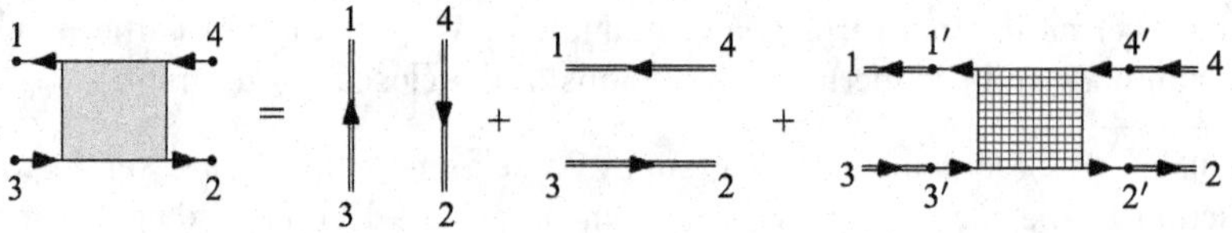

$$= \quad + \quad + \tag{14.13}$$

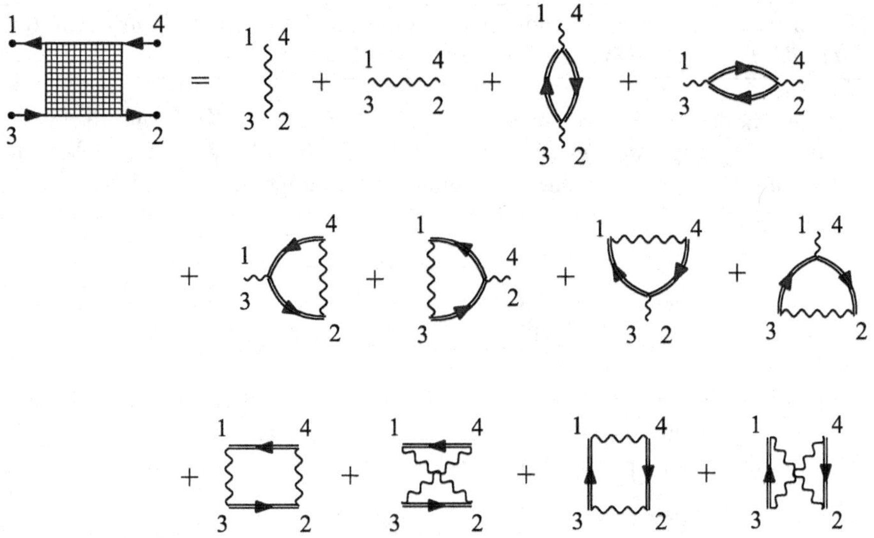

Figure 14.5 Diagrams for the kernel K_r up to second order in the interaction.

or, in formula,

$$G_2(1,2;3,4) = G(1;3)G(2;4) \pm G(1;4)G(2;3) + \int G(1;1')G(3';3)K_r(1',2';3',4')G(4';4)G(2;2'), \quad (14.14)$$

where the kernel K_r is represented by the square grid and accounts for all G_2-diagrams of order larger than zero. These diagrams are all connected since if the piece with external vertices, say, 1 and 3, was disconnected from the piece with external vertices 2 and 4 [like the first diagram in (14.13)], then there would certainly be a self-energy insertion. The rules to assign the variables to the square grid $K_r(1,2;3,4)$ are the same as those for G_2: The first two variables 1 and 2 label the opposite vertices with an outgoing line and the third variable 3 labels the vertex located to the right of an imaginary oriented line going from 1 to 2. The diagrams for K_r up to second order in the interaction are shown in Fig. 14.5. The Feynman rules to convert them into mathematical expressions are the same as those for G_2 (see the beginning of this section). The only extra rule is that if two external vertices i and j coincide, like in the first eight diagrams of Fig. 14.5, then we must multiply the diagram by $\delta(i;j)$. This is the same rule that we have introduced for the Hartree self-energy diagram, see discussion below (7.12).

From Fig. 14.5 we see that in the fourth, fifth, sixth, and ninth diagrams vertices 1 and 3 can be disconnected from vertices 2 and 4 by cutting two G-lines. The remaining diagrams are instead *two-particle irreducible* since $(1,3)$ and $(2,4)$ cannot be disjoint by cutting two G-lines. This is the reason for the subscript r in kernel K_r, which contains also two-particle *reducible* diagrams and can therefore be called a reducible kernel. Let us denote by K the *irreducible kernel.*

K is obtained from K_r by removing all two-particle reducible diagrams. The Feynman rules for K are identical to the Feynman rules for K_r.

A generic K_r-diagram has the structure $K^{(1)}GGK^{(2)}GG\ldots K^{(n)}$, with the $K^{(i)}$ two-particle irreducible diagrams. Let us check how the sign of a K_r-diagram is related to the sign of the constituents $K^{(i)}$ irreducible diagrams. Let l be the number of loops of a closed K_r-diagram with the structure $K^{(1)}GGK^{(2)}$:

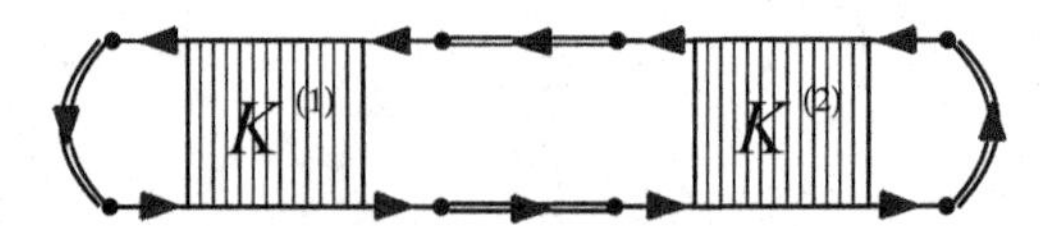

where the irreducible kernels are represented by a square with vertical stripes. We see that the closed $K^{(1)}$-diagram and $K^{(2)}$-diagram are obtained from the above diagram by an interchange of the starting points of the connecting GG-double-lines:

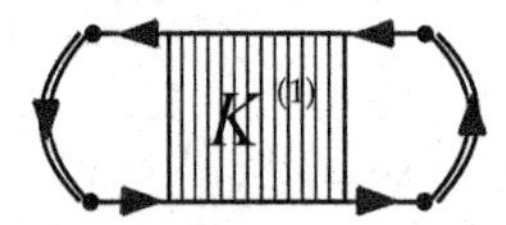

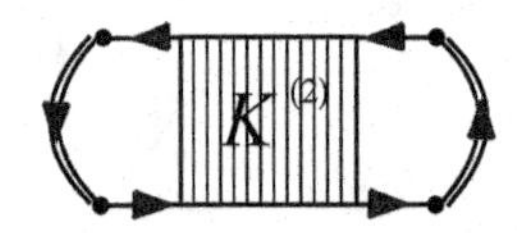

Since a single interchange adds or removes a loop, we conclude that $(\pm)^l = (\pm)^{l_1+l_2+1}$, where l_1 and l_2 are the number of loops in the closed diagrams $K^{(1)}$ and $K^{(2)}$. This property is readily seen to be general. If $K_r = K^{(1)}GGK^{(2)}GG\ldots K^{(n)}$ contains n irreducible kernels with signs $(\pm)^{l_i}$, then the sign of the K_r-diagram is $(\pm)^{l_1+\ldots+l_n+n-1}$. Thus we can alternatively formulate the Feynman rule for the prefactor of a K_r-diagram as:

- If $K_r = K^{(1)}GGK^{(2)}GG\ldots K^{(n)}$ with $K^{(i)}$ irreducible diagrams of order k_i and sign $(\pm)^{l_i}$, then the prefactor of K_r is

$$\mathrm{i}^{k_1+\ldots+k_n}(\pm)^{l_1+\ldots+l_n+n-1}.$$

In other words every GG-double-line contributes with a $(\pm)$ to the overall sign.

Let us consider a few examples. For the fourth K_r-diagram of Fig. 14.5 the prefactor is

$$\mathrm{i}^2(\pm)^l = \mathrm{i}^2(\pm)^3 = \pm\,\mathrm{i}^2. \tag{14.15}$$

The same diagram can be written as $K^{(1)}GGK^{(2)}$, where $K^{(1)}(1,2;3,4) = \text{prefactor} \times \delta(1;3)\delta(2;4)\, v(1;2)$ and $K^{(2)} = K^{(1)}$. Since the closed $K^{(1)}$-diagram has two loops, the prefactor of $K^{(1)}$ is

$$\mathrm{i}(\pm)^{l_1} = \mathrm{i}(\pm)^2 = \mathrm{i}.$$

The $K^{(2)}$-diagram has the same prefactor and since we have only one GG-double-line the overall prefactor is $\pm\,\mathrm{i}^2$, in agreement with (14.15). A second example is the fifth K_r-diagram of Fig. 14.5, whose prefactor is

$$\mathrm{i}^2(\pm)^l = \mathrm{i}^2(\pm)^2 = \mathrm{i}^2. \tag{14.16}$$

Also this diagram has the structure $K^{(1)}GGK^{(2)}$, where $K^{(1)}$ is the same as before while $K^{(2)}(1,2;3,4) = \text{prefactor} \times \delta(1;4)\delta(2;3)v(1;3)$. Since the closed $K^{(2)}$-diagram has one loop, the prefactor of $K^{(2)}$ is

$$\mathrm{i}(\pm)^{l_2} = \mathrm{i}(\pm)^1 = \pm\mathrm{i}.$$

Again the product of the prefactors of $K^{(1)}$ and $K^{(2)}$ times the $(\pm)$ sign coming from the GG-double-line agrees with the prefactor (14.16).

Bethe–Salpeter equation for the kernel According to the previous discussion we can write the following Dyson-like equation for the reducible kernel (integral over primed variables is understood):

$$K_r(1,2;3,4) = K(1,2;3,4) \pm \int K(1,2';3,4')G(4';1')G(3';2')K_r(1',2;3',4), \quad (14.17)$$

which is represented by the diagrammatic equation (remember that every GG-double-line contributes with a $\pm$):

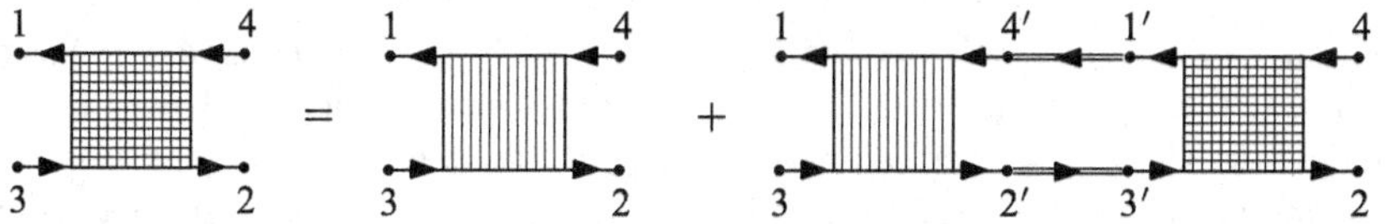

1 4 = 1 4 + 1 4′ 1′ 4
3 2 3 2 3 2′ 3′ 2 (14.18)

Equation (14.17) is known as the *Bethe–Salpeter equation* for the reducible kernel. Inserting (14.18) into (14.13), we find

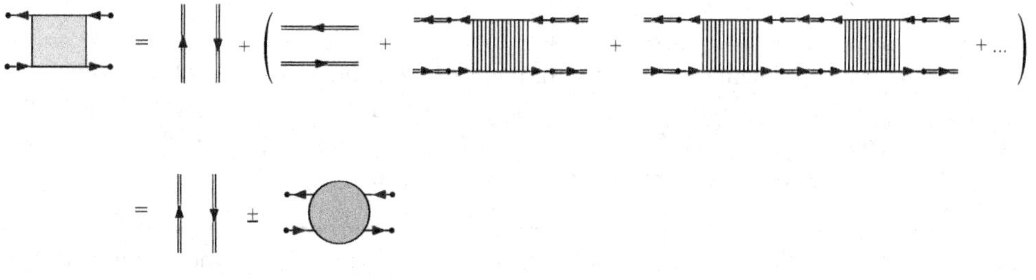

= + (+ + + ...)

= ±

where the gray blob is the two-particle XC function already encountered in (13.15):

$$L(1,2;3,4) \equiv \pm\big[G_2(1,2;3,4) - G(1;3)G(2;4)\big]. \quad (14.19)$$

The name "two-particle XC function" stems from the fact that L is obtained from G_2 by subtracting the Hartree diagram.

Bethe–Salpeter equation for the two-particle XC function The function L fulfills the diagrammatic equation

1 4 = ± 1 4 + 1 1′ 4′ 4
3 2 3 2 3 3′ 2′ 2

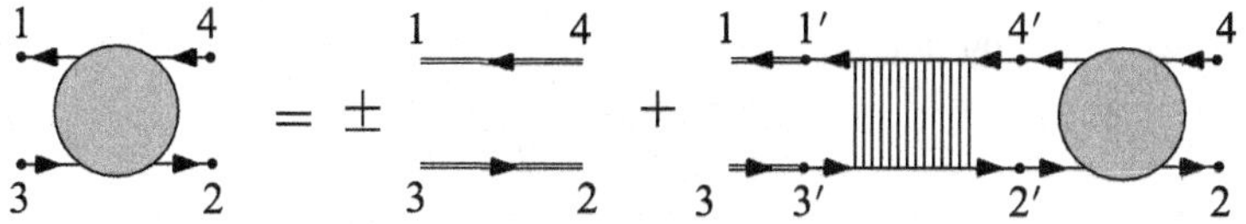

or, in formulas (integral over primed variables is understood),

$$\boxed{L(1,2;3,4) = G(1;4)G(2;3) \pm \int G(1;1')G(3';3)K(1',2';3',4')L(4',2;2',4)} \quad (14.20)$$

The reader can easily check the correctness of this equation by iterating it. Equation (14.20) is known as the *Bethe–Salpeter equation* for L.

There are several reasons for introducing yet another quantity like L. One reason is that by setting $4 = 2$ and $3 = 1$, the L-diagrams are the same as the χ-diagrams, with χ the density response function (or reducible polarization) defined in Section 11.7. In fact, the two quantities differ just by a prefactor that can be deduced by comparing the Feynman rules for polarization with the Feynman rules for G_2. A polarization diagram of order n has prefactor $\mathrm{i}^{n+1}(\pm)^l$, while the same diagram seen as an L-diagram in the limit $4 \to 2$ and $3 \to 1$ has prefactor $\mathrm{i}^n(\pm)^l$ times the sign $(\pm)$, which comes from the definition of L, see (14.19). Thus the relation between χ and L is

$$\boxed{\chi(1;2) = \pm\mathrm{i}\,L(1,2;1,2)} \quad (14.21)$$

We notice that (14.21) is not well defined if $L(1,2;1,2)$ is calculated as the difference between G_2 and GG. The reason is that in GG both Green's functions have the same time arguments. As already emphasized several times, these kind of ambiguities are always removed by shifting the arguments of the starting points. The precise way to write the relation between χ and L is $\chi(1;2) = \pm\mathrm{i}\,L(1,2;1^+,2^+)$. This is the same relation as (13.16) between χ_{00} and L. We conclude that

$$\chi_{00} = \chi.$$

If, on the other hand, L is defined as the solution of the Bethe–Salpeter equation or as the sum of all G_2-diagrams with the exception of the Hartree diagram, then (14.21) is well defined.

Another reason for introducing the two-particle XC function is that the change of the Green's function $G \to G + \delta G$ induced by a change of the single-particle Hamiltonian $h \to h + \delta h$ can be written in terms of L, see Section 14.6.

Exercise 14.2 Show that the vacuum diagrams in N are canceled by the denominator of (14.4).

Exercise 14.3 Show that (9.5) and (9.6) are both satisfied with G_2 given by (14.10) and Σ given by the second Born approximation.

Exercise 14.4 Show that the diagrams of K_r to second order in v are all those illustrated in Fig. 14.5.

14.5 Diagrammatic Proof of $K = \pm\delta\Sigma/\delta G$

In this section we prove the important relation

$$\boxed{K(1,2;3,4) = \pm\frac{\delta\Sigma(1;3)}{\delta G(4;2)}} \quad (14.22)$$

between the self-energy $\Sigma = \Sigma_s[G, v]$ and the kernel $K = K_s[G, v]$ of the Bethe–Salpeter equation. In (14.22) Σ_s and K_s are functionals of G and v (built using G-skeleton diagrams). The proof consists of showing that by cutting a G-line in all possible ways from every Σ-diagram,[2] we get the full set of K-diagrams each with the right prefactor. Let us first consider some examples. For the Hartree–Fock self-energy

$$\Sigma_{\rm HF}(1;3) =$$

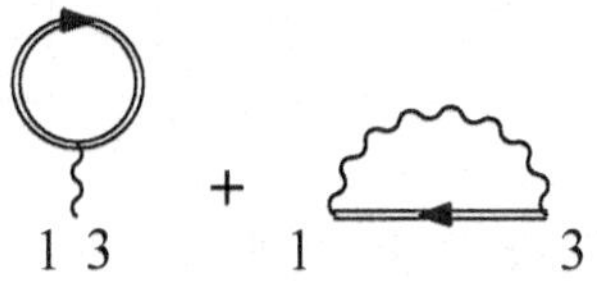

the prefactor is $(\pm\mathrm{i})$ for the first diagram and $(+\mathrm{i})$ for the second diagram. The functional derivative with respect to $G(4;2)$ yields the K-diagrams

$$\pm\frac{\delta\Sigma_{\rm HF}(1;3)}{\delta G(4;2)} = \qquad (14.23)$$

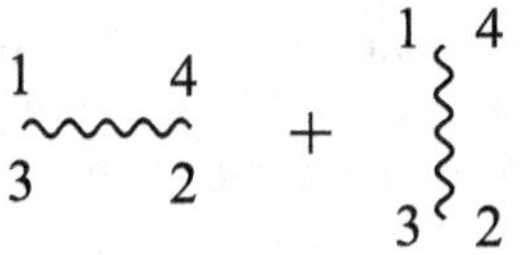

whose prefactors are correctly given by plus/minus the prefactors of the Σ-diagrams from which they originate. Another example is the first bubble diagram of the second Born approximation,

$$\Sigma_{\rm 2B,bubble}(1;3) =$$

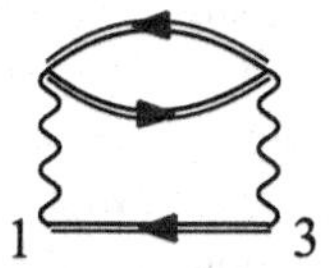

with prefactor $\pm\mathrm{i}^2$. Its functional derivative generates the following three diagrams for the kernel:

$$\pm\frac{\delta\Sigma_{\rm 2B,bubble}(1;3)}{\delta G(4;2)} =$$

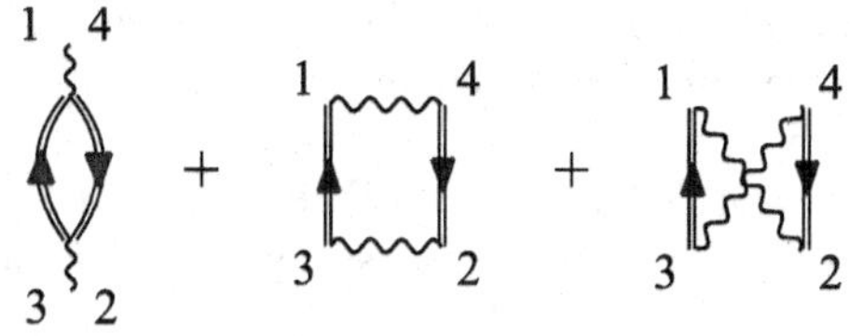

The prefactor of the first diagram is $\mathrm{i}^2(\pm)^l = \mathrm{i}^2(\pm)^2 = \mathrm{i}^2$. Similarly, one can calculate the prefactor of the other two diagrams and check that it is again given by i^2.

The general proof of (14.22) follows from a few statements:

- Closing a K-diagram $K^{(i)}(1,2;3,4)$ with a Green's function $G(4;2)$ leads to a skeletonic Σ-diagram up to a prefactor (we prove below that the prefactor is $\pm$).

---

[2] As already observed several times, to cut a G-line in all possible ways is equivalent of taking the functional derivative with respect to G.

Proof. Suppose that the self-energy diagram obtained from $K^{(i)}$ is not skeletonic. Then it must be of the form

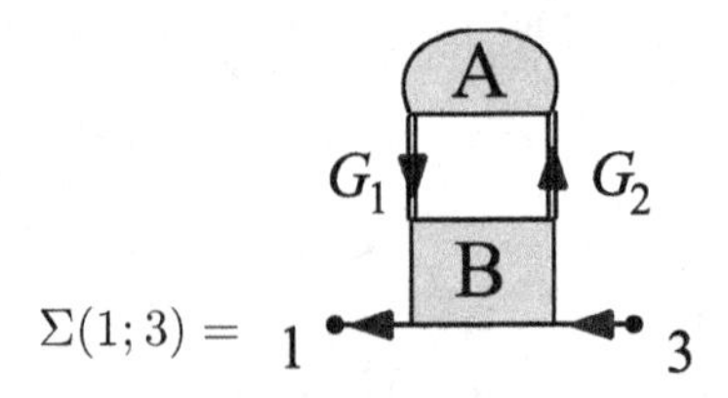

where A is a self-energy insertion. If the $G(4;2)$ that we used to close the K-diagram is in A, then the original K-diagram would be two-particle *reducible* since it could be divided into two disjoint pieces containing $(1,3)$ and $(2,4)$ by cutting two G-lines. This is in contradiction to the definition of the irreducible kernel K. The added line $G(4;2)$ cannot be G_1 or G_2 as otherwise the original K-diagram would contain a self-energy insertion and hence it would not be skeletonic. For the very same reason, the added line $G(4;2)$ cannot be in B either. We conclude that the self-energy diagram obtained by closing a K-diagram with a G-line is skeletonic.

- Every K-diagram is obtained from a unique Σ-diagram. Moreover, every Σ-diagram with n G-lines gives n topologically inequivalent K-diagrams.

  Proof. From the previous statement we know that by closing a K-diagram with a G-line we generate a Σ-diagram and hence there must be at least one Σ-diagram from which this K-diagram can be obtained. Furthermore, it is clear that the cutting of a G-line in two topologically inequivalent Σ-diagrams $\Sigma^{(1)}$ and $\Sigma^{(2)}$ cannot lead to topologically equivalent K-diagrams $K^{(1)}$ and $K^{(2)}$, since if we added back the G-line to $K^{(1)}$ and $K^{(2)}$ we would find that $\Sigma^{(1)}$ and $\Sigma^{(2)}$ have the same topology, contrary to the assumption. We thus conclude that every K-diagram comes from a unique Σ-diagram. The remaining question is then whether a given Σ-diagram can lead to two (or more) topologically equivalent K-diagrams $K^{(1)}$ and $K^{(2)}$ by cutting two different G-lines. In Section 11.3 we said that by cutting a G-line from a vacuum diagram we obtain N_S topological equivalent Σ-diagrams, where N_S is the number of symmetries of the vacuum diagram. Similarly, if $K^{(1)}$ and $K^{(2)}$ are topologically equivalent then there must be a symmetry s of the generating Σ-diagram that maps the set $\mathcal{G}$ of G-lines into itself. However, from the third statement of Section 11.3 it follows that the only possible symmetry is the identity permutation since the external vertices of Σ are fixed. We conclude that by removing two different G-lines we always obtain two topologically inequivalent K-diagrams.

- The prefactor of a K-diagram calculated from (14.22) agrees with the Feynman rules for the K-diagrams.

  Proof. The prefactor of a K-diagram calculated from (14.22) is plus/minus the prefactor of the generating Σ-diagram – that is, $\mathrm{i}^k(\pm)^{l+1}$, where k is the number of interaction lines and l is the number of loops in the Σ-diagram. Since the Σ- and K-diagrams have the same number of interaction lines, the factor i^k is certainly correct. We only need to check the sign. The sign of a K-diagram is, by definition, given by the number of loops in the diagram $K(1,2;3,4)$ closed with $G(4;2)$ and $G(3;1)$. When closing the K-diagram with $G(4;2)$ we get back $\Sigma(1;3)$. If we now close $\Sigma(1;3)$ with $G(3;1)$ we increase the number of loops in Σ by one. This proves the statement.

These three statements constitute the diagrammatic proof of (14.22). For Φ-derivable self-energies we can also write

$$K(1,2;3,4) = \pm\frac{\delta^2\Phi}{\delta G(3;1)\delta G(4;2)}. \tag{14.24}$$

Thus, K is obtained from Φ by differentiating twice with respect to G. Since the order of the functional derivatives does not matter, we have

$$K(1,2;3,4) = \pm\frac{\delta^2\Phi}{\delta G(3;1)\delta G(4;2)} = \pm\frac{\delta^2\Phi}{\delta G(4;2)\delta G(3;1)} = K(2,1;4,3), \tag{14.25}$$

which is indeed a symmetry of the two-particle Green's function, see (14.12). We can also regard (14.25) as a vanishing "curl" condition for $\Sigma = \Sigma_s[G,v]$,

$$\frac{\delta\Sigma(1;3)}{\delta G(4;2)} - \frac{\delta\Sigma(2;4)}{\delta G(3;1)} = 0, \tag{14.26}$$

which is a necessary condition for the existence of a functional Φ such that Σ is Φ-derivable. Thus, another way to establish whether a given Σ is Φ-derivable consists in checking the validity of (14.26).

14.6 Bethe–Salpeter Equation from the Variation of a Conserving G

In Section 14.4 we learned that L and hence $\chi_{\mu\nu}$ can be calculated by solving the Bethe–Salpeter equation. In practice this is done by approximating the kernel K in some way. We then inevitably come across the following question: How are the δJ_μ calculated from an approximate L via (13.14) and the δJ_μ calculated directly from an approximate G related? This question is of fundamental importance since we would like to export everything we have derived on conserving approximations into the linear response world, so that all basic conservation laws are automatically built in. In this section we show that the δJ_μ coming from a conserving G with self-energy Σ is the same as the δJ_μ coming from (13.14), where L is the solution of the Bethe–Salpeter equation with kernel $K = \pm\delta\Sigma/\delta G$. The proof is carried out in the general case of a time-dependent perturbation added to a pre-existing time-dependent electromagnetic field since no extra complications arise from this. Another motivation for keeping the formulas so general is to prove an important relation between L and δG, see Section 14.6.

We consider again a system of interacting particles of charge q subject to a time-dependent vector potential $\mathbf{A} + \delta\mathbf{A}$ and scalar potential $\phi + \delta\phi$. The interaction between the particles is accounted for by some Φ-derivable self-energy Σ. The equation of motion for the Green's function G_{tot} in the total fields is, to first order in $\delta\mathbf{A}$ and $\delta\phi$, given by

$$\left[\mathrm{i}\frac{d}{dz_1} + \frac{1}{2m}D_1^2 - q\phi(1) - q\left(\delta\phi(1) - \frac{1}{c}\frac{\mathbf{D}_1\cdot\delta\mathbf{A}(1) + \delta\mathbf{A}(1)\cdot\mathbf{D}_1}{2m\mathrm{i}}\right)\right] G_{\text{tot}}(1;3) - \int d2\, \Sigma[G_{\text{tot}}](1;2)G_{\text{tot}}(2;3) = \delta(1;3), \quad (14.27)$$

where the functional dependence of Σ on the Green's function is made explicit. In this equation the perturbing potentials with the time variable on the contour are defined similarly to (12.3) and (12.4) [and in agreement with the more general definition (4.6)]

$$\delta\mathbf{A}(\mathbf{x}, z = t_\pm) = \delta\mathbf{A}(\mathbf{x}, t), \qquad \delta\phi(\mathbf{x}, z = t_\pm) = \delta\phi(\mathbf{x}, t), \quad (14.28)$$

and $\delta\mathbf{A} = \delta\phi = 0$ for z on the vertical track. The Green's function G of the system with $\delta\mathbf{A} = \delta\phi = 0$ obeys the equation of motion

$$\left[\mathrm{i}\frac{d}{dz_1} + \frac{1}{2m}D_1^2 - q\phi(1)\right] G(1;3) - \int d3\, \Sigma[G](1;2)G(2;3) = \delta(1;3), \quad (14.29)$$

with the same functional form of the self-energy Σ (otherwise the two systems would be treated at a different level of approximation). For later convenience we introduce the shorthand notation [compare with (4.62)]

$$\overrightarrow{G}^{-1}(1;2) \equiv \left[\left(\mathrm{i}\frac{\overrightarrow{d}}{dz_1} + \frac{\overrightarrow{D}_1^2}{2m} - q\phi(1)\right)\delta(1;2) - \Sigma(1;2)\right],$$

$$\overleftarrow{G}^{-1}(1;2) \equiv \left[\delta(1;2)\left(-\mathrm{i}\frac{\overleftarrow{d}}{dz_2} + \frac{(\overleftarrow{D}_2^2)^*}{2m} - q\phi(2)\right) - \Sigma(1;2)\right],$$

where, as usual, the left/right arrows specify that the derivatives act on the quantity to their left/right. In terms of these differential operators, the equations of motion read

$$\int d2\, \overrightarrow{G}^{-1}(1;2)G(2;3) = \int d2\, G(1;2)\overleftarrow{G}^{-1}(2;3) = \delta(1;3). \quad (14.30)$$

We now derive an equation for the difference $\delta G = G_{\text{tot}} - G$. Adding and subtracting $\Sigma[G](1;2)G_{\text{tot}}(2;3)$ to and from the integrand of (14.29) and then subtracting (14.27), we find

$$\int d2\, \overrightarrow{G}^{-1}(1;2)\delta G(2;3) = \delta h(1)G_{\text{tot}}(1;3) + \int d2\, \delta\Sigma(1;2)G_{\text{tot}}(2;3), \quad (14.31)$$

where we define the perturbation

$$\delta h(1) \equiv q\left(\delta\phi(1) - \frac{1}{c}\frac{\mathbf{D}_1\cdot\delta\mathbf{A}(1) + \delta\mathbf{A}(1)\cdot\mathbf{D}_1}{2m\mathrm{i}}\right), \quad (14.32)$$

as well as the self-energy variation

$$\delta\Sigma(1;2) \equiv \Sigma[G_{\text{tot}}](1;2) - \Sigma[G](1;2).$$

Taking into account that G fulfills (14.30), we can rewrite (14.31) in the following integral form:

$$\delta G(1;3) = \int d2\, G(1;2)\delta h(2) G_{\rm tot}(2;3) + \int d2d4\, G(1;2)\delta\Sigma(2;4)G_{\rm tot}(4;3), \quad (14.33)$$

which can easily be verified by applying $\overrightarrow{G}^{-1}$ to both sides and by checking that both sides satisfy the KMS boundary conditions. We observe that if we had included the (discarded) term proportional to δA^2 in δh, then (14.33) would have been *exact* for all $\delta \mathbf{A}$ and $\delta\phi$. Since, however, we are interested in calculating δG to first order in the perturbing fields it is enough to consider the δh of (14.32). The next step is to evaluate the self-energy variation $\delta\Sigma$ to first order.

Consider a diagram for $\Sigma[G]$ containing n Green's functions. The variation of this diagram when $G \to G + \delta G$ is the sum of n diagrams obtained by replacing a G with a δG for all the G of the diagram. An example is the variation of the second-order bubble diagram illustrated below:

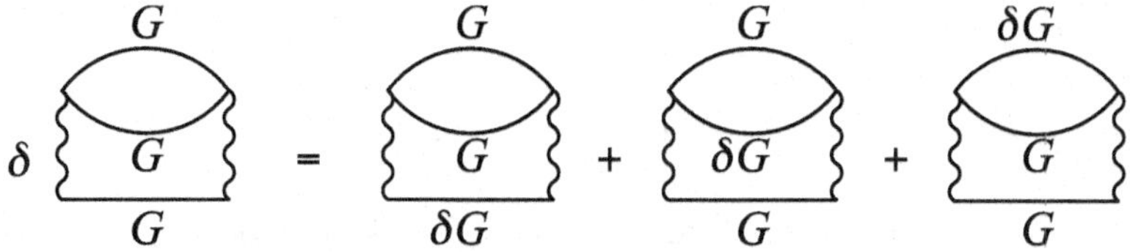

In accordance with the discussion on functional derivatives in Section 11.3 we see that the variation of Σ is the result of the operation $\delta\Sigma/\delta G$ (which cuts one Green's function line from Σ in all possible ways) multiplied by the variation of the Green's function. Thus, to first order we have

$$\delta\Sigma(2;4) = \Sigma[G+\delta G](2;4) - \Sigma[G](2;4) = \int d5d6\, \frac{\delta\Sigma(2;4)}{\delta G(5;6)}\delta G(5;6)$$

$$= \pm \int d5d6\, K(2,6;4,5)\delta G(5;6), \quad (14.34)$$

with K the kernel of the Bethe–Salpeter equation defined in Section 14.4. Equation (14.34) has the following diagrammatic representation (remember the rule for K: the first two variables label opposite vertices with outgoing lines and the third variable labels the vertex to the right of an imaginary oriented line connecting diagonally the first to the second variable):

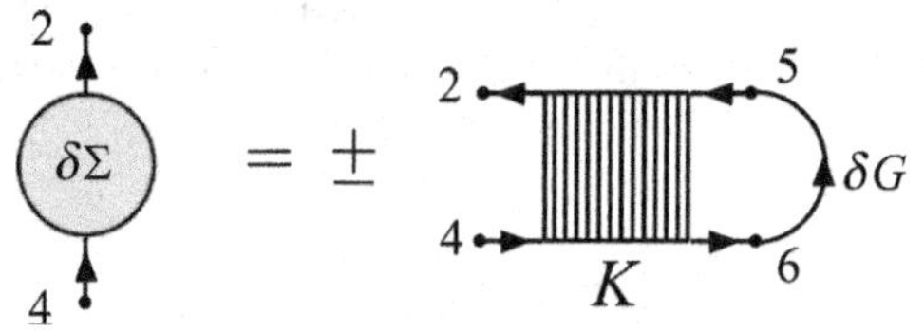

Since δh and $\delta\Sigma$ are of first order in the perturbing fields, we can replace $G_{\rm tot}$ with G in (14.33), thus obtaining

$$\delta G(1;3) = \int d2\; G(1;2)\delta h(2) G(2;3)$$

$$\pm \int d2d4d5d6\; G(1;2)G(4;3)K(2,6;4,5)\delta G(5;6). \quad (14.35)$$

This is a recursive equation for δG. Its solution provides the first-order change in G for a given Φ-derivable approximation to the self-energy.

To make contact with the discussion in Section 13.2, we manipulate the first term on the r.h.s. Using the explicit form of δh, we can write

$$\begin{aligned}\int d2\, G(1;2)\delta h(2)G(2;3) &= q\int d2\, G(1;2)\left(\delta\phi(2) - \frac{1}{c}\frac{-\overleftarrow{\mathbf{D}}_2^*\cdot\delta\mathbf{A}(2) + \delta\mathbf{A}(2)\cdot\mathbf{D}_2}{2mi}\right)G(2;3)\\ &= \int d2d4\left[\underbrace{q\delta(2;4)\left(\delta\phi(2) - \frac{\delta\mathbf{A}(2)}{c}\cdot\frac{\mathbf{D}_2 - \mathbf{D}_4^*}{2mi}\right)}_{\delta h(4;2)}G(1;4)G(2;3)\right],\end{aligned} \tag{14.36}$$

where in the first equality we perform an integration by parts (the arrow over $\mathbf{D}_2^*$ indicates that the derivative acts on the left) and use the identity (12.29). Substituting this result into (14.35), we find

$$\begin{aligned}\delta G(1;3) = &\int d2d4\ \delta h(4;2)\, G(1;4)G(2;3)\\ &\pm \int d2d4d5d6\ G(1;2)G(4;3)K(2,6;4,5)\delta G(5;6),\end{aligned} \tag{14.37}$$

where $\delta h(4;2)$ is the differential operator implicitly defined in (14.36). To visualize the structure of this equation, we give to $\delta h(4;2)$ the diagrammatic representation of a dashed line joining points 2 and 4. In this way, (14.37) is represented like this:

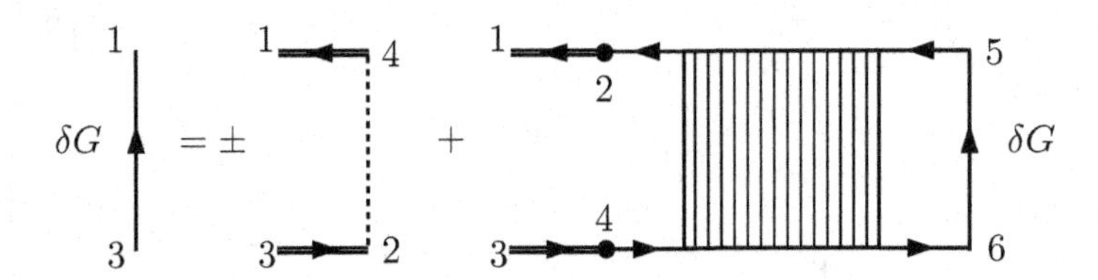

where every oriented line is a G-line and where the $\pm$ sign in (14.37) follows from the diagrammatic rule of the GG-double-line, see Section 14.4. Iterating this equation, we see that at every iteration the diagram with δG contains one more structure GGK. Therefore, the solution is

where the gray blob solves the Bethe–Salpeter equation (14.20) for the two-particle XC function with kernel $K = \pm\delta\Sigma/\delta G$, and hence it must be identified with L. Converting the diagrammatic solution into a formula, we obtain the following important result for the variation of a conserving Green's function:

$$\delta G(1;3) = \int d2d4\, \delta h(4;2)\, L(1,2;3,4)$$
$$= q\int d2 \left[\left(\delta\phi(2) - \frac{\delta \mathbf{A}(2)}{c}\frac{\mathbf{D}_2 - \mathbf{D}_4^*}{2mi}\right) L(1,2;3,4)\right]_{4=2}. \tag{14.38}$$

From this equation we can calculate the density and current variations,

$$\delta n(1) = \pm \mathrm{i}\delta G(1;1^+), \qquad \delta \mathbf{J}(1) = \pm \mathrm{i}\left(\frac{\mathbf{D}_1 - \mathbf{D}_2^*}{2mi}\delta G(1;2)\right)_{2=1^+} - \frac{q}{mc}n(1)\delta\mathbf{A}(1),$$

and it is a simple exercise to show that these variations can be written in the following compact form:[3]

$$\delta J_\mu(1) = q\int d2\, \chi_{\mu\nu}(1;2)\delta A^\nu(2) + \bar{\delta}_{0\mu}\frac{q}{m}n(1)\delta A^\mu(1), \tag{14.39}$$

with $\bar{\delta}_{0\mu} = 1 - \delta_{0\mu}$ and $\chi_{\mu\nu}$ the conserving response function defined, for a given L, as in (13.16) and (13.17). It is important to stress that in Section 13.2 the relations (13.16) and (13.17) were derived starting from the definitions of L and $\chi_{\mu\nu}$ in terms of the field operators. Here, instead, the same relations have been derived starting from the Dyson equation for G.[4] Equation (14.39) is an equation between quantities in Keldysh space and it contains an integral along the contour. To convert (14.39) into an equation on the real-time axis, we take, for example, $z_1 = t_{1-}$ on the forward branch and use the result of Exercise 5.5 to find

$$\delta J_\mu(\mathbf{x}_1,t_1) = q\int_{-\infty}^{\infty} dt_2 \int d\mathbf{x}_2\, \chi^{\mathrm{R}}_{\mu\nu}(\mathbf{x}_1,t_1;\mathbf{x}_2,t_2)\delta A^\nu(\mathbf{x}_2,t_2) + \bar{\delta}_{0\mu}\frac{q}{m}n(\mathbf{x}_1,t_1)\delta A^\mu(\mathbf{x}_1,t_1),$$

which is identical to (13.14), as it should be.

To summarize, we have shown that the current and the density variations produced by a conserving G with self-energy Σ are, to first order, obtained from (13.14) with a response function that satisfies the Bethe–Salpeter equation with kernel $K = \pm\delta\Sigma/\delta G$. In Section 14.10 we use (14.38) to prove an important identity between the vertex function and the self-energy.

14.7 Stationary Solutions of the Bethe–Salpeter Equation: Hartree plus Statically Screened Exchange Kernel

The Bethe–Salpeter equation has been successfully applied to the calculation of the photoabsorption spectra of several semiconductors and insulators, revealing excitonic features

[3] If L is defined as the solution of the Bethe–Salpeter equation, then $L(1,2;3,2) = L(1,2;3^+,2^+)$; this is due to the fact that in the Bethe–Salpeter equation the Hartree term in definition (13.15) is already subtracted and hence there is no ambiguity in setting $2^+ = 2$ and $3^+ = 3$.

[4] For an approximate response function there is no definition in terms of the field operators.

otherwise difficult to capture [206, 207, 208, 209]. In this section we show how to solve the Bethe–Salpeter equation for a system in a stationary state and for an approximation to the kernel that is suited to capture excitonic effects.

The lowest-order diagram of the Bethe–Salpeter equation (14.20) describes the propagation of a free particle–hole pair. In higher-order diagrams the particle interacts with the hole in many different ways, depending on the approximation to the kernel K. Thus, L can be interpreted as the interacting particle–hole propagator in a similar way as, for example, $G^<$ has been interpreted as a hole propagator, see Section 6.1.3. In Section 6.4.3 we learned that excitons are bound (or quasi-bound) particle–hole pairs. For L to describe the propagation of excitons, the kernel should be chosen in such a way that the particle and the hole interact directly through the Coulomb interaction. The lowest-order approximation to K in the skeletonic expansion with G and W is given by the first two diagrams of Fig. 14.5, where $v \to W$ in the first diagram.[5] Bearing in mind that we should use the lower sign for fermions, we have

$$K(1,2;3,4) =$$

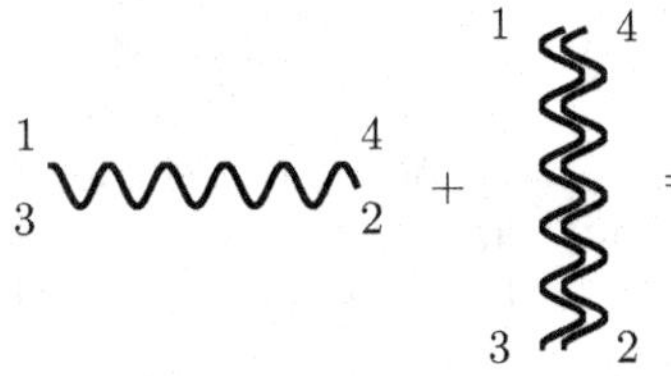

$$= \mathrm{i}\delta(1,3)\delta(2,4)v(1;2) - \mathrm{i}\delta(1,4)\delta(2,3)W(1;3) \qquad (14.40)$$

where the first diagram comes from the Hartree self-energy and the second diagram comes from the exchange (or Fock) self-energy with $v \to W$ (screened exchange). We refer to this approximation as the Hartree plus screened exchange (HSEX) approximation. Inserting this kernel into the Bethe–Salpeter equation, we find the diagrammatic equation for the two-particle XC function in Fig. 14.6. The exchange part of the kernel describes the direct interaction between the particle and the hole; we therefore expect the resulting L to contain the physics of excitons.

Statically screened approximation To calculate the response functions, we do not need L for all four different contour-time arguments – the knowledge of

$$L(\mathbf{x}_1, \mathbf{x}_2, \mathbf{x}_3, \mathbf{x}_4; z, z') \equiv L(\mathbf{x}_1, z, \mathbf{x}_2, z'; \mathbf{x}_3, z, \mathbf{x}_4, z') \qquad (14.41)$$

is enough, see again (13.16) and (13.17). It is generally not possible to close the Bethe–Salpeter equation for an L with this particular choice of times. However, if we make the so-called *statically screened approximation*[6]

$$W(1;2) \simeq \delta(z_1, z_2) W(\mathbf{r}_1, \mathbf{r}_2), \qquad (14.42)$$

then the Bethe–Salpeter equation becomes a closed equation for $L(\mathbf{x}_1, \mathbf{x}_2, \mathbf{x}_3, \mathbf{x}_4; z, z')$.

[5] In the second diagram of Fig. 14.5 the replacement $v \to W$ would generate two-particle reducible diagrams.

[6] This approximation amounts to ignoring the time nonlocality of the screened interaction. As we see in Chapter 15, the static screening is inadequate to capture an important collective excitation known as *plasmon*.

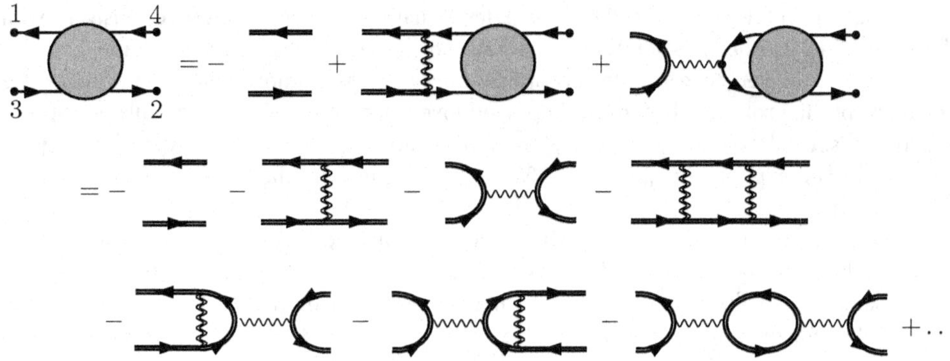

Figure 14.6 Diagrammatic representation of the XC response function in the HSEX approximation.

Using (14.42) in the diagrammatic equation of Fig. 14.6, we find

$$
\begin{aligned}
L(\mathbf{x}_1, \mathbf{x}_2, \mathbf{x}_3, \mathbf{x}_4; z, z') &= L^0(\mathbf{x}_1, \mathbf{x}_2, \mathbf{x}_3, \mathbf{x}_4; z, z') \\
&+ \mathrm{i} \int d\mathbf{x}_1' d\mathbf{x}_2' \int d\bar{z}\, L^0(\mathbf{x}_1, \mathbf{x}_2', \mathbf{x}_3, \mathbf{x}_1'; z, \bar{z}) W(\mathbf{x}_1', \mathbf{x}_2') L(\mathbf{x}_1', \mathbf{x}_2, \mathbf{x}_2', \mathbf{x}_4; \bar{z}, z') \\
&- \mathrm{i} \int d\mathbf{x}_1' d\mathbf{x}_2' \int d\bar{z}\, L^0(\mathbf{x}_1, \mathbf{x}_1', \mathbf{x}_3, \mathbf{x}_1'; z, \bar{z}) v(\mathbf{x}_1', \mathbf{x}_2') L(\mathbf{x}_2', \mathbf{x}_2, \mathbf{x}_2', \mathbf{x}_4; \bar{z}, z'),
\end{aligned}
$$

where

$$
L^0(\mathbf{x}_1, \mathbf{x}_2, \mathbf{x}_3, \mathbf{x}_4; z, z') = G(\mathbf{x}_1, z; \mathbf{x}_4, z') G(\mathbf{x}_2, z'; \mathbf{x}_3, z).
$$

In practical calculations the dressed Green's function G is approximated as a quasi-particle Green's function $G \simeq g$, where g has the form (13.38) and (13.39) with quasi-particle energies ϵ_n^{qp}.[7] Expanding L and L^0 as in (13.40), we can rewrite the Bethe–Salpeter equation in the static HSEX approximation as

$$
L_{\substack{ij\\mn}}(z, z') = \delta_{in}\delta_{jm}\, g_i(z, z') g_j(z', z) + \mathrm{i} \sum_{pq} \int d\bar{z}\; g_i(z, \bar{z}) g_j(\bar{z}, z) K_{\substack{ij\\pq}} L_{\substack{pq\\mn}}(\bar{z}, z'), \tag{14.43}
$$

where

$$
K_{\substack{ij\\pq}} \equiv W_{iqjp} - v_{iqpj} \tag{14.44}
$$

is the HSEX kernel. The four-index Coulomb integrals v_{iqpj} have been defined in (1.86). The screened tensor W_{iqjp} is defined in the same way with $v \to W$.

[7]Usually the quasi-particle energies are calculated from the self-consistent solution of $\epsilon_n^{\mathrm{qp}} = \mathrm{Re}\langle n|\hat{h} + \hat{\Sigma}(\epsilon_n)|n\rangle$, where the self-energy is often treated at the Hartree plus GW level.

At the time of writing, the Bethe–Salpeter equation has been solved exclusively with the kernel (14.44). The first solutions date back to the late 1970s, when the Bethe–Salpeter equation was applied to calculate the optical spectrum of diamond and silicon [212]. The accuracy of the solution has been improved over the years and the calculated optical spectra of several systems with strong excitonic features have been found to be in good agreement with experiments [207, 208, 209, 213]. Going beyond the static approximation for W represents a major computational challenge [214].

An alternative route to extract L with more sophisticated kernels consists in solving the Kadanoff–Baym equations for G [113, 215, 216]. In fact, the first-order change δG induced by a change δh in the single-particle Hamiltonian is given by (14.38), where L fulfills the Bethe–Salpeter equation with kernel $K(1,2;3,4) = \pm\delta\Sigma(1;3)/\delta G(4;2)$, see Section 14.6. Since the kernel (14.40) is just $\pm\delta\Sigma_{\rm HSEX}/\delta G$, it is clear that any approximation to the self-energy that goes beyond the HSEX approximation gives us access to L of higher degree of sophistication. Applications of this method are presented in Section 14.11.

14.7.1 Crystals: Bethe–Salpeter Equation at Finite Momentum Transfer

Let us specialize (14.43) to crystals. We write every one-particle label $i, j, \ldots$ in terms of a collective Greek index that specifies band and spin, and a Latin bold index that specifies the value of the quasi-momentum – for example, $i = \mathbf{k}\alpha$, $j = \mathbf{p}\beta$, see Section 2.3. The conservation of the quasi-momentum implies that the sum of the momenta in the pair (i, q) is the same as the sum of the momenta in the pair (j, p).[8] Therefore [217, 218],

$$K_{\substack{\mathbf{k}+\frac{\mathbf{q}}{2}\mu\ \mathbf{k}-\frac{\mathbf{q}}{2}\nu \\ \mathbf{k}''+\frac{\mathbf{q}''}{2}\alpha\ \mathbf{k}''-\frac{\mathbf{q}''}{2}\beta}} = \delta_{\mathbf{q}\mathbf{q}''} K^{\mathbf{q}}_{\substack{\mu\nu\mathbf{k} \\ \alpha\beta\mathbf{k}''}}, \tag{14.45}$$

which implicitly defines the tensor on the right-hand side. Taking into account that $v_{iqpj} = v^*_{jpqi}$ (and the like for W), see again (1.86), it is easy to show that $K^{\mathbf{q}}$ is self-adjoint:

$$K^{\mathbf{q}}_{\substack{\mu\nu\mathbf{k} \\ \alpha\beta\mathbf{k}''}} = K^{\mathbf{q}*}_{\substack{\alpha\beta\mathbf{k}'' \\ \mu\nu\mathbf{k}}}.$$

For a tensor K with the property in (14.45), the solution of (14.43) is a tensor L with the same property. Thus the Bethe–Salpeter equation reduces to

$$\begin{aligned} L_{\substack{\mathbf{k}+\frac{\mathbf{q}}{2}\mu\ \mathbf{k}-\frac{\mathbf{q}}{2}\nu \\ \mathbf{k}'+\frac{\mathbf{q}}{2}\rho\ \mathbf{k}'-\frac{\mathbf{q}}{2}\sigma}} &\equiv L^{\mathbf{q}}_{\substack{\mu\nu\mathbf{k} \\ \rho\sigma\mathbf{k}'}}(z,z') = \delta_{\mu\sigma}\delta_{\nu\rho}\delta_{\mathbf{k}\mathbf{k}'} g_{\mathbf{k}+\frac{\mathbf{q}}{2}\mu}(z,z') g_{\mathbf{k}-\frac{\mathbf{q}}{2}\nu}(z',z) \\ &+ \mathrm{i} \sum_{\alpha\beta\mathbf{k}''} \int d\bar{z}\, g_{\mathbf{k}+\frac{\mathbf{q}}{2}\mu}(z,\bar{z}) g_{\mathbf{k}-\frac{\mathbf{q}}{2}\nu}(\bar{z},z) K^{\mathbf{q}}_{\substack{\mu\nu\mathbf{k} \\ \alpha\beta\mathbf{k}''}} L^{\mathbf{q}}_{\substack{\alpha\beta\mathbf{k}'' \\ \rho\sigma\mathbf{k}'}}(\bar{z},z'). \end{aligned} \tag{14.46}$$

At this point we find it convenient to introduce the superindices $I = (\mu\nu\mathbf{k})$, $J = (\rho\sigma\mathbf{k}')$ etc. to rewrite (14.46) in the following compact form:

$$L^{\mathbf{q}}_{\substack{I \\ J}}(z,z') = \delta_{\substack{I \\ J}} \ell^{\mathbf{q}}_I(z,z') + \mathrm{i} \sum_M \int d\bar{z}\, \ell^{\mathbf{q}}_I(z,\bar{z}) K^{\mathbf{q}}_{\substack{I \\ M}} L^{\mathbf{q}}_{\substack{M \\ J}}(\bar{z},z'), \tag{14.47}$$

[8] Looking at the diagrammatic representation of v_{iqpj} in Section 7.8, we see that the pair (i, q) contains the quantum numbers of the outgoing particles, whereas the pair (j, p) contains the quantum numbers of the incoming particles.

where $\delta_{\substack{I\\J}} = \delta_{\substack{\mu\nu\mathbf{k}\\\rho\sigma\mathbf{k}'}} \equiv \delta_{\mu\sigma}\delta_{\nu\rho}\delta_{\mathbf{k}\mathbf{k}'}$ and

$$\ell^{\mathbf{q}}_I(z,z') = \ell^{\mathbf{q}}_{\mu\nu\mathbf{k}}(z,z') \equiv g_{\mathbf{k}+\frac{\mathbf{q}}{2}\mu}(z,z')g_{\mathbf{k}-\frac{\mathbf{q}}{2}\nu}(z',z)$$

is the quasi-particle electron–hole propagator. This quantity was already calculated in Section 13.3 for systems in a stationary state. Henceforth we discuss this special, yet relevant, case.

Taking the retarded/advanced component of (14.47) and Fourier transforming with respect to the time difference, we find the algebraic equation

$$L^{\mathbf{q},\mathrm{R/A}}_{\substack{I\\J}}(\omega) = \delta_{\substack{I\\J}}\ell^{\mathbf{q},\mathrm{R/A}}_I(\omega) + \mathrm{i}\sum_M \ell^{\mathbf{q},\mathrm{R/A}}_I(\omega)K^{\mathbf{q}}_{\substack{I\\M}}L^{\mathbf{q},\mathrm{R/A}}_{\substack{M\\J}}(\omega). \tag{14.48}$$

This equation depends parametrically on the transferred momentum $\mathbf{q}$ and should be solved for every $\mathbf{q}$ in the first Brillouin zone. Often the Bethe–Salpeter equation is solved only at zero-momentum transfer ($\mathbf{q} = 0$) since the momentum of light for typical perturbing fields (laser pulses with frequencies below 10 eV) can be discarded. The interest in the finite-$\mathbf{q}$ solutions has recently increased in order to study the exciton dynamics.[9] Below we solve (14.48) for any $\mathbf{q}$, but to lighten the notation we drop the dependence on $\mathbf{q}$ of the various quantities. In terms of the crystal quantum numbers, the quasi-particle electron–hole propagator (13.47) reads

$$\ell^{\mathrm{R/A}}_I(\omega) = \ell^{\mathrm{R/A}}_{\mu\nu\mathbf{k}}(\omega) = \mathrm{i}\frac{f_{\mathbf{k}-\frac{\mathbf{q}}{2}\nu} - f_{\mathbf{k}+\frac{\mathbf{q}}{2}\mu}}{\omega - \epsilon^{\mathrm{qp}}_{\mathbf{k}+\frac{\mathbf{q}}{2}\mu} + \epsilon^{\mathrm{qp}}_{\mathbf{k}-\frac{\mathbf{q}}{2}\nu} \pm \mathrm{i}\eta}. \tag{14.49}$$

Let us introduce two more quantities. We define the *particle–hole occupations*

$$f_I = f_{\mu\nu\mathbf{k}} \equiv f_{\mathbf{k}-\frac{\mathbf{q}}{2}\nu} - f_{\mathbf{k}+\frac{\mathbf{q}}{2}\mu}, \tag{14.50}$$

and the *particle–hole energies*

$$\Omega^0_I = \Omega^0_{\mu\nu\mathbf{k}} \equiv \epsilon^{\mathrm{qp}}_{\mathbf{k}+\frac{\mathbf{q}}{2}\mu} - \epsilon^{\mathrm{qp}}_{\mathbf{k}-\frac{\mathbf{q}}{2}\nu}.$$

Then (14.49) can be rewritten as

$$\ell^{\mathrm{R/A}}_I(\omega) = \mathrm{i}\frac{f_I}{\omega - \Omega^0_I \pm \mathrm{i}\eta}, \tag{14.51}$$

and substituting this result into (14.48) we find

$$(\omega - \Omega^0_I \pm \mathrm{i}\eta)L^{\mathrm{R/A}}_{\substack{I\\J}}(\omega) = \mathrm{i}f_I\delta_{\substack{I\\J}} - f_I\sum_M K_{\substack{I\\M}}L^{\mathrm{R/A}}_{\substack{M\\J}}(\omega). \tag{14.52}$$

Before solving this equation, we observe that $f_I = 0$ implies $L^{\mathrm{R/A}}_{\substack{I\\J}} = 0$. We then consider the subspace $\mathcal{S}$ of superindices I for which $f_I \neq 0$; the sum over M can be restricted to

[9] An optically excited exciton can scatter a phonon of finite momentum and turn into a finite-momentum exciton [219, 220, 221].

this subspace. We also observe that if $I \in \mathcal{S}$ and $J \notin \mathcal{S}$, then $\delta_{\substack{I\\J}} = 0$ and therefore (14.52) becomes a homogeneous system of equations. Consequently, $L^{\mathrm{R/A}}_{\substack{I\\J}}$ is nonvanishing only for $I, J \in \mathcal{S}$.

To proceed further we group the superindices into two classes, the class $\mathcal{R}$ of *resonant* excitations with $f_I > 0$, and the class $\mathcal{A}$ of *antiresonant* excitations with $f_I < 0$. These names originate from the notion that a resonant excitation in systems at zero temperature are associated with transitions from occupied to unoccupied states, hence $\Omega^0_I > 0$ and consequently $f_I > 0$. Antiresonant excitations describe the reverse process, hence $\Omega^0_I < 0$ and $f_I < 0$. We order all vectors and matrices in such a way that the first entries belong to $\mathcal{R}$ and the last entries belong to $\mathcal{A}$. Defining the matrices $\tilde{L}^{\mathrm{R/A}}$ and $\tilde{K}$ according to [222],

$$L^{\mathrm{R/A}}_{\substack{I\\J}}(\omega) \equiv \sqrt{|f_I|}\,\tilde{L}^{\mathrm{R/A}}_{\substack{I\\J}}(\omega)\sqrt{|f_J|} \quad ; \quad \tilde{K}_{\substack{I\\J}} \equiv \sqrt{|f_I|}\,K_{\substack{I\\J}}\sqrt{|f_J|}, \tag{14.53}$$

we can rewrite (14.52) as

$$\left[(\omega - \Omega^0 \pm \mathrm{i}\eta)\sigma_z + \tilde{K}\right]\tilde{L}^{\mathrm{R/A}}(\omega) = \mathrm{i}\mathbb{1}, \tag{14.54}$$

where $\mathbb{1}$ is the identity matrix, Ω^0 is the diagonal matrix with entries Ω^0_I, and

$$(\sigma_z)_{\substack{I\\J}} = \mathrm{sgn}(f_I)\delta_{\substack{I\\J}}\,.$$

Since $\tilde{K}$ is self-adjoint, see comment below (14.45), (14.54) implies that

$$[\tilde{L}^{\mathrm{R}}(\omega)]^\dagger = -\tilde{L}^{\mathrm{A}}(\omega). \tag{14.55}$$

The formal solution of (14.54) is obtained by solving the generalized eigenvalue problem

$$(\Omega^0\sigma_z - \tilde{K})\tilde{\mathbf{Y}}^\lambda = \Omega_\lambda \sigma_z \tilde{\mathbf{Y}}^\lambda, \tag{14.56}$$

where Ω_λ and $\tilde{\mathbf{Y}}^\lambda$ are the generalized eigenvalues and eigenvectors. The practical solution of this problem is discussed later for systems that are invariant under time-reversal. We here observe that for noninteracting electrons $\tilde{K} = 0$ and the generalized eigenvalues are the quasi-particle excitation energies Ω^0_I. For arbitrary self-adjoint matrices $\tilde{K}$, the Ω_λ can turn out to be complex, a situation that usually indicates the instability of the system. Below we assume that all Ω_λ are real (stability assumption). Then, (14.56) implies the generalized orthogonality condition $\tilde{\mathbf{Y}}^{\lambda'\dagger}\sigma_z\tilde{\mathbf{Y}}^\lambda = 0$ for $\Omega_{\lambda'} \neq \Omega_\lambda$.[10] On the other hand, if λ and λ' correspond to a degenerate eigenvalue, then we can always rotate the vectors of the degenerate multiplet to satisfy the generalized orthogonality condition. For the normalization of the generalized eigenvectors, we choose $\tilde{\mathbf{Y}}^{\lambda\dagger}\sigma_z\tilde{\mathbf{Y}}^\lambda = \sum_I \mathrm{sgn}(f_I)|\tilde{Y}^\lambda_I|^2 = s_\lambda$, where $s_\lambda = 1$ if the resonant component is larger than the antiresonant one and $s_\lambda = -1$ otherwise. In summary,

$$\tilde{\mathbf{Y}}^{\lambda'\dagger}\sigma_z\tilde{\mathbf{Y}}^\lambda = s_\lambda\delta_{\lambda\lambda'}. \tag{14.57}$$

[10] The adjoint of (14.56) reads $\tilde{\mathbf{Y}}^{\lambda\dagger}(\Omega^0\sigma_z - \tilde{K}) = \Omega_\lambda\tilde{\mathbf{Y}}^{\lambda\dagger}\sigma_z$ since $\tilde{K}$ is self-adjoint. Thus the multiplication of (14.56) from the right by $\tilde{\mathbf{Y}}^{\lambda'\dagger}$ can equivalently be written as $\Omega_\lambda\tilde{\mathbf{Y}}^{\lambda'\dagger}\sigma_z\tilde{\mathbf{Y}}^\lambda$ or $\Omega_{\lambda'}\tilde{\mathbf{Y}}^{\lambda'\dagger}\sigma_z\tilde{\mathbf{Y}}^\lambda$. Since $\Omega_{\lambda'} \neq \Omega_\lambda$, the result follows.

Let us now show that the identity matrix can be written as

$$\mathbb{1} = \sum_\lambda s_\lambda \tilde{\mathbf{Y}}^\lambda \tilde{\mathbf{Y}}^{\lambda\dagger} \sigma_z = \sum_\lambda s_\lambda \sigma_z \tilde{\mathbf{Y}}^\lambda \tilde{\mathbf{Y}}^{\lambda\dagger}. \tag{14.58}$$

We assume that the sets of vectors $\{\tilde{\mathbf{Y}}^\lambda\}$ and $\{\sigma_z \tilde{\mathbf{Y}}^\lambda\}$ are both complete sets. Multiplying the first equality by $\tilde{\mathbf{Y}}^{\lambda'}$ from the right or by $\tilde{\mathbf{Y}}^{\lambda'\dagger}\sigma_z$ from the left and using (14.57), we obtain an identity. Similarly, multiplying the second equality by $\sigma_z \tilde{\mathbf{Y}}^{\lambda'}$ from the right or by $\tilde{\mathbf{Y}}^{\lambda'\dagger}$ from the left and using again (14.57) we obtain an identity. As this is true for a complete set of vectors, then (14.58) follows. With the identity matrix written in terms of the generalized eigenvectors, it is straightforward to show that the solution of (14.54) is

$$\tilde{L}^{\mathrm{R/A}}(\omega) = \mathrm{i} \sum_\lambda \tilde{\mathbf{Y}}^\lambda \frac{s_\lambda}{\omega - \Omega_\lambda \pm \mathrm{i}\eta} \tilde{\mathbf{Y}}^{\lambda\dagger}. \tag{14.59}$$

In fact, (14.56) implies $\left[(\omega - \Omega^0 \pm \mathrm{i}\eta)\sigma_z + \tilde{K}\right] \tilde{\mathbf{Y}}^\lambda = \sigma_z(\omega - \Omega_\lambda \pm \mathrm{i}\eta)\tilde{\mathbf{Y}}^\lambda$, which together with (14.58) proves the statement. The matrices $\tilde{L}^{\mathrm{R/A}}$ satisfy (14.55) by inspection. The original matrices $L^{\mathrm{R/A}}$ are obtained from $\tilde{L}^{\mathrm{R/A}}(\omega)$ using (14.53). It is also easy to verify that in the noninteracting case (i.e., $K = 0$) we recover (14.51).

Time-reversal symmetry We now present a practical method to solve the generalized eigenvalue problem. For simplicity we confine our discussion to systems that are invariant under time-reversal. This means that the single-particle occupations and single-particle energies are even functions of $\mathbf{k}$: $f_{\mathbf{k}\mu} = f_{-\mathbf{k}\mu}$ and $\epsilon^{\mathrm{qp}}_{\mathbf{k}\mu} = \epsilon^{\mathrm{qp}}_{-\mathbf{k}\mu}$ [217, 218]. Then, for every index $I = (\mu\nu\mathbf{k}) \in \mathcal{R}$ there exists an index $\bar{I} = (\nu\mu - \mathbf{k}) \in \mathcal{A}$ since

$$f_{\bar{I}} = f_{-\mathbf{k}-\frac{\mathbf{q}}{2}\mu} - f_{-\mathbf{k}+\frac{\mathbf{q}}{2}\nu} = -(f_{\mathbf{k}-\frac{\mathbf{q}}{2}\nu} - f_{\mathbf{k}+\frac{\mathbf{q}}{2}\mu}) = -f_I.$$

Another interesting property of the index $\bar{I}$ is that

$$\Omega^0_{\bar{I}} = \epsilon^{\mathrm{qp}}_{-\mathbf{k}+\frac{\mathbf{q}}{2}\nu} - \epsilon^{\mathrm{qp}}_{-\mathbf{k}-\frac{\mathbf{q}}{2}\mu} = -(\epsilon^{\mathrm{qp}}_{\mathbf{k}+\frac{\mathbf{q}}{2}\mu} - \epsilon^{\mathrm{qp}}_{\mathbf{k}-\frac{\mathbf{q}}{2}\nu}) = -\Omega^0_I. \tag{14.60}$$

Let us analyze the matrix elements of the kernel K. From the definitions (14.44) and (14.45) we have for $I, J \in \mathcal{R}$

$$K_{\substack{I\\J}} = K_{\substack{\mu\nu\mathbf{k}\\ \rho\sigma\mathbf{k}'}} = W_{\mathbf{k}+\frac{\mathbf{q}}{2}\mu\,\mathbf{k}'-\frac{\mathbf{q}}{2}\sigma\,\mathbf{k}-\frac{\mathbf{q}}{2}\nu\,\mathbf{k}'+\frac{\mathbf{q}}{2}\rho} - v_{\mathbf{k}+\frac{\mathbf{q}}{2}\mu\,\mathbf{k}'-\frac{\mathbf{q}}{2}\sigma\,\mathbf{k}'+\frac{\mathbf{q}}{2}\rho\,\mathbf{k}-\frac{\mathbf{q}}{2}\nu} \tag{14.61}$$

while for $\bar{I}, \bar{J} \in \mathcal{A}$

$$K_{\substack{\bar{I}\\ \bar{J}}} = K_{\substack{\nu\mu-\mathbf{k}\\ \sigma\rho-\mathbf{k}'}} = W_{-\mathbf{k}+\frac{\mathbf{q}}{2}\nu\,-\mathbf{k}'-\frac{\mathbf{q}}{2}\rho\,-\mathbf{k}-\frac{\mathbf{q}}{2}\mu\,-\mathbf{k}'+\frac{\mathbf{q}}{2}\sigma} - v_{-\mathbf{k}+\frac{\mathbf{q}}{2}\nu\,-\mathbf{k}'-\frac{\mathbf{q}}{2}\rho\,-\mathbf{k}'+\frac{\mathbf{q}}{2}\sigma\,-\mathbf{k}-\frac{\mathbf{q}}{2}\mu}. \tag{14.62}$$

The Bloch wavefunctions of a system invariant under time-reversal have the property $\varphi_{\mathbf{k}\mu}(\mathbf{x}) = \langle \mathbf{x}|\mathbf{k}\mu\rangle = \varphi^*_{-\mathbf{k}\mu}(\mathbf{x})$. Recalling that $v_{iqpj} = v^*_{jpqi} = v^*_{pjiq}$ (and the like for W), see one more time (1.86), we conclude that

$$K_{\substack{\bar{I}\\ \bar{J}}} = K_{\substack{I\\J}}. \tag{14.63}$$

In a similar way, we can prove that for $I \in \mathcal{R}$ and $\bar{J} \in \mathcal{A}$ we have $K_{\substack{I\\ \bar{J}}} = K_{\substack{\bar{I}\\ J}}$. These properties of K transfer directly to $\tilde{K}$ since $|f_I| = |f_{\bar{I}}|$. It is then convenient to organize all matrices in a 2×2 block form where each block has indices either in the same (resonant or antiresonant) space or in different spaces. The 2×2 block form of the matrix Ω^0 contains the blocks $\Omega^0_{\mathcal{RR}} = -\Omega^0_{\mathcal{AA}}$ on the diagonal, see (14.60), and vanishing off-diagonal blocks. The kernel contains the blocks $\tilde{K}_{\mathcal{RR}} = \tilde{K}_{\mathcal{AA}}$ on the diagonal and the off-diagonal blocks $\tilde{K}_{\mathcal{RA}} = \tilde{K}_{\mathcal{AR}}$. We can then rewrite the generalized eigenvalue equation (14.56) as

$$\begin{pmatrix} \Omega^0_{\mathcal{RR}} - \tilde{K}_{\mathcal{RR}} & -\tilde{K}_{\mathcal{RA}} \\ -\tilde{K}_{\mathcal{RA}} & \Omega^0_{\mathcal{RR}} - \tilde{K}_{\mathcal{RR}} \end{pmatrix} \begin{pmatrix} \tilde{\mathbf{Y}}^{\lambda}_{\mathcal{R}} \\ \tilde{\mathbf{Y}}^{\lambda}_{\mathcal{A}} \end{pmatrix} = \Omega_\lambda \begin{pmatrix} \tilde{\mathbf{Y}}^{\lambda}_{\mathcal{R}} \\ -\tilde{\mathbf{Y}}^{\lambda}_{\mathcal{A}} \end{pmatrix}. \tag{14.64}$$

Notice that both $\tilde{K}_{\mathcal{RR}}$ and $\tilde{K}_{\mathcal{RA}}$ are self-adjoint. In solid-state physics calculations the off-diagonal blocks are often discarded; this is the so-called *Tamm–Dancoff approximation* [223, 224].

To shorten the equations and to highlight the mathematical structure, let us rename the diagonal and off-diagonal blocks of the matrix with the capital letters $D = \Omega^0_{\mathcal{RR}} - \tilde{K}_{\mathcal{RR}}$ and $O = -\tilde{K}_{\mathcal{RA}}$, respectively. We also rename the resonant and antiresonant components of the generalized eigenvectors with the letters $\mathbf{R} = \tilde{\mathbf{Y}}^{\lambda}_{\mathcal{R}}$ and $\mathbf{A} = \tilde{\mathbf{Y}}^{\lambda}_{\mathcal{A}}$. Then (14.64) is equivalent to the coupled system of equations,

$$\begin{array}{l} D\mathbf{R} + O\mathbf{A} = \Omega\mathbf{R} \\ O\mathbf{R} + D\mathbf{A} = -\Omega\mathbf{A} \end{array} \quad \Rightarrow \quad \begin{array}{l} (D+O)(\mathbf{R}+\mathbf{A}) = \Omega(\mathbf{R}-\mathbf{A}) \\ (D-O)(\mathbf{R}-\mathbf{A}) = \Omega(\mathbf{R}+\mathbf{A}) \end{array}. \tag{14.65}$$

Multiplying the first equation by $(D - O)$ and using the second equation, we find $(D - O)(D + O)(\mathbf{R} + \mathbf{A}) = \Omega^2(\mathbf{R} + \mathbf{A})$. Although this is now a standard eigenvalue equation, the matrix is not self-adjoint. We can easily remedy this problem provided that the matrix $(D - O)$ is positive definite. In the noninteracting case $(D - O) = \Omega^0_{\mathcal{RR}}$ and for systems in equilibrium at zero temperature, $\Omega^0_{\mathcal{RR}}$ is a diagonal matrix with all positive entries.[||] We assume that the interaction preserves the property of positive definiteness and define the matrix $S = (\frac{D-O}{|\Omega|})^{1/2}$. Then,

$$(D-O)(D+O)(\mathbf{R}+\mathbf{A}) = |\Omega| S \underbrace{S(D+O)S}_{Q} \underbrace{S^{-1}(\mathbf{R}+\mathbf{A})}_{\mathbf{Z}} = \Omega^2 S \underbrace{S^{-1}(\mathbf{R}+\mathbf{A})}_{\mathbf{Z}}.$$

The matrix $Q \equiv S(D + O)S$ is self-adjoint. If we define the vector $\mathbf{Z} \equiv S^{-1}(\mathbf{R} + \mathbf{A})$ and multiply the equation above from the left by S^{-1}, we obtain a Hermitian eigenvalue problem $Q\mathbf{Z} = |\Omega|\mathbf{Z}$. The eigenvalues of Q are therefore real and positive, and the corresponding eigenvectors can be chosen with unitary normalization: $\mathbf{Z}^\dagger\mathbf{Z} = 1$. This result also tells us that if Ω is a generalized eigenvalue of (14.64), then also $-\Omega$ is (the vector $\mathbf{Z}$ is of course the same for both). Let us denote by $\mathbf{R}^\pm$ and $\mathbf{A}^\pm$ the resonant and antiresonant components of the generalized eigenvectors with generalized eigenvalues $\pm|\Omega|$. From the definition of vector $\mathbf{Z}$ we have $(\mathbf{R}^\pm + \mathbf{A}^\pm) = S\mathbf{Z}$, and from the second equation of (14.65) we have $(\mathbf{R}^\pm - \mathbf{A}^\pm) = \pm S^{-2}(\mathbf{R}^\pm + \mathbf{A}^\pm) = \pm S^{-1}\mathbf{Z}$. We conclude that the generalized eigenvectors read

[||] For superindices I in the resonant space $\Omega^0_I > 0$.

$$\tilde{\mathbf{Y}}^{\pm} = \begin{pmatrix} \mathbf{R}^{\pm} \\ \mathbf{A}^{\pm} \end{pmatrix} = \frac{1}{2} \begin{pmatrix} (S \pm S^{-1})\mathbf{Z} \\ (S \mp S^{-1})\mathbf{Z} \end{pmatrix}. \tag{14.66}$$

The normalization of vector $\mathbf{Z}$ automatically guarantees the correct normalization (14.57) since

$$\tilde{\mathbf{Y}}^{\pm\dagger}\sigma_z\tilde{\mathbf{Y}}^{\pm} = \mathbf{R}^{\pm\dagger}\mathbf{R}^{\pm} - \mathbf{A}^{\pm\dagger}\mathbf{A}^{\pm} = \pm 1. \tag{14.67}$$

In a noninteracting system at zero temperature the resonant eigenvectors are those with positive energy, hence the $\tilde{\mathbf{Y}}^{+}$, and their antiresonant component is zero. The sign of (14.67) is consistently positive.

14.7.2 Excitons and Excitonic Mott Transition

To gain insight into the HSEX approximation for the kernel we solve (14.64) for the one-dimensional two-band model with Hamiltonian (6.105). We use the Hartree–Fock approximation for the quasi-particle Green's function g. The Hartree–Fock basis $\{|\lambda\rangle\}$ is the same as the $\{|\mu k\rangle\}$ basis used to write the Hamiltonian since the Hartree–Fock potential (8.31) is diagonal for $\{|\lambda\rangle\} = \{|\mu k\rangle\}$. In fact, in this case the one-particle density matrix $n_{p\alpha\, q\beta} = \delta_{\alpha\beta}\delta_{pq} f(\epsilon^{\mathrm{HF}}_{p\alpha} - \mu)$ and therefore (dropping the superscript "M" and using the lower sign for fermions)

$$V_{\mathrm{HF},k\mu\, k'\nu} = \sum_{p}\sum_{\alpha=c,v} f(\epsilon^{\mathrm{HF}}_{p\alpha} - \mu)(v_{k\mu\, p\alpha\, p\alpha\, k'\nu} - v_{k\mu\, p\alpha\, k'\nu\, p\alpha}). \tag{14.68}$$

From (6.105) we read that the four-index Coulomb tensor is[12]

$$v_{k\mu\, p\alpha\, q\beta\, k'\nu} = \delta_{\mu\nu}\delta_{\alpha\beta}(1 - \delta_{\mu\alpha})\delta_{k+p,k'+q}\frac{v_{k-k'}}{N}, \tag{14.69}$$

where N is the number of discretized quasi-momenta k, and substitution in (14.68) leads to the diagonal Hartree–Fock potential

$$V_{\mathrm{HF},k\mu\, k'\nu} = \delta_{\mu\nu}\delta_{kk'}\frac{v_0}{N}\sum_{p\alpha}(1 - \delta_{\mu\alpha}) f(\epsilon^{\mathrm{HF}}_{p\alpha} - \mu).$$

Interestingly, the Fock part does not contribute to V_{HF}.

Excitons In equilibrium at zero temperature, the occupations of the Bloch states are $f_{kc} = f(\epsilon^{\mathrm{HF}}_{kc} - \mu) = 0$ (empty conduction band) and $f_{kv} = f(\epsilon^{\mathrm{HF}}_{kv} - \mu) = 1$ (full valence band). Therefore the Hartree–Fock energies are

$$\begin{aligned} \epsilon^{\mathrm{HF}}_{kc} &= \epsilon_{kc} + V_{\mathrm{HF},kc\, kc} = \epsilon_{kc} + v_0, \\ \epsilon^{\mathrm{HF}}_{kv} &= \epsilon_{kv} + V_{\mathrm{HF},kv\, kv} = \epsilon_{kv}. \end{aligned}$$

Let us now construct the matrices Ω^0 and $\tilde{K}$. From (14.50) the resonant indices are $I = (cvk)$, whereas the antiresonant indices are $\bar{I} = (vc-k)$. Therefore,

$$\Omega^0_I = \Omega^0_{cvk} \equiv \epsilon^{\mathrm{HF}}_{k+\frac{q}{2}c} - \epsilon^{\mathrm{HF}}_{k-\frac{q}{2}v}.$$

[12] The interaction is nonzero only if $\mu = \nu = c$ and $\alpha = \beta = v$, or the other way around.

The blocks of the kernel have matrix elements, see (14.61) and (14.62),

$$K_{\substack{I\\J}} = K_{\substack{cvk\\cvk'}} = \frac{W_{k-k'}}{N}, \qquad K_{\substack{I\\\bar{J}}} = K_{\substack{cvk\\vc-k'}} = 0. \tag{14.70}$$

The off-diagonal blocks vanish since there are no Coulomb integrals for the scattering of two electrons in the same band. For this special case, the Tamm–Dancoff approximation is exact.[13] The problem is then reduced to the diagonalization of the self-adjoint matrix $\Omega^0_{\mathcal{R}\mathcal{R}} - K_{\mathcal{R}\mathcal{R}}$.[14] The eigenvalue equation reads

$$\sum_{k'}(\epsilon^{\rm HF}_{k+\frac{q}{2}c} - \epsilon^{\rm HF}_{k-\frac{q}{2}v})R_{cvk} - \frac{1}{N}\sum_{k'} W_{k-k'}R_{cvk'} = \Omega R_{cvk}. \tag{14.71}$$

Inserting the explicit form of the Hartree–Fock energies, we see that this result generalizes (6.107) to arbitrary momentum transfers and screened interactions, and it becomes identical to (6.107) for $q = 0$ and $W = v$. We conclude that the HSEX kernel is a good approximation to capture excitonic effects. In fact, (14.71) admits an exciton eigenvalue $\Omega_{\rm x}$ smaller than the gap, see Section 6.4.3, and therefore the photoabsorption spectrum (13.74) exhibits a peak at frequency $\omega = \Omega_{\rm x}$. Physically this means that photons of energy $\Omega_{\rm x}$ are absorbed via the excitation of a *bound* electron–hole pair (i.e., the exciton) and that density and current inside the semiconductor begin oscillating at the same frequency, in accordance with (13.14). The effect of the static screening is to reduce the effective electron–hole attraction, thereby reducing the exciton binding energy or, equivalently, increasing $\Omega_{\rm x}$.

Complex solutions Let us now solve the Bethe–Salpeter equation for the one-dimensional two-band model in a simple excited state. The excitation consists in promoting an electron from the top of the valence band to the bottom of the conduction band, yielding $f_{0v} = 0$ and $f_{0c} = 1$ (all other occupations remain the same as in the ground state). With these occupations the Hartree–Fock energies becomes

$$\epsilon^{\rm HF}_{kc} = \epsilon_{kc} + V_{{\rm HF},kc\,kc} = \epsilon_{kc} + \frac{N-1}{N}v_0,$$
$$\epsilon^{\rm HF}_{kv} = \epsilon_{kv} + V_{{\rm HF},kv\,kv} = \epsilon_{kv} + \frac{v_0}{N}.$$

To illustrate the emergence of complex eigenvalues in the Bethe–Salpeter equation it is enough to consider the $q = 0$ sector. Then the resonant indices are $I = (vc0)$ and $I = (cvk)$ for $k \neq 0$, and consequently the antiresonant indices are $\bar{I} = (cv0)$ and $\bar{I} = (vc-k)$ for $k \neq 0$.[15] The matrix Ω^0 has entries

$$\Omega^0_I = \begin{cases} \epsilon^{\rm HF}_{0v} - \epsilon^{\rm HF}_{0c} & I = (vc0) \\ \epsilon^{\rm HF}_{kc} - \epsilon^{\rm HF}_{kv} & I = (cvk),\ k \neq 0 \end{cases}.$$

The resonant–resonant block of the kernel has matrix elements

$$K_{\substack{vc0\\vc0}} = \frac{W_0}{N}, \quad K_{\substack{cvk\\cvk'}} = \frac{W_{k-k'}}{N}, \quad K_{\substack{vc0\\cvk'}} = K_{\substack{cvk\\vc0}} = 0,$$

[13] A posteriori we may say that the off-diagonal blocks are not responsible for the formation of an exciton; rather, they renormalize the exciton energy and wavefunction.

[14] We have used that $K = \tilde{K}$ since $|f_I| = 1$ for all I.

[15] For nonthermal stationary states $f_I > 0$ does not imply $\Omega_I > 0$.

whereas the resonant–antiresonant block of the kernel has matrix elements

$$K_{\substack{vc0\\cv0}} = K_{\substack{cvk\\vc-k'}} = 0, \quad K_{\substack{vc0\\vc-k'}} = \frac{W_{k'}}{N}, \quad K_{\substack{cvk\\cv0}} = \frac{W_k}{N}.$$

Thus the off-diagonal block O of the kernel acquires nonvanishing elements. Taking into account that $|f_I| = 1$ for all I, we also have $\tilde{K} = K$. The matrices D and O can then be written as

$$D = \begin{pmatrix} D_{00} & \mathbf{0}^\dagger \\ \mathbf{0} & \mathsf{D} \end{pmatrix}, \qquad \begin{aligned} D_{00} &= \epsilon^{\rm HF}_{0v} - \epsilon^{\rm HF}_{0c} - \tfrac{W_0}{N} \\ \mathsf{D}_{kk'} &= \delta_{kk'}(\epsilon^{\rm HF}_{kc} - \epsilon^{\rm HF}_{kv}) - \tfrac{W_{k-k'}}{N} \end{aligned},$$

where $\mathbf{0}$ is the null vector of dimension $(N-1)$, and

$$O = \begin{pmatrix} 0 & \mathbf{O}^\dagger \\ \mathbf{O} & \mathbb{O} \end{pmatrix}, \qquad (\mathbf{O})_k = -\frac{W_k}{N},$$

where $\mathbb{O}$ is the null matrix of dimension $(N-1) \times (N-1)$.

The solution of the generalized eigenvalue problem simplifies considerably if we take an interaction $W_k = W_0$ independent of k. The kth equation of the system $D\mathbf{R} + O\mathbf{A} = \Omega\mathbf{R}$ becomes

$$(\epsilon^{\rm HF}_{kc} - \epsilon^{\rm HF}_{kv})R_{cvk} - \frac{W_0}{N}(R + A_{cv0}) = \Omega R_{cvk}, \qquad R = \sum_{k \neq 0} R_{cvk}, \tag{14.72}$$

which implies $R = Q(\Omega)(R + A_{cv0})$ with

$$Q(\Omega) \equiv \frac{W_0}{N} \sum_{k \neq 0} \frac{1}{\epsilon^{\rm HF}_{kc} - \epsilon^{\rm HF}_{kv} - \Omega}.$$

We can easily express A_{cv0} in terms of R using the first equation of the system $O\mathbf{R} + D\mathbf{A} = -\Omega\mathbf{A}$ – that is, $D_{00}A_{cv0} - (W_0/N)R = -\Omega A_{cv0}$. Nontrivial solutions are therefore found from the zeros of the algebraic equation:

$$1 - Q(\Omega)\left(1 + \frac{W_0}{N}\frac{1}{\epsilon^{\rm HF}_{0v} - \epsilon^{\rm HF}_{0c} - \frac{W_0}{N} + \Omega}\right) = 0. \tag{14.73}$$

Let us discuss this result. We consider the same band dispersions as in Section 6.4.3 – that is, $\epsilon_{kv} = -\frac{\Delta}{2} - \Delta(1 - \cos k)$ and $\epsilon_{kc} = -\epsilon_{kv} - v_0$, with quasi-momentum $k = 2\pi m/(N+1)$ where $m = -N/2, \ldots, N/2$. We also take the same Hartree shift $v_0 = 0.65\Delta$. For noninteracting electrons (i.e., $v_0 = W_0 = 0$) the eigenvalues Ω are given by $\pm 2\epsilon_{kv}$ and have degeneracy 1 for $k = 0$ and degeneracy 2 otherwise, for a total of $(N+1)/2$ distinct eigenvalues of definite sign. In Fig. 14.7 we show the position of the noninteracting positive eigenvalues for $N = 9$ (five distinct eigenvalues) using vertical faint lines. The lowest eigenvalue corresponds to the excitation energy with the same value as the gap Δ, as it should. In Fig. 14.7(a) we plot the r.h.s. of (14.73) for a statically screened interaction $W_0 = 0.3\Delta < v_0$.

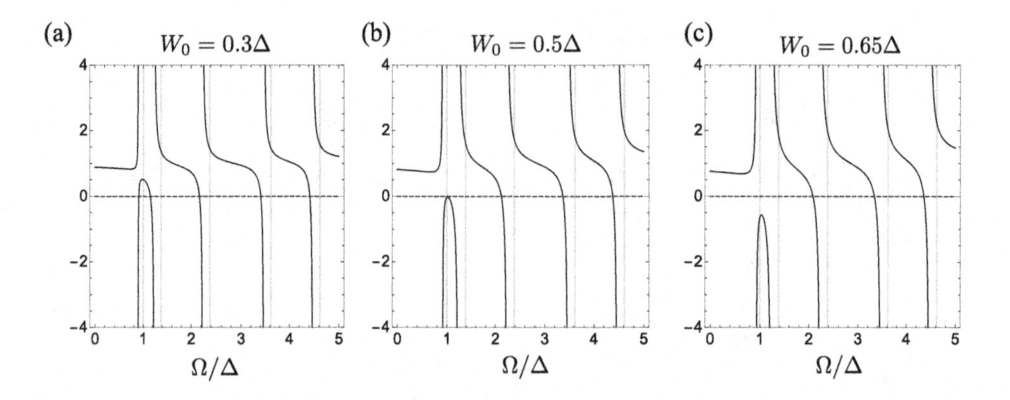

Figure 14.7 Graphical solution of (14.73) for different values of the statically screened interaction W_0. The faint vertical lines represent the positive eigenvalues of the noninteracting problem.

We observe as many zeros as the number of noninteracting eigenvalues. They are all shifted at a slightly lower energy, the lowest energy corresponding to a weakly bound exciton. Increasing $W_0 = 0.5\Delta$, the two lowest zeros become degenerate, see Fig. 14.7(b), and for the unscreened value $W_0 = 0.65\Delta$ the number of real zeros reduces from 5 to 3, see Fig. 14.7(c). Above the critical value $W_0 = 0.5\Delta$ the two lowest zeros cease to be real and become a complex conjugate pair. In general, the critical value depends on the density of electrons in the conduction band ($1/9$ in our example). In fact, it is more correct to speak in terms of *critical density* rather than critical value of W_0 as it is the density in the conduction band that is varied in an experiment. Above the critical density no excitonic solution is found, a phenomenon called the *excitonic Mott transition* [200, 225, 226]. We can understand the origin of this transition by looking at (14.53). If $|f_I| < 1$, then the effective interaction $\tilde{K}$ between the electron and the hole is reduced with respect to K, and for a too small effective interaction the electron and the hole do not bind.

Exercise 14.5 Show that the general interacting Hamiltonian in (1.83) with Coulomb integrals (14.69) leads to the interacting Hamiltonian in (6.105).

Exercise 14.6 Prove that for a Coloumb interaction as in (14.69) the kernel is given by (14.70).

Exercise 14.7 Solve numerically (14.71) for different q and study how the exciton energy changes by varying the momentum q.

14.8 T-Matrix Approximation in the Particle–Particle Channel

In this section we study the approximation to G_2 that leads to the T-matrix approximation in the particle–particle channel plus exchange (Tpp+X) for the self-energy. To illustrate the physics contained in this approximation we begin by considering a system with *zero* particles, and ask what physical processes G_2 should contain. Since there are no particles we can only add two particles (we cannot destroy particles) and the two added particles can only interact directly an arbitrary number of times. This is particularly clear in the Lippmann–Schwinger equation (M.6), see Appendix M, for the two-particle scattering state

$$|\psi\rangle = |\mathbf{k}\rangle + \frac{1}{E - \hat{\mathbf{p}}^2/m \pm i\eta} v(\hat{\mathbf{r}})|\psi\rangle, \tag{14.74}$$

of energy E and relative momentum $\mathbf{k}$. Iterating (14.74), we find the following expansion of $|\psi\rangle$ in powers of the interaction v:

$$|\psi\rangle = |\mathbf{k}\rangle + \frac{1}{E - \hat{\mathbf{p}}^2/m \pm i\eta} v(\hat{\mathbf{r}})|\mathbf{k}\rangle + \frac{1}{E - \hat{\mathbf{p}}^2/m \pm i\eta} v(\hat{\mathbf{r}}) \frac{1}{E - \hat{\mathbf{p}}^2/m \pm i\eta} v(\hat{\mathbf{r}})|\mathbf{k}\rangle + \ldots. \tag{14.75}$$

The first term is a free scattering state. In the second term the particles interact once and then continue to propagate freely.[16] In the third term we have a first interaction, then a free propagation followed by a second interaction, and then again a free propagation. In Section 14.3 we said that the second Born approximation for G_2 corresponds to truncate the expansion (14.75) to the term which is first order in v. We rewrite the second Born diagrammatic representation in a slightly different, yet topologically equivalent, way

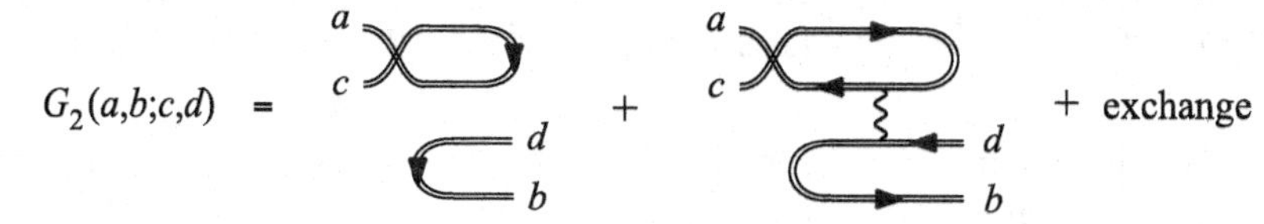

$G_2(a,b;c,d)$ = [diagram: a, c, d, b] + [diagram: a, c, d, b] + exchange (14.76)

where "exchange" represents the diagrams obtained from the explicitly drawn diagrams by exchanging c and d. The reader can easily verify that (14.76) is topologically equivalent to (14.10). The Tpp+X approximation for G_2 goes beyond the second Born approximation as it takes into account *all* multiple scatterings in the expansion (14.75). In the Tpp+X approximation the two added particles interact directly an arbitrary number of times. The

[16] The operator $(E - \hat{\mathbf{p}}^2/m \pm i\eta)^{-1}$ is the Fourier transform of the evolution operator of two noninteracting particles.

diagrammatic representation of this G_2 is then

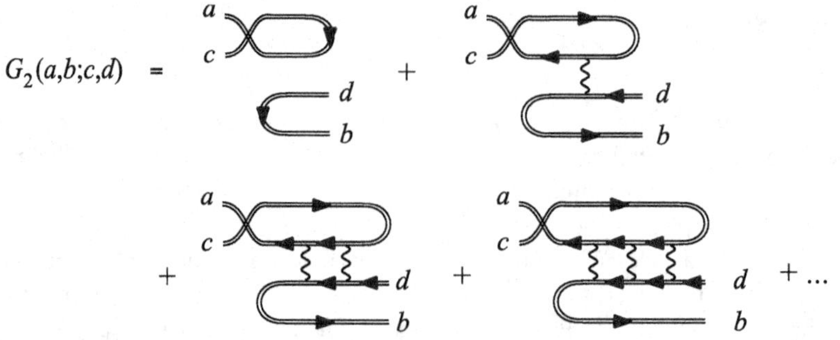

$G_2(a,b;c,d)$ = [diagram] + [diagram] + [diagram] + [diagram] + ...

+ **exchange** (14.77)

The name T-matrix approximation is given after a very useful quantity known as the *transfer matrix*. In scattering theory the transfer matrix $\hat{T}$ is the operator that transforms the free scattering state $|\mathbf{k}\rangle$ into the interacting scattering state $|\psi\rangle$: $\hat{T}|\mathbf{k}\rangle = v(\hat{\boldsymbol{r}})|\psi\rangle$. From (14.75) we see that

$$\hat{T} = v(\hat{\boldsymbol{r}}) + v(\hat{\boldsymbol{r}})\frac{1}{E - \hat{\mathbf{p}}^2/m \pm \mathrm{i}\eta}v(\hat{\boldsymbol{r}}) + \ldots = v(\hat{\boldsymbol{r}}) + \hat{T}\frac{1}{E - \hat{\mathbf{p}}^2/m \pm \mathrm{i}\eta}v(\hat{\boldsymbol{r}}). \quad (14.78)$$

In MBPT the transfer matrix is defined in a similar way:

$$T(1,2;1',2') = \delta(1;1')\delta(2;2')v(1';2') + \mathrm{i}\int d3d4\, T(1,2;3,4)G(3;1')G(4;2')v(1';2'). \quad (14.79)$$

The analogy between (14.79) and (14.78) is evident if we consider that the product $G(3;1')$ $G(4;2')$ is the propagator of two independent particles. The transfer matrix has already been encountered in the context of Φ-derivable approximations, see (12.36).

From a knowledge of the transfer matrix we can determine G_2 in the Tpp+X approximation since

$$v(1;2)G_2(1,2;1',2') = \int d3d4\, T(1,2;3,4)\left[G(3;1')G(4;2') \pm G(3;2')G(4;1')\right]. \quad (14.80)$$

We expect this approximation to be accurate in systems with low density of particles (since it is exact for vanishing densities) and short-range interactions v (since it originates from the ideas of scattering theory where two particles at large enough distance are essentially free). If these two requirements are met, then Tpp+X is expected to provide a reliable description of the many-particle system independently of the interaction strength. In this respect, Tpp+X is very different from GW. In Chapter 15 we show that, for systems like the electron gas, the GW approximation performs well in the high-density limit (Wigner–Seitz radius $r_s \to 0$) or, equivalently, in the weak coupling limit $e^2 \to 0$, see (15.28). The l.h.s. of (14.80) with $2' = 2^+$ also appears in definition (9.5) of the self-energy. Comparing these equations, it is straightforward to find the Tpp+X self-energy in (12.35).

Below we discuss a successful application of the Tpp+X approximation. The example is concerned with the formation of a Cooper pair in a superconductor.

14.8.1 Formation of a Cooper Pair

As a pedagogical example of the physics contained in the Tpp+X approximation, we consider the formation of a bound state in the BCS model. Consider the BCS Hamiltonian (2.46) and suppose that there is only one electron, say of spin down, in the system. Let us order the single-particle energies as $\epsilon_0 < \epsilon_1 < \ldots < \epsilon_N$ and assume no degeneracy. The ground state is $|\Psi_0\rangle = \hat{c}^\dagger_{0\downarrow}|0\rangle$ with ground-state energy ϵ_0. We are interested in calculating the spectral function for an electron of spin up. With only one spin-down electron in the ground state it is not possible to remove a spin-up electron, and therefore the spectral function contains only addition energies. Furthermore, the interaction couples pairs of electrons with the same k label and opposite spin. This implies that if we add an electron of momentum $k \neq 0$, the resulting two-particle state $\hat{c}^\dagger_{k\uparrow}|\Psi_0\rangle = \hat{c}^\dagger_{k\uparrow}\hat{c}^\dagger_{0\downarrow}|0\rangle$ is an eigenstate with energy $\epsilon_k + \epsilon_0$. Consequently the spectral function exhibits only one peak at the addition energy ϵ_k. The situation is much more interesting if we add a spin-up electron with $k = 0$ since $\hat{c}^\dagger_{0\uparrow}\hat{c}^\dagger_{0\downarrow}|0\rangle$ is not an eigenstate. Below we calculate the exact spectral function $A(\omega) \equiv A_{0\uparrow 0\uparrow}(\omega)$ and then compare the result with the spectral function in the Tpp+X and second Born approximation. As we shall see, the nonperturbative nature of the Tpp+X approximation gives very accurate results, while the second Born approximation performs rather poorly in this context.

Exact solution The spectral function is $A(\omega) = \mathrm{i}G^{>}_{0\uparrow 0\uparrow}(\omega)$ since the lesser Green's function vanishes. To obtain the greater Green's function, we have to calculate the energies $E_{2,m}$ of the two-electron eigenstates $|\Psi_{2,m}\rangle$ as well as the overlaps

$$P_m = \langle\Psi_0|\hat{c}_{0\uparrow}|\Psi_{2,m}\rangle = \langle 0|\hat{c}_{0\downarrow}\hat{c}_{0\uparrow}|\Psi_{2,m}\rangle$$

between the state resulting from the addition of a particle of spin up in level 0 and $|\Psi_{2,m}\rangle$, see (6.76). The two-electron eigenstates are either of the form $\hat{c}^\dagger_{k\uparrow}\hat{c}^\dagger_{p\downarrow}|0\rangle$ with $k \neq p$ (singly occupied levels are blocked) or linear combinations of doubly occupied levels. The former give zero overlaps, $P_m = 0$, and it is therefore sufficient to calculate the latter. In agreement with the notation of Section 2.7, we denote these eigenstates by $|\Psi^{(1)}_i\rangle$ where the superscript "(1)" indicates that there is one pair. The generic form of $|\Psi^{(1)}_i\rangle$ is

$$|\Psi^{(1)}_i\rangle = \sum_k \alpha^{(i)}_k b^\dagger_k|0\rangle,$$

where $b^\dagger_k = \hat{c}^\dagger_{k\uparrow}\hat{c}^\dagger_{k\downarrow}$ is the creation operator for a pair of electrons in level k. We have seen that the eigenvalues E_i of $|\Psi^{(1)}_i\rangle$ are the solutions of the algebraic equation

$$\sum_k \frac{1}{2\epsilon_k - E_i} = \frac{1}{v}, \tag{14.81}$$

and that the coefficients

$$\alpha^{(i)}_k = \frac{C}{2\epsilon_k - E_i},$$

with C a normalization constant. The l.h.s. of (14.81) approaches $+\infty$ for $E_i \to 2\epsilon_k$ from the left and $-\infty$ for $E_i \to 2\epsilon_k$ from the right. Thus, for $E_i = \epsilon_0 - \delta$, $\delta > 0$, the l.h.s. is

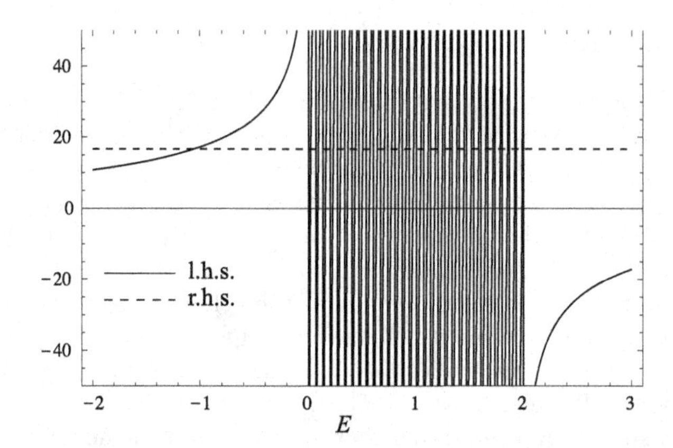

Figure 14.8 Graphical solution of the eigenvalue equation (14.81) for a single electron pair in the BCS model. The numerical parameters are $N = 30$ levels, equidistant energies $\epsilon_k = (\mathcal{E}/N)k$ with $k = 0, \ldots, N-1$, and $v = 1.8\mathcal{E}/N$ (all energies are measured in units of $\mathcal{E}$).

always positive, diverges when $\delta \to 0$, and goes to zero when $\delta \to \infty$. This means that for positive v (attractive interaction) we have a bound state. In the context of the BCS model this bound state is known as a *Cooper pair.* The graphical solution of (14.81) is displayed in Fig. 14.8 for $N = 30$ levels, equidistant energies $\epsilon_k = (\mathcal{E}/N)k$ with $k = 0, \ldots, N-1$, and $v = 1.8\mathcal{E}/N$. We clearly see an intersection at energy ~ -1 which corresponds to a Cooper pair. The full set of solutions E_i can easily be determined numerically and then used to construct the coefficients $\alpha_k^{(i)}$. We actually do not need these coefficients for all k since the overlap

$$P_i = \langle 0|\hat{c}_{0\downarrow}\hat{c}_{0\uparrow}|\Psi_i^{(1)}\rangle = \alpha_0^{(i)},$$

and it is therefore sufficient to calculate the coefficients with $k = 0$. Having the E_i and the $\alpha_0^{(i)}$, the exact spectral function reads

$$A(\omega) = 2\pi \sum_i |\alpha_0^{(i)}|^2 \, \delta(\omega - [E_i - \epsilon_0]). \tag{14.82}$$

This spectral function has an isolated peak in correspondence of the Cooper pair energy. We can then say that the addition of a spin-up electron in $k = 0$ excites the system in a combination of states of energy E_i, among which there is one state (the Cooper pair) with an infinitely long lifetime.

Tpp+X solution The first task is to rewrite the interaction Hamiltonian in a form suitable for MBPT:

$$\hat{H}_{\text{int}} = -v \sum_{kk'} \hat{c}^\dagger_{k\uparrow}\hat{c}^\dagger_{k\downarrow}\hat{c}_{k'\downarrow}\hat{c}_{k'\uparrow} = \frac{1}{2}\sum_{ijmn} v_{ijmn}\hat{c}^\dagger_i\hat{c}^\dagger_j\hat{c}_m\hat{c}_n,$$

where $i = (k_i, \sigma_i)$, $j = (k_j, \sigma_j)$, etc. are collective indices for the orbital and spin quantum numbers. The BCS interaction is nonvanishing only for $k_i = k_j$, $k_m = k_n$, and $\sigma_i = \bar{\sigma}_j =$

$\bar{\sigma}_m = \sigma_n$, where $\bar{\sigma}$ is the spin opposite to σ. Therefore,

$$v_{ijmn} = -v\,\delta_{k_i k_j}\delta_{k_m k_n}\delta_{\sigma_i \sigma_n}\delta_{\sigma_j \sigma_m}\delta_{\sigma_i \bar{\sigma}_j},$$

meaning that the only possible scattering processes are

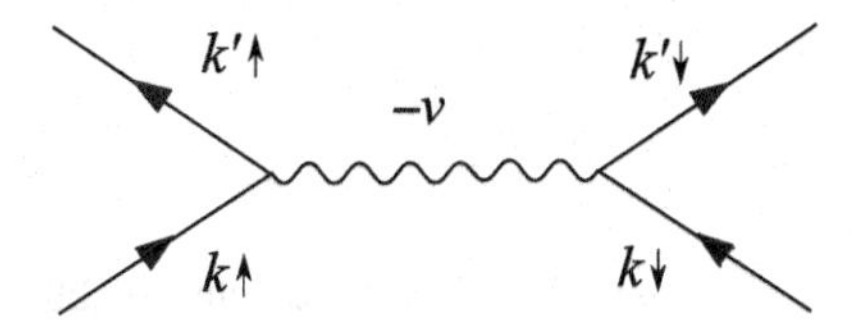

The interaction preserves the spin orientation of the scattered particles and the Green's function and the self-energy are diagonal in spin space. However, the spin-up and -down diagonal entries may differ since the ground state is not symmetric under a spin flip. Another interesting feature of the interaction is that if we add or remove an electron of label k, let the system evolve, and then remove or add an electron of label k' we find zero unless $k = k'$. Thus, the Green's function is also diagonal in k-space: $G_{k\sigma\, k'\sigma'} = \delta_{\sigma\sigma'}\delta_{kk'}G_{k\sigma}$. For a ground state with only one spin-down electron in level 0 the greater and lesser *noninteracting* Green's functions G_0 have diagonal matrix elements:

$$\begin{array}{ll} \text{spin up} & \text{spin down} \\ G^{>}_{0,k\uparrow}(\omega) = -2\pi\mathrm{i}\,\delta(\omega - \epsilon_k), & G^{>}_{0,k\downarrow}(\omega) = -2\pi\mathrm{i}\,(1 - \delta_{k0})\delta(\omega - \epsilon_k), \\ G^{<}_{0,k\uparrow}(\omega) = 0, & G^{<}_{0,k\downarrow}(\omega) = 2\pi\mathrm{i}\,\delta_{k0}\delta(\omega - \epsilon_0). \end{array} \tag{14.83}$$

Like the Green's function, the self-energy is diagonal in k- and spin-space $\Sigma_{k\sigma\, k'\sigma'} = \delta_{\sigma\sigma'}\delta_{kk'}\Sigma_{k\sigma}$. In the Tpp+X approximation Σ has the diagrammatic expansion of Fig. 12.3, and taking into account the explicit form of the BCS interaction we can easily assign the labels of each internal vertex. For instance, the self-energy for a spin-up electron looks like

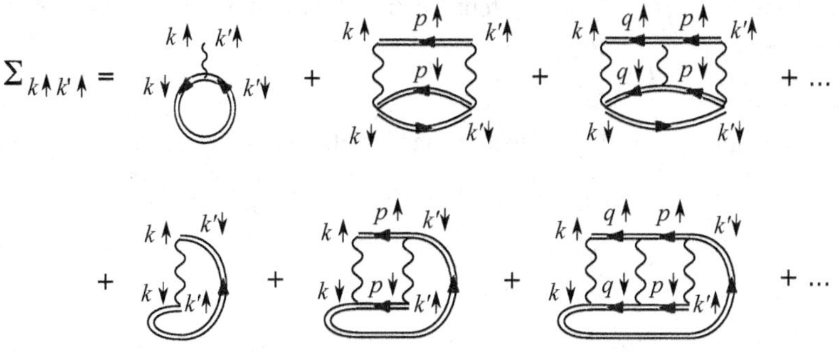

from which it is evident that the self-energy vanishes unless $k = k'$ (the Green's function in the lower part of all these diagrams starts with k and ends with k'). We also see that all exchange diagrams (second row) vanish since the spin orientation must be preserved. The mathematical expression of the self-energy is therefore (remember that the interaction is $-v$)

$$\Sigma_{k\uparrow}(z_1,z_2) = \mathrm{i}\underbrace{\left[v\delta(z_1,z_2) - \mathrm{i}v^2B(z_1,z_2) + \mathrm{i}^2v^3\int_\gamma d\bar{z}B(z_1,\bar{z})B(\bar{z},z_2) + \dots\right]}_{T(z_1,z_2)} G_{k\downarrow}(z_2,z_1), \tag{14.84}$$

where B is the Cooper pair propagator,

$$B(z_1,z_2) = \sum_p G_{p\uparrow}(z_1,z_2)G_{p\downarrow}(z_1,z_2).$$

From the self-energy we extract the spectral function using (10.20):

$$A(\omega) = \frac{\Gamma_{0\uparrow}(\omega)}{|\omega - \epsilon_0 - \Sigma^{\mathrm{R}}_{0\uparrow}(\omega)|^2} \qquad \text{where} \qquad \Gamma_{0\uparrow}(\omega) = -2\mathrm{Im}\left[\Sigma^{\mathrm{R}}_{0\uparrow}(\omega)\right].$$

To calculate $\Sigma_{0\uparrow}$ we observe that the reduced T-matrix defined as the quantity in the square brackets of (14.84) satisfies a Dyson-like equation,

$$T(z_1,z_2) = v\delta(z_1,z_2) - \mathrm{i}v\int_\gamma d\bar{z}\, T(z_1,\bar{z})B(\bar{z},z_2),$$

so that $\Sigma_{0\uparrow}(z_1,z_2) = \mathrm{i}T(z_1,z_2)G_{0\downarrow}(z_2,z_1)$. Using the Langreth rules of Table 5.1 the retarded component of the self-energy in frequency space reads

$$\Sigma^{\mathrm{R}}_{0\uparrow}(\omega) = \mathrm{i}\int\frac{d\omega'}{2\pi}\left[T^{\mathrm{R}}(\omega+\omega')G^{<}_{0\downarrow}(\omega') + T^{<}(\omega+\omega')G^{\mathrm{A}}_{0\downarrow}(\omega')\right]. \tag{14.85}$$

To extract the Keldysh components of the reduced T-matrix, we use the simplified Langreth rules for equilibrium systems (see the end of Section 10.5). Then (in frequency space) $T^{<} = -\mathrm{i}v[T^{\mathrm{R}}B^{<} + T^{<}B^{\mathrm{A}}]$, from which it follows that $T^{<} = -\mathrm{i}vT^{\mathrm{R}}B^{<}/(1+\mathrm{i}vB^{\mathrm{A}}) = 0$. Indeed, the lesser Cooper pair propagator $B^{<}(t_1,t_2) = \sum_p G^{<}_{p\uparrow}(t_1,t_2)G^{<}_{p\downarrow}(t_1,t_2) = 0$ since there are no electrons of spin up in the ground state (i.e., $G^{<}_{p\uparrow} = 0$). It remains to calculate the first term in (14.85). This can be done analytically if we use the noninteracting Green's function (14.83) (non self-consistent treatment). We find

$$\Sigma^{\mathrm{R}}_{0\uparrow}(\omega) = \mathrm{i}\int\frac{d\omega'}{2\pi}T^{\mathrm{R}}(\omega+\omega') \times 2\pi\mathrm{i}\,\delta(\omega'-\epsilon_0) = -T^{\mathrm{R}}(\omega+\epsilon_0).$$

The retarded component of the reduced T-matrix satisfies $T^{\mathrm{R}} = v - \mathrm{i}v\,T^{\mathrm{R}}B^{\mathrm{R}}$ and hence $T^{\mathrm{R}} = v/(1+\mathrm{i}vB^{\mathrm{R}})$. We are then left with the evaluation of the retarded Cooper pair propagator. Using again the noninteracting Green's function in (14.83),

$$\begin{aligned}
B^{>}(\omega) &= \sum_p\int\frac{d\omega'}{2\pi}G^{>}_{0,p\uparrow}(\omega-\omega')G^{>}_{0,p\downarrow}(\omega') \\
&= \sum_p\int\frac{d\omega'}{2\pi}\,2\pi\mathrm{i}\,\delta(\omega-\omega'-\epsilon_p)\times 2\pi\mathrm{i}\,(1-\delta_{p0})\delta(\omega'-\epsilon_p) \\
&= -2\pi\sum_{p\neq 0}\delta(\omega-2\epsilon_p)
\end{aligned}$$

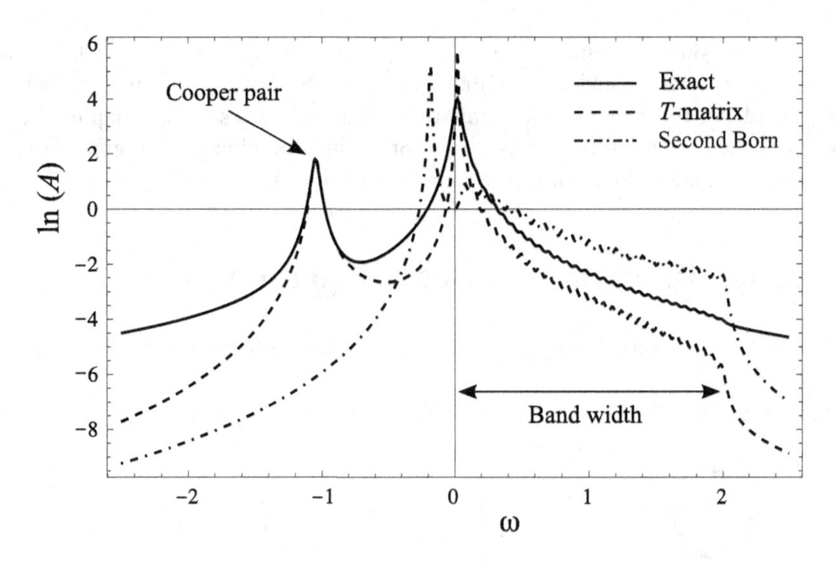

Figure 14.9 Logarithm of the spectral function $A(\omega)$ for the same parameters of Fig. 14.8 as obtained from the exact formula (14.82), the Tpp+X approximation (14.86), and the second Born approximation. The parameter $\eta = 0.3$ and for the exact formula we approximate $\delta(\omega) = \frac{1}{\pi}\frac{\eta}{\omega^2+\eta^2}$ (all energies are measured in units of $\mathcal{E}$).

and hence

$$B^{\mathrm{R}}(\omega) = \mathrm{i}\int \frac{d\omega'}{2\pi}\frac{B^{>}(\omega') - B^{<}(\omega')}{\omega - \omega' + \mathrm{i}\eta} = -\mathrm{i}\sum_{p\neq 0}\frac{1}{\omega - 2\epsilon_p + \mathrm{i}\eta}.$$

In conclusion, the spectral function in the nonself-consistent Tpp+X approximation reads

$$A(\omega) = 2\frac{\mathrm{Im}[T^{\mathrm{R}}(\omega+\epsilon_0)]}{|\omega - \epsilon_0 + T^{\mathrm{R}}(\omega+\epsilon_0)|^2}, \qquad T^{\mathrm{R}}(\omega) = \frac{v}{1 + v\sum_{p\neq 0}\frac{1}{\omega - 2\epsilon_p + \mathrm{i}\eta}}. \tag{14.86}$$

Comparison with exact solution In Fig. 14.9 we show the exact spectral function as well as the Tpp+X spectral function for the same parameters of Fig. 14.8. The position of the Cooper pair binding energy in the Tpp+X approximation is in excellent agreement with the exact result. The Tpp+X approximation overestimates the height of the onset of the continuum and predicts a too sharp transition at the band edge $\omega = 2$. Overall, however, the agreement is rather satisfactory. For comparison we also display the spectral function in the second Born approximation. The second Born self-energy is given by the first two terms of the T-matrix expansion and it is therefore perturbative in the interaction v. The second Born approximation severely underestimates the binding energy of the Cooper pair, and the continuum part of the spectrum decays far too slowly.

One final remark before concluding the section. It might look surprising that a second-order approximation like the second Born approximation is able to describe the formation of a bound state since a bound state is a highly nonperturbative (in v) solution of the eigenvalue problem. We should keep in mind, however, that what is perturbative here is the self-energy and *not* the Green's function. We have already stressed in Chapter 7 that the Green's function calculated from a self-energy with a finite number of diagrams corresponds to summing an infinite number of Green's function diagrams.

14.9 Vertex Function and Hedin Equations

In this section we derive a closed system of equations for the various many-body quantities introduced so far.

Vertex function Λ The self-energy $\Sigma = \Sigma_s[G, v]$ can be written as

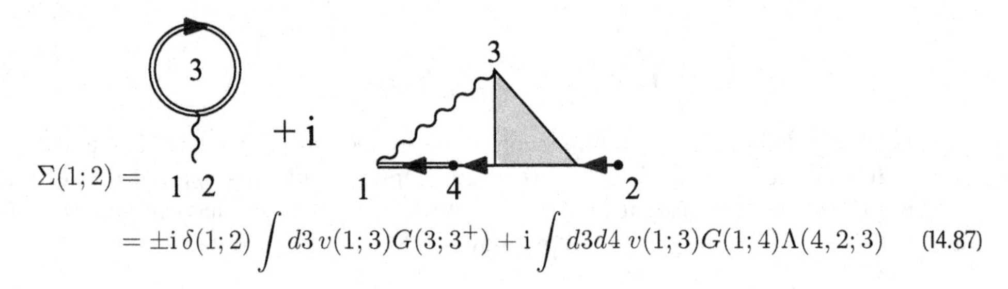

$$= \pm \mathrm{i}\,\delta(1;2) \int d3\, v(1;3)G(3;3^+) + \mathrm{i} \int d3d4\, v(1;3)G(1;4)\Lambda(4,2;3) \qquad (14.87)$$

The so-called *vertex function* (or just vertex)

$$\Lambda(1,2;3) =$$

contains all possible diagrams that we can enter into the second diagram of (14.87) to form a self-energy diagram. The zeroth-order vertex function is

$$\Lambda_0(1,2;3) = \delta(1;2^+)\delta(3;2) = \quad \bullet\,,$$

which is represented by a dot and yields the self-energy Fock diagram since

$$\mathrm{i} \int d3d4\, v(1;3)G(1;4)\delta(4;2^+)\delta(3;2) = \mathrm{i}\, v(1;2)G(1;2^+).$$

We can deduce an equation for the vertex by inserting in (9.5) the expression (14.87) for Σ and the expression (14.14) for G_2. It is a simple and instructive exercise for the reader to show that

$$\Lambda(1,2;3) = \delta(1;2^+)\delta(3;2) \pm \int d4d5\, K_r(1,4;2,5)G(5;3)G(3;4).$$

We can give to this formula the following diagrammatic representation:

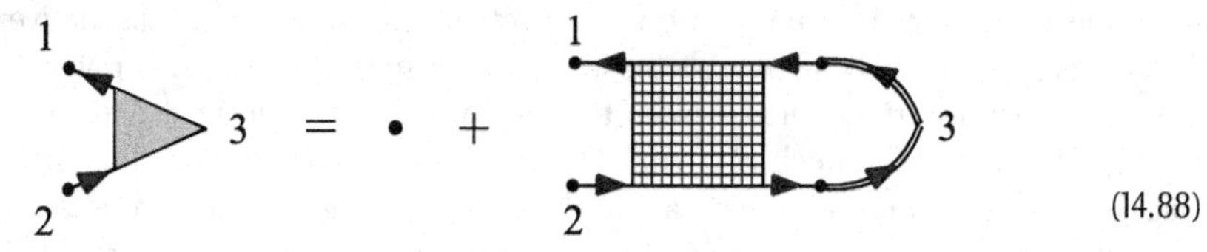

(14.88)

where we take into account that the GG-double-line gives a factor $(\pm)$ when the diagram is converted into a mathematical expression, see Section 14.4. Expanding the reducible kernel as in (14.17), we then obtain a Dyson equation for the vertex

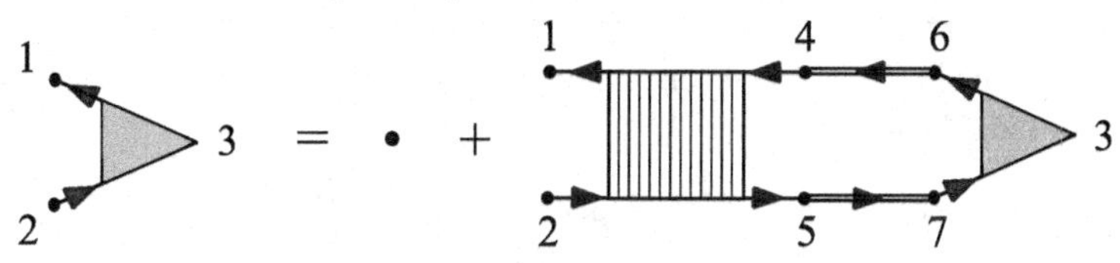

which in formulas reads (remember that $K = \pm\delta\Sigma/\delta G$):

$$\Lambda(1,2;3) = \delta(1;2^+)\delta(3;2) + \int d4d5d6d7 \frac{\delta\Sigma(1;2)}{\delta G(4;5)} G(4;6)G(7;5)\Lambda(6,7;3). \quad (14.89)$$

This is the Bethe–Salpeter equation for the vertex function.

Equations (14.87) and (14.89) provide a coupled system of equations from which to obtain the self-energy iteratively. If we start with the zeroth-order vertex function $\Lambda_0(1,2;3) = \delta(1;2^+)\delta(3;2)$, then (14.87) yields the Hartree–Fock self-energy whose functional derivative with respect to G yields the kernel (14.23). The insertion of this kernel into (14.89) generates the vertex function

(14.90)

The solution of (14.90) gives an infinite series of bubble and ladder diagrams similar to those in Fig. 14.6. The new vertex can now be inserted back into (14.87) to obtain a new self-energy. A subsequent differentiation with respect to G then gives yet another vertex and we can continue to iterate *ad infinitum*. We are not aware of any proof that such an iterative scheme converges nor that it generates all possible skeleton diagrams for the self-energy.

Vertex function Γ From the iterative solution of (14.87) and (14.89) we get the self-energy as a functional of G and v. The natural question then arises whether we can derive a similar system of equations from which to get the self-energy as a functional of G and W. To have Σ in terms of G and W, we must first replace $v \to W$ in the second diagram of (14.87), and then remove from Λ all diagrams containing a polarization insertion and again replace $v \to W$. In order to remove polarization insertions from Λ we need to introduce another kernel $K_{\mathrm{xc},r}$:

> $K_{\mathrm{xc},r}(1,2;3,4)$ *is obtained from* $K_r(1,2;3,4)$ *by discarding all those diagrams that are one-interaction-line reducible – that is, that can be broken into two disjoint pieces, one containing* $(1,3)$ *and the other containing* $(2,4)$*, by cutting an interaction line.*

In Fig. 14.5 the second, fourth, fifth, and sixth diagrams are one-interaction-line reducible while the remaining diagrams are one-interaction-line irreducible. Alternatively we can say that we are removing from K_r all those diagrams that give rise to a polarization insertion when we close the right vertices with two G-lines [so as to form a Λ-diagram, see (14.88)]. The reader can easily check that with the exception of the second, fourth, fifth, and sixth diagrams of Fig. 14.5 no other diagram in the same figure gives rise to a polarization insertion when it is closed with two G-lines. Since the kernel $K_{\mathrm{xc},r}$ is not two-particle irreducible we can introduce an irreducible kernel K_{xc} in the same fashion as we did for K_r.

The kernel K_{xc} is two-particle irreducible and one-interaction-line irreducible.

The only diagram in K which is one-interaction-line reducible is the Hartree diagram [the first diagram in (14.23)] and hence

$$K_{\mathrm{xc}}(1,2;3,4) = K(1,2;3,4) - \mathrm{i}\,\delta(1;3)\delta(2;4)v(1;2). \tag{14.91}$$

This equation is represented by the diagrammatic relation

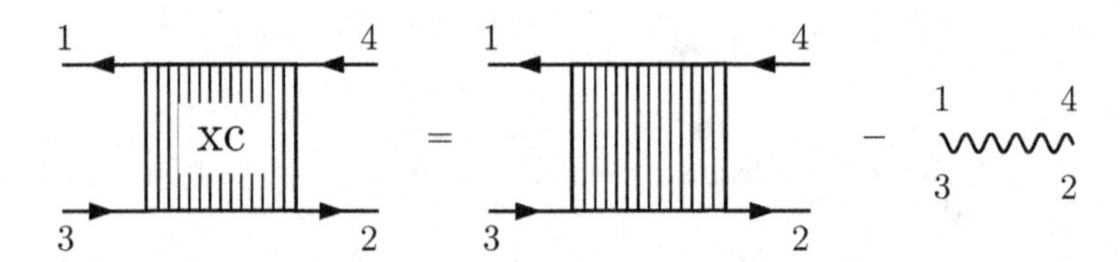

where the striped square with "xc" inside is K_{xc}. The reducible kernel $K_{\mathrm{xc},r}$ can then be expanded in terms of the irreducible kernel K_{xc} in a Bethe–Salpeter-like equation:

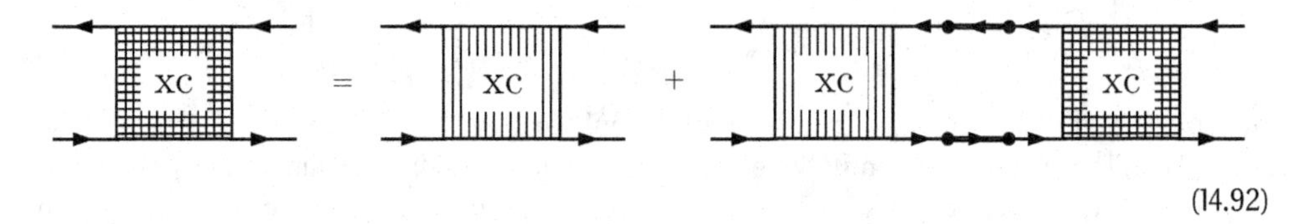

(14.92)

So far we regarded $K_{\mathrm{xc}} = K_{\mathrm{xc},s}[G,v]$ as a functional of G and v. If we remove from it all diagrams with a polarization insertion and subsequently replace $v \to W$ we obtain a new functional $K_{\mathrm{xc},ss}[G,W]$ which yields K_{xc} as a functional of G and W. Then, writing the self-energy as in (7.26) – that is,

$$\Sigma = \Sigma_{ss}[G,W] = \Sigma_{\mathrm{H}}[G,v] + \Sigma_{ss,\mathrm{xc}}[G,W],$$

we can establish the following relation between $K_{\mathrm{xc}} = K_{\mathrm{xc},ss}$ and $\Sigma_{\mathrm{xc}} = \Sigma_{ss,\mathrm{xc}}$:

$$\boxed{K_{\mathrm{xc}}(1,2;3,4) = \pm\frac{\delta\Sigma_{\mathrm{xc}}(1,3)}{\delta G(4;2)}} \tag{14.93}$$

The proof of (14.93) is identical to the proof of (14.22). We stress again that for (14.93) to be valid we must exclude from Σ_{xc} all self-energy diagrams with a polarization insertion and use W as the independent variable. We now have all ingredients to construct the vertex function $\Gamma = \Gamma_{ss}[G, W]$ without polarization insertions.[17] The new vertex function is obtained from $K_{\rm xc} = K_{{\rm xc},ss}[G, W]$ in a similar way as $\Lambda = \Lambda_s[G, v]$ is obtained from $K = K_s[G, v]$, see (14.89),

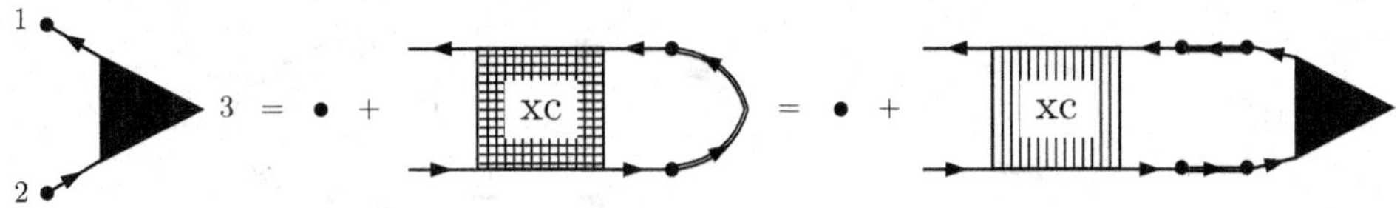

(14.94)

where Γ is represented by the black triangle. In the first equality $K_{{\rm xc},r} = K_{{\rm xc},r,ss}[G, W]$ is the sum of all the $K_{{\rm xc},r}$-diagrams which do not contain polarization insertions and in which $v \to W$; similarly in the second equality $K_{\rm xc} = K_{{\rm xc},ss}[G, W]$. The mathematical expression of this diagrammatic equation is

$$\Gamma(1,2;3) = \delta(1;2^+)\delta(3;2) + \int d4d5d6d7\, \frac{\delta\Sigma_{\rm xc}(1;2)}{\delta G(4;5)}\, G(4;6)G(7;5)\Gamma(6,7;3) \qquad (14.95)$$

Equation (14.95) is the Bethe–Salpeter equation for the G- and W-skeleton vertex function $\Gamma_{ss}[G, W]$. In the remainder of this section we regard all quantities as functional of the independent variables G and W; in other words, all quantities $Q = \Sigma$, Γ, $K_{\rm xc}$, P, etc. have to be understood as $Q_{ss}[G, W]$.

Hedin equations The vertex Γ allows us to write the self-energy as a functional of G and W according to

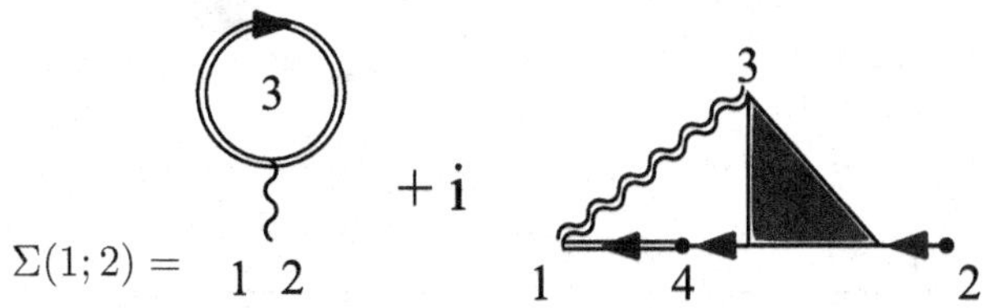

$\Sigma(1;2) =$ [diagram] $+\,\mathrm{i}$ [diagram]

or in formula

$$\Sigma(1;2) = \pm \mathrm{i}\,\delta(1;2)\int d3\, v(1;3)G(3;3^+) + \mathrm{i}\int d3d4\, W(1;3)G(1;4)\Gamma(4,2;3) \qquad (14.96)$$

[17]The reader may wonder why we did not use the symbol Λ_{ss} for this vertex. The vertex Γ is obtained from Λ by removing all diagrams containing a polarization insertion and then replacing $v \to W$. The point is that $\Lambda(1,2;3)$ can have a polarization insertion which is either internal or external. A Λ-diagram with an internal polarization insertion is, for example, the diagram resulting from the third term of Fig. 14.5, while a Λ-diagram with an external polarization insertion is, for example, the diagram resulting from the fourth, fifth, or sixth term of Fig. 14.5. The typical structure of a Λ-diagram with an external polarization insertion is $\Lambda(1,2;3) \propto \int d4d5\, \Lambda_i(1,2;4)v(4;5)P_j(5;3)$ with Λ_i some vertex diagram and P_j some polarization diagram. To construct Γ we have to remove from Λ the diagrams with internal and/or external polarization insertions. It is then clear that by expanding $W = v + vPv + \ldots$ in a Γ-diagram we get back only Λ-diagrams with *internal* polarization insertions – that is, we cannot recover the full Λ. In other words, $\Gamma_{ss}[G, W] \neq \Lambda_s[G, v]$. This is the motivation for using a different symbol. The situation is different for the self-energy and polarization since the expansion of W in $\Sigma_{ss}[G, W]$ and $P_{ss}[G, W]$ gives back $\Sigma_s[G, v]$ and $P_s[G, v]$.

We see that if we take a Γ-diagram and expand $W = v + vPv + \ldots$ in powers of v, we get Λ-diagrams with no external polarization insertions; hence the product $W\Gamma$ does not lead to double-counting.

The kernel $K_{\mathrm{xc},r}$ can also be used to write the polarization according to

$$P(1;2) = \text{[diagram]} + \mathrm{i}\ \text{[diagram with xc]} \tag{14.97}$$

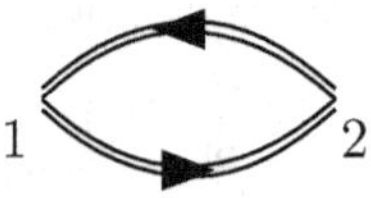

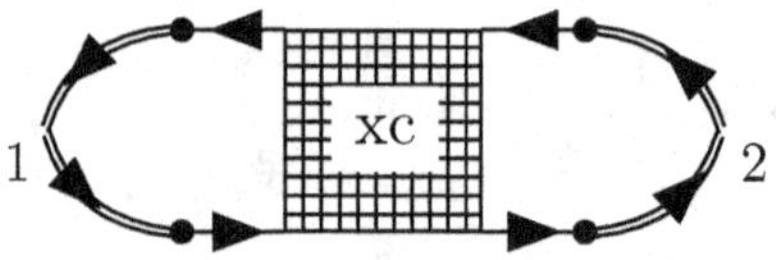

where the prefactor i in the second term on the r.h.s. is due to the different Feynman rules for P (m interactions give a prefactor i^{m+1}, see Section 7.8) and for $K_{\mathrm{xc},r}$ (m interactions give a prefactor i^m, see Section 14.4). The zeroth-order polarization diagram can also be seen as a closed Γ-diagram in which Γ is approximated by $\delta\delta = \bullet$:

$$\text{[diagram]} = \mathrm{i}\ \text{[diagram]}$$

It easy to check that this is an equality: The diagram on the l.h.s. is a polarization diagram and hence the prefactor is $\pm\mathrm{i}$, while the diagram on the r.h.s. is a closed Γ-diagram and hence the prefactor is $\pm$ due to the GG-double-line. This explains the presence of the factor i in front of it. We thus see that

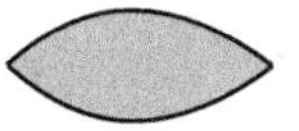

$$\text{[diagram]} = \mathrm{i}\ \text{[diagram]}$$

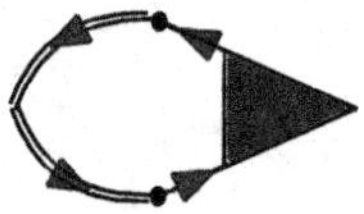

or in formulas

$$\boxed{P(1;2) = \pm\mathrm{i}\int d3d4\, G(1;3)G(4;1)\Gamma(3,4;2)} \tag{14.98}$$

Equations (14.95), (14.96), and (14.98) together with the Dyson equations $G = G_0 + G_0\Sigma G$ and $W = v + vPW$ are the famous *Hedin equations.* They form a set of coupled equations for G, Σ, P, W, and Γ whose solution yields the same G that solves the Martin–Schwinger hierarchy. We thus achieved a major reduction of the equations needed to determine the Green's function: from the infinite Martin–Schwinger hierarchy to the five Hedin equations. In Table 14.1 we list the Hedin equations and their diagrammatic representation.

Like the coupled equations (14.87) and (14.89), the Hedin equations can be iterated to obtain an expansion of Σ in terms of G and W. If we start with the zeroth-order vertex $\Gamma(1,2;3) = \delta(1;2^+)\delta(3;2)$, we find

$$\Sigma_{\mathrm{xc}}(1;2) = \mathrm{i}\,W(1;2)G(1;2) = \text{[diagram]} \tag{14.99}$$

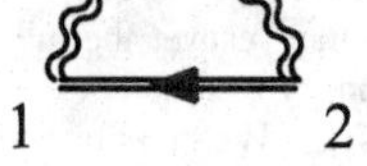

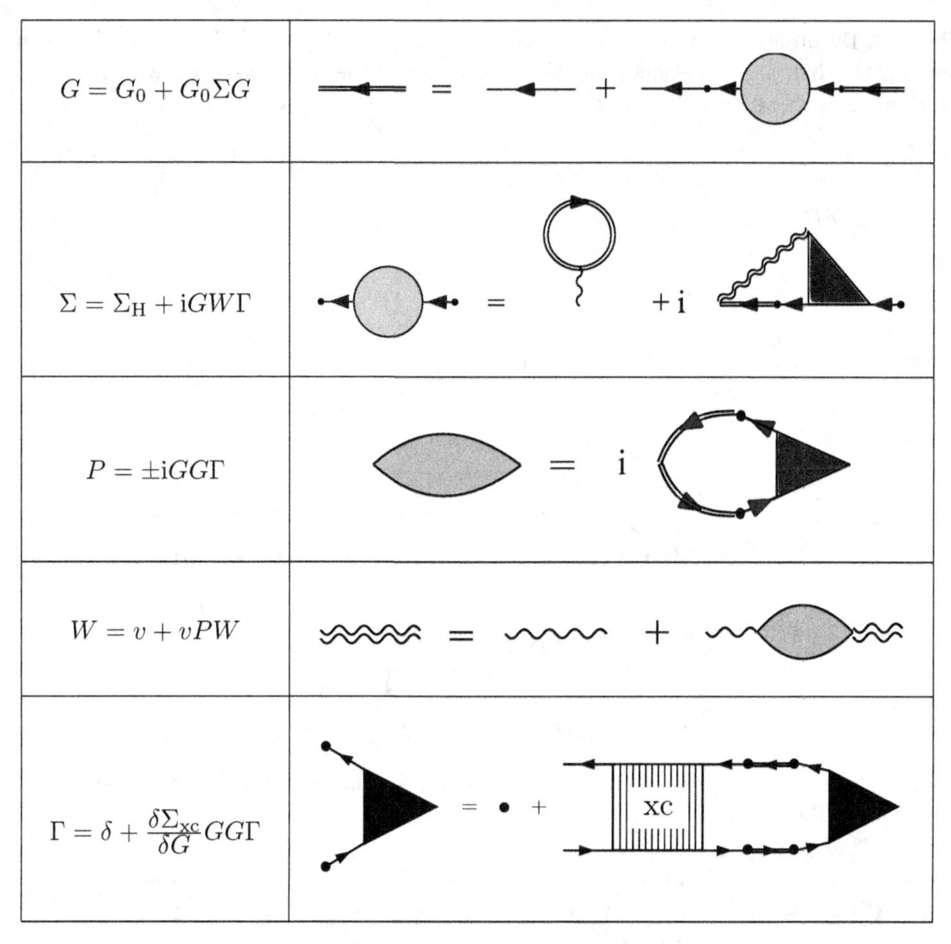

Expression	Diagram
$G = G_0 + G_0 \Sigma G$	
$\Sigma = \Sigma_{\mathrm{H}} + \mathrm{i} G W \Gamma$	$= \; + \mathrm{i}$
$P = \pm \mathrm{i} G G \Gamma$	$= \; \mathrm{i}$
$W = v + vPW$	
$\Gamma = \delta + \frac{\delta \Sigma_{\mathrm{xc}}}{\delta G} G G \Gamma$	xc

Table 14.1 Mathematical expression (left column) and diagrammatic representation (right column) of the Hedin equations.

and

$$P(1;2) = \pm \mathrm{i}\, G(1;2) G(2;1) = \quad 1 \quad 2 \tag{14.100}$$

This is the GW approximation already encountered in Section 7.7. Let us iterate further by calculating the kernel from the GW self-energy and then a new vertex Γ from (14.95). We have

$$\pm \frac{\delta \Sigma_{\mathrm{xc}}(1;3)}{\delta G(4;2)} = \pm \mathrm{i}\, \delta(1;4) \delta(2;3)\, W(1;3) \pm \mathrm{i}\, G(1;3) \frac{\delta W(1;3)}{\delta G(4;2)}.$$

The last term can be evaluated from the fourth Hedin equation $W = v + vPW$. We have

$$\frac{\delta W}{\delta G} = v \frac{\delta P}{\delta G} W + vP \frac{\delta W}{\delta G}.$$

This is a Dyson-like equation for $\delta W/\delta G$. We can expand $\delta W/\delta G$ in "powers" of v by iterations – that is, by replacing the $\delta W/\delta G$ in the last term of the r.h.s. with the whole r.h.s. In doing so we find the series

$$\frac{\delta W}{\delta G} = v\frac{\delta P}{\delta G}W + vPv\frac{\delta P}{\delta G}W + vPvPv\frac{\delta P}{\delta G}W + \ldots = (v + vPv + vPvPv + \ldots)\frac{\delta P}{\delta G}W$$
$$= W\frac{\delta P}{\delta G}W.$$

From (14.100) it follows that

$$\frac{\delta P(5;6)}{\delta G(4;2)} = \pm \mathrm{i}\delta(4;5)\delta(2;6)G(6;5) \pm \mathrm{i}\delta(4;6)\delta(2;5)G(5;6),$$

and therefore

$$\pm\frac{\delta\Sigma_{\mathrm{xc}}(1;3)}{\delta G(4;2)} = \pm\mathrm{i}\,\delta(1;4)\delta(2;3)\,W(1;3)$$
$$+\,\mathrm{i}^2\,G(1;3)W(1;4)G(2;4)W(2;3) + \mathrm{i}^2 G(1;3)W(1;2)G(2;4)W(4;3)$$

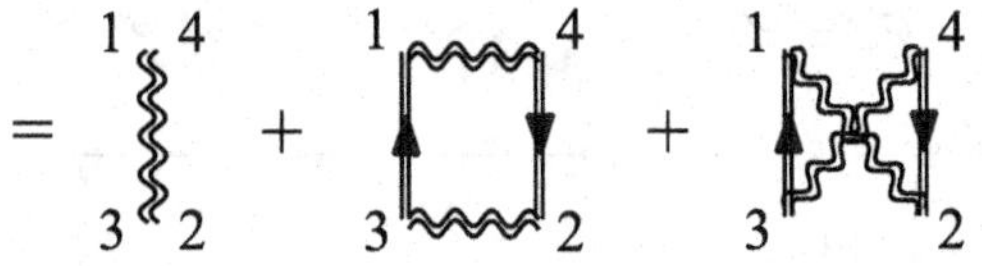

= [diagram: 1 4 / 3 2] + [diagram: 1 4 / 3 2] + [diagram: 1 4 / 3 2]

Inserting this kernel into the fifth Hedin equation, we obtain the following diagrammatic equation for the vertex:

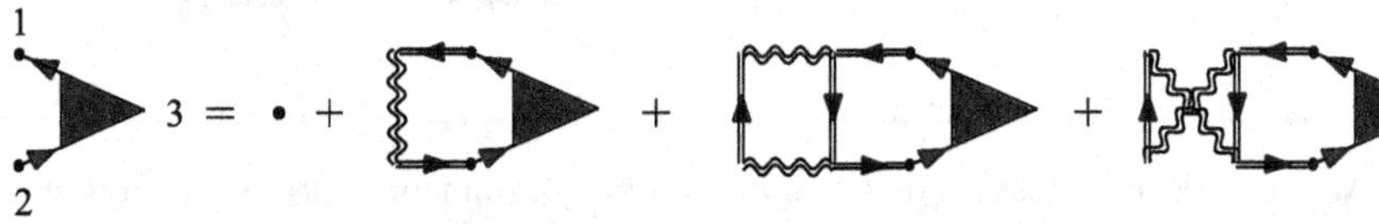

[diagram: 1, 2, 3 = • + … + … + …]

From this vertex we can then calculate a new self-energy, etc. This has actually been done in the context of the Hubbard model in Ref. [227]. As for the coupled equations (14.87) and (14.89), we are not aware of any proof that such an iterative scheme converges nor that it generates all possible skeleton diagrams for the self-energy.

In Appendix N we derive the Hedin equations using the so-called *source field* method introduced by Martin and Schwinger [50]. The source field method prescinds from any diagrammatic notion and allows us to arrive at the Hedin equations in a few steps. The source field method does not have the same physical appeal as the diagrammatic method since it does not tell us how to expand the various quantities Σ, P, Γ etc. Furthermore, every time the source field method generates an equation in which the Green's function has to be calculated at equal times, the most convenient way to resolve the ambiguity is to go back to the diagrammatic method.

Exercise 14.8 Prove (14.88).

14.10 Ward Identity and the f-sum Rule

There exist very special variations $\delta\mathbf{A}$ and $\delta\phi$ of the external potentials for which the variation of a conserving Green's function is trivial, and these are the variations induced by a gauge transformation. For an infinitesimal gauge transformation $\Lambda(\mathbf{x}, z)$ we have [see, e.g., (3.25)]

$$\delta\mathbf{A}(1) = \boldsymbol{\nabla}_1\Lambda(1), \qquad \delta\phi(1) = -\frac{1}{c}\frac{d}{dz_1}\Lambda(1). \tag{14.101}$$

Under this gauge transformation $h \to h + \delta h$ and consequently $G \to G + \delta G$. The transformed G is the Green's function that satisfies the Kadanoff–Baym equations with single-particle Hamiltonian $h + \delta h$. As already observed in (12.12), this Green's function is simply $G[\Lambda](1;2) = e^{\mathrm{i}\frac{q}{c}\Lambda(1)}G(1;2)e^{-\mathrm{i}\frac{q}{c}\Lambda(2)}$ for any Φ-derivable self-energy. Therefore, to first order in Λ

$$\delta G(1;2) = \left[e^{\mathrm{i}\frac{q}{c}\Lambda(1)}G(1;2)e^{-\mathrm{i}\frac{q}{c}\Lambda(2)} - G(1;2)\right] = \mathrm{i}\frac{q}{c}[\Lambda(1) - \Lambda(2)]G(1;2). \tag{14.102}$$

Inserting (14.101) and (14.102) into (14.38) we find

$$\mathrm{i}[\Lambda(1) - \Lambda(2)]G(1;2) = -\int d3\left[\left(\frac{d\Lambda(3)}{dz_3} + (\boldsymbol{\nabla}_3\Lambda(3))\cdot\frac{\mathbf{D}_3 - \mathbf{D}_4^*}{2m\mathrm{i}}\right)L(1,3;2,4^+)\right]_{4=3}.$$

We stress one more time that the replacement $L(1,4;2,3) \to L(1,4;2,3^+)$ on the r.h.s. is irrelevant if L comes from the solution of the Bethe–Salpeter equation, while it matters if L is calculated directly from (13.15) due to the presence of the Hartree term. If we now integrate by parts the r.h.s. and take into account that $\Lambda(\mathbf{x}, t_{0-}) = \Lambda(\mathbf{x}, t_0 - \mathrm{i}\beta) = 0$ (the perturbing field is switched on at times $t > t_0$), we get

$$\frac{d}{dz_3}L(1,3;2,3^+) + \boldsymbol{\nabla}_3\cdot\left(\frac{\mathbf{D}_3 - \mathbf{D}_4^*}{2m\mathrm{i}}L(1,3;2,4^+)\right)_{4=3} = \mathrm{i}\left[\delta(1;3) - \delta(2;3)\right]G(1;2), \tag{14.103}$$

which is valid for the exact L and G as well as for any conserving approximation to L and G. This relation is known as the *Ward identity* and it guarantees the gauge invariance of the theory.

The Ward identity is usually written in terms of the scalar vertex function Λ, defined in Section 14.9, and the vector vertex function $\boldsymbol{\Lambda}$:

$$\int d4d5\, G(1;4)G(5;2)\Lambda(4,5;3) \equiv L(1,3;2,3^+),$$

$$\int d4d5\, G(1;4)G(5;2)\boldsymbol{\Lambda}(4,5;3) \equiv \left(\frac{\mathbf{D}_3 - \mathbf{D}_4^*}{2m\mathrm{i}}L(1,3;2,4^+)\right)_{4=3}.$$

With these definitions (14.103) becomes

$$\int d4d5\, G(1;4)G(5;2)\left[\frac{d}{dz_3}\Lambda(4,5;3) + \boldsymbol{\nabla}_3\cdot\boldsymbol{\Lambda}(4,5;3)\right] = \mathrm{i}\left[\delta(1;3) - \delta(2;3)\right]G(1;2). \tag{14.104}$$

We can eliminate the Green's functions on the l.h.s. using the equations of motion. Acting on (14.104) with $\overrightarrow{G}^{-1}$ from the left and then with $\overleftarrow{G}^{-1}$ from the right, we arrive at the standard form of the Ward identity

$$\boxed{\frac{d}{dz_3}\Lambda(1,2;3) + \boldsymbol{\nabla}_3 \cdot \boldsymbol{\Lambda}(1,2;3) = \mathrm{i}\left[\overrightarrow{G}^{-1}(1;3)\delta(3;2) - \delta(1;3)\overleftarrow{G}^{-1}(3;2)\right]} \tag{14.105}$$

This equation relates the vertex to the self-energy and it is valid for any conserving approximation. It was derived in 1950 by Ward [228] (who published it in a letter of less than a column) in the context of quantum electrodynamics and it was used by the author to demonstrate an exact cancellation between divergent quantities. Later, in 1957, the Ward identity was generalized by Takahashi [229] to higher-order correlators and, especially in textbooks on quantum field theory [31, 32], these generalized identities are called the *Ward–Takahashi identities.* In the theory of Fermi liquids the Ward identity provides a relation between the vertex and the quasi-particle renormalization factor Z [185, 230], see Section 10.2. We emphasize that our derivation of the Ward identity does not require that the system is in the ground state; equation (14.105) is an identity between out-of-equilibrium correlators in the Keldysh space and it reduces to the standard Ward identity for systems in equilibrium.

***f*-sum rule** The satisfaction of the Ward identity implies that the response functions fulfill several exact relations. Among them is the f-sum rule discussed in Section 13.2.3. In the remainder of this section we prove the f-sum rule for conserving approximations. We start by taking the limit $2 \to 1$ in (14.103); using relations (13.16) and (13.17) we can write this limit as

$$\frac{d}{dz_3}\chi_{00}(1;3) + \partial_{3,k}\chi_{0k}(1;3) = 0, \tag{14.106}$$

where the sum over $k = 1, 2, 3$ is understood (Einstein convention). Similarly, we can apply the operator $(\mathbf{D}_1 - \mathbf{D}_2^*)/(2m\mathrm{i})$ to (14.103) and then take the limit $2 \to 1^+$. Using again the relations (13.16) and (13.17), we arrive at

$$\frac{d}{dz_3}\chi_{j0}(1;3) + \partial_{3,k}\chi_{jk}(1;3) = \frac{n(1)}{m}\partial_{1,j}\delta(1;3). \tag{14.107}$$

The response function $\chi_{\mu\nu}(1;3)$ belongs to the Keldysh space and it has a vanishing singular part. Therefore its structure is

$$\chi_{\mu\nu}(1;3) = \theta(z_1,z_3)\chi^>_{\mu\nu}(1;3) + \theta(z_3,z_1)\chi^<_{\mu\nu}(1;3). \tag{14.108}$$

Inserting (14.108) into (14.106), we find

$$\begin{aligned} -\delta(z_1,z_3)\left[\chi^>_{00}(1;3) - \chi^<_{00}(1;3)\right] &+ \theta(z_1,z_3)\left[\frac{d}{dz_3}\chi^>_{00}(1;3) + \partial_{3,k}\chi^>_{0k}(1;3)\right] \\ &+ \theta(z_3,z_1)\left[\frac{d}{dz_3}\chi^<_{00}(1;3) + \partial_{3,k}\chi^<_{0k}(1;3)\right] = 0, \end{aligned}$$

from which it follows that

$$\chi^>_{00}(\mathbf{x}_1,t_1;\mathbf{x}_3,t_1) - \chi^<_{00}(\mathbf{x}_1,t_1;\mathbf{x}_3,t_1) = 0, \tag{14.109}$$

and

$$\frac{d}{dt_3}\chi^{\lessgtr}_{00}(\mathbf{x}_1,t_1;\mathbf{x}_3,t_3)+\partial_{3,k}\chi^{\lessgtr}_{0k}(\mathbf{x}_1,t_1;\mathbf{x}_3,t_3)=0. \tag{14.110}$$

In a similar way, insertion of (14.108) into (14.107) leads to

$$\chi^{>}_{j0}(\mathbf{x}_1,t_1;\mathbf{x}_3,t_1)-\chi^{<}_{j0}(\mathbf{x}_1,t_1;\mathbf{x}_3,t_1)=-\frac{n(\mathbf{x}_1,t_1)}{m}\partial_{1,j}\delta(\mathbf{x}_1-\mathbf{x}_3) \tag{14.111}$$

and

$$\frac{d}{dt_3}\chi^{\lessgtr}_{j0}(\mathbf{x}_1,t_1;\mathbf{x}_3,t_3)+\partial_{3,k}\chi^{\lessgtr}_{jk}(\mathbf{x}_1,t_1;\mathbf{x}_3,t_3)=0. \tag{14.112}$$

Equations (14.110) and (14.112) are gauge conditions on the response functions. They guarantee that the switching of a pure gauge does not change the density and the current in the system. As for (14.109) and (14.111), we observe that in the exact case they are a direct consequence of the commutators $[\hat{n}(\mathbf{x}_1),\hat{n}(\mathbf{x}_3)]_-=0$ and of the commutator (13.33).

We now combine (14.110) and (14.111) using the symmetry property (13.10), which implies $\chi^{\lessgtr}_{0k}(\mathbf{x}_1,t_1;\mathbf{x}_3,t_3)=\chi^{\gtrless}_{k0}(\mathbf{x}_3,t_3;\mathbf{x}_1,t_1)$. We define the difference $\Delta_{\mu\nu}=\chi^{>}_{\mu\nu}-\chi^{<}_{\mu\nu}$ and write

$$\begin{aligned}\frac{d}{dt_3}\Delta_{00}(\mathbf{x}_1,t_1;\mathbf{x}_3,t_3)|_{t_3=t_1}&=-\partial_{3,k}\Delta_{0k}(\mathbf{x}_1,t_1;\mathbf{x}_3,t_1)=\partial_{3,k}\Delta_{k0}(\mathbf{x}_3,t_1;\mathbf{x}_1,t_1)\\&=-\frac{1}{m}\boldsymbol{\nabla}_3\cdot\left[n(\mathbf{x}_3,t_1)\boldsymbol{\nabla}_3\delta(\mathbf{x}_3-\mathbf{x}_1)\right].\end{aligned} \tag{14.113}$$

Let us specialize this formula to equilibrium situations: $\mathbf{A}(\mathbf{x},t)=\mathbf{A}(\mathbf{x})$ and $\phi(\mathbf{x},t)=\phi(\mathbf{x})$. Then the density $n(\mathbf{x}_3,t_1)=n(\mathbf{x}_3)$ is the equilibrium density and $\Delta_{\mu\nu}$ depends only on the time difference. Denoting by $\Delta_{\mu\nu}(\mathbf{x}_1,\mathbf{x}_3,\omega)$ its Fourier transform, (14.113) becomes

$$\mathrm{i}\int\frac{d\omega'}{2\pi}\,\omega'\Delta_{00}(\mathbf{x}_1,\mathbf{x}_3,\omega')=-\frac{1}{m}\boldsymbol{\nabla}_3\cdot\left[n(\mathbf{x}_3)\boldsymbol{\nabla}_3\delta(\mathbf{x}_3-\mathbf{x}_1)\right]. \tag{14.114}$$

Similarly, in equilibrium (14.109) can be rewritten as

$$\int\frac{d\omega'}{2\pi}\,\Delta_{00}(\mathbf{x}_1,\mathbf{x}_3,\omega')=0. \tag{14.115}$$

The relation between Δ_{00} and χ^{R} is provided by (13.22), which for large ω reads:

$$\chi^{\mathrm{R}}(\mathbf{x}_1,\mathbf{x}_3,\omega)=\frac{\mathrm{i}}{\omega}\int\frac{d\omega'}{2\pi}\Delta_{00}(\mathbf{x}_1,\mathbf{x}_3,\omega')+\frac{\mathrm{i}}{\omega^2}\int\frac{d\omega'}{2\pi}(\omega'-\mathrm{i}\eta)\Delta_{00}(\mathbf{x}_1,\mathbf{x}_3,\omega')+\mathcal{O}(\frac{1}{\omega^3}).$$

Using (14.114) and (14.115) we see that the large ω behavior of any *conserving* density response function is the same as the behavior (13.32) of the *exact* density response function. Furthermore, *assuming* that the conserving response function is analytic in the upper half of the complex ω-plane, we could apply the same trick as in Section 13.2.3 and integrate along contour C of Fig. 13.2 to find the f-sum rule again. The reason why we said "assuming" is that conserving approximations do not necessarily preserve the correct analytic structure of the response functions [231, 232].

14.11 Response Functions from Time-Propagation: Double Excitations

We saw in Section 13.2.1 that the spectral properties of neutral (particle-conserving) excitations can be read out from the two-particle XC function L or, equivalently, the response function χ. In Section 14.6 we derived the important result that the change δG of the Green's function induced by a change in the external potential can be obtained from the two-particle XC function that satisfies the Bethe–Salpeter equation with kernel $K = \pm\delta\Sigma/\delta G$, where $\Sigma = \Sigma[G]$ is the self-energy that determines G through $G = G_0 + G_0\Sigma G$. The direct implementation of the Bethe–Salpeter equation has proven to be computationally challenging and, in practice, often requires a number of additional approximations, such as neglect of self-consistency, kernel diagrams, and/or frequency dependence, see Section 14.7. Obtaining the response function by direct propagation of the Green's function does not require any of the aforementioned approximations [113, 215, 216, 233]. Moreover, the resulting response function automatically satisfies the f-sum rule. In this section we illustrate how to do this in practice.

Suppose that we are interested in calculating the change $\delta n_{ij}(t) \equiv -\mathrm{i}\,\delta G^{<}_{ij}(t,t)$ in the one-particle density matrix due to a change $\delta h_{mn}(t)$ of the single-particle Hamiltonian h. Then, from the first equality of (14.38), we have

$$\delta n_{ij}(t) = \pm\mathrm{i}\sum_{mn}\int_{-\infty}^{\infty} dt' L^{\mathrm{R}}_{\substack{ij\\mn}}(t,t')\,\delta h_{nm}(t'), \tag{14.116}$$

with response function

$$L_{\substack{ij\\mn}}(z,z') = -\mathrm{i}\langle\mathcal{T}\{\Delta\hat{n}_{ij,H}(z)\,\Delta\hat{n}_{mn,H}(z')\}\rangle$$

and $\Delta\hat{n}_{ij,H}(z) \equiv \hat{d}^{\dagger}_{j,H}(z)\hat{d}_{i,H}(z) - \langle\hat{d}^{\dagger}_{j,H}(z)\hat{d}_{i,H}(z)\rangle$.[18] Alternatively, (14.116) follows from the Kubo formula (13.5) with $\hat{O} = \hat{n}_{ij} = \hat{d}^{\dagger}_{j}\hat{d}_{i}$ and $\lambda\hat{H}'(t) = \sum_{mn}\delta h_{nm}(t)\hat{n}_{mn}$. Choosing to perturb the system with a δ-like time-dependent $\delta h_{nm}(t) = \delta(t)\delta h_{nm}$, we get

$$\frac{\delta n_{ij}(t)}{\delta h_{nm}} = \pm\mathrm{i}L^{\mathrm{R}}_{\substack{ij\\mn}}(t,0). \tag{14.117}$$

This equation tells us that if the amplitude δh_{nm} of the δ-function is small, then we can solve the Kadanoff–Baym equations with self-energy $\Sigma[G]$, calculate the one-particle density matrix $\delta n_{ij}(t) = n_{ij}(t) - n_{ij}(0)$, divide by δh_{nm}, and extract the retarded two-particle XC function $L^{\mathrm{R}}(t,0)$. Although nonlinear effects are always present, they are easily reduced by ensuring that the magnitude of the perturbation lies well in the linear response region. In practice, this is verified by doubling the amplitude δh_{nm} and checking that $\delta n_{ij}(t)$ doubles to a sufficient accuracy. The resulting two-particle XC function L solves the Bethe–Salpeter equation with kernel $K = \pm\delta\Sigma/\delta G$. Approximations to L that go beyond the HSEX approximation can in this way be studied with relatively simple self-energy approximations.

[18] For the expansion of L in arbitrary basis, see (13.40).

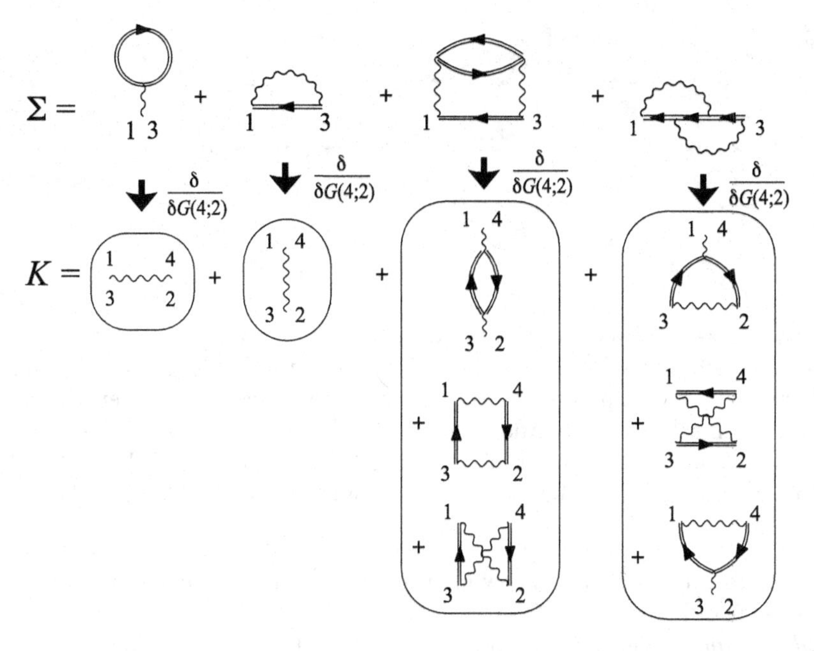

Figure 14.10 In the frame below every self-energy diagram $\Sigma^{(D)}$ we display the kernel diagrams generated from $\pm\delta\Sigma^{(D)}/\delta G$.

The static HSEX self-energy yields the *frequency-independent* kernel $K_{\rm HSEX}$, see Section 14.7, used in several state-of-the-art implementations of the Bethe–Salpeter equation. The diagrammatic expansion of the HSEX two-particle XC function is illustrated in Fig. 14.6, where the double lines are usually approximated with a quasi-particle Green's function. In Fig. 14.10 we illustrate the diagrammatic content of the second Born (2B) self-energy and kernel. In this figure the double lines are 2B Green's functions. Below each self-energy diagram $\Sigma^{(D)}$ we display the kernel diagrams generated from $\pm\delta\Sigma^{(D)}/\delta G$. It is instructive to expand the 2B L in terms of Hartree–Fock (HF) Green's functions, hence $G = G_{\rm HF} + G_{\rm HF}\Sigma_{\rm c}G$. To highlight the number of particle–hole excitations in every diagram we imagine that the direction of time is from left to right so that all v-lines are drawn vertical (the interaction is instantaneous). Some representative diagrams of the infinite series are displayed in Fig. 14.11. The series of bubbles and ladders in the first row corresponds to the HSEX approximation of Fig. 14.6 with $W \to v$; we clearly see that at every instant of time there is at most one particle–hole excitation. We infer that the HF (as well as the static HSEX) kernel renormalizes the single-particle excitations of the mean-field treatment but it cannot account for excitations of multiple-particle character, like the double-excitations discussed in the last paragraph of Section 13.3. These correlation-induced excitations call either for a Green's function beyond the Hartree–Fock approximation or for kernel diagrams beyond the Hartree–Fock (or static HSEX) approximation, hence *frequency-dependent* kernels. In the second row of Fig. 14.11 we show some of the diagrams with more than one particle–hole excitation. Every diagram contains a minimum of two particle–hole excitations and their

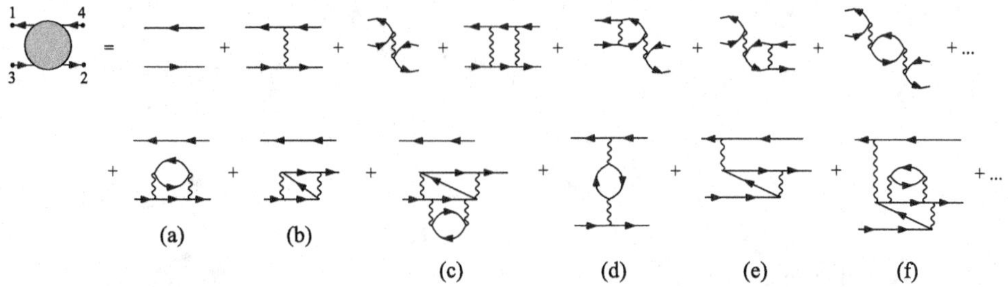

Figure 14.11 Expansion of the two-particle XC function in terms of the Hartree–Fock Green's function G_{HF} (oriented solid line). In the first row we have the diagrams of Fig. 14.6. In the second row we have examples of HF diagrams with one 2B self-energy insertion [(a) and (b)], and with two 2B self-energy insertions [(c)]. Then we have examples of diagrams with 2B kernel diagrams [(d) and (e)] and finally an example of a diagram with both 2B self-energy insertion and 2B kernel diagrams [(f)]. The direction of time is from left to right.

origin is due to either a 2B self-energy insertion [see (a), (b), and (c)] or 2B kernel diagrams [see (d) and (e)] or both [see (f)].

Hubbard ring We illustrate the above general considerations on a simple model system. We consider a Hubbard ring with Hamiltonian

$$\hat{H} = -T\sum_{s=1}^{N}\sum_{\sigma}\left(\hat{d}^{\dagger}_{s\sigma}\hat{d}_{s+1\sigma} + \hat{d}^{\dagger}_{s+1\sigma}\hat{d}_{s\sigma}\right) + U\sum_{s=1}^{N}\hat{n}_{s\uparrow}\hat{n}_{s\downarrow}, \tag{14.118}$$

where $\hat{n}_{s\sigma} = \hat{d}^{\dagger}_{s\sigma}\hat{d}_{s\sigma}$ is the occupation operator for site s and spin σ. In this equation the operator $\hat{d}_{N+1\sigma} \equiv \hat{d}_{1\sigma}$, so that the nearest-neighbor sites of 1 are 2 and N. In (14.118) the interaction is spin-dependent since if we want to rewrite the last term in the canonical form

$$\hat{H}_{\text{int}} = \frac{1}{2}\sum_{s\sigma,s'\sigma'} v_{s\sigma\,s'\sigma'}\,\hat{d}^{\dagger}_{s\sigma}\hat{d}^{\dagger}_{s'\sigma'}\hat{d}_{s'\sigma'}\hat{d}_{s\sigma}$$

then the Coulomb integral must be $v_{s\sigma\,s'\sigma'} = U\delta_{ss'}\delta_{\sigma\bar{\sigma}'}$ where $\bar{\sigma}'$ is the spin opposite to σ'. The reader can easily check that the exchange (Fock) self-energy diagram vanishes, whereas the Hartree self-energy is simply (see Exercise 8.5)

$$[\Sigma_{\text{H}}(z,z')]_{s\sigma\,s'\sigma'} = \delta_{ss'}\delta_{\sigma\sigma'}\delta(z,z')\,U\langle n_{s\bar{\sigma}}(z)\rangle.$$

The average occupation $\langle n_{s\sigma}(z)\rangle$ is independent of z in thermal equilibrium, independent of s due to the invariance of the ring under discrete rotations, and independent of σ if the number of spin-up and spin-down particles is the same. The HF eigenstates are then identical to the noninteracting eigenstates (like in the electron gas) and their energy is $\epsilon_k = -2T\cos(2\pi k/N) + Un/2$, with $n/2 = n_\sigma$ the average occupation per spin. In Fig. 14.12 we display the Hubbard ring with $N = 6$ sites as well as the HF energies and degeneracies at zero temperature for $n = 1/3$ and $U = T$. The particle–hole excitations of

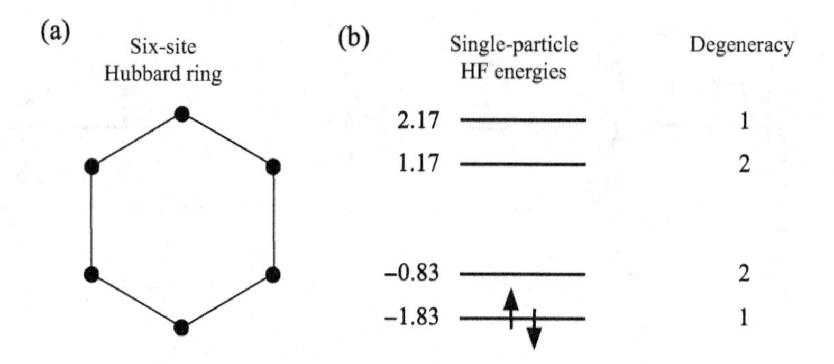

Figure 14.12 (a) Hubbard ring with 6 sites. (b) HF energy levels for $U = 1$. All energies are in units of $T > 0$.

$L^0 \equiv -\mathrm{i}G_{\mathrm{HF}}G_{\mathrm{HF}}$ are given by the energy differences between an unoccupied level and the only occupied level, see Section 13.3.

To assess the quality of the HF and 2B kernel we have calculated the eigenfunctions $|\Psi_k\rangle$ and eigenenergies E_k via exact diagonalization (ED) methods (Lanczos method [234, 235]), and then the excitation energies $\Omega_k = E_k - E_0$ for the transitions $|\Psi_0\rangle \leftrightarrow |\Psi_k\rangle$ between the ground state $|\Psi_0\rangle$ and the kth excited state $|\Psi_k\rangle$. Exact eigenstates and eigenenergies have also been used to simulate the time evolution of the system after a δ-like perturbation $\delta(t)\delta\epsilon_1$ of the onsite energy of site 1. According to (14.117), the ratio $\delta n_{11}(t)/\delta\epsilon_1$ is the response function $\chi^{\mathrm{R}}_{11}(t,0) \equiv -\mathrm{i}L^{\mathrm{R}}_{\substack{11\\11}}(t,0)$.

Numerical Fourier transform Both exact and Kadanoff–Baym simulations terminate after some finite time ΔT. We therefore approximate the Fourier transform with

$$\chi^{\mathrm{R}}_{11}(\omega) \sim \int_0^{\Delta T} dt\, e^{\mathrm{i}\omega t}\, \chi^{\mathrm{R}}_{11}(t,0). \tag{14.119}$$

The integral must be evaluated numerically as we know $\chi^{\mathrm{R}}_{11}(t,0)$ only at the discrete times $t = k\Delta_t$, with Δ_t the time step. To have an idea of how (14.119) looks, let us write $\chi^{\mathrm{R}}_{11}(t,0)$ in the form of (13.29):

$$\chi^{\mathrm{R}}_{11}(t,0) = -\mathrm{i}\theta(t)\sum_{k\neq 0}\left[e^{-\mathrm{i}\Omega_k t}|f_k(1)|^2 - e^{\mathrm{i}\Omega_k t}|f_k(1)|^2\right], \tag{14.120}$$

where $f_k(1) \equiv \sum_\sigma \langle\Psi_0|\hat{n}_{1\uparrow} + \hat{n}_{1\downarrow}|\Psi_k\rangle$ are the excitation amplitudes on site 1. Inserting (14.120) into (14.119) we get

$$\chi^{\mathrm{R}}_{11}(\omega) \sim -\mathrm{i}\sum_k |f_k(1)|^2\left[e^{\mathrm{i}(\omega-\Omega_k)\Delta T/2}\,\frac{\sin\frac{\omega-\Omega_k}{2}\Delta T}{\frac{\omega-\Omega_k}{2}} - e^{\mathrm{i}(\omega+\Omega_k)\Delta T/2}\,\frac{\sin\frac{\omega+\Omega_k}{2}\Delta T}{\frac{\omega+\Omega_k}{2}}\right].$$

The imaginary part of this function exhibits peaks of width $\sim 1/\Delta T$ instead of sharp δ-like peaks. These broadened peaks, however, are located exactly in $\pm\Omega_k$. We also observe that for any finite ΔT the imaginary part has an oscillatory background "noise" that goes to zero

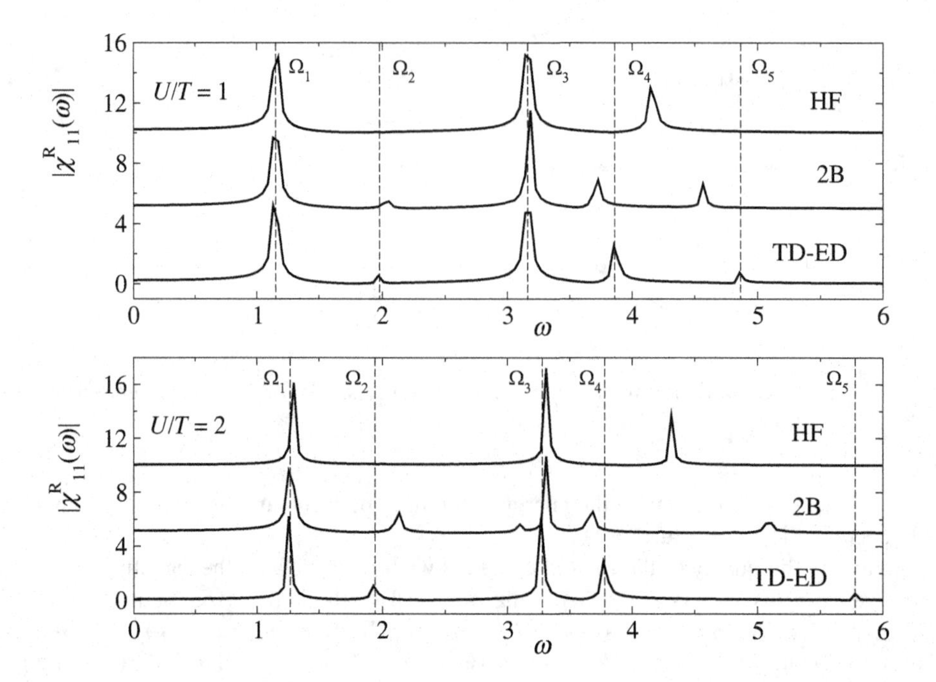

Figure 14.13 Modulus of the response function $\chi^{\rm R}_{11}(\omega)$ for $U/T = 1, 2$ as obtained from the exact time-propagation (TD-ED) and the solution of the Kadanoff–Baym equations in the HF and 2B approximations. The dashed vertical lines indicate the position of the ED excitations energies.

as $1/\Delta T$. Thus peaks of height less than $1/\Delta T$ are hardly visible. If we are only interested in the position and relative strength of the peaks, a better way to process the time-dependent information consists of performing a discrete Fourier transform:

$$\chi^{\rm R}_{11}(\omega_m) \sim \Delta_t \sum_{k=1}^{N_t} e^{{\rm i}\omega_m k\Delta_t} \chi^{\rm R}_{11}(k\Delta_t, 0),$$

where Δ_t is the time step, N_t is the total number of time steps, and the frequency is sampled according to $\omega_m = 2\pi m/(N_t\Delta_t)$. The discrete Fourier transform corresponds to extend periodically the time-dependent results between 0 and $\Delta T = N_t\Delta_t$ so that, for example, $\chi^{\rm R}_{11}(N_t\Delta_t + p\Delta_t, 0) = \chi^{\rm R}_{11}(p\Delta_t, 0)$. The imaginary part of $\chi^{\rm R}_{11}(\omega_m)$ can be negative for some ω_m due to the finite propagation time ΔT. It is aesthetically nicer to plot $|\chi^{\rm R}_{11}(\omega_m)|$: This quantity is always positive and it has maxima in $\pm\Omega_k$ whose height is proportional to the square of the excitation amplitudes.

Double excitations In Fig. 14.13 we plot the time-dependent ED (TD-ED) results for $|\chi^{\rm R}_{11}(\omega)|$. The position of the ED excitation energies (dashed vertical lines) matches exactly the position of the TD-ED peaks, in accordance with the previous discussion. In the same figure, the TD-ED response function is compared with the HF and 2B response functions as

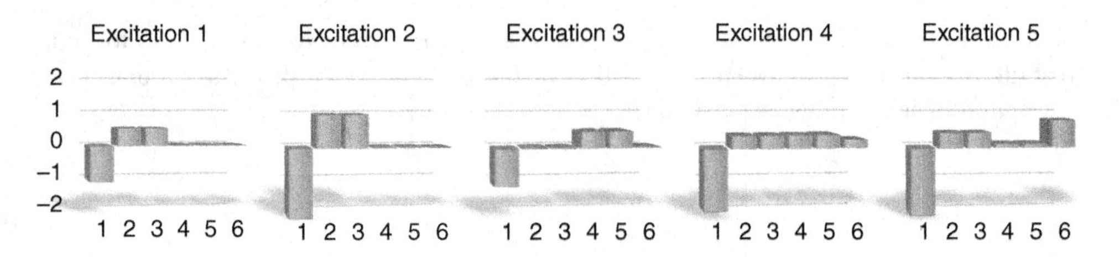

Figure 14.14 Histograms of the HF level occupation difference $\Delta n_l^{\mathrm{HF}}(\Omega)$ versus $l = 1, \ldots, 6$ for the excitation energies $\Omega = \Omega_1, .., \Omega_5$.

obtained from the solution of the Kadanoff–Baym equations for two different values $U = T$ and $U = 2T$ of the interaction.

As expected, not all excitations are reproduced within the HF approximation. The HF response function shows only three peaks corresponding to the (renormalized) single-excitation energies of the transition from the lowest HF level to one of the three excited HF levels of Fig. 14.12. Which of these HF excitations can be considered as an approximation to the exact excitations? Following Ref. [216], we calculate the average of the occupation operator $\hat{n}_l^{\mathrm{HF}} \equiv \sum_\sigma \hat{c}_{l\sigma}^\dagger \hat{c}_{l\sigma}$ over the excited states $|\Psi_k\rangle$. Here, $\hat{c}_{l\sigma}$ annihilates an electron of spin σ on the lth HF level. The difference

$$\Delta n_l^{\mathrm{HF}}(k) \equiv \langle \Psi_k | \hat{n}_l^{\mathrm{HF}} | \Psi_k \rangle - \langle \Psi_0 | \hat{n}_l^{\mathrm{HF}} | \Psi_0 \rangle$$

tells us how the occupation of the lth HF level changes in the transition $|\Psi_0\rangle \to |\Psi_k\rangle$. If we number the HF levels from the lowest to the highest in energy (thus, e.g., $l = 1$ is the lowest, $l = 2, 3$ the first excited, etc.), then single excitations are characterized by $\Delta n_1^{\mathrm{HF}}(k) \sim -1$, whereas double excitations by $\Delta n_1^{\mathrm{HF}}(k) \sim -2$. In Fig. 14.14 we show the histogram of

$$\Delta n_l^{\mathrm{HF}}(\Omega) = \frac{1}{d_\Omega} \sum_{k:\Omega_k = \Omega} \Delta n_l^{\mathrm{HF}}(k), \qquad d_\Omega = \text{degeneracy of excitation } \Omega$$

for the first five excitations $\Omega_1, \ldots, \Omega_5$ and for interaction $U = T$. We see that only Ω_1 and Ω_3 are single excitations and therefore only the first and the second HF peaks correspond to physical excitations. The third HF peak between Ω_4 and Ω_5 is instead unphysical and indeed quite far from the TD-ED peaks.

The qualitative agreement between TD-ED and MBPT results improves considerably in the 2B approximation. The 2B response function captures both single and double excitations! Unfortunately the position of the double-excitation peaks Ω_2, Ω_4, and Ω_5 is not as good as that of the single-excitation peaks and tends to worsen the higher we go in energy. Is this error due to an inaccurate estimation of the ground-state energy E_0, or of the excited-state energy E_k, or both? This question can be answered by calculating the ground-state energy either using the Galitskii–Migdal formula or the Luttinger–Ward functional,[19] and confronting the result with the exact ground-state energy E_0. What one finds is that the discrepancy between the 2B ground-state energy and E_0 is about 0.2% for $U = T$

[19] For any Φ-derivable approximation the two approaches yields the same result – see the discussion below (12.8).

and 1.2% for $U = 2T$. Thus the 2B ground-state energy is extremely accurate and we must attribute the mismatch between the position of the 2B and TD-ED peaks to an inaccurate description of the (doubly) excited states.

15

Electron Gas: Equilibrium and Nonequilibrium Correlation Effects

The electron gas introduced in Sections 8.2.2 and 8.3.2 is one of the most studied correlated systems of fermionic particles [108]. Over the past decades the electron gas has served as a training ground to test and understand the physical content of various approximations. Several useful notions such as dynamical screening, plasmons, and quasi-particles were originally introduced in this context and then exported to real materials. Despite the numerous papers dealing with the electron gas, we still lack full knowledge of its spectral properties. In this chapter we apply the nonequilibrium Green's function formalism to review some of the most important findings and take the opportunity to discuss some open issues.

15.1 Lifetime and Renormalization Factor

The simplest self-energy beyond the Hartree–Fock approximation is the second Born self-energy (7.16): $\Sigma = \Sigma_{\rm HF} + \Sigma_{\rm c}$. Using the Feynman rules of Section 7.8 for particles of spin S, it reads

$$\begin{aligned}\Sigma_{\rm c}(1;\bar{1}) &= \pm\, {\rm i}^2 \int d2d\bar{2}\; v(1;2)v(\bar{1};\bar{2})\big[G(1;\bar{1})G(2;\bar{2})G(\bar{2};2) \pm G(1;\bar{2})G(\bar{2};2)G(2;\bar{1})\big] \\ &= \pm\, {\rm i}^2\delta_{\sigma_1\bar{\sigma}_1}\int d\mathbf{r}_2 d\bar{\mathbf{r}}_2\; v(\mathbf{r}_1,\mathbf{r}_2)v(\bar{\mathbf{r}}_1,\bar{\mathbf{r}}_2) \\ &\quad\times\big[(2S+1)G(\mathbf{r}_1,z_1;\bar{\mathbf{r}}_1,\bar{z}_1)G(\mathbf{r}_2,z_1;\bar{\mathbf{r}}_2,\bar{z}_1)G(\bar{\mathbf{r}}_2,\bar{z}_1;\mathbf{r}_2,z_1) \\ &\quad\pm G(\mathbf{r}_1,z_1;\bar{\mathbf{r}}_2,\bar{z}_1)G(\bar{\mathbf{r}}_2,\bar{z}_1;\mathbf{r}_2,z_1)G(\mathbf{r}_2,z_1;\bar{\mathbf{r}}_1,\bar{z}_1)\big],\end{aligned} \tag{15.1}$$

where in the last equality we integrate over times and sum over spin, assuming that $v(1;2) = \delta(z_1,z_2)v(\mathbf{r}_1,\mathbf{r}_2)$ is spin-independent and that $G(1;2) = \delta_{\sigma_1\sigma_2}G(\mathbf{r}_1,z_1;\,\mathbf{r}_2,z_2)$ is spin-diagonal. The physical interpretation of (15.1) is particularly transparent in a system invariant under translations, like the electron gas. In this case all quantities depend only on the relative coordinate and can be Fourier transformed according to

$$G(1;2) = \delta_{\sigma_1\sigma_2}\int\frac{d\mathbf{p}}{(2\pi)^3}\, e^{{\rm i}\mathbf{p}\cdot(\mathbf{r}_1-\mathbf{r}_2)}G(\mathbf{p};z_1,z_2), \tag{15.2}$$

$$\Sigma_c(1;2) = \delta_{\sigma_1\sigma_2} \int \frac{d\mathbf{p}}{(2\pi)^3} e^{i\mathbf{p}\cdot(\mathbf{r}_1-\mathbf{r}_2)} \Sigma_c(\mathbf{p}; z_1, z_2). \tag{15.3}$$

It is a matter of simple algebra to show that in Fourier space (15.1) reads (lower sign for fermions):

$$\Sigma_c(\mathbf{p}; z_1, \bar{z}_1) = -i^2 \int \frac{d\mathbf{p}'}{(2\pi)^3} \frac{d\bar{\mathbf{p}}}{(2\pi)^3} \frac{d\bar{\mathbf{p}}'}{(2\pi)^3} (2\pi)^3 \delta(\mathbf{p} + \mathbf{p}' - \bar{\mathbf{p}} - \bar{\mathbf{p}}') B(\mathbf{p}, \bar{\mathbf{p}}, \bar{\mathbf{p}}')$$
$$\times G(\mathbf{p}'; \bar{z}_1, z_1) G(\bar{\mathbf{p}}; z_1, \bar{z}_1) G(\bar{\mathbf{p}}'; z_1, \bar{z}_1), \tag{15.4}$$

with B a quantity proportional to the differential cross section for the scattering $\mathbf{p}, \mathbf{p}' \to \bar{\mathbf{p}}, \bar{\mathbf{p}}'$ in the Born approximation (see Appendix M):

$$B(\mathbf{p}, \bar{\mathbf{p}}, \bar{\mathbf{p}}') = B(\mathbf{p}, \bar{\mathbf{p}}', \bar{\mathbf{p}}) = \frac{2S+1}{2} \left[\tilde{v}^2_{\mathbf{p}-\bar{\mathbf{p}}} + \tilde{v}^2_{\mathbf{p}-\bar{\mathbf{p}}'} \right] - \tilde{v}_{\mathbf{p}-\bar{\mathbf{p}}} \tilde{v}_{\mathbf{p}-\bar{\mathbf{p}}'} . \tag{15.5}$$

As usual, in this formula $\tilde{v}_{\mathbf{p}}$ is the Fourier transform of the interaction. From (15.4) we can easily extract the greater and lesser components of Σ_c. Going to frequency space, we get

$$\Sigma_c^{\gtrless}(\mathbf{p}, \omega) = -i^2 \int \frac{d\mathbf{p}' d\omega'}{(2\pi)^4} \frac{d\bar{\mathbf{p}} d\bar{\omega}}{(2\pi)^4} \frac{d\bar{\mathbf{p}}' d\bar{\omega}'}{(2\pi)^4} (2\pi)^4 \delta(\mathbf{p} + \mathbf{p}' - \bar{\mathbf{p}} - \bar{\mathbf{p}}') \delta(\omega + \omega' - \bar{\omega} - \bar{\omega}')$$
$$\times B(\mathbf{p}, \bar{\mathbf{p}}, \bar{\mathbf{p}}') G^{\lessgtr}(\mathbf{p}', \omega') G^{\gtrless}(\bar{\mathbf{p}}, \bar{\omega}) G^{\gtrless}(\bar{\mathbf{p}}', \bar{\omega}'). \tag{15.6}$$

In Section 10.2 we learned that the quasi-particle width $\Gamma(\mathbf{p}, \omega) = i[\Sigma_c^>(\mathbf{p}, \omega) - \Sigma_c^<(\mathbf{p}, \omega)]$ is a measure of the inverse lifetime of quasi-particles with momentum $\mathbf{p}$ and energy ω. We could interpret $\Sigma_c^>$ as the decay rate of an added particle and $\Sigma_c^<$ as the decay rate of a removed particle (hole). Then, according to the discussion below (14.3), $\Sigma_c^>$ describes a process in which a particle with momentum-energy $\mathbf{p}, \omega$ hits a particle with momentum-energy $\mathbf{p}', \omega'$, and after the scattering one particle goes into the state $\bar{\mathbf{p}}, \bar{\omega}$ and the other goes into the state $\bar{\mathbf{p}}', \bar{\omega}'$. The two δ-functions guarantee the conservation of momentum and energy. In Appendix M we show that the differential cross section for this process is proportional to B to lowest order in the interaction – that is, in the Born approximation. It is also intuitive to understand the appearance of the product of three G [109]. The probability to scatter off a particle with momentum-energy $\mathbf{p}', \omega'$ is given by the density of particles with this momentum-energy – that is, $f(\omega' - \mu) A(\mathbf{p}', \omega') = -iG^<(\mathbf{p}', \omega')$. The probability that after the scattering the particles end up in the states $\bar{\mathbf{p}}, \bar{\omega}$ and $\bar{\mathbf{p}}', \bar{\omega}'$ is given by the density of holes (available states): $\bar{f}(\bar{\omega} - \mu) A(\bar{\mathbf{p}}, \bar{\omega}) = iG^>(\bar{\mathbf{p}}, \bar{\omega})$ and $\bar{f}(\bar{\omega}' - \mu) A(\bar{\mathbf{p}}', \bar{\omega}') = iG^>(\bar{\mathbf{p}}', \bar{\omega}')$. In a similar way, we can discuss $\Sigma_c^<$. In the low-density limit $\beta\mu \to -\infty$ we see that the rate $\Sigma_c^<$ to scatter *into* the state $\mathbf{p}, \omega$ is negligible compared to the rate $\Sigma_c^>$ to scatter *out* of the same state [see also (10.5)]. On the contrary, at low temperatures the two rates are equally important.

Lifetime Let us estimate $\Sigma_c^{\lessgtr}(\mathbf{p}, \omega)$ at zero temperature for frequencies $\omega \sim \mu$ in a system of fermions. We assume that the integral over all momenta of the product of $\delta(\mathbf{p} + \mathbf{p}' - \bar{\mathbf{p}} - \bar{\mathbf{p}}')$, the differential cross section B, and the three spectral functions $A(\mathbf{p}', \omega') A(\bar{\mathbf{p}}, \bar{\omega}) A(\bar{\mathbf{p}}', \bar{\omega}')$ is a smooth function $F_{\mathbf{p}}(\omega', \bar{\omega}, \bar{\omega}')$ of the frequencies ω', $\bar{\omega}$ and $\bar{\omega}'$. This can rigorously be proven when $A(\mathbf{p}, \omega) = 2\pi\delta(\omega - p^2/2)$ is the noninteracting spectral function and the number of spatial dimensions is larger than one. At zero temperature the Fermi function $f(\omega-\mu)=\theta(\mu-\omega)$ and $\bar{f}(\omega-\mu)=1-\theta(\mu-\omega)=\theta(\omega-\mu)$, and therefore from (15.6) we have

$$\Sigma_{\rm c}^{>}(\mathbf{p},\omega)=-{\rm i}\int\frac{d\omega'}{2\pi}\frac{d\bar{\omega}}{2\pi}\frac{d\bar{\omega}'}{2\pi}F_{\mathbf{p}}(\omega',\bar{\omega},\bar{\omega}')\theta(\mu-\omega')\theta(\bar{\omega}-\mu)\theta(\bar{\omega}'-\mu)\delta(\omega+\omega'-\bar{\omega}-\bar{\omega}'). \tag{15.7}$$

Because of the θ the frequency $\omega'<\mu$ and $\bar{\omega}>\mu$, $\bar{\omega}'>\mu$. Since $\bar{\omega}+\bar{\omega}'>\omega'+\mu$ the frequency ω must be larger than μ, otherwise the argument of the δ-function cannot vanish. This agrees with the fluctuation-dissipation theorem for the self-energy according to which $\Sigma_{\rm c}^{>}(\mathbf{p},\omega)\propto 1-\theta(\mu-\omega)=\theta(\omega-\mu)$. For $\omega>\mu$ but very close to it the argument of the δ-function vanishes only in a region where ω', $\bar{\omega}$, and $\bar{\omega}'$ are very close to μ. Denoting by $u=\omega-\mu>0$ the distance of the frequency ω from the chemical potential, we can rewrite (15.7) as

$$\begin{aligned}\Sigma_{\rm c}^{>}(\mathbf{p},\omega)&=-{\rm i}\int_{\mu-u}^{\mu}\frac{d\omega'}{2\pi}\int_{\mu}^{\mu+u}\frac{d\bar{\omega}}{2\pi}\int_{\mu}^{\mu+u}\frac{d\bar{\omega}'}{2\pi}F_{\mathbf{p}}(\omega',\bar{\omega},\bar{\omega}')\delta(\omega+\omega'-\bar{\omega}-\bar{\omega}')\\&\sim-{\rm i}\frac{F_{\mathbf{p}}(\mu,\mu,\mu)}{u}\int_{-u}^{0}\frac{du'}{2\pi}\int_{0}^{u}\frac{d\bar{u}}{2\pi}\int_{0}^{u}\frac{d\bar{u}'}{2\pi}\delta(1+\frac{u'-\bar{u}-\bar{u}'}{u}),\end{aligned} \tag{15.8}$$

where in the last equality we approximate F with its value at the three frequencies equal to μ since F is smooth. Changing the integration variables $u'/u=x$, $\bar{u}/u=\bar{x}$ and $\bar{u}'/u=\bar{x}'$, we get a factor u^3 from $du'd\bar{u}d\bar{u}'$ and an integral which is independent of u. Thus we conclude that for ω very close to μ,

$$\Sigma_{\rm c}^{>}(\mathbf{p},\omega)=-{\rm i}\,C_{\mathbf{p}}\,\theta(\omega-\mu)\,(\omega-\mu)^2,$$

where $C_{\mathbf{p}}$ is a real constant. In a similar way we can derive that $\Sigma_{\rm c}^{<}(\mathbf{p},\omega)={\rm i}\,C_{\mathbf{p}}\,\theta(\mu-\omega)\,(\omega-\mu)^2$. Therefore, the quasi-particle width

$$\Gamma(\mathbf{p},\omega)=C_{\mathbf{p}}\,(\omega-\mu)^2 \tag{15.9}$$

vanishes quadratically as $\omega\to\mu$ *for all* $\mathbf{p}$.[1] This result is due to Luttinger [236], who also showed that self-energy diagrams with three or more interaction lines give contributions to the rates $\Sigma_{\rm c}^{\lessgtr}$ that vanish even faster as $\omega\to\mu$. In particular, these contributions vanish like $(\omega-\mu)^{2m}$ where m are integers larger than 1. Consequently (15.9) provides the *exact* leading term of the Taylor expansion of $\Gamma(\mathbf{p},\omega)$ in powers of $(\omega-\mu)$. The constant of proportionality

$$C_{\mathbf{p}}=\lim_{\omega\to\mu}\frac{\Gamma(\mathbf{p},\omega)}{(\omega-\mu)^2}>0,$$

since in fermionic systems Γ is positive, see the observation below (10.20).

From the Luttinger result we can understand why the low-energy excitations of metals are well described by an effective single-particle picture. Although the Coulomb interaction is strong, the scattering is very ineffective close to the Fermi surface due to phase space

[1] Equation (15.9) agrees with (10.17), according to which $\Gamma(\mathbf{p},\mu)=0$.

restrictions. This is not the case at higher energies. We emphasize that the details of the low-energy spectrum are still very much a many-body problem. In Section 15.6 we calculate $\Sigma_c^{\lessgtr}(\mathbf{p},\omega)$ as well as the quasi-particle width $\Gamma(\mathbf{p},\omega)$, the constant $C_{\mathbf{p}}$, the lifetime $\tau_{\mathbf{p}}$, etc. in the GW approximation.

Renormalization factor In Section 10.2 we define the quasi-particle energy $E_{\mathbf{p}}$ as the solution of $\omega - \epsilon_{\mathbf{p}}^{\mathrm{HF}} - \Lambda(\mathbf{p},\omega) = 0$. From the Luttinger–Ward theorem of Section 11.6 and from $\Gamma(\mathbf{p},\mu) = 0$ we infer that $E_{\mathbf{p}} \to \mu$ for $|\mathbf{p}| \to p_{\mathrm{F}}$. Thus the spectral function (10.24) becomes a δ-function of strength $Z = \lim_{|\mathbf{p}|\to p_{\mathrm{F}}} Z_{\mathbf{p}}$ centered at $\omega = \mu$ for $|\mathbf{p}| = p_{\mathrm{F}}$. The quasi-article renormalization factor Z, defined in (10.25), varies in the range

$$0 < Z \leq 1,$$

since the Hilbert transform relation (10.13) implies

$$\left.\frac{\partial\Lambda(\mathbf{p},\omega)}{\partial\omega}\right|_{\omega=\mu} = -\int \frac{d\omega'}{2\pi}\frac{\Gamma(\mathbf{p},\omega')}{(\mu-\omega')^2} < 0, \tag{15.10}$$

where we take into account that the rate function is nonnegative and vanishes quadratically as $\omega' \to \mu$. As the exact spectral function integrates to unity, see (6.82), this δ-function must be superimposed on a continuous function which integrates to $1 - Z$. The physical meaning of this result is that the removal or addition of a particle with momentum $|\mathbf{p}| = p_{\mathrm{F}}$ generates (with probability Z) an infinitely long-living excitation of energy μ and (with probability $1 - Z$) a continuum of other excitations at different energies. In the noninteracting case $Z = 1$; thus we can interpret Z as the renormalization of the square of the single-particle wavefunction due to interactions. From the viewpoint of a photoemission experiment (see Section 6.4), the spectral function $A(\mathbf{p},\omega)$ gives the probability that the system has changed its energy by ω when an electron of momentum $\mathbf{p}$ is photoemitted. For $|\mathbf{p}|$ values close to p_{F} this energy change is most likely equal to $E_{\mathbf{p}}$ and therefore the energy of the ejected electron is most likely equal to $\epsilon = \omega_0 - E_{\mathbf{p}}$, where ω_0 is the energy of the photon. In this way the quasi-particle energies can be measured experimentally in metals and the experiments have indeed confirmed the quasi-particle picture. As we see in Section 15.4.2, the removal or addition of a particle of momentum $\mathbf{p}$ *away from the Fermi momentum* p_{F} has large probability to excite electron–hole pairs as well as collective modes known as plasmons. These excitations introduce an uncertainty in the energy of the particle: The peak of the spectral function in $\omega = E_{\mathbf{p}}$ is broadened by an amount [see (10.26)] $\tau_{\mathbf{p}}^{-1} = Z_{\mathbf{p}}\Gamma(\mathbf{p},E_{\mathbf{p}})$ (mainly due to electron–hole excitations), and other prominent features appear in different spectral regions (mainly due to plasmons).

Luttinger liquids In a seminal footnote of the paper containing the proof of the result (15.9), see Ref. [236], Luttinger observed that the phase-space argument needs to be modified in one dimension. We can understand his observation by using the noninteracting spectral function $A(\mathbf{p},\omega) = 2\pi\delta(\omega - p^2/2)$. In this case the δ-function of momentum conservation appearing in (15.6) is sufficient to determine the energy, say, $\bar{\omega}' = \bar{p}'^2/2$ in terms of $\omega' = p'^2/2$ and $\bar{\omega} = \bar{p}^2/2$, and hence we can no longer treat $\bar{\omega}'$ as an independent variable. In other words, the function $F_{\mathbf{p}}(\omega',\bar{\omega},\bar{\omega}')$ is no longer smooth but rather it is proportional to a δ-function. As a consequence, (15.8) contains one integral less and $\Gamma(\mathbf{p},\omega) \propto |\omega - \mu|$. This fact has profound consequences for the nature of the interacting gas. For instance, the sharpness of the Fermi surface (discontinuity in the momentum distribution) requires that $\Gamma(\mathbf{p}, \omega)$ vanishes faster than $|\omega - \mu|$ as $\omega \to \mu$, see Section 11.6.1. In fact, in one dimension there is no Fermi surface and the interacting electron gas is said to be a *Luttinger liquid.* The interested reader can consult Ref. [237] for a detailed discussion on the properties of Luttinger liquids.

15.2 Correlation Energy in the GW Approximation

Preliminaries In the electron gas the interparticle interaction $v(\mathbf{x}_1, \mathbf{x}_2) = v(\mathbf{r}_1 - \mathbf{r}_2)$ is independent of spin and hence the screened interaction $W(\mathbf{r}_1, z_1; \mathbf{r}_2, z_2)$ is also independent of spin. The sum over spin in (7.23) gives

$$W(\mathbf{r}_1, z_1; \mathbf{r}_2, z_2) = v(\mathbf{r}_1, \mathbf{r}_2) + \int d\bar{\mathbf{r}}_1 d\bar{\mathbf{r}}_2 d\bar{z}\; v(\mathbf{r}_1, \bar{\mathbf{r}}_1) P(\bar{\mathbf{r}}_1, z_1; \bar{\mathbf{r}}_2, \bar{z}) W(\bar{\mathbf{r}}_2, \bar{z}; \mathbf{r}_2, z_2), \tag{15.11}$$

where the spin-independent polarization is defined as

$$P(\mathbf{r}_1, z_1; \mathbf{r}_2, z_2) \equiv \sum_{\sigma_1 \sigma_2} P(\mathbf{x}_1, z_1; \mathbf{x}_2, z_2).$$

For an electron gas all quantities in (15.11) depend only on the difference between the spatial coordinates. In momentum space (11.80) reads

$$\begin{aligned} \frac{E_{\text{int,c}}}{\mathrm{V}} &= \frac{\mathrm{i}}{2} \int \frac{d\mathbf{p} d\omega}{(2\pi)^4} \delta P^{\lessgtr}(\mathbf{p}, \omega) \tilde{v}_{\mathbf{p}} \\ &- \frac{1}{2} \int \frac{d\omega}{2\pi} \frac{d\omega'}{2\pi} \int \frac{d\mathbf{p}}{(2\pi)^3} \frac{P^<(\mathbf{p}, \omega) W^>(\mathbf{p}, \omega') - P^>(\mathbf{p}, \omega) W^<(\mathbf{p}, \omega')}{\omega - \omega'}, \end{aligned} \tag{15.12}$$

with V the system volume and $\tilde{v}_{\mathbf{p}}$ the Fourier transform of the interaction. This equation can be further manipulated by expressing W in terms of v and P. In momentum space the retarded/advanced component of the screened interaction can easily be derived from (15.11), and it reads:

$$W^{\mathrm{R/A}}(\mathbf{p}, \omega) = \frac{\tilde{v}_{\mathbf{p}}}{1 - \tilde{v}_{\mathbf{p}} P^{\mathrm{R/A}}(\mathbf{p}, \omega)}.$$

Using the fluctuation-dissipation theorem (10.38), for P and (10.39) for W, we have (omitting the explicit dependence on frequency and momentum)

$$W^> = \bar{f}\,[W^{\mathrm{R}} - W^{\mathrm{A}}] = |W^{\mathrm{R}}|^2\, P^>,$$

where

$$|W^{\mathrm{R}}|^2 = |W^{\mathrm{A}}|^2 = \frac{\tilde{v}^2}{(1 - \tilde{v} P^{\mathrm{R}})(1 - \tilde{v} P^{\mathrm{A}})}.$$

Similarly, for the lesser component of W we find

$$W^< = |W^{\mathrm{R}}|^2\, P^<. \tag{15.13}$$

Substituting these relations into (15.12) we obtain an equivalent formula for the correlation part of the interaction energy,

$$\frac{E_{\text{int,c}}}{\text{V}} = \frac{\text{i}}{2}\int \frac{d\mathbf{p}d\omega}{(2\pi)^4}\,\delta P^{\lessgtr}(\mathbf{p},\omega)\tilde{v}_{\mathbf{p}} - \frac{1}{2}\int \frac{d\omega}{2\pi}\frac{d\omega'}{2\pi}\int \frac{d\mathbf{p}}{(2\pi)^3}\,|W^{\text{R}}(\mathbf{p},\omega')|^2$$
$$\times \frac{P^{<}(\mathbf{p},\omega)P^{>}(\mathbf{p},\omega') - P^{>}(\mathbf{p},\omega)P^{<}(\mathbf{p},\omega')}{\omega-\omega'}. \tag{15.14}$$

We stress that no approximations have been done so far: Equation (15.14) is an exact rewriting of (15.12).

In the following we calculate the correlation energy (10.37) of an electron gas with Coulomb interaction $v(\mathbf{x}_1,\mathbf{x}_2) = 1/|\mathbf{r}_1-\mathbf{r}_2|$, and in the presence of an external potential $\phi(\mathbf{r}) = n\tilde{v}_0$ generated by a uniform positive background charge with the same density of the electrons. Let us start with the evaluation of the λ-integral of $E^{\text{HF}}_{\text{int}}(\lambda)$. This quantity is given in (10.32) with rescaled one-particle density matrix $n \to n_\lambda$ calculated in the Hartree-Fock approximation and, of course, with $v \to \lambda v$, see discussion below (10.37). Since the number of particles does not change when λ is varied, the density $n_\lambda(\mathbf{x}) = n/2$ is independent of λ. For the one-particle density matrix we have [see (8.36)]

$$n_\lambda(\mathbf{x}_1,\mathbf{x}_2) = -\text{i}G^{\text{M}}_\lambda(\mathbf{x}_1,\tau;\mathbf{x}_2,\tau^+) = \delta_{\sigma_1\sigma_2}\int \frac{d\mathbf{k}}{(2\pi)^3}e^{\text{i}\mathbf{k}\cdot(\mathbf{r}_1-\mathbf{r}_2)}\theta(p_{\text{F},\lambda}-k),$$

where $p_{\text{F},\lambda}$ is the Fermi momentum of the system with rescaled interaction. The Fermi momentum cannot depend on λ, otherwise $n_\lambda(\mathbf{x},\mathbf{x}) = n_\lambda(\mathbf{x})$ would be λ-dependent. Therefore the one-particle density matrix is independent of λ as well. We conclude that for an electron gas $E^{\text{HF}}_{\text{int}}(\lambda)$ is linear in λ and hence its λ-integral is trivial:

$$\int_0^1 \frac{d\lambda}{\lambda}E^{\text{HF}}_{\text{int}}(\lambda) = \text{V}\left[\frac{1}{2}\tilde{v}_0 n^2 - \int \frac{d\mathbf{p}}{(2\pi)^3}\frac{d\mathbf{k}}{(2\pi)^3}\tilde{v}_{\mathbf{p}-\mathbf{k}}\theta(p_{\text{F}}-k)\theta(p_{\text{F}}-p)\right]. \tag{15.15}$$

The reader can check that by adding E_0 (one-body part) to this energy one recovers the Hartree-Fock energy (8.39). We further encourage the reader to verify that we would get the same result by using the one-particle density matrix in the Hartree approximation. Indeed, both the Hartree and Hartree-Fock eigenstates are plane waves and since the density is constant in space $n_\lambda(\mathbf{x}_1,\mathbf{x}_2)$ is the same in both approximations (and independent of λ).

GW approximation To calculate the correlation energy we need an approximation to the self-energy. We here consider the GW approximation according to which $\Sigma_{\text{xc}}(1;2) = \text{i}G(1;2)W(1;2)$, with $W = v + vPW$ and $P = -\text{i}GG$. It is easy to show that $E_{\text{int,c}}$ evaluated with (10.35) using a GW self-energy or evaluated with (11.80) using $P = -\text{i}GG$ (equivalently $\delta P = 0$) yields the same result. The diagram that represent $\text{tr}_\gamma[\Sigma_{\text{xc}}G]$ in the GW approximation is

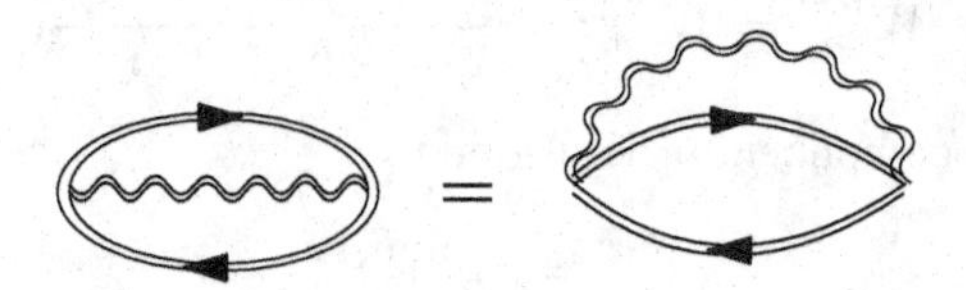

and it is topologically equivalent to the diagram on the r.h.s., which is represented by $\text{tr}_\gamma[-\text{i}GGW]$.

To carry on the calculation with pencil and paper we take the Green's function to be the Hartree Green's function.[2] Then, according to the observation below (10.37), we have $E_{\text{int,HF}}(\lambda) = E^{\text{HF}}_{\text{int}}(\lambda)$ since $n_\lambda(\mathbf{x}_1, \mathbf{x}_2)$ is the same in the Hartree and Hartree-Fock approximations. Thus the correlation energy (10.37) reduces to

$$E_{\text{corr}} = \int_0^1 \frac{d\lambda}{\lambda} E_{\text{int,c}}(\lambda). \tag{15.16}$$

The GW correlation energy corresponds to the resummation of the following (infinite) subset of diagrams (also called ring diagrams):

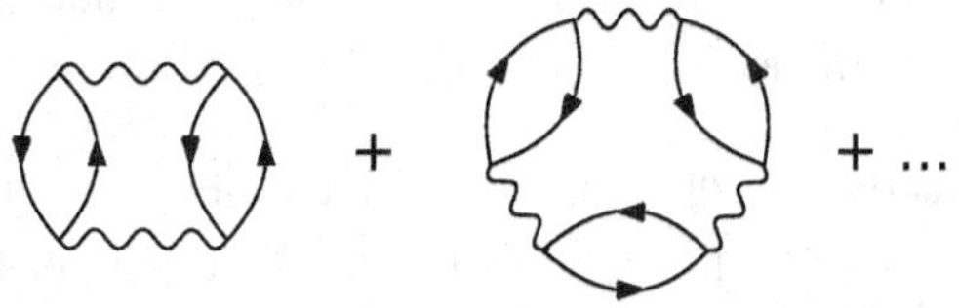

+ + ...

(15.17)

as can easily be checked from (11.78) with $P = -\text{i}GG$ and $W = v + vPW$.

In Fourier space the greater/lesser component of the spin-independent and noninteracting polarization $P^{\lessgtr}(\mathbf{r}_1, t_1; \mathbf{r}_2, t_2) = -\text{i}\sum_{\sigma_1\sigma_2} G^{\lessgtr}(1;2)G^{\gtrless}(2;1)$ reads

$$P^{\lessgtr}(\mathbf{p}, \omega) = -2\text{i} \int \frac{d\bar{\mathbf{p}}d\bar{\omega}}{(2\pi)^4} G^{\lessgtr}(\mathbf{p} + \bar{\mathbf{p}}, \omega + \bar{\omega})G^{\gtrless}(\bar{\mathbf{p}}, \bar{\omega}), \tag{15.18}$$

where we take into account that G is diagonal in spin space and hence the factor of 2 comes from spin. Inserting this polarization into (15.14), the formula for $E_{\text{int,c}}$ becomes (after some renaming of the integration variables)

$$\begin{aligned} \frac{E_{\text{int,c}}}{\text{V}} = -2 \int \frac{d\mathbf{p}d\omega}{(2\pi)^4} \frac{d\mathbf{p}'d\omega'}{(2\pi)^4} \frac{d\bar{\mathbf{p}}d\bar{\omega}}{(2\pi)^4} \int \frac{d\bar{\omega}'}{2\pi} \frac{|W^{\text{R}}(\mathbf{p} - \bar{\mathbf{p}}, \omega - \bar{\omega})|^2}{\omega + \omega' - \bar{\omega} - \bar{\omega}'} \\ \times \left[G^>(\mathbf{p}, \omega)G^>(\mathbf{p}', \omega')G^<(\bar{\mathbf{p}}, \bar{\omega})G^<(\mathbf{p} + \mathbf{p}' - \bar{\mathbf{p}}, \bar{\omega}')\right. \\ \left. -G^<(\mathbf{p}, \omega)G^<(\mathbf{p}', \omega')G^>(\bar{\mathbf{p}}, \bar{\omega})G^>(\mathbf{p} + \mathbf{p}' - \bar{\mathbf{p}}, \bar{\omega}')\right]. \end{aligned} \tag{15.19}$$

Next we use the explicit expression of the Hartree Green's function [see (6.45) and (6.46)]:

$$G^<(\mathbf{p}, \omega) = 2\pi\text{i}\,\theta(p_{\text{F}} - p)\delta(\omega - \frac{p^2}{2}),$$

$$G^>(\mathbf{p}, \omega) = -2\pi\text{i}\,\bar{\theta}(p_{\text{F}} - p)\delta(\omega - \frac{p^2}{2}),$$

[2]The Hartree Green's function of the electron gas in a uniform positive background charge is equal to the Green's function of a noninteracting electron gas *without* the positive background charge, see (8.40).

with $\bar{\theta}(x) = 1 - \theta(x)$, and perform the integral over all frequencies. Renaming the momenta as $\mathbf{p} = \mathbf{q} + \mathbf{k}/2$, $\mathbf{p}' = \mathbf{q}' - \mathbf{k}/2$, and $\bar{\mathbf{p}} = \mathbf{q} - \mathbf{k}/2$, we find

$$\frac{E_{\mathrm{int,c}}}{\mathrm{V}} = -4\int \frac{d\mathbf{q}}{(2\pi)^3}\frac{d\mathbf{q}'}{(2\pi)^3}\frac{d\mathbf{k}}{(2\pi)^3}\frac{|W^{\mathrm{R}}(\mathbf{k},\mathbf{q}\cdot\mathbf{k})|^2}{(\mathbf{q}-\mathbf{q}')\cdot\mathbf{k}} \times \bar{\theta}(p_{\mathrm{F}} - |\mathbf{q}+\frac{\mathbf{k}}{2}|)\bar{\theta}(p_{\mathrm{F}} - |\mathbf{q}'-\frac{\mathbf{k}}{2}|)\theta(p_{\mathrm{F}} - |\mathbf{q}-\frac{\mathbf{k}}{2}|)\theta(p_{\mathrm{F}} - |\mathbf{q}'+\frac{\mathbf{k}}{2}|). \tag{15.20}$$

In deriving (15.20) we use that the contributions coming from the four Green's functions in the second and third line of (15.19) are identical, so we only include the second line and multiply by 2. To prove that they are identical, one has to use the symmetry property (7.28), which implies $W^{\mathrm{R}}(\mathbf{k},\omega) = [W^{\mathrm{R}}(-\mathbf{k},-\omega)]^*$, and the invariance of the system under rotations, which implies that $W^{\mathrm{R}}(\mathbf{k},\omega)$ depends only on the modulus $k = |\mathbf{k}|$.

Before continuing with the evaluation of (15.20) an important observation is in order. By replacing $W(\mathbf{k},\omega)$ with $\tilde{v}_{\mathbf{k}} = 4\pi/k^2$, we get the $E_{\mathrm{int,c}}$ of the second-order ring diagram [the first term of the series (15.17)]. This quantity contains the integral $\int \frac{d\mathbf{k}}{(2\pi)^3}\tilde{v}_{\mathbf{k}}^2 \ldots = \int \frac{d\mathbf{k}}{(2\pi)^3}\frac{1}{k^4}\ldots$ and it is therefore divergent! In fact, each diagram of the series is divergent. As anticipated at the end of Section 5.3, this is a typical feature of bulk systems with long-range interparticle interactions. For these systems the resummation of the same class of diagrams to infinite order is absolutely essential to get meaningful results. We could compare this mathematical behavior to that of the Taylor expansion of the function $f(x) = -1 + 1/(1-x)$. This function is finite for $x \to \infty$, but each term of the Taylor expansion $f(x) = x + x^2 + \ldots$ diverges in the same limit. For bulk systems with long-range interactions MBPT can be seen as a Taylor series with vanishing convergence radius and hence we must resum the series (or pieces of it) to get a finite result. The GW approximation is the leading term in the G- and W-skeletonic formulation.

Let us go back to the evaluation of (15.20). In Section 15.4.1 we calculate the screened interaction $W = v + v\chi_0 W$ and show that at zero frequency and for small k,

$$W^{\mathrm{R}}(\mathbf{k},0) = \frac{4\pi}{k^2 + \lambda_{\mathrm{TF}}^{-2}}, \tag{15.21}$$

with $\lambda_{\mathrm{TF}} = \sqrt{\pi/(4p_{\mathrm{F}})}$ the so-called *Thomas–Fermi screening length.* Consequently, W^{R} does not diverge for small momenta and the integral (15.20) is finite. The effect of the screening is to transform the original (long-range) Coulomb interaction into a (short-range) Yukawa-type interaction since Fourier transforming (15.21) back to real space one finds

$$W^{\mathrm{R}}(\mathbf{r}_1,\mathbf{r}_2;\omega=0) = \frac{e^{-|\mathbf{r}_1-\mathbf{r}_2|/\lambda_{\mathrm{TF}}}}{|\mathbf{r}_1-\mathbf{r}_2|}. \tag{15.22}$$

We can move forward with the analytic calculations if we consider the high-density limit, $r_s \to 0$. Recalling that the Wigner–Seitz radius $r_s = (9\pi/4)^{1/3}/p_{\mathrm{F}}$, in this limit $\lambda_{\mathrm{TF}}^{-1} \ll p_{\mathrm{F}}$ and therefore the dominant contribution to the integral (15.20) comes from the region $k \ll p_{\mathrm{F}}$. For small k we can approximate $W^{\mathrm{R}}(\mathbf{k},\mathbf{q}\cdot\mathbf{k})$ with (15.21) since $W^{\mathrm{R}}(\mathbf{k},\omega)$

depends smoothly on ω for small ω, see Section 15.4. Then the integral (15.20) simplifies to

$$\frac{E_{\text{int,c}}}{\text{V}} = -4\int \frac{d\mathbf{q}}{(2\pi)^3}\frac{d\mathbf{q}'}{(2\pi)^3}\int_{k<p_{\text{F}}}\frac{d\mathbf{k}}{(2\pi)^3}\frac{(4\pi)^2}{(k^2+\lambda_{\text{TF}}^{-2})^2(\mathbf{q}-\mathbf{q}')\cdot\mathbf{k}}$$
$$\times\,\bar{\theta}(p_{\text{F}}-|\mathbf{q}+\frac{\mathbf{k}}{2}|)\bar{\theta}(p_{\text{F}}-|\mathbf{q}'-\frac{\mathbf{k}}{2}|)\theta(p_{\text{F}}-|\mathbf{q}-\frac{\mathbf{k}}{2}|)\theta(p_{\text{F}}-|\mathbf{q}'+\frac{\mathbf{k}}{2}|). \tag{15.23}$$

Next we observe that for $k \ll p_{\text{F}}$ the product $\bar{\theta}(p_{\text{F}}-|\mathbf{q}+\frac{\mathbf{k}}{2}|)\theta(p_{\text{F}}-|\mathbf{q}-\frac{\mathbf{k}}{2}|)$ is nonzero only for $\mathbf{q}$ close to the Fermi sphere. Denoting by c the direction cosine between $\mathbf{q}$ and $\mathbf{k}$, we then have $\mathbf{q}\cdot\mathbf{k}\sim p_{\text{F}}kc$ and $|\mathbf{q}\pm\frac{\mathbf{k}}{2}|\sim q\pm\frac{kc}{2}$. Further approximating

$$\int\frac{d\mathbf{q}}{(2\pi)^3}\sim\left(\frac{p_{\text{F}}}{2\pi}\right)^2\int_0^\infty dq\int_{-1}^{1}dc,$$

we get

$$\frac{E_{\text{int,c}}}{\text{V}} = -4\left(\frac{p_{\text{F}}}{2\pi}\right)^4\frac{(4\pi)^2}{2\pi^2}\frac{1}{p_{\text{F}}}\int_0^\infty dqdq'\int_{-1}^{1}dcdc'\int_0^{p_{\text{F}}}dk\frac{k}{(k^2+\lambda_{\text{TF}}^{-2})^2(c-c')}$$
$$\times\,\bar{\theta}(p_{\text{F}}-q-\frac{kc}{2})\bar{\theta}(p_{\text{F}}-q'+\frac{kc'}{2})\theta(p_{\text{F}}-q+\frac{kc}{2})\theta(p_{\text{F}}-q'-\frac{kc'}{2}). \tag{15.24}$$

The integral over q and q' is now straightforward since

$$\int dq\,\bar{\theta}(p_{\text{F}}-q-\frac{kc}{2})\theta(p_{\text{F}}-q+\frac{kc}{2}) = \int dx\,\bar{\theta}(x-\frac{kc}{2})\theta(x+\frac{kc}{2}) = \theta(c)kc.$$

Inserting this result into (15.24), we find

$$\frac{E_{\text{int,c}}}{\text{V}} = -2\left(\frac{p_{\text{F}}}{\pi^2}\right)^2 p_{\text{F}}\int_0^{p_{\text{F}}}dk\frac{k^3}{(k^2+\lambda_{\text{TF}}^{-2})^2}\underbrace{\int_0^1 dc\int_{-1}^0 dc'\frac{-cc'}{c-c'}}_{\frac{2}{3}(1-\ln 2)}. \tag{15.25}$$

The integral over k can be performed analytically since the integrand is a simple rational function. The result is

$$\int_0^{p_{\text{F}}}dk\frac{k^3}{(k^2+\lambda_{\text{TF}}^{-2})^2} = \frac{1}{2}\left(-1+\frac{\lambda_{\text{TF}}^{-2}}{\lambda_{\text{TF}}^{-2}+p_{\text{F}}^2}+\log(1+\lambda_{\text{TF}}^2p_{\text{F}}^2)\right)$$
$$\xrightarrow[r_s\to 0]{}\log(p_{\text{F}}\lambda_{\text{TF}})$$

The correlation energy (15.16) is given by the λ-integral of (15.25) with rescaled quantities. From (15.21) we see that $W^{\text{R}}\to\lambda W^{\text{R}}$, since p_{F} is independent of λ. Therefore, $E_{\text{int,c}}(\lambda)$ is given by the r.h.s. of (15.25) multiplied by λ^2. Then the λ-integral produces the extra factor $\int_0^1\frac{d\lambda}{\lambda}\lambda^2 = 1/2$ and the correlation energy reads

$$\boxed{E_{\text{corr}} = -n\text{V}\,\frac{2}{\pi^2}(1-\ln 2)\ln(p_{\text{F}}\lambda_{\text{TF}})} \tag{15.26}$$

where we use the relation $n = p_F^3/(3\pi^2)$ between the density and p_F. This result was first derived by Gell-Mann and Brueckner in 1957 [238]. In analogy with the Hartree–Fock energy (8.47), we can express (15.26) in terms of the Wigner-Seitz radius r_s.[3] Adding the resulting expression to the Hartree–Fock energy (8.48) we obtain

$$\boxed{\frac{E_{\text{tot}}}{n\text{V}} = \frac{1}{2}\left[\frac{2.21}{r_s^2} - \frac{0.916}{r_s} + 0.0622 \ln r_s\right] \quad \text{(in atomic units)}} \tag{15.27}$$

This formula contains the three leading contributions to E_{tot} in the high-density limit $r_s \to 0$. The second term is due to Fock. We conclude that the mean-field approximation becomes exact in the high-density limit $r_s \to 0$, in agreement with the discussion at the end of Section 8.1.

Asymptotic expansion At the end of Section 5.3 we mentioned that the MBPT expansion is an asymptotic expansion in the interaction strength.[4] Equation (15.27) provides a nice example of this statement. Reinserting all the fundamental constants in (15.27), the prefactor

$$\frac{1}{2} \to \frac{e^2}{2a_{\text{B}}},$$

with the Bohr radius $a_{\text{B}} = \frac{\hbar^2}{m_e e^2}$ and e and m_e the electron charge and mass, respectively. Taking into account that

$$r_s = \left(\frac{9\pi}{4}\right)^{\frac{1}{3}} \frac{1}{\hbar a_{\text{B}} p_{\text{F}}} \sim e^2, \tag{15.28}$$

we see that the first term $\propto e^0$, the second term $\propto e^2$, and the third term $\propto e^4 \ln e^2$. After many years of study of the electron gas the community feels reasonably confident that the correlation energy has the following asymptotic expansion:

$$\frac{E_{\text{corr}}}{n\text{V}} = a_0 \ln r_s + a_1 + a_2 r_s \ln r_s + a_3 r_s + a_4 r_s^2 \ln r_s + a_5 r_s^2 + \ldots \tag{15.29}$$

We have just shown that the coefficient a_0 originates from the sum of ring diagrams evaluated with the Hartree Green's function. This approximation is called the G_0W_0 approximation to distinguish it from the GW approximation, where G and W are self-consistently calculated from $G = G_0 + G_0(\text{i}GW)G$ and $W = v + v(-\text{i}GG)W$. The coefficients a_1 and a_2 can be extracted from the second-order exchange diagram and from a more precise evaluation of the sum of ring diagrams. The many-body origin of the coefficient a_3 is more complicated [239]. We write below the value of the first four coefficients in atomic units [240]:

$$\begin{aligned} a_0 &= +0.03109 \\ a_1 &= -0.04692 \\ a_2 &= +0.00923 \\ a_3 &= -0.010. \end{aligned}$$

[3] We remind the reader that r_s is related to the density n of the gas by $\frac{1}{n} = \frac{4\pi}{3}(a_{\text{B}} r_s)^3$, see Section 8.3.2.
[4] For a brief introduction to asymptotic expansions we refer the reader to Appendix F.

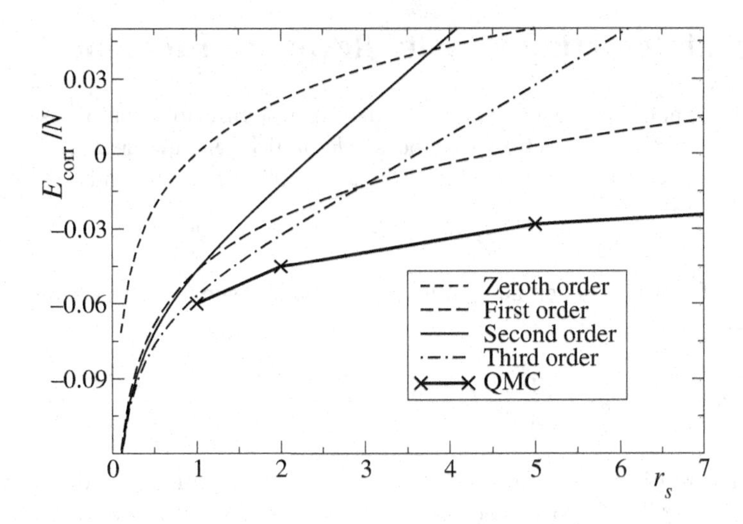

Figure 15.1 Correlation energy per particle (in atomic units) of the Coulombic electron gas.

In Fig. 15.1 we display the correlation energy per particle as obtained by truncating the sum (15.29) to the first m terms, with $m = 1, 2, 3, 4$. For comparison we also report the correlation energy per particle as obtained numerically [241] using a popular stochastic method, namely the Quantum Monte Carlo (QMC) method [242, 243]. A close inspection of the figure reveals the typical feature of an asymptotic expansion. We see that for $r_s = 1$ the third-order approximation is the most accurate. Going to larger values of r_s, however, the agreement deteriorates faster the higher is the order of the approximation; for example, the first-order approximation is more accurate than the second- and third-order approximations for $r_s = 5$.

Finally we would like to mention that the correlation energy calculated using the fully self-consistent GW approximation is as accurate as that from QMC calculations [149]. On the other hand, for spectral properties the G_0W_0 approximation seems to perform better than the GW approximation.[5] This is typically attributed to a lack of cancellations. We saw in Section 14.10 that conserving approximations in combination with gauge invariance lead to an equation between the vertex and the self-energy known as the *Ward identity*. The Ward identity suggests that self-consistent corrections are partially canceled by vertex corrections. Since the GW vertex is zero, this cancellation does not occur in the GW approximation [148, 169, 170, 171, 244, 245, 246, 247]. The fact that self-consistent calculations of total energies are very accurate even without these cancellations is still not completely understood, and is the topic of ongoing research.

Exercise 15.1 Prove (15.20).

[5] In systems with a gap, such as semiconductors, insulators, or molecules, the situation is much more unclear.

15.3 Noninteracting Density Response Function

The general formula for the noninteracting density response function χ^0 is given in, for example, (13.50). For an electron gas the single-particle energy eigenkets $|\mathbf{p}\tau\rangle$ are the momentum–spin kets, and therefore the excitation amplitudes (13.51) read:

$$f_{\mathbf{p}\tau\bar{\mathbf{p}}\bar{\tau}}(\mathbf{r}\sigma) = \langle \mathbf{r}\sigma|\bar{\mathbf{p}}\bar{\tau}\rangle\langle \mathbf{p}\tau|\mathbf{r}\sigma\rangle = \delta_{\sigma\bar{\tau}}\delta_{\sigma\tau}e^{\mathrm{i}(\bar{\mathbf{p}}-\mathbf{p})\cdot\mathbf{r}}.$$

Substitution of these excitation amplitudes into (13.50) and summation over the spin indices τ and $\bar{\tau}$ leads to

$$\chi^{0,\mathrm{R}}(\mathbf{x},\mathbf{x}';\omega) = \delta_{\sigma\sigma'}\int \frac{d\mathbf{p}d\bar{\mathbf{p}}}{(2\pi)^6} f_{\mathbf{p}}\bar{f}_{\bar{\mathbf{p}}}\left[\frac{e^{\mathrm{i}(\bar{\mathbf{p}}-\mathbf{p})\cdot(\mathbf{r}-\mathbf{r}')}}{\omega-(\epsilon_{\bar{\mathbf{p}}}-\epsilon_{\mathbf{p}})+\mathrm{i}\eta} - \frac{e^{-\mathrm{i}(\bar{\mathbf{p}}-\mathbf{p})\cdot(\mathbf{r}-\mathbf{r}')}}{\omega+(\epsilon_{\bar{\mathbf{p}}}-\epsilon_{\mathbf{p}})+\mathrm{i}\eta}\right],$$

with the standard notation $\mathbf{x} = \mathbf{r}\sigma$ and $\mathbf{x}' = \mathbf{r}'\sigma'$, and the zero-temperature Fermi function f and $\bar{f} = 1 - f$ for the occupied and unoccupied states, respectively. We rename the integration variables as $\mathbf{p} = \mathbf{k}$, $\bar{\mathbf{p}} = \mathbf{k}+\mathbf{q}$ in the first integral and $\bar{\mathbf{p}} = \mathbf{k}$, $\mathbf{p} = \mathbf{k}+\mathbf{q}$ in the second integral; in this way $\chi^{0,\mathrm{R}}$ becomes

$$\chi^{0,\mathrm{R}}(\mathbf{x},\mathbf{x}';\omega) = \delta_{\sigma\sigma'}\int \frac{d\mathbf{q}d\mathbf{k}}{(2\pi)^6} e^{\mathrm{i}\mathbf{q}\cdot(\mathbf{r}-\mathbf{r}')}\left[\frac{f_{\mathbf{k}}\bar{f}_{\mathbf{k}+\mathbf{q}} - f_{\mathbf{k}+\mathbf{q}}\bar{f}_{\mathbf{k}}}{\omega-(\epsilon_{\mathbf{k}+\mathbf{q}}-\epsilon_{\mathbf{k}})+\mathrm{i}\eta}\right]. \tag{15.30}$$

Due to the translational invariance of the system it is convenient to work with the Fourier transform $\chi^0(\mathbf{k},\omega)$, which we define similarly to (13.87):

$$\sum_{\sigma\sigma'}\chi^{0,\mathrm{R}}(\mathbf{x},\mathbf{x}';\omega) = \int \frac{d\mathbf{q}}{(2\pi)^3} e^{\mathrm{i}\mathbf{q}\cdot(\mathbf{r}-\mathbf{r}')}\chi^{0,\mathrm{R}}(\mathbf{q},\omega). \tag{15.31}$$

Comparing (15.31) with (15.30) and taking into account that the combination of Fermi functions $f_1\bar{f}_2 - f_2\bar{f}_1 = f_1 - f_2$, we obtain

$$\chi^{0,\mathrm{R}}(\mathbf{q},\omega) = 2\int \frac{d\mathbf{k}}{(2\pi)^3} f_{\mathbf{k}}\left[\frac{1}{\omega-(\epsilon_{\mathbf{k}+\mathbf{q}}-\epsilon_{\mathbf{k}})+\mathrm{i}\eta} - \frac{1}{\omega-(\epsilon_{\mathbf{k}}-\epsilon_{\mathbf{k}-\mathbf{q}})+\mathrm{i}\eta}\right]. \tag{15.32}$$

The evaluation of this integral is a bit tedious, but doable. For free electrons the energy dispersion is $\epsilon_{\mathbf{p}} = p^2/2$. Introducing the dimensionless variables

$$x = \frac{q}{p_{\mathrm{F}}}, \qquad y = \frac{k}{p_{\mathrm{F}}}, \qquad \nu = \frac{\omega}{\epsilon_{p_{\mathrm{F}}}} = \frac{2\omega}{p_{\mathrm{F}}^2}, \tag{15.33}$$

with $\epsilon_{p_{\mathrm{F}}}$ the noninteracting energy with Fermi momentum p_{F}, we can rewrite (15.32) as

$$\chi^{0,\mathrm{R}}(\mathbf{q},\omega) = \frac{4p_{\mathrm{F}}}{(2\pi)^2}\int_0^1 dy\, y^2\int_{-1}^1 dc\left[\frac{1}{\nu-x^2-2xyc+\mathrm{i}\eta} - \frac{1}{\nu+x^2-2xyc+\mathrm{i}\eta}\right], \tag{15.34}$$

where c is the cosine of the angle between the vectors $\mathbf{k}$ and $\mathbf{q}$. Due to the invariance of the system under rotations, $\chi^{0,\mathrm{R}}$ depends only on the modulus q of the vector $\mathbf{q}$. We now

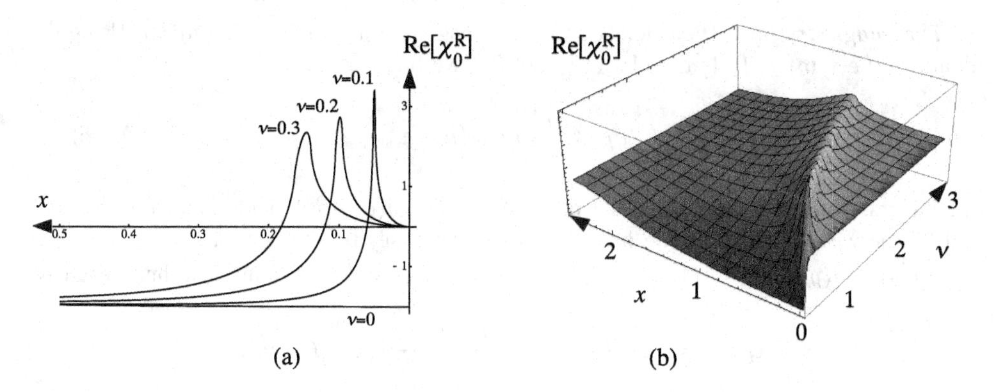

Figure 15.2 Real part of $\chi^{0,\mathrm{R}}$ in units of $2p_\mathrm{F}/(2\pi)^2$. (a) For frequencies $\nu = 0, 0.1, 0.2, 0.3$ as a function of x. (b) 3D plot as a function of ν and x.

calculate the real and the imaginary parts of $\chi^{0,\mathrm{R}}$. The resulting final form of $\mathrm{Re}\left[\chi^{0,\mathrm{R}}\right]$ and $\mathrm{Im}\left[\chi^{0,\mathrm{R}}\right]$ was for the first time worked out by Lindhard in a classic paper from 1954 [248]. Today it is common to refer to $\chi^{0,\mathrm{R}}$ as the *Lindhard function.*[6]

The real part To calculate the real part we simply set to zero the infinitesimal $i\eta$ in (15.34). Performing the integral over c, we are left with the evaluation of the logarithmic integral I defined in (8.43). It is then straightforward to arrive at

$$\mathrm{Re}\left[\chi^{0,\mathrm{R}}(\mathbf{q},\omega)\right] = \frac{2p_\mathrm{F}}{(2\pi)^2}\frac{I(2x,\nu - x^2) - I(2x,\nu + x^2)}{x}.$$

This function is discontinuous at the origin since the limit $x \to 0$ does not commute with the limit $\nu \to 0$. Using the explicit form of I, at zero frequency we find

$$\mathrm{Re}\left[\chi^{0,\mathrm{R}}(\mathbf{q},0)\right] = -\frac{4p_\mathrm{F}}{(2\pi)^2}F(x/2), \tag{15.35}$$

where F is the function (8.45) appearing in the Hartree-Fock eigenvalues of the electron gas. At any finite frequency ν the small-x behavior is instead parabolic,

$$\mathrm{Re}\left[\chi^{0,\mathrm{R}}(\mathbf{q}\to 0,\omega)\right] \to \frac{8p_\mathrm{F}}{(2\pi)^2}\frac{2}{3}\frac{x^2}{\nu^2}, \tag{15.36}$$

and the real part approaches zero as $x \to 0$. This is illustrated in Fig. 15.2(a), where the discontinuous behavior for $\nu = 0$ is clearly visible. For $\nu \to 0$ the function $\mathrm{Re}\left[\chi^{0,\mathrm{R}}\right]$ versus x exhibits a peak in $x_\mathrm{max} \sim \nu/2$ whose width decreases like $\sim \nu$ and whose height increases like $\sim \ln\nu$. The 3D plot of $\mathrm{Re}\left[\chi^{0,\mathrm{R}}\right]$ as a function of x and ν is displayed in Fig. 15.2(b); we see a maximum along the curve $\nu = x^2 + 2x$ and otherwise a rather smooth function away from the origin.

[6] In Appendix E we use the Lindhard function to evaluate the pair correlation function of the electron gas.

The imaginary part The calculation of the imaginary part of $\chi^{0,\mathrm{R}}$ is simpler. Using the identity $1/(x+\mathrm{i}\eta) = P(1/x) - \mathrm{i}\pi\delta(x)$ in (15.34), we find

$$\mathrm{Im}\left[\chi^{0,\mathrm{R}}(\mathbf{q},\omega)\right] = -\frac{4\pi p_\mathrm{F}}{(2\pi)^2}\int_0^1 dy\, y^2 \int_{-1}^1 dc \left[\delta(\nu - x^2 - 2xyc) - \delta(\nu + x^2 - 2xyc)\right].$$

The integral over c between -1 and 1 of the δ-function $\delta(\alpha - \beta c)$ yields $1/|\beta|$ if $|\alpha/\beta| < 1$ and 0 otherwise. Then, also the integral over y can easily be performed since it is of the form $\int_0^1 dy\, y\, \theta(y-\gamma) = \frac{1}{2}(1-\gamma^2)\,\theta(1-\gamma)$ with γ a positive constant. The final result is

$$\mathrm{Im}\left[\chi^{0,\mathrm{R}}(\mathbf{q},\omega)\right] = -\frac{2\pi p_\mathrm{F}}{(2\pi)^2}\frac{1}{x}\left[\frac{1-P_-^2(\nu,x)}{2}\,\theta(1-P_-(\nu,x)) - \frac{1-P_+^2(\nu,x)}{2}\,\theta(1-P_+(\nu,x))\right], \tag{15.37}$$

where we define

$$P_\pm(\nu,x) = \frac{|\nu \pm x^2|}{2x},$$

and where we take into account that both x and y are, by definition, positive quantities. In the quarter of the (ν,x) plane with $\nu > 0$ and $x > 0$ the first θ-function is nonvanishing in the area delimited by the parabolas $\nu = x^2 + 2x$, which we call p_1, and $\nu = x^2 - 2x$, which we call p_2, whereas the second θ-function is nonvanishing in the area below the parabola p_3 with equation $\nu = -x^2 + 2x$, see Fig. 15.3(a).[7] We define region I as the area where both the θ-functions are different from zero and region II as the area where the second θ-function does instead vanish. Then, from (15.37) we have

$$\mathrm{Im}\left[\chi^{0,\mathrm{R}}(\mathbf{q},\omega)\right] = -\frac{2\pi p_\mathrm{F}}{(2\pi)^2}\begin{cases}\frac{\nu}{2x} & \text{region I}\\ \frac{1}{2x}\left[1-\left(\frac{\nu - x^2}{2x}\right)^2\right] & \text{region II}\end{cases}. \tag{15.38}$$

The limits $\nu \to 0$ and $x \to 0$ must be done with care. If we approach the origin along the parabolas p_1 and p_3 that delimit region II, we find

$$\mathrm{Im}\left[\chi^{0,\mathrm{R}}(\mathbf{q},\omega)\right] \xrightarrow[\substack{\nu,x\to 0\\ \text{along } p_1}]{} 0, \qquad \mathrm{Im}\left[\chi^{0,\mathrm{R}}(\mathbf{q},\omega)\right] \xrightarrow[\substack{\nu,x\to 0\\ \text{along } p_3}]{} -\frac{2\pi p_\mathrm{F}}{(2\pi)^2}. \tag{15.39}$$

On the other hand, if we approach the origin from region I along the line $\nu = mx$ with $m < 2$, we find

$$\mathrm{Im}\left[\chi^{0,\mathrm{R}}(\mathbf{q},\omega)\right] \xrightarrow[\substack{\nu=mx,x\to 0\\ \text{(from region I)}}]{} -\frac{2\pi p_\mathrm{F}}{(2\pi)^2}\frac{m}{2}, \tag{15.40}$$

[7] It is intuitively clear that the imaginary part of $\chi^{0,\mathrm{R}}$ is nonvanishing in this region. For instance, at zero frequency the only electron-hole excitations which contribute to $\mathrm{Im}[\chi^{0,\mathrm{R}}]$ are those in which both the particle and the hole are on the Fermi surface since in this case their energy difference is zero. Then the maximum distance between the momentum $\mathbf{k}$ of the particle and the momentum $\mathbf{k}+\mathbf{q}$ of the hole is $q = 2p_\mathrm{F}$, which is obtained for $\mathbf{q} = -2\mathbf{k}$ (hole-momentum opposite to electron-momentum). Therefore at zero frequency, $\mathrm{Im}[\chi^{0,\mathrm{R}}]$ is nonzero for all $q \in (0, 2p_\mathrm{F})$. With similar geometric considerations one can generalize the above argument to finite frequencies.

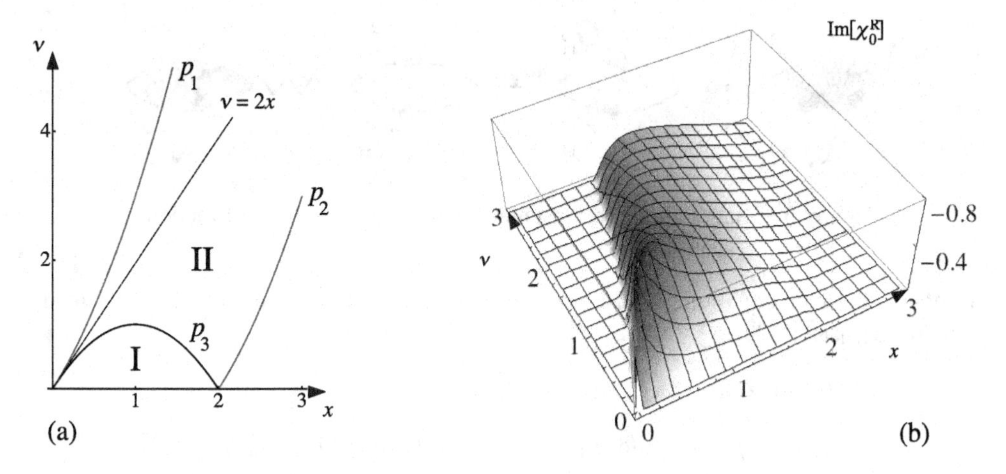

Figure 15.3 (a) Domain of the (ν, x) plane where the imaginary part of $\chi^{0,\mathrm{R}}$ is nonvanishing. (b) 3D plot of $\mathrm{Im}\left[\chi^{0,\mathrm{R}}\right]$ in units of $2\pi p_{\mathrm{F}}/(2\pi)^2$ as a function of ν and x.

which reduces to the value in the second limit of (15.39) for $m \to 2$. For any finite frequency ν the imaginary part of $\chi^{0,\mathrm{R}}$ is a continuous function of x with cusps (discontinuity of the first derivative) at the points where regions I and II start or end. The 3D plot of $\mathrm{Im}\left[\chi^{0,\mathrm{R}}\right]$ is shown in Fig. 15.3(b), where it is evident that the function is everywhere nonpositive, in agreement with our discussion on the energy-loss function in Section 13.5. The noninteracting electron gas can dissipate energy only via the creation or annihilation of electron–hole pairs. As we see in the next sections, the physics becomes more interesting in the interacting case.

Exercise 15.2 Show that in one dimension the Lindhard function at zero temperature and zero frequency is given by

$$\chi^{0,\mathrm{R}}(\mathbf{q}, 0) = -\frac{2}{\pi q} \ln \left| \frac{x+2}{x-2} \right|,$$

where $x = q/p_{\mathrm{F}}$.

Exercise 15.3 Show that in two dimensions the Lindhard function at zero temperature and zero frequency is given by

$$\chi^{0,\mathrm{R}}(\mathbf{q}, 0) = \begin{cases} -\frac{1}{\pi}\left(1 - \sqrt{1-(2/x)^2}\right) & x > 2 \\ -\frac{1}{\pi} & x < 2 \end{cases},$$

where $x = q/p_{\mathrm{F}}$.

15.4 RPA Density Response Function

In the electron gas the noninteracting response function χ^0 constitutes the basic ingredient to calculate the linear response density δn in the Hartree approximation. Let us consider

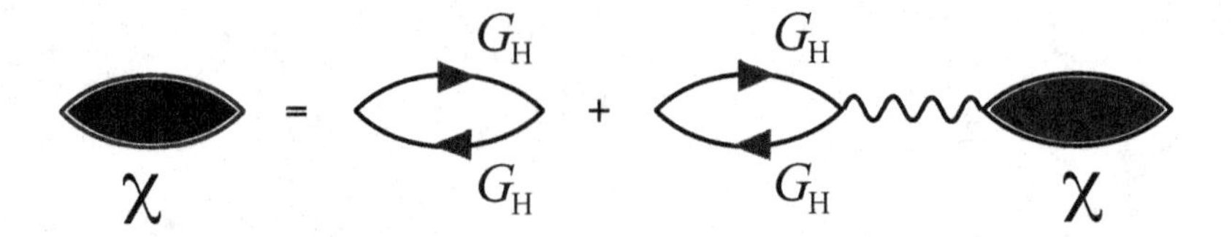

Figure 15.4 Diagrammatic representation of the RPA response function.

the Bethe–Salpeter equation for L corresponding to the variation of the Hartree Green's function $G_{\rm H}$ [see (14.20)]:

$$L_{\rm H}(1,2;3,4) = G_{\rm H}(1;4)G_{\rm H}(2;3) - \int d5d6d7d8\, G_{\rm H}(1;5)G_{\rm H}(7;3)K_{\rm H}(5,6;7,8)L_{\rm H}(8,2;6,4). \tag{15.41}$$

In this equation $K_{\rm H} \equiv -\delta\Sigma_{\rm H}/\delta G$ is the kernel that gives the linear response change of the Hartree self-energy $\Sigma_{\rm H}(1;2) = -{\rm i}\delta(1;2)\int d3\, v(1;3)G_{\rm H}(3;3^+)$ in accordance with (14.34). Equation (15.41) defines the two-particle XC function $L_{\rm H}$ in the Hartree approximation. The explicit form of the Hartree kernel is

$$K_{\rm H}(5,6;7,8) = -\frac{\delta\Sigma_{\rm H}(5;7)}{\delta G(8;6)} = {\rm i}\delta(5;7)\delta(6;8)v(5;8).$$

Inserting this result into (15.41) and taking the limit $3 \to 1$ and $4 \to 2$ we find

$$\chi(1;2) = -{\rm i}G_{\rm H}(1;2)G_{\rm H}(2;1) - {\rm i}\int d5d6\, G_{\rm H}(5;1)G_{\rm H}(1;5)v(5;6)\chi(6;4), \tag{15.42}$$

where $\chi(1;2) = -{\rm i}L_{\rm H}(1,2;1,2)$ is the density response function (we omit the subscript "H" in χ in order to simplify the notation), see (13.16). Thus, in the Hartree approximation, χ is given by a Dyson-like equation whose diagrammatic representation is shown in Fig. 15.4. For historical reasons this approximation is called the *Random Phase Approximation* (RPA). In 1953, Bohm and Pines [249] found an ingenious way to map the Hamiltonian of a gas of electrons interacting via the long-range Coulomb interaction into the Hamiltonian of a gas of electrons *plus* collective excitations interacting via a screened short-range Coulomb interaction. The RPA is equivalent to neglecting the interaction between the electrons and the collective excitations, and it becomes exact in the limit of very high densities (i.e., $r_s \to 0$). In Appendix O we present a simplified version of the original treatment by Bohm and Pines. Going beyond the Hartree-Fock approximation, the kernel K is no longer proportional to a product of δ-functions and if we take the limit $3 \to 1$ and $4 \to 2$ in the Bethe–Salpeter equation we find that L (under the integral sign) does not reduce to χ.

Let us now go back to (15.42) and consider the structure $-{\rm i}G_{\rm H}G_{\rm H}$ that appears in it. For an electron gas subject to a uniform potential ϕ_0 (positive background charge), the momentum–spin kets $|{\bf p}\sigma\rangle$ are eigenkets of the noninteracting Hamiltonian $\hat{h} = \hat{p}^2/2 - \phi_0\hat{\mathbb{1}}$ as well as of the Hartree Hamiltonian $\hat{h}_{\rm H} = \hat{p}^2/2 - \phi_0\hat{\mathbb{1}} + \hat{\mathcal{V}}_{\rm H}$ with $\hat{\mathcal{V}}_{\rm H} = n\tilde{v}_0\hat{\mathbb{1}}$, see (8.15). This means that for $\phi_0 = n\tilde{v}_0$ the Hartree Green's function $G_{\rm H}$ is the same as the Green's function G_0 of a system of noninteracting electrons with energy dispersion $\epsilon_{\mathbf{p}} = p^2/2$. In Section 13.3 we showed that $-\mathrm{i}G_0G_0$ is the noninteracting density response function χ^0, and in the previous section we calculated this χ^0 just for the case of interest - that is, electrons with energy dispersion $\epsilon_{\mathbf{p}} = p^2/2$. Therefore, in an electron gas the RPA density response function is the solution of

$$\chi(1;2) = \chi^0(1;2) + \int d3d4\, \chi^0(1;3)v(3;4)\chi(4;2), \tag{15.43}$$

with χ^0 given in (15.34). It is worth noticing that the Coulomb interaction v is spin independent and hence χ is not diagonal in spin space, although χ^0 is.[8] Equation (15.43) is an integral equation for two-point correlators in Keldysh space. As Keldysh functions, the Coulomb interaction $v(3;4) = \delta(z_3, z_4)/|\mathbf{r}_3 - \mathbf{r}_4|$ has only a singular part, whereas χ and χ^0 have the structure (14.108) and do not contain a singular part. Using the Langreth rules of Table 5.1 to extract the retarded component and Fourier transforming to frequency space, we obtain

$$\chi^{\mathrm{R}}(\mathbf{x}_1, \mathbf{x}_2; \omega) = \chi^{0,\mathrm{R}}(\mathbf{x}_1, \mathbf{x}_2; \omega) + \int d\mathbf{x}_3 d\mathbf{x}_4\, \chi^{0,\mathrm{R}}(\mathbf{x}_1, \mathbf{x}_3; \omega) v(\mathbf{x}_3, \mathbf{x}_4) \chi^{\mathrm{R}}(\mathbf{x}_4, \mathbf{x}_2; \omega).$$

Since both $\chi^{0,\mathrm{R}}$ and v depend only on the difference of the spatial arguments, χ^{R} depends only on the difference of the spatial arguments. If we define the Fourier transform of χ^{R} in analogy with (13.87),

$$\sum_{\sigma\sigma'} \chi^{\mathrm{R}}(\mathbf{x}, \mathbf{x}'; \omega) = \int \frac{d\mathbf{p}}{(2\pi)^3} e^{\mathrm{i}\mathbf{p}\cdot(\mathbf{r}-\mathbf{r}')} \chi^{\mathrm{R}}(\mathbf{p}, \omega),$$

then the RPA equation for χ^{R} becomes a simple algebraic equation whose solution is

$$\boxed{\chi^{\mathrm{R}}(\mathbf{q}, \omega) = \frac{\chi^{0,\mathrm{R}}(\mathbf{q}, \omega)}{1 - \tilde{v}_{\mathbf{q}} \chi^{0,\mathrm{R}}(\mathbf{q}, \omega)}, \qquad \tilde{v}_{\mathbf{q}} = \frac{4\pi}{q^2}} \tag{15.44}$$

In the remainder of this section we discuss this result.

15.4.1 Dynamical Screening and Static Thomas-Fermi Screening

The electron gas offers a simple and intuitive interpretation of the screened interaction $W^{\mathrm{R}}(t, t')$. We refer to this function away from the time diagonal as the *dynamically screened interaction* in order to distinguish it from the statically screened interaction introduced in (14.42).

Dynamical screening Let us investigate the space-time dependence of W^{R}. We consider the system in equilibrium, and at time $t = t_0$ we suddenly put and remove an electron of charge $q = -1$ in $\mathbf{r} = \mathbf{r}_0$. This particle generates an external potential

$$\phi(\mathbf{r}, t) = -\frac{1}{|\mathbf{r} - \mathbf{r}_0|}\delta(t - t_0) = -v(\mathbf{r}, \mathbf{r}_0)\delta(t - t_0). \tag{15.45}$$

[8] This is simply due to the fact that in the interacting gas a change of, say, the spin-up density affects both the spin-up and spin-down densities.

According to (13.14) the linear-response density change induced by (15.45) is

$$\delta n(\mathbf{r},t) = -\int_{-\infty}^{\infty} dt' \int d\mathbf{r}' \chi^{\mathrm{R}}(\mathbf{r},t;\mathbf{r}',t')\phi(\mathbf{r}',t') = \int d\mathbf{r}' \chi^{\mathrm{R}}(\mathbf{r},t;\mathbf{r}',t_0)v(\mathbf{r}',\mathbf{r}_0),$$

where $\chi^{\mathrm{R}}(\mathbf{r},t;\mathbf{r}',t') = \sum_{\sigma\sigma'} \chi^{\mathrm{R}}(\mathbf{x},t;\mathbf{x}',t')$ is the density response function summed over spin and $\delta n(\mathbf{r},t) = \sum_\sigma \delta n(\mathbf{x},t)$ is the total density change. This density change generates a classical (or Hartree) potential given by

$$\begin{aligned} V_{\mathrm{H}}(\mathbf{r},t) &= \int d\mathbf{r}'\, v(\mathbf{r},\mathbf{r}')\delta n(\mathbf{r}',t) \\ &= \int d\mathbf{r}' d\mathbf{r}''\, v(\mathbf{r},\mathbf{r}')\chi^{\mathrm{R}}(\mathbf{r}',t;\mathbf{r}'',t_0)v(\mathbf{r}'',\mathbf{r}_0) \\ &= \delta W^{\mathrm{R}}(\mathbf{r},t;\mathbf{r}_0,t_0), \end{aligned}$$

where in the last equality we define $\delta W = W - v = vPW = v\chi v$, see (7.25). Therefore $\delta W^{\mathrm{R}}(\mathbf{r},t;\mathbf{r}_0,t_0)$ can be seen as the classical interaction energy at time t between an electron in $\mathbf{r}$ and an electronic charge distribution $-\delta n(\mathbf{r}',t)$ induced by the sudden switch-on/switch-off of a charge $q = -1$ in $\mathbf{r}_0$ at time t_0.

This equivalence allows us to draw a few conclusions. At time $t = t_0^+$ the electron gas has no time to respond to the external perturbation and hence $\delta n(\mathbf{r}',t_0^+) = 0$, which implies

$$\delta W^{\mathrm{R}}(\mathbf{r},t_0^+;\mathbf{r}_0,t_0) = 0.$$

Let us denote by t_{resp} the timescale for the electron gas to respond to the external perturbation.[9] We expect that at times $t \sim t_{\mathrm{resp}}$ the electrons have had the time to move away from $\mathbf{r}_0$. Then the density change $\delta n(\mathbf{r}',t)$ is negative for $\mathbf{r}' \sim \mathbf{r}_0$ and positive away from this point since $\int d\mathbf{r}'\delta n(\mathbf{r}',t) = 0$ (conservation of the total charge). The dominant contribution to the interaction energy $V_{\mathrm{H}}(\mathbf{r}_0,t) = \int d\mathbf{r}'\delta n(\mathbf{r}',t)/|\mathbf{r}_0 - \mathbf{r}'|$ comes from the region $\mathbf{r}' \sim \mathbf{r}_0$ where $\delta n(\mathbf{r}',t) < 0$ for $t \sim t_{\mathrm{resp}}$. Therefore,

$$\delta W^{\mathrm{R}}(\mathbf{r},t;\mathbf{r}_0,t_0) < 0 \qquad \text{for } \mathbf{r} \sim \mathbf{r}_0 \text{ and } t \sim t_{\mathrm{resp}}.$$

We conclude that the retarded interaction is attractive at sufficiently short distances for times $t \sim t_{\mathrm{resp}}$. Finally we consider the limit of large times. In this case we expect that in a neighborhood of $\mathbf{r}_0$ the density has relaxed back to its equilibrium value: $\delta n(\mathbf{r}',t) \sim 0$ for $|\mathbf{r}' - \mathbf{r}_0| < R(t)$, where $R(t)$ is the extension of the relaxed region. $R(t)$ increases with time and the interaction energy $V_{\mathrm{H}}(\mathbf{r},t) = \int d\mathbf{r}'\delta n(\mathbf{r}',t)/|\mathbf{r} - \mathbf{r}'|$ approaches zero for any $|\mathbf{r} - \mathbf{r}_0|$ much smaller than $R(t)$:

$$\delta W^{\mathrm{R}}(\mathbf{r},t;\mathbf{r}_0,t_0) \sim 0 \qquad \text{for } |\mathbf{r} - \mathbf{r}_0| \ll R(t) \text{ and } t \gg t_{\mathrm{resp}}.$$

Static Thomas–Fermi screening Let us now focus on the static limit or, equivalently, the zero frequency limit. Approximating χ as in (15.44), the retarded component of W in Fourier space reads

$$W^{\mathrm{R}}(\mathbf{q},\omega) = \frac{\tilde{v}_{\mathbf{q}}}{1 - \tilde{v}_{\mathbf{q}}\,\chi^{0,\mathrm{R}}(\mathbf{q},\omega)}.$$

[9] In Section 15.5 we show that t_{resp} is roughly proportional to the inverse of the plasma frequency ω_{p}.

For $\omega = 0$ the imaginary part of $\chi^{0,\mathrm{R}}$ vanishes, see (15.40), while the real part approaches $-4p_{\mathrm{F}}/(2\pi)^2$ for $\mathbf{q} \to 0$, see (15.35). The small q behavior of the static effective interaction is therefore

$$W^{\mathrm{R}}(\mathbf{q} \to 0, 0) \to \frac{4\pi}{q^2 + \frac{4p_{\mathrm{F}}}{\pi}}. \tag{15.46}$$

This is exactly the screened interaction of the Thomas–Fermi theory [80, 250, 251]. If we Fourier transform $W^{\mathrm{R}}(\mathbf{q}, 0)$ back to real space, (15.46) implies that at large distances the effective interaction has the Yukawa form (15.22), with $\lambda_{\mathrm{TF}} = \sqrt{\pi/4p_{\mathrm{F}}}$ the *Thomas–Fermi screening length*. The precise physical interpretation of the static effective interaction is discussed in Section 15.5.

15.4.2 Plasmons and Landau Damping

We here show how the aforementioned collective excitations found by Bohm and Pines emerge from (15.44). For an electron gas subject to an external scalar potential $\delta\phi$, the Kubo formula in Fourier space reads $\delta n(\mathbf{q}, \omega) = -\chi^{\mathrm{R}}(\mathbf{q}, \omega)\delta\phi(\mathbf{q}, \omega)$ with $\delta n(\mathbf{q}, \omega)$ the Fourier transform of $\delta n(\mathbf{r}, t) = \sum_\sigma \delta n(\mathbf{x}, t)$ (for electrons the charge $q = -1$). Therefore,

$$\frac{\delta n(\mathbf{q}, \omega)}{\chi^{\mathrm{R}}(\mathbf{q}, \omega)} = -\delta\phi(\mathbf{q}, \omega). \tag{15.47}$$

Suppose that there exist points in the (q, ω) plane for which $1/\chi^{\mathrm{R}}(\mathbf{q}, \omega) = 0$. Then (15.47) is compatible with a scenario in which the electron density oscillates in space and time without any driving field: $\delta n \neq 0$ even though $\delta\phi = 0$. These persistent (undamped) density oscillations are the collective excitations of Bohm and Pines, who gave them the name of *plasmons*. From our previous analysis of the noninteracting response function it is evident that there are no points in the (q, ω) plane for which $1/\chi_0^{\mathrm{R}}(\mathbf{q}, \omega) = 0$. As we see below, the existence of plasmons is intimately connected to the long-range nature of the Coulomb interaction.

For an intuitive understanding of the self-sustained density oscillations in an electron gas we can use the following classical picture. Let $\mathbf{u}(\mathbf{r}, t)$ be the displacement of an electron from its equilibrium position in $\mathbf{r}$ at a certain time t. Consider the special case in which the displacement is of the form

$$\mathbf{u}(\mathbf{r}, t) = \mathbf{u}_0 \cos(\mathbf{q} \cdot \mathbf{r} - \omega t), \tag{15.48}$$

with $\mathbf{u}_0$ parallel to $\mathbf{q}$ (longitudinal displacement). The polarization associated with this displacement is $\mathbf{P}(\mathbf{r}, t) = -n\mathbf{u}(\mathbf{r}, t)$, with n the equilibrium density, and therefore the electric field $\mathbf{E} = -4\pi\mathbf{P} = 4\pi n\mathbf{u}$ is longitudinal as well. The classical equation of motion for the electron with equilibrium position in $\mathbf{r}$ reads:

$$\ddot{\mathbf{u}}(\mathbf{r}, t) = -\mathbf{E}(\mathbf{r}, t) = -4\pi n\mathbf{u}(\mathbf{r}, t).$$

For (15.48) to be a solution of this differential equation the frequency must be

$$\omega = \omega_{\mathrm{p}} = \sqrt{4\pi n} = \sqrt{\frac{3}{r_s^3}}. \tag{15.49}$$

This frequency is called the *plasma frequency*.[10]

[10]Expressing the plasma frequency in terms of the fundamental constants, we find $\omega_{\mathrm{p}} = \sqrt{\frac{4\pi e^2 n}{m_e}}$. Typical values of the plasma frequency are in the range $10 \div 20$ eV.

Let us investigate the possibility that the inverse of the RPA response function is zero. From (15.44) we see that $1/\chi^{\mathrm{R}}(\mathbf{q},\omega)=0$ implies

$$(a) \qquad \mathrm{Im}[\chi^{0,\mathrm{R}}(\mathbf{q},\omega)]=0, \tag{15.50}$$

$$(b) \qquad 1-\tilde{v}_{\mathbf{q}}\mathrm{Re}[\chi^{0,\mathrm{R}}(\mathbf{q},\omega)]=0. \tag{15.51}$$

We look for solutions with $x \ll 1$ - that is, to the left of regions I and II in Fig. 15.3(a) - so that $\mathrm{Im}[\chi^{0,\mathrm{R}}]=0$. We rewrite $\chi^{0,\mathrm{R}}$ in (15.34) in a slightly different form. Changing the variable $c \to -c$ in the first integral, we find the equivalent expression

$$\chi^{0,\mathrm{R}}(\mathbf{q},\omega)=\frac{8p_{\mathrm{F}}}{(2\pi)^2}\int_0^1 dy\, y^2\int_{-1}^{1} dc\,\frac{x^2-2xyc}{(\nu+\mathrm{i}\eta)^2-(x^2-2xyc)^2}.$$

To calculate the real part when $x \ll 1$ we set $\eta=0$ and expand the integrand in powers of (x^2-2xyc):

$$\mathrm{Re}\left[\chi^{0,\mathrm{R}}(\mathbf{q},\omega)\right]=\frac{8p_{\mathrm{F}}}{(2\pi)^2}\int_0^1 dy\, y^2\int_{-1}^{1} dc\,\frac{x^2-2xyc}{\nu^2}\left[1+\frac{(x^2-2xyc)^2}{\nu^2}+\dots\right].$$

Taking into account that terms with odd powers of c do not contribute, we obtain

$$\begin{aligned}\mathrm{Re}\left[\chi^{0,\mathrm{R}}(\mathbf{q},\omega)\right]&=\frac{8p_{\mathrm{F}}}{(2\pi)^2}\frac{2}{3}\left(\frac{x}{\nu}\right)^2\left[1+\frac{12}{5}\left(\frac{x}{\nu}\right)^2+\mathcal{O}(x^4)\right]\\&=\frac{p_{\mathrm{F}}^3}{3\pi^2}\left(\frac{q}{\omega}\right)^2\left[1+\frac{3}{5}\left(\frac{qp_{\mathrm{F}}}{\omega}\right)^2+\mathcal{O}(x^4)\right],\end{aligned}$$

where in the second equality we reintroduce the physical momentum and frequency. Inserting this result into (15.51) and taking into account the relation $n=p_{\mathrm{F}}^3/(3\pi^2)$ between the density and the Fermi momentum, as well as definition (15.49) of the plasma frequency, we get

$$1-\tilde{v}_{\mathbf{q}}\mathrm{Re}[\chi^{0,\mathrm{R}}(\mathbf{q},\omega)]=1-\frac{\omega_{\mathrm{p}}^2}{\omega^2}\left[1+\frac{3}{5}\left(\frac{qp_{\mathrm{F}}}{\omega}\right)^2+\mathcal{O}(x^4)\right]=0. \tag{15.52}$$

From this equation it follows that for $q=0$ the electron gas supports undamped density oscillations with frequency ω_{p} in agreement with the classical picture. For small q, however, the frequency changes according to

$$\omega=\omega_{\mathrm{p}}(q)\simeq\omega_{\mathrm{p}}\sqrt{1+\frac{3}{5}\left(\frac{qp_{\mathrm{F}}}{\omega_{\mathrm{p}}}\right)^2}, \tag{15.53}$$

which follows directly from (15.52) by replacing ω with ω_{p} in the square brackets.

An important consequence of the existence of plasmons is that the imaginary part of $\chi^{\mathrm{R}}(\mathbf{q},\omega)$ consists of a continuum of electron–hole excitations (as in the noninteracting electron gas) and a plasmon peak. To show this, let us introduce the real functions B_1, B_2 and A_1, A_2 that depend only on the modulus of $\mathbf{q}$ and on the frequency ω according to

$$B_1(q,\omega)+\mathrm{i}B_2(q,\omega)=\tilde{v}_{\mathbf{q}}\chi^{\mathrm{R}}(\mathbf{q},\omega), \tag{15.54a}$$

$$A_1(q,\omega)+\mathrm{i}A_2(q,\omega)=\tilde{v}_{\mathbf{q}}\chi^{0,\mathrm{R}}(\mathbf{q},\omega). \tag{15.54b}$$

If we multiply both sides of (15.44) by $\tilde{v}_{\mathbf{q}}$, we find that B_2 is

$$B_2 = \frac{A_2}{(1 - A_1)^2 + A_2^2},$$

where the dependence on (q, ω) has been dropped. From this equation we see that $A_2 \neq 0$ implies $B_2 \neq 0$ and therefore the imaginary part of the RPA response function is certainly nonvanishing in regions I and II of Fig. 15.3(a). Outside these regions A_2 is an infinitesimally small function proportional to η that vanishes only in the limit $\eta \to 0$. However, this does not imply that also B_2 vanishes since

$$\lim_{A_2 \to \pm 0} B_2 = \lim_{A_2 \to \pm 0} \frac{A_2}{(1 - A_1)^2 + A_2^2} = \pm\pi\delta(1 - A_1), \tag{15.55}$$

where $A_2 \to \pm 0$ signifies that A_2 approaches zero from negative/positive values. The r.h.s. of this equation is nonvanishing when the argument of the δ-function is zero. Taking into account that $1 - A_1(q, \omega) = 1 - \tilde{v}_{\mathbf{q}} \mathrm{Re}\left[\chi_0^{\mathrm{R}}(\mathbf{q}, \omega)\right]$ is exactly the function that establishes the existence of plasmons, see (15.51), we conclude that the imaginary part of the RPA response function has a δ-like peak along the plasmon curve in the (q, ω) plane.

According to the results derived in Section 13.5 we infer that the plasmon peak is related to a strong absorption of light at the plasma frequency, a result that has been experimentally confirmed in all simple metals. This absorption can also be deduced from the Maxwell equations. In an electron gas the Fourier transform of the electric field obeys the equation

$$q^2 \mathbf{E}(\mathbf{q}, \omega) - \frac{\omega^2}{c^2} \varepsilon^{\mathrm{R}}(\mathbf{q}, \omega) \mathbf{E}(\mathbf{q}, \omega) = 0,$$

with ε^{R} the dielectric function. Thus, an electromagnetic wave with wave vector $\mathbf{q}$ and energy ω can penetrate the medium provided that the dispersion relation $q^2 = \frac{\omega^2}{c^2} \varepsilon^{\mathrm{R}}(\mathbf{q}, \omega)$ is satisfied. The inverse of the dielectric function is $\varepsilon^{-1,\mathrm{R}}(\mathbf{q}, \omega) = 1 + \tilde{v}_{\mathbf{q}} \chi^{\mathrm{R}}(\mathbf{q}, \omega)$, see also Appendix N, and therefore $\varepsilon^{\mathrm{R}}(\mathbf{q}, \omega) = 1 - \tilde{v}_{\mathbf{q}} P^{\mathrm{R}}(\mathbf{q}, \omega)$ with P the polarization. In RPA $P = \chi^0$ and using (15.52) we deduce that for small q the dispersion relation becomes $q^2 = \frac{1}{c^2}(\omega^2 - \omega_{\mathrm{p}}^2)$. For energies larger than ω_{p} there always exist real wave vectors q for which light can penetrate and be absorbed. On the contrary, for $\omega < \omega_{\mathrm{p}}$ the wave vector is complex and the amplitude of the electric field decays exponentially in the medium. For such energies there cannot be any absorption by plasmons and hence the light is fully reflected. Visible light has energy below the typical plasma frequency and it is indeed common experience that it is reflected by most metals.

In Fig. 15.5 we display the domain of the (ν, x) plane where the imaginary part of χ^{R} is nonvanishing. The figure shows regions I and II, which are in common with the noninteracting response function as well as the numerical solution of the equation $1 - A_1(q, \omega) = 0$ for $r_s = 5$ (thick line). The plasmon branch is well approximated by (15.53) (thin line) for small x. Besides the plasmon curve, the function $1 - A_1$ vanishes also along a second curve (dashed line) which is, however, entirely contained in the region where $\mathrm{Im}\left[\chi^{\mathrm{R}}\right] \neq 0$. Inside region II the two solutions of $1 - A_1 = 0$ approach each other until they touch, and thereafter $1 - A_1$ is always different from zero. Just after the critical value x_c at which the plasmon branch crosses region II, the plasmon peak gets broadened by the electron–hole excitations

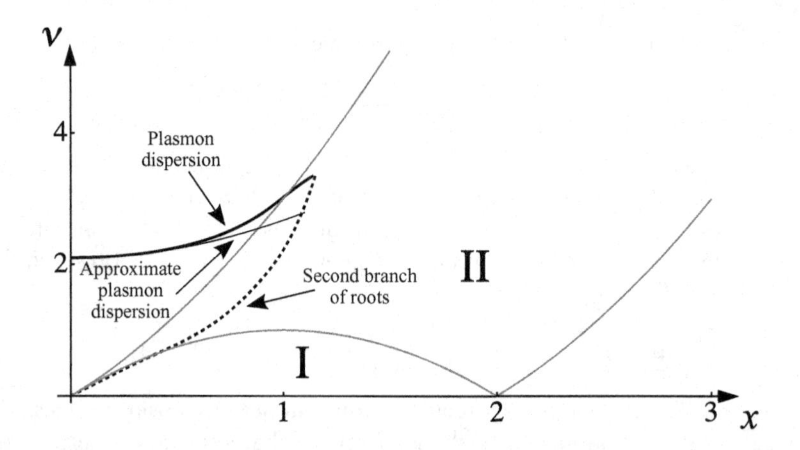

Figure 15.5 Domain of the (ν, x) plane where the imaginary part of the RPA response function is nonvanishing. Besides regions I and II we also have a δ-peak along the plasmon branch (thick line). The dashed line corresponds to a second solution of the equation $1 - A_1 = 0$. The thin line is the function in (15.53). In this plot $r_s = 5$.

and the lifetime of the corresponding plasmon excitation becomes finite. This phenomenon is known as *Landau damping.* As a matter of fact, the plasmon lifetime is finite at any finite temperature as well as at zero temperature if we go beyond RPA.

Exercise 15.4 Show that the RPA response function satisfies the f-sum rule.

15.5 Quench Dynamics of Plasmons and Electron–Hole Pairs

In this section we present a nice application of linear response theory in the electron gas. Suppose that a very high-energy photon is absorbed by the gas and that, as a consequence, an electron is instantaneously expelled by the system. How does the density of the gas rearrange in order to screen the suddenly created hole? And how long does it take? Do plasmons and electron–hole excitations contribute in a similar way to the post-quench dynamics? This study is relevant to the description of the transient screening of a core-hole in simple metals. We closely follow the derivation of Canright in Ref. [252].

The potential $\delta\phi$ generated by a charge Q suddenly created at time $t = 0$ in $\mathbf{r} = 0$ is

$$\delta\phi(\mathbf{x}, t) = \theta(t)\frac{Q}{r} = \int \frac{d\mathbf{q}}{(2\pi)^3} \int \frac{d\omega}{2\pi} e^{\mathrm{i}\mathbf{q}\cdot\mathbf{r} - \mathrm{i}\omega t}\, \delta\phi(\mathbf{q}, \omega),$$

with

$$\delta\phi(\mathbf{q}, \omega) = \frac{4\pi Q}{q^2}\frac{\mathrm{i}}{\omega + \mathrm{i}\eta} = \tilde{v}_{\mathbf{q}}\, Q \frac{\mathrm{i}}{\omega + \mathrm{i}\eta}.$$

From the linear response equation (15.47) we have

$$\delta n(\mathbf{r}, t) = \sum_\sigma \delta n(\mathbf{x}, t) = -\int \frac{d\mathbf{q}}{(2\pi)^3} \int \frac{d\omega}{2\pi} e^{\mathrm{i}\mathbf{q}\cdot\mathbf{r} - \mathrm{i}\omega t}\, B(q, \omega)\, Q \frac{\mathrm{i}}{\omega + \mathrm{i}\eta},$$

where the function $B = B_1 + \mathrm{i}B_2 = \tilde{v}\chi^{\mathrm{R}}$ is defined in (15.54). Since B depends only on the modulus q we can easily perform the angular integration and find

$$\delta n(\mathbf{r},t) = -\frac{4\pi Q}{(2\pi)^4}\frac{1}{r}\int_0^{\infty} dq\, q\sin(qr)\int_{-\infty}^{\infty} d\omega B(q,\omega)\frac{\mathrm{i}e^{-\mathrm{i}\omega t}}{\omega+\mathrm{i}\eta}. \tag{15.56}$$

The function B is analytic in the upper half of the complex ω-plane and goes to zero like $1/\omega^2$ when $\omega \to \infty$, see Section 14.10. Therefore, B has the required properties for the Kramers–Kronig relations (these relations are derived in Appendix P). Accordingly, we write B in terms the imaginary part B_2:

$$B(q,\omega) = -\frac{1}{\pi}\int_{-\infty}^{\infty} d\omega'\,\frac{B_2(q,\omega')}{\omega-\omega'+\mathrm{i}\eta}. \tag{15.57}$$

Inserting this relation into (15.56) and performing the integral over ω, we obtain

$$\delta n(\mathbf{r},t) = -\frac{16\pi Q}{(2\pi)^4}\frac{1}{r}\int_0^{\infty} dq\, q\sin(qr)\int_0^{\infty} d\omega B_2(q,\omega)\frac{1-\cos(\omega t)}{\omega}, \tag{15.58}$$

where we rename the integration variable ω' with ω and we use that B_2 is odd in ω, see Section 13.2.2.

From Fig. 15.5 we see that this integral naturally splits into two terms of physically different nature. At fixed q the function B_2 is a smooth function between the values $\omega_{\mathrm{min}}(q)$ and $\omega_{\mathrm{max}}(q)$ that delimit the electron–hole continuum (regions I+II) with

$$\omega_{\mathrm{min}}(q) = \epsilon_{p_{\mathrm{F}}}\begin{cases} 0 & \text{if } x<2 \\ x^2-2x & \text{if } x>2 \end{cases}, \qquad \omega_{\mathrm{max}}(q) = \epsilon_{p_{\mathrm{F}}}(x^2+2x),$$

and $x = q/p_{\mathrm{F}}$, like in (15.33). For $\omega \notin (\omega_{\mathrm{min}}(q), \omega_{\mathrm{max}}(q))$ the function B_2 is zero everywhere except along the plasmon branch $\omega_{\mathrm{p}}(q)$. Denoting by q_c the momentum at which the plasmon branch crosses region II, we have that for $q < q_c$ and for $\omega > \omega_{\mathrm{max}}(q)$ the function B_2 is given by (15.55):

$$B_2(q,\omega) = -\pi\delta(1-A_1(q,\omega)) = -\pi\frac{\delta(\omega-\omega_{\mathrm{p}}(q))}{\left|\frac{\partial A_1}{\partial\omega}(q,\omega_{\mathrm{p}}(q))\right|},$$

where we take into account that for positive frequencies A_2 is negative, see (13.89). From this analysis it is natural to write the density δn in (15.58) as the sum $\delta n_{\mathrm{eh}} + \delta n_{\mathrm{p}}$, where δn_{eh} is generated by the excitation of electron–hole pairs, whereas δn_{p} is generated by the excitation of plasmons. Specifically we have

$$\delta n_{\mathrm{eh}}(\mathbf{r},t) = -\frac{16\pi Q}{(2\pi)^4}\frac{1}{r}\int_0^{\infty} dq\, q\sin(qr)\int_{\omega_{\mathrm{min}}(q)}^{\omega_{\mathrm{max}}(q)} d\omega B_2(q,\omega)\frac{1-\cos(\omega t)}{\omega},$$

$$\delta n_{\mathrm{p}}(\mathbf{r},t) = \frac{16\pi^2 Q}{(2\pi)^4}\frac{1}{r}\int_0^{q_c} dq\, q\sin(qr)\,\frac{1-\cos(\omega_{\mathrm{p}}(q)t)}{\omega_{\mathrm{p}}(q)\left|\frac{\partial A_1}{\partial\omega}(q,\omega_{\mathrm{p}}(q))\right|}.$$

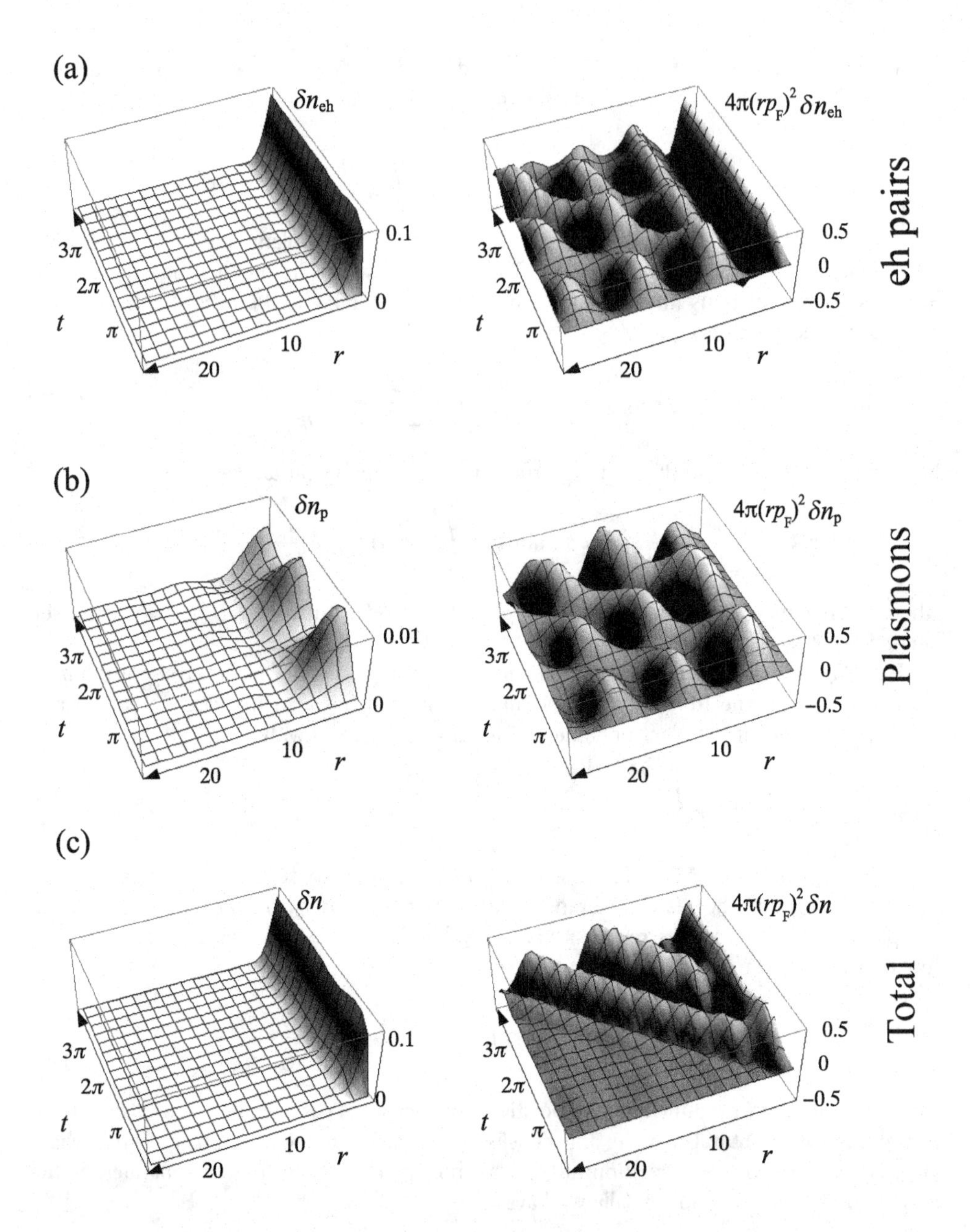

Figure 15.6 A 3D plot of the quench dynamics of an electron gas with $r_s = 3$ induced by the sudden creation of a point-like positive charge $Q = 1$ in the origin at $t = 0$. The contribution due to the excitation of electron–hole pairs (a) and plasmons (b) is, for clarity, multiplied by $4\pi(rp_{\rm F})^2$ in the plots to the right. Panel (c) is the sum of the two contributions. Units: r is in units of $1/p_{\rm F}$, t is in units of $1/\omega_{\rm p}$, and all densities are in units of $p_{\rm F}^3$.

In Fig. 15.6 we display the 3D plot of $\delta n_{\rm eh}$ (panel (a), left), $\delta n_{\rm p}$ (panel (b), left), and the total density δn (panel (c), left) as a function of time t and distance r from the origin. These plots have been generated by calculating the above integrals numerically for $Q = 1$ and $r_s = 3$. In the right part of the figure the densities are multiplied by $4\pi(rp_{\rm F})^2$ to highlight the large r behavior. As pointed out by Canright [252], the partitioning into electron–hole pairs and plasmons is "somewhat illuminating but also unphysical." The density $\delta n_{\rm eh}$, in contrast with the density $\delta n_{\rm p}$, clearly loses phase coherence at long times (panels (a) and (b), right). We also notice that at small r the screening is essentially due to $\delta n_{\rm eh}$, which builds up in time $t \sim 1/\omega_{\rm p}$ (panel (a), left); the plasmon contribution is an order of magnitude smaller (panel (b), left) and adds a small, damped ringing to the total density. The unphysical aspect of the partitioning is evident in the plots to the right of panels (a) and (b). The hole is suddenly created in $r = 0$ and, therefore, like for a stone tossed in a pond, the density in r should change only after the shockwave has had time to propagate till r. We instead see that at small t the densities $\delta n_{\rm eh}$ and $\delta n_{\rm p}$ decay very slowly for large r. This unphysical behavior is absent in the physical sum $\delta n = \delta n_{\rm eh} + \delta n_{\rm p}$ (panel (c), right), where an almost perfect cancellation occurs. We clearly see a density front that propagates at a speed of about the Fermi velocity $v_{\rm F} = p_{\rm F}$, which is $\simeq 0.65$ in our case. Thus the physical density $\delta n(r,t)$ is essentially zero up to a time $t \sim r/v_{\rm F}$; after this time δn changes and exhibits damped oscillations around its steady-state value.

The steady-state limit of the RPA density δn can easily be worked out from (15.58). Due to the Riemann–Lebesgue theorem, the term containing $\cos(\omega t)$ integrates to zero when $t \to \infty$ and hence

$$\delta n_s(\mathbf{r}) \equiv \lim_{t\to\infty} \delta n(\mathbf{r},t) = -\frac{Q}{2\pi^3}\frac{1}{r}\int_0^\infty dq\, q\sin(qr)\int_{-\infty}^{\infty} d\omega \frac{B_2(q,\omega)}{\omega}.$$

The integral over ω can be interpreted as a principal part since $B_2(q,0) = 0$, see (15.38). Then, from the Kramers–Kronig relation (15.57) we can replace the integral over ω with $B(q,0) = \tilde{v}_{\mathbf{q}}\chi^{\rm R}(\mathbf{q},0)$, and hence recover the well-known result

$$\delta n_s(\mathbf{r}) = -\frac{Q}{2\pi^2}\frac{1}{r}\int_0^\infty dq\, q\sin(qr)\tilde{v}_{\mathbf{q}}\,\chi^{\rm R}(\mathbf{q},0) = -Q\int\frac{d\mathbf{q}}{(2\pi)^3}\,e^{i\mathbf{q}\cdot\mathbf{r}}\,\tilde{v}_{\mathbf{q}}\,\chi^{\rm R}(\mathbf{q},0). \quad (15.59)$$

This formula yields the *screening density* in an electron gas. We defer the reader to the classic book of Fetter and Walecka [48] for a thorough analysis of (15.59), with $\chi^{\rm R}$ in the RPA.

Among the most interesting consequences of (15.59) with $\chi^{\rm R}$ in the RPA, we mention the following:

- The screening charge

$$\delta Q(R) \equiv -4\pi\int_0^R dr\, r^2 \delta n_s(\mathbf{r})$$

is close to minus the positive charge Q for $R \gtrsim \lambda_{\rm TF}$, meaning that the electron gas screens an external charge within a distance of a few $\lambda_{\rm TF}$. At very large distances the screening is perfect: $\delta Q(R\to\infty) = -Q$.

- For $r \gg 1/p_{\rm F}$ the density $\delta n_s(\mathbf{r})$ goes to zero as $\cos(2p_{\rm F}r)/r^3$, a results which was derived by Langer and Vosko in 1960 [253]. However, it was Friedel who first pointed out that these damped spatial oscillations are a general consequence of the discontinuity of the Fermi function at zero temperature [254]. For this reason, in the scientific literature they are known as the *Friedel oscillations*.

Equation (15.59) allows us to give a clear physical interpretation to the effective interaction of the Thomas–Fermi theory. The total change in the charge density is $\delta\rho_{\text{tot}}(\mathbf{r}) = Q\delta(\mathbf{r}) - \delta n_s(\mathbf{r})$, where $\delta(\mathbf{r})$ is the density of the suddenly created charge at the origin and the minus sign in front of $\delta n_s(\mathbf{r})$ originates from the fact that electrons have charge $q = -1$. The interaction energy between the charge distribution $\delta\rho_{\text{tot}}$ and a generic electron in position $\mathbf{r}$ is

$$e_{\text{int}}(\mathbf{r}) = -\int d\mathbf{r}' v(\mathbf{r}, \mathbf{r}')\delta\rho_{\text{tot}}(\mathbf{r}').$$

We see that $e_{\text{int}}(\mathbf{r})$ correctly reduces to $v(\mathbf{r}, 0) = -Q/r$ in an empty space since in this case the induced density $\delta n_s = 0$. Suppose now that the charge $Q = q = -1$ is the same as the electron charge. Then

$$\begin{aligned} e_{\text{int}}(\mathbf{r}) &= \int d\mathbf{r}' v(\mathbf{r}, \mathbf{r}') \left[\delta(\mathbf{r}') + \int \frac{d\mathbf{q}}{(2\pi)^3} e^{i\mathbf{q}\cdot\mathbf{r}'}\, \tilde{v}_{\mathbf{q}}\, \chi^{\text{R}}(\mathbf{q}, 0)\right] \\ &= \int \frac{d\mathbf{q}}{(2\pi)^3} e^{i\mathbf{q}\cdot\mathbf{r}} \left[\tilde{v}_{\mathbf{q}} + \tilde{v}_{\mathbf{q}}^2 \chi^{\text{R}}(\mathbf{q}, 0)\right] \\ &= \int \frac{d\mathbf{q}}{(2\pi)^3} e^{i\mathbf{q}\cdot\mathbf{r}}\, W^{\text{R}}(\mathbf{q}, 0) \xrightarrow[r\to\infty]{} \frac{e^{-r/\lambda_{\text{TF}}}}{r}. \end{aligned}$$

Thus the effective Yukawa interaction is the interaction between a "test" electron and a statically screened electron.

Exercise 15.5 Using (15.44) and the analytic result (15.35), show that

$$\delta n_s(\mathbf{r}) = \frac{Q}{\lambda_{\text{TF}}^2} \frac{e^{-r/\lambda_{\text{TF}}}}{4\pi r} - \frac{Q}{\pi} \frac{1}{(p_{\text{F}}\lambda_{\text{TF}})^2} \frac{1}{(4 + \frac{1}{2(p_{\text{F}}\lambda_{\text{TF}})^2})^2} \frac{\cos(2p_{\text{F}} r)}{r^3}.$$

15.6 Spectral Function and Momentum Distribution in the GW Approximation

In the previous section we have seen that a sudden creation of a positive charge in the electron gas yields a density response that can be physically interpreted as the sum of two contributions: It consists of a piece in which we excite electron–hole pairs and a piece in which we excite plasmons. We may therefore expect that the sudden addition or removal of an electron as described by the greater and lesser Green's functions induces similar excitations which appear as characteristic structures in the spectral function. This has indeed been found in photoemission and inverse photoemission experiments on metals such as sodium or aluminum [255], which strongly resemble the electron gas. To calculate the spectral function we must first determine the self-energy. Since $W = v + vPW = v + v\chi v$ already contains the physics of both single-particle excitations and plasmons, it is natural to consider the expansion of the self-energy $\Sigma_{ss,\text{xc}}[G, W]$ in terms of the dressed Green's function and screened interaction as described in Section 7.7. In the following we restrict ourselves to the lowest-order term of the expansion, which is the GW approximation

$\Sigma_{\rm xc}(1;2) = {\rm i}G(1;2)W(2;1)$ where the screened interaction is calculated from the polarization $P(1;2) = -{\rm i}G(1;2)G(2;1)$. This is the so-called RPA approximation for the screened interaction.

15.6.1 The G_0W_0 Self-Energy

In Chapter 8 we learned that the Hartree self-energy $\Sigma_{\rm H}$ cancels with the external potential of the positive background charge and can therefore be disregarded. In thermal equilibrium the electron gas is translationally invariant and we can Fourier transform to momentum-energy space. We write:

$$\Sigma^{\rm R}(\mathbf{p},\omega) = \Sigma_{\rm x}(\mathbf{p}) + \underbrace{{\rm i}\int \frac{d\omega'}{2\pi}\,\frac{\Sigma^{>}(\mathbf{p},\omega') - \Sigma^{<}(\mathbf{p},\omega')}{\omega - \omega' + {\rm i}\eta}}_{\Sigma^{\rm R}_{\rm c}(\mathbf{p},\omega)}, \tag{15.60}$$

where $\Sigma_{\rm x}(\mathbf{p}) = -\frac{2p_{\rm F}}{\pi}F(\frac{p}{p_{\rm F}})$ is the time-local exchange (or Fock) self-energy, see (8.46), and $\Sigma^{\rm R}_{\rm c}$ is the correlation self-energy, see (10.8). We remind the reader that in equilibrium the Green's function and the self-energy are diagonal in spin space since v, and hence W, are independent of spin. In accordance with our notation, the quantities $G(\mathbf{p},\omega)$ and $\Sigma(\mathbf{p},\omega)$ refer to the Fourier transform of the diagonal (and σ-independent) matrix elements $G(\mathbf{r}_1\sigma,t_1;\mathbf{r}_2\sigma,t_2)$ and $\Sigma(\mathbf{r}_1\sigma,t_1;\mathbf{r}_2\sigma,t_2)$. The electron gas is also rotationally invariant and all Fourier-transformed quantities depend only on the modulus of the momentum. To lighten the notation we denote $G(\mathbf{p},\omega)$ by $G(p,\omega)$, $\Sigma(\mathbf{p},\omega)$ by $\Sigma(p,\omega)$, $W(\mathbf{p},\omega)$ by $W(p,\omega)$, etc. In Fourier space the GW self-energy becomes a convolution between G and W:

$$\begin{aligned}\Sigma^{\lessgtr}(p,\omega) &= {\rm i}\int \frac{d\mathbf{k}d\omega'}{(2\pi)^4}\,G^{\lessgtr}(\mathbf{k},\omega')W^{\gtrless}(|\mathbf{k}-\mathbf{p}|,\omega'-\omega)\\ &= {\rm i}\int \frac{d\omega'}{(2\pi)^4}\,2\pi\int_{-1}^{1}dc\int_0^\infty dk\,k^2\,G^{\lessgtr}(k,\omega')W^{\gtrless}(\sqrt{k^2+p^2-2kpc}\,,\omega'-\omega),\end{aligned}$$

where we use that $\Sigma^{\lessgtr}(1;2) = {\rm i}G^{\lessgtr}(1;2)W^{\gtrless}(2;1)$. Instead of integrating over the cosine c we perform the substitution $q = \sqrt{k^2+p^2-2kpc}$ such that $qdq = -kpdc$, and rewrite the lesser/greater self-energy as

$$\Sigma^{\lessgtr}(p,\omega) = \frac{{\rm i}}{(2\pi)^3\,p}\int d\omega'\int_0^\infty dk\,k\,G^{\lessgtr}(k,\omega')\int_{|k-p|}^{k+p}dq\,q\,W^{\gtrless}(q,\omega'-\omega). \tag{15.61}$$

We have therefore reduced the calculation of the self-energy to a three-dimensional integral. Let us give a physical interpretation to $\Sigma^{\lessgtr}$ in the GW approximation. We have seen in Section 15.1 that $\Sigma^{>}(p,\omega)$ is the decay rate for an added particle with momentum p and energy ω. According to (15.61), this particle is scattered into momentum-energy state (k,ω'), thereby creating a electron–hole pair of momentum-energy $(|\mathbf{p}-\mathbf{k}|,\omega-\omega')$. If the energy ω is just above the chemical potential μ, then, since all states below the chemical potential are occupied, there is very little phase-space left to excite an electron–hole pair. We therefore expect that the scattering rate goes to zero as a function of $\omega - \mu$. According to our interpretation one factor $\omega - \mu$ comes from the phase-space requirement on $G^>$ (available states after the scattering) and another one from the phase-space requirement on $W^<$ (density of electron–hole pairs). Together this leads to the $(\omega - \mu)^2$ behavior that we have already deduced in Section 15.1 for the second Born approximation and that we derive more rigorously below for the approximation (15.61).[11] Similar considerations apply to $\Sigma^<$.

Equation (15.61) together with the equation for the screened interaction W and the Dyson equation,

$$G^{\mathrm{R}}(k,\omega) = \frac{1}{\omega - \epsilon_k - \Sigma^{\mathrm{R}}(k,\omega)} \quad , \quad \epsilon_k = k^2/2, \tag{15.62}$$

form a self-consistent set of equations which can only be solved with some considerable numerical effort [164]. Therefore, rather than solving the equations self-consistently we insert into them a physically motivated Green's function. The simplest Green's function we can consider is, of course, the noninteracting Green's function $G_0^{\mathrm{R}}(k,\omega) = (\omega - \epsilon_k + \mathrm{i}\eta)^{-1}$. However, this has some objections. First of all, the choice of the GW-diagram comes from a skeletonic expansion in dressed Green's functions. It would be preferable to include some kind of self-energy renormalization. The simplest of such a Green's function, which still allows for analytical manipulations, is given by

$$G^{\mathrm{R}}(k,\omega) = \frac{1}{\omega - \epsilon_k - \Delta + \mathrm{i}\eta}, \tag{15.63}$$

with Δ a real number [93]. In this poor man's choice we approximate the self-energy with a real number (as in the Hartree–Fock approximation, but this time k-independent). The next question is then how to choose Δ. We know from Section 15.1 that the exact and any approximate Green's function has a rate $\Gamma(k,\omega) \sim (\omega - \mu)^2$, and therefore that $G^{\mathrm{R}}(k,\mu)$ has a pole in $k = p_{\mathrm{F}}$ provided that p_{F} is chosen as the solution of

$$0 = \mu - \epsilon_{p_{\mathrm{F}}} - \Sigma^{\mathrm{R}}(p_{\mathrm{F}},\mu). \tag{15.64}$$

We then require that our approximate self-consistent $G^{\mathrm{R}}(k,\mu)$ in (15.63) has a pole at the same spot. This implies

$$0 = \mu - \epsilon_{p_{\mathrm{F}}} - \Delta,$$

and yields $\Delta = \mu - \epsilon_{p_{\mathrm{F}}}$. This shift simplifies considerably the analytical manipulations. In practice the Fermi momentum $p_{\mathrm{F}} = (3\pi^2 n)^{1/3}$ is determined by the density of the gas and, therefore, rather than fixing μ and calculating p_{F} we fix p_{F} and calculate μ. In this way our G^{R} has a pole at the desired p_{F}.[12] Thus the strategy is: First, evaluate Σ^{R} using the μ-dependent Green's function (15.63) and subsequently fix the chemical potential using (15.64). This is the so-called $\mathrm{G_0W_0}$ approximation for the self-energy.

The Green's function in (15.63) has the form of a noninteracting G and therefore using (6.68) and (6.69) we get

$$G^<(k,\omega) = 2\pi\mathrm{i} f(\omega - \mu)\delta(\omega - \epsilon_k - \Delta),$$
$$G^>(k,\omega) = -2\pi\mathrm{i} \bar{f}(\omega - \mu)\delta(\omega - \epsilon_k - \Delta).$$

[11] The fact that second Born and GW predict a rate proportional to $(\omega - \mu)^2$ is a direct consequence of the fact that both approximations contain the single bubble self-energy diagram.

[12] In general the comparison between Green's functions in different approximations makes sense only if these Green's functions have a pole at the same Fermi momentum. This means that different approximations to Σ yield different values of the chemical potential according to (15.64).

At zero temperature $f(\omega-\mu)=\theta(\mu-\omega)$ and due to the δ-function we can use $\theta(\mu-\epsilon_k-\Delta)=\theta(\epsilon_{p_F}-\epsilon_k)=\theta(p_F-k)$. Similarly, in $G^>$ we can replace $\bar{f}(\omega-\mu)$ with $\theta(k-p_F)$. Inserting $G^<$ in (15.61), we find

$$\Sigma^<(p,\omega)=\frac{-1}{4\pi^2 p}\int_0^{p_F} dk\,k\int_{|k-p|}^{k+p} dq\,q\,W^>(q,\epsilon_k+\Delta-\omega).$$

If we define the new variable $\omega'=\epsilon_k+\Delta-\omega$ so that $d\omega'=kdk$, then the integral becomes

$$\Sigma^<(p,\omega)=\frac{-1}{4\pi^2 p}\int_{\mu-\omega-\epsilon_{p_F}}^{\mu-\omega} d\omega'\int_{q_-}^{q_+} dq\,q\,W^>(q,\omega'),$$

where $q_-=|\sqrt{2(\omega'-\Delta+\omega)}-p|$ and $q_+=\sqrt{2(\omega'-\Delta+\omega)}+p$. In the same way, it is easy to show that

$$\Sigma^>(p,\omega)=\frac{1}{4\pi^2 p}\int_{\mu-\omega}^{\infty} d\omega'\int_{q_-}^{q_+} dq\,q\,W^<(q,\omega').$$

The calculation of the self-energy is now reduced to a two-dimensional integral. It only remains to calculate W from the polarization $P=-\mathrm{i}GG$. We have $W=v+vPW=v+v\chi v$, where $\chi=P+PvP+\ldots$. If we calculate P using the Green's function (15.63), then $P=\chi^0$ is the Lindhard function. Indeed, our approximate G only differs from the noninteracting G_0 by a constant shift of the one-particle energies, and hence the electron–hole excitation energies $\epsilon_{\mathbf{k}+\mathbf{q}}-\epsilon_{\mathbf{k}}$ in χ^0 are unchanged, see (15.32). Since $P=\chi^0$ then χ is exactly the RPA expression of Section 15.4. To extract the lesser/greater screened interaction we can use the fluctuation–dissipation theorem (10.39). Due to the property (13.26) we have $W^{\mathrm{A}}(q,\omega)=[W^{\mathrm{R}}(q,\omega)]^*$, and hence

$$W^<(q,\omega)=2\mathrm{i}f(\omega)\mathrm{Im}[W^{\mathrm{R}}(q,\omega)]\quad,\quad W^>(q,\omega)=2\mathrm{i}\bar{f}(\omega)\mathrm{Im}[W^{\mathrm{R}}(q,\omega)],$$

with f the Bose function and $\bar{f}=1+f$. At zero temperature, $f(\omega)=-\theta(-\omega)$ and $\bar{f}(\omega)=1+f(\omega)=1-\theta(-\omega)=\theta(\omega)$. Furthermore, $W^{\mathrm{R}}(q,\omega)=\tilde{v}_q+\tilde{v}_q^2\chi^{\mathrm{R}}(q,\omega)$ and therefore the lesser/greater self-energies assume the form

$$\mathrm{i}\Sigma^<(p,\omega)=\frac{1}{2\pi^2 p}\int_{\mu-\omega-\epsilon_{p_F}}^{\mu-\omega} d\omega'\,\theta(\omega')\int_{q_-}^{q_+} dq\,q\,\tilde{v}_q^2\,\mathrm{Im}[\chi^{\mathrm{R}}(q,\omega')],\tag{15.65}$$

$$\mathrm{i}\Sigma^>(p,\omega)=\frac{1}{2\pi^2 p}\int_{\mu-\omega}^{\infty} d\omega'\,\theta(-\omega')\int_{q_-}^{q_+} dq\,q\,\tilde{v}_q^2\,\mathrm{Im}[\chi^{\mathrm{R}}(q,\omega')].\tag{15.66}$$

Recalling that $\mathrm{Im}[\chi^{\mathrm{R}}(q,\omega')]$ is negative (positive) for positive (negative) frequencies ω', see (13.89), we infer that $\mathrm{i}\Sigma^>$ is nonvanishing only for $\omega>\mu$, in which case it is positive, and $\mathrm{i}\Sigma^<$ is nonvanishing only for $\omega<\mu$, in which case it is negative. Hence the rate function $\Gamma=\mathrm{i}(\Sigma^>-\Sigma^<)$ is positive as it should be. We also observe that both $\Sigma^<$ and $\Sigma^>$ depend only on $\omega-\mu$, and therefore (15.64) provides an explicit expression for μ:

$$\mu=\epsilon_{p_F}+\Sigma^{\mathrm{R}}(p_F,\mu)=\epsilon_{p_F}+\Sigma_{\mathrm{x}}(p_F)+\Sigma_{\mathrm{c}}^{\mathrm{R}}(p_F,\mu),\tag{15.67}$$

since the correlation self-energy $\Sigma_c^R(p_F, \mu)$ does not depend on μ. Any other shift Δ would have led to a more complicated equation for μ.

This is how far we can get with pencil and paper. What we have to do next is to calculate numerically $\Sigma^{\lessgtr}$, insert them into (15.60) to obtain Σ_c^R, and then extract the spectral function from (10.22). First, however, it is instructive to analyze the behavior of the self-energy for energies close to μ and for energies close to the plasma frequency ω_p. This analysis facilitates the interpretation of the numerical results and at the same time provides us with useful observations for the actual implementation.

15.6.2 G_0W_0 Self-Energy for Energies Close to the Chemical Potential

Let us analyze the integrals (15.65) and (15.66) for ω close to μ. We take, for example, the derivative of $i\Sigma^<$. For ω close to μ we have that $|\omega - \mu| < \epsilon_{p_F}$ and the lower limit of the integral in (15.65) can be set to zero. Differentiation with respect to ω then gives three terms, one term stemming from the differentiation of the upper limit of the ω' integral and two more terms stemming from the differentiation of q_- and q_+. These last two terms vanish when $\omega \to \mu^-$ since the interval of the ω' integration goes to zero in this limit. Therefore,

$$i\frac{\partial\Sigma^<(p,\omega)}{\partial\omega}\bigg|_{\omega=\mu^-} = -\frac{1}{2\pi^2 p}\int_{|p_F-p|}^{p_F+p} dq\, q\, \tilde{v}_q^2\, \mathrm{Im}[\chi^R(q,0)] = 0,$$

since $\mathrm{Im}[\chi^R(q,0)] = 0$. In a similar way one can show that the derivative of $\Sigma^>(p,\omega)$ with respect to ω vanishes for $\omega \to \mu^+$. Thus, for $\omega \sim \mu$ we have

$$i\Sigma^<(p,\omega) = -C_p^<\,\theta(\mu-\omega)\,(\omega-\mu)^2,$$
$$i\Sigma^>(p,\omega) = C_p^>\,\theta(\omega-\mu)\,(\omega-\mu)^2,$$

where $C_p^{\lessgtr} > 0$, see the discussion below (15.66). The constants $C_p^{\lessgtr}$ can easily be determined by taking the second derivative of $\Sigma^{\lessgtr}$ with respect to ω in $\omega = \mu^{\mp}$. It is a simple exercise for the reader to show that $C_p^> = C_p^< = C_p$ with

$$C_p = -\frac{1}{4\pi^2 p}\int_{|p_F-p|}^{p_F+p} dq\, q\, \frac{\tilde{v}_q^2}{|1-\tilde{v}_q\chi_0^R(q,0)|^2}\, \frac{\partial\, \mathrm{Im}[\chi_0^R(q,\omega)]}{\partial\omega}\bigg|_{\omega=0}, \tag{15.68}$$

where we use the RPA expression (15.44) for χ^R in terms of the Lindhard function $\chi^{0,R}$. The fact that $C_p^> = C_p^<$ agrees with the discussion in Section 15.1 and allows us to write the quasi-particle width as in (15.9).

Let us comment in more detail on the explicit formula of C_p in the G_0W_0 approximation. Since the derivative of $\mathrm{Im}[\chi^{0,R}(q,\omega)]$ with respect to ω is negative in $\omega = 0$ for $q \le 2p_F$ and zero otherwise, see Fig. 15.3(b), we find that the constant $C_p > 0$ for $|p_F - p| \le 2p_F$ (i.e., for $p \le 3p_F$) and zero otherwise. This is physically related to the fact that an injected particle with momentum $3p_F$ and energy μ can scatter into a particle with momentum p_F and energy μ and an electron–hole pair with momentum $2p_F$ and energy zero at the Fermi surface. Any particle with a momentum higher than $3p_F$ (and energy μ) can only excite electron–hole pairs with nonzero energy.

Lifetime From the knowledge of C_p we can calculate the lifetime τ_p of a quasi-particle (or quasi-hole) with momentum p. We recall that $\tau_p^{-1} = Z_p\Gamma(p, E_p)$ where Z_p is the quasi-particle renormalization factor and E_p is the quasi-particle energy defined as the solution of $\omega - \epsilon_p - \Sigma_x(p) - \Lambda(p,\omega) = 0$, see Section 10.2. For $p \sim p_F$ the quasi-particle energy $E_p \sim \mu$ and hence $\Gamma(p, E_p) \sim C_{p_F}(E_p - \mu)^2$. Expanding in powers of $p - p_F$ we then find

$$\frac{1}{\tau_p} = ZC_{p_F}\left(\frac{p_F}{m^*}\right)^2 (p-p_F)^2,$$

where $Z = Z_{p_F}$ and the *effective mass* m^* of the quasi-particle is defined by

$$\frac{1}{m^*} \equiv \frac{1}{p_F}\frac{dE_p}{dp}\bigg|_{p=p_F}.$$

Let us calculate a more explicit form of C_p. From (15.35) and (15.38) we have that $\chi^{0,R}(q,0) = -(p_F/\pi^2)F(q/2p_F)$ and $\partial\mathrm{Im}[\chi_0^R(q,\omega)]/\partial\omega|_{\omega=0} = -\theta(2p_F - q)/(2\pi q)$ (we use the expression in region I). Inserting these results into (15.68), we get

$$C_p = \frac{1}{8\pi^3 p}\int_{|p_F-p|}^{p_F+p} dq\,\theta(2p_F - q)W^R(q,0)^2, \quad \text{with} \quad W^R(q,0) = \frac{4\pi}{q^2 + \frac{4p_F}{\pi}F(\frac{q}{2p_F})}.$$

We now evaluate this formula for $p = p_F$:

$$C_{p_F} = \frac{1}{8\pi^3 p_F}\int_0^{2p_F} dq\left(\frac{4\pi}{q^2 + \frac{4p_F}{\pi}F(\frac{q}{2p_F})}\right)^2 = \frac{\pi}{4p_F^2}\xi(r_s),$$

where we define

$$\xi(r_s) = \int_0^1 dx\,\frac{1}{\left[F(x) + \left(\frac{9\pi}{4}\right)^{\frac{1}{3}}\frac{\pi}{r_s}x^2\right]^2}$$

and use that $p_F = (9\pi/4)^{\frac{1}{3}}/r_s$. Thus the quasi-particle lifetime becomes [108]

$$\frac{1}{\tau_p} = \frac{Z\pi}{4m^{*2}}\xi(r_s)\,(p-p_F)^2.$$

As an exercise we can analyze the high density $r_s \to 0$ behavior of $\xi(r_s)$. Since $F(0) = 1$ and $F(1) = 0$, the integrand of $\xi(r_s)$ is equal to 1 in $x = 0$ and it is proportional to r_s^2 in $x = 1$, which is small in the limit $r_s \to 0$. Therefore the largest contribution to the integral comes from the integration region around $x = 0$; we then approximate $F(x) \approx F(0) = 1$. Denoting by $\alpha = (9\pi/4)^{\frac{1}{6}}\sqrt{\pi/r_s}$ the square root of the constant which multiplies x^2, we have

$$\xi(r_s) = \frac{1}{\alpha}\int_0^\alpha dy\frac{1}{(1+y^2)^2} = \frac{1}{2\alpha}\arctan\alpha + \frac{1}{2(1+\alpha^2)} \sim \frac{\pi}{4\alpha} \quad (\alpha \to \infty).$$

So we see that $\xi(r_s) \sim \sqrt{r_s}$ for $r_s \to 0$ and hence the lifetime increases by increasing the density of the electron gas. Strictly speaking, also m^* and Z are functions of r_s, but they are weakly dependent on r_s and both approach 1 as $r_s \to 0$ [108]. This limiting behavior of m^* and Z is somehow intuitive since for $r_s \to 0$ the dominant self-energy diagram is the Hartree diagram, which is canceled by the external potential (the high-density interacting gas behaves like a noninteracting gas).

Quasi-particle energy The constant C_p pertains to the quasi-particle lifetime. What about the quasi-particle energy E_p? To answer this question we have to study the behavior of $\Lambda = \mathrm{Re}[\Sigma_c^R]$, since E_p is the solution of $\omega - \epsilon_p - \Sigma_x(p) - \Lambda(p,\omega) = 0$, see Section 10.2. This is not easily done on paper. As we discuss in more detail below, we find from numerical calculations for the electron gas at metallic densities that $E_p - \mu \approx \epsilon_p - \epsilon_{p_F}$. This implies that

$$\Sigma_x(p) + \Lambda(p, E_p) = E_p - \epsilon_p \approx \mu - \epsilon_{p_F}.$$

Therefore the real part of the self-energy does not change much when we move along the quasi-particle branch in the (p,ω)-plane. With hindsight our poor man's choice for G is more accurate than the Hartree–Fock G since in the Hartree–Fock approximation $\epsilon_p = p^2/2$ is renormalized by the exchange self-energy $\Sigma_x(p)$ and hence the Hartree–Fock $E_p = \epsilon_p^{\mathrm{HF}} = \epsilon_p + \Sigma_x(p)$ deviates considerably from a parabola, see Fig. 8.1(b).

15.6.3 G_0W_0 Self-Energy for Energies Close to the Plasma Frequency

So far we have only been looking at the behavior of the self-energy close to the Fermi surface. However, an added or removed particle at sufficiently high energy may also excite plasmons. For instance, for an added particle we would expect this to happen at energies $\omega \geq \mu + \omega_p$, where ω_p is the plasmon frequency. Let us study how these plasmon excitations appear from the equations. We observe that the function under the q-integral in (15.65) and (15.66) is $q\,\tilde{v}_q\,B_2(q,\omega')$, where B_2 is the imaginary part of the function $B = \tilde{v}\,\chi^{\mathrm{R}}$ defined in (15.54). As pointed out in (15.55), the function B_2 has δ-like singularities outside regions I and II [where $A_2(q,\omega') \sim -\eta\,\mathrm{sgn}(\omega')$] for $1 - A_1(q,\omega') = 0$ - that is, for $\omega' = \pm\omega_p(q)$. Thus, for $\omega' \sim \pm\omega_p(q)$ and for $q < q_c$ (q_c is the momentum at which the plasmon branch crosses the electron–hole continuum), we can write:

$$B_2(q,\omega') = -\pi \frac{\delta(\omega' - \omega_p(q))}{|\frac{\partial A_1}{\partial \omega'}(q,\omega_p(q))|} + \pi \frac{\delta(\omega' + \omega_p(q))}{|\frac{\partial A_1}{\partial \omega'}(q,\omega_p(q))|},$$

where we take into account that A_2 is negative (positive) for positive (negative) frequencies. The lesser and greater self-energy can then be split into an electron–hole part and a plasmon part $\Sigma^{\lessgtr} = \Sigma^{\lessgtr}_{\mathrm{eh}} + \Sigma^{\lessgtr}_{\mathrm{p}}$ where the electron–hole part is given by (15.65) and (15.66) integrated over regions I and II and the plasmon part is given by

$$\mathrm{i}\Sigma^<_{\mathrm{p}}(p,\omega) = -\frac{1}{2\pi p}\int_{\mu-\omega-\epsilon_{p_F}}^{\mu-\omega} d\omega'\,\theta(\omega')\int_{q_-}^{q_+} dq\,\frac{q\,\tilde{v}_q\,\theta(q_c - q)}{|\frac{\partial A_1}{\partial \omega'}(q,\omega_p(q))|}\,\delta(\omega' - \omega_p(q)),$$

$$\mathrm{i}\Sigma^>_{\mathrm{p}}(p,\omega) = \frac{1}{2\pi p}\int_{\mu-\omega}^{\infty} d\omega'\,\theta(-\omega')\int_{q_-}^{q_+} dq\,\frac{q\,\tilde{v}_q\,\theta(q_c - q)}{|\frac{\partial A_1}{\partial \omega'}(q,\omega_p(q))|}\,\delta(\omega' + \omega_p(q)).$$

When can we expect a contribution from the plasmon excitations? Let us first consider $\Sigma^<_{\mathrm{p}}$. Since the upper limit of the ω'-integration is given by $\mu - \omega$, we can only get a

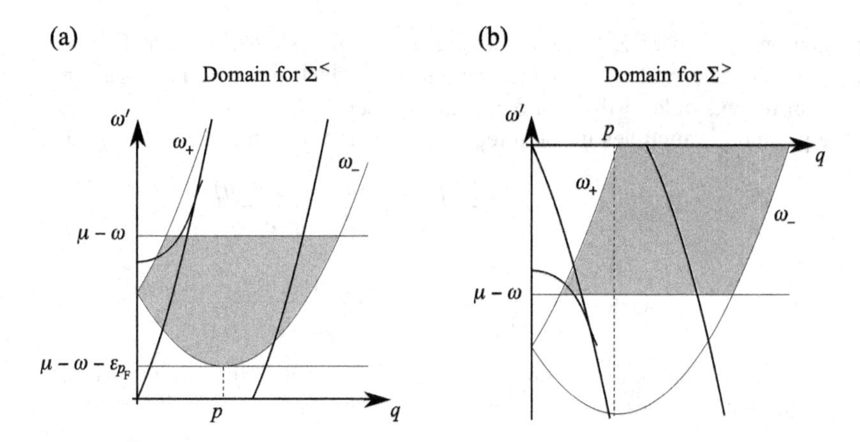

Figure 15.7 (a) Integration domain for $\Sigma^{<}(p,\omega')$ in the case that $\mu - \omega > \omega_{\rm p}$ and $\mu - \omega - \epsilon_{p_{\rm F}} > 0$ and $p < 2p_{\rm F}$. (b) Integration domain for $\Sigma^{>}(p,\omega')$ in the case that $\mu - \omega < -\omega_{\rm p}$ and $p < 2p_{\rm F}$. The figure further shows with thick lines regions I and II of the electron-hole continuum and the plasmon branch - compare with Fig. 15.5.

contribution from the integral when $\mu - \omega \geq \omega_{\rm p}(q) \geq \omega_{\rm p}$, with $\omega_{\rm p} = \omega_{\rm p}(0)$. The allowed q values are still determined by the functions q_+ and q_-, but it is at least clear that there can be no contribution unless $\omega \leq \mu - \omega_{\rm p}$. This means physically that a removed particle can only excite plasmons when its energy is at least $\omega_{\rm p}$ below the chemical potential. A similar story applies to $\Sigma^{>}_{\rm p}$: From the lower limit of the integral we see that there can be no contribution unless $\mu - \omega \leq -\omega_{\rm p}(q) \leq -\omega_{\rm p}$. Therefore an added particle can only excite plasmons when its energy is at least $\omega_{\rm p}$ above the chemical potential.

We can deduce the energies for which $\Sigma^{\lessgtr}_{\rm p}$ are maximal by analyzing in more detail the integration domain. For $\Sigma^{<}_{\rm p}$ the ω'-integral goes from $\max(0, \mu - \omega - \epsilon_{p_{\rm F}})$ to $\mu - \omega$ (which must be positive for otherwise $\Sigma^{<}_{\rm p} = 0$). Figure 15.7(a) shows an example of the ω'-domain (the region between the two horizontal lines) in the case that $\mu - \omega > \omega_{\rm p}$ and $\mu - \omega - \epsilon_{p_{\rm F}} > 0$. For any fixed ω' we have to integrate over q between q_- and q_+:

$$q_- = |\sqrt{2(\omega' - \mu + \omega + \epsilon_{p_{\rm F}})} - p| \leq q \leq \sqrt{2(\omega' - \mu + \omega + \epsilon_{p_{\rm F}})} + p = q_+,$$

where we use that $\Delta = \mu - \epsilon_{p_{\rm F}}$. This amounts to integrating over all q such that[13]

$$\omega_-(q) \leq \omega' \leq \omega_+(q) \quad , \quad \text{where} \quad \omega_\pm(q) = \frac{1}{2}(p \pm q)^2 + \mu - \omega - \epsilon_{p_{\rm F}}.$$

The parabolas $\omega_+(q)$ and $\omega_-(q)$ are displayed in Fig. 15.7, and together with the previous constraint lead to the integration domain represented by the gray-shaded area. Similarly,

[13] We can easily derive this equivalent form of the integration domain. Let $x = \omega' - \mu + \omega + \epsilon_{p_{\rm F}} > 0$. Then $|\sqrt{2x} - p| \leq q$ implies $-q \leq \sqrt{2x} - p \leq q$ or equivalently $p - q \leq \sqrt{2x} \leq p + q$. Denoting by $M_1 = \max(0, p-q)$ we can square this inequality and get $\frac{1}{2}M_1^2 \leq x \leq \frac{1}{2}(p+q)^2$. Next consider $q \leq \sqrt{2x} + p$, which implies $q - p \leq \sqrt{2x}$. Denoting by $M_2 = \max(0, q-p)$ we can square this inequality and get $\frac{1}{2}M_2^2 \leq x$. Since the maximum value between M_1^2 and M_2^2 is $(p-q)^2$, we find $\frac{1}{2}(p-q)^2 \leq x \leq \frac{1}{2}(p+q)^2$.

the integration domain of $\Sigma_{\rm p}^{>}$ is represented by the gray-shaded area in Fig. 15.7(b). Here the horizontal lines are at $\omega' = 0$ and $\omega' = \mu - \omega$ (which must be negative for otherwise $\Sigma_{\rm p}^{>} = 0$). In the example of the figure we have chosen $\mu - \omega < -\omega_{\rm p}$.

If the plasmon branch lies in the integration domain, then the integral over ω' leads to

$$\mathrm{i}\Sigma_{\rm p}^{<}(p,\omega) = -\frac{1}{2\pi p}\int_{q_0^<}^{q_1^<} dq\, \frac{q\,\tilde{v}_q\,\theta(q_{\rm c}-q)}{\left|\frac{\partial A_1}{\partial\omega'}(q,\omega_{\rm p}(q))\right|}, \tag{15.69}$$

$$\mathrm{i}\Sigma_{\rm p}^{>}(p,\omega) = \frac{1}{2\pi p}\int_{q_0^>}^{q_1^>} dq\, \frac{q\,\tilde{v}_q\,\theta(q_{\rm c}-q)}{\left|\frac{\partial A_1}{\partial\omega'}(q,\omega_{\rm p}(q))\right|}, \tag{15.70}$$

where $q_0^<$ and $q_1^<$ ($q_0^>$ and $q_1^>$) are the momenta at which the plasmon branch $\omega_{\rm p}(q)$ ($-\omega_{\rm p}(q)$) enters and leaves the gray-shaded area. Consider now the integrand for small q. We have

$$A_1(q,\omega') = \tilde{v}_q\,\mathrm{Re}[\chi_0^{\rm R}(q,\omega')] \xrightarrow[q\to 0]{} \tilde{v}_q\, n\,\frac{q^2}{\omega'^2} = \tilde{v}_q\,\frac{\omega_{\rm p}^2}{4\pi}\,\frac{q^2}{\omega'^2},$$

where we use (15.36) and the formula $\omega_{\rm p} = \sqrt{4\pi n}$ for the plasma frequency. Thus, for small momenta,

$$\frac{\partial A_1}{\partial\omega'}(q,\omega_{\rm p}(q)) = -\tilde{v}_q\,\frac{q^2}{2\pi\omega_{\rm p}}$$

and the integrand of $\Sigma_{\rm p}^{\lessgtr}$ becomes $2\pi\omega_{\rm p}/q$. This leads to a logarithmic behavior of $\Sigma_{\rm p}^{\lessgtr}$ when the lower integration limit $q_0^{\lessgtr}$ gets very small. Consequently, the quasi-particle width $\Gamma(p,\omega) = \mathrm{i}[\Sigma^>(p,\omega) - \Sigma^<(p,\omega)]$ also has a logarithmic behavior and its Hilbert transform $\Lambda(p,\omega)$ has a discontinuous jump at the same point.[14]

Let us see for which values of p and ω the logarithmic singularity occurs. For $q = 0$ the parabolas $\omega_+(0) = \omega_-(0) = \epsilon_p + \mu - \omega - \epsilon_{p_{\rm F}} \equiv \omega_0$ have the same value. For $\Sigma_{\rm p}^{<}$ the point $(0,\omega_0)$ is in the integration domain when $\omega_0 \le \mu - \omega$ (i.e., when $p \le p_{\rm F}$), whereas for $\Sigma_{\rm p}^{>}$ this point is in the integration domain when $\omega_0 \ge \mu - \omega$ (i.e., when $p \ge p_{\rm F}$). In the examples of Fig. 15.7 the point $(0,\omega_0)$ is inside the integration domain for $\Sigma_{\rm p}^{<}$ and outside for $\Sigma_{\rm p}^{>}$. The integrals in (15.69) and (15.70) become logarithmically divergent when the plasmon branches enter the integration domain exactly at the point $(0,\omega_0)$ (i.e., when $\omega_0 = \pm\omega_{\rm p}$) since in this case $q_0^{\lessgtr} = 0$. This happens at the frequency

$$\omega = \epsilon_p - \epsilon_{p_{\rm F}} + \mu - \omega_{\rm p} \qquad (\omega = \epsilon_p - \epsilon_{p_{\rm F}} + \mu + \omega_{\rm p})$$

in $\Sigma_{\rm p}^{<}$ ($\Sigma_{\rm p}^{>}$) for $p \le p_{\rm F}$ ($p \ge p_{\rm F}$). More generally, if ω_0 is close to $\pm\omega_{\rm p}$, the lower limit $q_0^{\lessgtr}$ is given by the solution of $\omega_-(q) = \pm\omega_{\rm p}(q) \sim \pm\omega_{\rm p}$ or $\omega_+(q) = \pm\omega_{\rm p}(q) \sim \pm\omega_{\rm p}$ (depending on whether ω_0 is larger or smaller than $\omega_{\rm p}$) and it is an easy exercise to check that $q_0^{\lessgtr} \sim \frac{1}{p}|\omega_0 \mp \omega_{\rm p}|$. Then the integrals (15.69) and (15.70) can be split into a singular contribution coming from the small momentum region and a remaining finite contribution. The singular part behaves as

$$\mathrm{i}\Sigma_{\rm p}^{\lessgtr}(p,\omega) = \pm\frac{\omega_{\rm p}}{p}\ln\frac{|\omega_0 \mp \omega_{\rm p}|}{p} = \pm\frac{\omega_{\rm p}}{p}\ln\frac{|\epsilon_p + \mu - \omega - \epsilon_{p_{\rm F}} \mp \omega_{\rm p}|}{p} \tag{15.71}$$

[14] If $\Lambda(p,\omega)$ has a discontinuity D in $\omega = \omega_D$ then $\Gamma(p,\omega) = -2P\int\frac{d\omega'}{\pi}\frac{\Lambda(p,\omega')}{\omega-\omega'} \sim -\frac{2D}{\pi}\ln|\omega-\omega_D|$.

for $p \leq p_\mathrm{F}$ ($p \geq p_\mathrm{F}$) for $\Sigma^<_\mathrm{p}$ ($\Sigma^>_\mathrm{p}$). In the limit $p \to 0$ this analysis must be done more carefully since the integration domain itself goes to zero as the two parabolas approach each other. The reader can work out this case. Equation (15.71) implies that the quasi-particle width $\Gamma(p,\omega)$ has a logarithmic singularity of strength $-\omega_\mathrm{p}/p$. Therefore $\Lambda(p,\omega)$ has a discontinuous jump of $\pi\omega_\mathrm{p}/(2p)$ at the same point. Strictly speaking, this is true only if p is smaller (larger) than p_F for $\Sigma^<_\mathrm{p}$ ($\Sigma^>_\mathrm{p}$). At exactly $p = p_\mathrm{F}$ the plasmon branch enters the integration domain at the critical point $(0, \pm\omega_\mathrm{p})$ when $\mu - \omega = \pm\omega_\mathrm{p}$, and it is outside the integration domains of $\Sigma^{\lessgtr}_\mathrm{p}$ for $|\mu - \omega| \leq \omega_\mathrm{p}$. Therefore the logarithmic singularity in (15.71) becomes one-sided and the Hilbert transform of the quasi-particle width induces a logarithmic singularity at $\mu - \omega = \pm\omega_\mathrm{p}$ in $\Lambda(p,\omega)$.

It is now time to interpret our results. We have seen that the quasi-particle energies satisfy the property $E_p - \mu \sim \epsilon_p - \epsilon_{p_\mathrm{F}}$. Therefore $\Sigma^<_\mathrm{p}$ exhibits a logarithmic peak at $\omega = E_p - \omega_\mathrm{p}$ for $p \leq p_\mathrm{F}$, whereas $\Sigma^>_\mathrm{p}$ exhibits a logarithmic peak at $\omega = E_p + \omega_\mathrm{p}$ for $p \geq p_\mathrm{F}$. As we show below, this leads to sharp features in the real part of Σ^R and peak structures at $\omega \sim E_p \pm \omega_\mathrm{p}$ in the spectral function, in addition to the quasi-particle peak at $\omega = E_p$. Physically this means that if we add a particle with momentum $p \geq p_\mathrm{F}$ then the created state evolves in time as a superposition of many-body eigenstates with energy components mainly centered around E_p (quasi-particle) and to a lesser extent around $E_p + \omega_\mathrm{p}$ (quasi-particle dressed by a plasmon). A similar interpretation can be given for the removal of a particle of momentum $p \leq p_\mathrm{F}$. After the removal the system is most likely to be found at the quasi-hole energy E_p and to a lesser extent at the energy $E_p - \omega_\mathrm{p}$ of a quasi-hole dressed by a plasmon.

15.6.4 Numerical Results

A practical calculation starts by specifying the density of the electron gas by choosing a value of r_s, and hence the Fermi momentum p_F. In our example we take an electron gas of density characterized by the Wigner–Seitz radius $r_s = 4$, which roughly corresponds to the density of valence electrons in the sodium crystal. Subsequently the chemical potential can be determined from (15.67) after the evaluation of $\Sigma^\mathrm{R}(p_\mathrm{F},\mu)$. We find that $\Sigma_\mathrm{x}(p_\mathrm{F}) = -1.33\,\epsilon_{p_\mathrm{F}}$ and $\Sigma^\mathrm{R}_\mathrm{c}(p_\mathrm{F},\mu) = -0.44\,\epsilon_{p_\mathrm{F}}$, yielding a chemical potential of $\mu = -0.77\epsilon_{p_\mathrm{F}}$, which is quite different from the value $\mu = \epsilon_{p_\mathrm{F}}$ of the noninteracting electron gas. The plasma frequency at $r_s = 4$ is $\omega_\mathrm{p} = 1.88\,\epsilon_{p_\mathrm{F}}$ and the plasmon branch enters the electron–hole continuum at the critical momentum $q_c = 0.945\,p_\mathrm{F}$. Then all parameters of the problem are determined and we can calculate the lesser/greater self-energy, the quasi-particle width, and the spectral function.

Quasi-particle width In Fig. 15.8 we display $-\mathrm{Im}[\Sigma^\mathrm{R}(p,\,\omega+\mu)] = \Gamma(p,\omega+\mu)/2$ relative to the chemical potential μ in the (p,ω)-plane. The first distinctive feature of this graph is the valley around the chemical potential where $\Gamma \sim (\omega-\mu)^2$ for $\omega \to \mu$ for all values of the momentum p. As discussed in this chapter and in Section 15.1, this behavior is completely determined by phase-space restrictions. Away from the chemical potential at energies $|\omega - \mu| \geq \omega_\mathrm{p}$ there is the possibility for the system with a particle of momentum p added or removed to decay into plasmons as well. As a consequence, the rate function is large for hole states with momentum $p < p_\mathrm{F}$ and for particle states with momentum $p > p_\mathrm{F}$. The boundary at $p = p_\mathrm{F}$ is not strict. Unlike the noninteracting electron gas for

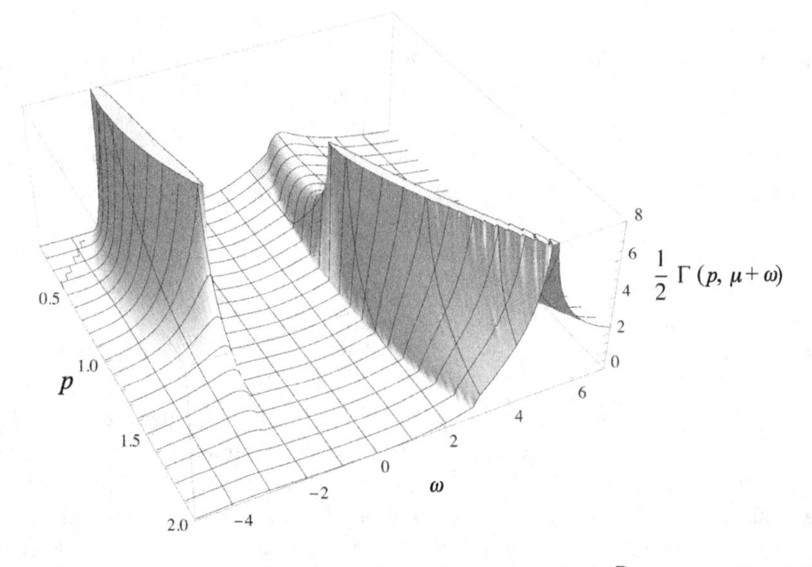

Figure 15.8 The imaginary part of the retarded self-energy $-\mathrm{Im}[\Sigma^{\mathrm{R}}(p,\omega+\mu)] = \Gamma(p,\omega+\mu)/2$ for an electron gas at $r_s = 4$ within the $\mathrm{G_0W_0}$ approximation as a function of the momentum and energy. The momentum p is measured in units of p_{F} and the energy ω and the self-energy in units of $\epsilon_{p_{\mathrm{F}}} = p_{\mathrm{F}}^2/2$.

an interacting gas it is possible to add a particle with momentum $p \leq p_{\mathrm{F}}$ or to remove a particle with momentum $p \geq p_{\mathrm{F}}$. As we see below the momentum distribution of the interacting gas is no longer a Heaviside step function.

Spectral function Let us now turn our attention to the real part of the self-energy and the spectral function. In Fig. 15.9 we show the real and imaginary parts of the self-energy and the spectral function as a function of the energy relative to μ for the momentum values $p/p_{\mathrm{F}} = 0.5, 1, 1.5$. For $p = 0.5\,p_{\mathrm{F}}$ (Fig. 15.9(a)) we see that $\mathrm{Im}[\Sigma^{\mathrm{R}}]$ (solid line) has a logarithmic singularity due to plasmons below the chemical potential. At the same point $\mathrm{Re}[\Sigma^{\mathrm{R}}]$ (dashed line) jumps discontinuously in accordance with the previous analysis. The structure of the spectral function is determined by the zeros of $\omega - \epsilon_p = \mathrm{Re}[\Sigma^{\mathrm{R}}(p,\omega)]$. These zeros occur at the crossings of the dotted line with the dashed line. At two of the crossings $\mathrm{Im}[\Sigma^{\mathrm{R}}]$ is small and as a consequence the spectral function has two pronounced peaks (Fig. 15.9(d)). The peak on the right close to μ represents the quasi-hole peak, whereas the peak to the left tells us that there is considerable probability to excite a plasmon. For $p = 1.5\,p_{\mathrm{F}}$ (Fig. 15.9(c)) the plasmon contribution occurs at positive energies. There are again three crossings between the dashed and dotted lines, but this time only one crossing (the quasi-particle one) occurs in an energy region where $-\mathrm{Im}[\Sigma^{\mathrm{R}}] \ll 1$. Thus the spectral function has only one main quasi-particle peak (Fig. 15.9(f)). There are still some plasmonic features visible at larger energies, but they are much less prominent than for the case of $p = 0.5\,p_{\mathrm{F}}$. This can be understood from the fact that plasmons are mainly excited at low momenta since for $p > q_c$ the plasmon branch enters the electron–hole continuum. For an added particle at high momentum to excite a plasmon at low momentum it should at the same time transfer a large momentum to an electron–hole excitation. However, if the particle thereby loses most of its energy to a plasmon, then the electron–hole pair should at the same time have large momentum and low energy, which is unlikely. Finally, in Fig. 15.9(b) we display the self-energy for $p = p_{\mathrm{F}}$. In this case $\mathrm{Im}[\Sigma^{\mathrm{R}}]$ displays a one-sided logarithmic singularity at $\omega - \mu = \pm\omega_{\mathrm{p}}$ (the numerics does not show this infinity exactly due to the finite energy grid), and as a consequence $\mathrm{Re}[\Sigma^{\mathrm{R}}]$ displays a logarithmic singularity at these points. These singularities would be smeared out to finite peaks if we had used a more advanced approximation for the screened interaction than the RPA in which we also allow for plasmon broadening. We further see that $\mathrm{Re}[\Sigma^{\mathrm{R}}]$ crosses the dotted line in only one point. At this point the derivative of $\mathrm{Re}[\Sigma^{\mathrm{R}}]$ is negative, in accordance with the general result (15.10), and $\mathrm{Im}[\Sigma^{\mathrm{R}}] = 0$. The quasi-particle peak becomes infinitely sharp and develops into a δ-function (vertical line Fig. 15.9(e)). For $r_s = 4$ the strength of the δ-function is $Z = 0.64$.

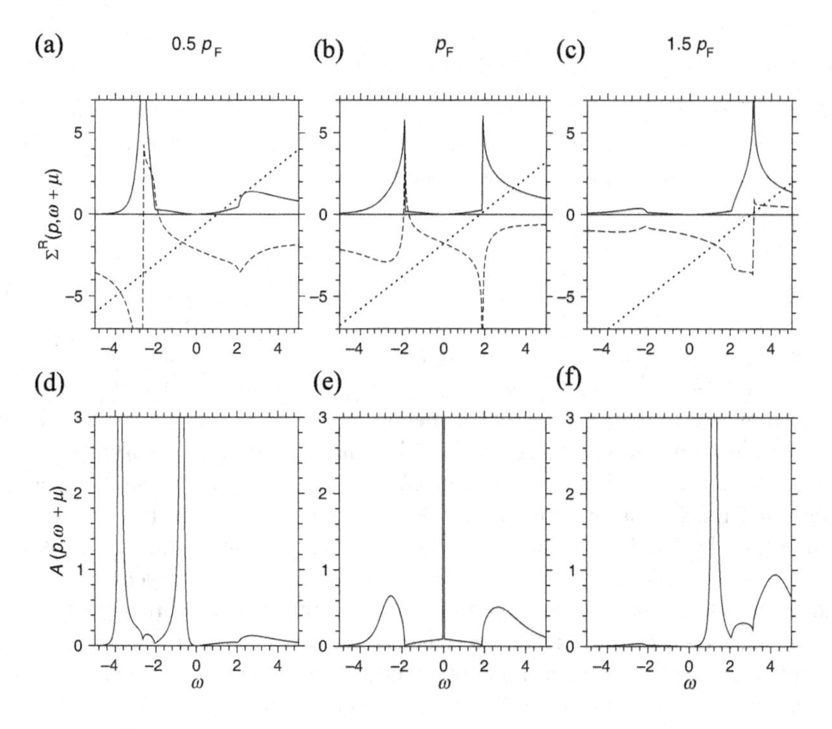

Figure 15.9 The self-energy and the spectral function in the $\mathrm{G_0W_0}$ approximation for $p/p_{\mathrm{F}} = 0.5,\ 1.0,\ 1.5$ at density $r_s = 4$ with ω and self-energies in units of $\epsilon_{p_{\mathrm{F}}} = p_{\mathrm{F}}^2/2$. In the top row the solid lines represent $-\mathrm{Im}[\Sigma^{\mathrm{R}}(p,\omega+\mu)] = \Gamma(p,\omega+\mu)/2$ and the dashed lines represent $\mathrm{Re}[\Sigma^{\mathrm{R}}(p,\omega+\mu)] = \Sigma_{\mathrm{x}}(p) + \Lambda(p,\omega+\mu)$. The dotted line represents the curve $\omega + \mu - \epsilon_p$. The crossings of this line with $\mathrm{Re}[\Sigma^{\mathrm{R}}(p,\omega+\mu)]$ determine the position of the peaks in the spectral function $A(p,\omega+\mu)$ displayed in the bottom row. The δ-function in the spectral function $A(p_{\mathrm{F}},\mu+\omega)$ at $\omega = 0$ is indicated by a vertical line and has strength $Z = 0.64$.

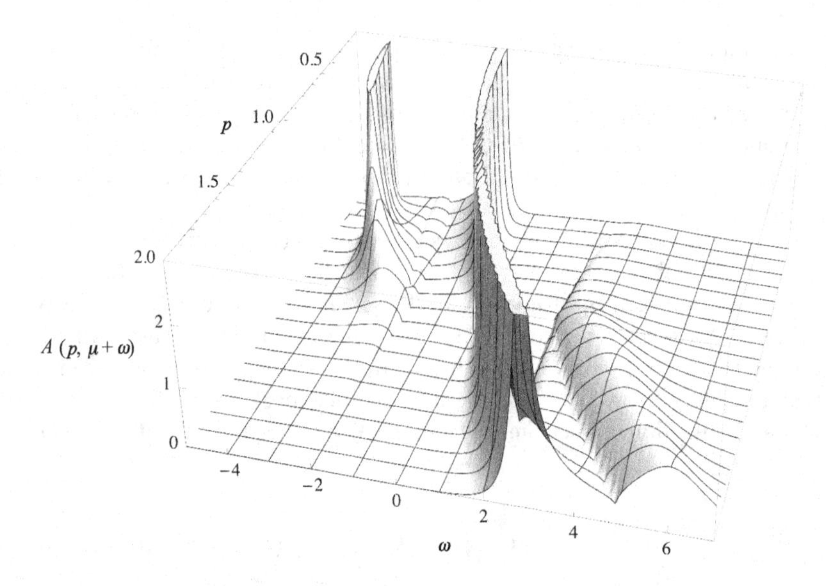

Figure 15.10 The spectral function $A(p, \mu + \omega)$ as a function of the momentum and energy for an electron gas at $r_s = 4$ within the G_0W_0 approximation. The momentum p is measured in units of p_F and the energy ω and the spectral function in units of $\epsilon_{p_F} = p_F^2/2$.

The spectral function in the (p, ω)-plane is displayed in Fig. 15.10. The quasi-particle peak appears in the middle of the figure and broadens when we move away from the Fermi momentum p_F. Its position at $E_p - \mu$ is approximately given by $p^2/2 - p_F^2/2$, or in units of the figure $(E_p - \mu)/\epsilon_{p_F} \approx (p/p_F)^2 - 1$. An important improvement of the dispersion curve E_p compared to the Hartree–Fock dispersion curve of Section 8.3.2 is that it does not exhibit the logarithmic divergence at $p = p_F$. The cancellation of the logarithmic divergence can be shown from a further analysis of the G_0W_0 equations. The interested reader can find the details and much more information on the spectral properties of the electron gas in a series of papers by Lundqvist [258, 259]. Physically, the logarithmic divergence is removed by the fact that the screened interaction W does not have the divergent behavior of the bare Coulomb interaction v at low momenta. Away from the quasi-particle structure we clearly see plasmonic features at energies below the chemical potential for $p \leq p_F$ (the addition of a hole can excite plasmons or, equivalently, the hole can decay into plasmon excitations). Similar plasmonic features are present for energies above the chemical potential for momenta $p \geq p_F$. These features are, however, less pronounced. As explained before, this is due to the fact that plasmons are mainly excited at low momenta.

Momentum distribution From the behavior of the spectral function we can also deduce the structure of the momentum distribution $n_p = \int_{-\infty}^{\mu} \frac{d\omega}{2\pi} A(p, \omega)$. For $p \leq p_F$ we integrate over the quasi-hole peak below the chemical potential. When p approaches p_F from below, the quasi-hole peak develops into a δ-function, while for p immediately above p_F the quasi-particle peak appears at an energy above μ and does not contribute to the integral. We thus expect that the momentum distribution suddenly jumps at $p = p_F$. This is indeed

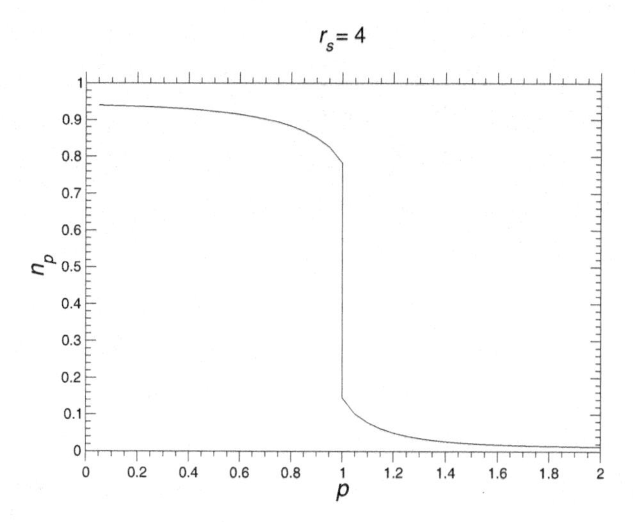

Figure 15.11 Momentum distribution n_p of an electron gas with $r_s = 4$ within the $\mathrm{G_0W_0}$ approximation. The momentum p is measured in units of p_F. The momentum distribution jumps with a value of $Z = 0.64$ at the Fermi surface.

shown rigorously in Section 11.6.1, where we prove that the discontinuity has exactly the size of the quasi-particle renormalization factor Z. The jump is directly related to the fact that the imaginary part of self-energy vanishes as $(\omega - \mu)^2$ close to the chemical potential. The formula for n_p in terms of the spectral function is awkward to use in practice since we need to integrate over a very spiky function which is not easily done numerically. We therefore use for the numerics the equivalent formula (11.49), where the self-energy for complex frequencies is calculated from the rate function according to

$$\Sigma^\mathrm{R}(p, \mu + \mathrm{i}\omega) = \Sigma_\mathrm{x}(p) + \int \frac{d\omega'}{2\pi} \frac{\Gamma(p, \omega')}{\mathrm{i}\omega + \mu - \omega' + \mathrm{i}\eta}.$$

The result is displayed in Fig. 15.11. We clearly see the sudden jump of magnitude Z at $p = p_\mathrm{F}$. Therefore the interacting electron gas preserves the sharp Fermi surface of the noninteracting system. The quasi-particle renormalization factor Z is smaller for lower densities, whereas for high densities it approaches unity [108]. The discontinuity of n_p and hence the sharpness of the Fermi surface is experimentally observable by means of Compton scattering on electron-gas-like metals such as sodium. For more details on this aspect we refer the reader to Ref. [260]. The sharpness of the Fermi surface applies to a large class of fermionic systems, not only metallic crystals but also other quantum systems such as liquid ^{3}He. In fact, theoretically it is used to define an important class of physical systems known as *Fermi liquid*, the low-energy behavior (excitations close to the Fermi surface) of which was first successfully described by Landau in a phenomenological way [263].

16

Green's Functions for Nonequilibrium Fermion–Boson Systems

So far we have considered systems of identical particles, either fermions or bosons. In this chapter we lay down the theory of multi-component systems, such as electrons and photons or electrons and phonons. We first revisit the quantization of light (photons) and nuclear vibrations (phonons), and then develop the MBPT for the interacting fermionic and bosonic nonequilibrium Green's functions. Special emphasis is given to systems of electrons and phonons for which we provide a pedagogical introduction to the ab initio NEGF theory [264].

16.1 Quantum Systems of Charged Particles and Photons

Let us begin by revisiting the classical theory of a system of N particles with mass m and charge q. In classical physics the particles are described by their positions $\mathbf{r}_i$ and velocities $\mathbf{v}_i = \dot{\mathbf{r}}_i$, $i = 1, \ldots, N$. The trajectory of the ith particle is determined by the Lorentz force, see Section 3.5,

$$m\dot{\mathbf{v}}_i(t) = q\mathbf{E}(\mathbf{r}_i, t) + \frac{q}{c}\mathbf{v}_i(t) \times \mathbf{B}(\mathbf{r}_i, t), \tag{16.1}$$

where the electric field $\mathbf{E}$ and magnetic field $\mathbf{B}$ solve the Maxwell equations:

$$\nabla \cdot \mathbf{E}(\mathbf{r}, t) = 4\pi q\, n(\mathbf{r}, t), \tag{16.2a}$$

$$\nabla \times \mathbf{E}(\mathbf{r}, t) = -\frac{1}{c}\frac{\partial \mathbf{B}(\mathbf{r}, t)}{\partial t}, \tag{16.2b}$$

$$\nabla \times \mathbf{B}(\mathbf{r}, t) = \frac{1}{c}\frac{\partial \mathbf{E}(\mathbf{r}, t)}{\partial t} + \frac{4\pi}{c} q\, \mathbf{J}(\mathbf{r}, t), \tag{16.2c}$$

$$\nabla \cdot \mathbf{B}(\mathbf{r}, t) = 0. \tag{16.2d}$$

In the Maxwell equations the density n and the current density $\mathbf{J}$ are given in terms of the positions and velocities of the particles:

$$n(\mathbf{r},t) = \sum_{i=1}^{N} \delta(\mathbf{r} - \mathbf{r}_i(t)), \tag{16.3a}$$

$$\mathbf{J}(\mathbf{r},t) = \sum_{i=1}^{N} \mathbf{v}_i(t)\delta(\mathbf{r} - \mathbf{r}_i(t)). \tag{16.3b}$$

The electromagnetic fields do therefore couple to the particles through the density and the current density. Equations (16.1), (16.2), and (16.3) are known as the *Maxwell–Lorentz equations*; they summarize the classical theory.[1] To lay down the quantum formulation of charged particles and electromagnetic fields, the route is: (1) determine the Lagrangian for which the Euler–Lagrange equations coincide with the classical equations; (2) calculate the conjugate momenta and impose the Dirac commutation relations between the Lagrangian coordinates and the conjugate momenta; and (3) construct the Hamiltonian to solve equilibrium and nonequilibrium problems.

We warn the reader about two common faults in the derivation of a quantized theory of charged particles and electromagnetic fields: (1) the use of the vector potential as a Lagrangian variable - this is not correct since the vector potential is constrained by a gauge condition (for instance, in the Coulomb gauge the vector potential is purely transverse); (2) the quantization of the electric field from the time derivative of the vector potential in *vacuum* - this is not correct since the presence of charged particles modify the time-dependence of the vector potential.

16.1.1 Lagrangian for Charged Particles and Electromagnetic Fields

The electric and magnetic fields can be obtained from the scalar potential ϕ and vector potential $\mathbf{A}$ in accordance with (3.2); in this way, two of the Maxwell equations, (16.2b) and (16.2d), are satisfied by construction.[2] However, the scalar and vector potentials are defined up to a gauge transformation. This means that the electric and magnetic fields calculated with ϕ and $\mathbf{A}$ are the same as those calculated with $\phi - \frac{1}{c}\frac{\partial \Lambda}{\partial t}$ and $\mathbf{A} + \boldsymbol{\nabla}\Lambda$ for any scalar function Λ, see (3.25). If we want to use ϕ and $\mathbf{A}$ as possible degrees of freedom, we must impose a gauge condition. A convenient condition to impose is the

[1]The Maxwell–Lorentz equations are affected by the self-interaction problem - that is, the particles feel the fields generated by themselves. This becomes most evident when considering the case of $N = 1$ particles. In this case the Lorentz equation (16.1) should simplify to $\dot{\mathbf{v}}_1 = 0$; however, it does not. The self-interaction problem is often ignored since we are either interested in solving the Maxwell equations for a given density and current or in solving the Lorentz equation for a given electromagnetic field. There have been several attempts to find a solution to the self-interaction problem, by Lorentz [265], Dirac [266], Wheeler and Feynman [267], and others [268, 269, 270]. Currently, however, a satisfactory theory from which to derive self-interaction free classical equations from a Lagrangian is missing. Here, we assume that $N \gg 1$ and that the particles' charges are infinitesimal, allowing us to neglect self-interactions. As we shall see, the quantization of the theory cures the self-interaction problem.

[2]This can easily be checked. We have $\boldsymbol{\nabla}\cdot\mathbf{B} = \boldsymbol{\nabla}\cdot(\boldsymbol{\nabla}\times\mathbf{A}) = 0$ since the divergence of a curl vanishes. We also have $\boldsymbol{\nabla}\times\mathbf{E} = \boldsymbol{\nabla}\times(-\boldsymbol{\nabla}\phi - \frac{1}{c}\partial\mathbf{A}/\partial t) = -\frac{1}{c}\frac{\partial}{\partial t}\boldsymbol{\nabla}\times\mathbf{A}$ since the curl of a gradient vanishes. Taking into account that $\mathbf{B} = \boldsymbol{\nabla}\times\mathbf{A}$, we recover (16.2b).

so-called *Coulomb gauge*,

$$\nabla \cdot \mathbf{A} = 0. \tag{16.4}$$

A possible choice of *Lagrangian coordinates* to formulate the quantum theory of matter and light are then the set of positions $\{\mathbf{r}_i(t)\}$, the scalar potential $\phi(\mathbf{r},t)$, and the vector potential $\mathbf{A}(\mathbf{r},t)$ *with* $\nabla \cdot \mathbf{A} = 0$. To shorten the notation we suppress the time argument during the calculations. The scalar and vector potentials can be seen as a *continuum* of Lagrangian coordinates as they are labeled by the continuum index $\mathbf{r}$.

We consider the Lagrangian [260]

$$\mathcal{L}[\{\mathbf{r}_i\},\{\mathbf{v}_i\},\phi,\dot{\phi},\mathbf{A},\dot{\mathbf{A}}] = \frac{m}{2}\sum_{i=1}^{N} v_i^2 + \int \frac{d\mathbf{r}'}{8\pi}\left[E^2(\mathbf{r}') - B^2(\mathbf{r}')\right]$$
$$- q\int d\mathbf{r}\Big[\underbrace{\sum_{i=1}^{N}\delta(\mathbf{r}-\mathbf{r}_i)}_{n(\mathbf{r})}\phi(\mathbf{r}) - \frac{1}{c}\underbrace{\sum_{i=1}^{N}\mathbf{v}_i\delta(\mathbf{r}-\mathbf{r}_i)}_{\mathbf{J}(\mathbf{r})}\cdot\mathbf{A}(\mathbf{r})\Big], \tag{16.5}$$

where $v_i^2 = \mathbf{v}_i \cdot \mathbf{v}_i$, $E^2 = \mathbf{E}\cdot\mathbf{E}$, and $B^2 = \mathbf{B}\cdot\mathbf{B}$. In (16.5) the electric and magnetic fields must be considered as functions of the independent variables ϕ, $\mathbf{A}$, and $\dot{\mathbf{A}} \equiv \partial\mathbf{A}/\partial t$ through (3.2), whereas the density n and current density $\mathbf{J}$ must be considered as functions of the independent variables $\{\mathbf{r}_i\}$ and $\{\mathbf{v}_i\}$.

According to Helmholtz theorem, any vector function $\mathbf{V}(\mathbf{r})$ (the electric and magnetic fields are two such functions) can uniquely be decomposed into a longitudinal part $\mathbf{V}^{\parallel}(\mathbf{r})$ which is curl-free – that is, $\nabla \times \mathbf{V}^{\parallel}(\mathbf{r}) = 0$ – and a transverse part $\mathbf{V}^{\perp}(\mathbf{r})$ which is divergenceless – that is, $\nabla \cdot \mathbf{V}^{\perp}(\mathbf{r}) = 0$. The magnetic field is purely transverse and so is the vector potential in the Coulomb gauge. In the same gauge the longitudinal and transverse part of the electric field are especially interesting. Writing $\mathbf{E} = \mathbf{E}^{\parallel} + \mathbf{E}^{\perp}$, we have

$$\mathbf{E}^{\parallel}(\mathbf{r},t) = -\nabla\phi(\mathbf{r},t), \qquad \mathbf{E}^{\perp}(\mathbf{r},t) = -\frac{1}{c}\frac{\partial\mathbf{A}(\mathbf{r},t)}{\partial t}. \tag{16.6}$$

As a consequence of this result, the cross product in $E^2 = E^{\parallel 2} + E^{\perp 2} + 2\mathbf{E}^{\parallel}\cdot\mathbf{E}^{\perp}$ does not contribute to the integral over space in (16.5). Indeed, $\mathbf{E}^{\parallel}\cdot\mathbf{E}^{\perp} \propto \nabla\phi\cdot\partial\mathbf{A}/\partial t$, which vanishes upon integration by parts (recall that in the Coulomb gauge $\nabla\cdot\mathbf{A} = 0$). Let us also inspect the coupling $\mathbf{J}\cdot\mathbf{A}$. Since $\mathbf{A}$ is purely transverse, there exists a vector function $\mathbf{V}$ such that $\mathbf{A} = \nabla \times \mathbf{V}$. The coupling between the current density and the vector potential can then be written as the integral over space of $\mathbf{J}\cdot(\nabla\times\mathbf{V})$ or, after an integration by parts, $-(\nabla\times\mathbf{J})\cdot\mathbf{V}$. Decomposing the current into longitudinal and transverse parts (i.e., $\mathbf{J} = \mathbf{J}^{\parallel} + \mathbf{J}^{\perp}$), we realize that only $\mathbf{J}^{\perp}$ couples to the vector potential (by definition $\nabla\times\mathbf{J}^{\parallel} = 0$).

To summarize, in the Coulomb gauge the Lagrangian (16.5) reads

$$\mathcal{L}[\{\mathbf{r}_i\},\{\mathbf{v}_i\},\phi,\dot{\phi},\mathbf{A},\dot{\mathbf{A}}] = \frac{m}{2}\sum_{i=1}^{N} v_i^2 - q\int d\mathbf{r}'\Big[n(\mathbf{r}')\phi(\mathbf{r}') - \frac{1}{c}\mathbf{J}^{\perp}(\mathbf{r}')\cdot\mathbf{A}(\mathbf{r}')\Big]$$
$$+ \int \frac{d\mathbf{r}'}{8\pi}\left[E^{\perp 2}(\mathbf{r}') + E^{\parallel 2}(\mathbf{r}') - B^2(\mathbf{r}')\right]. \tag{16.7}$$

The first Euler–Lagrange equation that we derive is the one for ϕ:

$$\frac{\delta \mathcal{L}}{\delta \phi(\mathbf{r})} = \frac{d}{dt}\frac{\delta \mathcal{L}}{\delta \dot{\phi}(\mathbf{r})}.$$

The r.h.s. vanishes since the Lagrangian does not depend on the time derivative of the scalar potential. For the l.h.s. we have

$$\frac{\delta \mathcal{L}}{\delta \phi(\mathbf{r})} = -qn(\mathbf{r}) + \int \frac{d\mathbf{r}'}{4\pi}\mathbf{E}^{\parallel}(\mathbf{r}')\cdot\underbrace{\frac{\delta \mathbf{E}^{\parallel}(\mathbf{r}')}{\delta\phi(\mathbf{r})}}_{-\boldsymbol{\nabla}'\delta(\mathbf{r}-\mathbf{r}')} = -qn(\mathbf{r},t) + \frac{\boldsymbol{\nabla}\cdot\mathbf{E}(\mathbf{r},t)}{4\pi},$$

where in the last equality we integrate by parts the second term and take into account that the divergence of $\mathbf{E}^{\perp}$ vanishes, hence $\boldsymbol{\nabla}\cdot\mathbf{E}^{\parallel} = \boldsymbol{\nabla}\cdot\mathbf{E}$. We also restore the time argument in the Lagrangian coordinates; the reader can easily recognize in the first Euler–Lagrange equation the Maxwell equation (16.2a).

Next we derive the Euler–Lagrange equation for $\mathbf{A}$:

$$\frac{\delta \mathcal{L}}{\delta A_\alpha(\mathbf{r})} = \frac{d}{dt}\frac{\delta \mathcal{L}}{\delta \dot{A}_\alpha(\mathbf{r})}, \qquad \alpha = x, y, z. \tag{16.8}$$

The r.h.s. is nonvanishing since $\mathbf{E}^{\perp}$ depends on $\dot{\mathbf{A}}$. We have

$$\frac{\delta \mathcal{L}}{\delta \dot{A}_\alpha(\mathbf{r})} = \int \frac{d\mathbf{r}'}{4\pi}\sum_\beta E^{\perp}_\beta(\mathbf{r}')\frac{\delta E^{\perp}_\beta(\mathbf{r}')}{\delta \dot{A}_\alpha(\mathbf{r})} = -\frac{1}{4\pi c}E^{\perp}_\alpha(\mathbf{r},t), \tag{16.9}$$

where we restore the time argument in the last equality. For the l.h.s. we have

$$\frac{\delta \mathcal{L}}{\delta A_\alpha(\mathbf{r})} = \frac{q}{c}J^{\perp}_\alpha(\mathbf{r}) - \int \frac{d\mathbf{r}'}{4\pi}\sum_\beta B_\beta(\mathbf{r}')\underbrace{\frac{\delta B_\beta(\mathbf{r}')}{\delta A_\alpha(\mathbf{r})}}_{\sum_\mu \varepsilon_{\beta\mu\alpha}\frac{\partial}{\partial r'_\mu}\delta(\mathbf{r}-\mathbf{r}')} = \frac{q}{c}J^{\perp}_\alpha(\mathbf{r},t) - \frac{1}{4\pi}\underbrace{\varepsilon_{\alpha\mu\beta}\frac{\partial}{\partial r_\mu}B_\beta(\mathbf{r},t)}_{[\boldsymbol{\nabla}\times\mathbf{B}(\mathbf{r},t)]_\alpha},$$

with $\varepsilon_{\alpha\mu\beta}$ the Levi-Civita tensor and $(r_x, r_y, r_z) \equiv (x, y, z)$. In the last equality we integrate by parts the second term and again restore the time arguments. Inserting these results into (16.8) we find

$$\boldsymbol{\nabla}\times\mathbf{B}(\mathbf{r},t) = \frac{1}{c}\frac{\partial \mathbf{E}^{\perp}(\mathbf{r},t)}{\partial t} + \frac{4\pi}{c}q\,\mathbf{J}^{\perp}(\mathbf{r},t).$$

This equation agrees with the Maxwell equation (16.2c) only provided that the longitudinal part of the r.h.s. of (16.2c) vanishes. This is indeed the case. The time derivative of the (already derived) Maxwell equation (16.2a) yields $4\pi q\frac{\partial n}{\partial t} = \boldsymbol{\nabla}\cdot\frac{\partial \mathbf{E}^{\parallel}}{\partial t}$, and from the continuity equation $\frac{\partial n}{\partial t} = -\boldsymbol{\nabla}\cdot\mathbf{J} = -\boldsymbol{\nabla}\cdot\mathbf{J}^{\parallel}$ we conclude that $\boldsymbol{\nabla}\cdot(\frac{\partial \mathbf{E}^{\parallel}}{\partial t} + 4\pi q\mathbf{J}^{\parallel}) = 0$. Since the electric field and the current density vanish at infinity, the parenthesis is zero.

The last Euler–Lagrange equation to consider is

$$\frac{\partial \mathcal{L}}{\partial r_{i,\alpha}} = \frac{d}{dt}\frac{\partial \mathcal{L}}{\partial v_{i,\alpha}}, \qquad \alpha = x, y, z. \tag{16.10}$$

To visualize the dependence on time of the various quantities we write the time argument explicitly. The r.h.s. gives

$$\frac{d}{dt}\frac{\partial \mathcal{L}}{\partial v_{i,\alpha}} = \frac{d}{dt}\Big[m v_{i,\alpha}(t) + \frac{q}{c}\int d\mathbf{r}\ \delta(\mathbf{r}-\mathbf{r}_i(t))A_\alpha(\mathbf{r},t)\Big]$$
$$= m\frac{dv_{i,\alpha}(t)}{dt} + \frac{q}{c}\sum_\mu \frac{\partial A_\alpha(\mathbf{r}_i,t)}{\partial r_{i,\mu}} v_{i,\mu}(t) + \frac{q}{c}\frac{\partial A_\alpha(\mathbf{r}_i,t)}{\partial t}, \tag{16.11}$$

where the time derivative of the delta function has been calculated using the chain rule: $\frac{d}{dt}\delta(\mathbf{r}-\mathbf{r}_i) = \sum_\mu \frac{\partial}{\partial r_{i,\mu}}\delta(\mathbf{r}-\mathbf{r}_i)v_{i,\mu}$. The l.h.s. can be calculated similarly and yields

$$\frac{\partial \mathcal{L}}{\partial r_{i,\alpha}} = -q\frac{\partial \phi(\mathbf{r}_i,t)}{\partial r_{i,\alpha}} + \frac{q}{c}\sum_\mu v_{i,\mu}(t)\frac{\partial A_\mu(\mathbf{r}_i,t)}{\partial r_{i,\alpha}}.$$

Inserting these results into (16.10) we get

$$m\frac{dv_{i,\alpha}(t)}{dt} = q\underbrace{\Big(-\frac{\partial \phi(\mathbf{r}_i,t)}{\partial r_{i,\alpha}} - \frac{1}{c}\frac{\partial A_\alpha(\mathbf{r}_i,t)}{\partial t}\Big)}_{E_\alpha(\mathbf{r}_i,t)} + \underbrace{\frac{q}{c}\sum_\mu v_{i,\mu}(t)\Big(\frac{\partial A_\mu(\mathbf{r}_i,t)}{\partial r_{i,\alpha}} - \frac{\partial A_\alpha(\mathbf{r}_i,t)}{\partial r_{i,\mu}}\Big)}_{[\mathbf{v}_i(t)\times(\boldsymbol{\nabla}\times\mathbf{A}(\mathbf{r}_i,t))]_\alpha},$$

which coincides with the equation of motion for the ith particle, see (16.1). The classical theory is therefore recovered by the Lagrangian (16.5) (or (16.7) if we work in the Coulomb gauge).

16.1.2 Quantization of the Electromagnetic Fields

The quantization of the classical theory is accomplished by transforming the Lagrangian coordinates and the conjugate momenta into self-adjoint operators satisfying the Dirac postulate. The conjugate momentum to ϕ is $P_\phi(\mathbf{r}) \equiv \delta\mathcal{L}/\delta\dot{\phi}(\mathbf{r}) = 0$, which implies that the scalar potential is *de facto* a classical field. The conjugate momentum to $\mathbf{r}_i$ is $\mathbf{p}_i \equiv \partial\mathcal{L}/\partial\mathbf{v}_i = m\mathbf{v}_i + (q/c)\mathbf{A}(\mathbf{r}_i)$, see first line in (16.11), and according to the Dirac postulate we have the well-known commutation relations $[\hat{r}_{i,\alpha}, \hat{p}_{j,\beta}]_- = \mathrm{i}\delta_{ij}\delta_{\alpha\beta}$.[3] The quantization of the vector potential is more subtle since $\mathbf{A}(\mathbf{r})$ is constrained to be purely transverse (Coulomb gauge), and this feature has to be transferred to the self-adjoint operator $\hat{\mathbf{A}}(\mathbf{r})$. Let us expand the vector potential in terms of sine and cosine functions:

$$\mathbf{A}(\mathbf{r}) = {\sum_{\mathbf{k}\lambda}}' \sqrt{\frac{8\pi c^2}{\mathrm{V}\omega_k}}\,\boldsymbol{\eta}_{\mathbf{k}\lambda}\big[\cos(\mathbf{k}\cdot\mathbf{r})u_{\mathbf{k}\lambda} - \sin(\mathbf{k}\cdot\mathbf{r})v_{\mathbf{k}\lambda}\big], \tag{16.12}$$

where $\lambda = 1, 2$ labels two orthonormal vectors $\boldsymbol{\eta}_{\mathbf{k}\lambda}$ which are in turn orthogonal to $\mathbf{k}$. We refer to the vectors $\boldsymbol{\eta}_{\mathbf{k}\lambda}$ as the *polarization vectors* since they specify the direction of the electromagnetic fields. The square root in (16.12) is a convenient prefactor; it contains

[3] We remind the reader that the operator $\hat{\mathbf{r}}_i$ ($\hat{\mathbf{p}}_i$) is defined as the position (momentum) operator acting on the ith particle and doing nothing to the other particles, see discussion below (1.16).

the volume V of the system and the light frequency $\omega_k = ck$ for waves of momentum $k = |\mathbf{k}|$. The amplitudes $u_{\mathbf{k}\lambda}$ and $v_{\mathbf{k}\lambda}$ are real [this guarantees the reality of $\mathbf{A}(\mathbf{r})$] and, more important, they are not constrained. It is easy to verify that $\nabla \cdot \mathbf{A} = 0$ for any choice of $u_{\mathbf{k}\lambda}$ and $v_{\mathbf{k}\lambda}$. The sum over $\mathbf{k}$ is primed since it runs over half of the $\mathbf{k}$-space (the cosine and sine functions with $\mathbf{k}$ and $-\mathbf{k}$ are not independent). Henceforth we use $u_{\mathbf{k}\lambda}$ and $v_{\mathbf{k}\lambda}$ instead of $\mathbf{A}(\mathbf{r})$ as Lagrangian coordinates.

Equation (16.6) implies that the transverse electric field can be expanded as

$$\mathbf{E}^{\perp}(\mathbf{r}) = -\sum_{\mathbf{k}\lambda}{}' \sqrt{\frac{8\pi}{V\omega_k}} \eta_{\mathbf{k}\lambda} \big[\cos(\mathbf{k}\cdot\mathbf{r})\dot{u}_{\mathbf{k}\lambda} - \sin(\mathbf{k}\cdot\mathbf{r})\dot{v}_{\mathbf{k}\lambda}\big]. \tag{16.13}$$

Accordingly, the only term in the Lagrangian that depends on the time derivative of the new Lagrangian coordinates can be written as[4]

$$\int \frac{d\mathbf{r}'}{8\pi} E^{\perp 2}(\mathbf{r}') = \sum_{\mathbf{k}\lambda}{}' \frac{1}{2\omega_k}(\dot{u}^2_{\mathbf{k}\lambda} + \dot{v}^2_{\mathbf{k}\lambda}).$$

The conjugate momentum to $u_{\mathbf{k}\lambda}$ is then $p_{u,\mathbf{k}\lambda} = \partial\mathcal{L}/\partial\dot{u}_{\mathbf{k}\lambda} = \dot{u}_{\mathbf{k}\lambda}/\omega_k$ and the conjugate momentum to $v_{\mathbf{k}\lambda}$ is $p_{v,\mathbf{k}\lambda} = \partial\mathcal{L}/\partial\dot{v}_{\mathbf{k}\lambda} = \dot{v}_{\mathbf{k}\lambda}/\omega_k$. We are now ready to quantize the electromagnetic field.

The Lagrangian coordinates and the conjugate momenta become self-adjoint operators satisfying the Dirac postulate,

$$[\hat{u}_{\mathbf{k}\lambda}, \hat{p}_{u,\mathbf{k}'\lambda'}]_- = \mathrm{i}\delta_{\mathbf{k},\mathbf{k}'}\delta_{\lambda\lambda'}, \qquad [\hat{v}_{\mathbf{k}\lambda}, \hat{p}_{v,\mathbf{k}'\lambda'}]_- = \mathrm{i}\delta_{\mathbf{k},\mathbf{k}'}\delta_{\lambda\lambda'}, \tag{16.14}$$

with all other commutators being zero. Let us introduce the nonself-adjoint operators

$$\hat{a}_{\mathbf{k}\lambda} \equiv \frac{1}{2}(\hat{u}_{\mathbf{k}\lambda} + \mathrm{i}\hat{v}_{\mathbf{k}\lambda} - \hat{p}_{v,\mathbf{k}\lambda} + \mathrm{i}\hat{p}_{u,\mathbf{k}\lambda}), \tag{16.15a}$$

$$\hat{a}_{-\mathbf{k}\lambda} \equiv \frac{1}{2}(\hat{u}_{\mathbf{k}\lambda} - \mathrm{i}\hat{v}_{\mathbf{k}\lambda} + \hat{p}_{v,\mathbf{k}\lambda} + \mathrm{i}\hat{p}_{u,\mathbf{k}\lambda}). \tag{16.15b}$$

The operator $\hat{a}_{\mathbf{k}\lambda}$ is in this way defined for all $\mathbf{k}$: If $\mathbf{k}$ belongs to the half-space of the primed sum (16.12), then $\hat{a}_{\mathbf{k}\lambda}$ is given by the first identity, otherwise it is given by the second identity. The reader can use the commutation relations (16.14) to verify that the bosonic commutation relations

$$[\hat{a}_{\mathbf{k}\lambda}, \hat{a}_{\mathbf{k}'\lambda'}] = 0, \qquad [\hat{a}_{\mathbf{k}\lambda}, \hat{a}^{\dagger}_{\mathbf{k}'\lambda'}] = \delta_{\lambda\lambda'}\delta_{\mathbf{k},\mathbf{k}'}$$

are satisfied for all $\mathbf{k}$ and $\mathbf{k}'$. From definition (16.15) we find $\hat{u}_{\mathbf{k}\lambda} + \mathrm{i}\hat{v}_{\mathbf{k}\lambda} = \hat{a}_{\mathbf{k}\lambda} + \hat{a}^{\dagger}_{-\mathbf{k}\lambda}$ and $\hat{u}_{\mathbf{k}\lambda} - \mathrm{i}\hat{v}_{\mathbf{k}\lambda} = \hat{a}_{-\mathbf{k}\lambda} + \hat{a}^{\dagger}_{\mathbf{k}\lambda}$. Using these identities in the expansion (16.12) along with the

[4]We use the orthonormality relations $\int d\mathbf{r}\cos(\mathbf{k}\cdot\mathbf{r})\cos(\mathbf{k}'\cdot\mathbf{r}) = \int d\mathbf{r}\sin(\mathbf{k}\cdot\mathbf{r})\sin(\mathbf{k}'\cdot\mathbf{r}) = \frac{1}{2}(2\pi^3)\delta(\mathbf{k}-\mathbf{k}') = \frac{1}{2}V\delta_{\mathbf{k},\mathbf{k}'}$ (the infinitesimal volume in $\mathbf{k}$-space is $d\mathbf{k} = (2\pi)^3/V$ and $\delta_{\mathbf{k},\mathbf{k}'}/d\mathbf{k} \to \delta(\mathbf{k}-\mathbf{k}')$ for $V \to \infty$) and $\int d\mathbf{r}\cos(\mathbf{k}\cdot\mathbf{r})\sin(\mathbf{k}'\cdot\mathbf{r}) = 0$.

definition[5] $\boldsymbol{\eta}_{-\mathbf{k}\lambda} \equiv \boldsymbol{\eta}_{\mathbf{k}\lambda}$, the vector potential operator becomes

$$\hat{\mathbf{A}}(\mathbf{r}) = \sum_{\mathbf{k}\lambda}{}' \sqrt{\frac{8\pi c^2}{\mathrm{V}\omega_k}} \boldsymbol{\eta}_{\mathbf{k}\lambda} \frac{1}{2} \left[e^{\mathrm{i}\mathbf{k}\cdot\mathbf{r}} (\hat{a}_{\mathbf{k}\lambda} + \hat{a}^\dagger_{-\mathbf{k}\lambda}) + e^{-\mathrm{i}\mathbf{k}\cdot\mathbf{r}} (\hat{a}_{-\mathbf{k}\lambda} + \hat{a}^\dagger_{\mathbf{k}\lambda}) \right]$$

$$= \sum_{\mathbf{k}\lambda} \sqrt{\frac{2\pi c^2}{\mathrm{V}\omega_k}} \boldsymbol{\eta}_{\mathbf{k}\lambda} e^{\mathrm{i}\mathbf{k}\cdot\mathbf{r}} (\hat{a}_{\mathbf{k}\lambda} + \hat{a}^\dagger_{-\mathbf{k}\lambda}), \tag{16.16}$$

where in the last equality the sum runs over the whole $\mathbf{k}$-space. Similarly, the expansion (16.13) of the transverse electric field operator becomes[6]

$$\hat{\mathbf{E}}^\perp(\mathbf{r}) = \mathrm{i} \sum_{\mathbf{k}\lambda} \sqrt{\frac{2\pi\omega_k}{\mathrm{V}}} \boldsymbol{\eta}_{\mathbf{k}\lambda} e^{\mathrm{i}\mathbf{k}\cdot\mathbf{r}} (\hat{a}_{\mathbf{k}\lambda} - \hat{a}^\dagger_{-\mathbf{k}\lambda}). \tag{16.17}$$

The quantization of particles and electromagnetic fields is in this way completed.[7]

The operators $\{\hat{\boldsymbol{r}}_i\}$ and $\{\hat{\boldsymbol{p}}_i\}$, $i = 1, \ldots, N$, act on the Hilbert space $\mathbb{H}_N$ of N identical particles, whereas the bosonic operators $\hat{a}_{\mathbf{k}\lambda}$ and $\hat{a}^\dagger_{\mathbf{k}\lambda}$ act on a bosonic Fock space $\mathbb{F}_b$. The bosonic particles in $\mathbb{F}_b$ are called *photons*, thus we can interpret $\hat{a}_{\mathbf{k}\lambda}$ ($\hat{a}^\dagger_{\mathbf{k}\lambda}$) as the operator that annihilates (creates) a photon with wavevector $\mathbf{k}$ and polarization $\boldsymbol{\eta}_{\mathbf{k}\lambda}$. A basis for the direct-product space $\mathbb{H}_N \otimes \mathbb{F}_b$ is, for example, the direct product of the position-spin kets in $\mathbb{H}_N$ and the momentum-polarization kets in $\mathbb{F}_b$:

$$|\mathbf{x}_1 \ldots \mathbf{x}_N; \mathbf{k}_1\lambda_1 \ldots \mathbf{k}_M\lambda_M\rangle \equiv |\mathbf{x}_1 \ldots \mathbf{x}_N\rangle \otimes \hat{a}^\dagger_{\mathbf{k}_1\lambda_1} \cdots \hat{a}^\dagger_{\mathbf{k}_M\lambda_M} |0\rangle, \tag{16.18}$$

where $M = 0, 1, \ldots, \infty$.

It is noteworthy to remark that the operator $\hat{\mathbf{A}}(\mathbf{r})$ is not the same as the operator $\mathbf{A}(\hat{\boldsymbol{r}})$ appearing in, for example, (3.1). In the *full quantum treatment* of particles and electromagnetic fields, $\hat{\mathbf{A}}(\mathbf{r})$ and $\hat{\mathbf{E}}^\perp(\mathbf{r})$ act on their own space, which is $\mathbb{F}_b$. In contrast, $\mathbf{A}(\hat{\boldsymbol{r}})$ acts on the one-particle Hilbert space $\mathbb{H}_1$, like the position operator $\hat{\boldsymbol{r}}$. The theory developed in the previous chapters treats the particles quantum mechanically and the electromagnetic fields classically. The full quantum treatment reduces to the one of the previous chapters when the electromagnetic fields are treated in the so-called Ehrenfest approximation, see Section 16.10.

[5] The $\boldsymbol{\eta}_{\mathbf{k}\lambda}$ are orthogonal to $\mathbf{k}$ and therefore they are also orthogonal to $-\mathbf{k}$.

[6] To obtain (16.17) we express $\dot{u}_{\mathbf{k}\lambda}$ and $\dot{v}_{\mathbf{k}\lambda}$ in terms of the conjugate momenta $p_{u,\mathbf{k}\lambda}$ and $p_{v,\mathbf{k}\lambda}$, make them become operators and then express these operators in terms of the bosonic operators $\hat{a}_{\mathbf{k}\lambda}$ and $\hat{a}^\dagger_{\mathbf{k}\lambda}$ using (16.15).

[7] The expansions (16.16) and (16.17) help us to appreciate the aforementioned complication in quantizing the theory using $\mathbf{A}(\mathbf{r})$ as the Lagrangian coordinate. The conjugate momentum to $\mathbf{A}(\mathbf{r})$ is $\mathbf{P}_{\mathrm{A}}(\mathbf{r}) \equiv \delta\mathcal{L}/\delta\dot{\mathbf{A}}(\mathbf{r}) = -\mathbf{E}^\perp(\mathbf{r})/(4\pi c)$, see (16.9), and the commutator between the two gives

$$[\hat{\mathbf{A}}(\mathbf{r}), -\frac{\hat{\mathbf{E}}^\perp(\mathbf{r}')}{4\pi c}]_- = \mathrm{i} \sum_{\mathbf{k}} \sum_{\lambda\lambda'} \frac{\boldsymbol{\eta}_{\mathbf{k}\lambda} \cdot \boldsymbol{\eta}_{\mathbf{k}\lambda'}}{\mathrm{V}} e^{\mathrm{i}\mathbf{k}\cdot(\mathbf{r}-\mathbf{r}')} = \mathrm{i}\delta^\perp(\mathbf{r} - \mathbf{r}'),$$

where $\delta^\perp(\mathbf{r} - \mathbf{r}')$ is the transverse delta function [199] guaranteeing that the divergence of the commutator with respect to $\mathbf{r}$ and $\mathbf{r}'$ vanishes.

16.1.3 Interacting Hamiltonian for Charged Particles and Photons

We are left with the derivation of the Hamiltonian $\mathcal{H}$. This is defined as the Legendre transform of the Lagrangian:

$$\mathcal{H} = \sum_{i=1}^{N} \mathbf{v}_i \cdot \mathbf{p}_i + \int d\mathbf{r}\, \dot{\mathbf{A}}(\mathbf{r}) \cdot \mathbf{P}_{\mathrm{A}}(\mathbf{r}) - \mathcal{L},$$

where $\mathbf{P}_{\mathrm{A}}(\mathbf{r}) \equiv \delta\mathcal{L}/\delta\dot{\mathbf{A}}(\mathbf{r}) = -\mathbf{E}^{\perp}(\mathbf{r})/(4\pi c) = \dot{\mathbf{A}}(\mathbf{r})/(4\pi c^2)$, see (16.9) and (16.6), is the conjugate momentum of the vector potential. The term proportional to $\dot{\phi}$ vanishes since $P_\phi = 0$. Expressing all velocities in terms of the conjugate momenta, we find

$$\begin{aligned}\mathcal{H} = &\sum_{i=1}^{N} \frac{1}{m}\big(\mathbf{p}_i - \tfrac{q}{c}\mathbf{A}(\mathbf{r}_i)\big) \cdot \mathbf{p}_i + \frac{1}{4\pi}\int d\mathbf{r}\, E^{\perp 2}(\mathbf{r}) - \frac{1}{2m}\sum_{i=1}^{N}\big(\mathbf{p}_i - \tfrac{q}{c}\mathbf{A}(\mathbf{r}_i)\big)^2 \\ &+ q\sum_{i=1}^{N}\Big[\phi(\mathbf{r}_i) - \frac{1}{mc}\big(\mathbf{p}_i - \tfrac{q}{c}\mathbf{A}(\mathbf{r}_i)\big)\cdot \mathbf{A}(\mathbf{r}_i)\Big] - \int \frac{d\mathbf{r}}{8\pi}\big[E^{\perp 2}(\mathbf{r}) + E^{\parallel 2}(\mathbf{r}) - B^2(\mathbf{r})\big],\end{aligned}$$

which correctly depends on the Lagrangian coordinate $\{\mathbf{r}_i\}$, ϕ, and $\mathbf{A}$, and the conjugate momenta $\{\mathbf{p}_i\}$ and $\mathbf{E}^{\perp}$ (the dependence on the velocities has been eliminated). The square of the longitudinal electric field is simply $(\boldsymbol{\nabla}\phi)\cdot(\boldsymbol{\nabla}\phi)$, see (16.6). Integrating by parts and using the Maxwell equation (16.2a) (i.e., $\boldsymbol{\nabla}\cdot\mathbf{E} = \boldsymbol{\nabla}\cdot\mathbf{E}^{\parallel} = -\nabla^2\phi = 4\pi q n$) we have

$$\int \frac{d\mathbf{r}}{8\pi} E^{\parallel 2}(\mathbf{r}) = \int \frac{d\mathbf{r}}{8\pi}\phi(\mathbf{r})(4\pi q)n(\mathbf{r}) = \frac{q}{2}\sum_{i=1}^{N}\phi(\mathbf{r}_i).$$

The solution of the *Poisson equation* $\nabla^2\phi = -4\pi q n$ for the density (16.3a) is $\phi(\mathbf{r}) = \sum_{j=1}^{N} q/|\mathbf{r} - \mathbf{r}_j|$. Collecting all terms, the Hamiltonian simplifies to

$$\mathcal{H}(N) = \frac{1}{2m}\sum_{i=1}^{N}\big(\mathbf{p}_i - \tfrac{q}{c}\mathbf{A}(\mathbf{r}_i)\big)^2 + \frac{1}{2}\sum_{ij=1}^{N}\frac{q^2}{|\mathbf{r}_i - \mathbf{r}_j|} + \int \frac{d\mathbf{r}}{8\pi}\big[E^{\perp 2}(\mathbf{r}) + B^2(\mathbf{r})\big],$$

where we emphasize that this form refers to a system of N particles by writing $\mathcal{H} = \mathcal{H}(N)$.

We are ready to transform the Hamiltonian into an operator acting on $\mathbb{H}_N \otimes \mathbb{F}_b$. Using the expansion (16.17) we get

$$\begin{aligned}\int \frac{d\mathbf{r}}{8\pi}\hat{E}^{\perp 2}(\mathbf{r}) &= -\frac{1}{4}\sum_{\mathbf{k}\lambda}\omega_k(\hat{a}_{\mathbf{k}\lambda} - \hat{a}^{\dagger}_{-\mathbf{k}\lambda})(\hat{a}_{-\mathbf{k}\lambda} - \hat{a}^{\dagger}_{\mathbf{k}\lambda}) \\ &= \frac{1}{4}\sum_{\mathbf{k}\lambda}\omega_k\big(2\hat{a}^{\dagger}_{\mathbf{k}\lambda}\hat{a}_{\mathbf{k}\lambda} + 1 - \hat{a}_{\mathbf{k}\lambda}\hat{a}_{-\mathbf{k}\lambda} - \hat{a}^{\dagger}_{-\mathbf{k}\lambda}\hat{a}^{\dagger}_{\mathbf{k}\lambda}\big),\end{aligned}$$

where we take into account that the frequency $\omega_k = ck$ depends only on the modulus of the vector $\mathbf{k}$. Using the expansion (16.16) we get

$$\int \frac{d\mathbf{r}}{8\pi}\hat{B}^2(\mathbf{r}) = \int \frac{d\mathbf{r}}{8\pi}|\nabla \times \hat{\mathbf{A}}(\mathbf{r})|^2 = \frac{1}{4}\sum_{\mathbf{k}\lambda} \frac{c^2|\mathbf{k}\times \boldsymbol{\eta}_{\mathbf{k}\lambda}|^2}{\omega_k}(\hat{a}_{\mathbf{k}\lambda} + \hat{a}^\dagger_{-\mathbf{k}\lambda})(\hat{a}_{-\mathbf{k}\lambda} + \hat{a}^\dagger_{\mathbf{k}\lambda})$$
$$= \frac{1}{4}\sum_{\mathbf{k}\lambda}\omega_k\big(2\hat{a}^\dagger_{\mathbf{k}\lambda}\hat{a}_{\mathbf{k}\lambda} + 1 + \hat{a}_{\mathbf{k}\lambda}\hat{a}_{-\mathbf{k}\lambda} + \hat{a}^\dagger_{-\mathbf{k}\lambda}\hat{a}^\dagger_{\mathbf{k}\lambda}\big),$$

where we take into account that the $\boldsymbol{\eta}_{\mathbf{k}\lambda}$ are unit vectors orthogonal to $\mathbf{k}$ and therefore $|\mathbf{k}\times\boldsymbol{\eta}_{\mathbf{k}\lambda}| = k$. The Hamiltonian operator $\hat{\mathcal{H}}(N)$ for N particles and an arbitrary number of photons is then

$$\hat{\mathcal{H}}(N) = \frac{1}{2m}\sum_{i=1}^{N}\big(\hat{\boldsymbol{p}}_i - \frac{q}{c}\hat{\mathbf{A}}(\hat{\boldsymbol{r}}_i)\big)^2 + \frac{1}{2}\sum_{i\neq j}^{N} v(\hat{\boldsymbol{r}}_i, \hat{\boldsymbol{r}}_j) + \sum_{\mathbf{k}\lambda}\omega_k(\hat{a}^\dagger_{\mathbf{k}\lambda}\hat{a}_{\mathbf{k}\lambda} + \frac{1}{2}), \tag{16.19}$$

where $v(\mathbf{r},\mathbf{r}') \equiv q^2/|\mathbf{r}-\mathbf{r}'|$. In (16.19) we also remove the constant (although infinite) shift arising from the contributions with $i = j$ in the second sum; this is how the quantum theory cures the self-interaction problem. The coupling between particles and photons is provided by the vector potential, which depends on both photonic operators and position operators, see again (16.16).

The Hamiltonian (16.19) is written in terms of operators $\{\hat{\boldsymbol{r}}_i\}$ and $\{\hat{\boldsymbol{p}}_i\}$ in first quantization. As pointed out in Section 1.3 this representation is problematic when dealing with identical particles. It is therefore useful to find the second-quantization form of the Hamiltonian that acts on the Fock space $\mathbb{F}\otimes\mathbb{F}_b$ of particles and photons. This is readily achieved by observing that for any ket $|\Psi\rangle \in \mathbb{H}_N$

$$\langle \mathbf{x}_1 \ldots \mathbf{x}_N|\sum_{i=1}^{N}\big(\hat{\boldsymbol{p}}_i - \frac{q}{c}\hat{\mathbf{A}}(\hat{\boldsymbol{r}}_i)\big)^2|\Psi\rangle = \sum_{i=1}^{N}(-\mathrm{i}\nabla_i - \frac{q}{c}\hat{\mathbf{A}}(\mathbf{r}_i))^2\langle \mathbf{x}_1\ldots\mathbf{x}_N|\Psi\rangle$$
$$= \langle \mathbf{x}_1\ldots\mathbf{x}_N|\int d\mathbf{x}\,\hat{\psi}^\dagger(\mathbf{x})\big(-\mathrm{i}\nabla - \frac{q}{c}\hat{\mathbf{A}}(\mathbf{r})\big)^2\hat{\psi}(\mathbf{x})|\Psi\rangle,$$

where in the last equality we use (1.80). Notice that in the second line $\hat{\mathbf{A}}(\mathbf{r})$ is still an operator depending on the photonic fields. As this identity is valid for any $\mathbf{x}_1,\ldots,\mathbf{x}_N$ and for any N, we conclude that the *particle-photon Hamiltonian* in second quantization is

$$\boxed{\begin{aligned}\hat{H} &= \frac{1}{2m}\int d\mathbf{x}\,\hat{\psi}^\dagger(\mathbf{x})\big(-\mathrm{i}\nabla - \frac{q}{c}\hat{\mathbf{A}}(\mathbf{r})\big)^2\hat{\psi}(\mathbf{x}) + \sum_{\mathbf{k}\lambda}\omega_k(\hat{a}^\dagger_{\mathbf{k}\lambda}\hat{a}_{\mathbf{k}\lambda} + \frac{1}{2})\\ &+ \frac{1}{2}\int d\mathbf{x}d\mathbf{x}'\hat{\psi}^\dagger(\mathbf{x})\hat{\psi}^\dagger(\mathbf{x}')v(\mathbf{r},\mathbf{r}')\hat{\psi}(\mathbf{x}')\hat{\psi}(\mathbf{x})\end{aligned}} \tag{16.20}$$

where we use (1.82) for the particle-particle interaction Hamiltonian.

It is instructive to separate the Hamiltonian as

$$\boxed{\hat{H} = \hat{H}_{\text{part}} + \hat{H}_{\text{phot}} + \hat{H}_{\text{part-phot}}} \tag{16.21}$$

where $\hat{H}_{\text{part}}$ is the Hamiltonian (1.83) for identical particles interacting through $v(\mathbf{r}, \mathbf{r}')$,

$$\hat{H}_{\text{phot}} = \sum_{\mathbf{k}\lambda} \omega_k (\hat{a}^\dagger_{\mathbf{k}\lambda} \hat{a}_{\mathbf{k}\lambda} + \frac{1}{2}) \tag{16.22}$$

is the free-photon Hamiltonian, and $\hat{H}_{\text{part-phot}}$ is the particle-photon interaction,[8]

$$\hat{H}_{\text{part-phot}} = -\frac{q}{c} \int d\mathbf{x}\, \hat{\mathbf{J}}(\mathbf{x}) \cdot \hat{\mathbf{A}}(\mathbf{r}), \tag{16.23}$$

where $\hat{\mathbf{J}} = \hat{\mathbf{j}} + \hat{\mathbf{j}}_d$ is the sum of the paramagnetic current density operator (3.27) and the diamagnetic current density operator (3.28). Notice that the vector potential in $\hat{\mathbf{j}}_d$ is not a vector function but the operator $\hat{\mathbf{A}}$ acting on $\mathbb{F}_b$. The particle-photon interaction (16.23) correctly agrees with the coupling between particles and classical electromagnetic fields derived in (3.30). However, the origin of the electromagnetic fields is different. In (3.30) ϕ and $\mathbf{A}$ are *external* potentials to be treated as inputs; they are generated by *external* charged particles. In (16.23) there are no external particles and hence no external potentials; the vector potential is generated by the particles of the system.

Expanding the vector potential according to (16.16) the Hamiltonian for the particle-photon interaction reads

$$\begin{aligned}\hat{H}_{\text{part-phot}} &= -\frac{q}{c} \sum_{\mathbf{k}\lambda} \sqrt{\frac{2\pi c^2}{\mathrm{V}\omega_k}} (\hat{a}_{\mathbf{k}\lambda} + \hat{a}^\dagger_{-\mathbf{k}\lambda}) \int d\mathbf{x}\, \boldsymbol{\eta}_{\mathbf{k}\lambda} \cdot \hat{\mathbf{j}}(\mathbf{x})\, e^{i\mathbf{k}\cdot\mathbf{r}} \\ &+ \frac{q^2}{m} \frac{2\pi}{\mathrm{V}} \sum_{\mathbf{k}\mathbf{k}'\lambda\lambda'} \frac{\boldsymbol{\eta}_{\mathbf{k}\lambda} \cdot \boldsymbol{\eta}_{\mathbf{k}'\lambda'}}{\sqrt{\omega_k \omega_{k'}}} (\hat{a}_{\mathbf{k}\lambda} + \hat{a}^\dagger_{-\mathbf{k}\lambda})(\hat{a}_{\mathbf{k}'\lambda'} + \hat{a}^\dagger_{-\mathbf{k}'\lambda'}) \int d\mathbf{x}\, \hat{n}(\mathbf{x})\, e^{i(\mathbf{k}-\mathbf{k}')\cdot\mathbf{r}}.\end{aligned} \tag{16.24}$$

The important message to take home from (16.24) is that $\hat{H}_{\text{part-phot}}$ does not commute with the total number of photon operator $\hat{N}_{\text{phot}} \equiv \sum_{\mathbf{k}\lambda} \hat{a}^\dagger_{\mathbf{k}\lambda} \hat{a}_{\mathbf{k}\lambda}$. This implies that the n-particle bosonic Green's functions G_n defined in (5.1) are not sufficient to close the Martin-Schwinger hierarchy for systems of interacting particles and photons since they contain an equal number of creation and annihilation operators. In Section 16.5 we introduce an extension of the bosonic G_n to deal with this situation. As we see, a Wick's theorem exists also for these new Green's functions and the whole NEGF formalism is not affected by this complication.

Exercise 16.1 Prove (16.23).

16.2 Quantum Systems of Electrons and Nuclei

So far we have been treating the nuclei as classical and infinitely heavy particles, frozen in their equilibrium positions, and only responsible for generating a static potential $\phi(\mathbf{r})$, see

[8] To derive (16.23) we integrate by parts and use the expansion (16.16).

for instance the Hamiltonian of the PPP model in Section 2.2, or the general Hamiltonian (3.1) and (L.1). However, the nuclei have a finite mass and, if perturbed, they either oscillate around the equilibrium positions (weak perturbations) or migrate far away from them (strong perturbations). Examples of the latter scenario are the breaking of chemical bonds in molecules or the melting of a crystal structure in solids.

In this section we lay down a quantum theory of electrons and nuclei which is suited whenever the average nuclear positions remain close to the equilibrium values. Under this "near-equilibrium hypothesis" the nuclei stay away from each other and can be treated as quantum *distinguishable* particles, see Section 1.2. In fact, we can distinguish them by means of techniques such as scanning tunneling microscopy or electron diffraction [271]. The following presentation is based on the work of Ref. [264].

The (nonrelativistic) Hamiltonian describing an unperturbed system of electrons interacting with N nuclei of charge Z_i and mass M_i, $i = 1, \dots N_{\rm n}$ reads (in our units the electron mass is $m = 1$ and the electric charge is $q = -1$)

$$\hat{H} = \int d\mathbf{x}d\mathbf{x}' \hat{\psi}^\dagger(\mathbf{x})\langle\mathbf{x}|\frac{\hat{p}^2}{2}|\mathbf{x}'\rangle\,\hat{\psi}(\mathbf{x}') + \hat{H}_{\rm int} + \sum_{i=1}^{N_{\rm n}} \frac{\hat{\mathcal{P}}_i^2}{2M_i} + \hat{H}_{\rm e-n} + \hat{H}_{\rm n-n}, \tag{16.25}$$

where the field operators $\hat{\psi}(\mathbf{x})$ annihilate an electron in $\mathbf{x} = (\mathbf{r}\sigma)$, hence they satisfy the anticommutation relations $[\hat{\psi}(\mathbf{x}), \hat{\psi}^\dagger(\mathbf{x}')]_+ = \delta(\mathbf{x} - \mathbf{x}')$, and the position and momentum operators of the nuclei satisfy the standard commutation relations $[\hat{\mathcal{R}}_{i,\kappa}, \hat{\mathcal{P}}_{j,\kappa'}]_- = \mathrm{i}\delta_{ij}\delta_{\kappa\kappa'}$, with κ and κ' running over the three components of the vectors. In accordance with our notation we use calligraphic letters for operators in first quantization, such as $\hat{\mathcal{R}}_{i,\kappa}$ and $\hat{\mathcal{P}}_{i,\kappa}$. We denote by $\hat{\boldsymbol{\mathcal{R}}}_i = (\hat{\mathcal{R}}_{i,x}, \hat{\mathcal{R}}_{i,y}, \hat{\mathcal{R}}_{i,z})$ the operator of the position vector and by $\hat{\boldsymbol{\mathcal{P}}}_i = (\hat{\mathcal{P}}_{i,x}, \hat{\mathcal{P}}_{i,y}, \hat{\mathcal{P}}_{i,z})$ the operator of the momentum vector of the ith nucleus. For the collection of all position and momentum vector operators we use the symbols $\hat{\boldsymbol{\mathcal{R}}} = (\hat{\boldsymbol{\mathcal{R}}}_1, \dots, \hat{\boldsymbol{\mathcal{R}}}_{N_{\rm n}})$ and $\hat{\boldsymbol{\mathcal{P}}} = (\hat{\boldsymbol{\mathcal{P}}}_1, \dots, \hat{\boldsymbol{\mathcal{P}}}_{N_{\rm n}})$. In (16.25) the first term is the electronic kinetic energy Hamiltonian, the second term is the electron–electron interaction Hamiltonian (1.82), the third term is the nuclear kinetic energy Hamiltonian ($\hat{\mathcal{P}}_i^2 = \hat{\boldsymbol{\mathcal{P}}}_i \cdot \hat{\boldsymbol{\mathcal{P}}}_i$ is the ith momentum operator squared), the fourth term is the electron–nucleus interaction Hamiltonian

$$\hat{H}_{\rm e-n} = -\int d\mathbf{x}\,\hat{n}(\mathbf{x}) \sum_{j=1}^{N_{\rm n}} Z_j v(\mathbf{r}, \hat{\boldsymbol{\mathcal{R}}}_j) = -\int d\mathbf{x}\,\hat{n}(\mathbf{x})\phi(\mathbf{r}, \hat{\boldsymbol{\mathcal{R}}}), \tag{16.26}$$

where $\hat{n}(\mathbf{x}) = \hat{\psi}^\dagger(\mathbf{x})\hat{\psi}(\mathbf{x})$ is the electron density operator coupled to the electrostatic potential operator $\phi(\mathbf{r}, \hat{\boldsymbol{\mathcal{R}}}) = \sum_j Z_j v(\mathbf{r}, \hat{\boldsymbol{\mathcal{R}}}_j)$, and the last term is the nucleus–nucleus interaction Hamiltonian

$$\hat{H}_{\rm n-n} = E_{\rm n-n}(\hat{\boldsymbol{\mathcal{R}}}) = \frac{1}{2} \sum_{i \neq j}^{N_{\rm n}} Z_i Z_j v(\hat{\boldsymbol{\mathcal{R}}}_i, \hat{\boldsymbol{\mathcal{R}}}_j).$$

All operators act on the direct-product space $\mathbb{F} \otimes \mathbb{D}_{N_{\rm n}}$, where $\mathbb{F}$ is the electronic Fock space and $\mathbb{D}_{N_{\rm n}}$ is the Hilbert space of the $N_{\rm n}$ distinguishable nuclei. Relativistic corrections [272] can be incorporated without any conceptual complication. In particular, the spin-orbit coupling emerges from the relativistic correction to the electron–nuclear interaction Hamiltonian in (16.26).

Expansion around thermal equilibrium Consider the interacting system of electrons and nuclei in thermal equilibrium at a certain temperature. Under the "near-equilibrium hypothesis" we can approximate the full Hamiltonian by its second-order Taylor expansion around the equilibrium values of the nuclear positions, which we name $\mathbf{R}^0 = (\mathbf{R}_1^0, \ldots, \mathbf{R}_{N_n}^0)$, and around the equilibrium value of the electronic density, which we name $n^0(\mathbf{x})$. In fact, also the electronic density must stay close to $n^0(\mathbf{x})$, otherwise the force acting on the nuclei would be strong enough to bring them away from the equilibrium positions. Let us introduce the displacement (or position fluctuation) operators $\hat{\mathcal{U}}_i$ and the density fluctuation operator $\Delta\hat{n}$ according to

$$\hat{\mathcal{U}}_i = \hat{\mathcal{R}}_i - \mathbf{R}_i^0, \qquad \Delta\hat{n}(\mathbf{x}) = \hat{n}(\mathbf{x}) - n^0(\mathbf{x}). \tag{16.27}$$

In the following we refer to these operators as *fluctuation operators.* Formally, the "near-equilibrium hypothesis" means that we can restrict the space $\mathbb{F} \otimes \mathbb{D}_{N_n}$ to the subspace of states giving a small average of $\hat{\mathcal{U}}_i$ and $\Delta\hat{n}$. The expansion of the electron–nucleus interaction Hamiltonian around the equilibrium nuclear positions yields

$$\hat{H}_{\text{e-n}} = -\int d\mathbf{x}\,\hat{n}(\mathbf{x})\Big[\phi(\mathbf{r}) - \sum_{i\kappa} g_{i,\kappa}(\mathbf{r})\hat{\mathcal{U}}_{i,\kappa} - \frac{1}{2}\sum_{i\kappa,j\kappa'} g^{\text{DW}}_{i,\kappa;j,\kappa'}(\mathbf{r})\hat{\mathcal{U}}_{i,\kappa}\hat{\mathcal{U}}_{j,\kappa'}\Big], \tag{16.28}$$

where we define

$$\phi(\mathbf{r}) \equiv \phi(\mathbf{r}, \mathbf{R}^0), \tag{16.29a}$$

$$g_{i,\kappa}(\mathbf{r}) \equiv -\left.\frac{\partial\phi(\mathbf{r},\mathbf{R})}{\partial R_{i,\kappa}}\right|_{\mathbf{R}=\mathbf{R}^0} = Z_i\frac{\partial}{\partial r_\kappa}v(\mathbf{r}, \mathbf{R}_i^0), \tag{16.29b}$$

$$g^{\text{DW}}_{i,\kappa;j,\kappa'}(\mathbf{r}) \equiv -\left.\frac{\partial^2\phi(\mathbf{r},\mathbf{R})}{\partial R_{i,\kappa}\partial R_{j,\kappa'}}\right|_{\mathbf{R}=\mathbf{R}^0} = -\delta_{ij}Z_i\frac{\partial^2}{\partial r_\kappa\partial r_{\kappa'}}v(\mathbf{r}, \mathbf{R}_i^0). \tag{16.29c}$$

In the last two equalities we take into account that the Coulomb potential $v(\mathbf{r},\mathbf{r}') = 1/|\mathbf{r}-\mathbf{r}'|$ depends only on the relative distance. The first term in (16.28) is a purely electronic operator; it is the potential energy operator for electrons in the classical field generated by a nuclear geometry $\mathbf{R}^0$. The second and third terms are new entries; they emerge when relaxing the infinite-mass approximation for the nuclei. Notice that the third term is already quadratic in the displacements and it can therefore be evaluated at the equilibrium density – that is, $\hat{n} \to n^0$ [273]. Going beyond the quadratic approximation, the replacement $\hat{n} \to n^0$ is no longer justified; in this case the third term is known as the *Debye–Waller interaction.* One of the most important effects of the Debye–Waller interaction is the renormalization of the electronic band structure [274].

The expansion of the nucleus–nucleus interaction Hamiltonian yields

$$\hat{H}_{\text{n-n}} = E_{\text{n-n}}(\mathbf{R}^0) + \sum_{i\kappa}\left.\frac{\partial E_{\text{n-n}}(\mathbf{R})}{\partial R_{i,\kappa}}\right|_{\mathbf{R}=\mathbf{R}^0}\hat{\mathcal{U}}_{i,\kappa} + \frac{1}{2}\sum_{i\kappa,j\kappa'}\left.\frac{\partial^2 E_{\text{n-n}}(\mathbf{R})}{\partial R_{i,\kappa}\partial R_{j,\kappa'}}\right|_{\mathbf{R}=\mathbf{R}^0}\hat{\mathcal{U}}_{i,\kappa}\hat{\mathcal{U}}_{j,\kappa'}. \tag{16.30}$$

In the first term we recognize the electrostatic energy of a nuclear geometry $\mathbf{R}^0$. As this term is only responsible for an overall energy shift we do not include it in the following.

The low-energy Hamiltonian Inserting the expansions of the electron–nucleus and nucleus–nucleus interaction Hamiltonians in (16.25) we can group all terms into three main contributions:

$$\boxed{\hat{H} = \hat{H}_{\rm e} + \hat{H}_{0,\rm phon} + \hat{H}_{\rm e-phon}} \tag{16.31}$$

The first term is a purely electronic Hamiltonian; it describes a quantum system of interacting electrons in the classical potential generated by a nuclear geometry $\mathbf{R}^0$

$$\boxed{\hat{H}_{\rm e} = \int d\mathbf{x}d\mathbf{x}'\hat{\psi}^\dagger(\mathbf{x})\langle\mathbf{x}|\frac{\hat{p}^2}{2}|\mathbf{x}'\rangle\,\hat{\psi}(\mathbf{x}') - \int d\mathbf{x}\,\hat{n}(\mathbf{x})\phi(\mathbf{r}) + \hat{H}_{\rm int}} \tag{16.32}$$

The reader can easily verify that $\hat{H}_{\rm e}$ coincides with the Hamiltonian in (1.83) with $\hat{h} = \frac{\hat{p}^2}{2} - \phi(\hat{\mathbf{r}})$, see also (3.1) and discussion below. The second term in (16.31) is a purely nuclear Hamiltonian; it describes a quantum system of N interacting nuclei in the electric field generated by a *frozen* electronic density $n^0(\mathbf{r})$

$$\boxed{\hat{H}_{0,\rm phon} = \sum_i \frac{\hat{\mathcal{P}}_i^2}{2M_i} + \frac{1}{2}\sum_{i\kappa,j\kappa'} K_{i,\kappa;j,\kappa'}\,\hat{\mathcal{U}}_{i,\kappa}\hat{\mathcal{U}}_{j,\kappa'}} \tag{16.33}$$

where the *elastic tensor*[9]

$$\boxed{K_{i,\kappa;j,\kappa'} \equiv \left.\frac{\partial^2 E_{\rm n-n}(\mathbf{R})}{\partial R_{i,\kappa}\partial R_{j,\kappa'}}\right|_{\mathbf{R}=\mathbf{R}^0} + \int d\mathbf{x}\,n^0(\mathbf{x})g^{\rm DW}_{i,\kappa;j,\kappa'}(\mathbf{r})} \tag{16.34}$$

is real and symmetric under the exchange $(i,\kappa) \leftrightarrow (j,\kappa')$. The last term in (16.31) describes the interaction between electrons and nuclei

$$\hat{H}_{\rm e-phon} = \sum_{i\kappa}\left[\left.\frac{\partial E_{\rm n-n}(\mathbf{R})}{\partial R_{i,\kappa}}\right|_{\mathbf{R}=\mathbf{R}^0} + \int d\mathbf{x}\;g_{i,\kappa}(\mathbf{r})\hat{n}(\mathbf{x})\right]\hat{\mathcal{U}}_{i,\kappa}. \tag{16.35}$$

The Hamiltonian (16.31) with the three contributions as in (16.32), (16.33), and (16.35) is identical to that of Ref. [273]. We here go a step further. Although it is not evident, the operator $\hat{H}_{\rm e-phon}$ is quadratic in the fluctuation operators. To show it, we consider the Heisenberg equation of motion (3.20) for the time-dependent average of the nuclear momentum operators $P_{i,\kappa}(t) = \mathrm{Tr}[\hat{\rho}\,\hat{U}(t_0,t)\hat{\mathcal{P}}_{i,\kappa}\hat{U}(t,t_0)]$. We find[10]

$$\frac{dP_{i,\kappa}(t)}{dt} = -\left.\frac{\partial E_{\rm n-n}(\mathbf{R})}{\partial R_{i,\kappa}}\right|_{\mathbf{R}=\mathbf{R}^0} - \int d\mathbf{x}\;g_{i,\kappa}(\mathbf{r})n(\mathbf{x},t) - \sum_{j\kappa'} K_{i,\kappa;j,\kappa'}U_{j,\kappa'}(t), \tag{16.36}$$

[9] According to the observation above (16.30) we make the replacement $\hat{n} \to n^0$.

[10] Use that the commutator $[\hat{\mathcal{U}}_{j,\kappa'}, \hat{\mathcal{P}}_{i,\kappa}] = \mathrm{i}\delta_{ij}\delta_{\kappa\kappa'}$.

where $n(\mathbf{x},t)$ and $U_{j,\kappa'}(t)$ are the time-dependent averages of the electronic density $\hat{n}(\mathbf{r})$ and of the nuclear displacement $\hat{\mathcal{U}}_{j,\kappa'}$. If the system is in thermal equilibrium and no external perturbing fields are present, then the l.h.s. vanishes, and $n(\mathbf{x},t) = n^0(\mathbf{x})$ and $U_{j,\kappa'} = 0$. We conclude that

$$\left.\frac{\partial E_{\text{n-n}}(\mathbf{R})}{\partial R_{i,\kappa}}\right|_{\mathbf{R}=\mathbf{R}^0} = -\int d\mathbf{x}\, g_{i,\kappa}(\mathbf{r}) n^0(\mathbf{x}). \tag{16.37}$$

Therefore (16.35) is equivalent to

$$\boxed{\hat{H}_{\text{e-phon}} = \sum_{i\kappa} \int d\mathbf{x}\, g_{i,\kappa}(\mathbf{r}) \Delta\hat{n}(\mathbf{x}) \hat{\mathcal{U}}_{i,\kappa}} \tag{16.38}$$

where $\Delta\hat{n}(\mathbf{r})$ is the density fluctuation operator defined in (16.27). In this form $\hat{H}_{\text{e-phon}}$ is manifestly quadratic in the fluctuation operators.

Inserting (16.37) in (16.36) we also see that the equation of motion for the momentum operators simplifies to

$$\boxed{\frac{dP_{i,\kappa}(t)}{dt} = -\int d\mathbf{x}\, g_{i,\kappa}(\mathbf{r}) \Delta n(\mathbf{x},t) - \sum_{j\kappa'} K_{i,\kappa;j,\kappa'} U_{j,\kappa'}(t)} \tag{16.39}$$

We can interpret the elastic tensor as the nuclear-force tensor of a system with frozen electronic density - that is, with $\Delta n(\mathbf{x},t) = 0$ - or alternatively with vanishing coupling $g_{i,\kappa}$. In general, $\Delta n(\mathbf{x},t) \neq 0$ and $g_{i,\kappa} \neq 0$, and the first term in (16.39) significantly contributes to the nuclear forces, see Section 16.2.2.

The Hamiltonian (16.31) is the low-energy approximation of the full Hamiltonian (16.25). However, the expansion around the equilibrium nuclear geometry and around the equilibrium density has inevitably made (16.31) depend on these quantities. The scalar potential ϕ and the electron–nuclear coupling g are determined from the sole knowledge of the equilibrium nuclear geometry $\mathbf{R}^0$, whereas the elastic tensor K depends on both $\mathbf{R}^0$ and n^0. Notice that the dependence of $\hat{H}$ on n^0 is through K as well as $\Delta\hat{n}$. In the following we assume that $\mathbf{R}^0$ is known and therefore that ϕ and g are given. Methods to obtain good approximations to the equilibrium nuclear geometry are indeed available (in Section 16.2.2 we describe one of them). Alternatively, $\mathbf{R}^0$ can be taken from X-ray crystallographic measurements. The value of the equilibrium electronic density must instead be determined, and the proper way of doing it is *self-consistently*. Let us expand on this point. We write the dependence of $\hat{H}$ on n^0 explicitly: $\hat{H} = \hat{H}[n^0]$. For any given many-body treatment (whether exact or approximate) a possible self-consistent strategy to obtain n^0 is: (1) make an initial guess n_1^0 and (2) use the chosen many-body treatment to calculate the equilibrium density n_2^0 of $\hat{H}[n_1^0]$, then the equilibrium density n_3^0 of $\hat{H}[n_2^0]$, the equilibrium density n_4^0 of $\hat{H}[n_3^0]$, and so on until convergence. If the initial guess n_1^0 already produces a good approximation to K, then a partial self-consistent scheme in which K is not updated is also conceivable. Self-consistency is, however, unavoidable to determine n^0 in $\Delta\hat{n}$. In fact, it is only at self-consistency that the equilibrium value $\Delta n = 0$, an essential requirement for the r.h.s. of (16.39) to vanish and hence for the nuclear geometry to remain still. In Section 16.11.3 we discuss how to implement the self-consistent strategy using NEGF. In particular, we show that n_0 can be obtained from the self-consistent solution of the Dyson equation for the Matsubara Green's functions.

16.2.1 Interacting Hamiltonian for Electrons and Phonons

Independently of the method chosen to find $\mathbf{R}^0$ and of the self-consistent many-body treatment chosen to determine n_0, the low-energy Hamiltonian of a system of electrons and nuclei is given by (16.31). Let us comment on the formal analogy between (16.31) and the Hamiltonian (16.21) for interacting particles and photons. The Hamiltonian $\hat{H}^0_{\mathrm{phon}}$ in (16.33) is quadratic like the photonic Hamiltonian $\hat{H}_{\mathrm{phot}}$ in (16.22), and the electron–nucleus interaction Hamiltonian in (16.38) is linear in the displacement operators like the first term in the electron–photon interaction Hamiltonian in (16.24). This analogy becomes even more evident in crystals.

In a crystal we can label the position of every nucleus with the vector (of integers) $\mathbf{n}$ of the unit cell it belongs to and with the position s relative to some point of the unit cell: $\mathbf{R}^0_{i=\mathbf{n},s} = \mathbf{R}_{\mathbf{n}} + \mathbf{R}_s$. If the unit cell contains N_u nuclei then $s = 1,\ldots,N_u$. By definition the vector $\mathbf{R}_s$ for the sth nucleus is the same in all unit cells, and the mass $M_{i=\mathbf{n},s} = M_s$ and charge $Z_{i=\mathbf{n},s} = Z_s$ of the sth nucleus in cell $\mathbf{n}$ is independent of $\mathbf{n}$. The invariance of the crystal under discrete translations implies that the elastic tensor depends only on the difference between unit cell vectors:

$$K_{\mathbf{n}s,\kappa;\mathbf{n}'s',\kappa'} = K_{s,\kappa;s',\kappa'}(\mathbf{n}-\mathbf{n}').$$

We thus see that $\hat{H}^0_{\mathrm{phon}}$ has the following form

$$\hat{H}_{0,\mathrm{phon}} = \sum_{\mathbf{n}s} \frac{\hat{\boldsymbol{\mathcal{P}}}^2_{\mathbf{n}s}}{2M_s} + \frac{1}{2}\sum_{\mathbf{n}s\kappa,\mathbf{n}'s'\kappa'} K_{s,\kappa;s',\kappa'}(\mathbf{n}-\mathbf{n}')\hat{\mathcal{U}}_{\mathbf{n}s,\kappa}\hat{\mathcal{U}}_{\mathbf{n}'s',\kappa'}. \tag{16.40}$$

Below we describe a procedure to bring $\hat{H}_{0,\mathrm{phon}}$ in a diagonal form. We may say that the final result is the nuclear counterpart of the Bloch theorem for electrons, see Section 2.3.

Bloch theorem for nuclear oscillators We define the reduced displacements and the reduced momenta operators according to

$$\hat{\overline{\boldsymbol{\mathcal{U}}}}_{\mathbf{n}s} \equiv \sqrt{M_s}\,\hat{\boldsymbol{\mathcal{U}}}_{\mathbf{n}s} \quad , \quad \hat{\overline{\boldsymbol{\mathcal{P}}}}_{\mathbf{n}s} \equiv \frac{\hat{\boldsymbol{\mathcal{P}}}_{\mathbf{n}s}}{\sqrt{M_s}}.$$

The reduced operators satisfy the same commutation relations as the original operators:

$$[\hat{\overline{\mathcal{U}}}_{\mathbf{n}s,\kappa}, \hat{\overline{\mathcal{P}}}_{\mathbf{n}'s',\kappa'}]_- = [\hat{\mathcal{U}}_{\mathbf{n}s,\kappa}, \hat{\mathcal{P}}_{\mathbf{n}'s',\kappa'}]_- = \mathrm{i}\delta_{\mathbf{n}\mathbf{n}'}\delta_{ss'}\delta_{\kappa\kappa'}. \tag{16.41}$$

In terms of the reduced operators the Hamiltonian (16.40) becomes

$$\hat{H}_{0,\mathrm{phon}} = \frac{1}{2}\sum_{\mathbf{n}s} \hat{\overline{\boldsymbol{\mathcal{P}}}}^2_{\mathbf{n}s} + \frac{1}{2}\sum_{\mathbf{n}s\kappa,\mathbf{n}'s'\kappa'} \overline{K}_{s,\kappa;s',\kappa'}(\mathbf{n}-\mathbf{n}')\,\hat{\overline{\mathcal{U}}}_{\mathbf{n}s,\kappa}\hat{\overline{\mathcal{U}}}_{\mathbf{n}'s',\kappa'}, \tag{16.42}$$

where

$$\overline{K}_{s,\kappa;s',\kappa'}(\mathbf{n}-\mathbf{n}') \equiv \frac{1}{\sqrt{M_s M_{s'}}} K_{s,\kappa;s',\kappa'}(\mathbf{n}-\mathbf{n}').$$

We now proceed along the same lines as in Section 2.3. We consider the set V_b of unit cells defined according to (2.13), impose the BvK boundary conditions, and expand the reduced operators according to [compare with the expansion (2.17)]

$$\hat{\overline{\mathcal{U}}}_{\mathbf{n}s} = \frac{1}{\sqrt{|V_b|}} \sum_{\mathbf{q}} e^{\mathrm{i}\mathbf{q}\cdot\mathbf{n}} \hat{\tilde{\mathcal{U}}}_{\mathbf{q}s} \quad , \quad \hat{\overline{\mathcal{P}}}_{\mathbf{n}s} = \frac{1}{\sqrt{|V_b|}} \sum_{\mathbf{q}} e^{\mathrm{i}\mathbf{q}\cdot\mathbf{n}} \hat{\tilde{\mathcal{P}}}_{\mathbf{q}s}. \tag{16.43}$$

The sums run over all $\mathbf{q}$ vectors defined in (2.16) and $|V_b|$ is the number of unit cells in V_b. Since the reduced operators are self-adjoint, the operators of the expansions must satisfy

$$\hat{\tilde{\mathcal{U}}}_{\mathbf{q}s} = \hat{\tilde{\mathcal{U}}}^{\dagger}_{-\mathbf{q}s} \quad , \quad \hat{\tilde{\mathcal{P}}}_{\mathbf{q}s} = \hat{\tilde{\mathcal{P}}}^{\dagger}_{-\mathbf{q}s}. \tag{16.44}$$

Taking into account that the matrix (2.15) is unitary we also infer that the commutation relations (16.41) lead to the commutation relations

$$[\hat{\tilde{\mathcal{U}}}_{\mathbf{q}s,\kappa}, \hat{\tilde{\mathcal{P}}}_{\mathbf{q}'s',\kappa'}]_- = \mathrm{i}\delta_{\mathbf{q},-\mathbf{q}'}\delta_{ss'}\delta_{\kappa\kappa'} \ ,$$

or equivalently [using (16.44)]

$$[\hat{\tilde{\mathcal{U}}}_{\mathbf{q}s,\kappa}, \hat{\tilde{\mathcal{P}}}^{\dagger}_{\mathbf{q}'s',\kappa'}]_- = \mathrm{i}\delta_{\mathbf{q},\mathbf{q}'}\delta_{ss'}\delta_{\kappa\kappa'} \ . \tag{16.45}$$

Inserting the expansion (16.43) in (16.42) we find

$$\hat{H}_{0,\mathrm{phon}} = \frac{1}{2}\sum_{\mathbf{q}s} \hat{\tilde{\mathcal{P}}}^{\dagger}_{\mathbf{q}s} \cdot \hat{\tilde{\mathcal{P}}}_{\mathbf{q}s} + \frac{1}{2}\sum_{\mathbf{q}}\sum_{s\kappa,s'\kappa'} \hat{\tilde{\mathcal{U}}}^{\dagger}_{\mathbf{q}s,\kappa} \tilde{K}_{s,\kappa;s',\kappa'}(\mathbf{q}) \hat{\tilde{\mathcal{U}}}_{\mathbf{q}s',\kappa'}, \tag{16.46}$$

where

$$\tilde{K}_{s,\kappa;s',\kappa'}(\mathbf{q}) \equiv \sum_{\mathbf{n}} e^{-\mathrm{i}\mathbf{q}\cdot\mathbf{n}} \overline{K}_{s,\kappa;s',\kappa'}(\mathbf{n}).$$

In a crystal the symmetry of the elastic tensor under an exchange of its indices reads $\overline{K}_{s,\kappa;s',\kappa'}(\mathbf{n}-\mathbf{n}') = \overline{K}_{s',\kappa';s,\kappa}(\mathbf{n}'-\mathbf{n})$. This property together with the fact that $\overline{K}$ is a real tensor can be used to show that the tensor $\tilde{K}(\mathbf{q}) = \tilde{K}^{\dagger}(\mathbf{q})$ for all $\mathbf{q}$:

$$\tilde{K}^{*}_{s',\kappa';s,\kappa}(\mathbf{q}) = \sum_{\mathbf{n}} e^{\mathrm{i}\mathbf{q}\cdot\mathbf{n}} \overline{K}_{s',\kappa';s,\kappa}(\mathbf{n}) = \sum_{\mathbf{n}} e^{\mathrm{i}\mathbf{q}\cdot\mathbf{n}} \overline{K}_{s,\kappa;s'\kappa'}(-\mathbf{n}) = \tilde{K}_{s,\kappa;s'\kappa'}(\mathbf{q}).$$

Similarly one can show that $\tilde{K}(-\mathbf{q}) = \tilde{K}^{T}(\mathbf{q})$, or equivalently $\tilde{K}^{*}(-\mathbf{q}) = \tilde{K}(\mathbf{q})$. Let $\mathbf{e}^{\nu}(\mathbf{q})$ and $\omega^2_{0\mathbf{q}\nu}$ be the normalized eigenvectors and eigenvalues of $\tilde{K}(\mathbf{q})$:

$$\tilde{K}(\mathbf{q})\mathbf{e}^{\nu}(\mathbf{q}) = \omega^2_{0\mathbf{q}\nu}\mathbf{e}^{\nu}(\mathbf{q}) \quad \overset{\text{component wise}}{\Longrightarrow} \quad \sum_{s'\kappa'} \tilde{K}_{s,\kappa;s'\kappa'}(\mathbf{q}) e^{\nu}_{s',\kappa'}(\mathbf{q}) = \omega^2_{0\mathbf{q}\nu} e^{\nu}_{s,\kappa}(\mathbf{q}).$$

We observe that the eigenvalues $\omega^2_{0\mathbf{q}\nu}$ of $\tilde{K}(\mathbf{q})$ are not guaranteed to be positive. Taking the conjugate of the eigenvalue equation, we deduce a useful property for eigenvalues and eigenvectors

$$\tilde{K}^*(-\mathbf{q})\mathbf{e}^{\nu*}(-\mathbf{q}) = \tilde{K}(\mathbf{q})\mathbf{e}^{\nu*}(-\mathbf{q}) = \omega^2_{0-\mathbf{q}\nu}\mathbf{e}^{\nu*}(-\mathbf{q}) \quad \Rightarrow \quad \begin{cases} \omega^2_{0-\mathbf{q}\nu} = \omega^2_{0\mathbf{q}\nu} \\ \mathbf{e}^{\nu*}(-\mathbf{q}) = \mathbf{e}^{\nu}(\mathbf{q}). \end{cases} \tag{16.47}$$

As the eigenvectors form an orthonormal basis for every $\mathbf{q}$ we can expand the nuclear operators according to

$$\hat{\tilde{\mathcal{U}}}_{\mathbf{q}s,\kappa} = \sum_{\nu} e^{\nu}_{s,\kappa}(\mathbf{q})\,\hat{\mathcal{U}}_{\mathbf{q}\nu} \quad , \quad \hat{\tilde{\mathcal{P}}}_{\mathbf{q}s,\kappa} = \sum_{\nu} e^{\nu}_{s,\kappa}(\mathbf{q})\,\hat{\mathcal{P}}_{\mathbf{q}\nu}. \tag{16.48}$$

The property (16.44) together with $\mathbf{e}^{\nu*}(-\mathbf{q}) = \mathbf{e}^{\nu}(\mathbf{q})$ imply that

$$\hat{\mathcal{U}}_{\mathbf{q}\nu} = \hat{\mathcal{U}}^{\dagger}_{-\mathbf{q}\nu} \quad , \quad \hat{\mathcal{P}}_{\mathbf{q}\nu} = \hat{\mathcal{P}}^{\dagger}_{-\mathbf{q}\nu}. \tag{16.49}$$

We can easily verify that the commutation relations (16.45) lead to the commutation relations[ll]

$$[\hat{\mathcal{U}}_{\mathbf{q}\nu}, \hat{\mathcal{P}}^{\dagger}_{\mathbf{q}'\nu'}]_- = \mathrm{i}\delta_{\mathbf{q},\mathbf{q}'}\delta_{\nu\nu'}. \tag{16.50}$$

Substituting the expansions (16.48) into (16.46) we eventually obtain the *Bloch theorem* for nuclear oscillators

$$\hat{H}_{0,\mathrm{phon}} = \frac{1}{2}\sum_{\mathbf{q}\nu} \hat{\mathcal{P}}^{\dagger}_{\mathbf{q}\nu}\hat{\mathcal{P}}_{\mathbf{q}\nu} + \frac{1}{2}\sum_{\mathbf{q}\nu} \omega^2_{0\mathbf{q}\nu}\hat{\mathcal{U}}^{\dagger}_{\mathbf{q}\nu}\hat{\mathcal{U}}_{\mathbf{q}\nu}. \tag{16.51}$$

Normal modes The eigenvectors $\mathbf{e}^{\nu}(\mathbf{q})$ are called the *normal modes* of the nuclear vibrations. In fact, the expansion of the original displacement operators in terms of the $\hat{\mathcal{U}}_{\mathbf{q}\nu}$ operators reads

$$\hat{\mathcal{U}}_{\mathbf{n}s,\kappa} = \frac{1}{\sqrt{M_s|\mathrm{V}_b|}}\sum_{\mathbf{q}} e^{\mathrm{i}\mathbf{q}\cdot\mathbf{n}} \sum_{\nu} e^{\nu}_{s,\kappa}(\mathbf{q})\,\hat{\mathcal{U}}_{\mathbf{q}\nu}. \tag{16.52}$$

According to this formula the nuclear vibrations can be decomposed into normal modes of quantum numbers $\mathbf{q}\nu$. For nonnegative eigenvalues $\omega^2_{0\mathbf{q}\nu}$ we can define the frequencies $\omega_{0\mathbf{q}\nu} = \sqrt{\omega^2_{0\mathbf{q}\nu}} \geq 0$, and the commutation relations (16.50) and the constraint (16.49) are both fulfilled at once if we write the displacement and momentum operators as

$$\hat{\mathcal{U}}_{\mathbf{q}\nu} = \frac{1}{\sqrt{2\omega_{0\mathbf{q}\nu}}}(\hat{b}_{0\mathbf{q}\nu} + \hat{b}^{\dagger}_{0-\mathbf{q}\nu}) \quad , \quad \hat{\mathcal{P}}_{\mathbf{q}\nu} = -\mathrm{i}\sqrt{\frac{\omega_{0\mathbf{q}\nu}}{2}}(\hat{b}_{0\mathbf{q}\nu} - \hat{b}^{\dagger}_{0-\mathbf{q}\nu}), \tag{16.53}$$

[ll]Let us calculate the commutator (16.45) using the expansion (16.48) and imposing (16.50). We have

$$[\hat{\tilde{\mathcal{U}}}_{\mathbf{q}s,\kappa}, \hat{\tilde{\mathcal{P}}}^{\dagger}_{\mathbf{q}'s',\kappa'}] = \sum_{\nu\nu'} e^{\nu}_{s,\kappa}(\mathbf{q})e^{\nu'*}_{s',\kappa'}(\mathbf{q}')[\hat{\mathcal{U}}_{\mathbf{q}\nu}, \hat{\mathcal{P}}^{\dagger}_{\mathbf{q}'\nu'}]_-$$
$$= \mathrm{i}\delta_{\mathbf{q},\mathbf{q}'}\sum_{\nu} e^{\nu}_{s,\kappa}(\mathbf{q})e^{\nu*}_{s',\kappa'}(\mathbf{q}) = \mathrm{i}\delta_{\mathbf{q},\mathbf{q}'}\delta_{ss'}\delta_{\kappa\kappa'}.$$

where the operators $\hat{b}_{0\mathbf{q}\nu}$ satisfy the bosonic commutation relations

$$[\hat{b}_{0\mathbf{q}\nu}, \hat{b}^{\dagger}_{0\mathbf{q}'\nu'}]_{-} = \delta_{\mathbf{q}\mathbf{q}'}\delta_{\nu\nu'}. \tag{16.54}$$

Using the property $\omega_{0\mathbf{q}\nu} = \omega_{0-\mathbf{q}\nu}$, the Hamiltonian (16.51) written in terms of the operators $\hat{b}_{0\mathbf{q}\nu}$ and $\hat{b}^{\dagger}_{0\mathbf{q}\nu}$ reads:

$$\hat{H}_{0,\mathrm{phon}} = \sum_{\mathbf{q}\nu} \omega_{0\mathbf{q}\nu}(\hat{b}^{\dagger}_{0\mathbf{q}\nu}\hat{b}_{0\mathbf{q}\nu} + \frac{1}{2}), \tag{16.55}$$

which is the Hamiltonian of a set of decoupled harmonic oscillators. We remark that bringing (16.33) in the form (16.51) is always possible, while the form (16.55) further requires that $\tilde{K}(\mathbf{q})$ is positive semidefinite.

Bare phonons versus dressed phonons We have already emphasized that the elastic tensor K yields the forces acting on the nuclei if the electron–nuclear coupling g vanishes, see (16.39). In this case electrons and nuclei are decoupled and for the solution $U_{i,\kappa} = 0$ to be a stable solution the eigenvalues of the tensor $\tilde{K}(\mathbf{q})$ must be nonnegative. The $\omega_{0\mathbf{q}\nu}$ can be interpreted as the *bare* frequencies of the normal modes. The operators $\hat{b}^{\dagger}_{0\mathbf{q}\nu}$ and $\hat{b}_{0\mathbf{q}\nu}$ change the energy of the $\mathbf{q}\nu$th normal mode by $\pm\omega_{0\mathbf{q}\nu}$. We can therefore interpret these operators as the creation and annihilation operators of a quantum of vibration, see also the discussion below (2.55). The quanta of vibrations are generally called *phonons*, and those of $\hat{H}_{0,\mathrm{phon}}$ are said to be *bare* phonons.[12] Phonons behave like bosonic particles, due to the commutation relations (16.54), and have the same kind of quantum numbers as electrons in a Bloch state – that is, a quasi-momentum $\mathbf{q}$ and a "band index" ν named the *phonon branch*. We can then say that the set of eigenvalues $\omega_{0\mathbf{q}\nu}$ for a fixed phonon branch is a bare *phononic band* and the set of all bands forms the bare *phononic band structure*. For phonons the band index runs from 1 to $3 \times N_u$, where 3 is the number of spatial dimensions and N_u is the number of nuclei in the unit cell. A phonon with these quantum numbers is a bosonic particle with a well-defined energy and hence an infinitely long lifetime.

In all known physical systems the electron–nuclear coupling g does not vanish and, consequently, the bare frequencies are not physical. In fact, the eigenvalues $\omega^2_{0\mathbf{q}\nu}$ are not even guaranteed to be positive, a circumstance that prevents the possibility of writing the displacement and momentum operators as in (16.53), and consequently to transform $\hat{H}_{0,\mathrm{phon}}$ from (16.51) to (16.55). This is by no means a problem for NEGF and MBPT; as we see, it is more convenient to work with the displacement and momentum operators than with the phonon creation and annihilation operators. In Section 16.11 we show that the electron–nuclear coupling changes the direction of the normal modes and renormalizes the eigenvalues $\omega^2_{0\mathbf{q}\nu}$, leading to the notion of *dressed phonons*. As the normal modes of $\tilde{K}$ are not physical we formulate the theory in an arbitrary basis – that is, we expand the nuclear operators as in (16.48), where the normal modes $\mathbf{e}^{\nu}(\mathbf{q})$ are left unspecified. The most suitable basis of normal modes is dictated by the choice of the many-body treatment. The only constraints we require on the normal modes are the properties of orthonormality and $\mathbf{e}^{\nu*}(-\mathbf{q}) = \mathbf{e}^{\nu}(\mathbf{q})$.

[12]The name "phonon" was originally coined by Frenkel in 1932 [275].

Electron–phonon Hamiltonian Let us expand the nuclear operators as in (16.48), where the unit vectors $\mathbf{e}^{\nu}(\mathbf{q})$ form some suitable basis of normal modes. Then the Hamiltonian of the bare phonons $\hat{H}_{0,\text{phon}}$ in (16.46) reads:

$$\boxed{\hat{H}_{0,\text{phon}} = \frac{1}{2}\sum_{\mathbf{q}\nu}\hat{\mathcal{P}}^{\dagger}_{\mathbf{q}\nu}\hat{\mathcal{P}}_{\mathbf{q}\nu} + \frac{1}{2}\sum_{\mathbf{q}\nu\nu'}\hat{\mathcal{U}}^{\dagger}_{\mathbf{q}\nu}K_{\nu\nu'}(\mathbf{q})\,\hat{\mathcal{U}}_{\mathbf{q}\nu'}} \tag{16.56}$$

where

$$K_{\nu\nu'}(\mathbf{q}) \equiv \sum_{s\kappa,s'\kappa'} e^{\nu*}_{s,\kappa}(\mathbf{q})\,\tilde{K}_{s,\kappa;s'\kappa'}(\mathbf{q})\,e^{\nu'}_{s'\kappa'}(\mathbf{q}) = K^{*}_{\nu\nu'}(-\mathbf{q}) = K^{*}_{\nu'\nu}(\mathbf{q}). \tag{16.57}$$

The expansion in normal modes can also be used to rewrite the electron–phonon interaction Hamiltonian (16.38) as

$$\boxed{\hat{H}_{\text{e-phon}} = \sum_{\mathbf{n}s\kappa}\int d\mathbf{x}\; g_{\mathbf{n}s,\kappa}(\mathbf{r})\Delta\hat{n}(\mathbf{x})\hat{\mathcal{U}}_{\mathbf{n}s,\kappa} = \sum_{\mathbf{q}\nu}\int d\mathbf{x}\; g_{-\mathbf{q}\nu}(\mathbf{r})\Delta\hat{n}(\mathbf{x})\,\hat{\mathcal{U}}_{\mathbf{q}\nu}} \tag{16.58}$$

where in the second equality we insert the expansion (16.52) and define

$$\boxed{g_{-\mathbf{q}\nu}(\mathbf{r}) \equiv \sum_{\mathbf{n}s\kappa}\frac{1}{\sqrt{M_s|\mathrm{V}_b|}}e^{\mathrm{i}\mathbf{q}\cdot\mathbf{n}}\,e^{\nu}_{s,\kappa}(\mathbf{q})\,g_{\mathbf{n}s,\kappa}(\mathbf{r}) = g^{*}_{\mathbf{q}\nu}(\mathbf{r})} \tag{16.59}$$

In crystals the electron–phonon coupling g satisfies an important property. According to the definition in (16.29) we have $g_{\mathbf{n}s,\kappa}(\mathbf{r}) = Z_s(\partial/\partial r_\kappa)v(\mathbf{r},\mathbf{R}^0_{\mathbf{n}s})$. The Coulomb interaction depends only on the relative coordinate and therefore $v(\mathbf{r},\mathbf{R}^0_{\mathbf{n}s}) = v(\mathbf{r}+\mathbf{R}_{\mathbf{n}'},\mathbf{R}^0_{\mathbf{n}s}+\mathbf{R}_{\mathbf{n}'}) = v(\mathbf{r}+\mathbf{R}_{\mathbf{n}'},\mathbf{R}^0_{\mathbf{n}+\mathbf{n}'s})$ for all vectors $\mathbf{R}_{\mathbf{n}'}$. This implies that $g_{\mathbf{n}s,\kappa}(\mathbf{r}) = g_{\mathbf{n}+\mathbf{n}'s,\kappa}(\mathbf{r}+\mathbf{R}_{\mathbf{n}'})$ and hence

$$g_{-\mathbf{q}\nu}(\mathbf{r}+\mathbf{R}_{\mathbf{n}}) = e^{\mathrm{i}\mathbf{q}\cdot\mathbf{n}}g_{-\mathbf{q}\nu}(\mathbf{r}). \tag{16.60}$$

We emphasize that no approximations have been made in writing (16.56) and (16.58), as we only change the basis from $\{\mathbf{n}s,\kappa\}$ to $\{\mathbf{q}\nu\}$. The *electron–phonon Hamiltonian* is given by (16.31) with $\hat{H}_{\text{el}}$, $\hat{H}_{0,\text{phon}}$, and $\hat{H}_{\text{el-phon}}$ defined in (16.32), (16.56), and (16.58), respectively. We also observe that the electron–phonon Hamiltonian can equivalently be expressed in terms of the bosonic operators $\hat{b}_{0\mathbf{q}\nu}$ and $\hat{b}^{\dagger}_{0\mathbf{q}\nu}$ using (16.53). In fact, the relation (16.53) guarantees that $\hat{\mathcal{U}}_{\mathbf{q}\nu}$ and $\hat{\mathcal{P}}_{\mathbf{q}\nu}$ satisfy (16.49) and (16.50) for *any* choice of the frequencies $\omega_{0\mathbf{q}\nu} \geq 0$.[13] We then see that the electron–phonon Hamiltonian, like the electron–photon Hamiltonian, does not commute with the total number of bosons operator $\hat{N}_{\text{bos}} \equiv \sum_{\mathbf{q}\nu}\hat{b}^{\dagger}_{0\mathbf{q}\nu}\hat{b}_{0\mathbf{q}\nu}$. This implies that the bosonic n-particle Green's functions G_n defined in (5.1) are not sufficient to close the Martin–Schwinger hierarchy. In Section 16.5 we introduce an extension of the bosonic G_n and show how to deal with this situation.

[13] Of course, an arbitrary choice of $\omega_{0\mathbf{q}\nu}$ does not allow us to write $\hat{H}_{0,\text{phon}}$ in the diagonal form (16.55).

16.2.2 Born-Oppenheimer Approximation

In this section we describe a popular and practical scheme to find an approximation to the equilibrium nuclear positions and to the equilibrium electronic density at *zero temperature*, thus providing an approximation to ϕ, g, and K. We refer to this scheme as the *Born-Oppenheimer approximation.* The starting point is the *Born-Oppenheimer Hamiltonian,*

$$\hat{H}^{\mathrm{BO}}(\hat{\boldsymbol{\mathcal{R}}}) \equiv \lim_{\{M_i\}\to\infty} \hat{H} = \int d\mathbf{x}d\mathbf{x}'\hat{\psi}^\dagger(\mathbf{x})\langle\mathbf{x}|\frac{\hat{p}^2}{2}|\mathbf{x}'\rangle\,\hat{\psi}(\mathbf{x}') + \hat{H}_{\mathrm{int}} - \int d\mathbf{x}\,\hat{n}(\mathbf{x})\phi(\mathbf{r},\hat{\boldsymbol{\mathcal{R}}}) + E_{\mathrm{n-n}}(\hat{\boldsymbol{\mathcal{R}}}),$$

where we use (16.25) to take the limit of infinite nuclear masses. The Born-Oppenheimer Hamiltonian commutes with all the nuclear position operators $\hat{\boldsymbol{\mathcal{R}}}_i$. Therefore the eigenkets of $\hat{H}^{\mathrm{BO}}(\hat{\boldsymbol{\mathcal{R}}})$ have the form $|\Psi(\mathbf{R})\rangle|\mathbf{R}_1\ldots\mathbf{R}_{N_\mathrm{n}}\rangle$ where the electronic ket $|\Psi(\mathbf{R})\rangle \in \mathbb{F}$ is an eigenket of $\hat{H}^{\mathrm{BO}}(\mathbf{R})$, and $|\mathbf{R}_1\ldots\mathbf{R}_{N_\mathrm{n}}\rangle \in \mathbb{D}_{N_\mathrm{n}}$ is a ket describing N_n distinguishable nuclei in positions $\mathbf{R} = (\mathbf{R}_1,\ldots,\mathbf{R}_{N_\mathrm{n}})$.

Let $|\Psi^{\mathrm{BO}}(\mathbf{R})\rangle$ be the ground state of $\hat{H}^{\mathrm{BO}}(\mathbf{R})$ with ground-state energy $E^{\mathrm{BO}}(\mathbf{R})$. An approximation for the equilibrium positions of the nuclei and for the equilibrium electronic density is found by minimizing $E_{\mathrm{BO}}(\mathbf{R})$ over all $\mathbf{R}$. Using the Hellmann-Feynman theorem, we have

$$F_{i,\kappa}(\mathbf{R}) \equiv \frac{\partial E^{\mathrm{BO}}(\mathbf{R})}{\partial R_{i,\kappa}} = \langle\Psi^{\mathrm{BO}}(\mathbf{R})|\frac{\partial \hat{H}^{\mathrm{BO}}(\mathbf{R})}{\partial R_{i,\kappa}}|\Psi^{\mathrm{BO}}(\mathbf{R})\rangle = -\int d\mathbf{x}\,n^{\mathrm{BO}}(\mathbf{x},\mathbf{R})\frac{\partial\phi(\mathbf{r},\mathbf{R})}{\partial R_{i,\kappa}} + \frac{\partial E_{\mathrm{n-n}}(\mathbf{R})}{\partial R_{i,\kappa}}, \tag{16.61}$$

where we define

$$n^{\mathrm{BO}}(\mathbf{x},\mathbf{R}) \equiv \langle\Psi^{\mathrm{BO}}(\mathbf{R})|\hat{n}(\mathbf{x})|\langle\Psi^{\mathrm{BO}}(\mathbf{R})\rangle.$$

The Born-Oppenheimer approximation consists of approximating the nuclear geometry with the solution of $F_{i,\kappa}(\mathbf{R}) = 0$ for all i,κ (i.e., $\mathbf{R}^0 \simeq \mathbf{R}^{0,\mathrm{BO}}$) and the equilibrium electronic density with $n^0(\mathbf{x}) \simeq n^{0,\mathrm{BO}}(\mathbf{x}) \equiv n^{\mathrm{BO}}(\mathbf{x},\mathbf{R}^{0,\mathrm{BO}})$. The approximated nuclear positions and density can then be used to obtain an approximation for $\phi \simeq \phi^{\mathrm{BO}}$, $g \simeq g^{\mathrm{BO}}$, and $K \simeq K^{\mathrm{BO}}$.

On the stability of the Born-Oppenheimer approximation As already pointed out at the end of Section 16.2, the equilibrium density n^0 enters the low-energy Hamiltonian through K *as well as* $\Delta\hat{n}$. The replacement $n^0 \simeq n^{0,\mathrm{BO}}$ in $\Delta\hat{n}$ leads to an inconsistency. In fact, any exact or even approximate treatment of the low-energy Hamiltonian (16.31) gives a ground-state density n^0 which is, in general, different from $n^{0,\mathrm{BO}}$. In equilibrium we then have $\Delta n = n^0 - n^{0,\mathrm{BO}} \neq 0$, and the nuclei move away from $\{\mathbf{R}\} = \{\mathbf{R}^{0,\mathrm{BO}}\}$ in accordance with the equation of motion (16.39). One way out of this impasse consists of treating n^0 in $\Delta\hat{n}$ self-consistently while keeping ϕ, g, and K fixed at the Born-Oppenheimer values. Whether this partial self-consistent scheme leads to a stable dynamics when the system is perturbed by weak external fields remains to be checked case by case.

Elastic tensor and Hessian of the Born–Oppenheimer energy The elastic tensor $K^{\rm BO}$ can alternatively be obtained from the Hessian of $E^{\rm BO}(\mathbf{R})$ and the density-density response function *at clamped nuclei*. Taking into account definitions (16.29), a second differentiation of (16.61) yields [276]

$$\begin{aligned}\mathrm{Hes}_{i,\kappa;j,\kappa'} \equiv \left.\frac{\partial^2 E^{\rm BO}(\mathbf{R})}{\partial R_{i,\kappa}\partial R_{j,\kappa'}}\right|_{\mathbf{R}=\mathbf{R}^{0,\rm BO}} &= \int d\mathbf{x}\, \left.\frac{\partial n^{\rm BO}(\mathbf{x},\mathbf{R})}{\partial R_{j,\kappa'}}\right|_{\mathbf{R}=\mathbf{R}^{0,\rm BO}} g^{\rm BO}_{i,\kappa}(\mathbf{r}) \\ &+ \int d\mathbf{x}\, n^{0,\rm BO}(\mathbf{x}) g^{\rm DW,BO}_{i,\kappa;j,\kappa'}(\mathbf{r}) + \left.\frac{\partial^2 E_{\rm n-n}(\mathbf{R})}{\partial R_{i,\kappa}\partial R_{j,\kappa'}}\right|_{\mathbf{R}=\mathbf{R}^{0,\rm BO}}. \end{aligned} \tag{16.62}$$

The first term on the r.h.s. can be rewritten in a more symmetric form using linear response theory. Let us imagine manually moving the infinitely heavy nuclei from $\mathbf{R}_j^{0,\rm BO}$ to $\mathbf{R}_j^{0,\rm BO} + \delta\mathbf{R}_j$. We indicate with $\delta\mathbf{R}_j(t)$ the extremely slow time-dependent function with the property that $\delta\mathbf{R}_j(t) = 0$ for $t = -\infty$ and $\delta\mathbf{R}_j(t) = \delta\mathbf{R}_j$ for $t = \infty$. This nuclear rearrangement induces a change in the electronic density. For the extremely slow (adiabatic) change considered here we expect the time-dependent electronic density $n(\mathbf{x}, t)$ to be the same as the ground-state electronic density $n^{\rm BO}(\mathbf{x}, \mathbf{R}(t))$ corresponding to a nuclear geometry $\mathbf{R}(t) = \mathbf{R}^{0,\rm BO} + \delta\mathbf{R}(t)$. This is the assumption of "instantaneous relaxation" of the electronic density. From (13.14) we then have (recall that the electronic charge is $q = -1$)

$$\begin{aligned}\delta n^{\rm BO}(\mathbf{x}, \mathbf{R}^{0,\rm BO}) &= -\lim_{t\to\infty}\int d\mathbf{x}'dt'\chi^{\rm R}_{\rm clamp}(\mathbf{x},t;\mathbf{x}';t')\delta\phi(\mathbf{r}',\mathbf{R}(t')) \\ &= \lim_{t\to\infty}\int d\mathbf{x}'dt'\chi^{\rm R}_{\rm clamp}(\mathbf{x},t;\mathbf{x}';t')\sum_{j\kappa'} g^{\rm BO}_{j,\kappa'}(\mathbf{r}')\delta R_{j,\kappa'}(t'),\end{aligned} \tag{16.63}$$

where $\chi^{\rm R}_{\rm clamp}$ is the equilibrium density-density response function of the purely electronic system with clamped nuclei in $\mathbf{R}^{0,\rm BO}$. For a response function with the property that $\chi^{\rm R}_{\rm clamp}(\mathbf{x},t;\mathbf{x}';t') \to 0$ for $|t-t'| \to \infty$ we can replace $\delta R_{j,\kappa'}(t') \to \delta R_{j,\kappa'}(t)$ since we have assumed extremely slow changes. Performing the integral over t', we get

$$\delta n^{\rm BO}(\mathbf{x}, \mathbf{R}^{0,\rm BO}) = \int d\mathbf{x}'\chi^{\rm R}_{\rm clamp}(\mathbf{x},\mathbf{x}';0)\sum_{j\kappa'} g^{\rm BO}_{j,\kappa'}(\mathbf{r}')\delta R_{j,\kappa'}, \tag{16.64}$$

where $\chi^{\rm R}_{\rm clamp}(\mathbf{x},\mathbf{x}';0)$ is the Fourier transform of the response function calculated at zero frequency. Using (16.64) to evaluate the derivative of the ground-state density in (16.62) and comparing with (16.34) we conclude that

$$K^{\rm BO}_{i,\kappa;j,\kappa'} = \mathrm{Hes}_{i,\kappa;j,\kappa'} - \int d\mathbf{x}d\mathbf{x}' g^{\rm BO}_{i,\kappa}(\mathbf{r})\chi^{\rm R}_{\rm clamp}(\mathbf{x},\mathbf{x}';0) g^{\rm BO}_{j,\kappa'}(\mathbf{r}'). \tag{16.65}$$

The elastic tensor in the Born–Oppenheimer approximation coincides with the Hessian of the Born–Oppenheimer energy only if the electron-nuclear coupling $g^{\rm BO}$ vanishes.

In crystals, the relation (16.65) reads:

$$K^{\rm BO}_{s,\kappa;s',\kappa'}(\mathbf{n}-\mathbf{n}') = \mathrm{Hes}_{s,\kappa;s',\kappa'}(\mathbf{n}-\mathbf{n}') - \int d\mathbf{x}d\mathbf{x}' g^{\rm BO}_{\mathbf{n}s,\kappa}(\mathbf{r})\chi^{\rm R}_{\rm clamp}(\mathbf{x},\mathbf{x}';0) g^{\rm BO}_{\mathbf{n}'s',\kappa'}(\mathbf{r}').$$

The Hessian tensor satisfies the same properties as the elastic tensor. Let $\omega^2_{\mathbf{q}\nu}$ and $\mathbf{e}^\nu(\mathbf{q})$ be the eigenvalues and normal modes of $\widetilde{\text{Hes}}(\mathbf{q})$. The eigenvalues $\omega^2_{\mathbf{q}\nu}$ are by construction nonnegative, since the Hessian of a function calculated in the global minimum is positive semidefinite. Multiplying by $e^{\nu*}_{s,\kappa}(\mathbf{q})e^{-i\mathbf{q}\cdot(\mathbf{n}-\mathbf{n}')}e^{\nu'}_{s'\kappa'}(\mathbf{q})/\sqrt{M_s M_{s'}}$ and summing over $\mathbf{n}s$, κ and $\mathbf{n}'s'$, κ', we obtain

$$\boxed{K^{\text{BO}}_{\nu\nu'}(\mathbf{q}) = \delta_{\nu\nu'}\omega^2_{\mathbf{q}\nu} - \int d\mathbf{x}d\mathbf{x}'g^{\text{BO}}_{\mathbf{q}\nu}(\mathbf{r})\,\chi^{\text{R}}_{\text{clamp}}(\mathbf{x},\mathbf{x}';0)\,g^{\text{BO}*}_{\mathbf{q}\nu'}(\mathbf{r}')} \tag{16.66}$$

Estimation of the nuclear forces We can use result (16.65) to estimate the forces acting on the nuclei. Let us go back to the equation of motion of the momentum operators. Substituting (16.65) into (16.39), we get the equation of motion in the Born–Oppenheimer approximation,

$$\begin{aligned}\frac{dP_{i,\kappa}(t)}{dt} &= -\sum_{j\kappa'}\text{Hes}_{\,i,\kappa;j,\kappa'}U_{j,\kappa'}(t)\\ &+\int d\mathbf{x}\,g^{\text{BO}}_{i,\kappa}(\mathbf{r})\Big[-\Delta n(\mathbf{x},t)+\int d\mathbf{x}'\chi^{\text{R}}_{\text{clamped}}(\mathbf{x},\mathbf{x}';0)\sum_{j\kappa'}g^{\text{BO}}_{j,\kappa'}(\mathbf{r}')U_{j,\kappa'}(t)\Big].\end{aligned}$$

The second term in the square brackets is the change $\delta n^{\text{BO}}(\mathbf{x},\mathbf{R}^{0,\text{BO}})$ due to a change $\{\mathbf{U}(t)\}$ of the nuclear geometry, see (16.64). For heavy nuclei the displacements $U_{j,\kappa'}(t)$ vary slowly in time. If no external perturbations are present then we can invoke the assumption of "instantaneous relaxation" to conclude that $\Delta n(\mathbf{x},t)\simeq\delta n^{\text{BO}}(\mathbf{x},\mathbf{R}^{0,\text{BO}})$. This approximate relation becomes exact only in the limit of infinite nuclear masses. We infer that the term in the square brackets is small and that the largest contribution to the nuclear forces comes from the Hessian of the Born–Oppenheimer energy. This result has led several authors to partition the low-energy Hamiltonian (16.31) in a slightly different way – see, for instance, Refs. [271, 276, 277, 278, 279]. The main difference consists of using the Hessian of E^{BO} to define the dressed phonon Hamiltonian,

$$\hat{H}_{\text{phon}} = \sum_{i=1}^{N}\frac{\hat{\mathcal{P}}_i^2}{2M_i}+\frac{1}{2}\sum_{i\kappa,j\kappa'}\text{Hes}_{\,i,\kappa;j,\kappa'}\hat{\mathcal{U}}_{i,\kappa}\hat{\mathcal{U}}_{j,\kappa'} \overset{\text{(for crystals)}}{=} \sum_{\mathbf{q}\nu}\omega_{\mathbf{q}\nu}(\hat{b}^\dagger_{\mathbf{q}\nu}\hat{b}_{\mathbf{q}\nu}+\frac{1}{2}),$$

where the possibility of writing the last equality comes from the fact that $\widetilde{\text{Hes}}(\mathbf{q})$ satisfies the same properties as $\widetilde{K}(\mathbf{q})$ *and* the eigenvalues $\omega^2_{\mathbf{q}\nu}\geq 0$. In the alternative partitioning the remainder (a quadratic form in the nuclear displacements),

$$\Delta\hat{H}_{\text{phon}} \equiv \hat{H}_{0,\text{phon}} - \hat{H}_{\text{phon}} = \frac{1}{2}\sum_{i\kappa,j\kappa'}(K_{i,\kappa;j,\kappa'}-\text{Hes}_{\,i,\kappa;j,\kappa'})\hat{\mathcal{U}}_{i,\kappa}\hat{\mathcal{U}}_{j,\kappa'}, \tag{16.67}$$

must be treated somehow. As we see, NEGF and related diagrammatic expansions are most easily formulated with the partitioning (16.31).

To summarize, the Born–Oppenheimer approximation offers a practical method to calculate the functions $\phi(\mathbf{r})$, $g_{i,\kappa}(\mathbf{r})$ as well as the tensor $K_{i,\kappa;j,\kappa'}$ entering the low-energy

Hamiltonian (16.31). The function $g_{i,\kappa}(\mathbf{r})$ provides the coupling between the electronic and nuclear degrees of freedom. If this coupling is zero then we have an uncoupled system of electrons and quantum nuclear oscillators.

16.3 General Hamiltonian for Interacting Identical Particles and Phosons

In this section we consider a system of *identical particles* (either fermions or bosons) interacting with N *distinguishable particles*. Photons and phonons can be seen as bosonic particles describing the elementary excitations of distinguishable particles. In fact, both the electron–photon Hamiltonian (16.20) and the electron–phonon Hamiltonian (16.31) can be written in terms of displacement and momentum operators.[14] We invent the name *phosons* for the bosonic particles describing the elementary excitations of the distinguishable particles. Photons and phonons are examples of phosons.

Let $\hat{\mathcal{U}}_\xi$ and $\hat{\mathcal{P}}_\xi$ be the displacement and momentum operators of the distinguishable particles where the index ξ runs over a suitable basis, for example, $\xi = \mathbf{k}\lambda$ for photons and $\xi = \mathbf{q}\nu$ for phonons. We also define the index ξ^* labeling the time-reversal state of the state labeled by ξ. For photons the time-reversal state of the state labeled by $\xi = \mathbf{k}\lambda$ has index $\xi^* = -\mathbf{k}\lambda$, and similarly for phonons the time-reversal state of the state labeled by $\xi = \mathbf{q}\nu$ has index $\xi^* = -\mathbf{q}\nu$. Of course, the time-reversal state of a time-reversal state is the original state: $\xi^{**} = \xi$. The operators with indices ξ and ξ^* are related by [compare with (16.49)]

$$\hat{\mathcal{U}}_\xi = \hat{\mathcal{U}}^\dagger_{\xi^*} \quad , \quad \hat{\mathcal{P}}_\xi = \hat{\mathcal{P}}^\dagger_{\xi^*}, \tag{16.68}$$

and satisfy the commutation relations [compare with (16.50)]

$$[\hat{\mathcal{U}}_\xi, \hat{\mathcal{P}}_{\xi'^*}]_- = [\hat{\mathcal{U}}_\xi, \hat{\mathcal{P}}^\dagger_{\xi'}]_- = \mathrm{i}\delta_{\xi\xi'}. \tag{16.69}$$

Particle–phoson Hamiltonian We consider a general particle–phoson Hamiltonian of the form

$$\boxed{\hat{H}(t) = \hat{H}_{\text{part}}(t) + \hat{H}_{0,\text{phos}}(t) + \hat{H}_{\text{part-phos}}(t)} \tag{16.70}$$

where $\hat{H}_{\text{part}}(t)$ is the Hamiltonian (1.83) for identical particles with single-particle Hamiltonian $\hat{h}(t)$ and interparticle interaction $v(\mathbf{x}, \mathbf{x}', t)$,

$$\hat{H}_{0,\text{phos}}(t) = \frac{1}{2}\sum_\xi \hat{\mathcal{P}}^\dagger_\xi \hat{\mathcal{P}}_\xi + \frac{1}{2}\sum_{\xi\xi'} \hat{\mathcal{U}}^\dagger_\xi Q_{\xi\xi'}(t)\hat{\mathcal{U}}_{\xi'} \tag{16.71}$$

[14] In the photonic case we have $\hat{\mathcal{U}}_{\mathbf{k}\lambda} = \frac{1}{\sqrt{2\omega_k}}(\hat{a}_{\mathbf{k}\lambda} + \hat{a}^\dagger_{-\mathbf{k}\lambda})$ and $\hat{\mathcal{P}}_{\mathbf{k}\lambda} = -\mathrm{i}\sqrt{\frac{\omega_k}{2}}(\hat{a}_{\mathbf{k}\lambda} - \hat{a}^\dagger_{-\mathbf{k}\lambda})$ and therefore

$$\hat{H}_{\text{phot}} = \sum_{\mathbf{k}\lambda} \omega_k(\hat{a}^\dagger_{\mathbf{k}\lambda}\hat{a}_{\mathbf{k}\lambda} + \frac{1}{2}) = \frac{1}{2}\sum_{\mathbf{k}\lambda} \hat{\mathcal{P}}^\dagger_{\mathbf{k}\lambda}\hat{\mathcal{P}}_{\mathbf{k}\lambda} + \frac{1}{2}\sum_{\mathbf{k}\lambda} \omega_k^2 \hat{\mathcal{U}}^\dagger_{\mathbf{k}\lambda}\hat{\mathcal{U}}_{\mathbf{k}\lambda}.$$

is the Hamiltonian describing noninteracting phosons with $Q_{\xi\xi'} = Q^*_{\xi^*\xi'^*} = Q^*_{\xi'\xi}$ [compare with (16.57)], and

$$\hat{H}_{\text{part-phos}}(t) = \sum_{\xi} \int d\mathbf{x}d\mathbf{x}' \mathfrak{a}_{\xi^*}(\mathbf{x}, \mathbf{x}', t)\hat{\psi}^\dagger(\mathbf{x})\hat{\psi}(\mathbf{x}')\hat{\mathcal{U}}_\xi + \sum_{\xi} \mathfrak{b}_{\xi^*}(t)\hat{\mathcal{U}}_\xi \tag{16.72}$$

is the particle–phoson interaction Hamiltonian with $\mathfrak{a}_{\xi^*}(\mathbf{x}, \mathbf{x}', t) = \mathfrak{a}^*_\xi(\mathbf{x}', \mathbf{x}, t)$ and $\mathfrak{b}_{\xi^*}(t) = \mathfrak{b}^*_\xi(t)$ [compare with (16.59)]. The time-dependence of the interparticle interaction v and particle–phoson couplings $\mathfrak{a}$ and $\mathfrak{b}$ could be due to, for example, an adiabatic switching protocol or a sudden quench of the interaction, whereas the time dependence of the one-particle Hamiltonian $\hat{h}(t)$ and matrix $Q_{\xi\xi'}(t)$ could be due to some external field, for example, laser fields, phonon drivings, etc. We consider systems that are initially described by a density matrix $\hat{\rho} = e^{-\beta\hat{H}^{\mathrm{M}}}/\mathrm{Tr}[e^{-\beta\hat{H}^{\mathrm{M}}}]$, where the Matsubara Hamiltonian has the same form as (16.70) but time-independent $\hat{h}(t) = \hat{h}^{\mathrm{M}}$, $v(\mathbf{x}, \mathbf{x}', t) = v^{\mathrm{M}}(\mathbf{x}, \mathbf{x}')$, $Q_{\xi\xi'}(t) = Q^{\mathrm{M}}_{\xi\xi'}$, $\mathfrak{a}_\xi(\mathbf{x}, \mathbf{x}', t) = \mathfrak{a}^{\mathrm{M}}_\xi(\mathbf{x}, \mathbf{x}')$, and $\mathfrak{b}_\xi(t) = \mathfrak{b}^{\mathrm{M}}_\xi$. Thermal equilibrium is recovered by choosing $\hat{H}^{\mathrm{M}} = \hat{H}(t_0) - \mu\hat{N}$, where μ is the chemical potential and $\hat{N} = \int d\mathbf{x}\,\hat{n}(\mathbf{x})$ is the total number of identical particles operator. The electron–phonon Hamiltonian and the electron–photon Hamiltonian *without* the last term in (16.24) are special cases of (16.70). In Table 16.1 we explicitly show how to choose the parameters of $\hat{H}(t)$ to recover these cases.[15] Although the last term in (16.24) does not represent a conceptual complication, we do not include it in the following treatment. However, we give enough details to make the reader able to extend the formalism to arbitrary Hamiltonians depending on $\hat{\mathcal{U}}_\xi$, $\hat{\mathcal{P}}_\xi$, $\hat{\psi}(\mathbf{x})$, and $\hat{\psi}^\dagger(\mathbf{x})$. Our priority here is to focus on the novel aspects brought about by a Hamiltonian that does not conserve the total number of phosons as well as to provide a pedagogical introduction to the basic ideas and mathematical tools required to deal with this kind of problem.

To shorten the equations we gather the displacement and momentum operators into a two-dimensional vector of operators having components

$$\hat{\phi}_{\xi,1} = \hat{\mathcal{U}}_\xi \quad , \quad \hat{\phi}_{\xi,2} = \hat{\mathcal{P}}_\xi. \tag{16.73}$$

[15] The $\mathfrak{a}_{\xi^*}(\mathbf{x}, \mathbf{x}')$ for photons may seem strange. However, we observe that

$$\frac{\mathrm{i}q}{m}\sqrt{\frac{\pi}{\mathrm{V}}} \int d\mathbf{x}d\mathbf{x}'\delta_{\sigma\sigma'} \sum_\alpha \eta_{\mathbf{k}\lambda,\alpha} \frac{e^{\mathrm{i}\mathbf{k}\cdot\mathbf{r}}\delta(\mathbf{r}' - \mathbf{r} - \epsilon\,\mathbf{e}_\alpha) - e^{\mathrm{i}\mathbf{k}\cdot\mathbf{r}'}\delta(\mathbf{r} - \mathbf{r}' - \epsilon\,\mathbf{e}_\alpha)}{\epsilon} \hat{\psi}^\dagger(\mathbf{x})\hat{\psi}(\mathbf{x}')$$

$$= \frac{\mathrm{i}q}{m}\sqrt{\frac{\pi}{\mathrm{V}}} \int d\mathbf{x}\, e^{\mathrm{i}\mathbf{k}\cdot\mathbf{r}} \sum_\alpha \eta_{\mathbf{k}\lambda,\alpha} \frac{\hat{\psi}^\dagger(\mathbf{x})\hat{\psi}(\mathbf{r} + \epsilon\,\mathbf{e}_\alpha, \sigma) - \hat{\psi}^\dagger(\mathbf{r} + \epsilon\,\mathbf{e}_\alpha, \sigma)\hat{\psi}(\mathbf{x})}{\epsilon}.$$

Adding and subtracting $\hat{\psi}^\dagger(\mathbf{x})\hat{\psi}(\mathbf{x})$ to the numerator and then taking the limit $\epsilon \to 0$, the fraction becomes $2m\mathrm{i}\,\hat{j}_\alpha(\mathbf{x})$, see (3.27), and therefore the whole quantity can be rewritten as $-q\sqrt{\frac{4\pi}{\mathrm{V}}}\int d\mathbf{x}\, e^{\mathrm{i}\mathbf{k}\cdot\mathbf{r}}\eta_{\mathbf{k}\lambda} \cdot \hat{\mathbf{j}}(\mathbf{x})$. Multiplying by $\hat{\mathcal{U}}_{\mathbf{k}\lambda} = \frac{1}{\sqrt{2\omega_k}}(\hat{a}_{\mathbf{k}\lambda} + \hat{a}^\dagger_{-\mathbf{k}\lambda})$ and summing over all $\mathbf{k}\lambda$, we recover the first term of (16.24).

Phosons	Photons	Phonons
ξ	$\mathbf{k}\lambda$	$\mathbf{q}\nu$
$Q_{\xi\xi'}$	$\delta_{\mathbf{kk}'}\delta_{\lambda\lambda'}\omega_k^2$	$\delta_{\mathbf{qq}'}K_{\nu\nu'}(\mathbf{q})$
$\mathfrak{a}_{\xi^*}(\mathbf{x},\mathbf{x}')$	$\frac{\mathrm{i}q\delta_{\sigma\sigma'}}{m}\sqrt{\frac{\pi}{\mathbb{V}}}\sum_\alpha \eta_{\mathbf{k}\lambda,\alpha}\frac{e^{\mathrm{i}\mathbf{k}\cdot\mathbf{r}}\delta(\mathbf{r}-\mathbf{r}'-\epsilon\,\mathbf{e}_\alpha)-e^{\mathrm{i}\mathbf{k}\cdot\mathbf{r}'}\delta(\mathbf{r}'-\mathbf{r}-\epsilon\,\mathbf{e}_\alpha)}{\epsilon}$	$\delta(\mathbf{x}-\mathbf{x}')g_{-\mathbf{q}\nu}(\mathbf{r})$
$\mathfrak{b}_{\xi^*}$	0	$-\int d\mathbf{x}\, g_{-\mathbf{q}\nu}(\mathbf{r})n^0(\mathbf{x})$

Table 16.1 Parameters of the general Hamiltonian for the case of photons (second column) and phonons (third column). In the $\mathfrak{a}_\xi(\mathbf{x},\mathbf{x}')$ of the second column the vectors $\mathbf{e}_\alpha$ are unit vectors along direction $\alpha = x, y, z$ and the limit $\epsilon \to 0$ is implicit.

We also introduce the composite index $\boldsymbol{\xi} = (\xi, i)$ with $i = 1, 2$ to compact the notation. Then the phoson Hamiltonian (16.71) takes the form

$$\boxed{\hat{H}_{0,\mathrm{phos}}(t) = \frac{1}{2}\sum_{\boldsymbol{\xi}\boldsymbol{\xi}'} Q_{\boldsymbol{\xi}\boldsymbol{\xi}'}(t)\hat{\phi}^\dagger_{\boldsymbol{\xi}}\hat{\phi}_{\boldsymbol{\xi}'}} \qquad \text{with } Q_{\boldsymbol{\xi}\boldsymbol{\xi}'} = Q_{\xi,i\,\xi',i'} = \begin{pmatrix} Q_{\xi\xi'} & 0 \\ 0 & \delta_{\xi\xi'} \end{pmatrix}_{ii'},$$

and similarly the particle–phoson interaction Hamiltonian (16.72) becomes

$$\boxed{\hat{H}_{\mathrm{part-phos}}(t) = \sum_{\boldsymbol{\xi}} \int d\mathbf{x}d\mathbf{x}'\,\mathfrak{a}_{\boldsymbol{\xi}^*}(\mathbf{x},\mathbf{x}',t)\hat{\psi}^\dagger(\mathbf{x})\hat{\psi}(\mathbf{x}')\,\hat{\phi}_{\boldsymbol{\xi}} + \sum_{\boldsymbol{\xi}} \mathfrak{b}_{\boldsymbol{\xi}^*}(t)\,\hat{\phi}_{\boldsymbol{\xi}}} \tag{16.74}$$

where

$$\mathfrak{a}_{\boldsymbol{\xi}=\xi,i} = \delta_{i,1}\mathfrak{a}_\xi \qquad , \qquad \mathfrak{b}_{\boldsymbol{\xi}=\xi,i} = \delta_{i,1}\mathfrak{b}_\xi .$$

In (16.74) we define the time-reversal index $\boldsymbol{\xi}^* \equiv (\xi^*, i)$. In the following treatment the explicit form of $Q_{\boldsymbol{\xi}\boldsymbol{\xi}'}$, $\mathfrak{a}_{\boldsymbol{\xi}}$, and $\mathfrak{b}_{\boldsymbol{\xi}}$ is not used. The only constraints we require on these quantities are the properties

$$\boxed{Q_{\boldsymbol{\xi}\boldsymbol{\xi}'} = Q^*_{\boldsymbol{\xi}^*\boldsymbol{\xi}'^*} = Q^*_{\boldsymbol{\xi}'\boldsymbol{\xi}} \quad , \quad \mathfrak{a}_{\boldsymbol{\xi}^*}(\mathbf{x},\mathbf{x}') = \mathfrak{a}^*_{\boldsymbol{\xi}}(\mathbf{x}',\mathbf{x}) \quad , \quad \mathfrak{b}_{\boldsymbol{\xi}^*} = \mathfrak{b}^*_{\boldsymbol{\xi}}} \tag{16.75}$$

which are of course fulfilled by the quantities of Table 16.1. Thus the formalism can deal with very general Hamiltonians such as a phoson Hamiltonian containing terms of the form $\hat{\mathcal{P}}^\dagger_\xi\hat{\mathcal{U}}_{\xi'}$ and $\hat{\mathcal{U}}^\dagger_\xi\hat{\mathcal{P}}_{\xi'}$ as well as interaction Hamiltonians where the particle degrees of freedom are coupled to the momentum operators.

16.4 Equations of Motion for Operators in the Contour Heisenberg Picture

The preliminary step to derive the Martin–Schwinger hierarchy for interacting particles and phosons is the derivation of the equations of motion for $\hat{\phi}_{\xi,H}(z)$ and $\hat{\psi}_H(\mathbf{x},z)$ in the contour Heisenberg picture, see (4.39).

Equation of motion for the displacement The commutation relations for the displacement operators follow directly from (16.69) and read

$$[\hat{\phi}_\xi, \hat{\phi}_{\xi'^*}]_- = [\hat{\phi}_\xi, \hat{\phi}^\dagger_{\xi'}]_- = J_{\xi\xi'} \quad , \quad J_{\xi\xi'} = J_{\xi,i\,\xi',i'} = \delta_{\xi\xi'} \begin{pmatrix} 0 & \mathrm{i} \\ -\mathrm{i} & 0 \end{pmatrix}_{ii'} . \tag{16.76}$$

The commutator between $\hat{\phi}_\xi$ and the Hamiltonian is therefore

$$[\hat{\phi}_\xi, \hat{H}(z)]_- = \frac{1}{2} \sum_{\xi'\xi''} \left(J_{\xi\xi'} Q_{\xi'\xi''}(z) \hat{\phi}_{\xi''} + J_{\xi\xi''^*} Q_{\xi'\xi''}(z) \hat{\phi}^\dagger_{\xi'} \right)$$
$$+ \sum_{\xi'} \int d\mathbf{x} d\mathbf{x}' J_{\xi\xi'^*} \mathfrak{a}_{\xi'^*}(\mathbf{x},\mathbf{x}',z) \hat{\psi}^\dagger(\mathbf{x}) \hat{\psi}(\mathbf{x}') + \sum_{\xi'} J_{\xi\xi'^*} \mathfrak{b}_{\xi'^*}(z),$$

where $\mathfrak{a}_\xi(\mathbf{x},\mathbf{x}',t_\pm) = \mathfrak{a}_\xi(\mathbf{x},\mathbf{x}',t)$ (independent of the branch), $\mathfrak{a}_\xi(\mathbf{x},\mathbf{x}',t_0 - \mathrm{i}\tau) = \mathfrak{a}^{\mathrm{M}}_\xi(\mathbf{x},\mathbf{x}')$ (independent of τ), and the like for $\mathfrak{b}_\xi(z)$. The first two terms on the r.h.s. are identical. Indeed, using the properties of (16.75) we have

$$\sum_{\xi'\xi''} J_{\xi\xi''^*} Q_{\xi'\xi''}(t) \hat{\phi}^\dagger_{\xi'} = \sum_{\xi'\xi''} J_{\xi\xi''} Q_{\xi'^*\xi''^*}(t) \hat{\phi}_{\xi'} = \sum_{\xi'\xi''} J_{\xi\xi''} Q_{\xi''\xi'}(t) \hat{\phi}_{\xi'} .$$

The equation of motion for the displacement operator is then

$$\mathrm{i}\frac{d}{dz} \hat{\phi}_{\xi,H}(z) = [\hat{\phi}_{\xi,H}(z), \hat{H}_H(z)]_-$$
$$= \sum_{\xi'} J_{\xi\xi'} \Big(\sum_{\xi''} Q_{\xi'\xi''}(z) \hat{\phi}_{\xi'',H}(z)$$
$$+ \int d\mathbf{x} d\mathbf{x}' \mathfrak{a}_{\xi'}(\mathbf{x},\mathbf{x}',z) \hat{\psi}^\dagger_H(\mathbf{x},z) \hat{\psi}_H(\mathbf{x}',z) + \mathfrak{b}_{\xi'}(z) \Big).$$

Using the property $\sum_{\xi_1} J_{\xi\xi_1} J_{\xi_1\xi'} = (J^2)_{\xi\xi'} = \delta_{\xi\xi'}$ the equation of motion can alternatively be written as

$$\sum_{\xi'} \left[\mathrm{i}\frac{d}{dz} J_{\xi\xi'} - Q_{\xi\xi'}(z) \right] \hat{\phi}_{\xi',H}(z) = \int d\mathbf{x} d\mathbf{x}' \mathfrak{a}_\xi(\mathbf{x},\mathbf{x}',z) \hat{\psi}^\dagger_H(\mathbf{x},z) \hat{\psi}_H(\mathbf{x}',z) + \mathfrak{b}_\xi(z). \tag{16.77}$$

The inhomogeneity due to the $\mathfrak{b}$-term prompts us to define a Green's function D_0 as the solution of

$$\sum_{\xi'} \left[\mathrm{i}\frac{d}{dz_1} J_{\xi_1\xi'} - Q_{\xi_1\xi'}(z_1) \right] D_{0,\xi'\xi_2}(z_1,z_2) = \delta_{\xi_1\xi_2^*} \delta(z_1,z_2), \tag{16.78}$$

with periodic KMS boundary conditions. The equation of motion for the *shifted displacement*,

$$\hat{\phi}^s_{\xi,H}(z) \equiv \hat{\phi}_{\xi,H}(z) - \underbrace{\sum_{\xi'} \int d\bar{z}\, D_{0,\xi\xi'^*}(z,\bar{z}) \mathfrak{b}_{\xi'}(\bar{z})}_{s_\xi(z)}, \tag{16.79}$$

reads:

$$\boxed{\sum_{\xi'} \left[\mathrm{i}\frac{d}{dz} J_{\xi\xi'} - Q_{\xi\xi'}(z) \right] \hat{\phi}^s_{\xi',H}(z) = \int d\mathbf{x} d\mathbf{x}' \mathfrak{a}_\xi(\mathbf{x},\mathbf{x}',z) \hat{\psi}^\dagger_H(\mathbf{x},z) \hat{\psi}_H(\mathbf{x}',z)} \tag{16.80}$$

which is a homogeneous equation in the field operators. The shift $s_\xi(z)$ depends only on the parameters Q and $\mathfrak{b}$ of the Hamiltonian. Notice that $\hat{\phi}^s_{\xi,H}(z)$ is the contour Heisenberg picture of the *time-dependent* operator,

$$\hat{\phi}^s_\xi(z) \equiv \hat{\phi}_\xi - s_\xi(z). \tag{16.81}$$

This follows from the fact that the shift is proportional to the identity operator and therefore

$$\hat{U}(z,z_\mathrm{i})\, \hat{\phi}^s_\xi\, \hat{U}(z_\mathrm{i},z) = \hat{U}(z,z_\mathrm{i}) \left(\hat{\phi}_\xi - s_\xi(z) \right) \hat{U}(z_\mathrm{i},z) = \hat{\phi}_{\xi,H}(z) - s_\xi(z) = \hat{\phi}^s_{\xi,H}(z),$$

where $\hat{U}$ is the contour evolution operator defined in Section 4.4. We encourage the reader to show that $s_\xi(t_\pm) = s_\xi(t)$ (independent of the branch) and $s_\xi(t_0 - \mathrm{i}\tau) = s^{\mathrm{M}}_\xi$ (independent of τ). The proportionality to the identity operator also implies that the operators $\hat{\phi}^s_\xi(z)$ and $\hat{\phi}_\xi(z)$ [these operators are not in the contour Heisenberg picture and in particular $\hat{\phi}_\xi(z)$ does not depend on z] satisfy the same commutation relations for any z and z':

$$[\hat{\phi}_\xi(z), \hat{\phi}^\dagger_{\xi'}(z')] = [\hat{\phi}_\xi(z), \hat{\phi}^{s\dagger}_{\xi'}(z')] = [\hat{\phi}^s_\xi(z), \hat{\phi}^{s\dagger}_{\xi'}(z')]. \tag{16.82}$$

Particle–phoson Hamiltonian in terms of the shifted displacement The shifted displacement operator can be used to write the full Hamiltonian (16.70) in a more suitable form for the subsequent analysis. According to definition (16.81), the particle–phoson Hamiltonian in (16.74) is the same as

$$\hat{H}_{\text{part-phos}}(t) = \sum_\xi \int d\mathbf{x} d\mathbf{x}' \langle \mathbf{x}|\hat{\mathfrak{a}}_{\xi^*}(t)|\mathbf{x}'\rangle \hat{\psi}^\dagger(\mathbf{x}) \hat{\psi}(\mathbf{x}') \left(\hat{\phi}^s_\xi + s_\xi(t) \right) + \sum_\xi \mathfrak{b}_{\xi^*}(t)\, \hat{\phi}_\xi,$$

where we define the operator in first quantization $\hat{\mathfrak{a}}_\xi(z)$ with matrix elements

$$\langle \mathbf{x}|\hat{\mathfrak{a}}_\xi(z)|\mathbf{x}'\rangle = \mathfrak{a}_\xi(\mathbf{x},\mathbf{x}',z) \quad \overset{(16.75)}{\Rightarrow} \quad \hat{\mathfrak{a}}_\xi(z) = \hat{\mathfrak{a}}^\dagger_{\xi^*}(z). \tag{16.83}$$

Let us separate the Hamiltonian for particles in the standard way – that is, $\hat{H}_{\text{part}} = \hat{H}_0 + \hat{H}_{\text{int}}$, where $\hat{H}_0$ is the noninteracting Hamiltonian (1.72) and $\hat{H}_{\text{int}}$ is the interaction Hamiltonian (1.82). We then have

$$\boxed{\hat{H}(t) = \hat{H}^s_0(t) + \hat{H}^s_{0,\text{phos}}(t) + \hat{H}_{\text{int}}(t) + \hat{H}_{\mathfrak{a}}(t)} \tag{16.84}$$

where the *shifted* noninteracting Hamiltonian for particles reads

$$\boxed{\hat{H}_0^s(t) \equiv \int d\mathbf{x}d\mathbf{x}' \langle \mathbf{x}|\hat{h}^s(t)|\mathbf{x}'\rangle \hat{\psi}^\dagger(\mathbf{x})\hat{\psi}(\mathbf{x}')} \qquad \hat{h}^s(t) \equiv \hat{h}(t) + \sum_\xi \hat{\mathfrak{a}}_{\xi^*}(t) s_\xi(t), \tag{16.85}$$

the *shifted* Hamiltonian for phosons reads

$$\boxed{\hat{H}_{0,\text{phos}}^s = \hat{H}_{0,\text{phos}} + \sum_\xi \mathfrak{b}_{\xi^*}(t)\,\hat{\phi}_\xi}$$

and the interaction Hamiltonian between particles and shifted phosons reads

$$\boxed{\hat{H}_{\mathfrak{a}}(t) \equiv \sum_\xi \int d\mathbf{x}d\mathbf{x}' \langle \mathbf{x}|\hat{\mathfrak{a}}_{\xi^*}(t)|\mathbf{x}'\rangle \hat{\psi}^\dagger(\mathbf{x})\hat{\psi}(\mathbf{x}')\,\hat{\phi}_\xi^s}$$

In this way, $\hat{H}_{0,\text{phos}}^s$ is written in terms of the original displacements $\hat{\phi}$, whereas $\hat{H}_{\mathfrak{a}}$ is written in terms of the shifted displacements $\hat{\phi}^s$.

Equation of motion for the field operator From the commutator $[\hat{\psi}(\mathbf{x}), \hat{H}(t)]$ we can easily derive the equation of motion

$$\boxed{\begin{aligned}\int d\mathbf{x}' \langle \mathbf{x}|\, \mathrm{i}\frac{d}{dz} - \hat{h}^s(z)|\mathbf{x}'\rangle \hat{\psi}_H(\mathbf{x}', z) &= \int d\mathbf{x}' v(\mathbf{x}, \mathbf{x}', z)\hat{n}_H(\mathbf{x}', z)\hat{\psi}_H(\mathbf{x}, z) \\ &+ \sum_\xi \int d\mathbf{x}' \mathfrak{a}_{\xi^*}(\mathbf{x}, \mathbf{x}', z)\hat{\psi}_H(\mathbf{x}', z)\,\hat{\phi}_{\xi,H}^s(z)\end{aligned}} \tag{16.86}$$

We observe that setting $v = \mathfrak{a} = 0$ in the r.h.s. of (16.80) and (16.86) we obtain the equations of motion of the field operators governed by the Hamiltonian $\hat{H}_0^s + \hat{H}_{0,\text{phos}}^s$. We use this observation in the next section.

16.5 Phosonic Green's Functions and Martin–Schwinger Hierarchy

The equations of motion suggest introducing an operator correlator made by shifted displacement operators. In analogy with the notation of Section 4.5, we use the short-hand notation

$$1 = \xi_1, z_1, \quad 2 = \xi_2, z_2, \quad \ldots; \quad 1^* = \xi_1^*, z_1, \quad 2^* = \xi_2^*, z_2, \quad \ldots$$

as well as

$$\tilde{1} = \tilde{\xi}_1, \tilde{z}_1, \quad \tilde{2} = \tilde{\xi}_2, \tilde{z}_2, \quad \ldots; \quad \tilde{1}^* = \tilde{\xi}_1^*, \tilde{z}_1, \quad \tilde{2}^* = \tilde{\xi}_2^*, \tilde{z}_2, \quad \ldots$$

etc., to assign the dependence of a displacement operator on the quantum number and contour time, hence $\hat{\phi}_{\xi_i,H}(z_i) = \hat{\phi}_H(i)$, $\hat{\phi}^s_{\xi_i,H}(z_i) = \hat{\phi}^s_H(i)$, etc.

Phosonic Green's functions We follow Ref. [280], but instead of using the displacement operators $\hat{\phi}$ we use the shifted displacement operators $\hat{\phi}^s$ to define the m-phoson Green's function[16] [compare with (5.1)]:

$$\widetilde{D}_m(1,2,\ldots,2m) \equiv \frac{1}{\mathrm{i}^m}\frac{\mathrm{Tr}\left[e^{-\beta\hat{H}^{\mathrm{M}}}\mathcal{T}\left\{\hat{\phi}^s_H(1)\hat{\phi}^s_H(2)\ldots\hat{\phi}^s_H(2m)\right\}\right]}{\mathrm{Tr}\left[e^{-\beta\hat{H}^{\mathrm{M}}}\right]}$$

$$= \frac{1}{\mathrm{i}^m}\frac{\mathrm{Tr}\left[\mathcal{T}\left\{e^{-\mathrm{i}\int_\gamma d\bar{z}\hat{H}(\bar{z})}\hat{\phi}^s(1)\hat{\phi}^s(2)\ldots\hat{\phi}^s(2m)\right\}\right]}{\mathrm{Tr}\left[\mathcal{T}\left\{e^{-\mathrm{i}\int_\gamma d\bar{z}\hat{H}(\bar{z})}\right\}\right]}, \tag{16.87}$$

where $\hat{H}(t_0 - \mathrm{i}\tau) = \hat{H}^{\mathrm{M}}$. In this definition m can also be a *half-integer.* We remark that the operators appearing in the second equality are not in the contour Heisenberg picture. An important property of the phosonic Green's function is the symmetry under an arbitrary permutation of its arguments:

$$\widetilde{D}_m(1,2,\ldots,2m) = \widetilde{D}_m(P(1),P(2),\ldots,P(2m))$$

for all permutations P of the integers $1,2,\ldots,2m$.

Martin–Schwinger hierarchy We proceed along the same lines as in Section 4.5 to find the equations of motion for the m-phoson Green's function $\widetilde{D}_m$ and the n-particle Green's function G_n. In analogy with definition (4.63) of the interparticle interaction, we define

$$\mathfrak{a}(\bar{1},\bar{1}';k) \equiv \delta(z_k,\bar{z}_1)\delta(z_k,\bar{z}_1')\mathfrak{a}_{\xi_k}(\bar{\mathbf{x}}_1,\bar{\mathbf{x}}_1',z_k), \tag{16.88}$$

and rewrite the equation of motion (16.80) as

$$\sum_{\xi'}\left[\mathrm{i}\frac{d}{dz_k}J_{\xi_k\xi'} - Q_{\xi_k\xi'}(z_k)\right]\hat{\phi}^s_{\xi',H}(z_k) = \int d\bar{1}d\bar{1}'\mathfrak{a}(\bar{1},\bar{1}';k)\hat{\psi}^\dagger_H(\bar{1})\hat{\psi}_H(\bar{1}').$$

Taking into account that the commutation relations for the shifted operators $\hat{\phi}^s$ are the same as for the original operators $\hat{\phi}$, see (16.82), we find for $m = 1/2, 1, 3/2, \ldots$

$$\sum_{\xi'}\left[\mathrm{i}\frac{d}{dz_k}J_{\xi_k\xi'} - Q_{\xi_k\xi'}(z_k)\right]\widetilde{D}_m(1,..,\xi',z_k,..,2m)$$
$$= \frac{1}{\mathrm{i}^m}\int d\bar{1}d\bar{1}'\mathfrak{a}(\bar{1},\bar{1}';k)\mathrm{Tr}\Big[\hat{\rho}\,\mathcal{T}\Big\{\hat{\phi}^s_H(1)\ldots\underbrace{\hat{\psi}^\dagger_H(\bar{1})\hat{\psi}_H(\bar{1}')}_{k\text{th place}}\ldots\hat{\phi}^s_H(2m)\Big\}\Big]$$
$$+\sum_{\substack{j=1\\ j\neq k}}^{2m}\delta_{\xi_k\xi_j^*}\delta(z_k,z_j)\widetilde{D}_{m-1}(1,..,\overset{\sqcap}{k},..,\overset{\sqcap}{j},..,2m)$$

[16] The phosonic Green's functions are the bosonic analog of the Nambu Green's functions in superconductivity.

where $\hat{\rho} = e^{-\beta \hat{H}^{\mathrm{M}}}/\mathrm{Tr}[e^{-\beta \hat{H}^{\mathrm{M}}}]$ is the density matrix operator and we define $\widetilde{D}_0 = 1$ and $\widetilde{D}_{-1/2} = 0$. We use the same notation introduced in Section 4.5, according to which the symbol $\sqcap$ above an index signifies that the index is missing from the list. The last term comes from the equal time commutator between displacement operators originating from the derivative of the θ-functions, see again Section 4.5. In a similar way we can work out the equation of motion for the n-particle Green's function and find for $n = 1, 2, 3, \ldots$ [compare with (4.64)]

$$\begin{aligned}
\int d\mathbf{x}' \langle \mathbf{x}_k | \mathrm{i}\frac{d}{dz_k} - \hat{h}^s(z_k) | \mathbf{x}' \rangle G_n(1, .., \mathbf{x}', z_k, .., n; 1', .., n') \\
= \frac{1}{\mathrm{i}^n} \int d\tilde{1} d\bar{1}' \mathfrak{a}(k, \bar{1}'; \tilde{1}^*) \mathrm{Tr}\Big[\hat{\rho}\, \mathcal{T}\Big\{\hat{\psi}_H(1) \ldots \underbrace{\hat{\phi}^s_H(\tilde{1}) \hat{\psi}_H(\bar{1}')}_{k\text{th place}} \ldots \hat{\psi}^\dagger_H(1')\Big\}\Big] \\
\pm \mathrm{i} \int d\bar{1}\, v(k; \bar{1})\, G_{n+1}(1, \ldots, n, \bar{1}; 1', \ldots, n', \bar{1}^+) \\
+ \sum_{j=1}^{n} (\pm)^{k+j}\, \delta(k; j')\, G_{n-1}(1, .., \overset{\sqcap}{k}, .., n; 1', .., \overset{\sqcap}{j'}, .., n')
\end{aligned}$$

where we define $G_0 = 1$. To distinguish the integration variables of the particle operators $\hat{\psi}$ from those of the phoson operators $\hat{\phi}^s$ we use a bar for the former and a tilde for the latter, thus $\int d\tilde{1} \equiv \int d\tilde{z}_1 \sum_{\tilde{\xi}_1}$. The equation of motion for G_n with derivative with respect to the primed arguments can be worked out similarly. All equations of motion must be solved with KMS boundary conditions – that is, $\widetilde{D}_m$ must be periodic on the contour with respect to all times $z_1, \ldots, z_{2m}$ and G_n must be periodic (antiperiodic) on the contour for bosons (fermions) with respect to all times $z_1, \ldots, z_n, z'_1, \ldots, z'_n$.

If the particle-phoson coupling $\mathfrak{a} = 0$ then $\widetilde{D}_m$ couples only to $\widetilde{D}_{m-1}$. The corresponding Green's function is denoted by $D_{0,m}$ and it is named the *noninteracting* phosonic Green's function. In the particle sector things are different. For $\mathfrak{a} = 0$ the equations of motion reduce to the Martin–Schwinger hierarchy (4.64) and (4.65), and G_n couples to G_{n-1} *and* G_{n+1} through the interparticle interaction v. For G_n to couple only to G_{n-1} the interparticle interaction has to vanish too.

When both interparticle and particle-phoson interactions are present, G_n couples to G_{n-1}, G_{n+1}, as well as to mixed Green's functions consisting of a mixed string of $\hat{\psi}$, $\hat{\psi}^\dagger$, and $\hat{\phi}^s$ operators, see the second line in the equations of motion. Likewise, $\widetilde{D}_m$ couples to $\widetilde{D}_{m-1}$ but also to mixed Green's functions, see the second line in the equations of motion. The equations of motion for the mixed Green's functions can be derived in precisely the same way, but it turns out that we do not need them for our purposes. For the derivation of these equations see Ref. [281]. The full set of equations is referred to as the Martin–Schwinger hierarchy. The exact solution is, of course, out of reach in most physical systems of relevance.

16.6 Wick's Theorem for the Phosonic Green's Functions

Wick's theorem provides the solution of the Martin–Schwinger hierarchy with r.h.s. evaluated at $\mathfrak{a} = v = 0$. This is the same as solving the Martin–Schwinger hierarchy for a system of electrons and phonons governed by the Hamiltonian $\hat{H}_0^s + \hat{H}_{0,\text{phos}}^s$, see comment below (16.86). As $\hat{H}_0^s$ depends on $\mathfrak{a}$ explicitly, setting $\mathfrak{a} = v = 0$ in the r.h.s. is not the same as solving the Martin–Schwinger hierarchy with $\hat{H}|_{\mathfrak{a}=v=0}$ (noninteracting hierarchy). Henceforth we name the Green's functions governed by the Hamiltonian $\hat{H}_0^s + \hat{H}_{0,\text{phos}}^s$ as the *independent* Green's functions and we denote them by $D_{0,m}$ and $G_{0,n}^s$. Wick's theorem turns out to be extremely useful to develop the diagrammatic theory. Let us consider the equation of motion for $D_{0,m}$. We have

$$\int d\tilde{1}\, \overrightarrow{D}_0^{-1}(k,\tilde{1}) D_{0,m}(1,..,\tilde{1},..,2m) = \sum_{\substack{j=1 \\ j\neq k}}^{2m} \delta(k,j^*) D_{0,m-1}(1,..,\sqcap\!\!\!\!k,..,\sqcap\!\!\!\!j,..,2m), \quad (16.89)$$

where the variable $\tilde{1}$ in the l.h.s. is at place k and we define

$$\overrightarrow{D}_0^{-1}(k,\tilde{1}) \equiv \left[\mathrm{i}\frac{d}{dz_k} J_{\boldsymbol{\xi}_k\tilde{\boldsymbol{\xi}}_1} - Q_{\boldsymbol{\xi}_k\tilde{\boldsymbol{\xi}}_1}(z_k)\right]\delta(z_k,\tilde{z}_1),$$

$$\delta(k,j) \equiv \delta_{\boldsymbol{\xi}_k\boldsymbol{\xi}_j}\delta(z_k,z_j).$$

The independent hierarchy (16.89) resembles the hierarchy for the noninteracting Green's functions, see (5.8) and (5.9). It is therefore not surprising that Wick's theorem for the independent $D_{0,m}$ and noninteracting $G_{0,n}$ is similar. The difference for the phosonic Green's functions is that all operators need to be contracted with each other, as there is no separation into creation and annihilation operators. Another feature that is different from the particle case is the appearance of an odd number of displacement operators when m is a half-integer. From (16.89) we see that the integer m connects with the integer $m-1$ and the half-integer m connects with the half-integer $m-1$. As such, we have two separate hierarchies of equations.

Wick's theorem The proof of Wick's theorem goes along the same lines as in Ref. [280]. The phosonic Green's functions $D_{0,m} = 0$ for all half-integers m. This can easily be proven by considering the ensemble average of the equation of motion (16.80) with $\mathfrak{a} = 0$. By definition this average yields the equation of motion for $D_{0,1/2}$ and the only solution satisfying the KMS boundary conditions is $D_{0,1/2} = 0$. Moreover, the Martin–Schwinger hierarchy couples half-integers to half-integers. Consider $m = 3/2$:

$$\int d\tilde{1}\, \overrightarrow{D}_0^{-1}(1,\tilde{1}) D_{0,3/2}(\tilde{1},2,3) = \delta(1,2^*)D_{0,1/2}(3) + \delta(1,3^*)D_{0,1/2}(2) = 0,$$

and the like for the variables 2 and 3. We see that $D_{0,3/2} = 0$ is a solution satisfying the KMS boundary conditions. By induction, $D_{0,m} = 0$ for half-integers. Henceforth we only consider integers m in the noninteracting case.

For $m = 1$ we have the equation of motion for $D_{0,1}$:

$$\int d\tilde{1}\, \overrightarrow{D}_0^{-1}(1,\tilde{1}) D_{0,1}(\tilde{1},2) = \delta(1,2^*). \quad (16.90)$$

Comparing with (16.78) we realize that $D_{0,1} = D_0$. Without any risk of ambiguity we denote $D_{0,1}(1,2)$ simply by $D_0(1,2)$ in the remainder of the book. For $D_{0,m}$ with $m > 1$ the solution of (16.89) is given by the so-called *hafnian* [31, 281, 282]. The hafnian can be defined recursively starting from any of the arguments in $D_{0,m}$. Choosing, for instance, the argument k, we have

$$D_{0,m}(1,..,2m) = \sum_{\substack{j=1 \\ j\neq k}}^{2m} D_0(k,j) D_{0,m-1}(1,..,\overset{\sqcap}{k},..,\overset{\sqcap}{j},..,2m). \tag{16.91}$$

As an example, consider $m = 3$. Then (16.91) with $k = 1$ yields

$$\begin{aligned} D_{0,3}(1,2,3,4,5,6) &= D_0(1,2)D_{0,2}(3,4,5,6) + D_0(1,3)D_{0,2}(2,4,5,6) \\ &+ D_0(1,4)D_{0,2}(2,3,5,6) + D_0(1,5)D_{0,2}(2,3,4,6) \\ &+ D_0(1,6)D_{0,2}(2,3,4,5), \end{aligned}$$

and using again (16.91) for $D_{0,2}$ we obtain an expansion of $D_{0,3}$ in terms of products of three D_0. The recursive form (16.91) makes it clear that $D_{0,m}$ satisfies (16.89) and the KMS boundary conditions.

We observe that the hafnian can be written in a more symmetric form using the symmetry $D_0(1,2) = D_0(2,1)$:

$$D_{0,m}(1,..,2m) = \frac{1}{2^m m!} \sum_P D_0(P(1),P(2)) \dots D_0(P(2m-1),P(2m)). \tag{16.92}$$

This rewriting highlights the main difference between Wick's theorem for $G_{0,n}$ and the one for $D_{0,m}$. For $G_{0,n}$ we need to connect unprimed arguments to primed arguments in all possible ways. For $D_{0,m}$ there is no such distinction, and we need to connect all arguments in all possible ways.

16.7 Exact Green's Functions from Wick's Theorem

The interacting Green's functions G_n and $\widetilde{D}_m$ can be expanded in powers of the interparticle interaction v and particle–phoson interaction $\mathfrak{a}$. In this section we focus on the one-particle Green's functions $G \equiv G_1$, the one-phoson Green's function $\widetilde{D} \equiv \widetilde{D}_1$, and the half-phoson Green's function $\widetilde{D}_{1/2}$. This section follows closely Ref. [280]. In Section 16.9 we show that the perturbative expansion leads to a closed system of equations for these quantities.

The starting point is the full Hamiltonian written in the form (16.84). Let us define the Hamiltonian for independent particles and phosons as

$$\hat{H}_{\rm ind}(t) = \hat{H}_0^s(t) + \hat{H}_{0,\rm phos}^s(t),$$

so that $\hat{H} = \hat{H}_{\rm ind} + \hat{H}_{\rm int} + \hat{H}_{\mathfrak{a}}$. Inside the contour-ordering the Hamiltonians $\hat{H}_{\rm ind}$, $\hat{H}_{\rm int}$, and $\hat{H}_{\mathfrak{a}}$ can be treated as commuting operators and hence the exponential of their sum can be separated into the product of three exponentials. It is then natural to define the independent averages as

$$\langle \mathcal{T}\{\ldots\}\rangle_{\text{ind}} \equiv \text{Tr}\Big[\mathcal{T}\{e^{-\text{i}\int_\gamma d\bar{z}\hat{H}_{\text{ind}}(\bar{z})}\ldots\}\Big].$$

It is worth stressing that the independent averages are not the same as the noninteracting averages – that is, the averages with $v = \mathfrak{a} = 0$ – since $\hat{H}_0^s$ contains the shifted Hamiltonian $\hat{h}^s$, see (16.85).

One-particle Green's function The one-particle Green's function $G \equiv G_1$ is defined in (5.1). Taking into account that $\hat{H} = \hat{H}_{\text{ind}} + \hat{H}_{\text{int}} + \hat{H}_{\mathfrak{a}}$, we have

$$G(a;b) = \frac{1}{\text{i}}\frac{\text{Tr}\left[\mathcal{T}\left\{e^{-\text{i}\int_\gamma d\bar{z}\hat{H}_{\text{ind}}(\bar{z})}e^{-\text{i}\int_\gamma d\bar{z}\hat{H}_{\text{int}}(\bar{z})}e^{-\text{i}\int_\gamma d\bar{z}\hat{H}_{\mathfrak{a}}(\bar{z})}\hat{\psi}(a)\hat{\psi}^\dagger(b)\right\}\right]}{\text{Tr}\left[\mathcal{T}\left\{e^{-\text{i}\int_\gamma d\bar{z}\hat{H}_{\text{ind}}(\bar{z})}e^{-\text{i}\int_\gamma d\bar{z}\hat{H}_{\text{int}}(\bar{z})}e^{-\text{i}\int_\gamma d\bar{z}\hat{H}_{\mathfrak{a}}(\bar{z})}\right\}\right]}. \tag{16.93}$$

The denominator in this equation is the interacting partition function Z. Expanding the exponentials containing $\hat{H}_{\text{int}}$ and $\hat{H}_{\mathfrak{a}}$, we find

$$G(a;b) = \frac{1}{\text{i}}\frac{\sum\limits_{k,p=0}^{\infty}\frac{(-\text{i})^{k+p}}{k!\,p!}\int_\gamma dz_1..d\tilde{z}_1..\langle\mathcal{T}\left\{\hat{H}_{\text{int}}(z_1)..\hat{H}_{\text{int}}(z_k)\hat{H}_{\mathfrak{a}}(\tilde{z}_1)..\hat{H}_{\mathfrak{a}}(\tilde{z}_p)\hat{\psi}(a)\hat{\psi}^\dagger(b)\right\}\rangle_{\text{ind}}}{\sum\limits_{k,p=0}^{\infty}\frac{(-\text{i})^{k+p}}{k!\,p!}\int_\gamma dz_1..d\tilde{z}_1..\langle\mathcal{T}\left\{\hat{H}_{\text{int}}(z_1)..\hat{H}_{\text{int}}(z_k)\hat{H}_{\mathfrak{a}}(\tilde{z}_1)..\hat{H}_{\mathfrak{a}}(\tilde{z}_p)\right\}\rangle_{\text{ind}}}.$$

To facilitate the identification of the expansion terms, we use a tilde for the contour times of the particle–phoson interaction Hamiltonian.

Let us write the integrated Hamiltonians in terms of $v(i;j)$ and $\mathfrak{a}(i,j;k)$. We have

$$\int dz_j\hat{H}_{\text{int}}(z_j) = \frac{1}{2}\int dj\,dj' v(j;j')\hat{\psi}^\dagger(j^+)\hat{\psi}^\dagger(j'^+)\hat{\psi}(j')\hat{\psi}(j), \tag{16.94a}$$

$$\int d\tilde{z}_j\hat{H}_{\mathfrak{a}}(\tilde{z}_j) = \int d\bar{j}d\bar{j}d\bar{j}'\mathfrak{a}(\bar{j},\bar{j}';\tilde{j}^*)\hat{\psi}^\dagger(\bar{j}^+)\hat{\psi}(\bar{j}')\hat{\phi}^s(\tilde{j}). \tag{16.94b}$$

We thus see that to evaluate the Green's function we need to be able to calculate contour-ordered strings like $\langle\mathcal{T}\{\hat{\psi}\ldots\hat{\psi}\hat{\psi}^\dagger\ldots\hat{\psi}^\dagger\hat{\phi}^s\ldots\hat{\phi}^s\}\rangle_{\text{ind}}$ with an arbitrary number of operators. To make progress, it is crucial to observe that $\hat{H}_0^s$ acts on the Fock space $\mathbb{F}$ of the particles and $\hat{H}_{0,\text{phos}}^s$ acts on the Hilbert space $\mathbb{D}_N$ of N distinguishable particles. As such, the eigenkets of $\hat{H}_{\text{ind}} = \hat{H}_0^s + \hat{H}_{0,\text{phos}}^s$ factorize into tensor products of kets in $\mathbb{F}$ and kets in $\mathbb{D}_N$. Therefore, the partition function for independent particles and phosons

$$\begin{aligned}Z_{\text{ind}} &= \text{Tr}\big[\mathcal{T}\{e^{-\text{i}\int_\gamma d\bar{z}\,\hat{H}_{\text{ind}}(\bar{z})}\}\big]\\&= \text{Tr}\big[\mathcal{T}\{e^{-\text{i}\int_\gamma d\bar{z}\,\hat{H}_0^s(\bar{z})}\}\big]\times\text{Tr}\big[\mathcal{T}\{e^{-\text{i}\int_\gamma d\bar{z}\,\hat{H}_{0,\text{phos}}^s(\bar{z})}\}\big]\\&= Z_0^s\times Z_{0,\text{phos}}^s\end{aligned}$$

factorizes into particle and phoson contributions. The same type of factorization allows us to simplify the independent average of any string of operators as

$$\langle\mathcal{T}\{\hat{\psi}\ldots\hat{\psi}\hat{\psi}^\dagger\ldots\hat{\psi}^\dagger\hat{\phi}^s\ldots\hat{\phi}^s\}\rangle_{\text{ind}} = \langle\mathcal{T}\{\hat{\psi}\ldots\hat{\psi}\hat{\psi}^\dagger\ldots\hat{\psi}^\dagger\}\rangle_0^s\times\langle\mathcal{T}\{\hat{\phi}^s\ldots\hat{\phi}^s\}\rangle_{0,\text{phos}}^s, \tag{16.95}$$

where the average $\langle\ldots\rangle_0^s$ is performed with $\hat{H}_0^s$ and the average $\langle\ldots\rangle_{0,\text{phos}}^s$ is performed with $\hat{H}_{0,\text{phos}}^s$.

To order k in v and to order p in $\mathfrak{a}$ the numerator of the Green's function contains the independent average of the following string:

$$
\begin{aligned}
&\langle\mathcal{T}\Big\{\underbrace{\hat{\psi}^\dagger(1^+)\hat{\psi}^\dagger(1'^+)\hat{\psi}(1')\hat{\psi}(1)\ldots}_{4k\text{ operators}}\underbrace{\hat{\psi}^\dagger(\bar{1}^+)\hat{\psi}(\bar{1}')\hat{\phi}^s(\tilde{1})\ldots}_{3p\text{ operators}}\hat{\psi}(a)\hat{\psi}^\dagger(b)\Big\}\rangle_{\text{ind}} \\
&= (\pm)^p\langle\mathcal{T}\Big\{\hat{\psi}(a)\underbrace{\hat{\psi}(1)\hat{\psi}(1')\ldots}_{2k\,\hat{\psi}}\underbrace{\hat{\psi}(\bar{1}')\ldots}_{p\,\hat{\psi}}\underbrace{\ldots\hat{\psi}^\dagger(\bar{1}^+)}_{p\,\hat{\psi}^\dagger}\underbrace{\ldots\hat{\psi}^\dagger(1'^+)\hat{\psi}^\dagger(1^+)}_{2k\,\hat{\psi}^\dagger}\hat{\psi}^\dagger(b)\Big\}\rangle_0^s \\
&\times\langle\mathcal{T}\Big\{\underbrace{\hat{\phi}^s(\tilde{1})\ldots}_{p\,\hat{\phi}^s}\Big\}\rangle_{0,\text{phos}}^s \\
&= (\pm)^p Z_0^s\,\mathrm{i}^{2k+p+1}G_{0,2k+p+1}^s(a,1,1',\ldots,\bar{1}',\ldots;b,1^+,1'^+,\ldots,\bar{1}^+,\ldots) \\
&\times Z_{0,\text{phos}}^s\,\mathrm{i}^{p/2}D_{0,p/2}(\tilde{1},\ldots),
\end{aligned}
$$

where in the last equality $G_{0,n}^s$ is the noninteracting n-particle Green's function *calculated with the shifted single-particle Hamiltonian* $\hat{h}^s$. With similar manipulations we can work out the expansion of the partition function. Taking into account that $D_{0,p/2}$ is nonvanishing only for even integers p, the expansion of the interacting Green's function reads

$$
\begin{aligned}
G(a;b) = \frac{Z_{\text{ind}}}{Z}\sum_{k,p=0}^{\infty}\frac{\mathrm{i}^{k+p}}{2^k k!\,(2p)!}\int d1d1'\ldots v(1;1')\ldots\int d\bar{1}d\bar{1}d\bar{1}'\ldots\mathfrak{a}(\bar{1},\bar{1}';\tilde{1}^*)\ldots \\
\times\, G_{0,2k+2p+1}^s(a,1,1',\ldots,\bar{1}',\ldots;b,1^+,1'^+,\ldots,\bar{1}^+,\ldots)D_{0,p}(\tilde{1},\ldots),
\end{aligned}
\tag{16.96}
$$

with

$$
\begin{aligned}
\frac{Z}{Z_{\text{ind}}} = \sum_{k,p=0}^{\infty}\frac{\mathrm{i}^{k+p}}{2^k k!\,(2p)!}\int d1d1'\ldots v(1;1')\ldots\int d\tilde{1}d\bar{1}d\bar{1}'\ldots\mathfrak{a}(\bar{1},\bar{1}';\tilde{1}^*)\ldots \\
\times\, G_{0,2k+2p}^s(1,1',\ldots,\bar{1}',\ldots;1^+,1'^+,\ldots,\bar{1}^+,\ldots)D_{0,p}(\tilde{1},\ldots).
\end{aligned}
\tag{16.97}
$$

We emphasize again that in expansion (16.96) the zeroth-order term ($k = p = 0$) is not the same as the noninteracting ($v = \mathfrak{a} = 0$) Green's function $G_{0,1}$ since $G_{0,1}^s$ is calculated with $\hat{h}^s$.

Using Wick's theorem for $G_{0,n}^s$ and $D_{0,m}$, results (16.96) and (16.97) provide an exact expansion in terms of the one-particle Green's function $G_{0,1}^s$ and the one-phoson Green's function $D_{0,1}$. Henceforth we shorten the notation and in addition to writing $D_0 \equiv D_{0,1}$ we also write $G_0^s \equiv G_{0,1}^s$.

One-phoson Green's function The one-phoson Green's function $\widetilde{D} \equiv \widetilde{D}_1$ is defined in (16.87). Splitting the Hamiltonian as in (16.93), expanding the exponentials containing $\hat{H}_{\text{int}}$

and $\hat{H}_{\mathfrak{a}}$, and using (16.94), we find

$$\widetilde{D}(\tilde{a},\tilde{b}) = \frac{Z_{\text{ind}}}{Z} \sum_{k,p=0}^{\infty} \frac{\mathrm{i}^{k+p}}{2^k k!\,(2p)!} \int d1d1' \ldots v(1;1') \ldots \int d\tilde{1}d\bar{1}d\bar{1}' \ldots \mathfrak{a}(\bar{1},\bar{1}';\tilde{1}^*;) \ldots$$
$$\times\, G^s_{0,2k+2p}(1,1',\ldots,\bar{1}',\ldots;1^+,1'^+,\ldots,\bar{1}^+,\ldots) D_{0,p+1}(\tilde{a},\tilde{b},\tilde{1},\ldots), \tag{16.98}$$

where we take into account that $D_{0,m}$ vanishes for half-integers m. Using Wick's theorem for $G^s_{0,n}$ and $D_{0,m}$ we have an exact expansion of the interacting one-phoson Green's function in terms of G^s_0 and D_0. The zeroth-order term ($k = p = 0$) yields the noninteracting ($v = \mathfrak{a} = 0$) result since $G_{0,0} = 1$.

Half-phoson Green's function The half-phoson Green's function $\widetilde{D}_{1/2}$ is proportional to the time-dependent ensemble average ϕ^s of the displacement operator $\hat{\phi}^s$:

$$\widetilde{D}_{1/2}(\tilde{a}) = \frac{1}{\mathrm{i}^{1/2}}\,\phi^s(\tilde{a}).$$

Proceeding along the same lines as for the derivation of the expansion (16.98), we find

$$\phi^s(\tilde{a}) = \frac{Z_{\text{ind}}}{Z} \sum_{k,p=0}^{\infty} \frac{(\pm)\mathrm{i}^{k+p+1}}{2^k k!\,(2p+1)!} \int d1d1' \ldots v(1;1') \ldots \int d\tilde{1}d\bar{1}d\bar{1}' \ldots \mathfrak{a}(\bar{1},\bar{1}';\tilde{1}^*) \ldots$$
$$\times\, G^s_{0,2k+2p+1}(1,1',\ldots,\bar{1}',\ldots;1^+,1'^+,\ldots,\bar{1}^+,\ldots) D_{0,p+1}(\tilde{a},\tilde{1},\ldots), \tag{16.99}$$

where we take into account that $D_{0,m}$ vanishes for half-integers m. The average ϕ^s vanishes for $\mathfrak{a} = 0$, in agreement with the equation of motion (16.80).

16.8 Diagrammatic Expansions

The expansions of G, $\widetilde{D}$, and ϕ^s contain the Green's functions G^s_0 and D_0, the interparticle interaction v and the particle-phoson interaction $\mathfrak{a}$. Let us assign a graphical object to these quantities. We continue using an oriented line from 2 to 1 to represent $G^s_0(1;2)$, and a wiggly line from 2 to 1 to represent $v(1;2) = v(2;1)$. For the noninteracting phosonic Green's function $D_0(1,2) = D_0(2,1)$ we use a spring from 1 to 2:

$$D_0(1,2) = 1 \quad 2$$

The particle-phoson coupling $\mathfrak{a}(1,2;3^*)$ is instead represented by the triangle

$$\mathfrak{a}(1,2;3^*) =$$

3

1 2

where the third argument is attached to the vertex with the white tip and the first argument is attached to the vertex on the right of an imaginary oriented line going from the white tip to the middle of the opposite edge.

We can now represent every term of the expansions with diagrams. As in the case of only particles, the diagrams for G are either connected or products of a connected diagram and a vacuum diagram. In a connected diagram for $G(a;b)$ all internal vertices are connected to *both* a and b through G_0^s, D_0, v, and $\mathfrak{a}$. Thus, a disconnected G-diagram is characterized by a subset of internal vertices that are not connected to either a or b and hence they form a vacuum diagram. Similarly, the diagrams for $\phi^s(\tilde{a})$ fall into two main classes: those with all internal vertices connected to $\tilde{a}$ and those where a subset of internal vertices is disconnected, thus forming a vacuum diagram. The diagrams for the one-phoson Green's function $\widetilde{D}(\tilde{a},\tilde{b})$ can instead be grouped into *three* different classes: (c1) all internal vertices connected to *both* $\tilde{a}$ and $\tilde{b}$; (c2) a subset of internal vertices connected only to $\tilde{a}$ and the complementary set connected only to $\tilde{b}$; and (c3) diagrams where a subset of internal vertices is not connected to either $\tilde{a}$ or $\tilde{b}$, thus forming a vacuum diagram. In all cases the contributions containing vacuum diagrams factorize and cancel with the expansion of Z. Furthermore, it turns out that many connected diagrams are topologically equivalent and as such we only need to sum over the topologically inequivalent ones. The number of topologically equivalent diagrams cancel the combinatorial factor $2^k k!\,(2p)!$ in (16.96) and (16.98) and $2^k k!\,(2p+1)!$ in (16.99). The proof of these statements goes along the same lines as the proof for only particles, see Chapter 7 and also Ref. [281]. The resulting formulas for G, $\widetilde{D}$, and ϕ^s become [280]:

$$G(a;b) = \sum_{k,p=0}^{\infty} \mathrm{i}^{k+p} \int d1d1' \ldots v(1;1') \ldots \int d\tilde{1}d\bar{1}d\bar{1}' \ldots \mathfrak{a}(\bar{1},\bar{1}';\tilde{1}^*) \ldots$$
$$\times\, G^s_{0,2k+2p+1}(a,1,1',\ldots,\bar{1}',\ldots;b,1^+,1'^+,\ldots,\bar{1}^+,\ldots) D_{0,p}(\tilde{1},\ldots)\Big|_{\substack{c\\t.i.}}, \quad (16.100)$$

$$\widetilde{D}(\tilde{a},\tilde{b}) = \sum_{k,p=0}^{\infty} \mathrm{i}^{k+p} \int d1d1' \ldots v(1;1') \ldots \int d\tilde{1}d\bar{1}d\bar{1}' \ldots \mathfrak{a}(\bar{1},\bar{1}';\tilde{1}^*) \ldots$$
$$\times\, G^s_{0,2k+2p}(1,1',\ldots,\bar{1}',\ldots;1^+,1'^+,\ldots,\bar{1}^+,\ldots) \mathrm{i}^{1+l} D_{0,p+1}(\tilde{a},\tilde{b},\tilde{1},\ldots)\Big|_{\substack{c\\t.i.}}, \quad (16.101)$$

and

$$\phi^s(\tilde{a}) = \sum_{k,p=0}^{\infty} (\pm)\mathrm{i}^{k+p+1} \int d1d1' \ldots v(1;1') \ldots \int d\tilde{1}d\bar{1}d\bar{1}' \ldots \mathfrak{a}(\bar{1},\bar{1}';\tilde{1}^*) \ldots$$
$$\times\, G^s_{0,2k+2p+1}(1,1',\ldots,\bar{1}',\ldots;1^+,1'^+,\ldots,\bar{1}^+,\ldots) D_{0,p+1}(\tilde{a},\tilde{1},\ldots)\Big|_{\substack{c\\t.i.}}, \quad (16.102)$$

where the abbreviations c and $t.i.$ indicate that we consider only *connected* and *topologically inequivalent* diagrams when expanding $G^s_{0,m}$ in permanents/determinants and $D_{0,m}$ in hafnians. In particular, the expansion (16.101) for $\widetilde{D}$ contains all diagrams in classes (c1) and (c2).

Diagrams for the one-particle Green's function In Fig. 16.1 we show a few low-order Feynman diagrams for G. The Feynman rules to convert the diagrams into a mathematical expression are

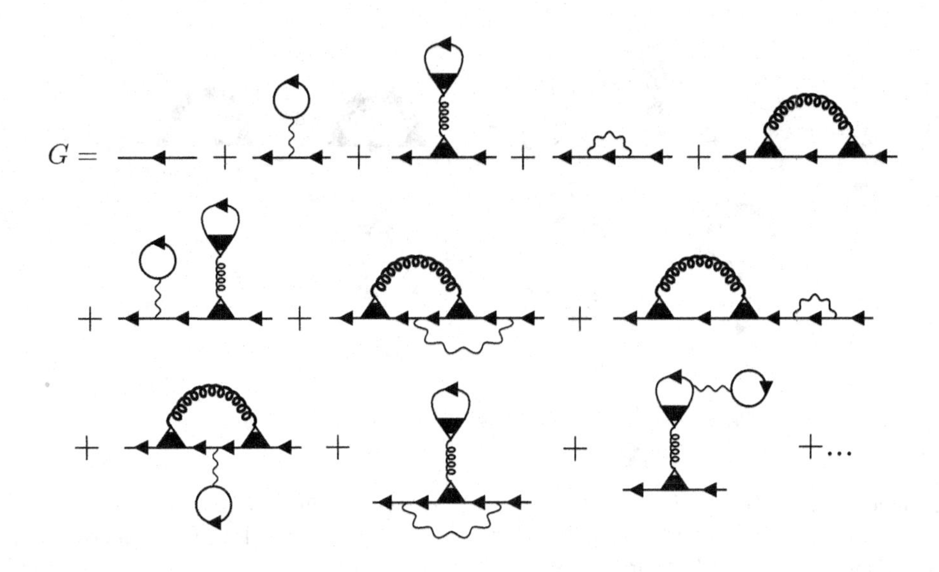

Figure 16.1 Low-order diagrams in the expansion of the interacting Green's function G.

- Number all vertices and assign an interaction $v(i;j)$ to a wiggly line connecting i and j, a Green's function $G_0^s(i;j^+)$ to an oriented line from j to i, a particle-phoson interaction $\mathfrak{a}(i,j;k^*)$ to a triangle with white tip in k and vertex i on the right of an imaginary oriented line going from the white tip to the middle of the opposite edge, and a phosonic Green's function $D_0(i,j)$ to a spring connecting i and j.
- Integrate over all internal vertices and multiply by $\mathrm{i}^{k+p}(\pm)^l$ where l is the number of *particle* loops, k is the number of wiggly lines, and $2p$ is the number of triangles.

We observe that there are diagrams (first and third diagrams in the second row of Fig. 16.1) containing repetitions of smaller blocks. As in the case of only particles, we define the irreducible self-energy Σ consisting of the set of diagrams contributing to G that, after the removal of the ingoing and outgoing G_0^s-lines, cannot be cut into two pieces by cutting an internal G_0^s-line. With this definition we can write G as a geometric series,

$$G = G_0^s + G_0^s \Sigma G_0^s + G_0^s \Sigma G_0^s \Sigma G_0^s + \ldots = G_0^s + G_0^s \Sigma G, \tag{16.103}$$

in which the products in this formula stand for convolutions. The self-energy $\Sigma = \Sigma[G_0^s, D_0, v, \mathfrak{a}]$ is itself expressed as an infinite sum of irreducible diagrams with G_0^s-lines and D_0-lines connected through v and $\mathfrak{a}$. Examples of diagrams for $\Sigma[G_0^s, D_0, v, \mathfrak{a}]$ are shown in Fig. 16.2.

Among the self-energy diagrams, there are some with self-energy insertions (those in the second line of Fig. 16.2) – that is, they contain pieces that can be made disjoint of the rest of the diagram by cutting two G_0^s-lines. The diagrams obtained by the removal of all self-energy insertions are called G-skeleton diagrams. The full set of Σ-diagrams can be

Figure 16.2 Low-order diagrams for the self-energy Σ.

obtained by dressing the skeleton diagrams with self-energy insertions in all possible ways, see Section 7.6. This amounts to the replacement of G_0^s by the full Green's function G in every diagram. We can therefore write

$$\Sigma = \Sigma[G_0^s, D_0, v, \mathfrak{a}] = \Sigma_s[G, D_0, v, \mathfrak{a}],$$

where we use the subscript "s" to denote that Σ_s is the sum of only G-skeleton diagrams.

Diagrams for the one-phoson and half-phoson Green's function Let us now proceed with the discussion of the phosonic Green's function $\widetilde{D}$. The diagrammatic structure comes from expanding $G_{0,n}^s$ and $D_{0,m}$ in (16.101), and example diagrams are shown in Fig. 16.3. The Feynman rules for the $\widetilde{D}$ diagrams are the same as for the G diagrams.

New types of diagrams are those belonging to class (c2). These are the double "tadpole" diagrams (second, fourth, fifth, and sixth diagrams) [280]. No such diagrams occur for G. The single tadpole diagrams constitute the diagrammatic expansion of the half-phoson Green's function $\widetilde{D}_{1/2} = \phi^s/\mathrm{i}^{1/2}$. They can be conveniently summed up, which is done in Fig. 16.4 where the oriented double line on the r.h.s. denotes the full Green's function G. The Feynman rules for the ϕ^s diagrams are the same as for the G diagrams and D diagrams, except that the prefactor is $\mathrm{i}^{k+p+1}(\pm)^l$, where l is the number of particle loops, k is the number of wiggly lines, and $2p+1$ is the number of triangles, see (16.102). We thus have

$$\phi^s(\tilde{a}) = \pm\mathrm{i}\int d\tilde{1}d\bar{1}d\bar{1}' D_0(\tilde{a}, \tilde{1})\mathfrak{a}(\bar{1}, \bar{1}'; \tilde{1}^*)G(\bar{1}'; \bar{1}^+), \tag{16.104}$$

a result that could have also been obtained by direct integration of the equation of motion (16.80). We can therefore write the expansion of $\widetilde{D}$ as

$$\widetilde{D}(\tilde{a}, \tilde{b}) = \widetilde{D}_{1/2}(\tilde{a})\widetilde{D}_{1/2}(\tilde{b}) + D_0(\tilde{a}, \tilde{b}) + (D_0\Pi D_0)(\tilde{a}, \tilde{b}) + (D_0\Pi D_0\Pi D_0)(\tilde{a}, \tilde{b}) + \ldots$$

in which the products in this formula stand for convolutions. The *phosonic self-energy* $\Pi = \Pi[G_0^s, D_0, v, \mathfrak{a}]$ is the collection of diagrams that cannot be made disjoint by cutting one D_0-line (one-D_0-line irreducible diagrams).

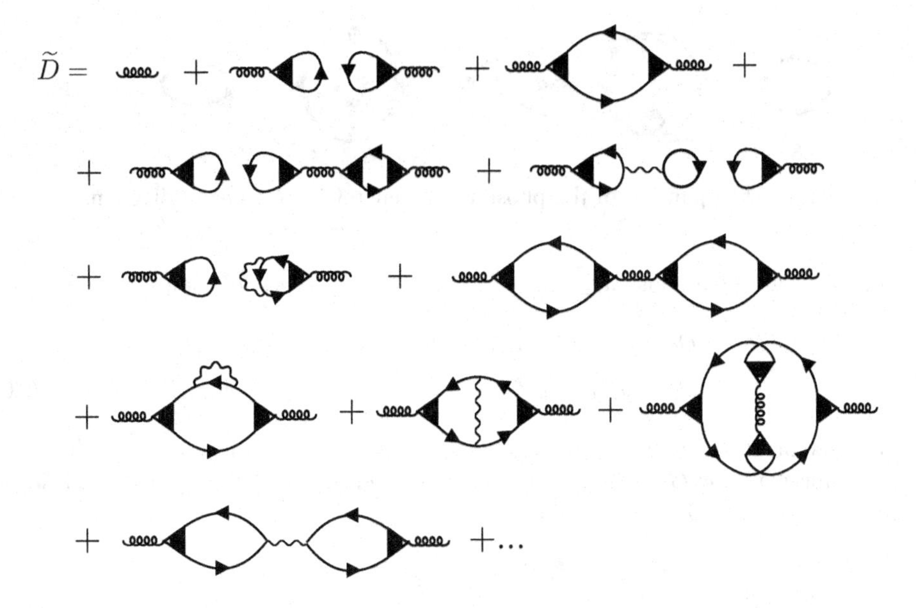

Figure 16.3 Low-order diagrams in the expansion of the interacting phosonic Green's function $\widetilde{D}$.

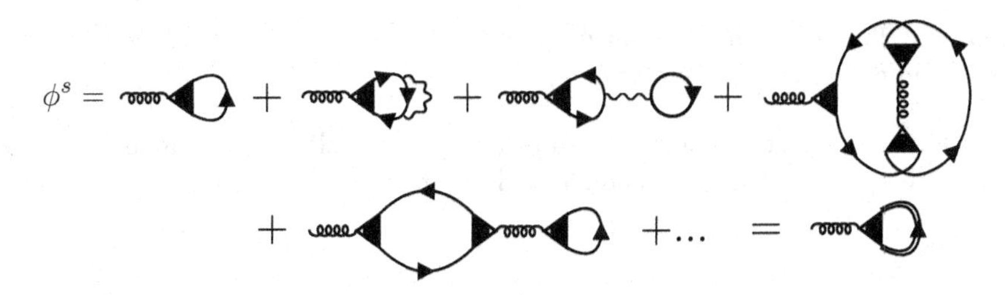

Figure 16.4 Resummation of the tadpole diagrams.

Differently from the particle case, $\widetilde{D}(\tilde{a},\tilde{b})$ does *not* fulfill a Dyson equation due to the presence of $\widetilde{D}_{1/2}(\tilde{a})\widetilde{D}_{1/2}(\tilde{b}) = \frac{1}{\mathrm{i}}\phi^s(\tilde{a})\phi^s(\tilde{b})$. This motivates us to define the fully connected phosonic Green's function [280]:

$$D(\tilde{a},\tilde{b}) \equiv \widetilde{D}(\tilde{a},\tilde{b}) - \frac{1}{\mathrm{i}}\phi^s(\tilde{a})\phi^s(\tilde{b}),$$

which has this term subtracted. $D(\tilde{a},\tilde{b})$ can alternatively be defined using the fluctuation operators $\Delta\hat{\phi}^s(\tilde{a}) \equiv \hat{\phi}^s(\tilde{a}) - \phi^s(\tilde{a}) = \hat{\phi}(\tilde{a}) - \phi(\tilde{a}) = \Delta\hat{\phi}(\tilde{a})$. We then have that

$$\boxed{D(\tilde{a},\tilde{b}) = \frac{1}{\mathrm{i}}\mathrm{Tr}\Big[\hat{\rho}\,\mathcal{T}\Big\{\Delta\hat{\phi}^s_H(\tilde{a})\Delta\hat{\phi}^s_H(\tilde{b})\Big\}\Big] = \frac{1}{\mathrm{i}}\mathrm{Tr}\Big[\hat{\rho}\,\mathcal{T}\Big\{\Delta\hat{\phi}_H(\tilde{a})\Delta\hat{\phi}_H(\tilde{b})\Big\}\Big]} \tag{16.105}$$

where $\hat{\rho} = e^{-\beta\hat{H}^{\mathrm{M}}}/Z$ is the density matrix operator.

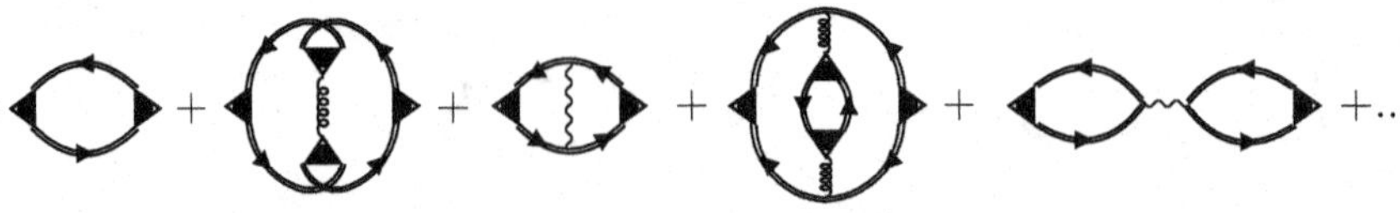

Figure 16.5 Expansion of the phosonic self-energy in G-skeleton diagrams.

The fully connected phosonic Green's function $D(\tilde{a}, \tilde{b})$ fulfills a Dyson equation since

$$\begin{aligned} D(\tilde{a}, \tilde{b}) &= D_0(\tilde{a}, \tilde{b}) + (D_0 \Pi D_0)(\tilde{a}, \tilde{b}) + (D_0 \Pi D_0 \Pi D_0)(\tilde{a}, \tilde{b}) + \ldots \\ &= D_0(\tilde{a}, \tilde{b}) + (D_0 \Pi D)(\tilde{a}, \tilde{b}). \end{aligned} \tag{16.106}$$

We can now proceed as before. We remove all diagrams with Σ-insertions inside the Π-diagrams (these are the G-skeleton diagrams for Π) and then replace G_0^s by the full Green's function G, thus obtaining the functional Π_s:

$$\Pi = \Pi[G_0^s, D_0, v, \mathfrak{a}] = \Pi_s[G, D_0, v, \mathfrak{a}].$$

Examples of Π_s diagrams are shown in Fig. 16.5.

Exercise 16.2 Prove that the vacuum diagrams in the expansion of G, $\widetilde{D}$, and ϕ^s can be factorized and cancel with the prefactor Z_{ind}/Z.

Exercise 16.3 Prove that the number of topologically equivalent diagrams for G and $\widetilde{D}$ to order k in the interparticle interaction v and $2p$ in the particle–phoson interaction $\mathfrak{a}$ is $2^k k!(2p)!$.

16.9 From the Skeletonic Expansion in D and W to the Hedin–Baym Equations

The topological idea of the skeletonic expansion in G can be further extended to the phosonic Green's function and the particle–particle interaction.

Skeletonic expansion in D We can remove all phosonic self-energy insertions inside the Π_s diagrams and then replace D_0 with D such that

$$\Pi = \Pi_s[G, D_0, v, \mathfrak{a}] = \Pi_{ss}[G, D, v, \mathfrak{a}].$$

Here, Π_{ss} contains all doubly skeletonic self-energy diagrams – that is, all those diagrams that do not contain either Σ-insertions or Π-insertions (the fourth diagram in Fig. 16.5 contains a Π-insertion and therefore is not doubly skeletonic). Examples of Π_{ss} diagrams are shown in Fig. 16.6, where the double spring represents D. A similar procedure can be carried out for the particle self-energy, except that for Σ we must exclude the time-local

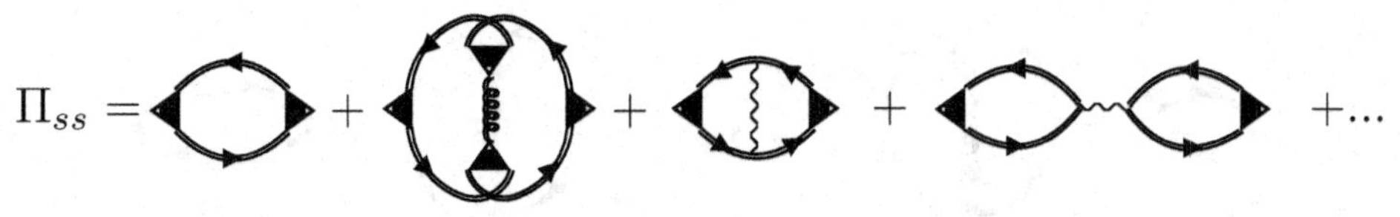

$\Pi_{ss} = \ldots + \ldots + \ldots + \ldots + \ldots$

Figure 16.6 Expansion of the phosonic self-energy in G-skeleton diagrams and D-skeleton diagrams.

$\Sigma_{ss} = \ldots + \ldots + \ldots + \ldots + \ldots + \ldots + \ldots$

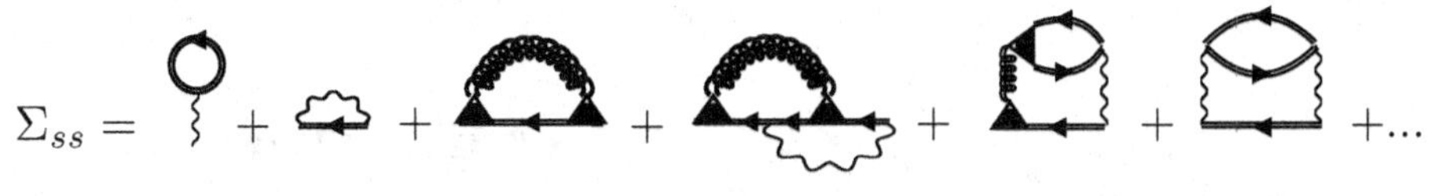

Figure 16.7 Expansion of the self-energy in G-skeleton diagrams and D-skeleton diagrams.

diagrams. In fact, a Π-insertion is here equivalent to a Σ-insertion and it would therefore lead to double-counting. Therefore,

$$\Sigma = \Sigma_s[G, D_0, v, \mathfrak{a}] = \Sigma_{\text{Eh}}[G, D_0, \mathfrak{a}] + \Sigma_{ss}[G, D, v, \mathfrak{a}],$$

where Σ_{Eh} is the second diagram in Fig. 16.2 with $G_0^s \to G$. Using the Feynman rules,

$$\begin{aligned} \Sigma_{\text{Eh}}[G, D_0, \mathfrak{a}](1;2) &= (\pm)\,\mathrm{i} \int d\tilde{1} d\tilde{2} d3 d4\, \mathfrak{a}(1,2;\tilde{1}^*) D_0(\tilde{1},\tilde{2}) \mathfrak{a}(3,4;\tilde{2}^*) G(4;3) \\ &= \int d\tilde{1}\, \mathfrak{a}(1,2;\tilde{1}^*) \phi^s(\tilde{1}), \end{aligned} \tag{16.107}$$

where in the last equality we use (16.104). The self-energy Σ_{Eh} is known as the *Ehrenfest* self-energy. In the context of electron–nuclear systems the *Ehrenfest approximation* consists of including the phononic feedback on the electrons through Σ_{Eh}, the electronic feedback on the phonons through the density matrix $G(\bar{1}'; \bar{1}^+)|_{\bar{z}_1' = \bar{z}_1} = \mathrm{i}n(\bar{\mathbf{x}}_1', \bar{\mathbf{x}}_1)$, see (16.104) and (8.20), and in setting $\Pi = 0$. In this approximation the nuclei are therefore treated as classical particles since they are described in terms of displacements and momenta only (i.e., the components of ϕ). Examples of Σ_{ss} diagrams are shown in Fig. 16.7. The analogy with the doubly skeletonic expansion in G and W discussed in Section 7.7 is evident; it has its origin in the same topological structure of W-diagrams and D-diagrams.

Skeletonic expansion in W Like in the case of only particles, we can further reduce the number of diagrams by removing all those diagrams containing a polarization insertion, which we here define as a piece that can be cut away by cutting two interaction lines v, *and at the same time* it does not break into two disjoint pieces by cutting one v-line or one D-line; an example is the sixth diagram in Fig. 16.7. The polarization diagrams are therefore one-v-line irreducible *and* one-D-line irreducible. In Fig. 16.8 we show a few low-order diagrams for the polarization $P = P_{ss}[G, D, v, \mathfrak{a}]$ which are both G-skeletonic and D-skeletonic. The polarization P contains both particle and phosons contributions. For later purposes we also define the one-v-line irreducible and one-D-line irreducible (and

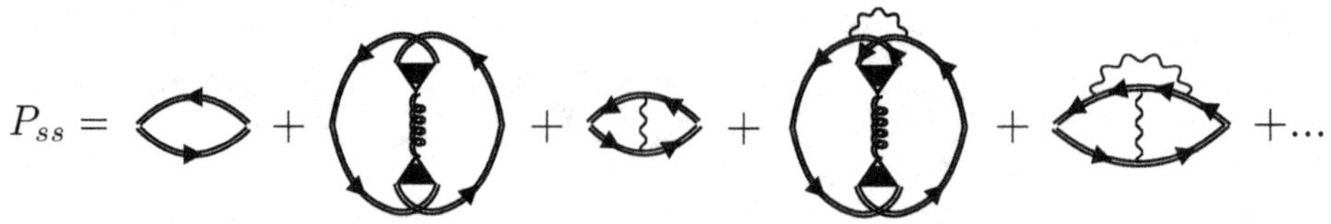

$P_{ss} =$ [diagram] + [diagram] + [diagram] + [diagram] + [diagram] + ...

Figure 16.8 Expansion of the polarization in G-skeleton diagrams and D-skeleton diagrams.

two-G-line reducible) kernel $K_{\mathrm{xc},r}$ in the same way as in (14.97),

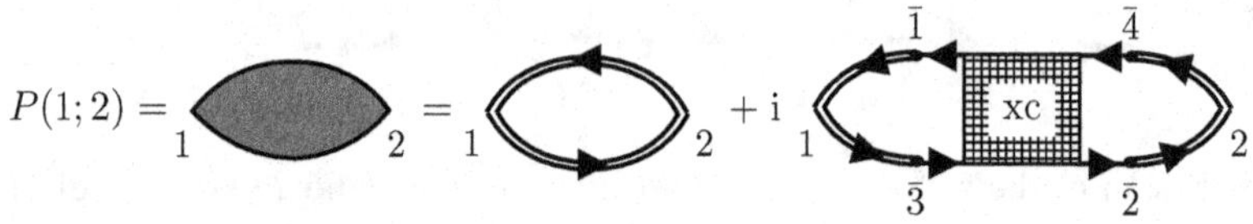

$$P(1;2) = \text{[diagram]} \qquad (16.108)$$

where the dark-gray bubble represents P and the square grid represents $K_{\mathrm{xc},r}(\bar{1},\bar{2};\bar{3},\bar{4}) = K_{\mathrm{xc},r}(\bar{2},\bar{1};\bar{4},\bar{3})$.

The polarization diagrams can be used to construct the *screened interaction* in the usual manner [compare with (7.22)]:

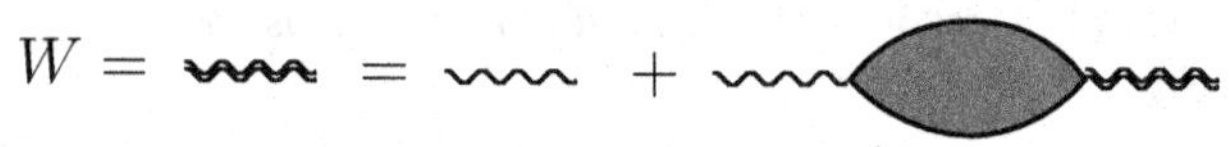

$$W = \text{[diagram]}$$

We say that a diagram is W-skeletonic if it does not contain P-insertions. Then the desired expression for Π and Σ is obtained by discarding all those diagrams that are not W-skeletonic, and then replacing v with W.

Phosonic self-energy For the phosonic self-energy we get

$$\Pi = \Pi_{ss}[G, D, v, \mathfrak{a}] = \Pi_{sss}[G, D, W, \mathfrak{a}],$$

where Π_{sss} is the sum of all the triply skeletonic self-energy diagrams – that is, all those diagrams that do not contain either Σ-insertions, Π-insertions, or P-insertions. In Fig. 16.9 we show a few low-order diagrams of the triply skeletonic expansion for Π_{sss}. They are all two-W-lines irreducible for otherwise they would contain a P-insertion. We then have either one-W-line reducible diagrams or one-W-line irreducible diagrams. We can write Π in a compact form using the one-W-line irreducible, one-D-line irreducible (and two-G-line reducible) kernel $K_{\mathrm{xc},r}$. We define the *dressed* (or *screened*) particle-phoson vertex $\mathfrak{a}^d$ as

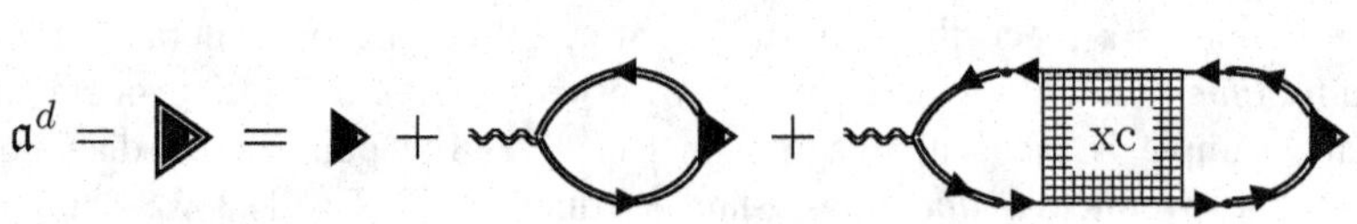

$$\mathfrak{a}^d = \text{[diagram]} \qquad (16.109)$$

By the same arguments as Section 14.9 one can show that $K_{\mathrm{xc},r}$ satisfies the Bethe–Salpeter equation (14.92) with $K_{\mathrm{xc}}(1,2;3,4) = \pm\frac{\delta\Sigma_{\mathrm{xc}}(1;3)}{\delta G(4;2)}$, see (14.93), and Σ_{xc} the XC self-energy

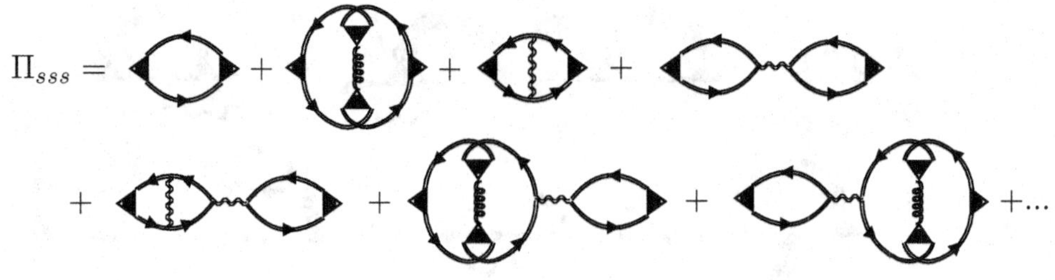

$\Pi_{sss} =$ + + + + + + +...

Figure 16.9 Expansion of the phosonic self-energy in G-, D-, and W-skeleton diagrams.

defined shortly in (16.110). In terms of the dressed particle-phoson vertex, the phosonic self-energy can be represented as [271]:

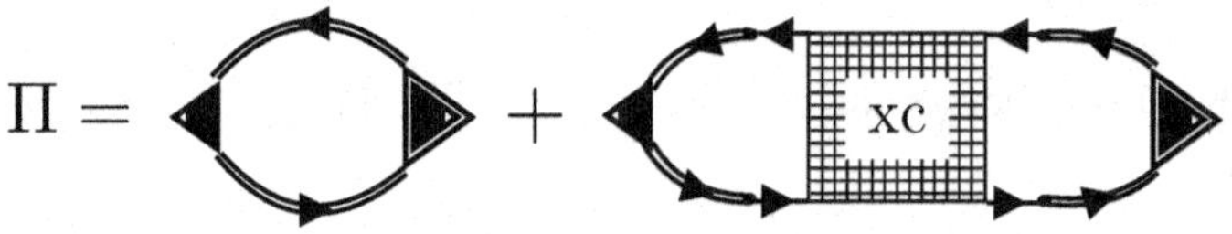

$\Pi =$ + xc

Particle self-energy For the particle self-energy Σ the only diagram for which we should not proceed with the replacement is the Hartree diagram (the first diagram in Fig. 16.7) since here every polarization insertion is equivalent to a self-energy insertion, hence the replacement $v \to W$ would lead to double-counting. Therefore,

$$\begin{aligned}\Sigma &= \Sigma_{\mathrm{Eh}}[G, D_0, \mathfrak{a}] + \Sigma_{ss}[G, D, v, \mathfrak{a}] \\ &= \Sigma_{\mathrm{Eh}}[G, D_0, \mathfrak{a}] + \Sigma_{\mathrm{H}}[G, v] + \Sigma_{sss,\mathrm{xc}}[G, D, W, \mathfrak{a}],\end{aligned} \tag{16.110}$$

where Σ_{H} is the Hartree diagram while the remaining part is the XC self-energy. We illustrate in Fig. 16.10 a few low-order diagrams of the triply skeletonic expansion of $\Sigma_{sss,\mathrm{xc}}$. Notice that the diagrams in the last row must be included since by cutting the two W-lines we get a piece that is one-D-line reducible and therefore it is not a polarization insertion.

Like the phosonic self-energy, the particle self-energy Σ can be written in a compact form using the kernel $K_{\mathrm{xc},r}$. Let us consider the admissible "effective" interactions that can sprout from, for example, the left vertex of Σ_{xc}. We symbolically represent the diagrammatic structure $\mathfrak{a}^d - \mathfrak{a}$ in (16.109) as $W\tilde{P}\mathfrak{a}$.[17] Keeping an eye on Fig. 16.10 we realize that we can have W, see first, fourth, and sixth diagrams, and $(\mathfrak{a}D\mathfrak{a})$, see second, third, and fifth diagrams. We can also have $W\tilde{P}(\mathfrak{a}D\mathfrak{a})$, see eighth diagram, but we cannot have $W\tilde{P}W$ since the corresponding diagram would contain a P-insertion. We can further have $(\mathfrak{a}D\mathfrak{a})\tilde{P}W$, see seventh diagram, but we cannot have $(\mathfrak{a}D\mathfrak{a})\tilde{P}(\mathfrak{a}D\mathfrak{a})$ since the corresponding diagram would contain a Π-insertion. We can finally have $W\tilde{P}(\mathfrak{a}D\mathfrak{a})\tilde{P}W$ (ninth and tenth diagrams). All other structures are nonskeletonic: $W\tilde{P}(\mathfrak{a}D\mathfrak{a})\tilde{P}(\mathfrak{a}D\mathfrak{a})$, $(\mathfrak{a}D\mathfrak{a})\tilde{P}(\mathfrak{a}D\mathfrak{a})\tilde{P}W$, and $(\mathfrak{a}D\mathfrak{a})\tilde{P}W\tilde{P}(\mathfrak{a}D\mathfrak{a})$ contain a Π-insertion, whereas $(\mathfrak{a}D\mathfrak{a})\tilde{P}W\tilde{P}W$ and $W\tilde{P}W\tilde{P}(\mathfrak{a}D\mathfrak{a})$

[17] In this symbolic representation, $\tilde{P}$ would be the polarization if the particle-phoson interaction $\mathfrak{a}(1,2;3) \propto \delta(1;2)$ were local in the first two arguments, compare (16.108) with (16.109). Notice that this is the case for the electron-phonon interaction, see Table 16.1.

$\Sigma_{sss,\text{xc}} =$ + + + + + + + + + +...

Figure 16.10 Expansion of the self-energy in G-, D-, and W-skeleton diagrams.

contain a polarization insertion. We conclude that the total "effective" interaction sprouting from the left vertex is

$$\begin{aligned}\widetilde{W} &= W + (\mathfrak{a}D\mathfrak{a}) + W\tilde{P}(\mathfrak{a}D\mathfrak{a}) + (\mathfrak{a}D\mathfrak{a})\tilde{P}W + W\tilde{P}(\mathfrak{a}D\mathfrak{a})\tilde{P}W \\ &= W + (\mathfrak{a} + W\tilde{P}\mathfrak{a})D(\mathfrak{a} + \mathfrak{a}\tilde{P}W) \\ &= W + \mathfrak{a}^d D\mathfrak{a}^d.\end{aligned}$$

To make these graphical considerations rigorous we define the phoson-induced particle–particle interaction from the dressed particle–phoson vertex according to

$$W_{\text{phos}}(1, 2; 4, 3) =$$

or in formulas

$$W_{\text{phos}}(1, 2; 4, 3) = \int d\tilde{1}d\tilde{2}\ \mathfrak{a}^d(1, 3; \tilde{1}^*)D(\tilde{1}, \tilde{2})\mathfrak{a}^d(2, 4; \tilde{2}^*),$$

and the total screened interaction $\widetilde{W} \equiv W + W_{\text{phos}}$ according to

$$\widetilde{W} = \qquad = \qquad +$$

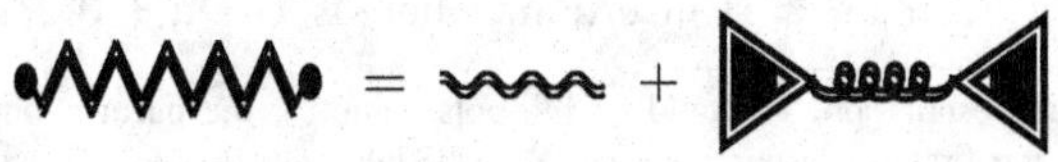

or in formulas

$$\widetilde{W}(1,2;4,3) = \delta(1;3)\delta(2;4)W(1;2) + W_{\rm phos}(1,2;4,3).$$

The self-energy can then be written as

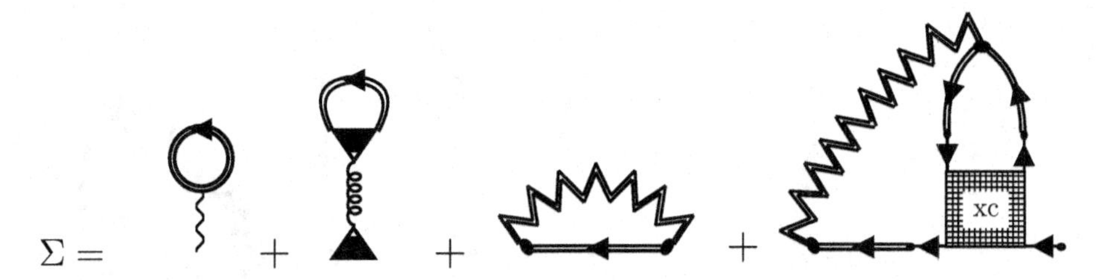

Hedin–Baym equations We summarize in Table 16.2 the fundamental equations that relate the various many-body quantities: G, D, Σ, Π, $\widetilde{W}$, W, P, $\mathfrak{a}^d$, and $K_{{\rm xc},r}$. For systems of electrons and phonons the equations for the phosonic Green's function and self-energy were derived by Baym [273], see also Ref. [283], whereas for systems of only electrons the equations for the electronic Green's function and self-energy were derived by Hedin [93], see Section 14.9 and Appendix N. We here align with Ref. [271] and name the full set of equations in Table 16.2 the *Hedin–Baym equations* for particles and phosons. These equations generalize those of Ref. [271] as they apply to systems in equilibrium at finite temperature as well as to systems driven away from equilibrium by external fields. They also generalize the equations of Ref. [264] as the electron–phoson coupling can be nonlocal in space. The Hedin–Baym equations provide a closed system of equations for any approximation to the XC self-energy through the irreducible kernel $K_{\rm xc} = \pm\delta\Sigma_{\rm xc}/\delta G$.

For systems in equilibrium at zero temperature the *adiabatic* assumption in conjunction with the assumption of a *nondegenerate ground state* allows for deforming the contour into a single branch going from $-\infty$ to $+\infty$ – that is, the real axis, see Section 5.4. In this case the contour Green's functions become the time-ordered Green's functions and the Hedin–Baym equations for a system of electrons and phonons reduce to those presented in Ref. [271]. We emphasize that no such shortcut is possible at finite temperature. One way to avoid the use of the Konstantinov–Perel's contour for equilibrium systems at finite temperature is the analytic continuation (from Matsubara to retarded to time-ordered), which may, however, be rather cumbersome in the presence of singularities or branch cuts, although notable progress has been made recently [284, 285, 286].

Like for the case of only particles, the Hedin–Baym equations can be iterated to obtain an expansion of Σ and Π in terms of G, W, D, and $\mathfrak{a}^d$. If we start with $K_{{\rm xc},r} = 0$, the self-energy Σ is approximated by

$$\Sigma = \Sigma_{\rm Eh} + \Sigma_{\rm H} + \Sigma_{\rm GW} + \Sigma_{\rm FM}, \tag{16.111}$$

where $\Sigma_{\rm GW} = {\rm i}GW$ is the GW self-energy with RPA screened interaction $W = v \pm {\rm i}vGGW$, and

$$\Sigma_{\rm FM} = {\rm i}\mathfrak{a}^d GD\mathfrak{a}^d$$

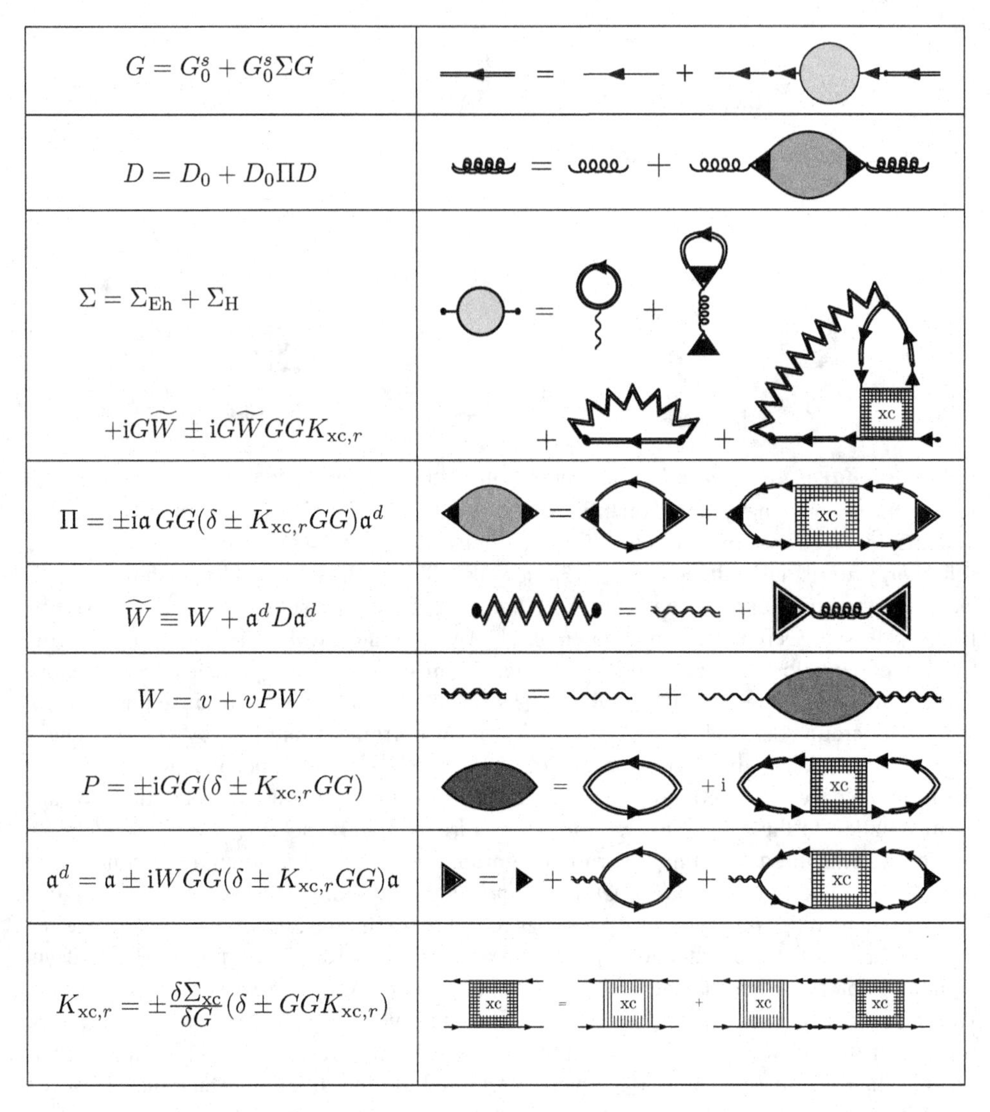

Mathematical expression	Diagrammatic representation
$G = G_0^s + G_0^s \Sigma G$	
$D = D_0 + D_0 \Pi D$	
$\Sigma = \Sigma_{\rm Eh} + \Sigma_{\rm H}$ $+ {\rm i}G\widetilde{W} \pm {\rm i}G\widetilde{W}GGK_{{\rm xc},r}$	
$\Pi = \pm {\rm i}\mathfrak{a}\, GG(\delta \pm K_{{\rm xc},r}GG)\mathfrak{a}^d$	
$\widetilde{W} \equiv W + \mathfrak{a}^d D \mathfrak{a}^d$	
$W = v + vPW$	
$P = \pm {\rm i}GG(\delta \pm K_{{\rm xc},r}GG)$	
$\mathfrak{a}^d = \mathfrak{a} \pm {\rm i}WGG(\delta \pm K_{{\rm xc},r}GG)\mathfrak{a}$	
$K_{{\rm xc},r} = \pm \frac{\delta \Sigma_{\rm xc}}{\delta G}(\delta \pm GGK_{{\rm xc},r})$	

Table 16.2 Mathematical expression (left column) and diagrammatic representation (right column) of the Hedin–Baym equations for systems of interacting particles and phosons.

is the so-called *Fan–Migdal self-energy* [287, 288] with dressed particle–phoson interaction

$$\mathfrak{a}^d = (\delta \pm {\rm i}WGG)\mathfrak{a}.$$

The phosonic self-energy for $K_{{\rm xc},r} = 0$ is simply

$$\Pi = \pm {\rm i}\,\mathfrak{a}\,GG\,\mathfrak{a}^d = \pm {\rm i}\,\mathfrak{a}(GG \pm {\rm i}GGWGG)\mathfrak{a}. \tag{16.112}$$

Inserting the approximated self-energies Σ and Π into the Dyson equations $G = G_0^s + G_0^s \Sigma G$ and $D = D_0 + D_0 \Pi D$, we obtain a closed system of equations for G and D. In Section 16.10 we show that the approximations (16.111) and (16.112) are conserving – that is, the resulting Green's functions satisfy all fundamental conservation laws.

The Hedin–Baym equations can alternatively be derived using the so-called *source field* method [271, 279, 289, 290]. We emphasize again that the source field method prescinds from any diagrammatic notion – that is, it does not tell us how to expand the various many-body quantities diagrammatically. We mention that in the context of electrons and phonons a partial diagrammatic analysis of the phosonic Green's function starting from a different independent Hamiltonian can be found in Refs. [277, 278].

Exercise 16.4 Prove that the irreducible kernel $K_{\rm xc}$ is the functional derivative of $\Sigma_{\rm xc}$ – that is, the relation (14.93) holds true also for systems of particles and phosons.

16.10 Kadanoff–Baym Equations and Conserving Approximations

More practical than the iterative solution of the Hedin–Baym equations is the solution of the Kadanoff–Baym equations. For simplicity we here consider the self-energies as functionals of G and D (doubly skeletonic expansion). Using the equations of motion for D_0, see (16.90), and for G_0^s, see (5.10a) with $G_0 \to G_0^s$ and $\overrightarrow{G}_0^{s,-1}(1;\bar{1}) \equiv \langle \mathbf{x}_1 | {\rm i}\frac{d}{dz_1} - \hat{h}^s(z_1) | \bar{\mathbf{x}}_1 \rangle \delta(z_1, \bar{z}_1)$ [compare with (4.62)], we can convert the Dyson equations (16.103) and (16.106) into integro-differential equations on the contour

$$\int d\bar{1}\, \overrightarrow{G}_0^{s,-1}(1;\bar{1}) G(\bar{1};2) = \delta(1;2) + \int d\bar{1}\, \Sigma(1;\bar{1}) G(\bar{1};2),$$

$$\int d\bar{1}\, \overrightarrow{D}_0^{-1}(1,\bar{1}) D(\bar{1},2^*) = \delta(1,2) + \int d\bar{1}\, \Pi(1^*,\bar{1}) D(\bar{1},2^*).$$

As in Section 9.1, we introduce operators in first quantization for the Green's functions and self-energies according to

$$\hat{\mathcal{G}}(z_1,z_2) = \int d\mathbf{x}_1 d\mathbf{x}_2 |\mathbf{x}_1\rangle G(1;2) \langle \mathbf{x}_2|, \tag{16.113a}$$

$$\hat{\Sigma}(z_1,z_2) = \int d\mathbf{x}_1 d\mathbf{x}_2 |\mathbf{x}_1\rangle \Sigma(1;2) \langle \mathbf{x}_2|, \tag{16.113b}$$

$$\hat{\mathcal{D}}(z_1,z_2) = \sum_{\boldsymbol{\xi}_1 \boldsymbol{\xi}_2} |\boldsymbol{\xi}_1\rangle D(1,2^*) \langle \boldsymbol{\xi}_2|, \tag{16.113c}$$

$$\hat{\Pi}(z_1,z_2) = \sum_{\boldsymbol{\xi}_1 \boldsymbol{\xi}_2} |\boldsymbol{\xi}_1\rangle \Pi(1^*,2) \langle \boldsymbol{\xi}_2|, \tag{16.113d}$$

as well as

$$\hat{\mathcal{Q}}(z) = \sum_{\xi_1\xi_2} |\xi_1\rangle Q_{\xi_1\xi_2}(z)\langle\xi_2|, \tag{16.114a}$$

$$\hat{\mathcal{J}} = \sum_{\xi_1\xi_2} |\xi_1\rangle J_{\xi_1\xi_2}\langle\xi_2|, \tag{16.114b}$$

where $\{|\xi\rangle\}$ is an orthonormal basis in the Hilbert space of one phoson. The integro-differential equations can then be seen as the representation in the $\{|\xi\rangle\}$ and $\{|\mathbf{x}\rangle\}$ bases of the following equations for operators in first quantization:

$$\left[\mathrm{i}\frac{d}{dz_1} - \hat{h}^s(z_1)\right]\hat{\mathcal{G}}(z_1,z_2) = \delta(z_1,z_2) + \int_\gamma d\bar{z}\,\hat{\Sigma}(z_1,\bar{z})\hat{\mathcal{G}}(\bar{z},z_2), \tag{16.115a}$$

$$\left[\mathrm{i}\frac{d}{dz_1}\hat{\mathcal{J}} - \hat{\mathcal{Q}}(z_1)\right]\hat{\mathcal{D}}(z_1,z_2) = \delta(z_1,z_2) + \int_\gamma d\bar{z}\,\hat{\Pi}(z_1,\bar{z})\hat{\mathcal{D}}(\bar{z},z_2). \tag{16.115b}$$

The properties of the matrices Q, see (16.75), and J, see (16.76), imply that the operators $\hat{\mathcal{Q}} = \hat{\mathcal{Q}}^\dagger$ and $\hat{\mathcal{J}} = \hat{\mathcal{J}}^\dagger$ are self-adjoint.

For actual implementations (see Appendix J) it is convenient to separate the particle self-energy into a time-local (singular) contribution $\hat{\Sigma}^\delta(z_1,z_2) \propto \delta(z_1,z_2)$ and a rest $\hat{\Sigma}_\mathrm{c}(z_1,z_2)$ which we continue calling the *correlation self-energy*:

$$\hat{\Sigma} = \hat{\Sigma}^\delta + \hat{\Sigma}_\mathrm{c}.$$

The singular contribution is given by the sum of the HF self-energy $\hat{\Sigma}_\mathrm{HF} = \delta(z_1,z_2)\hat{\mathcal{V}}_\mathrm{HF}(z)$, see Section 8.3, and the Ehrenfest self-energy:

$$\hat{\Sigma}^\delta = \hat{\Sigma}_\mathrm{HF} + \hat{\Sigma}_\mathrm{Eh}.$$

The Ehrenfest self-energy operator can be extracted from (16.107) by considering the definition (16.83) of the operator $\hat{\mathfrak{a}}$. We find

$$\hat{\Sigma}_\mathrm{Eh}(z_1,z_2) = \delta(z_1,z_2)\sum_{\xi} \hat{\mathfrak{a}}_{\xi^*}(z_1)\phi^s_\xi(z_1). \tag{16.116}$$

We bring the contribution containing the singular part of the self-energy on the l.h.s. and sum it to $\hat{h}^s$. Then the equation of motion for the Green's function becomes

$$\left[\mathrm{i}\frac{d}{dz_1} - \hat{h}_\mathrm{HF+Eh}(z_1)\right]\hat{\mathcal{G}}(z_1,z_2) = \delta(z_1,z_2) + \int_\gamma d\bar{z}\,\hat{\Sigma}_\mathrm{c}(z_1,\bar{z})\hat{\mathcal{G}}(\bar{z},z_2), \tag{16.117}$$

where

$$\begin{aligned}\hat{h}_\mathrm{HF+Eh}(z) &= \hat{h}^s(z) + \hat{\mathcal{V}}_\mathrm{HF}(z) + \sum_\xi \hat{\mathfrak{a}}_{\xi^*}(z_1)\phi^s_\xi(z_1)\\ &= \hat{h}_\mathrm{HF}(z) + \sum_\xi \hat{\mathfrak{a}}_{\xi^*}(z_1)\phi_\xi(z_1).\end{aligned}$$

In the last equality we use the definition (8.23) of the HF Hamiltonian, the explicit form (16.85) of the shifted Hamiltonian $\hat{h}^s$, and the definition (16.81) of the shifted displacement. Due to the appearance of the displacement we must complement the equations of motion for $\hat{\mathcal{G}}$ and $\hat{\mathcal{D}}$ with the equation of motion for $\phi_{\boldsymbol{\xi}}$:

$$\sum_{\xi'}\left[\mathrm{i}\frac{d}{dz}J_{\xi\xi'} - Q_{\xi\xi'}(z)\right]\phi_{\xi'}(z) = \pm\mathrm{i}\,\mathrm{Tr}\left[\hat{\mathfrak{a}}_{\xi}(z)\hat{\mathcal{G}}(z,z^+)\right] + \mathfrak{b}_{\xi}(z), \tag{16.118}$$

which follows directly from (16.77). Two more equations involving the time derivative with respect to z_2 can be derived along the same lines:

$$\hat{\mathcal{G}}(z_1,z_2)\left[-\mathrm{i}\frac{\overleftarrow{d}}{dz_2} - \hat{h}_{\mathrm{HF+Eh}}(z_2)\right] = \delta(z_1,z_2) + \int_\gamma d\bar{z}\,\hat{\mathcal{G}}(z_1,\bar{z})\hat{\Sigma}_{\mathrm{c}}(\bar{z},z_2), \tag{16.119a}$$

$$\hat{\mathcal{D}}(z_1,z_2)\left[-\mathrm{i}\frac{\overleftarrow{d}}{dz_2}\hat{\mathcal{J}} - \hat{\mathcal{Q}}(z_2)\right] = \delta(z_1,z_2) + \int_\gamma d\bar{z}\,\hat{\mathcal{D}}(z_1,\bar{z})\hat{\Pi}(\bar{z},z_2). \tag{16.119b}$$

Equations (16.115b), (16.117), (16.118), and (16.119) are closed equations once an approximation for Σ_{c} and Π in terms of G- and D-skeleton diagrams has been chosen.

Kadanoff–Baym equations Placing the time arguments on different branches of the contour and using the Langreth rules, we can convert the contour equations into a coupled system of equations for the Keldysh components of $\hat{\mathcal{G}}$ and $\hat{\mathcal{D}}$. These are the *Kadanoff–Baym equations* for systems of particles and phosons to be solved with KMS boundary conditions. As for the case of only particles, see Section 9.2, the equations for the Matsubara components decouple. The corresponding Green's functions $\hat{\mathcal{G}}^{\mathrm{M}}$ and $\hat{\mathcal{D}}^{\mathrm{M}}$ allow for calculating the initial ensemble average of any one-body operator for particles and of any quadratic operator in the displacements for phosons. In particular, for systems in thermal equilibrium we have access to finite-temperature quantities. The Kadanoff–Baym equations for the Green's functions with times on the horizontal branches (hence the left/right and lesser/greater components) allow for monitoring the system evolution. As already pointed out at the beginning of Section 16.3, we can study how the system responds to different kinds of external perturbations, such as interaction quenches, laser fields, phonon drivings, etc. To avoid making the presentation too abstract we discuss the Kadanoff–Baym equations for a system of electrons and phonons, see Section 16.11. The generalization to systems of particles and phosons with the most general Q, $\mathfrak{a}$, and $\mathfrak{b}$ does not present any extra complications.

Conserving approximations By a careful diagrammatic choice of the self-energies we can make sure that the observables calculated from the Green's functions satisfy the conservation laws for energy, angular momentum, momentum, and particle number. For this purpose, the self-energies must be chosen in a Φ-derivable manner. To define this properly in the context of particles and phosons we split off the Ehrenfest–Hartree part of the self-energy, as in (16.110). In the doubly skeletonic expansion the remainder is the XC self-energy $\Sigma_{ss,\mathrm{xc}}[G,D,v,g] = \Sigma_{sss,\mathrm{xc}}[G,D,W,g]$. We then construct the functional $\Phi_{\mathrm{xc}}[G,D,v,g]$ by closing each skeleton diagram for Σ_{xc} with a G-line, thereby producing a set of vacuum diagrams. The lowest-order diagrams of the expansion are shown in Fig. 16.11. Each of the topologically different vacuum diagrams is also multiplied by a symmetry factor $1/N_{\mathrm{S}}$,

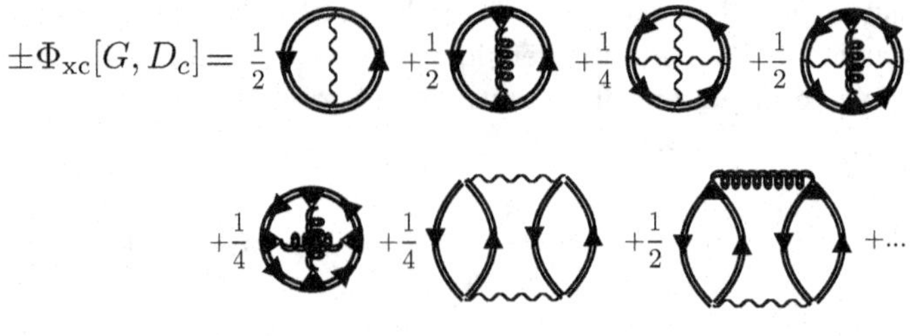

$$\pm\Phi_{\rm xc}[G, D_c] = \frac{1}{2} \;[\text{diagram}]\; + \frac{1}{2}\;[\text{diagram}]\; + \frac{1}{4}\;[\text{diagram}]\; + \frac{1}{2}\;[\text{diagram}]\; + \frac{1}{4}\;[\text{diagram}]\; + \frac{1}{4}\;[\text{diagram}]\; + \frac{1}{2}\;[\text{diagram}]\; + \ldots$$

Figure 16.11 Expansion of the Φ functional in G-skeleton and D-skeleton diagrams.

where N_S is the number of equivalent G-lines yielding the same self-energy diagram by their respective removal (the relevant fact that the different classes of equivalent lines in a vacuum diagram have the same number of elements can be proven as in Section 11.3). We furthermore introduce an additional plus/minus sign since the removal of a G-line from a vacuum diagram changes the number of particle loops by one. By construction, the $\Phi_{\rm xc}$ functional has the property that

$$\Sigma_{\rm xc}(1;2) = \frac{\delta\Phi_{\rm xc}}{\delta G(2;1)}. \tag{16.120}$$

Proceeding along the same lines of reasoning as in Section 11.7 we can also prove that

$$\frac{1}{2}\,\Pi(1,2) = \left.\frac{\delta\Phi_{\rm xc}}{\delta D(1,2)}\right|_S, \tag{16.121}$$

where the subscript "S" refers to the symmetrized derivative $[\delta/\delta D(1,2) + \delta/\delta D(2,1)]/2$. The Hartree self-energy is obtained from the functional derivative of the functional $\Phi_{\rm H}$, see first diagram in Fig. 11.4 and (11.63), whereas the Ehrenfest self-energy is obtained from the functional derivative of

$$\pm\Phi_{\rm Eh}[G, D_0] = \frac{1}{2}\;[\text{diagram}] \tag{16.122}$$

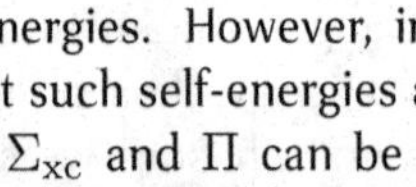

Therefore, the full self-energy Σ is the functional derivative of the Φ functional:

$$\Phi = \Phi_{\rm H}[G] + \Phi_{\rm Eh}[G, D_0] + \Phi_{\rm xc}[G, D].$$

So far we have dealt with the exact self-energies. However, in most cases we have to deal with approximate self-energies. We say that such self-energies are Φ-derivable whenever there exists a functional $\Phi_{\rm xc}[G, D]$ such that $\Sigma_{\rm xc}$ and Π can be written as in (16.120) and (16.121), respectively. In practice the approximate Φ functional is obtained by selecting an appropriate subset of Φ diagrams. The significance of this procedure lies in the fact that the Φ functional has invariance properties with regard to space–time and gauge transformations, implying the fulfillment of the corresponding conservation laws for the Green's functions that satisfy the Kadanoff–Baym equations with Φ-derivable self-energies, see Chapter 12.

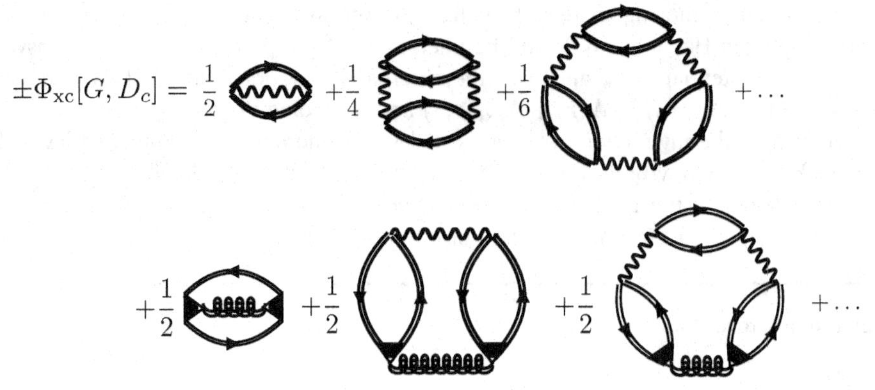

Figure 16.12 Diagrams of the $\Phi_{\rm xc}$ functional leading to the self-energies (16.111) and (16.112) through the functional derivatives (16.120) and (16.121).

We conclude by observing that the approximation to the self-energies discussed at the end of Section 16.9 (derived by setting $K_{{\rm xc},r} = 0$ in the Hedin–Baym equations) is Φ-derivable. The diagrams for $\Phi_{\rm xc}$ are those of the GW approximation, see Fig. 12.1, plus a second infinite sum of ring diagrams in which one v-line is replaced by $(\mathfrak{a}D\mathfrak{a})$, see Fig. 16.12.

16.10.1 Recovering the Classical Theory of Electromagnetism and Quantum Particles

For a system of charged particles and photons the equation of motion (16.117) is exactly the same as the equation of motion (12.5) with $w = 0$ and with vector potential given by the time-dependent ensemble average of the operator (16.16). Furthermore, the equation of motion (16.118) is equivalent to the Maxwell equation (16.2c) for $\boldsymbol{\xi} = (\xi, 1)$ and to the Maxwell equation (16.2b) for $\boldsymbol{\xi} = (\xi, 2)$, where the electric and magnetic fields are calculated from the time-dependent ensemble average of the operators (16.16) and (16.17). We encourage the reader to verify these results using the parameters of Table 16.1. If we treat the electron–photon interaction at the Ehrenfest level and the electron–electron interaction at the Hartree level then the equation of motion for the Green's function is the same as the equation of motion of a noninteracting Green's function G_0 where the vector potential $\mathbf{A}$ and the scalar potential ϕ of the single-particle Hamiltonian satisfy the Maxwell equations with density and current given by G_0.[18] With this in mind we can consider the (transverse) *external* vector potential and the *external* scalar potential of the previous chapters as the potentials generated by an *external* system of charged particles treated at the Ehrenfest–Hartree level. For instance, the nuclear potential felt by the electrons can be seen as the Hartree potential generated by an external system of infinitely heavy nuclei. Similarly, a (transverse) laser pulse that drives electrons out of equilibrium can be seen as the Ehrenfest potential generated by

[18] This is actually true only to first order in the vector potential as we have not included the quadratic term in the general Hamiltonian (16.70).

an external system of moving charges. If such an external system exists, then we must add to the particle–photon Hamiltonian (16.20) the interaction of the particles with the (transverse) external vector potential $\mathbf{A}_{\rm ext}$ and with the external scalar potential ϕ. This is done by replacing $\hat{\mathbf{A}}(\mathbf{r}) \to \hat{\mathbf{A}}_{\rm tot}(\mathbf{r}) = \hat{\mathbf{A}}(\mathbf{r}) + \mathbf{A}_{\rm ext}(\mathbf{r},t)$ and by adding the term $q\int d\mathbf{x}\phi(\mathbf{x},t)\hat{n}(\mathbf{x})$. If we further treat the interaction between particles and photons at the Ehrenfest level, then we are back to the many-body Hamiltonian (1.83) with one-particle Hamiltonian (3.1) where the vector potential is the *total* (and transverse) vector potential $\mathbf{A}_{\rm tot} = \langle\hat{\mathbf{A}}\rangle + \mathbf{A}_{\rm ext}$. We have in this way proven the statement in footnote 1 of Chapter 3.

Exercise 16.5 Prove (16.121).

16.11 Kadanoff–Baym Equations for Electron–Phonon Systems

In this section we apply the NEGF machinery developed so far to interacting systems of electrons and phonons. We consider the case of external perturbations that do not break the lattice periodicity of the crystal. The reader can consult Ref. [264] for a more extensive discussion.

16.11.1 Noninteracting Phononic Green's Function

We begin by calculating the noninteracting phononic Green's function. This is given by the solution of (16.90), where $\overrightarrow{D}_0^{-1}$ is defined below (16.89). Taking into account (16.76) and the definitions in Table 16.1, we have

$$J_{\boldsymbol{\xi}\boldsymbol{\xi}'} = J_{\mathbf{q}\nu,i\,\mathbf{q}'\nu',i'} = \delta_{\mathbf{q},\mathbf{q}'}\,J^{ii'}_{\nu\nu'}\,,$$
$$Q_{\boldsymbol{\xi}\boldsymbol{\xi}'}(z) = Q_{\mathbf{q}\nu,i\,\mathbf{q}'\nu',i'}(z) = \delta_{\mathbf{q},\mathbf{q}'}\,Q^{ii'}_{\nu\nu'}(\mathbf{q},z)\,,$$

where

$$J^{ii'}_{\nu\nu'} = \delta_{\nu\nu'}\begin{pmatrix} 0 & \mathrm{i} \\ -\mathrm{i} & 0 \end{pmatrix}_{ii'}, \qquad Q^{ii'}_{\nu\nu'}(\mathbf{q},z) = \begin{pmatrix} K_{\nu\nu'}(\mathbf{q},z) & 0 \\ 0 & \delta_{\nu\nu'} \end{pmatrix}_{ii'}. \tag{16.123}$$

As both J and Q are diagonal in $\mathbf{q}$-space the equation of motion (16.90) is solved by a D_0 which is proportional to $\delta_{\mathbf{q},-\mathbf{q}'}$:

$$D_0(\boldsymbol{\xi},z,\boldsymbol{\xi}',z') = D_0(\mathbf{q}\nu,i,z,\mathbf{q}'\nu',i',z') = \delta_{\mathbf{q},-\mathbf{q}'}\,D^{ii'}_{0,\mathbf{q}\nu\nu'}(z,z'), \tag{16.124}$$

where

$$\boxed{\left[\mathrm{i}\frac{d}{dz}J - Q(\mathbf{q},z)\right]D_{0,\mathbf{q}}(z,z') = \mathbb{1}\delta(z,z')} \tag{16.125}$$

Here, J, Q, and D_0 are 2×2 block matrices where the block (i, i') is given by the matrices $J^{ii'}$, $Q^{ii'}$, and $D_0^{ii'}$. Similarly, $\mathbb{1}$ is the identity matrix: $\mathbb{1}^{ii'}_{\nu\nu'} = \delta_{\nu\nu'}\delta_{ii'}$. According to definition (16.73) and result (16.105), the matrix elements of D_0 are correlators between fluctuation operators of displacements and momenta. For instance,

$$D^{11}_{0,\mathbf{q}\nu\nu'}(z, z') = \frac{1}{\mathrm{i}}\frac{1}{Z^s_{0,\mathrm{phon}}}\langle \mathcal{T}\Big\{ \Delta\hat{\mathcal{U}}_{\mathbf{q}\nu}(z)\Delta\hat{\mathcal{U}}_{-\mathbf{q}\nu'}(z')\Big\}\rangle^s_{0,\mathrm{phon}}, \tag{16.126}$$

where the noninteracting average $\langle \ldots \rangle^s_{0,\mathrm{phon}}$ is defined below (16.95). *Mutatis mutandis* we can derive the equation of motion with derivative with respect to z':

$$\boxed{D_{0,\mathbf{q}}(z, z')\Big[-\mathrm{i}\frac{\overleftarrow{d}}{dz'}J - Q(\mathbf{q}, z')\Big] = \mathbb{1}\delta(z, z')} \tag{16.127}$$

We now proceed along the same lines as in Section 6.2. We define the nonunitary matrices $W_{L\mathbf{q}}$ and $W_{R\mathbf{q}}$ as the solution of

$$\mathrm{i}\frac{d}{dz}W_{L\mathbf{q}}(z) = Q(\mathbf{q}, z)J\,W_{L\mathbf{q}}(z), \qquad -\mathrm{i}\frac{d}{dz}W_{R\mathbf{q}}(z) = W_{R\mathbf{q}}(z)Q(\mathbf{q}, z)J,$$

with boundary conditions $W_{L\mathbf{q}}(t_{0-}) = W_{R\mathbf{q}}(t_{0-}) = \mathbb{1}$. The explicit expression for these matrices is given in terms of the contour-ordering and anticontour-ordering operators

$$W_{L\mathbf{q}}(z) = \mathcal{T}\left\{e^{-\mathrm{i}\int_{t_{0-}}^{z} d\bar{z}\; Q(\mathbf{q},\bar{z})J}\right\}, \qquad W_{R\mathbf{q}}(z) = \bar{\mathcal{T}}\left\{e^{\mathrm{i}\int_{t_{0-}}^{z} d\bar{z}\; Q(\mathbf{q},\bar{z})J}\right\},$$

from which it follows that $W_{L\mathbf{q}}(z)W_{R\mathbf{q}}(z) = W_{R\mathbf{q}}(z)W_{L\mathbf{q}}(z) = \mathbb{1}$. We look for solutions of the form

$$D_{0,\mathbf{q}}(z, z') = -\mathrm{i}J\,W_{L\mathbf{q}}(z)F_{\mathbf{q}}(z, z')W_{R\mathbf{q}}(z').$$

Inserting this expression into (16.125) and (16.127), we obtain a couple of equations for the unknown matrix function $F_{\mathbf{q}}$:

$$\frac{d}{dz}F_{\mathbf{q}}(z, z') = -\frac{d}{dz'}F_{\mathbf{q}}(z, z') = \mathbb{1}\delta(z, z'),$$

which is solved by $F_{\mathbf{q}}(z, z') = \theta(z, z')F^>_{\mathbf{q}} + \theta(z', z)F^<_{\mathbf{q}}$, with

$$F^>_{\mathbf{q}} - F^<_{\mathbf{q}} = \mathbb{1}. \tag{16.128}$$

Imposing the KMS relations $D_{0,\mathbf{q}}(t_{0-}, z') = D_{0,\mathbf{q}}(t_0 - \mathrm{i}\beta, z')$, we find

$$F^<_{\mathbf{q}} = W_{L\mathbf{q}}(t_0 - \mathrm{i}\beta)F^>_{\mathbf{q}} = e^{-\beta Q^{\mathrm{M}}(\mathbf{q})J}F^>_{\mathbf{q}}, \tag{16.129}$$

where $Q^{\mathrm{M}}(\mathbf{q}) = Q(\mathbf{q}, t_0 - \mathrm{i}\tau)$. Equations (16.128) and (16.129) can be solved for $F^<_{\mathbf{q}}$ and $F^>_{\mathbf{q}}$, and the final expression for the noninteracting phononic Green's function reads

$$\boxed{D_{0,\mathbf{q}}(z, z') = -\mathrm{i}J\,W_{L\mathbf{q}}(z)\Big[\theta(z, z')\bar{f}(Q^{\mathrm{M}}(\mathbf{q})J) + \theta(z', z)f(Q^{\mathrm{M}}(\mathbf{q})J)\Big]W_{R\mathbf{q}}(z')} \tag{16.130}$$

where $f(\omega) = 1/(e^{\beta\omega} - 1)$ is the Bose function and $\bar{f}(\omega) = e^{\beta\omega} f(\omega)$. Having the Green's function on the contour, we can now extract all its Keldysh components.

Matsubara component The Matsubara component $D^{\mathrm{M}}_{0,\mathbf{q}}(\tau, \tau')$ is obtained by setting $z = t_0 - \mathrm{i}\tau$ and $z' = t_0 - \mathrm{i}\tau'$ in (16.130). Alternatively, we can set $z = t_0 - \mathrm{i}\tau$ and $z' = t_0 - \mathrm{i}\tau'$ in one of the equations of motion, and then solve for $D^{\mathrm{M}}_{0,\mathbf{q}}$. Choosing the equation of motion (16.125), we have

$$\boxed{\left[-\frac{d}{d\tau} J - Q^{\mathrm{M}}(\mathbf{q})\right] D^{\mathrm{M}}_{0,\mathbf{q}}(\tau, \tau') = \mathbb{1}\mathrm{i}\delta(\tau, \tau')}$$

Expanding the Matsubara Green's function in bosonic Matsubara frequencies according to (6.18) and the Dirac delta according to (6.20), we find an algebraic equation for the coefficients of the expansion

$$\left[\omega_m J - Q^{\mathrm{M}}(\mathbf{q})\right] D^{\mathrm{M}}_{0,\mathbf{q}}(\omega_m) = \mathbb{1}. \tag{16.131}$$

Taking into account Table 16.1 and the explicit form of the matrix J, (16.131) is converted into a set of algebraic equations for the four blocks:

$$\begin{pmatrix} -K(\mathbf{q}) & \mathrm{i}\omega_m \mathbb{1} \\ -\mathrm{i}\omega_m \mathbb{1} & -\mathbb{1} \end{pmatrix} \begin{pmatrix} D^{11,\mathrm{M}}_{0,\mathbf{q}}(\omega_m) & D^{12,\mathrm{M}}_{0,\mathbf{q}}(\omega_m) \\ D^{21,\mathrm{M}}_{0,\mathbf{q}}(\omega_m) & D^{22,\mathrm{M}}_{0,\mathbf{q}}(\omega_m) \end{pmatrix} = \begin{pmatrix} \mathbb{1} & 0 \\ 0 & \mathbb{1} \end{pmatrix}.$$

In this formula each of the entries of the 2×2 matrices is itself a matrix with indices ν, ν'. The reader can easily work out the expression of the four Green's function blocks. We report below the expression of the block $(1,1)$ corresponding to the displacement–displacement correlator in (16.126):

$$D^{11,\mathrm{M}}_{0,\mathbf{q}}(\omega_m) = \frac{1}{\omega_m^2 - K(\mathbf{q})}. \tag{16.132}$$

If we choose to work in the basis of the normal modes of K, then $K_{\nu\nu'}(\mathbf{q}) = \delta_{\nu\nu'}\omega^2_{0\mathbf{q}\nu}$ and the displacement–displacement correlator simplifies to

$$D^{11,\mathrm{M}}_{0,\mathbf{q}\nu\nu'}(\omega_m) = \frac{\delta_{\nu\nu'}}{\omega_m^2 - \omega^2_{0\mathbf{q}\nu}}.$$

We already emphasized in Section 16.2.1 that the eigenvalues $\omega^2_{0\mathbf{q}\nu}$ are not physical and can even be negative. Therefore, there is no particular convenience to work in the basis of the normal modes of K.

Lesser and greater components The lesser/greater component of the noninteracting phononic Green's function is obtained by setting $z = t_{-/+}$ and $z' = t'_{+/-}$ in (16.130). For times on the horizontal branches of the contour the matrices $W_{L\mathbf{q}}$ and $W_{R\mathbf{q}}$ are independent of the branch – that is, $W_{L\mathbf{q}}(t_\pm) = W_{L\mathbf{q}}(t)$ and $W_{R\mathbf{q}}(t_\pm) = W_{R\mathbf{q}}(t)$ with

$$W_{L\mathbf{q}}(t) = T\left\{e^{-\mathrm{i}\int_{t_0}^{t} d\bar{t}\, Q(\mathbf{q},\bar{t})J}\right\}, \qquad W_{R\mathbf{q}}(t) = \bar{T}\left\{e^{\mathrm{i}\int_{t_0}^{t} d\bar{t}\, Q(\mathbf{q},\bar{t})J}\right\},$$

and T and $\bar{T}$ the time-ordering and anti-time-ordering operators. We then have

$$D^{<}_{0,\mathbf{q}}(t,t') = -\mathrm{i}J\,W_{L\mathbf{q}}(t)f(Q^{\mathrm{M}}(\mathbf{q})J)W_{R\mathbf{q}}(t'), \tag{16.133a}$$

$$D^{>}_{0,\mathbf{q}}(t,t') = -\mathrm{i}J\,W_{L\mathbf{q}}(t)\bar{f}(Q^{\mathrm{M}}(\mathbf{q})J)W_{R\mathbf{q}}(t'). \tag{16.133b}$$

The greater component is obtained from the lesser component by replacing $f \to \bar{f}$. In the following we then consider only $D^{<}_{0,\mathbf{q}}$.

Let us discuss the special, yet most common, case of $Q(\mathbf{q},t) = Q^{\mathrm{M}}(\mathbf{q})$ independent of time (no phonon driving) and identical to its value along the Matsubara track. Then $W_{L\mathbf{q}}(t) = \exp[-\mathrm{i}Q^{\mathrm{M}}(\mathbf{q})J(t-t_0)]$ and $W_{R\mathbf{q}}(t) = \exp[\mathrm{i}Q^{\mathrm{M}}(\mathbf{q})J(t-t_0)]$ commute with the Bose function, and the lesser component can be written as

$$D^{<}_{0,\mathbf{q}}(t,t') = -\mathrm{i}Jf(Q^{\mathrm{M}}(\mathbf{q})J)e^{-\mathrm{i}Q^{\mathrm{M}}(\mathbf{q})J(t-t')}.$$

As this function depends only on the time difference, it can be Fourier transformed with respect to $t-t'$ according to (6.44). In frequency space the lesser component reads

$$D^{<}_{0,\mathbf{q}}(\omega) = -2\pi\mathrm{i}Jf(Q^{\mathrm{M}}(\mathbf{q})J)\,\delta(\omega - Q^{\mathrm{M}}(\mathbf{q})J) = -2\pi\mathrm{i}Jf(\omega)\delta(\omega - Q^{\mathrm{M}}(\mathbf{q})J). \tag{16.134}$$

We can easily extract the four blocks of $D^{<}_{0,\mathbf{q}}(\omega)$ using the Cauchy relation[19]

$$\begin{aligned}
-2\pi\mathrm{i}J\,\delta(\omega - Q^{\mathrm{M}}(\mathbf{q})J) &= J\left[\frac{1}{\omega - Q^{\mathrm{M}}(\mathbf{q})J + \mathrm{i}\eta} - \frac{1}{\omega - Q^{\mathrm{M}}(\mathbf{q})J - \mathrm{i}\eta}\right],\\
&= \frac{1}{(\omega+\mathrm{i}\eta)J - Q^{\mathrm{M}}(\mathbf{q})} - \frac{1}{(\omega-\mathrm{i}\eta)J - Q^{\mathrm{M}}(\mathbf{q})}.
\end{aligned}$$

We see that the fraction with $(\omega \pm \mathrm{i}\eta)$ is the same as $D^{\mathrm{M}}_{0,\mathbf{q}}(\omega_m)$ calculated in $\omega_m = \omega \pm \mathrm{i}\eta$, compare with (16.131). Therefore the block $(1,1)$ of $D^{<}_{0,\mathbf{q}}(\omega)$ can be read out from (16.132):

$$D^{11,<}_{0,\mathbf{q}}(\omega) = f(\omega)\left[\frac{1}{(\omega+\mathrm{i}\eta)^2 - K(\mathbf{q})} - \frac{1}{(\omega-\mathrm{i}\eta)^2 - K(\mathbf{q})}\right].$$

Using the same strategy, the reader can extract the explicit form of the remaining three blocks. The greater component has the same form as the lesser component with $f(\omega) \to \bar{f}(\omega)$.

Retarded and advanced components The retarded and advanced components can be calculated from the lesser and greater components, see Table 5.1. We have

$$D^{\mathrm{R}}_{0,\mathbf{q}}(t,t') = \theta(t-t')[D^{>}_{0,\mathbf{q}}(t,t') - D^{<}_{0,\mathbf{q}}(t,t')] = -\mathrm{i}\theta(t-t')J\,W_{L\mathbf{q}}(t)W_{R\mathbf{q}}(t'), \tag{16.135a}$$

$$D^{\mathrm{A}}_{0,\mathbf{q}}(t,t') = -\theta(t'-t)[D^{>}_{0,\mathbf{q}}(t,t') - D^{<}_{0,\mathbf{q}}(t,t')] = \mathrm{i}\theta(t'-t)J\,W_{L\mathbf{q}}(t)W_{R\mathbf{q}}(t'), \tag{16.135b}$$

from which it follows that we can write the lesser/greater components in (16.133) as [compare with (6.49)]

$$\boxed{D^{\lessgtr}_{0,\mathbf{q}}(t,t') = D^{\mathrm{R}}_{0,\mathbf{q}}(t,t_0)\,J\,D^{\lessgtr}_{0,\mathbf{q}}(t_0,t_0)\,J\,D^{\mathrm{A}}_{0,\mathbf{q}}(t_0,t')} \tag{16.136}$$

[19] Take into account that $J^{-1} = J$ and hence $JA^{-1} = J^{-1}A^{-1} = (AJ)^{-1}$ for any matrix A.

In the special case $Q(\mathbf{q},t) = Q^{\rm M}(\mathbf{q})$, the retarded/advanced components depend only on the time difference and they can be Fourier transformed. In Fourier space we have [compare with (6.58)]

$$D^{\rm R/A}_{0,\mathbf{q}}(\omega) = \mathrm{i}\int \frac{d\omega'}{2\pi}\frac{D^{>}_{0,\mathbf{q}}(\omega') - D^{<}_{0,\mathbf{q}}(\omega')}{\omega - \omega' \pm \mathrm{i}\eta} = \frac{1}{(\omega \pm \mathrm{i}\eta)J - Q^{\rm M}(\mathbf{q})}. \tag{16.137}$$

Notice that the advanced component is the adjoint of the retarded component. Also notice that $D^{\rm R/A}_{0,\mathbf{q}}(\omega) = D^{\rm M}_{0,\mathbf{q}}(\omega \pm \mathrm{i}\eta)$, see (16.131). Therefore, the block $(1,1)$ of the retarded/advanced Green's function is simply

$$D^{11,\rm R/A}_{0,\mathbf{q}}(\omega) = \frac{1}{(\omega \pm \mathrm{i}\eta)^2 - K(\mathbf{q})}.$$

Using the same strategy the reader can work out the expression of the remaining three blocks. We finally observe that from (16.134) and (16.137) we can derive the *fluctuation–dissipation theorem* for the noninteracting phononic Green's function,

$$D^{<}_{0,\mathbf{q}}(\omega) = f(\omega)\left[D^{\rm R}_{0,\mathbf{q}}(\omega) - D^{\rm A}_{0,\mathbf{q}}(\omega)\right],$$

and the like for the greater component.

16.11.2 Equations of Motion for the Contour Green's Functions

Let us write the equation of motion for the contour Green's functions G and D in the case of an interacting system of electrons and phonons.

Electronic Green's function The equation of motion for the contour Green's function G is given in (16.117). Let us calculate the explicit form of the term added to $\hat{h}_{\rm HF}$ in the l.h.s. According to Table 16.1, the coupling $\mathfrak{a}$ is proportional to a Dirac delta in position basis and the only nonvanishing component is the one that couples the density to the displacement $\phi_{\mathbf{q}\nu,1} = U_{\mathbf{q}\nu}$. Therefore,

$$\sum_{\xi}\langle \mathbf{x}_1|\hat{\mathfrak{a}}_{\xi^*}(z_1)|\mathbf{x}_2\rangle\phi_{\xi}(z_1) = \delta(\mathbf{x}_1 - \mathbf{x}_2)\sum_{\mathbf{q}\nu} g_{-\mathbf{q}\nu}(\mathbf{r}_1,z_1)U_{\mathbf{q}\nu}(z_1),$$

where $U_{\mathbf{q}\nu}$ is the average of $\hat{\mathcal{U}}_{\mathbf{q}\nu}$. This term acts like a local scalar potential. If we define the operator in first quantization

$$\hat{g}_{\mathbf{q}\nu}(z) \equiv \int d\mathbf{x}\,|\mathbf{x}\rangle\, g_{\mathbf{q}\nu}(\mathbf{r},z)\langle\mathbf{x}| \quad \Rightarrow \quad \langle\mathbf{x}_1|\hat{g}_{\mathbf{q}\nu}(z)|\mathbf{x}_2\rangle = \delta(\mathbf{x}_1 - \mathbf{x}_2)g_{\mathbf{q}\nu}(\mathbf{r}_1,z_1),$$

then $\sum_{\xi}\hat{\mathfrak{a}}_{\xi^*}(z_1)\phi_{\xi}(z_1) = \sum_{\mathbf{q}\nu}\hat{g}_{-\mathbf{q}\nu}(z_1)U_{\mathbf{q}\nu}(z_1)$ and (16.117) becomes

$$\left[\mathrm{i}\frac{d}{dz_1} - \hat{h}_{\rm HF}(z_1) - \sum_{\mathbf{q}\nu}\hat{g}_{-\mathbf{q}\nu}(z_1)U_{\mathbf{q}\nu}(z_1)\right]\hat{\mathcal{G}}(z_1,z_2) = \delta(z_1,z_2) + \int_{\gamma} d\bar{z}\,\hat{\Sigma}_{\rm c}(z_1,\bar{z})\hat{\mathcal{G}}(\bar{z},z_2). \tag{16.138}$$

We now use the working hypothesis that the external perturbations do not break the lattice periodicity of the crystal. It is then advantageous to represent the equation of motion (16.138) on some suitable one-electron Bloch basis $|\mathbf{k}\mu\rangle$, see Section 2.3, where μ is an index for the spin and band degrees of freedom. In fact, in such bases the Hartree–Fock Hamiltonian is diagonal in $\mathbf{k}$-space,

$$\langle\mathbf{k}\mu|\,\hat{h}_{\mathrm{HF}}\,|\mathbf{k}'\mu'\rangle = \delta_{\mathbf{k},\mathbf{k}'}h_{\mathrm{HF},\mu\mu'}(\mathbf{k}),$$

and consequently the Green's function and the self-energy are diagonal too, see Appendix C,

$$\langle\mathbf{k}\mu|\,\hat{\mathcal{G}}\,|\mathbf{k}'\mu'\rangle = \delta_{\mathbf{k},\mathbf{k}'}G_{\mathbf{k}\mu\mu'}, \qquad \langle\mathbf{k}\mu|\,\hat{\Sigma}_{\mathrm{c}}\,|\mathbf{k}'\mu'\rangle = \delta_{\mathbf{k},\mathbf{k}'}\Sigma_{\mathrm{c},\mathbf{k}\mu\mu'}.$$

The representation of $\hat{g}_{\mathbf{q}\nu}$ in the same basis yields

$$\langle\mathbf{k}\mu|\,\hat{g}_{\mathbf{q}\nu}(z)\,|\mathbf{k}'\mu'\rangle = \int d\bar{\mathbf{x}}\,\langle\mathbf{k}\mu|\bar{\mathbf{x}}\rangle\langle\bar{\mathbf{x}}|\mathbf{k}'\mu'\rangle\,g_{\mathbf{q}\nu}(\bar{\mathbf{r}},z).$$

Let us write $\bar{\mathbf{r}}$ as the sum of the vector $\mathbf{R}_{\bar{\mathbf{n}}}$ of the unit cell the point $\bar{\mathbf{r}}$ belongs to and a displacement $\bar{\mathbf{u}}$ spanning the unit cell centered at the origin: $\bar{\mathbf{r}} = \mathbf{R}_{\bar{\mathbf{n}}} + \bar{\mathbf{u}}$. Then the Bloch wavefunctions $\langle\bar{\mathbf{x}}|\mathbf{k}\mu\rangle = e^{\mathrm{i}\mathbf{k}\cdot\bar{\mathbf{n}}}u_{\mathbf{k}\mu}(\bar{\mathbf{u}}\bar{\sigma})/\sqrt{|\mathrm{V}_b|}$, see (2.18) and following discussion. Inserting this expression into the matrix element of $\hat{g}_{\mathbf{q}\nu}$ and writing the spatial integral over $\bar{\mathbf{r}}$ as $\int d\bar{\mathbf{r}} = \sum_{\bar{\mathbf{n}}}\int_{\mathrm{cell}} d\bar{\mathbf{u}}$, where $\int_{\mathrm{cell}}$ stands for an integral over the volume of the unit cell, we find

$$\begin{aligned}\langle\mathbf{k}\mu|\,\hat{g}_{\mathbf{q}\nu}(z)\,|\mathbf{k}'\mu'\rangle &= \sum_{\bar{\sigma}}\int_{\mathrm{cell}} d\bar{\mathbf{u}}\;u^*_{\mathbf{k}\mu}(\bar{\mathbf{u}}\bar{\sigma})u_{\mathbf{k}'\mu'}(\bar{\mathbf{u}}\bar{\sigma})\underbrace{\sum_{\bar{\mathbf{n}}}\frac{e^{-\mathrm{i}(\mathbf{k}-\mathbf{k}'+\mathbf{q})\cdot\bar{\mathbf{n}}}}{|\mathrm{V}_b|}}_{\delta_{\mathbf{k}-\mathbf{k}',-\mathbf{q}}}\;g_{\mathbf{q}\nu}(\bar{\mathbf{u}},z)\\ &\equiv \delta_{\mathbf{k}-\mathbf{k}',-\mathbf{q}}\;g_{\mathbf{q}\nu,\mu\mu'}(\mathbf{k},z),\end{aligned} \tag{16.139}$$

where in the second equality we use the property (16.60), whereas the last equality defines the matrix $g_{\mathbf{q}\nu}(\mathbf{k})$. We conclude that the sandwich of (16.138) between $\langle\mathbf{k}\mu|$ and $|\mathbf{k}\mu'\rangle$ gives the following equation of motion for the Green's function matrix $G_{\mathbf{k}}$:

$$\begin{aligned}\Big[\mathrm{i}\frac{d}{dz_1} - h_{\mathrm{HF}}(\mathbf{k},z_1) - \sum_{\nu} g_{\mathbf{0}\nu}(\mathbf{k},z_1)U_{\mathbf{0}\nu}(z_1)\Big]G_{\mathbf{k}}(z_1,z_2)\\ = \delta(z_1,z_2) + \int_\gamma d\bar{z}\,\Sigma_{\mathrm{c},\mathbf{k}}(z_1,\bar{z})G_{\mathbf{k}}(\bar{z},z_2)\end{aligned} \tag{16.140}$$

Displacements and momenta The average displacements and momenta satisfy the equation of motion (16.118). Taking into account Table 16.1, we find

$$\begin{aligned}\frac{dP_{\mathbf{q}\nu}(z)}{dz} &= -\int d\mathbf{x}\,g_{\mathbf{q}\nu}(\mathbf{r},z)\Delta n(\mathbf{x},z) - \sum_{\nu'}K_{\nu\nu'}(\mathbf{q},z)U_{\mathbf{q}\nu'}(z)\\ \frac{dU_{\mathbf{q}\nu}(z)}{dz} &= P_{\mathbf{q}\nu}(z)\end{aligned} \tag{16.141}$$

where $P_{\mathbf{q}\nu}$ is the average of $\hat{\mathcal{P}}_{\mathbf{q}\nu}$. The first of these equations agrees with (16.39), whereas the second equation establishes that $P_{\mathbf{q}\nu}$ is the conjugate momentum of $U_{\mathbf{q}\nu}$. Alternatively, $U_{\mathbf{q}\nu}$ and $P_{\mathbf{q}\nu}$ can be calculated from $\phi = \phi^s + s$, see (16.79), where ϕ^s is given by (16.104). Taking into account Table 16.1, it is straightforward to find

$$U_{\mathbf{q}\nu}(z) = \sum_{\nu'} \int d\bar{\mathbf{x}}d\bar{z}\, D^{11}_{0,\mathbf{q}\nu\nu'}(z,\bar{z}) g_{\mathbf{q}\nu'}(\bar{\mathbf{r}},\bar{z})\Delta n(\bar{\mathbf{x}},\bar{z}), \tag{16.142a}$$

$$P_{\mathbf{q}\nu}(z) = \sum_{\nu'} \int d\bar{\mathbf{x}}d\bar{z}\, D^{21}_{0,\mathbf{q}\nu\nu'}(z,\bar{z}) g_{\mathbf{q}\nu'}(\bar{\mathbf{r}},\bar{z})\Delta n(\bar{\mathbf{x}},\bar{z}). \tag{16.142b}$$

In equilibrium, $\Delta n = 0$ and therefore $U_{\mathbf{q}\nu} = P_{\mathbf{q}\nu} = 0$. Under the working hypothesis that the external perturbations do not break the lattice periodicity of the crystal, we also have $U_{\mathbf{q}\nu}(z) = P_{\mathbf{q}\nu}(z) = 0$ for all $\mathbf{q} \neq 0$.

Phononic Green's function Let us investigate the structure of the phononic self-energy Π. According to the fourth Hedin–Baym equation of Table 16.2, we preliminarily need to investigate the dressed electron–phonon coupling defined by the eighth Hedin–Baym equation. As the bare electron–phonon coupling is local in time and space, the dressed electron–phonon coupling can be expressed in terms of the polarization P, see the seventh Hedin–Baym equation. We have

$$\begin{aligned} \mathfrak{a}^d(\underbrace{\mathbf{x}_1, z_1}_{1}, \underbrace{\mathbf{x}_2, z_2}_{2}; \underbrace{-\mathbf{q}_3\nu_3, i_3, z_3}_{3^*}) &= \delta_{i_3,1}\delta(1,2)\delta(z_1,z_3) g_{-\mathbf{q}_3\nu_3}(\mathbf{r}_1, z_1) \\ &+ \delta_{i_3,1}\delta(1,2)\int d\bar{1}d\bar{2}\, W(1;\bar{1})P(\bar{1};\bar{2})\delta(\bar{z}_2,z_3) g_{-\mathbf{q}_3\nu_3}(\bar{\mathbf{r}}_2,\bar{z}_2). \end{aligned}$$

Let us define the time-local bare coupling and the time-nonlocal dressed coupling according to

$$g_{\mathbf{q}\nu}(1;z) = g_{\mathbf{q}\nu}(\mathbf{x}_1, z_1; z) \equiv \delta(z_1, z) g_{\mathbf{q}\nu}(\mathbf{r}_1, z_1), \tag{16.143a}$$

$$g^d_{\mathbf{q}\nu}(1;z) = g^d_{\mathbf{q}\nu}(\mathbf{x}_1, z_1; z) \equiv g_{\mathbf{q}\nu}(1;z) + \int d\bar{1}d\bar{2}\, W(1;\bar{1})P(\bar{1};\bar{2}) g_{\mathbf{q}\nu}(\bar{2};z). \tag{16.143b}$$

We then see that

$$\mathfrak{a}(1,2;3^*) = \delta_{i_3,1}\delta(1,2) g_{-\mathbf{q}_3\nu_3}(1;z_3) \quad , \quad \mathfrak{a}^d(1,2;3^*) = \delta_{i_3,1}\delta(1;2) g^d_{-\mathbf{q}_3\nu_3}(1;z_3).$$

The dressed coupling $\mathfrak{a}^d(1,2;3) \propto \delta(1;2)$ like the bare coupling $\mathfrak{a}$. Therefore, also the phononic self-energy can be written in terms of the polarization P. From the fourth Hedin–Baym equation we have

$$\begin{aligned} \Pi(1,2) &= \Pi(\mathbf{q}_1\nu_1, i_1, z_1, \mathbf{q}_2\nu_2, i_2, z_2) \\ &= \delta_{i_1,1}\delta_{i_2,1}\int d\bar{1}d\bar{2}\, g_{-\mathbf{q}_1\nu_1}(\bar{1};z_1) P(\bar{1};\bar{2}) g^d_{-\mathbf{q}_2\nu_2}(\bar{2};z_2). \end{aligned}$$

Notice that the only nonvanishing components of Π are those of the displacement–displacement block: $i_1 = i_2 = 1$. A more symmetric form for Π can be derived if we use the explicit expression (16.143b) of the dressed coupling. Indeed,

$$gPg^d = gP(g + WPg) = g(P + PWP)g = g\chi g,$$

where $\chi = P + PWP = P + Pv\chi$ is the density-density response function, see (7.24), and $W = v + vPW$ in accordance with the sixth Hedin-Baym equation of Table 16.2. Spelling out the space-spin-time convolutions,

$$\begin{aligned}\Pi(1,2) &= \Pi(\mathbf{q}_1\nu_1, i_1, z_1, \mathbf{q}_2\nu_2, i_2, z_2) \\ &= \delta_{i_1,1}\delta_{i_2,1} \int d\bar{1}d\bar{2}\, g_{-\mathbf{q}_1\nu_1}(\bar{1}; z_1)\chi(\bar{1};\bar{2})g_{-\mathbf{q}_2\nu_2}(\bar{2}; z_2). \end{aligned} \tag{16.144}$$

The periodicity of the crystal under discrete lattice translations can now be used to show that $\Pi(1,2)$ is nonvanishing only for $\mathbf{q}_1 = -\mathbf{q}_2$. Two-point correlators $c(\mathbf{r},\mathbf{r}') = c(\mathbf{R_n}+\mathbf{u}, \mathbf{R_{n'}}+\mathbf{u}')$ like the Green's function, the self-energy, or the polarization, depend only on the difference $\mathbf{n}-\mathbf{n}'$, see also Appendix C. We can then expand the response function according to

$$\begin{aligned}\chi(1,2) &= \chi(\mathbf{R}_{\mathbf{n}_1}+\mathbf{u}_1\sigma_1, z_1; \mathbf{R}_{\mathbf{n}_2}+\mathbf{u}_2\sigma_2, z_2) \\ &= \frac{1}{\sqrt{|\mathrm{V}_b|}}\sum_{\mathbf{q}} e^{\mathrm{i}\mathbf{q}\cdot(\mathbf{n}_1-\mathbf{n}_2)}\chi(\mathbf{u}_1\sigma_1, z_1; \mathbf{u}_2\sigma_2, z_2).\end{aligned}$$

Inserting this expansion in (16.144) and writing the spatial integral over $\bar{\mathbf{r}}_1 = \mathbf{R}_{\bar{\mathbf{n}}_1} + \bar{\mathbf{u}}_1$ as $\int d\bar{\mathbf{r}}_1 = \sum_{\bar{\mathbf{n}}_1}\int_{\text{cell}} d\bar{\mathbf{u}}_1$, we realize that the equation for Π involves the calculation of the sum

$$\sum_{\bar{\mathbf{n}}_1} g_{-\mathbf{q}_1\nu_1}(\mathbf{R}_{\bar{\mathbf{n}}_1}+\bar{\mathbf{u}}_1, z_1)e^{\mathrm{i}\mathbf{q}\cdot\bar{\mathbf{n}}_1} = \sum_{\bar{\mathbf{n}}_1} g_{-\mathbf{q}_1\nu_1}(\bar{\mathbf{u}}_1, z_1)e^{\mathrm{i}(\mathbf{q}+\mathbf{q}_1)\cdot\bar{\mathbf{n}}_1} \propto \delta_{\mathbf{q},-\mathbf{q}_1},$$

where in the second equality we use the property (16.60). Performing the same manipulations for the integral over $\bar{\mathbf{r}}_2$ in (16.144), we readily find that $\Pi(1,2)$ is proportional to $\delta_{\mathbf{q},\mathbf{q}_2}$. We conclude that

$$\Pi(1,2) = \delta_{\mathbf{q}_1,-\mathbf{q}_2}\, \Pi^{i_1i_2}_{-\mathbf{q}_1\nu_1\nu_2}(z_1,z_2).$$

As both $D_0(1,2)$ and $\Pi(1,2)$ are proportional to $\delta_{\mathbf{q}_1,-\mathbf{q}_2}$, the interacting phononic Green's function $D(1,2) = [D_0 + D_0\Pi D_0 + D_0\Pi D_0\Pi D_0 + \ldots](1,2)$ is also proportional to $\delta_{\mathbf{q}_1,-\mathbf{q}_2}$:

$$D(1,2) = \delta_{\mathbf{q}_1,-\mathbf{q}_2}\, D^{i_1i_2}_{\mathbf{q}_1\nu_1\nu_2}(z_1,z_2).$$

Notice that the phononic self-energy Π in the square brackets is defined with a $-\mathbf{q}_1$, whereas the phononic Green's function D in the square brackets is defined with a $\mathbf{q}_1$. These different definitions are chosen to have a more elegant Kadanoff-Baym equation. Indeed, if we sandwich (16.115b) between $\langle\mathbf{q}\nu, i|$ and $|\mathbf{q}\nu', i'\rangle$ we obtain (in matrix form)

$$\boxed{\left[\mathrm{i}\frac{d}{dz_1}J - Q(\mathbf{q},z_1)\right]D_{\mathbf{q}}(z_1,z_2) = \mathbb{1}\delta(z_1,z_2) + \int_\gamma d\bar{z}\,\Pi_{\mathbf{q}}(z_1,\bar{z})D_{\mathbf{q}}(\bar{z},z_2)} \tag{16.145}$$

where $\Pi_{\mathbf{q}}$ reads [see (16.144)]:

$$\Pi^{ii'}_{\mathbf{q}\nu\nu'}(z,z') = \delta_{i,1}\delta_{i',1}\int d\bar{1}d\bar{2}\, g_{\mathbf{q}\nu}(\bar{1};z)\chi(\bar{1};\bar{2})g_{-\mathbf{q}\nu'}(\bar{2};z')$$
$$= \delta_{i,1}\delta_{i',1}\int d\bar{\mathbf{x}}_1 d\bar{\mathbf{x}}_2\, g_{\mathbf{q}\nu}(\bar{\mathbf{r}}_1,z)\chi(\bar{\mathbf{x}}_1,z;\bar{\mathbf{x}}_2,z')g_{-\mathbf{q}\nu'}(\bar{\mathbf{r}}_2,z'). \quad (16.146)$$

Equations (16.140), (16.145), and their counterparts with derivatives with respect to z_2, along with the equations of motion (16.141) for the displacements and momenta, form a closed system of equations for any approximate functional $\Sigma_{c,ss}[G,D]$ and $\Pi_{ss}[G,D]$. Furthermore, if the self-energies are Φ-derivable then the resulting Green's functions satisfy all fundamental conservation laws. In the next sections we discuss the strategy to solve these equations.

16.11.3 Self-Consistent Matsubara Equations

Like in the case of only particles, the preliminary step to solve the equations of motion for the Green's functions on the contour consists of solving the Matsubara problem. The Matsubara self-energies do indeed depend only on the Matsubara Green's functions, see Section 9.2, and therefore the equations for the Matsubara components are closed. Expanding the Matsubara Green's functions and self-energies according to (6.18) (fermionic frequencies for G and Σ_c and bosonic frequencies for D and Π) and taking into account that along the vertical track $U_{\mathbf{q}\nu} = P_{\mathbf{q}\nu} = 0$ since $\Delta n = 0$, see (16.142), the equations of motion (16.140) and (16.145) yield

$$G^{\rm M}_{\mathbf{k}}(\omega_m) = \frac{1}{\omega_m - h^{\rm M}_{\rm HF}(\mathbf{k}) - \Sigma^{\rm M}_{c,\mathbf{k}}(\omega_m)}, \qquad h^{\rm M}_{\rm HF}(\mathbf{k}) = h_{\rm HF}(\mathbf{k}) - \mu\,, \quad (16.147a)$$

$$D^{\rm M}_{\mathbf{q}}(\omega_m) = \frac{1}{\omega_m J - Q^{\rm M}(\mathbf{q}) - \Pi^{\rm M}_{\mathbf{q}}(\omega_m)}, \qquad Q^{\rm M}(\mathbf{q}) = \begin{pmatrix} K(\mathbf{q}) & 0 \\ 0 & \mathbb{1} \end{pmatrix}. \quad (16.147b)$$

For any approximation to $\Sigma_c = \Sigma_{ss,c}[G,D]$ and $\Pi = \Pi_{ss}[G,D]$ these equations can be solved self-consistently. We recall that $K = K[n^0]$ depends on the equilibrium density,

$$n^0(\mathbf{x}) = -\mathrm{i}\langle\mathbf{x}|\hat{\mathcal{G}}^{\rm M}(\tau,\tau^+)|\mathbf{x}\rangle = -\mathrm{i}\sum_{\mathbf{k}}\sum_{\mu\mu'}\langle\mathbf{x}|\mathbf{k}\mu\rangle\, G^{\rm M}_{\mathbf{k}\mu\mu'}(\tau,\tau^+)\,\langle\mathbf{k}\mu'|\mathbf{x}\rangle, \quad (16.148)$$

see (16.34). Thus, the Matsubara equations are coupled even for noninteracting $v = g = 0$ systems. It is only in the partial self-consistent scheme discussed at the end of Section 16.2 that the noninteracting ($v = g = 0$) Matsubara equations decouple since the elastic tensor K is not updated in this case.

Phononic self-energy in the clamped+static approximation In all physical situations of relevance, v and g do not vanish and a diagrammatic approximation for the self-energies must be chosen. For the electronic self-energy the GW approximation is a "gold standard" for obtaining accurate or at least reasonable results [94, 95, 100, 291, 292, 293]. What about the phononic self-energy? Let us explore the physics of (16.146) when χ is calculated by summing all diagrams *without* electron–phonon vertices: $\chi \simeq \chi|_{g=0}$. This approximation for χ corresponds to the response function of a system of only electrons interacting through the Coulomb repulsion and feeling the potential ϕ generated by clamped nuclei in $\mathbf{R}^0$ (i.e., $\chi|_{g=0} = \chi_{\rm clamp}$), see Section 16.2.2. In such *clamped approximation* the phononic self-energy (16.146) reads

$$\Pi^{{\rm M},ii'}_{\mathbf{q}\nu\nu'}(\omega_m) \simeq \delta_{i,1}\delta_{i',1}\int d\bar{\mathbf{x}}_1 d\bar{\mathbf{x}}_2\, g_{\mathbf{q}\nu}(\bar{\mathbf{r}}_1)\chi^{\rm M}_{\rm clamp}(\bar{\mathbf{x}}_1,\bar{\mathbf{x}}_2;\omega_m)g_{-\mathbf{q}\nu'}(\bar{\mathbf{r}}_2),$$

where we use that along the vertical track $g_{\mathbf{q}\nu}(\bar{\mathbf{r}}_1, t_0 - {\rm i}\tau) = g_{\mathbf{q}\nu}(\bar{\mathbf{r}}_1)$ is independent of τ. Let us further approximate the r.h.s. with its value at $\omega_m = 0$ [283]. This is the so-called *static approximation* and it is similar in spirit to the statically screened approximation of W, see (14.42). According to (10.41) $\chi^{\rm M}(\bar{\mathbf{x}}_1,\bar{\mathbf{x}}_2;\pm{\rm i}\eta) = \chi^{\rm R/A}(\bar{\mathbf{x}}_1,\bar{\mathbf{x}}_2;0)$ and according to (13.27) the retarded/advanced response function is real at zero frequency. Therefore the function $\chi^{\rm M}(\bar{\mathbf{x}}_1,\bar{\mathbf{x}}_2;\zeta)$ of the complex frequency ζ has no discontinuity in $\zeta = 0$ and we can write $\chi^{\rm M}(\bar{\mathbf{x}}_1,\bar{\mathbf{x}}_2;0) = \chi^{\rm R/A}(\bar{\mathbf{x}}_1,\bar{\mathbf{x}}_2;0)$. In the clamped+static approximation the phononic self-energy can then be written as

$$\Pi^{{\rm M},ii'}_{\mathbf{q}\nu\nu'}(\omega_m) \simeq \delta_{i,1}\delta_{i',1}\Pi^{\rm clamp+stat}_{\mathbf{q}\nu\nu'}$$

with

$$\Pi^{\rm clamp+stat}_{\mathbf{q}\nu\nu'} \equiv \int d\bar{\mathbf{x}}_1 d\bar{\mathbf{x}}_2\, g_{\mathbf{q}\nu}(\bar{\mathbf{r}}_1)\chi^{\rm R}_{\rm clamp}(\bar{\mathbf{x}}_1,\bar{\mathbf{x}}_2;0)g_{-\mathbf{q}\nu'}(\bar{\mathbf{r}}_2). \tag{16.149}$$

Inserting this approximation into (16.147b) we see that the effect of Π is to renormalize the block $(1,1)$ of $Q^{\rm M}$. In other words, the interacting D in the clamped+static approximation has the same form as the noninteracting D_0 with a renormalized elastic tensor,

$$K_{\nu\nu'}(\mathbf{q}) \to K^{\rm renorm}_{\nu\nu'}(\mathbf{q}) = K_{\nu\nu'}(\mathbf{q}) + \Pi^{\rm clamp+stat}_{\mathbf{q}\nu\nu'}.$$

Connection with the Born–Oppenheimer approximation If we use the Born–Oppenheimer approximation of Section 16.2.2 to evaluate g and K, then the renormalized elastic tensor is exactly the Hessian, $K^{\rm renorm}_{\nu\nu'}(\mathbf{q}) = {\rm Hes}_{\nu\nu'}(\mathbf{q})$, compare with (16.66) and take into account the property (16.59). The Hessian has positive eigenvalues $\omega^2_{\mathbf{q}\nu}$ since $\mathbf{R}^{0,{\rm BO}}$ is the global minimum of the Born–Oppenheimer energy. The frequencies $\omega_{\mathbf{q}\nu}$ are called the phonon frequencies and they provide an excellent starting point already for a clamped response function $\chi_{\rm clamp}$ evaluated at the RPA level, see Section 15.4. The phononic band structure can indeed be measured using Raman spectroscopy. Although the agreement is often reasonable, the importance of going beyond the static approximation (especially for metallic systems) has been reported in the literature [294, 295, 296]. In the same way as the time-nonlocal electronic self-energy is responsible for renormalizing the band structure, producing a finite quasi-particle lifetime and giving rise to satellites such as plasmons, the nonstatic approximation for $\Pi_{\mathbf{q}}$ is expected to renormalize the phononic band structure, producing a finite phononic lifetime, see Section 16.11.4, and giving rise to phononic satellites.

In the basis of the normal modes of the Hessian we have ${\rm Hes}_{\nu\nu'}(\mathbf{q}) = \delta_{\nu\nu'}\omega^2_{\mathbf{q}\nu}$. Thus, the block $(1,1)$ of the interacting phononic Green's function in the clamped+static approximation for Π is simply [compare with (16.132)]

$$D^{11,{\rm M}}_{\mathbf{q}\nu\nu'}(\omega_m) = \frac{\delta_{\nu\nu'}}{\omega^2_m - \omega^2_{\mathbf{q}\nu}}.$$

As the phonon frequencies are all positive we can use them to construct the annihilation and creation operators $\hat{b}_{\mathbf{q}\nu}$ and $\hat{b}^\dagger_{\mathbf{q}\nu}$ according to (16.53), where $\hat{\mathcal{U}}_{\mathbf{q}\nu}$ and $\hat{\mathcal{P}}_{\mathbf{q}\nu}$ are the displacement and momentum operators in the basis of the normal modes of the Hessian. There is no particular convenience in rewriting the full electron–phonon Hamiltonian in terms of $\hat{b}_{\mathbf{q}\nu}$ and $\hat{b}^\dagger_{\mathbf{q}\nu}$ since $\hat{H}^0_{\text{phon}}$ is not diagonal in this basis. The real convenience of the Hessian basis is that the interacting phononic Green's function in the clamped+static approximation is diagonal. Therefore, we can interpret $\hat{b}_{\mathbf{q}\nu}$ and $\hat{b}^\dagger_{\mathbf{q}\nu}$ as the annihilation and creation operators of a phonon. In the clamped+static approximation these phonons have a well-defined energy and hence an infinitely long lifetime.

Conserving approximations Going beyond the clamped+static approximation, the frequencies and the direction of the normal modes change. Although we expect the changes to be small if the starting point is already a good one, the concept of phonons as infinitely long-lived lattice excitations becomes an approximate concept. Phonons become *quasi-phonons* or *dressed phonons* by acquiring a finite lifetime, see Section 16.11.4.

The results obtained in the clamped+static approximation suggest that the minimal meaningful phononic self-energy is $\Pi = g\chi g$ with the clamped RPA $\chi = \chi^0 + \chi^0 v \chi$, or equivalently $\Pi = g\chi^0 g^d$ with $g^d = (\delta + v\chi)g = (\delta + W\chi^0)g$ and $W = v + v\chi^0 W$. As pointed out in Section 16.10, this phononic self-energy is Φ-derivable, the XC functional Φ_{xc} being the sum of all diagrams in the second row of Fig. 16.12. For the theory to be conserving, the electronic correlation self-energy must be consistently obtained from the functional derivative of the same functional Φ_{xc}. Therefore, any calculation with $\Pi = g\chi g$ should be done with an electronic Fan–Migdal self-energy $\Sigma_{\text{FM}} = \mathrm{i} g^d G D g^d$. We can add to Σ_{FM} the GW self-energy and still be conserving since the diagrams in the first row of Fig. 16.12 do not contribute to Π.

16.11.4 Time-Dependent Evolution and Steady States

With the Matsubara Green's functions at our disposal we can proceed with the calculation of all other Keldysh components by time propagation. The equations for the right Green's functions are obtained by taking $z_1 = t_-$ or t_+ and $z_2 = t_0 - \mathrm{i}\tau$ in (16.140) and (16.145), and read:

$$\boxed{\left[\mathrm{i}\frac{d}{dt} - h_{\text{HF}}(\mathbf{k},t) - \sum_\nu g_{\mathbf{0}\nu}(\mathbf{k},t) U_{\mathbf{0}\nu}(t)\right] G^\rceil_{\mathbf{k}}(t,\tau) = \left[\Sigma^{\text{R}}_{\text{c},\mathbf{k}} \cdot G^\rceil_{\mathbf{k}} + \Sigma^\rceil_{\text{c},\mathbf{k}} \star G^{\text{M}}_{\mathbf{k}}\right](t,\tau)} \tag{16.150}$$

$$\boxed{\left[\mathrm{i}\frac{d}{dt} J - Q(\mathbf{q},t)\right] D^\rceil_{\mathbf{q}}(t,\tau) = \left[\Pi^{\text{R}}_{\mathbf{q}} \cdot D^\rceil_{\mathbf{q}} + \Pi^\rceil_{\mathbf{q}} \star D^{\text{M}}_{\mathbf{q}}\right](t,\tau)} \tag{16.151}$$

Similarly, for $z_1 = t_0 - \mathrm{i}\tau$ and $z_2 = t_-$ or t_+ the adjoint of the equations of motion (16.140) and (16.145) yield

$$\boxed{G^\lceil_{\mathbf{k}}(\tau,t)\left[-\mathrm{i}\frac{\overleftarrow{d}}{dt} - h_{\text{HF}}(\mathbf{k},t) - \sum_\nu g_{\mathbf{0}\nu}(\mathbf{k},t) U_{\mathbf{0}\nu}(t)\right] = \left[G^\lceil_{\mathbf{k}} \cdot \Sigma^{\text{A}}_{\text{c},\mathbf{k}} + G^{\text{M}}_{\mathbf{k}} \star \Sigma^\lceil_{\text{c},\mathbf{k}}\right](\tau,t)} \tag{16.152}$$

$$D^{\lceil}_{\mathbf{q}}(\tau,t)\Big[-\mathrm{i}\frac{\overleftarrow{d}}{dt}J-Q(\mathbf{q},t)\Big]=\Big[D^{\lceil}_{\mathbf{q}}\cdot\Pi^{\mathrm{A}}_{\mathbf{q}}+D^{\mathrm{M}}_{\mathbf{q}}\star\Pi^{\lceil}_{\mathbf{q}}\Big](\tau,t) \tag{16.153}$$

At fixed τ these equations are first-order integro-differential equations in t, which must be solved with initial conditions

$$\begin{aligned} G^{\rceil}_{\mathbf{k}}(t_0,\tau)&=G^{\mathrm{M}}_{\mathbf{k}}(0,\tau), & G^{\lceil}_{\mathbf{k}}(\tau,t_0)&=G^{\mathrm{M}}_{\mathbf{k}}(\tau,0),\\ D^{\rceil}_{\mathbf{q}}(t_0,\tau)&=D^{\mathrm{M}}_{\mathbf{q}}(0,\tau), & D^{\lceil}_{\mathbf{q}}(\tau,t_0)&=D^{\mathrm{M}}_{\mathbf{q}}(\tau,0). \end{aligned} \tag{16.154}$$

The dependence on time in $h_{\mathrm{HF}}(\mathbf{k},t)$ and $Q(\mathbf{q},t)$ may be due to some external laser field and/or phonon driving.

The retarded/advanced as well as the left/right components of the self-energies depend on the left and right Green's functions and on the lesser/greater Green's functions. Therefore, the four equations (16.150–16.153) do not form a closed set of equations for $G^{\rceil}_{\mathbf{k}}$, $G^{\lceil}_{\mathbf{k}}$, $D^{\rceil}_{\mathbf{q}}$, and $D^{\lceil}_{\mathbf{q}}$. To close the set we need the equations of motion for $G^{\lessgtr}_{\mathbf{k}}$ and $D^{\lessgtr}_{\mathbf{q}}$. These can easily be obtained by setting $z_1=t_{1\pm}$ and $z_2=t_{2\mp}$ in (16.140) and (16.145):

$$\begin{aligned} &\Big[\mathrm{i}\frac{d}{dt_1}-h_{\mathrm{HF}}(\mathbf{k},t_1)-\sum_\nu g_{0\nu}(\mathbf{k},t_1)U_{0\nu}(t_1)\Big]G^{\lessgtr}_{\mathbf{k}}(t_1,t_2)\\ &\qquad=\Big[\Sigma^{\lessgtr}_{\mathrm{c},\mathbf{k}}\cdot G^{\mathrm{A}}_{\mathbf{k}}+\Sigma^{\mathrm{R}}_{\mathrm{c},\mathbf{k}}\cdot G^{\lessgtr}_{\mathbf{k}}+\Sigma^{\rceil}_{\mathrm{c},\mathbf{k}}\star G^{\lceil}_{\mathbf{k}}\Big](t_1,t_2) \end{aligned} \tag{16.155}$$

$$\Big[\mathrm{i}\frac{d}{dt_1}J-Q(\mathbf{q},t_1)\Big]D^{\lessgtr}_{\mathbf{q}}(t_1,t_2)=\Big[\Pi^{\mathrm{R}}_{\mathbf{q}}\cdot D^{\lessgtr}_{\mathbf{q}}+\Pi^{\lessgtr}_{\mathbf{q}}\cdot D^{\mathrm{A}}_{\mathbf{q}}+\Pi^{\rceil}_{\mathbf{q}}\star D^{\lceil}_{\mathbf{q}}\Big](t_1,t_2) \tag{16.156}$$

$$\begin{aligned} &G^{\lessgtr}_{\mathbf{k}}(t_1,t_2)\Big[-\mathrm{i}\frac{\overleftarrow{d}}{dt_2}-h_{\mathrm{HF}}(\mathbf{k},t_2)-\sum_\nu g_{0\nu}(\mathbf{k},t_2)U_{0\nu}(t_2)\Big]\\ &\qquad=\Big[G^{\lessgtr}_{\mathbf{k}}\cdot\Sigma^{\mathrm{A}}_{\mathrm{c},\mathbf{k}}+G^{\mathrm{R}}_{\mathbf{k}}\cdot\Sigma^{\lessgtr}_{\mathrm{c},\mathbf{k}}+G^{\rceil}_{\mathbf{k}}\star\Sigma^{\lceil}_{\mathrm{c},\mathbf{k}}\Big](t_1,t_2) \end{aligned} \tag{16.157}$$

$$D^{\lessgtr}_{\mathbf{q}}(t_1,t_2)\Big[-\mathrm{i}\frac{\overleftarrow{d}}{dt_2}J-Q(\mathbf{q},t_2)\Big]=\Big[D^{\lessgtr}_{\mathbf{q}}\cdot\Pi^{\mathrm{A}}_{\mathbf{q}}+D^{\mathrm{R}}_{\mathbf{q}}\cdot\Pi^{\lessgtr}_{\mathbf{q}}++D^{\rceil}_{\mathbf{q}}\star\Pi^{\lceil}_{\mathbf{q}}\Big](t_1,t_2) \tag{16.158}$$

which must be solved with the initial conditions

$$\begin{aligned} G^{<}_{\mathbf{k}}(t_0,t_0)&=G^{\mathrm{M}}_{\mathbf{k}}(0,0^+), & G^{>}_{\mathbf{k}}(t_0,t_0)&=G^{\mathrm{M}}_{\mathbf{k}}(0^+,0),\\ D^{<}_{\mathbf{q}}(t_0,t_0)&=D^{\mathrm{M}}_{\mathbf{q}}(0,0^+), & D^{>}_{\mathbf{q}}(t_0,t_0)&=D^{\mathrm{M}}_{\mathbf{q}}(0^+,0). \end{aligned} \tag{16.159}$$

Finally, we need the equation of motion for the displacement and momentum at $\mathbf{q}=0$ since $U_{0\nu}(t)$ appears explicitly in the equations for the electronic Green's function. These are given by (16.141) when setting $z = t_\pm$, and read:

$$\boxed{\begin{aligned}\frac{dP_{\mathbf{0}\nu}(t)}{dt} &= -\int d\mathbf{x}\, g_{\mathbf{0}\nu}(\mathbf{r},t)\Delta n(\mathbf{x},t) - \sum_{\nu'} K_{\nu\nu'}(\mathbf{0},t)U_{\mathbf{0}\nu'}(t) \\ \frac{dU_{\mathbf{0}\nu}(t)}{dt} &= P_{\mathbf{0}\nu}(t)\end{aligned}} \tag{16.160}$$

where [compare with (16.148)]

$$\Delta n(\mathbf{x},t) = -\mathrm{i}\sum_{\mathbf{k}}\sum_{\mu\mu'}\langle\mathbf{x}|\mathbf{k}\mu\rangle\, G^<_{\mathbf{k}\mu\mu'}(t,t)\,\langle\mathbf{k}\mu'|\mathbf{x}\rangle - n^0(\mathbf{x}).$$

From these equations we see that the equilibrium density n^0 appearing in the electron–phonon Hamiltonian *must be* the same as the self-consistent density (16.148), otherwise the r.h.s. of the equation of motion for $P_{\mathbf{0}\nu}$ does not vanish at $U_{\mathbf{0}\nu} = 0$.

The set of equations (16.150–16.160) are the *Kadanoff–Baym equations* for systems of electrons and phonons.[20] The Kadanoff–Baym equations, together with the initial conditions for the Keldysh components of $G_{\mathbf{k}}$ and $D_{\mathbf{k}}$, completely determine the electronic and phononic Green's functions with one and two real times once a choice for the self-energies is made. These equations have been so far solved only in relatively simple model systems [119, 281, 297, 298]. In the clamped+static approximation for $\Pi_{\mathbf{q}}$ the Kadanoff–Baym equations for $D_{\mathbf{q}}$ simplify considerably since $\Pi^{\lessgtr}_{\mathbf{q}} = \Pi^{\rceil}_{\mathbf{q}} = \Pi^{\lceil}_{\mathbf{q}} = 0$ and

$$\Pi^{ii',\mathrm{R/A}}_{\mathbf{q}\nu\nu'}(t,t') = \delta_{i,1}\delta_{i',1}\delta(t-t')\Pi^{\text{clamp+stat}}_{\mathbf{q}\nu\nu'}, \tag{16.161}$$

in agreement with (16.149). We emphasize that in this approximation the resulting Green's functions are not conserving. In particular, the total energy of the unperturbed system does not remain constant in time.

Symmetry properties We have already pointed out in Section 9.2 that for the actual implementation of the Kadanoff–Baym equations the exact relations (9.30), (9.31), and (9.32) for the electronic Green's function are very useful. We can derive the analogous identities for the phononic Green's function. From definition (16.105) we find

$$D^{ii',<}_{\mathbf{q}\nu\nu'}(t,t') = D^{i'i,>}_{-\mathbf{q}\nu'\nu}(t',t), \tag{16.162a}$$

$$D^{ii',\gtrless}_{\mathbf{q}\nu\nu'}(t,t') = -\left[D^{i'i,\gtrless}_{\mathbf{q}\nu'\nu}(t',t)\right]^*, \tag{16.162b}$$

$$D^{ii',\lceil}_{\mathbf{q}\nu\nu'}(\tau,t) = -\left[D^{i'i,\rceil}_{\mathbf{q}\nu'\nu}(t,\beta-\tau)\right]^*, \tag{16.162c}$$

$$D^{ii',\mathrm{M}}_{\mathbf{q}\nu\nu'}(\tau,\tau') = -\left[D^{i'i,\mathrm{M}}_{\mathbf{q}\nu'\nu}(\tau',\tau)\right]^*. \tag{16.162d}$$

The phononic Green's function satisfies also another important property if the Hamiltonian is invariant under time-reversal. Let $\hat{\Theta}$ be the anti-unitary time-reversal operator.

[20] We observe that for vanishing electron–phonon coupling the Kadanoff–Baym equations for the electronic Green's function are the same as in Section 9.2 since $\hat{\Sigma}(z_1,z_2) = \delta(z_1,z_2)\hat{\mathcal{V}}_{\mathrm{HF}}(z_1) + \hat{\Sigma}_{\mathrm{c}}(z_1,z_2)$.

Time-reversal symmetry implies that we can choose the many-body eigenstates $|\Psi_A\rangle$ of $\hat{H}$ such that $|\Psi_A\rangle = \hat{\Theta}|\Psi_A\rangle$, and therefore we have the property[21] [299]

$$\langle\Psi_A|\hat{O}|\Psi_B\rangle = \langle\Psi_B|\hat{\Theta}\,\hat{O}^\dagger\hat{\Theta}^{-1}|\Psi_A\rangle, \tag{16.163}$$

for any operator $\hat{O}$. Under a time-reversal transformation the displacement operators are even, whereas the momentum operators are odd – that is, $\hat{\Theta}\hat{\mathcal{U}}_{\mathbf{n}s,\kappa}\hat{\Theta}^{-1} = \hat{\mathcal{U}}_{\mathbf{n}s,\kappa}$ and $\hat{\Theta}\,\hat{\mathcal{P}}_{\mathbf{n}s,\kappa}\hat{\Theta}^{-1} = -\hat{\mathcal{P}}_{\mathbf{n}s,\kappa}$. Expanding these operators as in (16.52), we find

$$\hat{\Theta}\,\hat{\mathcal{U}}_{\mathbf{q}\nu}\hat{\Theta}^{-1} = \hat{\mathcal{U}}_{-\mathbf{q}\nu} = \hat{\mathcal{U}}^\dagger_{\mathbf{q}\nu}, \tag{16.164a}$$

$$\hat{\Theta}\,\hat{\mathcal{P}}_{\mathbf{q}\nu}\hat{\Theta}^{-1} = -\hat{\mathcal{P}}_{-\mathbf{q}\nu} = -\hat{\mathcal{P}}^\dagger_{\mathbf{q}\nu}, \tag{16.164b}$$

where we take into account that $\hat{\Theta}c = c^*\hat{\Theta}$ for any complex number c and we use the property $\mathbf{e}^{\nu*}(-\mathbf{q}) = \mathbf{e}^{\nu}(\mathbf{q})$ of the normal modes. From Eq. (16.163) we then infer that

$$\langle\Psi_A|\hat{\mathcal{U}}_{\mathbf{q}\nu}|\Psi_B\rangle = \langle\Psi_B|\hat{\mathcal{U}}_{\mathbf{q}\nu}|\Psi_A\rangle, \tag{16.165a}$$

$$\langle\Psi_A|\hat{\mathcal{P}}_{\mathbf{q}\nu}|\Psi_B\rangle = -\langle\Psi_B|\hat{\mathcal{P}}_{\mathbf{q}\nu}|\Psi_A\rangle. \tag{16.165b}$$

Let E_A be the eigenenergy of $|\Psi_A\rangle$. If the system is in equilibrium, then $|\Psi_A\rangle$ is also an eigenket of the density matrix $\hat{\rho}$, and we denote by ρ_A its eigenvalue. Consider the block $(1,1)$ of the greater Green's function in Eq. (16.105). We have

$$D^{11,>}_{\mathbf{q}\nu\nu'}(t,t') = \frac{1}{\mathrm{i}}\sum_{A,B}\rho_A e^{\mathrm{i}(E_A-E_B)(t-t')} \times \langle\Psi_A|\Delta\hat{\mathcal{U}}_{\mathbf{q}\nu}|\Psi_B\rangle\langle\Psi_B|\Delta\hat{\mathcal{U}}_{-\mathbf{q}\nu'}|\Psi_A\rangle = D^{11,>}_{-\mathbf{q}\nu'\nu}(t,t'), \tag{16.166}$$

where in the second equality we use (16.165). We can analogously derive the relations for all other blocks and for the lesser Green's function. The final result is

$$D^{ii',\gtrless}_{\mathbf{q}\nu\nu'}(t,t') = (-)^{i+i'}D^{i'i,\gtrless}_{-\mathbf{q}\nu'\nu}(t,t') \quad [\hat{\Theta}\text{ invariance}]. \tag{16.167}$$

To highlight the mathematical structure of the derived relations we denote by D^T the transpose matrix of D: $[D^T]^{ii'}_{\nu\nu'} \equiv D^{i'i}_{\nu'\nu}$, and $D^\dagger = [D^T]^*$. Then, (16.162) and (16.167) take the following compact form:

$$D^<_{\mathbf{q}}(t,t') = [D^>_{-\mathbf{q}}(t',t)]^T, \tag{16.168a}$$

$$D^{\gtrless}_{\mathbf{q}}(t,t') = -[D^{\gtrless}_{\mathbf{q}}(t',t)]^\dagger, \tag{16.168b}$$

$$D^{\gtrless}_{\mathbf{q}}(t,t') = \sigma_z[D^{\gtrless}_{-\mathbf{q}}(t,t')]^T\sigma_z \quad [\hat{\Theta}\text{ invariance}], \tag{16.168c}$$

$$D^{\lceil}_{\mathbf{q}}(\tau,t) = -\left[D^{\rceil}_{\mathbf{q}}(t,\beta-\tau)\right]^\dagger, \tag{16.168d}$$

$$D^{\mathrm{M}}_{\mathbf{q}}(\tau,\tau') = -\left[D^{\mathrm{M}}_{\mathbf{q}}(\tau',\tau)\right]^\dagger, \tag{16.168e}$$

[21] Let $|\tilde{\Psi}\rangle = \hat{\Theta}|\Psi\rangle$ be the time-reversed state of $|\Psi\rangle$. By definition, $\langle\Psi_A|\Psi_B\rangle = \langle\tilde{\Psi}_B|\tilde{\Psi}_A\rangle$. Consider a generic operator $\hat{O}$ and the ket $|\Psi_C\rangle = \hat{O}^\dagger|\Psi_A\rangle$. We then have

$$\langle\Psi_A|\hat{O}|\Psi_B\rangle = \langle\Psi_C|\Psi_B\rangle = \langle\tilde{\Psi}_B|\tilde{\Psi}_C\rangle = \langle\tilde{\Psi}_B|\hat{\Theta}\hat{O}^\dagger|\Psi_A\rangle = \langle\tilde{\Psi}_B|\hat{\Theta}\hat{O}^\dagger\hat{\Theta}^{-1}\hat{\Theta}|\Psi_A\rangle = \langle\tilde{\Psi}_B|\hat{\Theta}\hat{O}^\dagger\hat{\Theta}^{-1}|\tilde{\Psi}_A\rangle.$$

Assuming that the states are invariant under time reversal (i.e., $|\tilde{\Psi}_{A,B}\rangle = |\Psi_{A,B}\rangle$), the property (16.163) follows.

where $[\sigma_z]^{ii'}_{\nu\nu'} = \delta_{\nu\nu'}\begin{pmatrix} 1 & 0 \\ 0 & -1 \end{pmatrix}_{ii'}$.

Steady-state solution We can formally solve the Dyson equation for the lesser/greater phononic Green's function following the same mathematical steps as in Section 9.3. Under the assumption that $\Pi_{\mathbf{q}}$ vanishes when the separation between its time arguments goes to infinity, we find the long-time limit solution

$$\lim_{t,t'\to\infty} D^{\lessgtr}_{\mathbf{q}}(t,t') = \left[D^{\rm R}_{\mathbf{q}} \cdot \Pi^{\lessgtr}_{\mathbf{q}} \cdot D^{\rm A}_{\mathbf{q}}\right](t,t').$$

Further assuming that in the long-time limit the system attains a steady state, we can Fourier transform with respect to the time difference every correlator and find

$$D^{\lessgtr}_{\mathbf{q}}(\omega) = D^{\rm R}_{\mathbf{q}}(\omega)\Pi^{\lessgtr}_{\mathbf{q}}(\omega)D^{\rm A}_{\mathbf{q}}(\omega),$$

where

$$D^{\rm R/A}_{\mathbf{q}}(\omega) = \frac{1}{(\omega \pm {\rm i}\eta)J - Q(\mathbf{q}) - \Pi^{\rm R/A}_{\mathbf{q}}(\omega)}, \tag{16.169}$$

and $Q(\mathbf{q}) = \lim_{t\to\infty} Q(\mathbf{q},t)$.[22]

In frequency space the first two equations in (16.168) read

$$D^{<}_{\mathbf{q}}(\omega) = \left[D^{>}_{-\mathbf{q}}(-\omega)\right]^T, \qquad D^{\lessgtr}_{\mathbf{q}}(\omega) = -\left[D^{\lessgtr}_{\mathbf{q}}(\omega)\right]^\dagger.$$

These properties imply that

$$D^{\rm R}_{\mathbf{q}}(\omega) = \left[D^{\rm A}_{-\mathbf{q}}(-\omega)\right]^T, \qquad D^{\rm R}_{\mathbf{q}}(\omega) = \left[D^{\rm A}_{\mathbf{q}}(\omega)\right]^\dagger,$$

which in turn lead to the following properties of the phononic self-energy[23] through (16.169):

$$\Pi^{\rm R}_{\mathbf{q}}(\omega) = \left[\Pi^{\rm A}_{-\mathbf{q}}(-\omega)\right]^T, \qquad \Pi^{\rm R}_{\mathbf{q}}(\omega) = \left[\Pi^{\rm A}_{\mathbf{q}}(\omega)\right]^\dagger.$$

We can then uniquely write the retarded/advanced phononic self-energy as

$$\boxed{\Pi^{\rm R/A}_{\mathbf{q}}(\omega) = \Lambda^{\rm phon}_{\mathbf{q}}(\omega) \mp \frac{\rm i}{2}\Gamma^{\rm phon}_{\mathbf{q}}(\omega)}$$

where $\Lambda^{\rm phon}_{\mathbf{q}}(\omega)$ and $\Gamma^{\rm phon}_{\mathbf{q}}(\omega)$ are self-adjoint. Then,

$$\Lambda^{\rm phon}_{\mathbf{q}}(\omega) = \left[\Lambda^{\rm phon}_{-\mathbf{q}}(-\omega)\right]^T, \qquad \Lambda^{\rm phon}_{\mathbf{q}}(\omega) = \left[\Lambda^{\rm phon}_{\mathbf{q}}(\omega)\right]^\dagger, \tag{16.170a}$$

$$\Gamma^{\rm phon}_{\mathbf{q}}(\omega) = -\left[\Gamma^{\rm phon}_{-\mathbf{q}}(-\omega)\right]^T, \qquad \Gamma^{\rm phon}_{\mathbf{q}}(\omega) = \left[\Gamma^{\rm phon}_{\mathbf{q}}(\omega)\right]^\dagger. \tag{16.170b}$$

[22]In most physical situations $Q(\mathbf{q},t) = Q^{\rm M}(\mathbf{q})$ is independent of time.
[23]Take into account that $J^T = -J$ and $Q^T(-\mathbf{q}) = Q(\mathbf{q})$.

For systems with time-reversal symmetry we also have the properties

$$D^{\lessgtr}_{\mathbf{q}}(\omega) = \sigma_z \big[D^{\lessgtr}_{-\mathbf{q}}(\omega)\big]^T \sigma_z \qquad \Rightarrow \qquad D^{\mathrm{R/A}}_{\mathbf{q}}(\omega) = \sigma_z \big[D^{\mathrm{R/A}}_{-\mathbf{q}}(\omega)\big]^T \sigma_z$$

and therefore

$$\Pi^{\mathrm{R/A}}_{\mathbf{q}}(\omega) = \sigma_z \big[\Pi^{\mathrm{R/A}}_{-\mathbf{q}}(\omega)\big]^T \sigma_z,$$

which imply

$$\Lambda^{\mathrm{phon}}_{\mathbf{q}}(\omega) = \sigma_z \big[\Lambda^{\mathrm{phon}}_{-\mathbf{q}}(\omega)\big]^T \sigma_z, \tag{16.171a}$$

$$\Gamma^{\mathrm{phon}}_{\mathbf{q}}(\omega) = \sigma_z \big[\Gamma^{\mathrm{phon}}_{-\mathbf{q}}(\omega)\big]^T \sigma_z. \tag{16.171b}$$

Thermal equilibrium In thermal equilibrium we can express all Keldysh components in terms of the *phononic spectral function*:

$$\begin{aligned} A^{\mathrm{phon}}_{\mathbf{q}}(\omega) &\equiv \mathrm{i}\Big[D^{\mathrm{R}}_{\mathbf{q}}(\omega) - D^{\mathrm{A}}_{\mathbf{q}}(\omega)\Big] = \mathrm{i}\Big[D^{>}_{\mathbf{q}}(\omega) - D^{<}_{\mathbf{q}}(\omega)\Big] \\ &= \mathrm{i}\, D^{\mathrm{R}}_{\mathbf{q}}(\omega)\Big[\Pi^{\mathrm{R}}_{\mathbf{q}}(\omega) - \Pi^{\mathrm{A}}_{\mathbf{q}}(\omega) - 2\mathrm{i}\eta J\Big] D^{\mathrm{A}}_{\mathbf{q}}(\omega) \\ &= D^{\mathrm{R}}_{\mathbf{q}}(\omega)\Big[\Gamma^{\mathrm{phon}}_{\mathbf{q}}(\omega) + 2\eta J\Big] D^{\mathrm{A}}_{\mathbf{q}}(\omega). \end{aligned} \tag{16.172}$$

The properties previously derived can be used to show that

$$A^{\mathrm{phon}}_{\mathbf{q}}(\omega) = -\big[A^{\mathrm{phon}}_{-\mathbf{q}}(-\omega)\big]^T, \qquad A^{\mathrm{phon}}_{\mathbf{q}}(\omega) = \big[A^{\mathrm{phon}}_{\mathbf{q}}(\omega)\big]^\dagger.$$

The fluctuation–dissipation theorem for the phononic Green's function was derived in Section 16.11.1 for the noninteracting case. The proof in the interacting case goes along the same lines as in Section 6.3.1, and we here write the final result:

$$\boxed{D^{<}_{\mathbf{q}}(\omega) = -\mathrm{i} f(\omega) A^{\mathrm{phon}}_{\mathbf{q}}(\omega), \qquad D^{>}_{\mathbf{q}}(\omega) = -\mathrm{i} \bar{f}(\omega) A^{\mathrm{phon}}_{\mathbf{q}}(\omega)} \tag{16.173}$$

Taking into account that $\bar{f}(\omega) - f(\omega) = 1$, we also have

$$D^{\mathrm{R/A}}_{\mathbf{q}}(\omega) = \mathrm{i}\int \frac{d\omega'}{2\pi} \frac{D^{>}_{\mathbf{q}}(\omega') - D^{<}_{\mathbf{q}}(\omega')}{\omega - \omega' \pm \mathrm{i}\eta} = \int \frac{d\omega'}{2\pi} \frac{A^{\mathrm{phon}}_{\mathbf{q}}(\omega')}{\omega - \omega' \pm \mathrm{i}\eta}.$$

Quasi-phonons The phononic self-energy has only one nonvanishing block, which is the block $(1,1)$. If $\Gamma^{11,\mathrm{phon}}_{\mathbf{q}}(\omega) \neq 0$ we can discard the infinitesimal in (16.172) and find for the block $(1,1)$ of the spectral function

$$A^{11,\mathrm{phon}}_{\mathbf{q}}(\omega) = D^{11,\mathrm{R}}_{\mathbf{q}}(\omega)\Gamma^{11,\mathrm{phon}}_{\mathbf{q}}(\omega) D^{11,\mathrm{A}}_{\mathbf{q}}(\omega), \tag{16.174}$$

with

$$D^{11,\mathrm{R/A}}_{\mathbf{q}}(\omega) = \frac{1}{(\omega \pm \mathrm{i}\eta)^2 - K(\mathbf{q}) - \Pi^{11,\mathrm{R/A}}_{\mathbf{q}}(\omega)}. \tag{16.175}$$

Equation (16.174) is particularly interesting. In the clamped+static Born–Oppenheimer approximation we have $\Pi^{11,\mathrm{R/A}}_{\mathbf{q}}(\omega) \to \Pi^{\mathrm{clam+stat}}_{\mathbf{q}}$ (independent of ω), see Section 16.11.3, and hence $D^{11,\mathrm{R/A}}_{\mathbf{q}}(\omega)$ has poles at the phonon frequencies $\pm\omega_{\mathbf{q}\nu}$ with $\omega_{\mathbf{q}\nu} \geq 0$. In accordance with the discussion in Section 10.2 we may say that in this approximation the phonons behave as infinitely long-lived bosonic particles of well-defined energy $\omega_{\mathbf{q}\nu}$. Going beyond the clamped+static approximation, phonons become *quasi-phonons* like particles become quasi-particles when going beyond the Hartree–Fock approximation. The phononic self-energy generally develops a finite imaginary part that is responsible for the damping of the lattice vibrations or, equivalently, for the finite lifetime of the quasi-phonons. The frequency dependence of the phononic self-energy is also responsible for a renormalization of the phonon frequencies and for the appearance of satellites.

We can estimate the frequency renormalization and the phononic lifetime assuming that the correction to the clamped+static Born–Oppenheimer approximation is small. Working in the basis of the normal modes of the Hessian, we then approximate

$$\Pi^{11,\mathrm{R/A}}_{\mathbf{q}\nu\nu'}(\omega) \simeq \Pi^{\mathrm{clam+stat}}_{\mathbf{q}\nu\nu'} + \delta_{\nu\nu'}\Big(\Lambda^{\mathrm{dyn}}_{\mathbf{q}\nu}(\omega) \mp \frac{\mathrm{i}}{2}\Gamma^{\mathrm{dyn}}_{\mathbf{q}\nu}(\omega)\Big),$$

where $\Lambda^{\mathrm{dyn}}_{\mathbf{q}\nu}(\omega)$ and $\Gamma^{\mathrm{dyn}}_{\mathbf{q}\nu}(\omega)$ are real functions, see (16.170). Bearing in mind that $K_{\nu\nu'}(\mathbf{q}) + \Pi^{\mathrm{clam+stat}}_{\mathbf{q}\nu\nu'} = \delta_{\nu\nu'}\omega^2_{\mathbf{q}\nu}$, see Section 16.11.3, we see that the block $(1,1)$ of the retarded/advanced phononic Green's function is diagonal in the chosen basis. To lowest order in $\Lambda^{\mathrm{dyn}}_{\mathbf{q}\nu}$ and $\Gamma^{\mathrm{dyn}}_{\mathbf{q}\nu}$ we can approximate (16.175) as

$$D^{11,\mathrm{R/A}}_{\mathbf{q}\nu\nu}(\omega) = \frac{1}{(\omega \pm \mathrm{i}\frac{\Gamma^{\mathrm{dyn}}_{\mathbf{q}\nu}(\omega)}{4\omega})^2 - (\omega_{\mathbf{q}\nu} + \frac{\Lambda^{\mathrm{dyn}}_{\mathbf{q}\nu}(\omega)}{2\omega_{\mathbf{q}\nu}})^2}. \tag{16.176}$$

To further manipulate this formula, we define

$$\lambda_{\mathbf{q}\nu}(\omega) \equiv \omega_{\mathbf{q}\nu} + \frac{\Lambda^{\mathrm{dyn}}_{\mathbf{q}\nu}(\omega)}{2\omega_{\mathbf{q}\nu}} \quad , \quad \gamma_{\mathbf{q}\nu}(\omega) \equiv \frac{\Gamma^{\mathrm{dyn}}_{\mathbf{q}\nu}(\omega)}{2\omega},$$

and rewrite (16.176) as

$$D^{11,\mathrm{R/A}}_{\mathbf{q}\nu\nu}(\omega) = \frac{1}{2\lambda_{\mathbf{q}\nu}(\omega)}\left[\frac{1}{\omega - \lambda_{\mathbf{q}\nu}(\omega) \pm \mathrm{i}\frac{\gamma_{\mathbf{q}\nu}(\omega)}{2}} - \frac{1}{\omega + \lambda_{\mathbf{q}\nu}(\omega) \pm \mathrm{i}\frac{\gamma_{\mathbf{q}\nu}(\omega)}{2}}\right]. \tag{16.177}$$

Let us define the *quasi-phonon frequencies* $\Omega^{\pm}_{\mathbf{q}\nu}$ as the solution of

$$\Omega^{\pm}_{\mathbf{q}\nu} \mp \lambda_{\mathbf{q}\nu}(\Omega^{\pm}_{\mathbf{q}\nu}) = \Omega^{\pm}_{\mathbf{q}\nu} \mp (\omega_{\mathbf{q}\nu} + \frac{\Lambda^{\mathrm{dyn}}_{\mathbf{q}\nu}(\Omega^{\pm}_{\mathbf{q}\nu})}{2\omega_{\mathbf{q}\nu}}) = 0. \tag{16.178}$$

To first order in $\omega - \Omega^{\pm}_{\mathbf{q}\nu}$ we can then write

$$\omega \mp \lambda_{\mathbf{q}\nu}(\omega) \simeq \left(1 \mp \frac{\partial\lambda_{\mathbf{q}\nu}}{\partial\omega}\right)_{\omega=\Omega^{\pm}_{\mathbf{q}\nu}} (\omega - \Omega^{\pm}_{\mathbf{q}\nu}).$$

If the function $\gamma_{\mathbf{q}\nu}(\omega)$ is small for $\omega \simeq \Omega^{\pm}_{\mathbf{q}\nu}$ and slowly varying in ω, then we can approximate $\gamma_{\mathbf{q}\nu}(\omega) \simeq \gamma^{\pm}_{\mathbf{q}\nu} \equiv \gamma_{\mathbf{q}\nu}(\Omega^{\pm}_{\mathbf{q}\nu})$ for $\omega \simeq \Omega^{\pm}_{\mathbf{q}\nu}$ and rewrite (16.177) as

$$D^{11,\mathrm{R/A}}_{\mathbf{q}\nu\nu}(\omega) = \frac{1}{2\lambda^{+}_{\mathbf{q}\nu}} \frac{Z^{+}_{\mathrm{phon},\mathbf{q}\nu}}{\omega - \Omega^{+}_{\mathbf{q}\nu} \pm \mathrm{i}/(2\tau^{+}_{\mathrm{phon},\mathbf{q}\nu})} - \frac{1}{2\lambda^{-}_{\mathbf{q}\nu}} \frac{Z^{-}_{\mathrm{phon},\mathbf{q}\nu}}{\omega - \Omega^{-}_{\mathbf{q}\nu} \pm \mathrm{i}/(2\tau^{-}_{\mathrm{phon},\mathbf{q}\nu})},$$

where

$$\lambda^{\pm}_{\mathbf{q}\nu} \equiv \lambda_{\mathbf{q}\nu}(\Omega^{\pm}_{\mathbf{q}\nu}) = \omega_{\mathbf{q}\nu} + \frac{\Lambda^{\mathrm{dyn}}_{\mathbf{q}\nu}(\Omega^{\pm}_{\mathbf{q}\nu})}{2\omega_{\mathbf{q}\nu}}, \tag{16.179a}$$

$$Z^{\pm}_{\mathrm{phon},\mathbf{q}\nu} \equiv \frac{1}{\left(1 \mp \frac{\partial \lambda_{\mathbf{q}\nu}}{\partial \omega}\right)_{\omega=\Omega^{\pm}_{\mathbf{q}\nu}}} = \frac{1}{\left(1 \mp \frac{1}{2\omega_{\mathbf{q}\nu}} \frac{\partial \Lambda^{\mathrm{dyn}}_{\mathbf{q}\nu}}{\partial \omega}\right)_{\omega=\Omega^{\pm}_{\mathbf{q}\nu}}}, \tag{16.179b}$$

$$\tau^{\pm}_{\mathrm{phon},\mathbf{q}\nu} \equiv Z^{\pm,\mathrm{phon}}_{\mathbf{q}\nu} \frac{\Gamma^{\mathrm{dyn}}_{\mathbf{q}\nu}(\Omega^{\pm}_{\mathbf{q}\nu})}{2\Omega^{\pm}_{\mathbf{q}\nu}}. \tag{16.179c}$$

Let us comment on these results. The *phononic quasi-particle renormalization factor* $Z^{\pm}_{\mathrm{phon},\mathbf{q}\nu}$ gives the probability that a lattice vibration with momentum $\mathbf{q}$ along the normal mode ν excites a phonon with quantum numbers $\mathbf{q}\nu$. The remaining spectral weight $(1 - Z^{\pm}_{\mathrm{phon},\mathbf{q}\nu})$ is absorbed by collective phononic excitations arising from correlation effects. In general, $\Omega^{+}_{\mathbf{q}\nu} \neq -\Omega^{-}_{\mathbf{q}\nu}$, and therefore the energy lost from the emission of a phonon $\mathbf{q}\nu$ is not the same as the energy gained by the absorption of the same phonon. Energy is, however, conserved since the same analysis carried for $D^{11,\mathrm{R/A}}_{-\mathbf{q}\nu\nu}(\omega)$ leads to

$$D^{11,\mathrm{R/A}}_{-\mathbf{q}\nu\nu}(\omega) = \frac{1}{2\lambda^{-}_{\mathbf{q}\nu}} \frac{Z^{-}_{\mathrm{phon},\mathbf{q}\nu}}{\omega + \Omega^{-}_{\mathbf{q}\nu} \pm \mathrm{i}/(2\tau^{-}_{\mathrm{phon},\mathbf{q}\nu})} - \frac{1}{2\lambda^{+}_{\mathbf{q}\nu}} \frac{Z^{+}_{\mathrm{phon},\mathbf{q}\nu}}{\omega + \Omega^{+}_{\mathbf{q}\nu} \pm \mathrm{i}/(2\tau^{+}_{\mathrm{phon},\mathbf{q}\nu})},$$

where we use the property $\omega_{\mathbf{q}\nu} = \omega_{-\mathbf{q}\nu}$, see (16.47), and the properties (16.170). To the best of our knowledge, this correlation-induced splitting of the phononic frequencies has not been observed yet. In crystals with time-reversal symmetry, $\Lambda^{\mathrm{dyn}}_{\mathbf{q}\nu}(\omega) = \Lambda^{\mathrm{dyn}}_{-\mathbf{q}\nu}(\omega)$, which combined with the properties (16.170) implies $\Lambda^{\mathrm{dyn}}_{\mathbf{q}\nu}(\omega) = \Lambda^{\mathrm{dyn}}_{\mathbf{q}\nu}(-\omega)$. In this case $\Omega^{+}_{\mathbf{q}\nu} = -\Omega^{-}_{\mathbf{q}\nu}$ and no correlation-induced splitting occurs. Finally, $\tau^{\pm}_{\mathrm{phon},\mathbf{q}\nu}$ dictates the timescale of the damping of the lattice vibration associated with the normal mode $\mathbf{q}\nu$, and it is called the *phononic lifetime*. The result (16.179c) agrees with the expression in Refs. [271, 274, 300] for $Z^{\pm}_{\mathrm{phon},\mathbf{q}\nu} = 1$.

Exercise 16.6 Find the expression for all blocks of the Matsubara, retarded, and advanced component of the noninteracting phononic Green's function using (16.131) and (16.137).

Exercise 16.7 Prove the fluctuation–dissipation theorem in (16.173).

17

From Green's Functions to Simplified Many-Body Approaches

The many-body diagrammatic theory represents a systematic way to deal with interacting systems of fermions and bosons. The Kadanoff–Baym equations provide information on the dynamical and spectral properties. However, the time nonlocality of the self-energy represents a numerical obstacle for the full two-times propagation. In fact, the two-times $G^{\gtrless}(t,t')$ involves the calculation of collision integrals with upper limits t or t', thereby making any time-stepping algorithm, see Appendix J, scale at least cubically (T^3) with the physical propagation time T. In this chapter we discuss simplifications of the Kadanoff–Baym equations at different levels. We anticipate that all simplified approaches remain approximate even for an exact self-energy, and that none of these approaches have access to the time off-diagonal Green's functions (needed for the spectral functions).

17.1 Generalized Kadanoff–Baym Ansatz and Its Mirrored Form

17.1.1 Fermions

Let us start our discussion by considering a system containing only one kind of identical particles, and let them be fermions. In Section 6.1 we have seen that the calculation of the time-dependent ensemble average of any one-body observable necessitates the use of the time-diagonal Green's function $G^<(t,t)$. It would therefore be desirable to generate an equation for such one-time quantity. The exact equation for $G^<(t,t)$ can be derived by subtracting (9.28) to (9.27) and then setting $t_1 = t_2 = t$:

$$\begin{aligned} \mathrm{i}\frac{d}{dt}\hat{\mathcal{G}}^<(t,t) - \left[\hat{h}(t), \hat{\mathcal{G}}^<(t,t)\right]_- &= \left[\hat{\Sigma}^< \cdot \hat{\mathcal{G}}^{\mathrm{A}} + \hat{\Sigma}^{\mathrm{R}} \cdot \hat{\mathcal{G}}^< + \hat{\Sigma}^{\rceil} \star \hat{\mathcal{G}}^{\lceil}\right](t,t) \\ &\quad - \left[\hat{\mathcal{G}}^< \cdot \hat{\Sigma}^{\mathrm{A}} + \hat{\mathcal{G}}^{\mathrm{R}} \cdot \hat{\Sigma}^< + \hat{\mathcal{G}}^{\rceil} \star \hat{\Sigma}^{\lceil}\right](t,t). \end{aligned} \quad (17.1)$$

Using the anti-hermicity properties of the Green's function, see (9.30) and (9.31), and self-energy, see (9.34) and (9.35), we easily realize that the second term in the r.h.s. is the Hermitian conjugate of the first term in the r.h.s.

For all derivations in this chapter we use the adiabatic contour of Section 4.3, and we assume that the adiabatic assumption is valid. We then have $\hat{\Sigma}^{\lceil} = \hat{\Sigma}^{\rceil} = 0$.[1] The retarded/advanced self-energy has the form, see (9.17),

$$\hat{\Sigma}^{\mathrm{R}}(t,t') = \left[\hat{\Sigma}^{\mathrm{A}}(t',t)\right]^{\dagger} = \delta(t-t')\hat{\Sigma}^{\delta}(t) + \theta(t-t')\left[\hat{\Sigma}^{>}(t,t') - \hat{\Sigma}^{<}(t,t')\right].$$

In an exact treatment the singular part $\hat{\Sigma}^{\delta}$ is the HF self-energy, see (9.17). We here keep open the possibility of making time-local approximations also for the retarded/advanced correlation self-energy.[2] Thus, for the time being, we do not provide an explicit expression for $\hat{\Sigma}^{\delta}$. We make the following two manipulations on (17.1):

1. Isolate the singular part of the self-energy and absorb it in $\hat{h}$.

2. Write the remainder of the collision integral as

$$\int_{-\infty}^{t} dt' \left[\hat{\Sigma}^{<}(t,t')\Big(-\hat{\mathcal{G}}^{>}(t',t) + \hat{\mathcal{G}}^{<}(t',t)\Big) + \Big(\hat{\Sigma}^{>}(t,t') - \hat{\Sigma}^{<}(t,t')\Big)\hat{\mathcal{G}}^{<}(t',t)\right]$$
$$= \int_{-\infty}^{t} dt' \left[\hat{\Sigma}^{>}(t,t')\hat{\mathcal{G}}^{<}(t',t) - \hat{\Sigma}^{<}(t,t')\hat{\mathcal{G}}^{>}(t',t)\right]. \quad (17.2)$$

Henceforth we do not write the lower integration limit, which we take as $-\infty$ unless otherwise stated. After manipulations (1) and (2) the equation of motion (17.1) becomes

$$\boxed{\frac{d}{dt}\hat{n}^{<}(t) + \mathrm{i}\left[\hat{h}_{\mathrm{qp}}(t), \hat{n}^{<}(t)\right]_{-} = -\int^{t} dt' \left[\hat{\Sigma}^{>}(t,t')\hat{\mathcal{G}}^{<}(t',t) - \hat{\Sigma}^{<}(t,t')\hat{\mathcal{G}}^{>}(t',t)\right] + \mathrm{h.c.}} \quad (17.3)$$

where "h.c." stands for the Hermitian conjugate of the first term in the r.h.s. To lighten the notation, in (17.3) we have defined the one-body operators

$$\hat{h}_{\mathrm{qp}}(t) = \hat{h}(t) + \hat{\Sigma}^{\delta}(t),$$

which plays the role of a quasi-particle Hamiltonian, see footnote 7 in Chapter 14, and

$$\boxed{\hat{n}^{\gtrless}(t) \equiv -\mathrm{i}\,\hat{\mathcal{G}}^{\lessgtr}(t,t)} \quad (17.4)$$

Using the fermionic anticommutation rules we can easily prove that

$$\hat{n}^{>}(t) = \hat{n}^{<}(t) - \hat{\mathbb{1}}. \quad (17.5)$$

It is worth noting that $\langle \mathbf{x}_1|\hat{n}^{<}(t)|\mathbf{x}_2\rangle = n(\mathbf{x}_1,\mathbf{x}_2,t)$ is the time-dependent one-particle density matrix defined in (8.20). Consequently, (17.3) serves as the equation of motion for

[1] Actually it is possible to avoid the use of the adiabatic contour by adding an extra term. We defer the interested reader to Ref. [301].

[2] An example of time-local approximations beyond HF is the HSEX self-energy, where the screened interaction of the GW diagram is calculated in the statically screened approximation, see (14.42).

the one-particle density matrix. This equation is closed on $\hat{n}^<$ for $\hat{\Sigma}^{\lessgtr} = 0$ since typical approximations to $\hat{\Sigma}^\delta$ are functionals of $\hat{n}^<$; an example is the HF self-energy, see (8.19) or (8.31). To close the equation even for nonvanishing $\hat{\Sigma}^{\lessgtr}$ we have to transform the time off-diagonal $\hat{\mathcal{G}}^{\gtrless}$ into a functional of $\hat{n}^<$. In this way the whole r.h.s. of (17.3) becomes a functional of $\hat{n}^<$ since $\hat{\Sigma}^{\lessgtr}$ is a functional of $\hat{\mathcal{G}}^{\gtrless}$.

GKBA for fermions In the mid 1980s Lipavský et al. [302] proposed an ansatz for the functional $\hat{\mathcal{G}}^{\gtrless}$. In essence the idea is to use the noninteracting (or mean-field) form (6.49) along with the group property of the noninteracting (or mean-field) retarded Green's function – that is, $\hat{\mathcal{G}}_0^{\rm R}(t,t_0) = {\rm i}\hat{\mathcal{G}}_0^{\rm R}(t,t_1)\hat{\mathcal{G}}_0^{\rm R}(t_1,t_0)$ for all $t > t_1 > t_0$ – and the like for the advanced component.[3] We have

$$\begin{aligned}
\hat{\mathcal{G}}_0^<(t,t') &= \hat{\mathcal{G}}_0^{\rm R}(t,t_0)\hat{\mathcal{G}}_0^<(t_0,t_0)\hat{\mathcal{G}}_0^{\rm A}(t_0,t') \\
&= {\rm i}\hat{\mathcal{G}}_0^{\rm R}(t,t')\underbrace{\hat{\mathcal{G}}_0^{\rm R}(t',t_0)\hat{\mathcal{G}}_0^<(t_0,t_0)\hat{\mathcal{G}}_0^{\rm A}(t_0,t')}_{\hat{\mathcal{G}}_0^<(t',t')} \\
&\quad - {\rm i}\underbrace{\hat{\mathcal{G}}_0^{\rm R}(t,t_0)\hat{\mathcal{G}}_0^<(t_0,t_0)\hat{\mathcal{G}}_0^{\rm A}(t_0,t)}_{\hat{\mathcal{G}}_0^<(t,t)}\hat{\mathcal{G}}_0^{\rm A}(t,t') \\
&= -\hat{\mathcal{G}}_0^{\rm R}(t,t')\hat{n}_0^<(t') + \hat{n}_0^<(t)\hat{\mathcal{G}}_0^{\rm A}(t,t'),
\end{aligned} \tag{17.6}$$

where $\hat{n}_0^<(t) \equiv -{\rm i}\hat{\mathcal{G}}_0^<(t,t)$. An identical relation holds for $\hat{\mathcal{G}}_0^>(t,t')$ provided that we replace $\hat{n}_0^<$ with $\hat{n}_0^>$. The *Generalized Kadanoff–Baym Ansatz* (GKBA) [302] amounts to approximating all interacting $\hat{\mathcal{G}}^{\gtrless}$ in the collision integral of (17.3) (including those in the self-energy) as

$$\boxed{\hat{\mathcal{G}}^{\gtrless}(t,t') \simeq -\hat{\mathcal{G}}^{\rm R}(t,t')\,\hat{n}^{\gtrless}(t') + \hat{n}^{\gtrless}(t)\,\hat{\mathcal{G}}^{\rm A}(t,t')} \tag{17.7}$$

where

$$\hat{\mathcal{G}}^{\rm R}(t,t') = -{\rm i}\,\theta(t-t')T\left\{e^{-{\rm i}\int_{t'}^{t} d\bar{t}\,\hat{h}_{\rm qp}(\bar{t})}\right\}, \qquad \hat{\mathcal{G}}^{\rm A}(t,t') = \left[\hat{\mathcal{G}}^{\rm R}(t',t)\right]^\dagger. \tag{17.8}$$

The r.h.s. of (17.7) is a functional of $\hat{n}^<$ since the Green's functions $\hat{\mathcal{G}}^{\rm R/A}$ are given in terms of $\hat{h}_{\rm qp}$, which is in turn a functional of $\hat{n}^<$. In this way (17.3) becomes a closed equation for $\hat{n}^<$ *for any diagrammatic approximation to the self-energy.* The GKBA is exact in the HF approximation and it is expected to be accurate when the average time between two consecutive collisions is longer than the quasi-particle decay time. Improved versions of the GKBA are associated with more sophisticated retarded/advanced Green's functions [303].

The simplification introduced by the GKBA comes with certain trade-offs that it is important to bear in mind. In a stationary state $\hat{n}^<(t) = \hat{n}^<$ is independent of time and $\hat{\mathcal{G}}^{\rm R/A}$ depends only on the time difference. Then, in Fourier space (17.7) reads

$$\hat{\mathcal{G}}^{\gtrless}(\omega) = -\hat{\mathcal{G}}^{\rm R}(\omega)\,\hat{n}^{\gtrless} + \hat{n}^{\gtrless}\,\hat{\mathcal{G}}^{\rm A}(\omega). \tag{17.9}$$

[3] In the HF approximation the retarded Green's function is given by (6.47) with $\hat{h} \to \hat{h}_{\rm HF}$.

In general $\hat{n}^{\gtrless}$ does not commute with $\hat{\mathcal{G}}^{\mathrm{R/A}}$, and therefore (17.9) does not guarantee the property of semidefinite positiveness of $\mathrm{i}\hat{\mathcal{G}}^{>}$ and $-\mathrm{i}\hat{\mathcal{G}}^{<}$, see Exercise 6.9. This shortcoming is at the origin of a possible violation of the Pauli exclusion principle, according to which the eigenvalues of $\hat{n}^{<}$ should fall within the range 0 to 1 [304]. Another important observation is that the GKBA equation does not offer any insights into the interacting spectral function, as it does not provide access to the time off-diagonal Green's function.

The GKBA equation of motion (17.3) has been successfully applied in a large variety of physical situations [305]. These include the nonequilibrium dynamics [306] and many-body localization [307] of Hubbard clusters, time-dependent quantum transport [303, 308], equilibrium absorption of sodium clusters [309], real-time dynamics of the Auger decay [310], transient absorption [311, 312, 313] and carrier dynamics [314, 315, 316, 317] of semiconductors, excitonic insulators out of equilibrium [318], and charge transfer [319] and charge migration [320, 321, 322, 323] in molecular systems.

Mirrored GKBA for fermions The GKBA is not the only ansatz to transform $\hat{\mathcal{G}}^{\lessgtr}$ into a functional of $\hat{n}^{<}$. An equally simple and legitimate ansatz can be obtained by observing that the group property of $\hat{\mathcal{G}}^{\mathrm{R/A}}$ as defined in (17.8) implies

$$\hat{\mathcal{G}}^{\mathrm{A}}(t_0,t') = \mathrm{i}\hat{\mathcal{G}}^{\mathrm{A}}(t_0,t)\hat{\mathcal{G}}^{\mathrm{R}}(t,t'), \qquad \forall t > t'$$
$$\hat{\mathcal{G}}^{\mathrm{R}}(t,t_0) = -\mathrm{i}\hat{\mathcal{G}}^{\mathrm{A}}(t,t')\hat{\mathcal{G}}^{\mathrm{R}}(t',t_0), \qquad \forall t < t'$$

Inserting these results into the first row of (17.6) we arrive at the *mirrored GKBA* (MGKBA) [324, 325]

$$\boxed{\hat{\mathcal{G}}^{\gtrless}(t,t') \simeq -\hat{n}^{\gtrless}(t)\hat{\mathcal{G}}^{\mathrm{R}}(t,t') + \hat{\mathcal{G}}^{\mathrm{A}}(t,t')\hat{n}^{\gtrless}(t')} \tag{17.10}$$

The GKBA and MGKBA can be obtained from one another by exchanging $\hat{\mathcal{G}}^{\mathrm{R}} \leftrightarrow -\hat{\mathcal{G}}^{\mathrm{A}}$. As pointed out in Ref. [201], any linear combination of the GKBA and MGKBA is also a legitimate ansatz. In MGKBA the one-particle density matrix is on the left (right) of $\hat{\mathcal{G}}^{\mathrm{R}}$ ($\hat{\mathcal{G}}^{\mathrm{A}}$) and it is calculated at time t (t'). Thus, the MGKBA equation of motion (17.3) can be written as

$$\boxed{\frac{d}{dt}\hat{n}^{<}(t) + \mathrm{i}\left[\hat{h}_{\mathrm{qp}}(t),\hat{n}^{<}(t)\right]_{-} = -\hat{\Gamma}^{\mathrm{el},>}(t)\hat{n}^{<}(t) - \hat{\Gamma}^{\mathrm{el},<}(t)\hat{n}^{>}(t) + \mathrm{h.c.}} \tag{17.11}$$

where

$$\hat{\Gamma}^{\mathrm{el},\gtrless}(t) = \pm\int^{t} dt'\,\hat{\Sigma}^{\gtrless}(t,t')\hat{\mathcal{G}}^{\mathrm{A}}(t',t)$$

can be interpreted as electronic scattering rates.

General observations An important feature of the GKBA and MGKBA is that the exact relation

$$\hat{\mathcal{G}}^{>}(t,t') - \hat{\mathcal{G}}^{<}(t,t') = \hat{\mathcal{G}}^{\mathrm{R}}(t,t') - \hat{\mathcal{G}}^{\mathrm{A}}(t,t')$$

is fulfilled independently of the choice of $\hat{\mathcal{G}}^{\mathrm{R/A}}$; this is a direct consequence of (17.5). In Section 17.2 we discuss another important merit of the (M)GKBA, namely the ability to uphold conservation laws for conserving self-energies.

17.1.2 Phosons

Through the (M)GKBA we obtain a closed equation of motion for the one-time quantity $n^<$. This represents a major simplification when compared to the original Kadanoff–Baym equations for the two-times Green's functions. However, the (M)GKBA for the fermionic Green's function alone does not help in problems with fermions and phosons (photons or phonons). The reason is that the fermionic and phosonic self-energies are functionals of G and D at different times and therefore we still need to solve the Kadanoff–Baym equations for D. In order to reduce the NEGF time-scaling of fermion–phoson systems we need an equation of motion for the time diagonal phosonic Green's function and extend the (M)GKBA to phosons.

Phosonic density matrix To lighten the notation we define the phosonic density matrix operators:

$$\boxed{\hat{\gamma}^{\lessgtr}(t) \equiv \mathrm{i}\hat{\mathcal{D}}^{\lessgtr}(t,t)} \tag{17.12}$$

We recall that the quantity $\hat{\mathcal{D}}$ is the phosonic Green's function operator in first quantization, see (16.113c) and (16.105). Using the commutation rules (16.76) between displacements and momenta we obtain the important relation

$$\hat{\gamma}^>(t) = \hat{\gamma}^<(t) + \hat{\mathcal{J}}. \tag{17.13}$$

The exact equation of motion for $\hat{\gamma}^<(t)$ can be derived by subtracting the lesser component of (16.119b) from the lesser component of (16.115b) and then setting equal times

$$\begin{aligned}\frac{d}{dt}\hat{\gamma}^<(t) + \mathrm{i}\Big(\hat{\mathcal{J}}\hat{\mathcal{Q}}(t)\hat{\gamma}^<(t) - \hat{\gamma}^<(t)\hat{\mathcal{Q}}(t)\hat{\mathcal{J}}\Big) &= \hat{\mathcal{J}}\left[\hat{\Pi}^< \cdot \hat{\mathcal{D}}^{\mathrm{A}} + \hat{\Pi}^{\mathrm{R}} \cdot \hat{\mathcal{D}}^<\right](t,t) \\ &\quad - \left[\hat{\mathcal{D}}^< \cdot \hat{\Pi}^{\mathrm{A}} + \hat{\mathcal{D}}^{\mathrm{R}} \cdot \hat{\Pi}^<\right](t,t)\hat{\mathcal{J}}.\end{aligned} \tag{17.14}$$

In (17.14) we use the adiabatic contour to be consistent with the fermionic case. Using the anti-hermicity properties of the Green's function and self-energy, see Section 16.11.4, we easily realize that the second term in the r.h.s. is the Hermitian conjugate of the first term in the r.h.s. We can now proceed as in the fermionic case. We write the retarded/advanced phosonic self-energy as

$$\hat{\Pi}^{\mathrm{R}}(t,t') = \left[\hat{\Pi}^{\mathrm{A}}(t',t)\right]^\dagger = \delta(t-t')\hat{\Pi}^{\delta}(t) + \theta(t-t')\left[\hat{\Pi}^>(t,t') - \hat{\Pi}^<(t,t')\right].$$

In an exact treatment the singular part $\hat{\Pi}^{\delta} = 0$. However, we are also interested in time-local approximations of the retarded/advanced correlation self-energy and, therefore, we keep open the possibility of having $\hat{\Pi}^{\delta} \neq 0$.[4] By employing the same manipulations that led to (17.3) on (17.14), we obtain:

$$\boxed{\frac{d}{dt}\hat{\gamma}^<(t) + \mathrm{i}\Big(\hat{\mathcal{J}}\hat{\mathcal{Q}}_{\mathrm{qp}}(t)\hat{\gamma}^<(t) - \hat{\gamma}^<(t)\hat{\mathcal{Q}}_{\mathrm{qp}}(t)\hat{\mathcal{J}}\Big) = \hat{\mathcal{J}}\left[\hat{\Pi}^> \cdot \hat{\mathcal{D}}^< - \hat{\Pi}^< \cdot \hat{\mathcal{D}}^>\right](t,t) + \mathrm{h.c.}} \tag{17.15}$$

[4] An example of time-local approximations is the clamped+static approximation, see (16.161).

where

$$\hat{\mathcal{Q}}_{\rm qp}(t) = \hat{\mathcal{Q}}(t) + \hat{\Pi}^{\delta}(t) \tag{17.16}$$

is the quasi-phoson, or dressed phoson, Hamiltonian.

GKBA for phosons Equations (17.3) and (17.15) form a closed system of equations for $\hat{n}^<$ and $\hat{\gamma}^<$ provided that we use the fermionic (M)GKBA for $\hat{\mathcal{G}}^{\gtrless}$ and find an expression of the time off-diagonal $\hat{\mathcal{D}}^{\gtrless}$ in terms of $\hat{\gamma}^{\gtrless}$. Such an expression has been found by Karlsson et al. [166]. The idea is again to consider the noninteracting form (16.136) of the phosonic Green's function and then take advantage of the group property of the noninteracting retarded component. For any $t > t' > t_0$ the result (16.135) implies

$$\begin{aligned}\hat{\mathcal{D}}_0^{\rm R}(t,t_0) &= -{\rm i}\hat{\mathcal{J}}\,\hat{\mathcal{W}}_L(t) = -{\rm i}\hat{\mathcal{J}}\,\hat{\mathcal{W}}_L(t)\underbrace{\hat{\mathcal{W}}_R(t')({\rm i}\hat{\mathcal{J}})(-{\rm i}\hat{\mathcal{J}})\hat{\mathcal{W}}_L(t')}_{\hat{\mathbb{1}}}\\ &= {\rm i}\hat{\mathcal{D}}_0^{\rm R}(t,t')\hat{\mathcal{J}}\,\hat{\mathcal{D}}_0^{\rm R}(t',t_0),\end{aligned} \tag{17.17}$$

and the like for the advanced component. In (17.17) all quantities are written in an operator form, for example, $\hat{\mathcal{W}}_L(t) = T\left\{e^{-{\rm i}\int_{t_0}^{t}d\bar{t}\;\hat{\mathcal{Q}}(\bar{t})\hat{\mathcal{J}}}\right\}$, see (16.114). Following the same steps leading to (17.6) we can rewrite (16.136) as

$$\hat{\mathcal{D}}_0^{\gtrless}(t,t') = {\rm i}\hat{\mathcal{D}}_0^{\rm R}(t,t')\,\hat{\mathcal{J}}\,\hat{\mathcal{D}}_0^{\gtrless}(t',t') - {\rm i}\hat{\mathcal{D}}_0^{\gtrless}(t,t)\,\hat{\mathcal{J}}\,\hat{\mathcal{D}}_0^{\rm A}(t,t'). \tag{17.18}$$

The *GKBA for phosons* [166] amounts to approximating all interacting $\hat{\mathcal{D}}^{\gtrless}$ in the r.h.s. of (17.15) as

$$\boxed{\hat{\mathcal{D}}^{\gtrless}(t,t') \simeq \hat{\mathcal{D}}^{\rm R}(t,t')\,\hat{\mathcal{J}}\,\hat{\gamma}^{\gtrless}(t') - \hat{\gamma}^{\gtrless}(t)\,\hat{\mathcal{J}}\,\hat{\mathcal{D}}^{\rm A}(t,t')} \tag{17.19}$$

with

$$\hat{\mathcal{D}}^{\rm R/A}(t,t') = -{\rm i}\theta(t-t')\hat{\mathcal{J}}\,T\left\{e^{-{\rm i}\int_{t'}^{t}d\bar{t}\;\hat{\mathcal{Q}}_{\rm qp}(\bar{t})\hat{\mathcal{J}}}\right\}. \tag{17.20}$$

Mirrored GKBA for phosons The phosonic GKBA is not the only ansatz to transform $\hat{\mathcal{D}}^{\lessgtr}$ into a functional of $\hat{\gamma}^<$. The definition in (17.20) implies

$$\begin{aligned}\hat{\mathcal{D}}^{\rm A}(t_0,t') &= \hat{\mathcal{D}}^{\rm A}(t_0,t)({\rm i}\hat{\mathcal{J}})\hat{\mathcal{D}}^{\rm R}(t,t'), &&\forall t > t',\\ \hat{\mathcal{D}}^{\rm R}(t,t_0) &= -\hat{\mathcal{D}}^{\rm A}(t,t')({\rm i}\hat{\mathcal{J}})\hat{\mathcal{D}}^{\rm R}(t',t_0), &&\forall t < t'.\end{aligned}$$

Inserting these results into (16.136), we obtain an equally simple and legitimate ansatz, which we refer to as the MGKBA for phosons:

$$\boxed{\hat{\mathcal{D}}^{\gtrless}(t,t') \simeq \hat{\gamma}^{\gtrless}(t)\,\hat{\mathcal{J}}\,\hat{\mathcal{D}}^{\rm R}(t,t') - \hat{\mathcal{D}}^{\rm A}(t,t')\,\hat{\mathcal{J}}\,\hat{\gamma}^{\gtrless}(t')} \tag{17.21}$$

In MGKBA, the equation of motion (17.15) can be written as

$$\boxed{\frac{d}{dt}\hat{\gamma}^<(t) + \mathrm{i}\Big(\hat{\mathcal{J}}\hat{\mathcal{Q}}_{\rm qp}(t)\hat{\gamma}^<(t) - \hat{\gamma}^<(t)\hat{\mathcal{Q}}_{\rm qp}(t)\hat{\mathcal{J}}\Big) = -\hat{\Gamma}^{\rm ph,>}(t)\hat{\gamma}^<(t) + \hat{\Gamma}^{\rm ph,<}(t)\hat{\gamma}^>(t)} \tag{17.22}$$

where

$$\hat{\Gamma}^{\rm ph,\lessgtr}(t) = \hat{\mathcal{J}}\int^t dt'\, \hat{\Pi}^{\lessgtr}(t,t')\hat{\mathcal{D}}^{\rm A}(t',t)\hat{\mathcal{J}}$$

can be interpreted as phononic scattering rates.

General observations Also for phosons the exact property

$$\hat{\mathcal{D}}^>(t,t') - \hat{\mathcal{D}}^<(t,t') = \hat{\mathcal{D}}^{\rm R}(t,t') - \hat{\mathcal{D}}^{\rm A}(t,t')$$

is fulfilled in (M)GKBA independently of the choice of $\hat{\mathcal{D}}^{\rm R/A}$; this is a direct consequence of (17.13) and $\hat{\mathcal{J}}^2 = \hat{\mathbb{1}}$.

Through the (M)GKBA for fermions and phosons the equations of motion (17.3) and (17.15) close on $\hat{n}^<$ and $\hat{\gamma}^<$ for any diagrammatic approximation to the self-energies Σ and Π. We observe that the presence of phosons leads to a slight modification of (17.3), where $\hat{h}_{\rm qp}(t)$ is replaced by $\hat{h}_{\rm qp}(t) + \sum_\xi \hat{\mathrm{a}}_{\xi^*}(t)\phi_\xi(t)$, see (16.117).

Phonons For phonons, a physically meaningful approximation to $\hat{\Pi}^\delta$ is the clamped+static approximation. As discussed in Section 16.11.3, in this approximation the $(1,1)$ block of the quasi-phonon Hamiltonian $\hat{\mathcal{Q}}_{\rm qp}$, see definition (17.16), is the same as the Hessian of the Born–Oppenheimer energy. In the basis $\{\mathbf{q}\alpha\}$ of the normal modes of the Hessian we can then write[5]

$$\langle \mathbf{q}\alpha, i|\hat{\mathcal{Q}}_{\rm qp}(t)|\mathbf{q}'\alpha', i'\rangle = \delta_{\mathbf{q},\mathbf{q}'}\delta_{\alpha\alpha'}\begin{pmatrix}\omega^2_{\mathbf{q}\alpha} & 0\\ 0 & 1\end{pmatrix}_{ii'}, \tag{17.23}$$

where $\omega^2_{\mathbf{q}\alpha} \geq 0$ are the eigenvalues of the Hessian. In the same basis we have

$$\langle \mathbf{q}\alpha, i|\hat{\mathcal{D}}^{\rm R}(t,t')|\mathbf{q}'\alpha', i'\rangle = D^{\rm R}(\mathbf{q}\alpha, i, t, -\mathbf{q}'\alpha', i', t') = \delta_{\mathbf{q},\mathbf{q}'}\delta_{\alpha\alpha'}D^{ii',\rm R}_{\mathbf{q}\alpha}(t,t')$$

with

$$\begin{aligned} D^{\rm R}_{\mathbf{q}\alpha}(t,t') &= -\mathrm{i}\theta(t-t')\begin{pmatrix}0 & \mathrm{i}\\ -\mathrm{i} & 0\end{pmatrix}\exp\left[-\mathrm{i}\begin{pmatrix}0 & \mathrm{i}\omega^2_{\mathbf{q}\alpha}\\ -\mathrm{i} & 0\end{pmatrix}(t-t')\right] \\ &= \mathrm{i}\frac{\theta(t-t')}{2\omega_{\mathbf{q}\alpha}}\left[e^{\mathrm{i}\omega_{\mathbf{q}\alpha}(t-t')}\begin{pmatrix}1 & -\mathrm{i}\omega_{\mathbf{q}\alpha}\\ \mathrm{i}\omega_{\mathbf{q}\alpha} & \omega^2_{\mathbf{q}\alpha}\end{pmatrix} - e^{-\mathrm{i}\omega_{\mathbf{q}\alpha}(t-t')}\begin{pmatrix}1 & \mathrm{i}\omega_{\mathbf{q}\alpha}\\ -\mathrm{i}\omega_{\mathbf{q}\alpha} & \omega^2_{\mathbf{q}\alpha}\end{pmatrix}\right]. \end{aligned} \tag{17.24}$$

This approximated form depends only on the time difference and can be Fourier transformed. We leave it as an exercise for the reader to show that $D^{11,\rm R}_{\mathbf{q}\alpha}(\omega) = 1/[(\omega + \mathrm{i}\eta)^2 -$

[5]We encourage the reader to have another glance at definitions (16.113) and (16.114).

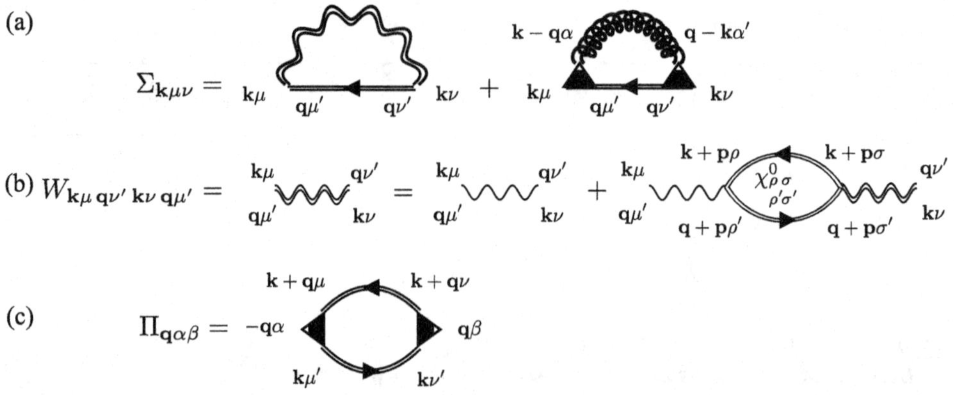

Figure 17.1 (a) Electronic self-energy in the GW plus Fan–Migdal approximation. (b) RPA equation for the screened interaction W. (c) Phononic self-energy.

$\omega^2_{\mathbf{q}\alpha}]$, in agreement with (16.175). The advanced component can easily be derived using the symmetry properties discussed in Section 16.11.4, and we find

$$D^{\rm A}_{\mathbf{q}\alpha}(t',t) = [D^{\rm R}_{\mathbf{q}\alpha}(t,t')]^\dagger.$$

In the next section we consider a popular approximation to the self-energies Σ and Π. This is relevant for the discussion of even more simplified approaches like the semiconductor Bloch equations and the Boltzmann equations.

Exercise 17.1 Calculate all matrix elements of the Fourier transform of (17.24).

17.1.3 Electron–Phonon Systems in the GW plus Fan–Migdal Approximation

Let us consider a crystal and assign a one-particle Bloch basis for electrons $\{\mathbf{k}\mu\}$ and phonons $\{\mathbf{q}\alpha\}$. We assume that the external fields do not break the lattice periodicity; hence the Green's functions and self-energies are diagonal in the quasi-momentum space. In this section we derive the expression of the electronic self-energy in the GW plus Fan–Migdal approximation and phononic self-energy in the bare bubble approximation, see Section 16.9.

The GW self-energy with RPA screened interaction W is shown in Fig. 17.1. In labeling the internal vertices we already take into account the conservation of the quasi-momentum. Since $\Sigma^{\lessgtr} = {\rm i}G^{\lessgtr}W^{\gtrless}$, we have to find an expression for $W^{\gtrless}$ in terms of $G^<$ and $G^>$. Starting from the Dyson equation $W = v + vPW$ and following the same steps as in Section 9.3 (with $G \to W$, $G_{\rm HF} \to v$, and $\Sigma \to P$), we find

$$W^{\lessgtr} = W^{\rm R}P^{\lessgtr}W^{\rm A}. \tag{17.25}$$

In deriving this result we use that $v^{\lessgtr} = 0$ and that the left, right, and Matsubara components of W vanish on the adiabatic contour. Equation (17.25) generalizes (15.13), which was obtained for the electron gas in equilibrium. To reduce the complexity of the equations, we evaluate $W^{\mathrm{R/A}}$ in the statically screened approximation, see (14.42). We then have [200, 326]

$$\begin{aligned}\Sigma^{\mathrm{GW},<}_{\mathbf{k}\mu\nu}(t,t') &= \mathrm{i}\sum_{\mathbf{q}\nu'\mu'} W^{<}_{\mathbf{k}\mu\,\mathbf{q}\nu'\,\mathbf{k}\nu\,\mathbf{q}\mu'}(t,t')G^{<}_{\mathbf{q}\mu'\nu'}(t,t') \\ &= \sum_{\mathbf{q}\nu'\mu'}\sum_{\mathbf{p}\rho\sigma\rho'\sigma'} W_{\mathbf{k}\mu\,\mathbf{q}+\mathbf{p}\rho'\,\mathbf{k}+\mathbf{p}\rho'\,\mathbf{q}\mu'}G^{>}_{\mathbf{q}+\mathbf{p}\sigma'\rho'}(t',t)G^{<}_{\mathbf{k}+\mathbf{p}\rho\sigma}(t,t') \\ &\quad\times W_{\mathbf{k}+\mathbf{p}\sigma\,\mathbf{q}\nu'\,\mathbf{k}\nu\,\mathbf{q}+\mathbf{p}\sigma'}G^{<}_{\mathbf{q}\mu'\nu'}(t,t'),\end{aligned} \quad (17.26)$$

where we use that $P = \chi^0 = -\mathrm{i}GG$ when W is evaluated at the RPA level. The greater component of the self-energy is obtained by exchanging $> \leftrightarrow <$. We can add to the GW self-energy the second-order exchange diagram with statically screened W lines [327]. This simply amounts to replacing $W_{\mathbf{k}_1\mu_1\,\mathbf{k}_2\mu_2\,\mathbf{k}_3\mu_3\,\mathbf{k}_4\mu_4} \to \frac{1}{2}[W_{\mathbf{k}_1\mu_1\,\mathbf{k}_2\mu_2\,\mathbf{k}_3\mu_3\,\mathbf{k}_4\mu_4} - W_{\mathbf{k}_1\mu_1\,\mathbf{k}_2\mu_2\,\mathbf{k}_4\mu_4\,\mathbf{k}_3\mu_3}]$ in (17.26). Notice that the unscreened version of this approximation, $W \to v$, is the same as the second Born approximation.

Next we consider the Fan–Migdal self-energy in Fig. 17.1. We take a bare electron–phonon coupling g and later discuss how to dress it. In accordance with the graphical representation in Section 16.8 we must evaluate the g in the time-reversal states $\mathbf{q}-\mathbf{k}\alpha$ and $\mathbf{k}-\mathbf{q}\alpha'$. Using the notation introduced in (16.139) and keeping an eye on Fig. 17.1, we have

$$\Sigma^{\mathrm{FM},<}_{\mathbf{k}\mu\nu}(t,t') = \mathrm{i}\sum_{\mathbf{q}\nu'\mu'\alpha\alpha'} g_{\mathbf{q}-\mathbf{k}\alpha,\mu\mu'}(\mathbf{k})D^{11,<}_{\mathbf{k}-\mathbf{q}\alpha\alpha'}(t,t')G^{<}_{\mathbf{q}\mu'\nu'}(t,t')g_{\mathbf{k}-\mathbf{q}\alpha',\nu'\nu}(\mathbf{q}).$$

This equation can be written in a more symmetric form if we use the property (16.59), which implies $\hat{g}^{\dagger}_{\mathbf{q}\alpha} = \hat{g}_{-\mathbf{q}\alpha}$. We have

$$\boxed{[g_{\mathbf{q}\alpha,\mu\nu}(\mathbf{k})]^* = \langle\mathbf{k}\mu|\hat{g}_{\mathbf{q}\alpha}|\mathbf{q}+\mathbf{k}\nu\rangle^* = \langle\mathbf{q}+\mathbf{k}\nu|\hat{g}_{-\mathbf{q}\alpha}|\mathbf{k}\mu\rangle = g_{-\mathbf{q}\alpha,\nu\mu}(\mathbf{q}+\mathbf{k})} \quad (17.27)$$

and accordingly

$$\Sigma^{\mathrm{FM},<}_{\mathbf{k}\mu\nu}(t,t') = \mathrm{i}\sum_{\mathbf{q}\nu'\mu'\alpha\alpha'} g_{\mathbf{q}-\mathbf{k}\alpha,\mu\mu'}(\mathbf{k})D^{11,<}_{\mathbf{k}-\mathbf{q}\alpha\alpha'}(t,t')G^{<}_{\mathbf{q}\mu'\nu'}(t,t')g^{*}_{\mathbf{q}-\mathbf{k}\alpha',\nu\nu'}(\mathbf{k}). \quad (17.28)$$

The Fan–Migdal self-energy calculated with a screened electron–phonon coupling $g^d = (1+WP)g = (1+v\chi)g$, see (16.143b), is more involved. To account for screening effects to some degree we can make the clamped+static approximation here too – that is, $g^d_{\mathbf{q}\alpha}(\mathbf{x},z;z') = \delta(z,z')g^d_{\mathbf{q}\alpha}(\mathbf{x})$ with

$$g^d_{\mathbf{q}\alpha}(\mathbf{x}) \simeq g_{\mathbf{q}\alpha}(\mathbf{r}) + \int d\mathbf{x}_1 d\mathbf{x}_2\, v(\mathbf{x},\mathbf{x}_1)\chi^{\mathrm{R}}_{\mathrm{clamp}}(\mathbf{x}_1,\mathbf{x}_2;\omega=0)g_{\mathbf{q}\alpha}(\mathbf{r}_2). \quad (17.29)$$

In this way the screened electron–phonon coupling is still time-local and the mathematical form of the Fan–Migdal lesser self-energy is identical to (17.28) with $g \to g^d$. The greater component is obtained by exchanging $> \leftrightarrow <$.

We finally consider the phononic self-energy. Keeping an eye on Fig. 17.1, we have

$$\begin{aligned}\Pi^{ii',<}_{\mathbf{q}\alpha\beta}(t,t') &= -\mathrm{i}\delta_{i1}\delta_{i'1}\sum_{\mathbf{k}\mu\nu\mu'\nu'} g_{\mathbf{q}\alpha,\mu'\mu}(\mathbf{k})G^{<}_{\mathbf{q}+\mathbf{k}\mu\nu}(t,t')G^{>}_{\mathbf{k}\nu'\mu'}(t',t)g_{-\mathbf{q}\beta,\nu\nu'}(\mathbf{q}+\mathbf{k})\\ &= -\mathrm{i}\delta_{i1}\delta_{i'1}\sum_{\mathbf{k}\mu\nu\mu'\nu'} g_{\mathbf{q}\alpha,\mu'\mu}(\mathbf{k})G^{<}_{\mathbf{q}+\mathbf{k}\mu\nu}(t,t')G^{>}_{\mathbf{k}\nu'\mu'}(t',t)g^{*}_{\mathbf{q}\beta,\nu'\nu}(\mathbf{k}), \end{aligned} \tag{17.30}$$

where in the second equality we use again (17.27). The greater component is obtained by exchanging $> \leftrightarrow <$. Dressing the electron–phonon coupling is here less obvious since only one g should be dressed [264, 271, 279, 328, 329], see (16.112). Replacing just one g in (17.30) with the clamped+static approximation (17.29) would lead to the violation of the properties (16.168). However, this is not the optimal way to proceed. The lesser component of the exact phononic self-energy is $\Pi^{<} = g\chi^{<}g$, see (16.144). Starting from the Dyson equation $\chi = P + Pv\chi$ and following the same steps as in Section 9.3 (with $G \to \chi$, $G_{\mathrm{HF}} \to P$ and $\Sigma \to v$), we find

$$\chi^{<} = (1+\chi^{\mathrm{R}}v)P^{<}(1+v\chi^{\mathrm{A}}), \tag{17.31}$$

and therefore

$$\Pi^{11,<}_{\mathbf{q}\alpha\beta} = g^{*}_{-\mathbf{q}\alpha}(1+\chi^{\mathrm{R}}v)P^{<}(1+v\chi^{\mathrm{A}})g_{-\mathbf{q}\beta} = g^{d,\mathrm{A}*}_{-\mathbf{q}\alpha}P^{<}g^{d,\mathrm{A}}_{-\mathbf{q}\alpha}. \tag{17.32}$$

We emphasize that (17.32) is an exact rewriting of the lesser phononic self-energy. By implementing the static approximation (17.29) at this stage, and approximating $P = -\mathrm{i}GG$, we obtain (17.30) with $g \to g^{d}$.

Through the GKBA for electrons and phonons the self-energies $\Sigma^{\mathrm{GW},\lessgtr}(t,t')$, $\Sigma^{\mathrm{FM},\lessgtr}(t,t')$, and $\Pi^{\lessgtr}(t,t')$ with $t > t'$ become functionals of $n^{<}(t')$ and $\gamma^{<}(t')$ with $t' < t$. Therefore, the GKBA equations (17.3) and (17.15) carry memory – that is, the density matrices at time t depend on the history. On the contrary, the MGKBA leads to equations of motion with no memory since the self-energies become functionals of $n^{<}(t)$ and $\gamma^{<}(t)$. The numerical solution of the (M)GKBA equations scales quadratically with the physical propagation time T since the calculation of the collision integrals scales linearly with T. The GW plus Fan–Migdal approximation has been implemented to study the relaxation of electrons and phonons in a photoexcited $\mathrm{MoS_2}$ monolayer [330].

As a general remark we note that the treatment of the Keldysh components of the self-energies lacks consistency in (M)GKBA. The retarded/advanced self-energies are usually evaluated in some quasi-particle approximation (hence they are time-local) and have the only purpose of improving the retarded/advanced Green's functions in the ansatzes. The lesser/greater self-energies, on the other hand, exhibit time nonlocality and originate from diagrammatic approximations that need not be the same as the one used for the retarded/advanced self-energies. Nonetheless, all fundamental conservation laws are preserved if the self-energies are Φ-derivable; this is the topic of the next section.

Exercise 17.2 Evaluate the self-energy in (17.26) for the two-band model of Section 6.4.3.

17.2 Enlarging the Class of Conserving Approximations

A conserving approximation yields observables that satisfy conservation laws, such as the conservation of particle number, momentum, angular momentum, and energy. Let us revisit the theory of conserving approximations presented in Chapter 12. Starting from the Green's function that solves the equations of motion (7.18) and (7.19), which we rewrite below for convenience

$$\int d3\, \overrightarrow{G}_0^{-1}(1;3)G(3;2) = \delta(1;2) + \int d3\, \Sigma(1;3)G(3;2),$$

$$\int d3\, G(1;3)\overleftarrow{G}_0^{-1}(3;2) = \delta(1;2) + \int d3\, G(1;3)\Sigma(3;2),$$

we have learned that the continuity equation is satisfied when (12.14) is fulfilled, the momentum conservation law is satisfied when the second row of (12.18) vanishes, etc. We can write the conditions for an approximation to be conserving in a slightly different form. Choosing $z_1 = t_{1+}$ or $z_1 = t_{1-}$ in (12.14), we recognize the lesser component of the convolutions ΣG and $G\Sigma$, both evaluated at equal times. On the adiabatic contour the Langreth rules yield

$$\int^{t_1} dt_3 \int d\mathbf{x}_3 \Big[\Sigma^>(1;3)G^<(3;1) - \Sigma^<(1;3)G^>(3;1)\Big] + \text{h.c.} = 0, \tag{17.33}$$

where we use (17.2). In the same way we can derive the condition for the conservation of the total momentum:

$$\int^{t_1} dt_3 \int d\mathbf{x}_1 d\mathbf{x}_3 \Big[\Sigma^>(1;3)\boldsymbol{\nabla}_1 G^<(3;1) - \Sigma^<(1;3)\boldsymbol{\nabla}_1 G^>(3;1)\Big] + \text{h.c.} = 0. \tag{17.34}$$

The conditions for the other conservation laws are derived similarly. Henceforth, we only discuss the continuity equation and the momentum conservation law for brevity.

We consider a Green's function G that satisfies the equations of motion,

$$\int d3\, \overrightarrow{G}_0^{-1}(1;3)G(3;2) = \delta(1;2) + \int d3\, \Sigma_A(1;3)G_A(3;2), \tag{17.35a}$$

$$\int d3\, G(1;3)\overleftarrow{G}_0^{-1}(3;2) = \delta(1;2) + \int d3\, G_A(1;3)\Sigma_A(3;2), \tag{17.35b}$$

where Σ_A and G_A are ansatzes for the self-energy and the Green's function. Following the approach outlined in Chapter 12, we can establish the conditions that guarantee the satisfaction of the conservation laws by this G. The unsurprising outcomes are (17.33) and (17.34) with replacements $\Sigma \to \Sigma_A$ and $G \to G_A$. The invariance of the Φ functional under gauge transformation and rigid shifts ensures that these conditions are automatically met for *any* ansatz of the form

$$\Sigma_A(1;2) = \Sigma[G_A](1;2) = \left.\frac{\delta\Phi[G]}{\delta G(2;1^+)}\right|_{G=G_A}.$$

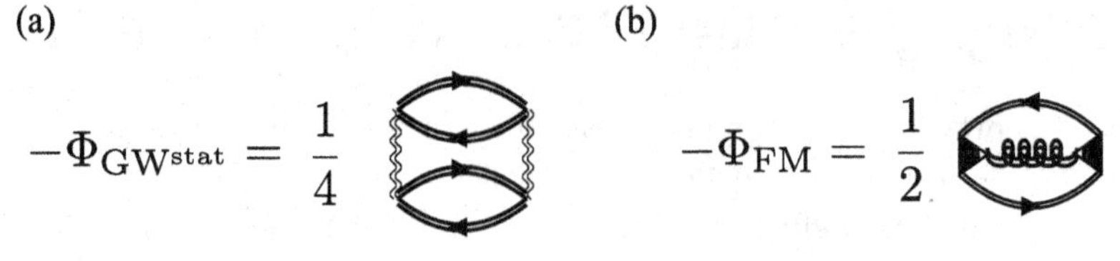

(a)

$$-\Phi_{\mathrm{GW^{stat}}} = \frac{1}{4}$$

(b)

$$-\Phi_{\mathrm{FM}} = \frac{1}{2}$$

Figure 17.2 Φ functional leading to the GW self-energy with a statically screened W (a) and to the Fan–Migdal self-energy (b).

We conclude that if the self-energy is Φ-derivable and evaluated at the same ansatz Green's function G_A with which it is multiplied then the Green's function G in (17.35) is conserving!

It remains to prove that the (M)GKBA equations preserve the conservation laws. Let us choose $G_A^{\lessgtr}$ to be the GKBA Green's function in (17.7) or the MGKBA Green's function in (17.10). Subtracting the second equation of (17.35) from the first and then setting the times $z_1 = t_-$ and $z_2 = t_+$, we find the (M)GKBA equation of motion for $n^<$. In other words, the (M)GKBA $n^<$ is the equal-time lesser component of the Green's function G in (17.35). As the particle number and the total momentum can be calculated from the equal-time $G^<$, we infer that the (M)GKBA equations preserve the conservation laws.

Although Baym's original derivation pertains to a self-consistent solution of the equations of motion, we have just learned that the whole proof goes through if the collision integral is evaluated at a Green's function G_A (and hence at a Φ-derivable self-energy $\Sigma_A = \Sigma[G_A]$) different from the G appearing in the l.h.s. of (17.35). In the context of particle conservation this fact was pointed out in Ref. [331] for G_A the one-shot Green's function of an electronic system. The (M)GKBA Green's function is another one among the infinitely many ansatzes for G. The argument is completely general and holds for all conservation laws, including energy conservation, as well as for fermion–phoson systems, thereby enlarging enormously the class of conserving approximations [166].

GW plus Fan–Migdal With such new perspective on conserving approximations, let us reexamine the GW plus Fan–Migdal approximation presented in Section 17.1.3. Although the GW self-energy in (17.26) contains two statically screened W, it is still Φ-derivable. The corresponding Φ functional is the one leading to the self-energy bubble diagram, with bare interactions v replaced by statically screened interactions W, see Fig. 17.2(a). Of course this Φ-diagram does not emerge from the diagrammatic expansion. Nonetheless, we can still define it and use it to prove the conserving nature of the self-energy. Next we consider the Fan–Migdal self-energy in (17.28). This self-energy too is Φ-derivable, the Φ-diagram being the one in Fig. 17.2(b). We observe that replacing the bare electron–phonon couplings with the statically screened ones, see (17.29), and then taking the functional derivative with respect to G leads to (17.28) with $g \to g^d$. Thus, the Fan–Migdal self-energy with two statically screened electron–phonon couplings is Φ-derivable as well. For the theory to be conserving, the phononic self-energy must be the functional derivative with respect to D of the *same* Φ functional. Thus, if the electron–phonon couplings of the Fan–Migdal self-energy are bare (dressed), then (17.30) needs to be evaluated with g (g^d). In all cases the (M)GKBA equations preserve all fundamental conservation laws.

17.3 Semiconductor Bloch Equations

The semiconductor Bloch equations [200, 332] apply to crystals with a finite gap in the one-particle spectrum, like semiconductors or insulators. Orthodox derivations of the semiconductor Bloch equations are based on the cluster expansion [333]. In this section we derive them from the MGKBA, highlighting all the underlying simplifications. We leave it as an exercise for the reader to go through the same mathematical steps using the GKBA. We have opted for the MGKBA formulation because it paves the way for even more simplified approaches, such as the Boltzmann equations, under the so-called Markovian approximation. In contrast, the Markovian approximation leads to unphysical divergencies in the GKBA formulation [325]. We discuss these issues in more detail in Section 17.4. The GKBA treatment (with no Markovian approximation) is presented in Section 17.6.

The **first simplification** is that phonons are assumed to remain in equilibrium at a certain temperature [326, 327, 334]. In fact, the semiconductor Bloch equations exclusively govern the dynamics of the electrons through the density matrix $\hat{n}^<$. For a given Bloch basis $\{\mathbf{k}\mu\}$ we define the electronic *occupations*

$$f^{\mathrm{el}}_{\mathbf{k}\mu}(t) \equiv \langle \mathbf{k}\mu|\hat{n}^<(t)|\mathbf{k}\mu\rangle$$

as the diagonal entries, and the *polarizations*

$$p_{\mathbf{k}\mu\nu}(t) \equiv \langle \mathbf{k}\mu|\hat{n}^<(t)|\mathbf{k}\nu\rangle, \quad \mu \neq \nu$$

as the off-diagonal entries of the one-particle density matrix. The polarizations carry information on the electronic coherence. The **second simplification** is that the quasi-particle Hamiltonian has matrix elements

$$\langle \mathbf{k}\mu|\hat{h}_{\mathrm{qp}}(t)|\mathbf{k}\nu\rangle \simeq \delta_{\mu\nu}\tilde{\epsilon}_{\mathbf{k}\mu}(t) + (1-\delta_{\mu\nu})\Omega_{\mathbf{k}\mu\nu}(t). \tag{17.36}$$

In this expression,

$$\tilde{\epsilon}_{\mathbf{k}\mu}(t) = \epsilon_{\mathbf{k}\mu} + \Omega_{\mathbf{k}\mu\mu}(t), \tag{17.37}$$

where $\epsilon_{\mathbf{k}\mu}$ is the equilibrium quasi-particle eigenvalue and the Rabi frequencies,

$$\begin{aligned}\Omega_{\mathbf{k}\mu\nu}(t) \equiv{}& \frac{1}{2c}\langle \mathbf{k}\mu|\hat{\mathbf{p}}\cdot\mathbf{A}(\hat{\mathbf{r}},t) + \mathbf{A}(\hat{\mathbf{r}},t)\cdot\hat{\mathbf{p}} + \frac{1}{c}A^2(\hat{\mathbf{r}},t)|\mathbf{k}\nu\rangle \\ &+ \sum_{\mathbf{k}'\mu'\nu'}(v_{\mathbf{k}\mu\mathbf{k}'\mu'\mathbf{k}'\nu'\mathbf{k}\nu} - W_{\mathbf{k}\mu\mathbf{k}'\mu'\mathbf{k}\nu\mathbf{k}'\nu'})\langle \mathbf{k}'\nu'|\hat{n}^<(t) - \hat{n}^<(0)|\mathbf{k}'\mu'\rangle,\end{aligned}$$

account for the coupling of the external vector potential (driving field) $\mathbf{A}$ with the electrons, see (3.1), and the change of the HSEX potential [compare with (8.31) or (8.37)]. We assume that the driving field has a finite duration T_{drive} and, without any loss of generality, we take $t_0 = 0$ so that $\mathbf{A}(t) = 0$ for $t < 0$ and for $t > T_{\mathrm{drive}}$.

Let us now go back to the equation of motion (17.3). Implementing the first and second simplifications, we find

$$\boxed{\frac{d}{dt}f^{\mathrm{el}}_{\mathbf{k}\mu} + \mathrm{i}\sum_{\alpha\neq\mu}\Big(\Omega_{\mathbf{k}\mu\alpha}\,p_{\mathbf{k}\alpha\mu} - p_{\mathbf{k}\mu\alpha}\,\Omega_{\mathbf{k}\alpha\mu}\Big) = S^{\mathrm{el}}_{\mathbf{k}\mu\mu}} \tag{17.38}$$

$$\boxed{\begin{aligned}\frac{d}{dt}p_{\mathbf{k}\mu\nu} + \mathrm{i}(\tilde{\epsilon}_{\mathbf{k}\mu} - \tilde{\epsilon}_{\mathbf{k}\nu})p_{\mathbf{k}\mu\nu}(t) + \mathrm{i}\Omega_{\mathbf{k}\mu\nu}(f^{\mathrm{el}}_{\mathbf{k}\nu} - f^{\mathrm{el}}_{\mathbf{k}\mu})& \\ + \mathrm{i}\sum_{\alpha\neq\nu}\Omega_{\mathbf{k}\mu\alpha}\,p_{\mathbf{k}\alpha\nu} - \mathrm{i}\sum_{\alpha\neq\mu}p_{\mathbf{k}\mu\alpha}\,\Omega_{\mathbf{k}\alpha\nu} &= S^{\mathrm{el}}_{\mathbf{k}\mu\nu}\end{aligned}} \tag{17.39}$$

where we define

$$S^{\mathrm{el}}_{\mathbf{k}\mu\nu}(t) \equiv \langle\mathbf{k}\mu| - \int^t dt'\left[\hat{\Sigma}^>(t,t')\hat{\mathcal{G}}^<(t',t) - \hat{\Sigma}^<(t,t')\hat{\mathcal{G}}^>(t',t)\right] + \text{h.c.}|\mathbf{k}\nu\rangle. \tag{17.40}$$

Henceforth we refer to S^{el} as the electronic scattering term. The equations of motion (17.38) and (17.39) with $S^{\mathrm{el}} = 0$ are equivalent to solving the time-dependent HSEX equations, leading to the Bethe–Salpeter equation in the linear response regime [233], see Sections 14.6 and 14.7; they have been recently implemented to investigate the dynamics of coherent excitons [87, 335, 336].

The **third simplification** consists in approximating the off-diagonal elements of the scattering term as

$$\boxed{S^{\mathrm{el}}_{\mathbf{k}\mu\nu} \simeq -\Gamma^{\mathrm{pol}}_{\mathbf{k}\mu\nu}p_{\mathbf{k}\mu\nu}, \qquad \forall\mu\neq\nu} \tag{17.41}$$

The polarization decay timescales typically range from tens to hundreds of femtoseconds, depending on the specific system and the type of photo-excitation. The reader can verify that (17.39) is solved by $p_{\mathbf{k}\mu\nu}(t) \propto e^{-\mathrm{i}(\tilde{\epsilon}_{\mathbf{k}\mu}-\tilde{\epsilon}_{\mathbf{k}\nu})t}e^{-\Gamma^{\mathrm{pol}}_{\mathbf{k}\mu\nu}t}$ for $t > T_{\mathrm{drive}}$ (recall that for $t > T_{\mathrm{drive}}$ the Rabi frequencies vanish). The polarization rates $\Gamma^{\mathrm{pol}}_{\mathbf{k}\mu\nu}$ can be calculated by different means [337, 338, 339], although they are often treated as fitting parameters. In Section 17.4.4 we discuss how to obtain a rough estimate of them.

We also simplify the diagonal part of the scattering term. As the polarizations are complex numbers with different phases it is reasonable to anticipate that their impact on the self-energies is secondary compared to the one of the occupations. The idea is then to set to zero the polarizations in the diagonal scattering term $S^{\mathrm{el}}_{\mathbf{k}\mu\mu}$. We consider an electronic MGKBA featuring diagonal retarded/advanced Green's functions

$$G^{\mathrm{R}}_{\mathbf{k}\mu\nu}(t,t') \simeq -\mathrm{i}\delta_{\mu\nu}\theta(t-t')e^{-\mathrm{i}\int_{t'}^{t}d\bar{t}\,\tilde{\epsilon}_{\mathbf{k}\mu}(\bar{t})} = [G^{\mathrm{A}}_{\mathbf{k}\nu\mu}(t',t)]^*,$$

and make a **fourth simplification**: $n^<_{\mathbf{k}\mu\nu} = \delta_{\mu\nu}f_{\mathbf{k}\mu}$. In this way (17.10) becomes

$$\boxed{G^<_{\mathbf{k}\mu\nu}(t,t') = \mathrm{i}\delta_{\mu\nu}e^{-\mathrm{i}\int_{t'}^{t}d\bar{t}\,\tilde{\epsilon}_{\mathbf{k}\mu}(\bar{t})}\Big(\theta(t-t')f^{\mathrm{el}}_{\mathbf{k}\mu}(t) + \theta(t'-t)f^{\mathrm{el}}_{\mathbf{k}\mu}(t')\Big)} \tag{17.42}$$

The expression for $G^>_{\mathbf{k}\mu\nu}(t,t')$ is identical with $f^{\mathrm{el}}_{\mathbf{k}\mu} \to f^{\mathrm{el}}_{\mathbf{k}\mu} - 1$, see (17.5).

***GW* scattering term** Let us analyze the GW self-energy in (17.26). Using the simplified

MGKBA in (17.42), we find for $t > t'$

$$\begin{aligned}
&\sum_{\nu} \Sigma^{\mathrm{GW},<}_{\mathbf{k}\mu\nu}(t,t') G^{>}_{\mathbf{k}\nu\mu}(t',t) \\
&= \sum_{\mathbf{qp}\mu'\nu'\nu} W_{\mathbf{k}\mu\,\mathbf{q+p}\nu'\,\mathbf{k+p}\nu\,\mathbf{q}\mu'} G^{>}_{\mathbf{q+p}\nu'\nu'}(t',t) G^{<}_{\mathbf{k+p}\nu\nu}(t,t') \\
&\quad \times W_{\mathbf{k+p}\nu\,\mathbf{q}\mu'\,\mathbf{k}\mu\,\mathbf{q+p}\nu'} G^{<}_{\mathbf{q}\mu'\mu'}(t,t') G^{>}_{\mathbf{k}\mu\mu}(t',t) \\
&= \sum_{\mathbf{qp}\mu'\nu'\nu} \left| W_{\mathbf{k}\mu\,\mathbf{q+p}\nu'\,\mathbf{k+p}\nu\,\mathbf{q}\mu'} \right|^2 e^{-\mathrm{i}\int_{t'}^{t} d\bar{t}\left(\tilde{\epsilon}_{\mathbf{k+p}\nu}(\bar{t}) + \tilde{\epsilon}_{\mathbf{q}\mu'}(\bar{t}) - \tilde{\epsilon}_{\mathbf{q+p}\nu'}(\bar{t}) - \tilde{\epsilon}_{\mathbf{k}\mu}(\bar{t})\right)} \\
&\quad \times \left(f^{\mathrm{el}}_{\mathbf{q+p}\nu'}(t) - 1\right)\left(f^{\mathrm{el}}_{\mathbf{k}\mu}(t) - 1\right) f^{\mathrm{el}}_{\mathbf{k+p}\nu}(t) f^{\mathrm{el}}_{\mathbf{q}\mu'}(t).
\end{aligned} \tag{17.43}$$

Analogously we can derive the expression for $\Sigma^{\mathrm{GW},>}G^{<}$ and construct the full scattering term in the GW approximation.

The **fifth simplification** originates from the fact that the system has a finite gap Δ in the ground state. This means that the zero-temperature equilibrium occupations $f^{\mathrm{el}}_{\mathbf{k}\mu} \simeq 1$ if μ is a valence band and $f^{\mathrm{el}}_{\mathbf{k}\mu} \simeq 0$ if μ is a conduction band. For such electronic state to be stationary in the absence of external fields the r.h.s. of (17.43) (and the analogous term arising from $\Sigma^{\mathrm{GW},>}G^{<}$) must vanish. We see by inspection that this is not the case when ν, μ' are valence bands and ν', μ are conduction bands. However, for this choice of band indices the exponential factor oscillates at a frequency $\gtrsim 2\Delta$ (for gaps $\Delta \gtrsim 1$ eV the period corresponding to a frequency 2Δ is $\lesssim 2$ fs) and average to zero when the integral over t' extends on a range of a few tens of femtoseconds or more. Since these contributions should be "physically" small we set them to zero from the outset – that is, all terms containing $W_{\mu\nu'\nu\mu'}$ with indices (ν, μ') in the conduction (valence) bands and indices (ν', μ) in the valence (conduction) bands are not included in the sum. We use the symbol $\sum'$ to denote a sum with the exclusion of the aforementioned terms. In this way the r.h.s. of (17.43) is rigorously zero for $f^{\mathrm{el}}_{\mathbf{k}\mu} = 1$ when μ is a valence band and $f^{\mathrm{el}}_{\mathbf{k}\mu} = 0$ when μ is a conduction band.[6] The final outcome for the scattering term in the GW approximation is

$$\boxed{\begin{aligned}
S^{\mathrm{GW}}_{\mathbf{k}\mu\mu}(t) = &\sum_{\mathbf{qp}\mu'\nu'\nu}{}' \int_0^t dt' e^{-\mathrm{i}\int_{t'}^{t} d\bar{t}\left(\tilde{\epsilon}_{\mathbf{k+p}\nu}(\bar{t}) + \tilde{\epsilon}_{\mathbf{q}\mu'}(\bar{t}) - \tilde{\epsilon}_{\mathbf{q+p}\nu'}(\bar{t}) - \tilde{\epsilon}_{\mathbf{k}\mu}(\bar{t})\right)} \\
&\times \left| W_{\mathbf{k}\mu\,\mathbf{q+p}\nu'\,\mathbf{k+p}\nu\,\mathbf{q}\mu'} \right|^2 \Big[\left(f^{\mathrm{el}}_{\mathbf{q+p}\nu'}(t) - 1\right)\left(f^{\mathrm{el}}_{\mathbf{k}\mu}(t) - 1\right) f^{\mathrm{el}}_{\mathbf{k+p}\nu}(t) f^{\mathrm{el}}_{\mathbf{q}\mu'}(t) \\
&- f^{\mathrm{el}}_{\mathbf{q+p}\nu'}(t) f^{\mathrm{el}}_{\mathbf{k}\mu}(t) \left(f^{\mathrm{el}}_{\mathbf{k+p}\nu}(t) - 1\right)\left(f^{\mathrm{el}}_{\mathbf{q}\mu'}(t) - 1\right)\Big] + \mathrm{h.c.}
\end{aligned}} \tag{17.44}$$

The lower integration limit has been set to zero to guarantee the continuity of the GW scattering term as t crosses zero.[7] This result has a transparent physical interpretation. The occupation $f^{\mathrm{el}}_{\mathbf{k}\mu}$ increases if two electrons in the occupied states $\mathbf{q}\mu'$ and $\mathbf{k+p}\nu$ interact and after the scattering end up in the unoccupied states $\mathbf{q+p}\nu'$ and $\mathbf{k}\mu$ (term in the second row). Conversely, the occupation $f^{\mathrm{el}}_{\mathbf{k}\mu}$ decreases if two electrons in the occupied

[6] The inclusion of the "physically" small terms may lead to severe instabilities at long times.

[7] In accordance with the fifth simplification, the GW scattering term vanishes for all times $t < 0$.

states $\mathbf{q}+\mathbf{p}\nu'$ and $\mathbf{k}\mu$ interact and after the scattering end up in the unoccupied states $\mathbf{q}\mu'$ and $\mathbf{k}+\mathbf{p}\nu$ (term in the third row).

Fan–Migdal scattering term Following a similar strategy we can add to the GW scattering term the contribution due to the Fan–Migdal self-energy. According to the first simplification, see beginning of the section, phonons are assumed to remain in thermal equilibrium. Then the lesser phononic Green's function can be calculated using the fluctuation–dissipation theorem (16.173). Fourier transforming (17.24), we find

$$D^{\mathrm{R}}_{\mathbf{q}\alpha}(\omega) = \frac{1}{2\omega_{\mathbf{q}\alpha}}\left[\frac{1}{\omega+\mathrm{i}\eta-\omega_{\mathbf{q}\alpha}}\begin{pmatrix}1 & \mathrm{i}\omega_{\mathbf{q}\alpha}\\ -\mathrm{i}\omega_{\mathbf{q}\alpha} & \omega^2_{\mathbf{q}\alpha}\end{pmatrix} - \frac{1}{\omega+\mathrm{i}\eta+\omega_{\mathbf{q}\alpha}}\begin{pmatrix}1 & -\mathrm{i}\omega_{\mathbf{q}\alpha}\\ \mathrm{i}\omega_{\mathbf{q}\alpha} & \omega^2_{\mathbf{q}\alpha}\end{pmatrix}\right],$$

and $D^{\mathrm{A}}_{\mathbf{q}\alpha}(\omega) = [D^{\mathrm{R}}_{\mathbf{q}\alpha}(\omega)]^\dagger$, leading to a spectral function, see (16.172),

$$\begin{aligned} A^{\mathrm{phon}}_{\mathbf{q}\alpha\beta}(\omega) &= \mathrm{i}\delta_{\alpha\beta}\left[D^{\mathrm{R}}_{\mathbf{q}\alpha}(\omega) - D^{\mathrm{A}}_{\mathbf{q}\alpha}(\omega)\right] \\ &= \delta_{\alpha\beta}\frac{2\pi\delta(\omega-\omega_{\mathbf{q}\alpha})}{2\omega_{\mathbf{q}\alpha}}\begin{pmatrix}1 & \mathrm{i}\omega_{\mathbf{q}\alpha}\\ -\mathrm{i}\omega_{\mathbf{q}\alpha} & \omega^2_{\mathbf{q}\alpha}\end{pmatrix} - \delta_{\alpha\beta}\frac{2\pi\delta(\omega+\omega_{\mathbf{q}\alpha})}{2\omega_{\mathbf{q}\alpha}}\begin{pmatrix}1 & -\mathrm{i}\omega_{\mathbf{q}\alpha}\\ \mathrm{i}\omega_{\mathbf{q}\alpha} & \omega^2_{\mathbf{q}\alpha}\end{pmatrix}. \end{aligned}$$

Therefore,

$$\begin{aligned} D^<_{\mathbf{q}\alpha\beta}(\omega) = -\mathrm{i}f_{\mathrm{B}}(\omega)A^{\mathrm{phon}}_{\mathbf{q}\alpha\beta}(\omega) = -\delta_{\alpha\beta}\frac{2\pi\mathrm{i}}{2\omega_{\mathbf{q}\alpha}}&\left[f_{\mathrm{B}}(\omega_{\mathbf{q}\alpha})\delta(\omega-\omega_{\mathbf{q}\alpha})\begin{pmatrix}1 & \mathrm{i}\omega_{\mathbf{q}\alpha}\\ -\mathrm{i}\omega_{\mathbf{q}\alpha} & \omega^2_{\mathbf{q}\alpha}\end{pmatrix}\right. \\ &\left. +\bar{f}_{\mathrm{B}}(\omega_{\mathbf{q}\alpha})\delta(\omega+\omega_{\mathbf{q}\alpha})\begin{pmatrix}1 & -\mathrm{i}\omega_{\mathbf{q}\alpha}\\ \mathrm{i}\omega_{\mathbf{q}\alpha} & \omega^2_{\mathbf{q}\alpha}\end{pmatrix}\right], \end{aligned}$$

where we rename the Bose function $f \to f_{\mathrm{B}}$ to aid the reader in distinguishing it from the electronic occupations, and we use the property $f_{\mathrm{B}}(-\omega) = -e^{\beta\omega}f_{\mathrm{B}}(\omega) = -\bar{f}_{\mathrm{B}}(\omega) = -[f_{\mathrm{B}}(\omega)+1]$. Fourier transforming back to real times,

$$\begin{aligned} D^<_{\mathbf{q}\alpha\beta}(t,t') &= \int\frac{d\omega}{2\pi}e^{-\mathrm{i}\omega(t-t')}D^<_{\mathbf{q}\alpha\beta}(\omega) = -\mathrm{i}\delta_{\alpha\beta}\frac{1}{2\omega_{\mathbf{q}\alpha}} \\ &\times\left[f_{\mathrm{B}}(\omega_{\mathbf{q}\alpha})e^{-\mathrm{i}\omega_{\mathbf{q}\alpha}(t-t')}\begin{pmatrix}1 & \mathrm{i}\omega_{\mathbf{q}\alpha}\\ -\mathrm{i}\omega_{\mathbf{q}\alpha} & \omega^2_{\mathbf{q}\alpha}\end{pmatrix} + \bar{f}_{\mathrm{B}}(\omega_{\mathbf{q}\alpha})e^{\mathrm{i}\omega_{\mathbf{q}\alpha}(t-t')}\begin{pmatrix}1 & -\mathrm{i}\omega_{\mathbf{q}\alpha}\\ \mathrm{i}\omega_{\mathbf{q}\alpha} & \omega^2_{\mathbf{q}\alpha}\end{pmatrix}\right]. \end{aligned} \tag{17.45}$$

The greater component is obtained by exchanging $f_{\mathrm{B}} \leftrightarrow \bar{f}_{\mathrm{B}}$.

We are now ready to evaluate the Fan–Migdal contribution to the scattering term. The Fan–Migdal self-energy is given in (17.28). Using (17.45) and (17.42), we find

$$
\begin{aligned}
\sum_{\nu} & \Sigma^{\mathrm{FM},<}_{\mathbf{k}\mu\nu}(t,t')G^{>}_{\mathbf{k}\nu\mu}(t',t) \\
&= \mathrm{i}\sum_{\mathbf{q}\nu\alpha} g_{\mathbf{q}-\mathbf{k}\alpha,\mu\nu}(\mathbf{k})D^{11,<}_{\mathbf{k}-\mathbf{q}\alpha\alpha}(t,t')G^{<}_{\mathbf{q}\nu\nu}(t,t')g^{*}_{\mathbf{q}-\mathbf{k}\alpha,\mu\nu}(\mathbf{k})G^{>}_{\mathbf{k}\mu\mu}(t',t) \\
&= -\sum_{\mathbf{q}\nu\alpha} \frac{\left|g_{\mathbf{q}-\mathbf{k}\alpha,\mu\nu}(\mathbf{k})\right|^2}{2\omega_{\mathbf{k}-\mathbf{q}\alpha}} \\
&\quad \times \Big\{ e^{-\mathrm{i}\int_{t'}^{t} d\bar{t}\left(\tilde{\epsilon}_{\mathbf{q}\nu}(\bar{t})-\tilde{\epsilon}_{\mathbf{k}\mu}(\bar{t})+\omega_{\mathbf{k}-\mathbf{q}\alpha}\right)} f^{\mathrm{el}}_{\mathbf{q}\nu}(t)\big(f^{\mathrm{el}}_{\mathbf{k}\mu}(t)-1\big) f_{\mathrm{B}}(\omega_{\mathbf{k}-\mathbf{q}\alpha}) \\
&\quad + e^{-\mathrm{i}\int_{t'}^{t} d\bar{t}\left(\tilde{\epsilon}_{\mathbf{q}\nu}(\bar{t})-\tilde{\epsilon}_{\mathbf{k}\mu}(\bar{t})-\omega_{\mathbf{k}-\mathbf{q}\alpha}\right)} f^{\mathrm{el}}_{\mathbf{q}\nu}(t)\big(f^{\mathrm{el}}_{\mathbf{k}\mu}(t)-1\big) \bar{f}_{\mathrm{B}}(\omega_{\mathbf{k}-\mathbf{q}\alpha}) \Big\}.
\end{aligned}
\tag{17.46}
$$

For the state with all electrons in the valence bands (hence no electrons in the conduction bands) to be a stationary state in the absence of external fields, the contribution with μ a conduction band, hence $f^{\mathrm{el}}_{\mathbf{k}\mu} = 0$, and ν a valence band, hence $f^{\mathrm{el}}_{\mathbf{q}\nu} = 1$, has to vanish. Failure to do so would result in a nonzero time derivative of the occupation $f^{\mathrm{el}}_{\mathbf{k}\mu}$. Inspecting (17.46), we see that, in general, this is not the case. An observation similar to the one made in the fifth simplification comes to the rescue. Phonon frequencies usually lie in the range $(0, 100)$ meV, whereas the values of the electronic gap Δ usually exceed 1 eV. For μ a conduction band and ν a valence band the exponential factors in (17.46) oscillate at a frequency $\simeq\Delta$ (corresponding to a period of $\simeq$4 fs) and average to zero when the integral over t' extends on a range of a few tens of femtoseconds or more. We leverage this physical consideration to construct a scattering term that ensures the stability of the zero-temperature electronic state. Let us split the electron–phonon coupling as

$$
g_{\mathbf{q}\alpha,\mu\nu}(\mathbf{k}) = g^{\mathrm{inter}}_{\mathbf{q}\alpha,\mu\nu}(\mathbf{k}) + g^{\mathrm{intra}}_{\mathbf{q}\alpha,\mu\nu}(\mathbf{k}),
\tag{17.47}
$$

where g^{inter} is nonvanishing only if μ is a valence (conduction) band and ν is a conduction (valence) band, whereas g^{intra} is nonvanishing only if μ and ν are both valence or conduction bands. The **sixth simplification** consists of replacing $g \to g^{\mathrm{intra}}$ in (17.46) since the contribution due to g^{inter} is "physically" small. In this way the r.h.s. of (17.46) vanishes in the absence of external fields. Analogously we can calculate $\Sigma^{\mathrm{FM},>}G^{<}$ and construct the full scattering term in the Fan-Migdal approximation. The final outcome is

$$
\boxed{
\begin{aligned}
S^{\mathrm{FM}}_{\mathbf{k}\mu\mu}(t) &= \sum_{\mathbf{q}\nu\alpha}\int_0^t dt' \frac{\left|g^{\mathrm{intra}}_{\mathbf{q}-\mathbf{k}\alpha,\mu\nu}(\mathbf{k})\right|^2}{2\omega_{\mathbf{k}-\mathbf{q}\alpha}} \Big\{ e^{-\mathrm{i}\int_{t'}^{t} d\bar{t}\left(\tilde{\epsilon}_{\mathbf{q}\nu}(\bar{t})-\tilde{\epsilon}_{\mathbf{k}\mu}(\bar{t})+\omega_{\mathbf{k}-\mathbf{q}\alpha}\right)} \\
&\times \Big[\big(f^{\mathrm{el}}_{\mathbf{q}\nu}(t)-1\big) f^{\mathrm{el}}_{\mathbf{k}\mu}(t)\bar{f}_{\mathrm{B}}(\omega_{\mathbf{k}-\mathbf{q}\alpha}) - f^{\mathrm{el}}_{\mathbf{q}\nu}(t)\big(f^{\mathrm{el}}_{\mathbf{k}\mu}(t)-1\big) f_{\mathrm{B}}(\omega_{\mathbf{k}-\mathbf{q}\alpha})\Big] \\
&\qquad + e^{-\mathrm{i}\int_{t'}^{t} d\bar{t}\left(\tilde{\epsilon}_{\mathbf{q}\nu}(\bar{t})-\tilde{\epsilon}_{\mathbf{k}\mu}(\bar{t})-\omega_{\mathbf{k}-\mathbf{q}\alpha}\right)} \\
&\times \Big[\big(f^{\mathrm{el}}_{\mathbf{q}\nu}(t)-1\big) f^{\mathrm{el}}_{\mathbf{k}\mu}(t) f_{\mathrm{B}}(\omega_{\mathbf{k}-\mathbf{q}\alpha}) - f^{\mathrm{el}}_{\mathbf{q}\nu}(t)\big(f^{\mathrm{el}}_{\mathbf{k}\mu}(t)-1\big) \bar{f}_{\mathrm{B}}(\omega_{\mathbf{k}-\mathbf{q}\alpha})\Big]\Big\} \\
&+ \mathrm{h.c.}
\end{aligned}
}
\tag{17.48}
$$

We set again the lower integration limit to zero to guarantee the continuity of the scattering term as t crosses zero. We can interpret this result in terms of electron–phonon scattering events. The occupation $f_{\mathbf{k}\mu}$ increases if the electron in the occupied state $\mathbf{q}\nu$ emits (absorbs) a phonon $\mathbf{q}-\mathbf{k}\alpha$ ($\mathbf{k}-\mathbf{q}\alpha$) and ends up in the unoccupied state $\mathbf{k}\mu$ (second and fourth terms). Conversely, the occupation $f_{\mathbf{k}\mu}$ decreases if the electron in the occupied state $\mathbf{k}\mu$ emits (absorbs) a phonon $\mathbf{k}-\mathbf{q}\alpha$ ($\mathbf{q}-\mathbf{k}\alpha$) and ends up in the unoccupied state $\mathbf{q}\nu$ (first and third terms).

Equations (17.38) and (17.39) with an off-diagonal scattering term given by (17.41) and a diagonal scattering term $S^{\rm el}_{\mathbf{k}\mu\mu}=S^{\rm GW}_{\mathbf{k}\mu\mu}+S^{\rm FM}_{\mathbf{k}\mu\mu}$ given by the sum of (17.44) and (17.48), together with the initial conditions $f_{\mathbf{k}\mu}(0)=0$ ($f_{\mathbf{k}\mu}(0)=1$) for μ a conduction (valence) band and $p_{\mathbf{k}\mu\nu}(0)=0$, are named the *semiconductor Bloch equations.*

17.4 Semiconductor Electron–Phonon Equations

An important limitation of the semiconductor Bloch equations is the neglect of the phonon dynamics, and in particular of the electronic feedback on the nuclear degrees of freedom. In this section we show that the phonon dynamics can be included at a small numerical effort. The first simplification of the previous section is replaced by a less drastic one; we take the phononic density matrix diagonal in the basis of the normal modes of the Born–Oppenheimer energy: $\gamma^<_{\mathbf{q}\alpha\beta}=\delta_{\alpha\beta}\gamma^<_{\mathbf{q}\alpha}$. Then the MGKBA (17.21) for the phononic Green's function becomes

$$D^{\gtrless}_{\mathbf{q}\alpha\beta}(t,t')\simeq\delta_{\alpha\beta}\Big[\gamma^{\gtrless}_{\mathbf{q}\alpha}(t)\,J\,D^{\rm R}_{\mathbf{q}\alpha}(t,t')-D^{\rm A}_{\mathbf{q}\alpha}(t,t')\,J\,\gamma^{\gtrless}_{\mathbf{q}\alpha}(t')\Big],\tag{17.49}$$

where $D^{\rm R/A}_{\mathbf{q}\alpha}$ has been explicitly calculated in (17.24).

17.4.1 Electrons

Let us get acquainted with (17.49) and extract the $(1,1)$ component, relevant for the calculation of the Fan–Migdal self-energy. The phononic density matrix has the following properties[8]

$$\gamma^{11,<}_{\mathbf{q}\alpha}(t)={\rm Tr}\big[\hat\rho\,\Delta\hat{\mathcal{U}}_{-\mathbf{q}\alpha,H}(t)\Delta\hat{\mathcal{U}}_{\mathbf{q}\alpha,H}(t)\big]=\gamma^{11,<}_{-\mathbf{q}\alpha}(t)=\gamma^{11,>}_{\mathbf{q}\alpha}(t)=\gamma^{11,>}_{-\mathbf{q}\alpha}(t),$$

$$\gamma^{22,<}_{\mathbf{q}\alpha}(t)={\rm Tr}\big[\hat\rho\,\Delta\hat{\mathcal{P}}_{-\mathbf{q}\alpha,H}(t)\Delta\hat{\mathcal{P}}_{\mathbf{q}\alpha,H}(t)\big]=\gamma^{22,<}_{-\mathbf{q}\alpha}(t)=\gamma^{22,>}_{\mathbf{q}\alpha}(t)=\gamma^{22,>}_{-\mathbf{q}\alpha}(t),$$

$$\gamma^{12,<}_{\mathbf{q}\alpha}(t)={\rm Tr}\big[\hat\rho\,\Delta\hat{\mathcal{P}}_{-\mathbf{q}\alpha,H}(t)\Delta\hat{\mathcal{U}}_{\mathbf{q}\alpha,H}(t)\big]=[\gamma^{21,<}_{\mathbf{q}\alpha}(t)]^*=\gamma^{21,>}_{-\mathbf{q}\alpha}(t)=[\gamma^{12,>}_{-\mathbf{q}\alpha}(t)]^*,$$

which, together with (17.13), can be summarized as follows

$$\begin{aligned}\gamma^{ij,\lessgtr}_{\mathbf{q}\alpha}(t)&=[\gamma^{ji,\lessgtr}_{\mathbf{q}\alpha}(t)]^*,\\ \gamma^{ij,>}_{\mathbf{q}\alpha}(t)&=\gamma^{ji,<}_{-\mathbf{q}\alpha}(t),\\ \gamma^{ij,>}_{\mathbf{q}\alpha}(t)&=\gamma^{ij,<}_{\mathbf{q}\alpha}(t)+\begin{pmatrix}0&{\rm i}\\-{\rm i}&0\end{pmatrix}_{ij}.\end{aligned}\tag{17.50}$$

[8]As usual, $\hat\rho$ is the thermal density matrix operator, see (4.14).

We then have

$$D^{11,\gtrless}_{\mathbf{q}\alpha\alpha}(t,t') = -\mathrm{i}\theta(t-t')\frac{1}{2\omega_{\mathbf{q}\alpha}}\left[B^{\gtrless *}_{\mathbf{q}\alpha}(t)e^{-\mathrm{i}\omega_{\mathbf{q}\alpha}(t-t')} + B^{\lessgtr}_{-\mathbf{q}\alpha}(t)e^{\mathrm{i}\omega_{\mathbf{q}\alpha}(t-t')}\right]$$
$$- \mathrm{i}\theta(t'-t)\frac{1}{2\omega_{\mathbf{q}\alpha}}\left[B^{\gtrless}_{\mathbf{q}\alpha}(t')e^{-\mathrm{i}\omega_{\mathbf{q}\alpha}(t-t')} + B^{\lessgtr *}_{-\mathbf{q}\alpha}(t')e^{\mathrm{i}\omega_{\mathbf{q}\alpha}(t-t')}\right], \quad (17.51)$$

where

$$B^{\gtrless}_{\mathbf{q}\alpha}(t) \equiv \omega_{\mathbf{q}\alpha}\gamma^{11,\gtrless}_{\mathbf{q}\alpha}(t) + \mathrm{i}\gamma^{21,\gtrless}_{\mathbf{q}\alpha}(t). \quad (17.52)$$

We observe that the hermiticity properties (16.168) are correctly satisfied. We also observe that $\lim_{t'\to t} D^{11,\gtrless}_{\mathbf{q}\alpha\alpha}(t,t') = -\mathrm{i}\gamma^{11,\gtrless}_{\mathbf{q}\alpha}(t)$, in agreement with (17.12). Had we used the GKBA, we would have found (17.51) with $[\cdot 1\cdot] \leftrightarrow [\cdot 2\cdot]$, where $[\cdot 1\cdot]$ and $[\cdot 2\cdot]$ are the terms in the square brackets of the first and second rows, respectively.

Inserting (17.51) in the first line of (17.46) and then in the analogous term with $< \leftrightarrow >$, we obtain a generalization of the Fan–Migdal scattering term. In this generalized form, the phononic density matrix is a dynamical variable to be determined by solving its own equation of motion. Comparing (17.51) with (17.45), the generalized form of the Fan–Migdal scattering term can easily be deduced. In (17.48) we must replace $f_{\mathrm{B}}(\omega_{\mathbf{q}\alpha}) \to B^{<*}_{\mathbf{q}\alpha}(t)$ and $\bar{f}_{\mathrm{B}}(\omega_{\mathbf{q}\alpha}) \to B^{>}_{-\mathbf{q}\alpha}(t)$ in the second and fourth terms, and $\bar{f}_{\mathrm{B}}(\omega_{\mathbf{q}\alpha}) \to B^{>*}_{\mathbf{q}\alpha}(t)$ and $f_{\mathrm{B}}(\omega_{\mathbf{q}\alpha}) \to B^{<}_{-\mathbf{q}\alpha}(t)$ in the first and third terms:

$$S^{\mathrm{FM}}_{\mathbf{k}\mu\mu}(t) = \int_0^t dt' \sum_{\mathbf{q}\nu\alpha} \frac{\left|g^{\mathrm{intra}}_{\mathbf{q}-\mathbf{k}\alpha,\mu\nu}(\mathbf{k})\right|^2}{2\omega_{\mathbf{k}-\mathbf{q}\alpha}} \Big\{ e^{-\mathrm{i}\int_{t'}^{t} d\bar{t}\left(\tilde{\epsilon}_{\mathbf{q}\nu}(\bar{t}) - \tilde{\epsilon}_{\mathbf{k}\mu}(\bar{t}) + \omega_{\mathbf{k}-\mathbf{q}\alpha}\right)}$$
$$\times \Big[\big(f_{\mathbf{q}\nu}(t)-1\big) f_{\mathbf{k}\mu}(t) B^{>*}_{\mathbf{k}-\mathbf{q}\alpha}(t) - f_{\mathbf{q}\nu}(t)\big(f_{\mathbf{k}\mu}(t)-1\big) B^{<*}_{\mathbf{k}-\mathbf{q}\alpha}(t)\Big]$$
$$+ e^{-\mathrm{i}\int_{t'}^{t} d\bar{t}\left(\tilde{\epsilon}_{\mathbf{q}\nu}(\bar{t}) - \tilde{\epsilon}_{\mathbf{k}\mu}(\bar{t}) - \omega_{\mathbf{k}-\mathbf{q}\alpha}\right)}$$
$$\times \Big[\big(f_{\mathbf{q}\nu}(t)-1\big) f_{\mathbf{k}\mu}(t) B^{<}_{\mathbf{q}-\mathbf{k}\alpha}(t) - f_{\mathbf{q}\nu}(t)\big(f_{\mathbf{k}\mu}(t)-1\big) B^{>}_{\mathbf{q}-\mathbf{k}\alpha}(t)\Big]\Big\} + \mathrm{h.c.} \quad (17.53)$$

To finalize the equations of motion of the electronic degrees of freedom, we also have to modify (17.36) by adding the Ehrenfest contribution, see (16.140). We have

$$\langle \mathbf{k}\mu|\hat{h}_{\mathrm{HF}}(t)|\mathbf{k}\nu\rangle \simeq \delta_{\mu\nu}\tilde{\epsilon}_{\mathbf{k}\mu}(t) + \sum_{\alpha} g_{\mathbf{0}\alpha,\mu\nu}(\mathbf{k})U_{\mathbf{0}\alpha}(t) + \Omega_{\mathbf{k}\mu\nu}(t). \quad (17.54)$$

The Ehrenfest contribution is off-diagonal, like the Rabi frequencies. The change in the electronic equations of motion (17.38) and (17.39) is a simple "renormalization" of the Rabi frequencies:

$$\Omega_{\mathbf{k}\mu\nu} \to \Omega^{\mathrm{ren}}_{\mathbf{k}\mu\nu} = \Omega_{\mathbf{k}\mu\nu} + \sum_{\alpha} g_{\mathbf{0}\alpha,\mu\nu}(\mathbf{k})U_{\mathbf{0}\alpha}. \quad (17.55)$$

For a full time-dependent framework of electrons and phonons, the equations of motion for the electronic occupations and polarizations must be coupled to the equations of motion for the phononic displacements, momenta, and density matrix.

Equilibrium response function The treatment of phonons necessitates a preliminary discussion on the equilibrium response function. We here consider the clamped-nuclei RPA response function $\chi_{\rm clamp} = \chi^0 + \chi^0 v \chi_{\rm clamp} = \chi^0 + \chi^0 W \chi^0$, see Section 15.4, and make a static approximation for W. Omitting the contour integrals and the momentum labels, and using the graphical rules in (7.30) and (7.31), we have

$$\chi_{{\rm clamp},\mu\nu'\mu'\nu} =$$

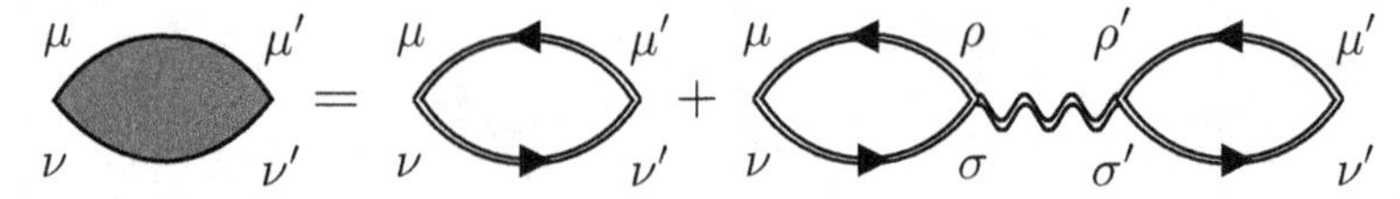

$$= \chi^0_{\mu\nu'\mu'\nu} + \sum_{\rho\rho'\sigma\sigma'} \chi^0_{\mu\sigma\rho\nu} W_{\rho\sigma'\rho'\sigma} \chi^0_{\rho'\nu'\mu'\sigma'}. \quad (17.56)$$

The MGKBA Green's functions are diagonal in the band indices, see (17.42), and therefore $\chi^0_{\mu\nu'\mu'\nu} = \delta_{\mu\mu'}\delta_{\nu\nu'}\chi^0_{\substack{\mu\\\nu}}$. We find it convenient to change the notation to better visualize which band indices appear on the left and which ones on the right of the bubble. In terms of the quantities

$$\chi_{{\rm clamp},\substack{\mu\mu'\\\nu\nu'}} = \chi_{{\rm clamp},\mu\nu'\mu'\nu} \quad , \quad W_{\substack{\rho\rho'\\\sigma\sigma'}} = W_{\rho\sigma'\rho'\sigma}, \quad (17.57)$$

the RPA equation (17.56) with a diagonal χ^0 reads

$$\chi_{{\rm clamp},\substack{\mu\mu'\\\nu\nu'}} = \delta_{\mu\mu'}\delta_{\nu\nu'}\chi^0_{\substack{\mu\\\nu}} + \chi^0_{\substack{\mu\\\nu}} W_{\substack{\mu\mu'\\\nu\nu'}} \chi^0_{\substack{\mu'\\\nu'}}.$$

Taking into account that at zero temperature $\chi^{0,\lessgtr}_{\substack{\mu\\\nu}} = 0$ if μ and ν are both conduction or both valence bands, we conclude that the only nonvanishing elements of the zero-temperature response function $\chi_{{\rm clamp},\substack{\mu\mu'\\\nu\nu'}}$ are those for which the indices of the pairs (μ,ν) and (μ',ν') are either conduction–valence or valence–conduction.

17.4.2 Nuclear Displacements

The equations of motion for the nuclear displacements and momenta are given in (16.160), and can be combined into a single second-order differential equation:

$$\frac{d^2 U_{\mathbf{0}\alpha}(t)}{dt^2} = -\sum_{\mathbf{k}\mu\nu} g_{\mathbf{0}\alpha,\nu\mu}(\mathbf{k})\Delta n^<_{\mathbf{k}\mu\nu}(t) - \sum_{\alpha'} K_{\alpha\alpha'}(\mathbf{0}) U_{\mathbf{0}\alpha'}(t). \quad (17.58)$$

In (17.58) the elastic tensor K appears. It would be desirable to generate a simplified equation where the elastic tensor is renormalized by the phononic self-energy, giving rise to the Born–Oppenheimer frequencies. Let us go back to the equations of motion (17.38) and (17.39). At $t = 0$ we have

$$\left.\frac{d}{dt} f^{\rm el}_{\mathbf{k}\mu}(t)\right|_{t=0} = 0 \quad , \quad \left.\frac{d}{dt} p_{\mathbf{k}\mu\nu}(t)\right|_{t=0} = -{\rm i}\Omega_{\mathbf{k}\mu\nu}\big(f^{\rm el}_{\mathbf{k}\nu}(0) - f^{\rm el}_{\mathbf{k}\mu}(0)\big).$$

We infer that only the polarizations $p_{\mathbf{k}\mu\nu}$ with indices (μ, ν) either conduction–valence or valence–conduction change linearly with the external driving field, since it is only in this case that $f_{\mathbf{k}\nu}(0) - f_{\mathbf{k}\mu}(0) \neq 0$. All other elements of the electronic density matrix change at least quadratically with the external field. We then split the polarizations like the electron–phonon coupling, see (17.47),

$$p_{\mathbf{k}\mu\nu} = p^{\rm inter}_{\mathbf{k}\mu\nu} + p^{\rm intra}_{\mathbf{k}\mu\nu},$$

where $p^{\rm inter}$ is nonvanishing only if μ is a valence (conduction) band and ν is a conduction (valence) band, whereas $p^{\rm intra}$ is nonvanishing only if μ and ν are both valence or conduction bands. Accordingly, we rewrite the first term in the r.h.s. of (17.58) as

$$\sum_{\mathbf{k}\mu\nu} g_{0\alpha,\nu\mu}(\mathbf{k})\Delta n^<_{\mathbf{k}\mu\nu} = \sum_{\mathbf{k}\mu\nu} g_{0\alpha,\nu\mu}(\mathbf{k}) p^{\rm inter}_{\mathbf{k}\mu\nu} + \sum_{\mathbf{k}\mu\nu} g_{0\alpha,\nu\mu}(\mathbf{k})\left[\delta_{\mu\nu}\Delta f^{\rm el}_{\mathbf{k}\mu} + p^{\rm intra}_{\mathbf{k}\mu\nu}\right], \quad (17.59)$$

where $\Delta f^{\rm el}_{\mathbf{k}\mu}(t) = f^{\rm el}_{\mathbf{k}\mu}(t) - f^{\rm el}_{\mathbf{k}\mu}(0)$. The **seventh simplification** consists in expressing $p^{\rm inter}_{\mathbf{k}\mu\nu}$ using the Kubo formula (13.5) of linear response theory, and in treating the nuclei in the Ehrenfest approximation. Omitting the dependence on momenta, the change in the density matrix to first order in the Rabi frequencies is

$$\Delta n^<_{\mu\nu}(t) = \int dt' \sum_{\mu'\nu'} \chi^{\rm R}_{\substack{\mu\mu' \\ \nu\nu'}}(t,t')\Omega_{\mu'\nu'}(t'). \quad (17.60)$$

The Ehrenfest approximation to the response function is

$$\chi_{\substack{\mu\mu' \\ \nu\nu'}}(z,z') = \chi_{{\rm clamp},\, \substack{\mu\mu' \\ \nu\nu'}}(z,z') + \sum_{\substack{\alpha\beta \\ \mu\mu' \\ \nu\nu'}} \int dz_1 dz_2\, \chi_{{\rm clamp},\, \substack{\mu\rho \\ \nu\sigma}}(z,z_1)$$
$$\times\, g_{\alpha,\rho\sigma} D^{11}_{\alpha\beta}(z_1,z_2) g_{\beta,\sigma'\rho'} \chi_{\substack{\rho'\mu' \\ \sigma'\nu'}}(z_2,z').$$

Taking the retarded component of this equation, substituting it into (17.60) and recalling that, see (16.142),

$$U_\alpha(t_1) = \sum_{\beta\rho'\sigma'} \int dt_2 D^{11,{\rm R}}_{\alpha\beta}(t_1,t_2) g_{\alpha,\sigma'\rho'} \Delta n^<_{\rho'\sigma'}(t_2),$$

we obtain

$$p^{\rm inter}_{\mu\nu}(t) = \int dt' \sum_{\mu'\nu'} \chi^{\rm R}_{{\rm clamp},\, \substack{\mu\mu' \\ \nu\nu'}}(t,t') \left[\sum_\alpha g_{\alpha,\mu'\nu'} U_\alpha(t') + \Omega_{\mu'\nu'}(t')\right]. \quad (17.61)$$

For a response function that decays fast as $|t-t'| \to \infty$, we can evaluate the slowly varying function U_α at time t instead of t', and hence perform the time integral $\int dt' \chi^{\rm R}_{\rm clamp}(t,t') =$

$\chi^{R}_{\text{clamp}}(\omega = 0)$. This is the **eighth simplification**, which allows us to derive the following important result

$$\sum_{\mu\nu} g_{\alpha,\nu\mu}\, p^{\text{inter}}_{\mu\nu} + \sum_{\alpha'} K_{\alpha\alpha'} U_{\alpha'} = \omega^2_\alpha U_\alpha + \int dt' \sum_{\substack{\mu\mu' \\ \nu\nu'}} g_{\alpha,\nu\mu} \chi^{R}_{\text{clamp},\ \substack{\mu\mu' \\ \nu\nu'}}(t,t')\Omega_{\mu'\nu'}(t'),$$

where we take into account that the clamped response function is nonvanishing only if (μ,ν) are either conduction-valence or valence-conduction and then use (16.66). In conclusion, the equation of motion (17.58) becomes (reintroducing the dependence on momenta)

$$\begin{aligned}\frac{d^2}{dt^2}U_{\mathbf{0}\alpha}(t) + \omega^2_{\mathbf{0}\alpha} U_{\mathbf{0}\alpha}(t) = &-\sum_{\mathbf{k}\mu\nu} g_{\mathbf{0}\alpha,\nu\mu}(\mathbf{k})\left[\delta_{\mu\nu}\Delta f^{\text{el}}_{\mathbf{k}\mu}(t) + p^{\text{intra}}_{\mathbf{k}\mu\nu}(t)\right] \\ &- \int dt' \sum_{\substack{\mathbf{k}\mathbf{k}' \\ \mu\mu' \\ \nu\nu'}} g_{\mathbf{0}\alpha,\nu\mu}(\mathbf{k}) \chi^{R}_{\text{clamp},\mathbf{k}\mathbf{k}'\ \substack{\mu\mu' \\ \nu\nu'}}(\mathbf{0};t,t')\Omega_{\mathbf{k}'\mu'\nu'}(t')\end{aligned} \tag{17.62}$$

We emphasize that in (17.62) the elecron-phonon coupling is the undressed one. We also observe that (17.62) differs from the equation of motion $\frac{d^2}{dt^2}U_{\mathbf{0}\alpha} + \omega^2_{\mathbf{0}\alpha}U_{\mathbf{0}\alpha}(t) = -\sum_{\mathbf{k}\mu\nu} g_{\mathbf{0}\alpha,\nu\mu}(\mathbf{k})\Delta\rho_{\mathbf{k}\mu\nu}(t)$, typical of electron-phonon model Hamiltonians. This equation involves the full density matrix, not just the intra-only elements; moreover, the last term in (17.62) is missing. The root cause of the discrepancy lies in the unjustified replacement of (16.56) with a set of harmonic oscillators: $\hat{H}_{0,\text{phon}} = \sum_{\mathbf{q}\alpha} \omega_{\mathbf{q}\alpha} \hat{b}^\dagger_{\mathbf{q}\alpha}\hat{b}_{\mathbf{q}\alpha}$. As pointed out in Chapter 16, see also Refs. [264, 278], model Hamiltonians suffer from a double renormalization of the phononic frequencies.

17.4.3 Phononic Density Matrix

The starting point is here the equation of motion (17.15). In the basis of the normal modes of the Born-Oppenheimer energy this equation is closed on the diagonal part of the phononic density matrix since the quasi-phonon Hamiltonian $\hat{Q}_{\text{qp}}$ and the MGKBA phononic Green's function are diagonal in such basis, see (17.23) and (17.49). We then have

$$\frac{d}{dt}\gamma^<_{\mathbf{q}\alpha}(t) + \begin{pmatrix} 0 & -1 \\ \omega^2_{\mathbf{q}\alpha} & 0 \end{pmatrix} \gamma^<_{\mathbf{q}\alpha}(t) - \gamma^<_{\mathbf{q}\alpha}(t)\begin{pmatrix} 0 & -\omega^2_{\mathbf{q}\alpha} \\ 1 & 0 \end{pmatrix} = S^{\text{phon}}_{\mathbf{q}\alpha\alpha}(t) \tag{17.63}$$

where

$$S^{\text{phon}}_{\mathbf{q}\alpha\alpha}(t) \equiv J \int^t dt' \left[\Pi^>_{\mathbf{q}\alpha\alpha}(t,t')D^<_{\mathbf{q}\alpha\alpha}(t',t) - \Pi^<_{\mathbf{q}\alpha\alpha}(t,t')D^>_{\mathbf{q}\alpha\alpha}(t',t)\right] + \text{h.c.} \tag{17.64}$$

is the phononic scattering term.

We already comment that S^{el} vanishes if the electronic occupations are the noninteracting zero-temperature ones. This fact assures the stationarity of the zero-temperature electronic state in the absence of external driving fields. What about the equilibrium value of the phononic density matrix? Is it also stationary? From definition (17.12) the value of $\gamma^<_{\mathbf{q}\alpha}(0)$ resulting from the equilibrium phononic Green's function (17.45) is

$$\gamma^<_{\mathbf{q}\alpha}(0) = \mathrm{i}D^<_{\mathbf{q}\alpha\alpha}(0,0) = \frac{1}{2\omega_{\mathbf{q}\alpha}}\begin{pmatrix} 1+2f_{\mathrm{B}}(\omega_{\mathbf{q}\alpha}) & -\mathrm{i}\omega_{\mathbf{q}\alpha} \\ \mathrm{i}\omega_{\mathbf{q}\alpha} & \omega^2_{\mathbf{q}\alpha}(1+2f_{\mathrm{B}}(\omega_{\mathbf{q}\alpha})) \end{pmatrix}. \tag{17.65}$$

The reader can easily verify that

$$\begin{pmatrix} 0 & -1 \\ \omega^2_{\mathbf{q}\alpha} & 0 \end{pmatrix}\gamma^<_{\mathbf{q}\alpha}(0) - \gamma^<_{\mathbf{q}\alpha}(0)\begin{pmatrix} 0 & -\omega^2_{\mathbf{q}\alpha} \\ 1 & 0 \end{pmatrix} = \begin{pmatrix} 0 & 0 \\ 0 & 0 \end{pmatrix}.$$

Thus, $\gamma^<_{\mathbf{q}\alpha}(0)$ in (17.65) is a solution of (17.63) provided that the phononic scattering term vanishes when evaluated at the equilibrium Green's functions. The idea behind the **ninth simplification** is to replace $g \to g^{\mathrm{intra}}$ in the bare bubble approximation (17.30), thereby guaranteeing that $S^{\mathrm{phon}}_{\mathbf{q}\alpha\alpha} = 0$ in equilibrium. In fact, using the diagonal MGKBA Green's function (17.42), we have

$$\Pi^{\lessgtr}_{\mathbf{q}\alpha\alpha}(t,t') = -\mathrm{i}\delta_{i1}\delta_{i'1}\sum_{\mathbf{k}\mu\nu} g^{\mathrm{intra}}_{\mathbf{q}\alpha,\mu\nu}(\mathbf{k})G^{\lessgtr}_{\mathbf{q}+\mathbf{k}\nu\nu}(t,t')G^{\gtrless}_{\mathbf{k}\mu\mu}(t',t)g^{\mathrm{intra}*}_{\mathbf{q}\alpha,\mu\nu}(\mathbf{k}). \tag{17.66}$$

If μ,ν are both conduction (valence) bands then the product $G^{\lessgtr}_{\mathbf{q}+\mathbf{k}\nu\nu}(t,t')G^{\gtrless}_{\mathbf{k}\mu\mu}(t',t) = 0$ for zero-temperature Green's functions since $f^{\mathrm{el}}_{\mathbf{k}\mu} = f^{\mathrm{el}}_{\mathbf{k}\nu} = 0$ ($f^{\mathrm{el}}_{\mathbf{k}\mu} = f^{\mathrm{el}}_{\mathbf{k}\nu} = 1$). Conversely, we may say that for our approximations to make sense the contribution with g^{inter} to the self-energy has to be small.[9] The continuity of $S^{\mathrm{phon}}_{\mathbf{q}\alpha\alpha}(t)$ as t crosses zero forces us to set the lower integration limit in (17.64) to zero.

Let us inspect the 2×2 matrix structure of the phononic scattering term. Naming the integral in (17.64) with the letter I and taking into account that the phononic self-energy has only one nonvanishing element, which is the $(1,1)$, we find

$$S^{\mathrm{phon}} = JI + \text{h.c.} = \begin{pmatrix} 0 & \mathrm{i} \\ -\mathrm{i} & 0 \end{pmatrix}\begin{pmatrix} I^{11} & I^{12} \\ 0 & 0 \end{pmatrix} + \text{h.c.} = \begin{pmatrix} 0 & \mathrm{i}I^{11*} \\ -\mathrm{i}I^{11} & -\mathrm{i}I^{12}+\mathrm{i}I^{12*} \end{pmatrix}.$$

To calculate the matrix elements of $S^{\mathrm{phon}}_{\mathbf{q}\alpha\alpha}$ we need the $(1,1)$ and $(1,2)$ elements of the phononic Green's function in (17.49). The $(1,1)$ element of $D^{\lessgtr}$ has been already derived, see (17.51). For the $(1,2)$ element we find[10]

$$\begin{aligned} D^{12,\gtrless}_{\mathbf{q}\alpha\alpha}(t,t') = &-\mathrm{i}\theta(t-t')\frac{1}{2\omega_{\mathbf{q}\alpha}}\left[\mathrm{i}\omega_{\mathbf{q}\alpha}B^{\gtrless *}_{\mathbf{q}\alpha}(t)e^{-\mathrm{i}\omega_{\mathbf{q}\alpha}(t-t')} - \mathrm{i}\omega_{\mathbf{q}\alpha}B^{\lessgtr}_{-\mathbf{q}\alpha}(t)e^{\mathrm{i}\omega_{\mathbf{q}\alpha}(t-t')}\right] \\ &-\mathrm{i}\theta(t'-t)\frac{1}{2\omega_{\mathbf{q}\alpha}}\left[\mathrm{i}C^{\gtrless}_{\mathbf{q}\alpha}(t')e^{-\mathrm{i}\omega_{\mathbf{q}\alpha}(t-t')} - \mathrm{i}C^{\lessgtr *}_{-\mathbf{q}\alpha}(t')e^{\mathrm{i}\omega_{\mathbf{q}\alpha}(t-t')}\right], \end{aligned} \tag{17.67}$$

where

$$C^{\gtrless}_{\mathbf{q}\alpha}(t) \equiv \gamma^{22,\gtrless}_{\mathbf{q}\alpha}(t) - \mathrm{i}\omega_{\mathbf{q}\alpha}\gamma^{12,\gtrless}_{\mathbf{q}\alpha}(t). \tag{17.68}$$

[9] This contribution is expected to be small from a physical standpoint, as discussed in more detail below (17.46).

[10] In the GKBA we would find (17.67) with $[\cdot 1\cdot] \leftrightarrow [\cdot 2\cdot]$, where $[\cdot 1\cdot]$ and $[\cdot 2\cdot]$ are the terms in the square brackets of the first and second rows, respectively

Evaluating the phononic self-energy (17.66) at the electronic MGKBA Green's function (17.42) and using (17.51) and (17.67) for the MGKBA phononic Green's function, we find

$$S^{21,\mathrm{phon}}_{\mathbf{q}\alpha\alpha}(t) = -\mathrm{i}\int_0^t dt' \sum_{\mathbf{k}\mu\nu} \frac{|g^{\mathrm{intra}}_{\mathbf{q}\alpha,\mu\nu}(\mathbf{k})|^2}{2\omega_{\mathbf{q}\alpha}} \Big\{ e^{-\mathrm{i}\int_{t'}^{t} d\bar{t}\left(\tilde{\epsilon}_{\mathbf{q}+\mathbf{k}\nu}(\bar{t})-\tilde{\epsilon}_{\mathbf{k}\mu}(\bar{t})-\omega_{\mathbf{q}\alpha}\right)}$$
$$\times\Big[\big(f^{\mathrm{el}}_{\mathbf{q}+\mathbf{k}\nu}(t)-1\big)f^{\mathrm{el}}_{\mathbf{k}\mu}(t)B^{<}_{\mathbf{q}\alpha}(t) - f^{\mathrm{el}}_{\mathbf{q}+\mathbf{k}\nu}(t)\big(f^{\mathrm{el}}_{\mathbf{k}\mu}(t)-1\big)B^{>}_{\mathbf{q}\alpha}(t)\Big]$$
$$+ e^{-\mathrm{i}\int_{t'}^{t} d\bar{t}\left(\tilde{\epsilon}_{\mathbf{q}+\mathbf{k}\nu}(\bar{t})-\tilde{\epsilon}_{\mathbf{k}\mu}(\bar{t})+\omega_{\mathbf{q}\alpha}\right)}$$
$$\times\Big[\big(f^{\mathrm{el}}_{\mathbf{q}+\mathbf{k}\nu}(t)-1\big)f^{\mathrm{el}}_{\mathbf{k}\mu}(t)B^{>*}_{-\mathbf{q}\alpha}(t) - f^{\mathrm{el}}_{\mathbf{q}+\mathbf{k}\nu}(t)\big(f^{\mathrm{el}}_{\mathbf{k}\mu}(t)-1\big)B^{<*}_{-\mathbf{q}\alpha}(t)\Big]\Big\},$$

$$S^{22,\mathrm{phon}}_{\mathbf{q}\alpha\alpha}(t) = -\mathrm{i}\int_0^t dt' \sum_{\mathbf{k}\mu\nu} \frac{|g^{\mathrm{intra}}_{\mathbf{q}\alpha,\mu\nu}(\mathbf{k})|^2}{2\omega_{\mathbf{q}\alpha}} \Big\{ \mathrm{i}e^{-\mathrm{i}\int_{t'}^{t} d\bar{t}\left(\tilde{\epsilon}_{\mathbf{q}+\mathbf{k}\nu}(\bar{t})-\tilde{\epsilon}_{\mathbf{k}\mu}(\bar{t})-\omega_{\mathbf{q}\alpha}\right)}$$
$$\times\Big[\big(f^{\mathrm{el}}_{\mathbf{q}+\mathbf{k}\nu}(t)-1\big)f^{\mathrm{el}}_{\mathbf{k}\mu}(t)C^{<}_{\mathbf{q}\alpha}(t) - f^{\mathrm{el}}_{\mathbf{q}+\mathbf{k}\nu}(t)\big(f^{\mathrm{el}}_{\mathbf{k}\mu}(t)-1\big)C^{>}_{\mathbf{q}\alpha}(t)\Big]$$
$$- \mathrm{i}e^{-\mathrm{i}\int_{t'}^{t} d\bar{t}\left(\tilde{\epsilon}_{\mathbf{q}+\mathbf{k}\nu}(\bar{t})-\tilde{\epsilon}_{\mathbf{k}\mu}(\bar{t})+\omega_{\mathbf{q}\alpha}\right)}$$
$$\times\Big[\big(f^{\mathrm{el}}_{\mathbf{q}+\mathbf{k}\nu}(t)-1\big)f^{\mathrm{el}}_{\mathbf{k}\mu}(t)C^{>*}_{-\mathbf{q}\alpha}(t) - f^{\mathrm{el}}_{\mathbf{q}+\mathbf{k}\nu}(t)\big(f^{\mathrm{el}}_{\mathbf{k}\mu}(t)-1\big)C^{<*}_{-\mathbf{q}\alpha}(t)\Big]\Big\} + \mathrm{h.c.},$$

and the more obvious ones

$$S^{11,\mathrm{phon}}_{\mathbf{q}\alpha\alpha}(t) = 0 \quad , \quad S^{12,\mathrm{phon}}_{\mathbf{q}\alpha\alpha}(t) = [S^{21,\mathrm{phon}}_{\mathbf{q}\alpha\alpha}(t)]^*.$$

Two observations are now in order. After the implementation of the nine simplifications, the equations of motion for the electronic and phononic density matrices at $\Omega_{\mathbf{k}\mu\nu} = 0$ are solved by a *noninteracting* $n^<$ and $\gamma^<$, a fact that makes physical sense only for systems featuring weak initial correlations. The second observation pertains to the temperature of the phononic and electronic subsystems. The most general initial value for the phononic density matrix is (17.65), where the temperature can be any. For the electronic density matrix, instead, only zero temperature is a permissible temperature (the occupations must be either 0 or 1 for the scattering terms to vanish). The restriction on the electronic temperature can be lifted if we implement the Markovian approximation.

17.4.4 Markovian Approximation

The **tenth simplification** is the *Markovian* approximation and consists in evaluating the scattering terms with quasi-particle energies $\tilde{\epsilon}_{\mathbf{k}\mu}(t)$ in equilibrium, hence $\tilde{\epsilon}_{\mathbf{k}\mu}(t) = \epsilon_{\mathbf{k}\mu}$, and in taking the limit $t \to \infty$ of the upper integration limit while keeping t finite in the electronic and phononic density matrices. In the Markovian approximation the integral of the exponential factors becomes

$$\lim_{t\to\infty}\int_0^t dt' e^{\mathrm{i}\omega(t-t')} = \lim_{\eta\to 0^+}\lim_{t\to\infty}\int_0^t d\tau e^{\mathrm{i}\omega\tau-\eta\tau} = \mathrm{i}P\frac{1}{\omega} + \pi\delta(\omega).$$

In the following we assume that the contribution of the principal part to the scattering terms is subdominant with respect to the Dirac delta, and we discard it.

In the Markovian approximation the GW scattering term (17.44) becomes[||]

$$\boxed{\begin{aligned} S^{\rm GW}_{\mathbf{k}\mu\mu}(t) &= 2\pi \sum_{\mathbf{q}\mathbf{p}\nu'\mu'\nu} \left|W_{\mathbf{k}\mu\,\mathbf{q}+\mathbf{p}\nu'\,\mathbf{k}+\mathbf{p}\nu\,\mathbf{q}\mu'}\right|^2 \delta\big(\epsilon_{\mathbf{k}+\mathbf{p}\nu} + \epsilon_{\mathbf{q}\mu'} - \epsilon_{\mathbf{q}+\mathbf{p}\nu'} - \epsilon_{\mathbf{k}\mu}\big) \\ &\times \Big[\big(f^{\rm el}_{\mathbf{q}+\mathbf{p}\nu'}(t) - 1\big)\big(f^{\rm el}_{\mathbf{k}\mu}(t) - 1\big) f^{\rm el}_{\mathbf{k}+\mathbf{p}\nu}(t) f^{\rm el}_{\mathbf{q}\mu'}(t) \\ &- f^{\rm el}_{\mathbf{q}+\mathbf{p}\nu'}(t) f^{\rm el}_{\mathbf{k}\mu}(t)\big(f^{\rm el}_{\mathbf{k}+\mathbf{p}\nu}(t) - 1\big)\big(f^{\rm el}_{\mathbf{q}\mu'}(t) - 1\big)\Big] \end{aligned}} \tag{17.69}$$

The Dirac delta leads to the conservation of energy in every scattering event. We can then perform an unrestricted sum over the band indices since the argument of the Dirac delta with indices (μ, ν') in the conduction (valence) bands and indices (ν, μ') in the valence (conduction) bands exceeds 2Δ, see discussion below (17.43).

The Markovian approximation of the Fan–Migdal scattering term (17.53) leads to

$$\boxed{\begin{aligned} S^{\rm FM}_{\mathbf{k}\mu\mu}(t) &= 2\pi \sum_{\mathbf{q}\nu\alpha} \frac{\left|g_{\mathbf{q}-\mathbf{k}\alpha,\mu\nu}(\mathbf{k})\right|^2}{2\omega_{\mathbf{k}-\mathbf{q}\alpha}} \Big\{\delta\big(\epsilon_{\mathbf{q}\nu} - \epsilon_{\mathbf{k}\mu} + \omega_{\mathbf{k}-\mathbf{q}\alpha}\big) \\ &\times \Big[\big(f^{\rm el}_{\mathbf{q}\nu}(t) - 1\big) f^{\rm el}_{\mathbf{k}\mu}(t) \mathrm{Re}\big[B^{>}_{\mathbf{k}-\mathbf{q}\alpha}(t)\big] - f^{\rm el}_{\mathbf{q}\nu}(t)\big(f^{\rm el}_{\mathbf{k}\mu}(t) - 1\big)\mathrm{Re}\big[B^{<}_{\mathbf{k}-\mathbf{q}\alpha}(t)\big]\Big] \\ &\qquad + \delta\big(\epsilon_{\mathbf{q}\nu} - \epsilon_{\mathbf{k}\mu} - \omega_{\mathbf{k}-\mathbf{q}\alpha}\big) \\ &\times \Big[\big(f^{\rm el}_{\mathbf{q}\nu}(t) - 1\big) f^{\rm el}_{\mathbf{k}\mu}(t) \mathrm{Re}\big[B^{<}_{\mathbf{q}-\mathbf{k}\alpha}(t)\big] - f^{\rm el}_{\mathbf{q}\nu}(t)\big(f^{\rm el}_{\mathbf{k}\mu}(t) - 1\big)\mathrm{Re}\big[B^{>}_{\mathbf{q}-\mathbf{k}\alpha}(t)\big]\Big]\Big\} \end{aligned}} \tag{17.70}$$

where we replace $g^{\rm intra} \to g$ since the argument of the Dirac delta evaluated for μ and ν either conduction–valence bands or valence–conduction bands is of order Δ, and therefore the corresponding contribution vanishes. The same consideration applies to the phononic scattering terms which take the form

$$\begin{aligned} S^{21,\rm phon}_{\mathbf{q}\alpha\alpha}(t) &= -\mathrm{i}\pi \sum_{\mathbf{k}\mu\nu} \frac{|g_{\mathbf{q}\alpha,\mu\nu}(\mathbf{k})|^2}{2\omega_{\mathbf{q}\alpha}} \Big\{\delta\big(\epsilon_{\mathbf{q}+\mathbf{k}\nu} - \epsilon_{\mathbf{k}\mu} - \omega_{\mathbf{q}\alpha}\big) \\ &\times \Big[\big(f^{\rm el}_{\mathbf{q}+\mathbf{k}\nu}(t) - 1\big) f^{\rm el}_{\mathbf{k}\mu}(t) B^{<}_{\mathbf{q}\alpha}(t) - f^{\rm el}_{\mathbf{q}+\mathbf{k}\nu}(t)\big(f^{\rm el}_{\mathbf{k}\mu}(t) - 1\big) B^{>}_{\mathbf{q}\alpha}(t)\Big] \\ &\qquad + \delta\big(\epsilon_{\mathbf{q}+\mathbf{k}\nu} - \epsilon_{\mathbf{k}\mu} + \omega_{\mathbf{q}\alpha}\big) \\ &\times \Big[\big(f^{\rm el}_{\mathbf{q}+\mathbf{k}\nu}(t) - 1\big) f^{\rm el}_{\mathbf{k}\mu}(t) B^{>*}_{-\mathbf{q}\alpha}(t) - f^{\rm el}_{\mathbf{q}+\mathbf{k}\nu}(t)\big(f^{\rm el}_{\mathbf{k}\mu}(t) - 1\big) B^{<*}_{-\mathbf{q}\alpha}(t)\Big]\Big\}, \end{aligned}$$

$$\begin{aligned} S^{22,\rm phon}_{\mathbf{q}\alpha\alpha}(t) &= \pi \sum_{\mathbf{k}\mu\nu} \frac{|g_{\mathbf{q}\alpha,\mu\nu}(\mathbf{k})|^2}{2} \Big\{\delta\big(\epsilon_{\mathbf{q}+\mathbf{k}\nu} - \epsilon_{\mathbf{k}\mu} - \omega_{\mathbf{q}\alpha}\big) \\ &\times \Big[\big(f^{\rm el}_{\mathbf{q}+\mathbf{k}\nu}(t) - 1\big) f^{\rm el}_{\mathbf{k}\mu}(t) C^{<}_{\mathbf{q}\alpha}(t) - f^{\rm el}_{\mathbf{q}+\mathbf{k}\nu}(t)\big(f^{\rm el}_{\mathbf{k}\mu}(t) - 1\big) C^{>}_{\mathbf{q}\alpha}(t)\Big] \\ &\qquad - \delta\big(\epsilon_{\mathbf{q}+\mathbf{k}\nu} - \epsilon_{\mathbf{k}\mu} + \omega_{\mathbf{q}\alpha}\big) \\ &\times \Big[\big(f^{\rm el}_{\mathbf{q}+\mathbf{k}\nu}(t) - 1\big) f^{\rm el}_{\mathbf{k}\mu}(t) C^{>*}_{-\mathbf{q}\alpha}(t) - f^{\rm el}_{\mathbf{q}+\mathbf{k}\nu}(t)\big(f^{\rm el}_{\mathbf{k}\mu}(t) - 1\big) C^{<*}_{-\mathbf{q}\alpha}(t)\Big]\Big\} + \mathrm{h.c.} \end{aligned}$$

[||]The reader can easily show that the contribution of the principal part of the GW scattering term vanishes.

Phononic occupations and coherences To gain some more physical intuition on the equation of motion for $\gamma^<$, let us write the displacements and momenta in terms of the dressed phononic operators [compare with (16.53)]:

$$\hat{\mathcal{U}}_{\mathbf{q}\alpha} = \frac{1}{\sqrt{2\omega_{\mathbf{q}\alpha}}}(\hat{b}_{\mathbf{q}\alpha} + \hat{b}^\dagger_{-\mathbf{q}\alpha}) \quad , \quad \hat{\mathcal{P}}_{\mathbf{q}\alpha} = -\mathrm{i}\sqrt{\frac{\omega_{\mathbf{q}\alpha}}{2}}(\hat{b}_{\mathbf{q}\alpha} - \hat{b}^\dagger_{-\mathbf{q}\alpha}). \tag{17.71}$$

Taking into account that $U_{\mathbf{q}\alpha} = P_{\mathbf{q}\alpha} = 0$ for all $\mathbf{q} \neq 0$ (the driving field does not break the lattice periodicity), we have

$$\begin{aligned}\gamma^{11,<}_{\mathbf{q}\alpha}(t) &= \mathrm{Tr}\big[\hat{\rho}\,\Delta\hat{\mathcal{U}}_{-\mathbf{q}\alpha,H}(t)\Delta\hat{\mathcal{U}}_{\mathbf{q}\alpha,H}(t)\big]\\ &= \frac{1}{2\omega_{\mathbf{q}\alpha}}\Big(f^{\mathrm{phon}}_{\mathbf{q}\alpha}(t) + f^{\mathrm{phon}}_{-\mathbf{q}\alpha}(t) + 1 + \Theta_{\mathbf{q}\alpha}(t) + \Theta^*_{\mathbf{q}\alpha}(t)\Big) - \delta_{\mathbf{q},\mathbf{0}}|U_{\mathbf{0}\alpha}(t)|^2,\end{aligned} \tag{17.72a}$$

$$\begin{aligned}\gamma^{22,<}_{\mathbf{q}\alpha}(t) &= \mathrm{Tr}\big[\hat{\rho}\,\Delta\hat{\mathcal{P}}_{-\mathbf{q}\alpha,H}(t)\Delta\hat{\mathcal{P}}_{\mathbf{q}\alpha,H}(t)\big]\\ &= \frac{\omega_{\mathbf{q}\alpha}}{2}\Big(f^{\mathrm{phon}}_{\mathbf{q}\alpha}(t) + f^{\mathrm{phon}}_{-\mathbf{q}\alpha}(t) + 1 - \Theta_{\mathbf{q}\alpha}(t) - \Theta^*_{\mathbf{q}\alpha}(t)\Big) - \delta_{\mathbf{q},\mathbf{0}}|P_{\mathbf{0}\alpha}(t)|^2,\end{aligned} \tag{17.72b}$$

$$\begin{aligned}\gamma^{12,<}_{\mathbf{q}\alpha}(t) &= \mathrm{Tr}\big[\hat{\rho}\,\Delta\hat{\mathcal{P}}_{-\mathbf{q}\alpha,H}(t)\Delta\hat{\mathcal{U}}_{\mathbf{q}\alpha,H}(t)\big]\\ &= \frac{\mathrm{i}}{2}\Big(f^{\mathrm{phon}}_{\mathbf{q}\alpha}(t) - f^{\mathrm{phon}}_{-\mathbf{q}\alpha}(t) - 1 - \Theta_{\mathbf{q}\alpha}(t) + \Theta^*_{\mathbf{q}\alpha}(t)\Big) - \delta_{\mathbf{q},\mathbf{0}}P^*_{\mathbf{0}\alpha}(t)U_{\mathbf{0}\alpha}(t),\end{aligned} \tag{17.72c}$$

where we introduce the *phononic occupations*

$$\boxed{f^{\mathrm{phon}}_{\mathbf{q}\alpha}(t) \equiv \mathrm{Tr}\big[\hat{\rho}\,b^\dagger_{\mathbf{q}\alpha,H}(t)b_{\mathbf{q}\alpha,H}(t)\big]} \tag{17.73}$$

and the *phononic coherences*

$$\boxed{\Theta_{\mathbf{q}\alpha}(t) = \Theta_{-\mathbf{q}\alpha}(t) \equiv \mathrm{Tr}\big[\hat{\rho}\,b_{\mathbf{q}\alpha,H}(t)b_{-\mathbf{q}\alpha,H}(t)\big]} \tag{17.74}$$

In the subsequent derivations we consider $\mathbf{q} \neq \mathbf{0}$. Notice that $\mathbf{q} = 0$ is a set of zero measure in the first Brillouin zone, and we can safely ignore the contribution with $\mathbf{q} = \mathbf{0}$ in the Fan–Migdal scattering term (17.70). Expressing the B-like and C-like combinations defined in (17.52) and (17.68) in terms of the phononic occupations and coherences, we find

$$B^<_{\mathbf{q}\alpha}(t) = f^{\mathrm{phon}}_{\mathbf{q}\alpha}(t) + \Theta_{\mathbf{q}\alpha}(t), \tag{17.75a}$$

$$B^>_{\mathbf{q}\alpha}(t) = f^{\mathrm{phon}}_{\mathbf{q}\alpha}(t) + 1 + \Theta_{\mathbf{q}\alpha}(t), \tag{17.75b}$$

$$C^<_{\mathbf{q}\alpha}(t) = \omega_{\mathbf{q}\alpha}\big[f^{\mathrm{phon}}_{\mathbf{q}\alpha}(t) - \Theta_{\mathbf{q}\alpha}(t)\big], \tag{17.75c}$$

$$C^>_{\mathbf{q}\alpha}(t) = \omega_{\mathbf{q}\alpha}\big[f^{\mathrm{phon}}_{\mathbf{q}\alpha}(t) + 1 - \Theta_{\mathbf{q}\alpha}(t)\big]. \tag{17.75d}$$

With these preliminary steps, we can now formulate an equation of motion for the phononic occupations and coherences. From (17.72) we have

$$f^{\mathrm{phon}}_{\mathbf{q}\alpha} = \frac{1}{2}\Big[\omega_{\mathbf{q}\alpha}\gamma^{11,<}_{\mathbf{q}\alpha} + \frac{1}{\omega_{\mathbf{q}\alpha}}\gamma^{22,<}_{\mathbf{q}\alpha} - \mathrm{i}\big(\gamma^{12,<}_{\mathbf{q}\alpha} - \gamma^{21,<}_{\mathbf{q}\alpha}\big)\Big].$$

Let $T_{\mathbf{q}\alpha} = \begin{pmatrix} 0 & -1 \\ \omega^2_{\mathbf{q}\alpha} & 0 \end{pmatrix} \gamma^<_{\mathbf{q}\alpha} - \gamma^<_{\mathbf{q}\alpha} \begin{pmatrix} 0 & -\omega^2_{\mathbf{q}\alpha} \\ 1 & 0 \end{pmatrix}$ be the 2×2 matrix appearing on the l.h.s. of (17.63). Taking into account the explicit form of $\gamma^<_{\mathbf{q}\alpha}$ in (17.72), we find

$$\omega_{\mathbf{q}\alpha} T^{11}_{\mathbf{q}\alpha} + \frac{1}{\omega_{\mathbf{q}\alpha}} T^{22}_{\mathbf{q}\alpha} - \mathrm{i}\big(T^{12}_{\mathbf{q}\alpha} - T^{21}_{\mathbf{q}\alpha}\big) = 0,$$

and therefore

$$\boxed{\frac{d}{dt} f^{\mathrm{phon}}_{\mathbf{q}\alpha}(t) = S^{\mathrm{phon-occ}}_{\mathbf{q}\alpha\alpha}(t)} \tag{17.76}$$

where the scattering term for the phononic occupations is $S^{\mathrm{phon-occ}}_{\mathbf{q}\alpha\alpha} = \frac{1}{2}\Big[\omega_{\mathbf{q}\alpha} S^{11,\mathrm{phon}}_{\mathbf{q}\alpha\alpha} + \frac{1}{\omega_{\mathbf{q}\alpha}} S^{22,\mathrm{phon}}_{\mathbf{q}\alpha\alpha} - \mathrm{i}\big(S^{12,\mathrm{phon}}_{\mathbf{q}\alpha\alpha} - S^{21,\mathrm{phon}}_{\mathbf{q}\alpha\alpha}\big)\Big]$ and reads:

$$\boxed{\begin{aligned} S^{\mathrm{phon-occ}}_{\mathbf{q}\alpha\alpha}(t) &= 2\pi \sum_{\mathbf{k}\mu\nu} \frac{|g_{\mathbf{q}\alpha,\mu\nu}(\mathbf{k})|^2}{2\omega_{\mathbf{q}\alpha}} \Big\{\delta\big(\epsilon_{\mathbf{q}+\mathbf{k}\nu} - \epsilon_{\mathbf{k}\mu} - \omega_{\mathbf{q}\alpha}\big) \\ &\times \Big[\big(f^{\mathrm{el}}_{\mathbf{q}+\mathbf{k}\nu}(t) - 1\big) f^{\mathrm{el}}_{\mathbf{k}\mu}(t) f^{\mathrm{phon}}_{\mathbf{q}\alpha}(t) - f^{\mathrm{el}}_{\mathbf{q}+\mathbf{k}\nu}(t)\big(f^{\mathrm{el}}_{\mathbf{k}\mu}(t) - 1\big)\big(f^{\mathrm{phon}}_{\mathbf{q}\alpha}(t) + 1\big)\Big] \\ &\quad + \delta\big(\epsilon_{\mathbf{q}+\mathbf{k}\nu} - \epsilon_{\mathbf{k}\mu} + \omega_{\mathbf{q}\alpha}\big) \\ &\times \Big[\big(f^{\mathrm{el}}_{\mathbf{q}+\mathbf{k}\nu}(t) - 1\big) f^{\mathrm{el}}_{\mathbf{k}\mu}(t) - f^{\mathrm{el}}_{\mathbf{q}+\mathbf{k}\nu}(t)\big(f^{\mathrm{el}}_{\mathbf{k}\mu}(t) - 1\big)\Big] \mathrm{Re}\big[\Theta_{\mathbf{q}\alpha}(t)\big]\Big\} \end{aligned}} \tag{17.77}$$

The equation of motion for the phononic coherences can be derived similarly. We have

$$\Theta_{\mathbf{q}\alpha} = \frac{1}{2}\Big[\omega_{\mathbf{q}\alpha} \gamma^{11,<}_{\mathbf{q}\alpha} - \frac{1}{\omega_{\mathbf{q}\alpha}} \gamma^{22,<}_{\mathbf{q}\alpha} + \mathrm{i}\big(\gamma^{12,<}_{\mathbf{q}\alpha} + \gamma^{21,<}_{\mathbf{q}\alpha}\big)\Big].$$

Taking into account the explicit form of $\gamma^<_{\mathbf{q}\alpha}$ in (17.72), we find

$$\omega_{\mathbf{q}\alpha} T^{11}_{\mathbf{q}\alpha} - \frac{1}{\omega_{\mathbf{q}\alpha}} T^{22}_{\mathbf{q}\alpha} + \mathrm{i}\big(T^{12}_{\mathbf{q}\alpha} + T^{21}_{\mathbf{q}\alpha}\big) = 4\mathrm{i}\omega_{\mathbf{q}\alpha}\Theta_{\mathbf{q}\alpha},$$

and therefore

$$\boxed{\frac{d}{dt}\Theta_{\mathbf{q}\alpha}(t) + 2\mathrm{i}\omega_{\mathbf{q}\alpha}\Theta_{\mathbf{q}\alpha} = S^{\mathrm{phon-coh}}_{\mathbf{q}\alpha\alpha}(t)} \tag{17.78}$$

where the scattering term for the phononic coherences is $S^{\mathrm{phon-coh}}_{\mathbf{q}\alpha\alpha} = \frac{1}{2}\big[\omega_{\mathbf{q}\alpha} S^{11,\mathrm{phon}}_{\mathbf{q}\alpha\alpha} -$

$\frac{1}{\omega_{\mathbf{q}\alpha}} S^{22,\text{phon}}_{\mathbf{q}\alpha\alpha} + \mathrm{i}(S^{12,\text{phon}}_{\mathbf{q}\alpha\alpha} + S^{21,\text{phon}}_{\mathbf{q}\alpha\alpha})]$ and reads

$$\boxed{\begin{aligned} S^{\text{phon}-\text{coh}}_{\mathbf{q}\alpha\alpha}(t) &= \pi \sum_{\mathbf{k}\mu\nu} \frac{|g_{\mathbf{q}\alpha,\mu\nu}(\mathbf{k})|^2}{2\omega_{\mathbf{q}\alpha}} \\ &\times \Big\{ \delta\big(\epsilon_{\mathbf{q}+\mathbf{k}\nu} - \epsilon_{\mathbf{k}\mu} - \omega_{\mathbf{q}\alpha}\big) \Big[\big(f^{\text{el}}_{\mathbf{q}+\mathbf{k}\nu}(t) - 1\big) f^{\text{el}}_{\mathbf{k}\mu}(t) \big(\Theta_{\mathbf{q}\alpha}(t) - f^{\text{phon}}_{\mathbf{q}\alpha}(t)\big) \\ &\qquad - f^{\text{el}}_{\mathbf{q}+\mathbf{k}\nu}(t)\big(f^{\text{el}}_{\mathbf{k}\mu}(t) - 1\big)\big(\Theta_{\mathbf{q}\alpha}(t) - f^{\text{phon}}_{\mathbf{q}\alpha}(t) - 1\big)\Big] \\ &+ \delta\big(\epsilon_{\mathbf{q}+\mathbf{k}\nu} - \epsilon_{\mathbf{k}\mu} + \omega_{\mathbf{q}\alpha}\big) \Big[\big(f^{\text{el}}_{\mathbf{q}+\mathbf{k}\nu}(t) - 1\big) f^{\text{el}}_{\mathbf{k}\mu}(t) \big(f^{\text{phon}}_{-\mathbf{q}\alpha}(t) + 1 - \Theta_{\mathbf{q}\alpha}(t)\big) \\ &\qquad - f^{\text{el}}_{\mathbf{q}+\mathbf{k}\nu}(t)\big(f^{\text{el}}_{\mathbf{k}\mu}(t) - 1\big)\big(f^{\text{phon}}_{-\mathbf{q}\alpha}(t) - \Theta_{\mathbf{q}\alpha}(t)\big)\Big]\Big\} \end{aligned}} \tag{17.79}$$

The equation of motion for the phononic coherence deserves further investigation. Let us write $\Theta_{\mathbf{q}\alpha} = \Theta^{(r)}_{\mathbf{q}\alpha} + \mathrm{i}\Theta^{(i)}_{\mathbf{q}\alpha}$ as the sum of its real and imaginary parts. Then (17.78) and (17.79) imply

$$\frac{d}{dt}\Theta^{(i)}_{\mathbf{q}\alpha}(t) + 2\omega_{\mathbf{q}\alpha}\Theta^{(r)}_{\mathbf{q}\alpha} = -\Gamma^{\text{coh}}_{\mathbf{q}\alpha}(t)\Theta^{(i)}_{\mathbf{q}\alpha}(t), \tag{17.80a}$$

$$\frac{d}{dt}\Theta^{(r)}_{\mathbf{q}\alpha}(t) - 2\omega_{\mathbf{q}\alpha}\Theta^{(i)}_{\mathbf{q}\alpha} = -\Gamma^{\text{coh}}_{\mathbf{q}\alpha}(t)\Theta^{(r)}_{\mathbf{q}\alpha}(t) + S^{\text{phon}-\text{coh}}_{\mathbf{q}\alpha}(t)\big|_{\Theta_{\mathbf{q}\alpha}=0}, \tag{17.80b}$$

where

$$\Gamma^{\text{coh}}_{\mathbf{q}\alpha}(t) = \pi \sum_{\mathbf{k}\mu\nu} \frac{|g_{\mathbf{q}\alpha,\mu\nu}(\mathbf{k})|^2}{2\omega_{\mathbf{q}\alpha}} \big[\delta\big(\epsilon_{\mathbf{q}+\mathbf{k}\nu} - \epsilon_{\mathbf{k}\mu} - \omega_{\mathbf{q}\alpha}\big) - \delta\big(\epsilon_{\mathbf{q}+\mathbf{k}\nu} - \epsilon_{\mathbf{k}\mu} + \omega_{\mathbf{q}\alpha}\big)\big] \times \big[f^{\text{el}}_{\mathbf{k}\mu}(t) - f^{\text{el}}_{\mathbf{q}+\mathbf{k}\nu}(t)\big].$$

The *coherence rate* $\Gamma^{\text{coh}}_{\mathbf{q}\alpha}$ is positive if the electronic occupations satisfy the inequality $f_{\mathbf{k}\mu} > f_{\mathbf{q}+\mathbf{k}\nu}$ for $\epsilon_{\mathbf{q}+\mathbf{k}\nu} > \epsilon_{\mathbf{k}\mu}$, and indices (μ,ν) that are both conduction or both valence bands (contributions with μ a valence index and ν a conduction index or vice versa vanish due to the Dirac delta); this condition holds true for a quasi-thermalized distribution of carriers. We prove in the next section that $S^{\text{phon}-\text{coh}}_{\mathbf{q}\alpha}(t)\big|_{\Theta_{\mathbf{q}\alpha}=0}$ vanishes for thermal electronic and phononic occupations, and it is therefore small for occupations close to thermal ones. Ignoring this term in (17.80b) and assuming $\Gamma^{\text{coh}}_{\mathbf{q}\alpha}(t)$ weakly dependent on time, the most general solution for the phononic coherence is

$$\Theta^{(i)}_{\mathbf{q}\alpha}(t) = \Theta_{0,\mathbf{q}\alpha}\cos\big(2\omega_{\mathbf{q}\alpha}t + \phi_{0,\mathbf{q}\alpha}\big)e^{-\Gamma^{\text{coh}}_{\mathbf{q}\alpha}t},$$
$$\Theta^{(r)}_{\mathbf{q}\alpha}(t) = \Theta_{0,\mathbf{q}\alpha}\sin\big(2\omega_{\mathbf{q}\alpha}t + \phi_{0,\mathbf{q}\alpha}\big)e^{-\Gamma^{\text{coh}}_{\mathbf{q}\alpha}t}.$$

It is worth noting that the Markovian approximation of the GKBA equations of motion for phononic coherences differs from the MGKBA equation (17.80) in that the sign of $\Gamma^{\text{coh}}_{\mathbf{q}\alpha}$ is reversed. This implies that the equilibrium solution is unstable.

Short and long driving In Table 17.1 we summarize the equations of motion for the electronic and phononic degrees of freedom, which we refer to as the *semiconductor*

Definition	Equations
Electronic occupation	$\frac{d}{dt}f^{\mathrm{el}}_{\mathbf{k}\mu} + \mathrm{i}\sum_{\nu\neq\mu}\left(\Omega^{\mathrm{ren}}_{\mathbf{k}\mu\nu}\,p_{\mathbf{k}\nu\mu} - p_{\mathbf{k}\mu\nu}\,\Omega^{\mathrm{ren}}_{\mathbf{k}\nu\mu}\right) = S^{\mathrm{el}}_{\mathbf{k}\mu\mu}$
Electronic polarization	$\frac{d}{dt}p_{\mathbf{k}\mu\nu} + \mathrm{i}(\tilde{\epsilon}_{\mathbf{k}\mu} - \tilde{\epsilon}_{\mathbf{k}\nu})p_{\mathbf{k}\mu\nu} + \mathrm{i}\Omega^{\mathrm{ren}}_{\mathbf{k}\mu\nu}(f^{\mathrm{el}}_{\mathbf{k}\nu} - f^{\mathrm{el}}_{\mathbf{k}\mu})$ $+\mathrm{i}\sum_{\nu'\neq\nu}\Omega^{\mathrm{ren}}_{\mathbf{k}\mu\nu'}\,p_{\mathbf{k}\nu'\nu} - \mathrm{i}\sum_{\nu'\neq\mu}p_{\mathbf{k}\mu\nu'}\,\Omega^{\mathrm{ren}}_{\mathbf{k}\nu'\nu} = -\Gamma^{\mathrm{pol}}_{\mathbf{k}\mu\nu}p_{\mathbf{k}\mu\nu}$
Nuclear displacement	$\frac{d^2}{dt^2}U_{\mathbf{0}\alpha} + \omega^2_{\mathbf{0}\alpha}U_{\mathbf{0}\alpha} = -\sum_{\mathbf{k}\mu\nu}g_{\mathbf{0}\alpha,\nu\mu}(\mathbf{k})\left[\delta_{\mu\nu}\Delta f^{\mathrm{el}}_{\mathbf{k}\mu} + p^{\mathrm{intra}}_{\mathbf{k}\mu\nu}\right]$ $-\int dt'\sum_{\substack{\mathbf{k}\mathbf{k}'\\ \mu\mu'\\ \nu\nu'}}g_{\mathbf{0}\alpha,\nu\mu}(\mathbf{k})\chi^{\mathrm{R}}_{\mathrm{clamp},\mathbf{k}\mathbf{k}'\,\substack{\mu\mu'\\ \nu\nu'}}(\mathbf{0};t,t')\Omega_{\mathbf{k}'\mu'\nu'}(t')$
Phonon occupation	$\frac{d}{dt}f^{\mathrm{phon}}_{\mathbf{q}\alpha}(t) = S^{\mathrm{phon-occ}}_{\mathbf{q}\alpha\alpha}(t)$
Phonon coherence	$\frac{d}{dt}\Theta_{\mathbf{q}\alpha}(t) + 2\mathrm{i}\omega_{\mathbf{q}\alpha}\Theta_{\mathbf{q}\alpha} = S^{\mathrm{phon-coh}}_{\mathbf{q}\alpha\alpha}(t)$
Quasi-particle energy	$\tilde{\epsilon}_{\mathbf{k}\mu}(t) = \epsilon_{\mathbf{k}\mu} + \sum_{\mathbf{k}'\mu'}(v_{\mathbf{k}\mu\mathbf{k}'\mu'\mathbf{k}'\mu'\mathbf{k}\mu} - W_{\mathbf{k}\mu\mathbf{k}'\mu'\mathbf{k}\mu\mathbf{k}'\mu'})\Delta f^{\mathrm{el}}_{\mathbf{q}\mu'}$
Renorm. Rabi frequency	$\Omega^{\mathrm{ren}}_{\mathbf{k}\mu\nu} = \Omega_{\mathbf{k}\mu\nu} + \sum_{\alpha}g_{\mathbf{0}\alpha,\mu\nu}(\mathbf{k})U_{\mathbf{0}\alpha}$

Table 17.1 Table summarizing the semiconductor eletron-phonon equations.

electron-phonon equations. The electronic scattering term is given by the sum of (17.69) - GW - and (17.70) - Fan-Migdal - whereas the phononic scattering terms are given by (17.77) and (17.79). All scattering rates except the GW one depend on both phononic occupations and coherences. The unique feature of the semiconductor electron-phonon equations is a consistent treatment of phononic occupations and coherences as well as the inclusion of the renormalization of the electronic quasi-particle energies induced by the nuclear displacements. The former aspect opens the door to studies of phonon squeezed states in optically excited semiconductors [340, 341, 342, 343] and time-dependent Debye-Waller factors [344, 345, 346, 347].

If the duration T_{drive} of the driving field is much shorter than a typical phonon period (for a phonon frequency $\lesssim 100$ meV the phonon period is $\gtrsim 40$ fs), then the nuclei remain essentially still while the field is on. For such short drivings, and for a response function $\chi^{\mathrm{R}}(t,t')$ that decays rapidly as $|t-t'|$ increases, we can neglect the last term in the third equation of Table 17.1, as it grows linearly with T_{drive}. The energy shift $\sum_{\alpha}g_{\mathbf{0}\alpha,\mu\nu}(\mathbf{k})U_{\mathbf{0}\alpha}$ in $\Omega^{\mathrm{ren}}_{\mathbf{k}\mu\nu}$, see (17.55), is typically of the order of a few meV and is responsible for time-dependent modulations in the photoabsorption spectra [348, 349, 350, 351, 352]. To capture this effect it is crucial to use $\Omega^{\mathrm{ren}}_{\mathbf{k}\mu\nu}$ in the two terms that multiply the polarizations in the second equation of Table 17.1 [400]. All other $\Omega^{\mathrm{ren}}_{\mathbf{k}\mu\nu}$ can be approximated with $\Omega_{\mathbf{k}\mu\nu}$.

For long driving fields – that is, $T_{\rm drive}$ a few hundreds of femtoseconds or longer – the last term in the third equation of Table 17.1 cannot be discarded. In this case the nuclei are expected to slowly attain new positions and no time-dependent modulations of the photoabsorption spectra are to be expected. The nuclear shifts are mainly responsible for a few meV renormalization of the quasi-particle energies. Thus, if our focus is solely on occupations and coherences, or if we do not require an meV resolution of the spectrum, then we can solve the semiconductor electron–phonon equations with $U_{\mathbf{0}\alpha} = 0$.

Polarization rates The electronic scattering term has the following mathematical structure

$$S^{\rm el}_{\mathbf{k}\mu\mu}(t) = S^{\rm GW}_{\mathbf{k}\mu\mu}(t) + S^{\rm FM}_{\mathbf{k}\mu\mu}(t) = -2\Gamma^{{\rm el},>}_{\mathbf{k}\mu\mu}(t) f^{\rm el}_{\mathbf{k}\mu}(t) - 2\Gamma^{{\rm el},<}_{\mathbf{k}\mu\mu}(t)\big(f^{\rm el}_{\mathbf{k}\mu}(t) - 1\big).$$

The reader can check that $\Gamma^{{\rm el},\gtrless}_{\mathbf{k}\mu\mu}(t) > 0$ for vanishing phononic coherences (i.e., $\Theta_{\mathbf{q}\alpha} = 0$). Comparing this equation with (17.11), we infer that our simplifications have led to diagonal one-body operators $\hat{\Gamma}^{{\rm el},\gtrless}$ in the Bloch basis: $\langle \mathbf{k}\mu|\hat{\Gamma}^{{\rm el},\gtrless}|\mathbf{k}\nu\rangle = \delta_{\mu\nu}\Gamma^{{\rm el},\gtrless}_{\mathbf{k}\mu\mu}$. Bracketing (17.11) with $\langle \mathbf{k}\mu|$ and $|\mathbf{k}\nu\rangle$, we then obtain an expression for the polarization rates in (17.41):

$$\Gamma^{\rm pol}_{\mathbf{k}\mu\nu} = \Gamma^{{\rm el},>}_{\mathbf{k}\mu\mu} + \Gamma^{{\rm el},<}_{\mathbf{k}\mu\mu} + \Gamma^{{\rm el},>}_{\mathbf{k}\nu\nu} + \Gamma^{{\rm el},<}_{\mathbf{k}\nu\nu}.$$

It is worth remarking that the Markovian approximation of the GKBA equations of motion leads to unphysical polarization rates.

17.4.5 Equilibrium and Steady-State Solutions

Let us discuss the stationary solutions of the semiconductor electron–phonon equations. We assign a common chemical potential μ_c to all conduction bands and a common chemical potential μ_v to all valence bands, and show that the scattering terms vanish if the electronic occupations $f^{\rm el}_{\mathbf{k}\nu} = 1/[e^{\beta(\epsilon_{\mathbf{k}\nu}-\mu_\nu)} + 1]$ (noninteracting finite-temperature electrons) and the phononic density matrix $\gamma_{\mathbf{q}\alpha}$ is given by (17.65) with $f_{\rm B}(\omega_{\mathbf{q}\alpha}) = 1/[e^{\beta\omega_{\mathbf{q}\alpha}} - 1]$ (noninteracting finite-temperature phonons). In this case the definitions in (17.73) and (17.74) yield

$$f^{\rm phon}_{\mathbf{q}\alpha} = f_{\rm B}(\omega_{\mathbf{q}\alpha}) \quad , \quad \Theta_{\mathbf{q}\alpha} = 0.$$

The reader can verify that the equilibrium phononic Green's function (17.45) is recovered by inserting these values into (17.51), as it should be. Let us consider the GW scattering term in (17.69). Taking into account that

$$\frac{f^{\rm el}_{\mathbf{k}\nu}}{1 - f^{\rm el}_{\mathbf{k}\nu}} = e^{-\beta(\epsilon_{\mathbf{k}\nu}-\mu_\nu)},$$

the energy conservation enforced by the Dirac delta implies

$$\frac{f^{\rm el}_{\mathbf{k}+\mathbf{p}\nu} f^{\rm el}_{\mathbf{q}\mu'}}{(f^{\rm el}_{\mathbf{k}+\mathbf{p}\nu} - 1)(f^{\rm el}_{\mathbf{q}\mu'} - 1)} e^{\beta(\mu_{\mu'}+\mu_\nu)} = \frac{f^{\rm el}_{\mathbf{q}+\mathbf{p}\nu'} f^{\rm el}_{\mathbf{k}\mu}}{(f^{\rm el}_{\mathbf{q}+\mathbf{p}\nu'} - 1)(f^{\rm el}_{\mathbf{k}\mu}(t) - 1)} e^{\beta(\mu_{\nu'}+\mu_\mu)},$$

and therefore all terms in $S^{\rm GW}_{\mathbf{k}\mu\mu}$ with $\mu_{\mu'}+\mu_\nu=\mu_{\nu'}+\mu_\mu$ vanish. The only case for which $\mu_{\mu'}+\mu_\nu\neq\mu_{\nu'}+\mu_\mu$ is when (μ',ν) are both conduction (valence) bands and (ν',μ) are both valence (conduction) bands. However, for this choice of indices the argument of the Dirac delta is at least 2Δ. We conclude that $S^{\rm GW}_{\mathbf{k}\mu\mu}=0$.

Similarly, for the Fan–Migdal scattering term in (17.70) we have

$$\begin{aligned}\frac{f^{\rm el}_{\mathbf{k}\mu}}{f^{\rm el}_{\mathbf{k}\mu}-1}\bar{f}_{\rm B}(\omega_{\mathbf{k}-\mathbf{q}\alpha}) &= -e^{-\beta(\epsilon_{\mathbf{k}\mu}-\mu_\mu)}\bar{f}_{\rm B}(\omega_{\mathbf{k}-\mathbf{q}\alpha})\\ &= -e^{-\beta(\mu_\nu-\mu_\mu)}e^{-\beta(\epsilon_{\mathbf{q}\nu}-\mu_\nu)}e^{-\beta\omega_{\mathbf{k}-\mathbf{q}\alpha}}\bar{f}_{\rm B}(\omega_{\mathbf{k}-\mathbf{q}\alpha})\\ &= e^{-\beta(\mu_\nu-\mu_\mu)}\frac{f^{\rm el}_{\mathbf{q}\nu}}{f^{\rm el}_{\mathbf{q}\nu}-1}f_{\rm B}(\omega_{\mathbf{k}-\mathbf{q}\alpha}),\end{aligned}$$

where in the second row we enforce the energy conservation $\epsilon_{\mathbf{k}\mu}=\epsilon_{\mathbf{q}\nu}+\omega_{\mathbf{k}-\mathbf{q}\alpha}$. A similar relation can be derived for the term with $\epsilon_{\mathbf{k}\mu}=\epsilon_{\mathbf{q}\nu}-\omega_{\mathbf{k}-\mathbf{q}\alpha}$. Therefore, all terms in $S^{\rm FM}_{\mathbf{k}\mu\mu}$ with $\mu_\nu=\mu_\mu$ vanish. The only case for which $\mu_\nu\neq\mu_\mu$ is when ν is a conduction (valence) band and μ is a valence (conduction) band. However, for this choice of indices the argument of the Dirac delta is of order Δ. We conclude that $S^{\rm FM}_{\mathbf{k}\mu\mu}=0$. The same arguments can be used to show that $S^{\rm phon-occ}_{\mathbf{q}\alpha\alpha}=S^{\rm phon-coh}_{\mathbf{q}\alpha\alpha}=0$.

The most general steady-state solution of the semiconductor electron–phonon equations with $\Omega^{\rm ren}_{\mathbf{k}\mu\nu}=0$ is given by noninteracting electronic and phononic occupations *at the same temperature*, and vanishing electronic polarizations and phononic coherences. Among all these solutions there exists the equilibrium one, where β is the equilibrium inverse temperature and $\mu_c=\mu_v=\mu$ is the equilibrium chemical potential. The steady-state solution, if attained, is expected to have a temperature higher than the equilibrium temperature since the driving field injects energy into the system. This means that $\Delta f^{\rm el}_{\mathbf{k}\mu}=f^{\rm el}_{\mathbf{k}\mu}(t\to\infty)-f^{\rm el}_{\mathbf{k}\mu}(t=0)$ is, in general, different from zero, and therefore the nuclei attain new positions $U_{\mathbf{0}\alpha}(t\to\infty)=-\frac{1}{\omega^2_{\mathbf{0}\alpha}}\sum_{\mathbf{k}\mu}g_{\mathbf{0}\alpha,\mu\mu}(\mathbf{k})\Delta f^{\rm el}_{\mathbf{k}\mu}$, see (17.62). These displacements can be either negative or positive, depending on the sign and magnitude of the electron–phonon couplings. Because of the nonzero nuclear displacements, $\Omega^{\rm ren}_{\mathbf{k}\mu\nu}(t\to\infty)$ is small but not exactly zero. Consequently, the steady-state occupations, polarizations, and coherences differ slightly from the thermal values.

Exercise 17.3 Derive the expression of $D^{ij,\lessgtr}_{\mathbf{q}\alpha\beta}(t,t')$ in GKBA.

Exercise 17.4 Derive the equation of motion for the phononic occupation and coherence with $\mathbf{q}=0$.

Exercise 17.5 Derive the equation of motion for the phononic coherences in GKBA.

Exercise 17.6 Write (17.60) without omitting the dependence on momenta and prove (17.62) by following the same steps.

17.5 Boltzmann Equations

The semiconductor electron–phonon equations simplify in the so-called *incoherent regime*. For times long after the perturbation caused by the external field it is reasonable to assume that the system is well described by a many-body density matrix of the form

$$\hat{\rho}(t) = \hat{U}(t, t_0)\, \hat{\rho}\, \hat{U}(t_0, t) = \sum_k w_k(t) |\Psi_k\rangle\langle\Psi_k|,$$

where the many-body states $|\Psi_k\rangle$ have a well-defined number of electrons and phonons or, equivalently, are eigenstates of the electronic and phononic number operators,

$$\hat{n}^{\mathrm{el}}_{\mathbf{k}\mu}|\Psi_k\rangle = \hat{d}^{\dagger}_{\mathbf{k}\mu}\hat{d}_{\mathbf{k}\mu}|\Psi_k\rangle = n^{\mathrm{el}}_{\mathbf{k}\mu}|\Psi_k\rangle, \qquad n^{\mathrm{el}}_{\mathbf{k}\mu} = 0, 1, \tag{17.81a}$$

$$\hat{n}^{\mathrm{phon}}_{\mathbf{q}\alpha}|\Psi_k\rangle = \hat{b}^{\dagger}_{\mathbf{q}\alpha}\hat{b}_{\mathbf{q}\alpha}|\Psi_k\rangle = n^{\mathrm{phon}}_{\mathbf{q}\alpha}|\Psi_k\rangle, \qquad n^{\mathrm{phon}}_{\mathbf{q}\alpha} = 0, 1, 2, \ldots. \tag{17.81b}$$

Equations (17.81) constitute the **eleventh simplification**. In the incoherent regime we have

$$\boxed{p_{\mathbf{k}\mu\nu}(t) = 0} \tag{17.82}$$

$$\boxed{U_{\mathbf{0}\alpha}(t) = P_{\mathbf{0}\alpha}(t) = 0} \tag{17.83}$$

and

$$\boxed{\Theta_{\mathbf{q}\alpha}(t) = 0}$$

The semiconductor electron–phonon equations reduce to only two coupled equations for the electronic and phononic occupations, known as the *Boltzmann equations* [109, 353], which are the first and fourth equations in Table 17.1:

$$\frac{d}{dt} f^{\mathrm{el}}_{\mathbf{k}\mu}(t) = S^{\mathrm{el}}_{\mathbf{k}\mu\mu}(t),$$

$$\frac{d}{dt} f^{\mathrm{phon}}_{\mathbf{q}\alpha}(t) = S^{\mathrm{phon-occ}}_{\mathbf{q}\alpha\alpha}(t).$$

The Boltzmann equations do not account for the interaction between electrons and transverse fields.[12] The nonequilibrium distribution of electrons and phonons enters as the initial value of $f^{\mathrm{el}}_{\mathbf{k}\mu}$ and $f^{\mathrm{ph}}_{\mathbf{q}\alpha}$. The scattering terms of the Boltzmann equations are obtained from (17.69), (17.70), and (17.77) by setting $\Theta_{\mathbf{q}\alpha} = 0$. In particular, $S^{\mathrm{GW}}_{\mathbf{k}\mu\mu}$ does not change, see (17.69),

[12] Historically the Boltzmann equations contain the coupling between a longitudinal electric field $\mathbf{E} = -\boldsymbol{\nabla} V$ and the $\mathbf{k}$-gradient of the occupations.

whereas $S^{\rm FM}_{\mathbf{k}\mu\mu}$ and $S^{\rm phon-occ}_{\mathbf{q}\alpha\alpha}$ simplify to

$$S^{\rm FM}_{\mathbf{k}\mu\mu}(t) = 2\pi\sum_{\mathbf{q}\nu\alpha}\frac{\left|g_{\mathbf{q}-\mathbf{k}\alpha,\mu\nu}(\mathbf{k})\right|^2}{2\omega_{\mathbf{k}-\mathbf{q}\alpha}}\Big\{\delta\big(\epsilon_{\mathbf{q}\nu}-\epsilon_{\mathbf{k}\mu}+\omega_{\mathbf{k}-\mathbf{q}\alpha}\big)$$
$$\times\Big[\big(f^{\rm el}_{\mathbf{q}\nu}(t)-1\big)f^{\rm el}_{\mathbf{k}\mu}(t)\big(f^{\rm phon}_{\mathbf{k}-\mathbf{q}\alpha}(t)+1\big)-f^{\rm el}_{\mathbf{q}\nu}(t)\big(f^{\rm el}_{\mathbf{k}\mu}(t)-1\big)f^{\rm phon}_{\mathbf{k}-\mathbf{q}\alpha}(t)\Big]$$
$$+\,\delta\big(\epsilon_{\mathbf{q}\nu}-\epsilon_{\mathbf{k}\mu}-\omega_{\mathbf{k}-\mathbf{q}\alpha}\big)$$
$$\times\Big[\big(f^{\rm el}_{\mathbf{q}\nu}(t)-1\big)f^{\rm el}_{\mathbf{k}\mu}(t)f^{\rm phon}_{\mathbf{q}-\mathbf{k}\alpha}(t)-f^{\rm el}_{\mathbf{q}\nu}(t)\big(f^{\rm el}_{\mathbf{k}\mu}(t)-1\big)\big(f^{\rm phon}_{\mathbf{q}-\mathbf{k}\alpha}(t)+1\big)\Big]\Big\}, \quad (17.84)$$

$$S^{\rm phon-occ}_{\mathbf{q}\alpha\alpha}(t) = 2\pi\sum_{\mathbf{k}\mu\nu}\frac{|g_{\mathbf{q}\alpha,\mu\nu}(\mathbf{k})|^2}{2\omega_{\mathbf{q}\alpha}}\delta\big(\epsilon_{\mathbf{q}+\mathbf{k}\nu}-\epsilon_{\mathbf{k}\mu}-\omega_{\mathbf{q}\alpha}\big)$$
$$\times\Big[\big(f^{\rm el}_{\mathbf{q}+\mathbf{k}\nu}(t)-1\big)f^{\rm el}_{\mathbf{k}\mu}(t)f^{\rm phon}_{\mathbf{q}\alpha}(t)-f^{\rm el}_{\mathbf{q}+\mathbf{k}\nu}(t)\big(f^{\rm el}_{\mathbf{k}\mu}(t)-1\big)\big(f^{\rm phon}_{\mathbf{q}\alpha}(t)+1\big)\Big]. \quad (17.85)$$

All contributions to the scattering terms have a clear physical interpretation in terms of two fundamental scattering events:

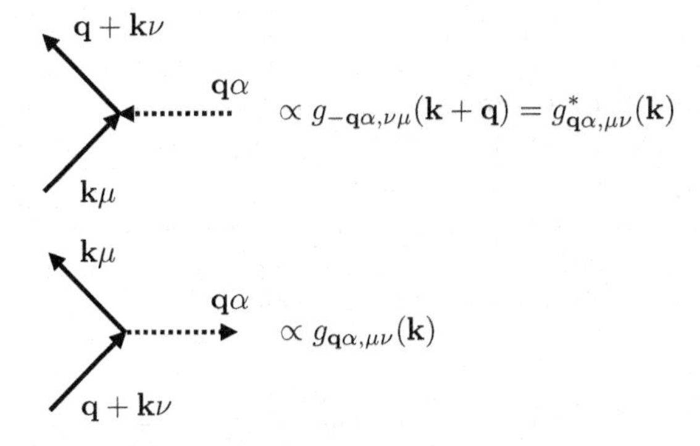

both having probability $|g_{\mathbf{q}\alpha,\mu\nu}(\mathbf{k})|^2$. We mention that the Boltzmann equations can alternatively be derived using the Wigner function [43, 109] or the Fermi golden rule [354, 355, 356].

17.5.1 Irreversibility and the H-theorem

The dynamics governed by the Boltzmann equations is not reversible. In fact, the eleven simplifications needed to arrive at the Boltzmann equations have transformed the original Kadanoff–Baym equations, which maintain time reversal invariance for Hamiltonians that are invariant under time reversal, into equations with a "time arrow." Put differently, looking at the dynamics we can establish whether the system is moving forward or backward in time. This statement in known as the *H-theorem* and can be mathematically formulated as follows. Consider the noninteracting form of the total entropy of the system: $\mathrm{S} = \mathrm{S}^{\rm el} + \mathrm{S}^{\rm phon}$, where, see (D.16),

$$\mathrm{S}^{\mathrm{el}} = -K_{\mathrm{B}} \sum_{\mathbf{k}\mu} \Big[f^{\mathrm{el}}_{\mathbf{k}\mu} \ln f^{\mathrm{el}}_{\mathbf{k}\mu} + (1 - f^{\mathrm{el}}_{\mathbf{k}\mu}) \ln(1 - f^{\mathrm{el}}_{\mathbf{k}\mu}) \Big], \tag{17.86a}$$

$$\mathrm{S}^{\mathrm{phon}} = -K_{\mathrm{B}} \sum_{\mathbf{q}\alpha} \Big[f^{\mathrm{phon}}_{\mathbf{q}\alpha} \ln f^{\mathrm{phon}}_{\mathbf{q}\alpha} - (1 + f^{\mathrm{phon}}_{\mathbf{q}\alpha}) \ln(1 + f^{\mathrm{phon}}_{\mathbf{q}\alpha}) \Big]. \tag{17.86b}$$

The H-theorem establishes that the quantity $\mathrm{H} = -\mathrm{S}/K_{\mathrm{B}}$ has a nonpositive time derivative, $d\mathrm{H}/dt \leq 0$. The remainder of this section contains the proof of the H-theorem [325].

Let $\mathrm{H}^{\mathrm{el}} = -\mathrm{S}^{\mathrm{el}}/K_{\mathrm{B}}$ and $\mathrm{H}^{\mathrm{phon}} = -\mathrm{S}^{\mathrm{phon}}/K_{\mathrm{B}}$, so that $\mathrm{H} = \mathrm{H}^{\mathrm{el}} + \mathrm{H}^{\mathrm{phon}}$. We have

$$\frac{d\mathrm{H}^{\mathrm{el}}}{dt} = \sum_{\mathbf{k}\mu} \frac{df^{\mathrm{el}}_{\mathbf{k}\mu}}{dt} \ln \frac{f^{\mathrm{el}}_{\mathbf{k}\mu}}{1 - f^{\mathrm{el}}_{\mathbf{k}\mu}} = \sum_{\mathbf{k}\mu} (S^{\mathrm{GW}}_{\mathbf{k}\mu\mu} + S^{\mathrm{FM}}_{\mathbf{k}\mu\mu}) \ln \frac{f^{\mathrm{el}}_{\mathbf{k}\mu}}{1 - f^{\mathrm{el}}_{\mathbf{k}\mu}}. \tag{17.87}$$

The GW scattering term is given in (17.69) and can be rewritten as

$$S^{\mathrm{GW}}_{\mathbf{k}\mu\mu} = 2\pi \sum_{\mu'\nu\nu'} \sum_{\mathbf{k}'\mathbf{p}\mathbf{p}'} \big| W_{\mathbf{k}\mu\,\mathbf{k}'\nu'\,\mathbf{p}\nu\,\mathbf{p}'\mu'} \big|^2 \delta\big(\epsilon_{\mathbf{p}\nu} + \epsilon_{\mathbf{p}'\mu'} - \epsilon_{\mathbf{k}'\nu'} - \epsilon_{\mathbf{k}\mu}\big) \delta_{\mathbf{k}+\mathbf{k}',\mathbf{p}+\mathbf{p}'}$$
$$\times \Big[\bar{f}^{\mathrm{el}}_{\mathbf{k}'\nu'} \bar{f}^{\mathrm{el}}_{\mathbf{k}\mu} f^{\mathrm{el}}_{\mathbf{p}\nu} f^{\mathrm{el}}_{\mathbf{p}'\mu'} - f^{\mathrm{el}}_{\mathbf{k}'\nu'} f^{\mathrm{el}}_{\mathbf{k}\mu} \bar{f}^{\mathrm{el}}_{\mathbf{p}\nu} \bar{f}^{\mathrm{el}}_{\mathbf{p}'\mu'} \Big],$$

where we define $\bar{f}^{\mathrm{el}}_{\mathbf{k}\mu} = 1 - f^{\mathrm{el}}_{\mathbf{k}\mu} \geq 0$.[13] To recognize the mathematical structure of this equation we find it convenient to introduce the superindices $i = \mathbf{k}\mu$, $j = \mathbf{k}'\nu'$, $m = \mathbf{p}\nu$, and $n = \mathbf{p}'\mu'$. Then, the GW contribution to (17.87) reads

$$\left.\frac{d\mathrm{H}^{\mathrm{el}}}{dt}\right|_{\mathrm{GW}} = \sum_{ijmn} \Gamma_{ijmn} \Big[\bar{f}^{\mathrm{el}}_i \bar{f}^{\mathrm{el}}_j f^{\mathrm{el}}_m f^{\mathrm{el}}_n - f^{\mathrm{el}}_i f^{\mathrm{el}}_j \bar{f}^{\mathrm{el}}_m \bar{f}^{\mathrm{el}}_n \Big] \ln \frac{f^{\mathrm{el}}_i}{\bar{f}^{\mathrm{el}}_i}, \tag{17.88}$$

with

$$\Gamma_{ijmn} = 2\pi \big| W_{ijmn} \big|^2 \delta\big(\epsilon_i + \epsilon_j - \epsilon_m - \epsilon_n\big) \delta_{\mathbf{k}_i + \mathbf{k}_j, \mathbf{k}_m + \mathbf{k}_n} = \Gamma_{jinm} = \Gamma_{nmji} \geq 0,$$

and $\mathbf{k}_i$ is the momentum carried by the superindex i, $\mathbf{k}_j$ the momentum carried by the superindex j, etc. The term in the square brackets of (17.88) is symmetric under the exchange $i \leftrightarrow j$ and $m \leftrightarrow n$, whereas it is antisymmetric under the exchange $i \leftrightarrow n$ and $j \leftrightarrow m$. Therefore, we can rewrite (17.88) as

$$\left.\frac{d\mathrm{H}^{\mathrm{el}}}{dt}\right|_{\mathrm{GW}} = \sum_{ijmn} \frac{\Gamma_{ijmn}}{4} \Big[\bar{f}^{\mathrm{el}}_i \bar{f}^{\mathrm{el}}_j f^{\mathrm{el}}_m f^{\mathrm{el}}_n - f^{\mathrm{el}}_i f^{\mathrm{el}}_j \bar{f}^{\mathrm{el}}_m \bar{f}^{\mathrm{el}}_n \Big] \Big[\ln \frac{f^{\mathrm{el}}_i}{\bar{f}^{\mathrm{el}}_i} + \ln \frac{f^{\mathrm{el}}_j}{\bar{f}^{\mathrm{el}}_j} - \ln \frac{f^{\mathrm{el}}_m}{\bar{f}^{\mathrm{el}}_m} - \ln \frac{f^{\mathrm{el}}_n}{\bar{f}^{\mathrm{el}}_n} \Big]$$
$$= \sum_{ijmn} \frac{\Gamma_{ijmn}}{4} \bar{f}^{\mathrm{el}}_i \bar{f}^{\mathrm{el}}_j f^{\mathrm{el}}_m f^{\mathrm{el}}_n \Big[1 - \frac{f^{\mathrm{el}}_i f^{\mathrm{el}}_j \bar{f}^{\mathrm{el}}_m \bar{f}^{\mathrm{el}}_n}{\bar{f}^{\mathrm{el}}_i \bar{f}^{\mathrm{el}}_j f^{\mathrm{el}}_m f^{\mathrm{el}}_n} \Big] \ln \frac{f^{\mathrm{el}}_i f^{\mathrm{el}}_j \bar{f}^{\mathrm{el}}_m \bar{f}^{\mathrm{el}}_n}{\bar{f}^{\mathrm{el}}_i \bar{f}^{\mathrm{el}}_j f^{\mathrm{el}}_m f^{\mathrm{el}}_n}.$$

[13] To recover (17.69) we perform the sum over $\mathbf{k}'$, which fixes $\mathbf{k}' = \mathbf{p} + \mathbf{p}' - \mathbf{k}$, and then change variables according to: $\mathbf{p}' \to \mathbf{q}$ and $\mathbf{p} \to \mathbf{k} + \mathbf{p}$.

Taking into account that $\Gamma_{ijmn}\bar{f}_i^{\rm el}\bar{f}_j^{\rm el}f_m^{\rm el}f_n^{\rm el} \geq 0$ and that the function $(1-x)\ln x \leq 0$ for all $x > 0$, we conclude that

$$\left.\frac{d\mathrm{H}^{\rm el}}{dt}\right|_{\rm GW} \leq 0. \tag{17.89}$$

Let us now come to the Fan–Migdal contribution. The Fan–Migdal scattering term is given in (17.84) and can be rewritten as

$$\begin{aligned} S^{\rm FM}_{\mathbf{k}\mu\mu} &= -2\pi\sum_{\mathbf{q}\alpha\mathbf{k}'\nu}\frac{|g_{-\mathbf{q}\alpha,\mu\nu}(\mathbf{k})|^2}{2\omega_{\mathbf{q}\alpha}}\delta(\epsilon_{\mathbf{k}'\nu}-\epsilon_{\mathbf{k}\mu}+\omega_{\mathbf{q}\alpha})\delta_{\mathbf{q},\mathbf{k}-\mathbf{k}'} \\ &\times\left[\bar{f}^{\rm el}_{\mathbf{k}'\nu}f^{\rm el}_{\mathbf{k}\mu}\bar{f}^{\rm phon}_{\mathbf{q}\alpha}-f^{\rm el}_{\mathbf{k}'\nu}\bar{f}^{\rm el}_{\mathbf{k}\mu}f^{\rm phon}_{\mathbf{q}\alpha}\right] \\ &-2\pi\sum_{\mathbf{q}\alpha\mathbf{k}'\nu}\frac{|g_{\mathbf{q}\alpha,\mu\nu}(\mathbf{k})|^2}{2\omega_{\mathbf{q}\alpha}}\delta(\epsilon_{\mathbf{k}'\nu}-\epsilon_{\mathbf{k}\mu}-\omega_{\mathbf{q}\alpha})\delta_{\mathbf{q},\mathbf{k}'-\mathbf{k}} \\ &\times\left[\bar{f}^{\rm el}_{\mathbf{k}'\nu}f^{\rm el}_{\mathbf{k}\mu}f^{\rm phon}_{\mathbf{q}\alpha}-f^{\rm el}_{\mathbf{k}'\nu}\bar{f}^{\rm el}_{\mathbf{k}\mu}\bar{f}^{\rm phon}_{\mathbf{q}\alpha}\right], \end{aligned} \tag{17.90}$$

where we define $\bar{f}^{\rm phon}_{\mathbf{q}\alpha} = f^{\rm phon}_{\mathbf{q}\alpha} + 1 \geq 0$. Let us introduce the quantity

$$\Gamma_{\mathbf{k}\mu\,\mathbf{k}'\nu\,\mathbf{q}\alpha} \equiv 2\pi\frac{|g_{-\mathbf{q}\alpha,\mu\nu}(\mathbf{k})|^2}{2\omega_{\mathbf{q}\alpha}}\delta(\epsilon_{\mathbf{k}'\nu}-\epsilon_{\mathbf{k}\mu}+\omega_{\mathbf{q}\alpha})\delta_{\mathbf{q},\mathbf{k}-\mathbf{k}'} \geq 0.$$

Using the property (17.27) we can easily prove that the third line of (17.90) is $-\Gamma_{\mathbf{k}'\nu\,\mathbf{k}\mu\,\mathbf{q}\alpha}$, and therefore the Fan–Migdal scattering term takes the following compact form

$$\begin{aligned} S^{\rm FM}_{\mathbf{k}\mu\mu} &= -\sum_{\mathbf{q}\alpha\mathbf{k}'\nu}\Gamma_{\mathbf{k}\mu\,\mathbf{k}'\nu\,\mathbf{q}\alpha}\left[\bar{f}^{\rm el}_{\mathbf{k}'\nu}f^{\rm el}_{\mathbf{k}\mu}\bar{f}^{\rm phon}_{\mathbf{q}\alpha}-f^{\rm el}_{\mathbf{k}'\nu}\bar{f}^{\rm el}_{\mathbf{k}\mu}f^{\rm phon}_{\mathbf{q}\alpha}\right] \\ &-\sum_{\mathbf{q}\alpha\mathbf{k}'\nu}\Gamma_{\mathbf{k}'\nu\,\mathbf{k}\mu\,\mathbf{q}\alpha}\left[\bar{f}^{\rm el}_{\mathbf{k}'\nu}f^{\rm el}_{\mathbf{k}\mu}f^{\rm phon}_{\mathbf{q}\alpha}-f^{\rm el}_{\mathbf{k}'\nu}\bar{f}^{\rm el}_{\mathbf{k}\mu}\bar{f}^{\rm phon}_{\mathbf{q}\alpha}\right]. \end{aligned}$$

Again, we highlight the mathematical structure by introducing the superindices $i = \mathbf{k}\mu$, $j = \mathbf{k}'\nu'$, and $q = \mathbf{q}\alpha$. Let us define

$$R_{ijq} \equiv \Gamma_{ijq}\left[f_i^{\rm el}\bar{f}_j^{\rm el}\bar{f}_q^{\rm phon} - \bar{f}_i^{\rm el}f_j^{\rm el}f_q^{\rm phon}\right] = -\Gamma_{ijq}\bar{f}_i^{\rm el}f_j^{\rm el}f_q^{\rm phon}\left[1-\frac{f_i^{\rm el}\bar{f}_j^{\rm el}\bar{f}_q^{\rm phon}}{\bar{f}_i^{\rm el}f_j^{\rm el}f_q^{\rm phon}}\right]. \tag{17.91}$$

Then $S^{\rm FM}_{\mathbf{k}\mu\mu} = -\sum_{jq}(R_{ijq}-R_{jiq})$ and the Fan–Migdal contribution to (17.87) reads

$$\begin{aligned} \left.\frac{d\mathrm{H}^{\rm el}}{dt}\right|_{\rm FM} &= -\sum_{ijq}(R_{ijq}-R_{jiq})\ln\frac{f_i^{\rm el}}{\bar{f}_i^{\rm el}} = -\sum_{ijq}R_{ijq}\left[\ln\frac{f_i^{\rm el}}{\bar{f}_i^{\rm el}}-\ln\frac{f_j^{\rm el}}{\bar{f}_j^{\rm el}}\right] \\ &= -\sum_{ijq}R_{ijq}\left[\ln\frac{f_i^{\rm el}\bar{f}_j^{\rm el}\bar{f}_q^{\rm phon}}{\bar{f}_i^{\rm el}f_j^{\rm el}f_q^{\rm phon}}-\ln\frac{\bar{f}_q^{\rm phon}}{f_q^{\rm phon}}\right]. \end{aligned} \tag{17.92}$$

Next we calculate $d\mathrm{H}^{\mathrm{phon}}/dt$. Using (17.86b) we find

$$\frac{d\mathrm{H}^{\mathrm{phon}}}{dt} = \sum_{\mathbf{q}\alpha} \frac{df^{\mathrm{phon}}_{\mathbf{q}\alpha}}{dt} \ln \frac{f^{\mathrm{phon}}_{\mathbf{q}\alpha}}{\bar{f}^{\mathrm{phon}}_{\mathbf{q}\alpha}} = \sum_{\mathbf{q}\alpha} S^{\mathrm{phon-occ}}_{\mathbf{q}\alpha\alpha} \ln \frac{f^{\mathrm{phon}}_{\mathbf{q}\alpha}}{\bar{f}^{\mathrm{phon}}_{\mathbf{q}\alpha}}. \tag{17.93}$$

The phononic scattering term is given in (17.85) and can be rewritten as

$$\begin{aligned} S^{\mathrm{phon-occ}}_{\mathbf{q}\alpha\alpha}(t) &= -2\pi \sum_{\mathbf{k}\mu\mathbf{k}'\nu} \frac{|g_{\mathbf{q}\alpha,\mu\nu}(\mathbf{k})|^2}{2\omega_{\mathbf{q}\alpha}} \delta\big(\epsilon_{\mathbf{k}'\nu} - \epsilon_{\mathbf{k}\mu} - \omega_{\mathbf{q}\alpha}\big)\delta_{\mathbf{q},\mathbf{k}'-\mathbf{k}} \\ &\quad \times \left[\bar{f}^{\mathrm{el}}_{\mathbf{k}'\nu} f^{\mathrm{el}}_{\mathbf{k}\mu} f^{\mathrm{phon}}_{\mathbf{q}\alpha} - f^{\mathrm{el}}_{\mathbf{k}'\nu} \bar{f}^{\mathrm{el}}_{\mathbf{k}\mu} \bar{f}^{\mathrm{phon}}_{\mathbf{q}\alpha}\right] \\ &= -\sum_{\mathbf{k}\mu\mathbf{k}'\nu} \Gamma_{\mathbf{k}'\nu\,\mathbf{k}\mu\,\mathbf{q}\alpha} \left[\bar{f}^{\mathrm{el}}_{\mathbf{k}'\nu} f^{\mathrm{el}}_{\mathbf{k}\mu} f^{\mathrm{phon}}_{\mathbf{q}\alpha} - f^{\mathrm{el}}_{\mathbf{k}'\nu} \bar{f}^{\mathrm{el}}_{\mathbf{k}\mu} \bar{f}^{\mathrm{phon}}_{\mathbf{q}\alpha}\right] = \sum_{ij} R_{jiq}, \end{aligned}$$

where in the second equality we recognize the same term as in the third line of (17.90), and in the last equality we use the definition (17.91). Inserting this result into (17.93) and using superindices we obtain the following compact expression:

$$\frac{d\mathrm{H}^{\mathrm{phon}}}{dt} = \sum_{ijq} R_{ijq} \ln \frac{f^{\mathrm{phon}}_q}{\bar{f}^{\mathrm{phon}}_q}, \tag{17.94}$$

which is the negative of the last term in (17.92). We conclude that

$$\begin{aligned} \left.\frac{d\mathrm{H}^{\mathrm{el}}}{dt}\right|_{\mathrm{FM}} + \frac{d\mathrm{H}^{\mathrm{phon}}}{dt} &= -\sum_{ijq} R_{ijq} \ln \frac{f^{\mathrm{el}}_i \bar{f}^{\mathrm{el}}_j \bar{f}^{\mathrm{phon}}_q}{\bar{f}^{\mathrm{el}}_i f^{\mathrm{el}}_j f^{\mathrm{phon}}_q} \\ &= \sum_{ijq} \Gamma_{ijq} \bar{f}^{\mathrm{el}}_i f^{\mathrm{el}}_j f^{\mathrm{phon}}_q \left[1 - \frac{f^{\mathrm{el}}_i \bar{f}^{\mathrm{el}}_j \bar{f}^{\mathrm{phon}}_q}{\bar{f}^{\mathrm{el}}_i f^{\mathrm{el}}_j f^{\mathrm{phon}}_q}\right] \ln \frac{f^{\mathrm{el}}_i \bar{f}^{\mathrm{el}}_j \bar{f}^{\mathrm{phon}}_q}{\bar{f}^{\mathrm{el}}_i f^{\mathrm{el}}_j f^{\mathrm{phon}}_q}. \end{aligned} \tag{17.95}$$

Taking into account that $\Gamma_{ijq}\bar{f}^{\mathrm{el}}_i f^{\mathrm{el}}_j f^{\mathrm{phon}}_q \geq 0$ and that the function $(1-x)\ln x \leq 0$ for all $x > 0$, we infer that the r.h.s. of (17.95) is nonpositive. This result together with (17.89) concludes the proof of the H-theorem.

Exercise 17.7 Write the Boltzmann equations for the Hubbard model, see Section 2.5, and for the Holstein model, see Section 2.8.

17.6 Time-Linear Scheme for the GKBA Equations

In this section we go back to the equations of motion for the density matrices, see Section 17.1, and reexamine the (M)GKBA *without* making any additional simplifications. For pedagogical reasons we consider purely fermionic systems and defer the interested reader to Refs. [166, 357, 358] for the inclusion of phosonic degrees of freedom. We also confine the discussion to the GKBA, leaving the MGKBA formulation as an exercise for the reader.

Our aim here is to show that the timescaling of the GKBA equations can be lowered from quadratic to linear by introducing an extra equation of motion for the equal-time two-particle Green's function. This idea was put forward in Refs. [359, 360]. The derivation below closely follows Ref. [361].

The equation of motion (17.3) for the electronic density matrix in a general one-particle basis reads

$$\frac{d}{dt}n^{<}_{il}(t) + \mathrm{i}\sum_{j}\Big(h_{\mathrm{qp},ij}(t)n^{<}_{jl}(t) - n^{<}_{ij}(t)h_{\mathrm{qp},jl}(t)\Big) = S^{\mathrm{el}}_{il}(t), \tag{17.96}$$

with electronic scattering term [compare with (17.40)]

$$\begin{aligned} S^{\mathrm{el}}_{il}(t) &= -\int^{t} dt'\Big[\Sigma^{>}(t,t')G^{<}(t',t) - \Sigma^{<}(t,t')G^{>}(t',t)\Big]_{il} + (i \leftrightarrow l)^{*} \\ &= \mathrm{i}\sum_{rpn} v_{irpn}\,\mathcal{G}_{\substack{pl\\rn}}(t) + (i \leftrightarrow l)^{*}. \end{aligned} \tag{17.97}$$

In the last equality we recognize the correlated part of the equal-time two-particle Green's function, see (9.5),

$$\mathcal{G}_{\substack{pl\\rn}}(t) \equiv -\langle \hat{d}^{\dagger}_{l,H}(t)\hat{d}^{\dagger}_{r,H}(t)\hat{d}_{p,H}(t)\hat{d}_{n,H}(t)\rangle_{\mathrm{c}}, \tag{17.98}$$

where the subscript "c" signifies that the HF part has been subtracted.[14]

17.6.1 Second Born Approximation

Let us begin by considering the second Born approximation. From (7.29) we have

$$\begin{aligned} \Big[\Sigma^{>}(t,t')G^{<}(t',t)\Big]_{il} = \sum_{j}\Sigma^{>}_{ij}(t,t')G^{<}_{jl}(t',t) = \mathrm{i}^{2}\sum_{rpn}v_{irpn}\sum_{jmqs}(v_{qmsj} - v_{qmjs}) \\ \times\, G^{>}_{nm}(t,t')G^{<}_{sr}(t',t)G^{>}_{pq}(t,t')G^{<}_{jl}(t',t). \end{aligned} \tag{17.99}$$

The matrix elements of the product $\Sigma^{<}(t,t')G^{>}(t',t)$ are obtained from (17.99) by interchanging $> \leftrightarrow <$. The mathematical structure underlying these expressions emerges clearly if we introduce the Coulomb tensor [compare with (17.57)]

$$w_{\substack{qj\\sm}} \equiv v_{qmjs} - v_{qmsj} = w^{*}_{\substack{jq\\ms}}, \tag{17.100}$$

and the response function

$$\chi^{0,\gtrless}_{\substack{pq\\rs}}(t,t') \equiv -\mathrm{i}G^{\gtrless}_{pq}(t,t')G^{\lessgtr}_{sr}(t',t). \tag{17.101}$$

Expressing S^{el} in terms of these quantities, we infer that

$$\mathcal{G}_{\substack{pl\\rn}}(t) = -\mathrm{i}\int^{t} dt' \sum_{qsjm}\Big[\chi^{0,>}_{\substack{pq\\rs}}(t,t')w_{\substack{qj\\sm}}\chi^{0,<}_{\substack{jl\\mn}}(t',t) - \chi^{0,<}_{\substack{pq\\rs}}(t,t')w_{\substack{qj\\sm}}\chi^{0,>}_{\substack{jl\\mn}}(t',t)\Big]. \tag{17.102}$$

[14] The HF part of the two-particle Green's function corresponds to the first two diagrams in the r.h.s. of (14.13).

The square bracket in (17.102) is the sum of simple products between matrices in the two-particle space. Using capital letters for superindices composed by pairs of one-particle indices, for example, $A = (p, r)$, $B = (l, n)$, etc., the matrix elements of $\mathcal{G}$ can also be written as

$$\mathcal{G}_{AB}(t) = -\mathrm{i} \int^t dt' \sum_{A'B'} \left[\chi^{0,>}_{AA'}(t,t') w_{A'B'} \chi^{0,<}_{B'B}(t',t) - \chi^{0,<}_{AA'}(t,t') w_{A'B'} \chi^{0,>}_{B'B}(t',t) \right]. \tag{17.103}$$

The GKBA form of $\mathcal{G}$ Up to this point we have not yet used the GKBA to transform the scattering term, or equivalently $\mathcal{G}$, into a functional of the density matrix. Evaluating the greater response function for $t > t'$ using the GKBA in (17.7), we get

$$\chi^{0,>}_{\substack{pq\\rs}}(t,t') = \mathrm{i} \underbrace{\sum_a G^{\mathrm{R}}_{pa}(t,t') n^>_{aq}(t')}_{-G^>_{pq}(t,t')} \underbrace{\sum_b n^<_{sb}(t') G^{\mathrm{A}}_{br}(t',t)}_{G^<_{sr}(t',t)} = \sum_{ab} P^{\mathrm{R}}_{\substack{pa\\rb}}(t,t') n^{(2)>}_{\substack{aq\\bs}}(t'),$$

where we define

$$P^{0,\mathrm{R}}_{\substack{pa\\rb}}(t,t') \equiv \mathrm{i}\, G^{\mathrm{R}}_{pa}(t,t') G^{\mathrm{A}}_{br}(t',t), \qquad n^{(2)>}_{\substack{aq\\bs}}(t') \equiv n^>_{aq}(t') n^<_{sb}(t').$$

The two-time function $P^{0,\mathrm{R}}$ can be interpreted as the propagator of an electron–hole pair. Notice that $G^{\mathrm{R}}(t^+, t) = -\mathrm{i}$ and $G^{\mathrm{A}}(t, t^+) = \mathrm{i}$, and hence

$$P^{0,\mathrm{R}}_{\substack{pa\\rb}}(t^+,t) = \mathrm{i}\delta_{pa}\delta_{rb} \qquad \text{or equivalently} \qquad P^{0,\mathrm{R}}_{AB}(t^+,t) = \mathrm{i}\delta_{AB}. \tag{17.104}$$

For $t < t'$ the GKBA implies

$$\chi^{0,>}_{\substack{pq\\rs}}(t,t') = \mathrm{i} \underbrace{\sum_a n^>_{pa}(t) G^{\mathrm{A}}_{aq}(t,t')}_{G^>_{pq}(t,t')} \underbrace{\sum_b G^{\mathrm{R}}_{sb}(t',t) n^<_{br}(t)}_{-G^<_{sr}(t',t)} = -\sum_{ab} n^{(2)>}_{\substack{pa\\rb}}(t) P^{0,\mathrm{A}}_{\substack{aq\\bs}}(t,t'),$$

where

$$P^{0,\mathrm{A}}_{\substack{aq\\bs}}(t,t') \equiv -\mathrm{i}\, G^{\mathrm{A}}_{aq}(t,t') G^{\mathrm{R}}_{sb}(t',t) = \left[P^{0,\mathrm{R}}_{\substack{qa\\sb}}(t',t) \right]^*. \tag{17.105}$$

Using the superindex notation, the greater response function in GKBA for any t and t' then reads:

$$\chi^{0,>}_{AB}(t,t') = \sum_C \left[P^{0,\mathrm{R}}_{AC}(t,t') n^{(2)>}_{CB}(t') - n^{(2)>}_{AC}(t) P^{0,\mathrm{A}}_{CB}(t,t') \right]. \tag{17.106}$$

An analogous derivation can be performed for the lesser response function. The final result is identical to (17.106), but the matrix $n^{(2)>}$ is replaced by the matrix

$$n^{(2)<}_{\substack{aq\\bs}}(t) \equiv n^<_{aq}(t) n^>_{sb}(t).$$

Hence,

$$\chi^{0,\lessgtr}_{AB}(t,t') = \sum_C \left[P^{0,\mathrm{R}}_{AC}(t,t')n^{(2)\lessgtr}_{CB}(t') - n^{(2)\lessgtr}_{AC}(t)P^{0,\mathrm{A}}_{CB}(t,t')\right]. \tag{17.107}$$

We are now ready to transform $\mathcal{G}$ into an explicit functional of the one-particle density matrix. Taking into account that in (17.103) the time $t' < t$ we obtain (in matrix form)

$$\mathcal{G}(t) = \mathrm{i}\int^t dt'\, P^{0,\mathrm{R}}(t,t')\left[n^{(2)>}(t')w\, n^{(2)<}(t') - n^{(2)<}(t')w\, n^{(2)>}(t')\right]P^{0,\mathrm{A}}(t',t). \tag{17.108}$$

The quadratic timescaling of the GKBA equations of motion is evident from (17.108): A time step from t to $t+\Delta_t$ necessitates the calculation of $\mathcal{G}(t)$, and $\mathcal{G}(t)$ contains an integral with an upper limit that grows like t. It is worth observing that the evaluation of $\mathcal{G}(t)$ necessitates a knowledge of the density matrix for all times $t' < t$ - that is, the GKBA equations are history-dependent. The same approximation in MGKBA leads to a $\mathcal{G}(t)$ depending only on the density matrix at time t, see (17.43).

Time-linear scaling The key insight to transition from quadratic to linear timescaling is that the particle–hole propagator $P^{0,\mathrm{R}}$ satisfies an elementary equation of motion. From (17.8) we have for any $t > t'$:

$$\mathrm{i}\frac{d}{dt}P^{0,\mathrm{R}}_{\substack{pa\\rb}}(t,t') = \sum_c h_{\mathrm{qp},pc}(t)P^{0,\mathrm{R}}_{\substack{ca\\rb}}(t,t') - \sum_d h_{\mathrm{qp},dr}(t)P^{0,\mathrm{R}}_{\substack{pa\\db}}(t,t').$$

Introducing the quasi-particle Hamiltonian in the two-particle space,

$$h^{(2)}_{\mathrm{qp},\substack{pc\\rd}}(t) = h_{\mathrm{qp},pc}(t)\delta_{rd} - \delta_{pc}h_{\mathrm{qp},dr}(t),$$

we can rewrite the equation of motion for $P^{0,\mathrm{R}}$ in matrix form as follows:

$$\mathrm{i}\frac{d}{dt}P^{0,\mathrm{R}}(t,t') = h^{(2)}_{\mathrm{qp}}(t)P^{0,\mathrm{R}}(t,t'), \qquad t > t'. \tag{17.109}$$

Let us now come to $P^{0,\mathrm{A}}$. In matrix form (17.105) reads $P^{0,\mathrm{A}}(t',t) = [P^{0,\mathrm{R}}(t,t')]^\dagger$. Taking the adjoint of (17.109) we then get

$$-\mathrm{i}\frac{d}{dt}P^{0,\mathrm{A}}(t',t) = P^{0,\mathrm{A}}(t',t)h^{(2)}_{\mathrm{qp}}(t), \qquad t > t', \tag{17.110}$$

where we use that the matrix $h^{(2)}_{\mathrm{qp}}$ is self-adjoint in the two-particle space: $h^{(2)}_{\mathrm{qp},AB} = h^{(2)*}_{\mathrm{qp},BA}$. The equations of motion for $P^{0,\mathrm{R}}$ and $P^{0,\mathrm{A}}$ can be used to construct an equation of motion for $\mathcal{G}$. Recalling the differentiation rule,

$$\frac{d}{dt}\int^t dt'\, f(t,t') = f(t^+,t) + \int^t dt'\, \frac{d}{dt}f(t,t'), \tag{17.111}$$

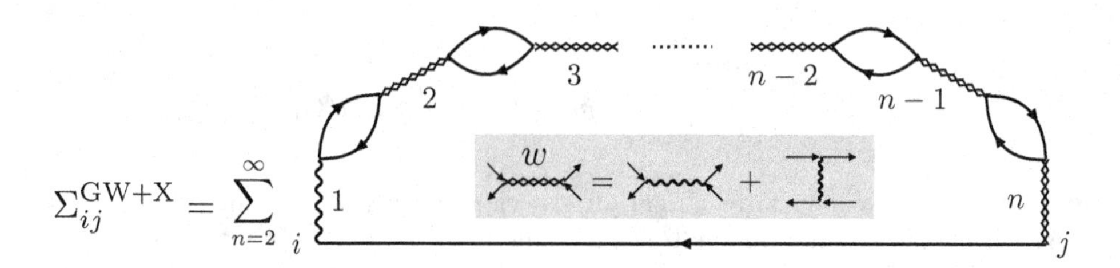

Figure 17.3 Diagrams for the GW+X self-energy.

where f is an arbitrary function of two times, we find from (17.108)

$$\boxed{\mathrm{i}\frac{d}{dt}\mathcal{G}(t) = -\Psi(t) + h^{(2)}_{\text{qp}}(t)\mathcal{G}(t) - \mathcal{G}(t)h^{(2)}_{\text{qp}}(t)} \tag{17.112}$$

with

$$\boxed{\Psi(t) \equiv n^{(2)>}(t)w\,n^{(2)<}(t) - n^{(2)<}(t)w\,n^{(2)>}(t)} \tag{17.113}$$

Equation (17.112) together with the equation of motion (17.96) for $n^<$ form a closed system of ordinary differential equations whose numerical solution scales linearly with the maximum propagation time. We also observe that replacing $v \to W$ in the two-particle tensor w, the same equation of motion corresponds to the approximation (17.26) plus second-order exchange with statically screened W lines.

17.6.2 GW plus Exchange

The time-linear scaling holds for the second Born approximation as well as for state-of-the-art diagrammatic methods like GW and T-matrix (both in the particle–hole and particle–particle channels) [360], see Section 12.6. In Ref. [361] the time-linear scaling formulation has been extended to GW plus exchange, T-matrix plus exchange, and self-energies with three-particle correlations. In Refs. [166, 357, 358] a time-linear scaling formulation has been established also for systems of interacting electrons and phosons. In this section we derive the GKBA equations of motion in the GW plus exchange (GW+X) approximation. As we see, the difference with the second Born approximation in (17.112) is only minor. Readers who wish to immediately compare second Born with GW+X or who prefer to go through the derivation in a second reading can jump directly to (17.126).

The GW+X self-energy is displayed in Fig. 17.3, where the double zig-zag lines represent the tensor w defined in (17.100), see also the diagrammatic representation of w in the gray box of Fig. 17.3. The GW self-energy is recovered by neglecting the exchange contribution in w - that is, by setting $w_{\substack{qj\\sm}} = v_{qmjs}$ (the first term on the right-hand side of the diagrammatic equation in the gray box). The figure shows the diagrams of order n in the interaction strength. The first term of the sum has $n = 2$ and it coincides with the correlated part of the second Born self-energy, see (7.16). Thus, the GW+X self-energy is obtained from the second Born self-energy by replacing the bare response function $\chi^0_{\substack{pq\\rs}}(z,z') \equiv -\mathrm{i}G_{pq}(z,z')G_{sr}(z',z)$ with the RPA+X response function (in matrix form)

$$\chi(z,z') = \chi^0(z,z') + \int_\gamma d\bar{z}\,\chi^0(z,\bar{z})\,w\,\chi(\bar{z},z'). \tag{17.114}$$

This implies that $\mathcal{G}$ in (17.102) changes into

$$\mathcal{G}(t) = -\mathrm{i}\int^t dt' \left[\chi^>(t,t')\,w\,\chi^{0,<}(t',t) - \chi^<(t,t')\,w\,\chi^{0,>}(t',t)\right]. \tag{17.115}$$

Up to this point we have not yet used the GKBA to transform $\mathcal{G}$ into a functional of the density matrix.

Before using the GKBA, we need to extract the lesser and greater components of the RPA+X response function. These components follow from (17.114) when setting $z = t_\mp$ and $z' = t_\pm$. Using the Langreth rules derived in Section 5.5 and taking into account that we are working on the adiabatic contour, we find

$$\chi^\lessgtr(t,t') = \chi^{0,\lessgtr}(t,t') + \left[\chi^{0,\mathrm{R}}\cdot w\cdot\chi^\lessgtr + \chi^{0,\lessgtr}\cdot w\cdot\chi^{\mathrm{A}}\right](t,t'). \tag{17.116}$$

Similarly, for the retarded component we have

$$\chi^{\mathrm{R}}(t,t') = \chi^{0,\mathrm{R}}(t,t') + \left[\chi^{0,\mathrm{R}}\cdot w\cdot\chi^{\mathrm{R}}\right](t,t'). \tag{17.117}$$

The advanced response function can be derived similarly; the result is the same as in (17.117) with $\mathrm{R}\to\mathrm{A}$. The equations for $\chi^{\mathrm{R/A}}$ are useful to isolate $\chi^\lessgtr$. We rewrite (17.116) as

$$(\delta - \chi^{0,\mathrm{R}}\cdot w)\cdot\chi^\lessgtr = \chi^{0,\lessgtr}\cdot(\delta + w\cdot\chi^{\mathrm{A}}), \tag{17.118}$$

where δ stands for the Dirac delta in time and the Kronecker delta in the two-particle space - that is, for any two-times correlator C we have $[\delta\cdot C](t,t') = \int d\bar{t}\,\delta(t-\bar{t})C(\bar{t},t') = C(t,t')$. Next we observe that

$$(\delta + \chi^{\mathrm{R}}\cdot w)\cdot(\delta - \chi^{0,\mathrm{R}}\cdot w) = \delta + (\chi^{\mathrm{R}} - \chi^{0,\mathrm{R}} - \chi^{\mathrm{R}}\cdot w\cdot\chi^{0,\mathrm{R}})\cdot w = \delta, \tag{17.119}$$

since the term in parentheses vanishes according to (17.117). Convoluting (17.118) with $(\delta + \chi^{\mathrm{R}}\cdot w)$ on the right, we then find

$$\chi^\lessgtr = (\delta + \chi^{\mathrm{R}}\cdot w)\cdot\chi^{0,\lessgtr}\cdot(\delta + w\cdot\chi^{\mathrm{A}}). \tag{17.120}$$

The GKBA form of $\mathcal{G}$ At this point we have all ingredients to evaluate $\mathcal{G}$ using the GKBA. From (17.106) and (17.107) we first obtain the GKBA form of the retarded/advanced response functions

$$\chi^{0,\mathrm{R}}(t,t') = P^{0,\mathrm{R}}(t,t')n^{(2)\Delta}(t'), \qquad \chi^{0,\mathrm{A}}(t,t') = n^{(2)\Delta}(t)P^{0,\mathrm{A}}(t,t'), \tag{17.121}$$

where we define

$$n^{(2)\Delta}(t) \equiv n^{(2)>}(t) - n^{(2)<}(t).$$

Inserting (17.121) into (17.117) and in the analogous equation for χ^{A}, we obtain the GKBA expression for the retarded/advanced RPA+X response functions

$$\chi^{\mathrm{R}}(t,t') = P^{\mathrm{R}}(t,t')n^{(2)\Delta}(t'), \qquad \chi^{\mathrm{A}}(t,t') = n^{(2)\Delta}(t)P^{\mathrm{A}}(t,t'), \tag{17.122}$$

where the dressed electron–hole propagators satisfy the integral equation:

$$P^{\mathrm{R}} = P^{0,\mathrm{R}} + P^{0,\mathrm{R}}n^{(2)\Delta}\cdot w\cdot P^{\mathrm{R}} = P^{0,\mathrm{R}} + P^{\mathrm{R}}n^{(2)\Delta}\cdot w\cdot P^{0,\mathrm{R}}, \tag{17.123a}$$

$$P^{\mathrm{A}} = P^{0,\mathrm{A}} + P^{0,\mathrm{A}}\cdot w\cdot n^{(2)\Delta}P^{\mathrm{A}} = P^{0,\mathrm{A}} + P^{\mathrm{A}}\cdot w\cdot n^{(2)\Delta}P^{0,\mathrm{A}}. \tag{17.123b}$$

We observe that w is time-local and therefore the convolution dot to its left and/or right can be removed, see Exercise 5.2. Using the GKBA expression for $\chi^{0,\lessgtr}$ in (17.106) and (17.107) as well as the GKBA expression for $\chi^{\mathrm{R/A}}$ in (17.122), the lesser/greater RPA+X response function in (17.120) becomes

$$\begin{aligned}\chi^{\lessgtr} &= (\delta + P^{\mathrm{R}}n^{(2)\Delta}w)\cdot(P^{0,\mathrm{R}}n^{(2)\lessgtr} - n^{(2)\lessgtr}P^{0,\mathrm{A}})\cdot(\delta + wn^{(2)\Delta}P^{\mathrm{A}})\\ &= P^{\mathrm{R}}n^{(2)\lessgtr}\cdot(\delta + wn^{(2)\Delta}P^{\mathrm{A}}) - (\delta + P^{\mathrm{R}}n^{(2)\Delta}w)\cdot n^{(2)\lessgtr}P^{\mathrm{A}}.\end{aligned}$$

This result allows for rewriting $\mathcal{G}$ in (17.115) in a very elegant form. Taking into account that for any $\bar{t} < t$ the results in (17.106) and (17.107) yield $\chi^{0,\lessgtr}(\bar{t},t) = -[n^{(2)\lessgtr}P^{0,\mathrm{A}}](\bar{t},t)$, we find

$$\begin{aligned}\mathcal{G}(t) &= \mathrm{i}\Big[\underbrace{\Big(P^{\mathrm{R}}n^{(2)>}\cdot(\delta + wn^{(2)\Delta}P^{\mathrm{A}}) - (\delta + P^{\mathrm{R}}n^{(2)\Delta}w)\cdot n^{(2)>}P^{\mathrm{A}}\Big)}_{\chi^{>}}\cdot w\cdot\underbrace{\Big(n^{(2)<}P^{0,\mathrm{A}}\Big)}_{-\chi^{0,<}}\Big](t,t)\\ &\quad -\Big[>\leftrightarrow<\Big]\\ &= \mathrm{i}\Big[P^{\mathrm{R}}\cdot\Big(n^{(2)>}w\,n^{(2)<} + \underbrace{(n^{(2)>}\,w\,n^{(2)\Delta} - n^{(2)\Delta}wn^{(2)>})}_{-\Psi}P^{\mathrm{A}}\cdot wn^{(2)<}\Big)\cdot P^{0,\mathrm{A}}\Big](t,t)\\ &\quad -\mathrm{i}\Big[P^{\mathrm{R}}\cdot\Big(n^{(2)<}w\,n^{(2)>} + \underbrace{(n^{(2)<}\,w\,n^{(2)\Delta} - n^{(2)\Delta}wn^{(2)<})}_{-\Psi}P^{\mathrm{A}}\cdot wn^{(2)>}\Big)\cdot P^{0,\mathrm{A}}\Big](t,t)\\ &= \mathrm{i}\Big[P^{\mathrm{R}}\cdot\Psi\cdot P^{0,\mathrm{A}} + P^{\mathrm{R}}\cdot\Psi\cdot P^{\mathrm{A}}\cdot wn^{(2)\Delta}P^{0,\mathrm{A}}\Big](t,t)\\ &= \mathrm{i}\Big[P^{\mathrm{R}}\cdot\Psi\cdot P^{\mathrm{A}}\Big](t,t),\end{aligned} \tag{17.124}$$

where in the second equality we observe that $[n^{(2)\gtrless}P^{\mathrm{A}}\cdot wn^{(2)\lessgtr}P^{0,\mathrm{A}}](t,t) = 0$ since P^{A} contains a $\theta(\bar{t}-t)$ and $P^{0,\mathrm{A}}$ contains a $\theta(t-\bar{t})$. We also recognize the quantity Ψ defined in (17.113).

Time-linear scaling It is now easy to prove also that the GW+X method scales linearly in time. Taking into account the equation of motion (17.109) for $P^{0,\mathrm{R}}$ and the differentiation rule in (17.111), we find from (17.123a):

$$\begin{aligned}\mathrm{i}\frac{d}{dt}P^{\mathrm{R}}(t,t') &= h^{(2)}_{\mathrm{qp}}(t)P^{0,\mathrm{R}}(t,t') + \mathrm{i}\frac{d}{dt}\int^{t} d\bar{t}\,P^{0,\mathrm{R}}(t,\bar{t})n^{(2)\Delta}(\bar{t})wP^{\mathrm{R}}(\bar{t},t')\\ &= \Big(h^{(2)}_{\mathrm{qp}}(t) - n^{(2)\Delta}(t)w\Big)P^{\mathrm{R}}(t,t').\end{aligned} \tag{17.125}$$

Equation (17.123a) also implies that $P^{\rm R}(t^+,t) = {\rm i}$ and that $P^{\rm A}(t',t) = [P^{\rm R}(t,t')]^\dagger$. Thus, using again the differentiation rule in (17.111),

$$\boxed{{\rm i}\frac{d}{dt}\mathcal{G}(t) = -\Psi(t) + h^{(2)}_{\rm eff}(t)\mathcal{G}(t) - \mathcal{G}(t)h^{(2)\dagger}_{\rm eff}(t)} \tag{17.126}$$

where

$$\boxed{h^{(2)}_{\rm eff}(t) \equiv h^{(2)}_{\rm qp}(t) - n^{(2)\Delta}(t)w} \tag{17.127}$$

Comparing this result with the second Born equation of motion (17.112), we conclude that the only change brought about by the GW+X approximation is the replacement of $h^{(2)}_{\rm qp}$ with the effective Hamiltonian $h^{(2)}_{\rm eff}$. The effective Hamiltonian is not self-adjoint and therefore the last two terms in (17.126) cannot be grouped to form a commutator.

17.6.3 Open Systems and Quantum Transport

The theory of quantum transport began with the pioneering works of Landauer and Büttiker [137, 138, 139] and became a mature field after the works of Meir, Wingreen, and Jauho [130, 131], who provided a general formula for the time-dependent current through correlated junctions in terms of NEGF, see Section 9.5. Nonetheless, the ability to harness the full power of the Meir–Wingreen formula is hampered by the underlying two-time structure of the NEGF, a feature that makes real-time simulations computationally challenging. In this section we rewrite the Meir–Wingreen formula in terms of an *embedding correlator* and show how the use of the GKBA allows for evaluating the time-dependent current at a *time-linear* cost. Such development has been put forward in Ref. [362]; we here closely follow the original derivation.

The starting point is the equation of motion of the electronic density matrix (17.96). If the system is open, we have to add the embedding self-energy $\Sigma_{\rm em}$ to the many-body self-energy Σ, see (9.53). Hence, for open systems

$$\frac{d}{dt}n^<_{il}(t) + {\rm i}\sum_j \Big(h_{{\rm qp},ij}(t)n^<_{jl}(t) - n^<_{ij}(t)h_{{\rm qp},jl}(t)\Big) = S^{\rm el}_{il}(t) + S^{\rm em}_{il}(t), \tag{17.128}$$

where the embedding scattering term is defined as

$$\begin{aligned} S^{\rm em}_{il}(t) &= -\int^t dt' \Big[\Sigma^>_{\rm em}(t,t')G^<(t',t) - \Sigma^<_{\rm em}(t,t')G^>(t',t)\Big]_{il} + (i \leftrightarrow l)^* \\ &= -\int^\infty dt' \Big[\Sigma^{\rm R}_{\rm em}(t,t')G^<(t',t) + \Sigma^<_{\rm em}(t,t')G^{\rm A}(t',t)\Big]_{il} + (i \leftrightarrow l)^*. \end{aligned} \tag{17.129}$$

Let us write $\Sigma_{\rm em} = \sum_\alpha \Sigma_\alpha$, where the sum runs over all electrodes. As we are using the adiabatic contour, we introduce a switch-on function $s_\alpha(t)$ for the contact between the system and electrode α: $T_{m\,k\alpha} \to s_\alpha(t)T_{m\,k\alpha}$, see discussion below (9.48). The functions

$s_\alpha(t)$ are zero for $t < 0$ and slowly attain the value 1 in a finite time window. In Appendix K we show that the Keldysh components of Σ_α in WBLA are

$$\Sigma^{\rm R}_{\alpha,mn}(t,t') = -\frac{\rm i}{2}s^2_\alpha(t)\Gamma_{\alpha,mn}\delta(t-t'), \tag{17.130a}$$

$$\Sigma^<_{\alpha,mn}(t,t') = {\rm i}\Gamma_{\alpha,mn}s_\alpha(t)s_\alpha(t')e^{-{\rm i}\phi_\alpha(t,t')}\int\frac{d\omega}{2\pi}f(\omega-\mu)e^{-{\rm i}\omega(t-t')}, \tag{17.130b}$$

where $\phi_\alpha(t,t') \equiv \int_{t'}^t d\bar{t}\,V_\alpha(\bar{t})$ is a time-dependent phase generated by the time-dependent voltage V_α in electrode α [363]. Accordingly, the embedding scattering term in (17.129) simplifies to (in matrix form)

$$S^{\rm em}(t) = -\int^t dt'\Sigma^<_{\rm em}(t,t')G^{\rm A}(t',t) - \frac{\Gamma(t)}{2}n^<(t) + {\rm h.c.}, \tag{17.131}$$

where $\Gamma(t) = \sum_\alpha s^2_\alpha(t)\Gamma_\alpha$.

Before proceeding further, a discussion on the retarded/advanced Green's functions appearing in the GKBA is needed. In noninteracting but open systems the only self-energy is the embedding one. From (17.130a) we see that

$$\Sigma^{\rm R}_{\rm em}(t,t') = -\frac{\rm i}{2}\Gamma(t)\delta(t-t'),$$

which is local in time. Therefore, the noninteracting retarded Green's function of an open system is given by (6.47), with $h \to h - \frac{\rm i}{2}\Gamma$. As the GKBA Green's function must reduce to the noninteracting one in the absence of interactions, we use (17.8) with $h_{\rm qp} \to \tilde{h}_{\rm qp} = h_{\rm qp} - \frac{\rm i}{2}\Gamma$:

$$G^{\rm R}(t,t') = -{\rm i}\,\theta(t-t')T\left\{e^{-{\rm i}\int_{t'}^t d\bar{t}\,\tilde{h}_{\rm qp}(\bar{t})}\right\}, \qquad G^{\rm A}(t,t') = \left[G^{\rm R}(t',t)\right]^\dagger. \tag{17.132}$$

Notice that $\tilde{h}_{\rm qp}$ is not self-adjoint. In Ref. [303] it was shown that at the Hartree–Fock level the equation of motion (17.128) is exactly reproduced in GKBA, provided that the retarded/advanced Green's functions are chosen as in (17.132).

To construct the time-linear scheme we use an efficient pole expansion of the Fermi function $f(\omega) = \frac{1}{2} - \sum_l \eta_l\left(\frac{1}{\beta\omega+{\rm i}\zeta_l} + \frac{1}{\beta\omega-{\rm i}\zeta_l}\right)$ with residues η_l and poles ${\rm i}\zeta_l$, ${\rm Re}[\zeta_l] > 0$, obtained from the solution of the eigenvalue problem of a tridiagonal matrix [364]. We rewrite the lesser self-energy in (17.130b) for $t > t'$ as

$$\Sigma^<_\alpha(t,t') = \frac{\rm i}{2}\delta(t-t')s^2_\alpha(t)\Gamma_\alpha - s_\alpha(t)\sum_l\frac{\eta_l}{\beta}F_{l\alpha}(t,t')\Gamma_\alpha,$$

with

$$F_{l\alpha}(t,t') = s_\alpha(t')e^{-{\rm i}\phi_\alpha(t,t')}e^{-{\rm i}(\mu-{\rm i}\frac{\zeta_l}{\beta})(t-t')}. \tag{17.133}$$

Inserting the result into (17.131), the equation of motion (17.128) for the electronic density matrix becomes (in matrix form)

$$\frac{d}{dt}n^<(t) + {\rm i}\tilde{h}_{\rm qp}(t)n^<(t) - {\rm i}n^<(t)\tilde{h}^\dagger_{\rm qp}(t) = S^{\rm el}(t) + \frac{\Gamma(t)}{2} + \sum_{l\alpha}s_\alpha(t)\frac{\eta_l}{\beta}\Gamma_\alpha\mathcal{G}^{\rm em}_{l\alpha}(t) + {\rm h.c.}, \tag{17.134}$$

where

$$\mathcal{G}^{\rm em}_{l\alpha}(t) \equiv \int d\bar{t}\; F_{l\alpha}(t,\bar{t})G^{\rm A}(\bar{t},t)$$

is the embedding correlator. Taking into account the explicit expressions in (17.132) and (17.133), it is straightforward to find

$$\mathrm{i}\frac{d}{dt}\mathcal{G}^{\rm em}_{l\alpha}(t) = -s_\alpha(t) - \mathcal{G}^{\rm em}_{l\alpha}(t)\Big(\tilde{h}^{\dagger}_{\rm qp}(t) - V_\alpha(t) - \mu + \mathrm{i}\frac{\zeta_l}{\beta}\Big). \tag{17.135}$$

Equations (17.134) and (17.135), together with the equations of motion of $\mathcal{G}$ [necessary to construct $S^{\rm el}$, see (17.97)], form a coupled system of ordinary differential equations for correlated real-time simulations of open systems. This time-linear scheme becomes similar to the one of Refs. [365, 366, 367, 368, 369, 370] for $S^{\rm el} = 0$. The scaling with the system size of (17.135) grows like $N^3_{\rm sys} \times N_{\rm electrodes} \times N_p$, where $N_{\rm sys}$ is the number of one-particle basis functions, $N_{\rm electrodes}$ is the number of electrodes, and N_p is the number of poles for the expansion of $f(\omega)$.

Exercise 17.8 Derive the expression of $\mathcal{G}$ in the second Born approximation using the MGKBA. How would you formulate the time-linear scheme?

17.7 Systems Coupled to Baths: Redfield and Lindblad Equations

Both the Hamiltonian for electrons and photons and for electrons and phonons have the general structure

$$\hat{H}(t) = \hat{H}_s(t) + \hat{H}_b(t) + \sum_\alpha \hat{S}_\alpha(t)\hat{B}_\alpha(t), \tag{17.136}$$

where $\hat{H}_s$ contains only electronic operators, $\hat{H}_b$ contains only phosonic operators, and the interaction between the two subsystems is mediated by the last term, where $\hat{S}_\alpha$ contains only electronic operators and $\hat{B}_\alpha$ only phosonic operators. The time dependence is usually due to some time-dependent coupling, which is here absorbed in the operators. In this section we consider the more general problem of operators acting either on a Hilbert space $\mathbb{H}_s$ (which is the Fock space $\mathbb{F}$ in the case of the electrons) or on a Hilbert space $\mathbb{H}_b$ (which is the Fock space $\mathbb{F}_b$ in the case of the phosons). We refer to $\hat{H}_s$ as the *system* Hamiltonian and to $\hat{H}_b$ as the *bath* Hamiltonian. The operators of the system–bath interaction do not have to be self-adjoint; however, it must hold

$$\sum_\alpha \hat{S}_\alpha(t)\hat{B}_\alpha(t) = \sum_\alpha \hat{S}^\dagger_\alpha(t)\hat{B}^\dagger_\alpha(t). \tag{17.137}$$

Let $\hat{\rho}$ be the many-body density matrix of the system coupled to the bath and $\hat{U}(t,t')$ be the evolution operator for the Hamiltonian $\hat{H}$. The time-dependent ensemble average of an operator $\hat{O}_s$ acting only on $\mathbb{H}_s$ is given by

$$\begin{aligned} O_s(t) &= \mathrm{Tr}\big[\hat{\rho}\hat{U}(t_0,t)\hat{O}_s(t)\hat{U}(t,t_0)\big] \\ &= \mathrm{Tr}_s\Big[\underbrace{\mathrm{Tr}_b\big[\hat{U}(t,t_0)\hat{\rho}\hat{U}(t_0,t)\big]}_{\hat{\rho}_s(t)}\hat{O}_s(t)\Big]. \end{aligned} \tag{17.138}$$

In the second equality we have introduced the partial traces over the system and bath degrees of freedom. Mathematically the partial traces can be defined as follows. Let $|\Psi_{s,i}\rangle|\Psi_{b,j}\rangle$ be a complete set of states in the direct product Hilbert space $\mathbb{H}_s \otimes \mathbb{H}_b$. Given an operator $\hat{O}$ acting on $\mathbb{H}_s \otimes \mathbb{H}_b$, we have

$$\mathrm{Tr}_s\big[\hat{O}\big] = \sum_i \langle\Psi_{s,i}|\hat{O}|\Psi_{s,i}\rangle, \qquad \mathrm{Tr}_b\big[\hat{O}\big] = \sum_j \langle\Psi_{b,j}|\hat{O}|\Psi_{b,j}\rangle.$$

Thus, $\mathrm{Tr}_s\big[\hat{O}\big]$ is an operator acting only on $\mathbb{H}_b$, whereas $\mathrm{Tr}_b\big[\hat{O}\big]$ is an operator acting only on $\mathbb{H}_s$, and $\mathrm{Tr}\big[\hat{O}\big] = \mathrm{Tr}_b\big[\mathrm{Tr}_s\big[\hat{O}\big]\big] = \mathrm{Tr}_s\big[\mathrm{Tr}_b\big[\hat{O}\big]\big]$. If $\hat{O} = \hat{O}_s$ acts only on the Hilbert space $\mathbb{H}_s$, then $\mathrm{Tr}\big[\hat{O}_s\big] = \mathrm{Tr}_s\big[\hat{O}_s\big]$. Similarly, if $\hat{O} = \hat{O}_b$ acts only on the Hilbert space $\mathbb{H}_b$ then $\mathrm{Tr}\big[\hat{O}_b\big] = \mathrm{Tr}_b\big[\hat{O}_b\big]$, and $\mathrm{Tr}\big[\hat{O}_b\hat{O}_s\big] = \mathrm{Tr}_b\big[\hat{O}_b\big]\mathrm{Tr}_s\big[\hat{O}_s\big]$.

The goal of this section is to obtain a closed equation of motion for the reduced density matrix $\hat{\rho}_s(t) = \mathrm{Tr}_b\big[\hat{U}(t,t_0)\hat{\rho}\hat{U}(t_0,t)\big]$, thus eliminating the bath degrees of freedom. We closely follow the presentation of Ref. [371]. Let us define

$$\hat{\rho}(t) = \hat{U}(t,t_0)\hat{\rho}\hat{U}(t_0,t) \quad \Rightarrow \quad \frac{d\hat{\rho}(t)}{dt} = -\mathrm{i}\big[\hat{H}(t),\hat{\rho}(t)\big]_-, \tag{17.139}$$

where we use that $\mathrm{i}\frac{d}{dt}\hat{U}(t,t_0) = \hat{H}(t)\hat{U}(t,t_0)$ and the like for $\hat{U}(t_0,t)$, see (3.13). The equation of motion for $\hat{\rho}(t)$ is known as the *Liouville-von Neumann equation.* We write the full Hamiltonian in (17.136) as $\hat{H} = \hat{H}_0 + \hat{H}_{eb}$, with $\hat{H}_0 = \hat{H}_e + \hat{H}_b$ and $\hat{H}_{eb} = \sum_\alpha \hat{S}_\alpha\hat{B}_\alpha$, and define $\hat{U}_0(t,t')$ as the evolution operator for the Hamiltonian $\hat{H}_0$. The evolution operator $\hat{U}_0(t,t') = \hat{U}_s(t,t')\hat{U}_b(t,t')$ since $\hat{H}_0$ is the direct sum of the operator $\hat{H}_s$ acting only on $\mathbb{H}_s$ and the operator $\hat{H}_b$ acting only on $\mathbb{H}_b$. We further introduce the following operators in the Heisenberg representation with respect to $\hat{H}_0$:

$$\begin{aligned} \hat{\rho}_H(t) &= U_0(t_0,t)\hat{\rho}(t)U_0(t,t_0), \\ \hat{H}_{eb,H}(t) &= U_0(t_0,t)\hat{H}_{eb}(t)U_0(t,t_0), \\ \hat{H}_{0,H}(t) &= U_0(t_0,t)\hat{H}_0(t)U_0(t,t_0). \end{aligned}$$

We have

$$\begin{aligned} \frac{d\hat{\rho}_H(t)}{dt} &= \mathrm{i}\big[\hat{H}_{0,H}(t),\hat{\rho}_H(t)\big]_- + U_0(t_0,t)\frac{d\hat{\rho}(t)}{dt}U_0(t,t_0) \\ &= -\mathrm{i}\big[\hat{H}_{eb,H}(t),\hat{\rho}_H(t)\big]_-, \end{aligned} \tag{17.140}$$

where in the last equality we use (17.139). Integrating (17.140) between 0 and t, we find a recursive formula for $\hat{\rho}_H(t)$. To second order in the system–bath interaction we get

$$\hat{\rho}_H(t) = \hat{\rho}(t_0) - \mathrm{i}\int_{t_0}^{t} dt_1 \big[\hat{H}_{eb,H}(t_1), \hat{\rho}(t_0)\big]_- \\ - \int_{t_0}^{t} dt_1 \int_{t_0}^{t_1} dt_2 \big[\hat{H}_{eb,H}(t_1), \big[\hat{H}_{eb,H}(t_2), \hat{\rho}(t_0)\big]_-\big]_-. \tag{17.141}$$

To proceed further we have to make some assumptions on the initial many-body density matrix and on the Hamiltonian $\hat{H}$. We assume that at time $t = t_0$ the system and the bath are decoupled so that

$$\hat{\rho}(t_0) = \hat{\rho}_{s,0}\,\hat{\rho}_{b,0},$$

where $\hat{\rho}_{s,0}$ acts only on $\mathbb{H}_s$ and $\hat{\rho}_{b,0}$ acts only on $\mathbb{H}_b$. We further assume that $\hat{H}_b$ is independent of time, hence

$$\hat{U}_0(t,t') = \hat{U}_s(t,t')e^{-\mathrm{i}\hat{H}_b(t-t')},$$

and that the bath is initially in a stationary ensemble:

$$\big[\hat{H}_b, \rho_{b,0}\big]_- = 0. \tag{17.142}$$

We also make a reasonable assumption for the bath operators $\hat{B}_\alpha(t)$ - that is, $\hat{B}_\alpha(t) = \hat{B}_\alpha$ and

$$\mathrm{Tr}_b[\hat{B}_\alpha \rho_{b,0}] = 0. \tag{17.143}$$

This assumption is reasonable since in systems of electrons and phosons the $\hat{B}_\alpha$ operators are the displacement operators, which do not have an explicit time dependence and have zero ensemble average in equilibrium. For this class of problems we have

$$\mathrm{Tr}_b\big[\hat{H}_{eb,H}(t)\hat{\rho}(t_0)\big] = \sum_\alpha \hat{S}_{\alpha,H}(t)\hat{\rho}_{s,0}\mathrm{Tr}_b\big[e^{\mathrm{i}\hat{H}_b(t-t_0)}\hat{B}_\alpha e^{-\mathrm{i}\hat{H}_b(t-t_0)}\hat{\rho}_{b,0}\big] \\ = \sum_\alpha \hat{S}_{\alpha,H}(t)\hat{\rho}_{s,0}\mathrm{Tr}_b\big[e^{\mathrm{i}\hat{H}_b(t-t_0)}\hat{B}_\alpha\hat{\rho}_{b,0}e^{-\mathrm{i}\hat{H}_b(t-t_0)}\big] = 0, \tag{17.144}$$

where in the second equality we use the assumption (17.142), and in the last equality we use the cyclic property of the trace and the assumption (17.143). The system operators in the Heisenberg picture are defined similarly to all other operators:

$$\hat{S}_{\alpha,H}(t) = \hat{U}_0(t_0,t)\hat{S}_\alpha(t)\hat{U}_0(t,t_0) = \hat{U}_s(t_0,t)\hat{S}_\alpha(t)\hat{U}_s(t,t_0).$$

In the same way, we can show that $\mathrm{Tr}_b\big[\hat{\rho}(t_0)\hat{H}_{eb,H}(t)\big] = 0$.

Taking into account (17.144), we see that if we trace (17.141) over the bath degrees of freedom then the term that is of first order in $\hat{H}_{eb}$ vanishes and we find a formula for

$\hat{\rho}_{s,H}(t) \equiv \mathrm{Tr}_b[\hat{\rho}_H(t)]$:

$$\begin{aligned}
\hat{\rho}_{s,H}(t) &= \hat{\rho}_{s,0} - \int_{t_0}^{t} dt_1 \int_{t_0}^{t_1} dt_2 \mathrm{Tr}_b\Big[\big[\hat{H}_{eb,H}(t_1), \big[\hat{H}_{eb,H}(t_2), \hat{\rho}(t_0)\big]_-\big]_-\Big] \\
&= \hat{\rho}_{s,0} - \sum_{\alpha\beta} \int_{t_0}^{t} dt_1 \int_{t_0}^{t_1} dt_2 \\
&\times \Big\{ C_{\alpha\beta}(t_1,t_2)\Big(\hat{S}_{\alpha,H}(t_1)\hat{S}_{\beta,H}(t_2)\hat{\rho}_{s,0} - \hat{S}_{\beta,H}(t_2)\hat{\rho}_{s,0}\hat{S}_{\alpha,H}(t_1)\Big) \\
&- C^*_{\alpha\beta}(t_1,t_2)\Big(\hat{S}^\dagger_{\alpha,H}(t_1)\hat{\rho}_{s,0}\hat{S}^\dagger_{\beta,H}(t_2) - \hat{\rho}_{s,0}\hat{S}^\dagger_{\beta,H}(t_2)\hat{S}^\dagger_{\alpha,H}(t_1)\Big)\Big\},
\end{aligned} \tag{17.145}$$

where we use (17.137) for the terms in the last line, and we define the *bath correlators*[15]

$$C_{\alpha\beta}(t_1,t_2) \equiv \mathrm{Tr}_b\Big[\hat{B}_{\alpha,H}(t_1)\hat{B}_{\beta,H}(t_2)\rho_{b,0}\Big] = \mathrm{Tr}_b\Big[e^{\mathrm{i}\hat{H}_b(t_1-t_2)}\hat{B}_\alpha e^{-\mathrm{i}\hat{H}_b(t_1-t_2)}\hat{B}_\beta\rho_{b,0}\Big]. \tag{17.146}$$

Taking the time derivative of (17.145), we obtain the important result

$$\begin{aligned}
\frac{d\hat{\rho}_{s,H}(t)}{dt} &= -\sum_{\alpha\beta}\int_{t_0}^{t} dt' \\
&\times \Big\{ C_{\alpha\beta}(t,t')\Big(\hat{S}_{\alpha,H}(t)\hat{S}_{\beta,H}(t')\hat{\rho}_{s,H}(t) - \hat{S}_{\beta,H}(t')\hat{\rho}_{s,H}(t)\hat{S}_{\alpha,H}(t)\Big) \\
&- C^*_{\alpha\beta}(t,t')\Big(\hat{S}^\dagger_{\alpha,H}(t)\hat{\rho}_{s,H}(t)\hat{S}^\dagger_{\beta,H}(t') - \hat{\rho}_{s,H}(t)\hat{S}^\dagger_{\beta,H}(t')\hat{S}^\dagger_{\alpha,H}(t)\Big)\Big\},
\end{aligned} \tag{17.147}$$

where in the r.h.s. we replace $\hat{\rho}_{s,0} \to \hat{\rho}_{s,H}(t)$ everywhere. Such replacement introduces an error which is of fourth order in the system–bath Hamiltonian, see again (17.145); therefore, (17.147) is still correct to second order.

The equation of motion (17.147) allows us to obtain the equations of motion for the quantity of interest, see (17.138):

$$\begin{aligned}
\hat{\rho}_s(t) = \mathrm{Tr}_b\big[\hat{\rho}(t)\big] &= \mathrm{Tr}_b\big[\hat{U}_0(t,t_0)\hat{\rho}_H(t)\hat{U}_0(t_0,t)\big] \\
&= \hat{U}_s(t,t_0)\mathrm{Tr}_b\big[e^{-\mathrm{i}\hat{H}_b(t-t_0)}\hat{\rho}_H(t)e^{\mathrm{i}\hat{H}_b(t-t_0)}\big] \\
&= \hat{U}_s(t,t_0)\hat{\rho}_{s,H}(t)\hat{U}_s(t_0,t),
\end{aligned} \tag{17.148}$$

where in the last equality we use the cyclic property of the trace. We have

$$\begin{aligned}
\frac{d\hat{\rho}_s(t)}{dt} &= -\mathrm{i}\big[\hat{H}_s(t),\hat{\rho}_s(t)\big]_- + \hat{U}_s(t,t_0)\frac{d\hat{\rho}_{s,H}(t)}{dt}\hat{U}_s(t_0,t) \\
&= -\mathrm{i}\big[\hat{H}_s(t),\hat{\rho}_s(t)\big]_- - \sum_{\alpha\beta}\int_{t_0}^{t} dt' \\
&\times \Big\{ C_{\alpha\beta}(t,t')\Big(\hat{S}_\alpha(t)\hat{U}_s(t,t')\hat{S}_\beta(t')\hat{U}_s(t',t)\hat{\rho}_s(t) - \hat{U}_s(t,t')\hat{S}_\beta(t')\hat{U}_s(t',t)\hat{\rho}_s(t)\hat{S}_\alpha(t)\Big) \\
&- C^*_{\alpha\beta}(t,t')\Big(\hat{S}^\dagger_\alpha(t)\hat{\rho}_s(t)\hat{U}_s(t,t')\hat{S}^\dagger_\beta(t')\hat{U}_s(t',t) - \hat{\rho}_s(t)\hat{U}_s(t,t')\hat{S}^\dagger_\beta(t')\hat{U}_s(t',t)\hat{S}^\dagger_\alpha(t)\Big)\Big\}.
\end{aligned}$$

[15] Notice that $\mathrm{Tr}_b\Big[\hat{B}^\dagger_{\beta,H}(t_2)\hat{B}^\dagger_{\alpha,H}(t_1)\rho_{b,0}\Big] = \Big(\mathrm{Tr}_b\Big[\hat{B}_{\alpha,H}(t_1)\hat{B}_{\beta,H}(t_2)\rho_{b,0}\Big]\Big)^* = C^*_{\alpha\beta}(t_1,t_2)$.

Redfield equation The equation of motion for $\hat{\rho}_s$ simplifies further if we make the Markovian approximation, which consists of letting the upper integration limit $t \to \infty$, see Section 17.4.4. Let us define the system operators

$$\hat{T}_\alpha(t) \equiv \sum_\beta \int_{t_0}^{\infty} dt' C_{\alpha\beta}(t,t') \hat{U}_s(t,t') \hat{S}_\beta(t') \hat{U}_s(t',t). \tag{17.149}$$

In terms of these operators the equation of motion for $\hat{\rho}_s$ in the Markovian approximation can be written as

$$\frac{d\hat{\rho}_s(t)}{dt} = -\mathrm{i}\big[\hat{H}_s(t), \hat{\rho}_s(t)\big]_- - \sum_\alpha \Big[\hat{S}_\alpha(t)\hat{T}_\alpha(t)\hat{\rho}_s(t) - \hat{T}_\alpha(t)\hat{\rho}_s(t)\hat{S}_\alpha(t) - \hat{S}^\dagger_\alpha(t)\hat{\rho}_s(t)\hat{T}^\dagger_\alpha(t) + \hat{\rho}_s(t)\hat{T}^\dagger_\alpha(t)\hat{S}^\dagger_\alpha(t)\Big]. \tag{17.150}$$

This result is known as the *Redfield equation* [372]. The r.h.s. is self-adjoint and therefore the Redfield equation guarantees the hermiticity of $\hat{\rho}_s(t)$. Another important merit of (17.150) is the trace-preserving property - that is, $\mathrm{Tr}_s[\hat{\rho}_s(t)] = \mathrm{Tr}_s[\hat{\rho}_s(t_0)]$ is independent of time. This follows directly from the observation that the trace over the system degrees of freedom of the r.h.s. vanishes. Unfortunately, however, the Redfield equation does not guarantee the positivity of the eigenvalues of $\hat{\rho}_s(t)$ [373, 374]. It has been observed that the positivity property can be violated when the eigenvalues of $\hat{\rho}_s(t_0)$ are either too close to zero or too close to unity.

Lindblad equation Without any loss of generality we consider self-adjoint operators $\hat{S}_\alpha = \hat{S}^\dagger_\alpha$ and $\hat{B}_\alpha = \hat{B}^\dagger_\alpha$.[16] Then, the bath correlators have the property

$$C_{\alpha\beta}(t_1,t_2) = C^*_{\beta\alpha}(t_2,t_1). \tag{17.152}$$

We now show that the positivity property is recovered when the bath correlators are time-local:

$$C_{\alpha\beta}(t,t') = 2c_{\alpha\beta}\delta(t-t'),$$

according to which

$$\hat{T}_\alpha = \sum_\beta c_{\alpha\beta}\hat{S}_\beta(t).$$

[16] In fact, these operators can always be chosen self-adjoint by a proper redefinition:

$$\hat{S}'_\alpha = \frac{1}{2}(\hat{S}_\alpha + \hat{S}^\dagger_\alpha), \qquad \hat{S}''_\alpha = \frac{\mathrm{i}}{2}(\hat{S}_\alpha - \hat{S}^\dagger_\alpha),$$

$$\hat{B}'_\alpha = \frac{1}{2}(\hat{B}_\alpha + \hat{B}^\dagger_\alpha), \qquad \hat{B}''_\alpha = -\frac{\mathrm{i}}{2}(\hat{B}_\alpha - \hat{B}^\dagger_\alpha),$$

so that

$$\sum_\alpha \hat{S}_\alpha\hat{B}_\alpha = \sum_\alpha(\hat{S}'_\alpha\hat{B}'_\alpha + \hat{S}''_\alpha\hat{B}''_\alpha). \tag{17.151}$$

The property (17.152) implies that $c_{\alpha\beta} = c^*_{\beta\alpha}$, and therefore there exists a unitary matrix U such that

$$c_{\alpha\beta} = \sum_\gamma \mathrm{U}^*_{\gamma\alpha}\lambda_\gamma \mathrm{U}_{\gamma\beta}.$$

From definition (17.146) we also infer that the matrix $C_{\alpha\beta}(t,t)$ is positive semidefinite, and therefore the matrix $c_{\alpha\beta}$ is positive semidefinite as well. Consequently the eigenvalues $\lambda_\gamma \geq 0$, and we can define the *Lindblad operators*

$$\hat{L}_\gamma(t) = \sqrt{\lambda_\gamma}\sum_\beta \mathrm{U}_{\gamma\beta}\hat{S}_\beta(t). \tag{17.153}$$

Taking into account that $\sum_\alpha \hat{S}^\dagger_\alpha \hat{T}^\dagger_\alpha = \sum_{\alpha\beta}\hat{S}_\alpha c^*_{\alpha\beta}\hat{S}_\beta = \sum_{\alpha\beta}\hat{S}_\alpha c_{\beta\alpha}\hat{S}_\beta = \sum_\alpha \hat{T}_\alpha\hat{S}_\alpha$, the Redfield equation (17.150) can be written in terms of the Lindblad operators as follows:

$$\frac{d\hat{\rho}_s(t)}{dt} = -\mathrm{i}\big[\hat{H}_s(t),\hat{\rho}_s(t)\big]_- + \sum_\gamma \Big[2\hat{L}_\gamma(t)\hat{\rho}_s(t)\hat{L}^\dagger_\gamma(t) - \hat{L}^\dagger_\gamma(t)\hat{L}_\gamma(t)\hat{\rho}_s(t) - \hat{\rho}_s(t)\hat{L}^\dagger_\gamma(t)\hat{L}_\gamma(t)\Big]. \tag{17.154}$$

This is the so-called *Lindblad equation* [375]. It has the same nice properties as the Redfield equations – that is, the hermiticity and the trace-preserving properties. In addition, it preserves the positivity of $\hat{\rho}_s(t)$. The positivity property follows from the explicit solution of (17.154). Let us define the positive semidefinite self-adjoint operator $\hat{\Gamma}(t) = 2\sum_\gamma \hat{L}^\dagger_\gamma(t)\hat{L}_\gamma(t)$ and rewrite the Lindblad equation as (omitting the dependence on time)

$$\frac{d\hat{\rho}_s}{dt} = -\mathrm{i}(\hat{H}_s - \frac{\mathrm{i}}{2}\hat{\Gamma})\hat{\rho}_s + \mathrm{i}\hat{\rho}_s(\hat{H}_s + \frac{\mathrm{i}}{2}\hat{\Gamma}) + 2\sum_\gamma \hat{L}_\gamma\hat{\rho}_s\hat{L}^\dagger_\gamma. \tag{17.155}$$

We further define the open system (non-Hermitian) Hamiltonian $\hat{H}_o \equiv \hat{H}_s - \frac{\mathrm{i}}{2}\hat{\Gamma}$ and the open system (nonunitary) evolution operator

$$\hat{U}_o(t) \equiv T\{e^{-\mathrm{i}\int_{t_0}^t dt'\hat{H}_o(t')}\} \quad \Rightarrow \quad \frac{d}{dt}\hat{U}_o(t) = -\mathrm{i}\hat{H}_o(t)\hat{U}_o(t).$$

The reader can easily check that (17.155) is solved by

$$\hat{\rho}_s(t) = \hat{U}_o(t)\Big[\hat{\rho}_{s,0} + \int_{t_0}^t dt'\hat{U}_o^{-1}(t')\Big(2\sum_\gamma \hat{L}_\gamma(t')\hat{\rho}_s(t')\hat{L}^\dagger_\gamma(t')\Big)\hat{U}_o^{\dagger-1}(t')\Big]\hat{U}_o^\dagger(t). \tag{17.156}$$

The quantity on the r.h.s. is a positive semidefinite operator for a positive semidefinite $\hat{\rho}_s(t')$ with $t' \leq t$.

Connection to Green's functions Through the Lindblad equation the problem of calculating the time-dependent ensemble average of an operator $\hat{O}_s$ acting on a system coupled to a bath is reduced to calculate $O_s(t) = \mathrm{Tr}[\hat{O}_s\hat{\rho}_s(t)]$, where $\hat{\rho}_s(t)$ satisfies (17.154). Merging the Lindblad idea with the NEGF formalism poses a nontrivial challenge. The first issue has to do with the Lindblad operators. Considering $\rho_{s,\infty} = \lim_{t\to\infty} \hat{\rho}_s(t) = e^{-\beta \hat{H}^{\mathrm{M}}}/Z$, the critical query arises: What is the relationship between $\beta\hat{H}^{\mathrm{M}}$ and the Lindblad operators? Even more relevant is the question: Can the time-dependent Lindblad operators be chosen in a manner that allows the recovery of the thermal Hamiltonian $\hat{H}^{\mathrm{M}} = \hat{H} - \mu\hat{N}$?

Another important issue pertains to the equations of motion of the field operators, see (4.44) and (4.45), which play a crucial role in the development of the NEGF formalism. How should these equations change when the dynamics is governed by the Lindblad equation? A possibility consists of defining a "Lindblad" representation $\hat{O}_{s,L}(t)$ of the operator $\hat{O}_s$ with the property

$$\mathrm{Tr}[\hat{O}_{s,L}(t)\hat{\rho}_{s,0}] = \mathrm{Tr}[\hat{O}_s\hat{\rho}_s(t)]. \tag{17.157}$$

Using (17.155) we find

$$\begin{aligned}\mathrm{Tr}\Big[\frac{d\hat{O}_{s,L}(t)}{dt}\hat{\rho}_{s,0}\Big] &= \mathrm{Tr}\Big[\hat{O}_s\frac{d\hat{\rho}_s(t)}{dt}\Big]\\ &= \mathrm{Tr}\Big[\Big(-\mathrm{i}\hat{O}_s\hat{H}_o(t) + \mathrm{i}\hat{H}_o^\dagger(t)\hat{O}_s + 2\sum_\gamma \hat{L}_\gamma^\dagger(t)\hat{O}_s\hat{L}_\gamma(t)\Big)\hat{\rho}_s(t)\Big],\end{aligned} \tag{17.158}$$

where the cyclic property of the trace has been used. Taking into account that (17.157) and (17.158) must be satisfied for any $\hat{O}_s$ and $\hat{\rho}_{s,0}$, we conclude that

$$\frac{d\hat{O}_{s,L}(t)}{dt} = \Big(-\mathrm{i}\hat{O}_s\hat{H}_o + \mathrm{i}\hat{H}_o^\dagger\hat{O}_s + 2\sum_\gamma \hat{L}_\gamma^\dagger\hat{O}_s\hat{L}_\gamma\Big)_L(t).$$

The main difficulty that arises with this equation is that the Lindblad representation of the product of two operators need not be the same as the product of the two operators in the Lindblad representation: $(\hat{A}\hat{B})_L(t) \neq \hat{A}_L(t)\hat{B}_L(t)$. The Lindblad representation reduces to the Heisenberg representation for vanishing Lindblad operators; in this case $(\hat{A}\hat{B})_L(t) = (\hat{A}\hat{B})_H(t) = \hat{A}_H(t)\hat{B}_H(t)$, and for $\hat{A}$ and $\hat{B}$ field operators we recover (4.44) and (4.45). Although not obvious, the nonequilibrium Green's function theory can nevertheless be extended to systems with Lindbladian dynamics. Interested readers may wish to consult Refs. [376, 377, 378, 379] for a field theory formulation and Ref. [46] for a formulation based on second quantization.

Appendix A: From the N Roots of 1 to the Dirac δ-Function

The roots of the equation

$$z = \sqrt[N]{1} = 1^{1/N}$$

are the complex numbers $z_k = \exp[\frac{2\pi i}{N} k]$ with integers $k = 1, \ldots, N$, as can be readily verified by taking the Nth power of z_k. In the complex plane these roots lie at the vertices of a regular polygon inscribed in a unit circle. In Fig. A.1(a) we show the location of the roots for $N = 3$.

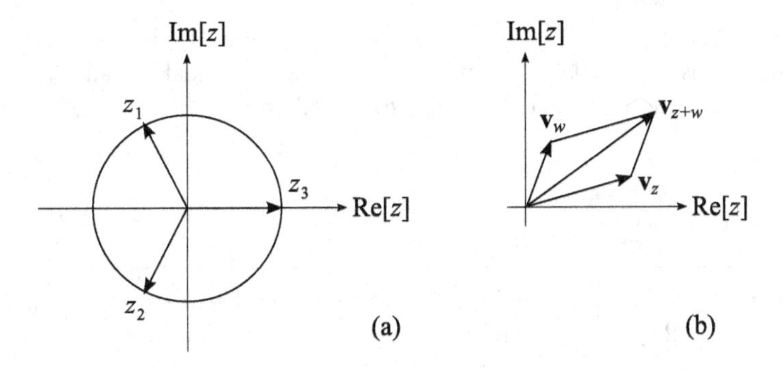

Figure A.1 (a) Location of the cubic roots of 1 in the complex plane. (b) Graphical representation of the sum of two complex numbers.

Which complex number do we get when we sum all roots? The addition of complex numbers is equivalent to the addition of vectors. If $\mathbf{v}_z$ is the vector going from the origin to z, then $\mathbf{v}_z + \mathbf{v}_w = \mathbf{v}_{z+w}$, as shown in Fig. A.1(b). It is then clear that

$$\sum_{k=1}^{N} z_k = \sum_{k=1}^{N} e^{\frac{2\pi i}{N} k} = 0.$$

This is a special case of a more general result. Instead of considering the sum of the roots, let us consider the sum of the nth power of the roots. The sum is a geometric sum since

$z_k^n = (z_1^n)^k$, and hence

$$\frac{1}{N}\sum_{k=1}^{N} z_k^n = \frac{1}{N}\sum_{k=1}^{N}(z_1^n)^k = \frac{(z_1^n)}{N}\,\frac{1-(z_1^n)^N}{1-(z_1^n)}. \tag{A.1}$$

The numerator in the r.h.s. is always zero since $(z_1^n)^N = (z_1^N)^n = 1$, while the denominator is zero only provided that n is a multiple of N. In this case, $z_k^n = 1$ for all k and we conclude that

$$\frac{1}{N}\sum_{k=1}^{N} z_k^n = \frac{1}{N}\sum_{k=1}^{N} e^{\frac{2\pi \mathrm{i}}{N}kn} = \ldots + \delta_{n,-N} + \delta_{n,0} + \delta_{n,N} + \delta_{n,2N} + \ldots. \tag{A.2}$$

Next we observe that result (A.2) does not change if we sum over k between M and $M+N$ - that is, over N arbitrary consecutive integers, rather than between 1 and N. This is obvious from the geometrical interpretation of (A.2), since it amounts to summing over all roots starting from M instead of 1. Let us prove it also with some algebra. We have

$$\frac{1}{N}\sum_{k=M}^{M+N-1} z_k^n = \frac{1}{N}\sum_{k=1}^{N} z_{k+M-1}^n = \frac{z_{M-1}^n}{N}\sum_{k=1}^{N} z_k^n,$$

where we use that $z_{k_1+k_2} = z_{k_1}z_{k_2}$. The statement follows from the fact that the sum vanishes unless n is a multiple of N, in which case the sum equals N and $z_{M-1}^n = 1$.

Let us consider the case that N is, say, even and choose $M = -N/2$. Then we can write

$$\boxed{\frac{1}{N}\sum_{k=-N/2}^{N/2-1} e^{\frac{2\pi \mathrm{i}}{N}kn} = \ldots + \delta_{n,-N} + \delta_{n,0} + \delta_{n,N} + \delta_{n,2N} + \ldots} \tag{A.3}$$

Now we define the variable $y_k = 2\pi k/N$. The distance between two consecutive y_k is $\Delta_y = y_{k+1} - y_k = 2\pi/N$. Therefore, taking the limit $N \to \infty$ we get

$$\lim_{N\to\infty}\frac{1}{N}\sum_{k=-N/2}^{N/2-1} e^{\frac{2\pi \mathrm{i}}{N}kn} = \lim_{\Delta_y\to 0}\sum_{k=-N/2}^{N/2-1}\frac{\Delta_y}{2\pi}\, e^{\mathrm{i}y_k n} = \int_{-\pi}^{\pi}\frac{dy}{2\pi}\, e^{\mathrm{i}yn}, \tag{A.4}$$

where in the last equality we transform the sum into an integral in accordance with the standard definition of integrals as the limit of a Riemann sum. Setting $n = m - m'$ and taking into account that for $N \to \infty$ only the Kronecker $\delta_{n,0}$ remains in the r.h.s. of (A.3), we obtain the following important identity:

$$\boxed{\int_{-\pi}^{\pi}\frac{dy}{2\pi}\, e^{\mathrm{i}y(m-m')} = \delta_{m,m'}} \tag{A.5}$$

In (A.3) we could, alternatively, think of the variable $n/N = y_n$ as a continuous variable when $N \to \infty$. The infinitesimal increment is then $\Delta_y = y_{n+1} - y_n = 1/N$, and taking the limit $N \to \infty$ we find

$$\sum_{k=-\infty}^{\infty} e^{2\pi i k y} = \lim_{N\to\infty} \frac{1}{\Delta_y}(\ldots + \delta_{n,-N} + \delta_{n,0} + \delta_{n,N} + \delta_{n,2N} + \ldots). \tag{A.6}$$

For $N \to \infty$ and hence $\Delta_y \to 0$, the quantity $\delta_{n,mN}/\Delta_y$ is zero unless $n = y_n N = mN$ (i.e., $y_n = m$), in which case it diverges like $1/\Delta_y$. Since for any function $f(y)$ we have

$$\begin{aligned} f(m) = \sum_{n=-\infty}^{\infty} \delta_{n,mN} f(y_n) &= \sum_{n=-\infty}^{\infty} \Delta_y \frac{\delta_{n,mN}}{\Delta_y} f(y_n) \\ &\xrightarrow[\Delta_y \to 0]{} \int_{-\infty}^{\infty} dy \left(\lim_{\Delta_y \to 0} \frac{\delta_{n,mN}}{\Delta_y} \right) f(y), \end{aligned}$$

we can identify the Dirac δ-function

$$\lim_{\Delta_y \to 0} \frac{\delta_{n,mN}}{\Delta_y} = \delta(y - m). \tag{A.7}$$

Therefore, taking the limit $\Delta_y \to 0$ in (A.6), we obtain a second important identity:

$$\boxed{\sum_{k=-\infty}^{\infty} e^{2\pi i k y} = \ldots + \delta(y+1) + \delta(y) + \delta(y-1) + \delta(y-2) + \ldots}$$

Lastly, we consider again (A.5) and divide both sides by an infinitesimal Δ_p

$$\frac{1}{\Delta_p} \int_{-\pi}^{\pi} \frac{dy}{2\pi}\, e^{iy(m-m')} = \frac{\delta_{m,m'}}{\Delta_p}.$$

In the limit $\Delta_p \to 0$ we can define the continuous variables $p = m\Delta_p$ and $p' = m'\Delta_p$, and using (A.7) we get

$$\lim_{\Delta_p \to 0} \frac{1}{\Delta_p} \int_{-\pi}^{\pi} \frac{dy}{2\pi}\, e^{iy(m-m')} = \delta(p - p').$$

The product in the exponential can be rewritten as $y(m - m') = y(p - p')/\Delta_p$. Thus, if we change variable $x = y/\Delta_p$ the above equation becomes

$$\boxed{\int_{-\infty}^{\infty} \frac{dx}{2\pi}\, e^{ix(p-p')} = \delta(p - p')}$$

which is one of the possible representations of the Dirac δ-function. In conclusion, the Dirac δ-function is intimately related to the sum of the N roots of 1 when $N \to \infty$.

Appendix B: Graphical Approach to Permanents and Determinants

The quantum states of a set of identical particles are either symmetric or antisymmetric in the single-particle labels. An orthonormal basis for the space spanned by these quantum states is formed by the complete set of (anti)symmetric products of single-particle orthonormal states, and their manipulation leads naturally to the consideration of permanents and determinants. The standard algebraic derivation of identities for permanents and determinants usually involves several steps with much relabeling of permutations that makes the derivations often long and not very insightful. In this appendix we instead give a simple and intuitive graphical derivation of several of those identities. The basic ingredient of the graphical approach is the *permutation graph* that keeps track of how the permutation moves around the elements on which it acts. A permutation P is defined as a one-to-one mapping from the set of integers $(1, \dots, n)$ to itself,

$$P(1, \dots, n) = (P(1), \dots, P(n)) = (1', \dots, n'),$$

where $j' = P(j)$ denotes the image of j under permutation P. The permutation graph is then defined by drawing a figure with the numbers $(1, \dots, n)$ as dots ordered from top to bottom along a vertical line on the left and with the images $(1', \dots, n')$ as dots also ordered from top to bottom along a vertical line on the right, and by connecting the dots j with the images j'. For example, if $n = 5$, we can consider the permutation

$$P(1, 2, 3, 4, 5) = (2, 1, 5, 3, 4) = (1', 2', 3', 4', 5'). \tag{B.1}$$

The permutation graph is then given by

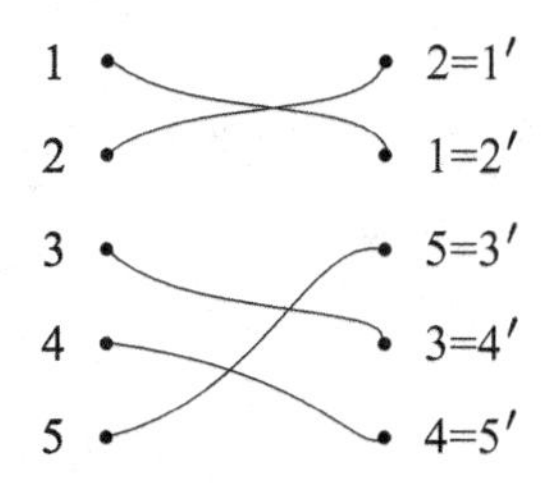

The permutation graph thus simply provides a graphical representation of how the numbers $(1, \dots, n)$ are moved around by the permutation. An important feature of the permutation graph is that, for a given permutation, the number of crossings of the lines is always even or odd, no matter how we deform the lines in the diagram, provided that we do not deform the lines outside the left and right boundaries. For example, we have:

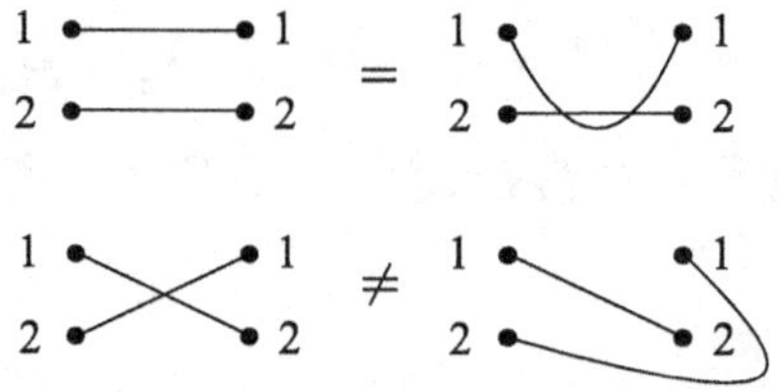

It should be noted that for this to be true we also need to exclude drawings in which the lines touch each other in one point and drawings in which multiple lines cross in exactly the same point. The crossing rule is intuitively clear since any deformation of the lines that creates new crossing points always creates two of them. Hence the parity of the number of crossings is preserved. We can therefore divide the set of permutations into two classes, those with an even number of crossings and those with an odd number of crossings. These are naturally referred to as even and odd permutations. Another way to characterize this property is by defining the *sign* of a permutation P to be

$$\text{sgn}\, P \equiv (-)^{n_c}, \tag{B.2}$$

where n_c is the number of crossings in a permutation graph and $(-)^{n_c}$ is a short-hand notation for $(-1)^{n_c}$. So even permutations have sign 1 and odd permutations have sign -1. It is also intuitively clear that any permutation can be built up from successive interchanges of two numbers. This is what one, for instance, would do when given the task of reordering numbered balls by each time swapping two of them. Each such swap is called a transposition. If a transposition interchanges labels i and j, then we write it as $(i\, j)$. For instance, permutation (B.1) can be constructed from the identity permutation by the subsequent transpositions $(4\,5)$, $(3\,5)$, and $(1\,2)$. This is described by the following four permutation graphs (to be read from left to right):

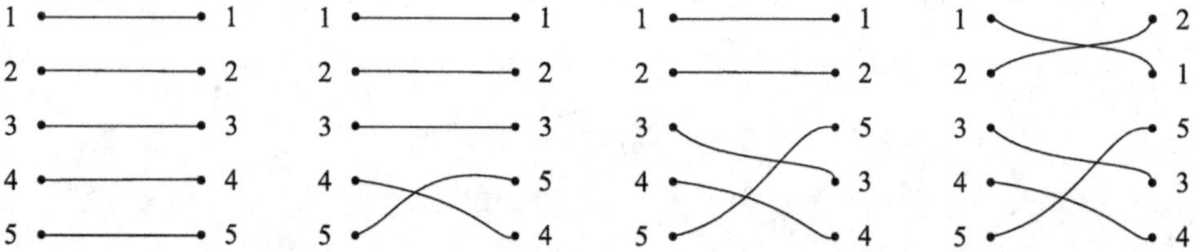

We can thus write

$$P = (1\,2)\,(3\,5)\,(4\,5), \tag{B.3}$$

in which the subsequent transpositions in this expression are carried out from right to left. We therefore need three transpositions to reorder numbered balls from $(1,2,3,4,5)$ to $(2,1,5,3,4)$. From these graphs we further see the interesting fact that every transposition changes the number of crossings by an odd number (which is equal to 1 in the example above). This fact is easily deduced from the following graph that displays the very right-hand side of a permutation graph:

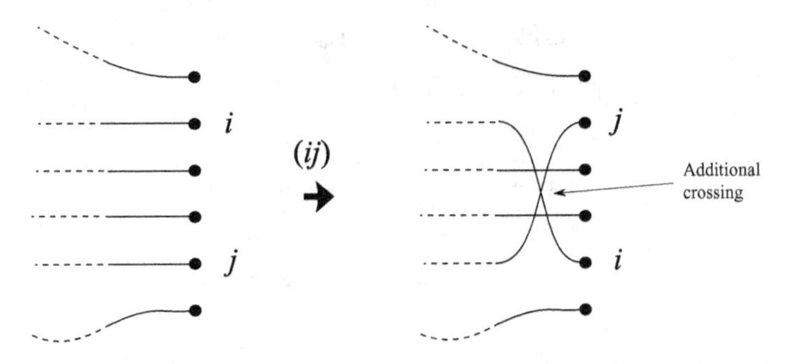

By interchanging i and j on the right-hand side we introduce one additional crossing plus an even number of crossings due to the upward and downward running lines induced by the swap. Hence a transposition always generates an odd number of crossings. It therefore follows that even permutations are built up from an even number of transpositions and odd permutations are built up from an odd number of transpositions. For example, permutation P of our example (B.3) is odd and indeed built up from three transpositions. As experience with reordering objects tells us, the way to achieve a given reordering is not unique. However, as we just proved, the parity of the number of transpositions is unique, and therefore the permutation in our example can only be decomposed using an odd number of transpositions. For instance, it is not difficult to check that permutation (B.3) can also be written as

$$P = (2\,3)\,(2\,5)\,(4\,5)\,(1\,3)\,(2\,3),$$

which indeed again has an odd number of transpositions, as it should. From our considerations we thus conclude that if $|P|$ is the number of transpositions in a given decomposition of a permutation P, then the sign of this permutation is given by $(-)^{|P|}$. By comparing with (B.2) we find the useful relation

$$(-)^{n_c} = (-)^{|P|}.$$

Alternatively, this relation is usually written as

$$(-)^{n_c} = (-)^{P},$$

where $(-)^P$ simply signifies the sign of the permutation. Before continuing let us illustrate another useful way of looking at the permutation graphs. In the graphs above we thought of the labels j as numbered balls and imaged the balls being moved around. This image is, for instance, very useful when thinking about operator orderings, as we did in Chapter 4. However, for deriving identities for permanents and determinants it is useful to think about the permutation graphs in a slightly different way. In permutation P, the ball numbered $P(i)$ acquires the new position i. For instance, example (B.1) tells us that after permutation P the balls labeled $(2,1,5,3,4)$ can be found at positions $(1,2,3,4,5)$. We can express this information in a new permutation graph where, on the left-hand side, we put the positions i and, on the right-hand side, we put the ball numbers $P(i)$ directly opposite to i and then connect them with horizontal lines. In our example this gives:

Position	Ball number
1	2
2	1
3	5
4	3
5	4

For instance, we have the line $(3,5)$ which tells us that ball 5 is found at position 3 after the permutation. Now we can reorder the balls on the right-hand side of the graph to their original positions by subsequent transpositions:

Position	Ball number		Position	Ball number
1	2		1	1
2	1		2	2
3	5	→	3	3
4	3		4	4
5	4		5	5

Since each transposition leads to an odd number of crossings, the sign of the permutation is again given by $(-1)^{n_c}$, with n_c the number of crossings, but now for a graph in which both sides are labeled 1 to n from top to bottom and where we connect i to $P(i)$. This is an alternative way to calculate the sign of a permutation from a permutation graph, and it proves to be very useful in deriving some important identities for permanents and determinants.

Permanent and determinant The permanent or determinant of an $n \times n$ matrix A_{ij} is defined as

$$|A|_{\pm} \equiv \sum_{P} (\pm)^P \prod_{i=1}^{n} A_{i\,P(i)},$$

where the "+" sign refers to the permanent and the "−" sign to the determinant [in this formula $(+)^P = (+1)^{|P|} = 1$ for all P]. For instance, for a 2×2 matrix we only have the even permutation $P(1,2) = (1,2)$ and the odd permutation $P(1,2) = (2,1)$, and therefore we have that the permanent/determinant is

$$|A|_\pm = A_{11}A_{22} \pm A_{12}A_{21}. \tag{B.4}$$

The permanent or determinant can be visualized using permutation graphs. First draw the permutation graph that connects with lines points i to points $P(i)$, and then associate with every line the matrix element $A_{i\,P(i)}$. For instance, the permanent/determinant in (B.4) can be written as

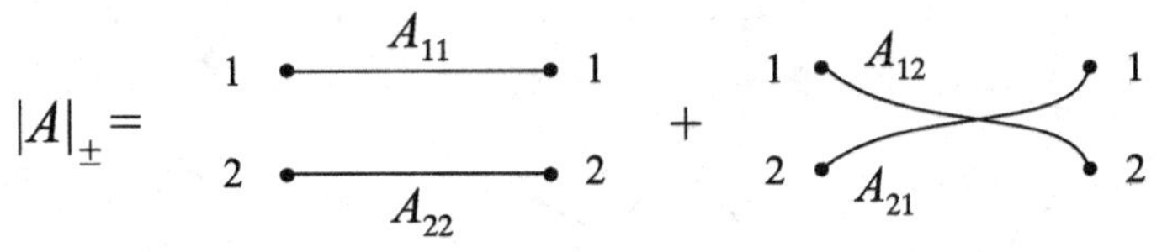

$$|A|_\pm = A_{11}A_{22} + A_{12}A_{21}$$

and for a 6×6 permanent/determinant we have a term of the form:

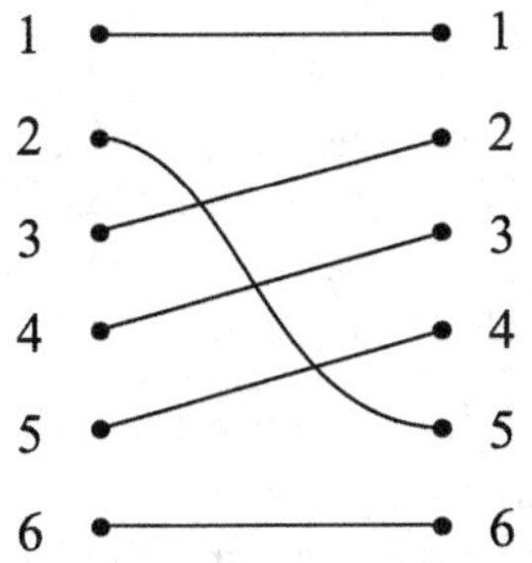

$$= (\pm)^3 A_{11}A_{25}A_{32}A_{43}A_{54}A_{66}$$

For permanents and determinants we can now prove the following Laplace expansion theorems:

$$|A|_\pm = \sum_{j=1}^{n} (\pm)^{i+j} A_{ij}\, \tilde{D}_{ij}$$

$$|A|_\pm = \sum_{i=1}^{n} (\pm)^{i+j} A_{ij}\, \tilde{D}_{ij}$$

in which $\tilde{D}_{ij}$ is the minor of $|A|_\pm$ - that is, the permanent/determinant of matrix A in which row i and column j are removed. The first expression gives the expansion of $|A|_\pm$ along row i, whereas the second one gives its expansion along column j. These identities can be proven almost immediately from drawing a single diagram. We consider a particular permutation that moves the ball numbered j in $(1,\ldots,n)$ just before the ball numbered i. If $j > i$, this permutation is explicitly given by

$$P(1,\ldots,n) = (1,\ldots,i-1,j,i,\ldots,j-1,j+1,\ldots,n),$$

with a very similar expression when $j < i$. Since there are $|j-i|$ crossings in the permutation graph, the sign of this permutation graph is $(\pm)^{i-j} = (\pm)^{i+j}$. For example, for $n = 6$ we can consider the permutation with $j = 5$ and $i = 2$:

$$P(1,2,3,4,5,6) = (1,5,2,3,4,6),$$

which is graphically given by the permutation graph in the previous figure. Let us now fix the line $(i,j) = (2,5)$ and subsequently connect the remaining lines in all possible ways. In this way we are exactly constructing the minor $\tilde{D}_{25}$:

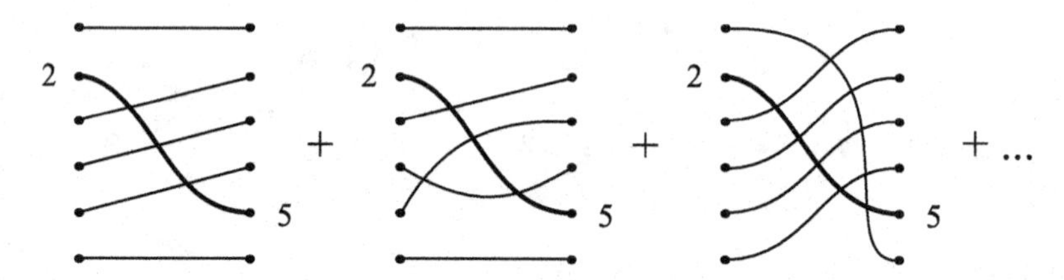

$$= (\pm)^{2+5} A_{25} \tilde{D}_{25} = (\pm)^{2+5} A_{25} \begin{vmatrix} A_{11} & A_{12} & A_{13} & A_{14} & A_{16} \\ A_{31} & A_{32} & A_{33} & A_{34} & A_{36} \\ A_{41} & A_{42} & A_{43} & A_{44} & A_{46} \\ A_{51} & A_{52} & A_{53} & A_{54} & A_{56} \\ A_{61} & A_{62} & A_{63} & A_{64} & A_{66} \end{vmatrix}_{\pm}$$

Indeed, the sign of the generic permutation graph in the above expansion is given by $(\pm)^{2+5} \times (\pm)^{n_c^{(25)}}$, where $n_c^{(25)}$ is the number of crossings *without* the line $(2,5)$.[1] It is clear that we obtain the full permanent/determinant by doing the same after singling out the lines $(2,j)$ for $j = 1,2,3,4,6$ and adding the results, since in that case we have summed over all possible connections exactly once. This yields

$$|A|_{\pm} = \sum_{j=1}^{6} (\pm)^{2+j} A_{2j} \tilde{D}_{2j}.$$

This is the Laplace formula for expanding the permanent/determinant along row 2. Alternatively we could have singled out the lines $(i,5)$ for $i = 1,\ldots,6$ and obtained

$$|A|_{\pm} = \sum_{i=1}^{6} (\pm)^{i+5} A_{i5} \tilde{D}_{i5},$$

which gives the expansion along row 5. It is clear that there is nothing special about this example and that the general proof goes the same way.

Generalization of the Laplace formula This graphical proof also gives an idea on how to generalize the Laplace formula. Consider a permutation graph in which we single out

[1] The reader can easily check that the line $(2,5)$ is always crossed an odd number of times (three times in the second graph and five times in the third graph).

three lines, (i_1, j_1), (i_2, j_2), and (i_3, j_3), where $i_i < i_2 < i_3$ and $j_1 < j_2 < j_3$, and where the remaining lines are drawn in a noncrossing way:

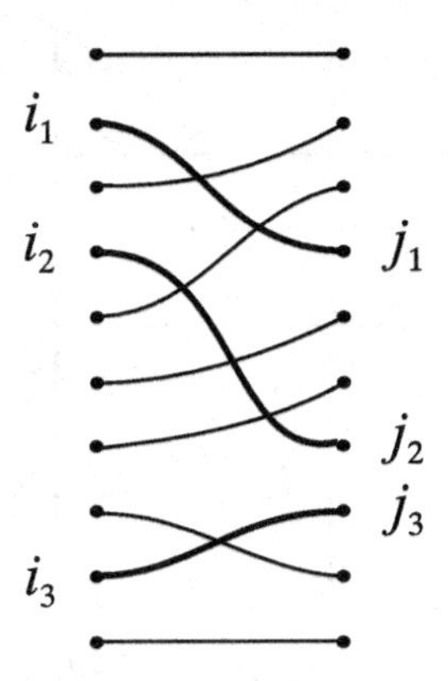

The sign of this permutation graph is clearly given by

$$(\pm)^{j_1-i_1}(\pm)^{j_2-i_2}(\pm)^{j_3-i_3} = (\pm)^{i_1+i_2+i_3+j_1+j_2+j_3}.$$

If we now fix the lines (i_1, j_1), (i_2, j_2), and (i_3, j_3) – that is, we fix the position of the ball numbered j_1 to be i_1, etc. – and make a permutation Q of the remaining balls (so that their positions are different from the position they have in the above permutation graph), we get a new permutation graph. As in the case of a single line, the total number of crossings for the lines (i_1, j_1), (i_2, j_2), and (i_3, j_3) always has the same parity (even or odd) for all Q and therefore the sign of this new permutation graph is

$$(\pm)^{i_1+i_2+i_3+j_1+j_2+j_3} \times (\pm)^{n_c^{(i_1j_1),(i_2j_2),(i_3j_3)}},$$

where $n_c^{(i_1j_1),(i_2j_2),(i_3j_3)}$ is the number of crossings with the lines (i_1, j_1), (i_2, j_2), and (i_3, j_3) removed from the graph. This observation allows us to derive an important identity. Let $\Pi_{i_1i_2i_3,j_1j_2j_3}(Q)$ be this new permutation graph. Then we can write

$$\sum_Q \Pi_{i_1i_2i_3,j_1j_2j_3}(Q) = (\pm)^{i_1+i_2+i_3+j_1+j_2+j_3} A_{i_1j_1} A_{i_2j_2} A_{i_3j_3} \tilde{D}_{i_1i_2i_3,j_1j_2j_3}. \tag{B.5}$$

In this equation $\tilde{D}_{i_1i_2i_3,j_1j_2j_3}$ is a generalized minor – that is, the permanent/determinant of matrix A in which we remove rows i_1, i_2, i_3 and columns j_1, j_2, j_3:

$$\tilde{D}_{i_1 i_2 i_3, j_1 j_2 j_3} = \begin{array}{c|ccccccc|}
 & & j_1 & & j_2 & & j_3 & \\
 & \vdots & & \vdots & & \vdots & & \vdots \\
i_1 & & A_{i_1 j_1} & & A_{i_1 j_2} & & A_{i_1 j_3} & \\
 & \vdots & & \vdots & & & & \vdots \\
i_2 & & A_{i_2 j_1} & & A_{i_2 j_2} & & A_{i_2 j_3} & \\
 & \vdots & & \vdots & & & & \vdots \\
i_3 & & A_{i_3 j_1} & & A_{i_3 j_2} & & A_{i_3 j_3} & \\
 & \vdots & & \vdots & & \vdots & & \vdots
\end{array}_{\pm}$$

Subsequently, in (B.5) we sum over all the permutations P of the integers j_1, j_2, and j_3. The sign of the permutation graph $\Pi_{i_1 i_2 i_3, P(j_1)P(j_2)P(j_3)}(Q)$ differs from the sign of $\Pi_{i_1 i_2 i_3, j_1 j_2 j_3}(Q)$ by a factor $(\pm)^{n_c}$, with n_c the number of crossings of the lines $(i_1, P(j_1))$, $(i_2, P(j_2))$, and $(i_3, P(j_3))$. Therefore we have

$$\sum_P \sum_Q \Pi_{i_1 i_2 i_3, P(j_1)P(j_2)P(j_3)}(Q) = (\pm)^{i_1+i_2+i_3+j_1+j_2+j_3} D_{i_1 i_2 i_3, j_1 j_2 j_3} \tilde{D}_{i_1 i_2 i_3, j_1 j_2 j_3},$$

where $D_{i_1 i_2 i_3, j_1 j_2 j_3}$ is the permanent/determinant consisting of the rows i_1, i_2, i_3 and columns j_1, j_2, j_3 of matrix A:

$$D_{i_1 i_2 i_3, j_1 j_2 j_3} = \left| \begin{array}{ccc} A_{i_1 j_1} & A_{i_1 j_2} & A_{i_1 j_3} \\ A_{i_2 j_1} & A_{i_2 j_2} & A_{i_2 j_3} \\ A_{i_3 j_1} & A_{i_3 j_2} & A_{i_3 j_3} \end{array} \right|_{\pm}.$$

If we finally sum over all triples (j_1, j_2, j_3) with $j_1 < j_2 < j_3$, then we clearly obtain the full permanent/determinant of matrix A since we have summed exactly once over all possible connections of lines in the permutation graph. We therefore find that

$$|A|_{\pm} = \sum_{j_1 < j_2 < j_3}^{n} (\pm)^{i_1+i_2+i_3+j_1+j_2+j_3} D_{i_1 i_2 i_3, j_1 j_2 j_3} \tilde{D}_{i_1 i_2 i_3, j_1 j_2 j_3}.$$

It is clear that the choice of the initial three lines in this example was arbitrary. We might as well have chosen $m < n$ lines instead. Then nothing essential would change in the derivation. In that case, we obtain the formula

$$\boxed{|A|_{\pm} = \sum_{J}^{n} (\pm)^{|I+J|} D_{IJ} \tilde{D}_{IJ}} \tag{B.6}$$

where we sum over the ordered m-tuple,

$$J = (j_1, \ldots, j_m) \quad \text{with} \quad j_1 < \ldots < j_m,$$

for a fixed-order m-tuple

$$I = (i_1, \ldots, i_m) \quad \text{with} \quad i_1 < \ldots < i_m,$$

and define

$$|I + J| = i_1 + \ldots + i_m + j_1 + \ldots + j_m.$$

D_{IJ} is the permanent/determinant containing rows I and columns J, whereas $\tilde{D}_{IJ}$ is the generalized minor, not containing the rows I and columns J. Result (B.6) is known as the *generalized Laplace formula* for permanents/determinants. Alternatively we could in the derivation have summed over the rows rather than the columns, and write:

$$\boxed{|A|_\pm = \sum_I^n (\pm)^{|I+J|} D_{IJ}\, \tilde{D}_{IJ}} \tag{B.7}$$

The number of terms in the summation in the generalized Laplace formula is equal to the number of ordered m-tuples that can be chosen from n indices and is therefore equal to $\binom{n}{m}$. The use of permutation graphs leads to a very compact derivation of the identities (B.6) and (B.7). As an example of (B.6) we can expand the permanent/determinant of a 4×4 matrix into products of 2×2 permanents/determinants. Taking, for example, $(i_1, i_2) = (1, 2)$, we have

$$\begin{aligned}
|A|_\pm &= \sum_{j_1<j_2} (\pm)^{1+2+j_1+j_2} D_{12,j_1j_2} \tilde{D}_{12,j_1j_2} \\
&= \begin{vmatrix} A_{11} & A_{12} \\ A_{21} & A_{22} \end{vmatrix}_\pm \begin{vmatrix} A_{33} & A_{34} \\ A_{43} & A_{44} \end{vmatrix}_\pm \pm \begin{vmatrix} A_{11} & A_{13} \\ A_{21} & A_{23} \end{vmatrix}_\pm \begin{vmatrix} A_{32} & A_{34} \\ A_{42} & A_{44} \end{vmatrix}_\pm \\
&+ \begin{vmatrix} A_{11} & A_{14} \\ A_{21} & A_{24} \end{vmatrix}_\pm \begin{vmatrix} A_{32} & A_{33} \\ A_{42} & A_{43} \end{vmatrix}_\pm + \begin{vmatrix} A_{12} & A_{13} \\ A_{22} & A_{23} \end{vmatrix}_\pm \begin{vmatrix} A_{31} & A_{34} \\ A_{41} & A_{44} \end{vmatrix}_\pm \\
&\pm \begin{vmatrix} A_{12} & A_{14} \\ A_{22} & A_{24} \end{vmatrix}_\pm \begin{vmatrix} A_{31} & A_{33} \\ A_{41} & A_{43} \end{vmatrix}_\pm + \begin{vmatrix} A_{13} & A_{14} \\ A_{23} & A_{24} \end{vmatrix}_\pm \begin{vmatrix} A_{31} & A_{32} \\ A_{41} & A_{42} \end{vmatrix}_\pm .
\end{aligned}$$

Permanent and determinant for the sum of matrices Finally, we derive a useful formula for the permanent/determinant of the sum of two matrices. Let us start with an example. If A and B are two 2×2 matrices, then we can readily calculate that

$$\begin{aligned}
&\begin{vmatrix} A_{11}+B_{11} & A_{12}+B_{12} \\ A_{21}+B_{21} & A_{22}+B_{22} \end{vmatrix}_\pm = \\
&\quad \begin{vmatrix} A_{11} & A_{12} \\ A_{21} & A_{22} \end{vmatrix}_\pm + A_{11}B_{22} \pm A_{12}B_{21} \pm A_{21}B_{12} + A_{22}B_{11} + \begin{vmatrix} B_{11} & B_{12} \\ B_{21} & B_{22} \end{vmatrix}_\pm .
\end{aligned}$$

This can be written in a compact way as

$$|A + B|_\pm = |A|_\pm + \sum_{i_1,j_1}^{2} (\pm)^{i_1+j_1} A_{i_1j_1} B_{\check{i}_1,\check{j}_1} + |B|_\pm,$$

where $\check{i}$ is the element complementary to i in the set $(1,2)$ - that is, if $i=1$ then $\check{i}=2$, and vice versa. We now derive the following generalization of this equation:

$$\boxed{|A+B|_\pm = |A|_\pm + \sum_{l=1}^{n-1}\sum_{IJ}(\pm)^{|I+J|}|A|_{l,\pm}(I,J)|B|_{n-l,\pm}(\check{I},\check{J}) + |B|_\pm} \tag{B.8}$$

where A and B are $n\times n$ matrices. In (B.8) I is an ordered l-tuple $(i_1,\ldots,i_l)$ with $i_1<\ldots<i_l$ and similarly for J. Further, $\check{I}$ is the set of ordered complementary indices to I in the set $(1,\ldots,n)$ and similarly for $\check{J}$. For example, if $n=5$ and $I=(1,3,4)$, then $\check{I}=(2,5)$. The quantity $|A|_{l,\pm}(I,J)$ denotes the $l\times l$ permanent/determinant of A with rows I and columns J. If $l=n$, the only possible n-tuple is $I=J=(1,\ldots,n)$ and therefore $|A|_{n,\pm}(I,J)=|A|_\pm$. Analogously, $|B|_{n-l,\pm}(\check{I},\check{J})$ denotes the $(n-l)\times(n-l)$ permanent/determinant of B with rows $\check{I}$ and columns $\check{J}$. To prove (B.8) we start with the example of the 2×2 matrix $C=A+B$. Let the thick line (i,j) represents the matrix element $C_{ij}=A_{ij}+B_{ij}$:

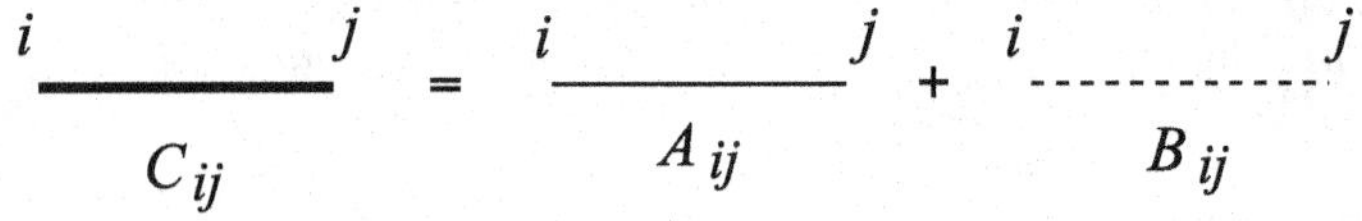

$$C_{ij} = A_{ij} + B_{ij}$$

where the thin line represents matrix element A_{ij} and the dashed line the matrix element B_{ij}. The graphic expression of the 2×2 permanent/determinant $|C|_\pm$ is then given by

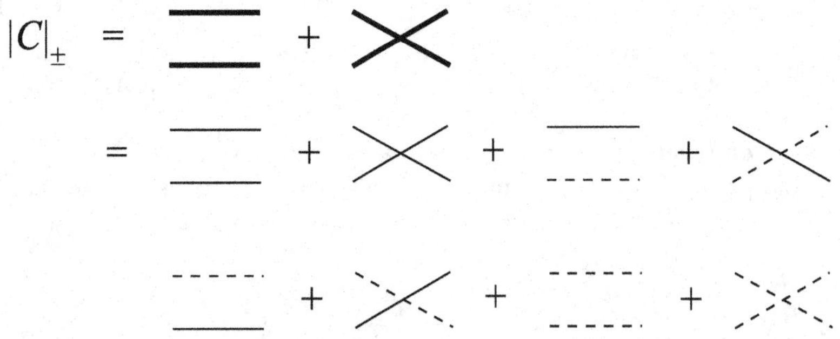

$$|C|_\pm = \;=\; + \;\times$$

$$= \;=\; + \;\times\; + \;=\; + \;\times$$

$$+ \;=\; + \;\times\; + \;=\; + \;\times$$

In the general case that C is an $n\times n$ matrix we can consider a given graph containing l solid lines and $n-l$ dashed lines. Let the thin lines run from the ordered set I to the ordered set J and the remaining dashed lines from $\check{I}$ to $\check{J}$. Since the sets are ordered, there is no crossing between two or more thin lines and between two or more dashed lines. The graph representing this situation looks like this:

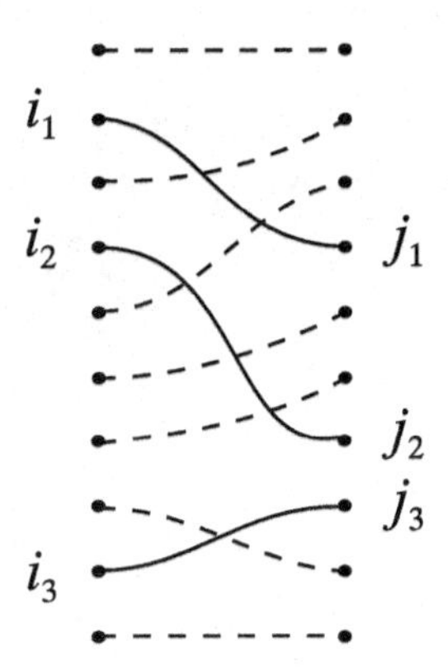

The sign of the graph is $(\pm)^{|I+J|}$. If we sum over all the dashed connections obtained by considering all permutations of the elements in the $(n-l)$-tuple $\breve{J}$ and multiply by the prefactor $(\pm)^{n_c}$ for each crossing of the dashed lines, we build the permanent/determinant $|B|_{n-l,\pm}(\breve{I}, \breve{J})$. The number of additional crossings of the dashed lines with the thin lines that are caused by these permutations is always even and we therefore do not need to add additional signs. Subsequently summing over all permutations in the l-tuples J and multiplying by the prefactor $(\pm)^{n_c}$ for each crossing of the thin lines, we build the permanent/determinant $|A|_{l,\pm}(I, J)$. As before, the number of additional crossings of the solid lines with the dashed lines caused by these permutations is always even, so we do not need to count these crossings to get the correct sign. When we finally sum over all ordered l-tuples I and J, then we sum over all graphs that have l solid lines and $n-l$ dashed lines. It is then clear that all possible graphs are obtained by summing over l from 1 to $n-1$ and adding $|A|_\pm$ and $|B|_\pm$. This proves (B.8).

To make the formula (B.8) explicit, we give an example for the case of 3×3 matrices A and B:

$$
\begin{aligned}
|A+B|_\pm &= |A|_\pm + \sum_{i_1, j_1}^{3} (\pm)^{i_1+j_1} |A|_{1,\pm}(i_1, j_1) |B|_{2,\pm}(\breve{i}_1, \breve{j}_1) \\
&+ \sum_{\substack{i_1<i_2 \\ j_1<j_2}} (\pm)^{i_1+i_2+j_1+j_2} |A|_{2,\pm}(i_{12}, j_{12}) |B|_{1,\pm}(\breve{i}_{12}, \breve{j}_{12}) + |B|_\pm,
\end{aligned}
$$

where we use the short-hand notation $i_{12} = (i_1, i_2)$ and $j_{12} = (j_1, j_2)$. As a final remark we note that for the generalized Laplace formulas and the formula for the permanent/determinant of the sum of two matrices, the symmetric (permanent) and antisymmetric (determinant) case can be treated on equal footings. This is no longer the case for the product of matrices. While it is true that for a product AB of two matrices one has $|AB|_- = |A|_-|B|_-$ (the determinant of a product is the product of the determinants) we have that $|AB|_+ \neq |A|_+|B|_+$ in general.

Appendix C: Green's Functions and Lattice Symmetry

In Section 2.3 we introduced several important physical systems with periodic symmetry, such as carbon nanotubes and the graphene lattice. The discussion was simplified by the fact that we dealt with noninteracting electrons. In this appendix we work out the consequence of lattice symmetry for interacting systems and ask ourselves in which way the many-body states can be labeled by the crystal momentum vectors. The fact that this should be possible is suggested by (inverse) photoemission experiments on crystals which clearly measure the band structure. The deeper physical meaning of band structure of interacting systems should therefore be apparent from a calculation of the spectral function see Section 6.4. To elucidate these aspects we must study the symmetry properties of the interacting Green's function under lattice translations. For concreteness we consider the electronic system of a three-dimensional crystal with lattice vectors $\mathbf{v}_1$, $\mathbf{v}_2$, and $\mathbf{v}_3$. This means that every unit cell is repeated periodically along $\mathbf{v}_1$, $\mathbf{v}_2$, and $\mathbf{v}_3$.

The starting point of our discussion is the Hamiltonian (the electron charge is $q = -1$)

$$\hat{H} = \int d\mathbf{x}\, \hat{\psi}^\dagger(\mathbf{x}) \Big(-\frac{\nabla^2}{2} - \phi(\mathbf{r}) \Big) \hat{\psi}(\mathbf{x}) + \frac{1}{2} \int d\mathbf{x} d\mathbf{x}'\, v(\mathbf{x}, \mathbf{x}') \hat{\psi}^\dagger(\mathbf{x}) \hat{\psi}^\dagger(\mathbf{x}') \hat{\psi}(\mathbf{x}') \hat{\psi}(\mathbf{x}), \quad \text{(C.1)}$$

where we take the interaction to be $v(\mathbf{x}, \mathbf{x}') = v(\mathbf{r} - \mathbf{r}')$ and the potential ϕ generated by the atomic nuclei in the crystal to be spin-independent and lattice-periodic:

$$\phi(\mathbf{r} + \mathbf{v}_j) = \phi(\mathbf{r}) \qquad j = 1, 2, 3.$$

This symmetry is, strictly speaking, only valid in a truly infinite system. The mathematical treatment of a Hamiltonian with an infinite number of particles is ill-defined as the Schrödinger equation becomes a differential equation for a wavefunction with infinitely many coordinates. A real piece of solid is, of course, finite and the symmetry is an approximation that is very good for bulk electrons. Since for the theoretical treatment the full periodic symmetry is a clear advantage, we impose periodic boundary conditions on a large box (BvK boundary conditions), as we did in Section 2.3. For simplicity we choose the box with edges given by the vectors $\mathbf{V}_j = N_j \mathbf{v}_j$, each radiating from, say, the origin of our reference frame. In the language of Section 2.3 this choice corresponds to taking $\mathbf{N}_1 = (N_1, 0, 0)$, $\mathbf{N}_2 = (0, N_2, 0)$, and $\mathbf{N}_3 = (0, 0, N_3)$. The requirement of periodicity for a many-body state $|\Psi\rangle$ with N particles implies that $|\Psi\rangle$ satisfies

$$\langle \mathbf{x}_1 \ldots (\mathbf{r}_i + \mathbf{V}_j)\sigma_i \ldots \mathbf{x}_N | \Psi \rangle = \langle \mathbf{x}_1 \ldots \mathbf{x}_i \ldots \mathbf{x}_N | \Psi \rangle \quad \text{(C.2)}$$

for all $j = 1, 2, 3$. In this way we have as many conditions as spatial differential operators in the Schrödinger equation (since the kinetic energy operator is a second-order differential operator we also need to put periodic boundary conditions on the derivatives of the many-body wavefunctions). Although these conditions seem natural, they are not trivial in the presence of many-body interactions. This is because the many-body interaction is invariant under the simultaneous translation of all particles [i.e., $v(\mathbf{r}-\mathbf{r}') = v((\mathbf{r}+\mathbf{V}_j)-(\mathbf{r}'+\mathbf{V}_j))$] but not under the translation of a single particle. The Hamiltonian does not therefore have a symmetry compatible with the boundary conditions (C.2). To solve this problem we replace the two-body interaction by

$$v(\mathbf{r}-\mathbf{r}') = \frac{1}{\mathrm{V}} \sum_{\mathbf{k}} \tilde{v}_{\mathbf{k}}\, e^{\mathrm{i}\mathbf{k}\cdot(\mathbf{r}-\mathbf{r}')}, \tag{C.3}$$

where V is the volume of the box, $\tilde{v}_{\mathbf{k}}$ is the Fourier transform of v and the sum runs over all vectors such that $\mathbf{k}\cdot\mathbf{V}_j = 2\pi m_j$ for $j = 1, 2, 3$ and m_j integers. The interaction (C.3) satisfies $v(\mathbf{r}+\mathbf{V}_j-\mathbf{r}') = v(\mathbf{r}-\mathbf{r}')$ and becomes equal to the original interaction when we take the limit $\mathrm{V}\to\infty$. With this replacement and the BvK boundary conditions (C.2) the eigenvalue equation for the Hamiltonian $\hat{H}$ becomes well defined (of course the spatial integrations in (C.1) must be restricted to the box).

Let us now explore the lattice symmetry. We consider the total momentum operator of the system defined in (3.33):

$$\hat{\mathbf{P}} \equiv \frac{1}{2\mathrm{i}} \int d\mathbf{x} \left[\hat{\psi}^\dagger(\mathbf{x})\left(\boldsymbol{\nabla}\hat{\psi}(\mathbf{x})\right) - \left(\boldsymbol{\nabla}\hat{\psi}^\dagger(\mathbf{x})\right)\hat{\psi}(\mathbf{x})\right].$$

This operator has the property that

$$-\mathrm{i}\boldsymbol{\nabla}\hat{\psi}(\mathbf{x}) = \left[\hat{\psi}(\mathbf{x}), \hat{\mathbf{P}}\right]_- \quad ; \quad -\mathrm{i}\boldsymbol{\nabla}\hat{\psi}^\dagger(\mathbf{x}) = \left[\hat{\psi}^\dagger(\mathbf{x}), \hat{\mathbf{P}}\right]_- .$$

From these equations we deduce that

$$\hat{\psi}(\mathbf{r}+\mathbf{r}'\sigma) = e^{-\mathrm{i}\hat{\mathbf{P}}\cdot\mathbf{r}}\,\hat{\psi}(\mathbf{r}'\sigma)\,e^{\mathrm{i}\hat{\mathbf{P}}\cdot\mathbf{r}}, \tag{C.4}$$

with a similar equation with $\hat{\psi}^\dagger$ replacing $\hat{\psi}$. Equation (C.4) is readily checked by taking the gradient with respect to $\mathbf{r}$ on both side and verifying the condition that both sides are equal for $\mathbf{r} = \mathbf{r}'$. It is now readily seen that the unitary operator $e^{-\mathrm{i}\hat{\mathbf{P}}\cdot\mathbf{r}}$ is the operator that shifts all particle coordinates over the vector $\mathbf{r}$. This follows from the adjoint of (C.4) if we evaluate

$$\begin{aligned}
|\mathbf{r}_1+\mathbf{r}\,\sigma_1,\ldots,\mathbf{r}_N+\mathbf{r}\,\sigma_N\rangle &= \hat{\psi}^\dagger(\mathbf{r}_N+\mathbf{r}\,\sigma_N)\ldots\hat{\psi}^\dagger(\mathbf{r}_1+\mathbf{r}\,\sigma_1)|0\rangle \\
&= e^{-\mathrm{i}\hat{\mathbf{P}}\cdot\mathbf{r}}\hat{\psi}^\dagger(\mathbf{r}_N\sigma_N)\ldots\hat{\psi}^\dagger(\mathbf{r}_1\sigma_1)e^{\mathrm{i}\hat{\mathbf{P}}\cdot\mathbf{r}}|0\rangle \\
&= e^{-\mathrm{i}\hat{\mathbf{P}}\cdot\mathbf{r}}|\mathbf{x}_1,\ldots,\mathbf{x}_N\rangle,
\end{aligned}$$

where we use that the action of $\hat{\mathbf{P}}$ on the empty ket $|0\rangle$ is zero. It therefore follows in particular that the many-body states that satisfy the BvK boundary conditions (C.2) also satisfy

$$e^{\mathrm{i}\hat{\mathbf{P}}\cdot\mathbf{V}_j}|\Psi\rangle = |\Psi\rangle. \tag{C.5}$$

Let us now consider the Hamiltonian (C.1) with the spatial integrals restricted to the box. We first demonstrate that

$$\hat{H} = e^{-i\hat{\mathbf{P}}\cdot\mathbf{v}_j}\hat{H}e^{i\hat{\mathbf{P}}\cdot\mathbf{v}_j}, \qquad j = 1, 2, 3. \tag{C.6}$$

The one-body operator $\hat{H}_{\rm pot}$ of the potential energy is

$$\begin{aligned}\hat{H}_{\rm pot} &= \int_V d\mathbf{x}\, \phi(\mathbf{r})\hat{\psi}^\dagger(\mathbf{x})\hat{\psi}(\mathbf{x}) = \int_V d\mathbf{x}\, \phi(\mathbf{r})\hat{\psi}^\dagger(\mathbf{r}+\mathbf{v}_j\,\sigma)\hat{\psi}(\mathbf{r}+\mathbf{v}_j\,\sigma)\\ &= \int_V d\mathbf{x}\, e^{-i\hat{\mathbf{P}}\cdot\mathbf{v}_j}\hat{\psi}^\dagger(\mathbf{x})\hat{\psi}(\mathbf{x})e^{i\hat{\mathbf{P}}\cdot\mathbf{v}_j}\phi(\mathbf{r}) = e^{-i\hat{\mathbf{P}}\cdot\mathbf{v}_j}\hat{H}_{\rm pot}\, e^{i\hat{\mathbf{P}}\cdot\mathbf{v}_j}.\end{aligned}$$

Similarly we can check the same transform law for the kinetic energy operator and the interaction operator with the interaction (C.3). We have therefore proven (C.6) or equivalently

$$\left[\hat{H}, e^{i\hat{\mathbf{P}}\cdot\mathbf{v}_j}\right]_- = 0, \qquad j = 1, 2, 3.$$

The set of unitary operators $e^{i\hat{\mathbf{P}}\cdot\mathbf{v}}$ with $\mathbf{v} = n_1\mathbf{v}_1 + n_2\mathbf{v}_2 + n_3\mathbf{v}_3$ all commute with the Hamiltonian and we can thus find a common set of eigenstates. These eigenstates satisfy

$$e^{i\hat{\mathbf{P}}\cdot\mathbf{v}_j}|\Psi\rangle = e^{i\alpha_j}|\Psi\rangle,$$

where the eigenvalues are a pure phase factor. The conditions (C.5) on the eigenstates imply that (recall that $\mathbf{V}_j = N_j\mathbf{v}_j$)

$$\left(e^{i\hat{\mathbf{P}}\cdot\mathbf{v}_j}\right)^{N_j}|\Psi\rangle = |\Psi\rangle. \tag{C.7}$$

Then we see that the phases α_j are not arbitrary but have to fulfill $e^{i\alpha_j N_j} = 1$ and hence $\alpha_j = 2\pi m_j/N_j$ with m_j an integer. Notice that the dimensionless vectors $\tilde{\mathbf{k}} = (\alpha_1, \alpha_2, \alpha_3)$ are exactly the vectors of Section 2.3, see (2.16). Equation (C.7) tells us that we can label the eigenstates of $\hat{H}$ with a crystal momentum $\mathbf{k}$ and a remaining quantum number l, since for an arbitrary translation $\mathbf{v} = n_1\mathbf{v}_1 + n_2\mathbf{v}_2 + n_3\mathbf{v}_3$ we have

$$e^{i\hat{\mathbf{P}}\cdot\mathbf{v}}|\Psi_{\mathbf{k}\,l}\rangle = e^{in_1\alpha_1+n_2\alpha_2+n_3\alpha_3}|\Psi_{\mathbf{k}\,l}\rangle = e^{i\tilde{\mathbf{k}}\cdot\mathbf{n}}|\Psi_{\mathbf{k}\,l}\rangle = e^{i\mathbf{k}\cdot\mathbf{v}}|\Psi_{\mathbf{k}\,l}\rangle, \tag{C.8}$$

where $\mathbf{k}$ is such that $\mathbf{k}\cdot\mathbf{v}_j = \alpha_j$.[1] Similarly to the $\tilde{\mathbf{k}}$ vectors, the $\mathbf{k}$ vectors that differ by a vector $\mathbf{K}$ or multiples thereof with the property that $\mathbf{K}\cdot\mathbf{v}_j = 2\pi$ must be identified. The set of inequivalent $\mathbf{k}$ vectors is called the *first Brillouin zone*. The vectors $\mathbf{K}$ are known as reciprocal lattice vectors. With this convention (C.8) represents the many-body generalization

[1]The vectors $\mathbf{k}$ have the physical dimension of the inverse of a length and should not be confused with the dimensionless vectors $\tilde{\mathbf{k}}$ of Section 2.3. Let us derive the relation between the two. The vectors $\tilde{\mathbf{k}}$ can be written as $\tilde{\mathbf{k}} = \sum_j \alpha_j\mathbf{e}_j$. On the other hand, the solution of $\mathbf{k}\cdot\mathbf{v}_j = \alpha_j$ is $\mathbf{k} = \sum_j \alpha_j\mathbf{b}_j$, where

$$\mathbf{b}_1 = \frac{\mathbf{v}_2\times\mathbf{v}_3}{\mathbf{v}_1\cdot(\mathbf{v}_2\times\mathbf{v}_3)}, \qquad \mathbf{b}_2 = \frac{\mathbf{v}_3\times\mathbf{v}_1}{\mathbf{v}_1\cdot(\mathbf{v}_2\times\mathbf{v}_3)}, \qquad \mathbf{b}_3 = \frac{\mathbf{v}_1\times\mathbf{v}_2}{\mathbf{v}_1\cdot(\mathbf{v}_2\times\mathbf{v}_3)}.$$

In the special case that $\mathbf{v}_1 = (a_1, 0, 0)$, $\mathbf{v}_2 = (0, a_2, 0)$, and $\mathbf{v}_3 = (0, 0, a_3)$ are orthogonal we simply have $\mathbf{b}_j = \mathbf{e}_j/a_j$.

of the Bloch theorem for single-particle states. It tells us that a simultaneous translation of all particles over a lattice vector $\mathbf{v}$ changes the many-body states by a phase factor.

We are now ready to study the symmetry properties of the Green's function. Let us, for example, consider the expression (6.76) for $G^>$ in position basis. The labels m in our case are $\mathbf{k}\,l$ and we have

$$G^>(\mathbf{x},\mathbf{x}';\omega) = -2\pi\mathrm{i}\sum_{\mathbf{k}\,l} P_{\mathbf{k}\,l}(\mathbf{x})P^*_{\mathbf{k}\,l}(\mathbf{x}')\delta(\omega - [E_{N+1,\mathbf{k}\,l} - E_{N,0}])$$

with quasi-particle wavefunctions

$$P_{\mathbf{k}\,l}(\mathbf{x}) = \langle\Psi_{N,0}|\hat{\psi}(\mathbf{x})|\Psi_{N+1,\mathbf{k}\,l}\rangle.$$

Let us now see how the quasi-particle wavefunctions change under a lattice translation:

$$\begin{aligned} P_{\mathbf{k}\,l}(\mathbf{x}+\mathbf{v}) &= \langle\Psi_{N,0}|\hat{\psi}(\mathbf{x}+\mathbf{v})|\Psi_{N+1,\mathbf{k}\,l}\rangle \\ &= \langle\Psi_{N,0}|e^{-\mathrm{i}\hat{\mathbf{P}}\cdot\mathbf{v}}\hat{\psi}(\mathbf{x})e^{\mathrm{i}\hat{\mathbf{P}}\cdot\mathbf{v}}|\Psi_{N+1,\mathbf{k}\,l}\rangle = e^{\mathrm{i}\mathbf{k}\cdot\mathbf{v}}P_{\mathbf{k}\,l}(\mathbf{x}), \end{aligned} \tag{C.9}$$

where we assume that $e^{\mathrm{i}\hat{\mathbf{P}}\cdot\mathbf{v}}|\Psi_{N,0}\rangle = |\Psi_{N,0}\rangle$ – that is, we assume that the ground state with N particles has $\mathbf{k}=0$. In the limit of vanishing interactions the quasi-particle wavefunctions become equal to the single-particle eigenstates and we recover the well-known symmetry property of the single-particle Bloch orbitals under lattice translations. From (C.9) we can derive

$$G^>(\mathbf{x}+\mathbf{v},\mathbf{x}'+\mathbf{v}';\omega) = \sum_{\mathbf{k}} e^{\mathrm{i}\mathbf{k}\cdot(\mathbf{v}-\mathbf{v}')}G^>(\mathbf{x},\mathbf{x}';\mathbf{k},\omega),$$

where we define

$$G^>(\mathbf{x},\mathbf{x}';\mathbf{k},\omega) = -2\pi\mathrm{i}\sum_{l} P_{\mathbf{k}\,l}(\mathbf{x})P^*_{\mathbf{k}\,l}(\mathbf{x}')\delta(\omega - [E_{N+1,\mathbf{k}\,l} - E_{N,0}]).$$

Thus if we know the Green's function $G^>(\mathbf{x},\mathbf{x}';\mathbf{k},\omega)$ for $\mathbf{x}$ and $\mathbf{x}'$ in a given unit cell, then the Green's function $G^>(\mathbf{x},\mathbf{x}';\omega)$ in all units cells is readily calculated. The problem is thus reduced to calculating $G^>(\mathbf{x},\mathbf{x}';\mathbf{k},\omega)$ in a single unit cell. So far we worked in the position basis, but in general we can use any basis we like inside this unit cell. If we expand the field operators in a unit cell in a localized basis as

$$\hat{\psi}(\mathbf{x}) = \sum_{s\tau}\varphi_{s\tau}(\mathbf{x})\hat{d}_{s\tau},$$

where s is a label for basis functions in the unit cell and τ is a spin index, then in the new basis we have

$$G^>_{s\tau\,s'\tau'}(\mathbf{k},\omega) = -2\pi\mathrm{i}\sum_{l} P_{\mathbf{k}\,l}(s\tau)P^*_{\mathbf{k}\,l}(s'\tau')\delta(\omega - [E_{N+1,\mathbf{k}\,l} - E_{N,0}]).$$

Here,

$$P_{\mathbf{k}\,l}(s\tau) = \langle\Psi_{N,0}|\hat{d}_{s\tau}|\Psi_{N+1,\mathbf{k}\,l}\rangle = \int_{\mathrm{V}} d\mathbf{x}\,P_{\mathbf{k}\,l}(\mathbf{x})\varphi^*_{s\tau}(\mathbf{x}).$$

A commonly used orthonormal basis in the unit cell is the set of functions

$$\varphi_{\mathbf{K}\tau}(\mathbf{x}) = \frac{\delta_{\sigma\tau}}{\sqrt{\mathrm{v}}}\, e^{\mathrm{i}\mathbf{K}\cdot\mathbf{r}}, \tag{C.10}$$

where v is the volume of the unit cell and $\mathbf{K}$ is a reciprocal lattice vector. An advantage of using this basis is that the set of plane waves $e^{\mathrm{i}(\mathbf{k}+\mathbf{K})\cdot\mathbf{r}}$ with $\mathbf{k}$ in the first Brillouin zone and $\mathbf{K}$ a reciprocal lattice vector is the same set used to expand the interaction in (C.3). Therefore the Coulomb integrals are especially simple in this basis.

Appendix D: Thermodynamics and Quantum Statistical Mechanics

Thermodynamics We consider a system with a given number N of particles in a box of volume $\mathtt{V}$ and at temperature T. The first law of thermodynamics establishes a relation between the change of energy dE of the system, the work $\delta\mathcal{L}$ done by the system, and the heat δQ absorbed by the system:

$$dE = \delta Q - \delta\mathcal{L}. \tag{D.1}$$

In (D.1) the infinitesimals δQ and $\delta\mathcal{L}$ are not exact differentials since the total absorbed heat or the total work done depend on the path of the transformation, which brings the system from one state to another and not only on the initial and final states. The second law of thermodynamics establishes that

$$\delta Q \leq Td\mathrm{S},$$

where S is the entropy and the equality holds only for reversible processes. In what follows we consider only reversible processes and write $\delta Q = Td\mathrm{S}$. As the only work that the system can make is to change its volume, we have $\delta\mathcal{L} = Pd\mathtt{V}$ where P is the internal pressure. Therefore, (D.1) takes the form

$$dE = Td\mathrm{S} - Pd\mathtt{V}.$$

From this relation we see that the internal energy $E = E(\mathrm{S}, \mathtt{V}, N)$ is a function of the entropy, the volume, and the number of particles.

If we now allow also the number of particles to change, then dE acquires the extra term $(\partial E/\partial N)_{\mathrm{S},T}\, dN$. The quantity

$$\mu \equiv \left(\frac{\partial E}{\partial N}\right)_{\mathrm{S},T}$$

is known as the chemical potential and represents the energy cost of adding a particle to the system. In conclusion, the change in energy of a system that can exchange heat and particles and that can expand its volume reads

$$dE = Td\mathrm{S} - Pd\mathtt{V} + \mu dN. \tag{D.2}$$

The internal energy E is not the most convenient quantity to work with due to its explicit dependence on S, a difficult quantity to control and measure. The only situation for which we do not need to bother about S is that of a system at zero temperature, since then $Td\mathrm{S} = 0$. In fact, the energy is mostly used in this circumstance. Experimentally, it is much easier to control and measure the temperature, the pressure, and the volume. We can then introduce more convenient quantities by a Legendre transformation. The *free energy* or *Helmholtz energy* is defined as

$$F = E - T\mathrm{S} \qquad \Rightarrow \qquad dF = -\mathrm{S}dT - Pd\mathtt{V} + \mu dN \tag{D.3}$$

and depends on T, $\mathtt{V}$, and N. From it we can also define the *Gibbs energy,*

$$G = F + P\mathtt{V} \qquad \Rightarrow \qquad dG = -\mathrm{S}dT + \mathtt{V}dP + \mu dN, \tag{D.4}$$

which instead depends on T, P, and N. We see that in G the only extensive variable is the number of particles, and therefore

$$G(T, P, N) = N\mu(T, P), \tag{D.5}$$

according to which the chemical potential is the Gibbs energy per particle. For later purposes we also define another important thermodynamic quantity known as the *grand potential,*

$$\Omega = F - \mu N \qquad \Rightarrow \qquad d\Omega = -\mathrm{S}dT - Pd\mathtt{V} - Nd\mu, \tag{D.6}$$

which depends explicitly on the chemical potential μ. From the knowledge of these thermodynamic energies we can extract all thermodynamic quantities by differentiation. For instance, from (D.2) we have

$$\left(\frac{\partial E}{\partial \mathrm{S}}\right)_{\mathtt{V},N} = T, \qquad \left(\frac{\partial E}{\partial \mathtt{V}}\right)_{\mathrm{S},N} = -P, \qquad \left(\frac{\partial E}{\partial N}\right)_{\mathrm{S},\mathtt{V}} = \mu,$$

which can be conveniently shortened as

$$(\partial_{\mathrm{S}}, \partial_{\mathtt{V}}, \partial_N)\, E = (T, -P, \mu).$$

In this compact notation it is evident that E depends on the independent variables $(\mathrm{S}, \mathtt{V}, N)$. Similarly we can write

$$\begin{aligned}
(\partial_T, \partial_{\mathtt{V}}, \partial_N)\, F &= (-\mathrm{S}, -P, \mu), \\
(\partial_T, \partial_P, \partial_N)\, G &= (-\mathrm{S}, \mathtt{V}, \mu), \\
(\partial_T, \partial_{\mathtt{V}}, \partial_\mu)\, \Omega &= (-\mathrm{S}, -P, -N).
\end{aligned}$$

Let us briefly summarize what we have seen so far. By combining the first and second laws of thermodynamics we are able to express the *differential* of the various energies in terms of the basic thermodynamic variables (S, T, P, $\mathtt{V}$, N, μ). Using (D.5) we can *integrate dG* and obtain $G = \mu N$. This result is particularly important since we can now give an explicit form to all other energies. If we insert (D.5) into (D.4) we find

$$F = -P\mathtt{V} + \mu N$$

and similarly we can construct the energy and the grand potential

$$E = TS - P\mathrm{V} + \mu N, \qquad \Omega = -P\mathrm{V}. \tag{D.7}$$

The grand potential is simply the product of pressure and volume, and recalling that its independent variables are T, V, μ, we have

$$d\Omega = -\left(\frac{\partial P}{\partial T}\right)_{\mathrm{V},\mu} \mathrm{V}dT - \left(\frac{\partial P}{\partial \mathrm{V}}\right)_{T,\mu} \mathrm{V}d\mathrm{V} - \left(\frac{\partial P}{\partial \mu}\right)_{T,\mathrm{V}} \mathrm{V}d\mu - Pd\mathrm{V}.$$

Comparing this result with (D.6), we find the important relations

$$\mathrm{S} = \left(\frac{\partial P}{\partial T}\right)_{\mathrm{V},\mu} \mathrm{V}, \qquad N = \left(\frac{\partial P}{\partial \mu}\right)_{T,\mathrm{V}} \mathrm{V}, \tag{D.8}$$

along with the obvious one $(\partial P/\partial \mathrm{V})_{T,\mu} = 0$ (the pressure of a system with a given temperature and density is the same for all volumes). The second part of (D.8) was used in Section 8.2.2 to calculate the pressure of an electron gas in the Hartree approximation.

Statistical approach All relations found so far are useless without a microscopic way to calculate the energies. Statistical mechanics, in its classical or quantum version, provides the bridge between the microscopic laws governing the motion of particles and the macroscopic quantities of thermodynamics. The connection between statistical mechanics and thermodynamics is due to Boltzmann. Let E_n be one of the possible energies of the system with N_n particles and let us consider $\mathcal{M}$ identical copies of the same system. This hypersystem is usually referred to as an *ensemble*. Suppose that in the ensemble there are M_1 systems with energy E_1 and number of particles N_1, M_2 systems with energy E_2 and number of particles N_2, etc. with, of course, $M_1 + M_2 + \ldots = \mathcal{M}$. The total energy of the ensemble is therefore $\mathcal{E} = \sum_n M_n E_n$, while the total number of particles is $\mathcal{N} = \sum_n M_n N_n$. Let us calculate the degeneracy of level $\mathcal{E}$. This is a simple combinatorial problem. Consider, for instance, an ensemble of 12 copies like this:

E_1	E_2	E_3	E_2
E_3	E_3	E_1	E_2
E_2	E_1	E_2	E_3

We have $M_1 = 3$, $M_2 = 5$, and $M_3 = 4$, and the total energy of the ensemble is $\mathcal{E} = 3E_1 + 5E_2 + 4E_3$. The degeneracy $d_{\mathcal{E}}$ of level $\mathcal{E}$ is

$$d_{\mathcal{E}} = \frac{12!}{3!\,5!\,4!} = 27720.$$

In general, the number of ways to have M_1 systems with energy E_1, M_2 systems with energy E_2, etc. over an ensemble of $M_1 + M_2 + \ldots = \mathcal{M}$ copies is given by

$$d_{\mathcal{E}} = \frac{\mathcal{M}!}{M_1!\,M_2!\ldots}.$$

If the ensemble is formed by a very large number of copies, $\mathcal{M} \gg 1$, then also $M_k \gg 1$ and it is more convenient to work with the logarithm of the degeneracy. For a large number P we can use the *Stirling formula*:

$$\ln P! \sim P \ln P - P$$

so that

$$\begin{aligned}\ln d_{\mathcal{E}} &\sim \mathcal{M} \ln \mathcal{M} - \mathcal{M} - \sum_n (M_n \ln M_n - M_n) \\ &= -\mathcal{M} \sum_n w_n \ln w_n,\end{aligned}$$

where $w_n \equiv M_n/\mathcal{M}$ can be interpreted as the probability of finding one of the systems of the ensemble in a state of energy E_n. It is now reasonable to expect that an ensemble in equilibrium is an ensemble in the most probable configuration – that is, an ensemble that maximizes the degeneracy $d_{\mathcal{E}}$. Let us then study what we get if we maximize the quantity

$$\mathrm{S}[\{w_n\}] \equiv -K_{\mathrm{B}} \sum_n w_n \ln w_n, \tag{D.9}$$

under the constraints that the probabilities w_n sum up to 1, $\sum_n w_n = 1$, that the average energy is E, $\sum_n w_n E_n = E$, and that the average number of particles is N, $\sum_n w_n N_n = N$. The constant K_{B} in (D.9) is completely irrelevant for our purpose and its presence is justified below. With the help of three Lagrange multipliers we must find the unconstrained maximum of the function

$$\begin{aligned}\frac{1}{K_{\mathrm{B}}} \tilde{\mathrm{S}}[\{w_n\}, \lambda_1, \lambda_2, \lambda_3] = &-\sum_n w_n \ln w_n - \lambda_1 \Big(\sum_n w_n - 1\Big) \\ &- \lambda_2 \Big(\sum_n w_n E_n - E\Big) - \lambda_3 \Big(\sum_n w_n N_n - N\Big)\end{aligned}$$

with respect to all w_n and the three λ. Setting to zero the derivative of $\tilde{\mathrm{S}}$ with respect to w_n, we find

$$\frac{\partial \tilde{\mathrm{S}}}{\partial w_n} = -\ln w_n - 1 - \lambda_1 - \lambda_2 E_n - \lambda_3 N_n = 0,$$

from which it follows that

$$w_n = e^{-(1+\lambda_1) - \lambda_2 E_n - \lambda_3 N_n}.$$

This solution is a maximum since $\frac{\partial^2 \tilde{\mathrm{S}}}{\partial w_n \partial w_m} = -\frac{\delta_{mn}}{w_n} < 0$. To find the Lagrange multipliers we use the constraints. The first constraint $\sum_n w_n = 1$ yields

$$e^{1+\lambda_1} = \sum_n e^{-\lambda_2 E_n - \lambda_3 N_n} \equiv Z, \qquad \Rightarrow \qquad w_n = \frac{e^{-\lambda_2 E_n - \lambda_3 N_n}}{Z}. \tag{D.10}$$

Choosing λ_2 and λ_3 so as to satisfy the constraints

$$\sum_n \underbrace{\frac{e^{-\lambda_2 E_n - \lambda_3 N_n}}{Z}}_{w_n} E_n = E, \qquad \sum_n \underbrace{\frac{e^{-\lambda_2 E_n - \lambda_3 N_n}}{Z}}_{w_n} N_n = N,$$

then $\tilde{\mathrm{S}}$ evaluated at the w_n of (D.10) is equal to

$$\mathrm{S} = K_{\mathrm{B}} \left(\lambda_2 \sum_n w_n E_n + \lambda_3 \sum_n w_n N_n + \ln Z \right) = K_{\mathrm{B}}(\lambda_2 E + \lambda_3 N + \ln Z). \tag{D.11}$$

The groundbreaking idea of Boltzmann was to identify this quantity with the entropy. The constant $K_{\mathrm{B}} = 8.3 \times 10^{-5}$ eV/K, duly called the *Boltzmann constant,* was chosen to fit the first thermodynamic relation in (D.7). In fact, from the comparison between (D.11) and (D.7) we see that the Boltzmann idea is very sound and sensible. The physical meaning of λ_2, λ_3, and Z can be deduced by equating these two formulas; what we find is

$$\lambda_2 = \beta = \frac{1}{K_{\mathrm{B}}T}, \qquad \lambda_3 = -\beta\mu, \qquad \ln Z = \beta P \mathrm{V} = -\beta\Omega. \tag{D.12}$$

Connection to quantum mechanics In quantum mechanics the possible values of the energies E_k are the eigenvalues of the Hamiltonian operator $\hat{H}$. Then the quantity Z, also called the partition function, can be written as

$$Z = \sum_k e^{-\beta(E_k - \mu N_k)} = \mathrm{Tr}\left[e^{-\beta(\hat{H} - \mu\hat{N})} \right].$$

It is possible to show that all the thermodynamics derivatives agree with the Boltzmann idea. For instance, from (D.6) we have

$$-\left(\frac{\partial}{\partial\beta}\beta\Omega \right)_{\mathrm{V},\mu} = -\Omega - \beta \left(\frac{\partial\Omega}{\partial T} \right)_{\mathrm{V},\mu} \frac{\partial T}{\partial\beta} = P\mathrm{V} - T\mathrm{S} = -E + \mu N,$$

and the same result follows from (D.12) since

$$-\left(\frac{\partial}{\partial\beta}\beta\Omega \right)_{\mathrm{V},\mu} = \left(\frac{\partial \ln Z}{\partial\beta} \right)_{\mathrm{V},\mu} = -\frac{\mathrm{Tr}\left[e^{-\beta(\hat{H} - \mu\hat{N})} (\hat{H} - \mu\hat{N}) \right]}{\mathrm{Tr}\left[e^{-\beta(\hat{H} - \mu\hat{N})} \right]} = -E + \mu N.$$

Similarly, from (D.6) we have

$$\left(\frac{\partial\Omega}{\partial\mu} \right)_{\mathrm{V},T} = -N,$$

and the same result follows from (D.12) since

$$\left(\frac{\partial\Omega}{\partial\mu} \right)_{\mathrm{V},T} = -\frac{1}{\beta} \left(\frac{\partial \ln Z}{\partial\mu} \right)_{\mathrm{V},T} = -\frac{\mathrm{Tr}\left[e^{-\beta(\hat{H} - \mu\hat{N})} \hat{N} \right]}{\mathrm{Tr}\left[e^{-\beta(\hat{H} - \mu\hat{N})} \right]} = -N.$$

The statistical approach to thermodynamics also allows us to understand the positivity of the temperature from a microscopic point of view. By increasing the average energy E, the degeneracy $d_{\mathcal{E}}$ increases and hence also the entropy increases, $\partial\mathrm{S}/\partial E > 0$. From (D.2) we see that $\partial\mathrm{S}/\partial E = 1/T$ and hence the temperature must be positive.

In conclusion, the procedure to extract thermodynamic quantities from a system described by a Hamiltonian $\hat{H}$ with eigenkets $|\Psi_k\rangle$ of energy E_k and number of particles N_k is:

- Calculate the partition function

$$Z = \sum_k e^{-\beta(E_k - \mu N_k)},$$

which depends on μ, T, V (the dependence on V comes from the Hamiltonian).

- Calculate the grand potential

$$\Omega = -\frac{1}{\beta} \ln Z.$$

- To calculate the energy we can use

$$E - \mu N = -\frac{\partial}{\partial \beta} \ln Z. \tag{D.13}$$

In this way, however, the energy depends on μ, T, V. To eliminate μ in favor of N we must invert

$$N = \frac{1}{\beta}\frac{\partial}{\partial \mu} \ln Z. \tag{D.14}$$

- To find the equation of state of the system, we can use the third of (D.12) and eliminate μ in favor of N as described above.

- Other thermodynamics quantities of interest to the reader can be obtained with similar techniques.

Free fermions and bosons Let us apply the procedure to systems of noninteracting fermions and bosons. We denote by ϵ_i the one-particle energy levels. The total energy E_k of a generic many-body state with $N_k = N[\{n_i\}] = \sum_i n_i$ particles can be written as

$$E[\{n_i\}] = \sum_i \epsilon_i n_i,$$

where $n_i = 0, 1$ for fermions and $n_i = 0, 1, 2, \ldots, \infty$ for bosons. Then, the partition function is given by

$$Z = \sum_{n_1, n_2, \ldots} e^{-\beta(E[\{n_i\}] - \mu N[\{n_i\}])} = \prod_i \sum_{n_i} e^{-\beta(\epsilon_i - \mu) n_i} = \prod_i (1 \mp e^{-\beta(\epsilon_i - \mu)})^{\mp 1},$$

where the upper sign refers to bosons and the lower sign to fermions. The average number of particles is obtained from (D.14) and reads

$$N = \sum_i f_i, \quad f_i = \frac{1}{e^{\beta(\epsilon_i - \mu)} \mp 1}.$$

We see that in thermal equilibrium we can express the one-particle energies in terms of the average occupations:

$$\epsilon_i - \mu = \frac{1}{\beta} \ln \frac{1 \pm f_i}{f_i}.$$

Using (D.13) we find

$$E - \mu N = \sum_i (\epsilon_i - \mu) f_i = \frac{1}{\beta} \sum_i f_i \ln \frac{1 \pm f_i}{f_i}. \tag{D.15}$$

Therefore the entropy, see (D.11), can be written as

$$\mathrm{S} = K_{\mathrm{B}} \beta (E - \mu N) + K_{\mathrm{B}} \ln Z = -K_{\mathrm{B}} \sum_i \Big[f_i \ln f_i \mp (1 \pm f_i) \ln(1 \pm f_i) \Big]. \tag{D.16}$$

Appendix E: Density Matrices and Pair Correlation Function

In Section 5.1 we observed that from a knowledge of the *equal-time* n-particle Green's function we can calculate the time-dependent ensemble average of any n-body operator. Let us introduce the time-dependent n-particle density matrix Γ_n:

$$\begin{aligned}\Gamma_n(\mathbf{x}_1,...,\mathbf{x}_n;\mathbf{x}'_1,...,\mathbf{x}'_n|t) &= (\pm\mathrm{i})^n \lim_{\substack{z_i,z'_i\to t_- \\ z'_1>...>z'_n>z_n>...>z_1}} G_n(1,...,n;1',...,n') \\ &= \mathrm{Tr}\left[\hat{\rho}\,\hat{\psi}^\dagger_H(\mathbf{x}'_1,t)...\hat{\psi}^\dagger_H(\mathbf{x}'_n,t)\;\hat{\psi}_H(\mathbf{x}_n,t)...\hat{\psi}_H(\mathbf{x}_1,t)\right] \\ &= \mathrm{Tr}\Big[\underbrace{\hat{U}(t,t_0)\,\hat{\rho}\,\hat{U}(t_0,t)}_{\hat{\rho}(t)}\,\hat{\psi}^\dagger(\mathbf{x}'_1)...\hat{\psi}^\dagger(\mathbf{x}'_n)\;\hat{\psi}(\mathbf{x}_n)...\hat{\psi}(\mathbf{x}_1)\Big],\end{aligned} \tag{E.1}$$

where in the last equality we use the cyclic property of the trace. The reader can check that the same result would follow by choosing t_+ instead of t_- in the first line of (E.1). From Γ_n we can calculate the time-dependent ensemble average of any n-body operator. The one-particle density matrix $\Gamma_1(\mathbf{x};\mathbf{x}'|t)$ is the same as the quantity $n(\mathbf{x},\mathbf{x}',t)$ defined in Section 8.3. From the one-particle density matrix we can, for instance, calculate the time-dependent ensemble average of the noninteracting Hamiltonian $\hat{H}_0$

$$E_{\mathrm{one}}(t) = \mathrm{Tr}[\hat{\rho}\,\hat{H}_{0,H}(t)] = \mathrm{Tr}[\hat{\rho}(t)\,\hat{H}_0(t)] = \int d\mathbf{x}d\mathbf{y}\langle\mathbf{x}|\hat{h}(t)|\mathbf{y}\rangle\Gamma_1(\mathbf{y};\mathbf{x}|t).$$

Taking, for example, the Hamiltonian $\hat{h}$ with matrix elements (1.13), we find

$$E_{\mathrm{one}}(t) = \sum_{\sigma'\sigma}\int d\mathbf{r}\; h_{\sigma'\sigma}(\mathbf{r},-\mathrm{i}\boldsymbol{\nabla},\mathbf{S})\;\Gamma_1(\mathbf{r}\sigma;\mathbf{r}'\sigma'|t)\Big|_{\mathbf{r}'=\mathbf{r}}.$$

Similarly, from the *two-particle density matrix* we can, for instance, calculate the time-dependent ensemble average of the interaction Hamiltonian:

$$E_{\mathrm{int}}(t) = \mathrm{Tr}[\hat{\rho}\,\hat{H}_{\mathrm{int},H}(t)] = \mathrm{Tr}[\hat{\rho}(t)\,\hat{H}_{\mathrm{int}}] = \frac{1}{2}\int d\mathbf{x}d\mathbf{x}'\,v(\mathbf{x},\mathbf{x}')\Gamma_2(\mathbf{x},\mathbf{x}';\mathbf{x},\mathbf{x}'|t). \tag{E.2}$$

Properties of the one-particle density matrix For any time t the one-particle density matrix is Hermitian in the position–spin indices:

$$\boxed{\Gamma_1(\mathbf{y};\mathbf{x}|t) = \Gamma_1^*(\mathbf{x};\mathbf{y}|t)}$$

and its trace equals the average number of particles:

$$\boxed{\int d\mathbf{x}\, \Gamma_1(\mathbf{x},\mathbf{x}|t) = \int d\mathbf{x}\, n(\mathbf{x},t) = N(t)} \tag{E.3}$$

The eigenfunctions $\phi_k(t)$ of $\Gamma_1(t)$ are defined as the solution of the secular problem

$$\int d\mathbf{x}\, \Gamma_1(\mathbf{y};\mathbf{x}|t)\phi_k(\mathbf{x},t) = n_k(t)\phi_k(\mathbf{y},t), \tag{E.4}$$

and they are called the *natural orbitals.* The natural orbitals can always be chosen orthonormal and their set forms a basis in the one-particle Hilbert space. The eigenvalues $n_k(t)$ can be interpreted as the occupation of the natural orbitals for the following reasons. In the first place, they sum up to $N(t)$ due to (E.3):

$$\sum_k n_k(t) = N(t).$$

Second, if we multiply both sides of (E.4) by $\phi_k^*(\mathbf{y},t)$ and integrate over $\mathbf{y}$ we find

$$n_k(t) = \int d\mathbf{x}d\mathbf{y}\, \phi_k^*(\mathbf{y},t) \underbrace{\mathrm{Tr}\left[\hat{\rho}(t)\, \hat{\psi}^\dagger(\mathbf{x})\hat{\psi}(\mathbf{y})\right]}_{\Gamma_1(\mathbf{y};\mathbf{x}|t)} \phi_k(\mathbf{x},t) = \mathrm{Tr}\left[\hat{\rho}(t)\, \hat{d}_k^\dagger(t)\hat{d}_k(t)\right],$$

where the $\hat{d}$-operators are defined as in (1.52) and (1.53), the only difference being that the wavefunctions depend parametrically on time. Thus, $n_k(t)$ is the quantum average of the occupation operator of the time-dependent kth natural orbital. In fermionic systems the eigenvalues of $\hat{d}_k^\dagger \hat{d}_k$ are either 0 or 1, see (1.91), and therefore[1]

$$0 \le n_k(t) \le 1.$$

If at a certain time t the time-dependent density matrix $\hat{\rho}(t) = |\Psi\rangle\langle\Psi|$ is a pure state, hence $\mathrm{Tr}\left[\hat{\rho}(t)\, \hat{\psi}^\dagger(\mathbf{x})\hat{\psi}(\mathbf{y})\right] = \langle\Psi|\hat{\psi}^\dagger(\mathbf{x})\hat{\psi}(\mathbf{y})|\Psi\rangle$, and $|\Psi\rangle = |m_1 \ldots m_N\rangle$ is the permanent/determinant of N functions belonging to some orthonormal basis $\varphi_m(\mathbf{x}) = \langle\mathbf{x}|m\rangle$, then the natural orbitals are simply the functions of the basis (i.e., $\phi_k(t) = \varphi_k$), as can easily be checked. If the quantum number k appears N_k times in the string $m_1 \ldots m_N$, then the eigenvalues $n_k(t)$ are equal to N_k.

[1] Any fermionic ket $|\Psi\rangle$ can be written as the linear combination of two eigenkets $|\Psi_0\rangle$, $|\Psi_1\rangle$ of the occupation operator $\hat{d}_k^\dagger \hat{d}_k$ with eigenvalues 0 and 1, respectively: $|\Psi\rangle = \alpha|\Psi_0\rangle + \beta|\Psi_1\rangle$. Then, $\langle\Psi|\hat{d}_k^\dagger \hat{d}_k|\Psi\rangle = |\beta|^2 \le 1$ since the normalization of $|\Psi\rangle$ requires that $|\alpha|^2 + |\beta|^2 = 1$.

Properties of the n-particle density matrices for ensembles with fixed number of particles Let us consider a density matrix operator $\hat{\rho}(t) = \sum_q w_q|\chi_q\rangle\langle\chi_q|$, where each state of the ensemble is an eigenstate of the total number of particle operator with eigenvalue N. Then $\Gamma_n = 0$ for all $n > N$. The $N-1$-particle density matrix can be obtained from Γ_N by integrating out one coordinate. Setting $\mathbf{y}_N = \mathbf{x}_N$, we have

$$\begin{aligned}
&\int d\mathbf{x}_N\Gamma_N(\mathbf{y}_1,\dots,\mathbf{y}_{N-1},\mathbf{x}_N;\mathbf{x}_1,\dots,\mathbf{x}_N|t)\\
&\quad= \sum_q w_q\langle\chi_q|\,\hat{\psi}^\dagger(\mathbf{x}_1)\dots\hat{\psi}^\dagger(\mathbf{x}_{N-1})\underbrace{\left(\int d\mathbf{x}_N\,\hat{\psi}^\dagger(\mathbf{x}_N)\,\hat{\psi}(\mathbf{x}_N)\right)}_{\hat{N}}\hat{\psi}(\mathbf{y}_{N-1})\dots\hat{\psi}(\mathbf{y}_1)|\chi_q\rangle\\
&\quad= \sum_q w_q\langle\chi_q|\hat{\psi}^\dagger(\mathbf{x}_1)\dots\hat{\psi}^\dagger(\mathbf{x}_{N-1})\,\hat{\psi}(\mathbf{y}_{N-1})\dots\hat{\psi}(\mathbf{y}_1)|\chi_q\rangle\\
&\quad= \Gamma_{N-1}(\mathbf{y}_1,\dots,\mathbf{y}_{N-1};\mathbf{x}_1,\dots,\mathbf{x}_{N-1}|t),
\end{aligned}$$

where we use that the operator $\hat{N}$ acts on a one-particle state. We can continue this procedure and integrate out coordinate $\mathbf{y}_{N-1} = \mathbf{x}_{N-1}$. We then obtain again an expression involving the operator $\hat{N}$ that now acts on a two-particle state finally yielding $2\Gamma_{N-2}$. Subsequently integrating out more coordinates, we find

$$\begin{aligned}
&\Gamma_k(\mathbf{y}_1,\dots,\mathbf{y}_k;\mathbf{x}_1,\dots,\mathbf{x}_k|t)\\
&\quad= \frac{1}{(N-k)!}\int d\mathbf{x}_{k+1}\dots d\mathbf{x}_N\,\Gamma_N(\mathbf{y}_1,\dots,\mathbf{y}_k,\mathbf{x}_{k+1},\dots,\mathbf{x}_N;\mathbf{x}_1,\dots,\mathbf{x}_N|t). \quad \text{(E.5)}
\end{aligned}$$

Thus, the k-particle density matrix is obtained by integrating out $N-k$ coordinates out of the N-particle density matrix.

Probability interpretation We now discuss how the density matrices are related to the conditional probability of finding a particle in some point in space when a certain number of other particles are in given points in space. Let us, for simplicity, consider a system of spinless bosons in one dimension and in a pure state, hence $\hat{\rho}(t) = |\Psi\rangle\langle\Psi|$, with $|\Psi\rangle$ belonging to the N-particle Hilbert space $\mathbb{H}_N$. Then the states $|x_1\dots x_N\rangle$ with $x_1 \geq \dots \geq x_N$ form a basis in $\mathbb{H}_N$. Any state $|\Psi\rangle \in \mathbb{H}_N$ can be expanded as[2]

$$|\Psi\rangle = \int_{x_1\geq\dots\geq x_N} dx_1\dots dx_N|x_1\dots x_N\rangle\,\underbrace{\langle x_1\dots x_N|\Psi\rangle}_{\Psi(x_1,\dots,x_N)}$$

and the normalization condition reads

$$\begin{aligned}
1 = \langle\Psi|\Psi\rangle &= \int_{x_1\geq\dots\geq x_N} dx_1\dots dx_N\langle\Psi|x_1\dots x_N\rangle\langle x_1\dots x_N|\Psi\rangle\\
&= \frac{1}{N!}\int dx_1\dots dx_N\;\underbrace{|\Psi(x_1,\dots,x_N)|^2}_{\Gamma_N(x_1,\dots,x_N;x_1,\dots,x_N|t)}\;. \quad \text{(E.6)}
\end{aligned}$$

[2]The domain of integration with two or more bosons in the same position has zero measure and we therefore ignore that these states are not normalized to unity.

Let us now ask what is the probability density $p(z)$ for finding a particle at position z. We start with the first nontrivial case of $N = 2$ particles. The configuration space is given by the gray area in the figure:

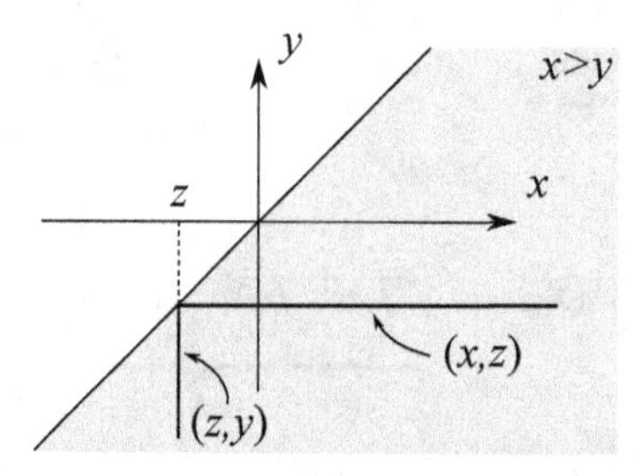

We then have

$$p(z) = \int_z^\infty dx\, |\Psi(x,z)|^2 + \int_{-\infty}^z dy\, |\Psi(z,y)|^2 = \int_{-\infty}^\infty dy |\Psi(z,y)|^2,$$

where in the last equality we use the symmetry of the wavefunction. We observe, however, that

$$\int_{-\infty}^\infty dz\, p(z) = \int_{-\infty}^\infty dz dy\, |\Psi(z,y)|^2 = 2,$$

as follows directly from (E.6). Mathematically the fact that $p(z)$ does not integrate to unity is not surprising since, according to (E.5) with $N = 2$ and $k = 1$, $p(z) = \Gamma_1(z;z|t)$ is the density of particles and hence it correctly integrates to 2. Intuitively, however, it seems strange that a probability does not integrate to unity and therefore an explanation is needed.

The point is that the event of finding a particle in z and the event of finding a particle in z' are not independent events. To calculate $p(z)$ we have to integrate $|\Psi(z,y)|^2$ over all y and this includes the point $y = z'$. Thus, among the events with a particle in z there is an event in which there is a particle in z'. This fact can most easily be illustrated by a simple discrete example. Consider two bosons that can either occupy state $|1\rangle$ or state $|2\rangle$. Then, $\mathbb{H}_2 = \{|11\rangle,\ |12\rangle,\ |22\rangle\}$. A general state describing these two particles is a linear combination of the basis states in $\mathbb{H}_2$:

$$|\Psi\rangle = \Psi(1,1)|11\rangle + \Psi(1,2)|12\rangle + \Psi(2,2)|22\rangle,$$

and if $|\Psi\rangle$ is normalized we have[3]

$$2|\Psi(1,1)|^2 + |\Psi(1,2)|^2 + 2|\Psi(2,2)|^2 = 1.$$

Now the probability to find a particle in state 1 is

$$p(1) = \left|\frac{\langle 11|\Psi\rangle}{\sqrt{2}}\right|^2 + |\langle 12|\Psi\rangle|^2 = 2|\Psi(1,1)|^2 + |\Psi(1,2)|^2.$$

[3]According to our convention, the states $|11\rangle$ and $|22\rangle$ are normalized to 2, see Section 1.3.

Similarly, the probability to find a particle in state 2 is

$$p(2) = |\Psi(1,2)|^2 + 2|\Psi(2,2)|^2.$$

However, the probability to find a particle in state 1 or in state 2 is not the sum of the two probabilities, since

$$p(1) + p(2) = 2|\Psi(1,1)|^2 + 2|\Psi(1,2)|^2 + 2|\Psi(2,2)|^2 = 1 + |\Psi(1,2)|^2.$$

In this way we double-count the state $|12\rangle$. The correct probability formula for overlapping event sets A and B is $p(A \cup B) = p(A) + p(B) - p(A \cap B)$. The fact that we do find a particle in state 1 does not exclude the fact that another particle can be found in state 2. This joint probability is $|\Psi(1,2)|^2$ and needs to be subtracted from $p(1) + p(2)$. In this way the probability to find a particle either in 1 or 2 is unity, as it should be.

Let us now calculate the probability density $p(z)$ for N bosons in one dimension. We have

$$\begin{aligned}
p(z) &= \int_{z \geq x_2 \geq \ldots \geq x_N} dx_2 \ldots dx_N |\Psi(z, x_2, \ldots, x_N)|^2 \\
&\quad + \int_{x_1 \geq z \geq x_3 \geq \ldots \geq x_N} dx_1 dx_3 \ldots dx_N |\Psi(x_1, z, x_3, \ldots, x_N)|^2 + \ldots \\
&= \int_{x_2 \geq \ldots \geq x_N} dx_2 \ldots dx_N |\Psi(z, x_2, \ldots, x_N)|^2 \\
&= \frac{1}{(N-1)!} \int dx_2 \ldots dx_N |\Psi(z, x_2, \ldots, x_N)|^2,
\end{aligned}$$

where we use the symmetry of the wavefunction. As in the previous case $p(z) = \Gamma_1(z; z|t)$ is the density of particles in z and it does not integrate to unity but to the total number of particles N. The reason is the same as before. Clearly the above derivation can readily be generalized to particles with spin, dimensions higher than 1, and to the fermionic case. Thus the physical interpretation of the density $n(\mathbf{x})$ is the probability of finding a particle in $\mathbf{x}$.

Coming back to our one-dimensional system of bosons, we now consider the joint probability $p(z, z')$ of finding a particle at z and another at z'. Since $p(z, z') = p(z', z)$, we can assume that $z > z'$. Then,

$$\begin{aligned}
p(z, z') &= \int_{z \geq z' \geq x_3 \geq \ldots \geq x_N} dx_3 \ldots dx_N |\Psi(z, z', x_3, \ldots, x_N)|^2 \\
&\quad + \int_{z \geq x_2 \geq z' \geq \ldots \geq x_N} dx_2 dx_4 \ldots dx_N |\Psi(z, x_2, z', \ldots, x_N)|^2 \\
&\quad + \ldots + \int_{x_1 \geq z \geq z' \geq \ldots \geq x_N} dx_1 dx_4 \ldots dx_N |\Psi(x_1, z, z', \ldots, x_N)|^2 + \ldots \\
&= \int_{x_3 \geq \ldots \geq x_N} dx_3 \ldots dx_N |\Psi(z, z', x_3, \ldots, x_N)|^2 \\
&= \frac{1}{(N-2)!} \int dx_3 \ldots dx_N |\Psi(z, z', x_3, \ldots, x_N)|^2,
\end{aligned}$$

where again the symmetry of the wavefunction has been used. We thus see that

$$p(z, z') = \Gamma_2(z, z'; z, z'|t)$$

is the two-particle density matrix. The same derivation is readily seen to be valid also for fermions and also in the case of arbitrary spin and spatial dimensions. Thus, more generally, we have $p(\mathbf{x}, \mathbf{x}') = \Gamma_2(\mathbf{x}, \mathbf{x}'; \mathbf{x}, \mathbf{x}'|t)$. The probability interpretation of Γ_2 gives a nice interpretation of the interaction energy (E.2):

$$E_{\text{int}}(t) = \frac{1}{2} \int d\mathbf{x} d\mathbf{x}' v(\mathbf{x}, \mathbf{x}') p(\mathbf{x}, \mathbf{x}').$$

The interaction energy is simply the product of the probability of finding a particle at $\mathbf{x}$ and another at $\mathbf{x}'$ times their interaction; the factor $1/2$ makes sure that we count each pair only once, see discussion before (1.81).

Pair correlation function Henceforth we do not explicitly write the time variable in the various quantities as time plays no role in the derivations. If points $\mathbf{r}$ and $\mathbf{r}'$ are far from each other, then it is reasonable to expect that the probability $p(\mathbf{x}, \mathbf{x}')$ becomes the independent product $n(\mathbf{x})n(\mathbf{x}')$ for any physically relevant state $|\Psi\rangle$. We therefore define the *pair correlation function*,

$$g(\mathbf{x}, \mathbf{x}') = \frac{p(\mathbf{x}, \mathbf{x}')}{n(\mathbf{x})n(\mathbf{x}')}. \tag{E.7}$$

It is easy to show that $g \to 1$ for $|\mathbf{r} - \mathbf{r}'| \to \infty$ when the state $|\Psi\rangle$ is the ground state of some homogeneous system so that $n(\mathbf{x}) = n(\mathbf{r}\sigma) = n_\sigma$ is independent of position and $p(\mathbf{x}, \mathbf{x}') = p_{\sigma\sigma'}(\mathbf{r} - \mathbf{r}')$ depends only on the coordinate difference. If the system is in a cubic box of volume $\mathrm{V} = L^3$, then

$$p_{\sigma\sigma'}(\mathbf{r} - \mathbf{r}') = \frac{1}{\mathrm{V}} \sum_{\mathbf{k}} e^{\mathrm{i}\mathbf{k}\cdot(\mathbf{r}-\mathbf{r}')} \tilde{p}_{\sigma\sigma'}(\mathbf{k}).$$

For large $|\mathbf{r} - \mathbf{r}'|$ we can restrict the sum over $\mathbf{k}$ to those wavevectors with $|\mathbf{k}| \lesssim 2\pi/|\mathbf{r} - \mathbf{r}'|$ since the sum over the wavevectors with $|\mathbf{k}| \gtrsim 2\pi/|\mathbf{r} - \mathbf{r}'|$ gives approximately zero. Thus, in the limit $|\mathbf{r} - \mathbf{r}'| \to \infty$ only the term with $\mathbf{k} = 0$ survives

$$\lim_{|\mathbf{r}-\mathbf{r}'| \to \infty} p_{\sigma\sigma'}(\mathbf{r} - \mathbf{r}') = \frac{1}{\mathrm{V}} \tilde{p}_{\sigma\sigma'}(0). \tag{E.8}$$

To calculate $\tilde{p}_{\sigma\sigma'}(0)$ we observe that by definition

$$p_{\sigma\sigma'}(\mathbf{r} - \mathbf{r}') = \langle\Psi|\hat{\psi}^\dagger(\mathbf{x})\hat{\psi}^\dagger(\mathbf{x}')\hat{\psi}(\mathbf{x}')\hat{\psi}(\mathbf{x})|\Psi\rangle = \langle\Psi|\hat{n}(\mathbf{x})\hat{n}(\mathbf{x}')|\Psi\rangle \pm \delta(\mathbf{x} - \mathbf{x}')n(\mathbf{x}). \tag{E.9}$$

Therefore,

$$\tilde{p}_{\sigma\sigma'}(0) = \int d\mathbf{r}\, p_{\sigma\sigma'}(\mathbf{r}) = N_\sigma n_{\sigma'} \pm \delta_{\sigma\sigma'} n_\sigma,$$

where we use that $|\Psi\rangle$ is an eigenstate of the total number of particles of spin σ operator $\hat{N}_\sigma = \int d\mathbf{r}\; n(\mathbf{x})$ with eigenvalue N_σ. Inserting this result into (E.8), taking the limit $\mathrm{V} \to \infty$, and using that $N_\sigma/\mathrm{V} = n_\sigma$, we get

$$\lim_{|\mathbf{r}-\mathbf{r}'| \to \infty} p_{\sigma\sigma'}(\mathbf{r} - \mathbf{r}') = n_\sigma n_{\sigma'},$$

which implies that $g \to 1$ in the same limit.

A formula that is commonly found in the literature is

$$E_{\rm int} = \frac{1}{2}\int d\mathbf{x}d\mathbf{x}' v(\mathbf{x},\mathbf{x}')n(\mathbf{x})n(\mathbf{x}') + \frac{1}{2}\int d\mathbf{x}d\mathbf{x}' v(\mathbf{x},\mathbf{x}')n(\mathbf{x})n(\mathbf{x}')\left[g(\mathbf{x},\mathbf{x}')-1\right].$$

The first term represents the classical (Hartree) interaction between two densities $n(\mathbf{x})$ and $n(\mathbf{x}')$, whereas the second term denotes the exchange-correlation part of the interaction.

In fact, g is important also because it can be measured using X-ray scattering [260]. Theoretically the quantity $p(\mathbf{x},\mathbf{x}')$ is closely related to the density response function, since in equilibrium systems we can rewrite (E.9) as

$$p(\mathbf{x},\mathbf{x}') = \langle\Psi|\hat{n}_H(\mathbf{x},t)\hat{n}_H(\mathbf{x}',t)|\Psi\rangle \pm \delta(\mathbf{x}-\mathbf{x}')n(\mathbf{x}).$$

In the first term we recognize the greater part of the density response function $\chi(\mathbf{x},z;\mathbf{x}',z')$ – that is,

$$\begin{aligned}\langle\Psi|\hat{n}_H(\mathbf{x},t)\hat{n}_H(\mathbf{x}',t)|\Psi\rangle &= n(\mathbf{x})n(\mathbf{x}') + \mathrm{i}\chi^>(\mathbf{x},t;\mathbf{x}',t)\\ &= n(\mathbf{x})n(\mathbf{x}') + \mathrm{i}\int\frac{d\omega}{2\pi}\chi^>(\mathbf{x},\mathbf{x}';\omega),\end{aligned}$$

where we use (13.16). This means that $p(\mathbf{x},\mathbf{x}')$ can be calculated from diagrammatic perturbation theory, as explained in Chapter 14.

Pair correlation function in the electron gas As an example we consider the electron gas. In momentum space the relation between p and χ reads

$$\tilde{p}_{\sigma\sigma'}(\mathbf{k}) = n_\sigma n_{\sigma'}(2\pi)^3\delta(\mathbf{k}) - \delta_{\sigma\sigma'}n_\sigma + \mathrm{i}\int\frac{d\omega}{2\pi}\chi^>_{\sigma\sigma'}(\mathbf{k},\omega). \tag{E.10}$$

We define $p=\sum_{\sigma\sigma'}p_{\sigma\sigma'}$ and similarly $\chi=\sum_{\sigma\sigma'}\chi_{\sigma\sigma'}$ and $n=\sum_\sigma n_\sigma$. Then, summing (E.10) over σ and σ' and using the fluctuation–dissipation theorem (13.20) for $\chi^>$, we get

$$\tilde{p}(\mathbf{k}) = n^2(2\pi)^3\delta(\mathbf{k}) - n + \mathrm{i}\int\frac{d\omega}{2\pi}\bar{f}(\omega)\left[\chi^{\rm R}(\mathbf{k},\omega)-\chi^{\rm A}(\mathbf{k},\omega)\right].$$

In this equation the difference inside the square brackets is $2\mathrm{i}\,\mathrm{Im}[\chi^{\rm R}(\mathbf{k},\omega)]$ since the property (13.26) implies that $\chi^{\rm A}(\mathbf{k},\omega)=[\chi^{\rm R}(\mathbf{k},\omega)]^*$. In the zero-temperature limit $\beta\to\infty$, the Bose function $\bar{f}(\omega)=1/(1-e^{-\beta\omega})$ vanishes for negative ω and is unity for positive ω. Therefore,

$$\tilde{p}(\mathbf{k}) = n^2(2\pi)^3\delta(\mathbf{k}) - n - \frac{1}{\pi}\int_0^\infty d\omega\,\mathrm{Im}\left[\chi^{\rm R}(\mathbf{k},\omega)\right].$$

Let us evaluate the frequency integral when χ is the Lindhard density response function discussed in Section 15.3. According to Fig. 15.3(a) and to equation (15.38) we have for $k=|\mathbf{k}|<2p_{\rm F}$

$$\begin{aligned}\int_0^\infty d\omega\,\mathrm{Im}\left[\chi^{\rm R}(\mathbf{k},\omega)\right] &= \epsilon_{p_{\rm F}}\int_0^{-x^2+2x}d\nu\left(-\frac{p_{\rm F}}{2\pi}\right)\frac{\nu}{2x}\\ &\quad+\epsilon_{p_{\rm F}}\int_{-x^2+2x}^{x^2+2x}d\nu\left(-\frac{p_{\rm F}}{2\pi}\right)\frac{1}{2x}\left[1-\left(\frac{\nu-x^2}{2x}\right)^2\right]\\ &= -\frac{p_{\rm F}^3}{4\pi}\left(x-\frac{x^3}{12}\right),\end{aligned}$$

whereas for $k > 2p_{\rm F}$

$$\int_0^\infty d\omega\ {\rm Im}\left[\chi^{\rm R}(\mathbf{k},\omega)\right] = \epsilon_{p_{\rm F}} \int_{x^2-2x}^{x^2+2x} d\nu \left(-\frac{p_{\rm F}}{2\pi}\right)\frac{1}{2x}\left[1-\left(\frac{\nu-x^2}{2x}\right)^2\right]$$
$$= -\frac{p_{\rm F}^3}{4\pi}\frac{4}{3}.$$

In these equations $x = k/p_{\rm F}$, $\nu = \omega/\epsilon_{p_{\rm F}}$, and $\epsilon_{p_{\rm F}} = p_{\rm F}^2/2$, see (15.33). Taking into account that $p_{\rm F}^3 = 3\pi^2 n$, see (8.42), we conclude that

$$\tilde{p}(\mathbf{k}) = \begin{cases} n^2(2\pi)^3\delta(\mathbf{k}) + n(-1+\frac{3x}{4}-\frac{x^3}{16}) & |\mathbf{k}| < 2p_{\rm F} \\ 0 & |\mathbf{k}| > 2p_{\rm F} \end{cases}.$$

Fourier transforming back to real space we find

$$\begin{aligned} p(\mathbf{r}) &= \int \frac{d\mathbf{k}}{(2\pi)^3} e^{i\mathbf{k}\cdot\mathbf{r}}\,\tilde{p}(\mathbf{k}) \\ &= n^2 + \frac{n}{2\pi^2 r}\int_0^{2p_{\rm F}} dk \left(-1+\frac{3x}{4}-\frac{x^3}{16}\right) k\sin kr \qquad (\text{setting } \alpha = p_{\rm F} r) \\ &= n^2 + \frac{3n^2}{2\alpha}\int_0^2 dx\left(-x+\frac{3x^2}{4}-\frac{x^4}{16}\right)\sin x\alpha \\ &= n^2 - \frac{9n^2}{2}\left(\frac{\sin\alpha - \alpha\cos\alpha}{\alpha^3}\right)^2. \end{aligned} \tag{E.11}$$

In the electron gas the pair correlation function (E.7) reads

$$g(\mathbf{x},\mathbf{x}') = g_{\sigma\sigma'}(\mathbf{r}-\mathbf{r}') = \frac{p_{\sigma\sigma'}(\mathbf{r}-\mathbf{r}')}{n_\sigma n_{\sigma'}}.$$

If we define $g = \frac{1}{4}\sum_{\sigma\sigma'} g_{\sigma\sigma'}$ and use that $n_\sigma = n/2$, then (E.11) implies

$$g(\mathbf{r}) = \frac{p(\mathbf{r})}{n^2} = 1 - \frac{9}{2}\left(\frac{\sin p_{\rm F}r - p_{\rm F}r\,\cos p_{\rm F}r}{(p_{\rm F}r)^3}\right)^2.$$

As expected, $g \to 1$ for $r \to \infty$. For small r we have $\sin\alpha - \alpha\cos\alpha = \alpha^3/3 + \ldots$ and hence $g(0) = 1/2$. This result can easily be interpreted. Since we use the noninteracting density response function we have only incorporated antisymmetry. Therefore, like-spin electrons are correlated through the Pauli exclusion principle, but unlike-spin electrons are not. If we go beyond the simple noninteracting case, then we find that $g(0) < 1/2$. In fact, if we use the RPA approximation for χ we would find that $g(0)$ can even become negative at small enough densities. Going beyond RPA repairs this situations. For a thorough discussion on these topics, see Refs. [108, 260, 380].

Variational principle with density matrices Before concluding the appendix we would like to draw the attention of the reader to an important point. In many physical situations one is interested in calculating the ground-state energy of the Hamiltonian $\hat{H} = \hat{H}_0 + \hat{H}_{\rm int}$.

From basic courses of quantum mechanics we know that this energy can be obtained, in principle, by minimizing the quantity $\langle\Psi|\hat{H}|\Psi\rangle$ over *all* possible (normalized) N-particle states $|\Psi\rangle$. This is in general a formidable task since the wavefunction $\Psi(\mathbf{x}_1,\ldots,\mathbf{x}_N)$ depends on N coordinates. We have just learned, however, that

$$\langle\Psi|\hat{H}|\Psi\rangle = \int d\mathbf{x}d\mathbf{y}\,\langle\mathbf{x}|\hat{h}|\mathbf{y}\rangle\,\Gamma_1(\mathbf{y};\mathbf{x}) + \frac{1}{2}\int d\mathbf{x}d\mathbf{x}'\,v(\mathbf{x},\mathbf{x}')\Gamma_2(\mathbf{x},\mathbf{x}';\mathbf{x},\mathbf{x}').$$

Couldn't we minimize the quantum average of the Hamiltonian with respect to Γ_1 and Γ_2? This would be a great achievement since Γ_1 and Γ_2 depend only on 2 and 4 coordinates respectively. Unfortunately the answer to the question is, at present, negative. While we know the constraints to construct physical one-particle density matrices (for instance, for fermions Γ_1 must be Hermitian with eigenvalues between 0 and 1 that sum up to N) we still do not know the constraints for Γ_2 [381].

Appendix F: Asymptotic Expansions

Very often we are interested in calculating functions $f(x)$ depending on a variable x when x is close to some value x_0 or when x approaches x_0 if $x_0 = \infty$. If no exact solution exists (or if the exact solution is exceedingly complicated), it is useful to devise "perturbative" methods to approximate f. These perturbative methods lead to an expansion of $f(x)$ in terms of functions $\varphi_n(x)$ that for $x \to x_0$ approach zero faster and faster with increasing n. MBPT is one such method. The set of functions $\varphi_n(x)$, $n = 0, 1, 2, \ldots$, are called an asymptotic set if

$$\lim_{x \to x_0} \frac{\varphi_{n+1}(x)}{\varphi_n(x)} = 0,$$

and the expansion

$$f(x) \sim \sum_{n=0}^{\infty} a_n \varphi_n(x) \tag{F.1}$$

is called an *asymptotic expansion* if

$$\lim_{x \to x_0} \frac{f(x) - \sum_{n=0}^{N} a_n \varphi_n(x)}{\varphi_N(x)} = 0 \quad \text{for all } N. \tag{F.2}$$

Obviously a Taylor series is an asymptotic expansion while a Fourier series is not. In (F.1) we use the symbol "$\sim$" because the series need not be convergent! In many interesting physical situations the first terms of an asymptotic expansion decrease rapidly for small $|x - x_0|$, while higher-order terms increase wildly with increasing n (at fixed $|x - x_0|$). In these cases we can get a good approximation to f by summing just the first few terms of the expansion. This is why asymptotic expansions, even when divergent, are very useful in practice.

To summarize, an asymptotic expansion need not be convergent and a convergent series need not be an asymptotic expansion (a Fourier series is not asymptotic). Distinguishing between these two concepts – that is, convergence versus asymptoticity – is crucial to appreciate the power of asymptotic expansions. In mathematical terms we can say that convergence pertains to the behavior of the partial sum

$$S_N(x) = \sum_{n=0}^{N} a_n \varphi_n(x)$$

for $N \to \infty$, while asymptoticity pertains to the behavior of S_N for $x \to x_0$.

From the above definitions it follows that the coefficient a_{N+1} is given by

$$a_{N+1} = \lim_{x \to x_0} \frac{f(x) - \sum_{n=0}^{N} a_n \varphi_n(x)}{\varphi_{N+1}(x)}.$$

Thus, if a function has an asymptotic expansion then this expansion is unique given the φ_n. The inverse is not true since different functions can have the same asymptotic expansion. For example, for any constant c

$$\frac{1}{1-x} + c\, e^{-1/x^2} = 1 + x + x^2 + \ldots .$$

The r.h.s. is an asymptotic expansion around $x_0 = 0$ and is independent of c since $\lim_{x\to 0} e^{-1/x^2}/x^n = 0$ for all n. We wish to conclude this appendix with a classical example of asymptotic expansions.

Let us consider the error function

$$\text{erf}(x) = \frac{2}{\sqrt{\pi}} \int_0^x dt\, e^{-t^2} = 1 - \frac{2}{\sqrt{\pi}} \int_x^\infty dt\, e^{-t^2} = 1 - \frac{1}{\sqrt{\pi}} \int_{x^2}^\infty ds\, \frac{e^{-s}}{\sqrt{s}}.$$

For nonnegative integers n we define the function

$$\begin{aligned} F_n(x) &= \int_{x^2}^\infty ds\, s^{-n-1/2} e^{-s} = \frac{e^{-x^2}}{x^{2n+1}} - (n + \frac{1}{2}) \int_{x^2}^\infty ds\, s^{-n-1-1/2} e^{-s} \\ &= \frac{e^{-x^2}}{x^{2n+1}} - (n + \frac{1}{2}) F_{n+1}(x). \end{aligned}$$

The asymptotic expansion of erf(x) around $x_0 = \infty$ follows from the repeated use of the above recursive relation

$$\begin{aligned} \text{erf}(x) &= 1 - \frac{1}{\sqrt{\pi}} F_0(x) \\ &= 1 - \frac{1}{\sqrt{\pi}} \left[\frac{e^{-x^2}}{x} - \frac{1}{2} F_1(x) \right] \\ &= 1 - \frac{1}{\sqrt{\pi}} \left[\frac{e^{-x^2}}{x} - \frac{1}{2} \frac{e^{-x^2}}{x^3} + \frac{1}{2}\frac{3}{2} F_2(x) \right] \\ &= 1 - \frac{e^{-x^2}}{\sqrt{\pi}} \sum_{n=0}^{\infty} (-)^n \frac{(2n-1)!!}{2^n x^{2n+1}}, \end{aligned} \tag{F.3}$$

which is clearly a divergent series for all finite x. To show that this is an asymptotic expansion we have to prove that the functions $\varphi_n \sim e^{-x^2}/x^{2n+1}$ form an asymptotic set (which is obvious) and that (F.2) is fulfilled. Let us define

$$\begin{aligned} R_{N+1}(x) &= \text{erf}(x) - 1 + \frac{e^{-x^2}}{\sqrt{\pi}} \sum_{n=0}^{N} (-)^n \frac{(2n-1)!!}{2^n x^{2n+1}} \\ &= -\frac{(-)^{N+1}}{\sqrt{\pi}} \frac{(2N+1)!!}{2^{N+1}} F_{N+1}(x). \end{aligned}$$

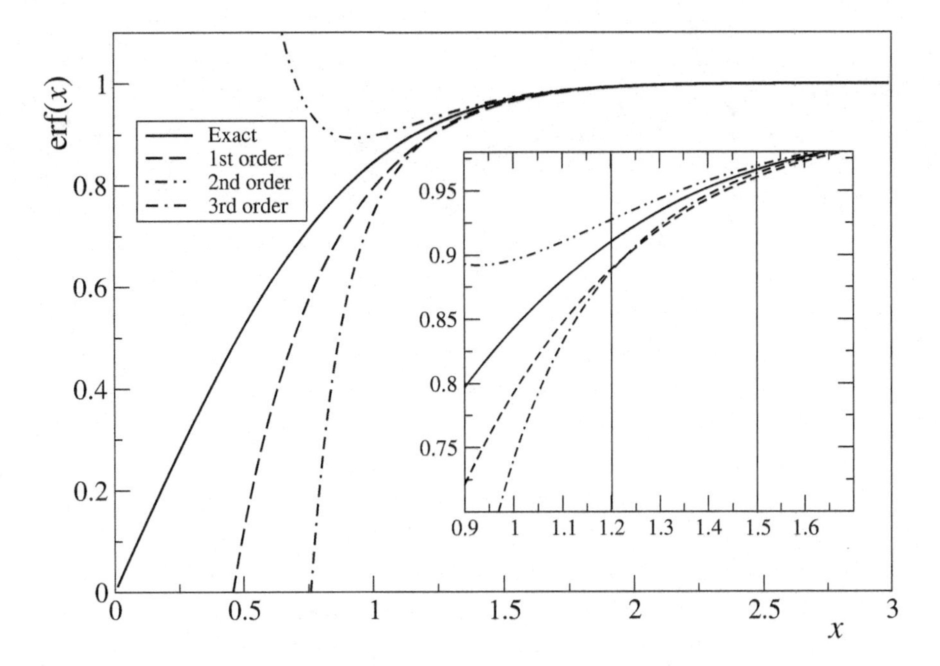

Figure F.1 Error function erf(x) and the partial sums $S_N(x)$ of the asymptotic expansion (F.3) for $N = 1, 2, 3$. The inset is a magnification of the same functions in a restricted domain of the x variable.

We have

$$|F_{N+1}(x)| = \left| \int_{x^2}^{\infty} ds\, s^{-N-3/2} e^{-s} \right| \le \frac{1}{x^{2N+3}} \int_{x^2}^{\infty} ds\, e^{-s} = \frac{e^{-x^2}}{x^{2N+3}},$$

which goes to zero faster than $\varphi_n(x) \sim e^{-x^2}/x^{2N+1}$ for $x \to \infty$. Therefore (F.3) is an asymptotic expansion.

In Fig. F.1 we display the plot of erf(x) as well as the partial sums of the asymptotic expansion with $N = 0,\ 1,\ 2$, which corresponds to the first-, second-, and third-order approximation. We see that for $x < 1$ the best approximation is the first-order one. The inset shows a magnification of the same curves in a narrower window of the x variable. For $x = 1.2$ we get closer to the exact curve by adding to the term with $n = 0$ the term with $n = 1$ (second order); the further addition of the term with $n = 3$ worsens the approximation. Increasing x even further, e.g., for $x = 1.5$, the third-order approximation is superior to the first- and second-order approximations. Thus, given x there exists an optimal N for which the partial sum performs best, whereas the inclusion of higher-order terms brings the partial sum further away from the exact value. In this example the optimal value of N increases with increasing x.

Appendix G: BBGKY Hierarchy

In Appendix E we established the connection between the equal-time n-particle Green's function G_n and the n-particle density matrix Γ_n, see (E.1). It would be valuable if we could generate a closed system of equations for the Γ_n as all time-dependent ensemble averages can be calculated from them. Let us write (E.1) again:

$$\Gamma_n(\mathbf{x}_1, ..., \mathbf{x}_n; \mathbf{x}'_1, ..., \mathbf{x}'_n|t) = (\pm \mathrm{i})^n \lim_{\substack{z_i, z'_i \to t_- \\ z'_1 > ... > z'_n > z_n > ... > z_1}} G_n(1, ..., n; 1', ..., n'). \tag{G.1}$$

A hierarchy of equations for the Γ_n can be derived starting from the Martin–Schwinger hierarchy. For notational convenience we introduce the collective coordinates $X_N = (\mathbf{x}_1, ..., \mathbf{x}_n)$ and $X'_N = (\mathbf{x}'_1, ..., \mathbf{x}'_n)$:

$$\Gamma_n(X_N; X'_N|t) \equiv \Gamma_n(\mathbf{x}_1, ..., \mathbf{x}_n; \mathbf{x}'_1, ..., \mathbf{x}'_n|t).$$

For simplicity we also consider a one-particle Hamiltonian which is diagonal in spin space [see (1.13)]:

$$\langle \mathbf{x}_1|\hat{h}(z_1)|\bar{\mathbf{x}}\rangle = \delta(\mathbf{x}_1, \bar{\mathbf{x}})h(1). \tag{G.2}$$

Then (5.2) and (5.3) can be rewritten as

$$\left[\mathrm{i}\frac{d}{dz_k} - h(k)\right] G_n(1, ..., n; 1', ..., n') = \pm \mathrm{i} \int d\bar{1}\, v(k; \bar{1})\, G_{n+1}(1, .., n, \bar{1}; 1', .., n', \bar{1}^+)$$
$$+ \sum_{j=1}^{n} (\pm)^{k+j}\, \delta(k; j')\, G_{n-1}(1, .., \overset{\sqcap}{k}, .., n; 1', .., \overset{\sqcap}{j'}, .., n'), \tag{G.3}$$

and

$$\left[-\mathrm{i}\frac{d}{dz'_k} - h(k')\right] G_n(1, ..., n; 1', ..., n') = \pm \mathrm{i} \int d\bar{1}\, v(k'; \bar{1})\, G_{n+1}(1, ..., n, \bar{1}^-; 1', ..., n', \bar{1})$$
$$+ \sum_{j=1}^{n} (\pm)^{k+j}\, \delta(j; k')\, G_{n-1}(1, .., \overset{\sqcap}{j}, .., n; 1', .., \overset{\sqcap}{k'}, .., n'). \tag{G.4}$$

Now we subtract (G.4) from (G.3), multiply by $(\pm \mathrm{i})^n$, sum over k between 1 and n, and take the limit $z_i, z_i' \to t_-$ as in (G.1). The sum of all the time derivatives of G_n yields the time derivative of Γ_n:

$$(\pm \mathrm{i})^n \lim_{z_i, z_i' \to t_-} \sum_k \left[\mathrm{i}\frac{d}{dz_k} + \mathrm{i}\frac{d}{dz_k'} \right] G_n(1, ..., n; 1', ..., n') = \mathrm{i}\frac{d}{dt}\Gamma_n(X_N; X_N'|t).$$

The terms with the δ-functions cancel out since

$$\sum_{k,j=1}^{n} (\pm)^{k+j} \left[\delta(k; j') G_{n-1}(1, ... \overset{\sqcap}{k} ..., n; 1', ... \overset{\sqcap}{j'} ..., n') \right.$$
$$\left. - \, \delta(j; k') \, G_{n-1}(1, ... \overset{\sqcap}{j} ..., n; 1', ... \overset{\sqcap}{k'} ..., n') \right] = 0,$$

as follows directly by renaming the indices $k \leftrightarrow j$ in, say, the second term. Interestingly, this cancellation occurs independently of the choice of the contour arguments z_i and z_i'. The terms with the single-particle Hamiltonian h are also easy to handle since

$$(\pm \mathrm{i})^n \lim_{z_i, z_i' \to t_-} h(k) G_n(1, ..., n; 1', ..., n') = h(\mathbf{x}_k, t) \Gamma_n(X_N; X_N'|t),$$

and similarly for $h(k')G_n(1, ..., n; 1', ..., n')$, and hence

$$(\pm \mathrm{i})^n \lim_{z_i, z_i' \to t_-} \left[\sum_{k=1}^{n} h(k) - \sum_{k=1}^{n} h(k') \right] G_n(1, ..., n; 1', ..., n')$$
$$= \left[\sum_{k=1}^{n} h(\mathbf{x}_k, t) - \sum_{k=1}^{n} h(\mathbf{x}_k', t) \right] \Gamma_n(X_N; X_N'|t).$$

We are left with the integrals $\int v \, G_{n+1}$. Consider, for instance, (G.3). By definition we have

$$G_{n+1}(1,..., n, \bar{1}; 1', ..., n', \bar{1}^+)$$
$$= \frac{(\pm)}{\mathrm{i}^{n+1}} \mathrm{Tr} \left[\hat{\rho} \mathcal{T} \left\{ \hat{\psi}_H(1) ... \hat{n}_H(\bar{1}) \hat{\psi}_H(k) ... \hat{\psi}_H(n) \hat{\psi}_H^\dagger(n') ... \hat{\psi}_H^\dagger(1') \right\} \right].$$

As the interaction is local in time we have to evaluate this G_{n+1} when the time $\bar{z}_1 = z_k$. For these times the operator $\hat{n}_H(\bar{1})\hat{\psi}_H(k) = \hat{n}_H(\bar{\mathbf{x}}_1, z_k)\hat{\psi}_H(\mathbf{x}_k, z_k)$ should be treated as a composite operator when we take the limit $z_i, z_i' \to t_-$. Consequently we can write

$$\lim_{z_i, z_i' \to t_-} G_{n+1}(1, ..., n, \bar{\mathbf{x}}_1, z_k; 1', ..., n', \bar{\mathbf{x}}_1, z_k^+)$$
$$= \frac{(\pm)^{n+1}}{\mathrm{i}^{n+1}} \mathrm{Tr} \left[\hat{\rho} \, \hat{\psi}_H^\dagger(\mathbf{x}_1', t) ... \hat{\psi}_H^\dagger(\mathbf{x}_n', t) \; \hat{\psi}_H(\mathbf{x}_n, t) ... \hat{n}_H(\bar{\mathbf{x}}_1, t) \hat{\psi}_H(\mathbf{x}_k, t) ... \hat{\psi}_H(\mathbf{x}_1, t) \right].$$

Next we want to move the density operator $\hat{n}_H(\bar{\mathbf{x}}_1, t)$ between $\hat{\psi}_H^\dagger(\mathbf{x}_n', t)$ and $\hat{\psi}_H(\mathbf{x}_n, t)$ so as to form the $(n+1)$-particle density matrix. This can easily be done since operators in the Heisenberg picture at equal time satisfy the same (anti)commutation relations as the original operators. From (1.48) we have

$$\hat{\psi}(\mathbf{x}_j)\hat{n}(\bar{\mathbf{x}}_1) = \hat{n}(\bar{\mathbf{x}}_1)\hat{\psi}(\mathbf{x}_j) + \delta(\mathbf{x}_j - \bar{\mathbf{x}}_1)\hat{\psi}(\mathbf{x}_j)$$

and therefore

$$(\pm\mathrm{i})^{n+1} \lim_{z_i, z_i' \to t_-} G_{n+1}(1, ..., n, \bar{\mathbf{x}}_1, z_k; 1', ..., n', \bar{\mathbf{x}}_1, z_k^+)$$
$$= \Gamma_{n+1}(X_N, \bar{\mathbf{x}}_1; X_N', \bar{\mathbf{x}}_1|t) + \sum_{j>k} \delta(\mathbf{x}_j - \bar{\mathbf{x}}_1)\Gamma_n(X_N; X_N'|t).$$

If we multiply this equation by $v(\mathbf{x}_k, \bar{\mathbf{x}}_1)$ and integrate over $\bar{\mathbf{x}}_1$, we get exactly the integral that we need since in the Martin–Schwinger hierarchy $\int v\, G_{n+1}$ is multiplied by $\pm\mathrm{i}$ and we said before that we are multiplying the nth equation of the hierarchy by $(\pm\mathrm{i})^n$. Thus,

$$(\pm\mathrm{i})^{n+1} \lim_{z_i, z_i' \to t_-} \int d\bar{1}\, v(k; \bar{1})\, G_{n+1}(1, ..., n, \bar{1}; 1', ..., n', \bar{1}^+)$$
$$= \int d\bar{\mathbf{x}}_1\, v(\mathbf{x}_k, \bar{\mathbf{x}}_1)\, \Gamma_{n+1}(X_N, \bar{\mathbf{x}}_1; X_N', \bar{\mathbf{x}}_1|t) + \sum_{j>k} v(\mathbf{x}_k, \mathbf{x}_j)\, \Gamma_n(X_N; X_N'|t).$$

In a similar way we can prove that

$$(\pm\mathrm{i})^{n+1} \lim_{z_i, z_i' \to t_-} \int d\bar{1}\, v(k'; \bar{1})\, G_{n+1}(1, ..., n, \bar{1}^-; 1', ..., n', \bar{1})$$
$$= \int d\bar{\mathbf{x}}_1\, v(\mathbf{x}_k', \bar{\mathbf{x}}_1)\, \Gamma_{n+1}(X_N, \bar{\mathbf{x}}_1; X_N', \bar{\mathbf{x}}_1|t) + \sum_{j>k} v(\mathbf{x}_k', \mathbf{x}_j')\, \Gamma_n(X_N; X_N'|t).$$

Thus the interaction produces terms proportional to Γ_{n+1} and terms proportional to Γ_n. Collecting all the pieces, we find the following equation:

$$\left[\mathrm{i}\frac{d}{dt} - \sum_{k=1}^{n}\left(h(\mathbf{x}_k, t) + \sum_{j>k} v(\mathbf{x}_k, \mathbf{x}_j)\right) + \sum_{k=1}^{n}\left(h(\mathbf{x}_k', t) + \sum_{j>k} v(\mathbf{x}_k', \mathbf{x}_j')\right)\right]\Gamma_n(X_N; X_N'|t)$$
$$= \sum_{k=1}^{n} \int d\bar{\mathbf{x}}_1 \left(v(\mathbf{x}_k, \bar{\mathbf{x}}_1) - v(\mathbf{x}_k', \bar{\mathbf{x}}_1)\right) \Gamma_{n+1}(X_N, \bar{\mathbf{x}}_1; X_N', \bar{\mathbf{x}}_1|t)$$

This hierarchy of equations for the density matrices is known as the Born–Bogoliubov–Green–Kirkwood–Yvon (BBGKY) hierarchy [382, 383, 384, 385].

Let us spell out the lowest-order equations. For $n = 1$ we have

$$\left[\mathrm{i}\frac{d}{dt} - h(\mathbf{x}_1, t) + h(\mathbf{x}_1', t)\right]\Gamma_1(\mathbf{x}_1; \mathbf{x}_1'|t)$$
$$= \int d\bar{\mathbf{x}}_1 \left(v(\mathbf{x}_1, \bar{\mathbf{x}}_1) - v(\mathbf{x}_1', \bar{\mathbf{x}}_1)\right) \Gamma_2(\mathbf{x}_1, \bar{\mathbf{x}}_1; \mathbf{x}_1', \bar{\mathbf{x}}_1|t),$$

while for $n = 2$ we have

$$\left[\mathrm{i}\frac{d}{dt} - h(\mathbf{x}_1,t) - h(\mathbf{x}_2,t) - v(\mathbf{x}_1,\mathbf{x}_2) + h(\mathbf{x}_1',t) + h(\mathbf{x}_2',t) + v(\mathbf{x}_1',\mathbf{x}_2')\right]\Gamma_2(\mathbf{x}_1,\mathbf{x}_2;\mathbf{x}_1',\mathbf{x}_2'|t)$$
$$= \int d\bar{\mathbf{x}}_1 \left(v(\mathbf{x}_1,\bar{\mathbf{x}}_1) + v(\mathbf{x}_2,\bar{\mathbf{x}}_1) - v(\mathbf{x}_1',\bar{\mathbf{x}}_1) - v(\mathbf{x}_2',\bar{\mathbf{x}}_1)\right)\Gamma_3(\mathbf{x}_1,\mathbf{x}_2,\bar{\mathbf{x}}_1;\mathbf{x}_1',\mathbf{x}_2',\bar{\mathbf{x}}_1|t).$$

For the BBGKY hierarchy to be useful, one should devise suitable truncation schemes to express Γ_m in terms of Γ_n with $n < m$. Standard truncation schemes give rise to several problems in the time domain - see Ref. [386] and references therein - although progress has been made recently [387, 388, 389].

Appendix H: From δ-like Peaks to Continuous Spectral Functions

Photocurrent experiments on bulk systems like metals and semiconductors reveal that the photocurrent $I_{ph}(\epsilon)$ and hence the spectral function $A(\epsilon)$, see Section 6.4, are continuous functions of ϵ. How can a sum of δ-like peaks transform into a continuous function? Understanding the transition from δ-like peaks to continuous functions does not only satisfy a mathematical curiosity, but is also important to make connection with experiments. In this appendix we explain the "mystery" with an example.

Consider a one-dimensional crystal like the one in Fig. 2.5, with zero on-site energy and nearest-neighbor hoppings T. The atoms are labeled from 1 to N and 1 is the nearest neighbor of N (BvK boundary conditions). The single-particle eigenkets $|k\rangle$ are Bloch functions with amplitude on site j given by

$$\langle j|k\rangle = \frac{e^{\mathrm{i}kj}}{\sqrt{N}},$$

and with eigenenergy $\epsilon_k = 2T\cos k$, where $k = 2\pi m/N$ and $m = 1, \ldots, N$. We assume that the electrons are noninteracting and calculate the spectral function $A_{11}(\omega)$ projected on atom 1.[1] By definition,

$$\begin{aligned} A_{11}(\omega) &= \mathrm{i}\left[G^{\mathrm{R}}_{11}(\omega) - G^{\mathrm{A}}_{11}(\omega)\right] = -2\mathrm{Im}\left[G^{\mathrm{R}}_{11}(\omega)\right] = -2\mathrm{Im}\,\langle 1|\frac{1}{\omega - \hat{h} + \mathrm{i}\eta}|1\rangle \\ &= \frac{2}{N}\sum_{m=1}^{N}\frac{\eta}{(\omega - 2T\cos\frac{2\pi m}{N})^2 + \eta^2}, \end{aligned}$$

where in the last equality we insert the completeness relation $\sum_k |k\rangle\langle k| = \hat{\mathbb{1}}$. Since the eigenvalues with m and $m' = N - m$ are degenerate, we can rewrite the sum over m as

$$A_{11}(\omega) = \frac{4}{N}\sum_{m=1}^{N/2}\frac{\eta}{(\omega - 2T\cos\frac{2\pi m}{N})^2 + \eta^2}. \tag{H.1}$$

[1] Since $|1\rangle$ is not an eigenket we expect that $A_{11}(\omega)$ has peaks at different energies. Furthermore, due to the discrete translational invariance of the Hamiltonian, $A_{jj}(\omega)$ is independent of the atomic site j.

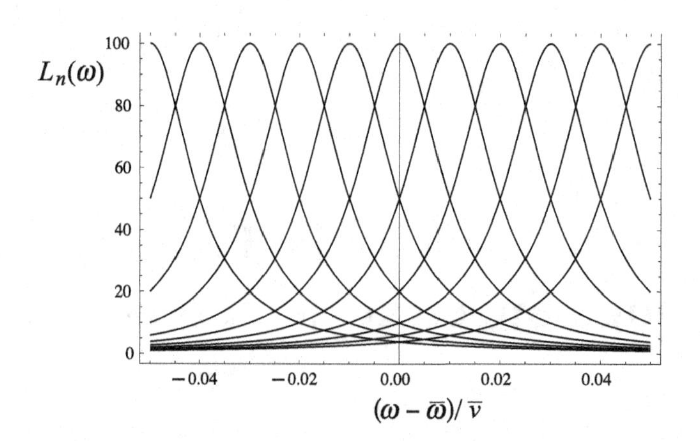

Figure H.1 Lorentzian functions $L_n(\omega) = \frac{\eta}{(\omega-\bar{\omega}-2\pi\bar{v}n/N)^2+\eta^2}$ for $n = -50, -40, \ldots, 40, 50$ in units of $1/(2\pi\bar{v})$. The other parameters are $\eta/(2\pi\bar{v}) = 0.01$ and $N = 1000$. We see that $L_{\pm 50}(\bar{\omega})$ is already very small and can be discarded.

Restricting the sum between 1 and $N/2$ guarantees that different m correspond to different peaks. For every finite N the spectral function becomes a sum of δ-functions when the limit $\eta \to 0$ is taken. However, if we first take the limit $N \to \infty$ and then $\eta \to 0$, the situation is very different. For any arbitrary small but finite η the height of the peaks goes to zero like the prefactor $1/N$, while the position of the peaks moves closer and closer. When the energy spacing between two consecutive eigenvalues becomes much smaller than η, many Lorentzians contribute to the sum for a given ω. Let us study the value of the sum for a frequency $\bar{\omega} = 2T\cos\frac{2\pi\bar{m}}{N}$ corresponding to the position of one of the peaks. For $m = \bar{m} + n$ with $n \ll N$, we can write

$$\epsilon_k = 2T\cos\frac{2\pi m}{N} = \bar{\omega} + 2\pi\bar{v}\frac{n}{N} + \mathcal{O}(n^2/N^2), \tag{H.2}$$

where $\bar{v} = -2T\sin\frac{2\pi\bar{m}}{N}$. We see that for $n = \pm N\eta/(2\pi\bar{v}) \ll N$ the r.h.s. of (H.2) is equal to $\bar{\omega} \pm \eta$. We can use this approximate expression of ϵ_k in $A_{11}(\bar{\omega})$ since if the eigenvalues are much larger than $\bar{\omega} + \eta$ or much smaller than $\bar{\omega} - \eta$ the corresponding contribution is negligible. In Fig. H.1 we show the plot of several Lorentzians of width η centered in $\bar{\omega} + 2\pi\bar{v}n/N$ for $\eta/(2\pi\bar{v}) = 0.01$ and $N = 1000$. The value that they assume in $\bar{\omega}$ is rather small already for $2\pi\bar{v}n/N = \pm 5\eta$ (i.e., $|n| = 5N\eta/(2\pi\bar{v}) = 50 \ll N$). Inserting the approximate expression of the eigenvalues in A_{11} and extending the sum over n between $-\infty$ and ∞, we find

$$A_{11}(\bar{\omega}) = \lim_{N\to\infty}\frac{4}{N}\sum_{n=-\infty}^{\infty}\frac{\eta}{(2\pi\bar{v}n/N)^2 + \eta^2} = 4\int_{-\infty}^{\infty}dx\frac{\eta}{(2\pi\bar{v}x)^2 + \eta^2} = \frac{2}{|\bar{v}|}.$$

The final result is *finite and independent of* η. The limiting function can easily be expressed in terms of $\bar{\omega}$. Defining $\bar{k} = 2\pi\bar{m}/N$, we have $\bar{v} = -2T\sin\bar{k}$ and $\bar{\omega} = 2T\cos\bar{k}$. Solving

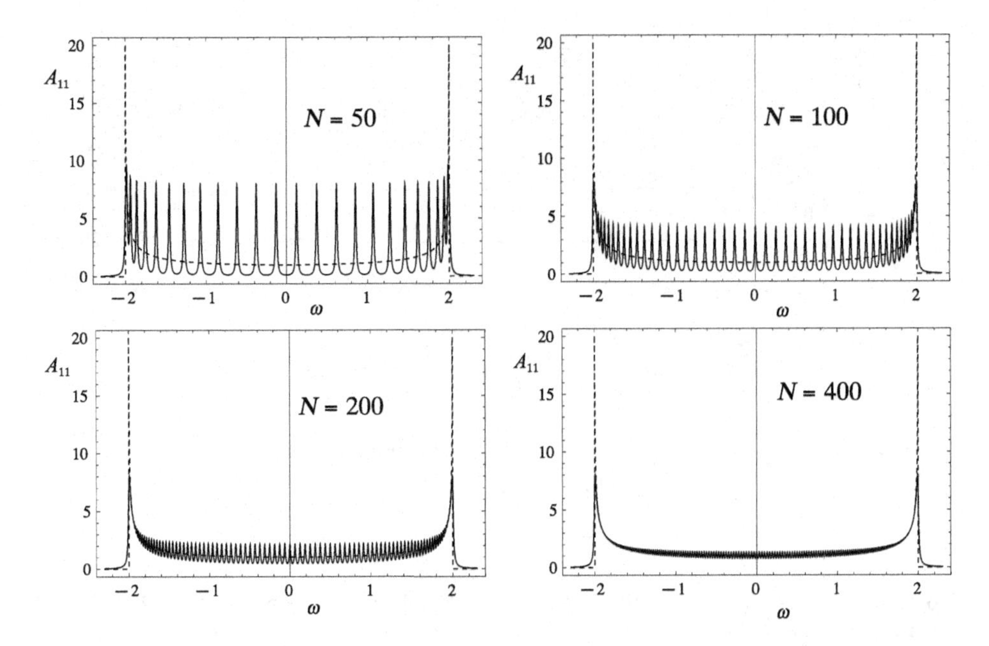

Figure H.2 Comparison between the limiting spectral function (dashed) and the spectral function (H.1) for $N = 50, 100, 200, 400$ (solid). The parameter $\eta = 0.01$ and all energies are in units of $|T|$.

the latter equation for $\bar{k}$ we get

$$A_{11}(\bar{\omega}) = \frac{2}{\left|2T \sin\left(\arccos \frac{\bar{\omega}-\epsilon}{2T}\right)\right|}.$$

In Fig. H.2 we show how fast the discrete spectral function (H.1) converges to the limiting spectral function as N increases. We can see that N must be larger than $\sim 5|T|/\eta$ for the peaks to merge and form a smooth function.

It is instructive to look at the transition $N \to \infty$ also in the time domain. For every finite N and at zero temperature the Green's function, say, $G^{<}_{11}(t, t')$ is an oscillatory function of the form

$$G^{<}_{11}(t, t') = \pm \mathrm{i} \int \frac{d\omega}{2\pi} e^{-\mathrm{i}\omega(t-t')} f(\omega - \mu) A_{11}(\omega) = \pm \mathrm{i} \frac{1}{N} \sum_{\epsilon_k < \mu} e^{-\mathrm{i}\epsilon_k (t-t')}.$$

In Fig. H.3 we plot the absolute value squared of $G^{<}_{11}(t, 0)$, which corresponds to the probability of finding the system unchanged when removing a particle from site 1 at time 0 and then putting the particle back in the same site at time t. The probability initially decreases but for times $t \sim N/(2T)$ (roughly the inverse of the average energy spacing) there is a revival. The revival time becomes longer when N becomes larger. In the continuum limit $N \to \infty$ the revival time becomes infinite and the probability decays to zero.[2] If the

[2]This follows from the fact that we can replace the sum by an integral and use the Riemann–Lebesgue theorem.

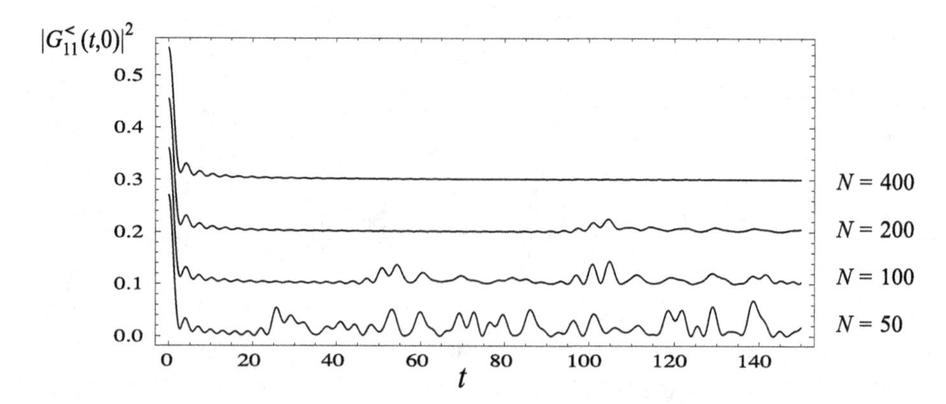

Figure H.3 Absolute value squared of $G^{<}_{11}(t, 0)$ at zero temperature and $\mu = 0$. The curves with $N = 100, 200, 400$ are shifted upward by $0.1,\ 0.2,\ 0.3$ respectively. Times are in units of $1/|T|$.

decay is exponential, $e^{-\Gamma t}$, then the spectral function in frequency space is a Lorentzian of width Γ. This means that the limiting spectral function can have poles (or branch cuts) in the complex plane even though its discrete version has poles just above and below the real axis and is analytic everywhere else, see again (H.1). The appearance of poles in the complex plane is related to the infinite revival time which, in turn, is due to the negative interference of many oscillatory functions with different but very close energies.

Appendix I: Fermi Golden Rule and Shortcomings of Linear Response Theory

In this appendix we revisit some basic concepts of the first-order time-dependent perturbation theory or, more briefly, *linear response theory.* Let us denote by $|\Psi_k\rangle$ the orthonormal eigenkets of the Hamiltonian $\hat{H}$ with eigenvalues E_k:

$$\hat{H} = \sum_k E_k |\Psi_k\rangle\langle\Psi_k|, \qquad \langle\Psi_k|\Psi_{k'}\rangle = \delta_{kk'}.$$

The most general form of the perturbing Hamiltonian in the basis $\{|\Psi_k\rangle\}$ is

$$\hat{H}'(t) = \lambda \sum_{kk'} T_{kk'}(t) |\Psi_k\rangle\langle\Psi_{k'}|, \tag{I.1}$$

with $T_{kk'} = T^*_{k'k}$ and λ a parameter controlling the strength of the perturbation. If we expand the time-evolved ket $|\Psi(t)\rangle$ as

$$|\Psi(t)\rangle = \sum_k c_k(t) e^{-\mathrm{i}E_k t} |\Psi_k\rangle, \tag{I.2}$$

the problem of solving the time-dependent Schrödinger equation is equivalent to solving the following linear system of coupled differential equations:

$$\mathrm{i}\frac{d}{dt} c_k(t) = \lambda \sum_{k'} T_{kk'}(t) e^{\mathrm{i}\omega_{kk'}t} c_{k'}(t), \tag{I.3}$$

with $\omega_{kk'} \equiv E_k - E_{k'}$ the *Bohr frequencies.* We emphasize that (I.3) is an exact reformulation of the original Schrödinger equation. It is easy to show that the coefficients $c_k(t)$ which solve (I.3) preserve the normalization of the state:[1]

$$\sum_k |c_k(t)|^2 = \sum_k |c_k(t_0)|^2 = \langle\Psi(t_0)|\Psi(t_0)\rangle. \tag{I.4}$$

[1]To prove (I.4) we multiply both sides of (I.3) by $c^*_k(t)$, take the complex conjugate of the resulting equation and subtract it from the original one. Then, summing over all k, we find that the time derivative of the l.h.s. of (I.4) is zero.

From this exact result it follows that if the sum over k is restricted to a subset $\mathcal{S}$ of eigenkets, then

$$\sum_{k\in\mathcal{S}} |c_k(t)|^2 \leq \langle\Psi(t_0)|\Psi(t_0)\rangle. \tag{I.5}$$

Equation (I.5) provides a useful benchmark for assessing the accuracy of linear response theory. Let us denote by $c_k^{(n)}(t)$ the nth-order coefficient of the expansion of $c_k(t)$ in powers of λ. In most cases the initial state is an eigenstate of $\hat{H}$, say $|\Psi(t_0)\rangle = |\Psi_i\rangle$, and therefore $c_k^{(0)}(t) = 0$ for all $k \neq i$. If the subset $\mathcal{S}$ includes all states with the exception of $|\Psi_i\rangle$, then the lowest-order contribution to the sum in (I.5) is obtained by replacing $c_k(t)$ with $c_k^{(1)}(t)$. This leads to the following upper bound for the linear response coefficients:

$$\sum_{k\neq i} |c_k^{(1)}(t)|^2 \leq \langle\Psi(t_0)|\Psi(t_0)\rangle. \tag{I.6}$$

This inequality is obviously fulfilled at $t = t_0$ since $c_k^{(1)}(t_0) = 0$ for $k \neq i$. As we see, however, the inequality (I.6) may be violated at sufficiently long times. In these cases the probabilistic interpretation of the coefficients $|c_k^{(1)}(t)|^2$ breaks down and the results of linear response theory become unreliable.

Let $|\Psi_j\rangle$ be the discrete eigenkets of $\hat{H}$ with eigenenergies E_j and $|\Psi_\alpha\rangle$ be the continuum eigenkets of $\hat{H}$; the collective index $\alpha = (E, Q)$ comprises the continuum eigenenergy E and a set of discrete or continuum quantum numbers Q which uniquely identify the state. In order to highlight the typical structure of the linear response formulas, we find it convenient to introduce a notation that treats discrete and continuum states on an equal footing. We define the label a which runs over both discrete states, $a = j$, and continuum states, $a = \alpha$, and the short-hand notation $\int da \equiv \sum_j + \int d\alpha$. The eigenkets of the Hamiltonian $\hat{H}$ are therefore represented by the set $\{|\Psi_a\rangle\}$ with inner product

$$\langle\Psi_a|\Psi_{a'}\rangle = \delta(a - a'), \tag{I.7}$$

where $\delta(a-a') = \delta_{jj'}$ if $a = j$ and $a' = j'$, and $\delta(a-a') = \delta(\alpha-\alpha') = \delta(E-E')\delta(Q-Q')$ if $a = \alpha$ and $a' = \alpha'$. The expansion of a generic ket reads

$$|\Psi\rangle = \sum_j c_j|\Psi_j\rangle + \int d\alpha\; c_\alpha|\Psi_\alpha\rangle = \int da\; c_a|\Psi_a\rangle \tag{I.8}$$

and the normalization reads

$$\langle\Psi|\Psi\rangle = \sum_j |c_j|^2 + \int d\alpha\; |c_\alpha|^2 = \int da\; |c_a|^2. \tag{I.9}$$

If $|\Psi(0)\rangle = |\Psi_{a_0}\rangle$ is an eigenket at time $t_0 = 0$ then $c_a(0) = \delta(a - a_0)$ and to first order in λ the solution of (I.3) reads

$$c_a(t) \sim c_a^{(0)}(t) + c_a^{(1)}(t) = \delta(a - a_0) - \mathrm{i}\lambda \int_0^t dt'\, T_{aa_0}(t')e^{\mathrm{i}\omega_{aa_0}t'}. \tag{I.10}$$

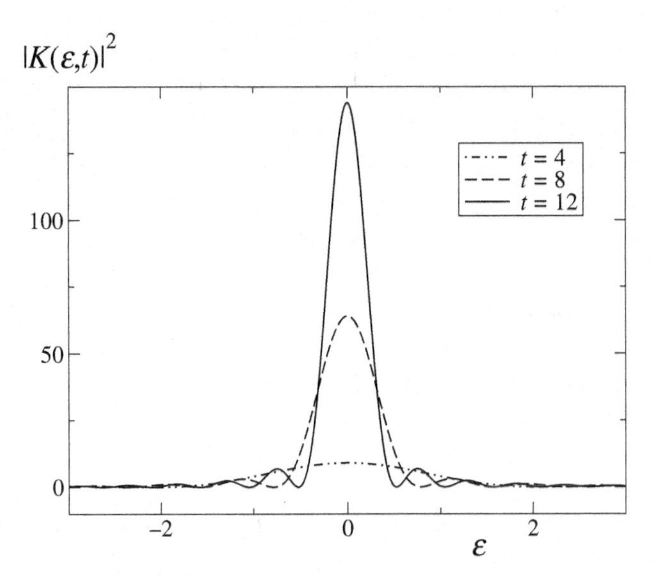

Figure I.1 Square modulus $|K(\epsilon,t)|^2$ as a function of ϵ (arbitrary units) for different values of the time parameter t (arbitrary units).

The analytic evaluation of the r.h.s. is rather simple for monochromatic perturbations. This is not a serious restriction since any perturbation can be written as a linear combination (Fourier transform) of monochromatic perturbations. For simplicity we set the diagonal element $T_{a_0a_0}(t) = 0$; its inclusion is straightforward and is left as an exercise for the reader. As for the off-diagonal elements we take the monochromatic dependence $T_{aa_0}(t) = T_{aa_0}e^{-\mathrm{i}\omega t}$. Inserting this form into (I.10), we obtain for $a \neq a_0$

$$c_a^{(1)}(t) = -\mathrm{i}\lambda T_{aa_0} K(\omega - \omega_{aa_0}, t),$$

where we introduce the function

$$K(\epsilon,t) = e^{-\mathrm{i}\epsilon t/2}\frac{\sin\left(\frac{\epsilon}{2}t\right)}{\left(\frac{\epsilon}{2}\right)}. \tag{I.11}$$

The square modulus $|K(\epsilon,t)|^2$ is displayed in Fig. I.1 as a function of ϵ for different values of t. It has a peak in $\epsilon = 0$, whose height increases like t^2 and whose width decreases like $2\pi/t$. For $t \to \infty$ we can approximate $|K(\epsilon,t)|^2$ as a "square barrier" of height t^2, width $2\pi/t$, centered in $\epsilon = 0$:[2]

$$\lim_{t\to\infty} |K(\epsilon,t)|^2 \sim \lim_{t\to\infty} t^2\,\theta\left(\frac{\pi}{t} - |\epsilon|\right) = \lim_{t\to\infty} 2\pi t\,\delta(\epsilon), \tag{I.12}$$

where in the last step we use the representation of the δ-function,

$$\delta(\epsilon) = \lim_{\xi\to 0}\theta(\xi - |\epsilon|)/(2\xi).$$

[2]A rigorous derivation of (I.12) can be found in any textbook of quantum mechanics, see, e.g., Ref. [390].

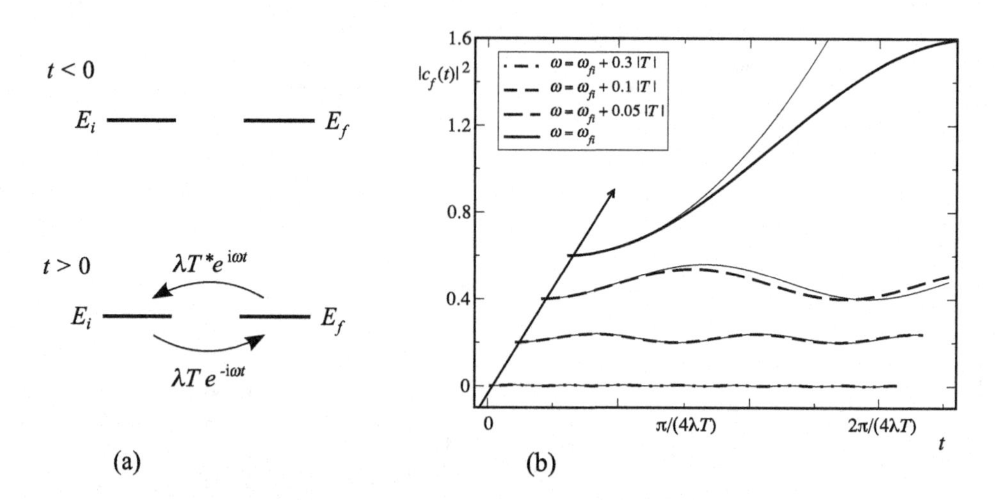

Figure I.2 (a) Schematic representation of the time-dependent perturbation. (b) The coefficient $|c_f(t)|^2$ for $\lambda = 0.01$ and different values of ω. The thin curves are the linear response results.

The result (I.12) allows us to extract the long-time limit of the probability (if $a = j$) or the probability density (if $a = \alpha$) for the transition $a_0 \to a$:

$$\lim_{t\to\infty} |c_a^{(1)}(t)|^2 = |\lambda T_{aa_0}|^2 2\pi\, t\, \delta(\omega - \omega_{aa_0}). \tag{I.13}$$

Below we study the solution (I.13) for $|\Psi(t_0)\rangle = |\Psi_{a_0}\rangle$, which is either a discrete state or a continuum state.

First case: $|\Psi(t_0)\rangle$ is a discrete state. Let $|\Psi(0)\rangle = |\Psi_i\rangle$ be one of the discrete eigenkets of $\hat{H}$. Then the normalization is $\langle\Psi(0)|\Psi(0)\rangle = 1$. From (I.13) the probability for the transition to another discrete state $j \neq i$ is

$$\lim_{t\to\infty} |c_j^{(1)}(t)|^2 = |\lambda T_{ji}|^2 2\pi\, t\, \delta(\omega - \omega_{ji}).$$

This result tells us that for $\omega = \omega_{ji}$ the probability $|c_j^{(1)}(t)|^2$ grows like $t\delta(0) \sim t^2$ (unless $T_{ji} = 0$) and eventually becomes larger than unity.[3] Consequently the probability overcomes the upper bound (I.6) and the theory of linear response breaks down. To better appreciate in what way linear response theory breaks down, let us consider the simplest possible case: a perturbation that couples only two states, see Fig. I.2(a). Let $k = i$ and $k = f$ be the label of these states. The perturbing Hamiltonian can then be written as

$$\hat{H}'(t) = Te^{-i\omega t}|\Psi_f\rangle\langle\Psi_i| + T^* e^{i\omega t}|\Psi_i\rangle\langle\Psi_f|.$$

[3]Taking the limit $\epsilon \to 0$ in (I.12) before the limit $t \to \infty$, we see that

$$\lim_{\epsilon\to 0}\delta(\epsilon) = \delta(0) = \lim_{t\to\infty}\lim_{\epsilon\to 0}\frac{t}{2\pi}\,\theta\left(\frac{\pi}{t} - |\epsilon|\right) = \lim_{t\to\infty}\frac{t}{2\pi}.$$

This Hamiltonian is a special case of (I.1) where all $T_{kk'}(t) = 0$ except for $T_{fi}(t) = T^*_{if}(t) = Te^{-\mathrm{i}\omega t}$. If the system is in $|\Psi_i\rangle$ at time t_0 the differential equations (I.3) must be solved with boundary conditions $c_k(t_0) = 0$ if $k \neq i$ and $c_i(t_0) = 1$. Without loss of generality we take $t_0 = 0$. As the perturbation couples $|\Psi_i\rangle$ to $|\Psi_f\rangle$ only, the coefficients $c_k(t)$ with $k \neq i, f$ remain zero at all times. On the other hand, the coefficients c_i and c_f are time-dependent since they are solutions of

$$\mathrm{i}\frac{d}{dt}c_i(t) = \lambda T^* e^{\mathrm{i}(\omega-\omega_{fi})t}c_f(t), \tag{I.14}$$

$$\mathrm{i}\frac{d}{dt}c_f(t) = \lambda T e^{-\mathrm{i}(\omega-\omega_{fi})t}c_i(t). \tag{I.15}$$

This coupled system of equations can easily be solved. We define the frequency $\bar{\omega} = \omega - \omega_{fi}$. Multiplying (I.15) by $e^{\mathrm{i}\bar{\omega}t}$, differentiating with respect to t, and using (I.14), we find

$$\frac{d^2}{dt^2}c_f(t) + \mathrm{i}\bar{\omega}\frac{d}{dt}c_f(t) + \lambda^2|T|^2 c_f(t) = 0,$$

the general solution of which is $c_f(t) = Ae^{\mathrm{i}\alpha_+ t} + Be^{\mathrm{i}\alpha_- t}$ with

$$\alpha_\pm = \frac{1}{2}\left(-\bar{\omega} \pm \sqrt{\bar{\omega}^2 + 4\lambda^2|T|^2}\right).$$

The boundary condition $c_f(0) = 0$ implies $A = -B$. To determine the constant A we calculate $c_i(t)$ from (I.15) and impose that $c_i(0) = 1$ (boundary condition). The final result is

$$c_f(t) = -\frac{2\mathrm{i}\lambda T}{\sqrt{\bar{\omega}^2 + 4\lambda^2|T|^2}}e^{-\mathrm{i}\bar{\omega}t/2}\sin\left(\frac{\sqrt{\bar{\omega}^2 + 4\lambda^2|T|^2}}{2}t\right).$$

The probability $|c_f(t)|^2$ of finding the system in $|\Psi_f\rangle$ at time t oscillates in time and the amplitude of the oscillations is correctly bounded between 0 and 1; it is easy to verify that $|c_i(t)|^2 + |c_f(t)|^2 = 1$ for all t. Notice that the amplitude reaches the maximum value of 1 when the frequency of the perturbing field is equal to the Bohr frequency ω_{fi}, in which case $\bar{\omega} = 0$. This is the so-called *resonance phenomenon*: $|c_f(t)|^2$ increases *quadratically* up to a time $t_{\rm max} \sim \pi/(4\lambda|T|)$ when it reaches the value $\sim 1/2$. For frequencies ω much larger or smaller than ω_{fi} the amplitude of the oscillations is of the order of $(\lambda T/\bar{\omega})^2$, see Fig. I.2(b), and hence the eigenket $|\Psi_f\rangle$ has a very small probability of being excited. The linear response result is obtained by approximating $c_f(t)$ to first order in λ and reads:

$$c_f(t) \sim c_f^{(1)}(t) = -\mathrm{i}\lambda T e^{-\mathrm{i}\bar{\omega}t/2}\frac{\sin\left(\frac{\bar{\omega}}{2}t\right)}{\left(\frac{\bar{\omega}}{2}\right)} = -\mathrm{i}\lambda T K(\bar{\omega}, t). \tag{I.16}$$

The comparison between the exact and the linear response solution is displayed in Fig. I.2(b). We can distinguish two regimes:

- $|\bar{\omega}| \gg \lambda|T|$: The linear response theory is accurate for short times but it eventually breaks down at times $t > |\bar{\omega}|/(\lambda|T|)^2$ since $c_f(t)$ and $c_f^{(1)}(t)$ start to oscillate out of phase. The amplitude of the oscillations of $|c_f^{(1)}(t)|^2$ is, however, in good agreement with the exact solution.

- $|\bar{\omega}| \ll \lambda|T|$: At the frequency of the resonance phenomenon ($\bar{\omega} = 0$) we have $c_f^{(1)}(t) = -\mathrm{i}\lambda T\, t$, which steadily increases with time. The square modulus $|c_f^{(1)}(t)|^2$ becomes larger than 1 for $t > 1/(\lambda T)$, see Fig. I.2(b), and the inequality (I.6) breaks down. More generally, close to a resonance neither the frequency nor the amplitude of the oscillations are well reproduced in linear response.

For the probability density of the transition to a continuum state α we have

$$\lim_{t\to\infty} |c_\alpha^{(1)}(t)|^2 = |\lambda T_{\alpha i}|^2 2\pi\, t\, \delta(\omega - \omega_{\alpha i}). \tag{I.17}$$

Even though the probability density can vary between 0 and ∞, the continuum states can lead to the breakdown of the upper bound (I.6). Indeed, (I.17) implies

$$\lim_{t\to\infty} \int d\alpha\, |c_\alpha^{(1)}(t)|^2 = \lim_{t\to\infty} 2\pi\, t\, \lambda^2 \int dQ\, |T_{\alpha i}|^2_{E=E_i+\omega} \to \infty,$$

where in the last step we split the integral over $\alpha = (E, Q)$ into an integral over energy E and quantum number Q. It is noteworthy, however, that the divergence of the continuum states is milder than that of the discrete states. The former goes like $\sim t$ independent of the value of ω, while the latter goes like $t\delta(0) \sim t^2$ when the frequency ω matches a Bohr frequency.

Second case: $|\Psi(t_0)\rangle$ is a continuum state. Let us study the linear response solution when $|\Psi(0)\rangle = |\Psi_{\alpha_0}\rangle$ is one of the continuum eigenkets of $\hat{H}$. In this case the normalization is $\langle\Psi(0)|\Psi(0)\rangle = \delta_\alpha(0)$. The notation $\delta_\alpha(0) = \delta(\alpha - \alpha)$ is used to distinguish the δ-function in α-space from the δ-function in energy space $\delta(0) = \delta(E - E)$. We may write that $\delta_\alpha(0) = \delta(0)\delta_Q(0)$ with $\delta_Q(0) = \delta(Q - Q)$ the δ-function in Q space. The quantity $\delta_Q(0)$ is unity only provided that Q is a discrete quantum number; if Q is an angle (or a set of angles) or a momentum (or a set of momenta) or any other kind of continuous quantum number then the quantity $\delta_Q(0)$ is infinite.

The probability for the transition to a discrete state j can be extracted from (I.13) by setting $a = j$ and $a_0 = \alpha_0$:

$$\lim_{t\to\infty} |c_j^{(1)}(t)|^2 = |\lambda T_{j\alpha_0}|^2 2\pi\, t\, \delta(\omega - \omega_{j\alpha_0}). \tag{I.18}$$

For $\omega = \omega_{j\alpha_0}$ we obtain the (by now) familiar divergence $t\delta(0)$. This divergence, however, does not necessarily imply that the inequality (I.6) is violated when $t \to \infty$. Indeed, for large t we can write $t\delta(0) \sim \delta(0)\delta(0)$, which must be compared with the initial normalization $\delta_\alpha(0) = \delta(0)\delta_Q(0)$. Only if Q is a discrete quantum number or if $\delta(0)$ "diverges faster" than $\delta_Q(0)$ when Q is a continuum quantum number then the probability $|c_j^{(1)}(t \to \infty)|^2$ exceeds the upper bound in (I.6). In all other cases (I.6) is fulfilled and other criteria must be found to assess the quality of the linear response approximation.

Next we consider the probability density for the transition to another continuum state, $\alpha \neq \alpha_0$. From (I.10) and (I.13) we find

$$\lim_{t\to\infty} |c_\alpha^{(1)}(t)|^2 = |\lambda T_{\alpha\alpha_0}|^2 2\pi\, t\, \delta(\omega - \omega_{\alpha\alpha_0}). \tag{I.19}$$

We can check the accuracy of this result by replacing the sum over k in (I.6) with the integral over $\alpha \neq \alpha_0$. We have

$$\lim_{t\to\infty} \int_{\alpha\neq\alpha_0} d\alpha \, |c_\alpha^{(1)}(t)|^2 = \left[(2\pi\lambda)^2 \int dQ \, |T_{\alpha\alpha_0}|^2_{E=E_0+\omega} \right] \times \delta(0), \tag{I.20}$$

where we replace t with $2\pi\delta(0)$ and define E_0 as the eigenenergy of $|\Psi_{\alpha_0}\rangle$. Excluding pathological perturbations, the integral over Q is either finite (when Q is a discrete quantum number[4]) or at most divergent like $\delta_Q(0)$ [when the coupling is extremely localized in Q-space like, e.g., $T_{\alpha\alpha_0} = T_{EE_0}\delta(Q - Q_0)$]. Thus, for sufficiently small λ the r.h.s. of (I.20) is always smaller than $\delta_\alpha(0) = \delta(0)\delta_Q(0)$ and the inequality (I.6) is fulfilled.

Fermi golden rule From the previous analysis we can conclude that if the frequency of the perturbation is far from the Bohr frequencies $\omega_{j\alpha_0} = E_j - E_0$, then the linear response coefficients fulfill the upper bound (I.6). We stress, however, that (I.6) is not the only condition to fulfill for the long-time results to be accurate and reliable. Things considerably improve when there are only continuum states. In most of these cases the linear response formulas are accurate also when $t \to \infty$ and, admittedly, very elegant. Considering again (I.19) we see that the probability density $|c_\alpha(t)|^2$ grows linearly in time provided that the energy $E = E_0 + \omega$. Therefore, it makes sense to define a probability density per unit time and cast (I.19) in the form

$$\boxed{\lim_{t\to\infty} \frac{d}{dt}|c_\alpha(t)|^2 \sim 2\pi|\lambda T_{\alpha\alpha_0}|^2 \delta(E - E_0 - \omega), \qquad \alpha \neq \alpha_0} \tag{I.21}$$

This equation is known as the *Fermi golden rule* and it is of great practical use.[5] For example, the power P fed into the system is equal to the integral of the probability density per unit time to excite a state of energy E times the energy $E - E_0$ of the transition. In the long-time limit we can write

$$\lim_{t\to\infty} P(t) = \lim_{t\to\infty} \int d\alpha \, (E - E_0) \frac{d}{dt}|c_\alpha(t)|^2 \sim 2\pi\omega \int dQ \, |\lambda T_{\alpha\alpha_0}|^2_{E=E_0+\omega}.$$

Thus to lowest order in λ the system can absorb energy only via transitions to states with energy $E = E_0 + \omega$. This implies that the linear response theory can only describe the absorption or emission of a single photon.

[4]In this case the integral over Q is actually a sum.

[5]Despite its name, the Fermi golden rule was derived the first time by Dirac in 1927 [391]. In 1949 Fermi gave a course on nuclear physics at the University of Chicago (his lectures have been collected in a book entitled *Nuclear Physics*, see Ref. [392]) in which he rederived (I.21) and coined for it the name "Golden rule No. 2." This name was so appealing that people started to refer to (I.21) initially as "Golden rule No. 2, by E. Fermi" and eventually as the "Fermi golden rule," see also Ref. [393].

Appendix J: How to Solve the Kadanoff–Baym Equations in Practice

In this appendix we discuss how to implement the Kadanoff–Baym equations for the Green's function. The presentation follows very closely that of Ref. [118]. Before entering into the details of the procedure we observe that the Kadanoff–Baym equations are first-order differential equations in time. The numerical technique to solve these kinds of equations is the so-called *time-stepping technique*. Consider the differential equation

$$\frac{d}{dt}f(t) = g(t),$$

where $g(t)$ is a known function. From the value of f at the initial time t_0 we can calculate the value of f one time step later, at $t_0 + \Delta_t$, by approximating the derivative with a finite difference:

$$\frac{f(t_0 + \Delta_t) - f(t_0)}{\Delta_t} = g(t_0).$$

Of course, $f(t_0 + \Delta_t)$ calculated in this way provides a good approximation if the time step Δ_t is much smaller than the typical timescale over which $g(t)$ varies. From $f(t_0 + \Delta_t)$ we can then calculate $f(t_0 + 2\Delta_t)$ by approximating again the derivative with a finite difference. In general, from the value of f at time $t_0 + n\Delta_t$ we can calculate the value of f at time $t_0 + (n+1)\Delta_t$ from

$$\frac{f(t_0 + (n+1)\Delta_t) - f(t_0 + n\Delta_t)}{\Delta_t} = g(t_0 + n\Delta_t). \tag{J.1}$$

This is the basic idea of the time-stepping technique. There are of course dozens (or maybe hundreds) of refinements and variants to make the iterative procedure more stable and accurate. For example, one can replace $g(t_0 + n\Delta_t)$ in the r.h.s. of (J.1) with the average $\frac{1}{2}[g(t_0 + n\Delta_t) + g(t_0 + (n+1)\Delta_t)]$ or with the value of g at half the time step – that is, $g(t_0 + (n + \frac{1}{2})\Delta_t))$. The advantage of these variants is clear: If we use the same scheme to go backward in time then we recover exactly $f(t_0)$ from $f(t_0 + n\Delta_t)$. The variant of choice depends on the physical problem at hand. Below we explain how to solve the Kadanoff–Baym equations using a generic time-stepping technique and then give details on how to implement the second Born and GW approximations.

General strategy Let us represent the Green's function and self-energy operators in some suitable one-particle basis and denote by G and Σ the corresponding matrices. We begin by writing the self-energy Σ as the sum of the Hartree–Fock self-energy and the correlation self-energy

$$\Sigma_{\text{tot}} = \Sigma_{\text{HF}} + \Sigma_{\text{c}}.$$

Then the Kadanoff–Baym equations with at least one real-time argument can be rewritten with $h \to h_{\text{HF}}$ and $\Sigma \to \Sigma_{\text{c}}$ since $\Sigma^{\text{R/A}}(t,t') = \delta(t,t')V_{\text{HF}}(t) + \Sigma_{\text{c}}^{\text{R/A}}(t,t')$, while for all other Keldysh components $\Sigma = \Sigma_{\text{c}}$. Let us assume that we have calculated the Matsubara Green's function, and hence the Matsubara self-energy, using some suitable self-consistent method. We now show that the lesser and greater Green's functions can be obtained by solving the following four coupled equations:

$$\left[\mathrm{i}\frac{d}{dt} - h_{\text{HF}}(t)\right] G^{\rceil}(t,\tau) = \tilde{\mathcal{I}}_L^{\rceil}(t,\tau), \tag{J.2a}$$

$$\left[\mathrm{i}\frac{d}{dt} - h_{\text{HF}}(t)\right] G^{>}(t,t') = \tilde{\mathcal{I}}_L^{>}(t,t'), \tag{J.2b}$$

$$G^{<}(t,t')\left[-\mathrm{i}\frac{\overleftarrow{d}}{dt'} - h_{\text{HF}}(t')\right] = \tilde{\mathcal{I}}_R^{<}(t,t'), \tag{J.2c}$$

$$\mathrm{i}\frac{d}{dt}G^{<}(t,t) - \left[h_{\text{HF}}(t), G^{<}(t,t)\right]_{-} = \tilde{\mathcal{I}}_L^{<}(t,t) - \tilde{\mathcal{I}}_R^{<}(t,t), \tag{J.2d}$$

where[1]

$$\begin{aligned}
\tilde{\mathcal{I}}_L^{\rceil}(t,\tau) &= \left[\Sigma_{\text{c}}^{\text{R}} \cdot G^{\rceil} + \Sigma_{\text{c}}^{\rceil} \star G^{\text{M}}\right](t,\tau), \\
\tilde{\mathcal{I}}_L^{\lessgtr}(t,t') &= \left[\Sigma_{\text{c}}^{\text{R}} \cdot G^{\lessgtr} + \Sigma_{\text{c}}^{\lessgtr} \cdot G^{\text{A}} + \Sigma_{\text{c}}^{\rceil} \star G^{\lceil}\right](t,t'), \\
\tilde{\mathcal{I}}_R^{\lessgtr}(t,t') &= \left[G^{\text{R}} \cdot \Sigma_{\text{c}}^{\lessgtr} + G^{\lessgtr} \cdot \Sigma_{\text{c}}^{\text{A}} + G^{\rceil} \star \Sigma_{\text{c}}^{\lceil}\right](t,t').
\end{aligned} \tag{J.3a}$$

In other words, we do not need the equation for $G^{\lceil}$ nor the equation for $G^{>}(t,t')$ with the derivative with respect to t' and neither the equation for $G^{<}(t,t')$ with the derivative with respect to t. The fourth equation (J.2d) for the time-diagonal Green's function can be derived by subtracting the Kadanoff–Baym equations (9.27) and (9.28) for $G^{<}(t_1,t_2)$ and then setting $t_1 = t_2 = t$. For open systems $\Sigma_{\text{c}} \to \Sigma_{\text{c}} + \Sigma_{\text{em}}$, see Section 9.4.

Without loss of generality we choose $t_0 = 0$. If we use the simple time-stepping method (J.1) then the collision integral (i.e., the r.h.s.) of the first equation for $t = 0$ is $[\Sigma^{\rceil} \star G^{\text{M}}](0,\tau) = [\Sigma^{\text{M}} \star G^{\text{M}}](0,\tau)$, which is known; we can make one time step to calculate $G^{\rceil}(\Delta_t,\tau)$ since $G^{\rceil}(0,\tau) = G^{\text{M}}(0,\tau)$, which is also known. Similarly, the collision integral of the second and third equation is known for $t = t' = 0$ and we can make one time step to calculate $G^{>}(\Delta_t, 0)$ and $G^{<}(0,\Delta_t)$ since we know $G^{>}(0,0) = G^{\text{M}}(0^+,0)$ and $G^{<}(0,0) = G^{\text{M}}(0,0^+)$. Using the symmetry properties (9.30) and (9.31), we then also know

[1] To distinguish the collision integral $\mathcal{I}$ of Section 9.2 calculated with Σ from the collision integral calculated with Σ_{c}, we use the symbol $\tilde{\mathcal{I}}$.

$G^{\lceil}(\tau, \Delta_t)$ as well as $G^{>}(0, \Delta_t)$ and $G^{<}(\Delta_t, 0)$. In order to calculate $G^{<}(\Delta_t, \Delta_t)$ we use the fourth equation. The value of $G^{>}(\Delta_t, \Delta_t)$ can be computed from

$$G^{>}(t,t) = \mathrm{i}\hat{\mathbb{1}} + G^{<}(t,t),$$

which is a direct consequence of the (anti)commutation rules of the field operators. In more refined time-stepping techniques one needs the collision integrals not only at $t = t' = 0$ but also at the first time step, see again the discussion below (J.1). In these cases one typically implements a *predictor corrector* scheme. The idea is essentially to make the first time step as described above, use the result to improve the collision integrals (for instance by taking an average), and to make again the first time step. In principle we can do more than one predictor corrector – that is, we can use the result of the first predictor corrector to improve further the collision integrals and make again the first time step. In most cases, however, one predictor corrector is enough.

It should now be clear how to proceed. Having the Green's functions with real times up to Δ_t, we can compute the collision integrals up to the same time. Then, we can use the first equation to calculate $G^{\rceil}(2\Delta_t, \tau)$ and the second and third equation to calculate $G^{>}(2\Delta_t, n\Delta_t)$ and $G^{<}(n\Delta_t, 2\Delta_t)$ with $n = 0, 1$. The quantities $G^{\lceil}(\tau, 2\Delta_t)$, $G^{>}(n\Delta_t, 2\Delta_t)$, and $G^{<}(2\Delta_t, n\Delta_t)$ follow directly from the symmetry properties. Finally we use the fourth equation to calculate $G^{<}(2\Delta_t, 2\Delta_t)$. At the end of the second time step we have the Green's function with real times smaller or equal to $2\Delta_t$. In Fig. J.1 we illustrate how the procedure works from time step m to time step $m+1$. The first equation is used to evaluate $G^{\rceil}((m+1)\Delta_t, \tau)$, the second and third equations to evaluate $G^{>}((m+1)\Delta_t, n\Delta_t)$ and $G^{<}(n\Delta_t, (m+1)\Delta_t)$ for all $n \leq m$. The symmetry properties are used to extract the Green's functions with interchanged time arguments and finally the fourth equation is used to calculate $G^{<}((m+1)\Delta_t, (m+1)\Delta_t)$. In the following we describe in detail a practical propagation algorithm to implement this strategy. This algorithm has been applied to finite systems [113, 118] as well as to open systems [114, 115].

Time-stepping algorithm Since the integral of $h_{\mathrm{HF}}(t)$ can attain large values, it is favorable to eliminate this term from the time-stepping equations. For each time step $T \to T + \Delta_t$ we therefore absorb the term in a time evolution operator of the form

$$U(t) = e^{-\mathrm{i}\bar{h}_{\mathrm{HF}}(T)t}, \qquad 0 \leq t \leq \Delta_t,$$

where $\bar{h}_{\mathrm{HF}}(T) = h(T + \Delta_t/2) + V_{\mathrm{HF}}(T)$. The Hamiltonian $h(t)$ is explicitly known as a function of time and can be evaluated at half the time step. The term V_{HF} is only known at time T and it is recalculated in the repeated time step (predictor corrector). In terms of the operator $U(t)$ we define new Green's functions g^{x} ($\mathrm{x} = \rceil, \lceil, >, <$) as

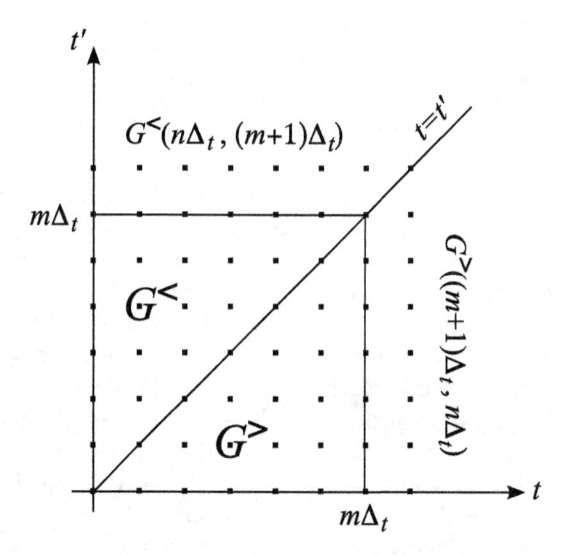

Figure J.1 Time-stepping technique in the (t, t') plane. $G^>(t, t')$ is calculated for $t > t'$ and $G^<(t, t')$ is calculated for $t \le t'$.

$$
\begin{aligned}
G^{\rceil}(T+t, \tau) &= U(t) g^{\rceil}(t, \tau), \\
G^{>}(T+t, t') &= U(t) g^{>}(t, t'), \\
G^{<}(t', T+t) &= g^{<}(t', t) U^{\dagger}(t), \\
G^{<}(T+t, T+t) &= U(t) g^{<}(t, t) U^{\dagger}(t),
\end{aligned}
$$

where $t' < T$ and $0 < t < \Delta_t$. We can now transform the Kadanoff–Baym equations into equations for g. For instance, $g^{\rceil}(t, \tau)$ satisfies the equation

$$
\mathrm{i}\frac{d}{dt} g^{\rceil}(t, \tau) = U^{\dagger}(t) \left[h_{\mathrm{HF}}(T+t) - \bar{h}_{\mathrm{HF}}(T) \right] G^{\rceil}(T+t, \tau) + U^{\dagger}(t) \tilde{\mathcal{I}}_L^{\rceil}(T+t, \tau).
$$

Since $\bar{h}_{\mathrm{HF}}(T) \sim h_{\mathrm{HF}}(T+t)$ for times $0 \le t \le \Delta_t$, we can neglect for these times the first term on the r.h.s. We then find

$$
\begin{aligned}
G^{\rceil}(T+\Delta_t, \tau) &= U(\Delta_t) \left[G^{\rceil}(T, \tau) + \int_0^{\Delta_t} dt\, \frac{d}{dt} g^{\rceil}(t, \tau) \right] \\
&\sim U(\Delta_t) G^{\rceil}(T, \tau) - \mathrm{i} U(\Delta_t) \left[\int_0^{\Delta_t} dt\, e^{\mathrm{i}\bar{h}_{\mathrm{HF}}(T) t} \right] \tilde{\mathcal{I}}_L^{\rceil}(T, \tau) \\
&= U(\Delta_t) G^{\rceil}(T, \tau) - V(\Delta_t) \tilde{\mathcal{I}}_L^{\rceil}(T, \tau),
\end{aligned} \tag{J.4}
$$

where $V(\Delta_t)$ is defined according to

$$
V(\Delta_t) = \frac{1 - e^{-\mathrm{i}\bar{h}_{\mathrm{HF}}(T)\Delta_t}}{\bar{h}_{\mathrm{HF}}(T)}.
$$

In a similar way we can use (J.2b) and (J.2c) to propagate the greater and lesser Green's functions and we find

$$G^{>}(T+\Delta_t,t') = U(\Delta_t)G^{>}(T,t') - V(\Delta_t)\tilde{\mathcal{I}}_L^{>}(T,t'), \tag{J.5}$$

$$G^{<}(t',T+\Delta_t) = G^{<}(t',T)U^{\dagger}(\Delta_t) - \tilde{\mathcal{I}}_R^{<}(t',T)V^{\dagger}(\Delta_t). \tag{J.6}$$

For time-stepping along the diagonal, the equation for $g^{<}(t,t)$ reads:

$$\mathrm{i}\frac{d}{dt}g^{<}(t,t) = U^{\dagger}(t)\left[\tilde{\mathcal{I}}_L^{<}(t,t) - \tilde{\mathcal{I}}_R^{<}(t,t)\right]U(t),$$

where again we approximate the difference $\bar{h}_{\mathrm{HF}}(T) - h_{\mathrm{HF}}(T+t) \sim 0$. Integrating over t between 0 and Δ_t we then find

$$G^{<}(T+\Delta_t,T+\Delta_t) \sim U(\Delta_t)G^{<}(T,T)U^{\dagger}(\Delta_t) - \mathrm{i}U(\Delta_t)\left[\int_0^{\Delta_t} dt\, U^{\dagger}(t)\left(\tilde{\mathcal{I}}_L^{<}(T,T) - \tilde{\mathcal{I}}_R^{<}(T,T)\right)U(t)\right]U^{\dagger}(\Delta_t).$$

By using the operator expansion

$$e^{A}Be^{-A} = B + [A,B]_- + \frac{1}{2!}\left[A,[A,B]_-\right]_- + \frac{1}{3!}\left[A,\left[A,[A,B]_-\right]_-\right]_- + \ldots,$$

it follows that

$$-\mathrm{i}\int_0^{\Delta_t} dt\, U^{\dagger}(t)\left(\tilde{\mathcal{I}}_L^{<}(T,T) - \tilde{\mathcal{I}}_R^{<}(T,T)\right)U(t) = \sum_{n=0}^{\infty} C_n,$$

where

$$C_{n+1} = \frac{\mathrm{i}\Delta_t}{n+2}\left[\bar{h}_{\mathrm{HF}}(T), C_n\right]_-$$

and $C_0 = -\mathrm{i}\Delta_t\left(\tilde{\mathcal{I}}_L^{<}(T,T) - \tilde{\mathcal{I}}_R^{<}(T,T)\right)$. Inserting this result into the formula for the equal time $G^{<}$, we finally obtain

$$G^{<}(T+\Delta_t,T+\Delta_t) = U(\Delta_t)\left[G^{<}(T,T) + \sum_{n=0}^{\infty} C_n\right]U^{\dagger}(\Delta_t). \tag{J.7}$$

Typically, keeping terms for $n \leq 3$ yields sufficient accuracy. Equations (J.4), (J.5), (J.6), and (J.7) together with the symmetry properties discussed in Section 9.2 form the basis of the time-stepping algorithm. At each time step, it requires the construction of the operators $U(\Delta_t)$ and $V(\Delta_t)$ and therefore the diagonalization of $\bar{h}_{\mathrm{HF}}(T)$. The implementation of the predictor corrector consists of carrying out the time step $T \to T + \Delta_t$ as many times as needed to get the desired accuracy. According to our experience, repeating every time step twice is already enough to obtain accurate results (provided that the time step Δ_t is small enough). To summarize, the procedure is as follows:

- The collision integrals and $\bar{h}_{\rm HF}$ at time T are calculated from the Green's functions with real times up to T.
- A step in the Green's function is taken according to (J.4), (J.5), (J.6), and (J.7).
- A new $\bar{h}_{\rm HF}$ and new collision integrals $\tilde{\mathcal{I}}^{\rceil}_L(T+\Delta_t,\tau)$, $I^{>}_L(T+\Delta_t,t')$, and $I^{<}_R(t',T+\Delta_t)$ are calculated from the new Green's functions with real times up to $T+\Delta_t$.
- The arithmetic average of $\bar{h}_{\rm HF}$ and of the collision integrals for times T and $T+\Delta_t$ is calculated.
- The time step $T \to T+\Delta_t$ is then repeated using the arithmetic average values of $\bar{h}_{\rm HF}$ and of the collision integrals according to (J.4), (J.5), (J.6), and (J.7).

This concludes the general time-stepping procedure for the Green's function.

In the following we work out explicitly the self-energy in terms of the Green's function. The derivation is very instructive since it confirms with two examples the validity of the observation made in the discussion on the Matsubara formalism in Section 5.4 [see also discussion below (9.21)], according to which in order to perform the $(m+1)$th time step we do not need the self-energy with time arguments larger than $m\Delta_t$.

Second Born self-energy We study how the Keldysh components of the self-energy are expressed in terms of the Keldysh components of the Green's function and highlight the dependence on the time variables. No matter what the approximation to the self-energy is, the four differential equations (J.2) tell us that in order to evaluate the time derivative with respect to T of $G^{\rceil}(T,\tau)$, $G^{>}(T,\bar{t})$, and $G^{<}(\bar{t},T)$ with $\bar{t}\leq T$, and of $G^{<}(T,T)$ we only need to know $\Sigma^{\rceil}(\bar{t},\tau)$ for $\bar{t}\leq T$, $\Sigma^{>}(\bar{t},\bar{t}')$ for $\bar{t}'\leq\bar{t}\leq T$, and $\Sigma^{<}(\bar{t},\bar{t}')$ for $\bar{t}\leq\bar{t}'\leq T$. We now show that to calculate these self-energies in the second Born approximation it is enough to know the Green's function with real times up to T.

The second Born self-energy is the sum of the Hartree–Fock self-energy, $\Sigma_{\rm HF}$, and the second-order correlation self-energy $\Sigma_{\rm c}$. The Hartree–Fock self-energy is local in time and can be calculated from the sole knowledge of $G^{<}(\mathbf{x},t;\mathbf{x}',t)$. The second-order correlation self-energy is given in (7.16). Using the Feynman rules [see also (15.1)] we convert the diagrams into a mathematical expression and find for fermions

$$\Sigma_{\rm c}(1;2) = -{\rm i}^2\int d3d4\; v(1;3)v(2;4)\left[G(1;2)G(3;4)G(4;3) - G(1;4)G(4;3)G(3;2)\right].$$

For convenience of notation we suppress the position–spin variables. Taking into account that v is local in time, the right component of $\Sigma_{\rm c}$ is

$$\Sigma^{\rceil}_{\rm c}(t,\tau) = -{\rm i}^2\int d\mathbf{x}_3 d\mathbf{x}_4\, v\, v\left[G^{\rceil}(t,\tau)G^{\rceil}(t,\tau)G^{\lceil}(\tau,t) - G^{\rceil}(t,\tau)G^{\lceil}(\tau,t)G^{\rceil}(t,\tau)\right].$$

The left component $\Sigma^{\lceil}_{\rm c}$ can be worked out similarly and, like $\Sigma^{\rceil}_{\rm c}$, contains only $G^{\rceil}(t,\tau)$ and $G^{\lceil}(\tau,t)$. The lesser and greater components read

$$\Sigma^{\lessgtr}_{\rm c}(t,t') = -{\rm i}^2\int d\mathbf{x}_3 d\mathbf{x}_4\, v\, v\left[G^{\lessgtr}(t,t')G^{\lessgtr}(t,t')G^{\gtrless}(t',t) - G^{\lessgtr}(t,t')G^{\gtrless}(t',t)G^{\lessgtr}(t,t')\right].$$

Thus we see that in order to calculate the self-energy with real times up to T we only need the Green's function with real times up to T.

GW self-energy Let us show that also the GW self-energy with times up to a maximum time T can be calculated from Green's functions with times less than T. In the GW approximation the XC part of the self-energy is the product of the Green's function G and the screened interaction W, see the end of Section 7.7. The screened interaction satisfies the Dyson equation,

$$W(1;2) = v(1;2) + \int d3d4\, v(1;3)P(3;4)W(4;2),$$

where the polarization is approximated as $P(1;2) = -\mathrm{i}G(1;2)G(2;1)$. In order to split the self-energy into a Hartree–Fock part and a correlation part we write $W = v + \delta W$ so that the GW self-energy reads $\Sigma = \Sigma_{\mathrm{HF}} + \Sigma_{\mathrm{c}}$ with

$$\Sigma_{\mathrm{c}}(1;2) = -\mathrm{i}G(1;2)\delta W(2;1).$$

Suppressing again the dependence on the position–spin variables, we have

$$\Sigma_{\mathrm{c}}^{\rceil}(t,\tau) = -\mathrm{i}G^{\rceil}(t,\tau)\delta W^{\lceil}(\tau,t) \quad ; \quad \Sigma_{\mathrm{c}}^{\lceil}(\tau,t) = -\mathrm{i}G^{\lceil}(\tau,t)\delta W^{\rceil}(t,\tau)$$

and

$$\Sigma_{\mathrm{c}}^{\lessgtr}(t,t') = -\mathrm{i}G^{\lessgtr}(t,t')\delta W^{\gtrless}(t',t).$$

As $W(1;2) = W(2;1)$ is symmetric, see property (7.28), we only need to calculate $\delta W^{\rceil}(t,\tau)$ and $\delta W^{<}(t,t')$ for $t \le t'$ since[2]

$$\delta W^{\lceil}(\tau,t) = \delta W^{\rceil}(t,\tau) \quad \text{and} \quad \delta W^{>}(t,t') = \delta W^{<}(t',t).$$

From the Dyson equation for the screened interaction we find

$$\delta W^{\rceil}(t,\tau) = vP^{\rceil}(t,\tau)v + vX^{\rceil}(t,\tau)$$

and

$$\delta W^{<}(t,t') = vP^{<}(t,t')v + vX^{<}(t,t'),$$

where

$$P^{\rceil}(t,\tau) = -\mathrm{i}G^{\rceil}(t,\tau)G^{\lceil}(\tau,t) \quad \text{and} \quad P^{<}(t,t') = -\mathrm{i}G^{<}(t,t')G^{>}(t',t),$$

and

$$X^{\rceil}(t,\tau) = \int_0^t d\bar{t}\, P^{\mathrm{R}}(t,\bar{t})\delta W^{\rceil}(\bar{t},\tau) - \mathrm{i}\int_0^{\beta} d\bar{\tau}\, P^{\rceil}(t,\bar{\tau})\delta W^{\mathrm{M}}(\bar{\tau},\tau)$$

$$X^{<}(t,t') = \int_0^t d\bar{t}\, P^{\mathrm{R}}(t,\bar{t})\delta W^{<}(\bar{t},\tau) + \int_0^{t'} d\bar{t}\, P^{<}(t,\bar{t})\delta W^{\mathrm{A}}(\bar{t},\tau) - \mathrm{i}\int_0^{\beta} d\bar{\tau}\, P^{\rceil}(t,\bar{\tau})\delta W^{\lceil}(\bar{\tau},t').$$

[2]We recall that we only need $\Sigma^{>}(t,t')$ for $t \ge t'$ and $\Sigma^{<}(t,t')$ for $t \le t'$.

Suppose that we have propagated up to a maximum time T and let us see whether we can calculate $\delta W^{\rceil}$ and $\delta W^{<}$ with real times up to T. The first term in $\delta W^{\rceil}$ and $\delta W^{<}$ is known since the polarization is given in terms of Green's functions with real times smaller than T. The same is true for the second term since the time integrals in X have an upper limit which never exceeds T. For instance, for the first time step we need

$$X^{\rceil}(0,\tau) = -\mathrm{i}\int_0^\beta d\bar{\tau}\, P^{\mathrm{M}}(0,\bar{\tau})\delta W^{\mathrm{M}}(\bar{\tau},\tau)$$

and $X^{<}(0;0) = X^{\rceil}(0,0^{+})$. These quantities are known from the solution of the equilibrium problem. After the first time step we can calculate P and δW with real times up to Δ_t and hence X with real times up to Δ_t. We can then perform the second time step and so on. In more refined time-stepping techniques we also need to implement a predictor corrector. This means that to go from time step m to time step $m+1$ we also need δW with real times equal to $(m+1)\Delta_t$. In this case we can use an iterative scheme similar to the one previously described. As a first guess for, for example, $\delta W^{<}(n\Delta_t,(m+1)\Delta_t)$ we take $\delta W^{<}(n\Delta_t, m\Delta_t)$. Similarly, we approximate $\delta W^{\rceil}((m+1)\Delta_t,\tau) \sim \delta W^{\rceil}(m\Delta_t,\tau)$ and $\delta W^{<}((m+1)\Delta_t,(m+1)\Delta_t) \sim \delta W^{<}(m\Delta_t, m\Delta_t)$. We then calculate X with real times equal to $(m+1)\Delta_t$ and use it to obtain a new value of δW with real times equal to $(m+1)\Delta_t$. This new value can then be used to improve X and in turn δW, and the process is repeated until convergence is reached.

Appendix K: Time-Dependent Landauer-Büttiker Formula

In this appendix we consider the system of Fig. 9.1 and solve the Kadanoff-Baym equations for noninteracting electrons, hence the correlation self-energy Σ_{MM} in (9.53) is vanishing. We are interested in the dynamics induced by the switch-on of an external voltage at times $t > t_0$ - that is, $\epsilon_{k\alpha}(t) = \epsilon_{k\alpha} + V_\alpha(t)$. To calculate the time-dependent ensemble average of any one-body operator, such as the occupation of the molecular levels or the current flowing through the molecule, we must determine $G^<_{MM}(t,t)$. To lighten the notation we drop the subscript "MM" since all quantities in (9.53) are matrices with indices in the molecular region and therefore there is no risk of misunderstandings.

We can generate an equation for $G^<(t,t)$ by subtracting (9.28) from (9.27) and then setting $z = t_-$, $z' = t_+$. The result is

$$\mathrm{i}\frac{d}{dt}G^<(t,t) - \left[h(t), G^<(t,t)\right]_- = \left[\Sigma^<_{\rm em}\cdot G^{\rm A} - G^{\rm R}\cdot\Sigma^<_{\rm em} + \Sigma^{\rm R}_{\rm em}\cdot G^< - G^<\cdot\Sigma^{\rm A}_{\rm em}\right](t,t) + \left[\Sigma^\rceil_{\rm em}\star G^\lceil - G^\rceil\star\Sigma^\lceil_{\rm em}\right](t,t). \tag{K.1}$$

It is easy to see that the terms with the plus sign on the r.h.s. are the transpose conjugate of the terms with the minus sign. For instance, from (6.28) and (9.34)

$$\int_{t_0}^\infty d\bar{t}\,\Sigma^<_{\rm em}(t,\bar{t})G^{\rm A}(\bar{t},t) = -\int_{t_0}^\infty d\bar{t}\,[\Sigma^<_{\rm em}(\bar{t},t)]^\dagger[G^{\rm R}(t,\bar{t})]^\dagger = -\left[G^{\rm R}\cdot\Sigma^<_{\rm em}\right]^\dagger(t,t),$$

and similarly from (9.31) and (9.35)

$$-\mathrm{i}\int_0^\beta d\tau\,\Sigma^\rceil_{\rm em}(t,\tau)G^\lceil(\tau,t) = -\mathrm{i}\int_0^\beta d\tau\,[\Sigma^\lceil_{\rm em}(\beta-\tau,t)]^\dagger[G^\rceil(t,\beta-\tau)]^\dagger = -\left[G^\rceil\star\Sigma^\lceil_{\rm em}\right]^\dagger(t,t).$$

Therefore we can rewrite (K.1) as

$$\mathrm{i}\frac{d}{dt}G^<(t,t) - \left[h(t), G^<(t,t)\right]_- = -\left[G^{\rm R}\cdot\Sigma^<_{\rm em} + G^<\cdot\Sigma^{\rm A}_{\rm em} + G^\rceil\star\Sigma^\lceil_{\rm em}\right](t,t) + \text{H.c.}, \tag{K.2}$$

where H.c. stands for Hermitian conjugate. In order to solve this differential equation we must calculate the Keldysh components of the embedding self-energy as well as the Green's function $G^{\rm R}$ and $G^{\rceil}$. $G^{\rm R}$ can easily be derived from the sole knowledge of $\Sigma^{\rm R}_{\rm em}$. To calculate $G^{\rceil}$ we must first calculate the Matsubara Green's function $G^{\rm M}$, see (9.24). We then proceed as follows. First we determine the embedding self-energy, then we calculate $G^{\rm M}$ from (9.23) and use it to calculate $G^{\rceil}$, which we then use to solve (K.2). This way of proceeding is completely general. One always starts from the Matsubara Green's function since it is the only Kadanoff–Baym equation which is not coupled to the others. The difference in the interacting case is that the Matsubara self-energy is a functional of $G^{\rm M}$ and therefore cannot be determined a priori like the embedding self-energy. In the interacting case the Matsubara Kadanoff–Baym equation must be solved self-consistently.

Embedding self-energy For simplicity we take $V_\alpha(t) = V_\alpha$ independent of time – that is, we consider a sudden switch-on of the bias. Then the Hamiltonian is time independent on the horizontal branches of the contour, and according to the results of Section 6.2 the retarded and advanced Green's functions $g^{\rm R/A}$ depend only on the time difference. The advanced embedding self-energy for reservoir α is then

$$\Sigma^{\rm A}_{\alpha,mn}(t,t') = \int \frac{d\omega}{2\pi} e^{-{\rm i}\omega(t-t')} \underbrace{\sum_k T_{m\,k\alpha}\, g^{\rm A}_{k\alpha}(\omega)\, T_{k\alpha\,n}}_{\Sigma^{\rm A}_{\alpha,mn}(\omega)} .$$

To simplify the analytic calculations we take the eigenvalues of the molecular Hamiltonian h well inside the continuum spectrum of reservoir α and make the WBLA according to which

$$\Sigma^{\rm A}_{\alpha,mn}(\omega) = \sum_k T_{m\,k\alpha} \frac{1}{\omega - \epsilon_{k\alpha} - V_\alpha - {\rm i}\eta} T_{k\alpha\,n} = \frac{\rm i}{2}\Gamma_{\alpha,mn}$$

is a purely imaginary constant. The quantity

$$\Gamma_{\alpha,mn} = 2\pi \sum_k T_{m\,k\alpha}\, \delta(\omega - \epsilon_{k\alpha} - V_\alpha)\, T_{k\alpha\,n} = \Gamma^*_{\alpha,nm}$$

can be seen as the (m,n) matrix element of a self-adjoint matrix Γ_α. Consequently,

$$\Sigma^{\rm A}_\alpha(t,t') = \frac{\rm i}{2}\Gamma_\alpha \delta(t-t').$$

In the WBLA, $\Sigma^{\rm A}_\alpha(\omega)$, and hence $\Sigma^{\rm R}_\alpha(\omega) = [\Sigma^{\rm A}_\alpha(\omega)]^\dagger$, is the same as in equilibrium since the dependence on V_α drops out. Then we can use (10.15) and write

$$\Sigma^{\rm M}_\alpha(\omega_q) = \frac{\rm i}{2}\begin{cases} -\Gamma_\alpha & \text{if } {\rm Im}[\omega_q] > 0 \\ +\Gamma_\alpha & \text{if } {\rm Im}[\omega_q] < 0. \end{cases}$$

The right and left embedding self-energy can now be easily derived. Without loss of generality we choose the time at which the bias is switched on to be $t_0 = 0$. Using the relation (6.50),

$$g^{\rceil}_{k\alpha}(t,\tau) = {\rm i} g^{\rm R}_{k\alpha}(t,0) g^{\rm M}_{k\alpha}(0,\tau) = e^{-{\rm i}(\epsilon_{k\alpha}+V_\alpha)t} g^{\rm M}_{k\alpha}(0,\tau).$$

Expanding $g^{\rm M}_{k\alpha}(0,\tau)$ in Matsubara frequencies, see (6.18), we find

$$\Sigma^{\rceil}_{\alpha,mn}(t,\tau) = \frac{1}{-\mathrm{i}\beta}\sum_q e^{\omega_q\tau}\sum_k T_{m\,k\alpha}\frac{e^{-\mathrm{i}(\epsilon_{k\alpha}+V_\alpha)t}}{\omega_q-\epsilon_{k\alpha}+\mu}T_{k\alpha\,n}$$
$$= \frac{1}{-\mathrm{i}\beta}\sum_q e^{\omega_q\tau}\int\frac{d\omega}{2\pi}\underbrace{2\pi\sum_k T_{m\,k\alpha}\delta(\omega-\epsilon_{k\alpha})T_{k\alpha\,n}}_{\Gamma_{\alpha,mn}}\frac{e^{-\mathrm{i}(\omega+V_\alpha)t}}{\omega_q-\omega+\mu}.$$

The trick employed to transform the summation over k into an integration over ω is frequently utilized in the subsequent derivations. We observe that in the original definition of Γ_α the argument of the δ-function is $\omega-\epsilon_{k\alpha}-V_\alpha$. However, since Γ_α is independent of ω we can shift ω as we like without changing the value of the sum. In conclusion,

$$\Sigma^{\rceil}_{\alpha}(t,\tau) = \Gamma_\alpha\times\frac{1}{-\mathrm{i}\beta}\sum_q e^{\omega_q\tau}\int\frac{d\omega}{2\pi}\frac{e^{-\mathrm{i}(\omega+V_\alpha)t}}{\omega_q-\omega+\mu}.$$

In a completely analogous way the reader can check that the expression for $\Sigma^{\lceil}_{\alpha}$ is

$$\Sigma^{\lceil}_{\alpha}(\tau,t) = \Gamma_\alpha\times\frac{1}{-\mathrm{i}\beta}\sum_q e^{-\omega_q\tau}\int\frac{d\omega}{2\pi}\frac{e^{\mathrm{i}(\omega+V_\alpha)t}}{\omega_q-\omega+\mu}.$$

It is also straightforward to verify that $\Sigma^{\rceil}_{\alpha}(t,\tau)=[\Sigma^{\lceil}_{\alpha}(\beta-\tau,t)]^\dagger$, in agreement with (9.35). Finally, the lesser self-energy is obtained from

$$\Sigma^{<}_{\alpha,mn}(t,t') = \sum_k T_{m\,k\alpha}\underbrace{\mathrm{i}f(\epsilon_{k\alpha}-\mu)e^{-\mathrm{i}(\epsilon_{k\alpha}+V_\alpha)(t-t')}}_{g^{<}_{k\alpha}(t,t')}T_{k\alpha\,n}$$
$$= \mathrm{i}\,\Gamma_{\alpha,mn}\int\frac{d\omega}{2\pi}f(\omega-\mu)e^{-\mathrm{i}(\omega+V_\alpha)(t-t')}.$$

Matsubara Green's function We are now in the position to solve the Kadanoff–Baym equations. The Matsubara Green's function is simply

$$G^{\rm M}(\omega_q) = \frac{1}{\omega_q-h-\Sigma^{\rm M}_{\rm em}(\omega_q)+\mu} = \begin{cases}\dfrac{1}{\omega_q-h+\mathrm{i}\Gamma/2+\mu} & \text{if Im}[\omega_q]>0\\ \dfrac{1}{\omega_q-h-\mathrm{i}\Gamma/2+\mu} & \text{if Im}[\omega_q]<0\end{cases}, \tag{K.3}$$

with $\Gamma=\sum_\alpha\Gamma_\alpha$. The combination $h\pm\mathrm{i}\Gamma/2$ appears very often below and it is therefore convenient to introduce an effective *non-Hermitian* Hamiltonian,

$$h_{\rm eff} = h-\frac{\mathrm{i}}{2}\Gamma \quad\Rightarrow\quad h^\dagger_{\rm eff} = h+\frac{\mathrm{i}}{2}\Gamma,$$

to shorten the formulas.

Right Green's function To calculate $G^{\rceil}$ we use its equation of motion

$$\left[\mathrm{i}\frac{d}{dt} - h\right] G^{\rceil}(t,\tau) = \int_0^\infty d\bar{t}\, \Sigma^{\mathrm{R}}_{\mathrm{em}}(t,\bar{t})G^{\rceil}(\bar{t},\tau) - \mathrm{i}\int_0^\beta d\bar{\tau}\, \Sigma^{\rceil}_{\mathrm{em}}(t,\bar{\tau})G^{\mathrm{M}}(\bar{\tau},\tau).$$

Taking into account that $\Sigma^{\mathrm{R}}_{\mathrm{em}}(t,\bar{t}) = [\Sigma^{\mathrm{A}}_{\mathrm{em}}(\bar{t},t)]^\dagger = -\frac{\mathrm{i}}{2}\Gamma\delta(\bar{t}-t)$, the first term on the r.h.s. is simply $-\frac{\mathrm{i}}{2}\Gamma G^{\rceil}(t,\tau)$. Moving this term to the l.h.s., we form the combination $h - \frac{\mathrm{i}}{2}\Gamma = h_{\mathrm{eff}}$ and hence

$$G^{\rceil}(t,\tau) = e^{-\mathrm{i}h_{\mathrm{eff}}t}\left[G^{\mathrm{M}}(0,\tau) - \int_0^t dt' e^{\mathrm{i}h_{\mathrm{eff}}t'}\int_0^\beta d\bar{\tau}\,\Sigma^{\rceil}_{\mathrm{em}}(t',\bar{\tau})G^{\mathrm{M}}(\bar{\tau},\tau)\right], \tag{K.4}$$

where we take into account that $G^{\mathrm{M}}(0,\tau) = G^{\rceil}(0,\tau)$.

Retarded Green's function In order to solve (K.2) we must calculate one remaining quantity, which is the retarded Green's function. Again we observe that the Hamiltonian is time-independent on the horizontal branches of the contour and therefore $G^{\mathrm{R}}(t,t')$ depends only on the time difference $t-t'$:

$$G^{\mathrm{R}}(t,t') = \int \frac{d\omega}{2\pi} e^{-\mathrm{i}\omega(t-t')}G^{\mathrm{R}}(\omega).$$

Using the retarded Dyson equation we find

$$G^{\mathrm{R}}(\omega) = \frac{1}{\omega - h + \mathrm{i}\eta}\left[1 + \Sigma^{\mathrm{R}}_{\mathrm{em}}(\omega)G^{\mathrm{R}}(\omega)\right] \quad\Rightarrow\quad G^{\mathrm{R}}(\omega) = \frac{1}{\omega - h_{\mathrm{eff}}}, \tag{K.5}$$

and hence

$$G^{\mathrm{R}}(t,t') = -\mathrm{i}\theta(t-t')e^{-\mathrm{i}h_{\mathrm{eff}}(t-t')}.$$

Collision integral We can now evaluate the three terms in the square brackets of (K.2). Let us start with the first term:

$$\begin{aligned}\left[G^{\mathrm{R}}\cdot\Sigma^{<}_{\mathrm{em}}\right](t,t) &= -\mathrm{i}e^{-\mathrm{i}h_{\mathrm{eff}}t}\int_0^t dt' e^{\mathrm{i}h_{\mathrm{eff}}t'}\,\mathrm{i}\sum_\alpha \Gamma_\alpha \int\frac{d\omega}{2\pi}f(\omega-\mu)e^{-\mathrm{i}(\omega+V_\alpha)t'}e^{\mathrm{i}(\omega+V_\alpha)t}\\ &= \mathrm{i}\sum_\alpha\int\frac{d\omega}{2\pi}f(\omega-\mu)\left(1 - e^{\mathrm{i}(\omega+V_\alpha-h_{\mathrm{eff}})t}\right)G^{\mathrm{R}}(\omega+V_\alpha)\Gamma_\alpha.\end{aligned} \tag{K.6}$$

The second term involves Σ^{A}, which is proportional to a δ-function and therefore

$$\left[G^{<}\cdot\Sigma^{\mathrm{A}}\right](t,t) = \frac{\mathrm{i}}{2}G^{<}(t,t)\Gamma. \tag{K.7}$$

The third term is more complicated even though the final result is rather simple. Using (K.4) we find

$$\left[G^{\rceil}\star\Sigma^{\lceil}_{\mathrm{em}}\right](t,t) = e^{-\mathrm{i}h_{\mathrm{eff}}t}\left\{\left[G^{\mathrm{M}}\star\Sigma^{\lceil}_{\mathrm{em}}\right](0,t) - \mathrm{i}\int_0^t dt' e^{\mathrm{i}h_{\mathrm{eff}}t'}\left[\Sigma^{\rceil}_{\mathrm{em}}\star G^{\mathrm{M}}\star\Sigma^{\lceil}_{\mathrm{em}}\right](t',t)\right\}.$$

We now show that the second term on the r.h.s. vanishes. Taking into account the explicit form of the left and right embedding self-energies,

$$\left[\Sigma^{\rceil}_{\rm em}\star G^{\rm M}\star\Sigma^{\lceil}_{\rm em}\right](t',t)=\int\frac{d\omega}{2\pi}\frac{d\omega'}{2\pi}\sum_{\alpha\alpha'}\Gamma_\alpha\frac{1}{-{\rm i}\beta}\sum_q\frac{e^{-{\rm i}(\omega+V_\alpha)t'}}{\omega_q-\omega+\mu}G^{\rm M}(\omega_q)\frac{e^{{\rm i}(\omega'+V_{\alpha'})t}}{\omega_q-\omega'+\mu}\Gamma_{\alpha'},$$

where we use $\int_0^\beta d\tau\, e^{(\omega_q-\omega_{q'})\tau}=\beta\delta_{qq'}$, see Appendix A. Since both t and t' are positive, the integral over ω can be performed by closing the contour in the lower half of the complex ω-plane, whereas the integral over ω' can be performed by closing the contour in the upper half of the complex ω'-plane. We then see that the integral over ω is nonzero only for $\mathrm{Im}[\omega_q]<0$, whereas the integral over ω' is nonzero only for $\mathrm{Im}[\omega_q]>0$. This means that for every ω_q the double integral vanishes. We are left with the calculation of the convolution between $G^{\rm M}$ and $\Sigma^{\lceil}$. We have

$$\left[G^{\rm M}\star\Sigma^{\lceil}_{\rm em}\right](0,t)=\int\frac{d\omega}{2\pi}\frac{1}{-{\rm i}\beta}\sum_q\frac{G^{\rm M}(\omega_q)e^{\eta\omega_q}}{\omega_q-\omega+\mu}\sum_\alpha\Gamma_\alpha e^{{\rm i}(\omega+V_\alpha)t},$$

where the factor $e^{\eta\omega_q}$ stems from the fact that in the limit $t\to 0$ the quantity on the l.h.s. must be equal to $\left[G^{\rm M}\star\Sigma^{\rm M}_{\rm em}\right](0,0^+)$. To evaluate the sum over the Matsubara frequencies we use the identity (6.23) and then deform the contour from Γ_a to Γ_b since $G^{\rm M}(\zeta)$ is analytic in the upper and lower halves of the complex ζ-plane, see (K.3). We then have

$$\left[G^{\rm M}\star\Sigma^{\lceil}_{\rm em}\right](0,t)=\int\frac{d\omega}{2\pi}\frac{d\omega'}{2\pi}f(\omega')\left[\frac{G^{\rm M}(\omega'-{\rm i}\delta)}{\omega'-\omega+\mu-{\rm i}\delta}-\frac{G^{\rm M}(\omega'+{\rm i}\delta)}{\omega'-\omega+\mu+{\rm i}\delta}\right]\sum_\alpha\Gamma_\alpha e^{{\rm i}(\omega+V_\alpha)t},$$

where, as usual, the limit $\delta\to 0$ is implicit. As $t>0$ we can close the integral over ω in the upper half of the complex ω-plane. Then only the second term in the square brackets contributes and we find

$$\begin{aligned}\left[G^{\rm M}\star\Sigma^{\lceil}_{\rm em}\right](0,t)&={\rm i}\int\frac{d\omega'}{2\pi}f(\omega')G^{\rm M}(\omega'+{\rm i}\delta)\sum_\alpha\Gamma_\alpha e^{{\rm i}(\omega'+\mu+V_\alpha)t}\\&={\rm i}\int\frac{d\omega}{2\pi}f(\omega-\mu)G^{\rm R}(\omega)\sum_\alpha\Gamma_\alpha e^{{\rm i}(\omega+V_\alpha)t},\end{aligned}$$

where in the last equality we change the integration variable $\omega'=\omega-\mu$ and observe that $G^{\rm M}(\omega-\mu+{\rm i}\delta)=G^{\rm R}(\omega)$. Therefore the last term on the r.h.s. of (K.2) reads

$$\left[G^{\rceil}\star\Sigma^{\lceil}_{\rm em}\right](t,t)={\rm i}\int\frac{d\omega}{2\pi}f(\omega-\mu)\sum_\alpha e^{{\rm i}(\omega+V_\alpha-h_{\rm eff})t}G^{\rm R}(\omega)\Gamma_\alpha. \tag{K.8}$$

Equal-time lesser Green's function We now have all the ingredients to solve the equation of motion for $G^<(t,t)$. Using the results (K.6), (K.7), and (K.8),

$$\begin{aligned}{\rm i}\frac{d}{dt}G^<(t,t)-h_{\rm eff}G^<(t,t)+G^<(t,t)h^\dagger_{\rm eff}&=-{\rm i}\int\frac{d\omega}{2\pi}f(\omega-\mu)\sum_\alpha\\&\times\left\{G^{\rm R}(\omega+V_\alpha)+e^{{\rm i}(\omega+V_\alpha-h_{\rm eff})t}\left[G^{\rm R}(\omega)-G^{\rm R}(\omega+V_\alpha)\right]\right\}\Gamma_\alpha+{\rm H.c.}\end{aligned}$$

It is natural to make the transformation

$$G^{<}(t,t) = e^{-\mathrm{i}h_{\mathrm{eff}}t}\tilde{G}^{<}(t,t)e^{\mathrm{i}h^{\dagger}_{\mathrm{eff}}t}, \tag{K.9}$$

so that the differential equation for $\tilde{G}^{<}(t,t)$ reads

$$\begin{aligned}
\mathrm{i}\frac{d}{dt}\tilde{G}^{<}(t,t) =& -\mathrm{i}\int\frac{d\omega}{2\pi}f(\omega-\mu)\sum_{\alpha}e^{\mathrm{i}h_{\mathrm{eff}}t}\left[G^{\mathrm{R}}(\omega+V_\alpha)\Gamma_\alpha-\Gamma_\alpha G^{\mathrm{A}}(\omega+V_\alpha)\right]e^{-\mathrm{i}h^{\dagger}_{\mathrm{eff}}t}\\
&-\mathrm{i}\int\frac{d\omega}{2\pi}f(\omega-\mu)\sum_{\alpha}V_\alpha\left[G^{\mathrm{R}}(\omega)G^{\mathrm{R}}(\omega+V_\alpha)\Gamma_\alpha e^{\mathrm{i}(\omega+V_\alpha-h^{\dagger}_{\mathrm{eff}})t}-\mathrm{H.c.}\right],
\end{aligned}$$

where in the second line we use the Dyson-like identity

$$G^{\mathrm{R}}(\omega+V_\alpha) = G^{\mathrm{R}}(\omega) - V_\alpha G^{\mathrm{R}}(\omega)G^{\mathrm{R}}(\omega+V_\alpha) \tag{K.10}$$

and the symbol "H.c." now refers to the transpose conjugate of the quantity in the square bracket: $\mathrm{H.c.} = e^{-\mathrm{i}(\omega+V_\alpha-h_{\mathrm{eff}})t}\Gamma_\alpha G^{\mathrm{A}}(\omega+V_\alpha)G^{\mathrm{A}}(\omega)$. The quantity $\tilde{G}^{<}(t,t)$ can be calculated by direct integration of the r.h.s. The integral over time between 0 and t of the second line is easy since the dependence on t is all contained in the exponential. To integrate the first line, we use the identity

$$\int_0^t dt'\,e^{\mathrm{i}At'}\left[\frac{1}{x-A}B-B\frac{1}{x-A^{\dagger}}\right]e^{-\mathrm{i}A^{\dagger}t'} = -\mathrm{i}e^{\mathrm{i}At'}\frac{1}{x-A}B\frac{1}{x-A^{\dagger}}e^{-\mathrm{i}A^{\dagger}t'}\Big|_0^t,$$

which is valid for arbitrary matrices A and B as it can be verified by direct differentiation of the r.h.s. after the replacements $e^{\mathrm{i}At'}\to e^{\mathrm{i}(A-x)t'}$ and $e^{-\mathrm{i}A^{\dagger}t'}\to e^{-\mathrm{i}(A^{\dagger}-x)t'}$ (clearly this replacement does not change the r.h.s.). Finally the integration between 0 and t of the l.h.s. yields the difference $\tilde{G}^{<}(t,t)-\tilde{G}^{<}(0,0)$. The matrix $\tilde{G}^{<}(0,0)=G^{<}(0,0)=G^{\mathrm{M}}(0,0^{+})$ and

$$\begin{aligned}
G^{\mathrm{M}}(0,0^{+}) &= \frac{1}{-\mathrm{i}\beta}\sum_q e^{\eta\omega_q}G^{\mathrm{M}}(\omega_q) = \int\frac{d\omega}{2\pi}f(\omega)[G^{\mathrm{M}}(\omega-\mathrm{i}\delta)-G^{\mathrm{M}}(\omega+\mathrm{i}\delta)]\\
&= \mathrm{i}\int\frac{d\omega}{2\pi}f(\omega-\mu)G^{\mathrm{R}}(\omega)\Gamma G^{\mathrm{A}}(\omega).
\end{aligned} \tag{K.11}$$

This result generalizes (6.24) to molecules with an arbitrary number of levels. Equation (K.11) can also be derived from the fluctuation-dissipation theorem since $G^{\mathrm{M}}(0,0^{+}) = G^{<}(0,0) = \int\frac{d\omega}{2\pi}G^{<}(\omega)$ with $G^{<}(\omega)=\mathrm{i}f(\omega-\mu)A(\omega)$ and the spectral function

$$A(\omega) = \mathrm{i}\left[G^{\mathrm{R}}(\omega)-G^{\mathrm{A}}(\omega)\right] = G^{\mathrm{R}}(\omega)\Gamma G^{\mathrm{A}}(\omega).$$

Interestingly, the spectral function can be written as the sum of partial spectral functions corresponding to different reservoirs:

$$A(\omega) = \sum_\alpha A_\alpha(\omega), \quad \text{where} \quad A_\alpha(\omega) = G^{\mathrm{R}}(\omega)\Gamma_\alpha G^{\mathrm{A}}(\omega).$$

Collecting the terms coming from the integration and taking into account (K.9) and (K.10), after some algebra we arrive at

$$\boxed{\begin{aligned}-\mathrm{i}G^<(t,t) = \int \frac{d\omega}{2\pi} f(\omega-\mu)\sum_\alpha \Big\{ & A_\alpha(\omega+V_\alpha) \\ & + V_\alpha \left[e^{\mathrm{i}(\omega+V_\alpha-h_{\rm eff})t}\, G^{\rm R}(\omega)A_\alpha(\omega+V_\alpha) + \text{H.c.} \right] \\ & + V_\alpha^2\, e^{-\mathrm{i}h_{\rm eff}t}\, G^{\rm R}(\omega)A_\alpha(\omega+V_\alpha)G^{\rm A}(\omega)\, e^{\mathrm{i}h^\dagger_{\rm eff}t}\Big\}\end{aligned}} \tag{K.12}$$

Considering the original complication of the problem, equation (K.12) is an extremely compact result and, as we see below, contains a lot of physics. The derivation can easily be generalized to situations where the molecular Hamiltonian has terms that flip the spin [394], for example, $\sum_{mn}(T^{\rm sf}_{mn}d^\dagger_{m\uparrow}d_{n\downarrow} + T^{\rm sf}_{nm}d^\dagger_{n\uparrow}d_{m\downarrow})$, ferromagnetic leads where $\epsilon_{k\alpha\uparrow} \neq \epsilon_{k\alpha\downarrow}$, and arbitrary time-independent perturbations in the molecular region, $T_{mn} \to T_{mn}(t) = T'_{mn}$ [129]. The derivation has also been generalized to deal with arbitrary time-dependent biases [363, 395]. The only crucial ingredient to obtain a close analytic formula for $G^<(t,t)$ is the WBLA.

Let us come back to (K.12). We can easily check that $G^<(0,0) = G^{\rm M}(0,0^+)$ and that for zero bias, $V_\alpha = 0$, one has $G^<(t,t) = G^<(0,0)$ at all times, as it should be. Due to the nonhermiticity of $h_{\rm eff} = h - \mathrm{i}\Gamma/2$, the terms in the second and third line vanish exponentially fast in the limit $t \to \infty$. Consequently, $G^<(t,t)$ approaches the steady-state value

$$\lim_{t\to\infty} G^<(t,t) = \mathrm{i}\int \frac{d\omega}{2\pi} f(\omega-\mu)\sum_\alpha A_\alpha(\omega+V_\alpha).$$

This result should be compared with the equilibrium result (K.11), in which the partial spectral functions are all calculated at the same frequency ω. Instead, out of equilibrium the partial spectral functions are calculated at frequencies shifted by the applied bias, $\omega \to \omega + V_\alpha$.

The transient behavior of $G^<(t,t)$ is described by the last two lines of (K.12). These terms have different physical origins. At sufficiently low temperatures the Fermi function has a sharp step at $\omega = \mu$ and the second line gives rise to transient oscillations with frequencies $\omega_j \sim |\mu + V_\alpha - h_j|$ with damping times τ_j, where $h_j - \mathrm{i}\tau_j^{-1}/2$ are the eigenvalues of the effective Hamiltonian $h - \mathrm{i}\Gamma/2$. These oscillations originate from virtual transitions between the resonant levels of the molecule and the Fermi level of the biased reservoirs. The third line of (K.12) describes intramolecular transitions. If the effective Hamiltonian $h_{\rm eff}$ does not commute with Γ_α (and hence with A_α), this term produces oscillations of frequency $\omega_{ij} = h_i - h_j$ which are exponentially damped over a timescale $\tau_{ij} = 1/(\tau_i^{-1} + \tau_j^{-1})$. It is worth stressing that if $h_{\rm eff}$ and Γ_α commute, no intramolecular frequencies are observed in the transient behavior of $G^<(t,t)$ since $e^{-\mathrm{i}h_{\rm eff}t + \mathrm{i}h^\dagger_{\rm eff}t} = e^{-\Gamma t}$. It is also interesting to observe that intramolecular transitions are damped on a faster timescale than the transitions between the Fermi level of the reservoirs and the molecular levels.

Time-dependent current Having all the self-energies and the Green's functions, we can also calculate the time-dependent current $I_\alpha(t)$ in (9.57). The mathematical steps are very similar to the ones leading to (K.12). We leave it as an exercise for the reader to prove that (take into account that the electron charge is $q = -1$):

$$
\boxed{
\begin{aligned}
I_\alpha(t) = 2 \int \frac{d\omega}{2\pi} f(\omega - \mu) \sum_\beta \mathrm{Tr}_M \Big\{ \\
& + \Gamma_\alpha G^{\mathrm{R}}(\omega + V_\alpha)\Gamma_\beta G^{\mathrm{A}}(\omega + V_\alpha) - \Gamma_\alpha G^{\mathrm{R}}(\omega + V_\beta)\Gamma_\beta G^{\mathrm{A}}(\omega + V_\beta) \\
& - V_\beta \Big[\Gamma_\alpha e^{\mathrm{i}(\omega + V_\beta - h_{\mathrm{eff}})t} G^{\mathrm{R}}(\omega)\big(- \mathrm{i}\delta_{\alpha\beta} G^{\mathrm{R}}(\omega + V_\beta) + A_\beta(\omega + V_\beta)\big) + \mathrm{H.c.} \Big] \\
& - V_\beta^2 \, \Gamma_\alpha e^{-\mathrm{i}h_{\mathrm{eff}}t} \, G^{\mathrm{R}}(\omega) A_\beta(\omega + V_\beta) G^{\mathrm{A}}(\omega) \, e^{\mathrm{i}h^\dagger_{\mathrm{eff}}t} \Big\}
\end{aligned}
}
\tag{K.13}
$$

where the sum over β is a sum over all reservoirs. This is the time-dependent generalization of the Landauer–Büttiker formula (9.62).

The analysis of (K.13) is very similar to the analysis for the time-dependent behavior of $G^<(t,t)$. The last two lines vanish exponentially fast in the long-time limit and the steady-state value of the current is given by the first two terms in the curly bracket. The reader can easily verify that the long-time limit agrees with (9.62). Let us evaluate (K.13) for the simple case of a 2×2 Hamiltonian h with matrix elements

$$
h = \begin{pmatrix} \epsilon_0 + \Delta & 0 \\ 0 & \epsilon_0 - \Delta \end{pmatrix}
$$

and two reservoirs that we call left, $\alpha = L$, and right, $\alpha = R$. For simplicity we take $\Gamma_L = \Gamma_R$ (proportionate coupling). We study two different scenarios: In the first scenario the matrix Γ_L is proportional to the 2×2 identity matrix, while in the second scenario Γ_L is proportional to a 2×2 matrix with all identical entries:

$$
\text{first scenario}: \quad \Gamma_L = \Gamma_0 \begin{pmatrix} 1 & 0 \\ 0 & 1 \end{pmatrix} \quad \Bigg| \quad \text{second scenario}: \quad \Gamma_L = \Gamma_0 \begin{pmatrix} 1 & 1 \\ 1 & 1 \end{pmatrix}. \tag{K.14}
$$

The system is initially in equilibrium at zero temperature and chemical potential $\mu = 0$. At time $t = 0$ we suddenly switch on a bias $V_L = 2\epsilon_0$ in the left reservoir while we keep the right reservoir at zero bias, $V_R = 0$. If we take $\epsilon_0 > 0$ and $\Delta \ll \epsilon_0$, and if the molecule is weakly coupled to the reservoirs, $\Gamma_0 \ll \epsilon_0$, then the initial number of electrons on the molecule is approximately zero. The effect of the bias is to raise the occupied levels of the left reservoir above the energy of the molecular levels – see schematic representation below:

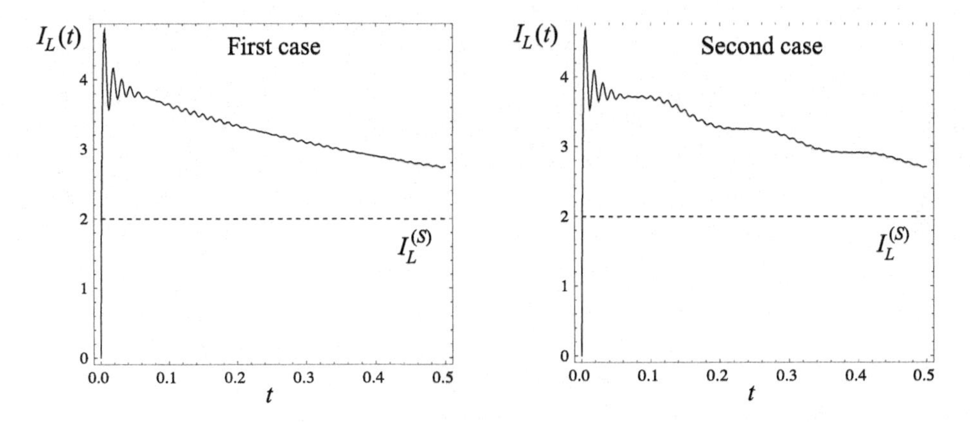

Figure K.1 Left current $I_L(t)$ for the two-level molecular system described in the main text with parameters $\epsilon_0 = 500$, $\Delta = 20$, $V_L = 2\epsilon_0$, and $V_R = 0$. The left (right) panel corresponds to the first (second) scenario, see (K.14). Energies and current are in units of Γ_0 and time is in units of Γ_0^{-1}.

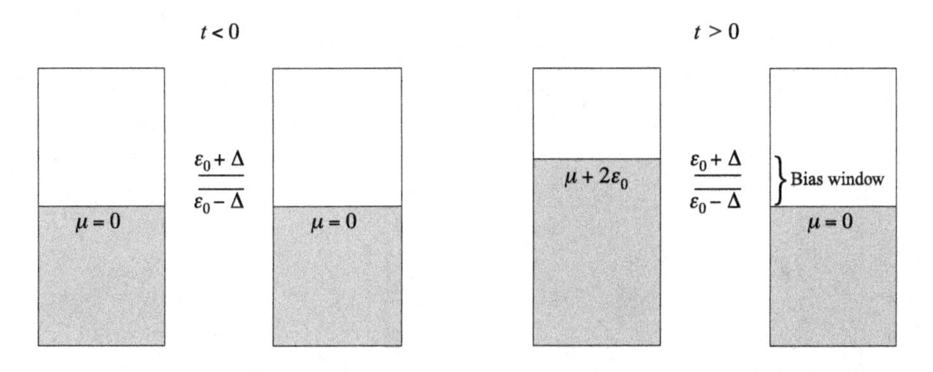

In this situation both molecular levels are inside the bias window and a net current starts flowing. In the first scenario the transient current exhibits only oscillations at frequency $\omega = |V_L - \epsilon_0 \pm \Delta| = |\epsilon_0 \pm \Delta|$ and $\omega = |V_R - \epsilon_0 \pm \Delta| = |\epsilon_0 \mp \Delta|$ corresponding to transitions between the molecular levels and the Fermi energy of the biased reservoirs. No intramolecular transitions are possible since Γ_α commutes with h. This is nicely illustrated in the left panel of Fig. K.1, where we plot the function $I_L(t)$ of (K.13). The only two visible frequencies $\omega_\pm = |\epsilon_0 \pm \Delta|$ produce coherent quantum beats in the current. This phenomenon has been observed in the context of spin transport, where the energy spacing Δ should be seen as the Zeeman splitting induced by an external magnetic field [394, 396, 397, 398]. In the second scenario the matrices Γ_α do not commute with h and according to our previous analysis we should observe an additional frequency $\omega = 2\Delta$ corresponding to the intramolecular transition. In the right panel of Fig. K.1 we display the left current $I_L(t)$ for the second scenario. We clearly see an additional oscillation of frequency $\omega = 2\Delta$ superimposed on the oscillations of the first scenario.

Appendix L: Virial Theorem for Conserving Approximations

We consider a system of identical particles with charge q and mass m, mutually interacting with a Coulomb interaction $v(\mathbf{r}_1, \mathbf{r}_2) = q^2/|\mathbf{r}_1 - \mathbf{r}_2|$ and subject to the static potential generated by nuclei of charge $\{-Z_i q\}$ in position $\{\mathbf{R}_i\}$. The single-particle Hamiltonian operator $\hat{h}$ reads

$$\hat{h} = \frac{\hat{p}^2}{2m} - q^2 \sum_i \frac{Z_i}{|\hat{\boldsymbol{r}} - \mathbf{R}_i|} \equiv \frac{\hat{p}^2}{2m} + q\phi(\hat{\boldsymbol{r}}; \{\mathbf{R}_i\}). \tag{L.1}$$

The total energy of the system,

$$E_{\text{tot}} = E + E_{\text{n-n}}, \tag{L.2}$$

is the sum of the energy defined in (10.27),

$$E = E_{\text{one}} + E_{\text{int}},$$

and the classical nuclear energy,

$$E_{\text{n-n}} = \frac{1}{2} q^2 \sum_{i \neq j} \frac{Z_j Z_i}{|\mathbf{R}_j - \mathbf{R}_i|}.$$

For later purposes it is convenient to split the one-body part of the energy into a kinetic and potential part,

$$E_{\text{one}} = E_{\text{kin}} + E_{\text{e-n}}(\{\mathbf{R}_i\}),$$

with

$$E_{\text{kin}} = \pm \mathrm{i} \int d\mathbf{x}_1 \left[-\frac{\nabla_1^2}{2m} G(1;2) \right]_{2=1^+},$$

and

$$E_{\text{e-n}}(\{\mathbf{R}_i\}) = \pm \mathrm{i} \int d\mathbf{x}_1 q\phi(\mathbf{r}_1; \{\mathbf{R}_i\}) G(1;1^+) = \int d\mathbf{x}_1 q\phi(\mathbf{r}_1; \{\mathbf{R}_i\}) n(1), \tag{L.3}$$

where $n(1) = \pm \mathrm{i} G(1;1^+)$ is the density. In (L.3) the dependence of $E_{\text{e-n}}$ on the nuclear coordinates is explicitly reported since this quantity depends on $\{\mathbf{R}_i\}$ implicitly through G and explicitly through ϕ.

In this appendix we show that the virial theorem

$$\boxed{2E_{\text{kin}} + E_{\text{e-n}}(\{\mathbf{R}_i\}) + E_{\text{int}} + E_{\text{n-n}} = 0}$$

is satisfied provided that

Condition (1): G is self-consistently calculated from a conserving self-energy.
Condition (2): The nuclei are in their equilibrium configuration:

$$\frac{dE_{\text{tot}}}{d\mathbf{R}_i} = 0 \quad \text{for all } i. \tag{L.4}$$

This result was first derived in Ref. [103].

We begin by exploiting the important stationary property of the Klein functional, see Section 11.5,

$$\frac{\delta \Omega_{\text{K}}[G, v]}{\delta G} = 0,$$

when G equals the self-consistent Green's function corresponding to a conserving self-energy [condition (1)]. We then define a Green's function G^λ which depends on the real parameter λ according to

$$G^\lambda(1; 2) = \lambda^3 G(\lambda \mathbf{r}_1 \, \sigma_1, z_1; \lambda \mathbf{r}_2 \, \sigma_2, z_2).$$

The Green's function $G^{\lambda=1} = G$ is the self-consistent Green's function and therefore

$$\left. \frac{d\Omega_{\text{K}}[G^\lambda, v]}{d\lambda} \right|_{\lambda=1} = 0. \tag{L.5}$$

Let us calculate this derivative from the expression (11.38) of the Klein functional. The term $\text{tr}_\gamma \left[\ln(-G^\lambda) \right] = \text{tr}_\gamma \left[\ln(-G) \right]$ is independent of λ since

$$\text{tr}_\gamma \left[(G^\lambda)^m \right] = \int d1 \ldots dm \, G^\lambda(1; 2) \ldots G^\lambda(m, 1^+) = \text{tr}_\gamma \left[G^m \right].$$

As a consequence, (L.5) can be written as

$$\left. \frac{d\Phi[G^\lambda, v]}{d\lambda} \right|_{\lambda=1} - \frac{d}{d\lambda} \text{tr}_\gamma \left[G^\lambda \overleftarrow{G}_0^{-1} \right]_{\lambda=1} = 0. \tag{L.6}$$

To evaluate the derivative of Φ, consider the expansion (11.22). The nth-order term consists of n interaction lines, $2n$ Green's function lines, and integration over $2n$ spatial coordinates. Taking into account that the interaction v is Coulombic, we have

$$\int d\mathbf{r}_1 \ldots d\mathbf{r}_{2n} \underbrace{v \ldots v}_{n \text{ times}} \underbrace{G^\lambda \ldots G^\lambda}_{2n \text{ times}} = \lambda^n \int d\mathbf{r}_1 \ldots d\mathbf{r}_{2n} \underbrace{v \ldots v}_{n \text{ times}} \underbrace{G \ldots G}_{2n \text{ times}}$$

and therefore

$$\Phi[G^\lambda, v] = \sum_{n=1}^{\infty} \frac{\lambda^n}{2n} \, \text{tr}_\gamma \left[\Sigma_s^{(n)} G \right].$$

The λ-derivative of this quantity is proportional to the interaction energy,

$$\left.\frac{d\Phi[G^\lambda, v]}{d\lambda}\right|_{\lambda=1} = \frac{1}{2}\mathrm{tr}_\gamma\,[\Sigma G] = \frac{-\mathrm{i}\beta}{2}\int d\mathbf{x}_1 d2\,\Sigma(1;2)G(2;1^+) = \mp\beta E_{\mathrm{int}},$$

where in the last equality we use (10.29). Next we consider the second term of (L.6). We have

$$\mathrm{tr}_\gamma\left[G^\lambda \overleftarrow{G}_0^{-1}\right] = \int d1\; G^\lambda(1;2)\left[-\mathrm{i}\frac{\overleftarrow{d}}{dz_2} + \frac{\overleftarrow{\nabla}_2^2}{2m} - q\phi(\mathbf{r}_2;\{\mathbf{R}_i\})\right]_{2=1^+}$$
$$= \int d1\left[-\mathrm{i}\frac{d}{dz_2}G^\lambda(1;2)\right]_{2=1^+} + \underbrace{\int d1\left[\frac{\nabla_2^2}{2m} - q\phi(\mathbf{r}_2;\{\mathbf{R}_i\})\right]G^\lambda(1;2)\Big|_{2=1^+}}_{\pm\beta[\lambda^2 E_{\mathrm{kin}} + \lambda E_{\mathrm{e-n}}(\{\lambda\mathbf{R}_i\})]}$$

and hence

$$\frac{d}{d\lambda}\mathrm{tr}_\gamma\left[G^\lambda \overleftarrow{G}_0^{-1}\right]_{\lambda=1} = \pm\beta\left[2E_{\mathrm{kin}} + E_{\mathrm{e-n}}(\{\mathbf{R}_i\}) + \int d\mathbf{x}_1 q\left.\frac{d\phi(\mathbf{r}_1;\{\lambda\mathbf{R}_i\})}{d\lambda}\right|_{\lambda=1} n(\mathbf{x}_1)\right].$$

Substituting these results into (L.6), we find

$$2E_{\mathrm{kin}} + E_{\mathrm{e-n}}(\{\mathbf{R}_i\}) + E_{\mathrm{int}} + \int d\mathbf{x}_1 q\left.\frac{d\phi(\mathbf{r}_1;\{\lambda\mathbf{R}_i\})}{d\lambda}\right|_{\lambda=1} n(\mathbf{x}_1) = 0, \tag{L.7}$$

which is true for any conserving approximation.

The next step consists in evaluating the λ-derivative in (L.7). For this purpose we consider the variation of the energy E induced by a variation $\delta\mathbf{R}_j$ of the nuclear coordinates. In the zero-temperature limit $\Omega = E - \mu N$ [see (D.7)] and the variations $\delta\Omega$ and δE are equal since the total number of particles is independent of the nuclear position, provided that there is a gap between the ground state and the first excited state. Then we can write

$$\delta E = \mathrm{tr}_\gamma\left[\frac{\delta\Omega_{\mathrm{K}}}{\delta G}\delta G\right]_{\{\mathbf{R}_i\}} + \left.\frac{d\Omega_{\mathrm{K}}}{d\mathbf{R}_i}\right|_G \cdot \delta\mathbf{R}_i,$$

where we take into account that Ω_{K} depends on $\{\mathbf{R}_i\}$ implicitly through G and explicitly through G_0. The first term on the r.h.s. vanishes for a conserving Green's function (stationary property). In the Klein functional the position of the nuclei only enters as parameters in the term G_0^{-1}, while the Green's function is an independent variable. This means that

$$\left.\frac{d\Omega_{\mathrm{K}}}{d\mathbf{R}_j}\right|_G = \int d\mathbf{x}_1 q\frac{d\phi(\mathbf{r}_1;\{\mathbf{R}_i\})}{d\mathbf{R}_j}n(\mathbf{x}_1).$$

From these considerations we conclude that the variation of E induced by the variation $\{\mathbf{R}_i\} \to \{\lambda\mathbf{R}_i\}$ is

$$\left.\frac{dE}{d\lambda}\right|_{\lambda=1} = \int d\mathbf{x}_1 q\left.\frac{d\phi(\mathbf{r}_1;\{\lambda\mathbf{R}_i\})}{d\lambda}\right|_{\lambda=1} n(\mathbf{x}_1).$$

This is an important result since it establishes a relation between the derivative appearing in (L.7) and the derivative of E. This latter derivative can easily be worked out using (L.4), which can alternatively be written as $dE/d\mathbf{R}_i = -dE_{\text{n-n}}/d\mathbf{R}_i$, see (L.2). Hence,

$$\left.\frac{dE}{d\lambda}\right|_{\lambda=1} = -\left.\frac{d}{d\lambda}\frac{1}{2}q^2\sum_{ij}\frac{Z_j Z_i}{\lambda|\mathbf{R}_j - \mathbf{R}_i|}\right|_{\lambda=1} = E_{\text{n-n}}.$$

The virial theorem simply follows from (L.7) and the combination of these last two equations.

Appendix M: Lippmann–Schwinger Equation and Cross Section

Let us consider two nonidentical particles with, for simplicity, the same mass m and described by the Hamiltonian (in first quantization):

$$\hat{\mathcal{H}} = \frac{\hat{\boldsymbol{p}}_1^2}{2m} + \frac{\hat{\boldsymbol{p}}_2^2}{2m} + v(\hat{\boldsymbol{r}}_1 - \hat{\boldsymbol{r}}_2), \tag{M.1}$$

where we use the standard notation $\hat{\boldsymbol{p}}_1 = \hat{\boldsymbol{p}} \otimes \hat{\mathbb{1}}$ for the momentum of the first particle, $\hat{\boldsymbol{p}}_2 = \hat{\mathbb{1}} \otimes \hat{\boldsymbol{p}}$ for the momentum of the second particle, and similarly for the position operators. For the time being we ignore the spin of the particles. In the absence of the interaction v, the eigenkets of $\hat{\mathcal{H}}$ are the momentum eigenkets $|\mathbf{k}_1\rangle|\mathbf{k}_2\rangle$. We are interested in calculating how the unperturbed eigenkets change due to the presence of the interaction. Since the interaction preserves the total momentum, it is convenient to work in the reference frame of the center of mass. Let us then introduce a slightly different basis. The wavefunction

$$\Psi_{\mathbf{k}_1,\mathbf{k}_2}(\mathbf{r}_1, \mathbf{r}_2) = \langle\mathbf{r}_2|\langle\mathbf{r}_1|\;|\mathbf{k}_1\rangle|\mathbf{k}_2\rangle = e^{\mathrm{i}\mathbf{k}_1\cdot\mathbf{r}_1 + \mathrm{i}\mathbf{k}_2\cdot\mathbf{r}_2} \tag{M.2}$$

can also be written as

$$\Psi_{\mathbf{k}_1,\mathbf{k}_2}(\mathbf{r}_1, \mathbf{r}_2) = e^{\mathrm{i}\mathbf{K}\cdot\mathbf{R} + \mathrm{i}\mathbf{k}\cdot\mathbf{r}}, \tag{M.3}$$

with $\mathbf{K} = \mathbf{k}_1 + \mathbf{k}_2$ the total momentum, $\mathbf{k} = \frac{1}{2}(\mathbf{k}_1 - \mathbf{k}_2)$ the relative momentum, $\mathbf{R} = \frac{1}{2}(\mathbf{r}_1 + \mathbf{r}_2)$ the center of mass coordinate, and $\mathbf{r} = \mathbf{r}_1 - \mathbf{r}_2$ the relative coordinate. We then see that we can define an equivalent basis $|\mathbf{K}\rangle|\mathbf{k}\rangle$ in which the first ket refers to the center of mass degree of freedom and the second ket to the relative coordinate degree of freedom. The ket $|\mathbf{K}\rangle|\mathbf{k}\rangle$ is an eigenket of the total momentum operator $\hat{\boldsymbol{\mathcal{P}}} = \hat{\boldsymbol{p}}_1 + \hat{\boldsymbol{p}}_2$ with eigenvalue $\mathbf{K}$ and of the relative momentum operator $\hat{\boldsymbol{p}} = \frac{1}{2}(\hat{\boldsymbol{p}}_1 - \hat{\boldsymbol{p}}_2)$ with eigenvalue $\mathbf{k}$. Similarly, the ket $|\mathbf{R}\rangle|\mathbf{r}\rangle$ is an eigenket of the center-of-mass position operator $\hat{\boldsymbol{\mathcal{R}}} = \frac{1}{2}(\hat{\boldsymbol{r}}_1 + \hat{\boldsymbol{r}}_2)$ with eigenvalue $\mathbf{R}$ and of the relative coordinate operator $\hat{\boldsymbol{r}} = \hat{\boldsymbol{r}}_1 - \hat{\boldsymbol{r}}_2$ with eigenvalue $\mathbf{r}$. The inner products of these kets are

$$\begin{aligned}
\langle\mathbf{r}|\langle\mathbf{R}|\;|\mathbf{K}\rangle|\mathbf{k}\rangle &= e^{\mathrm{i}\mathbf{K}\cdot\mathbf{R} + \mathrm{i}\mathbf{k}\cdot\mathbf{r}}, \\
\langle\mathbf{r}|\langle\mathbf{K}'|\;|\mathbf{K}\rangle|\mathbf{k}\rangle &= (2\pi)^3\delta(\mathbf{K} - \mathbf{K}')e^{\mathrm{i}\mathbf{k}\cdot\mathbf{r}},
\end{aligned}$$

etc. In terms of the operators $\hat{\mathcal{P}}$, $\hat{p}$, $\hat{\mathcal{R}}$, and $\hat{r}$ the Hamiltonian in (M.1) takes the form

$$\hat{\mathcal{H}} = \frac{\hat{\mathcal{P}}^2}{4m} + \frac{\hat{\mathbf{p}}^2}{m} + v(\hat{\mathbf{r}}), \tag{M.4}$$

which is independent of the center-of-mass position operator $\hat{\mathcal{R}}$. We use this observation to write a generic eigenket of $\hat{\mathcal{H}}$ as $|\mathbf{K}\rangle|\psi\rangle$, where $|\psi\rangle$ fulfills

$$\hat{h}|\psi\rangle = \left[\frac{\hat{\mathbf{p}}^2}{m} + v(\hat{\mathbf{r}})\right]|\psi\rangle = E|\psi\rangle. \tag{M.5}$$

The eigenenergy of $|\mathbf{K}\rangle|\psi\rangle$ is then $E + K^2/(4m)$. Let us take E in the continuum part of the spectrum of the operator $\hat{h}$. Assuming that $v(\mathbf{r})$ vanishes when $r \to \infty$, we have that $E \geq 0$. Then the eigenfunction $\langle \mathbf{r}|\psi\rangle$ is a plane wave in this region of space – that is, $\lim_{r\to\infty}\langle \mathbf{r}|\psi\rangle = \langle \mathbf{r}|\mathbf{k}\rangle = e^{i\mathbf{k}\cdot\mathbf{r}}$, with $E = k^2/m$. We calculate $|\psi\rangle$ using the *Lippmann–Schwinger equation*:

$$|\psi\rangle = |\mathbf{k}\rangle + \frac{1}{E - \hat{\mathbf{p}}^2/m \pm i\eta} v(\hat{\mathbf{r}})|\psi\rangle. \tag{M.6}$$

The equivalence between (M.5) and (M.6) can easily be verified by multiplying both sides with $(E - \hat{\mathbf{p}}^2/m \pm i\eta)$ and by letting $\eta \to 0$. The ket $|\mathbf{k}\rangle$ on the r.h.s. of (M.6) guarantees that for $v = 0$ we recover the noninteracting solution. In order to choose the sign in front of $i\eta$ we must understand what physical state $|\psi\rangle$ is. For this purpose we suppose that we prepare the system at time $t = 0$ in the state $|\mathbf{k}\rangle$ and then evolve the system in time according to the Hamiltonian $\hat{h}$. The state of the system at time t is therefore

$$|\psi(t)\rangle = e^{-i(\hat{h}-E)t}|\mathbf{k}\rangle, \tag{M.7}$$

where, for convenience, we subtract from $\hat{h}$ the eigenvalue E so that the noninteracting ket $|\psi(t)\rangle = |\mathbf{k}\rangle$ is independent of t (in the noninteracting case $v = 0$ and hence $|\mathbf{k}\rangle$ is an eigenket of $\hat{h}$ with eigenvalue $E = k^2/m$). Now for any $t > 0$ we can write

$$\begin{aligned} e^{-i(\hat{h}-E)t} &= i\int \frac{d\omega}{2\pi} e^{-i(\omega-E)t} \frac{1}{\omega - \hat{h} + i\eta} \\ &= i\int \frac{d\omega}{2\pi} e^{-i(\omega-E)t} \left[\frac{1}{\omega - \hat{\mathbf{p}}^2/m + i\eta} + \frac{1}{\omega - \hat{\mathbf{p}}^2/m + i\eta} v(\hat{\boldsymbol{r}}) \frac{1}{\omega - \hat{h} + i\eta}\right], \end{aligned}$$

where in the last equality we expand $1/(\omega - \hat{h} + i\eta)$ in a Dyson-like equation. Inserting this result into (M.7), we find

$$|\psi(t)\rangle = |\mathbf{k}\rangle + i\int \frac{d\omega}{2\pi} e^{-i(\omega-E)t} \frac{1}{\omega - \hat{\mathbf{p}}^2/m + i\eta} v(\hat{\boldsymbol{r}}) \frac{1}{\omega - \hat{h} + i\eta}|\mathbf{k}\rangle. \tag{M.8}$$

In the limit $t \to \infty$ the dominant contribution to the ω-integral comes from the region $\omega \sim E$ (Riemann–Lebesgue theorem). If we replace $\omega \to E$ in the first denominator, (M.8) becomes

$$\lim_{t\to\infty}|\psi(t)\rangle = |\mathbf{k}\rangle + \lim_{t\to\infty} \frac{1}{E - \hat{\mathbf{p}}^2/m + i\eta} v(\hat{\boldsymbol{r}})|\psi(t)\rangle.$$

Comparing this with the Lippmann–Schwinger equation, we see that $|\psi\rangle = \lim_{t\to\infty} |\psi(t)\rangle$ when the sign of $i\eta$ in (M.6) is plus. Similarly, it is easy to show that $|\psi\rangle = \lim_{t\to-\infty} |\psi(t)\rangle$ when the sign of $i\eta$ in (M.6) is minus. As in an experiment we can only study the forward evolution, we take the sign of $i\eta$ in (M.6) to be plus in the following discussion.

We multiply (M.6) from the left with $\langle \mathbf{r}|$ and insert the completeness relation $\hat{\mathbb{1}} = \int d\mathbf{r}' |\mathbf{r}'\rangle\langle\mathbf{r}'|$ to the left of v:

$$\psi(\mathbf{r}) = \langle \mathbf{r}|\psi\rangle = e^{\mathrm{i}\mathbf{k}\cdot\mathbf{r}} + \int d\mathbf{r}' \langle \mathbf{r}| \frac{1}{E - \hat{\mathbf{p}}^2/m + \mathrm{i}\eta} |\mathbf{r}'\rangle v(\mathbf{r}')\psi(\mathbf{r}'). \tag{M.9}$$

The kernel of this integral equation can be evaluated as follows:

$$\begin{aligned}
\langle \mathbf{r}| \frac{1}{E - \hat{\mathbf{p}}^2/m + \mathrm{i}\eta} |\mathbf{r}'\rangle &= \int \frac{d\mathbf{q}}{(2\pi)^3} \frac{e^{\mathrm{i}\mathbf{q}\cdot(\mathbf{r}-\mathbf{r}')}}{E - q^2/m + \mathrm{i}\eta} \\
&= \int_0^\infty \frac{dq}{(2\pi)^2} q^2 \int_{-1}^1 dc \, \frac{e^{\mathrm{i}q|\mathbf{r}-\mathbf{r}'|c}}{E - q^2/m + \mathrm{i}\eta} \\
&= -\frac{m}{8\pi^2} \frac{1}{\mathrm{i}|\mathbf{r}-\mathbf{r}'|} \int_{-\infty}^\infty dq \, q \, \frac{e^{\mathrm{i}q|\mathbf{r}-\mathbf{r}'|} - e^{-\mathrm{i}q|\mathbf{r}-\mathbf{r}'|}}{q^2 - k^2 - \mathrm{i}\eta} \\
&= -\frac{m}{4\pi} \frac{e^{\mathrm{i}k|\mathbf{r}-\mathbf{r}'|}}{|\mathbf{r}-\mathbf{r}'|},
\end{aligned}$$

where in the last equality we take into account that the denominator has simple poles in $q = \pm(k + \mathrm{i}\eta)$. If the interaction v is short-ranged and if we are only interested in calculating the wavefunction far away from the interacting region, we can approximate the kernel in (M.9) with its expansion for $r \gg r'$. We have $|\mathbf{r} - \mathbf{r}'| = (r^2 + r'^2 - 2\mathbf{r}\cdot\mathbf{r}')^{1/2} \sim r - \mathbf{r}\cdot\mathbf{r}'/r$. Let $\mathbf{k}' = k\mathbf{r}/r$ be the propagation vector pointing in the same direction as the vector $\mathbf{r}$ (where we want to compute ψ) and having the same modulus of the relative momentum. Then, we can approximate (M.9) as

$$\psi(\mathbf{r}) = e^{\mathrm{i}\mathbf{k}\cdot\mathbf{r}} - \frac{m}{4\pi} \frac{e^{ikr}}{r} \int d\mathbf{r}' e^{-\mathrm{i}\mathbf{k}'\cdot\mathbf{r}'} v(\mathbf{r}')\psi(\mathbf{r}') \equiv e^{\mathrm{i}\mathbf{k}\cdot\mathbf{r}} + \frac{e^{ikr}}{r} f(\mathbf{k}', \mathbf{k}). \tag{M.10}$$

We thus see that $\psi(\mathbf{r})$ is written as the sum of an incident wave and an outgoing spherical wave. This wavefunction has a similar structure to the continuum eigenfunction of a one-dimensional system with a potential barrier: an incident wave e^{ikx} on which we superimposed a reflected wave Re^{-ikx} for $x \to -\infty$, and a transmitted wave Te^{ikx} for $x \to \infty$. In three dimensions the reflected and transmitted waves are both incorporated into the outgoing spherical wave. The probability of being reflected (transmitted) is obtained by choosing $\mathbf{r}$, or equivalently $\mathbf{k}'$, in the opposite (same) direction of the incident wavevector $\mathbf{k}$. In three dimensions, however, there are infinitely many more possibilities and the outgoing wave yields the probability of measuring a relative momentum $\mathbf{k}'$ for two particles scattering with relative momentum $\mathbf{k}$:

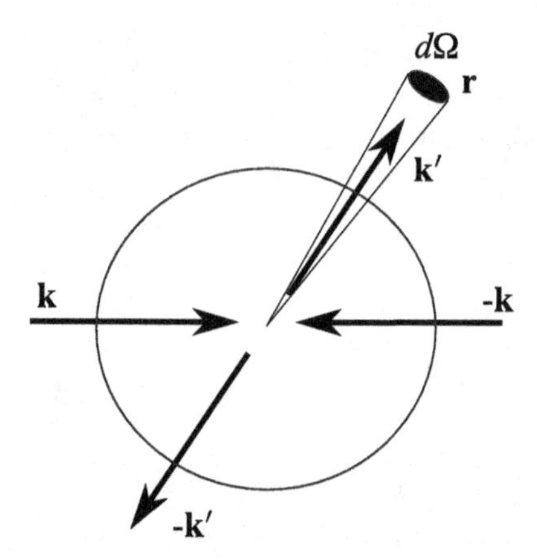

In an experiment we can measure the flux of incident particles as well as how many particles per unit time hit a detector in $\mathbf{r}$ with surface area $r^2 d\Omega$. The flux of incident particles is just the modulus of the current density of the initial plane wave, which in our case is k. The current density of the scattered particles in the direction $\mathbf{k}'$ is, to lowest order in $1/r$, given by $k|f(\mathbf{k}',\mathbf{k})/r|^2$. Therefore, the number of particles that hit the detector per unit time is $k|f(\mathbf{k}',\mathbf{k})|^2 d\Omega$. The ratio between these two quantities has the physical dimension of an area and it is called the *differential cross section*:

$$d\sigma \equiv \frac{k|f(\mathbf{k}',\mathbf{k})|^2 d\Omega}{k} \quad \Rightarrow \quad \frac{d\sigma}{d\Omega} = |f(\mathbf{k}',\mathbf{k})|^2. \tag{M.11}$$

We can calculate f by iterating (M.10) – that is, by replacing ψ in the r.h.s. with the whole r.h.s. *ad infinitum*. The first iteration corresponds to the second Born approximation and gives

$$f(\mathbf{k}',\mathbf{k}) = -\frac{m}{4\pi}\tilde{v}_{\mathbf{k}'-\mathbf{k}} = -\frac{m}{4\pi}\tilde{v}_{\mathbf{k}-\mathbf{k}'}, \tag{M.12}$$

where in the last equality we use that $v(\mathbf{r}) = v(-\mathbf{r})$ and hence the Fourier transform is real and symmetric.

So far we have considered two distinguishable particles. If the particles are identical, their orbital wavefunction is either symmetric or antisymmetric. This means that the initial ket is $|\mathbf{k}\rangle \pm |-\mathbf{k}\rangle$. Consequently, we must replace (M.10) with

$$\begin{aligned}\psi(\mathbf{r}) &= \left(e^{i\mathbf{k}\cdot\mathbf{r}} \pm e^{-i\mathbf{k}\cdot\mathbf{r}}\right) - \frac{m}{4\pi}\frac{e^{ikr}}{r}\int d\mathbf{r}' e^{-i\mathbf{k}'\cdot\mathbf{r}'} v(\mathbf{r}')\psi(\mathbf{r}') \\ &= \left(e^{i\mathbf{k}\cdot\mathbf{r}} \pm e^{-i\mathbf{k}\cdot\mathbf{r}}\right) + \frac{e^{ikr}}{r} f(\mathbf{k}',\mathbf{k}).\end{aligned} \tag{M.13}$$

In the second Born approximation we then have $f(\mathbf{k}',\mathbf{k}) = -\frac{m}{4\pi}(\tilde{v}_{\mathbf{k}-\mathbf{k}'} \pm \tilde{v}_{\mathbf{k}+\mathbf{k}'})$ and the differential cross section becomes

$$\frac{d\sigma_\pm}{d\Omega} = \left(\frac{m}{4\pi}\right)^2 \left(\tilde{v}^2_{\mathbf{k}-\mathbf{k}'} + \tilde{v}^2_{\mathbf{k}+\mathbf{k}'} \pm 2\tilde{v}_{\mathbf{k}-\mathbf{k}'}\tilde{v}_{\mathbf{k}+\mathbf{k}'}\right).$$

We can rewrite this formula using the momentum of the particles in the original reference frame. We use the same notation as in Section 15.1 and call $\mathbf{p} = \mathbf{K}/2 + \mathbf{k}$, $\mathbf{p}' = \mathbf{K}/2 - \mathbf{k}$ the momenta of the incident particles, and $\bar{\mathbf{p}} = \mathbf{K}/2 + \mathbf{k}'$, $\bar{\mathbf{p}}' = \mathbf{K}/2 - \mathbf{k}'$ the momenta of the particles after the scattering. Then we see that $\mathbf{k} - \mathbf{k}' = \mathbf{p} - \bar{\mathbf{p}}$ and $\mathbf{k} + \mathbf{k}' = \mathbf{p} - \bar{\mathbf{p}}'$, and hence

$$\frac{d\sigma_\pm}{d\Omega} = \left(\frac{m}{4\pi}\right)^2 \left(\tilde{v}^2_{\mathbf{p}-\bar{\mathbf{p}}} + \tilde{v}^2_{\mathbf{p}-\bar{\mathbf{p}}'} \pm 2\tilde{v}_{\mathbf{p}-\bar{\mathbf{p}}}\tilde{v}_{\mathbf{p}-\bar{\mathbf{p}}'}\right). \tag{M.14}$$

To complete the discussion we must combine this result with the spin degrees of freedom. Let us start by considering two bosons of spin S (hence S is an integer). In the basis of the total spin they can form multiplets of spin $2S$, $2S - 1, \ldots, 0$. The multiplets with even total spin are symmetric under an exchange of the particles, while those with odd total spin are antisymmetric. We define N_+ and N_- as the total number of states in the symmetric and antisymmetric multiplets, respectively. We then have $N_+ + N_- = (2S+1)^2$. Furthermore, since the multiplet with spin $S' \leq 2S$ has always two states more than the multiplet with spin $S' - 1$ (the eigenvalues of the z-component of the total spin go from $-S'$ to S') we also have $N_+ = N_- + 2S + 1$. These two relations allow us to express N_+ and N_- in terms of S, and what we find is

$$N_+ = \frac{(2S+1)(2S+2)}{2}, \qquad N_- = \frac{(2S+1)2S}{2}. \tag{M.15}$$

In the absence of any information about the spin of the particles we must calculate the differential cross section using a statistical weight $N_+/(2S+1)^2$ for the symmetric orbital part and $N_-/(2S+1)^2$ for the antisymmetric orbital part:

$$\begin{aligned}\frac{d\sigma}{d\Omega} &= \frac{N_+}{(2S+1)^2}\frac{d\sigma_+}{d\Omega} + \frac{N_-}{(2S+1)^2}\frac{d\sigma_-}{d\Omega} \\ &= \left(\frac{m}{4\pi}\right)^2 \left(\tilde{v}^2_{\mathbf{p}-\bar{\mathbf{p}}} + \tilde{v}^2_{\mathbf{p}-\bar{\mathbf{p}}'} + \frac{2}{2S+1}\tilde{v}_{\mathbf{p}-\bar{\mathbf{p}}}\tilde{v}_{\mathbf{p}-\bar{\mathbf{p}}'}\right),\end{aligned} \tag{M.16}$$

which is proportional to the bosonic $B(\mathbf{p}, \bar{\mathbf{p}}, \bar{\mathbf{p}}')$ in (15.5). In a similar way, we can proceed for two fermions. It is easy to show that N_+ and N_- are still given by (M.15). This time, however, we must combine the symmetric multiplets with the antisymmetric wavefunction and the antisymmetric multiplets with the symmetric wavefunction. We find

$$\begin{aligned}\frac{d\sigma}{d\Omega} &= \frac{N_-}{(2S+1)^2}\frac{d\sigma_+}{d\Omega} + \frac{N_+}{(2S+1)^2}\frac{d\sigma_-}{d\Omega} \\ &= \left(\frac{m}{4\pi}\right)^2 \left(\tilde{v}^2_{\mathbf{p}-\bar{\mathbf{p}}} + \tilde{v}^2_{\mathbf{p}-\bar{\mathbf{p}}'} - \frac{2}{2S+1}\tilde{v}_{\mathbf{p}-\bar{\mathbf{p}}}\tilde{v}_{\mathbf{p}-\bar{\mathbf{p}}'}\right),\end{aligned} \tag{M.17}$$

which is proportional to the fermionic $B(\mathbf{p}, \bar{\mathbf{p}}, \bar{\mathbf{p}}')$ in (15.5).

Appendix N: Hedin Equations from a Generating Functional

The idea behind the *source field method* (also named the *field-theoretic approach*) is to construct a generating functional from which to obtain the quantities of interest by functional differentiation with respect to some "source" field S. Classical examples of generating functionals are the exponential e^{2xt-t^2} whose nth derivative with respect to the "source" variable t calculated in $t = 0$ gives the nth Hermite polynomial, or the function $1/\sqrt{1 - 2xt + t^2}$ whose nth derivative with respect to the "source" variable t calculated in $t = 0$ gives the nth Legendre polynomial multiplied by $n!$. Clearly there exist infinitely many formulations that are more or less complicated depending on which are the quantities of interest. In this appendix we follow the presentation of Strinati in Ref. [185], see also Ref. [283]. We define a generating functional which is an extension of definition (5.1) of the Green's function:[1]

$$G_S(1; 2) = \frac{1}{\mathrm{i}} \frac{\mathrm{Tr}\left[\mathcal{T}\left\{e^{-\mathrm{i}\int_\gamma d\bar{z}\hat{H}_S(\bar{z})}\hat{\psi}(1)\hat{\psi}^\dagger(2)\right\}\right]}{\mathrm{Tr}\left[\mathcal{T}\left\{e^{-\mathrm{i}\int_\gamma d\bar{z}\hat{H}_S(\bar{z})}\right\}\right]}, \tag{N.1}$$

where $\hat{H}_S$ is given by the sum of the Hamiltonian $\hat{H}$ of the system and the coupling between a source field S and the density $\hat{n}(\mathbf{x}) = \hat{\psi}^\dagger(\mathbf{x})\hat{\psi}(\mathbf{x})$:

$$\hat{H}_S(z) = \hat{H}(z) + \int d\mathbf{x}\, S(\mathbf{x}, z)\hat{n}(\mathbf{x}).$$

The special feature of the source field S is that it can take different values on the two horizontal branches of the contour and, furthermore, it can vary along the imaginary track. This freedom allows for unconstrained variations of S in G_S, a necessary property if we want to take functional derivatives. In the special case $S(\mathbf{x}, t_+) = S(\mathbf{x}, t_-)$ and $S(\mathbf{x}, t_0 - \mathrm{i}\tau) = S^{\mathrm{M}}(\mathbf{x})$, the addition of the source field is equivalent to the addition of a scalar potential. In particular, the generating functional G_S reduces to the standard Green's function G for $S = 0$. We also observe that G_S satisfies the KMS relations for any S.

[1]In (N.1) the field operators are *not* in the Heisenberg picture, otherwise we would have added the subscript "H"; their dependence on the contour time is fictitious and it has only the purpose of specifying where the operators lie on the contour, see Section 4.2.

Let us derive the equations of motion for G_S.[2] We write the numerator in (N.1) as

$$\mathcal{T}\left\{e^{-i\int_\gamma d\bar{z}\hat{H}_S(\bar{z})}\hat{\psi}(1)\hat{\psi}^\dagger(2)\right\}$$
$$= \theta(z_1, z_2)\mathcal{T}\left\{e^{-i\int_{z_1}^{t_0-i\beta} d\bar{z}\hat{H}_S(\bar{z})}\right\}\hat{\psi}(1)\mathcal{T}\left\{e^{-i\int_{t_{0-}}^{z_1} d\bar{z}\hat{H}_S(\bar{z})}\hat{\psi}^\dagger(2)\right\}$$
$$\pm\, \theta(z_2, z_1)\mathcal{T}\left\{e^{-i\int_{z_1}^{t_0-i\beta} d\bar{z}\hat{H}_S(\bar{z})}\hat{\psi}^\dagger(2)\right\}\hat{\psi}(1)\mathcal{T}\left\{e^{-i\int_{t_{0-}}^{z_1} d\bar{z}\hat{H}_S(\bar{z})}\right\}.$$

Taking the derivative with respect to z_1, we find

$$i\frac{d}{dz_1}G_S(1;2) = \delta(1;2) + \frac{1}{i}\frac{\mathrm{Tr}\left[\mathcal{T}\left\{e^{-i\int_\gamma d\bar{z}\hat{H}_S(\bar{z})}\left[\hat{\psi}(1), \hat{H}_S(z_1)\right]\hat{\psi}^\dagger(2)\right\}\right]}{\mathrm{Tr}\left[\mathcal{T}\left\{e^{-i\int_\gamma d\bar{z}\hat{H}_S(\bar{z})}\right\}\right]}. \tag{N.2}$$

The commutator between the field operators and the Hamiltonian has been encountered several times (see, for instance, Section 3.3) and the reader should have no problems in rewriting (N.2) as

$$i\frac{d}{dz_1}G_S(1;2) - \int d3[h(1;3) + \delta(1;3)S(3)]G_S(3;1)$$
$$= \delta(1;2) \pm i\int d3\, v(1;3)G_{2,S}(1,3;2,3^+),$$

where we define the generalized two-particle Green's function

$$G_{2,S}(1,2;3,4) = \frac{1}{i^2}\frac{\mathrm{Tr}\left[\mathcal{T}\left\{e^{-i\int_\gamma d\bar{z}\hat{H}_S(\bar{z})}\hat{\psi}(1)\hat{\psi}(2)\hat{\psi}^\dagger(4)\hat{\psi}^\dagger(3)\right\}\right]}{\mathrm{Tr}\left[\mathcal{T}\left\{e^{-i\int_\gamma d\bar{z}\hat{H}_S(\bar{z})}\right\}\right]},$$

which reduces to G_2 for $S = 0$. We can also introduce a self-energy Σ_S as we have done in (9.5):

$$\int d3\, \Sigma_S(1;3)G_S(3;2) = \pm i\int d3\, v(1;3)G_{2,S}(1,3;2,3^+), \tag{N.3}$$

and rewrite the equation of motion as

$$\int d3\, \overrightarrow{G}_S^{-1}(1;3)G_S(3;2) = \delta(1;2), \tag{N.4}$$

where the operator $\overrightarrow{G}_S^{-1}$ (as usual the arrow signifies that it acts on quantities to its right) is defined according to

$$\overrightarrow{G}_S^{-1}(1;3) = i\frac{\overrightarrow{d}}{dz_1}\delta(1;3) - S(1)\delta(1;3) - h(1;3) - \Sigma_S(1;3).$$

[2]In the special case $G_S = G$ the derivation below provides an alternative proof of the equations of motion for the Green's function.

Proceeding along the same lines, it is possible to derive the adjoint equation of motion

$$\int d3\, G_S(1;3)\overleftarrow{G}_S^{-1}(3;2) = \delta(1;2), \tag{N.5}$$

with

$$\overleftarrow{G}_S^{-1}(3;2) = -\delta(3;2)\mathrm{i}\frac{\overleftarrow{d}}{dz_2} - \delta(3;2)S(2) - h(3;2) - \Sigma_S(3;2). \tag{N.6}$$

Thus, no extra complication arises in deriving the equations of motion of the new Green's function provided that we properly extend the definition of the two-particle Green's function and self-energy.

We make extensive use of the operators $\overrightarrow{G}_S^{-1}$ and $\overleftarrow{G}_S^{-1}$ in what follows. These operators have the property that for any two functions $A(1;2)$ and $B(1;2)$ which fulfill the KMS boundary conditions

$$\int d3d4\, A(1;3)\overrightarrow{G}_S^{-1}(3;4)B(4;2) = \int d3d4\, A(1;3)\overleftarrow{G}_S^{-1}(3;4)B(4;2),$$

since an integration by parts with respect to the contour time does not produce any boundary term as a consequence of the KMS boundary conditions. Using (N.5) we can also solve (N.3) for Σ_S:

$$\Sigma_S(1;4) = \pm\mathrm{i}\int d2d3\, v(1;3)G_{2,S}(1,3;2,3^+)\overleftarrow{G}_S^{-1}(2;4), \tag{N.7}$$

a result that we need later.

The starting point of our new derivation of the Hedin equations is a relation between $G_{2,S}$ and $\delta G_S/\delta S$. The generating functional G_S depends on S through the dependence on $\hat{H}_S$, which appears in the exponent of both the numerator and the denominator. Therefore,

$$\begin{aligned}
\frac{\delta G_S(1;2)}{\delta S(3)} &= \frac{1}{\mathrm{i}}\frac{\mathrm{Tr}\left[\mathcal{T}\left\{e^{-\mathrm{i}\int_\gamma d\bar{z}\hat{H}_S(\bar{z})}(-\mathrm{i})\hat{n}(3)\hat{\psi}(1)\hat{\psi}^\dagger(2)\right\}\right]}{\mathrm{Tr}\left[\mathcal{T}\left\{e^{-\mathrm{i}\int_\gamma d\bar{z}\hat{H}_S(\bar{z})}\right\}\right]} \\
&\quad - \frac{1}{\mathrm{i}}\frac{\mathrm{Tr}\left[\mathcal{T}\left\{e^{-\mathrm{i}\int_\gamma d\bar{z}\hat{H}_S(\bar{z})}\hat{\psi}(1)\hat{\psi}^\dagger(2)\right\}\right]}{\left(\mathrm{Tr}\left[\mathcal{T}\left\{e^{-\mathrm{i}\int_\gamma d\bar{z}\hat{H}_S(\bar{z})}\right\}\right]\right)^2} \times \mathrm{Tr}\left[\mathcal{T}\left\{e^{-\mathrm{i}\int_\gamma d\bar{z}\hat{H}_S(\bar{z})}(-\mathrm{i})\hat{n}(3)\right\}\right] \\
&= \pm\left[G_{2,S}(1;3;2,3^+) - G_S(1;2)G_S(3;3^+)\right].
\end{aligned} \tag{N.8}$$

Inserting this result into (N.7), we find

$$\Sigma_S(1;4) = \underbrace{\pm\mathrm{i}\delta(1;4)\int d3\, v(1;3)G_S(3;3^+)}_{\Sigma_{\mathrm{H},S}(1;4)} + \underbrace{\mathrm{i}\int d2d3\, v(1;3)\frac{\delta G_S(1;2)}{\delta S(3)}\overleftarrow{G}_S^{-1}(2;4)}_{\Sigma_{\mathrm{xc},S}(1;4)}. \tag{N.9}$$

We recognize in the first term on the r.h.s. the Hartree self-energy $\Sigma_{\mathrm{H},S}(1;4) = \delta(1;4)V_{\mathrm{H},S}(1)$ with Hartree potential

$$V_{\mathrm{H},S}(1) = \pm\mathrm{i}\int d3\, v(1;3)G_S(3;3^+),$$

in agreement with (8.5). The sum of the source field S and the Hartree energy $V_{\mathrm{H},S}$ defines the so-called classical (or total) energy,

$$C(1) = S(1) + V_{\mathrm{H},S}(1). \tag{N.10}$$

We recall that in extended systems, such as an electron gas, the external potential and the Hartree potential are separately infinite while their sum is finite and meaningful, see Section 8.3.2. It is therefore more natural to study variations with respect to C rather than with respect to S. Regarding G_S as a functional of C instead of S we can apply the chain rule to the XC self-energy of (N.9) and obtain

$$\Sigma_{\mathrm{xc},S}(1;4) = \mathrm{i}\int d2d3d5\, v(1;3)\frac{\delta G_S(1;2)}{\delta C(5)}\frac{\delta C(5)}{\delta S(3)}\overleftarrow{G}_S^{-1}(2;4). \tag{N.11}$$

The derivative of the total field with respect to the source field in $S = 0$ is the *inverse (longitudinal) dielectric function* ε^{-1} defined as

$$\varepsilon^{-1}(1;2) = \left.\frac{\delta C(1)}{\delta S(2)}\right|_{S=0},$$

and it is calculated shortly. The derivative $\delta G_S/\delta C$ is most easily worked out from the derivative $\delta\overleftarrow{G}_S^{-1}/\delta C$. Using the equations of motions (N.4) and (N.5) we have (omitting the arguments)

$$\int \frac{\delta G_S}{\delta C}\overleftarrow{G}_S^{-1} + \int G_S\frac{\delta \overleftarrow{G}_S^{-1}}{\delta C} = 0 \quad\Rightarrow\quad \frac{\delta G_S}{\delta C} = -\int G_S\frac{\delta\overleftarrow{G}_S^{-1}}{\delta C}G_S, \tag{N.12}$$

and hence (N.11) can also be written as[3]

$$\Sigma_{\mathrm{xc},S}(1;4) = -\mathrm{i}\int d2d3d5\, v(1;3)G_S(1;2)\frac{\delta\overleftarrow{G}_S^{-1}(2;4)}{\delta C(5)}\frac{\delta C(5)}{\delta S(3)}. \tag{N.13}$$

This result leads to the second Hedin equation. In fact, all Hedin equations naturally follow from (N.13) when forcing on this equation the structure $\mathrm{i}GW\Gamma$.

Let us start by proving that the vertex function

$$\Gamma(1,2;3) \equiv -\left.\frac{\delta\overleftarrow{G}_S^{-1}(1;2)}{\delta C(3)}\right|_{S=0} \tag{N.14}$$

appearing in (N.13) obeys the fifth Hedin equation. From (N.6) and the definition (N.10) of the classical energy we can rewrite $\overleftarrow{G}_S^{-1}$ as

$$\overleftarrow{G}_S^{-1}(1;2) = \delta(1;2)\left[-\mathrm{i}\frac{\overleftarrow{d}}{dz_2} - C(2)\right] - h(1;2) - \Sigma_{\mathrm{xc},S}(1;2).$$

[3] In principle we could have added to $\delta G_S/\delta C$ any function F such that $\int F\,\overleftarrow{G}_S^{-1} = 0$. However, one can show that this possibility is excluded due to the satisfaction of the KMS relations.

We thus see that

$$\Gamma(1,2;3) = \delta(1;2)\delta(1;3) + \left.\frac{\delta\Sigma_{\mathrm{xc},S}(1;2)}{\delta C(3)}\right|_{S=0}.$$

The XC self-energy depends on C only through its dependence on the generating functional G_S; using the chain rule we then find

$$\begin{aligned}\Gamma(1,2;3) &= \delta(1;2)\delta(2;3) + \int d4d5\, \left.\frac{\delta\Sigma_{\mathrm{xc},S}(1;2)}{\delta G_S(4;5)}\frac{\delta G_S(4;5)}{\delta C(3)}\right|_{S=0} \\ &= \delta(1;2)\delta(2;3) + \int d4d5d6d7\, \frac{\delta\Sigma_{\mathrm{xc}}(1;2)}{\delta G(4;5)} G(4,6)G(7;5)\Gamma(6,7;3),\end{aligned}$$

where in the last step we use (N.12). This is exactly the Bethe–Salpeter equation for Γ, see (14.95), provided that we manually shift the contour time $z_2 \to z_2^+$ in $\delta(1;2)$.[4]

Next we observe that (N.8) implies

$$\begin{aligned}\left.\frac{\delta G_S(1;1^+)}{\delta S(3)}\right|_{S=0} &= \pm\left[G_2(1;3;1^+,3^+) - G(1;1^+)G(3;3^+)\right] \\ &= L(1,3;1,3) \\ &= \mp \mathrm{i}\chi(1;3). \end{aligned} \tag{N.15}$$

If we define the polarization P as

$$P(1;2) = \pm\mathrm{i}\left.\frac{\delta G_S(1;1^+)}{\delta C(2)}\right|_{S=0}, \tag{N.16}$$

then (N.15) provides the following relation between χ and P:

$$\chi(1;2) = \int d3\, P(1;3)\left.\frac{\delta C(3)}{\delta S(2)}\right|_{S=0} = \int d3\, P(1;3)\varepsilon^{-1}(3;2). \tag{N.17}$$

Let us calculate the inverse dielectric function. Using the definition of the classical energy (N.10), we have

$$\begin{aligned}\varepsilon^{-1}(3;2) &= \delta(3;2) \pm \mathrm{i}\int d4\, v(3;4)\left.\frac{\delta G_S(4;4^+)}{\delta S(2)}\right|_{S=0} \\ &= \delta(3;2) + \int d4\, v(3;4)\chi(4;2). \end{aligned} \tag{N.18}$$

Inserting this result into (N.17) we get the familiar relation, see (7.24), of the density response function in terms of the polarization: $\chi = P + Pv\chi$. This suggests that we are on the right track in defining P according to (N.16). To make this argument more stringent we evaluate

[4]This shift is crucial in order to get the correct Fock (or exchange) self-energy.

the functional derivative (N.16). Using (N.12) and the definition (N.14) of the vertex function we find

$$P(1;2) = \mp \mathrm{i} \int d3d4\, G(1;3) \left.\frac{\delta \overleftarrow{G}_S^{-1}(3;4)}{\delta C(2)}\right|_{S=0} G(4;1^+)$$
$$= \pm \mathrm{i} \int d3d4\, G(1;3)G(4;1)\Gamma(3,4;2),$$

where in the last step we (safely) replaced 1^+ with 1 in the second Green's function. We have just obtained the third Hedin equation.

The only remaining quantity to evaluate is $v \times \delta C/\delta S$, which appears in the XC self-energy (N.13). From (N.18) we see that

$$W(1;2) \equiv \int d3\, v(1;3) \left.\frac{\delta C(2)}{\delta S(3)}\right|_{S=0} = v(1;2) + \int d3d4\, v(1;3)\chi(3;4)v(2;4).$$

Expanding χ in powers of the polarization, we get $W = v + vPv + vPvPv + \ldots$ which is solved by the fourth Hedin equation,

$$W(1;2) = v(1;2) + \int d3d4\, v(1;3)P(3;4)W(4;2).$$

Finally, taking into account the definition of Γ and W, we can write the self-energy (N.13) in $S = 0$ as

$$\Sigma_{\rm xc}(1;2) \equiv \Sigma_{{\rm xc},S=0}(1;2) = \mathrm{i} \int d3d4\, W(1;3)G(1;4)\Gamma(4,2;3),$$

which coincides with the second Hedin equation. This concludes the alternative derivation of the Hedin equations using a generating functional.

Appendix O: Why the Name Random Phase Approximation?

In Chapter 13 we discussed the density response function χ in the time-dependent Hartree approximation, and we have called this approximation the Random Phase Approximation (RPA). The origin of this name is based on an idea of Bohm and Pines, who derived the RPA response function without using any diagrammatic technique [249, 399]. In this appendix we do not present the original derivation of Bohm and Pines since it is rather lengthy (a pedagogical discussion of the original derivation can be found in Ref. [401]). We instead use the idea of Bohm and Pines in a derivation based on the equation of motion for χ, thus still providing a justification for the name RPA. A similar derivation can be found in Refs. [49, 260].

Let us consider the Hamiltonian of an electron gas in a large three-dimensional box of volume $\mathrm{V} = L^3$:

$$\begin{aligned}\hat{H} &= -\frac{1}{2}\int d\mathbf{x}\, \hat{\psi}^\dagger(\mathbf{x})\nabla^2\hat{\psi}(\mathbf{x}) + \frac{1}{2}\int d\mathbf{x}d\mathbf{x}'\, v(\mathbf{r}-\mathbf{r}')\hat{\psi}^\dagger(\mathbf{x})\hat{\psi}^\dagger(\mathbf{x}')\hat{\psi}(\mathbf{x}')\hat{\psi}(\mathbf{x})\\ &= -\frac{1}{2}\int d\mathbf{x}\, \hat{\psi}^\dagger(\mathbf{x})\left(\nabla^2 + v(0)\right)\hat{\psi}(\mathbf{x}) + \frac{1}{2}\int d\mathbf{x}d\mathbf{x}'\, v(\mathbf{r}-\mathbf{r}')\hat{n}(\mathbf{x})\hat{n}(\mathbf{x}'),\end{aligned}$$

where the space integral is restricted to the box. Imposing the BvK periodic boundary conditions along x, y, and z, we can expand the field operators according to

$$\hat{\psi}(\mathbf{x}) = \frac{1}{\mathrm{V}}\sum_{\mathbf{p}} e^{\mathrm{i}\mathbf{p}\cdot\mathbf{r}}\,\hat{d}_{\mathbf{p}\sigma}, \qquad \Rightarrow \qquad \hat{\psi}^\dagger(\mathbf{x}) = \frac{1}{\mathrm{V}}\sum_{\mathbf{p}} e^{-\mathrm{i}\mathbf{p}\cdot\mathbf{r}}\,\hat{d}^\dagger_{\mathbf{p}\sigma},$$

where $\mathbf{p} = \frac{2\pi}{L}(n_x, n_y, n_z)$. The inverse relations read

$$\hat{d}_{\mathbf{p}\sigma} = \int d\mathbf{r}\, e^{-\mathrm{i}\mathbf{p}\cdot\mathbf{r}}\,\hat{\psi}(\mathbf{x}) \qquad \Rightarrow \qquad \hat{d}^\dagger_{\mathbf{p}\sigma} = \int d\mathbf{r}\, e^{\mathrm{i}\mathbf{p}\cdot\mathbf{r}}\,\hat{\psi}^\dagger(\mathbf{x}).$$

It is easy to check that the $\hat{d}$-operators satisfy the anticommutation relations

$$\left[\hat{d}_{\mathbf{p}\sigma}, \hat{d}^\dagger_{\mathbf{p}'\sigma'}\right]_- = \mathrm{V}\delta_{\sigma\sigma'}\delta_{\mathbf{p}\mathbf{p}'}, \tag{O.1}$$

and that in the limit $\mathrm{V} \to \infty$ they become the creation/annihilation operators of the momentum–spin kets, see (1.65). In terms of the $\hat{d}$-operators the density operator reads

$$\hat{n}(\mathbf{x}) = \frac{1}{\mathrm{V}^2}\sum_{\mathbf{p}\mathbf{p}'} e^{-\mathrm{i}(\mathbf{p}-\mathbf{p}')\cdot\mathbf{r}}\, \hat{d}^{\dagger}_{\mathbf{p}\sigma}\hat{d}_{\mathbf{p}'\sigma} = \frac{1}{\mathrm{V}^2}\sum_{\mathbf{p}\mathbf{q}} e^{\mathrm{i}\mathbf{q}\cdot\mathbf{r}}\, \hat{d}^{\dagger}_{\mathbf{p}\sigma}\hat{d}_{\mathbf{p}+\mathbf{q}\sigma}. \tag{O.2}$$

We define the density fluctuation operator as

$$\hat{\rho}_{\mathbf{q}} \equiv \frac{1}{\mathrm{V}}\sum_{\mathbf{p}\sigma} \hat{d}^{\dagger}_{\mathbf{p}\sigma}\hat{d}_{\mathbf{p}+\mathbf{q}\sigma}. \tag{O.3}$$

This operator can be seen as the Fourier transform of $\hat{n}(\mathbf{r})$, since from (O.2)

$$\hat{n}(\mathbf{r}) = \sum_{\sigma}\hat{n}(\mathbf{x}) = \frac{1}{\mathrm{V}}\sum_{\mathbf{q}} e^{\mathrm{i}\mathbf{q}\cdot\mathbf{r}}\hat{\rho}_{\mathbf{q}} \qquad \Rightarrow \qquad \hat{\rho}_{\mathbf{q}} = \int d\mathbf{r}\; e^{-\mathrm{i}\mathbf{q}\cdot\mathbf{r}}\,\hat{n}(\mathbf{r}).$$

The $\hat{d}$-operators and the density fluctuation operator can be used to rewrite the Hamiltonian of the electron gas in the following form:

$$\hat{H} = \frac{1}{\mathrm{V}}\sum_{\mathbf{p}\sigma}\left(\epsilon_{\mathbf{p}} - \frac{v(0)}{2}\right)\hat{d}^{\dagger}_{\mathbf{p}\sigma}\hat{d}_{\mathbf{p}\sigma} + \frac{1}{2\mathrm{V}}\sum_{\mathbf{q}}\tilde{v}_{\mathbf{q}}\,\hat{\rho}_{\mathbf{q}}\hat{\rho}_{-\mathbf{q}},$$

where $\epsilon_{\mathbf{p}} = p^2/2$ and $\tilde{v}_{\mathbf{q}} = \int d\mathbf{r}\; e^{-\mathrm{i}\mathbf{q}\cdot\mathbf{r}}v(\mathbf{r})$.

Let us now consider the density response function:

$$\begin{aligned}\chi^{\mathrm{R}}(\mathbf{r},t;\mathbf{r}',t') &= \sum_{\sigma\sigma'} -\mathrm{i}\theta(t-t')\langle\,[\hat{n}_H(\mathbf{x},t),\hat{n}_H(\mathbf{x}',t')]_-\rangle \\ &= \frac{1}{\mathrm{V}^2}\sum_{\mathbf{q}\mathbf{q}'} -\mathrm{i}\theta(t-t')e^{\mathrm{i}\mathbf{q}\cdot\mathbf{r}+\mathrm{i}\mathbf{q}'\cdot\mathbf{r}'}\langle\,[\hat{\rho}_{\mathbf{q},H}(t),\hat{\rho}_{\mathbf{q}',H}(t')]_-\rangle,\end{aligned} \tag{O.4}$$

where $\langle\ldots\rangle$ denotes the equilibrium ensemble average. Due to translational invariance the response function depends only on the difference $\mathbf{r}-\mathbf{r}'$ and hence the only nonvanishing averages in (O.4) are those for which $\mathbf{q}' = -\mathbf{q}$. If we define

$$\chi^{\mathrm{R}}_{\mathbf{p}}(\mathbf{q},t-t') \equiv \frac{1}{\mathrm{V}}\sum_{\sigma} -\mathrm{i}\theta(t-t')\langle\,[\hat{d}^{\dagger}_{\mathbf{p}\sigma,H}(t)\hat{d}_{\mathbf{p}+\mathbf{q}\,\sigma,H}(t),\hat{\rho}_{-\mathbf{q},H}(t')]_-\rangle,$$

then we can rewrite the density response function as

$$\chi^{\mathrm{R}}(\mathbf{r},t;\mathbf{r}',t') = \frac{1}{\mathrm{V}^2}\sum_{\mathbf{q}}\sum_{\mathbf{p}} e^{\mathrm{i}\mathbf{q}\cdot(\mathbf{r}-\mathbf{r}')}\,\chi^{\mathrm{R}}_{\mathbf{p}}(\mathbf{q},t-t'). \tag{O.5}$$

We now derive the equation of motion for $\chi^{\mathrm{R}}_{\mathbf{p}}(\mathbf{q},t)$. The derivative of this quantity with respect to t contains a term in which we differentiate the Heaviside function and a term in which we differentiate the operators in the Heisenberg picture. We have

$$\begin{aligned}\mathrm{i}\frac{d}{dt}\chi^{\mathrm{R}}_{\mathbf{p}}(\mathbf{q},t) &= \frac{1}{\mathrm{V}}\sum_{\sigma}\delta(t)\,\langle\,[\hat{d}^{\dagger}_{\mathbf{p}\sigma,H}(t)\hat{d}_{\mathbf{p}+\mathbf{q}\,\sigma,H}(t),\hat{\rho}_{-\mathbf{q},H}(t)]_-\,\rangle \\ &+ \frac{1}{\mathrm{V}}\sum_{\sigma} -\mathrm{i}\theta(t)\,\langle\,\Big[[\hat{d}^{\dagger}_{\mathbf{p}\sigma,H}(t)\hat{d}_{\mathbf{p}+\mathbf{q}\,\sigma,H}(t),\hat{H}_H(t)]_-,\hat{\rho}_{-\mathbf{q},H}(0)\Big]_-\,\rangle.\end{aligned} \tag{O.6}$$

To evaluate the r.h.s. of this equation we need the basic commutator

$$\sum_{\sigma\sigma'}\left[\hat{d}^{\dagger}_{\mathbf{p}\sigma}\hat{d}_{\mathbf{p}+\mathbf{q}\,\sigma},\hat{d}^{\dagger}_{\mathbf{p}'\sigma'}\hat{d}_{\mathbf{p}'+\mathbf{q}'\,\sigma'}\right]_{-}=\mathrm{V}\sum_{\sigma}\left(\delta_{\mathbf{p}+\mathbf{q}\,\mathbf{p}'}\,\hat{d}^{\dagger}_{\mathbf{p}\sigma}\hat{d}_{\mathbf{p}+\mathbf{q}+\mathbf{q}'\,\sigma}-\delta_{\mathbf{p}\,\mathbf{p}'+\mathbf{q}'}\,\hat{d}^{\dagger}_{\mathbf{p}-\mathbf{q}'\sigma}\hat{d}_{\mathbf{p}+\mathbf{q}\,\sigma}\right),$$

where we explicitly use the anticommutation rules (O.1). The first commutator in the r.h.s. of (O.6) is then

$$\sum_{\sigma}\left[\hat{d}^{\dagger}_{\mathbf{p}\sigma}\hat{d}_{\mathbf{p}+\mathbf{q}\,\sigma},\hat{\rho}_{-\mathbf{q}}\right]_{-}=\sum_{\sigma}\left(\hat{d}^{\dagger}_{\mathbf{p}\sigma}\hat{d}_{\mathbf{p}\sigma}-\hat{d}^{\dagger}_{\mathbf{p}+\mathbf{q}\,\sigma}\hat{d}_{\mathbf{p}+\mathbf{q}\,\sigma}\right). \tag{O.7}$$

The commutator with the Hamiltonian is the sum of two terms. The first term involves the one-body part of $\hat{H}$ and yields

$$\sum_{\sigma}\left[\hat{d}^{\dagger}_{\mathbf{p}\sigma}\hat{d}_{\mathbf{p}+\mathbf{q}\,\sigma},\frac{1}{\mathrm{V}}\sum_{\mathbf{p}'\sigma'}(\epsilon_{\mathbf{p}'}-\frac{1}{2}v(0))\hat{d}^{\dagger}_{\mathbf{p}'\sigma'}\hat{d}_{\mathbf{p}'\sigma'}\right]_{-}=(\epsilon_{\mathbf{p}+\mathbf{q}}-\epsilon_{\mathbf{p}})\sum_{\sigma}\hat{d}^{\dagger}_{\mathbf{p}\sigma}\hat{d}_{\mathbf{p}+\mathbf{q}\,\sigma}.$$

We thus see that the dependence on $v(0)$ disappears. The second term involves the commutator with $\hat{\rho}_{\mathbf{q}}\hat{\rho}_{-\mathbf{q}}$ and must be approximated. Here is where the idea of Bohm and Pines comes into play. It is easy to find

$$\begin{aligned}\sum_{\sigma}&\left[\hat{d}^{\dagger}_{\mathbf{p}\sigma}\hat{d}_{\mathbf{p}+\mathbf{q}\,\sigma},\frac{1}{2\mathrm{V}}\sum_{\mathbf{q}'}\tilde{v}_{\mathbf{q}'}\,\hat{\rho}_{\mathbf{q}'}\hat{\rho}_{-\mathbf{q}'}\right]_{-}\\&=\frac{1}{2\mathrm{V}}\sum_{\mathbf{q}'\sigma}\tilde{v}_{\mathbf{q}'}\left[\left(\hat{d}^{\dagger}_{\mathbf{p}\sigma}\hat{d}_{\mathbf{p}+\mathbf{q}+\mathbf{q}'\,\sigma}-\hat{d}^{\dagger}_{\mathbf{p}-\mathbf{q}'\,\sigma}\hat{d}_{\mathbf{p}+\mathbf{q}\sigma}\right),\hat{\rho}_{-\mathbf{q}'}\right]_{+}.\end{aligned} \tag{O.8}$$

To cast the r.h.s. of this equation in terms of an anticommutator we use $\tilde{v}_{\mathbf{q}}=\tilde{v}_{-\mathbf{q}}$. Now we observe that the low-energy states of an electron gas are states with very delocalized electrons. This means that the expansion

$$|\Psi\rangle=\frac{1}{N!}\int d\mathbf{x}_1\ldots d\mathbf{x}_N\Psi(\mathbf{x}_1,\ldots,\mathbf{x}_N)|\mathbf{x}_1\ldots\mathbf{x}_N\rangle \tag{O.9}$$

of a low-energy state has a wavefunction that is a smooth function of the coordinates. Let us consider the action of $\sum_{\sigma}\hat{d}^{\dagger}_{\mathbf{p}\sigma}\hat{d}_{\mathbf{p}+\mathbf{q}\,\sigma}$ on a position-spin ket,

$$\begin{aligned}\sum_{\sigma}\hat{d}^{\dagger}_{\mathbf{p}\sigma}\hat{d}_{\mathbf{p}+\mathbf{q}\,\sigma}|\mathbf{x}_1\ldots\mathbf{x}_N\rangle&=\sum_{\sigma}\int d\mathbf{r}d\mathbf{r}'e^{\mathrm{i}\mathbf{p}\cdot(\mathbf{r}-\mathbf{r}')}e^{-\mathrm{i}\mathbf{q}\cdot\mathbf{r}'}\;\hat{\psi}^{\dagger}(\mathbf{r}\sigma)\hat{\psi}(\mathbf{r}'\sigma)|\mathbf{x}_1\ldots\mathbf{x}_N\rangle\\&=\int d\mathbf{r}\sum_{i=1}^{N}e^{\mathrm{i}\mathbf{p}\cdot(\mathbf{r}-\mathbf{r}_i)}e^{-\mathrm{i}\mathbf{q}\cdot\mathbf{r}_i}|\mathbf{x}_1\ldots\mathbf{x}_{i-1}\,\mathbf{r}\sigma_i\,\mathbf{x}_{i+1}\mathbf{x}_N\rangle,\end{aligned}$$

where we use (1.45). Multiplying this equation by the wavefunction of a low-energy state and integrating over all coordinates, we see that the dominant contribution of the integral over $\mathbf{r}$ comes from the region $\mathbf{r}\sim\mathbf{r}_i$ since for large $|\mathbf{r}-\mathbf{r}_i|$ the exponential $e^{\mathrm{i}\mathbf{p}\cdot(\mathbf{r}-\mathbf{r}_i)}$

becomes a highly oscillating function. Under the integral over $\mathbf{x}_1 \ldots \mathbf{x}_N$ we then make the approximation

$$\sum_\sigma \hat{d}^\dagger_{\mathbf{p}\sigma} \hat{d}_{\mathbf{p}+\mathbf{q}\,\sigma} |\mathbf{x}_1 \ldots \mathbf{x}_N\rangle \sim \Delta \mathrm{V} \sum_{i=1}^{N} e^{-\mathrm{i}\mathbf{q}\cdot\mathbf{r}_i} |\mathbf{x}_1 \ldots \mathbf{x}_N\rangle, \tag{O.10}$$

where $\Delta \mathrm{V}$ is some volume element proportional to $1/p$. The r.h.s. in (O.10) is the position–spin ket multiplied by the sum of exponentials with "randomly" varying phases. In the large N limit, or equivalently in the limit of Wigner–Seitz radius $r_s \to 0$, this sum approaches zero for all $\mathbf{q} \neq 0$. The idea of Bohm and Pines was then to say that every operator $\hat{d}^\dagger_{\mathbf{p}\sigma} \hat{d}_{\mathbf{p}+\mathbf{q}\,\sigma}$ appearing in (O.8) generates small contributions unless $\mathbf{q} = 0$. This means that the dominant contributions in (O.8) are those with either $\mathbf{q}' = -\mathbf{q}$ (in this case the operator in parenthesis becomes $\hat{d}^\dagger_{\mathbf{p}\sigma} \hat{d}_{\mathbf{p}\sigma} - \hat{d}^\dagger_{\mathbf{p}+\mathbf{q}\,\sigma} \hat{d}_{\mathbf{p}+\mathbf{q}\sigma}$) or those with $\mathbf{q}' = 0$ since according to (O.3) the main contribution to the density fluctuation operator comes from $\hat{\rho}_{\mathbf{q}'}$ with $\mathbf{q}' = 0$ (in this case $\hat{\rho}_0 = \frac{1}{\mathrm{V}} \sum_{\mathbf{p}'\sigma} \hat{d}^\dagger_{\mathbf{p}'\sigma} \hat{d}_{\mathbf{p}'\sigma}$). For $\mathbf{q}' = 0$, however, the r.h.s. of (O.8) vanishes and therefore the Bohm–Pines approximation reduces to

$$\sum_\sigma \Big[\hat{d}^\dagger_{\mathbf{p}\sigma} \hat{d}_{\mathbf{p}+\mathbf{q}\,\sigma}, \frac{1}{2\mathrm{V}} \sum_{\mathbf{q}'} \tilde{v}_{\mathbf{q}'}\, \hat{\rho}_{\mathbf{q}'} \hat{\rho}_{-\mathbf{q}'}\Big]_- \sim \frac{1}{2\mathrm{V}} \sum_\sigma \tilde{v}_{\mathbf{q}} \left[\left(\hat{d}^\dagger_{\mathbf{p}\sigma} \hat{d}_{\mathbf{p}\sigma} - \hat{d}^\dagger_{\mathbf{p}+\mathbf{q}\,\sigma} \hat{d}_{\mathbf{p}+\mathbf{q}\sigma}\right), \hat{\rho}_{\mathbf{q}}\right]_+ .$$

Next we observe that $\hat{d}^\dagger_{\mathbf{p}\sigma} \hat{d}_{\mathbf{p}\sigma}$ is the occupation operator for electrons of momentum $\mathbf{p}$ and spin σ. The noninteracting ground state of the electron gas has N electrons in the lowest-energy $\mathbf{p}$-levels and therefore is an eigenstate of $\hat{d}^\dagger_{\mathbf{p}\sigma} \hat{d}_{\mathbf{p}\sigma}$ with eigenvalue $\mathrm{V} f_{\mathbf{p}}$ where $f_{\mathbf{p}} = 1$ if $\mathbf{p}$ is occupied and $f_{\mathbf{p}} = 0$ otherwise. The second approximation of Bohm and Pines was to assume that things are not so different in the interacting case and hence that we can replace $\hat{d}^\dagger_{\mathbf{p}\sigma} \hat{d}_{\mathbf{p}\sigma}$ with $\mathrm{V} f_{\mathbf{p}}$. In conclusion,

$$\sum_\sigma \Big[\hat{d}^\dagger_{\mathbf{p}\sigma} \hat{d}_{\mathbf{p}+\mathbf{q}\,\sigma}, \frac{1}{2\mathrm{V}} \sum_{\mathbf{q}'} \tilde{v}_{\mathbf{q}'}\, \hat{\rho}_{\mathbf{q}'} \hat{\rho}_{-\mathbf{q}'}\Big]_- \sim 2(f_{\mathbf{p}} - f_{\mathbf{p}+\mathbf{q}})\, \tilde{v}_{\mathbf{q}}\, \hat{\rho}_{\mathbf{q}}$$

where the factor of 2 comes from spin. Putting together all these results, the equation of motion (O.6) becomes

$$\mathrm{i}\frac{d}{dt} \chi^{\mathrm{R}}_{\mathbf{p}}(\mathbf{q},t) = 2\delta(t)(f_{\mathbf{p}} - f_{\mathbf{p}+\mathbf{q}}) + (\epsilon_{\mathbf{p}+\mathbf{q}} - \epsilon_{\mathbf{p}})\, \chi^{\mathrm{R}}_{\mathbf{p}}(\mathbf{q},t) + 2\tilde{v}_{\mathbf{q}}(f_{\mathbf{p}} - f_{\mathbf{p}+\mathbf{q}}) \frac{1}{\mathrm{V}} \sum_{\mathbf{p}'} \chi^{\mathrm{R}}_{\mathbf{p}'}(\mathbf{q},t).$$

Fourier transforming as

$$\chi^{\mathrm{R}}_{\mathbf{p}}(\mathbf{q},t) = \int \frac{d\omega}{2\pi} e^{-\mathrm{i}\omega t} \chi^{\mathrm{R}}_{\mathbf{p}}(\mathbf{q},\omega),$$

and rearranging the terms, we find the solution

$$\chi^{\mathrm{R}}_{\mathbf{p}}(\mathbf{q},\omega) = 2\frac{f_{\mathbf{p}} - f_{\mathbf{p}+\mathbf{q}}}{\omega - \epsilon_{\mathbf{p}+\mathbf{q}} - \epsilon_{\mathbf{p}} + \mathrm{i}\eta} \left(1 + \tilde{v}_{\mathbf{q}} \frac{1}{\mathrm{V}} \sum_{\mathbf{p}'} \chi^{\mathrm{R}}_{\mathbf{p}'}(\mathbf{q},\omega)\right), \tag{O.11}$$

where we add the infinitesimal $i\eta$ to ensure that $\chi^{\mathrm{R}}_{\mathbf{p}}(\mathbf{q}, t < 0) = 0$. From (O.5) we see that the Fourier transform of the density response function is

$$\chi^{\mathrm{R}}(\mathbf{q}, \omega) = \frac{1}{\mathrm{V}} \sum_{\mathbf{p}} \chi^{\mathrm{R}}_{\mathbf{p}}(\mathbf{q}, \omega).$$

Thus, summing (O.11) over $\mathbf{p}$ and dividing by the volume we get

$$\chi^{\mathrm{R}}(\mathbf{q}, \omega) = \frac{2}{\mathrm{V}} \sum_{\mathbf{p}} \frac{f_{\mathbf{p}} - f_{\mathbf{p}+\mathbf{q}}}{\omega - \epsilon_{\mathbf{p}+\mathbf{q}} - \epsilon_{\mathbf{p}} + i\eta} \left[1 + \tilde{v}_{\mathbf{q}} \chi^{\mathrm{R}}(\mathbf{q}, \omega)\right].$$

In the limit $\mathrm{V} \to \infty$ we have $\frac{1}{\mathrm{V}} \sum_{\mathbf{p}} \to \int \frac{d\mathbf{p}}{(2\pi)^3}$ and we recognize in this equation the noninteracting response function (15.32). Thus,

$$\chi^{\mathrm{R}}(\mathbf{q}, \omega) = \chi^{\mathrm{R}}_{0}(\mathbf{q}, \omega) \left[1 + \tilde{v}_{\mathbf{q}} \chi^{\mathrm{R}}(\mathbf{q}, \omega)\right]. \tag{O.12}$$

Solving (O.12) for χ^{R} we recover the RPA response function (15.44).

Appendix P: Kramers–Kronig Relations

Let us consider a function $A(z)$ of the complex variable $z = x + \mathrm{i}y$ which is analytic in the upper half-plane and with the property that

$$\lim_{z\to\infty} zA(z) = 0, \qquad \text{with } \mathrm{Arg}(z) \in (0, \pi). \tag{P.1}$$

For z in the upper half-plane we can use the Cauchy residue theorem to write this function as

$$A(z) = \oint_C \frac{dz'}{2\pi\mathrm{i}} \frac{A(z')}{z' - z}, \tag{P.2}$$

where the integral is along an anticlockwise-oriented curve C that is entirely contained in the upper half-plane and inside which there is the point z. Let C be the curve in the figure below and $z = x + \mathrm{i}\eta$ a point infinitesimally above the real axis.

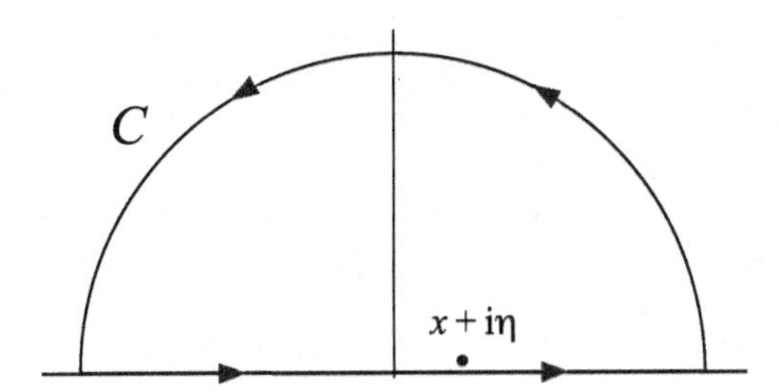

Due to property (P.1) the contribution of the integral along the arc vanishes in the limit of infinite radius and therefore (P.2) implies that

$$A(x + \mathrm{i}\eta) = \int_{-\infty}^{\infty} \frac{dx'}{2\pi\mathrm{i}} \frac{A(x')}{x' - (x + \mathrm{i}\eta)} = P\int_{-\infty}^{\infty} \frac{dx'}{2\pi\mathrm{i}} \frac{A(x')}{x' - x} + \frac{1}{2}A(x), \tag{P.3}$$

where in the second equality we use the Cauchy relation

$$\frac{1}{x' - x - \mathrm{i}\eta} = P\frac{1}{x' - x} + \mathrm{i}\pi\delta(x' - x), \tag{P.4}$$

with P the principal part. Since η is an infinitesimal positive constant we can safely replace $A(x + \mathrm{i}\eta)$ with $A(x)$ in the l.h.s. of (P.3), thereby obtaining the following important identity:

$$A(x) = \frac{\mathrm{i}}{\pi} P \int_{-\infty}^{\infty} dx' \frac{A(x')}{x - x'}.$$

If we separate $A(x) = A_1(x) + \mathrm{i}A_2(x)$ into a real and an imaginary part, then we see that A_1 and A_2 are related through a Hilbert transformation,

$$\boxed{A_1(x) = -\frac{1}{\pi} P \int_{-\infty}^{\infty} dx' \, \frac{A_2(x')}{x - x'}} \tag{P.5}$$

$$\boxed{A_2(x) = \frac{1}{\pi} P \int_{-\infty}^{\infty} dx' \, \frac{A_1(x')}{x - x'}}$$

These relations are known as the *Kramers–Kronig relations* and allow us to express the real/imaginary part of the function A in terms of its imaginary/real part. Consequently, the full function A can be written solely in terms of its real or imaginary part since

$$A(x) = A_1(x) + \mathrm{i}\frac{1}{\pi} P \int_{-\infty}^{\infty} dx' \, \frac{A_1(x')}{x - x'} = \frac{\mathrm{i}}{\pi} \int_{-\infty}^{\infty} dx' \, \frac{A_1(x')}{x - x' + \mathrm{i}\eta},$$

and also

$$A(x) = -\frac{1}{\pi} P \int_{-\infty}^{\infty} dx' \, \frac{A_2(x')}{x - x'} + \mathrm{i}A_2(x) = -\frac{1}{\pi} \int_{-\infty}^{\infty} dx' \, \frac{A_2(x')}{x - x' + \mathrm{i}\eta}.$$

References

[1] P. A. M. Dirac, *The Principles of Quantum Mechanics*, 4th edition (Oxford University Press, 1958).

[2] J. J. Sakurai and J. Napolitano, *Modern Quantum Mechanics* (Addison-Wesley, 1994).

[3] A. Bohm, *Quantum Mechanics: Foundations and Applications* (Springer, 1994).

[4] R. Pariser and R. G. Parr, *J. Chem. Phys* **21**, 466 (1953).

[5] J. A. Pople, *Trans. Faraday Soc.* **42**, 1375 (1953).

[6] J. Linderberg and Y. Öhrn, *Propagators in Quantum Chemistry* (Wiley, 2000).

[7] P. Atkins, *Physical Chemistry* (W. H. Freeman, 1998).

[8] K. J. H. Giesbertz, A.-M. Uimonen, and R. van Leeuwen, *Eur. Phys. J. B* **91**, 282 (2018).

[9] K. S. Novoselov, A. K. Geim, S. V. Morozov, et al., *Science* **306**, 666 (2004).

[10] A. H. C. Neto, F. Guinea, N. M. R. Peres, K. S. Novoselov, and A. K. Geim, *Rev. Mod. Phys.* **81**, 109 (2009).

[11] A. Mielke, *J. Phys. A* **24**, L73 (1991).

[12] H. Tasaki, *Phys. Rev. Lett.* **69**, 1608 (1992).

[13] A. Mielke and H. Tasaki, *Commun. Math. Phys.* **158**, 341 (1993).

[14] P. W. Anderson, *Phys. Rev.* **124**, 41 (1961).

[15] J. Kondo, *Progr. Theoret. Phys.* **32**, 37 (1964).

[16] A. C. Hewson, *The Kondo Problem to Heavy Fermions* (Cambridge University Press, 2007).

[17] J. Hubbard, *Proc. Roy. Soc. A* **276**, 238 (1963).

[18] H. Tasaki, *J. Phys. Condens. Matter* **10**, 4353 (1998).

[19] D. Baeriswyl, D. K. Campbell, J. M. P. Carmelo, F. Guinea and E. Louis (eds.), *The Hubbard Model: Its Physics and Mathematical Physics* (Springer, 1995).

[20] A. Montorsi (ed.), *The Hubbard Model: A Reprint Volume* (World Scientific, 1992).

[21] P. W. Anderson, *Phys. Rev.* **115**, 2 (1959).

[22] J. Bardeen, L. N. Cooper, and J. R. Schrieffer, *Phys. Rev.* **106**, 162 (1957).

[23] D. C. Ralph, C. T. Black, and M. Tinkham, *Phys. Rev. Lett.* **76**, 688 (1996).

[24] R. W. Richardson, *Phys. Lett.* **3**, 277 (1963).

[25] L. N. Cooper, *Phys. Rev.* **104**, 1189 (1956).

[26] H. Frölich, *Adv. Phys.* **3**, 325 (1954).

[27] R. P. Feynman, *Phys. Rev.* **97**, 660 (1955).

[28] W. P. Su, J. R. Schrieffer, and A. J. Heeger, *Phys. Rev. B* **22**, 2099 (1980).

[29] T. Holstein, *Ann. Phys.* **8**, 325 (1959).

[30] P. A. M. Dirac, *Proc. R. Soc. Lond. A* **117**, 610 (1928).

[31] C. Itzykson and J.-B. Zuber *Quantum Field Theory* (McGraw-Hill, 1980).

[32] L. H. Ryder, *Quantum Field Theory* (Cambridge University Press, 1985).

[33] F. Mandl and G. Shaw, *Quantum Field Theory* (Wiley, 1984).

[34] S. Weinberg, *The Quantum Theory of Fields – Volume I: Foundations* (Cambridge University Press, 1995).

[35] P. Ramond, *Field Theory: A Modern Primer* (Benjamin/Cummings, Inc., 1981).

[36] M. E. Peskin and D. V. Schroeder, *An Introduction to Quantum Field Theory* (Addison-Wesley, 1995).

[37] R. E. Peierls, *Quantum Theory of Solids* (Clarendon, 1955).

[38] I. G. Lang and Y. A. Firsov, *Zh. Eksp. Teor. Fiz.* **43**, 1843 (1962) [*Sov. Phys. JETP* 16, 1301 (1962)].

[39] L. V. Keldysh, *Sov. Phys. JETP* **20**, 1018 (1965).

[40] J. Schwinger, *J. Math. Phys.* **2**, 407 (1961).

[41] L. V. Keldysh, in *Progress in Nonequilibrium Green's Functions II*, edited by M. Bonitz and D. Semkat (World Scientific, 2003).

[42] O. V. Konstantinov and V. I. Perel', *Sov. Phys. JETP* **12**, 142 (1961).

[43] P. Danielewicz, *Ann. Phys.* **152**, 239 (1984).

[44] M. Wagner, *Phys. Rev. B* **44**, 6104 (1991).

[45] P. Gaspard and M. Nagaoka, *J. Chem. Phys.* **111**, 5676 (1999).

[46] G. Stefanucci, *Phys. Rev. Lett.* **133**, 066901 (2024).

[47] M. Gell-Mann and F. Low, *Phys. Rev.* **84**, 350 (1951).

[48] A. L. Fetter and J. D. Walecka, *Quantum Theory of Many-Particle Systems* (McGraw-Hill, 1971).

[49] H. Bruus and K. Flensberg, *Many-Body Quantum Theory in Condensed Matter Physics: An Introduction* (Oxford University Press, 2004).

[50] P. C. Martin and J. Schwinger, *Phys. Rev.* **115**, 1342 (1959).

[51] R. Kubo, *J. Phys. Soc. Jpn.* **12**, 570 (1957).

[52] R. Haag, N. M. Hugenholtz, and M. Winnink, *Comm. Math. Phys.* **5**, 215 (1967).

[53] G. C. Wick, *Phys. Rev.* **80**, 268 (1950).

[54] D. Kaiser, *Phil. Mag.* **43**, 153 (1952).

[55] J. Olsen, P. Jørgensen, T. Helgaker, and O. Christiansen, *J. Chem. Phys.* **112**, 9736 (2000).

[56] D. C. Langreth, in *Linear and Nonlinear Electron Transport in Solids*, edited by J. T. Devreese and E. van Doren (Plenum, 1976), pp. 3–32.

[57] M. J. Hyrkäs, D. Karlsson, and R. van Leeuwen, *J. Phys. A Math. Theoret.* **52**, 215303 (2019).

[58] M. J. Hyrkäs, D. Karlsson, and R. van Leeuwen, *J. Phys. Math. Theoret.* **55**, 335301 (2022).

[59] G. Stefanucci and C.-O. Almbladh, *Phys. Rev. B* **69**, 195318 (2004).

[60] M. Knap, E. Arrigoni, and W. von der Linden, *Phys. Rev. B* **81**, 024301 (2010).

[61] S. Ejima, H. Fehske, and F. Gebhard, *Europhys. Lett.* **93**, 30002 (2011).

[62] C. W. J. Beenakker, *Phys. Rev. B* **44**, 1646 (1991).

[63] H. Haug and A.-P. Jauho, *Quantum Kinetics in Transport and Optics of Semiconductors* (Springer, 2008).

[64] C. Stampfl, K. Kambe, J. D. Riley, and D. F. Lynch, *J. Phys. Condens. Matter* **5**, 8211 (1993).

[65] N. V. V. Stojić, A. Dal Corso, B. Zhou, and S. Baroni, *Phys. Rev. B* **77**, 195116 (2008).

[66] J. Braun, R. Rausch, M. Potthoff, J. Minár, and H. Ebert, *Phys. Rev. B* **91**, 035119 (2015).

[67] H. Joon Choi and J. Ihm, *Phys. Rev. B* **59**, 2267 (1999).

[68] A. Smogunov, A. Dal Corso, and E. Tosatti, *Phys. Rev. B* **70**, 045417 (2004).

[69] S. Moser, *J. Electron Spectros. Relat. Phenomena* **214**, 29 (2017).

[70] E. Perfetto, D. Sangalli, A. Marini, and G. Stefanucci, *Phys. Rev. B* **94**, 245303 (2016).

[71] J. K. Freericks, H. R. Krishnamurthy, and T. Pruschke, *Phys. Rev. Lett.* **102**, 136401 (2009).

[72] W. Schattke, M. A. Van Hove, F. J. García de Abajo, R. Díez Muino, and N. Mannella, in *Solid-State Photoemission and Related Methods: Theory and Experiment*, edited by W. Schattke and M. A. Van Hove (Wiley, 2003).

[73] A. Damascelli, Z. Hussain, and Z.-X. Shen, *Rev. Mod. Phys.* **75**, 473 (2003).

[74] G. D. Mahan, *Phys. Rev. B* **2**, 4334 (1970).

[75] W. L. Schaich and N. W. Ashcroft, *Phys. Rev. B* **3**, 2452 (1971).

[76] C. Caroli, D. Lederer-Rozenblatt, B. Roulet, and D. Saint-James, *Phys. Rev. B* **8**, 4552 (1973).

[77] C.-O. Almbladh, *Physica Scripta* **32**, 341 (1985).

[78] C.-O. Almbladh and L. Hedin, in *Handbook on Synchrotron Radiation* Vol. 1, edited by E. E. Koch (North-Holland Publishing, 1983).

[79] W. Bardyszewski and L. Hedin, *Physica Scripta* **32**, 439 (1985).

[80] G. Grosso and G. Pastori Parravicini, *Solid State Physics* (Academic Press, 2000).

[81] H. Ohnishi, N. Tomita, and K. Nasu, *Int. J. Mod. Phys. B* **32**, 1850094 (2018).

[82] A. Rustagi and A. F. Kemper, *Phys. Rev. B* **97**, 235310 (2018).

[83] G. Stefanucci and E. Perfetto, *Phys. Rev. B* **103**, 245103 (2021).

[84] A. Steinhoff, M. Florian, M. Rösner, et al., *Nat. Commun.* **8**, 1166 (2017).

[85] D. Christiansen, M. Selig, E. Malic, R. Ernstorfer, and A. Knorr, *Phys. Rev. B* **100**, 205401 (2019).

[86] E. Perfetto and G. Stefanucci, *Phys. Rev. B* **103**, L241404 (2021).

[87] E. Perfetto, D. Sangalli, A. Marini, and G. Stefanucci, *Phys. Rev. Mater.* **3**, 124601 (2019).

[88] E. Perfetto and G. Stefanucci, *Phys. Rev. Lett.* **125**, 106401 (2020).

[89] E. Perfetto, S. Bianchi, and G. Stefanucci, *Phys. Rev. B* **101**, 041201 (2020).

[90] M. K. L. Man, J. Madeo, C. Sahoo, et al., *Sci. Adv.* **7**, eabg0192 (2021).

[91] S. Dong, M. Puppin, T. Pincelli, et al., *Natural Sci.* **1**, e10010 (2021).

[92] L. G. Molinari and N. Manini, *Eur. Phys. J. B* **51**, 331–336 (2006).

[93] L. Hedin, *Phys. Rev.* **139**, A796 (1965).

[94] M. S. Hybertsen and S. G. Louie, *Phys. Rev. B* **34**, 5390 (1986).

[95] F. Aryasetiawan and O. Gunnarsson, *Rep. Prog. Phys.* **61**, 237 (1998).

[96] W. G. Aulbur, L. Jönsson, and J. W. Wilkins, in *Solid State Physics*, edited by H. Ehrenreich and F. Spaepen (Academic Press, 1999).

[97] A. Marini, C. Hogan, M. Grüning, and D. Varsano, *Comput. Phys. Commun.* **180**, 1392 (2009).

[98] J. Deslippe, G. Samsonidze, D. A. Strubbe, et al., *Comput. Phys. Commun.* **183**, 1269 (2012).

[99] L. Reining, *WIREs Compu. Molec. Sci.* **8**, e1344 (2018).

[100] D. Golze, M. Dvorak, and P. Rinke, *Fron. Chem.* **7** (2019).

[101] C.-O. Almbladh, U. von Barth, and R. van Leeuwen, *Int. J. Mod. Phys. B* **13**, 535 (1999).

[102] P. García-González and R. W. Godby, *Phys. Rev. B* **63**, 075112 (2001).

[103] N. E. Dahlen and R. van Leeuwen, *J. Chem. Phys.* **122**, 164102 (2005).

[104] D. R. Hartree, *Proc. Cambridge Phil. Soc.* **24**, 89 (1928).

[105] E. P. Gross, *Il Nuovo Cimento* **20**, 454 (1961).

[106] L. P. Pitaevskii, *JETP* **13**, 451 (1961).

[107] F. Dalfovo, S. Giorgini, L. P. Pitaevskii, and S. Stringari, *Rev. Mod. Phys.* **71**, 463 (1999).

[108] G. F. Giuliani and G. Vignale, *Quantum Theory of the Electron Liquid* (Cambridge University Press, 2005).

[109] L. P. Kadanoff and G. Baym, *Quantum Statistical Mechanics* (W. A. Benjamin, Inc., 1962).

[110] R. Rajaraman, *Solitons and Instantons* (North-Holland, 1982).

[111] J. C. Slater, *Phys. Rev.* **35**, 210 (1930).

[112] V. Fock, *Z. Physik* **61**, 126 (1930).

[113] N. E. Dahlen and R. van Leeuwen, *Phys. Rev. Lett.* **98**, 153004 (2007).

[114] P. Myöhänen, A. Stan, G. Stefanucci, and R. van Leeuwen, *Europhys. Lett.* **84**, 67001 (2008).

[115] P. Myöhänen, A. Stan, G. Stefanucci, and R. van Leeuwen, *Phys. Rev. B* **80**, 115107 (2009).

[116] M. Puig von Friesen, C. Verdozzi, and C.-O. Almbladh, *Phys. Rev. Lett.* **103**, 176404 (2009).

[117] M. Puig von Friesen, C. Verdozzi, and C.-O. Almbladh, *Phys. Rev. B* **82**, 155108 (2010).

[118] A. Stan, N. E. Dahlen, and R. van Leeuwen, *J. Chem. Phys.* **130**, 224101 (2009).

[119] M. Schüler, J. Berakdar, and Y. Pavlyukh, *Phys. Rev. B* **93**, 054303 (2016).

[120] M. Schüler, D. Golez, Y. Murakami, et al., *Comp. Phys. Commun.* **257**, 107484 (2020).

[121] J. Kaye and D. Golež, *SciPost Phys.* **10**, 091 (2021).

[122] F. Meirinhos, M. Kajan, J. Kroha, and T. Bode, *SciPost Phys. Core* **5**, 030 (2022).

[123] X. Dong, E. Gull, and H. U. R. Strand, *Phys. Rev. B* **106**, 125153 (2022).

[124] M. A. Reed, C. Zhou, C. J. Muller, T. P. Burgin, and J. M. Tour, *Science* **278**, 252 (1997).

[125] A. I. Yanson, G. R. Bollinger, H. E. van den Brom, N. Agrait, and J. M. van Ruitenbeek, *Nature* **395**, 783 (1998).

[126] C. Joachim and M. A. Ratner, *Proc. Nat. Acad. Sci.* **102**, 8801 (2005).

[127] G. Cuniberti, G. Fagas, and K. Richter, *Introducing Molecular Electronics* (Springer, 2005).

[128] J. Cuevas and E. Scheer, *Molecular Electronics: An Introduction to Theory and Experiment* (World Scientific, 2010).

[129] R. Tuovinen, E. Perfetto, G. Stefanucci, and R. van Leeuwen, *Phys. Rev. B* **89**, 085131 (2014).

[130] Y. Meir and N. S. Wingreen, *Phys. Rev. Lett.* **68**, 2512 (1992).

[131] A.-P. Jauho, N. S. Wingreen, and Y. Meir, *Phys. Rev. B* **50**, 5528 (1994).

[132] G. Stefanucci, *Phys. Rev. B* **75**, 195115 (2007).

[133] E. Khosravi, G. Stefanucci, S. Kurth, E. K. U. Gross, *App. Phys. A* **93** 355 (2008).

[134] E. Khosravi, G. Stefanucci, S. Kurth, E. K. U. Gross, *Phys. Chem. Chem. Phys.* **11**, 4535 (2009).

[135] P. Myöhänen, R. Tuovinen, T. Korhonen, G. Stefanucci, and R. van Leeuwen, *Phys. Rev. B* **85**, 075105 (2012).

[136] E. Khosravi, A.-M. Uimonen, A. Stan, et al., *Phys. Rev. B* **85**, 075103 (2012).

[137] R. Landauer, *IBM J. Res. Dev.* **1**, 233 (1957).

[138] M. Büttiker, Y. Imry, R. Landauer, and S. Pinhas, *Phys. Rev. B* **31**, 6207 (1985).

[139] M. Büttiker, *Phys. Rev. Lett.* **57**, 1761 (1986).

[140] N. Bushong, N. Sai, and M. Di Ventra, *Nano Lett.* **5**, 2569 (2005).

[141] J. M. Luttinger and J. C. Ward, *Phys. Rev.* **118**, 1417 (1960).

[142] N. E. Dahlen, R. van Leeuwen, and U. von Barth, *Phys. Rev. A* **73**, 012511 (2006).

[143] N. E. Dahlen and U. von Barth, *Phys. Rev. B* **69**, 195102 (2004).

[144] N. E. Dahlen and U. von Barth, *J. Chem. Phys.* **120**, 6826 (2004).

[145] U. von Barth, N. E. Dahlen, R. van Leeuwen, and G. Stefanucci, *Phys. Rev. B* **72**, 235109 (2005).

[146] A. Klein, *Phys. Rev.* **121**, 950 (1961).

[147] G. Baym, *Phys. Rev.* **171**, 1391 (1962).

[148] G. Stefanucci, Y. Pavlyukh, A.-M. Uimonen, and R. van Leeuwen, *Phys. Rev. B* **90**, 115134 (2014).

[149] B. Holm and U. von Barth, *Phys. Rev. B* **57**, 2108 (1998).

[150] S. Tomonaga, *Progr. Theoret. Phys.* **5**, 544 (1950).

[151] J. M. Luttinger, *J. Math. Phys.* **4**, 1154 (1963).

[152] D. C. Mattis and E. H. Lieb, *J. Math. Phys.* **6**, 304 (1963).

[153] J. Friedel, *Il Nuovo Cimento* **7**, 287 (1958).

[154] J. Friedel, *Adv. Phys.* **50**, 539 (2001).

[155] D. C. Langreth, *Phys. Rev.* **150**, 516 (1966).

[156] A. Kawabata, *J. Phys. Soc. Japan* **60**, 3222 (1991).

[157] H. Mera, K. Kaasbjerg, Y. M. Niquet, and G. Stefanucci, *Phys. Rev. B* **81**, 035110 (2010).

[158] G. Stefanucci and S. Kurth, *Phys. Rev. Lett.* **107**, 216401 (2011).

[159] J. P. Bergfield, Z.-F. Liu, K. Burke, and C. A. Stafford, *Phys. Rev. Lett.* **108**, 066801 (2012).

[160] P. Tröster, P. Schmitteckert, and F. Evers, *Phys. Rev. B* **85**, 115409 (2012).

[161] I. Aleiner, P. Brouwer, and L. Glazman, *Phys. Rep.* **358**, 309 (2002).

[162] C. T. De Dominicis, *J. Math. Phys.* **4**, 255 (1963).

[163] C. T. De Dominicis and P. C. Martin, *J. Math. Phys.* **5**, 14 (1964).

[164] C. T. De Dominicis and P. C. Martin, *J. Math. Phys.* **5**, 31 (1964).

[165] M. Hindgren, *Model Vertices Beyond the GW Approximation*, PhD thesis (1997).

[166] D. Karlsson, R. van Leeuwen, Y. Pavlyukh, E. Perfetto, and G. Stefanucci, *Phys. Rev. Lett.* **127**, 036402 (2021).

[167] D. Karlsson and R. van Leeuwen, *Phys. Rev. B* **94**, 125124 (2016).

[168] G. Baym and L. P. Kadanoff, *Phys. Rev.* **124**, 287 (1961).

[169] Y. Pavlyukh, A.-M. Uimonen, G. Stefanucci, and R. van Leeuwen, *Phys. Rev. Lett.* **117**, 206402 (2016).

[170] M. Holzmann, B. Bernu, C. Pierleoni, et al., *Phys. Rev. Lett.* **107**, 110402 (2011).

[171] Y. Pavlyukh, G. Stefanucci, and R. van Leeuwen, *Phys. Rev. B* **102**, 045121 (2020).

[172] M. J. van Setten, F. Caruso, S. Sharifzadeh, et al., *J. Chem. Theor. Comput.* **11**, 5665 (2015).

[173] H. Stolz and R. Zimmermann, *Physica Status Solidi (B)* **94**, 135 (1979).

[174] D. Kremp, W. Kraeft, and A. Lambert, *Physica A Stat. Mech. Appl.* **127**, 72 (1984).

[175] R. Zimmermann, *Many-Particle Theory of Highly Excited Semiconductors* (Teubner, 1987).

[176] W. D. Kraeft, D. Kremp, W. Ebeling, and G. Röpke, *Quantum Statistics of Charged Particle Systems* (Akademie-Verlag, 1986).

[177] D. O. Gericke, S. Kosse, M. Schlanges, and M. Bonitz, *Phys. Rev. B* **59**, 10639 (1999).

[178] C. Piermarocchi and F. Tassone, *Phys. Rev. B* **63**, 245308 (2001).

[179] N. H. Kwong, G. Rupper, and R. Binder, *Phys. Rev. B* **79**, 155205 (2009).

[180] M. F. Pereira and K. Henneberger, *Phys. Rev. B* **58**, 2064 (1998).

[181] D. J. Thouless, *Ann. Phys.* **10**, 553 (1960).

[182] S. Schmitt-Rink, C. M. Varma, and A. E. Ruckenstein, *Phys. Rev. Lett.* **63**, 445 (1989).

[183] L. P. Kadanoff and P. C. Martin, *Phys. Rev.* **124**, 670 (1961).

[184] A.-M. Uimonen, E. Khosravi, A. Stan, et al., *Phys. Rev. B* **84**, 115103 (2011).

[185] G. Strinati, *Riv. Nuovo Cim.* **11**, 1 (1986).

[186] U. Fano, *Phys. Rev.* **124**, 1866 (1961).

[187] W. Thomas, *Naturwissenschaften* **13**, 627 (1925).

[188] W. Kuhn, *Z. Phys.* **33**, 408 (1925).

[189] F. Reiche and W. Thomas, *Z. Phys.* **34**, 510 (1925).

[190] R. W. Ziolkowski, J. M. Arnold, and D. M. Gogny, *Phys. Rev. A* **52**, 3082 (1995).

[191] R. Santra, V. S. Yakovlev, T. Pfeifer, and Z.-H. Loh, *Phys. Rev. A* **83**, 033405 (2011).

[192] J. C. Baggesen, E. Lindroth, and L. B. Madsen, *Phys. Rev. A* **85**, 013415 (2012).

[193] M. Wu, S. Chen, K. J. Schafer, and M. B. Gaarde, *Phys. Rev. A* **87**, 013828 (2013).

[194] A. N. Pfeiffer, M. J. Bell, A. R. Beck, et al., *Phys. Rev. A* **88**, 051402 (2013).

[195] E. Perfetto and G. Stefanucci, *Phys. Rev. A* **91**, 033416 (2015).

[196] S. L. Adler, *Phys. Rev.* **126**, 413 (1962).

[197] N. Wiser, *Phys. Rev.* **129**, 62 (1963).

[198] R. Del Sole and E. Fiorino, *Phys. Rev. B* **29**, 4631 (1984).

[199] S. Mukamel, *Principles of Nonlinear Optics and Spectroscopy* (Oxford University Press, 1995).

[200] H. Haug and S. W. Koch, *Quantum Theory of the Optical and Electronic Properties of Semiconductors* (World Scientific, 1994).

[201] W. Schäfer and M. Wegener, *Semiconductor Optics and Transport Phenomena* (Springer, 2002).

[202] A. C. Phillips, *Phys. Rev.* **142**, 984 (1966).

[203] M. Cini, E. Perfetto, G. Stefanucci, and S. Ugenti, *Phys. Rev. B* **76**, 205412 (2007).

[204] L. D. Faddeev, *JETP* **12**, 1014 (1961).

[205] K. Held, in *DMFT at 25: Infinite Dimensions*, Lecture Notes of the Autumn School on Correlated Electrons, edited by E. Pavarini, E. Koch, and S. Zhang. (Forschungszentrum Jülich, 2014).

[206] M. Rohlfing and S. G. Louie, *Phys. Rev. B* **62**, 4927 (2000).

[207] G. Onida, L. Reining, and A. Rubio, *Rev. Mod. Phys.* **74**, 601 (2002).

[208] F. Bechstedt, *Many-Body Approach to Electronic Excitations* (Springer, 2015).

[209] R. M. Martin, L. Reining, and D. M. Ceperley, *Interacting Electrons* (Cambridge University Press, 2016).

[210] W. Hanke and L. J. Sham, *Phys. Rev. Lett.* **33**, 582 (1974).

[211] W. Hanke and L. J. Sham, *Phys. Rev. B* **12**, 4501 (1975).

[212] W. Hanke and L. J. Sham, *Phys Rev. B* **21**, 4656 (1980).

[213] C. Vorwerk, B. Aurich, C. Cocchi, and C. Draxl, *Electronic Struct.* **1**, 037001 (2019).

[214] D. Sangalli, P. Romaniello, G. Onida, and A. Marini, *J. Chem. Phys.* **134**, 034115 (2011).

[215] N.-H. Kwong and M. Bonitz, *Phys. Rev. Lett.* **84**, 1768 (2000).

[216] N. Säkkinen, M. Manninen, and R. van Leeuwen, *New J. Phys.* **14**, 013032 (2012).

[217] F. Furche, *J. Chem. Phys.* **114**, 5982 (2001).

[218] T. Sander, E. Maggio, and G. Kresse, *Phys. Rev. B* **92**, 045209 (2015).

[219] P. Cudazzo, *Phys. Rev. B* **102**, 045136 (2020).

[220] H.-Y. Chen, D. Sangalli, and M. Bernardi, *Phys. Rev. Lett.* **125**, 107401 (2020).

[221] F. Paleari and A. Marini, *Phys. Rev. B* **106**, 125403 (2022).

[222] A. Schleife, C. Rödl, F. Fuchs, K. Hannewald, and F. Bechstedt, *Phys. Rev. Lett.* **107**, 236405 (2011).

[223] I. Tamm, in *Selected Papers*, edited by B. Bolotovskii, V. Frenkel, and R. Peierls (Springer, 1991), pp. 157–174.

[224] S. M. Dancoff, *Phys. Rev.* **78**, 382 (1950).

[225] L. V. Keldysh and Y. U. Kopaev, *Sov. Phys. Solid State* **6**, 2219 (1965).

[226] S. W. Koch, W. Hoyer, M. Kira, and V. S. Filinov, *Physica Status Solidi (B)* **238**, 404 (2003).

[227] A. Schindlmayr and R. W. Godby, *Phys. Rev. Lett.* **80**, 1702 (1998).

[228] J. C. Ward, *Phys. Rev.* **79**, 182 (1950).

[229] Y. Takahashi, *Nuovo Cim.* **6**, 371 (1957).

[230] P. Nozières, *Theory of Interacting Fermi Systems* (Benjamin, 1964).

[231] M. Hellgren and U. von Barth, *J. Chem. Phys.* **131**, 044110 (2009).

[232] A.-M. Uimonen, G. Stefanucci, Y. Pavlyukh, and R. van Leeuwen, *Phys. Rev. B* **91**, 115104 (2015).

[233] C. Attaccalite, M. Grüning, and A. Marini, *Phys. Rev. B* **84**, 245110 (2011).

[234] T. J. Park and J. C. Light, *J. Chem. Phys.* **85**, 5870 (1986).

[235] H. Q. Lin and J. E. Gubernatis, *Comput. Phys.* **7**, 400 (1993).

[236] J. M. Luttinger, *Phys. Rev.* **121**, 942 (1961).

[237] T. Giamarchi, *Quantum Physics in One Dimension* (Clarendon, 2004).

[238] M. Gell-Mann and K. Brueckner, *Phys. Rev.* **106**, 364 (1957).

[239] T. Endo, M. Horiuchi, Y. Takada, and H. Yasuhara, *Phys. Rev. B* **59**, 7367 (1999).

[240] J. Sun, J. P. Perdew, and M. Seidl, *Phys. Rev. B* **81**, 085123 (2010).

[241] D. M. Ceperly and B. J. Alder, *Phys. Rev. Lett.* **45**, 566 (1980).

[242] W. M. C. Foulkes, L. Mitas, R. J. Needs, and G. Rajagopal, *Rev. Mod. Phys.* **73**, 33 (2001).

[243] K. Binder and D. W. Heermann, *Monte Carlo Simulations in Statistical Physics: An Introduction* (Springer, 2010).

[244] G. D. Mahan and B. E. Sernelius, *Phys. Rev. Lett.* **62**, 2718 (1989).

[245] S. Hong and G. D. Mahan, *Phys. Rev. B* **50**, 8182 (1994).

[246] P. A. Bobbert and W. van Haeringen, *Phys. Rev. B* **49**, 10326 (1994).

[247] R. Del Sole, L. Reining, and R. W. Godby, *Phys. Rev. B* **49**, 8024 (1994).

[248] J. Lindhard, *Det Kgl. Danske Vid. Selskab, Matematisk-fysiske Meddelelser* **28** (1954).

[249] D. Bohm and D. Pines, *Phys. Rev.* **92**, 609 (1953).

[250] L. H. Thomas, *Proc. Cambridge Phil. Soc.* **23**, 542 (1927).

[251] E. Fermi, *Rend. Accad. Naz. Lincei* **6**, 602 (1927).

[252] G. S. Canright, *Phys. Rev. B* **38**, 1647 (1988).

[253] J. S. Langer and S. H. Vosko, *J. Phys. Chem. Solids* **12**, 196 (1960).

[254] J. Friedel, *Am. Sci.* **93**, 156 (2005).

[255] F. Aryasetiawan, L. Hedin, and K. Karlsson, *Phys. Rev. Lett.* **77**, 2268 (1996).

[256] B. I. Lundqvist, *Phys. Kondens. Materie* **6**, 193 (1967).

[257] B. I. Lundqvist, *Phys. Kondens. Materie* **6**, 206 (1967).

[258] B. I. Lundqvist, *Phys. Kondens. Materie* **7**, 117 (1968).

[259] B. I. Lundqvist and V. Samathiyakanit, *Phys. Kondens. Materie* **9**, 231 (1969).

[260] G. D. Mahan, *Many-Particle Physics*, 2nd edition (Plenum Press, 1990).

[261] L. Landau, *Sov. Phys. JETP* **3**, 920 (1957).

[262] L. Landau, *Sov. Phys. JETP* **5**, 101 (1957).

[263] L. Landau, *Sov. Phys. JETP* **8**, 70 (1959).

[264] G. Stefanucci, R. van Leeuwen, and E. Perfetto, *Phys. Rev. X* **13**, 031026 (2023).

[265] H. Lorentz, *Weiterbildung der Maxwell'schen Theorie: Elektronentheorie.* In Enzyklopa die der Mathematischen Wissenschaften, volume 5 T.2, pp. 145–280 (1904).

[266] P. A. M. Dirac, *Proc. Roy. Soc. Lond. A Math. Phys. Sci.* **167**, 148 (1938).

[267] J. A. Wheeler and R. P. Feynman, *Rev. Mod. Phys.* **21**, 425 (1949).

[268] H. Spohn, *Dynamics of Charged Particles and Their Radiation Field* (Cambridge University Press, 2004).

[269] G. Bauer, D.-A. Deckert, and D. Dürr, *Commun. Partial Diff. Equations.* **38**, 1519 (2013).

[270] D. Lazarovici, *Eu. J. Philos. Sci.* **8**, 145 (2018).

[271] F. Giustino, *Rev. Mod. Phys.* **89**, 015003 (2017).

[272] T. Itoh, *Rev. Mod. Phys.* **37**, 159 (1965).

[273] G. Baym, *Ann. Phys.* **14**, 1 (1961).

[274] P. B. Allen and V. Heine, *J. Phys. C Solid State Phys.* **9**, 2305 (1976).

[275] J. Frenkel, *Wave Mechanics: Elementary Theory* (Oxford University Press, 1932).

[276] S. Baroni, S. de Gironcoli, A. Dal Corso, and P. Giannozzi, *Rev. Mod. Phys.* **73**, 515 (2001).

[277] P. N. Keating, *Phys. Rev.* **175**, 1171 (1968).

[278] A. Marini, S. Poncé, and X. Gonze, *Phys. Rev. B* **91**, 224310 (2015).

[279] A. Marini, *Phys. Rev. B* **107**, 024305 (2023).

[280] D. Karlsson and R. van Leeuwen, in *Handbook of Materials Modeling*, edited by W. Andreoni and S. Yip (Springer, 2020), pp. 367–395.

[281] N. Säkkinen, *Application of Time-Dependent Many-Body Perturbation Theory to Excitation Spectra of Selected Finite Model Systems*, PhD thesis (2016).

[282] E. R. Caianiello, *Combinatorics and Renormalization in Quantum Field Theory* (Benjamin, 1973).

[283] L. Hedin and S. Lundqvist, *Solid State Phys.* **23**, 1 (1970).

[284] J. Fei, C.-N. Yeh, and E. Gull, *Phys. Rev. Lett.* **126**, 056402 (2021).

[285] J. Fei, C.-N. Yeh, D. Zgid, and E. Gull, *Phys. Rev. B* **104**, 165111 (2021).

[286] K. Nogaki and H. Shinaoka, *J. Phys. Soc. Japan* **92**, 035001 (2023).

[287] H. Y. Fan, *Phys. Rev.* **82**, 900 (1951).

[288] A. B. Migdal, *Sov. Phys. JETP* **7**, 996 (1958).

[289] R. van Leeuwen, *Phys. Rev. B* **69**, 115110 (2004).

[290] P. M. M. C. de Melo and A. Marini, *Phys. Rev. B* **93**, 155102 (2016).

[291] M. Shishkin and G. Kresse, *Phys. Rev. B* **75**, 235102 (2007).

[292] D. Nabok, A. Gulans, and C. Draxl, *Phys. Rev. B* **94**, 035118 (2016).

[293] A. Rasmussen, T. Deilmann, and K. S. Thygesen, *NPJ Comput. Mater.* **7**, 22 (2021).

[294] E. Maksimov and S. Shulga, *Solid State Commun.* **97**, 553 (1996).

[295] M. Lazzeri and F. Mauri, *Phys. Rev. Lett.* **97**, 266407 (2006).

[296] S. Pisana et al., *Nat. Mater.* **6**, 198 (2007).

[297] N. Säkkinen, Y. Peng, H. Appel, and R. van Leeuwen, *J. Chem. Phys.* **143**, 234101 (2015).

[298] N. Säkkinen, Y. Peng, H. Appel, and R. van Leeuwen, *J. Chem. Phys.* **143**, 234102 (2015).

[299] J. J. Sakurai and J. Napolitano, *Modern Quantum Mechanics*, 2nd edition (Cambridge University Press, 2017).

[300] G. Grimvall, *The Electron–Phonon Interaction in Metals* (North-Holland, 1981).

[301] D. Karlsson, R. van Leeuwen, E. Perfetto, and G. Stefanucci, *Phys. Rev. B* **98**, 115148 (2018).

[302] P. Lipavský, V. Špička, and B. Velický, *Phys. Rev. B* **34**, 6933 (1986).

[303] S. Latini, E. Perfetto, A.-M. Uimonen, R. van Leeuwen, and G. Stefanucci, *Phys. Rev. B* **89**, 075306 (2014).

[304] Y. Pavlyukh, E. Perfetto, D. Karlsson, R. van Leeuwen, and G. Stefanucci, *Phys. Rev. B* **105**, 125135 (2022).

[305] E. Perfetto and G. Stefanucci, *J. Phys. Condens. Matter* **30**, 465901 (2018).

[306] S. Hermanns, N. Schlünzen, and M. Bonitz, *Phys. Rev. B* **90**, 125111 (2014).

[307] Y. Bar Lev and D. R. Reichman, *Phys. Rev. B* **89**, 220201 (2014).

[308] R. Tuovinen, R. van Leeuwen, E. Perfetto, and G. Stefanucci, *J. Chem. Phys.* **154**, 094104 (2021).

[309] G. Pal, Y. Pavlyukh, W. Hübner, and H. C. Schneider, *Eur. Phys. J. B* **79**, 327 (2011).

[310] F. Covito, E. Perfetto, A. Rubio, and G. Stefanucci, *Phys. Rev. A* **97**, 061401 (2018).

[311] E. Perfetto, A.-M. Uimonen, R. van Leeuwen, and G. Stefanucci, *Phys. Rev. A* **92**, 033419 (2015).

[312] D. Sangalli, S. Dal Conte, C. Manzoni, G. Cerullo, and A. Marini, *Phys. Rev. B* **93**, 195205 (2016).

[313] E. A. A. Pogna, M. Marsili, D. De Fazio, et al., *ACS Nano* **10**, 1182 (2016).

[314] L. Bányai, Q. T. Vu, B. Mieck, and H. Haug, *Phys. Rev. Lett.* **81**, 882 (1998).

[315] Q. T. Vu, H. Haug, W. A. Hügel, S. Chatterjee, and M. Wegener, *Phys. Rev. Lett.* **85**, 3508 (2000).

[316] Q. T. Vu and H. Haug, *Phys. Rev. B* **62**, 7179 (2000).

[317] D. Sangalli and A. Marini, *EPL* **110**, 47004 (2015).

[318] R. Tuovinen, D. Golež, M. Eckstein, and M. A. Sentef, *Phys. Rev. B* **102**, 115157 (2020).

[319] E. V. Boström, A. Mikkelsen, C. Verdozzi, E. Perfetto, and G. Stefanucci, *Nano Lett.* **18**, 785 (2018).

[320] E. Perfetto, D. Sangalli, A. Marini, and G. Stefanucci, *J. Phys. Chem. Lett.* **9**, 1353 (2018).

[321] E. Perfetto, D. Sangalli, M. Palummo, A. Marini, and G. Stefanucci, *J. Chem. Theor. Comput.* **15**, 4526 (2019).

[322] E. Perfetto, A. Trabattoni, F. Calegari, et al., *J. Phys. Chem. Lett.* **11**, 891 (2020).

[323] E. P. Månsson, S. Latini, F. Covita, et al., *Commun. Chem.* **4**, 73 (2021).

[324] A. Kalvová, V.Špička, B. Velický, and P. Lipavský, *Europhysics Letters* **141**, 16002 (2023).

[325] G. Stefanucci and E. Perfetto, *SciPost Phys.* **16**, 073 (2024).

[326] A. Marini, *J Phys. Conf. Ser.* **427**, 012003 (2013).

[327] A. Steinhoff et al., *2D Mater.* **3**, 031006 (2016).

[328] M. Calandra, G. Profeta, and F. Mauri, *Phys. Rev. B* **82**, 165111 (2010).

[329] J. Berges, N. Girotto, T. Wehling, N. Marzari, and S. Poncé, *Phys. Rev. X* **13**, 041009 (2023).

[330] E. Perfetto and G. Stefanucci, *Nano Lett.* **23**, 7029 (2023).

[331] H. Mera, M. Lannoo, C. Li, N. Cavassilas, and M. Bescond, *Phys. Rev. B* **86**, 161404 (2012).

[332] O. Schubert et al., *Nat. Photonics* **8**, 119 (2014).

[333] M. Kira and S. Koch, *Progr. Quantum Electron.* **30**, 155 (2006).

[334] A. Molina-Sánchez, D. Sangalli, L. Wirtz, and A. Marini, *Nano Lett.* **17**, 4549 (2017).

[335] D. Sangalli, *Phys. Rev. Mater.* **5**, 083803 (2021).

[336] Y.-H. Chan, D. Y. Qiu, F. H. da Jornada, and S. G. Louie, *Proc. Nat. Acad. Sci.* **120**, e2301957120 (2023).

[337] Y. Toyozawa, *J. Phys. Chem. Solids* **25**, 59 (1964).

[338] A. Marini, *Phys. Rev. Lett.* **101**, 106405 (2008).

[339] Y.-H. Chan, J. B. Haber, M. H. Naik, et al., *Nano Lett.* **23**, 3971 (2023).

[340] G. A. Garrett, A. G. Rojo, A. K. Sood, J. F. Whitaker, and R. Merlin, *Science* **275**, 1638 (1997).

[341] S. L. Johnson, P. Neaud, E. Vorobeva, et al., *Phys. Rev. Lett.* **102**, 175503 (2009).

[342] F. Benatti, M. Esposito, D. Fausti, et al., *New J. Phys.* **19**, 023032 (2017).

[343] M. Lakehal, M. Schiró, I. M. Eremin, and I. Paul, *Phys. Rev. B* **102**, 174316 (2020).

[344] T. Chase, M. Trigo, A. Reid, et al., *App. Phys. Lett.* **108**, 041909 (2016).

[345] L. Waldecker, T. Vasileiadis, R. Bertoni, et al., *Phys. Rev. B* **95**, 054302 (2017).

[346] D. Filippetto, P. Musumeci, R. Li, et al., *Rev. Mod. Phys.* **94**, 045004 (2022).

[347] T. L. Britt, Q. Li, P. Rene de Cotret, et al., *Nano Lett.* **22**, 4718 (2022).

[348] C. Trovatello, H. Miranda, A. Molina-Sanchez, et al., *ACS Nano* **14**, 5700 (2020).

[349] D. Li, C. Trovatello, S. Dal Conte, et al., *Nat. Commun.* **12**, 954 (2021).

[350] S. Mor, V. Gosetti, A. Molina-Sanchez, et al., *Phys. Rev. Res.* **3**, 043175 (2021).

[351] T. Y. Jeong, B. Moon Jin, S. Rhim, et al., *ACS Nano* **10**, 5560 (2016).

[352] C. J. Sayers, A. Genco, C. Trovatello, et al., *Nano Lett.* **23**, 9235 (2023).

[353] L. P. Pitaevskii and E. M. Lifshitz, *Physical Kinetics: Volume 10* (Butterworth-Heinemann, 1981).

[354] J. M. Ziman, *Electrons and Phonons: The Theory of Transport Phenomena in Solids* (Clarendon, 1960).

[355] S. Poncé, E. R. Margine, and F. Giustino, *Phys. Rev. B* **97**, 121201 (2018).

[356] S. Sadasivam, M. K. Y. Chan, and P. Darancet, *Phys. Rev. Lett.* **119**, 136602 (2017).

[357] Y. Pavlyukh, E. Perfetto, D. Karlsson, R. van Leeuwen, and G. Stefanucci, *Phys. Rev. B* **105**, 125134 (2022).

[358] Y. Pavlyukh, E. Perfetto, and G. Stefanucci, *Phys. Rev. B* **106**, L201408 (2022).

[359] N. Schlünzen, J.-P. Joost, and M. Bonitz, *Phys. Rev. Lett.* **124**, 076601 (2020).

[360] J.-P. Joost, N. Schlünzen, and M. Bonitz, *Phys. Rev. B* **101**, 245101 (2020).

[361] Y. Pavlyukh, E. Perfetto, and G. Stefanucci, *Phys. Rev. B* **104**, 035124 (2021).

[362] R. Tuovinen, Y. Pavlyukh, E. Perfetto, and G. Stefanucci, *Phys. Rev. Lett.* **130**, 246301 (2023).

[363] M. Ridley, A. MacKinnon, and L. Kantorovich, *Phys. Rev. B* **91**, 125433 (2015).

[364] J. Hu, R.-X. Xu, and Y. Yan, *J. Chem. Phys.* **133**, 101106 (2010).

[365] A. Croy and U. Saalmann, *Phys. Rev. B* **80**, 245311 (2009).

[366] X. Zheng, G. Chen, Y. Mo, et al., *J. Chem. Phys.* **133**, 114101 (2010).

[367] Y. Zhang, S. Chen, and G. Chen, *Phys. Rev. B* **87**, 085110 (2013).

[368] Y. H. Kwok, H. Xie, C. Y. Yam, X. Zheng, and G. H. Chen, *J. Chem. Phys.* **139**, 224111 (2013).

[369] R. Wang, D. Hou, and X. Zheng, *Phys. Rev. B* **88**, 205126 (2013).

[370] Y. Kwok, G. Chen, and S. Mukamel, *Nano Lett.* **19**, 7006 (2019).

[371] P. Gaspard and M. Nagaoka, *J. Chem. Phys.* **111**, 5668 (1999).

[372] A. G. Redfield, *IBM J. Res. Dev.* **1**, 19 (1957).

[373] A. Suárez, R. Silbey, and I. Oppenheim, *J. Chem. Phys.* **97**, 5101 (1992).

[374] P. Pechukas, *Phys. Rev. Lett.* **73**, 1060 (1994).

[375] G. Lindblad, *Commun. Math. Phys.* **48**, 119 (1976).

[376] L. M. Sieberer, M. Buchhold, and S. Diehl, *Rep. Progr. Phys.* **79**, 096001 (2016).

[377] H. C. Fogedby, *Phys. Rev. A* **106**, 022205 (2022).

[378] A. McDonald and A. A. Clerk, *Phys. Rev. Res.* **5**, 033107 (2023).

[379] F. Thompson and A. Kamenev, *Ann. Phys.* **455**, 169385 (2023).

[380] E. Lipparini, *Modern Many-Particle Physics: Atomic Gases, Quantum Dots and Quantum Fluids* (World Scientific, 2003).

[381] R. McWeeny, *Rev. Mod. Phys.* **32**, 335 (1960).

[382] M. Born and H. S. Green, *Proc. R. Soc. Lond. A* **191**, 168 (1947).

[383] N. N. Bogoliubov, *The Dynamical Theory in Statistical Physics* (Hindustan Pub. Corp., 1965).

[384] J. Kirkwood, *J. Chem. Phys.* **14**, 180 (1946).

[385] J. Yvon, *Nucl. Phys.* **4**, 1 (1957).

[386] A. Akbari, M. J. Hashemi, R. M. Nieminen, R. van Leeuwen, and A. Rubio, *Phys. Rev. B* **85**, 235121 (2012).

[387] F. Lackner, I. Březinová, T. Sato, K. L. Ishikawa, and J. Burgdörfer, *Phys. Rev. A* **91**, 023412 (2015).

[388] F. Lackner, I. Březinová, T. Sato, K. L. Ishikawa, and J. Burgdörfer, *Phys. Rev. A* **95**, 033414 (2017).

[389] J.-P. Joost, N. Schlunzen, H. Ohldag, et al., *Phys. Rev. B* **105**, 165155 (2022).

[390] C. Cohen-Tannoudji, B. Diu, and F. Laloë, *Quantum Mechanics* (Wiley, 1977).

[391] P. A. M. Dirac, *Proc. R. Soc. Lond. A* **114**, 243 (1927).

[392] E. Fermi, *Nuclear Physics* (Univerisity of Chicago, 1950).

[393] T. D. Visser, *Am. J. Phys.* **77**, 487 (2009).

[394] E. Perfetto, G. Stefanucci, and M. Cini, *Phys. Rev. B* **78**, 155301 (2008).

[395] M. Ridley, N. W. Talarico, D. Karlsson, N. L. Gullo, and R. Tuovinen, *J. Phys. A Math. Theoret.* **55**, 273001 (2022).

[396] J. A. Gupta, D. D. Awschalom, X. Peng, and A. P. Alivisatos, *Phys. Rev. B* **59**, R10421 (1999).

[397] A. Greilich, R. Oulton, E. Zhukov, et al., *Phys. Rev. Lett.* **96**, 227401 (2006).

[398] F. M. Souza, *Phys. Rev. B* **76**, 205315 (2007).

[399] D. Pines, *Elementary Excitations in Solids* (Benjamin, 1964).

[400] E. Perfetto, K. Wu, and G. Stefanucci, *npj 2D Mater Appl* **8**, 40 (2024).

[401] E. K. U. Gross, E. Runge and O. Heinonen, *Many-Particle Theory* (Adam Hilger, 1991).

Index

Printed in the United States
by Baker & Taylor Publisher Services